Australian Ecosystems: 200 years of Utilization, Degradation and Reconstruction

Proceedings of a Symposium held in Geraldton, Western Australia
28 August-2 September, 1988

Edited by

D. A. Saunders
CSIRO Division of Wildlife and Ecology
LMB No. 4, P.O. Midland, Western Australia 6056

A. J. M. Hopkins
Department of Conservation and Land Management
P.O. Box 51, Wanneroo, Western Australia 6065

R. A. How
Western Australian Museum,
Francis Street, Perth, Western Australia 6000

1990

Proceedings of the Ecological Society of Australia
Volume 16

The National Library of Australia
Cataloguing-in-Publication entry:

Australian Ecosystems: 200 years of utilization, degradation and reconstruction.

Bibliography.
Includes index.
ISBN 0 949324 26 4.
ISSN 0077-834

1. Ecology — Australia — Congresses. 2. Biotic communities — Australia — Congresses. 3. Conservation of natural resources — Australia — Congresses. I. Saunders, D. A. (Denis Allan), 1947- . II. Hopkins, A. J. M. III. How, R. A. IV. Ecological Society of Australia. (Series: Proceedings of the Ecological Society of Australia; v. 16).

574.50994

PRINTED AND PUBLISHED IN AUSTRALIA
FOR THE ECOLOGICAL SOCIETY OF AUSTRALIA BY
SURREY BEATTY & SONS PTY LTD
43 Rickard Road, Chipping Norton, NSW 2170 Australia

Contents

Foreword v
Preface vii

EVOLUTIONARY AND LAND USE HISTORY

Ecological history for environmental management. R. L. Clark 1-21

An ecological study of traditional Aboriginal use of "country": Martu in the Great and Little Sandy Deserts, Western Australia. F. J. Walsh 23-37

USES OF AND IMPACTS ON AQUATIC SYSTEMS

The impact of nutrient enrichment on nearshore and estuarine ecosystems in Western Australia. K. Hillman, R. J. Lukatelich & A. J. McComb 39-53

Resolving the conflict between conservation values and extractive use of the Abrolhos coral reefs. A. I. Hatcher, G. D. Wright & B. G. Hatcher 55-70

Ecological changes due to our use of the coast: research needs versus effort. P. G. Fairweather 71-77

Australian upland streams: ecological degradation and possible restoration. P. S. Lake & R. Marchant 79-91

MANAGEMENT OF AUSTRALIAN VEGETATION

From frontier to fragments: European impact on Australia's vegetation. R. J. Hobbs & A. J. M. Hopkins 93-114

Sydney's vegetation 1788-1988: utilization, degradation and rehabilitation. D. H. Benson & J. Howell 115-27

The role of synthetic vegetation in present and future landscapes of Australia. P. B. Bridgewater 129-34

The revegetation of Victoria: the first stages. A description of the revegetation movement in Victoria. S. Burke & R. Youl 135-40

Interactions of native and introduced species in new habitats. M. D. Fox 141-47

Tree dieback and insect dynamics in remnants of native woodlands on farms. J. Landsberg, J. Morse & P. Khanna 149-65

The European honeybee in Western Australian kwongan: foraging preferences and some implications for management. R. T. Wills, M. N. Lyons & D. T. Bell 167-76

Fire and management of south-western Australian banksias. R. M. Cowling, B. B. Lamont & N. J. Enright ... 177-83

MANAGEMENT OF THE ARID ZONE

Where the creeks run dry or ten feet high: pastoral management in arid Australia. M. H. Friedel, B. D. Foran & D. M. Stafford Smith 185-94

Pattern and production in arid lands. D. M. Stafford Smith & G. Pickup 195-200

The impact of European settlement on the vertebrate animals of arid Australia: a conceptual model. S. R. Morton 201-13

Uluru (Ayers Rock-Mt Olga) National Park: an assessment of a fire management programme. G. Allan & L. Baker 215-20

Application of fire ecology to fire management: a manager's perspective. N. Preece 221-33

INTERACTION BETWEEN GRAZING AND NATURAL ECOSYSTEMS

The effect of grazing on Australian ecosystems. A. D. Wilson 235-44

Analysis of stocking records from 1884 to 1988 during the subdivision of *Momba*, the largest property in semi-arid New South Wales. John Pickard 245-53

Cattle grazing within subalpine heathland and grassland communities on the Bogong High Plains: disturbance, regeneration and the shrub-grass balance. R. J. Williams 255-65

Regeneration in populations of the arid zone plants *Acacia carnei* and *A. oswaldii*. T. D. Auld 267-72

Soil and vegetation characteristics of dry tropical rangelands for predicting regeneration in exclosures. J. G. McIvor & C. J. Gardener 273-77

Dry tropical rangelands: solving one problem and creating another. G. J. Gardener, J. G. McIvor & J. Williams 279-86

AUSTRALIAN FAUNA — ITS PAST AND FUTURE

A review of current ideas of the extinction, conservation and management of Australia's terrestrial vertebrate fauna. H. F. Recher & L. Lim 287-301

The impact of agricultural and pastoral industries on birds in the southern half of Western Australia: past, present and future. D. A. Saunders & P. J. Curry 303-21

State forests in Australia and their role in wildlife conservation. S. M. Davey & T. W. Norton 323-45

The use of ant communities to evaluate change in Australian terrestrial ecosystems: a review and a recipe. A. N. Andersen 347-57

The value of road reserves to the avifauna of the central wheatbelt of Western Australia. P. Cale 359-67

The numbat *Myrmecobius fasciatus* (Myrmecobiidae): history of decline and potential for recovery. J. A. Friend 369-77

Factors affecting small mammal distribution and abundance in the Eastern Otway Ranges, Victoria. B. A. Wilson, D. Robertson, D. J. Moloney, G. R. Newell & W. S. Laidlaw 379-96

Restoration of biological scenarios: the role of museum collections. Barbara York Main 397-409

PLANNING, LANDUSE AND MANAGEMENT AT A REGIONAL AND LOCAL LEVEL

Naturalness: the concept and its application to Australian ecosystems. S. G. Taylor 411-18

Major issues in land management in Australia with particular reference to the role of educators. V. R. Squires 419-25

Some management implications for the conservation of vegetation remnants and associated fauna in the central western region of New South Wales. D. C. Goldney & I. J. S. Bowie 427-40

Satellite monitoring of soils, vegetation and grazing impacts in the semi-arid region of western New South Wales. A. L. O'Neill & D. J. Eldridge 441-47

Integration of satellite imagery and geographic information systems for habitat mapping and the prediction of fauna distributions: preliminary results and future directions. J. K. Hibberd, A. L. O'Neill, J. Marthick & R. L. Sim 449-57

Reconstruction of South Australia's arid lands: the conservation option. B. Cohen 459-65

Natural regions of western New South Wales and their use for environmental management. G. Morgan & J. Terrey 467-73

Land evaluation: the Victorian land capability aassessment approach. A. Lindsay and K. Rowe 475-90

Identifying the National Estate: issues and directions in the assessment of significance of places. M. O'Brien & R. Purdie 491-504

Institutional processes affecting the management of natural resources. R. D. McKellar 505-07

RECONSTRUCTION OF EXTENSIVELY MODIFIED ECOSYSTEMS

Zoological considerations

Rehabilitation of disturbed land: long-term prospects for the recolonization of fauna. J. D. Major 509-17

Two hundred years of disturbance: how has it aided our understanding of succession in Australia? B. J. Fox 521-29

Botanical considerations

Seed ecology in relation to reclamation: lessons from mined-lands in Western Australia. D. T. Bell, S. Vlahos & S. M. Bellairs 531-35

Weed management and revegetation programmes. F. D. Panetta & R. H. Groves 537-43

Saltland reclamation: selection of superior Australian tree genotypes for discharge sites. P. G. van der Moezel & D. T. Bell 545-49

Rehabilitation agronomy: guidelines for revegetating degraded land. C. V. Malcolm 551-56

Bauxite mine rehabilitation in the Darling Range, Western Australia. S. C. Ward, J. M. Koch & O. G. Nichols 557-65

THE GENETIC RESOURCE BASE FOR THE FUTURE

Conservation of genetic resources in Australia's flora and fauna. S. D. Hopper & D. J. Coates 567-77

Managing Australia's environment: directions for the future. A. J. M. Hopkins, R. A. How and D. A. Saunders 579-83

Index 585-602

Foreword

As we approach the 21st century, Australians face some major decisions about our management of the environment.

Concern for the environment is not, as it has been painted by some detractors, a middle class trendy issue. Nor is it a subversive notion working to undermine the economic development of Australia. The Australian Government is tackling a wide range of environmental issues; from those involving preservation of national and world heritage values, prevention of land and water degeneration to containing the spread of hazardous wastes and combatting further depletion of the ozone layer.

The days have gone when there was always more land, more trees, more minerals, more resources. We have too many examples of our diminishing resources. One example is the Huon Pine, once mainstay of a shipbuilding industry, and now very limited in distribution. We must evaluate our role on this continent over the last 200 years and look to our responsibilities in the next. While the national economy depends substantially on the export earning of rural and mining industries and increasingly on the tourist industry, continued economic success is more likely if long-term gain is not sacrificed for short-term profit.

Australia has been subjected to intense development for less than 200 years, and yet in that time: 70% of the plant communities which existed in 1788 have been modified in some way; over 65% of the original tree cover has been removed; and up to 75% of our rainforest has been cleared for grazing and agriculture.

As a result of wholesale exploitation of a fragile continent: 55% of arid lands and 45% of non-arid areas require treatment for land degradation; 4.2 million hectares are affected by salting; 78 species of plants and 17 species of mammals are believed to be extinct; and land degradation in the Murray-Darling basin alone, reduces agricultural productivity by at least $200 million annually.

Development interests have dominated land use decision making in this country over the past 200 years, often at the expense of the environment. This is reflected in costs to the primary industry sector. For example, the costs of restoring degraded lands and costs to individual graziers from pesticide residue problems.

In Australia today there are three big environmental issues which have to be faced by us all. They are: identification and rehabilitation of land degraded by past land use practices and the development of guidelines for the future protection of our land and its resources; identification and protection of those natural areas which are particularly fragile, unique or of special importance to Australians and the world; and minimizing climate change but at the same time preparing ourselves for the inevitable effects on our environment, and consequently on our economy.

There are no quick fixes for these problems. They all require long-term commitment and a change in our attitude to the use of our resources. They all require considerable financial commitment over time. Whilst issues such as the Franklin River Dam, woodchipping, mining in Kakadu, logging of the North Queensland rainforests and the southern forests of Tasmania all attract publicity, very little is heard of perhaps the most critical environmental issue; land degradation.

The National Conservation Strategy for Australia provides nationally agreed guidelines for the use of living resources by Australians so that the reasonable needs and aspirations of society can be sustained. Since endorsement of the strategy by Commonwealth and State Governments many programmes consistent with its objectives have been introduced involving all levels of government and the community at large. Action plans are currently being developed to implement the principles adopted in the strategy.

The Government has maintained and expanded the national soil conservation programme, increasing funding from $6 million in 1987-88 to $10.8 million in 1988-89. The

Australian Soil Conservation Council has endorsed a national soil conservation strategy which was released in 1989. It provides a framework for nationally agreed policies and priority actions for the prevention and control of land degradation and for the rehabilitation of affected areas.

A major further increase in the National Soil Conservation Programme was announced in the Prime Minister's statement "Our Country, Our Future" with funding for 1989-90 set at $23.4 million. The National Soil Conservation Programme funding forms part of a major landcare package, with $320 million to be spent during the next decade for land care, related tree planting and remnant vegetation conservation.

There is no doubt, however, that any soil conservation programme has to look at ways of changing the management practices of farmers. This is something farmers are already aware of and implementing. It is clearing and overgrazing that cause most of the problems. This is where ecologists can be of particular help.

Specialist scientific advice is always helpful to the Government, but never as widely available as industry advocacy, or the lobbying of the conservation movement. Governments at all levels need objective scientific advice and it is the responsibility of societies such as the Ecological Society of Australia to be proactive with advice for Government.

At the same time simple requests for more funds for basic research will not be helpful. These days most governments are aware that we know all too little about most of our ecosystems, but we do know something of general principles which will be helpful. Too many scientists are ready to hide behind statements like "we don't really know enough about . . ." rather than say "we don't know it all, but we can tell that certain consequences will flow from these actions."

Those sentiments were echoed by Prince Charles in 1987 when addressing a conference on the North Sea. He dismissed the persistent excuse that there is not enough proof of ecological damage to justify stricter environmental controls by saying "while we wait for the doctor's diagnosis the patient may easily die."

Let me urge you to make your views known to government, but in positive terms. That way governments will be more inclined to listen when you inevitably ask for more research funds! I am grateful to the Ecological Society of Australia for providing their perspective in this volume. I welcome the constructive suggestions that are made. It is obvious that we are working to the common goal of making a more ecologically sustainable environment in which to live.

Senator Graham Richardson
Parliament House Canberra

Preface

The objective of the 1988 Symposium of the Ecological Society of Australia was to provide a perspective for Australian ecosystems in the next century. The theme, AUSTRALIAN ECOSYSTEMS: 200 YEARS OF UTILIZATION, DEGRADATION AND RECONSTRUCTION, required an assessment of past changes to and the present status of Australian ecosystems. Early circulars emphasized the desire to bring together contributions from scientists, managers and others involved in exploiting Australian ecosystems in order to examine important unifying principles and establish priorities for research and management during the next century.

The initial response to the call for contributions was most encouraging and over 60 papers and posters were offered, necessitating an extension of the Symposium to four days. It is a measure of the diversity of the responses that only two contributions were solicited for subjects not covered by the offerings generated by the call for papers. These were contributions on Aboriginal use of Australian ecosystems and the use and future of Australian soils. The chapters that follow in these proceedings represent those that have been submitted to review by two referees and accepted by the editors. Several papers presented at the Symposium are not included here because they were not submitted for refereeing, they were rejected by the referees as not relevant to the theme or as not of sufficient standard to warrant publication.

It is apparent that dramatic changes have occurred in Australian ecosystems over the last 200 years, particularly in the latter half. The extensive modifications to both aquatic and terrestrial ecosystems since European occupation were highlighted in all of the review papers and mentioned in the majority of case studies.

At the Plenary Session, held at the conclusion of the Symposium, there was consensus amongst the 190 participants that there should be immediate action to arrest the degradation of ecosystems and reverse deleterious trends. Specific resolutions were adopted that called on governments to take the lead in planning for and management of all aspects of the environment, and outlining the need for a high level of public involvement in this process. The adopted resolutions were sent to all relevant State and Commonwealth ministers as well as heads of departments responsible for conservation and land management. In keeping with the warmth of the spring sunshine in the Queens Park Amphitheatre and the mood of the Symposium, the Plenary Session also concurred with the view that ecologists must communicate more to both the press and public on issues of environmental importance. As part of this process, it was suggested that the Society establish a communication committee.

The final day of the meeting in Geraldton was devoted to the Open Forum where thirty papers were read in three pairs of concurrent sessions. A distinguished panel sat in judgment on the 17 papers by students in order to award the prize for the best presented paper. Our congratulations to Craig James and Heather Keith on winning the honour and sharing the generous prize donated by the Hammersley Iron Pty Ltd.

During the planning stages of the Symposium the organizing committee deliberated on the choice of venue for the 1988 Symposium. It was decided that a regional centre would be preferable to Perth as it would provide a sequestered environment where all participants could indulge in lengthy hours of informal discussion. It was decided that the city of Geraldton, 420 km north of Perth, provided an ideal venue for such an environment and had excellent facilities to cater for around 200 delegates. A consequence of the selection of Geraldton as the venue was the paucity of media attendance and coverage at the Symposium and the limited public involvement in sessions and discussions.

A Public Forum was held on the penultimate evening of the Symposium to which participants, press and public alike were invited to attend and to provide the opportunity for the local community to participate. This Public Forum was chaired by the local MLA and Minister for Mines and Local Government, the Hon. Geoff Carr, and had as distinguished speakers Ralph Slatyer, Bert Main, Harry Recher and Bill Hare. The organizers are grateful to these five for providing a most informative and stimulating evening in which many of the contrasting views held by ecologists were aired and in which the public entered into lively debate.

The formal social aspects of the Conference centred around the Civic Reception by the City of Geraldton, where the delegates were made to feel welcome through the generosity of the Mayor Faye Simpson and her fellow councillors. The Annual Dinner was held in the Catholic Community Centre where Bert Main was presented with the Gold Medal of the Ecological Society of Australia and gave a most enlightening talk to a well preserved audience.

The selection of Geraldton as the venue dictated the choice of pre- or post-conference tours. The pre-conference tour was designed to take in major developmental and research sites on the floristically rich northern Sandplains (kwongan) of the Perth Basin and nearly half the delegates took advantage of this to travel from Perth to Geraldton. The four-day post-conference tour to Kalbarri, Shark Bay, Carnarvon and Perth was conducted by Greg Keighery and examined many areas of ecological and biological interest on the west coast of Western Australia.

It is axiomatic to mention that with such a large influx of scientists to a small provincial city, there needs to be great support from the local community. The organizing committee would like to extend their appreciation to the following for great patience and assistance during the course of the Symposium: Narelle Taylor of the Geraldton Tourist Bureau; Faye Simpson, Mayor of Geraldton; Rod Wilson and Greg Wallace of the Geraldton Branch of the Western Australian Museum; Geoff Mercer, Geraldton Regional Manager of the Department of Conservation and Land Management; Peter Clough, Secretary of the local MLA; and Judy Kennington of the Hampton Park Inn, Greenough. We would also like to thank: Brian Tobin of the Catholic Community Centre; Guy Hooper of the Queens Park Theatre; John Beard, the organizer of the poster session; Dinky Gobble-Garrett who organized the accommodation; and our fellow councillors, Arthur McComb, Kieran McNamara, John Majer and Greg Beeston, for support behind the scenes and the extensive time spent in discussions and achieving common goals. We would also like to pay tribute to those who went beyond the call of duty; Libby Mattiske, John Dell, Judith Harvey and Dennis King who worked tirelessly behind the scenes to help make the Symposium the success that it was.

We would like to thank the following for their willing help in acting as referees: G. W. Arnold, K. J. Atkins, T. D. Auld, A. N. Andersen, M. P. Austin, P. Baverstock, J. A. Beck, G. Beeston, D. T. Bell, P. F. Berry, J. Blyth, R. W. Braithwaite, L. W. Braithwaite, P. B. Bridgewater, M. G. Brooker, A. A. Burbidge, S. Bunn, S. Burke, R. Clay, K. D. Cocks, B. Cohen, P. J. Curry, P. G. Fairweather, M. D. Fox, M. Friedal, G. R. Friend, J. A. Friend, D. G. Goldney, R. D. Graetz, G. Griffin, T. Griffin, R. H. Groves, G. N. Harrington, J. M. Harvey, J. K. Hibberd, K. Hillman, R. J. Hobbs, K. C. Hodgkinson, A. J. M. Hopkins, R. A. How, J. Howell, W. F. Humphreys, G. Keighery, A. Kinnear, J. Koch, D. J. Kitchener, P. S. Lake, D. Lamb, B. B. Lamont, A. Lindsay, J. F. Lynch, A. R. Main, J. D. Majer, C. V. Malcolm, J. Marshall, E. M. Mattiske, A. J. McComb, N. L. McKenzie, S. A. Moore, S. R. Morton, O. G. Nichols, J. C. Noble, R. A. Nulsen, M. O'Brien, A. L. O'Neill, J. Pickard, N. Preece, R. Purdie, H. F. Recher, D. A. Saunders, J. C. Short, G. T. Smith, S. G. Taylor, P. G. van der Moezel, D. I. Walker, S. C. Ward, R. J. Williams and A. D. Wilson.

We are extremely grateful to Lesley Brooker who carried out the enormous task of collating the index. Ivor Beatty and his staff at Surrey Beatty & Sons have done a marvellous job in producing this volume and we are grateful for their support.

Last but not least we would like to thank the following for generous support towards the running of the Symposium: Alcoa of Aust.; Ansett Airlines of Aust.; Association Minerals Consolidated Ltd.; Australian Groundwater Consultants; CSIRO Division of Wildlife and Ecology; Curtin University of Technology; CRA Exploration Pty. Ltd.; Dames and Moore Pty. Ltd.; Western Australian Department of Conservation and Land Management; Western Australian Environmental Protection Authority; Geography Teachers' Association of Western Australia; Geraldton Museum; Geraldton Tourist Bureau; Hammersley Iron Pty. Ltd.; Kinhill Sterns Pty. Ltd.; Le Provost, Semeniuk and Chalmer Pty Ltd.; Maunsell and Partners Pty. Ltd.; Western Australian Museum; Woodside Offshore Petroleum Pty. Ltd.; and Worsly Alumina Pty. Ltd.

R. A. HOW
A. J. M. HOPKINS
D. A. SAUNDERS

Proc. Ecol. Soc. Aust. 1990 — 16:1-21

Ecological history for environmental management

R. L. CLARK

CSIRO, Division of Water Resources
P.O. Box 1666, Canberra, A.C.T. 2601 Australia

Abstract

Management of ecosystems requires knowledge of their history. Description and observation are inadequate for detecting underlying trends or understanding and predicting rates, directions and magnitudes of change in complex and dynamic systems. The application of ecological history to environmental management depends on the scale and resolution both available from the historical evidence and required by a management problem. Evidence from the environment itself, particularly that contained in sediments or peat, complements and extends the written historical and instrumental records. The temporal resolution obtained from the environmental evidence is usually one to two orders of magnitude smaller than the time scale covered. Managers need a similar or greater degree of resolution, but are rarely concerned with a future beyond the next few decades. Knowledge of the long-term development of the environment is a necessary background for management, but not directly applicable. Of greater value is the opportunity medium- to long-term environmental history provides for testing ecological theories. By contrast, studies of the last few hundred years, at appropriate resolutions, are critical for understanding the dynamics of Australian ecosystems before and since European settlement in 1788 and for predicting effects of local, regional and global environmental changes over the next hundred years or so. This kind of research bridges the gap between traditional palaeoecology on the one hand and monitoring, observation and experiment on the other, and can answer specific management questions. Examples are given of the application of environmental history on several time-scales to ecological theory and to practical questions of environmental management.

Introduction

Managing ecosystems without any knowledge of their history may well invite future disaster. Simple description of the environment or observation of environmental variables over a few years are essential, but inadequate, for detecting longer-term trends and understanding rates, directions and magnitudes of change in complex and dynamic systems. This is particularly so in Australia where climatic variability is high and extreme climatic events may be more important than means in determining what actually is at any one place at a particular time. Any period of observation is likely to be unrepresentative of the longer-term and may or may not contain rare events. Shorter-term change or stability must be placed in the context of longer-term trends or variability.

In order to predict the outcomes of management actions, studies of present-day pattern must be complemented by studies of current processes (Hopkins & Saunders 1987), but both kinds of study need to be extended by investigations of pattern and process in the past, leading up to the present-day. Predictions are dependent on explanations of how the present came to be, on understanding the complex interactions over time between processes and the patterns they create (Wasson & Clark 1985). Environmental pattern at any one time is both the outcome of preceding interactions between processes and what was there, and the template on which processes act in the immediate future.

This volume is concerned with the effects on Australian ecosystems of European land use since the establishment of the first British settlement at Sydney Cove in 1788. It is not intended to be simply a catalogue of environmental degradation and disaster, but a positive attempt to learn from the mistakes and successes of the past and to combine this knowledge with current expertise for improved environmental management, sustained yield from agricultural lands and rehabilitation of degraded country. The present chapter has two aims: (i) to provide the longer-term context of the changes that have taken place over the last 200 years and that are discussed in subsequent chapters; and (ii) to show how palaeoecological methods can be applied on time scales and with resolutions appropriate to management questions, thus providing information critical to our understanding of enviromental change over the next hundred years or so. No attempt has been made to comprehensively review palaeoecological studies in Australia. Rather, what follows is a partisan argument for the application of enviromental history to management questions, and an attempt to persuade managers and ecologists that some of their questions can be answered by looking to the past.

Cainozoic history

The development of Australian environments, particularly of vegetation, over the last 65 million years has been outlined or reviewed recently by several authors (Kershaw 1981; Walker & Singh 1981; Bowler 1982; Hope 1982; Truswell & Harris 1982; Hope 1984; Wasson & Clark 1987). An extensive review by Kershaw (1988) considers the evidence for, and questions assumptions about, the course of environmental change over the past 20 million years.

About 54 million years ago, at the beginning of the Eocene, the Australian continental plate separated from Antarctica and began drifting north, carrying with it a Gondwanic flora and fauna. Indomalaysian rainforest taxa migrated into the north from the early Tertiary onwards (Kershaw 1988). Throughout the Tertiary, most of the continent was covered by closed forests, but they were gradually replaced by more open vegetation over the last twenty million years (Truswell & Harris 1982; Kershaw 1988), possibly in response to increasing seasonality and greater variability in thermal and water regimes (Nix 1982). By 2.5 million years ago, a winter rainfall regime was established in the south-east, replacing the summer rainfall regime that had previously covered the entire continent (Bowler 1982).

The tendency towards seasonal aridity became even more marked through the Quaternary (the last 1.8 million years), particularly over the last 700 000 years as the amplitude of glacial cycles increased (Bowler 1982). The course of climatic change over the last 500 000 years, as reflected in the deep-sea record of $\delta^{18}O$ and sea surface temperatures, derived from foraminiferal assemblages, is shown in Figure 1. Glacial and interglacial cycles are evident, probably resulting from changes in insolation at the earth's surface (Milankovich 1941). It is apparent that interglacials, such as we are experiencing at the moment, are atypical of the Quaternary and have occupied only about 10% of the time. Through most of the Quaternary, terrestrial environments have been colder and drier than at present. Australian vegetation now is predominantly grasslands or shrublands and the extent of treeless or non-tree dominated vegetation has been even greater in the past (Hope 1984, 1989). In the drier forests and woodlands, *Casuarina* was more abundant than *Eucalyptus* until the last hundred or so thousand years (Singh & Geissler 1985; Kershaw 1988).

From the long-term record three points, at least, should be made.

(a) Almost all of the Australian continent remained unglaciated through the Quaternary, so soils and landforms are very different from those of Europe and much of North America. With Australia's unique flora, fauna and geological history, ecological theories and land management practices developed in the northern hemisphere may not apply to the Australian environment. The last 200 years have made this abundantly clear.

(b) Recent evolution of the biota has been, for most of the time, in environments very different from those we experience today. Present-day distributions may reflect adjustment to past

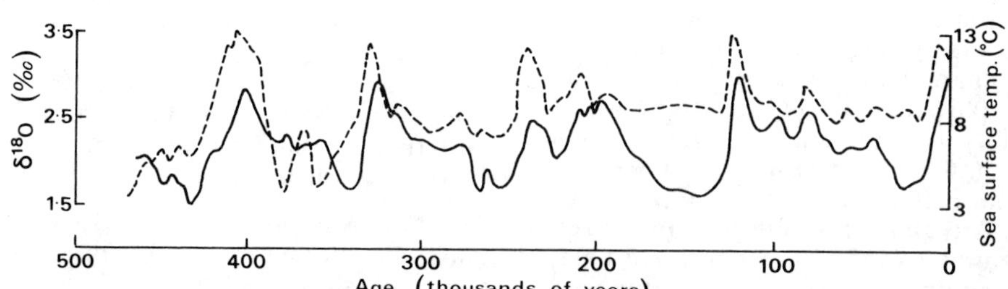

FIG. 1. Glacial and interglacial periods over the last 500 000 years as reflected in variations in $\delta^{18}O$ (solid line) and sea surface temperatures (dashed line), derived from foraminiferal assemblages, recorded in Indian Ocean cores. Peaks of the curves indicate interglacials and troughs full glacial conditions. Adapted from Hays *et al.* (1976).

environments more than to present ones. Some adaptations to present conditions may be fortuitous; for example, adaptations to fire may be traits evolved to survive other stresses such as droughts (Gill 1981).

(c) comparing the present interglacial with previous ones, and taking into account insolation cycles, it is likely that cooling towards the next glacial period will not begin until about 5 000 years hence (Berger 1981), but anthropogenic effects on the atmosphere could alter the normal course of events.

Aboriginal settlement of Australia

The earliest unequivocal evidence for Aboriginal presence in Australia comes from Lake Mungo in western New South Wales where skeletal remains have been dated to between 30 000 and 40 000 years BP (Bowler 1976; Thorne 1981). The age of these skeletons may be considerably greater (Bowler 1989). Occupation sites around the coast range in age to about 30 000 BP (Bowdler 1977), with possibly older dates from near Perth (Pearce & Barbetti 1981) and west of Sydney (Nanson *et al.* 1987). The predominantly coastal distribution of older sites led Bowdler (1977) to suggest that Aboriginal colonization of Australia was from the sea, around the coast and up rivers. More recent discovery of evidence for occupation in Central Australia around 22 000 BP (Smith 1987) challenges this hypothesis, but may only change ideas of how quickly people occupied the whole continent.

The initial arrival of Aborigines at any place must have affected the environment, particularly through changes in fire regime. While there has been much speculation, wild or well-founded, on the effects of past Aboriginal burning (Jones 1975; Hughes & Sullivan 1981; Horton 1982; Clark 1983a; Bowman & Brown 1986; Head 1988), there is no certain evidence at present for vegetation changes resulting from the first impact of Aborigines. This may be only because there are so few sites spanning the last hundred thousand years or more. The best evidence comes from Lynch's Crater on the Atherton Tableland in north-east Queensland, where Kershaw (1986) found an unprecedented increase in charcoal about 40 000 years ago that may well have resulted from increased incidence of fire due to Aborigines. At the same time, rainforest Gymnosperms almost disappeared, some, for example *Dacrydium*, becoming extinct in Australia.

The last 40 000 years

The last 40 000 years have seen major changes worldwide as the environment became colder and drier towards the last glacial maximum, about 20 000 to 16 000 years ago, and then warmed to

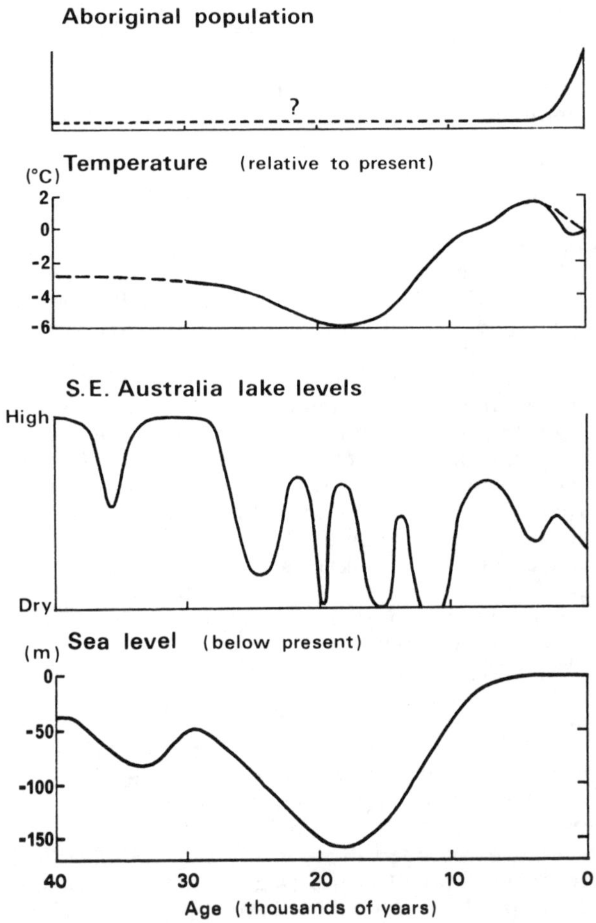

FIG. 2. Relative changes over the last 40 000 years in Australian Aboriginal population, temperature, lake levels and sea level. Adapted from Wasson (1988).

the present interglacial (Fig. 2). Changes through the last 10 000 years, the Holocene period, have been small compared with those which came before (Bowler *et al.* 1976). Changes in the environment since European settlement, with the exception of urban areas and lands used for intensive agriculture, are not of the order of magnitude of those in the longer term, although the rate of change is greater and many recent changes are of a different kind and probably irreversible.

Sea level fell to its lowest about 18 000 BP, then rose to its present level by 6 000 BP (Thom & Chappell 1975). As the climate ameliorated following the last glacial, rainforests expanded into areas occupied by drier vegetation types (Ash 1983, 1988; Kershaw 1986) and, in southeastern Australia, trees spread into grasslands and shrublands (Hope 1984, 1989). The first postglacial woodlands in some cases were dominated by *Casuarina* or *Callitris*, *Eucalyptus* replacing them as late as 6 000 BP (Clark 1983b; Luly, pers. comm.). Most of

Australia's forests and woodlands were established, or re-established, within the last 10 000 years and are not *in situ* ancient ecosystems.

Figure 2 highlights our ignorance of Aboriginal population changes over the past 40 000 years, apart from a widespread increase in population in mid- to late-Holocene times (Lourandos 1983; Ross 1985). We can speculate about how climatic change, and consequent changes in resources, might have affected the size and distribution of Aboriginal populations (Head 1987), but there is little hard evidence beyond the last 10 000 years. We know virtually nothing of the dynamics of Aboriginal social organization, technologies, use of the land and effects on the environment. We cannot assume that Aboriginal society remained static over tens of thousands of years, nor that it was uniform throughout the continent. The longer-term historical record suggests that climate remained the major determinant of vegetation distribution and change (Clark 1983a) but, as Macphail (1980) suggests, Aboriginal land use, particularly burning, could affect the rate of change by opposing or reinforcing the climatically determined direction of change.

The rapid increase in Aboriginal populations over the last few thousand years has been attributed to favourable climatic change and the resultant increase in resources, or to social change, such as a shift to small tools or a higher level of social organization (Lampert & Ross 1989). As with the arrival of Aborigines, the environmental evidence for intensification of land use resulting from rapid population increase is not immediately obvious (Head 1983). One basic difficulty is distinguishing anthropogenic change from other kinds of change in the environmental record. What is certain is that the landscape the first Europeans encountered was a "cultural" landscape, if not as affected by human use as European landscapes were.

European settlement

Europeans arriving in Australia over the last few hundred years could be regarded as another wave of boat people, the first of whom were Aborigines and the most recent Vietnamese (Thorne 1980). But European settlement of the continent was different. A whole civilization with all its trappings was rapidly imposed on an entire landscape and people. The extent of the changes might not, as yet, be as great as those caused by past major climatic changes through glacial and interglacial cycles, but the rates of change are unprecedented, as are many of the types of change. Further, many of the land-use techniques developed over centuries in Europe were inappropriate to the Australian environment, leading to failure of many early agricultural enterprises and, in many instances, rapid and possibly irreversible environmental damage. This process continues; we are still learning and still making mistakes. Two factors in particular have been difficult to accept, accommodate and adjust to. Over most of the continent, soils are old, shallow and infertile and rates of soil formation at present are extremely low; and climate in most regions is harsh, extremely variable and largely unpredictable from season to season and year to year. In particular, rainfall is markedly seasonal over much of Australia and, in most years, there is less than the mean annual rainfall. These points are elaborated by other authors in this volume.

While rates of environmental change have altered since European settlement and new factors, such as weeds, hoofed domesticated animals and rabbits have been introduced (Adamson & Fox 1982), the processes acting in the environment remain unchanged, as do chains of causation and interaction in complex and dynamic ecosystems. The impact of Europeans on the landscape has varied enormously as has the timing of those impacts. Only the Port Jackson region was affected in the early years of settlement, from 1788 onwards, and the greatest impact over most of the continent did not occur until the latter half of the nineteenth century or well into the twentieth century. Parts of the environment have been totally transformed, while others have been, so far, relatively little touched. For the future, in areas not alienated for towns and cities, we need to distinguish between management of natural ecosystems, as reserves or with non-intensive exploitation, and management of agricultural and plantation lands (see Hobbs & Hopkins, this volume).

What use is history?

All this is essential background for environmental managers if they are to understand the context in which they are working, but, beyond that, is it useful for making decisions affecting land management now and over the next two hundred years?

Before this question is addressed, some definition of general terms applied to time-scales in this chapter is necessary. The long-term record refers here to a record longer than the last thousand years; the medium-term record to one covering the last two hundred to one thousand years; and the short-term record to one that spans less than the last two hundred years, the period of European settlement of Australia. This is a nested series in time, beginning with the present-day and extending for different distances into the past. For management, the continuity of past, present and future is most important, although windows in the past can throw light on some questions or provide past analogues of the present or possible futures. It must be said that most of the long-term record is not directly and immediately applicable, for reasons that will be dealt with subsequently, but, indirectly, it can be of great value. In recent years many palaeoecologists have reversed the uniformitarian principle of using present environments to interpret the past and, instead, use past environments to throw light on the present and future (Doe 1983; Wasson & Clark 1985). Secondly, the past can tell us how ecological processes work, and ecological theories can be tested by studying past environmental changes, particularly vegetation changes (Walker 1970; Davis 1981; Walker & Chen 1987; Dodson 1988). It is essential to note here Passmore's (1974) distinction between problems in ecology and ecological problems. Problems in ecology are those defined within or by the discipline itself, while ecological problems are those that we perceive in the environment and that may require action to resolve or remedy. As Oldfield (1987) observed, what environmental history can tell us about the former may not be directly applicable, but what it can tell us about the latter may be critical for our understanding of the present and future.

To deal with ecological problems, the most compelling reasons for studying environmental history arise from the following considerations.

(a) Many problems are inherently historical, for example, the impact of European settlement on pre-existing Australian environments. While written and ethnographic records can provide much information about what was here when Europeans first arrived, reconstructions of the medium-term past are needed in order to perceive the kind and magnitude of induced changes.

(b) Instrumental records are invariably too short to predict the return periods of less frequent or rare events, particularly those which are of greatest magnitude. This applies to disturbances such as floods, fires or cyclones, as well as to stresses applied over longer periods such as droughts.

(c) Many phenomena have been recorded rarely, or discontinuously in space and time, or not at all; for example, stream sediment loads or species composition changes in aquatic ecosystems. Often, monitoring is only initiated when something is perceived as a problem or becomes a political issue. The environmental record can be used to reconstruct the history of such variables or events, both quickly and economically. An example is provided by current research in Britain, which demonstrates that power station emissions have been responsible for acidification of lakes (Battarbee *et al.* 1988).

(d) The environment is a complex and dynamic system, with many interacting processes. The past can provide a record of rare, and often critical, concatenations of variables or events, such as droughts, frosts, fires, floods, vegetation, land use and soil erosion. The historical record can be used to distinguish events, or sequences of events, that may not be of the greatest magnitude but may be of most significance in their effects on species, communities or ecosystems. For example, a single frost in a drought might have a far greater effect than frost or drought alone.

(e) While ecosystems might remain stable for long periods, there may be considerable shorter-term variation within those periods. To manage ecosystems, we need to know the limits of reversible variation and, if possible, the thresholds for irreversible change and the likely agents of change. The time-span required to cover the range of variability depends on the longevity of dominant or critical species. For forests with trees that live for hundreds of years, only the long-term environmental record can provide this information. Adaptive and, in particular, experimental environmental management (Holling 1978; Hopkins & Saunders 1987) can provide some information on limits and thresholds in ecosystems, but to rely only on the effects of management is to ignore the vast storehouse of experiments that have already been conducted in the past (Deevey 1969), on time-scales appropriate to the ecosystems being managed.

(f) Short-term variability may mask long-term trends and only the historical record can provide this information, particularly in an environment such as Australia where short-term climatic variability is so high.

(g) Predictions from modelling and time series analysis will prove inadequate, or even dangerously misleading, if boundary conditions are changing. We need evidence for rates and directions of change and how these have altered over time in, for example, climate, water quality and species composition of terrestrial ecosystems.

(h) If, as predicted, the earth experiences rapid, global environmental change over the next century, any changes will be superimposed on pre-existing dynamics, not a static world. To predict effects of global change, we need to know what the dynamics are, including longer-term trends, limits of variability and the frequency and magnitude of rare events.

Historical methodology

Figure 3, adapted from Oldfield (1983) suggests the time spans of: (a) levels of organization within the environment, from individuals to the biosphere; (b) affective times of major agents of environmental change; and (c) methods of study. Delcourt *et al.* (1983) present a more comprehensive and complex diagram of temporal and spatial scales of disturbance regimes, responses and vegetation patterns, but this complexity is unnecessary for the present argument.

Of the levels of organization, the ecosystem is the most difficult to circumscribe because its limits are defined by the researcher. The time span of

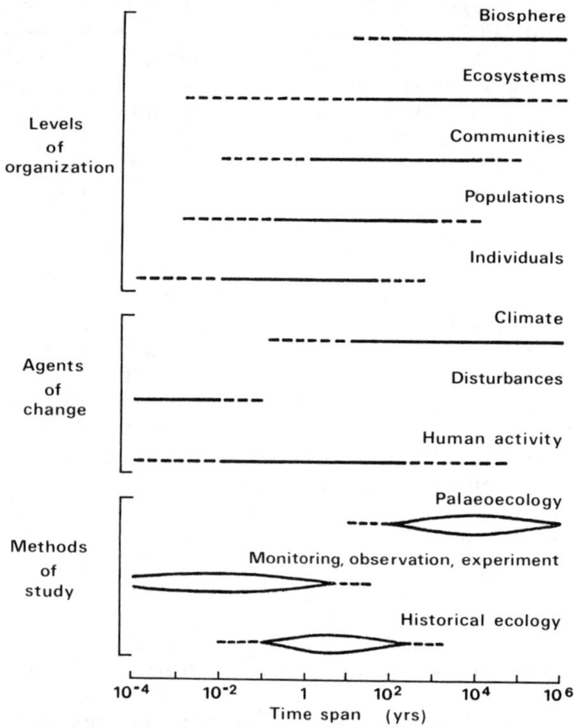

FIG. 3. The time spans of stability at different levels of ecological organization down to individuals and of affective times of the main agents of ecological change as compared with the time spans covered by different methods of study. Based on Oldfield (1983).

greatest human environmental impact overlaps with those of long-term climatic change and short-term disturbances such as fires, floods or frosts. Of the methods of study, the two most commonly used, traditional palaeoecology on the one hand and monitoring, observation and experiment on the other, hardly overlap. More importantly, the gap between them is at just the time span covered by the most intensive recent human impact worldwide. The method of study needed, then, is historical ecology, reconstructing the history of the environment on a time scale of years to centuries, with the appropriate resolution. Figure 4 adds a spatial scale to the temporal and suggests approximate boundaries between the different fields of study. Ideally, there should be no boundaries. Historical ecology, or recent palaeoecology, completes a continuum from experiment, observation and monitoring through to long-term palaeoecology. This needs stressing while palaeoecologists seem to believe that anything that has happened in the recent past is of little interest compared with what went before and while ecologists tend to overlook the processes that have shaped the present environment.

The three types of evidence used to reconstruct the past are written and oral records, instrumental records and the record contained or preserved within the present environment. Each kind of evidence has its strengths and limitations, so all three should be used, where possible, to give the most complete and accurate reconstruction (Oldfield & Clark 1989). For Australia, the known written record began in the early sixteenth century, but until 1788, remained extremely spasmodic. Unlike parts of Asia, north Africa and Europe, Australia has a very short written history. The instrumental record of climatic and hydrological variables is even shorter, with the exception of the Sydney region, and extremely patchy in both time and space.

Less easily interpreted is the environmental record, but a variety of techniques can be used from a range of scientific disciplines. The landscape of today, and the vegetation in particular, can tell us much about the past that shaped it, but the most complete record comes from undisturbed sediments or peats in lakes, bogs, swamps, estuaries, reservoirs and dams. If the sediments are unmixed, they can provide a continuous record from the oldest at the base to the most recent at the top. It is the sedimentary record that will be emphasized here.

Scales and resolution

The critical problem in applying palaeoecology to management is with spatial and temporal scales and resolution. At the extremes, we could ask how to reconcile patch dynamics with global environmental change, or the one to two million years of

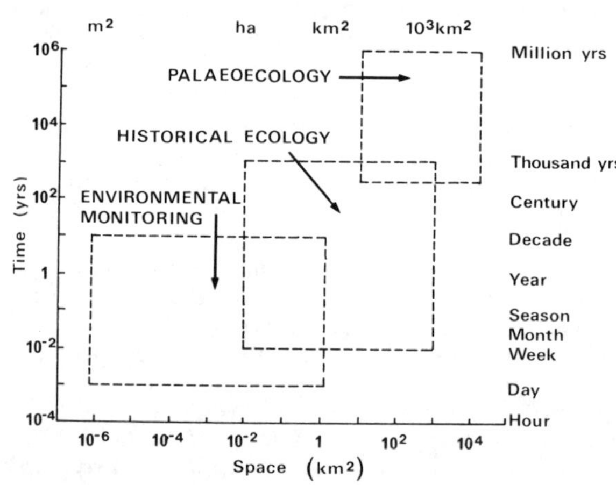

FIG. 4. The temporal and spatial scales over which methods of ecological study usually operate. In this chapter, the short-term environmental record is defined as one within the last 200 years, the period of European settlement of Australia; the medium-term record covers the last 200 to 1 000 years; and the long-term record is one that spans more than the last 1 000 years.

the Quaternary with the next one or two centuries. Questions of scale are discussed by di Castri and Hadley (1988), who point out that technological, societal and conceptual developments over the last few decades have led to increased concern with larger spatial, but shorter temporal scales. Interest in global problems necessitates scaling up in both space and time from the more usual levels of enquiry, but the profound and far-reaching effects of human activity over the last century have so altered the environment that, for many questions, only the shorter time scales are directly relevant. Di Castri and Hadley also stress the usefulness of switching scales, including those of time and space, when studying multi-layered natural systems, moving up one level from that in which a phenomenon occurs to understand constraints and down one level to understand causality.

Figure 5 plots time scale against the resolution obtainable from the environmental and, particularly, sedimentary record. Resolution is usually between one and two orders of magnitude less than the time span covered, with finer resolution sometimes obtainable. The limit to the resolution that can be achieved depends on sedimentation rate and whether mixing has occurred or not. For very low sedimentation rates, the minimum sample size needed to obtain statistically significant results may contain sediment deposited over hundreds or thousands of years. For environmental managers interested in change or stability on a scale of decades to one or two centuries, deposition sites are needed with a resolution of from months to decades, or finer if possible, over the same time span in which they are interested. Because of the lack of resolution at an appropriate scale, most of the long-term environmental histories, dealing with thousands to millions of years, are of no direct relevance to environmental management, while fine-resolution records of the more recent past are. Often it will not be possible to obtain sediments with the required resolution over a long time span, but there are many sites that can yield at least part of the record, usually the most recent.

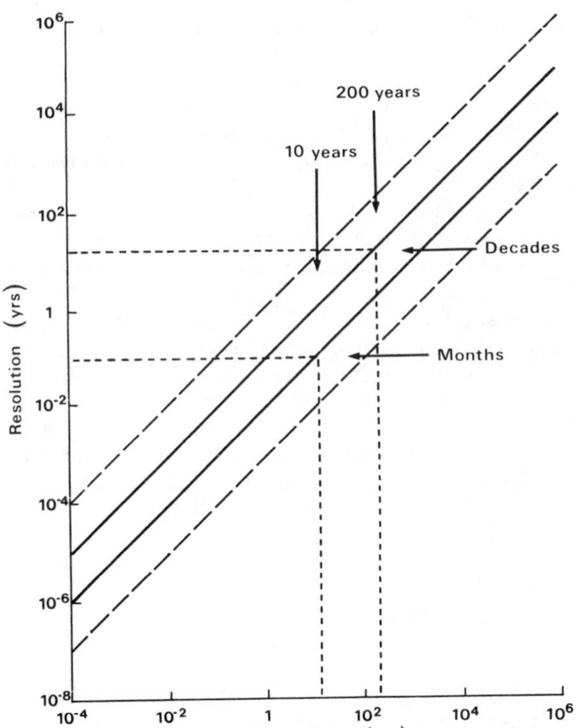

FIG. 5. The relationship between resolution and time scale. In the environmental or sedimentary record, the resolution obtainable is usually one to two, and occasionally three, orders of magnitude smaller than the period covered. For management, the required resolution is usually one to two orders of magnitude less than the period of interest.

The long-term record and environmental management

The long-term environmental record, on a time scale of several thousand years, provides the context for recent history, a temporal sequence of interactions and effects of environmental processes, and analogues for present and future environmental conditions. Ecological theory can be tested by looking at experiments already conducted in the past (Deevey 1969). This time scale falls within the concern of traditional palaeoecology, but scientists often ignore the implications of historical reconstructions for managing present and future landscapes. By asking management questions at the outset and by developing appropriate methodology, palaeoecology can go far beyond mere description of past environments and past environmental change. The history of the Magela floodplain, in Kakadu National Park in the Northern Territory, provides an example.

The Magela floodplain

The main aim of the research on the Magela floodplain (12°25′S., 132°50′E.), was to predict the likely fate of uranium mill tailings should they be stored above ground when the Ranger mine ceases operation (Pickup *et al.* 1987). Environmental history can contribute to decisions about storage of tailings in two principal ways. The long-term record can be used to improve predictions of environmental conditions over the next thousand years or so, so that some allowance for likely environmental change could be built into the tailings containment structure, and the medium-term record can be used to predict where any escaped tailings might be deposited, and in what concentration. Analysis of the sedimentary record provides information relevant to other environmental problems because the results shed light on ecological theory and can be used for management of wetlands in the monsoon tropics.

Sediments were sampled over the length and breadth of the Magela floodplain (Fig. 6) and many of their physical and chemical properties have been determined, along with analyses of inclusions such as pollen and diatoms (Clark & Guppy 1988). Radiocarbon dating has been used extensively. Before about 7 000 years BP, the present floodplain was a broad shallow valley with the Magela flowing through in a sandy channel. As the sea level rose following the last glaciation, the valley was gradually inundated by the sea, tidal limits reaching as far as a kilometre above Mudginberri (Fig. 6.). The Magela embayment became a backwater swamp as the levee of the East Alligator River built up and an extensive mangrove forest, consisting mainly of *Rhizophora* spp., was established.

About 3 000 BP, other mangroves, particularly *Avicennia*, began to replace *Rhizophora* and, over most of the floodplain, all mangroves disappeared by about 1 500 BP as freshwater grasses, sedges and paperbarks became established (Fig. 7). The change from *Rhizophora* to dominance of other mangroves and the shift to freshwater vegetation occurred at about the present-day tidal limits of *Rhizophora stylosa* and *Avicennia marina* respectively, but the change from *Rhizophora* to several other mangrove species suggests the creation of a range of new tidal environments, or areal extension of existing ones, by channel cutting and channel shifts (Clark & Guppy 1988). Lees and Clements (1987; Lees, pers. comm.) have evidence from across the north of Australia that suggests decreased wet season precipitation between about 2 800 BP and 1 600 BP and this might well explain the downcutting of channels in the Magela sediments.

Several conclusions for ecological theory can be drawn from this long-term record, on a scale of thousands of years and resolved to decades.

(a) Mutually exclusive ecosystems replaced each other, but changes in the physical environment were necessary before the vegetation changed.

(b) Each stage of the transition from *Rhizophora* forest to freshwater wetland could be reversed, and usually was once or twice, before the final shift. This is argued more fully in Clark and Guppy (1988).

(c) Long-term ecosystem stability contained short-term variability; the concept of stability is dependent on spatial and temporal scale and resolution.

(d) The period of least environmental stability contained the greatest species diversity, due to the creation of a small-scale mosaic of different environments supporting distinct vegetation types.

(e) Transitions from one ecosystem to another occurred relatively rapidly in what might be termed punctuated succession. The model that best fits the sequence of ecosystem change on the Magela floodplain is that of Laszlo (1972; Oldfield 1983) with variation around a mean in a stable state and change to a new and different stable state triggered by some external factor, in this case changes in the physical environment related to sea level and climate (Oldfield & Clark 1989).

Changes in the dominant pollen taxa through the freshwater wetland over the past 1 500 years at one site are plotted on a ternary diagram in Figure 8a. This shows that the present-day vegetation at the site is very similar to that at about 1 500 BP and for much of the time since. There have been fluctuations in the proportion of Myrtaceae pollen and one period when Cyperaceae increased markedly. From a grid of surface samples distributed over the floodplain it was found that the proportion of Myrtaceae pollen is significantly correlated with distance to the nearest paperbark (*Melaleuca* spp.). This means that most Myrtaceae pollen in the floodplain sediments is from *Melaleuca* spp. and that local pollen production dominates, with minimal input from terrestrial environments surrounding and upstream of the floodplain. The transfer function derived from the relationship between *Melaleuca* pollen and distance (Fig. 8b) can be applied to the ternary diagram and suggests that the distance to the nearest *Melaleuca* at this site has varied between about 50 m and 750 m over the past 1 500 years.

There are several conclusions from the floodplain history that are important for management, as well as those drawn above from the longer term record.

(a) Observed and recorded present-day freshwater wetland variability between seasons, years and decades, is within the range experienced over the past 1 500 years.

(b) A sea level rise of 50-100 cm would inundate the floodplain with sea water at high tides and allow *Avicennia* to grow, but a rise of 200 cm would be needed to re-establish *Rhizophora* forest. Predicted sea level rise due to the greenhouse effect over the next century is between 20 cm and 140 cm (Bolin *et al.* 1986); if this occurs, then many of the coastal freshwater wetlands in northern Australia will revert to mangroves.

(c) Buffalo damage to banks of tidal channels towards the mouth of the Magela may lead, or may have led, to local salinization, but cannot cause widespread salt water incursions throughout the Magela floodplain.

(d) A change in geometry of the Magela outflow, due to extreme floods, tidal surges or other effects of cyclones, might result in increased drainage and a new phase of channel incision or increased tidal influence, but this would only affect the downstream reaches of the floodplain.

For management of uranium mill tailings, the local long-term history needs to be supplemented by climatic histories of other sites in the region, in

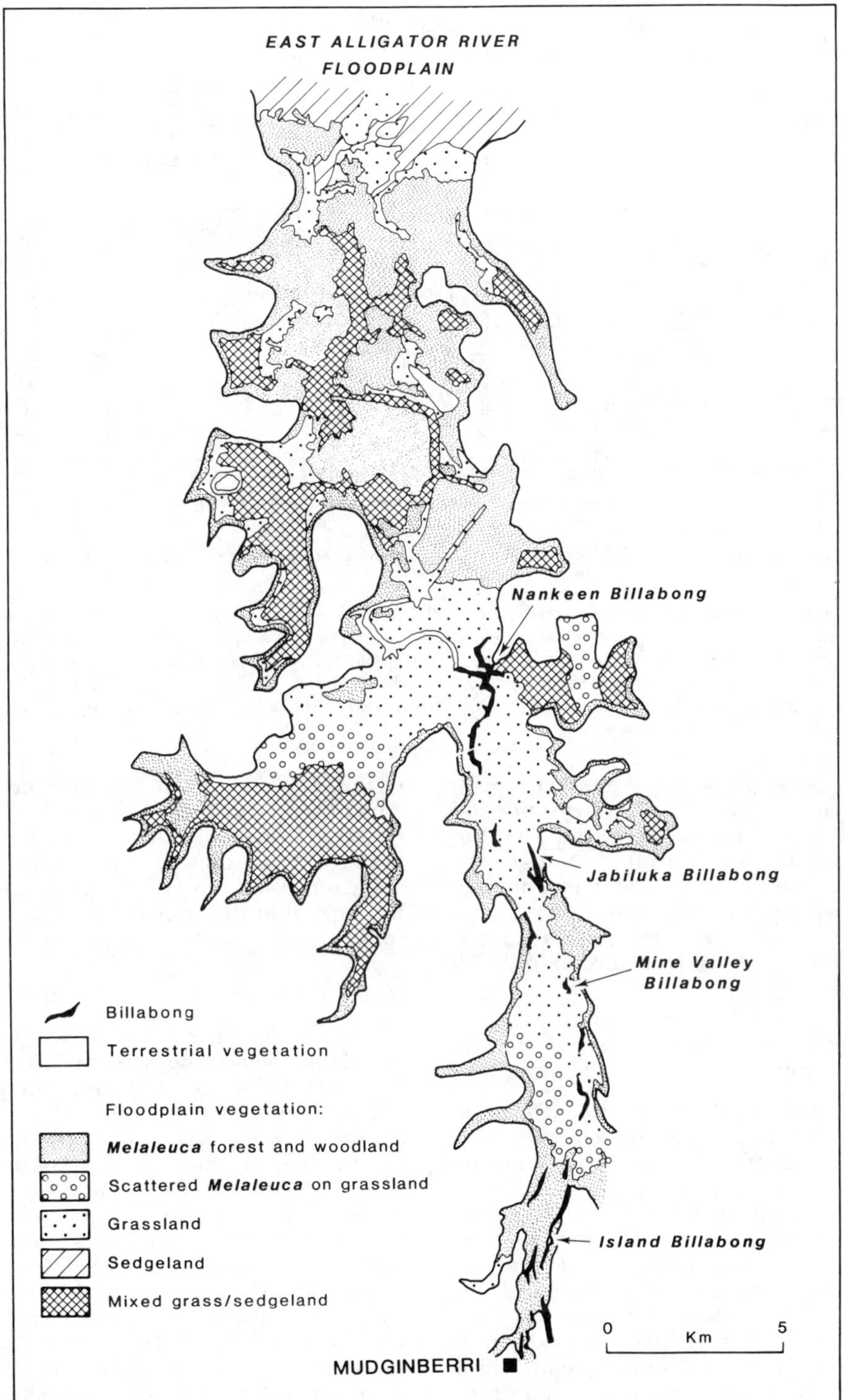

FIG. 6. A map of the Magela floodplain showing the main billabongs and the distribution of vegetation. Adapted from Finlayson *et al.* (1985).

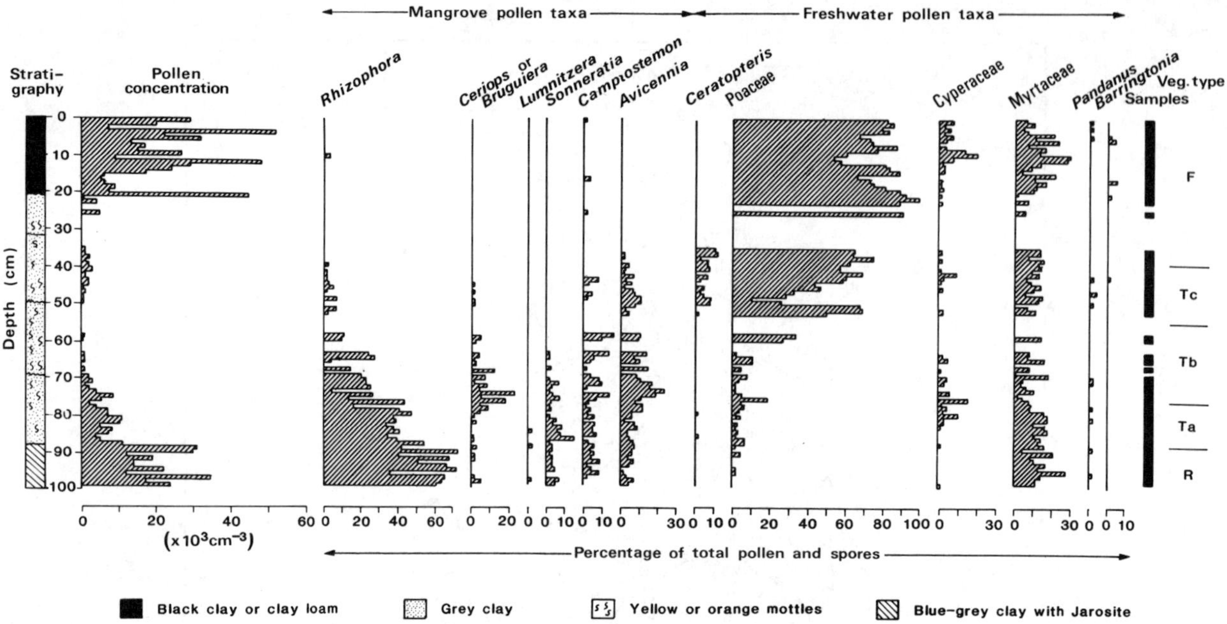

FIG. 7. The change from mangrove forest to freshwater wetland on the Magela floodplain as shown by the relative pollen content of the sediment. Where there are no samples, the sediment was oxidized and too few pollen grains remained. Vegetation types are, from the earliest at the base of the section upward: R, *Rhizophora* forest (to *ca.* 3 000 BP); T, transition, dominated by mangroves other than *Rhizophora* and divided into three phases (*ca.* 3 000-1 500 BP); and F, freshwater wetland (*ca.* 1 500 BP to present). This is the upper part of a core taken near Mine Valley Billabong (Fig. 6). Adapted from Clark and Guppy (1988).

Australia and around the world. Then better estimates of probabilities of future environmental changes can be made and of likely effects on the landscape and on any above-ground tailings containment structure. When the research is complete, changes in sediment influx and deposition patterns through the freshwater wetland period should allow the prediction of likely future deposition sites.

Comparisons before and after European settlement

The Magela Creek study seeks to determine longer-term trends and provide a baseline description of the environment and, more importantly, its dynamics. Against this background, better predictions can be made of the effects of future land uses, and changes can be measured as they occur. Magela Creek is in an area where the greatest impact of European settlement is yet to come, from tourism, mining and townships. Other parts of the country have already been changed radically and, in these areas, the questions asked of the historical record might be about detecting long term trends, as on the Magela; looking for past analogues of present-day conditions, events or concatenations of events; or measuring changes in rates of processes before and after European disturbances. As an example of the second, a fire history from the Australian Capital Territory will be used, and of the third, erosion rates in western New South Wales. These examples lead back into questions of scale and resolution, which are further exemplified by vegetation histories from Tasmanian rainforest soils.

Rotten Swamp

An intense bushfire swept across Rotten Swamp in the Brindabella ranges in the Australian Capital Territory (35°45′S., 148°54′E.) in January 1983, following a severe drought. Many hummocks of *Sphagnum cristatum*, which formed an ombrotrophic bog on part of the swamp, were burned. *Sphagnum* bogs occur in the headwaters of many water supply catchments in the mountains of south-eastern Australia and, by impeding and regulating flow, they minimize downstream channel erosion and the consequent deterioration in water quality in reservoirs. Rotten Swamp is one of several such bogs in the Cotter River catchment for Canberra's water supply, and the catchment managers wanted to know whether such a devastating fire had occurred before and, if so, how long it took for the bog to recover.

Figure 9 shows charcoal and pollen from a section through the peat; methods, inferences and

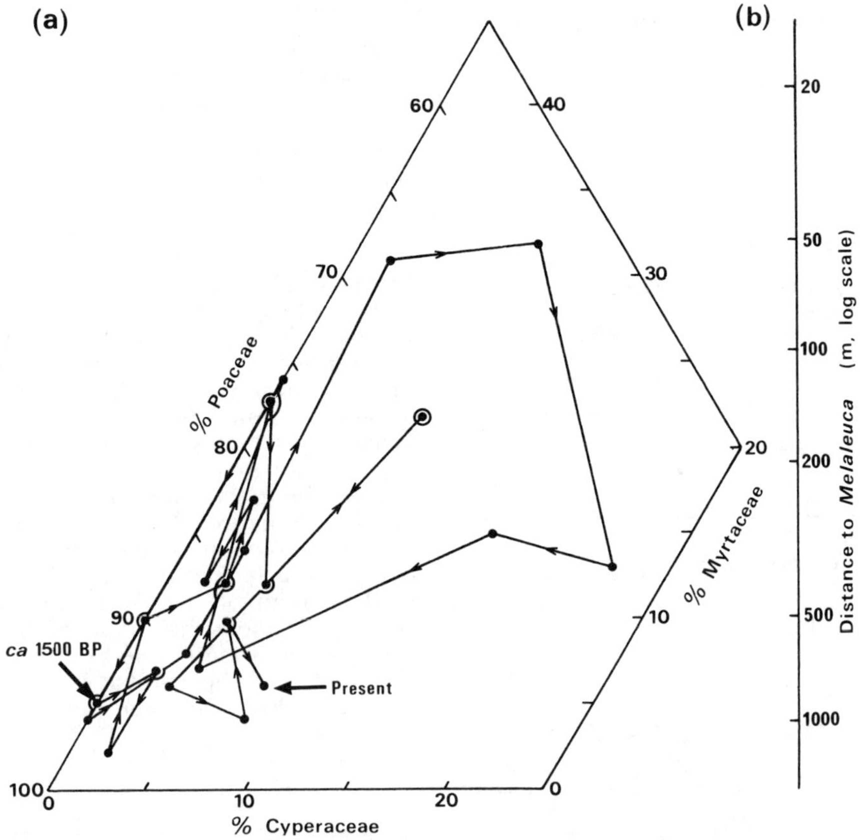

FIG. 8. (a) A ternary diagram showing relative changes in the three dominant pollen taxa through the freshwater period (*ca.* 1 500 BP to present) in the core shown in Figure 7. (b) A scale of the transfer function relating percentage Myrtaceae pollen to distance to the nearest *Melaleuca*, derived from analysis of a grid of surface samples taken over the whole floodplain (n=105; r=0.77).

conclusions are described by Clark (1986a, 1987). The charcoal curve shows a peak late last century, corresponding with the time when European stockmen used the high country for summer pasture, burning the vegetation when and where they could to provide fresh green feed (Banks 1982). As fire prevention and suppression measures were introduced early this century, the incidence of fires decreased and this is reflected in the charcoal. There are no charcoal bands or obvious discontinuities that would suggest that the peat itself had ever burned, nor high concentrations of charcoal before European settlement that would indicate fires burning *Sphagnum* hummocks. The pollen and spores show the bog becoming drier towards the surface, after a wet period through the last thousand years.

While it is likely that there have been droughts and fires at Rotten Swamp over the last 10 000 years, the critical post-settlement change was to the hydrology of the bog, allowing it to dry out and burn. The changed hydrology was probably caused by the destruction by stock of stream-bank vegetation, including *Sphagnum* hummocks, which resulted in increased drainage from the bog. Wimbush and Costin (1979) have documented similar effects of grazing on vegetation and hydrology in the Kosciusko area. The combination of drought, intense fire and dry *Sphagnum* hummocks appears unprecedented before European settlement. To prevent recurrence, action can be taken to restore stream-bank vegetation and regenerate the bog.

The level of resolution of the pollen and charcoal records from Rotten Swamp differ because of the sampling techniques used. For pollen, samples were taken at intervals down the core, while charcoal was sampled continuously. This sampling scheme was chosen to minimize the time required to answer the questions asked, but other questions about vegetation change or succession after fire would require closer sampling for pollen analysis. The number of years within each sample of equal volume increases with depth or with age, due to

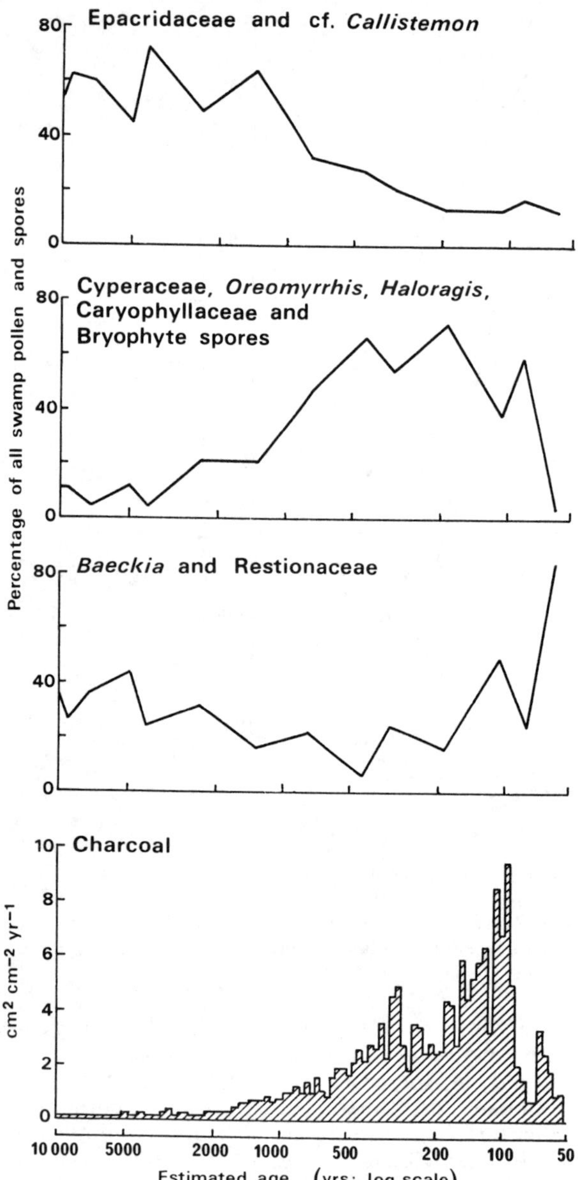

FIG. 9. Changes through time in the proportions of swamp pollen and spore types in a section of *Sphagnum* peat at Rotten Swamp, with mean annual charcoal influx to the same peat section. All samples shown are from peat and not from the partly humified *Sphagnum* grading into the hummock above. No *Sphagnum* spores are included with the Bryophytes as none were found, but *Sphagnum* tissue was identified throughout. Adapted from Clark (1986a).

compression and differential preservation of the various organic and inorganic components of the peat. Thus, the resolution obtainable for the recent past at this site is far greater than that for the more distant past.

Umberumberka Reservoir

One of the worst consequences of inappropriate European land use has been wind and water erosion of soil. Australian soils are usually thin, infertile and forming at very slow rates. Wasson and Galloway (1986) calculated pre- and post-settlement water erosion rates in the Barrier Range near Broken Hill. They made their estimates from the amount of sediment deposited in Umberumberka Reservoir (31°50′S., 141°15′E.), built between 1913 and 1915, and in dated pre-reservoir alluvial deposits where Umberumberka Creek formerly disgorged onto the Mundi Mundi Plain. Post-settlement erosion rates were about 50 times higher than in the previous 3 000 years and twice as high as in the 3 000 years before then (Fig. 10). While erosion rates have dropped since the 1940s, this is of little comfort in a region where most of the remaining soil mantle is from aeolian accession and soil formation in the catchment is negligible.

The resolution obtainable in pre-reservoir deposits was limited to two periods of 3 000 years by the scarcity of material for radiocarbon dating. By contrast, sediment surveys of the reservoir have been made several times since 1915 and deep coring in 1982 showed that 14 m of sediment had built up behind the dam wall, allowing very fine

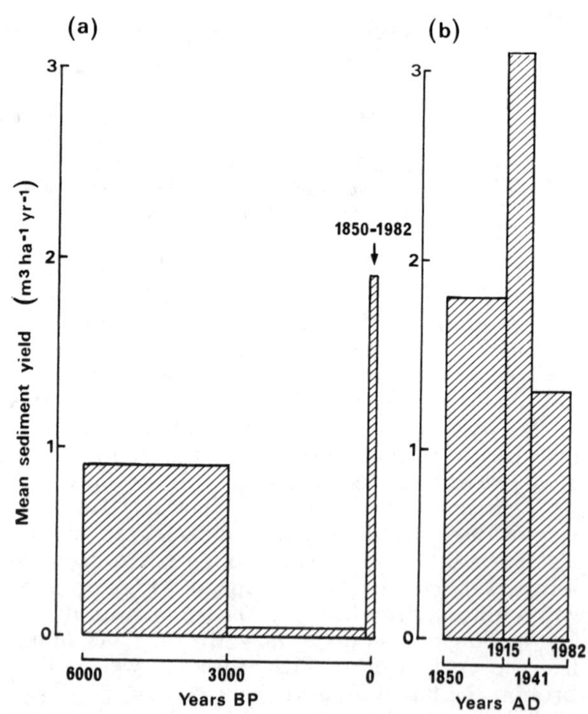

FIG. 10. Changes over time in erosion rates in the Umberumberka Creek catchment: (a) 6 000 BP to the present; (b) 1850-1982 AD. Adapted from Wasson & Galloway (1986).

resolution studies. Most of the sediment consists of the aeolian fraction of the catchment soils, the silt-sized clay pellets, or parna, blown in at different times during the last 30 000 years or more (Chartres 1983; Wasson & Galloway 1986). The sediment is deposited in units of centimetres to tens of centimetres thick, probably representing individual inflow events, with small, highly organic layers between, containing autochthonous organic matter and aeolian deposits of inorganic particles and pollen grains. Figure 11 shows changes in water content, anhysteretic remanent magnetization (ARM; an indicator of fine-grained magnetite and maghemite), pollen concentration and the relative frequencies of major pollen types through one of these inflows. Water content, ARM and pollen concentration indicate the input of fine silt early in the event, followed by deposition of clay. Where deposition basins have inflowing streams, waterborne pollen from the catchment area is far more abundant than airborne pollen (Clark 1986b). Pollen composition may therefore indicate the source of sediment within the catchment, slugs of sediment coming from different distances, different areas, different vegetation types, or different depths in the soil profile. Particle size and pollen concentration and composition may well distinguish between erosion of surface soils and of channel banks. In this inflow event (Fig. 11), tentatively dated to February, 1928, changes in the proportions of different pollen types suggest changes in sediment source areas, early input being from the edges of the reservoir and its exposed sediments, and later input from areas upstream. The later sediments are probably mixtures of surface soils, with abundant pollen, and old material eroded from channel banks, with no pollen.

The resolution through single events obtainable at such sites as Umberumberka is greater than that usually aimed for or achieved by environmental monitoring programmes. At Umberumberka, there is a record of 80-100 sedimentation events in sequence, which can be analysed at several levels of resolution, from within events to seasons, years or decades, showing both short-term changes in response to rainfall and longer-term trends due to climate and stocking rates.

Tasmanian rainforest peats

It is also possible at some sites to achieve a fine spatial resolution. For aquatic ecosystems, sediments in shallow lakes, reservoirs or estuaries can be used. In ombrotrophic bogs such as Rotten Swamp, the peat preserves pollen and other organic remains produced at the site as well as atmospheric inputs from short and long distances. In wet forests, mor humus soils contain local pollen (Iversen 1964; O'Sullivan 1973) and sometimes consist almost solely of pollen grains from the plant that was growing on site. Figure 12 summarizes the relative pollen content of samples from two cores through rainforest peats in southern Tasmania, from the Celerytop Islands (43°22′S., 146°09′E.) and near Hastings (43°20′S., 146°50′E.). Both cores show marked changes that reflect the

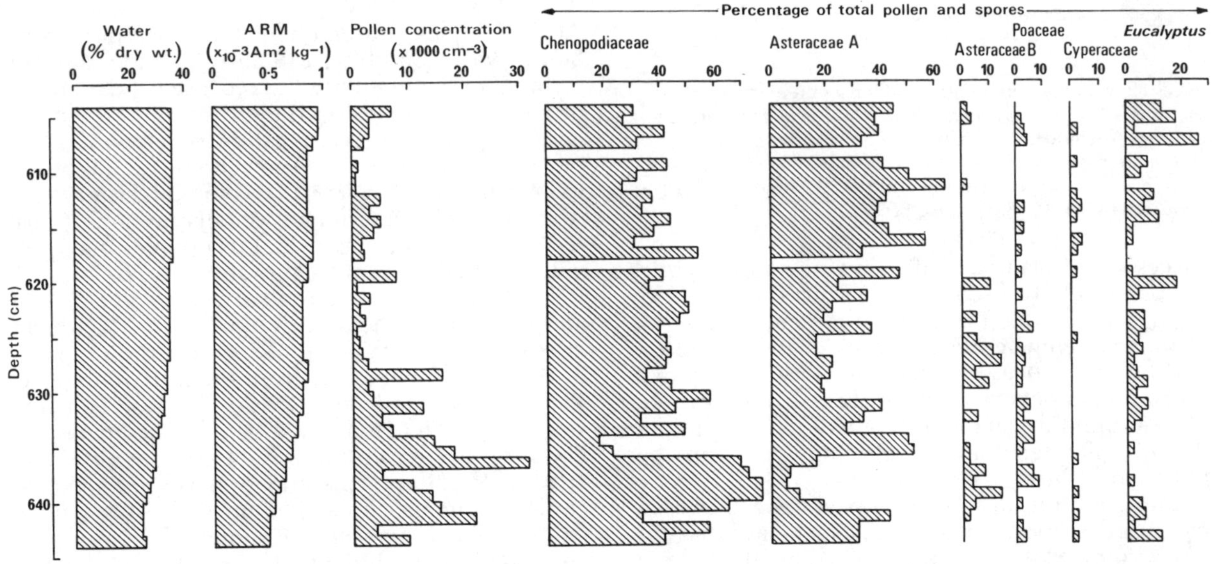

FIG. 11. Properties of the sediment in contiguous samples through a stratigraphic unit from Umberumberka Reservoir, deposited in a single inflow event, possibly in February 1928. The first sediment deposited is at the base and the last at the top. ARM is a magnetic measurement that reflects iron mineralogy and magnetic domain size, here related to particle size.

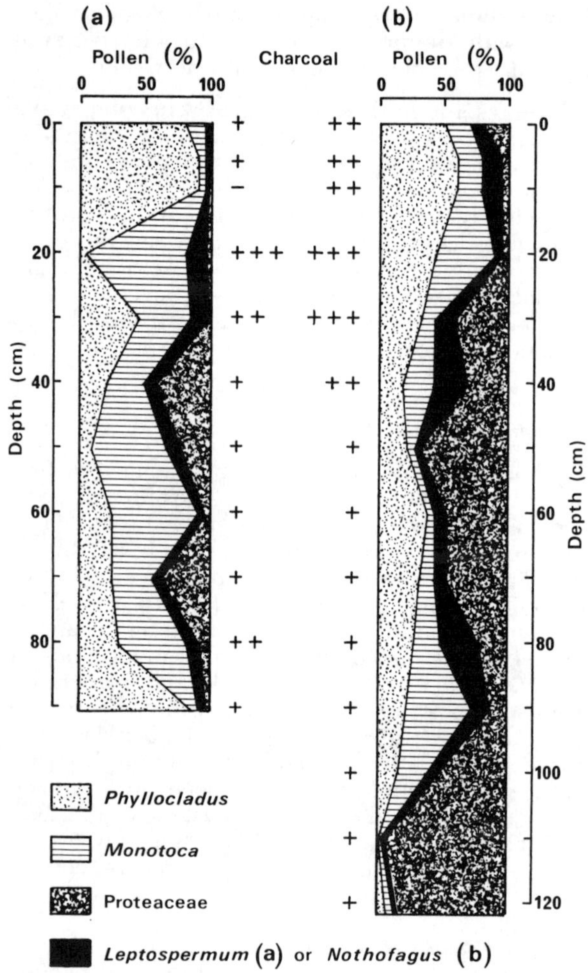

FIG. 12. Changes in relative pollen content through two sections of rainforest peat from southern Tasmania: (a) Celerytop Islands; (b) Hastings. The relative abundance of charcoal in the samples is indicated by crosses. A radiocarbon date from 120 cm deep in the Hastings section was about 12 000 B.P. (F. Podger, pers. comm.).

succession of plants actually growing on site. A grid of cores could be used to reconstruct the dynamics of a patch of rainforest over time and relate the changes to climate, disturbances, plant succession and human activity. In this case, very high spatial resolution is possible, but, because of the low rate of peat build-up (about 100 years for a centimetre depth) the temporal resolution that can be obtained is fairly coarse.

The questions one would expect to be answered by fine temporal resolution such as that at Umberumberka are different from those one would ask of a site such as those in Tasmanian rainforests with fine spatial resolution. At the ideal site, both would be possible.

The last 200 years

Reservoirs and other water storages such as farm dams and tanks, throughout the country, are sites of sedimentation of waterborne or airborne particles (Clark & Wasson 1986). These man-made water storages supplement the record available from natural lakes, swamps, bogs, estuaries and alluvium. Coastal swamps and estuaries, in particular, may have high sedimentation rates, allowing fine resolution, although spatial deposition patterns may be complex due to the interaction of river flow and tides. Some inland swamps also have the potential to provide fine resolution records over the short-term (Green *et al.* 1988). In Australia, many reservoirs are sited in areas where there are few or no permanent water bodies and reservoir sedimentation rates are often higher than those of lakes or swamps. No existing dam covers the whole 200 years of European settlement, but there are many built more than a century ago, such as Parramatta Reservoir, used by Dolman (1987; Green & Dolman 1988) to reconstruct changes in vegetation in an urban reserve. Umberumberka Reservoir, discussed in the last section, provides a rare opportunity to study past changes in a range of environmental variables over many decades in the arid zone. As a further example, some results of ongoing research at Burrinjuck Reservoir will be given. The potential of sediment studies of both large and small water storages for application to management is discussed by Clark and Wasson (1986).

Burrinjuck Reservoir

At Burrinjuck Reservoir in south-eastern New South Wales (35°00′S., 148°35′E.), a wide range of techniques has been used to reconstruct the effects of land use changes on the environment, particularly on soil erosion and vegetation. The techniques are listed in Table 1 and highlight the interdisciplinary nature of ecological history. No single field of study can answer all questions (Clark & Wasson 1986), but several techniques may provide mutually reinforcing answers. Burrinjuck Reservoir, when full, has a surface area of 57 km^2 and maximum water depth of 67 m. Its catchment area is 13 500 km^2 and water enters the reservoir in three principal arms, from the Murrumbidgee, Yass and Goodradigbee Rivers. The reservoir first stored water in 1912 but it was not filled until a flood of record and unpredicted magnitude in 1925. This flood appears to have resuspended earlier sediments, mixed them with soils eroded from old river terraces, the shores of the reservoir and from the catchment, and redeposited them in a thick layer on the floor of the reservoir. Surviving sediments cover the 60 or more years since 1925 and intermediate dates have been determined by a

TABLE 1. Methods used to reconstruct the history of Burrinjuck Reservoir and its catchment.

A. Historical records
- Land use history
- Documentation of the environment
- Records of events

B. Instrumental records
- Meteorological
- Hydrological

C. Sediment studies
- Seismic sounding
- Mapping
- Sedimentology
- Sampling and stratigraphy
- Dating (^{137}Cs, ^{210}Pb, fires, floods, pollen)
- Chemistry and pollutants
- Radionuclides ($^{210}Pb/^{226}Ra$, ^{137}Cs)
- Mineral magnetics
- Pollen
- Diatoms

D. Catchment studies
- Mapping
- Remote sensing
- Sediment tracing:
  - radionuclides
  - mineral magnetics
  - physical and chemical properties
  - pollen

variety of methods (Clark 1986b; Wasson *et al.* 1987; and Table 1). The resolution obtained varies from seasons to several years at different depths of the sediment. The record from Burrinjuck Reservoir, as well as throwing light on recent environmental history, exemplifies several points made earlier.

Written sources record a major bushfire around the main basin of the reservoir on 24 December 1972. Figure 13 shows a thin section through part of a charcoal band in nearby sediments that was attributed to this bushfire. From the instrumental record, the first significant subsequent rainfall was in February 1973, and this would have washed in the silt and large charcoal particles evident in the lower part of Figure 13. An algal monitoring programme was operating at the time and, in May 1973, the first flush of the diatom *Melosira granulata* was recorded at the same site from which the core was taken (May 1978). Dead *Melosira* sank to the sediment surface and their siliceous shells were preserved, as shown in the upper part of Figure 13. Information from written and instrumental records and from monitoring thus verified the dating of one level in the sediment and gave the dating detail and precision otherwise unobtainable. From the dated sedimentary record we can calculate the amount of deposition and, hence, erosion resulting from the bushfire. Because the sedimentary record extends beyond the monitoring period (1972-1975; May 1978), both before and after, we can place this single event in the context of longer-term trends and of similar events in the past, as well as establish its antecedent conditions and follow its consequences, if any, for years afterwards. The complementarity of historical and instrumental records, monitoring and several kinds of palaeoecological studies is clearly evident.

Figure 14 shows the relative abundances of some diatom taxa in the Burrinjuck Reservoir sediments from 1925 to 1982. Many more were identified (Holland & Clark 1989), but these are the most abundant and demonstrate the limitations of monitoring. If living diatom populations had been monitored for any one, five or ten year period, for example, over the 60 years recorded, it would not have been possible to predict the composition of future diatom populations. There is no obvious simple explanation for the changes over time in the composition of the diatom flora. While the core from which the diatoms were extracted has not yet been dated in detail, it is evident that widespread bushfires in 1939, 1944, 1951-52, 1956-58, 1964-65, 1967-68 and 1972-73 had either no effect on the diatom flora or a different effect each time. This might be related to differences in subsequent rainfall and nutrient input. The increase in *Melosira distans* may reflect the rapid rise in Canberra's population in the 1960s, but so far we have no explanation for the change in dominance from *Melosira distans* to *Melosira granulata*. If the variability and unpredictability evident in Figure 14 results from sequential colonization of a new water body by different diatom species, then even the 60 years of sedimentary record is inadequate for predicting future changes. This will only be possible when, and if, a dynamic equilibrium is reached.

Like other sedimentary sites with a significant inflow of water, waterborne pollen grains washed in from the Burrinjuck catchment with other silt-sized particles predominate over airborne pollen grains blowing in from the region (Clark 1986b). Traditionally, palynologists looked for sites with little or no inflow in order to reconstruct the regional picture, but a great deal more information, and different kinds of information, can be obtained from waterborne pollen.

An example is given in Figure 15, which shows three pollen taxa, Myrtaceae, Poaceae and *Pinus*, expressed as a percentage of all pollen in the sediments for the years 1935 to 1957. The sediments were sampled at contiguous 1 cm intervals and data from all samples within each year were combined to give mean annual figures. Trends over the whole period are apparent, with a decrease in Myrtaceae pollen, mainly from native forests, and increases in Poaceae pollen, largely from pastures, and in *Pinus* as the area of plantation forests increased. As most of the pollen is waterborne, washing in with fine silt and clay, then its composition in any one sample or year should

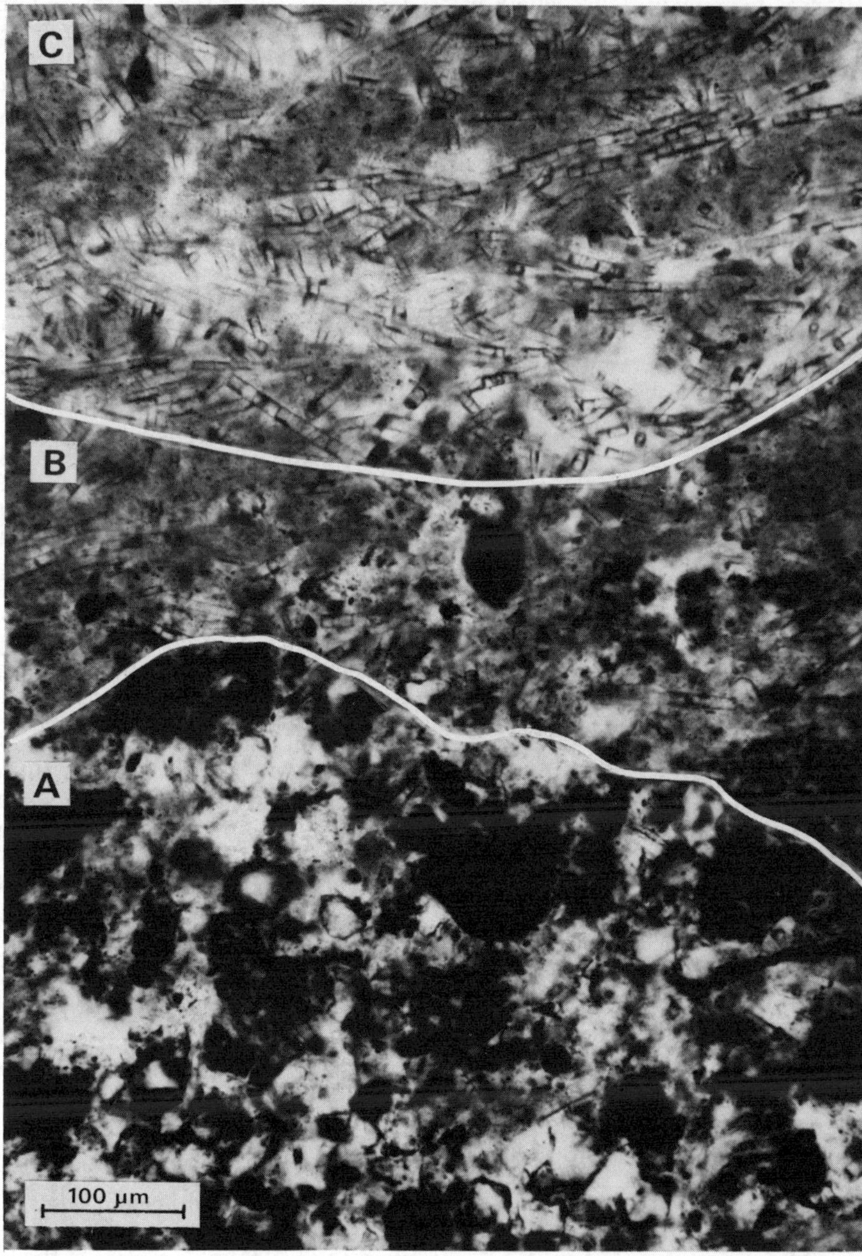

FIG. 13. Part of a thin section through sediments from Burrinjuck Reservoir showing: A, charcoal and silt washed in during February 1973 following a bushfire in December 1972; B, subsequent settling of finer particles; C, diatoms (*Melosira granulata*) from a population increase recorded in May 1973, and probably resulting from the post-fire input of nutrients.

indicate the sediment source. A year-by-year comparison of the three pollen taxa in Figure 15 suggests that there was little sediment from pine forests in the years 1945-1947 but a lot in 1942 and 1956. Pastures were eroded in 1946, 1953 and 1956, and native forests in 1943 and 1955. Most of the sediment deposited in the flood of 1956 came from pastures and pine forests, and not from native forests. Burrinjuck has three major tributary areas: the Yass River catchment, which is predominantly pastures; the Goodradigbee River catchment, which is almost all native forest; and

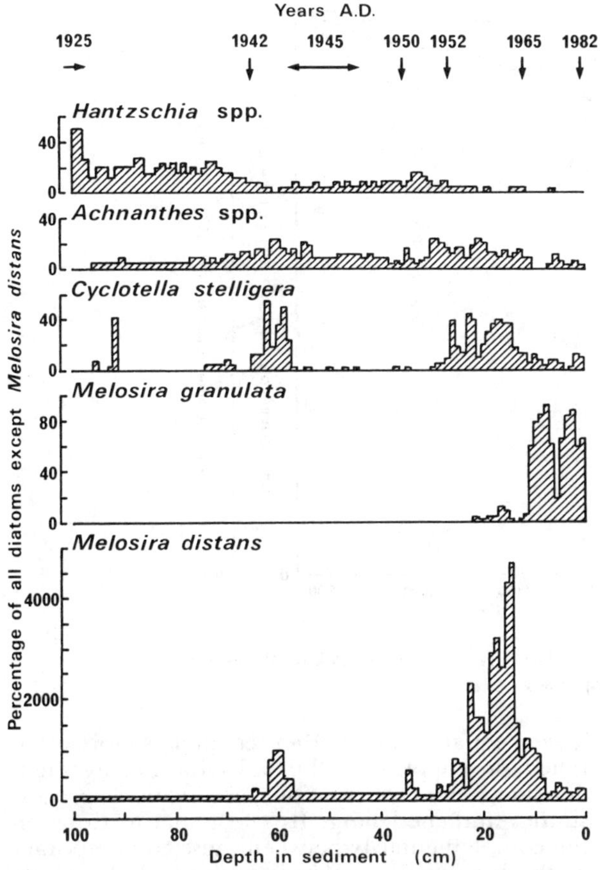

FIG. 14. Changes over time in relative diatom abundances in Burrinjuck Reservoir as recorded in the sediments. Note the change of scale for *Melosira distans*. Data provided by John Holland.

the Murrumbidgee River catchment, which is about half forest and half pasture, but also contains most of the pine plantations and the Canberra urban area.

The link between pollen, vegetation and sediment sources means that the climatic and hydrological record can be searched for identified parts of the catchment, large or small, in specified years to find the most significant rainfall and erosion events and determine their antecedent conditions in terms of climate. Land use and condition at the time can be determined from historical records. Recommendations can then be made for better management of the land, given specified uses and sequences of climatic or hydrologic conditions.

Cumulative sedimentation near the dam wall in Burrinjuck Reservoir up to 1982 is shown in Figure 16. A further 30 g/cm^2 of sediment was deposited in 1983. Three major events brought in 57% of the sediment in the main basin in 5% of the years of deposition since 1925. By far the largest amount was deposited in the major flood of 1925; the two later events, in 1945 and 1983, were both intense, local rain at the end of severe droughts. Extreme erosion in 1945 and 1983 resulted from exceptional conditions antecedent to unexceptional rainfall events. The sediment deposited in the reservoir's main basin was a mixture of soil freshly eroded from bare pastures and ploughed fields in the catchment and resuspended sediment from the dry, exposed upper reaches of the reservoir. If a frequency/magnitude relationship between stream flow and suspended sediment were used to predict sedimentation in the reservoir, it would not have predicted two of the three main deposition events. Once the risks of combinations of droughts, stocking rates, fallowing practices and intense rainfall are better known, the farmer can adjust land management practices to minimize the risk of further soil loss.

Spatial and temporal patterns

Sedimentation rates at Burrinjuck Reservoir have been falling over time, paticularly since the mid 1940s (Fig. 16). This may be due to a number of factors, including better land management, with improved pastures and large numbers of farm

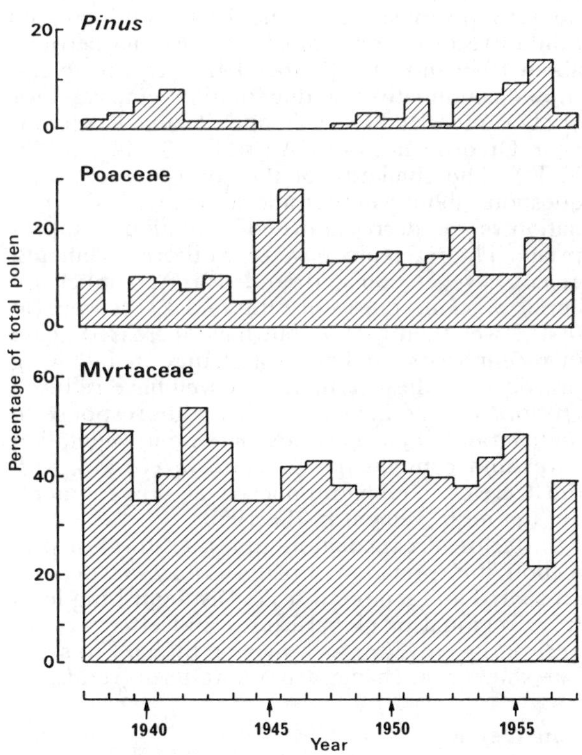

FIG. 15. Relative changes in abundance of three major pollen taxa in sediments from Burrinjuck Reservoir for each year from 1938 to 1957. Adapted from Clark (1986b).

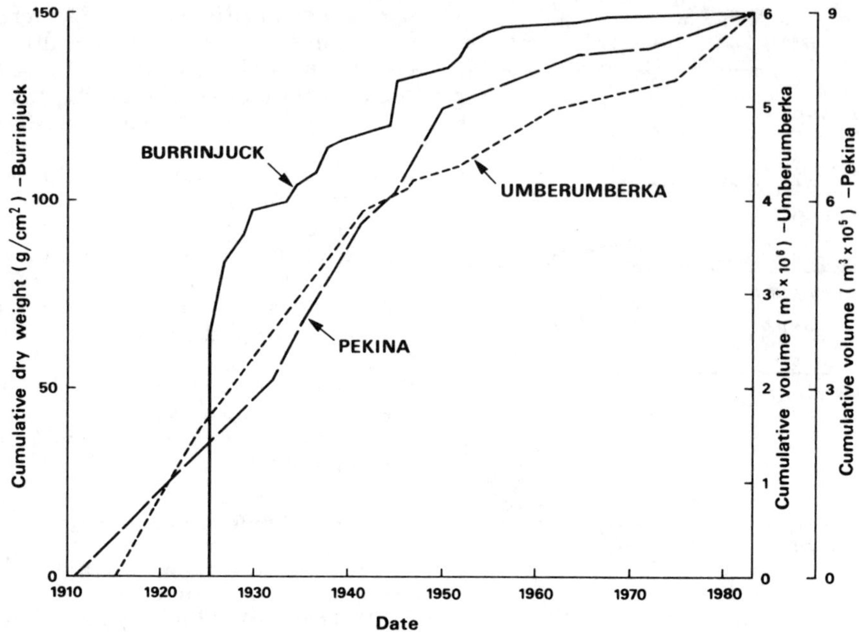

FIG. 16. Cumulative sedimentation in Burrinjuck, Umberumberka and Pekina Reservoirs up to 1982. Scales have been adjusted for comparison.

dams, widespread destruction of rabbit populations by myxomatosis in the 1950s, or increased rainfall recorded over much of Australia between about 1945 and 1975 (Pittock 1975). Figure 16 also shows cumulative sedimentation curves for Umberumberka Reservoir and Pekina Reservoir, near Orroroo in South Australia (32°44′S., 138° 37′E.). The similarity of the three curves raises questions about whether a local or regional explanation of the decrease in sedimentation is appropriate. The reservoirs have very different climates, geology, vegetation and land use. At Umberumberka, stocking rates since the mid 1940s were, at first, lower than before, but have increased again in recent years. At Pekina a change in fallowing practices by wheat farmers may well have reduced erosion. But perhaps at all three the response to higher rainfall has been an increase in vegetation cover and consequent reduction in erosion, with the decrease in rabbit numbers at all sites augmenting the improvement in the 1950s.

In terms of explanation and, therefore, of prediction, interpretation of patterns may well depend on spatial and temporal scale. Distinguishing local from regional or global causes and effects is not necessarily easy, just as it may be difficult to separate short-term change from long-term trends.

Conclusion

Studies of the past can illuminate environmental dynamics in a unique way, but can never provide the whole of an answer to management questions. This is equally true of other techniques more often relied on to provide the basis for management decisions: experimentation, observation, process studies and modelling. It is essential to draw on the complementarity between these contemporary methods and ecological history, particularly in fully integrated studies of the present and the past in catchments with a deposition basin at their outlet (Oldfield 1977). Lake catchments provide a natural unit for study and the presence of a deposition basin should be one of the main criteria used to select sites for long-term environmental monitoring.

Predicting future effects of natural, accidental or imposed environmental changes requires knowledge of the past and of how the present came to be, and palaeoecological research can be tailored to answer specific environmental questions. The time span and resolution of the historical record appropriate to any management question depends on the time span in the future with which that question is concerned. Studies of the recent past, on a time-scale of decades to centuries, are critical for understanding the dynamics of Australian ecosystems before and since European settlement in 1788 and for predicting effects of local, regional and global environmental changes over the next few decades and centuries.

Much of Australia's unique environment remains relatively intact, despite the destruction, degradation and alterations of the last two hundred years. This gives us an opportunity few countries now have, or can afford, to learn from the past and apply present knowledge to improve both management of productive land and maintenance of relatively natural ecosystems. Perhaps Australia is fortunate that European settlement was so recent.

Acknowledgements

I am indebted to all those with whom I have worked and talked over the last few years, particularly Julian Ash and Frank Oldfield, and to Angas Hopkins for his valuable comments on the original manuscript. Above all, I am grateful to Bob Wasson, with whom many of these ideas, approaches and research projects were developed.

References

Adamson D. A. & Fox M. D. (1982) Change in Australasian vegetation since European settlement. In: *A History of Australasian Vegetation* (ed. J. M. B. Smith) pp. 109-46. McGraw Hill, Sydney.

Ash J. (1983) Rainfall patterns in northeastern Queensland: 7±2KA. In: *CLIMANZ, A symposium of results and discussions concerned with Late Quaternary climatic history of Australia, New Zealand and surrounding seas* (eds J. M. A. Chappell & A. Grindrod) vol. 1, p.90 and vol. 2, fig.76. Department of Biogeography and Geomorphology, Australian National University, Canberra.

Ash J. (1988) The location and stability of rainforest boundaries in north-eastern Queensland, Australia. *J. Biogeog.* **15,** 619-30.

Banks J. C. G. (1982) *The use of dendrology in interpreting the dynamics of snow gum forest.* Ph.D. Thesis, Australian National University, Canberra.

Battarbee R. W., Anderson N. J., Appleby P. G., Flower R. J., Fritz S. C., Haworth E. Y., Higgitt S., Jones V. J., Kreiser A., Munro M. A. R., Natkanski J., Oldfield F., Patrick S. T., Richardson N. G., Rippey B. & Stevenson A. C. (1988) *Lake acidification in the United Kingdom 1800-1986. Evidence from analysis of lake sediments.* Report to the Department of the Environment. Palaeoecology Research Unit, University College, London.

Berger A. L. (1981) The astronomical theory of paleoclimates. In: *Climatic Variations and Variability: Facts and Theories* (ed. A. L. Berger) pp. 501-25. Reidel, Dordrecht.

Bolin B., Doos B. R., Jager J. & Warrick R. A. (eds) (1986) *The greenhouse effect, climatic change and ecosystems.* Wiley, Chichester.

Bowdler S. (1977) The coastal colonisation of Australia. In: *Sunda and Sahul. Prehistoric Studies in Southeast Asia, Melanesia and Australia* (eds J. Allen, J. Golson & R. Jones) pp. 205-46. Academic Press, London.

Bowler J. M. (1976) Recent developments in reconstructing late Quaternary environments in Australia. In: *The Origin of the Australians* (eds R. L. Kirk & A. G. Thorne) pp. 55-77. Australian Institute of Aboriginal Studies, Canberra.

Bowler J. M. (1982) Aridity in the late Tertiary and Quaternary of Australia. In: *Evolution of the flora and fauna of arid Australia* (eds W. R. Barker & P. J. M. Greenslade) pp. 35-46. Peacock, Frewville.

Bowler J. M. (1989) Late Pleistocene high rainfall: dating and seasonality implications. In: *CLIMANZ III, Proceedings of a conference on quantification and modelling of palaeoclimates* (eds R. J. Wasson & T. H. Donnelly). CSIRO, Division of Water Resources, Canberra.

Bowler J. M., Hope G. S., Jennings J. N., Singh G. & Walker D. (1976) Late Quaternary climates of Australia and New Guinea. *Quat. Res.* **6,** 359-94.

Bowman D. M. J. S. & Brown M. J. (1986) Bushfires in Tasmania: a botanical approach to anthropological questions. *Archaeol. Oceania* **31,** 166-71.

Chartres C. J. (1983) The pedogenesis of desert loam soils in the Barrier Range, western New South Wales. II Weathering and soil formation. *Aust. J. Soil Res.* **21,** 1-13.

Clark R. L. (1983a) Pollen and charcoal evidence for the effects of Aboriginal burning on the vegetation of Australia. *Archaeol. Oceania* **28,** 32-7.

Clark R. L. (1983b) *Fire history from fossil charcoal in lake and swamp sediments.* Ph.D. Thesis, Australian National University, Canberra.

Clark R. L. (1986a) *The fire history of Rotten Swamp, A.C.T.* Report to the A.C.T. Parks and Conservation Service. CSIRO, Division of Water and Land Resources, Canberra.

Clark R. L. (1986b) Pollen as a chronometer and sediment tracer, Burrinjuck Reservoir, Australia. *Hydrobiologia* **143,** 63-9.

Clark R. L. (1987) Fire history from charcoal: effects of sampling on interpretation. In: *Archaeometry: Further Australasian Studies* (eds W. R. Ambrose & J. M. J. Mummery) pp.135-42. Prehistory Department, Research School of Pacific Studies, Australian National University, and Australian National Gallery, Canberra.

Clark R. L. & Guppy J. C. (1988) A transition from mangrove forest to freshwater wetland in the monsoon tropics of Australia. *J. Biogeog.* **15,** 665-84.

Clark R. L. & Wasson R. J. (1986) Reservoir sediments. In: *Limnology in Australia* (eds P. De Deckker & W. D. Williams) pp. 497-507. CSIRO, Melbourne and Junk, Dordrecht.

Davis M. B. (1981) Quaternary history and the stability of forest communities. In: *Forest succession: Concepts and Application* (eds D. C. West, H. H. Shugart & D. B. Botkin) pp. 132-53. Springer-Verlag, New York.

Deevey E. S. (1969) Coaxing history to conduct experiments. *Bioscience* **19,** 40-3.

Delcourt H. R., Delcourt P. A. & Webb T. III (1983) Dynamic plant ecology: the spectrum of vegetational change in space and time. *Quat. Sci. Rev.* **1,** 153-75.

di Castri F. & Hadley M. (1988) Enhancing the credibility of ecology: interacting along and across hierarchical scales. *GeoJournal* **17,** 5-35.

Dodson J. R. (1988) The perspective of pollen records to study response, competition and resilience in vegetation on Barrington Tops, Australia. *Prog. Physical Geogr.* **12,** 183-208.

Doe B. R. (1983) The past is the key to the future. *Geochim. Cosmochim. Acta* **47,** 1341-54.

Dolman G. S. (1987) *Dynamics of Sydney's urban bushland from fine resolution pollen analysis.* Ph.D. Thesis, Australian National University, Canberra.

Finlayson C. M., Cowie I., Bailey B. & Armstrong M. (1985). Recycling of nutrients and heavy metals by floodplain vegetation. In: *Alligator Rivers Region Research Institute, Annual Research Summary 1984-85* pp. 35-46. Australian Government Publishing Service, Canberra.

Gill A. M. (1981) Adaptive responses of Australian vascular plant species to fire. In: *Fire and the Australian Biota* (eds A. M. Gill, R. H. Groves & I. R. Noble) pp. 243-72. Australian Academy of Science, Canberra.

Green D. G. & Dolman G. S. (1988) Fire resolution pollen analysis. *J. Biogeog.* **15,** 685-701.

Green D. G., Singh G., Polach H. A., Moss D., Banks J. & Geissler E. (1988). A fine resolution palaeoecology and palaeoclimatology from southeast Australia. *J. Ecol.* **76,** 790-806.

Hays J. D., Imbrie J. & Shackleton N. J. (1976). Variations in the earth's orbit: pacemaker of the Ice Ages. *Science* **194,** 1121.

Head L. (1983) Environment as artefact: a geographic perspective on the Holocene occupation of southwestern Victoria. *Archaeol. Oceania* **18,** 73-80.

Head L. (1987) The Holocene prehistory of a coastal wetland system: Discovery Bay, south-eastern Australia. *Human Ecol.* **15,** 435-62.

Head L. (1988) Holocene vegetation, fire and environmental history of the Discovery Bay region, south-western Victoria. *Aust. J. Ecol.* **13,** 21-49.

Holland J. & Clark R. L. (1989) *Diatoms of Burrinjuck Reservoir, New South Wales, Australia.* Divisional Report 89/1, CSIRO, Division of Water Resources, Canberra.

Holling C. S. (ed.) (1978) *Adaptive environmental assessment and management.* Wiley, Chichester.

Hope G. S. (1984) Australian environmental change: timing, directions, magnitudes and rates. In: *Quaternary Extinctions: A Prehistoric Revolution* (eds P. S. Martin & R. G. Klein) pp. 681-90. University of Arizona Press, Tucson.

Hope G. S. (1989) The climatic implications of timberline changes in Australasia, from 30 000 BP to present. In: *CLIMANZ III, Proceedings of a conference on quantification and modelling of palaeoclimates* (eds R. J. Wasson & T. H. Donnelly). CSIRO, Division of Water Resources, Canberra.

Hope J. H. (1982) Late Cainozoic vertebrate faunas and the development of aridity in Australia. In: *Evolution of the flora and fauna of arid Australia* (eds W. R. Barker & P. J. M. Greenslade) pp. 85-100. Peacock, Frewville.

Hopkins A. J. M. & Saunders D. A. (1987) Ecological studies as the basis for management. In: *Nature conservation: the role of remnants of native vegetation* (eds D. A. Saunders, G. W. Arnold, A. A. Burbidge & A. J. M. Hopkins) pp. 15-28. Surrey Beatty, Chipping Norton.

Horton D. R. (1982) The burning question: Aborigines, fire and Australian ecosystems. *Mankind* **13,** 237-51.

Hughes P. J. & Sullivan M. E. (1981) Aboriginal burning and late Holocene geomorphic events in eastern NSW. *Search* **21,** 277-8.

Iversen J. (1964) Retrogressive vegetational succession in the Post-Glacial. *J. Ecol.* **52 (supplement),** 59-70.

Jones R. (1975) The Neolithic Palaeolithic and hunting gardeners: Man and land in the Antipodes. In: *Quaternary Studies* (eds R. P. Suggate & M. M. Cresswell) pp. 21-34. Royal Society of New Zealand, Wellington.

Kershaw A. P. (1981) Quaternary vegetation and environments. In: *Ecological Biogeography of Australia* (ed. A. Keast) pp. 81-101. Junk, The Hague.

Kershaw A. P. (1986) Climatic change and Aboriginal burning in north-east Australia during the last two glacial/interglacial cycles. *Nature* **322,** 47-49.

Kershaw A. P. (1988) The Late Cainozoic history of Australasia: 20 million to 20 thousand years B.P. In: *Vegetation History* (eds B. Huntly & T. Webb III). Kluwer, Dordrecht.

Lampert R. J. & Ross A. (1989) A question of numbers: the interaction of people and environment in the Holocene. In: *CLIMANZ III, Proceedings of a conference on quantification and modelling of palaeoclimates* (eds R. J. Wasson & T. H. Donnelly). CSIRO, Division of Water Resources, Canberra.

Laszlo E. (1972) *Introduction to systems philosophy.* Harper and Row, New York.

Lees B. G. & Clements A. (1987) Climatic implications of chenier dates in northern Australia. *Radiocarbon* **29,** 311-7.

Lourandos H. (1983) Intensification: A Late Pleistocene-Holocene archaeological sequence from southwestern Victoria. *Archaeol. Oceania* **16,** 145-54.

Macphail M. K. (1980) Regeneration processes in Tasmanian forests. *Search* **11,** 184-90.

May V. (1978) Areas of recurrence of toxic algae within Burrinjuck Dam, New South Wales, Australia. *Telopea* **1,** 295-313.

Milankovitch M. (1941) Canon of insolation and the Ice Age problem. Royal Serbian Academy, Belgrade. (Translation, Israel Program for Scientific Translation, Jerusalem, 1969).

Nanson G. C., Young R. W. & Stockton E. (1987) Chronology and palaeoenvironment of the Cranbrook Terrace (near Sydney) containing artefacts more than 40 000 years old. *Archael. Oceania* **22,** 72-8.

Nix H. (1982) Environmental determinants of biogeography and evolution in Terra Australis. In: *Evolution of the flora and fauna of arid Australia* (eds W. R. Barker & P. J. M. Greenslade) pp. 47-66. Peacock, Frewville.

Oldfield F. (1977) Lakes and their drainage basins as units of sediment-based ecological study. *Prog. Physical Geogr.* **1,** 460-504.

Oldfield F. (1983) Man's impact on the environment: some recent perspectives. *Geography* **68,** 245-56.

Oldfield F. (1987) The future of the past — a perspective on palaeoenvironmental study. In: *Horizons in Physical Geography* (eds M. J. Clark, K. J. Gregory & A. M. Gurnell) pp. 10-26. Macmillan, Basingstoke.

Oldfield F. & Clark R. L. (1989) Environmental history: the environmental evidence. In: *European Environmental History* (eds P. Brimblecombe & C. Pfister) Springer-Verlag, Berlin.

O'Sullivan P. E. (1973) Pollen analysis of mor humus layers from a native Scots pine ecosystem, interpreted with surface samples. *Oikos* **24,** 259-72.

Passmore J. (1974) *Man's Responsibility for Nature.* Duckworth, London.

Pearce R. H. & Barbetti M. (1981) A 38 000 year old archaeological site at Upper Swan, Western Australia. *Archaeol. Oceania* **16,** 173-8.

Pickup G., Wasson R. J., Warner R. F., Tongway D. & Clark R. L. (1987) *A feasibility study of geomorphic research for the long term management of uranium mill tailings.* CSIRO, Institute of Natural Resources and Environment, Division of Water Resources Research, Divisional Report 87/2, Canberra.

Pittock A. B. (1975) Climatic change and the patterns of variation in Australian rainfall. *Search* **6,** 498-504.

Ross A. (1985) Archaeological evidence for population change in the middle to late Holocene in southeastern Australia. *Archaeol. Oceania* **20,** 81-9.

Singh G. & Geissler E. A. (1985) Late Cainozoic history of vegetation, fire, lake levels and climate at Lake George, New South Wales, Australia. *Phil. Trans Roy. Soc. Lond.* **B 311,** 379-447.

Smith M. A. (1987) Pleistocene occupation in arid Central Australia. *Nature* **328,** 710-1.

Thom B. G. & Chappell J. (1975) Holocene sea levels relative to Australia. *Search* **6,** 90-3.

Thorne A. G. (1980) The longest link: human evolution in Southeast Asia and the settlement of Australia. In: *Indonesia: Australian Perspectives* (eds J. J. Fox, R. G. Garnaut, P. T. McCawley & J. A. C. Mackie) pp. 35-43. Research School of Pacific Studies, Australian National University, Canberra.

Thorne A. G. (1981) The arrival and adaptation of Australian aborigines. In: *Ecological Biogeography of Australia* (ed. A. Keast) pp. 1751-60. Junk, The Hague.

Truswell E. M. & Harris W. K. (1982) The Cainozoic palaeobotanical record in arid Australia: fossil evidence for the origin of an arid-adapted flora. In: *Evolution of the flora and fauna of arid Australia* (eds W. R. Barker & P. J. M. Greenslade) pp. 67-76. Peacock, Frewville.

Walker D. (1970) Direction and rate in some British Postglacial hydroseres. In: *Studies in the vegetational history of the British Isles* (eds D. Walker & R. G. West) pp. 117-39. Cambridge University Press, Cambridge.

Walker D. & Chen Y. (1987) Palynological light on tropical rainforest dynamics. *Quat. Sci. Rev.* **6,** 77-92.

Walker D. & Singh G. (1981) Vegetation history. In: *Australian Vegetation* (ed. R. H. Groves) pp. 26-43. Cambridge University Press, Cambridge.

Wasson R. J. (1988) Landscape denudation in Australia in the Late Holocene. In: *Global Change* pp. 56-68. Australian Academy of Science, Canberra.

Wasson R. J. & Clark R. L. (1985) Environmental history for explanation and prediction. *Search* **16,** 258-63.

Wasson R. J. & Clark R. L. (1987) The Quaternary in Australia — past, present and future. *Australian Geoscience 1986-87* pp. 29-34. Annual Report of the Australian Geoscience Council. Bureau of Mineral Resources Report 282. Australian Government Publishing Service, Canberra.

Wasson R. J., Clark R. L., Nanninga P. M. & Waters J. (1987) ^{210}Pb as a chronometer and tracer, Burrinjuck Reservoir, Australia. *Earth Surf. Proc. Landf.* **12,** 399-414.

Wasson R. J. & Galloway R. W. (1986) Sediment yield in the Barrier Range before and after European settlement. *Aust. Rangel. J.* **8,** 79-90.

Wimbush D. J. & Costin A. B. (1979) Trends in vegetation at Kosciusko. I Grazing trials in the subalpine zone, 1957-1971. *Aust. J. Bot.* **27,** 741-87.

Proc. Ecol. Soc. Aust. 1990 — 16:23-37

An ecological study of traditional Aboriginal use of "country": Martu in the Great and Little Sandy Deserts, Western Australia

F. J. WALSH

Department of Botany, The University of Western Australia, Nedlands, W.A. 6009, Australia

Abstract

An ecological framework was used to reconstruct aspects of the traditional lifestyle of Martu Aborigines of the Great Sandy and Little Sandy Deserts. Their subsistence strategies were directed toward ensuring significant plant food species could be selected from the broad resource base. This was achieved by targeting resource rich plant communities and occupying sites that ensured access to a variety of plant communities and a reliable water source. Movement pathways in proximity to and between occupation sites were determined by the prevailing topography. The Martu distinguished temporal and spatial patterns of change on a perceptual basis that differed from European notions of a desert environment. This perception of availability affected their use of resources. The need for ecologists and land managers to consider the implications of Aboriginal resource use and management within the wider context of their subsistence lifestyle is emphasized.

Introduction

Aboriginal people are recognized as having unique and intimate associations with the landscape. Their experiences are interpreted and expressed through a body of traditional ecological knowledge (Baines 1986). In areas of arid Australia, Aboriginal people play a pivotal role, through their interaction with past and present environments. In the past, as hunter-gatherers, they were components of arid ecosystems. At present, they are restating and reasserting their affiliations with the land. For the future, they have identified a desire to be involved in nature conservation and land management (Kean *et al.* 1988).

The habitats and ecosystems occupied by Aboriginal people varied throughout the continent; consequently, the material basis of their lifestyle varied. There is a large amount of ethnobiological literature that documents the species, particularly plants, that were used by Aboriginal people. Most of this research has been in tropical and arid environments where in Aboriginal communities a modified traditional lifestyle is maintained. Increasingly, social scientists are utilizing ethnobiological material to describe the economic, foraging or subsistence strategies of Aboriginal societies. For example, detailed studies have been conducted in Cape York (Chase & Sutton 1981), Arnhem Land (Meehan 1982; Altman 1987; Rose 1987), Dampier Peninsula (Smith & Kalotas 1985), the northern Great Sandy Desert (Cane 1984), Central Australia (Latz 1982) and the southern Western Desert (Gould 1980).

Some biologists have documented the knowledge of Aboriginal elders and used ethnohistorical records to describe the biology of extinct, rare or endangered species, or those with restricted distributions (Gott 1983; Burbidge *et al.* 1988). Newsome (1980) has interpreted Arunda mythology and used it to complement scientific knowledge of the migratory behaviour of red kangaroos. In Australia, other examples of collaboration between biologists and Aboriginal people are limited. The use of traditional ecological knowledge by ecologists and land managers has generally been restricted to the documentation of Aboriginal burning patterns (see Allen & Baker this volume). However, Kimber (1984) recorded examples of additional resource management practices and Stevenson (1985) discussed concepts that were employed by the Tiwi in making resource management decisions. Few ecologists have referred to subsistence studies as a means of assessing the interactions between Aboriginal people and their environment, and the consequences of this interaction.

In Aboriginal culture there is a continuity between the past and present. Many traditional practices continue to be used and applied by people, although frequently these have been modified by technological, material and lifestyle changes. Information from observations of contemporary lifetyle, historical and ethnographic records and the interpretation of archaeological and palaeontological material can be used to reconstruct precontact subsistence patterns. The most complete reconstructions result from an approach that integrates information from various disciplines.

Reconstructions of traditional subsistence patterns and strategies ideally attempt to account for components of the material, social and spiritual culture that exerted an influence on where, when and why people moved and camped. In the past, sociobiologists have adopted aspects of biological theory and applied it to their studies of subsistence strategies and foraging behaviour in hunter-gatherer groups (see Winterhalder & Smith 1981). Some anthropologists have argued against this approach, largely on the basis that it does not account for the decisions made by individuals in response to their social and spiritual responsibilities (see Rappaport 1984). It is now considered that people's activities were *constrained* rather than *determined* by their environment. These constraints varied according to the relative richness of the environment. In the arid zone they were particularly severe; Tindale (1981:1855) referred to the Western Desert as "the most challenging environment".

This chapter utilizes material compiled from research with the Martu who live in the Great and Little Sandy Deserts of Western Australia. First, it aims to show how a combination of ecological and ethnographic techniques may be used to determine the spatial and temporal distribution of resources used by the Martu. Second, it aims to describe the subsistence activities that were influenced by this distribution. Finally, it relates the implications of these activities to the landscape in which the Martu live.

Background to the region and its people

The Martu continue to interact with and utilize country that was occupied by them or their predecessors. Until precontact times they lived as hunter-gatherers in a region that lies approximately 300 km east of Newman in the Pilbara, Western Australia (Fig. 1). Telfer is situated approximately 100 km north-west of Rudall River and the study area. Its average annual rainfall (1974-1987) was 306 mm; the bioclimate was classified by Goble-Garratt (1987) as semi-desertic tropical. Rainfall volumes declined to the east of Telfer, the research area lies within the 150-200 mm isohyet (Anon. 1979). Rains from cyclonic depressions and convectional thunderstorms resulted in localized falls of high intensity in the hotter months. On average, January to March were the wettest months, monthly rainfalls declined until August to October which were the driest months. In July and August a small increase in rainfall resulted from widespread, low intensity rains associated with weather systems that originated in the south.

The region occupied by the Martu straddles the Little Sandy and Great Sandy Deserts. Dissected hills rise above the plains to form range systems. In the south-west of the region the Broadhurst Range comprises the watershed of the Rudall River. The primary river channel flows north-east through dunefields. In the south-east the McKay and Harbutt Ranges and smaller isolated ranges are drained by ephemeral headwater streams that dissipate onto the surrounding sandplains and dunefields. Expansive dunefields dominate the region. Sand dunes generally form parallel north-west ridges, except where sand and wind movement is interrupted by topographical features. The Percival palaeoriver persists as a series of disconnected saline playas.

The region lies within the Canning and Keartland Botanical Districts described by Beard and Webb (1974). The shrub steppe of stony plateaux and ranges is dominated by *Acacia pyrifolia*. *A. coriacea* and *Hakea suberea* shrubs characterize the sandy plains that surround the uplands. On these shrub steppe hummock grasses, *Triodia basedowii* and *T. pungens* are abundant. Rudall River is typified by a tree savannah of *Eucalyptus camaldulensis* and *E. microtheca* on river flats with mixed grasses. *Acacia aneura* and *Melaleuca glomerata* and *M. lasiandra* occur on less permanent wetlands and in valleys and depressions. A mixed shrub (*Acacia, Hakea* and *Grevillea*) steppe occurs on the dune slopes and swales with *T. pungens* and *Plectrachne schinzii* hummock grasses. The saline playas are fringed by samphire.

It is necessary to consider reconstructions of hunter-gatherer subsistence in a time frame that accounts for changes in the landscape and its use. Archaeological investigations have identified patterns of colonization and occupation by Aborigines of the arid zone. The excavation of an occupation site in the Central Australian Ranges has shown the region was occupied 22 000 years ago, during the Pleistocene (Smith 1988). In contrast, Veth (1989) suggested the dunefields of the Simpson Desert, the Little and Great Sandy Deserts and possibly the Great Victoria Desert were occupied more recently, in the mid-Holocene or later. The Sandy Deserts were continuously occupied from 5 000 B.P. to Present (Veth 1987). The time of colonization coincided with a period of climatic amelioration. Since this time there were changes in the material basis of occupation. Veth and Hiscock (1988) postulated that the adoption

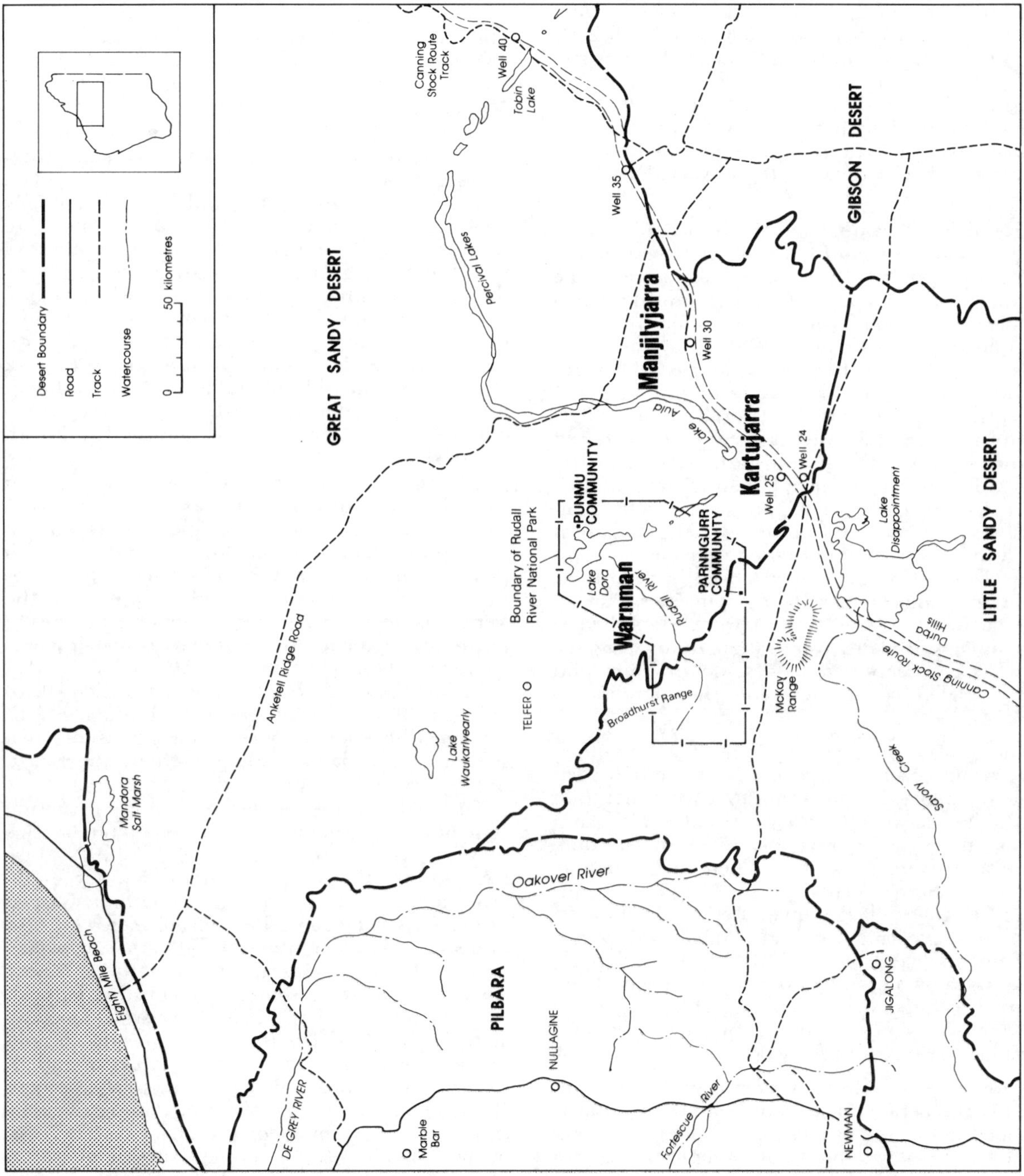

Cartography by David Buchan

FIG. 1. Location of research area.

of small stone tools enabled people to make wooden artefacts. These included bowls suited to carrying food and water. Thus, the use of these tools may have facilitated the expansion of populations into marginal desert areas. Seed grinding activity intensified between 3 000 and 1 500 B.P., indicating a change in the subsistence base. A rise in the rate of artefact discard suggested population densities increased from 1 500 to 700 years ago (Veth 1987). Changes in resource use and land management practices are expected to have occurred associated with these variations over time.

Five dialectal groups traditionally occupied lands in the Great Sandy and Little Sandy Deserts in the vicinity of Rudall River and Lake Auld. The Manjilyjarra, Kartujarra, Kuratjarra and Putijarra people have been collectively referred to as Mardudjara (Tonkinson 1978); they and the Warnman now call themselves "Martu". They are included amongst groups belonging to the Western Desert culture bloc (Berndt 1959). Their hunter-gatherer lifestyle was irreversibly altered following contact with Europeans in the early 1900s. By 1965 these people had all migrated or been forced from their lands and settled in missions, stations and towns. In the northern Western Desert the outstation movement commenced in 1983 and Martu moved back to their traditional homelands. Two communities, Punmu and Parnngurr (Cotten Creek), were established in what is now Rudall River National Park (Fig. 1). Martu have continued to resettle in these and other smaller communities on their traditional lands.

The lifestyle of these people has been modified since contact with Europeans. They are now relatively sedentary. People prefer to drive rather than walk. Their movements are restricted to roads and tracks and areas in proximity to the community camp and overnight camps. Their diet is mainly comprised of Western foods, although they continue to collect large quantities of bushfoods. Men now use rifles to hunt. Metal and plastic implements have replaced the wooden digging sticks and bowls used by women. There have been dramatic changes in their material culture. Despite these changes many aspects of the social and spiritual cultures are retained.

Knowledge of Martu history can be used to identify the period during which land use patterns changed. After the establishment of the Canning Stock Route, Martu land management techniques were applied with decreasing intensity as people moved away from their country; they ceased in the late 1950s when the last groups left. Subsequently, there was a 20 year lapse; then over the last decade some of those techniques have been reapplied, although spatially and temporally modified.

Methods

Ethnographic research

Ethnographic field research was conducted over an area ranging from Punmu in the north to Durba Hills in the south, Jigalong in the west and the Canning Stock Route in the east. A total of five months field research was conducted between March 1987 and August 1988.

Ethnographic research was conducted with a core group of seven Martu women from Parnngurr. In addition, at various times, I worked with a wider range of people that eventually included 15 individuals. In the core group were women from three generation levels. Interaction with these age groups was necessary to overcome the language problems experienced when talking to the older women and to balance the substantial loss of traditional knowledge in younger generations. On several occasions the opportunity to work in collaboration with a male linguist and a fluent English-Pitjantjatjarra speaker was very productive. These men also facilitated my work with Martu men.

Ethnographic material was compiled during formal and informal interviews and discussions. Interviews were recorded on tape cassette when it was appropriate. To establish the data base on plant resource species a standard set of questions was asked. Consequently, information on the Martu names for each species, its use, processing technique, habitat and seasonal availability was compiled. The question list was used only as a guideline; this ensured there was sufficient flexibility to follow up interesting comments with further discussion. Voucher specimens of each species and variants distinguished by the Martu were collected and preserved.

To make observations on food gathering activities, I accompanied Martu women on the trips they regularly made from Parnngurr. More frequently, a group of women and I would establish camps away from the main community for two to five days. The camp locations chosen were also precontact occupation sites. Food gathering excursions were then undertaken from them. These excursions were essentially reconstructions of traditional activities. They focussed on the activities associated with finding bushfoods and processing them so they were suitable for eating. Decisions to be made were referred to the women. Generally my role as an observer and passive participant was maintained. Excursions were conducted on foot. If the women suggested we go places more than eight kilometres from the outlying camps then we would drive. When working with Martu a considerable amount of flexibility, patience and time was required.

On each excursion the following was recorded: the route taken; the plant community that was traversed; the activities of the women; the time devoted to these activities; and, the resource species that were observed or harvested. In the later stages of the project the edible materials that had been collected were weighed and/or measured. Some material was preserved for nutritional analyses.

During interviews and discussions specific topics were explored with the Martu. These topics included: seasons and weather; descriptions of the landscape and its components; precontact activities; custodianship for sites and areas of land; and, fires and burning. This resulted in a large, fragmented array of notes from which material can be drawn to compare and contrast the field observations, the results of archaeological research and ethnohistorical literature.

Vegetation surveys

Vegetation surveys were conducted after accompanying Martu women on gathering excursions. Survey quadrats were situated in plant communities:

a) where resources had been harvested;
b) that had been traversed in search of resources; and,
c) that had not been visited at all.

Quadrats were distinguished as burnt (less than five years since a fire) or unburnt (more than five years since a fire). This was a subjective assessment based on the relative size of spinifex hummocks. Where possible this distinction was corroborated by ethnographic information; Martu have been reapplying fire regimes since 1983. There was no control for the influence of rainfall on vegetation patches.

The sampling quadrats were 400 m^2. The percent cover of each species in a total of 36 quadrats was determined. Ordination and classification techniques were used to distinguish vegetation types. The ecological parameters of species richness and Shannon-Weiner diversity indices were calculated and used as tools for comparing the resources available at different locations and times. The initial data set included all plant species that occurred in the quadrats. This was reduced to include only resource species. Data were later combined and averaged for burnt and unburnt quadrats that were located on the major topographical units.

A discussion of Martu subsistence patterns

Senft *et al.* (1987) suggested an ecological hierarchy which provided a useful model for investigating the behaviour of species seeking widely dispersed, relatively low quality food. These factors characterized the plant resource base on which Martu were reliant for a large proportion of their energy, nutrient and bulk food requirements. The hierarchy proposed by Senft *et al.* (1987) consisted of selection at increasing spatial scales. These were: the selection of preferred plant species from within micropatches; the targeting of certain plant communities; the use of particular landscape systems; and finally, the habitation of general regions. This model was utilized as the basis for this ethnoecological study. The approach does not assume hunter-gatherer activities were environmentally determined, it merely provides a framework for describing them. The following section notes the role of plant foods relative to water and meat sources, it then identifies the plant species that were preferred by the Martu. The next section briefly discusses the spatial and temporal availability of species from an ecological and ethnographic perspective. This is then related to the movement of people in the landscape and through the region. The information presented herein is mainly descriptive, some of it is inferred from available data. Further data analysis and field work are required to corroborate the material presented. In the future, detailed comparisons with archaeological and ethnohistorical literature will be made.

The resource base

Water was the constraining resource of the Sandy Deserts. Within the region, the Rudall River was the only continuous primary river channel; it was over 120 km in length. Elsewhere, drainage lines were less than 40 km long. Surface waters were scattered and varied in their degree of persistence. The wetland classes of the region ranged from permanent springs to ephemeral claypans (after Masini 1988). In addition to using surface waters, Martu drew on relatively reliable subterranean supplies by digging wells and soaks. At least 650 water sources have been named and located by Martu in an area of approximately 150 000 km^2 (Gallagher, pers. comm.).

The location and degree of permanency of water sources limited or expanded the tracts of country that Martu exploited. Their distribution was a key factor that influenced the movement of people. The distribution of surface waters was altered by the prevailing weather and varied throughout the year.

The proportion of animal to plant food eaten by the Martu has not been determined. Plant foods comprised the bulk of their diet. However, meat was highly prized and contained nutrients and energy that did not occur in plant foods. Gould (1969) stated that in the Western Desert plant food contributed 70-80% of the diet's bulk. This figure may have been an overestimate; it represented the food resources recorded after periods of high rainfall, when the availability of plant food was relatively greater after periods of average rainfall.

Veth and Walsh (1988) suggested Gould's data were relevant to a geographically restricted area, not the entire Western Desert.

At least 12 species of native mammals were once hunted by Martu (see Burbidge *et al.* 1988). Adult birds and nestlings were killed and eaten. Reptiles were an important component of the diet. The information from older Martu people has indicated that reptiles were a reliable food source. Statements that alluded to their reliability were common: "always get 'em pula pula (small lizards)" was frequently repeated. This has been supported by indications of the resilience and diversity of reptile populations in the arid zone (Morton & James 1988).

Vegetation surveys in the Sandy Deserts have recorded at least 330 plant species (Beard & Webb 1974; George & Mitchell 1983; Walsh, unpubl.). To date, the use of 148 species has been documented. These provided the raw materials for food, medicine, tobacco, ash, tools, implements and other uses. Of these species, 106 were potential sources of food. This represented the use of approximately 35% of the known flora for food.

Several variables have been proposed to determine the significance or the relative contribution of a species to the total diet. Importantly, they were dependant on variations in space and time; they are:

a) the abundance, predictability and accessibility of each species;
b) the energy costs of processing each species;
c) the energy and nutrient returns of each species;
d) cultural preferences (Veth & Walsh 1988:19).

Between 20 and 40 species used by Martu may be referred to as significant species or "staples". The maintenance of a broad resource base was a vital factor that contributed to desert survival. It prevented the reliance on any one species that may have been unproductive under adverse conditions. Also, it may have ensured Martu avoided the over-exploitation of a particular resource. Importantly, it permitted flexibility and selectivity in their diet.

Plant foods may be conveniently separated into categories (Fig. 2) according to the plant part consumed (see Latz 1982 for a detailed discussion of plant use in these categories). The categories are constructed from a Eurocentric background. They do not always correspond to divisions used by Martu. Aboriginal people use systems of ethnotaxonomy that have some similarity and many differences to the Linnaean system (see Sutton 1976).

The seeds of 51 grass, sedge, herb and acacia species were food sources. This is the largest number of seed species recorded for use by an Aboriginal group. The apparent importance of this food group has been supported by ethnographic and historical records. Species that were used by the Martu included: *Acacia ampliceps, A. dictyophleba, A. eriopoda, Amphipogon caricinus, Eragrostis eriopoda, Fimbristylis oxystachya, Portulaca oleracea* and *Tecticornia verrucosa.* In the hummock grasslands and acacia shrublands of the region, seed producing species were dominant. Following suitable climatic conditions these provided an abundant resource. They were suited to exploitation when large wooden bowls, grindstones and water were available.

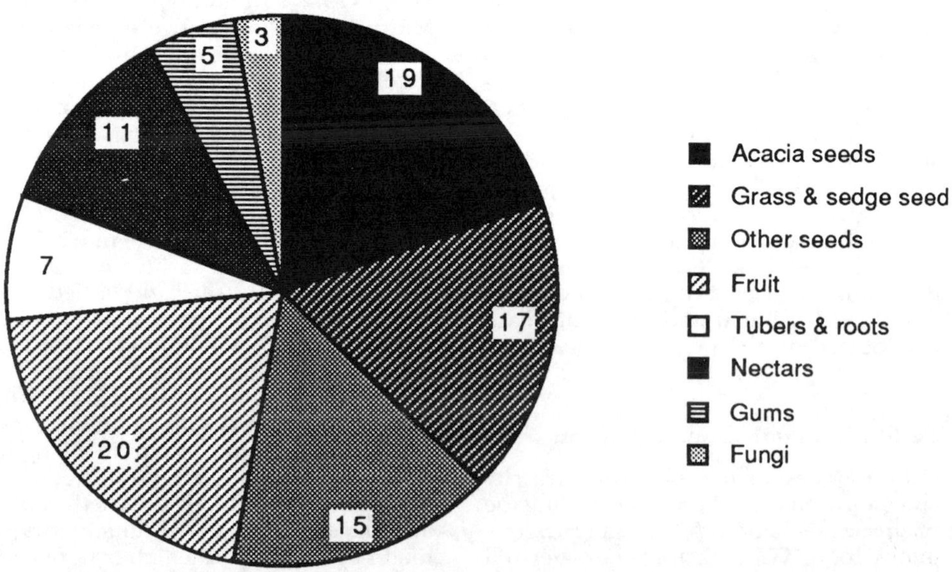

FIG. 2. Proportion of plant species used in major food groups (from a total of 91 species).

Seeds were preferentially used when large quantities could be harvested, therefore increasing the efficiency of processing. *Yakirra australiensis* was a favoured seed source; it could be harvested and cleaned at a rate of at least 490 g/hr. Therefore, in four hours of preparation approximately 2 kg of seed could be procured. Under widespread dry conditions the productivity of many seed producing species could have been expected to decline. Consequently, other foods were sought or alternative subsistence strategies were implemented.

Fruits were highly ranked as preferable food items. Six species of *Solanum* were eaten. Other species gathered included: *Amyema quandang, Canthium latifolium, Cucumis melo* and *Santalum acuminatum.* Most fruits were easily picked, cleaned and eaten; they required little preparation. They were prized for their succulence and sweetness. Several species were collected in large amounts, for example, in an hour 50 fresh fruit and 100 dried fruit of *Solanum diversiflorum* were harvested by one woman, providing 1 010 g of fruit. Ecological reasons also influenced their significance; several species continued to be productive under low rainfall conditions, these included *Solanum diversiflorum, S. chippendalei, Ficus platypoda* and *Santalum lanceolatum* (Pate 1986). Furthermore, the distribution of *Solanum* and several other species could be manipulated by fire regimes and the spreading of seed. This provided predictable resource patches that could then be exploited.

The tubers of *Vigna lanceolata* and *Cyperus bulbosus* were also important components of the traditional diet. They were moist and starchy to eat; commonly grew in association with wetlands that were frequently visited; large quantities could be collected; they continued to be available in periods when other food resources were limited; and, women could continue to talk to each other when digging them. The roots of five other species were edible. Although less preferred, they may be referred to as drought foods as they were used only in dry periods.

Nectars and gums were available from sixteen species. They were highly prized. Twenty-five plant species hosted edible insects. Cossid moth larvae (witchetty grubs), wasp galls and the scale of lerp insects were also eaten. They are often overlooked as important dietary items, yet they were highly nutritious and tasty, were available in dry periods and could be readily collected. Other insects eaten included termite aeletes and ant eggs.

It may be noted that many of the foods discussed above have been described as significant. This reflected the value Martu placed on having a varied diet; foods were considered to be complementary and important for different reasons. It also reflected the array of species that were productive under different environmental conditions; as these altered so the choice of species had to change.

Climate and seasons

The entire Western Desert has been characterized as an aseasonal environment, there apparently being "no predictable season when plants are expected to ripen" (Gould 1980:66). Yet early in the research Martu indicated that there was an appropriate time to look for specific resources and undertake certain activities. This identified the need for further consideration of climatic patterns and seasonality of the region.

Average rainfall figures obscure the large variation in yearly totals. This variation was reflected in long-term data recorded over 73 years at Roy Hill, 260 km west of the study area. Although the average annual rainfall was 265 mm, the standard error was 141 mm. Furthermore, comparison of rainfall data between two weather stations indicated there were large differences in the volume of rainfall received in the summer months. This was a consequence of the highly localized nature of summer falls. Although there was a predictable pattern to the time when rain was expected to fall, the total amount of rainfall was far less predictable.

As a result of the rainfall pattern the distribution of surface waters in the warmer months and cooler months was very different. In the warmer months localized falls rapidly filled wetlands with impermeable substrates such as claypans. Sufficient rainfall to create runoff was required to replenish semi-permanent pools and rockholes. Generally, in the warmer periods surface waters were ephemeral and widely dispersed; in the cooler months they were less abundant but lasted for longer.

Further research is required to determine seasonal patterns of resource productivity. The establishment and phenology of species depended on the timing, duration and intensity of climatic conditions. In the arid zone these responses were variable, ranging from the rapid seeding of ephemeral species following rainfall events to the eventual seeding of perennial species such as *Acacia aneura* after a series of rainfall events that induced flowering and then seeding (Maconochie 1982). In general terms it was apparent that there were predictable periods when some resources were expected, other species were available throughout the year in good seasons, whilst others were in season at no predictable time. It will be necessary to determine resource availability on the basis of individual species phenology.

Ethnographic questions were pursued to determine whether Martu recognized and followed patterns of environmental change. Relevant information was recorded in many separate pieces.

Martu named periods of time when specific resources were expected to be most suitable for collection. They identified three major periods that followed in succession. These progressed from

"Tulparra" to "Yalijarra" to "Wantajarra" (using Kartujarra terminology). These may be equated to the seasons of spring, summer and winter respectively. Changes in temperature were the basis for distinguishing the transition from one "season" to another. Within each period were "sub-seasons" when certain climatic events were expected to occur. The Martu had a detailed vocabulary for various types of rain, cloud, wind and other weather phenomena (Tonkinson 1978). Biotic responses to these climatic events were identified and equate to "sub-seasons". These frequently overlapped the progression from one season to another, due to the delay in response. For example, "yukurigara" or "green grass time" commenced late in "Yalijarra", after the summer rains fell, it then continued through much of "Wantajarra".

Biological indicators conveyed to Martu information on specific events. These were interpreted to predict forthcoming seasons and other events. For example, the calling of the cricket "urata ura-ta" signalled the beginning of "Tulparra". Biological indicators were also used to specify the availability of particular resources. To give another example, when budgerigar chicks were heard, then grass seeds were expected to be ready for harvest.

Martu identified the season when a potential food producing species was expected to be suitable for harvesting. This information was sometimes conveyed by reciting the sequence of species that supplied a particular food source. The availability of a "wama" (nectar) was expressed in this way: "Wantajarra you have jakata, then jalpinpa then yumpalpa, last one, wirarajartu, in Tulparra". This identified the flowering succession of *Grevillea wickhamii, G.* aff. *eriostachya, Hakea rhombales* and *H. suberea* respectively.

Temperature was the most predictable component of the climate; perhaps for this reason it was used as the basis for the main seasonal distinctions by Martu. The "sub-seasons" were reliant on rainfall to initiate the chain of biological events associated with them. This was disrupted when perturbations from low rainfall or drought occurred. In "Yalijarra", rains were expected to fall in the period from December to March. However, there was considerable variation in the actual timing of these rainfall events. This indicates why it was inappropriate to use a refined time scale, such as calendar months, in such an unpredictable environment. Martu referred to the position of the Pleiades (Seven Sisters) constellation while discussing when seasons were expected to occur. These stars provided them with a fixed means of monitoring the variation in events.

Martu did not identify a specific cycle of seasons. Their concept of seasonality was primarily focused on a sequence of events that did not have a recognized heirarchy; these events were spaced by intervening periods of time. They placed little emphasis on the long-term chronology of these events.

The distribution and location of resources at the community level

At the next scale in an ecological hierarchy it may be considered necessary for Martu to determine where in the landscape the plant communities or habitats of resource species were expected to occur. This process will be documented in this section.

The spatial distribution of plant communities in the East Pilbara and Great Sandy Desert was closely related to the prevailing landform type (Beard & Webb 1974; Atkins 1985; Goble-Garratt 1987). Generally, this was supported by ordination and classification of the vegetation sites surveyed in this study. Within the ordination space, sites that were situated on the same landform unit were most similar; those located on hills and ranges were most distinct from those adjacent to wetlands; and, sites situated on sand dunes and sandplains had characteristics intermediate to uplands and wetlands. On sandplains and sand dunes there was considerable variation in vegetation type but, so far, no discernible patterns have been observed. The vegetation associated with wetlands was also variable. In the region the fixed distribution of plant species that was influenced by landform was superimposed with dynamic variations that resulted from localized events, notably burning and rainfall. This may explain the variation observed in communities on sandy substrates. The average species richness and diversity of resource species occurring on the major landform units were calculated (Fig. 3). These measures serve as comparisons between sites and are only valid for the time when they were sampled.

In some seasons when collecting plant foods, Martu women selectively targeted certain plant communities. This was determined using estimates of the time spent in them. In "Wantajarra", food resources were most frequently sought and harvested from communities that had high levels of species diversity and richness. Generally, women targeted plant communities where the probability of encountering resources was greatest. This pattern of selective gathering indicated they were aware of the different spatial and temporal productivity of plant communities.

Burnt wetland communities (Fig. 3) had a high species richness (12.8 species) and diversity index (2.0). In "Wantajarra", wetland systems were preferentially selected as areas where plant resources could be gathered in large quantities. Associated foods included: the seeds of *Brachiaria miliiformis, Panicum decompositum, Phragmites* sp. FW 142 and other annual grasses; the seeds of *Acacia eriopoda* and *Eucalyptus microtheca;* tubers of *Cyperus bulbosus*

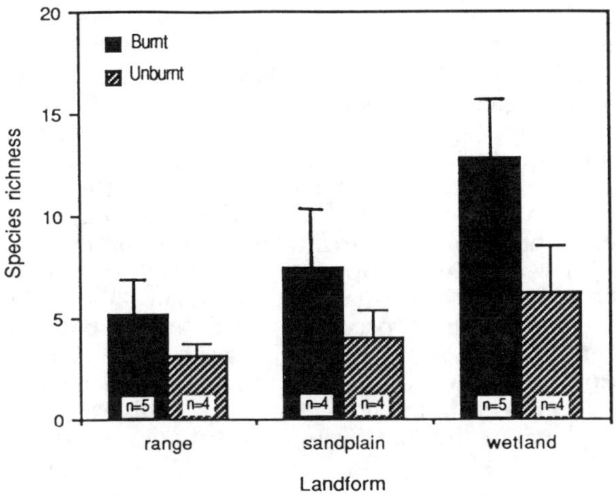

FIG. 3A. Species richness of food plants that occurred on major landform units. Mean and standard error.

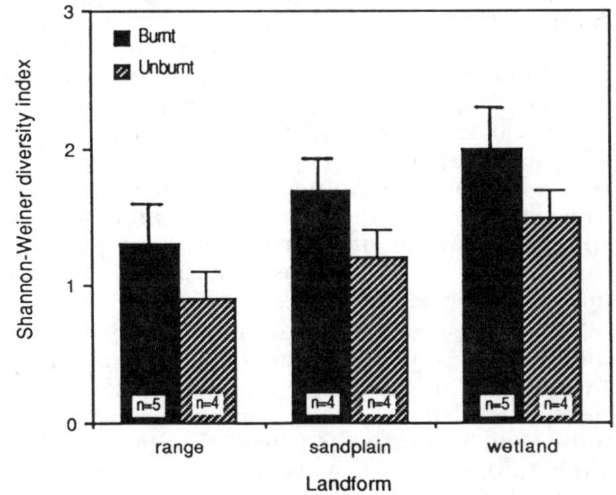

FIG. 3B. Diversity indices of food plants that occurred on major landform units. Mean and standard error.

and *Vigna lanceolata;* and the fruit of *Santalum lanceolatum.* Watercourses were also important sources of shade and often water. The variety and significance of resources they provided increased their importance.

Although not identified in Figure 3, ecotonal communities had high levels of species richness and diversity. Their productivity was apparently enhanced by the surface runoff they received from adjacent rocky surfaces or the overflow and seepage from adjacent wetlands. The ecotones between range and sandplain and wetlands and sandplain were often used as areas for resource collection. They were targeted less specifically than wetlands but were frequently traversed by women on routes connecting one gathering patch to another or one camp site to another.

Patches of burnt spinifex sandplain were specifically targeted by Martu women. Frequently these had a high diversity and richness of resource species. Latz (1982) noted in Central Australia that 70% of resource species regenerated after a fire. The extent of resprouting and reseeding of arid zone species depended on rainfall after the fire (Griffin & Friedel 1984a,b). Potential resource species associated with early seral communities included: *Eragrostis eriopoda, Sida fibulifera* and *Yakirra australiensis* with edible seeds; five species of *Solanum* with edible fruits; and *Petalostylis cassioides* which hosted a sugary lerp. Species such as these persisted for several years after a fire. A paired t-test indicated that on all landform units the richness and diversity index of areas burnt less than five years ago was significantly greater ($t=1.9$, $n=36$, $0.05<p<0.1$) than that burnt more than five years ago. The diversity of these burnt patches declined as successional species became established.

Later seral communities were low in species diversity and richness. However, several significant resource species occurred in them. These were often in monospecific stands, for example the nectar producing *Grevillea* aff. *eriostachya.* When these species were in flower, women would make long diversions over relatively unproductive country to collect these prized resources.

The specificity with which Martu women targeted plant communities varied throughout the seasonal cycle. This was because of the changes in species availability. Data recorded in "Tulparra" indicated women spent up to 30% of their time on collecting excursions tracking and excavating reptiles. In precontact times they may have also hunted small marsupials and other game. It is difficult to determine the spatial extent and frequency of these activities due to the widespread decline of medium-sized marsupials (Burbidge *et al.* 1988). Hunting excursions were conducted over a wider range of habitats than gathering excursions. When collecting reptiles, women spent relatively more time on sandplains and interdunal areas.

Generally, men were less selective of the plant communities they visited and ranged over larger areas than women. This was partially because, when hunting game, they traversed the array of habitats frequented by animals that occurred at low densities. It was mainly when finding the raw materials for tools and implements that men focussed their activities on certain plant communities.

Quantitative survey methods provided a scientific approach to identifying the characteristics of the plant communities where Martu women harvested resources. But how did they distinguish the distribution of resources and then find these in the vast, apparently homogeneous landscape of the Sandy Deserts? The vegetation surveys indicated there was considerable heterogeneity in the distribution of plant communities. The answer is complex. It reflects the detailed body of traditional ecological knowledge held by Aboriginal people. This has been accumulated by long-term association with their environment. It also reflects the interplay between the mythological and material spheres of Martu life. First, it was necessary to determine how Martu classified their landscape and its components.

At the broader level of landscape systems, Martu identified country according to the dominant landform type. For example, in Manjilyjarra language: sandplain ("lannga") was distinguished from sand dune ("tali") and claypan ("linji") was distinguished from saltlake ("warla"). Within these landforms, soil and rock types were named to refine the reference. Again in Manjilyjarra: red rock ("miji miji puli") and white rock ("pilya puli") were named; these distinguished sandstone rock from quartz rock.

Within these landform units, Martu named plant associations by reference to the species that was dominant. Commonly, this was refined by naming a group of plants according to the presence of resource species. The suffix "-kurru" was added to signify dominance of the reference species. In Manjilyjarra, an *Acacia coriacea-Hakea suberea-Triodia basedowii* association was referred to as a "janparrakurru" (after the dominant *T. basedowii*). However, within this association a stand of *Acacia jensenii* (a preferred spear tree) was precisely identified as a "mulyartikurru".

Some *a priori* knowledge of a region was transmitted through the oral literature. In song cycles and stories recited by Martu women the tracts of country associated with camping grounds and geographic features were described. Information on the resources to be found near these areas was conveyed in some verses and lines. The experiences based on previous visits to areas were also drawn upon. The location and productivity of certain species was predictable. Some sites were specifically described as being "the place" for a species. For example, a major occupation site situated on a sandplain by an ephemeral watercourse was referred to as a "kunyjamurra (bush yam) place". Furthermore, when people had visited tracts of country they frequently returned to camp and recounted the location, abundance and condition of resource species they had encountered. Such information was commonly exchanged between men and women, particularly when they had found or sighted evidence of species generally collected by the opposite sex. For example, women described to men the species, condition and ages of animals whose tracks they had encountered and the direction the tracks took, thereby, providing information that was useful to the hunter. The environment and its components was, and continues to be, a regular topic of conversation.

When questioned about where a plant was found Martu generally identified the landform type where the species was expected to grow. They often qualified this by naming a place where it was abundant, such as "Winakarujunu" — the yam place described above. Rarely did they detail the description by referring to the plant community with which the species was commonly associated. Occasionally individual plants or small stands of a species did not occur in their usual habitat. Interestingly, some of these plants were of significance to the Martu; they identified them as having a special role in mythological events. This was their explanation for the disjunct distribution of these plants.

Movements in the landscape

The next scale in the ecological hierarchy involved the selection of routes and sites in the landscape; these ensured productive plant communities and other resources were encountered.

Traditional camping grounds were situated at locations where water was readily accessible, usually less than 2 km away and in locations that ensured access to a variety of plant communities. These sites varied from simple camps occupied by small groups for a few days to extensive habitation sites where large groups aggregated (Veth 1987). Occupation sites were the focus of peoples activities; from them people ventured through the landscape. Two types of movement may be distinguished in relation to camp sites (after Binford 1980). First, those in proximity to a camp, and second, those that involved travel from one camp to another.

Gould (1969) suggested that in proximity to their camp, Ngatatjarra consumed staples collected from a radius of eight to sixteen km. The distances covered by Martu women were far less. The farthest they walked from camp was within a radius of 5.6 km. The size of this discrepancy suggested that different mobility strategies were practised by these groups. The application of a standard foraging radius can be misleading due to the heterogeneous use of the landscape. However, it does help to indicate the size of the area from where plant resources were intensively collected. The reconstructions have shown that in proximity to the camp, plant food gathering focussed on communities associated with wetland systems and ecotones. Hunting and seasonal gathering excursions diverged from these communities. This pattern may be extrapolated to identify travel routes between camps. Martu have described some of the routes taken from one site to another: these followed watercourses and the base of hills and ranges. Inevitably, this was not possible in dunefields where watercourses and uplands were present. It was apparent that within a group's range, the prevailing topography influenced movement pathways.

Broad comparisons were made between the mobility patterns of three language groups in the region (see Fig. 1). Rudall River represented a relative corridor of reliable food and water resources along its 120 km length. The Warnman utilized the riverine system as a major movement pathway and occupied a series of camping grounds along its length. Sites that ensured access to both range and sandplain habitats, such as along the north-west section, were preferred.

To the south-east, the Kartujarra inhabited land characterized by a series of headwater and sandplain streams that drained from the ranges and hills onto the sandplains. These streams were up to 25 km long and were areas of resource concentration. Surface water was available in rockholes, soaks, claypans and ephemeral pools. In this area, movement pathways paralleled the streams that flowed from the ranges, then crossed sandplains, before connecting to streams associated with the next range system.

Extensive dunefields and saltlakes dominated Manjilyjarra country to the north-east of Rudall River. Surface waters were limited and there were no continuous watercourses. A series of saline playas influenced the direction of movements. Resources and soaks along their shorelines were utilized, whilst the vast saline expanses were not accessible. It would have been necessary for the Manjilyjarra to make more regular traverses of sandplains and dunefields than Warnman or Kartujarra people.

Regional movements in space and time

The Martu had a dynamic pattern of group aggregation and dispersal and lived in a changing environment. A preliminary descriptive model of resource availability in the region and subsistence patterns adopted by them has been described (Fig. 4). It documents variables that affected the intensity of resource and land use. It assumed an "average" rainfall year. The relative significance of resource groups is presented. The plant communities targeted are suggested. Mobility was described as the relative frequency of moves from one camp to another. Group size indicated the relative number

KARTUJARRA SEASON	YALIJARRA	WANTAJARRA	TULPARRA	YALIJARRA
WEATHER				
Temperature	hot	cool	warm	hot
Rainfall	high intensity showers	widespread rains	occasional rains	occasional thunderstorms
WATER SOURCES				
Permamency	intermittent & ephemeral	ephemeral & semi-permament	semi-pemament	permament
Classification	claypans	claypans, pools	rockholes, pools	rockholes, springs, soaks
ANIMAL RESOURCES				
Mammals and birds	***	***	***	***
e.g.	emus, red kangaroos	bustards, nestlings, pups	bustards	euros
Reptiles	***	*	**	**
Insects	***	***	***	**
e.g.	termite aeletes	witchetty grubs	witchetty grubs	lerps
PLANT RESOURCES				
Seeds	*	***	***	**
Fruits	*	**	***	**
Tubers	**	***	***	**
Stores	**	*	-	***
SELECTED PLANT GATHERING AREAS	claypans & surrounding sandplain/dune country	primary & headwater streams	sandplain & continue to target watercourses	variety of habitats
GROUP MOBILITY	high	moderate	low	relatively sedentary
GROUP SIZE	small	medium	medium to large	small
PARTS OF RANGE UTILISED	marginal areas of range	outer favoured areas of range	inner favoured areas of range	dry reserve areas, probably of estate

APPROXIMATING MONTHS: ----J--------F--------M--------A--------M--------J--------J--------A--------S--------O--------N--------D--------J---

Relative importance of food groups: *** = high ** = moderate * = low - = not available

FIG. 4. Preliminary descriptive model of resource availability and subsistence patterns adopted by the Martu.

of individuals who were based at a single water source. Commonly, the basic foraging or economic unit (the band) comprised two or more family groups of 6 to 30 individuals (Tonkinson 1978), the size of these groups varied in relation to the scarcity or abundance of resources. The bands collected resources over an area referred to as the range which varied according to prevailing environmental conditions. In the future this model will be refined by the relative quantification of additional variables, such as the extent of a group's range, the frequency with which areas were traversed and the length of time an area was occupied. Presently, it is possible only to provide a general descriptive account of when and where Martu were utilizing resources; this is presented below and in Figure 4.

Late in "Yalijarra" (approximately January to March) summer rains fell. Two to three weeks after the rains commenced termite aeletes were collected. Surface waters were widely scattered, game and plant resources occurred at low densities. Consequently, small groups were highly mobile. The scattered ephemeral waters permitted short-term access to marginal areas and the use of associated unharvested resources. Preceding the cool period, varanid and other reptiles were heavily hunted because of the relatively large fat reserves that they carried. Toward the end of the rainy season, semi-permanent rockholes and pools replenished. Ephemeral waters evaporated. Groups of people retracted to areas within their estate and the spatial extent of their movements contracted.

"Wantajarra" (approximately March to August) was the cool period of the year. Plants were important food sources, particularly seeds, tubers and early ripening fruits. Reptiles were less important as they hibernated. Women targeted resource areas associated with wetlands most specifically in this period. Excess seed foods were harvested and stored. Bands moved between semi-permanent water sources. Gatherings and meetings occurred at reliable waters in areas with abundant resources.

"Tulparra" (approximately August to October) was considered a pleasant period of the year. A diverse array of plant foods was available. Foraging pathways extended over sandplains and sand dunes, patches of burnt spinifex grassland produced *Solanum* fruit. Seeds continued to be harvested along watercourses. Reptiles re-emerged and were again added to the menu.

In the early dry months of "Yalijarra" temperatures increased. Bands made their way to their "yinta" (focal area). This included a permanent spring or soak where the water persisted through dry periods. Within the estate of all family groups there was at least one "yinta". The array of available plant food declined. Acacia seeds, tubers and some fruits were used and stored reserves were drawn upon. Animal foods increased in significance as dietary components. Men and women were less specific in the habitats they targeted. Camp sites were occupied for relatively long periods of time as the distance between potable sources was large. The depletion of food resources was a major factor that influenced the need to relocate camps. With the commencement of rains later in "Yalijarra" groups fragmented and dispersed; so the cycle continued.

Management strategies used by Martu

Harvesting and cultivation were activities of the Martu that potentially affected the availability of plant species. The specific impact of these activities was not explored in this study. There is ethnographic evidence that localized depletions of preferred plant foods occurred when areas in proximity to occupation sites and along movement pathways were intensively foraged. Martu recounted times when "mai (plant food) all finished, we go (somewhere else)". This was reinforced by several statements that groups moved in response to women complaining it was difficult to find plant foods. Species producing fruits and tubers were likely to be prone to short supply because they were restricted to specific habitats, were not widespread and were preferentially harvested. There is ethnographic evidence that Martu managed some of these species. Specific techniques were used to harvest *Vigna lanceolata* and *Cyperus bulbosus* tubers, efforts were made to leave older *V. lanceolata* "mother tubers" and avoid damaging those that remained. Veth and Walsh (1988) suggested the propagation of *V. lanceolata* may have been enhanced by digging that fragmented its tubers and roots, thereby enhancing the dispersal and reproduction of the species.

There is other evidence that Martu cultivated preferred species. An elderly Martu man described the deliberate process of how, prior to rains in "Yalijarra", bush tomato *(Solanum diversiflorum)* seed was scattered on burnt areas near the places where he had camped as a young man. Although less conclusive, Martu had stated that after Contact their relatives spread the seed of bitter melon *(Citrullus lanatus)* and buffel grass *(Cenchurus ciliaris)* along sections of the Rudall River. Other authors have described the consequences of incipient cultivation practices. For example: Hynes and Chase (1982) suggested Aboriginal activities may have created some plant communities of Cape York Peninsula; and, Gott (1983) described how the abundance and productivity of the daisy yam *(Microseris scapigera)* was likely to have been increased by Aboriginal harvesting. Ethnohistorical and ethnographic evidence collated by Kimber (1984) indicated Aboriginal people in central Australia took measures to enhance the productivity of certain species. He suggested these aided in making "life more secure in a drought-prone area" (Kimber 1984:21).

The introduction of species to sites beyond their "normal" range or habitat may have been a consequence of trade, exchange and storage. Martu stated that plant foods, particularly seeds and fruits, were exchanged at social events and through trade practices. They dried and preserved a range of fruit species for use in the short-term, and cached large quantities of seed and *Cyperus bulbosus* tubers in anticipation of drought and large meetings. Kimber (1984) also emphasized that storage was a necessary practice.

The use of fire by Aborigines in arid Australia has been well documented (see Kimber 1983). Latz and Griffin (1978) proposed that in central Australia fire effectively increased the diversity of habitats to form a patch mosaic. Furthermore, Latz (1982) suggested fire was utilized to promote the regeneration of preferred resource species. In the Great Sandy Desert, post-burn seral stages are frequently observed along the roads and tracks and radiating from the camps Martu now use. Within this study, vegetation surveys identified the relatively greater diversity and richness of resource species growing on recently burnt patches. From this evidence it may be inferred Martu also used fire as a resource management tool.

The above paragraphs address the tangible consequences of Aboriginal resource management activities that may be readily recognized by ecologists. In contrast, anthropologists discuss active resource management in a far broader sense, looking at the implications of "strategic ecological or economic courses of action via social controls and political manoeuvre . . . or by symbol and ritual" (Hunn & Williams 1982).

Stevenson (1985) provided a more integrated approach in his description of resource management by the Tiwi of Melville and Bathurst Islands. He considered management beyond the use of tools and addressed the basis of their management goals.

The maintenance of a broad resource base was a critical determinant of Martu subsistence strategies. Within this resource base some species were preferred to others. Their management goals were aimed at maintaining this resource base (Veth & Walsh 1988) and increasing the productivity of preferred species. This was achieved by the adoption of subsistence strategies that were dynamic thereby ensuring that they had access to resources that were productive under the prevailing conditions.

In summary, Martu focussed their activities on discrete camping grounds that occurred in close proximity to water sources and resource-rich areas. The numbers of people at these sites and the length of time they were occupied for varied according to the permanency of water, abundance of food sources and social demands. Generally, foraging excursions radiated from these sites; the women focussed their gathering on plant communities associated with wetlands and areas of recently burnt spinifex sandplain, while avoiding hills and ranges. When people moved from one camp site to another they generally continued to utilize these communities, taking routes that followed watercourses, the base of hills and traversed sandplains. It was over this spatial framework that management tools, burning, harvesting etc., were applied.

Collaboration in the future

There is a large body of traditional ecological knowledge that is relevant to the pursuits of ecologists. This includes information applicable to the reconstruction of pre-contact environments, identification of post-contact environmental changes, studies of species biology and community interactions and alternative resource management practices. The preceding sections presented an overview of traditional patterns of resource use and management by Martu. This section proposes some ways in which these may be interpreted and applied by Western ecologists and land managers.

There are fundamental differences in Western and Aboriginal concepts of time and seasons (see Rose 1988). Furthermore, there are variations in how seasons are described in different environments. Aboriginal people living in tropical environments used biological indicators to distinguish seasons; these combined to form cycles of well-defined periods of time (see Chase & Sutton 1981; Smith & Kalotas 1985). In contrast, Martu never expressed seasons in terms of a cycle, instead, they described continuous periods that were named according to the prevailing weather. There was a probability that climatic events (usually rainfall) would punctuate these periods; these events would then initiate a chain of biological responses. There was a seasonal pattern to the probability of this rainfall. Management practices should be implemented as appropriate to the prevailing season; given that these may vary in relation to a Roman calender, these practices should be applied on a flexible time schedule.

The use of management tools was briefly discussed above. Attempts to recapitulate traditional burning patterns have been trialled in contemporary management situations (Allan & Baker, this volume). Some details relevant to the frequency, seasonal timing and habitat specificity of fires continue to elude researchers. It was a common practice for Aboriginal people to burn patches of country as they travelled. Griffin and Allan (1986) presented a diagram of a hypothetical fire mosaic imposed on a route traversed by

Aborigines moving from one waterhole to another. By using a reconstruction of spatial and temporal patterns of resource and land use, fire ecologists may be better equipped to predict when and where it is appropriate to burn. This may be achieved by the establishment of a pilot study in a restricted and defined region; alternatively modelling procedures may be trialled for a hypothetical landscape. Firstly, it would be necessary to identify the distribution and location of different types of habitation sites on topographical map sheets. Secondly, the movement pathways that linked these could be plotted in relation to plant communities of the region. Finally, the temporal patterns, when and how frequently sites were occupied or pathways traversed, could be superimposed.

This procedure would be more readily facilitated if a system of land management units were adopted. Latz (1982) proposed the abundance of food and water resources and the presence of mythological sites formed the basis of a system of land classification. He suggested six different classes of country were defined; the use of these by people differed according to prevailing conditions. There is no conclusive evidence that in the arid zone Aboriginal people perceived such a system, however, this possibility should be explored in the future. An integrated approach to mapping anthropological and environmental data on western Cape York was developed by Cribb (1986); this method has the potential to be utilized for management and land use planning procedures in the arid zone.

It was noted above, that even within Martu "country" there were variations in subsistence and mobility patterns in relation to the prevailing topography, species availability and water sources. This emphasises that it is inappropriate to consider these patterns to have been the same throughout the Western Desert and central Australia. For example, general comparisons show that the use of resources by the Ngatatjara was likely to have been less intensive than resource use by Martu. The Ngatatjara were less sedentary than Martu, being highly mobile between widespread points in their landscape (Veth 1987). Alternatively, the use of plant communities varies amongst language groups. In contrast to Martu, the Alyawara visited riverine floodplain communities less frequently than sandhill communities disturbed by fire; their activities were primarily focused on mulga communities (O'Connell & Hawkes 1981). This was largely because of the absence of mulga communities and preferred species, particularly *Ipomoea costata,* in Martu country; this apparently small difference may be manifested in different mobility patterns. It is important that land managers recognize the spatial heterogeneity of desert ecosystems, particularly in localized regions.

The opportunities for collaboration between ecologists and Aboriginal people are present. However, a sense of urgency invariably pervades those already involved in such work. Inevitably detailed knowledge is lost when people are no longer living a hunter-gatherer lifestyle. Despite the interest shown by younger Aboriginal people in traditional ways; there are numerous distractions to the learning process. With the passing of elders also goes knowledge that may be irretrievable.

Increasingly, Aboriginal people are land managers operating independently or in conjunction with management agencies. Furthermore, they continue to interact with, utilize and exert an influence on their environment. To effectively communicate it is necessary to appreciate and understand the different perceptions of Aboriginal and non-Aboriginal cultures. This understanding may be achieved by working in collaboration with Aboriginal people and social scientists, to ensure that, as ecologists, we maintain, interpret and apply a body of traditional ecological knowledge that is relevant to future environments.

Acknowledgements

The Department of Botany, the Australian Institute of Aboriginal Studies and the Western Desert Puntukurnuparna have supported this research with the provision of funds and facilities. The project is supervised by Dr David Bell, Dr William Loneragan and Prof. Robert Tonkinson. Field assistance was provided by Tim Foley, John Walsh, Louis Warren and Matthew Wrigley. Aspects of the work involved collaboration with Arpad Kalotas, Nick Thieberger and Peter Veth. Attendance at a Workshop on Traditional Ecological Knowledge increased my understanding of this field of study. I thank all these people for their contribution. Finally, I hope this accurately conveys the perceptions and some of the aspirations of the Martu who have been the focal point of this work.

References

Altman J. C. (1987) *Hunter-gatherers today: an Aboriginal economy in north Australia.* Australian Institute of Aboriginal Studies, Canberra.

Anon. (1979) *Average annual rainfall map of Western Australia.* Bureau of Meteorology, Perth.

Atkins K. J. (1985) *Studies into the ecological regeneration of an iron ore mine site in the arid Pilbara region of Western Australia.* Ph.D. Thesis. University of Western Australia, Perth.

Baines G. (1986) Ecologists and anthropologists, unite. *Tradition, conservation and development.* Occasional newsletter of the Commission on Ecology's Working Group on Traditional Ecological Knowledge, **4,** I.U.C.N./C.O.E.

Beard J. S. & Webb M. J. (1974) *Vegetation survey of Western Australia, Great Sandy Desert 1:1 000 000 Vegetation series.* University of Western Australia Press, Perth.

Berndt R. M. (1959) The concept of "the tribe" in the Western Desert of Australia. *Oceania* **50,** 118-32.

Binford L. R. (1980) Willow smoke and dogs' tails: hunter-gatherer settlement systems and archaeological site formation. *Am. Antiquity* **1,** 5-31.

Burbidge A. A., Johnson K., Fuller P. J. & Southgate R. I. (1988) Aboriginal knowledge of the mammals of the Central Deserts of Australia. *Aust. Wildl. Res.* **15,** 9-39.

Cane S. B. (1984) *Desert camps: a case study of stone artefacts and Aboriginal behaviour in the Western Desert.* Ph.D. Thesis. Australian National University, Canberra.

Chase A. & Sutton P. (1981) Hunters and gatherers in a rich environment: Aboriginal coastal exploitation in Cape York Peninsula. In: *Ecological biogeography of Australia* (ed. A. Keast). Dr Junk, London.

Cribb R. (1986) A graphics system for site-based anthropological data. *Aust. Ab. St.* **1986/2,** 24-30.

George A. S. & Mitchell A. S. (1983) Part II Flora. In: Wildlife of the Great Sandy Desert, Western Australia (eds A. A. Burbidge and N. L. McKenzie). *Wildl. Res. Bull. West. Aust.* **12,** 38-61.

Goble-Garratt E. M. (1987) *Phytosociology of the Telfer area of the Great Sandy Desert, Western Australia.* M.Sc. Thesis. University of Western Australia, Perth.

Gott B. (1983) Murnong- *Microseris scapigera:* a study of staple food of Victorian Aborigines. *Aust. Ab. St.* 2-14.

Gould R. A. (1969) Subsistence behaviour among the Western Desert Aborigines of Australia. *Oceania* **39,** 253-74.

Gould R. A. (1980) *Living archaeology.* Cambridge University Press, Cambridge.

Griffin G. F. & Allan G. (1986) Fire and management of Aboriginal owned lands in central Australia. In: *The application of science and technology to Aboriginal development in central Australia* (eds B. Foran & B. Walker) CSIRO Division of Wildlife and Rangelands Research, Project Report 3. CSIRO, Melbourne.

Griffin G. F. & Friedel M. H. (1984a) Effects of fire on central Australian rangelands. I Fire and fuel characteristics and changes in herbage and nutrients. *Aust. J. Ecol.* **9,** 381-93.

Griffin G. F. & Friedel M. H. (1984b) Effects of fire on central Australian rangelands. II Changes in tree and shrub populations. *Aust. J. Ecol.* **9,** 395-403.

Hunn E. S. & Williams N. M. (1982) Introduction. In: *Resource managers: North American and Australian hunter-gatherers* (eds N. M. Williams & E. S. Hunn). Australian Institute of Aboriginal Studies, Canberra.

Hynes R. A. & Chase A. R. (1982) Plants, sites and domiculture: Aboriginal influence upon plant communities in Cape York Peninsula. *Archaeol. Oceania* **17,** 38-50.

Kean J. S., Richardson G. & Trueman N. (1988) Aboriginal role in nature conservation. *Emu Conference.* June 7-9, 1989, Emu, South Australia.

Kimber R. G. (1983) Black lightning: Aborigines and fire in central Australia and the Western Desert. *Archaeol. Oceania* **18,** 38-45.

Kimber R. G. (1984) Resource use and management in central Australia. *Aust. Ab. St.* **1984/2,** 12-23.

Latz P. K. (1982) *Bushfires and bushtucker: Aborigines and plants in central Australia.* M.A. Thesis. University of New England, Armidale.

Latz P. K. & Griffin G. F. (1978) Changes in Aboriginal land management in relation to fire and food plants in central Australia. In: *The nutrition of Aborigines in relation to the ecosystems of Central Australia* (eds B. S. Hetzel & H. J. Frith). CSIRO Melbourne.

Maconochie J. R. (1982) Regeneration of arid zone plants: a floristic survey. In: *Evolution of the fauna and flora of arid Australia* (eds W. R. Barker & P. J. M. Greenslade). Peacock Publications, Frewville.

Masini R. J. (1988) *Inland waters of the Pilbara Part 1.* Environmental Protection Authority Technical Series **10,** Environmental Protection Authority, Perth.

Meehan B. (1982) *Shell bed to shell midden.* Australian Institute of Aboriginal Studies, Canberra.

Morton S. R. & James C. D. (1988) The diversity and abundance of lizards in arid Australia. *Am. Nat.* **132,** 237-56.

Newsome A. E. (1980) The eco-mythology of the Red Kangaroo in central Australia. *Mankind* **12,** 327-33.

O'Connell J. F. & Hawkes K. (1981) Alyawara plant use and optimal foraging theory. In: *Hunter-gatherer foraging strategies: ethnographic and archaeological analyses* (eds B. Winterhalder & E. A. Smith). University of Chicago Press, Chicago.

Pate D. (1986) The effects of drought on Ngatatjara plant use: an evaluation of optimal foraging theory. *Hum. Ecol.* **14,** 95-115.

Rappaport R. A. (1984) *Pigs for the ancestors: ritual in the ecology of a New Guinea people.* Yale University Press, New Haven.

Rose D. B. (1988) Exploring an Aboriginal land ethic. *Meanjin* **4,** 378-87.

Rose F. G. G. (1987) *The traditional mode of production of the Australian Aborigines.* Angus & Robertson, Sydney.

Senft R. L., Coughenour M. B., Bailey D. W., Rittenhouse L. R., Sala O. E., & Swift D. M. (1987) Large herbivore foraging and ecological hierarchies. *BioScience* **37,** 789-99.

Smith M. & Kalotas A. C. (1985) Bardi plants: an annotated list of plants and their use by the Bardi Aborigines of Dampierland, in north-western Australia. *Rec. West. Aust. Mus.* **12,** 317-59.

Smith M. A. (1988) Pleistocene occupation in central arid Australia. *Nature* **328,** 710-1.

Stevenson P. M. (1985) Traditional Aboriginal resource management in the wet-dry tropics: Tiwi case study. *Proc. Ecol. Soc. Aust.* **13,** 309-15.

Sutton P. (1976) Linguistic aspects of ethnobotanical research. Unpubl. manuscript, Australian Institute of Aboriginal Studies, Canberra.

Tindale N. B. (1981) Desert Aborigines and the southern coastal peoples: some comparisons. In: *Ecological biogeography of Australia* (ed. A. Keast). Dr Junk, London.

Tonkinson R. (1978) *The Mardudjara Aborigines: living the dream in Australia's desert.* Holt, Rinehart and Winston, New York.

Veth P. M. (1987) Martujarra prehistory: variations in arid zone adaptations. *Aust. Archaeol.* **25,** 102-11.

Veth P. M. (1989) The relevance of the desert to Australian prehistory. Seminar presented to the Department of Archaeology, University of Western Australia, March 1989.

Veth P. M. & Hiscock P. (1988) The tula adzes ain't: a reanalysis of the adzes from Puntutjarpa. Paper presented to the Technology Conference, Armidale, Nov. 1989.

Veth P. M. & Walsh F. J. (1988) The concept of "staple" plant foods in the Western Desert of Western Australia. *Aust. Ab. St.* **1988/2,** 19-25.

Winterhalder B. & Smith E. A. (eds) (1981) *Hunter-gatherer foraging strategies: ethnographic and archaeological analyses.* University of Chicago Press, Chicago.

Proc. Ecol. Soc. Aust. 1990 — 16:39-53

The impact of nutrient enrichment on nearshore and estuarine ecosystems in Western Australia

K. HILLMAN, R. J. LUKATELICH and A. J. McCOMB

Centre for Water Research, Department of Botany, The University of Western Australia, Nedlands, W.A. 6009

Abstract

This chapter describes how industrial development and agricultural practices have caused the nutrient enrichment of nearshore and estuarine ecosystems. It discusses management strategies that have been used or are being developed to overcome the resulting eutrophication problems, using two Western Australian ecosystems as case studies. These are the Peel-Harvey estuarine system and Cockburn Sound.

The Peel-Harvey estuarine system is a broad shallow water body with restricted exchange with the ocean. Clearing of native vegetation and drainage in the catchment (especially the coastal plain portion) have greatly increased the amount of nutrients entering the system since the 1950s. Environmental deterioration of the marine embayment of Cockburn Sound also dates from the 1950s, but has been caused by nutrient enrichment from industrial outfalls following development of the area as a major industrial complex and harbour.

Emphasis is given to two major principles underlying the management of aquatic ecosystems: that such ecosystems cannot be studied as isolated water bodies, and that a multi-disciplinary approach is essential for successful management.

Introduction

Human settlement has usually centred around bodies of water. Seven of the ten largest metropolitan areas in the world border estuaries (Cronin 1967). This pattern of settlement is particularly noticeable in Australia, where the aridity of the centre of the continent has reinforced development in coastal areas around rivers and estuaries. Such areas offered advantages of easy transport, easy waste disposal, water for industry, farming and domestic use, a ready food supply and, of course, aesthetic surroundings. This settlement pattern has resulted in water bodies being subject to the combined pressures of residential, industrial and farming landuse. There are now many examples both within Australia (Lake Illawarra in New South Wales and the Gippsland Lakes in Victoria) and overseas (Chesapeake Bay, the Thames estuary) where these landuse pressures have resulted in the environmental degradation of water bodies. Major management issues have arisen because waterscapes are also preferred by most people, both local and tourists, for their passive and/or active recreation.

In developed countries, political awareness of the need for sound management of aquatic ecosystems has been helped by the growing importance of tourism. Tourism is now regarded as Australia's largest industry based on percentage of G.N.P. (1981) and Western Australia's second largest (Hitchen cited in Thurlow *et al.* 1986). In addition, the public expresses concern when aquatic recreational areas are degraded and, in this respect, managers of water bodies have perhaps been more fortunate than those involved in other ecosystems. In Western Australia for example, public and governmental concern has led to studies being carried out on a number of water bodies, including the Peel-Harvey estuarine system (Hodgkin *et al.* 1985), Cockburn Sound (Anon. 1979), the Blackwood River estuary (Hodgkin 1978), the Swan/Canning estuary (Thurlow *et al.* 1986), Wilson Inlet (Lukatelich *et al.* 1984) and Princess Royal and Oyster Harbours, Albany (Anon. 1988).

In this chapter, a review is presented of the problems, causes, management strategies and responses to management of two aquatic ecosystems in Western Australia that are affected detrimentally by nutrient enrichment: the Peel-Harvey estuarine system and Cockburn Sound (Figs 1 and 6). These water bodies, the two most intensively studied aquatic ecosystems in Western Australia, illustrate the differences in management approach to nutrient enrichment from a point source (Cockburn Sound) versus a diffuse source (the Peel-Harvey).

Cockburn Sound

The nature of the problem. Cockburn Sound is a sheltered marine embayment situated some 20 km south of the Perth-Fremantle area. It is 16 km long by 9 km wide, and consists of a deep central basin (17-22 m deep) surrounded by shallow sandy platforms which range in width from 50 m to 3 km (Fig. 1). The Sound has long been considered a valuable recreational asset of the Perth region because of its proximity, its sheltered waters and beaches, and its abundant supply of fish, crustaceans and molluscs.

Prior to 1954, Cockburn Sound was used mainly for recreational purposes and commercial fishing, but its relatively protected deep waters also constituted an excellent anchorage, hence its choice as the outer harbour for the Perth-Fremantle area. Industrial development of the eastern shore of Cockburn Sound commenced with the establishment of an oil refinery in 1955 and the next 25 years saw the addition of iron, steel, alumina and nickel refining or processing plants, chemical and fertilizer production plants and a bulk grain terminal. With the development of the industrial complex, wharves and groynes were built and channels dredged for shipping access. In addition to industrial development, a waste water treatment plant was commissioned in 1966 at the northern end of the Sound, whilst at the southern entrance a rockfill causeway was built in 1971-73 connecting Garden Island with the mainland (Fig. 1). The causeway is broken by two bridges (one 305 m long, and one 610 m long), through which limited oceanic exchange occurs.

The 25 years of industrial expansion also saw an increase in population, an increase in leisure time and an upsurge in popularity of water-oriented activities, leading to inevitable conflict between recreational and industrial, land and water use. During the late 1960s and early 1970s, public concern at the deterioration of the environment resulted in a range of baseline environmental studies being carried out. Two major environmental problems were identified, both of which attracted considerable public attention.

(1) The initial and continuing dieback of subtidal seagrass meadows. This loss was of concern from both physical and biological viewpoints. As the name implies, seagrasses are very like terrestrial grasses, and their roots bind and therefore stabilize sediments, whilst their dense leaf canopy acts to baffle water movement, thereby reducing erosion (Fonseca & Kenworthy 1987). Seagrass meadows also provide a sheltered habitat for aquatic fauna (Pollard 1984), and constitute a ready food supply; seagrass epiphytes (flora and fauna) are the preferred food source of many species of fish and crustacea (Lenanton *et al.* 1982), whilst seagrass leaves themselves become a considerable detrital food source (Walker & McComb 1985).
(2) The deterioration of water quality, as seen by the marked colour changes and increased turbidity associated with phytoplankton blooms (Anon. 1979; Chiffings 1979). On occasions the quality of water in popular swimming areas deteriorated to the extent that it became unattractive to swimmers and beach users. There was also public concern about contact with effluents whilst in the water, and contamination of local seafood.

The cause of the problem. In 1975 a review of the baseline studies was carried out and the Western Australian Government funded a three-year (1976-1979) environmental study. The objective of the study was to obtain the information necessary to manage the Sound for multipurpose use, accommodating recreational and fishing activities as well as the port and industry. To achieve this objective seven major topics were identified as requiring detailed investigation.

(1) The social and recreational uses and value of the Sound.
(2) The fisheries production and potential for amateur and professional fishermen.
(3) Description and distribution of urban and industrial effluents.
(4) Assessment of the cause of seagrass dieback.
(5) The cause and possible cures of the phytoplankton blooms.
(6) Appraisal of water movement in the Sound and therefore the transportation, dispersion and dilution of effluents.
(7) Sand movement, including the sedimentation of the Sound, state of sand dunes and recent changes in the eastern shoreline.

Studies of the first two topics documented the present and potential future importance of Cockburn Sound for recreation and fishing, and the concern of community groups about pollution and the future availability and quality of beaches if industrial growth continued (Feilman Planning Consultants 1978; Dybdahl 1979; Marsh & Devaney 1978; Wells 1978; Wilson *et al.* 1978). The remaining five topics are interrelated, and discussed below.

Urban and industrial effluents reaching Cockburn Sound by both direct discharge (Murphy 1979) and groundwater flow (Layton Groundwater Consultants 1979) were documented; the contribution of air emissions was not studied initially but

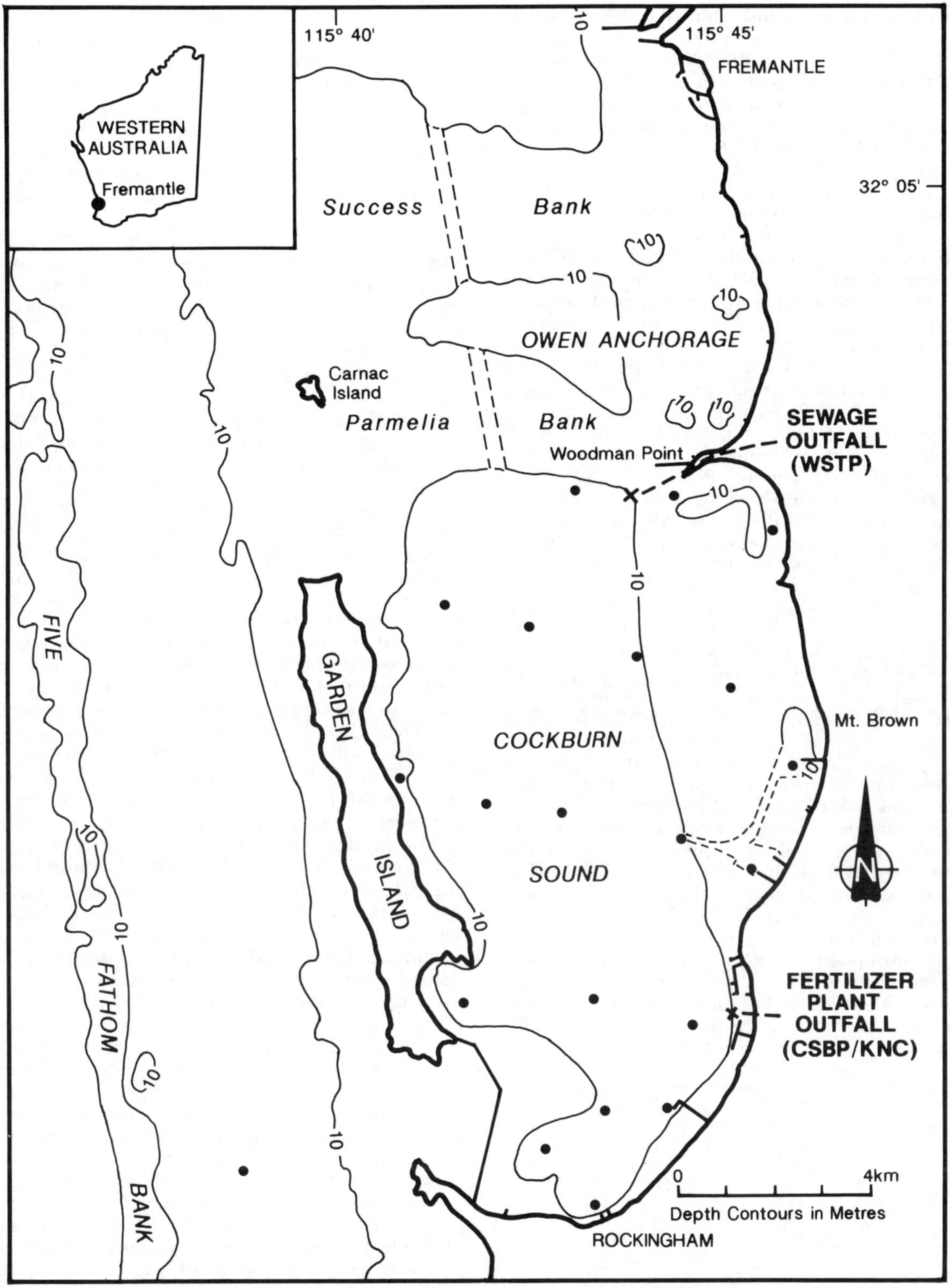

FIG. 1. Cockburn Sound, Western Australia. The 10 m contour is shown; the depth of the basin reaches 20 m. Outfalls from the sewage works and a fertilizer plant are indicated. Redrawn from Chiffings and McComb (1981).

subsequent estimates indicated that this contribution was slight compared to that of the first two sources (Chiffings 1979). Whilst the phytoplankton blooms in Cockburn Sound were caused clearly by nutrient enrichment, a variety of substances could have been responsible for the seagrass dieback. Thus effluents and groundwater were sampled for metals, nitrogen and phosphorous, "other chemicals" [petroleum hydrocarbons from the oil refinery, sulphides from the wastewater treatment plant and oil refinery, soluble fluoride from the Cumming Smith British Petro (CSBP) fertilizer plant], and pH and temperature. In response to public concern, levels of contaminants in fish and shellfish were also sampled (Chegwidden 1979; Iveson 1979) and concentrations of cadmium, lead, chromium and zinc were found to exceed accepted health standards in certain species of edible shellfish. The CSBP discharge was identified as the major source of cadmium, and with the British Petroleum (BP) oil refinery, Woodman Point Waste Water Treatment Plant (WPTP), State Energy Commission power plant (SEC) and Australian Iron and Steel Pty Ltd (AIS), made up most of the input of other heavy metals. Unsafe levels of the bacteria *Escherischia coli* and *Salmonella* serotypes were also found in mussels in the vicinity of the WPTP sewage works. There was some evidence of occasional petroleum-derived hydrocarbons tainting the flesh of various species of fish, which was probably related to the discharge of hydrocarbons from the BP refinery and to oil spills. A wide variety of contaminants was also found in the groundwater of the unconfined aquifers which underlie the shoreline of Cockburn Sound. Common contaminants included nitrogen, chromium, nickel, aluminium oxide, sulphate, chloride and hydrocarbons and, since groundwater flow is essentially seawards, these would eventually end up in the Sound. However, of the range of substances being discharged into the Sound, those found to have the most widespread harmful effects on the environment were nitrogen and phosphorous.

The inputs of nitrogen and phosphorus to Cockburn Sound from the four main sources, as determined in the 1976-79 study, are presented in Table 1. Nearly 80% of the nitrogen and over 95% of the phosphorus were contributed by industrial effluents and sewage plant waste water (Murphy 1979). It was also found that over 90% of both nutrients contributed by industrial effluents and wastewater came from two sources; the outfall shared by the CSBP superphosphate plant and Kwinana Nitrogen Company (KNC), and the WPTP outfall (Table 2). CSBP was found to contribute almost all of the phosphorus in the outfall, and KNC 98% of the nitrogen (Murphy 1979).

TABLE 1. Estimates of nutrient loads to Cockburn Sound waters from external sources.

Source	Total Nitrogen (kg/day)	(%)	Total Phosphorus (kg/day)	(%)
Groundwater	447	7	1	<1
Coastal waters	218	3	145	4
Industrial Effluents and Sewage	4 979	79	3 766	96
Air Emissions	667	10	1	<1

Algal bioassays and investigations of the N:P ratios of the available forms of these elements in the water determined that of the two nutrients, nitrogen was the one limiting phytoplankton growth in the Sound (Chiffings 1979), and therefore that nitrogen input controlled phytoplankton biomass. Studies of the seawater circulation in the Sound concluded that exchange with the ocean was limited, therefore effluents do not move out of the Sound readily. In effect the Sound behaves like a tidal "lake" for most of the time (Steedman & Craig 1979) and phytoplankton blooms can be related to the history of nitrogen loading (Fig. 2). When the sewage plant (WPTP) commenced operation in 1966 there was a large increase in nitrogen loading, and an even larger increase when KNC commenced production in late 1968. Noticeable phytoplankton blooms started occurring when nitrogen loads exceeded approximately 2 000 kg day^{-1}. Since water movement in the Sound is largely wind driven and under most conditions an anticlockwise gyre is present (Steedman & Craig 1979), typical phytoplankton distribution patterns (Fig. 3) also tended to follow nutrient dispersal patterns radiating out from the KNC/CSBP outfall (just south of James Point — see Fig. 1) and WPTP outfall, under the influence of the gyre.

The sudden increase in nitrogen loading due to discharge from the KNC/CSBP outfall also coincided with the beginning of seagrass dieback. As can be seen in Figure 4, the main dieback of seagrass occurred between 1968 and 1972, and commenced just south of James Point, then proceeded northwards along the eastern shore. An investigation into the seagrass dieback indicated that it was due largely to the increased nitrogen supply, which stimulated excessive growth of algal epiphytes on the seagrass leaves, thereby blocking out the supply of light to the seagrass (Cambridge *et al.* 1986; Silberstein *et al.* 1986). In addition, excessive growth of unattached algae, and increased turbidity due to phytoplankton blooms further reduced the amount of light received by the seagrass. This light reduction led to a fall in seagrass productivity, which in turn led to a thinning of the leaf canopy. Once the canopy became thin and/or patchy, there was a reduction in the capacity of the meadows to act as sediment stabilizers, and the plants were abraded by sand winnowed out from around their bases. Thin canopies were also

TABLE 2. Loads of nitrogen and phosphorus (kg/day) in industrial outfalls and sewage discharged to Cockburn Sound.

Outfall	Ammonia (kg/day)	(%)	Nitrate (kg/day)	(%)	Total Inorganic Nitrogen (kg/day)	(%)	Total Nitrogen (kg/day)	(%)	Total Phosphorus (kg/day)	(%)
AIS	51	(1)	–		51	(1)	49	(1)	2	(<1)
BHP	0.2	(<1)	0.6	(<1)	0.8	(<1)	0.9	(<1)	<1	(<1)
BP	271	(7)	4.2	(1)	275	(6)	322	(6)	8	(<1)
KNC/CSBP	2 350	(62)	665	(98)	3 015	(68)	3 075	(62)	3 275	(87)
WPTP	1 114	(29)	10	(1)	1 124	(25)	1 422	(29)	261	(7)
SEC	–		–		–		110	(2)	220	(6)
Total	3 786		679.8		4 465.8		4 979		3 766	

more open to attack by sea urchins, and this combination of factors eventually lead to the complete loss of the meadows (Cambridge *et al.* 1986).

Management Strategy. The Cockburn Sound Study (Anon. 1979) concluded by making a number of management recommendations to overcome the problems of groundwater contamination, microbial contamination of waters and heavy metal contamination of seafood, most of which have since been implemented by the industries concerned. The main recommendation was that nitrogen loads to the Sound be reduced substantially to improve water quality and halt the seagrass dieback. In late December 1982, KNC responded by commissioning a steam scrubber designed to remove a large proportion of nitrogen from the plant's effluent and, in July 1984, waste water originally discharged from the sewage works outfall was diverted to a new pipeline which now discharges into deep water outside the Sound. Although the contribution of nitrogen from the sewage works outfall was much less

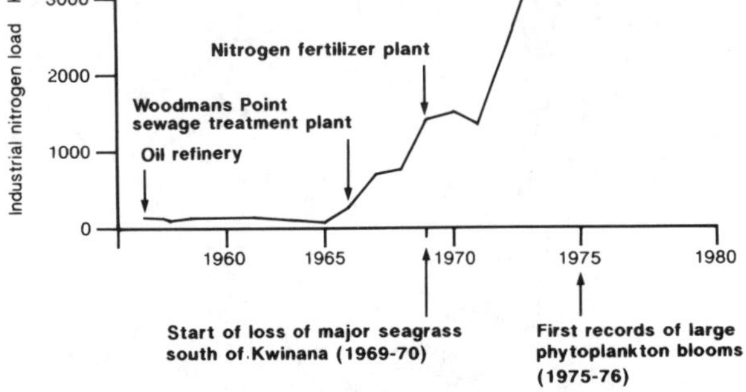

FIG. 2. Estimated nitrogen loads entering Cockburn Sound showing the commencement of discharge from the oil refinery, sewage treatment plant and fertilizer works, together with the major time of seagrass loss and the first record of marked phytoplankton blooms. Redrawn from Cambridge *et al.* (1986).

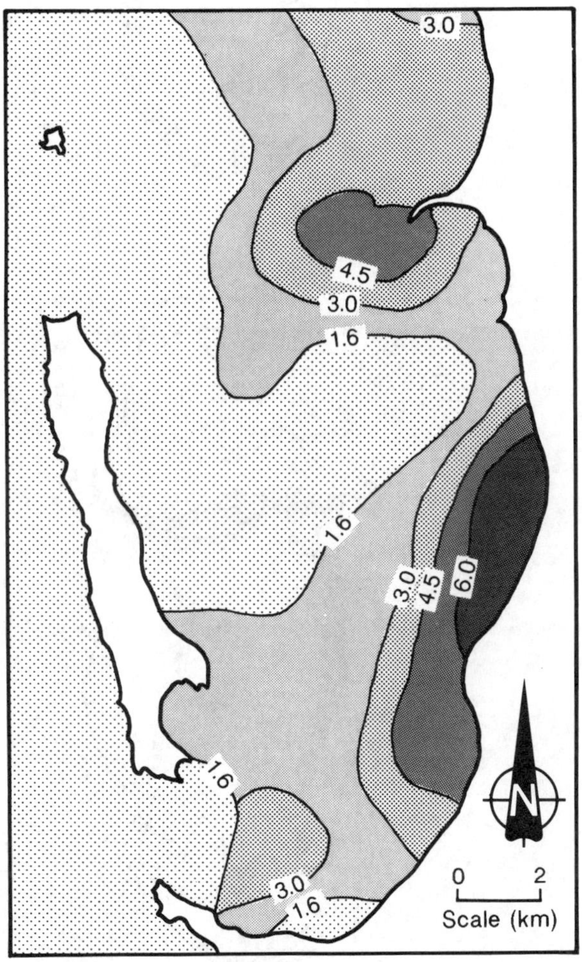

FIG. 3. Typical distribution of chlorophyll "a" (mg m^{-3}) in Cockburn Sound during spring 1977. Redrawn from Chiffings and McComb (1981).

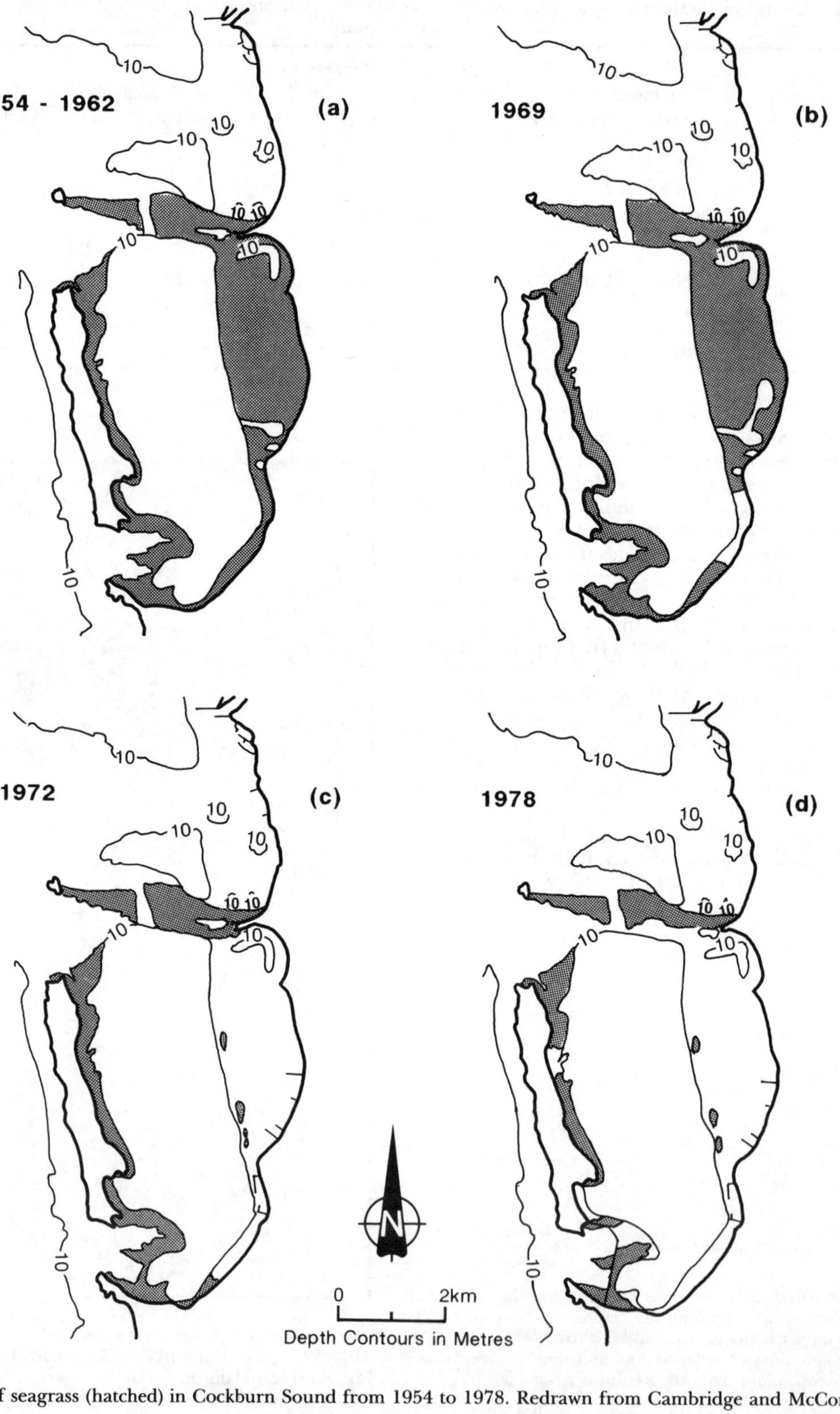

FIG. 4. Loss of seagrass (hatched) in Cockburn Sound from 1954 to 1978. Redrawn from Cambridge and McComb (1984).

TABLE 3. Changes in total nitrogen loads (kg/day) to Cockburn Sound from Woodman Point treatment plant and KNC/CSBP outfalls. Adapted from Hillman and Bastyan (1988).

	WPTP	KNC/CSBP	TOTAL
1977/1978	2 456	3 015	5 561
1981	1 560	1 593	3 153
1982/1983	2 004	563*	2 567
1984/1985	0	974*	974
1986/1987	0	1 128	1 128

*From KNC only.

than that of KNC, this was deemed a necessary management measure because of increased loads predicted from an unexpected population increase.

In order to assess the effect of these management measures on the environmental quality of the Sound, a series of monitoring exercises were carried out over the summers of 1982/83 (Chiffings & McComb 1983), 1984/85 (Hillman 1986) and 1986/87 (Hillman & Bastyan 1988). Table 3 shows the considerable decrease in nitrogen loads from the KNC/CSBP and WPTP outfalls since the 1976-79 study and these improvements were reflected in the environmental response of the ecosystem. In 1984/85, seagrass dieback was found to have halted (Hillman 1986), and phytoplankton levels and water turbidity had decreased significantly (Fig. 5). The findings of the most recent study (1986/87) illustrate the importance of continuous monitoring in ecosystem management. Evidence was found of a slight increase in nitrogen loading from the KNC/CSBP outfall, and a slight decrease in water quality. There was also evidence that because there has been such a marked fall in phosphate loading from CSBP, there could be a shift towards phosphorus limitation of phytoplankton growth. Current monitoring programmes are now designed to determine quickly whether the ecosystem is nitrogen-limited or phosphorus-limited, since this may affect management practices.

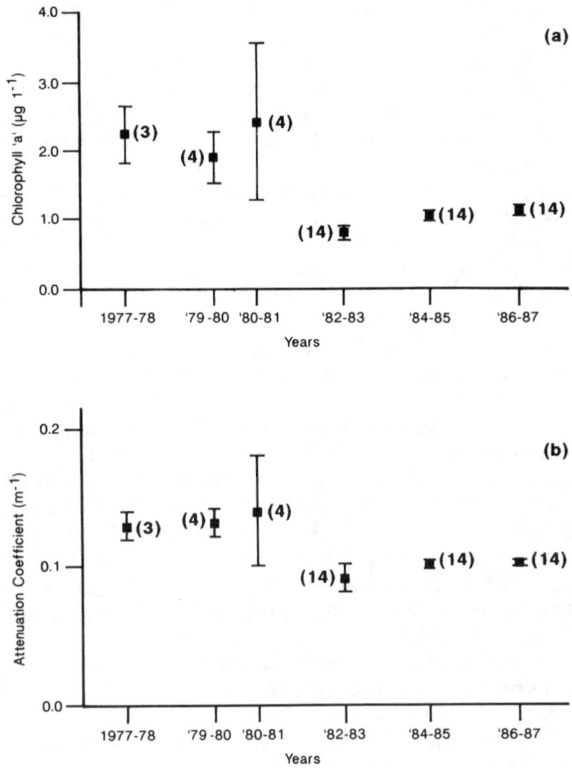

FIG. 5. (a) Mean chlorophyll "a" concentrations for sampling periods (monthly 1977-78, 1979-80, 1980-81; weekly 1982-83, 1984-85, 1986-87) during summer in Cockburn Sound. Standard error bars are shown, and the number in brackets represents the number of sampling occasions. Redrawn from Hillman and Bastyan (1988). (b) Mean light attenuation coefficients for sampling periods during summer in Cockburn Sound. Details as for Figure 5a.

The Peel-Harvey Estuarine System

The nature of the problem. The Peel-Harvey system is an important recreational waterway for local residents and visitors from Perth, and also supports the largest estuarine fishery on the Western Australian coast (Hodgkin *et al.* 1985). The two linked estuaries show two major symptoms of eutrophication (Hodgkin *et al.* 1981; McComb *et al.* 1981). In Peel Inlet, large quantities of rotting macroalgae (mainly *Cladophora* in earlier years, but more recently *Chaetomorpha* and *Ulva*) accumulate on what were, until the late 1960s, clean sandy beaches, becoming foetid masses with a nauseating stench of hydrogen sulphide. Measures to ameliorate the worst features of the macroalgal problem by removal of accumulating algae, by onshore weed clearing and offshore weed harvesting have been only partially successful in keeping beaches near inhabited areas free from decomposing algae. The current cost (1986-87) of the programme is A$260 000 per year. The increase in macroalgal growth has been accompanied by the increase in the commercial fish catch (Lenanton *et al.* 1984).

The macroalgae are largely benthic species and their distribution is mainly confined to Peel Inlet; their growth is inhibited in Harvey Estuary because of the higher turbidity and consequent poor light penetration through the water (Gordon *et al.* 1981a). Resuspension rates are much higher in Harvey Estuary compared to most of Peel Inlet; Harvey Estuary lies parallel and exposed to the prevailing winds (S-SW and N-NW), while most of Peel Inlet is sheltered in large embayments and by shallow marginal platforms (Gabrielson & Lukatelich 1985).

Although there is little macroalgal biomass in Harvey Estuary, in late spring and early summer there are massive blooms of blue-green alga *Nodularia spumigena* Mert. Blooms have occurred each year since 1978, except 1979 and 1987. The alga has a distinctive and nauseating smell and may accumulate and rot on the shore, similar to that described by Francis (1878) in his work on Lake Alexandrina in South Australia. Fish and crabs avoid the estuaries during *Nodularia* blooms and commercial catches are reduced dramatically (Hodgkin *et al.* 1985; Potter *et al.* 1983). The odours from decomposing *Nodularia* have been blamed for serious illness in local residents (Hodgkin *et al.* 1985).

As a result of public concern of the cost of macroalgal clearing, the Estuarine and Marine Advisory Committee of the Western Australian Environmental Protection Authority requested the Department of Conservation and Environment to commission a study of the algal problem. The multidisciplinary study was initiated in 1976 with two main objectives: to determine the cause of excessive algal growth, and more generally to gain an understanding of this particular estuarine system. A bibliography of the 32 technical reports and 73 papers published to 1985 is listed in Hodgkin *et al.* (1985).

The Peel-Harvey System. Harvey Estuary occupies 56 km^2 and is a long, shallow (less than 2.5 m deep) coastal lagoon in an interdunal depression. Peel Inlet occupies 75 km^2 and is a shallow (less than 2 m deep), nearly circular basin in a sandy coastal plain; it has wide peripheral platforms (<0.5 m deep) which occupy more than half the total area. The basins are of similar volume; Peel Inlet, 61×10^6 m^3, and Harvey Estuary, 56×10^6 m^3. Peel Inlet is connected to the ocean by a 5 km narrow inlet channel (Fig. 6).

The total catchment area of the system is 11 378 km^2. It is principally agricultural and 1 720 km^2 has been dammed. Three rivers enter the system. The Murray River drains forest as well as agricultural land and in an average year contributes a volume of water equivalent to about four times than of Peel Inlet. The Serpentine River, which enters Peel Inlet at the mouth of the Murray, has been dammed and contributes a relatively small volume of water compared to the Murray. The Harvey River contributes a volume of water equivalent to about four times the volume of Harvey Estuary in an average year. The upper reaches of the Harvey River have been dammed and the lower reaches drain sandy coastal-plain soils of low relief, which have been extensively cleared and drained for grazing. For example, in 1978 (an average year) the total freshwater inflow to the system was 627 million cubic metres. Forty-eight percent of this total came from the Murray, 15% from the Serpentine, and 37% from the Harvey (Hodgkin *et al.* 1981). River flow is highly seasonal, with 90% occurring in the four winter months from June to September.

The astronomic tidal amplitude in the system is usually less than 10 cm, with a maximum daily range of 20 cm, whereas the variation in water level brought about by longer-term forcing, such as changes in barometric pressure, may provide a range of up to 50 cm over periods of five to ten days (Black *et al.* 1981).

The cause of the problem. Both nitrogen and phosphorus are essential to plant growth and a shortage of either can limit the amount of algal biomass. However, it is generally phosphorus rather than nitrogen that is the limiting nutrient which determines the biomass of macroalgae (Gordon *et al.* 1981b; Lukatelich & McComb 1985). Phosphorus is also the nutrient which limits the growth of the nitrogen-fixing blue-green alga *Nodularia* in Harvey Estuary. Potential sources of phorphorus input to the estuary are: river and drain flow, groundwater flow, urban sewage, surface run-off, seawater and the atmosphere. However, the rivers and drains are by far the greatest source, contributing over 90% of net input of phosphorus (Birch 1982).

The principal source of phosphorus to the rivers and drains is the sandy soils of agricultural land on the coastal plain. Some 125 000 ha (81%) of the coastal plain catchment which have been cleared for agriculture are sandy soils. These sandy soils are very porous, naturally deficient in phosphorus and have a low capacity to bind and retain phosphorus. Phosphorus has been applied as superphosphate fertilizer to these soils for the last 40 years. During the last decade an average of about 1 500 tonnes (18 kg P ha^{-1}) has been applied annually to the coastal plain catchment (Hodgkin *et al.* 1985).

Phosphorus input from riverflow has increased greatly since the 1950s when it was measured by the CSIRO Division of Fisheries and Oceanography (CSIRO 1952-1957). At median flow rates the Serpentine and Murray Rivers delivered, respectively, nine and 50 times more phosphorus in 1972-78 than in 1949-56 (Hodgkin *et al.* 1981).

The amount of phosphorus entering the system each year varies substantially, the variation largely reflecting the volume of riverflow. Diatom blooms follow almost immediately the onset of significant riverflow each winter (Fig. 7) and available data suggest that 50-70% of the winter phosphorus load

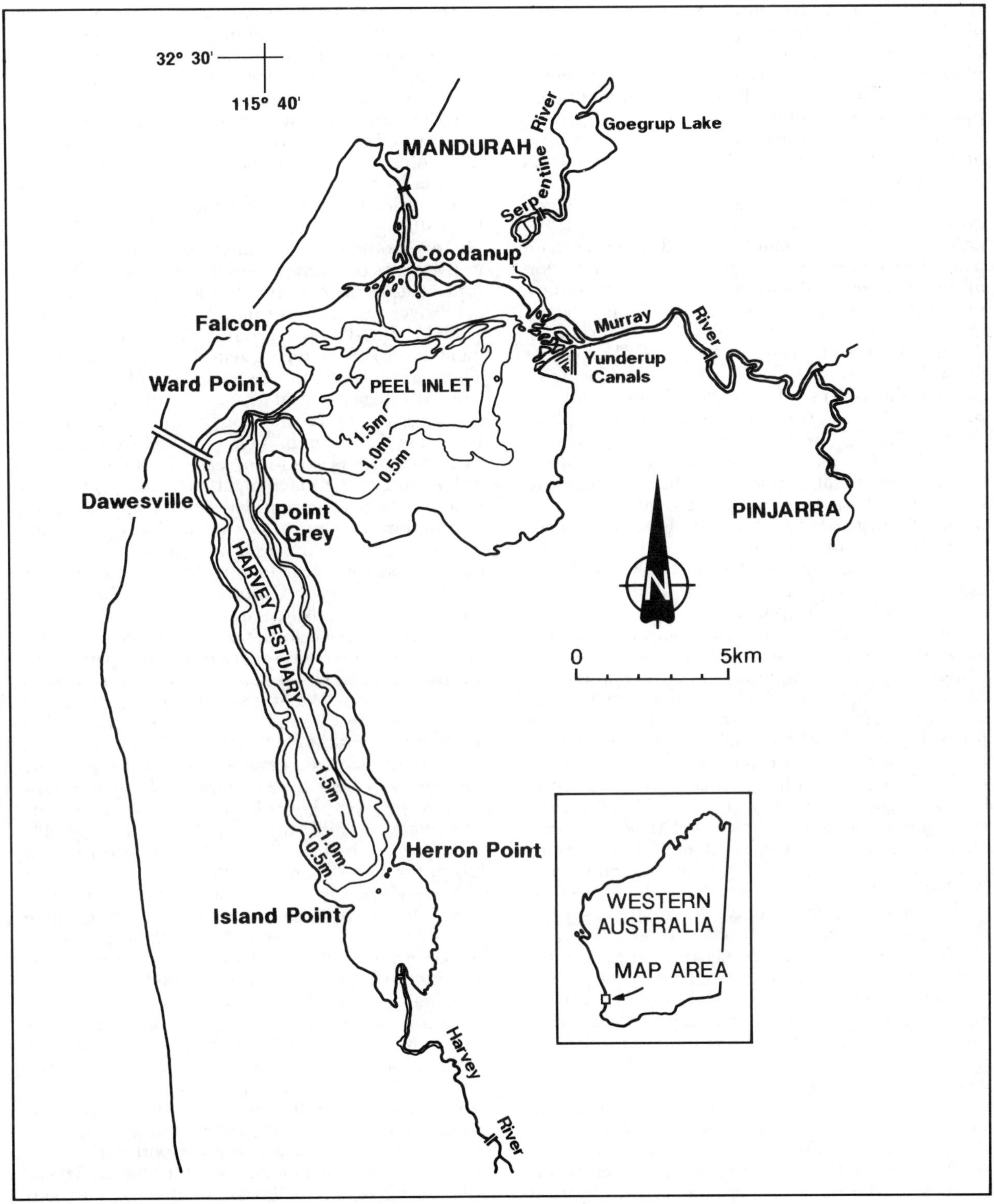

FIG. 6. The Peel-Harvey Estuarine System, Western Australia. The 0.5 m, 1.0 m and 1.5 m depth contours are shown. Mandurah is a town near the channel connecting Peel Inlet to the ocean. The location of the proposed Dawesville Channel is shown. Insert shows location of the system.

is trapped by these blooms. In river water, 70-80% of the total phosphorus is in the form of phosphate. There is little colloidal or particulate phosphorus in the river water, so there is little potential for the flocculation of colloids and sedimentation upon mixing with saline estuary water. Phosphorus trapped by the diatoms can settle out in the form of senescent cells or faecal pellets derived from grazing zooplankton.

At the time of peak phosphorus input, *Nodularia* growth is limited by temperature (Huber 1981) and there is approximately a 100 day lag between peak riverine phosphorus input and the initiation of the *Nodularia* bloom (Fig. 7). Water column phosphorus concentrations are low immediately prior to the initiation of the *Nodularia* blooms (Fig. 7) and the blooms rely on recycled phosphorus from the sediments as their major source of phosphorus. The collapse of the diatom blooms loads the surface sediments with organic matter and phosphorus. The decomposition of this material results in the depletion of oxygen, due to a high biological oxygen demand at a time when the water column is usually stratified, so that there is little replenishment of oxygen below the halocline. This leads to high rates of phosphorus release from the sediments (Lukatelich & McComb 1985) and provides the developing *Nodularia* bloom with an ample supply of phosphorus.

There is a quantitative relationship between the amount of phosphorus entering the system in winter and the magnitude of the subsequent *Nodularia* bloom (Fig. 8). The relationship between minimum salinity each winter and the size of the following *Nodularia* bloom in summer is explained by the relationship between river volume (minimum salinity) and phosphorus load. No *Nodularia* blooms were recorded in 1977, 1979 and 1987 when minimum salinities exceeded 10‰, *Nodularia* akinetes require salinities less than 15‰ and temperatures greater than 18°C for mass germination (Huber 1981). In those years when there were no *Nodularia* blooms, the salinity exceeded 15‰ by the time water temperatures reached 18°C.

The macroalgae presumably rely on the winter diatom blooms to trap and precipitate phosphorus and, at the time of peak phosphorus input, light and temperature limit macroalgal growth (Gordon *et al.* 1981a). Subsequent recycling from the sediments provides phosphorus for macroalgal growth when conditions become favourable. Lukatelich and McComb (1985) found that light was the only variable correlated with macroalgal biomass (Fig. 9). Biomass was significantly negatively correlated ($r=0.82$, $p<0.05$) with the mean attenuation coefficient during the summer growth period (December-March). Tissue nitrogen and phosphorus concentrations showed no evidence of increased nutrient limitation during periods of low biomass (Lukatelich & McComb 1985).

The proposed management strategy. In reviewing the Peel-Harvey eutrophication problem on behalf of the Department of Conservation and Environment, Black and Hodgkin (1984) stated that the principal objective of a management programme should be to reduce the algal nuisance to acceptable levels without further damage to the estuarine environment. This objective was to be accomplished without causing loss of production of the estuarine fishery or of agriculture in the catchment. An acceptable level of algal nuisance was defined as: *Nodularia* blooms should not occur more frequently than once in five years on average and macroalgae should not foul beaches near populated areas. Based on the extensive research results, Black and Hodgkin (1984) defined a "preferred strategy" for management of the Peel-Harvey system. This had three parts. These were the continued harvesting of macroalgae from the beaches and from offshore, agronomic measures to reduce the input of phosphorus to the system, and creation of a new channel from Harvey Estuary to the ocean to increase flushing.

The aim of the macroalgal harvesting is to keep beaches clean near populated areas to ameliorate a situation that is offensive to local residents. Harvesting only removes a small proportion of the macroalgae each year (about 2% of peak biomass in 1984). The loss of nutrients from the system due to harvesting is insignificant.

Fertilizer applied to pasture on the coastal plain catchment is the main source of phosphorus entering the Peel-Harvey system (see above). The main aim of the catchment programme has been to devise new fertilizers and fertilizer application methods more suitable to sandy soils of the coastal plain. In the past, superphosphate has been widely applied each year at the rate of 18 kg of phosphorus per ha. Phosphorus applied as superphosphate is 84% water-soluble and is readily leached to drainage. A new slow release form of superphosphate (Coastal Superphosphate) has been developed for use on the coastal plain. Only 20% of the phosphorus in Coastal Superphosphate is water-soluble and a much smaller proportion is lost direct to drainage. More of the phosphorus remains to supply plant requirements in subsequent years and therefore less needs to be applied.

A "soil phosphorus store" has accumulated as the result of continued application of phosphatic fertilizer in excess of the requirements of the pastures. Soil surveys made in 1983 and 1984 have shown that about 70% of the coastal plain could support high crop production for at least one year, and in most cases up to five years, without further application of phosphorus (Hodgkin *et al.* 1985). Soil testing of all paddocks on the coastal plain catchment has been carried out since 1984 and individual farmers are advised as to the appropriate type and rate of fertilizer to be applied to their paddocks.

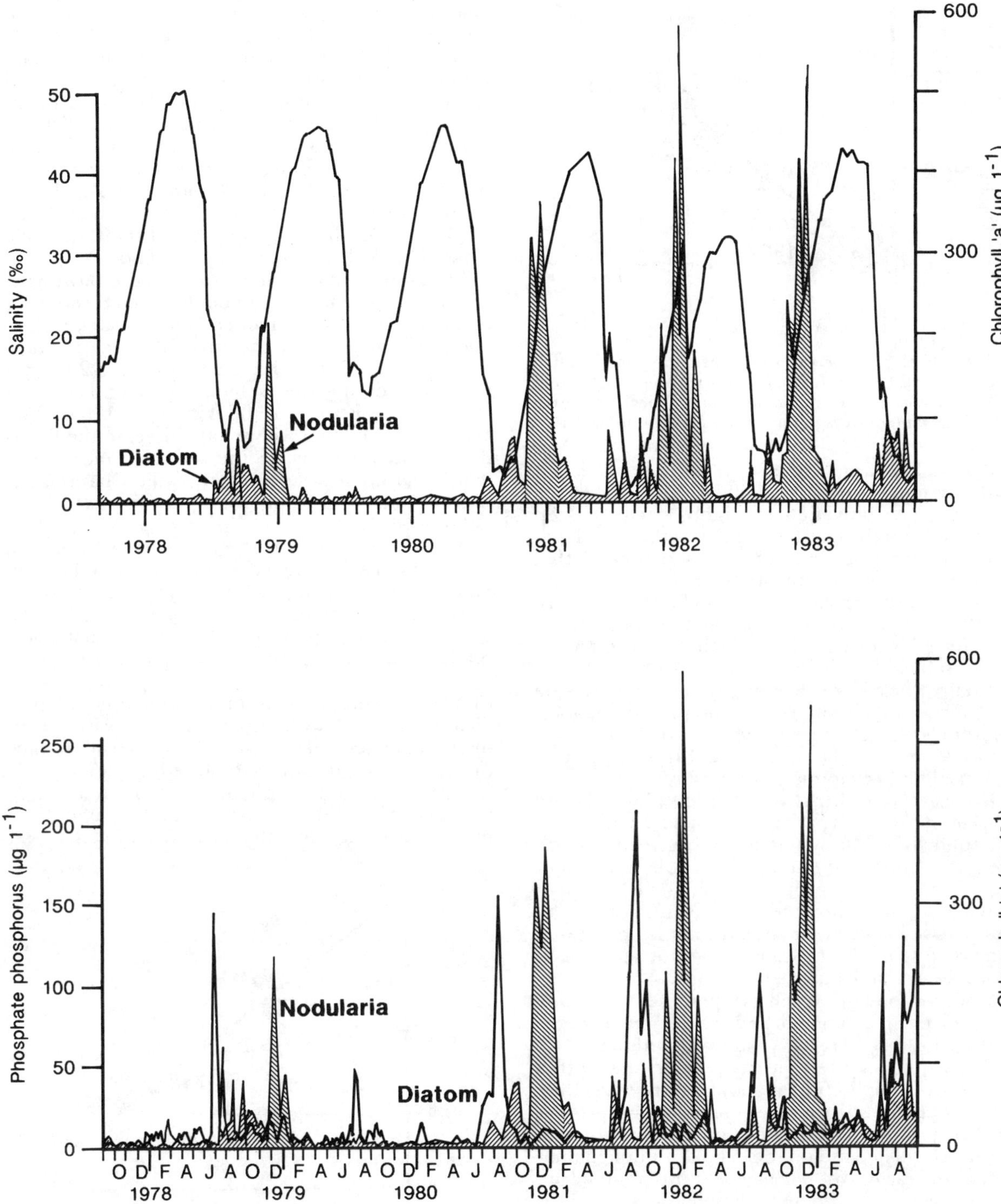

FIG. 7. Changes in salinity and chlorophyll "a" (hatched) (above), and phosphate and chlorophyll "a" (below) with time (1978-1983) in Harvey Estuary. Periods of diatom and *Nodularia* dominance of the phytoplankton (chlorophyll "a") are shown. Redrawn from Lukatelich (1984).

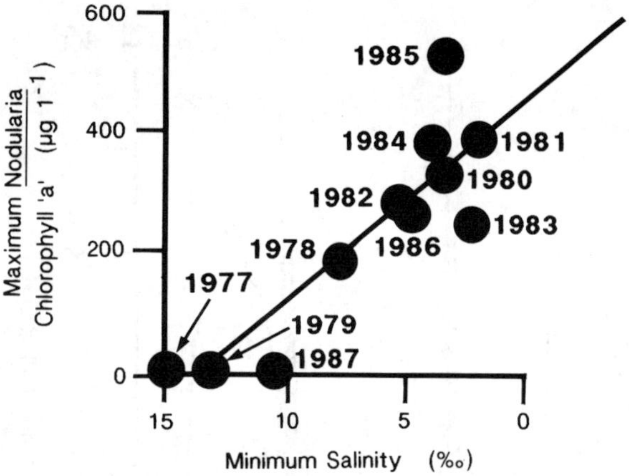

FIG. 8. Relationship between the maximum concentration of chlorophyll "a" during each summer *Nodularia* bloom in Harvey Estuary and minimum salinity measured each preceding winter (average for surface and bottom water).

Part of the large increase in phosphorus input to the system since about 1960 has been due to the clearing and cultivation of highly leaching deep grey sands on the coastal plain, and the associated improved drainage network. It is proposed that there be a moratorium on further clearing and drainage until policies and guidelines for future land clearing and drainage are developed.

Conversion of agricultural land to forestry on sandy soils also provides a means to reduce substantially phosphorus loading to the estuary in the long-term. Preliminary economic evaluation (Mattinson & Morrison 1985) shows that this measure warrants large-scale field trials to test its technical and economic practicability.

Intensive agricultural activities like piggeries and market gardens are potential major sources of phosphorus for the system in the future and hence are an important component of catchment management. It is proposed that further expansion of intensive agricultural activities within the Peel-Harvey catchment should only be permitted subject to stringent restrictions on nutrient loss to groundwater and surface drainage.

Modification of fertilizer practices in 1984 resulted in an estimated 25% reduction in phosphorus input to Harvey Estuary (Hodgkin *et al.* 1985). In the long term it has been estimated that phosphorus application at about half the pre-1984 rate should be sufficient to maintain the present level of pasture production. Long-term reduction in inputs will depend on the extent of implementation of the catchment management programme and the depletion of the soil phosphorus store; it is anticipated that there will be a progressive reduction in input of 30-40% over 10 years (Hodgkin *et al.* 1985). The success of the catchment management programme depends on a high level of farmer compliance with advice given.

In order to reduce the algal nuisance to acceptable levels in the short-term (within five years) it has been proposed that a new channel (Dawesville Channel) be constructed to the sea, to increase phosphorus export. The location of the proposed channel is near the north end of Harvey Estuary (Fig. 6). The selection of the site was dictated mainly by the practical considerations of length, topography and existing development. The channel has been designed with a depth of 4.5 m below mean sea level, a bottom width of 200 m and side slopes of 1:5. The estimated cost of the Dawesville Channel is $31 million (Hodgkin *et al.* 1985).

The volume of exchange through the Dawesville Channel will be more than double the present flow through the existing inlet channel (Hodgkin *et al.* 1985). It will also have the effect of increasing flow through the existing channel by some 15-20% and will nearly double the exchange between Peel Inlet and Harvey Estuary (Hodgkin *et al.* 1985). The Dawesville Channel will greatly increase the flushing of estuary water to the sea. The increase in daily tidal range with the Dawesville Channel has been estimated to be about 0.2 m (Tong 1985).

Predicted outcome of the proposed strategy. Mathematical modelling suggests that the Dawesville Channel will at least halve flushing times in the Peel-Harvey system (Tong 1985). Phosphorus retention in Harvey Estuary in winter may be reduced by up to half that at present. It is anticipated that phosphorus will mainly be lost in particulate form (phytoplankton cells) because phytoplankton uptake of phosphate is rapid compared to predicted flushing rates. In combination with catchment management measures, which will reduce phosphorus inputs, this should reduce significantly the size of the winter phytoplankton blooms, and lead to a marked reduction in the amount of organic material settling onto the surface sediments. Not

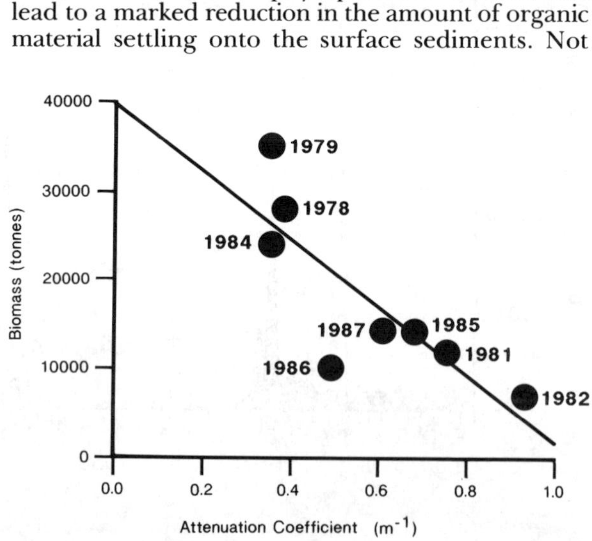

FIG. 9. Relationship between mean macroalgal biomass and mean attenuation coefficient for the period December to March.

only will the phosphorus load to the sediments decrease, the potential for this and stored phosphorus to be released to the water column will be reduced because of the lower organic loading.

The most significant effect of the Dawesville Channel on the microalgae will be the elimination of the massive *Nodularia* blooms. It is expected that mean Harvey Estuary salinities, except in extremely wet winters, will reach 30‰ by the end of October (Hearn & Hunter 1986). *Nodularia* blooms collapse at the present time when salinities reach 30‰ (Fig. 7). The initiation of the *Nodularia* blooms normally occurs when temperatures have risen in late September. Akinete germination is severely depressed at salinities above 15‰ (Huber 1981) and it is expected that after Channel construction salinities will exceed 20‰ by the end of September, even at the southern end of the Harvey (Hearn & Hunter 1986). The effect of the Dawesville Channel on salinities in Peel Inlet has not been modelled, but it is expected that salinities will be too high at the end of winter for initiation of *Nodularia* blooms.

The critical factors most likely to affect macroalgae are light and nutrient concentrations. Salinity changes due to the Dawesville Channel can be ignored in this context, because the nuisance algae in the system appear to have a broad salinity tolerance (Gordon *et al.* 1981a). The more marine salinity regime may enable some marine macroalgae to become established but it is not expected that they would affect the present nuisance species.

Macroalgal biomass in different years is significantly, negatively correlated with light attenuation during the main period of macroalgal growth. Turbidity should fall in Peel Inlet to levels similar to those recorded in 1977-78, and there should be a concomitant increase in macroalgal biomass. It is difficult to predict how quickly the biologically available phosphorus in the sediments would be reduced sufficiently to cause a decline in macroalgal growth. However, in time, after construction of the Dawesville Channel, nuisance weed growth should be reduced eventually to acceptable levels. It is unlikely that macroalgal accumulations comparable with those in Peel Inlet will occur in Harvey Estuary because most of the turbidity in the Harvey, except during phytoplankton blooms, is due to the resuspension of the surface sediments.

General Discussion

The environmental studies carried out on the Peel-Harvey estuarine system and Cockburn Sound have yielded valuable information for the future management of aquatic ecosystems in Western Australia, and possibly elsewhere as well. For both these waterbodies, the detrimental effects of high nutrient loading were exacerbated by poor flushing characteristics, and therefore the determination of water retention times and circulation patterns has become a high priority for current and future studies on aquatic ecosystems. The Peel-Harvey study has also highlighted the need to restrict nutrient-generating activities on poor sandy soils in estuarine catchments, whilst studies on Cockburn Sound have emphasized the need for rigorous measurement and control of the inputs of industrial and urban wastes to water bodies.

In a more general management sense, four main points are highlighted by both studies. Firstly, that aquatic ecosystems cannot be treated as isolated water bodies. Potential sources of nutrients and contaminants must be traced and the inputs measured. Secondly, a multidisciplinary approach is essential for successful management. Biological responses cannot be correctly interpreted without knowing the physical and chemical processes occurring in the ecosystem. Thirdly, ongoing monitoring is necessary to make sure management objectives are being achieved, and for the early detection of changes in the ecosystem. The fourth point is an automatic consequence of the third. Management plans must be constantly reviewed and updated to respond to changing conditions.

Acknowledgements

The Cockburn Sound Study and The Peel-Harvey Estuarine System Study, which supported this research financially, were funded by the Department of Conservation and Environment (now Environmental Protection Authority), Western Australia.

References

Anon. (1979) *Cockburn Sound Environmental Study 1976-1979.* Report No. 9, Department of Conservation and Environment, Perth, Western Australia. 103pp.

Anon. (1988) *What's happening in our harbours? Princess Royal and Oyster Harbours, Albany.* Bulletin No. 341, Environmental Protection Authority, Perth, Western Australia. 17pp.

Birch P. B. (1982) Phosphorus export from coastal plain drainage into the Peel-Harvey Estuarine System of Western Australia. *Aust. J. Mar. Freshw. Res.* **33,** 23-32.

Black R. E. & Hodgkin E. P. (1984) *Management of Peel Inlet and Harvey Estuary.* Bulletin No. 170, Department of Conservation and Environment, Perth, Western Australia.

Black R. E., Lukatelich R. J., McComb A. J. & Rosher J. E. (1981) Exchange of water, salt, nutrients and phytoplankton between Peel Inlet, Western Australia and the Indian Ocean. *Aust. J. Mar. Freshw. Res.* **32,** 709-20.

Cambridge M. L., Chiffings A. W., Brittan C., Moore L. & McComb A. J. (1986) The loss of seagrass in Cockburn Sound, Western Australia. II. Possible causes of seagrass decline. *Aquat. Bot.* **24,** 269-85.

Cambridge M. L. & McComb A. J. (1984) The loss of seagrass in Cockburn Sound, Western Australia. I. The time course and magnitude of seagrass decline in relation to industrial development. *Aquat. Bot.* **20,** 229-43.

Chegwidden A. (1979) Cockburn Sound Environmental Study. Technical Report on Distribution of Contaminants. Report No. 5, Department of Conservation and Environment, Perth, Western Australia.

Chiffings A. W. (1979) *Cockburn Sound Environmental Study. Technical Report on Nutrient Enrichment and Phytoplankton.* Report No. 3, Department of Conservation and Environment, Perth, Western Australia. 59pp.

Chiffings A. W. & McComb A. J. (1981) Boundaries in phytoplankton populations. *Proc. Ecol. Soc. Aust.* **11,** 27-38.

Chiffings A. W. & McComb A. J. (1983) *The effects of nutrient load reduction on water quality in Cockburn Sound. An interim report to the Fremantle Port Authority Water Quality Advisory Committee.* Department of Conservation and Environment, Perth, Western Australia. 20pp.

Cronin L. E. (1967) The role of man in estuarine processes. In *Estuaries* (ed. G. H. Lauff). *Am. Assoc. Adv. Sci.* **83,** 667-89.

Dybdahl R. E. (1979) *Cockburn Sound Environmental Study. Technical Report on Fish Productivity.* Report No. 4, Department of Conservation and Environment, Perth, Western Australia. 96pp.

Feilman Planning Consultants Pty. Ltd. (1978) *Cockburn Sound Recreation Survey: A Report to the Cockburn Sound Study Group.* Department of Conservation and Environment, Perth, Western Australia.

Fonseca M. S. & Kenworthy W. J. (1987) Effects of current on photosynthesis and distribution of seagrasses. *Aquat. Bot.* **27,** 59-78.

Francis G. (1878) Poisonous Australian Lake. *Nature* **2,** 11-2.

Gabrielson J. O. & Lukatelich R. J. (1985) Wind-related resuspension of sediments in the Peel-Harvey Estuarine system. *Est. Coast. Shelf Sci.* **20,** 135-45.

Gordon D. M., Birch P. B. & McComb A. J. (1981a) *The ecology of* Cladophora *in the Peel-Harvey Estuarine System.* Bulletin No. 91, Department of Conservation and Environment, Perth, Western Australia. 204pp.

Gordon D. M., Birch P. B. & McComb A. J. (1981b) Effects of inorganic phosphorus and nitrogen on the growth of an estuarine *Cladophora* in culture. *Bot. Mar.* **24,** 93-106.

Hearn C. J. & Hunter J. R. (1986) *Estimates of barotrophic flushing time for Harvey Estuary.* Report to Kinhill Stearns, Perth, Western Australia. 25pp.

Hillman K. (1986) *Nutrient Load Reduction, Water Quality and Seagrass Dieback in Cockburn Sound 1984-85.* Technical Series No. 5, Department of Conservation and Environment, Perth, Western Australia. 25pp.

Hillman K. & Bastyan G. (1988) *Nutrient Load and Water Quality in Cockburn Sound 1986-87.* Technical Series No. 26, Department of Conservation and Environment, Perth, Western Australia. 11pp.

Hodgkin E. P. (1978) *An Environmental Study of the Blackwood River Estuary.* Report No. 1, Department of Conservation and Environment, Perth, Western Australia. 78pp.

Hodgkin E. P., Birch P. B., Black R. E. & Hillman K. (1985) *The Peel-Harvey Estuarine System Proposals for Management.* Report No. 14, Department of Conservation and Environment, Perth, Western Australia. 54pp.

Hodgkin E. P., Birch P. B., Black R. E. & Humphries R. B. (1981) *The Peel-Harvey Estuarine System Study (1976-80),* Report No. 9, Department of Conservation and Environment, Perth, Western Australia. 71pp.

Huber A. L. (1981) *Cyanobacteria and nitrogen fixation in the Peel-Harvey estuarine system.* Bulletin No. 94, Department of Conservation and Environment, Perth, Western Australia.

Iveson J. B. (1979) *Salmonella* and *E. coli* in coastal waters, effluents and fauna in the Owen Anchorage Area including Carnac Island. A report on the Department of Health and Medical Services and Cockburn Sound Study Team Investigations.

Layton Groundwater Consultants (1979) *Cockburn Sound Groundwater Survey for Cockburn Sound Study Group.* Department of Conservation and Environment, Perth, Western Australia.

Lenanton R. C. J., Robertson A. I. & Hansen J. A. (1982) Nearshore accumulation of detached macrophytes as nursery areas for fish. *Mar. Ecol. Prog. Ser.* **9,** 51-7.

Lenanton R. C. J., Potter I. C., Loneragan N. R. & Chrystal P. J. (1984) Age structure and changes in the abundance of three important species of teleosts in the eutrophic estuary. *J. Zool.* (Lond.). **203,** 311-27.

Lukatelich R. J. (1984) Macroalgae, phytophankton and nutrients. In: *Potential for Management of the Peel-Harvey Estuary.* Pp. 167-85. Bulletin No. 160, Department of Conservation and Environment, Western Australia.

Lukatelich R. J. & McComb A. J. (1985) Nutrient recycling and the growth of phytoplankton and macroalgae in the Peel-Harvey estuarine system. Report No. ED-85-118, Western Australia. 80pp.

Lukatelich R. J., Schofield N. J. & McComb A. J. (1984) *The Nutrient Status of Wilson inlet 1982-1983.* Bulletin No. 159, Department of Conservation and Environment, Perth, Western Australia. 50pp.

Marsh L. M. & Devaney D. M. (1978) *The Benthic Fauna of Cockburn Sound, Western Australia. Part II: Coelenterata. Part III: Echinodermata: Ophiuroidea. Part IV: Echinodermata: Crinoidea, Asteroidea, Echinoidea and Holothuroidea.* Western Australian Museum, Perth, Western Australia.

Mattinson B. C. & Morrison D. A. (1985) A cost-benefit study of alternative strategies for reducing the algae nuisance in the Peel-Harvey estuary. In: *Peel-Harvey Estuarine System Study — Management of the Estuary* (ed. E. P. Hodgkin) pp. 251-84. Bulletin No. 195, Department of Conservation and Environment, Perth, Western Australia.

McComb A. J., Atkins R. P., Birch P. B., Gordon D. M. & Lukatelich R. J. (1981) Eutrophication in the Peel-Harvey Estuarine System, Western Australia. In: *Estuaries and Nutrients* (eds B. J. Neilson & L. E. Cronin) pp. 323-42. Humana Press, New Jersey.

Murphy P. J. (1979) *Cockburn Sound Environmental Study. Technical Report on Industrial Effluents.* Report No. 6, Department of Conservation and Environment, Perth, Western Australia. 279pp.

Pollard D. A. (1984) A review of ecological studies on seagrass-fish communities with particular reference to recent studies in Australia. *Aquat. Bot.* **18,** 3-42.

Potter I., Loneragan N. R., Lenanton R. C. J. & Chrystal P. J. (1983) Blue-green algae and fish population changes in a eutrophic estuary. *Mar. Poll. Bull.* **14,** 228-33.

Silberstein K., Chiffings A. W. & McComb A. J. (1986) The loss of seagrass in Cockburn Sound, Western Australia. III. The effect of epiphytes on productivity of *Posidonia australis* Hook f. *Aquat. Bot.* **24,** 355-71.

Steedman R. K. & Craig P. D. (1979) *Numerical model Study of Circulation and other Oceanographic Aspects of Cockburn Sound.* Prepared for Department of Conservation and Environment, Perth, Western Australia. R. K. Steedman and Asociates.

Thurlow B. A., Chambers J. & Klemm V. V. (1986) *Swan-Canning Estuarine System. Environment, Use and the Future.* Report No. 9, Waterways Commission, Perth, Western Australia. 463pp.

Tong G. D. (1985) *Modelling studies of Dawesville Cut-Harvey Estuary*. Bulletin No. 195, Department of Conservation and Environment, Perth, Western Australia. Pp. 139-63.

Walker D. I. & McComb A. J. (1985) Decomposition of leaves from *Amphibolis antarctica* (Labill.) Sonder ex Aschers. and *Posidonia australis* Hook. f., the major seagrass species of Shark Bay, Western Australia. *Bot. Mar.* **28,** 407-13.

Wells F. E. (1978) *A Quantitative Examination of the Benthic Molluscs of Cockburn Sound, Western Australia.* Western Australian Museum, Perth, Western Australia.

Wilson B. R., Kendrick G. W. & Brearley A. (1978) *The Benthic Fauna of Cockburn Sound, Western Australia. Part I: Prosobranch Gastropod and Bivalve Molluscs.* Western Australian Museum, Perth, Western Australia.

Proc. Ecol. Soc. Aust. 1990 — 16:55-70

Resolving the conflict between conservation values and extractive use of the Abrolhos coral reefs

A. I. HATCHER[1], G. D. WRIGHT[2] and B. G. HATCHER[3]

[1]*Hatcher Research Associates, 52a Marine Terrace, Marmion, W.A. 6020, Australia*

[2]*Department of Anthropology, University of Western Australia, Nedlands, W.A. 6009, Australia*

[3]*Department of Zoology, University of Western Australia, Nedlands, W.A. 6009, Australia (requests for reprints to this address only)*

Abstract

The Houtman Abrolhos reefs of Western Australia are the southernmost coral reef platforms in the Indian Ocean. They are not in pristine state, having supported the activities of miners, fisherman, tourists and scientists for many years. Until recently, the sustained yield of the lucrative fishery for the Western rock lobster was the primary goal of management. Increasing pressure from conservation groups has led to the consideration of more equitable management goals, which acknowledge the non-commercial values of the entire reef ecosystems. Using a combined ecological and anthropological approach to management-oriented research, we attempted to match the scales of social and ecological data collection. Benthic community structure was classified at spatial scales determined by patterns of human use of the reefs' resources. An extreme range of benthic communities occurs in each of the three reef groups along an approximately 15 km gradient, from dense kelp beds in the SW regions to rich living coral assemblages in the NE regions. The unusual oceanographic conditions and juxtaposition of marine communities at the Abrolhos results in a dynamic and diverse biotic mosaic which is highly productive, educational, attractive, and in the case of coral communities, fragile. By overlapping maps of ecological units and fishing effort intensity, specific areas of potential conflict between user groups were identified, providing managers with information relevant to zoning decisions. The majority of fishing effort was found to occur in ecological units with low densities of live coral, but significant exceptions to this generalization exist. The cause of the inverse relationship between fishing activity and live coral abundance has not been determined. This study, which was sensitive to the human component of the ecosystem, can provide a basis for equitable management of the Abrolhos reefs, at least partially defusing the confrontation between two major interest groups.

Introduction

Coral reefs, like most other ecosystems, are increasingly used by humans to provide food and other raw materials through extractive industries (Hatcher *et al.* 1989). Such activities usually alter components of the reef ecosystem (Dahl & Salvat 1988), and the benefits of exploitation must be weighed against the costs of environmental degradation. Inevitably, conflicts arise between user groups which place different values on the reef's resources (Kenchington 1988). Like rainforests, coral reefs have been identified as having particularly high conservation values (Hatcher *et al.* 1989) because they are biotically diverse, aesthetically pleasing, and serve as major reservoirs of genetic variability. The goal of maintaining these values may not always be compatible with the goal of maximizing sustainable yields of reef resources.

The difficulty for managers of human activities in complex ecosystems lies in determining the type and level of exploitation which is possible without jeopardizing the poorly understood processes that maintain biological diversity and functional complexity. While there is no substitute for experimental investigation of causal relationships between human activities and community structure, a necessary first step is to describe the patterns of usage and resources. More often than not these descriptions are quite independent in terms of methodologies, and spatial and temporal scales.

For example, fisheries catch statistics may be categorized on a latitudinal grid, while the habitats of fish communities may be classified in terms of geomorphological zonation. The incompatibility of such measures makes the identification of specific areas of conflict between extraction and conservation goals virtually impossible, thereby hindering the development of sound zoning and management strategy.

Here we present a case study of a set of heavily exploited, high latitude coral reef ecosystems, emphasizing the methods by which we interfaced measures of natural resource and human usage. Our basic precept was that the pattern of fishing should dictate the scales at which ecological data was collected. The work is part of a broader study of the Houtman Abrolhos reefs (Hatcher 1985; Hatcher *et al.* 1988) and will contribute to the development of management strategy for the area.

Study Area

The Houtman Abrolhos Islands and their associated reefs (Fig. 1) comprise three geographically distinct carbonate platforms separated by deep channels, located near the edge of the continental shelf off Western Australia. In the northern group (Wallabi), one island (North Is.) is separated from the others by an extensive submerged ridge. The islands cover an area of approximately 18.5 km^2 to the low tide mark, and the associated reefs an area of approximately 1100 km^2 to the 40 m contour line. The reefs consist of emergent perimeter and central platform reefs defining lagoons which range in mean depth from 4 m to 25 m.

Most types of reef structures are represented at the Abrolhos. Several of these are relicts of Pleistocene origin, while others are the result of recent accretion (Teichert 1947). Because the current

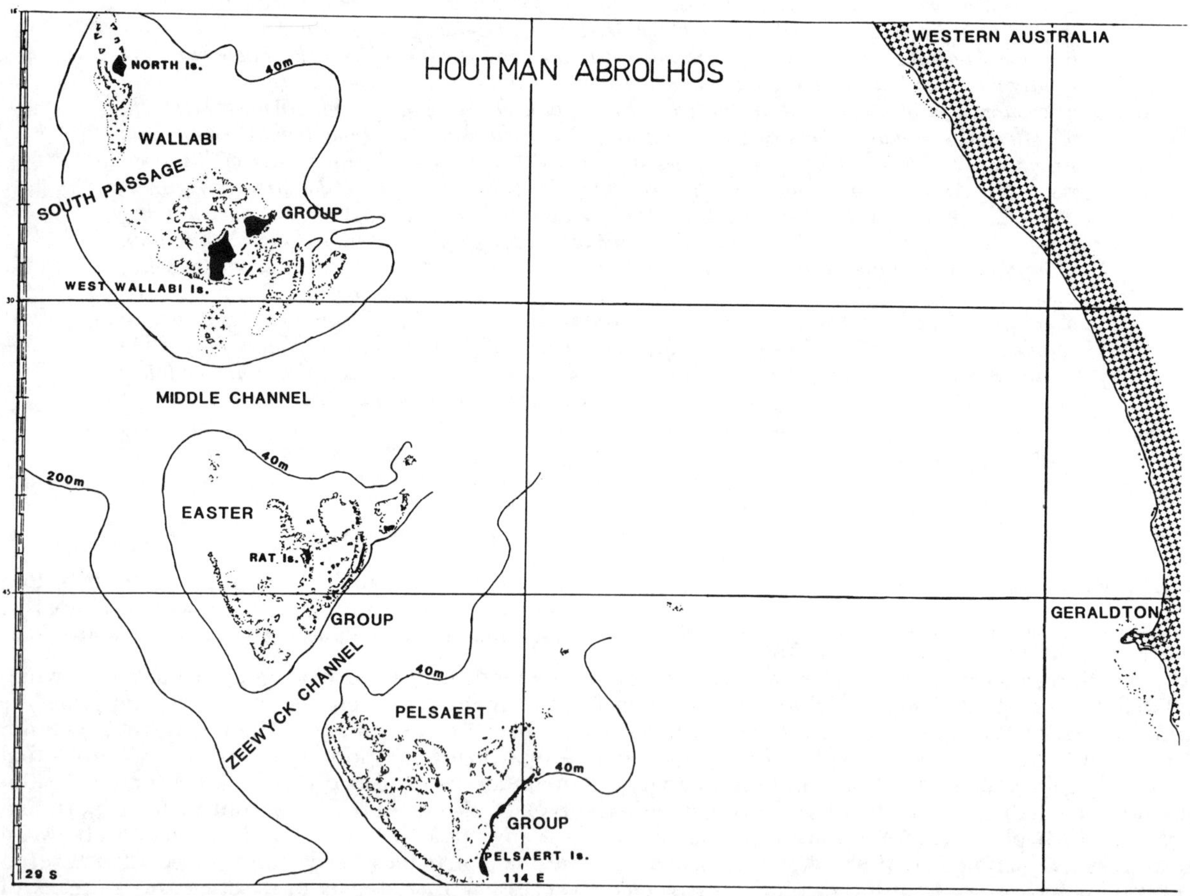

FIG. 1. General location of the Houtman Abrolhos relative to the coast of Western Australia. The three reef platforms and four island groups are labelled.

controversy over conservation and extractive uses of the Abrolhos reefs centres on Holocene coral assemblages, we concentrated on those structures in this study. The most prolific coral growth occurs on the leeward reef slopes, some of the shallow reef perimeters and lagoon patch reefs in the north-eastern sections of each reef platform (Wilson & Marsh 1979). Other areas of hard substratum are characterized by extensive stands of macrophytes (including kelp beds), or a mixture of macroalgae, corals and other sessile invertebrates (Hatcher 1985; Hatcher *et al.* 1987).

The marine environment of the Abrolhos has sustained nearly 100 years of extractive use by people (Hatcher & Walker 1984). Since the 1940s the major human activity at the Abrolhos has been the fishing of the Western rock lobster *Panulirus cygnus*, and the related activities of supply and residency associated with that fishery (Sheard 1962). Because of the long history of extensive exploitation, the human elements of the ecosystem must be an integral part of any examination of the natural environment of the Abrolhos. Social and ecological science have been combined in a tandem approach in this study.

Coral diversity and growth

The coral diversity at the Abrolhos is very high for reefs at this latitude (Wilson & Marsh 1979; Veron 1986). There are about 180 species of hermatypic corals in 40 genera (Veron & Marsh 1988), plus a further 10 species of ahermatypes in eight genera described to date. All but two coral species are tropical. This diversity contrasts with only about 60 species in 30 genera at Lord Howe Is. (Veron & Done 1979), and only eight to 15 genera on other coral reefs at similar latitude around the world (Wells 1955; Veron 1986).

Extensive marine habitats of the Abrolhos reefs support prolific and exceptionally diverse coral communities. An observation directly relevant to this study is that the growth rates of at least two abundant species of coral (*Acropora formosa* and *Pocillopora damicornis*) measured at the Abrolhos (Crossland 1981; Simpson 1988; Hatcher, unpubl. data) are 20 to 90% slower than rates measured at other locations in the tropics (Crossland 1981; Oliver 1984; Simpson 1988; T. Done, pers. comm.).

Unlike their tropical counterparts, corals at the Abrolhos compete with macroalgae for space, light and nutrients. Macroalgae are able to overgrow live coral in the absence of large herbivores, reducing growth rates by as much as 60%, and survivorship by up to 50% in one year (Crossland 1981; Hatcher *et al.* 1988; Hatcher, unpubl. data). Reduced light, coupled with low temperatures cause at least one of the dominant corals at the Abrolhos to put most of its carbonate production into the building of skeletal framework (which results in elongation) at the expense of skeletal infilling (which solidifies the skeletal framework) (Crossland 1984).

The tendency to build delicate structure makes the corals of the Abrolhos highly susceptible to physical damage. The slow growth rates and macroalgal competitors hinder recovery from such damage. These factors have been implicated in large-scale considerations of coral reef distribution (Johannes *et al.* 1983), and have serious implications for the assessment and management of activities causing physical disturbance in the Abrolhos coral communities.

It is primarily the beauty and diversity of the coral-dominated communities that has drawn the interest of the conservationist lobby to the marine environment of the Abrolhos. The threat of physical damage to corals there as a result of commercial trap fishing for the Western rock lobster was identified as a priority issue (Anon. 1986), and prompted this comparative study of the spatial overlap of rock lobster fishing and sensitive benthic communities.

Methods

Collection of data on fishing activity

Information was initially collected from the Western Australia Department of Fisheries (voluntary log books completed by approximately 20% of the fishermen), the Western Australian Fisheries Industry Council (WAFIC), and the Geraldton Professional Fishermen's Association (GPFA) to determine the extent of fishing around the Abrolhos reefs. These three organizations, while willing to contribute information, had little data which would allow spatial comparisons between fishing activity and benthic community structure. The scale of the fisheries data (latitudinal blocks) is not one at which ecologically distinct areas are identifiable.

To get relevant information, we asked a sample of fishermen to define their use of the reefs in personal interviews. Lack of time and money restricted the sample to 40 of the 201 fishermen in the Abrolhos rock lobster industry, and excluded fin-fish and scallop fishermen. The sample of rock lobster fishermen interviewed was restricted to those setting the majority of their traps (hereafter called "pots" using the fishermen's vernacular) in depths of less than 20 fathoms (36 m), and reflected the three distinct island groups, the recommendations of other fishermen and the availability of those willing to be interviewed. We focused on shallow water fishermen because they work those areas of the reef which are of most obvious interest to other user groups.

These fishermen supplied information on the spatial and temporal extent and intensity of their fishing, which was plotted on individual sheets of tracing paper laid over 1:25 000 depth-contoured orthophotomosaics of each reef platform. They were also asked for their observations of marine life, and in particular, about their knowledge of macroalgal and coral distribution and growth. The interviews were conducted between 19 October and 13 November, 1987.

Quantification and integration of the interview data: All interviewed fishermen were asked to draw the geographical extent of their fishing activity during a typical season. They did so at levels of precision ranging from exact points representing individual pot drops for a typical day of fishing, to more sweeping indications of the general areas in which they fished. Those who intensely fished small areas offered the most precise data, while more extensive fishermen were less able to relate their fishing ambit to finely detailed geographic features. For each geographically distinct area fished, the fisherman was asked to provide an estimate of the proportion of his total seasonal effort deployed there.

Unlike most ecological data, the data on intensity of fishing effort had the potential to be biased by the perceptions or misconception of the subjects. The individual drawings by the fishermen were compiled by tracing them onto master charts with each trace given a distinct colour and number. Thus, individual's plots overlapped each other in areas that were heavily fished. Claims by fishermen of fishing areas which are not productive of rock lobsters, and hence not fished by other fishermen, would have become obvious when compared with the general pattern. Such data would have been deleted if an adequate explanation was not available (i.e., the discovery of a new, productive ground). There were, however, no cases which warranted distrust on this basis. The honesty of the fishermen may be attributed to the fact that they recognized this research as something which would be critical in the future of their fishery. We feel that this attitude was largely due to justification given by us to the fishermen through their professional associations, and at the interviews.

The master charts of fishing effort were used in conjunction with the orthophotomosaics to define fishing zones for each island group (the reefs around North Island were analyzed separately from the rest of the Wallabi Group because of the distinct fishing community there). The fishing effort falling into each fishing zone was calculated by summation. A value for effort intensity was estimated as the total effort in a zone divided by its area. Total effort was calculated as the proportion of the pots lifted by the sampled fishermen who fish in the zone multiplied by the number of fishermen in the island group. Corrections were applied to this value of total effort to account for the proportion of pot drops in less than 36 m and the proportion of potential effort deployed (the average of seven seasons spanning 1980 to 1986). These correction factors were obtained from the independent sample of similar size (approximately 20% of all Abrolhos rock lobster fishermen) collected by the Western Australian Department of Fisheries.

The comparative variable used to assess variation in intensity of fishing effort across fishing zones was the *total number of pots dropped per hectare of seabed per fishing season.* For comparative analysis and graphical presentation, six categories of fishing effort intensity were identified, ranging from 0 to >100 pots ha^{-1} yr^{-1}.

Basis and method of ecological classification

No two ecologists view nature in exactly the same way. Classifying (i.e., simplifying) natural systems always involves partitioning continua. Other ecologists would undoubtedly have chosen different criteria for defining ecological units at the Abrolhos. Paramount in our scheme were considerations of the time and space scales at which humans utilize the resources of the reefs. This reflects the major role of humans as components of the Abrolhos ecosystems, and the management orientation of this research.

The charts of ecological units were derived in tandem with the charts of fishing zones. Spatial criteria, used in the definition of ecological units, were determined by the precision with which the fishermen perceived the environment. Little effort was expended on describing ecological characteristics at a finer level of precision than that required to describe the fishing zones. This is the key to the effective meshing of these sets of ecological and anthropological data.

An ecological classification scheme was derived *a priori* based on our knowledge of the Abrolhos reefs. Visually apparent boundaries between ecologically distinct areas (e.g., lagoon sand and patch reefs) were then identified and traced from the same 1:25 000 orthophotomosaics on which the fishermen had identified the fishing zones. The enclosed areas (termed "Ecological units") were then classified according to the scheme depicted schematically in Figure 2. The classification scheme is hierarchical, with ecological units as the lowest strata. Higher levels of classification are used to group sets of ecological units on the basis of their similarities.

Areas which are unknown to us, or unidentifiable from the aerial photos, were examined before classification by towing skin divers over shallow regions, and by scuba divers at deeper sites. For morphologically complex ecological units such as those in dissected limestone platforms or complex karst platforms (see Table 1), the proportional

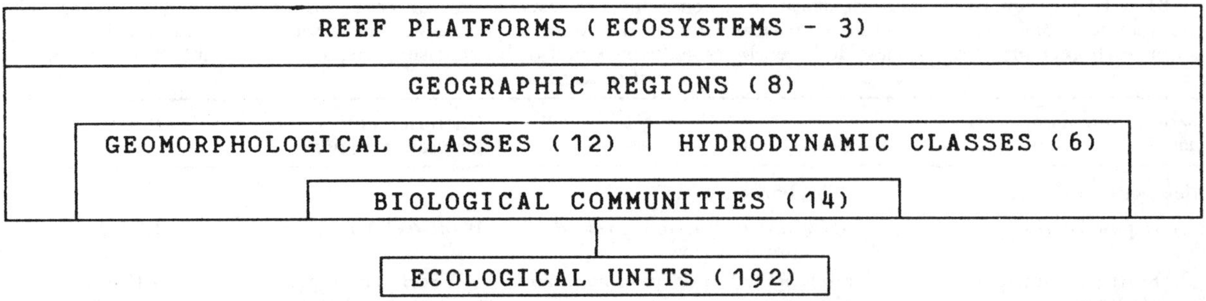

FIG. 2. A classification scheme for the marine ecological units of the Houtman Abrolhos. The scheme is essentially a spatial hierarchy, with parallel elements at certain strata. All regions, classes and communities are represented in (i.e., orthogonal to) each of the three ecosystems. The twelve geomorphological and six hydrodynamic classes are nested within (i.e., not orthogonal to) the eight geographic regions, and the fourteen biological communities are nested within the geomorphological-hydrodynamic stratum. The 192 unique ecological units identified throughout the three ecosystems are fully specified according to the higher-level classifications.

cover of component biological communities within each unit was measured from the photomosaics using the method of repeated grid sampling (Pielou 1969).

Assessment of the sensitivity of the benthos to fishing

For the purposes of this study, sensitivity is defined as the susceptibility of the benthic biological communities to physical damage caused by rock lobster pots and hauling ropes, impact from jet boat hulls and wash, and the deployment of anchors. This was assessed subjectively for each biological community from a consideration of its diversity, and the relative fragility and ability for repair and regrowth of the dominant organisms comprising it. Three broad levels of sensitivity were recognized: high, medium and low, and the benthic community classifications were ranked across them.

An index of the sensitivity of the large areas of seabed within fishing zones was formulated by ranking geomorphological classes according to the relative mix and sensitivity of their characteristic biological communities. This provided an index which was spatially and statistically compared to the distribution of fishing effort. The potential for physical damage to the benthos caused by rock lobster fishing within areas of different fishing effort intensity was estimated by an application of the index to the ecological units (classified according to geomorphology) which were overlaid by each fishing zone. If ecological units of more than one geomorphological class fell within a fishing zone, the sensitivity indices were averaged, because it was not possible to resolve the spatial distribution of fishing effort intensity below the scale of geomorphological classes. This method does not take into account the areal extent of different ecological units within each fishing zone, only their presence or absence. Areal weighting would add little to the usefulness of the analysis because occurrence, rather than abundance, is the primary criterion for conservation.

Results

Ecological description and mapping

The three geographically distinct reef platforms were considered to be separate ecosystems: the Pelsaert (southern) Group, the Easter (middle) Group, and the Wallabi Group (the Wallabi Islands and North Island). Within each group, eight geographic regions were defined which have clearly discernible counterparts in the other groups such as western lagoon, SW reef perimeter, etc. The geomorphological classes are listed in Table 1, which also shows the characteristic geographic regions in which they occur, and the biological communities occurring within them.

Six hydrodynamic classes were recognized on the basis of the flow characteristics (stability, uniformity, velocity), and forcing (wind, wave, tide). The hydrodynamic and geomorphological classifications are neither hierarchical nor orthogonal: some categories of one are found only in certain categories of the other. Because classification in one scheme is a fairly good predictor of the other, only the geomorphological scheme is used for the purposes of this report.

Biological communities: The 14 biological communities identified at the Abrolhos (Table 2) were classified according to the fundamental group of organisms (e.g., plants, corals, filter feeders, infauna, plankton) which dominate their structure. Sub-divisions indicate the structural-functional group of plant (e.g., macroalgae, seagrass), the nature of the coral assemblage (e.g., adapted to high or low wave energy) and the density of the infauna.

TABLE 1. A geomorphological classification scheme for the marine environments of the Houtman Abrolhos. The codes under "Biological Communities" indicate the most common type of benthic assemblage (underlined) occuring within each geomorphological class, followed by types represented with decreasing frequency (see Table 2 for description of codes).

Geomorphological class	Description	Characteristic geographic region(s)	Biological communities
Pleistocene structures:	relict, hardened surfaces.		
1) Exposed reef slope	windward, compacted, grooved; to approx. 40 m.	W off-reef margins	1,5,6,4,2, 10,7,12
2) Emergent limestone platform	low relief, intertidal pavements and reef crests.	S-W reef perimeters Central reef platforms	5,12,11,7 5,12,11,4
3) Submerged limestone platform	low-medium relief pavements with seds.	N-W reef perimeters W lagoons	4,1,5,11 5,12,4,11
Holocene Structures:	recent friable surfaces		
4) Sheltered reef slope	leeward, uncompacted, steep; to approx. 40 m.	N-E off-reef margins	7,10,8,2,9,11
5) Storm rubble field	leeward, unconsolidated, reef crests and flats.	E reef perimeters	5,6,4,12,2
6) Back reef	unconsolidated rubble and seds. behind reef crests.	S-W reef perimeters	10,4,8,12
7) Mobile sediment sheets	coarse sands in shallows and prograding sheets.	S-W reef perimeters W lagoons	12,13,4 13,12,3,4
8) Static sediment deposits	fine sands in sheltered and deep lagoon basins	E lagoons	13,12,3
Combined Structures:	thin Holocene veneers on Pleistocene basements		
9) Dissected limestone platform	spur/groove and elongate segments in reef passes.	W reef perimeters	4,7,8,10,6
10) Complex karst platform	high relief, deep pits in extensive shallows.	W lagoons and perimeters	13,4,10,5,8,6,12
11) Drowned doline field	high relief, submerged, reticulated reef lines	W lagoons	9,8,7,4,11,13
12) Isolated patch reefs	discrete, rounded reefs surrounded by sediments.	W lagoons E lagoons	4,10,2,1,5 10,9,4,11

The occurrence of biological communities across the geomorphological-hydrodynamic stratum is highly variable (Table 1). For example meso-macro algal communities (No. 4 in Table 2) can be found in virtually every geomorphological class, while high diversity coral communities (No. 9 in Table 2) are restricted to only a few. This variability means that higher level classifications are not always accurate predictors of the biological community.

Ecological units: Individual ecological units are the finest scale of resolution provided in the mapping of the marine resources of the Abrolhos. Each of the 192 units identified is a unique combination of all the classifications discussed above. Hence, it occurs at only one set of global co-ordinates, and is unambiguously located and described in terms of the higher-level classifications.

Ecological units were plotted on charts of the reef groups using a unique code for each unit. Their area was determined, and a master table prepared which specified the geomorphological, hydrodynamic and biological community classes describing them. Euclidian cluster analysis (Wilkinson 1986) demonstrated that geomorphological classes, grouped according to geographic regions provided the most realistic grouping of benthic biological communities (Table 1). Where possible, we used ecological units classified according to geomorphological class for comparisons with fishing zones (Fig. 3).

Distribution of ecological units containing living coral: Sheltered reef slopes form a continuum with exposed reef slopes as one progresses from western and southern to eastern and northern reef margins of the Abrolhos reefs. The transition is rarely clear-cut, but is usually marked by spur and groove topography, an increase in the steepness of the outer reef slope, and an increase in cover of live coral. The resulting assemblages of robust plating

TABLE 2. A classification scheme for the marine biological communities of the Houtman Abrolhos. Thirteen benthic communities and one planktonic community are identified.

Class	Description	Characteristic geographical regions
Plant Communities:		
1) mixed macrophyte assemblage	multiple canopies of kelps, corticated algae and seagrass.	W off-reef margins and N-W reef margins
2) *Sargassum* spp.	>50% cover, perennial stands with crustose understorey.	N-W reef margins and W lagoons
3) Seagrass bed	>50% cover, perennial stands rooted in sediments.	N-W reef margins and W lagoons
4) Meso-Macro algal assemblage	monolayers of turf and foliose algae; scattered large fronds.	W lagoons and Central reef platforms
5) Coralgal pavement	endo-epilithic, crustose and turf algae; encrusting invertebrates.	S-W reef perimeters and Central reef platforms
6) Calcareous algal field	unattached branching coralline algae and rhodoliths.	E lagoons and S-W reef parimeters
Coral Communities:		
7) High energy assemblage	>50% cover robust forms (thick branching, tabulate and encrusting).	reef perimeters and lagoons (shallow)
8) Low energy assemblage	>50% cover delicate forms (thin branching, foliose and plating).	E lagoons
9) High diversity assemblage	species-rich mixtures including massive and solitary forms.	E off-reef margins and E lagoons (deep)
Coral-Macroalgal Communities		
10) Coral-Macroalgae assemblages	*Acropora* and *Sargassum, Eucheuma* or *Lobophora* (20%-50% of each).	lagoons and E off-reef margins
Filter-feeding Communities:		
11) Dense invertebrate filter-feeders	sessile sponges, ascidians, anthozoans and sediment-tolerant corals.	lagoons (and channels)
Sedimentary Communities:		
12) Depauperate	low organics; sparse microflora and macrofauna.	S-W reef perimeters and lagoons (shallow)
13) Infaunal	high organics; rich microflora and macrofauna.	lagoons (sheltered)
Planktonic Communities:		
14) Suspended microbial assemblages	particulate and dissolved organic material and associated microbes.	ubiquitous

corals and macroalgae are visually appealing, and scientifically interesting.

Further north-east the transition to coral-rich communities is complete. Steep slopes exhibit a depth zonation from tabulate and plating corals with seasonal *Sargassum* cover, through branching corals (often with 100% cover) to high diversity coral and filter-feeder assemblages at the slope base. These leeward slope communities are most like the coral communities of moderately exposed to sheltered tropical reef habitats (see Veron 1986).

Karst features dominate the eastern portions of the Abrolhos platforms. They are characterized by deep (maximum 28 m), circular "pot holes" (drowned dolines, see Backshaw *et al.* 1979) ranging in size from tens to hundreds of metres in diameter. Thin margins of live coral surround the edges of the holes. These structures grade into drowned doline fields as the ratio of deep reef to shallow platform increases. Live coral cover increases with this ratio (25%-49% of total area). The coral communities in these ecological units are extremely rich, particularly on the deeper sides of the dolines.

Distribution of fishing effort over time and space

Fishing strategies are individual responses to a set of environmental, economic and social circumstances. The Abrolhos rock lobster fishing season has three distinct periods, each with a characteristic style of fishing effort.

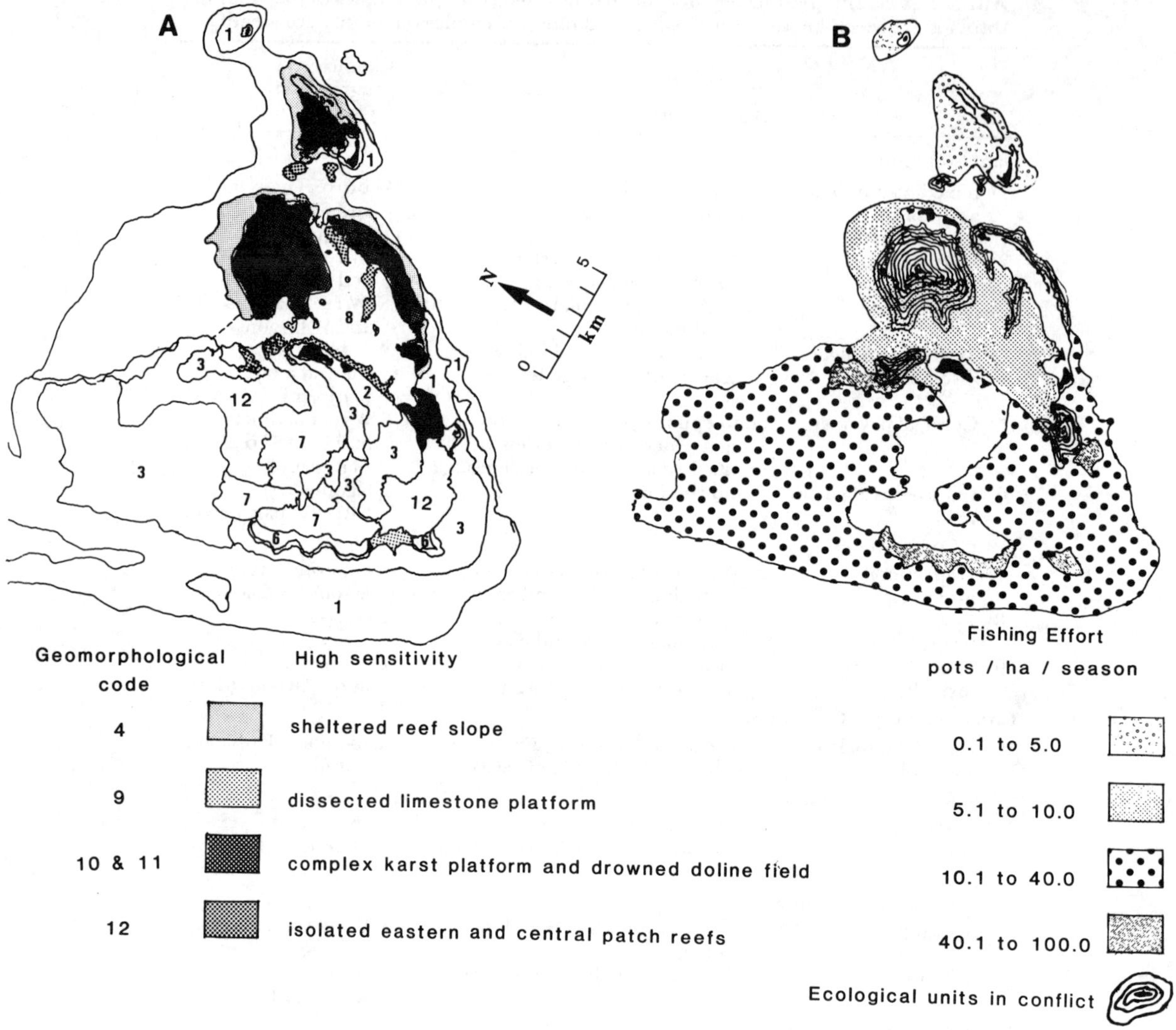

FIG. 3. Ecological units (A) and fishing zones (B) in the Easter Group of the Houtman Abrolhos. Islands are black, and the charts correspond spatially. The ecological units are numerically coded according to geomorphological class (Table 1), with those classes of high sensitivity (Table 5) shaded. The classifications of fishing effort intensity in fishing zones are indicated by the type of shading. Areas of potential conflict between user groups were determined by overlaying maps (a) and (b) (see text), and are identified on map (B) by concentric lines.

The Abrolhos season opens on 15 March. Rock lobsters which have grown to legal size in the eight months since the close of the last season, as well as survivors from previous cohorts, are eligible catch. During the first week or so, nearly all boats fish within the lagoon or close to the main reef structures. Catches are usually very high in this brief opening period as accumulated local stocks of high catchability are fished out.

The opening period is followed by a period of progressively more extensive fishing which lasts through the remainder of March and into April. This is the portion of the season in which the fishermen expend the bulk of their effort. Nearly all fishermen work every day of this middle period. Effort is distributed widely across the western portions of the reefs at this time, extending well out beyond the windward perimeters.

When the fishermen perceive that the summer weather pattern is being replaced by the winter pattern, they move to more sheltered fishing areas in the eastern portions of the reefs. The western fishing areas, which have sustained a great deal of fishing pressure, decline steadily in catch rates during the middle period. The strong southerly winds of summer change to a more variable pattern late in the season, bringing oceanic swell from the north-western quadrant. Areas in the eastern portions of the lagoons are said to "fire" as the deep swells disturb resident rock lobsters, which then become catchable. Fishermen tend to utilize more specialized fishing areas within the eastern grounds, and may have long histories of fishing a particular spot and obtaining a predictable yield from it.

From the charts of fishing intensity (e.g., Fig. 3) it is clear that the major focus of fishing effort is on reef areas west of the central island platforms in each group. Four of the five categories of effort intensity used to divide effort amongst zones are shown in Figure 3. The sixth category of greater than 100 pots per hectare per season was documented only in the Wallabi Group.

The best spatial matching of ecological units to fishing zones occurs at the scales of geomorphological classes. Rock lobster fishing in the western regions of the reefs during the opening and middle periods is concentrated on exposed reef slopes, back reefs, submerged limestone platforms and lagoon patch reefs. Rock lobster fishing in the eastern regions of the reefs during the closing period generally occurs on lagoon patch reefs, drowned doline fields and sheltered reef slopes.

Patterns of fishing also reflect the diversity of benthic biological communities within a fishing zone. For example the geomorphological classes containing coral communities in which rock lobster fishing occurs (e.g., dissected limestone platforms, back reefs, eastern lagoon patch reefs and sheltered reef slopes) also include plant and sediment communities in a fine scale mosaic. Fishermen generally place pots along the interfaces of these communities.

Although the boundaries of fishing zones were often coincident with boundaries of geomorphologically-defined ecological units identified in this study (reflecting the importance of physiography to fishermen), fishing intensities within units of a given geomorphological class varied among Island groups, as did the relative abundance of different biological communities contained within those units.

Overall, the eastern portions of the lagoons are fished much less heavily than the western portions, but they contain biological communities which are potentially more susceptible to physical damage by pots because of the proliferation of living coral communities there (Table 1, Fig. 3).

Integrating ecological data with fishing effort intensity

By overlaying the charts of fishing zones and ecological units, the distribution and intensity of fishing effort was compared to the distribution of geomorphological structures and biological communities of the submarine environment at the Abrolhos (Fig. 3). The relationship was evaluated with the aim of assessing the relative value of ecological units to the fishery, and the potential negative effects of fishing and related activities on those units (particularly those which contain living coral communities). By defining the realized and potential values of ecological units (e.g., in terms of fishery's yield, or coral cover and diversity) to users of the marine resources of the Abrolhos, and the susceptibility of those units to damage from that usage, conceptual and geographical areas of potential conflict between user groups were identified.

Rock lobster fishing effort and the composition of the benthic communities in the marine environment of the Abrolhos: The attractiveness of a fishing site at the Abrolhos to most rock lobster fishermen is largely due to its return in rock lobster flesh. This yield is undoubtedly related to a number of ecological factors including the rugosity of the substratum (which influences the availablity of shelter sites), the form and magnitude of benthic production (which influences the availability of suitable food), and the behaviour of the lobsters (which influences catchability). However, other factors such as weather, sea-state, depth, time of year, size of boat and personality come into play when fishermen choose their fishing spots.

With the methods used in this study, the spatial resolution of fishing effort (fishing zones) was generally coarser than the resolution of benthic community structure. The available data do not allow unequivocal statements as to why certain areas are favoured as fishing grounds, nor whether differences in yield among fishing zones are related to biological productivity of ecological units within those zones. Our aim here was to document the spatial relationship between effort and benthic community structure, and assess the sensitivity of the ecological units to fishing activity.

The reported rock lobster fishing effort ranged from 0 to 266 pot drops per hectare of reef surface per season. The zero value does not indicate that rock lobster pots are *never* set in an area, but that none were acknowledged to have been set there by fishermen in our sample.

An average of over 200 rock lobster pots were set on every hectare of the most intensely fished grounds during each of the seven fishing seasons comprising this sample. These intense fishing zones generally correspond to submerged limestone platforms. In one such fishing zone in the Wallabi group, there is a mixture of biological communities including meso/macro algal assemblages, seagrass, and sediment. In another near North Island, there is abundant cover of mixed macrophytes and meso/macro algal assemblages, as well as extensive areas of coralgal pavement.

The most intensely fished zone in the Easter Group (Fig. 3) is a reef crest (emergent platform) and back reef area supporting a mixture of coralgal pavements, meso/macro algal assemblages with scattered coral colonies and mobile sediments. This shallow region was fished exclusively by jet boats.

Only a few of the intensely-fished zones have a significant (>10%) cover of live coral. The most intensely fished zone in the Pelsaert Group consists of submerged and dissected limestone platforms which support mainly coralgal pavements, meso/macro algal assemblages and high energy coral assemblages.

Not all submerged limestone platforms experienced intense fishing effort; some attracted very little. The submerged limestone platforms in the two fishing zones with highest effort intensity are characterized by a moderate to high degree of topographic complexity, and extensive seagrass and mixed macrophyte communities. The submerged limestone platforms which attracted little fishing effort have low relief, support some filter feeding communities, scattered tabulate coral colonies, and extensive areas of virtually bare pavements covered by a thin layer of sediments. The hydrodynamic characteristics in the fishing zones encompassing submerged limestone platforms include a uniform, moderate to high velocity flow in two fishing zones with high effort intensity, and shallow, wind-driven flow in two zones with low effort intensity. Thus, the high rock lobster yields from submerged limestone platforms appear to be related to the presence of topographic relief, macrophytes and a rapid turnover of water.

Fishing zones with the lowest intensities of effort mainly overlay mobile and static sediments, storm rubble fields, emergent limestone platforms, complex karst platforms and drowned doline fields. The latter two geomorphological classes contain areas of coral-macroalgal and coral communities (Table 1). When data were pooled among all the island groups, the intensity of fishing effort was found to be significantly lower in fishing zones that contain extensive live coral or filter feeding communities (Table 3).

Fishing effort and the sensitivity of Ecological Units

Among the biological communities, the low energy coral assemblages are the most susceptible to physical damage (Table 4), because of their dominance by fragile branching corals. Other communities containing high cover of live coral or benthic filter feeders were also classified as having high sensitivity. Communities dominated by macrophytes are much less sensitive to physical damage because of their flexible structure and relatively high specific growth rates. Turf and coralline algal pavements, and sediments were judged to be relatively insensitive to physical damage in the context of this study.

TABLE 4. Ranking of benthic biological community classes of the marine ecological units at the Houtman Abrolhos according to their relative sensitivity to anthropogenic physical damage.

Biological community class	Sensitivity	Rank
8) Low energy coral assemblage	high	1
9) High diversity coral assemblage	high	2
7) High energy coral assemblage	high	2
11) Filter-feeding community	high	3
10) Coral-macroalgae assemblage	moderate	4
3) Seagrass	moderate	4
1) Mixed macrophytes	moderate	5
2) *Sargassum* spp.	moderate	6
4) Meso-macro algae	low	6
5) Coralgal pavement	low	7
6) Calcareous algal field	low	7
12) Depauperate sediments	low	7
13) Infaunal sediments	low	7

The sensitivity of geomorphological classes to anthropogenic physical damage (Table 5) is largely a function of the sensitivity of the biological communities which characterize them (Tables 4 and 1). For example, it is due to the presence of low energy coral assemblages that the drowned doline fields and leeward reef slopes were classified as highly sensitive. The sensitivity of specific ecological zones within these broad geomorphological

TABLE 3. Statistical comparison of rock lobster fishing effort intensity in Abrolhos fishing zones with and without coral asemblages (i.e., biological community codes 7-9, Table 2) and/or, filter-feeding communities (i.e., biological community code 11, Table 2).

Ranking variable	Number of fishing zones	Fishing effort intensity (pot drops ha^{-1} yr^{-1}) Maximum	Mean	S.E.
Sensitive communities present	49	111.13	13.49	3.43
Sensitive communities absent	28	265.80	38.79	11.79

Mann-Whitney U=976; P=0.002**

classes varies according to the relative mix of biological communities. An important example of this is the distinction made in Table 5 between the isolated patch reefs in the western regions of the reef groups and those in the eastern and central regions. Sensitivity in the former areas is lower because of the low ratio of reef area to unconsolidated sediments, and the predominance of algal-dominated communities in many of these extensive ecological units.

TABLE 5. Ranking of geomorphological classes of the marine ecological units at the Houtman Abrolhos according to their relative sensitivity to anthropogenic physical damage.

Geomorphological class	Sensitivity	Rank
11) Drowned doline field	high	1
4) Sheltered reef slope	high	2
10) Complex karst platform	high	2
12) Isolated patch reefs (eastern, central)	high	3
9) Dissected limestone platform	high	3
6) Back reef	moderate	4
12) Isolated patch reefs (western)	moderate	4
1) Exposed reef slope	moderate	5
3) Submerged limestone platform	moderate	5
2) Emergent limestone platform	low	6
5) Storm rubble field	low	7
7) Mobile sediment sheet	low	7
8) Static sediment deposit	low	7

There is a significant relationship between the sensitivity index of ecological units and the intensity of fishing effort within them (Table 6, Fig. 4). The highest intensities of effort occurred in fishing zones overlaying ecological units with sensitivity indices between 3 and 6; habitats which range from moderately sensitive to relatively insensitive to physical damage. Lower effort intensity is associated with fishing zones which cover either highly sensitive, or relatively insensitive ecological units.

Fishing effort intensity was generally lower in fishing zones containing significant cover of coral or filter-feeding communities than in those lacking these communities (Table 3). When fishing zones were categorized according to the presence of low energy coral assemblages (the most sensitive, Table 4), the mean effort intensity tended to be greater in zones from which they were absent (Table 7). This difference, however, is statistically significant only in the Easter Group. In the other island groups there was a more even distribution of mean fishing effort among fishing zones with and without these assemblages of fragile corals.

Low energy coral assemblages occur in a larger proportion of the fishing zones in the Wallabi Group than they do in the other island groups. The mean fishing effort intensity in zones containing these coral communities was also higher in the

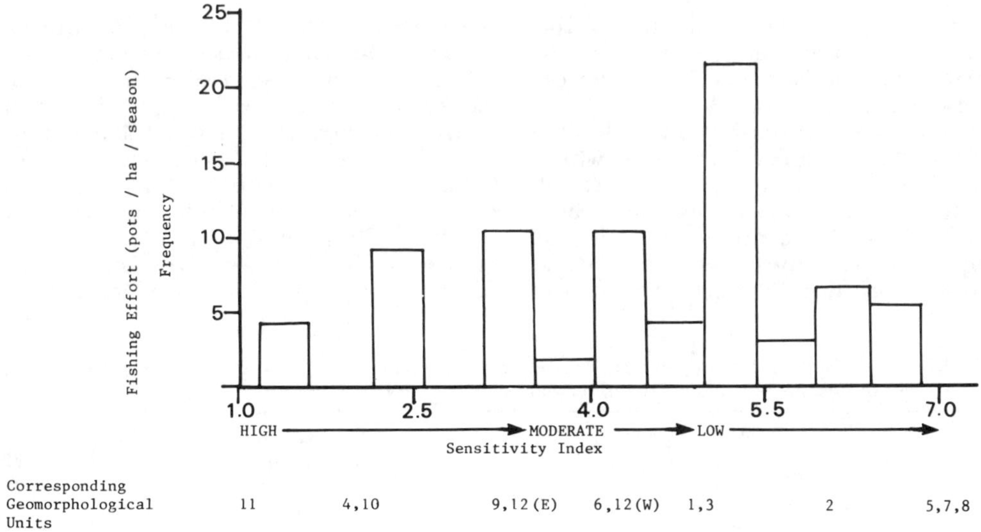

FIG. 4. The relationship between the sensitivity of the ecological units (classified according to geomorphological class) to anthropogenic physical damage, and fishing effort intensity within fishing zones on the Houtman Abrolhos reefs of Western Australia. Data are pooled for all four island groups. The codes 12 (E) and 12 (W) refer to the sensitivity index (Table 5) of isolated patch reefs (Table 1) in the eastern and western regions of the reef platforms respectively.

TABLE 6. The relationship between the relative sensitivity of ecological units and fishing effort intensity within overlaying fishing zones at the Houtman Abrolhos.

Sensitivity	Index value[1]	No. of zones[2]	Mean effort[3]	Standard error	Maximum effort[4]
High	1.0	4	2.99	1.49	7.05
	1.5	1	6.22	—	6.22
	2.0	9	6.65	0.71	9.37
	2.5	1	0.30	—	0.30
	3.0	8	19.18	5.12	46.84
Moderate	3.3	2	0.99	0.99	1.97
	3.5	2	57.92	53.21	111.13
	4.0	7	27.39	11.07	84.45
	4.3	3	0.95	0.81	2.56
	4.5	2	0.81	0.81	1.62
	4.7	1	18.12	—	18.12
	4.8	1	10.46	—	10.46
	5.0	21	40.35	15.54	265.80
	5.5	2	37.35	31.15	68.48
	5.7	1	96.03	—	96.03
Low	6.0	7	21.44	8.55	55.52
	6.5	1	0.00	—	0.00
	7.0	4	1.15	0.84	3.55

Comparison amongst Index-value groups:
Kruskal-Wallis non-parametric multiple comparison statistic,
$K=30.13$; $P=0.025$

[1]Average sensitivity rank of all geomorphological classes of ecological units within the fishing zone.
[2]Includes fishing zones in all Island groups.
[3]Average number of rock lobster pots dropped per hectare per season within fishing zones.
[4]Calculated number of rock lobster pots dropped per hectare per season in the most intensely fished zone.

Wallabi Group (Table 7). Over all groups, the mean intensity of effort averaged over all fishing zones was highest around North Island and lowest in the Pelsaert Group.

Ecological units of high sensitivity to physical damage (ranks 1,2,3 in Tables 4,5, Fig. 4) which overlap with fishing zones were defined as areas of potential conflict between conservation values and extractive use. These areas are shown for the Easter Group in Figure 3. Because of the incomplete sample on which fishing effort intensity was based, highly senitive units which partially overlap, or are immediately adjacent to zones where fishing effort intensity was greater than zero were also considered to be areas of potential conflict. Conversely, ecological units which fall within a highly sensitive geomorphological class (Table 5) but which lack highly sensitive biological communities (Table 4) were not included as areas of potential conflict. These exceptions explain apparent discrepancies between sensitivity and fishing effort in the areas of potential conflict identified in Figure 3.

TABLE 7. The distribution of fishing effort intensity related to the presence of low energy coral communities (see Table 2) in the fishing zones pooled within four island groups of the Houtman Abrolhos.

Island group	Low energy coral community present: Mean effort	Standard error	n^1	Low energy coral community absent: Mean effort	Standard error	n^2	Mann-Whitney test: U	P
Pelsaert	5.21	3.45	5	11.94	0.94	18	47	0.910
Easter	2.64	1.33	5	30.18	9.29	13	51	*0.067*
Wallabi	20.20	9.78	12	30.45	15.31	14	94	0.607
North Island	12.73	5.84	2	50.81	31.98	8	8	1.000

n^1 is the total number of fishing zones within each island group that contain low energy coral communities.
n^2 is the total number of fishing zones within each island group that do not contain low energy coral communities.

Discussion

The conclusions from this research depend largely on the interview data collected from a sample of 20% of the Abrolhos rock lobster fishermen. While the intended bias towards shallow-water fishermen does not jeopardize the validity of the data as applied to the questions posed, other factors had the potential to make the sample unrepresentative of that group of fishermen. Concerns about confidentiality or censure may have influenced the willingness of some fishermen to be interviewed, given the competition for good fishing ground at the Abrolhos, and the sensitivity of the issues involved. We have addressed these problems elsewhere (Wright *et al.* 1988), and have received comment from fishermen and representatives of the relevant fishermen's associations (K. Altham, pers. comm.) which lead us to conclude that such biases in the sample, if they exist, are small.

On average, commercial fishing for rock lobsters at the Abrolhos during the period covered by this study was concentrated in benthic habitats which were only moderately sensitive, or insensitive, to the type of physical damage associated with the activity. This finding corroborates the fishermen's claims that they do not often set pots on coral.

The differences among island groups in the relationship between fishing effort intensity and the presence of sensitive coral or filter-feeding communities (Table 7) is noteworthy. The significant difference in fishing effort associated with the presence or absence of these communities in the Easter Group may be due to the fact that there is very little live coral on the western reef perimeters and back reefs of this group, while there is somewhat more live coral in these geographic regions in the other two groups. In the Pelsaert Group, the most intense fishing occurred in the same geographic regions, but these contain several ecological units having high live coral cover. Another hypothesis which may explain these differences is that the fishing zones containing low energy coral assemblages are less accessible in the Easter Group than in the other two groups. For example, in the Wallabi Group, the karst and doline fields are closer to the western portions of the reef where the bulk of the fishing takes place, and fishermen can conveniently work these areas throughout the season in favourable conditions. The data necessary to test these hypotheses are lacking.

Despite the average trend of fishermen to avoid areas of high coral cover, many ecological units can be identified in which rock lobster fishing occurred, and where there is significant cover of live coral. Most of these are doline fields, lagoon patch reefs and leeward reef slopes in the north-eastern regions of each group (e.g., Fig. 3), although some back reefs and dissected limestone platforms in the Pelsaert Group provide examples from western reef margins.

Living coral colonies can be found in almost every area of the Abrolhos, but in many they occur as isolated organisms rather than communities. Their existence provides structural variety, biotic diversity and microhabitat for associated fauna. Areas of low coral cover predominate on the west sides of the reef groups, but again, the dissected limestone platforms and back reefs of the Pelsaert Group are exceptions to this generalization.

The coral communities on sheltered reef slopes, in drowned doline fields and on isolated patch reefs in eastern lagoons are diverse and extensive. The particular assemblages of coral species in these units are unique to the Abrolhos. Two of the most common species, *Acropora spicifera* and *A. abrolhosensis* are not found in significant densities elsewhere, and the abundant species *Alveopora gigas,* has been observed nowhere else (Veron 1985a). The proliferation of coral in these ecological units rivals the best-developed tropical communities both in terms of species diversity and percent living cover (Veron 1985b).

As well as being spectacular as individual colonies or assemblages, the coral communities of the Abrolhos provide structure to the submarine environment at two temporal scales. In the short term, they provide shelter and habitat for many species of invertebrates and fish. For example, the diversity of invertebrates inhabiting *Pocillopora damicornis* colonies at the Abrolhos is significantly higher than in colonies of the same species at the tropical Ningaloo Reef tract (Fitzpatrick 1987). On a long-term basis, coral communities are the major producers of the calcium carbonate framework at the Abrolhos reefs, effecting, along with coralline algae, the construction necessary to counteract the erosional forces of waves on the reef structure. Hence the coral communities are crucial to the continued existence of the marine ecosystems of the Abrolhos.

The ecological units which are of greatest interest to user groups other than fishermen are those with high proportions of living coral communities. It is the perceived destruction of these coral communities at the Abrolhos that has stirred the conservationist lobby to action (e.g., Anon. 1986; Bell 1986; Hatcher 1987 a,b). It is the luxuriant coral communities and their associated invertebrates and fish which attract non-extractive users (e.g., tourists) to the reefs, while the diversity and abundance of coral amidst kelp beds in high latitude habitats attract the interest of scientists (e.g., Hatcher & Walker 1984). Accretion by the coral communities maintains the diversity of shallow water habitats at the Abrolhos, which in turn maintain the great genetic diversity of the flora and fauna, and the visual attractiveness of the marine environment there.

There are no data on the response of the coral communities at the Abrolhos to high frequency, low intensity anthropogenic physical damage. Relevant ecological investigations elsewhere indicate that chronic, localized abrasion or breakage of coral can have disproportionate effects on the surrounding community because it does not allow adequate time for regrowth (e.g., Alino *et al.* 1985; Tilmant 1987), alters competitive interactions (e.g., Dustan 1987), or fosters pathogenic infection (e.g., Segel & Ducklow 1982).

The fragility of Abrolhos corals and their slow rates of growth (Crossland 1981, 1984), the presence of macroalgae within coral communities (Johannes *et al.* 1983), the steep slopes of the reef surfaces where coral growth is most prolific, and the common depth zonation of the corals all serve to enhance the susceptibility of Abrolhos coral communities to anthropogenic physical damage. For example tabulate corals dislodged from the upper surfaces of reef slopes may tumble downslope, leaving a swathe of crushed branching coral far more extensive than the area of the original (causal) damage (Hatcher, pers. obs.).

The presence of competent macroalgal competitors and the slow growth rates of Abrolhos corals also serve to hinder the recovery of coral communities from human-induced damage. Regrowth to pre-damage condition of experimentally damaged areas of live coral took over four years (Hatcher, unpubl. data). The presence of a common macroalga *(Eucheuma striatum)* in experimental plots the size of rock lobster pots has effectively prevented coral recovery to pre-damage abundance for a period of almost five years, due to overgrowth of the coral by the alga. This process, by which human-induced damage alters competitive heirarchies, has been implicated in the replacement of large areas of tabulate corals by branching corals in Abrolhos habitats, and more generally, the replacement of living coral communities by macroalgal communities there (Anon. 1986). No data exist to prove or disaprove these hypotheses, but macroalgal-dominated communities appear to be viable alternative stable states for benthic communities on reefs (e.g., Johannes *et al.* 1983; Hatcher 1984; Carpenter 1988).

If anthropogenic effects on benthic coral communities are significant at the Abrolhos, they will undoubtedly be highly site-specific, because the abundance and colonizing ability of various corals and macroalgae varies across reef habitats. On other coral reefs subjected to intense human activity, point-source disturbance of coral as a result of physical damage caused by recreational activities (e.g., boat anchors and diving) has rarely caused extensive degradation of coral communities (Tilmant 1987). On the other hand, physical damage to benthic coral communities caused by fishermen's use of traps and small explosive charges has resulted in the virtual denudation of large ($\approx$10 km^2) areas of coral reef in the Philippines (Alcala & Gomez 1987).

The present study was not designed to determine whether the fishing practices at the Abrolhos have caused degradation of the coral communities there. In the absence of historical data, the resolution of this issue is not possible without quantitative measurement of fishing impacts and experimental testing of hypotheses concerning their effects on benthic community structure. However, we have demonstrated that a tandem collection of ecological and anthropological data can allow an assessment of the potential for chronic damage to sensitive areas.

There are two aspects to the assessment of potential damage to benthic communities. Firstly, it is necessary to define the type, and delineate the spatial boundaries and average intensity of usage relative to the type and sensitivity of the biotic assemblages. The present study suggests ways to do this effectively. Secondly, it is necessary to assess the abilities of benthic communities to recover from damage caused by this usage, as well as from acute damage incurred outside of the average pattern. This study has not addressed the second aspect.

Because of the unusual characteristics of the coral communities at the Abrolhos, conservationists and scientists have drawn attention to the potential for damage which is associated with the deployment of rock lobster pots. The unique environmental conditions of high latitude reefs mean that their corals are physically fragile and relatively susceptible to physical damage and overgrowth by competitors such as macroalgae when man alters the benthic community structure. Thus, even small numbers of pot drops in a sensitive area could produce long-term degradation of coral colonies there. Definitive answers to these questions cannot be gleaned from the patterns of average usage. The next step is to measure community-specific coral recruitment, growth and mortality under controlled damage regimes.

The present study has partially defused the confrontation among interest groups at the Abrolhos (Anon. 1989). This has been done by simultaneously collecting ecological and anthropological data at similar spatial scales. Thus, we have identified the areas of overlap between fishing and conservation interest groups. The significant result of this exercise is not the fact that most fishing occurs in areas where corals are sparsely distributed. The study sheds no light on the causality of the historical relationship between fishing intensity and coral abundance. Its significance lies in identifying areas where there is significant potential for damage, and where few fishermen will be disadvantaged by the selective preservation of these areas for the use of many other individuals in the present and future.

Acknowledgements

This study was funded by the Government of Western Australia through the Abrolhos Islands Task Force. B.G.H. was supported by grants from the Australian Marine Sciences and Technologies Grants Scheme, and the Australian Research Council. Useful assistance was provided by the Australian Survey and Land Information Group, the Marine Laboratories of the Commonwealth Scientific and Industrial Research Organization, the Geraldton Professional Fishermen's Association, the Western Australian Departments of Conservation and Land Management and Fisheries, the Western Australian Fisheries Industry Council, and the Western Australian Museum. In particular, P. Baker, R. Black, N. Caputi, C. Chubb, B. Gilligan, J. Henry, M. Johnson, L. Joll, R. Owens, J. Penn, C. Simpson, I. Sutherland, J. Vernon, G. Wallace and J. Watson contributed valuable scientific and/or logistic input. We especially wish to thank K. Altham, W. Bailey, T. Beavor, F. Dewer, G. Leyland and forty Abrolhos rock lobster fishermen for their co-operation.

References

Alcala A. C. & Gomez E. D. (1987) Dynamiting coral reefs for fish: a resource-destructive fishing method. In: *Human Impacts on Coral Reefs: Facts and Recommendations* (ed. B. Salvat) pp. 51-60. Antenne Museum-EPHE, Tahiti, French Polynesia.

Alino P. M., Banzon P. V., Yap H. T., Gomes E. D., Morales J. T. & Boyonete R. P. (1985) Recovery and recolonization on a damaged back reef at Cangalayan Is. (Northern Philippines). In: *Proc. Fifth Int. Coral Reef Cong. Vol. 4* (ed. B. Delesalle *et al.*) pp. 279-84. Antenne Museum-EPHE, Moorea, French Polynesia.

Anon. (1986) *One million years of growth reduced to rubble.* 5 pp. The Conservation Council, Perth, Western Australia.

Anon. (1989) *Abrolhos Islands planning strategy — final report.* 29 pp. Abrolhos Islands Consultative Committee, Geraldton, Western Australia.

Backshaw D. G., Barnett J., Davies P. J., Duncan D. S., Harvey N., Hopley D., Isdale P. J., Jennings J. N. & Moss R. (1979) Drowned dolines — blue hole of the Pompey Reefs, Great Barrier Reef. *Bureau Mineral Res. J. Aust. Geol. Geophys.* **4,** 99-109.

Bell J. (1986) Abrolhos Islands in danger. *Skindiving Asia & South Pacific* **16/5,** 104-6.

Carpenter R. C. (1988) Mass mortality of a Carribean sea urchin: immediate effects on community metabolism and other herbivores. *Proc. Nat. Acad. Sci. U.S.A.* **85,** 511-4.

Crossland C. J. (1981) Seasonal growth of *Acropora* cf. *formosa* and *Pocillopora damicornis* on a high latitude reef (Houtman Abrolhos, Western Australia). In: *Proc. Fourth Int. Coral Reef Symp. Vol. 1,* (ed. E. D. Gomez *et al.*) pp. 663-7. University of the Philippines, Manila.

Crossland C. J. (1984) Seasonal variations in the rates of calcification and productivity in the coral *Acropora formosa* on a high latitude reef. *Mar. Ecol. Prog. Ser.* **15,** 135-40.

Dahl A. L. & Salvat B. (1988) Are human impacts, either through traditional or contemporary uses, stabilizing or destabilizing to reef community structure? In: *Proc. Sixth Int. Coral Reef Symp.* Vol. 1 (eds J. H. Choat *et al.*) pp. 63-9. James Cook University, Townsville.

Dustan P. (1987) Preliminary observations on the vitality of reef corals in San Salvador, Bahamas. In: *Proc. Third Symp. Geol. Bahamas* (ed. H. A. Curran) pp. 57-65. CCFL Bahamian Field Station, Fort Lauderdale, Florida.

Fitzpatrick J. (1987) Species richness and abundance in the community associated with Colonies of the coral *Pocillopora damicornis,* in Western Australia: An investigation into the relative influence of local environment and latitudinal position. B.Sc. (Hon.) Thesis, University of Western Australia.

Hatcher A. I., Hatcher B. G. & Wright G. D. (1988) *A preliminary report on the interaction between the major human activities and the marine environments at the Houtman Abrolhos Islands of Western Australia.* Report to The Abrolhos Islands Task Force, Department of Conservation and Land Management, Perth, Western Australia, 132 pp.

Hatcher B. G. (1984) A maritime accident provides evidence for alternate stable states in benthic communities on coral reefs. *Coral Reefs* **3,** 199-204.

Hatcher B. G. (1985) Ecological research at the Houtman Abrolhos: high latitude reefs of Western Australia. In: *Proc. Fifth Int. Coral Reef Cong. Vol. 5* (eds B. Delesalle *et al.*) pp. 291-7. Antenne Museum-EPHE, Moorea, French Polynesia.

Hatcher B. G. (1987a) The Houtman Abrolhos: research and conservation status. *Aust. Coral Reef Soc. Newslett.* **7,** 23-6.

Hatcher B. G. (1987b) The Houtman Abrolhos: More reefs in need of management. *Bull. Aust. Littoral Soc.* **10(6),** 5-6.

Hatcher B. G., Johannes R. E. & Robertson A. I. (1989) Review of research relevant to the conservation of shallow tropical marine ecosystems. *Oceanogr. Mar. Biol. Ann. Rev.* **27,** 337-414.

Hatcher B. G., Kirkman H. & Wood W. F. (1987) The growth of the kelp *Ecklonia radiata* near the northern limit of its range in Western Australia. *Mar. Biol.* **95,** 63-73.

Hatcher B. G. & Walker D. (eds) (1984) *Proceedings of a workshop on the Houtman Abrolhos.* Australian Marine Sciences Association and Western Australian Department of Fisheries & Wildlife. Special Publication No. 8401, 51 pp. Fisheries Department, Perth, Western Australia.

Johannes R. E., Wiebe W. J., Crossland C. J., Rimmer D. W. & Smith S. V. (1983) Latitudinal limits of coral reef growth. *Mar. Ecol. Prog. Ser.* **11,** 105-11.

Kenchington R. A. (1988) Managing reefs and inter-reefal environments and resources for sustained exploitive, extractive and recreational uses. In: *Proc. Sixth Int. Coral Reef. Symp.* Vol. 1 (eds J. H. Choat *et al.*) pp. 81-7. James Cook University, Townsville, Queensland.

Oliver J. K. (1984) Intra-colony variation in the growth of *Acropora formosa:* Extension rates and skeletal structure of white (zooxanthellae-free) and brown-tipped branches. *Coral Reefs* **3,** 139-47.

Pielou E. C. (1969) *Introduction to mathematical ecology.* Wiley Inter-Science, New York, 286 pp.

Segel C. A. & Ducklow H. (1982) A theoretical investigation into the influence of sublethal stress on coral — bacterial ecosystem dynamics. *Bull. Mar. Sci.* **32,** 919-35.

Sheard K. (1962) *The Western Australian Crayfishery 1944-1961.* Paterson Brokensha, Perth, 107 pp.

Simpson C. J. (1988) *Ecology of scleractinian coral in the Dampier Archipelago, Western Australia.* Environmental Protection Authority Technical Series No. 23, 238 pp. Environmental Protection Authority, Perth, Western Australia.

Teichert C. (1947) Contributions to the geology of the Houtman's Abrolhos, Western Australia. *Proc. Linn. Soc. N.S.W.* **71,** 145-96.

Tilmant J. T. (1987) Impacts of recreational activities on coral reefs. In: *Human Impacts on Coral Reefs: Facts and Recommendations* (ed. B. Salvat) pp. 192-214. Antenne Museum-EPHE, Tahiti, French Polynesia.

Veron J. E. . (1985a) New Scleractinia from Australian coral reefs. *Rec. West. Aust. Mus.* **12,** 147-83.

Veron J. E. N. (1985b) Aspects of the biogeography of hermatypic corals. In: *Proc. Fifth Int. Coral Reef Cong. Vol. 4* (eds B. Delesalle *et al.*) pp. 83-8. Antenne Museum-EPHE, Moorea, French Polynesia.

Veron J. E. N. (1986) *Corals of Australia and the Indo-Pacific.* Angus & Robertson, Sydney, 644 pp.

Veron J. E. N. & Done T. J. (1979) Corals and coral communities of Lord Howe Island. *Aust. J. Mar. Freshw. Res.* **30,** 203-36.

Veron J. E. N. & Marsh L. M. (1988) Hermatypic Corals of Western Australia. *Rec. West. Aust. Museum Suppl. No. 29,* 136 pp.

Wells J. W. (1955) A survey of the distribution of reef coral genera in the Great Barrier Reef region. *Repts. Great Barrier Reef Comm.* **6(2),** 21-9.

Wilkinson L. (1986) *SYSTAT — The system for statistics.* SYSTAT Inc., Evanston, Illinois, 682 pp.

Wilson B. R. & Marsh L. M. (1979) Coral reef communities at the Houtman Abrolhos, Western Australia in a zone of biogeographic overlap. In: *Proc. Int. Symp. Mar. Biogeogr. Evol. South. Hemi.,* pp. 259-78. D.S.I.R., Auckland, New Zealand.

Wright G. D., Hatcher A. I. & Hatcher B. G. (1988) Clarifying the impact of fishing activity on the reefs of the Houtman Abrolhos: results of a tandem approach between anthropology and marine science. In: *Proc. Sixth Int. Coral Reef Symp.* Vol. 2. (eds J. H. Choat *et al.*) pp. 433-7. James Cook University, Townsville.

Proc. Ecol. Soc. Aust. 1990 — 16:71-77

Ecological changes due to our use of the coast: research needs versus effort

P. G. FAIRWEATHER

Graduate School of the Environment, Macquarie University, N.S.W. 2019

Abstract

Australian society uses coastal, estuarine and nearshore marine ecosystems for a variety of recreational, industrial and social ends. The ocean and its ecosystems are used as receptacles for sewage and other pollution, as reservoirs of organisms for food or bait, as sites for recreational activities like playing, fishing or boating, and are subject to reclamation for development of residences, tourism and industry next to the sea. These uses have increased greatly during the 200 years of white settlement and have brought good and bad changes.

Central to determining limits of impacts and appropriate management is our understanding of the ecosystems involved. Much of our understanding of impacts lies in the "grey" literature, and therefore beyond the assessment of peer review. Thus the scientific value of this information is very uneven. In this chapter, the scant ecological information on the impacts of the above uses on marine flora and fauna in Australia is reviewed via searches of published literature in the 1980s. Some coastal habitats, such as saltmarshes and sandy beaches, are grossly under-represented in our published work. Little is known about how communities function along sandy and muddy stretches of open coast and on rocky shores of our estuaries. Attention is drawn to these and other poorly understood areas. The type of information needed to plan for the siting of marine reserves is highlighted.

Introduction

Living on an island, Australians are very aware of the sea. Attention is often drawn to the intensely urbanized population, with approximately 80% of our people living within 160 km of the ocean and some 60% in cities sited on estuaries. Linked with these urban areas are industry, tourism and recreational opportunities. Each use made of coastal habitats can present the ecosystem involved with some potential threat (Table 1). Most of these uses are increasing in magnitude, especially those resulting from recreation due to changes in work practices, increased leisure time and growing reliance on tourist revenue.

This chapter discusses the ecological research required for informed decisions about management and conservation to be made in relation to these potential threats. Such basic research is necessary: environmental management is based on an understanding of the ecosystems involved. This understanding begins with taxonomic and biogeographical surveys, but must extend to functioning within communities (e.g., successional change, role of disturbances, interactions among species) and ecosystems (e.g., nutrient cycling, energy flow). Functional understanding of ecological phenomena is often best gained using experimental protocols, and within Australian marine ecosystems, much experimental research has been done on rocky shores (e.g., see Underwood & Fairweather 1986) and coral reefs (see Doherty & Williams 1988). If we wish to manage ecological units like particular saltmarshes or sandy beaches, then research is needed in these habitats. This chapter asks: are we doing that research and is it readily available to interested parties?

Methods

This review of marine, estuarine and nearshore ecology was done in three parts. In the first phase the few published reviews of impacts on marine ecosystems in Australia were consulted. In the second phase, the computerised DIALOG system was searched to find out more about the subject. This was done to check the dearth of material unearthed in the first search. The search included four databases: BIOSIS, ENVIROLINE, Oceanic Abstracts and Aquatic Science Abstracts. The

TABLE 1. List of some potential impacts and threats to coastal ecosystems, and some of the habitats most affected.

Development	Habitat
Reclamation and removal of vegetation	Mangroves, saltmarsh, sand dunes
Port works and dredging	Sandy, muddy & rocky subtidal areas
Industrial discharges	Estuarine habitats generally, beaches
Alterations to drainage	Saltmarshes, mangroves, estuaries
Exploitation	
Fisheries	Estuaries, open sea, coral reefs, sandy subtidal
Bait, food, materials	Rocky shores, estuaries
Oil, gas, ore mining	Beaches, open sea
Sand mining	Beaches and dunes
Recreation	
Fishing	Rocky shores, estuaries, beaches
Offroad vehicles	Saltmarsh, dunes
Passive (=tourism)	Coral reefs, beaches
Sea bathing	Beaches, estuaries
Pollution e.g., TBT	Estuaries
Public Amenities and Utilities	
Sewage	Rocky shores, beaches, estuaries, sandy subtidal
Urban runoff	Rocky shores, beaches, estuaries, sandy subtidal
Agricultural discharges	Estuaries
Sea-level rise from Greenhouse effect	All shallow water/intertidal/supratidal habitats

keywords searched for included Australia, marine, coastal, estuary, wetland, ecology, ecosystem, species, degradation, impact, exploitation, harvest, conservation, regeneration and their derivatives. This yielded some 154 references, of which the most recent 100 were abstracted and viewed.

The last phase focussed on refereed publications about marine ecology in Australian ecosystems to look at the concentration of research effort among habitats and types of studies (Fairweather 1988). All issues published in 1980 to 1987 (inclusive) of some 27 marine and ecological journals (see Table 2) were manually searched. This covered recent research, especially that done in the period of the Australian Marine Sciences and Technology Advisory Committee and Marine Science and Technology Grants. The selection of journals to contribute papers for this sample could influence any such review: I attempted to include all journals that I expected would contain a substantial portion of this published research. Thus, this reflects my own experience. Specifically omitted were taxonomically restricted (e.g., fish) journals, those which yielded proportionally few, if any, papers upon perusal (e.g., *Canadian Journal of Fisheries and Aquatic Sciences, Hydrobiologia* and journals of the Royal and Linnean Societies), and some which were not available (e.g., *Journal of Plankton Research).* Additions made by interested readers to the journals covered in this database would be welcome.

All papers found in this sample that dealt with Australian marine and estuarine ecosystems (i.e., up to supratidal heights), and that were empirical and field-based, were read. Taxonomic or natural history papers without ecological data were omitted from this database. The following data were extracted from each paper and used in subsequent analyses: title, reference, author(s), institution(s) where the work was done (not current address), geographic locality and geomorphology of study

TABLE 2. List of the 27 journals examined (with the volume numbers in parentheses) and the number of papers on the ecology (in the broadest sense) of Australian marine/estuarine/coastal ecosystems.

Type of journal, title and volumes	No. of relevant papers
Australian/Marine	
Aust. J. Mar. Freshwat. Res. (31-38)	145
Wetlands (1-7, Vol. 2 incomplete)	27
International/Marine	
Mar. Ecol. Progr. Ser. (2-41)	106
J. Exp. Mar. Biol. Ecol. (42-114)	99
Mar. Biol. (55-96)	90
Aquat. Bot. (8-29)	37
Coral Reefs (1-6)	36
Bull. Mar. Sci. (30-41)	24
Est. Coast. Shelf Sci. (10-25)	23
Mar. Pollut. Bull. (11-18)	21
Limnol. Oceanogr. (25-32)	17
Bot. Mar. (23-29, Vol. 30 unavailable)	9
Biol. Bull. (158-173)	6
Phycologia (19-26)	4
J. Phycol. (16-23)	3
Estuaries (3-10)	2
PSZNI Mar. Ecol. (1-8)	0
Australian/Ecological	
Aust. J. Ecol. (5-12)	16
International/Ecological	
Oecologia (44-74)	41
Ecology (61-68)	9
Biological Conservation (17-42)	5
Ecological Monographs (50-57)	4
J. Appl. Ecol. (17-24)	2
J. Anim. Ecol. (49-56)	1
J. Ecol. (68-75)	1
Oikos (34-50)	1
Environ. Manage. (4-11)	0
Total	729

site, habitat-type, tidal height and the methodology used in the study. This body of research (729 papers) is discussed in the light of what we need to know to manage these ecosystems, recommendations are made for filling gaps in this research record, and the issue of the design, siting and function of marine reserves is taken as a case study.

Results and Discussion

Reviews

Two main points were suggested by the results of the first two phases of the research. First, although there have been many reviews of coastal impacts written in the last 18 years (see e.g., Nix & Elliot 1975), most do not take a wide view of the situation. For example, authors have reviewed changes and problems in particular locales (see e.g., Burgess & Woolmington 1981; Hopley 1988; McGuinness 1988), for particular habitats (see e.g., Allaway 1987; McComb & Lake 1988), or a combination of place and habitat (see e.g., Buckney 1987; McLoughlin 1987).

The second point is that much of the relevant information resides in the "grey" literature. This includes environmental impact statements (EISs), proceedings of conferences, reports of working parties, technical memoranda, governmental papers, etc. Most reviews did not appear in the usual scientific journals. Although the grey literature is directly involved in guiding management decisions, reviews appearing therein are not part of science in one sense. That is, they may not have been subjected to the scrutiny of peer review, which is known as one of the self-correcting mechanisms of scientific practice (and part of what sets it apart from other forms of knowledge). Much that is published in the scientific literature is of uneven quality, and work in the grey literature may, on the whole, be worse (e.g., see review by McGuinness 1988). The old adage "It isn't finished until it's published" applies here, and few detailed studies of impacts on coastal Australia (especially in the long-term) have been published. Perhaps this is due to a lack of self-consciousness (*sensu* McIntosh 1985) amongst Australian marine ecologists. For example, the first comprehensive text on Australian marine biology (Hammond & Synnot 1989) is only now nearing completion, despite our long-standing and significant contribution to the global body of marine ecological research (Fairweather 1988).

Literature search

The manual survey of refereed publications revealed much about the concentration of marine and related research on ecology in the 1980s. More than 60% of the 729 papers found in the survey were published in only four journals (see Table 2). Most studies (quantitatively measured by output of papers) were done subtidally in open coast environments (Table 3), whereas there were very few publications from studies in the open sea or supratidal (mostly saltmarsh) habitats. Intertidal areas, where the public often visit, were relatively well represented considering their small area. Coral and rocky reefs were the most popular habitat-types for study, but there were also considerable numbers of studies of seagrass and mangroves. Studies of sandy areas, waters and muddy bottoms were fewer.

TABLE 3. Numbers of papers dealing with each of several habitat features. Total numbers vary because more than one habitat may be investigated in some papers. Studies were not distributed evenly across any set of categories (χ^2 tests, significant at $P<0.05$).

Geomorphology	
Open coasts and embayments[a]	537
Estuaries[b]	182
Open sea[c]	11
Tidal Heights	
Subtidal	523
Intertidal	204
Supratidal	20
Habitat-type	
Coral reef	229
Rocky reef[d]	160
Seagrass	115
Mangal	94
Sandy bottoms	70
Sea surface and watercolumn[e]	49
Muddy bottoms	41
Saltmarsh	25

a — includes lagoons of coral reefs.
b — includes saline coastal lagoons.
c — pelagic communities (especially plankton) are probably quite under-represented in this sample because the *Journal of Plankton Research* could not be included.
d — includes other hard substrata such as pier pilings.
e — includes estuarine waterbodies, and studies on fish or plankton where no benthic habitat was specified.

It was also possible to analyse geographically the research effort (Table 4). There have been several attempts by geographers to classify the coastline of Australia into different dominant substrata. For instance, sand makes up nearly half of our coastline, but it makes up much more in some States (Table 4a). Rock is the next most common, followed by mud and coral. Mangroves and the mouths of estuaries account for very little of the overall coastline, but much of the coast of our northernmost States. This breakdown suggests that the analysis of habitat-types in Table 3 can be improved upon by considering the availability of the different substrata on a State by State basis. The null hypothesis might be that studies should be distributed as a pro rata representation of the

TABLE 4. (a) Types of coastline in each State of Australia expressed as percentages of the coastline (modified after Table 3 in Barson & Heatwole 1985), based on a total coastline of 35 299 km. I estimated the coral reefs of Queensland as an additional 40% of the coastline and for Western Australia as 9%; these may be underestimates.

	NT	Q	NSW	V	T	SA	WA	Total
Coast-type								
Mangrove	42	18	<1	2	0	7	17	2
Mud	6	<1	<1	7	2	<1	2	17
Sand	40	36	65	64	68	59	49	47
Rock	9	4	33	26	29	33	19	18
Estuary mouth	3	2	1	<1	1	<1	4	2
Coral	?	40	0	0	0	0	9	14
Total	14	29	5	5	6	9	32	

(b) Studies in the survey split by State or Territory and by habitat (data are the percentages of papers from each State in a given habitat class, for comparison with the types of coastline in each State given above).

	Percentage of all studies in each State							Australia Total	
Habitat	NT	Q	NSW	V	T	SA	WA	No. of papers	%
Mangrove	26	11	13	5	0	8	6	81	10
Muddy bottoms	26	4	6	14	3	2	6	44	6
Sandy bottoms	26	7	7	3	5	14	13	67	8
Rocky reef[a]	0	2	43	32	40	45	20	176	22
Estuary[b]	16	9	27	34	17	31	35	171	21
Coral reef	5	65	0	0	0	0	15	229	28
Pelagic[c]	0	2	5	12	35	0	4	44	5
Total no. of papers	19	321	200	59	40	51	124		
%	2	39	25	7	5	6	15		

a — includes all hard substrata except coral and trunks of mangroves;

b — includes seagrass, saltmarsh, some mangrove and estuarine waterbodies;

c — includes open sea, continental shelf and watercolumn.

types of coastline. Table 4b highlights the uneven publication record among States and among different habitats. Clearly our research effort does not match the availability of different habitats within each State, but we might not expect it to. Perhaps better criteria would be the number of scientific institutions or scientists per State, or the total population size of each State. The concentration on coral reef research in Queensland is defensible, at least in part, on the grounds that the Great Barrier Reef has global significance and Australians are its custodians on behalf of the rest of the world. Even for the best known habitats, the coverage across Australia is very uneven (Table 4b).

There were especially few studies of saltmarsh or intertidal areas of either sand or mud. This lack of knowledge is surprising on the grounds of possible but unknown threats, and hence there is a need for research. Saltmarsh has disappeared at an alarming rate since white settlement (McLoughlin 1987; Adam *et al.* 1988) and it was recently identified as a crucial habitat in New South Wales by the State Environmental Planning Policy No. 14. Supratidal areas like saltmarsh are actually easier to work in than intertidal or subtidal areas (pers. obs.). The question of the values of wetlands such as saltmarsh to our coastal fisheries and recreation has not yet been answered conclusively. For instance, are saltmarshes net importers or exporters of nutrients? What part do estuarine mudflats play in nutrient exchanges? Possibly even more alarming is our lack of understanding of sandy beaches as ecosystems. The Australian contribution to the major book on this topic (MacLachlan & Erasmus 1984) was made by geographers! Yet beaches are an intimate part of the Australian lifestyle and culture, especially in summer. What threats, if any, are posed by such recreation and development to their ecology or to public health? There is not enough information to answer such questions, because relatively few of the studies of sandy bottoms in Tables 3 and 4 are of intertidal sites. It is a sad indictment of Australian ecologists that visiting scientists or geographers were responsible for most of the few and the best ecological studies of sandy beaches in New South Wales (e.g., Dexter 1984) and Western Australia (e.g., McLachlan & Hesp 1984). On the whole this research has not gone beyond descriptions of taxonomy, distribution and abundance or organisms.

Different institutions were involved to different degrees with this published research effort (Table 5). Again this is probably to be expected, but these numbers are useful in identifying the most vigourous centres of research (Fairweather 1988). It is interesting that some institutions involved only in research like the Commonwealth Scientific and Industrial Research Organization and the Australian Institute of Marine Science were matched in output by some universities (which have the added responsibility of teaching), while management organizations (e.g., fisheries departments in each State) contributed fewer papers. Perhaps better criteria here would include the output relative to the size of the institution (e.g., the number of personnel involved or the amount of funding), or involve other outputs (such as the number of higher degrees awarded for universities). From this sample of papers, very active centres of research based in Sydney (e.g., rocky shores, coral reefs), Townsville (e.g., coral reefs, mangroves) and Perth (e.g., seagrasses, estuaries) were easily identified (unpublished data).

The studies were also scored as to their location on a methodological continuum. Much has been made recently of the need for field experiments in ecology, in particular the progression from descriptions of ecological patterns through postulating and correlating explanatory processes with these, and then to test these processes experimentally (see e.g., Underwood & Fairweather 1986).

TABLE 5. Numbers of papers emanating from work done at different scientific institutions. The favoured journals (i.e., ≥5 papers) are also listed, in order of their frequency.

Institution	Papers	Journals*						
University of Sydney	111	1	2	3	4	5	6	7
CSIRO	108	4	5	2	1	8		
AIMS	93	2	1	9	10	8		
James Cook University	64	5	2	9	4	13		
University of Western Australia	47	5	2	12	14			
University of Melbourne	38	4	1					
New South Wales Fisheries	29	4	12	11	1			
University of Queensland	29	5						
University of Adelaide	22	1	3					
Macquarie University	21	1						
Australian Museum	19	4						
University of New South Wales	18	11	4					
University of Tasmania	17	5	4	1				
Western Australian Marine Science Labs	17	13						
Griffith University	15	5						
Murdoch University	14	4						
Victorian Marine Science Labs	12							
Queensland Fisheries	12							
Tasmanian Fisheries	9							
ANU	7							
Latrobe University	6							
University of Technology, Sydney	5							
University of Wollongong	4							
Monash University	3							
University of Newcastle	3							
Flinders University	2							
University of New England	2							
Deakin University	1							
Other Australian[a]	42							
Overseas	43	1	4					

a — 19 governmental agencies (including fisheries departments from other States), and 14 papers from private companies.

* 1—J. Exp. Mar. Biol. Ecol.
2—Mar. Ecol. Progr. Ser.
3—Oecologia
4—Aust. J. Mar. Freshwat. Res.
5—Mar. Biol.
6—Ecology
7—Bull. Mar. Sci.
8—Est. Coast. Shelf Sci.
9—Coral Reefs
10—Limnol. Oceanogr.
11—Wetlands
12—Aquat. Bot.
13—Mar. Pollut. Bull.
14—Bot. Mar.

It is evident from Table 6 that ecologists from the government and university sectors approach their research problems with different methodological tools, viz. a greater proportion of university scientists take their studies to an experimental stage. It is unclear whether experimental manipulations are more difficult in some ecosystems (although each habitat defined in Table 3 saw at least some experimentation) or whether scientists in different work environments differ in their willingness or opportunities to put their ideas to some critical test.

TABLE 6. Cross-classification of type of institution involved in each study and type of methodology employed in the study. Studies with both governmental and educational (>99% from universities) scientists were scored in each category. There were 21 educational, 27 governmental and nine private institutions represented in the complete sample of papers. The methodological categories are viewed as a progressive continuum (descriptive – process – experimental, see text) and each study is scored for the most advanced type of methodology employed. The experimental category represents any independent test of a hypothesis and includes lab tests, but most were done in the field.

	Government	Education	Total
Descriptive	178	198	376
Process	95	100	195
Experimental	61	456	517

Case Study: Marine reserves

Why do we need ecological knowledge and an understanding of how ecosystems function? Why is such information integral to management decisions? For example, it has long been realized that a system or network of nature reserves in Australia is very desirable (Fenner 1975). The proclamation and management of reserves and protected areas in marine habitats has lagged behind the terrestrial scene, especially in New South Wales (Ivanovici 1984). There has been a dearth of Australian publications about the principles for, and outcomes of, the design of marine reserves (although the need for good inventory data is addressed by Margules *et al.* 1988). This is despite much discussion overseas about the proper units to reserve and manage (watersheds, Ray & Hayden 1987), the role of natural ecological boundaries in limiting areas (Salm 1984) and the effectiveness and ecological effects of such protection (White 1986).

Mainstream ecological research might suggest new roles and aspects of marine reserve design. For example, the part played by variation in recruitment of larvae in marine populations and communities has been recently emphasized (Underwood & Fairweather 1989). This recognition that most marine ecosystems are open systems offers much weight to arguments for a system of marine reserves. Such a system, including rich and diverse communities and located along a coastline, would provide refuges and growth centres (Cairns & Elliot 1987) for the unprotected coast by providing a recurrent source of potential recruits (Underwood & Fairweather 1989). This should allow the replenishment of areas after severe impacts and so form part of an integrated management scheme for any coastline. This effect could be especially so where the effects of currents and coastal landforms on larval transport (Wolanski & Hamner 1988) are taken into account. After reserves are declared, there are very real problems of management (e.g., White & Bratton 1980) to be faced. This can only be done with a thorough understanding of the "natural" range of variation and the "normal" dynamics within the communities being preserved. This suggests a need for long-term studies in ecosystems of concern. Ecologists have a clear role and indeed a duty here. This potential of marine reserves to function as sources of recolonization as well as refuges requires more research.

Recommendations for research

Based on this review, the following research topics are highlighted as urgent priorities to aid the management of our coast:

1) little-studied habitats, e.g., saltmarshes, sandy beaches and mudflats almost anywhere in Australia;
2) any marine or estuarine habitat in Northern Territory and the north generally (apart from the Great Barrier Reef);
3) seagrass in States other than Western Australia and New South Wales;
4) rocky shores other than in New South Wales;
5) coral reefs in Western Australia;
6) mangroves elsewhere than in north Queensland;
7) some habitats in estuaries e.g., rocky shores;
8) consequences of sea-level change from the Greenhouse Effect for intertidal and supratidal habitats;
9) more focus on ecological processes so that we can understand how these ecosystems function: description is a necessary first step, but is not enough alone;
10) addressing the hard questions of the direct economic benefits (e.g., value of wetlands to fisheries, recreational usage of the coast, etc.) or the conservation value of marine ecosystems;
11) input or collaboration with other disciplines, such as physical oceanography, biogeography or geomorphology, to solve environmental problems on our coasts, including the most useful design of marine reserves; and,
12) publication in the open scientific literature of any relevant data presently in hand concerning the above areas or for specific environmental threats.

Conclusions

The ecological changes due to our use of the Australian coast, and attempts to correct them, have not been summarized in any coherent fashion. This is in part due to little credit being given for such reviews but also probably results from a lack of self-consciousness among Australian marine ecologists, despite their major contribution to work that is significant on the world stage. This contribution should be brought home to politicians, scientific administrators and managers of natural resources. Judged by published studies in the last eight years, our understanding of coastal ecosystems and habitats is very uneven. Much work has been done on coral reefs and rocky shores (which, could be argued, are significant in terms of global attraction and links to ecological theory). Seagrasses, mangroves and estuaries more generally have received some attention (but perhaps we do not know as much as we think). Sandy beaches, mudflats and saltmarshes have had little beyond initial descriptive studies published about them. If management does stem from knowledge then much more research is needed for the protection and wise use of coastal environments in Australia.

Acknowledgements

Thanks are due to V. Nelson for help with references and reading a draft. I especially thank L. Handley for many hours in the library searching the literature and M. Glass of the Library for the DIALOG search. Many helpful comments were made by the reviewers, and A. Underwood, A. McComb, G. Grigg and others at talks given at the ESA Symposium, the AMSA Silver Jubilee Conference and the University of Sydney. The preparation of this paper was supported by a Marine Sciences and Technologies Grant and a Macquarie University Research Grant.

References

Adam P., Wilson N. C. & Huntley B. (1988) The phytosociology of coastal saltmarsh vegetation in New South Wales. *Wetlands (Australia)* **7,** 35-85.

Allaway W. G. (1987) Exploitation and destruction of mangroves in Australia. In: *Mangrove Ecosystems of Asia and the Pacific: status, exploitation and management* (eds C. D. Field & A. J. Dartnall) pp. 183-92. AIMS, Townsville.

Barson M. & Heatwole H. (1985) Seashore Vegetation. In: *Seashore Ecology* (T. Carefoot & R. D. Simpson) pp. 195-216. University of Queensland Press, Brisbane.

Buckney R. T. (1987) Three decades of habitat change: Kooragang Island, New South Wales. In: *Nature Conservation: the role of remnants of native vegetation* (eds D. A. Saunders, G. W. Arnold, A. A. Burbidge & A. J. M. Hopkins) pp. 227-32. Surrey Beatty & Sons, Chipping Norton.

Burgess J. S. & Woolmington E. (1981) Threat and stress in the Clarence River estuary of northern New South Wales. *Human Ecol.* **9,** 419-32.

Cairns D. K. & Elliot R. D. (1987) Oil spill impact assessment for seabirds: the role of refugia and growth centres. *Biol. Conserv.* **40,** 1-9.

Dexter D. (1984) Temporal and spatial variability in the community structure of the fauna of four sandy beaches in southeastern New South Wales. *Aust. J. Mar. Freshwat. Res.* **35,** 663-72.

Doherty P. J. & Williams D. McB. (1988) The replenishment of coral reef fish populations. *Oceanogr. Mar. Biol. Ann. Rev.* **26,** 487-551.

Fairweather P. G. (1988) Australian marine ecology in the '80s: the scientometrics of who publishes what where. In: *Australian Marine Sciences Association Silver Jubilee Conference Commemorative Volume,* pp. 6-10. Australian Marine Sciences Association, Melbourne.

Fenner F. (1975) A national system of ecological reserves in Australia. *Search* **6,** 108-11.

Hammond L. & Synnot R. (1989) *Australian Marine Biology.* Longman Cheshire, Melbourne (in press).

Hopley D. (1988) Anthropogenic influences on Australia's Great Barrier Reef. *Aust. Geographer* **19,** 26-45.

Ivanovici A. (ed.) (1984) *Inventory of Declared Marine and Estuarine Protected Areas in Australian Waters, Volume 2,* Australian National Parks & Wildlife Special Publication 12.

MacLachlan A. & Erasmus T. (1984) *Sandy Beaches as Ecosystems.* Dr W. Junk, The Hague.

MacLachlan A. & Hesp P. (1984) Faunal response to morphology and water circulation of a sandy beach with cusps. *Mar. Ecol. Progr. Ser.* **19,** 133-44.

Margules C. R., Nicholls A. O. & Pressey R. L. (1988) Selecting networks of reserves to maximise biological diversity. *Biol. Conserv.* **43,** 63-76.

McComb A. J. & Lake P. S. (eds) (1988) *The Conservation of Australian Wetlands.* Surrey Beatty & Sons, Chipping Norton.

McGuinness K. A. (1988) *The Ecology of Botany Bay and the Effects of Man's Activities: a Critical Synthesis.* Institute of Marine Ecology, University of Sydney.

McIntosh R. P. (1985) *The Background of Ecology.* Cambridge University Press, Cambridge.

McLoughlin L. (1987) Mangroves and grass swamps: changes in the shoreline vegetation of the middle Lane Cove River, Sydney, 1780's-1980's. *Wetlands (Australia)* **7,** 13-24.

Nix H. A. & Elliot M. A. (eds) (1975) *Managing Aquatic Ecosystems. Proc. Ecol. Soc. Aust. 8.*

Ray G. C. & Hayden B. P. (1987) Biogeographic differentiation of the coastal zone. In: *Symposium on the Ecological Management of Coastal Zones* (ed. J. I. Furtado), Dr W. Junk, The Hague.

Salm R. V. (1984) Ecological boundaries for coral-reef reserves: principles and guidelines. *Environ. Conserv.* **11,** 209-15.

Underwood A. J. & Fairweather P. G. (1986) Intertidal communities: do they have different ecologies or different ecologists? *Proc. Ecol. Soc. Aust.* **14,** 7-16.

Underwood A. J. & Fairweather P. G. (1989) Supply-side ecology and benthic marine assemblages. *Trends in Ecol. and Evol.* **4,** 16-20.

White A. T. (1986) Marine reserves: how effective as management strategies for Philippine, Indonesian and Malaysian coral reef environments? *Ocean Management* **10,** 137-59.

White P. S. & Bratton S. P. (1980) After preservation: philosophical and practical problems of change. *Biol. Conserv.* **18,** 241-55.

Wolanski E. & Hamner W. M. (1988) Topographically controlled fronts in the ocean and their biological influence. *Science* **241:** 177-81.

Proc. Ecol. Soc. Aust. 1990 — 16:79-91

Australian upland streams: ecological degradation and possible restoration

P. S. LAKE[1] and R. MARCHANT[2]

[1]*Department of Botany and Zoology and Centre for Stream Ecology, Monash University, Clayton, Victoria 3168*

[2]*Museum of Victoria and Centre for Stream Ecology, 71 Victoria Crescent, Abbotsford, Victoria 3067*

Abstract

During Australia's development by Europeans, upland streams and their catchments have been subjected to various forms of degradation. Catchments have been cleared for agriculture or for timber extraction resulting in sedimentation of and nutrient leakage into streams. Riparian vegetation has been damaged disrupting a critical component of the linkage between streams and their catchments. In certain areas, stream salinization has occurred following land clearance because reduced evapotranspiration has enabled saline groundwater to rise. Mining has contaminated stretches of some upland rivers with persistent pollutants such as heavy metals. Construction of dams has altered lotic faunal communities by changing thermal regimes downstream and by interfering with migration. Introduced exotic fish, particularly trout, have become well established in upland streams to the detriment of some native fish and invertebrates.

Australian upland streams have distinctive biological and hydrological characteristics. For adequate repair of streams and their catchments both restoration and rehabilitation will be required. Lack of detailed ecological knowledge of the structure and functioning of upland streams hampers any attempt at restoration and rehabilitation of this habitat. As a first step towards repair, inventories that survey the extent of damage to these streams and their catchments should be started and water authorities should be encouraged to implement integrated management policies for whole catchments.

Introduction

During the course of Australia's economic development there has been little concern for maintaining the ecological integrity of streams and rivers. Concerns over widespread soil erosion has only rarely considered the state of the rivers. Until recently streams and rivers were predominantly regarded as channels to carry the economic commodity of water. Instances of reported damage to streams and rivers due to mining, sluicing, pollution and land clearance were few and far between and were generally regarded as being part of a litany of special cases and the price of progress. Needless to say, in an era of rampant development, stream ecological research in Australia, apart from investigations of freshwater fisheries, was minimal.

With the rise of environmentalism in the 1970s and the realization of the amount of degradation inflicted upon Australia by European development, a strong and popular concern arose for the conservation of aquatic ecosystems including streams and rivers (see Lake 1978). This concern gave rise to several important stocktaking exercises, the most influential being the report "Water 2000: A perspective on Australia's Water Resources to the Year 2000" (Department of Resources and Energy 1983). This report detailed the ecological damage that has been wrought upon Australia's running waters and pointed out the lack of ecological knowledge about them.

In this chapter we will outline some of the major forms of degradation that have been wrought upon Australian upland streams and their catchments. Some of these forms of degradation, such as damage to catchment vegetation and consequent erosion may be capable of arrest and possible restoration, some such as the lingering effects of heavy metal contamination may not. Lack of knowledge about the ecological effects of such degradation is obvious and this lack is a major obstacle to the successful repair of past damage and to the prevention of future damage.

Streams and their catchments

From its source to its mouth, a natural stream is a continuum. This continuity applies for water, dissolved material and much of the suspended solids. More debatable is the suggestion that there is a continuum of ecological communities and of organic matter input and utilization — a suggestion enunciated in the River Continuum Concept (Vannote *et al.* 1980; Minshall *et al.* 1983). This concept has drawn critics (e.g., Winterbourn *et al.* 1981; Barmuta & Lake 1982) largely because of its deterministic statements and its inapplicability to a range of river types. Nevertheless it is a crucial property of running waters that upstream processes influence those downstream. For nutrients this property is embodied in the Nutrient Spiralling Concept (Elwood *et al.* 1983). River management, especially in this country, has greatly exploited the continuity of water transport in streams but has neglected or disregarded the continuity required by other stream components such as the biota.

Streams are inseparable from their catchments from which they derive water, dissolved solids including nutrients, particulate solids, some dissolved gases (e.g., CO_2,) and allochthonous organic matter (see Hynes 1975). The inputs of all these entities are spatially variable and, in upland streams, the direction of their export is down the slope whereas in depositional rivers in times of flood these entities (e.g., nutrients, organic matter) may be exported onto the floodplain. In upland streams draining forested catchments, allochthonous organic matter is a major energy source for the stream's consumers and decomposers (Vannote *et al.* 1980; Cummins *et al.* 1983). The input of dissolved organic matter may be of the same magnitude as that of particulate organic matter (e.g., Cummins *et al.* 1983) and can be incorporated into bacteria and fungi which may in turn be consumed (e.g., Rounick & Winterbourn 1983). Particulate organic matter from the catchment is used as both food and shelter by steam decomposers and consumers. In upland streams without shading by a forested canopy, primary production largely by periphyton may be the important energy source (Minshall 1978).

The riparian zone

Recently it has become clear that the riparian zone is a critical component in the linkage between streams and their catchments (Cummins 1986). This zone may shade streams, lowering primary production and decreasing temperature fluctuations. It may also stabilize the stream channel and its banks, contribute organic matter vital to the functioning of the detritivore communities, moderate sediment, nutrient and water inputs and contribute large woody debris to the stream (Lowrance *et al.* 1984; Barton *et al.* 1985; Cummins 1986; Fitzpatrick 1986; Petersen *et al.* 1987). Riparian vegetation harbours arthropods some of which fall into the stream and may be consumed. This source of food is seasonally important for salmonids (Hunt 1975) and for native fish, such as *Galaxias olidus,* in Victorian upland streams (Cadwallader *et al.* 1980).

Upland streams

Whereas running waters change their characteristics continuously downstream, there is a discernible division between upland and lowland sections. Upland sections are erosional whereas lowland sections are depositional (Cummins 1972). Upland streams are generally characterized by steep gradients, coarse substrata, low levels of dissolved and suspended solids, high dissolved oxygen levels, low nutrient levels and in many instances relatively low light availability and water temperatures. This distinction in abiotic variables is also clearly shown in the biological communities (Hynes 1970; Statzner & Higler 1986). Upland streams are low-order streams and it must be pointed out, especially in an Australian context, that upland streams do not have to be at high elevations; many, for example, run directly into the sea.

Australian upland streams and their biota

Although upland streams are widely distributed in Australia, to date most of the ecological knowledge of such streams has come from work in south-eastern Australia including Tasmania (see Lake *et al.* 1985; Lake 1987). For tropical Australia, ecological information on rainforest upland streams in Queensland is available from the studies of Pearson *et al.* (1986) and Benson and Pearson (1987a,b 1988) and on an upland section of Magela Creek, Northern Territory from Outridge (1988). South-western Australia and southern parts of South Australia have a Mediterranean climate with wet winters and dry summers. Ecological information on streams in south-western Australia is available from the studies of Bunn (1986a), Bunn *et al.* (1986) and Bunn and Davies (1989), whereas for South Australia there is the study of Towns (1985) of an intermittent stream near Adelaide.

Australian upland streams are distinctive both in terms of their fauna (Lake 1982; Williams 1983; Lake *et al.* 1985; Williams & Campbell 1987) and their hydrological behaviour (McMahon 1982; Finlayson *et al.* 1986). Compared with the general world trend, Australian streams have low mean annual discharges and a high variability of flow. This variability, in direct contrast to the world trend, increases as the catchments and streams get smaller. The mean peak annual floods are an order of magnitude greater and more variable than they are for world average streams (McMahon 1982). This variability in flow makes management of Australian rivers and their catchments more difficult than elsewhere. This difficulty it seems is only now being realized.

The upland stream fauna of eastern Australia (Queensland to Tasmania) (especially the invertebrates) has a high level of endemism and a relatively high species richness (Lake *et al.* 1985; Williams & Campbell 1987). In contrast the upland stream fauna of south-western Australia has a relatively low species richness (Bunn & Davies 1989). In comparison with many other regions of the world e.g., Europe, North America, the invertebrate fauna of Australian upland intermittent streams is relatively high in species richness and overlaps substantially with that in nearby permanent streams (Lake *et al.* 1985; Boulton & Suter 1986).

Australian temperate upland streams differ from Northern Hemisphere temperate streams in gaining a large amount of their leaf litter in summer (Lake 1982; Lake *et al.* 1985; Bunn 1986b). In Queensland rainforest streams the peak litter availability is in the late dry season (Benson & Pearson 1988). Many temperate Australian upland streams differ from Northern Hemisphere temperate streams in receiving throughout the year a large amount of woody debris, especially bark, from decorticating eucalypts in the catchment (Lake 1982; Bunn 1986b; Williams & Campbell 1987).

Catchment land use and upland streams

In the natural state in upland catchments, the transport of water, dissolved solids including nutrients, and particulate solids into streams is regulated substantially by the vegetation (Likens & Bormann 1974). Damage to the vegetation cover and soil surface of catchments by activities, ranging from logging to urbanization, disrupts the linkages between streams and their catchments. The catchments may then leak, in an unregulated way, water, dissolved solids and particulate solids into streams (Likens & Bormann 1974; Bormann & Likens 1979; Williams 1980).

The degree of disruption of the land surface of a catchment and the intensity of consequent land use are reflected in changes in input into the stream of dissolved and particulate solids (Likens & Bormann 1974). With a reduction in evapotranspiration due to loss of vegetation, runoff into streams may rise (Hewlett 1982). The potential for soil erosion rises with disruption of vegetation. Soil erosion by water is a two-phase process involving the detachment of particles from the soil and their subsequent transport by water (Smith & Finlayson 1988); both of these phases are exacerbated by vegetation damage. With increased runoff and loss of the retentive capacity of the vegetation, nutrients may leak into the stream (Petersen *et al.* 1987) which may trigger undesirable eutrophic effects downstream.

Soil erosion and stream sedimentation

Upland catchments in Australia have been cleared for pasture and crop production, for grazing, for forestry operations and urbanization. It is very evident from both popular (e.g., Standing Consultative Committee on River Improvement 1983) and technical accounts (e.g., Boughton 1970; Williams 1980; O'Brien *et al.* 1983; Woods 1983; Warner 1984) that large amounts of soil and sediments have been lost from upland catchments and have ended up in rivers. Their persistence in the river is largely dependent on the retentive capacity and erosive power of the stream. It is surprising considering the economic value of soil that data on soil loss in Australia are so scanty (Olive & Walker 1982; Woods 1983; Smith & Finlayson 1988). Given the high variability in discharge of Australian streams and the small particle size of many Australian soils (Olive & Walker 1982), soil losses may be very great and have occurred over a large area. For example of the 1.34 million square kilometres of non-arid grazing areas in Australia in 1975, 364 000 square kilometres were degraded by water erosion (Woods 1983).

Increased loads of suspended sediment downstream from disturbed catchments and increased streambed siltation and sedimentation have been shown in numerous overseas studies, especially in North America, to affect stream periphyton, invertebrate and fish communities (e.g., Hynes 1973; Alabaster & Lloyd 1980; Milner *et al.* 1981; Lemly 1982; Webster *et al.* 1983; Wiederholm 1984; Alexander & Hansen 1986). In eastern Australia, siltation and sedimentation have been seen as contributing to the population decline in some stream fish such as Macquarie perch, *Macquaria australasica*, (Cadwallader 1978) and Australian bass, *M. novemaculeata*, (Pollard *et al.* 1980) but the effects on plant and invertebrate communities have not been studied; except in relation to the construction of dams (see later) and to decomposition of leaf litter. In a Western Australian stream with high levels of sediment due to strip-mining for bauxite, Bunn (1988) found sediment smothered leaves on the bottom of the stream disrupting litter decomposition, inhibiting the buildup of invertebrates and micro-organisms and even rendering the litter anoxic.

The potential for modern forestry operations, which involve clear felling and roading, to degrade upland streams and their biota is controversial (Michaelis 1984). Given the great expansion of this industry this uncertainty is an unnecessary state of affairs and one that should be corrected by research. Studies in North America and New Zealand have shown that forestry operations which result in increased sediment deposition and increased water temperatures in streams have adversely affected both invertebrate (e.g., Tebo 1955; Graynoth 1979; Newbold *et al.* 1980; Lemly 1982; Silsbee & Larson 1983; Webster *et al.* 1983) and fish communities (Burns 1972; Morgan & Graynoth 1978; Murphy & Hall 1981; Moring 1982). It is also evident from detailed studies such as those at Coweeta in North Carolina, that the effects of logging are persistent (Webster *et al.* 1983). In Australia only two studies of limited scope have investigated the ecological effects of forestry operations on stream biota. Robinson (1977) in the Coranderrk experimental area in southern Victoria observed a different and less abundant macroinvertebrate fauna in a creek draining a clearfelled area compared with the fauna of an adacent creek draining an unlogged control area. In a coastal stream in a woodchipping area of southern New South Wales, Richardson (1985) produced evidence to suggest that increased sedimentation from road construction adversely affected both invertebrates and fish. Much more research needs to be done if the timber industry intends to reduce damage to upland streams effectively.

Nutrient retention and leakage

Degradation of the vegetation cover of catchments reduces the capacity of the catchment to retain nutrients and leads to nutrient leakage into streams (Bormann & Likens 1979; Webster *et al.* 1983). Logging operations consistently increase nutrient levels in streams and these levels have been shown to increase periphyton production (Webster *et al.* 1983). Australian upland streams generally have low total dissolved solid levels and low nutrient levels (e.g., Cullen 1983; Campbell *et al.* 1986; Marchant *et al.* 1985; Bunn *et al.* 1986; Pearson *et al.* 1986; Chessman 1986) and may thus be particularly susceptible to increased primary production as the result of nutrient addition from such activities as logging (Michaelis 1984; Campbell & Doeg 1984). That small additions of nutrients may have this effect on Australian upland streams is evident from the work of Cullen (1983, 1985) who observed that small amounts of nutrient addition to the Thredbo River, New South Wales, stimulated algal production in summer.

Riparian zone management

As previously mentioned, riparian zones constitute a critical component to safeguarding stream integrity. Stream surveys, such as those carried out in Victoria, (e.g., Macmillan *et al.* 1987; Kunert & Macmillan 1988) have revealed very extensive damage to the riparian zone of numerous streams by land clearance and grazing. It is vital for effective stream management that riparian zones, where undamaged, remain intact and, where they are damaged, effective programmes should be undertaken for restoration. In stream restoration the riparian zone must initially receive particular attention (Petersen *et al.* 1987) and failure to give it may negate effective restoration efforts elsewhere in the catchment. For example, in the Kosciusko National Park, Wimbush and Costin (1983) observed that streams in peaty valleys with damaged riparian zones continued to erode even though surrounding catchment vegetation had largely recovered from the effects of grazing. They concluded that "such hydro-ecological ruts . . . may be self-perpetuating even when the initiating agents are removed".

When upland catchments are cleared either on a temporary basis, such as by logging, or on a permanent basis, such as by pasture development, riparian zones, especially those of first and second-order streams must be left intact. The extent of such zones must be sufficient to shield the stream from the effects of land clearance and should be determined by research. Overseas values cannot be readily used, recommended widths determined by research vary with the forest ecosystem involved and the type of stream (e.g., Newbold *et al.* 1980; Barton *et al.* 1985).

Stream salinization

In some upland areas of Australia, notably in south-western Western Australia, the groundwater of the water table is saline. Reduction of evapotranspiration by land clearance may cause saline groundwater to rise and contribute to stream flow (Wood 1924; Shea *et al.* 1975; Loh & Stokes 1981; Smith & Finlayson 1988) inducing stream salinization (Williams 1987). This form of dryland salinity which may occur after clearing of upland areas, may even reverse the normal salinity profiles of rivers, for example, the Blackwood River, Western Australia, (Morrissy 1974). Stream salinization has important economic implications (see Peck *et al.* 1983) but it is depressing to reflect that to date there have been no investigations of the effects such stream salinization has on the biota of Australian upland streams. In Western Australia where there has been considerable work on the hydrology and economic effects of stream salinization (e.g., Peck *et al.* 1983; Steering Committee for Research on Land Use and Water Supply 1987), knowledge of the ecological effects is scanty. The reduction in the fauna of the Avon River may be partly due to salinization (Kendrick 1976) and marron, *Cherax tenuimanus*, have disappeared from salinized rivers (Morrissy 1978). We agree with Williams (1987) that there must be "a better understanding of the ecological effects occasioned by salinization of rivers and streams". Salinity in streams may rise with logging or land clearance and the higher salinity values may only occur for a short period (Steering Committee for Research on Land Use and Water Supply 1987), but such a change may have considerable ecological impact, a point stressed by Williams (1987).

Mining

Gravel and sand extraction, gold dredging and tin sluicing churn up the stream bed and can create sedimentation problems. While this form of damage has been recognized as posing a threat to upland stream biota in Australia (Wharton 1969, O'Brien *et al.* 1983), the effects have been poorly investigated. A short study of the effects of eductor dredging for gold (Doeg 1985) on the benthic invertebrates of the Upper Goulburn River, Victoria, showed this activity caused no long-term degradation of the fauna. However, large mining operations may have physical effects on the stream channel long after mining ceases (e.g., Warner 1984; Knighton 1987).

In this era of mine development, environment impact statements and environmental consultants, it seems to have been forgotten that past mining has already left Australia with a legacy of pollution hazards. We will give two types of examples to stress this point. Both involve heavy metals.

In goldmining, before the advent of cyanide treatment of the ores, mercury amalgamation was a widely used method of gold extraction (Coller 1983). The amount of mercury consumed in amalgamation and consequently lost to the environment was about equal to the amount of gold recovered. This could mean very substantial environmental pollution. For example, the amount of mercury released to the environment from the Bendigo field could be up to 900 tonnes (Bycroft *et al.* 1982). Mercury in tailings dumps poses a pollution hazard for adjacent streams.

In the catchment of the upper Lerderderg River, southern Victoria, in the vicinity of Blackwood, gold-mining operations involving amalgamation were carried out until about 50 years ago. In 1980 mercury concentrations up to 130 μg/g were found in sediments of the Lerderderg River and the pattern of sediment mercury concentration is clearly related to past mining activities (Bycroft *et al.* 1982). The sediment mercury has not been transported very far downstream. Both blackfish, *Gadopsis marmoratus*, and brown trout, *Salmo trutta*, were found to be contaminated by mercury (Bycroft *et al.* 1982).

In the Brisbane Ranges, southern Victoria, centred on Stieglitz, substantial alluvial and quartz mining operation for gold occurred in the 1870s and 1890s (Lake & Sokol 1986). Sediments in an upland stream, Sutherland Creek, and a pond near the mining operations are contaminated with mercury and this contamination was reflected in elevated mercury concentrations in the abdominal muscles of yabbies, *Cherax destructor* (Lake & Sokol 1986).

Sediments along a substantial length of the upper Goulburn River have been contaminated with mercury from gold mining operations (Ealey *et al.* 1983). This sediment contamination is reflected in high concentrations of mercury in English perch, *Perca fluviatilis,* and trout in the river and Lake Eildon (Ealey *et al.* 1983) and such contamination poses a hazard to human health. The operation of eductor dredging in the river by agitating mercury-enriched sediments may exacerbate the problem. Other rivers affected by mercury from past goldmining operations include the Thomson and La Trobe Rivers (Glover *et al.* 1980).

Near Canberra, there is a well-known case of persistent heavy metal contamination (Weatherley *et al.* 1980). The upper Molonglo River has been contaminated with heavy metals (zinc, copper, lead and cadmium) from mine workings at Captains Flat, New South Wales (Weatherley *et al.* 1967, 1980; Norris 1986). Mining ceased in 1962 and subsequently four major sources of heavy metal contamination were identified (Anon 1974); mine waste dumps, mine water and sediments in the river and sediments along its banks and on the flood plain. A survey (Lake 1963; Weatherley *et al.* 1967) revealed reduced invertebrate species richness and abundance and an absence of fish for 40 km downstream of Captains Flat. The pollution was sufficient to affect adversely plankton and fish populations downstream in Lake Burley Griffin (Weatherley & Dawson 1973; Hillman 1980). Subsequent surveys of the river from 1974 to 1977 by Nicholas and Thomas (1978) revealed the same trend in invertebrate species richness and abundance as found by Lake (1963) and Weatherley *et al.* (1967).

As the result of an inquiry, recommendations were made for the abatement of the pollution by specific engineering works (Anon 1974). At Captains Flat, tailings dumps were reshaped with gentler slopes, capped with impervious material and revegetated (Craze 1977). Slime dumps were removed and small streams diverted from the contaminated areas. It appears however that such restoration has been in vain. In 1982 the river was again surveyed (Norris 1986) using the original sites of Weatherley *et al.* (1967). The river was still badly contaminated with heavy metals and there had been little discernible change in the distribution and abundance of invertebrates. Norris (1986) concluded that "the remedial works conducted at Captains Flat have not been successful in rehabilitating the Molonglo River, as the mine seeps and probably sedimented material remain as major sources for the addition of trace metals to the river". Effective restoration of the Molonglo River and its valley would be a very costly operation, possibly more than the value of the metals commercially extracted. Other rivers damaged by heavy metals from past mining operations include the King River (Lake *et al.* 1977) and the South Esk River in Tasmania (Norris *et al.* 1982). Rehabilitation of old mines can be expensive and difficult, witness Rum Jungle (Ryan 1987). Even with elaborate safeguards, heavy metal mining particularly in upland catchments must always constitute a pollution hazard. Accidents can give rise to irreversible ecological damage. Even in upland streams with considerable erosive power and hence ability to move sediments, heavy metals in sediments persist and move only very slowly downstream.

The effect of dams

Many dams have been built on upland rivers in Australia, e.g., the Snowy Mountains Scheme, yet little ecological work has been carried out on the consequences of these impoundments for the lotic biota. Only in Tasmania and Victoria has there been any attempt to determine the ecological impact of the construction and operation of dams. Much of this work concerns the effects of dams on the benthic invertebrate communities of the disturbed rivers, though there is also some information on fish.

Coleman (1978) sampled the benthic invertebrates downstream of the Gordon Dam on the Gordon River in south-west Tasmania. His first samples were taken in summer approximately two and a half years after dam closure (April 1974) and before any water had been released. He found that the benthic fauna at four downstream sites within 5 km of the dam was clearly depauperate (4-10 taxa) compared with that further downstream 20-48 taxa) or with that from tributaries (17-44 taxa). A second set of samples taken a year later gave the same pattern of species richness, although by this stage the dam was discharging. He attributed the reduction in species richness close to the dam to the combined effects of reduction in flow and siltation from cement dust and noted that the release of water had not obviously changed the pattern of distribution of the fauna.

Water was released from the hypolimnetic zone of the dam and in summer this resulted in a 5°C lowering of downstream water temperatures (from approximately 16°C). In addition, oxygen saturation levels at the site nearest the dam declined to 20-25% (compared with values of 80-100% at other sites) and there was a strong smell of hydrogen sulphide; these conditions were due to decomposition of recently inundated organic matter. Coleman commented that if such poor water quality persisted then recolonization of benthic communities close to the dam even by cold-adapted species would not occur.

The benthic invertebrates below the Dartmouth Dam on the Mitta Mitta River in north-eastern Victoria were sampled over eight years both during and after the construction of the dam (Blyth *et al.* 1984; Doeg 1984). Species richness at three downstream sites within 20 km of the dam decreased from a mean of 45 taxa per site to a mean of 39 taxa during the construction period (1974-77) and the communities became dominated by oligochaetes and chironomids. Together these formed over 80% of the total number of individuals at these sites. The major impact on the river during this period was sedimentation from the construction site which produced a thick (1-10 mm) silt and algal cover at these sites. Such a habitat is favoured by worms and worm-like larvae but it is well known to be unsuitable for organisms such as mayfly and stonefly nymphs and caddisfly larvae (Milner *et al.* 1981) as silt clogs their gills and fills the interstices between rocks that they inhabit.

After closure of the dam (in 1977) no discharge occurred for two weeks. Massive drift of aquatic insects was observed near the dam wall (even during daylight) when flow stopped. However, community composition changed little over this period at the three sites. While the dam was filling, the river immediately downstream was subjected to reduced summer (by 8-10°C) and elevated winter water temperatures, low oxygen levels (down to 10% saturation) and high levels of hydrogen sulphide. These effects were due to hypolimnetic release of water from the dam. During this period the benthic community at the site closest to the dam was reduced to 26 taxa while the communities at the other two sites were less severely affected. A subsequent large release of hypolimnetic water (in summer for three months) for irrigation caused a reduction in species richness at the lowest site (the only site within 20 km of the dam that was sampled in this instance), which lasted until the following summer. This reduction was attributed to the disruption of life cycles of various species by the high flow and the depressed water temperature.

The benthic communities near the Thomson Dam on the Thomson River in southern Victoria have been studied for a decade (Malipatil & Blyth 1982; Doeg *et al.* 1987; Marchant 1989) before, during and after dam construction (i.e., 1977-87).

During construction there was a clear change in community composition at two sites (9 km and 35 km) below the dam, but oligochaetes and chironomids did not come to dominate the fauna as they did below the Dartmouth Dam nor was there an obvious reduction in species richness. This faunal change was also attributed to the effects of sedimentation, although a thick silt cover did not develop. Some time after dam closure (in 1983) hypolimnetic water was discharged for approximately fifteen months. This resulted in colder (by about 8°C) summer water temperatures, but had little effect on other aspects of water quality. A decrease in species richness (from 59 to 50 taxa) at the site closest to the dam coincided with the release of this water; temperatures at the other site were unaffected. When surface discharge from the dam began and summer water temperatures returned to normal a steady increase in species richness occurred at the site nearest the dam (from 50 to 87 taxa). Measurements of the levels of fine sediment at both sites indicated that levels had decreased since dam closure at the site closest to the dam, but had increased at the lower site. Species richness at the lower site began to decrease before dam closure and continued to do so as sediment accumulated at this site (from a mean of 60 taxa to a mean of 46).

It is evident that the benthic fauna in the Thomson River was not as severely affected by dam construction and operation as the fauna in the other two rivers. The increase in species richness after the temporary release of cold hypolimnetic water indicated that the benthic community showed some resilience. The effects of sedimentation were still evident at the lowest site. However, the fact that species richness at the upper site by the end of the study exceeded that recorded during dam construction and the fact that sediment levels had fallen at this site both suggest that the species richness at the lower site may also increase if sediment levels fall.

Chessman *et al.* (1987) studied the effects on the benthic fauna of the construction and operation of a much smaller dam than the previous three. It was the Blue Rock Dam on the Tanjil River, a tributary of the La Trobe River, in southern Victoria. Particular care was taken during construction to minimize sediment erosion by grassing disturbed areas, using silt traps and constructing retarding and settlement basins. Little noticeable change occurred in benthic communities during the construction phase at two sites (3 and 6 kms) below the dam and this was directly attributed to the success of the erosion controls: only light sedimentation occurred. After dam closure, decreases in species richness occurred over the following year and oligochaetes and chironomids increased in abundance. These groups were apparently favoured by a build up of fine organic matter that occurred when discharge decreased. The decline in species richness was associated with periodic episodes of very low flow when discharge from the dam was curtailed (for maintenance) and with the release of deoxygenated (and sulphide laden) water during the first six months of filling. When sampling stopped 18 months after dam closure there were indications that the benthic community was reverting to its original composition.

Thus is seems that dams that have surface rather than hypolimnetic discharge cause less degradation to the downstream benthic fauna and that it is possible to prevent damaging sedimentation during dam construction. It does not seem possible, however, to rehabilitate upland rivers that receive continual hypolimnetic drainage from dams. In these cases a cold-adapted but permanently impoverished benthic fauna will probably develop immediately downstream.

Dams on rivers are also barriers to fish movement. The barrier effect may not only divide populations but prevent migrations necessary for life cycle processes. Many Australian freshwater fish are diadromous; they need to move between marine environments and rivers to complete their life cycle. Most of these fish are catadromous; they mature in freshwater and move downstream to breed in brackish or sea water (e.g., Australian bass, eels and *Galaxias maculatus*). Some are anadromous [e.g., lampreys (Harris 1986)]. In inland rivers there are potamodromous fish that migrate considerable distances [e.g., Macquarie perch, and Golden perch, *Macquaria ambigua* (Harris 1986)]. If flow is stopped by a dam, even for short periods, fish kills may result. Below the Tallowa Dam on the Shoalhaven River. New South Wales, large numbers of fish, including Australian grayling, *Prototroctes maraena,* were killed on four separate occasions when discharge was stopped (Bishop & Bell 1978). Dams have undoubtedly contributed to the decline of stocks of migratory fish species (Pollard *et al.* 1980; O'Brien *et al.* 1983; Harris 1986).

The construction of barriers (dams and causeways) is in direct proportion to the level of water exploitation and thus it is not surprising to find that the barrier problem is particularly evident in south-eastern Australia (O'Brien *et al.* 1983). Harris (1984a) surveyed the eastern-flowing drainage systems from the Mary River in Queensland to Lakes Entrance, Victoria, and found, across the 22 separate drainage systems, between one-third to one-half of the habitat potentially usable by migratory fish had been affected by the presence of barriers. Over the same area, Harris (1984b) found 29 fishways (on only 9.2% of the barriers surveyed), but 23 of these failed to provide suitable conditions for fish migration at the time of survey. There is thus a clear need for research into the appropriate design and mode of operation of adequate fishways. Construction of appropriate fishways must then follow.

Impact of exotic biota

Compared with lowland depositional sections of rivers and lentic water bodies, upland streams in Australia have not suffered from the introduction of undesirable aquatic plants (Arthington & Mitchell 1986). The riparian zones have suffered from the introduction of vigorously growing weeds, blackberries, *Rubus* spp., and willows, *Salix* spp., being obvious examples. Willows may stabilize damaged banks but may also clog channels with their prodigious root growth.

Lowland rivers and many lentic water bodies have been invaded by a wide range of exotic fish including carp, *Cyprinus carpio,* goldfish, *Carassius auratus,* English perch, poeciliids especially the mosquito fish, *Gambusia affinis,* and cichlids (e.g., *Oreochromis mossambica*) (Tilzey 1980; Merrick & Schmida 1984; Fletcher 1986; Arthington 1986; Arthington & Mitchell 1986). In upland streams most of the exotic fish are salmonid with brown and rainbow trout, *Oncorhynchus mykiss,* being the most important (Tilzey 1980; Fletcher 1986). Trout have been very successful, and Tilzey (1977) ascribes this to favourable physico-chemical conditions, minimal competition from native fish, the abundance and availability of food, especially of certain native food species, and the virtual absence of parasites and disease. To this list could be added the absence of effective aquatic piscivores.

Trout eat a wide range of native aquatic and terrestrial invertebrates and fish and thus may be expected to have both competitive and predatory influences on other stream fauna. In Tasmania, the syncarid shrimp, *Anaspides tasmaniae,* is absent or rare in waters, both lotic and lentic, which contain trout (Lake & Knott 1973; Williams 1974; Knott *et al.* 1978). It has been suggested that redfin and trout have contributed to the decline of Macquarie perch populations (Cadwallader & Rogan 1977; Jackson 1981). In the course of work on the taxonomy and distribution of galaxiids, Frankenberg (1966) noted that the mountain galaxias, *Galaxias olidus,* had a fragmentary distribution. It was found in sections of headwater streams inaccessible to trout. Such a pattern of parapatric distribution involving *G. olidus* and trout has been confirmed by Tilzey (1976), Cadwallader (1979), Fletcher (1979) and Jackson and Williams (1980). In 1971 in Four Mile Creek, a tributary of the Eucumbene River, Tilzey (1976), found only *G. olidus.* Between 1971 and 1974 trout invaded the creek but their invasion upstream was limited by a waterfall. Below the waterfall the *G. olidus* population was eliminated whereas above the waterfall the galaxiids survived in abundance. Although competition may have been partly responsible for the decline of *G. olidus* populations, predation by trout seems a more potent force, especially as it is known that trout will readily eat *G. olidus* (Fletcher 1979; Tilzey 1977).

There has been one attempt to understand the interaction between trout and *G. olidus* by experiment (Fletcher 1979). Watchbox Creek in the Strathbogie Ranges has a downstream trout section and upstream galaxiid section separated by waterfalls. Trout were temporarily introduced into a stretch of the galaxiid section for four months in the summer of 1978. Galaxiid numbers in the trout stretch were greatly reduced, presumably through predation. There were detectable differences in the invertebrate communities between the trout stretch and an adjacent galaxiid stretch at the end of the experiment and some of these changes were similar to those found between the trout and galaxiid sections of the stream. Although this experiment is preliminary, it suggests that trout, in eliminating *G. olidus,* may also alter the composition of benthic invertebrate communities. This suggestion needs to be tested with rigorous long-term experiments.

It appears that trout and blackfish, *Gadopsis marmoratus,* may co-exist even though they have similar diets (Jackson 1978, 1981; Fletcher 1979, 1986). Jackson (1978) suggested that this was possible because the two species partitioned the habitat. It may also be due to temporal separation in feeding, with trout being diurnally active and blackfish nocturnally active.

Thus trout can adversely affect both invertebrate and native fish populations and this form of *species pollution* (Cadwallader 1978) must be restricted. Trout should not be introduced into new upland rivers, marginal populations should not be maintained by hatchery operations and upland streams free of introduced fish should be identified and conserved as habitats free of exotic fishes.

What needs to be done

It is evident that of the stresses that can be applied to ecosystems by human activity (see Rapport *et al.* 1985), Australian upland streams have been exposed to all four. Stream catchments have been harvested and mined (*sensu* Brinck *et al.* 1988) leading to a depletion of renewable resources (soil, nutrients, vegetation). Catchments and streams have been physically re-structured leading to alterations in their transport and storage abilities. For example, dams have altered downstream movements of organic matter, nutrients and biota whereas catchment clearance has lead to great changes in the dynamics of water and salt transport. Both short-lived (e.g., organic wastes) and long-lived pollutants (e.g., heavy metals) have been put into streams. Some of these pollutants may persist for a very long period and affect considerable lengths of streams. Finally exotic species, many of which have become pests, have been introduced deliberately into upland streams.

As a first step to redressing this state of affairs we urge water management authorities, preferably at a State level, to carry out surveys of the rivers and their catchments. An inventory of what is there and its state is not available and is clearly required (viz. McComb & Lake 1988). Such an exercise has been begun in Victoria by the Department of Water Resources and the Commissioner for the Environment. Two things should arise from such an inventory; an assessment of the type and extent of damage to stream systems and conversely, an identification of those rivers that have escaped damage. Firm moves could then be made to protect such rivers of high conservation value.

In the area of research, it is evident that our current understanding of Australian stream ecosystems is poor to rudimentary (Lake 1987). Thus we need research on the energetics, nutrient dynamics and community structure and function of representative streams and their catchments. This research effort must be long-term and is necessary if we are to understand the ecological impact of human activities. The effort spent attempting to understand the basic structure and functioning of stream ecosystems should be carried out concurrently with applied research on the effect of human activities on stream ecosystems. For example, the effects of modern forestry operations need to be understood adequately so that sound management guidelines can be implemented.

In our opinion, it is not necessary to wait for scientific research to be completed, to make major improvements in the management of Australian streams. At the State level, moves should be made to implement catchment-wide management policies. Such management should aim to prevent loss of soil and leakage of nutrients from catchments, to prevent the introduction of persistent pollutants and to lessen the barrier and flow regulatory effects of dams. As regards the latter, there is a need for minimal flow requirements for important ecological components of rivers (e.g., fish) to be devised by research and for these requirements to be met by altering the operation of current dams and incorporating the requirements into the design of new dams. The need for integrated management of the whole catchment is much greater for developed and developing catchments than for remote pristine catchments.

It is clear from the above account that many Australian streams and their catchments have been damaged by human activities. Some may be abated (e.g., cessation of nutrient leakage), while some of this damage may be irreversible (the release of heavy metals, hypolimnetic drainage from dams). There has been very little research in restoration ecology in Australia and it is worthwhile indicating the type of strategies which might be used to overcome or at least offset ecological damage. Brinck *et al.* (1988) outline some of these strategies. Active restoration uses specific repair activities to recreate the stream and catchment conditions that existed before human disturbance. Passive restoration, which is applicable to many forested catchments requires some specific repair activities but is allowed to occur without large-scale human intervention. It is aimed at achieving a similar state as active restoration. Rehabilitation, which is applicable to many catchments affected by clearing and grazing, requires extensive repair using ecologically sound procedures. It should be noted that after rehabilitation the communities both on the catchment and in the stream may be quite different from those that existed prior to disturbance. For Australian streams and their catchments we have very little knowledge to guide restoration and rehabilitation. It is a high priority to gain this knowledge and to provide the resources for effective restoration and rehabilitation.

Acknowledgements

We are grateful for the great assistance given by Pat Coates and Dave Morton in preparation of the paper.

References

Alabaster J. S. & Lloyd R. (1980) *Water Quality Criteria for Freshwater Fish.* Butterworths, London.

Alexander G. R. & Hansen E. A. (1986) Sand bed load in a Brook Trout stream. *N. Am. J. Fish. Manage.* **6,** 9-23.

Anon. (1974) *Mine Waste Pollution of the Molonglo River. Report by the Joint Government Technical Committee on Mine Waste Pollution of the Molonglo River.* Australian Government Publishing Service, Canberra, A.C.T.

Arthington A. H. (1986) Introduced cichlid fish in Australian inland waters. In: *Limnology in Australia* (eds P. De Deckker & W. D. Williams) pp. 239-48. CSIRO, Melbourne.

Arthington A. H. & Mitchell D. (1986) Aquatic invaders. In: *The Ecology of Biological Invasions* (eds R. Groom & J. Burdon) pp. 34-53. Australian Academy of Science, Canberra.

Barmuta L. & Lake P. S. (1982) On the value of the River Continuum Concept. *N.Z. J. Mar. Freshwat. Res.* **16,** 227-31.

Barton D. R., Taylor W. D. & Biette R. M. (1985) Dimensions of riparian buffer strips required to maintain trout habitat in southern Ontario streams. *N. Am. J. Fish. Manage.* **5,** 364-78.

Benson L. J. & Pearson R. G. (1987a) Drift and upstream movement in Yuccabine Creek, an Australian tropical stream. *Hydrobiologia* **153,** 225-39.

Benson L. J. & Pearson R. G. (1987b) The role of drift and effect of season in macroinvertebrate colonization of implanted substrata in a tropical Australian stream. *Freshwat. Biol.* **18,** 109-16.

Benson L. J. & Pearson R. G. (1988) Diversity and seasonality of adult Trichoptera captured in a night trap at Yuccabine Creek, a tropical Australian rainforest stream. *Aust. J. Ecol.* **13,** 337-44.

Bishop K. A. & Bell J. D. (1978) Observations on the fish fauna below Tallowa Dam (Shoalhaven River, N.S.W.) during river flow stoppages. *Aust. J. Mar. Freshwat. Res.* **29,** 543-9.

Blyth J. D., Doeg T. J. & St Clair R. M. (1984) Response of the macroinvertebrate fauna of the Mitta Mitta River, Victoria, to the construction and operation of Dartmouth Dam. I. Construction and initial filling period. *Occ. Pap. Mus. Vict.* **1,** 83-100.

Bormann F. H. & Likens G. E. (1979) *Pattern and Process in a Forested Ecosystem.* Springer-Verlag, New York.

Boughton W. C. (1970) Effects of land management on quantity and quality of available water. *Australian Water Resources Council Research Project 68/2. Report No.* 120, 1-330.

Boulton A. J. & Suter P. J. (1986) Ecology of temporary streams — an Australian perspective. In: *Limnology in Australia* (eds P. De Deckker & W. D. Williams) pp. 313-27. CSIRO, Melbourne.

Brinck P., Nilsson L. M. & Svedin U. (1988) Ecosystem redevelopment. *Ambio* **17,** 84-9.

Bunn S. W. (1986a) Spatial and temporal variation in the macroinvertebrate fauna of streams of the northern jarrah forest, Western Australia: functional organization. *Freshwat. Biol.* **16,** 621-32.

Bunn S. W. (1986b) Origin and fate of organic matter in Australian upland streams. In: *Limnology in Australia* (eds P. De Deckker & W. D. Williams) pp. 277-91. CSIRO, Melbourne.

Bunn S. E. (1988) Processing of leaf litter in two northern jarrah forest streams, Western Australia: II. The role of macroinvertebrates and the influence of soluble polyphenols and inorganic sediment. *Hydrobiologia* **162,** 211-23.

Bunn S. W. & Davies P. M. (1989) Why is the stream fauna of south-western Australia so impoverished? *Hydrobiologia* (in press).

Bunn S. E., Edward D. H. & Loneragan N. R. (1986) Spatial and temporal variation in the macroinvertebrate fauna of streams of the northern jarrah forest, Western Australia: community structure. *Freshwat. Biol.* **16,** 67-91.

Burns J. W. (1972) Some effects of logging and associated road construction on northern Californian streams. *Trans. Am. Fish. Soc.* **101,** 1-17.

Bycroft B. M., Coller B. A. W., Deacon G. B., Coleman D. J. & Lake P. S. (1982) Mercury contamination of the Lerderderg River, Victoria, Australia, from an abandoned gold field. *Environ. Pollut. Ser. A.* **28,** 135-47.

Cadwallader P. L. (1978) Some causes of the decline in range and abundance of native fish in the Murray-Darling River system. *Proc. Roy. Soc. Vict.* **90,** 211-24.

Cadwallader P. L. (1979) Distribution of native and introduced fish in the Seven Creek River System, Victoria. *Aust. J. Ecol.* **4,** 361-85.

Cadwallader P. L. & Rogan P. L. (1977) The Macquarie Perch, *Macquaria australasica* (Pisces: Percichthyidae) of Lake Eildon, Victoria. *Aust. J. Ecol.* **2,** 409-18.

Cadwallader P. L., Eden A. K. & Hook R. A. (1980) Role of streamside vegetation as a food source for *Galaxias olidus* Gunther (Pisces: Galaxidae). *Aust. J. Mar. Freshwat. Res.* **31,** 257-62.

Campbell I. C. & Doeg T. J. (1984) *The impact of the Timber Industry on Aquatic Values. Report to the Timber Industry Inquiry, Victoria.* Water Studies Centre, Chisholm Institute of Technology, Caulfield, Victoria.

Campbell I. C., McKaige M. E. & Lake P. S. (1986) The fauna of Australian high mountain streams: ecology, zoogeography and evolution. In: *Flora and Fauna of Alpine Australasia. Ages and Origins* (ed. B. A. Barlow) pp. 83-104. CSIRO in association with Australian Systematic Botany Society, Melbourne.

Chessman B. C. (1986) Diatom flora of an Australian river system: spatial patterns and environmental relationships. *Freshwat. Biol.* **16,** 805-19.

Chessman B. C., Robinson D. P. & Hortle K. G. (1987) Changes in the riffle macroinvertebrate fauna of the Tanjil River, southeastern Australia, during construction of Blue Rock Dam. *Regulated Rivers* **1,** 317-29.

Coleman D. (1978) Downstream effects of the Gordon Dam. Lower Gordon River Scientific Survey No. 10. The Hydro-Electric Commission, Hobart.

Coller B. A. W. (1983) Sources of mercury. In: *Mercury in the Environment* (eds G. B. Deacon & J. D. Smith). Royal Australian Chemical Institute, Victorian Branch, Melbourne.

Craze B. (1977) Restoration of the Captains Flat mining area. *J. Soil Conserv. N.S.W.* **33,** 98-105.

Cullen P. (1983) Sewage effluent disposal in the Crackenback River. The assimilation capacity of an upland stream. Report to Kosciusko Thredbo Pty. Ltd., Canberra.

Cullen P. (1985) Effluent disposal to rivers: the concept of assimilation capacity. *Search* **16,** 149-51.

Cummins K. W. (1972) What is a river? — zoological description. In: *River Ecology and Man* (eds R. T. Oglesby, C. A. Carlson & A. McCann) pp. 33-52. Academic Press, London.

Cummins K. W. (1986) Riparian influence on stream ecosystems. In: *Stream Protection. The Management of Rivers for Instream Uses* (ed. I. C. Campbell) pp. 45-55. Water Studies Centre, Chisholm Institute of Technology, Caulfield, Victoria.

Cummins K. W., Sedell J. R., Swanson F. J., Minshall G. W., Fisher S. G., Cushing C. E., Petersen R. C. & Vannote R. L. (1983) Organic matter budgets for stream ecosystems: problems in their evaluation. In: *Stream Ecology. Application and Testing of General Ecological Theory* (eds J. R. Barnes & G. W. Minshall) pp. 299-353. Plenum Press, New York.

Department of Resources and Energy (1983) *Water 2000: A Perspective on Australia's Water Resources to the Year 2000.* Australian Government Publishing Service, Canberra.

Doeg T. J. (1984) Response of the macroinvertebrate fauna of the Mitta Mitta River, Victoria, to the construction and operation of Dartmouth Dam. 2. Irrigation Release. *Occ. Pap. Mus. Vic.* **1,** 101-8.

Doeg T. J. (1985) Effect of eductor dredging operations on benthic macroinvertebrate communities in four rivers draining into Lake Eildon, Victoria. Report to Victorian Department of Minerals and Energy, Melbourne.

Doeg T. J., Davey G. W. & Blyth J. D. (1987) Response of the aquatic macroinvertebrate communities to dam construction on the Thomson River, southeastern Australia. *Regulated Rivers* **1,** 195-209.

Ealey E. H. M., Deacon G. B., Coller B. A. W., Bird G. J., Bos-van der Zalm C. H., Raper W. G. C. & State College of Victoria, Rusden (1983) Mercury in the Food Web of Raspberry Creek. Victorian Environment Protection Authority, Melbourne. Publication No. 153.

Elwood J. W., Newbold J. D., O'Neill R. V. & Van Winkle W. (1983) Resource spiralling: an operational paradigm for analyzing lotic ecosystems. In: *Dynamics of Lotic Ecosystems* (eds T. D. Fontaine III & S. M. Bartell) pp. 3-27. Ann Arbor Science Publishers, Ann Arbor, Michigan, U.S.A.

Finlayson B. L., McMahon T. A., Srikanthan R. & Haines A. (1986) World hydrology: A new database for comparative analysis. In: *Hydrology and Water Resources* pp. 288-96. Institution of Engineers, Australia, National Conference Publication, No. 86/13.

Fitzpatrick C. (1986) The use of buffer strips in controlling agricultural runoff. In: *Stream Protection. The Management of Rivers for Instream Uses* (ed. I. C. Campbell) pp. 75-84. Water Studies Centre, Chisholm Institute of Technology, Caulfield, Victoria.

Fletcher A. R. (1979) Effects of *Salmo trutta* on *Galaxias olidus* and macroinvertebrates in stream communities. M.Sc. Thesis, Department of Zoology, Monash University, Clayton, Victoria.

Fletcher A. R. (1986) Effects of introduced fish in Australia. In: *Limnology in Australia* (eds P. De Deckker & W. D. Williams) pp. 231-38. CSIRO, Melbourne, Victoria.

Frankenberg R. S. (1966) Fishes of the family Galaxiidae. *Aust. Nat. Hist.* **15,** 161-4.

Glover J. W., Bacher G. J. & Pearce T. S. (1980) Gippsland regional environmental study. Heavy metals in biota and sediments of the Gippsland lakes. Environmental Studies Report No. 279, Ministry for Conservation, Melbourne, Victoria.

Graynoth F. (1979) Effects of logging on stream environments and faunas in Nelson. *N. Z. J. Mar. Freshwat. Res.* **13,** 79-109.

Harris J. H. (1984a) A survey of fishways in streams of coastal south-eastern Australia. *Aust. Zool.* **21,** 219-33.

Harris J. H. (1984b) Impoundment of coastal drainages of south-eastern Australia, and a review of its relevance to fish migrations. *Aust. Zool.* **21,** 235-50.

Harris J. H. (1986) Fish passage in Australia. In: *Stream Protection. The Management of Rivers for Instream Uses* (ed. I. C. Campbell) pp. 135-41. Water Studies Centre, Chisholm Institute of Technology, Caulfield, Victoria.

Hewlett J. D. (1982) *Principles of Forest Hydrology.* University of Georgia Press, Athens, Georgia.

Hillman T. J. (1980) Lake Burley Griffin. In: *An Ecological Basis for Water Resource Management* (ed. W. D. Williams) pp. 317-23. Australian National University Press, Canberra, A.C.T.

Hunt R. L. (1975) Use of terrestrial invertebrates as food by salmonids. In: *Coupling of Land and Water Systems* (ed. A. D. Hasler) pp. 137-51. Springer-Verlag, Berlin.

Hynes H. B. N. (1970) *The Ecology of Running Waters.* Liverpool University Press, Liverpool.

Hynes H. B. N. (1973) The influence of sediment on the biota in running water. In: *Fluvial Processes and Sedimentation.* Proceedings of Hydrology Symposium, University of Alberta, Edmonton, 1973 pp. 653-63. Canada Department of the Environment, Ottawa.

Hynes H. B. N. (1975) The stream and its valley. *Verh. Internat. Verein. Limnol.* **19,** 1-15.

Jackson P. D. (1978) Benthic invertebrate fauna and feeding relationships of brown trout, *Salmo trutta* Linnaeus, and river blackfish, *Gadopsis marmoratus* Richardson in the Aberfeldy River, Victoria. *Aust. J. Mar. Freshwat. Res.* **29,** 725-42.

Jackson P. D. (1981) Trout introduced into south-eastern Australia: their interaction with native fishes. *Vic. Nat.* **98,** 18-24.

Jackson P. D. & Williams W. D. (1980) Effects of brown trout, *Salmo trutta* Linnaeus, on the distribution of some native fishes in three areas of southern Victoria, Australia. *Aust. J. Mar. Freshwat. Res.* **31,** 61-67.

Kendrick G. W. (1976) The Avon: faunal and other notes on a dying river in south-western Australia. *West. Aust. Nat.* **13,** 97-114.

Knighton A. D. (1987) Tin mining and sediment supply to the Ringarooma River, Tasmania, 1875-1979. *Aust. Geogr. Stud.* **25,** 83-97.

Knott B., Suter P. J. & Richardson A. M. M. (1978) A preliminary observation on the littoral rock fauna of Hartz Lake and Hartz Creek, southern Tasmania, with notes on the water chemistry of some neighbouring lakes. *Aust. J. Mar. Freshwat. Res.* **29,** 703-15.

Kunert C. & Macmillan L. (1988) Conservation value and status of Victorian rivers. Part III. The Wimmera River and its catchment. Faculty of Environmental Design and Construction, Royal Melbourne Institute of Technology, Melbourne, Victoria.

Lake P. S. (1963) A study of the macroinvertebrate fauna of the zinc polluted Molonglo River. Honours Thesis, Department of Zoology. Australian National University, Canberra, A.C.T.

Lake P. S. (1978) On the conservation of rivers in Australia. *Aust. Soc. Limnol. Newsl.* **16,** 24-8.

Lake P. S. (1982) The 1981 Jolly Award Address — Ecology of the macroinvertebrates of Australian upland streams: a review of current knowledge. *Bull. Aust. Soc. Limnol.* **8,** 1-15.

Lake P. S. (1987) Stream ecosystem research in Australia: problems and strategies. In: *National Water Research Seminar. Discussion Papers* pp. 49-80. Australian Water Research Advisory Council, Department of Resources and Energy, Canberra.

Lake P. S. & Knott B. (1973) On the freshwater crustaceans of the Central Plateau. In: *The Lake Country of Tasmania* (ed. M. R. Banks) pp. 95-9. Royal Society of Tasmania, Hobart.

Lake P. S. & Sokol A. (1986) Ecology of the yabby *Cherax destructor* Clark (Crustacea: Decapoda: Parastacidae) and its potential as a sentinel animal for mercury and lead pollution. *Australian Water Resource Council Technical Paper* **87,** 1-186.

Lake P. S., Coleman D., Mills B. & Norris R. (1977) A reconnaissance of pollution of the King River in the Comstock-Crotty area, western Tasmania. In: *Landscape and Man* (eds M. R. Banks & J. Kirkpatrick) pp. 157-73. Royal Society of Tasmania, Hobart.

Lake P. S., Barmuta L. A., Boulton A. J., Campbell I. C. & St Clair R. M. (1985) Australian streams and Northern Hemisphere stream ecology: comparisons and problems. *Proc. Ecol. Soc. Aust.* **14,** 61-82.

Lemly A. D. (1982) Modification of benthic insect communities in polluted streams: combined effects of sedimentation and nutrient enrichment. *Hydrobiologia* **87,** 229-45.

Likens G. & Bormann F. H. (1974) Linkages between terrestrial and aquatic ecosystems. *Bioscience* **24,** 447-56.

Loh I. C. & Stokes R. A. (1981) Predicting stream salinity changes in south-western Australia. *Agricultural Water Management* **4,** 227-54.

Lowrance R., Todd R., Fail J., Hendrickson Jr. O., Leonard R. & Asmussen L. (1984) Riparian forests as nutrient filters in agricultural watersheds. *Bioscience* **34,** 374-77.

McComb A. J. & Lake P. S. (Eds) (1988) *The Conservation of Australian Wetlands.* Surrey Beatty & Sons, Chipping Norton, New South Wales.

McMahon T. A. (1982) World hydrology: does Australia fit? In: *Hydrology and Water Resources Symposium 1982.* The Institution of Engineers, Australia. 7 pp.

Macmillan L. A., Kunert C. & Blakers M. (1987) Nature Conservation value and status of rivers in the South Western Region, Victoria. Report to the Victorian Department of Water Resources. Department of Planning, Policy and Landscape, Royal Melbourne Institute of Technology, Melbourne.

Malipatil M. B. & Blyth J. D. (1982) A qualitative study of the macroinvertebrate fauna of the Thomson River and its major tributaries, Gippsland, Victoria. *Rep. Nat. Mus. Vict.* **1,** 1-95.

Marchant R. (1989) Changes in the benthic invertebrate communities of the Thomson River south-eastern Australia, after dam construction. *Regulated Rivers* **4,** 71-88.

Marchant R., Metzeling L., Graesser A. & Suter P. (1985) The organization of macroinvertebrate communities on the major tributaries of the La Trobe River, Victoria, Australia. *Freshwat. Biol.* **15,** 315-31.

Merrick J. R. & Schmida G. E. (1984) Australian Freshwater Fishes. Biology and Management. John R. Merrick, Macquarie University, North Ryde, New South Wales.

Michaelis F. B. (1984) Possible effects of forestry on inland waters of Tasmania: a review. *Environ. Conserv.* **11,** 331-43.

Milner N. J., Scullion J., Carling P. A. & Crisp D. T. (1981) The effects of discharge on sediment dynamics and consequent effects on invertebrates and salmonids in upland rivers. *Adv. Appl. Biol.* **6,** 153-220.

Minshall G. W. (1978) Autotrophy in stream ecosystems. *Bioscience* **28,** 767-71.

Minshall G. W., Petersen R. C., Cummins K. W., Bott T. L., Sedell J. R., Cushing C. E. & Vannote R. L. (1983) Interbiome comparisons of stream ecosystem dynamics. *Ecol. Monogr.* **53,** 1-25.

Morgan D. R. & Graynoth E. (1978) The influence of forestry practices on the ecology of freshwater fish in New Zealand. An introduction to the literature. Fisheries Research Division Occasional Publication No. 14. New Zealand Ministry of Agriculture and Fisheries, Wellington.

Moring J. R. (1982) Decrease in stream gravel permeability after clear-cut logging: an indication of intragravel conditions for developing Salmonid eggs and alevins. *Hydrobiologia* **88,** 295-8.

Morrissy N. M. (1974) Reversed longitudinal salinity profile of a major river in the south-west of Western Australia. *Aust. J. Mar. Freshwat. Res.* **25,** 327-35.

Morrissy N. M. (1978) The amateur marron fishery in Western Australia. *Fish. Res. Bull. West. Aust.* **21,** 1-44.

Murphy M. L. & Hall J. D. (1981) Varied effects of clear-cut logging on predators and their habitat in small streams of the Cascade Mountains, Oregon. *Can. J. Fish. Aquat. Sci.* **38,** 137-45.

Newbold J. D., Erman D. C. & Roby K. B. (1980) Effects of logging on macroinvertebrates in streams with and without buffer strips. *Can. J. Fish. Aquat. Sci.* **37,** 1076-85.

Nicholas W. L. & Thomas M. (1978) Biological release and recycling of toxic metals from lake and river sediments. *Australian Water Resources Council Tech. Pap. No.* **33,** 1-99.

Norris R. H. (1986) Mine waste pollution of the Molonglo River, New South Wales and the Australian Capital Territory: effectiveness of remedial works at Captains Flat mining area. *Aust. J. Mar. Freshwat. Res.* **37,** 147-57.

Norris R. H., Lake P. S. & Swain R. (1982) Ecological effects of mine effluents on the South Esk River, north-eastern Tasmania. III. Benthic macroinvertebrates. *Aust. J. Mar. Freshwat. Res.* **33,** 789-809.

O'Brien W. T., McGregor A. M. & Crawshaw B. W. (1983) Instream uses of water in Australia. In: *In-stream and Environmental Issues. Water 2000. Consultants Report No. 9.* Australian Government Publishing Services, Canberra.

Olive L. J. & Walker P. H. (1982) Processes in overland flow-erosion and production of suspended material. In: *Prediction in Water Quality* (eds E. M. O'Loughlin & P. Cullen) pp. 87-119. Australian Academy of Science, Canberra.

Outridge P. M. (1988) Seasonal and spatial variations in benthic macroinvertebrate communities of Magela Creek, Northern Territory. *Aust. J. Mar. Freshwat. Res.* **39,** 211-23.

Pearson R. G., Benson L. J. & Smith R. E. W. (1986) Diversity and abundance of the fauna in Yuccabine Creek, a tropical rainforest stream. In: *Limnology in Australia* (eds P. De Deckker & W. D. Williams) pp. 329-42. CSIRO, Melbourne.

Peck A. J., Thomas J. F. & Williamson D. R. (1983) Salinity issues. In: *Effects of men on salinity in Australia. Water 2000: Consultants Report No. 8.* Australian Government Publishing Service, Canberra.

Petersen Jr. R. C., Madsen B. L., Wilzbach M. A., Magadza C. H. D., Paarlberg A., Kullberg A. & Cummins K. W. (1987) Stream management: emerging global similarities. *Ambio* **16,** 166-79.

Pollard D. A., Llewellyn L. C. & Tilzey R. D. J. (1980) Management of freshwater fish and fisheries. In: *An Ecological Basis for Water Resource Management* (ed. W. D. Williams) pp. 227-70. Australian National University Press, Canberra.

Rapport D. J., Regier H. A. & Hutchinson T. C. (1985) Ecosystem behavior under stress. *Am. Nat.* **125,** 617-40.

Richardson B. A. (1985) The impact of forest road construction on the benthic invertebrate and fish fauna of a coastal stream in southern New South Wales. *Bull. Aust. Soc. Limnol.* **10,** 65-88.

Robinson D. P. (1977) The study of the invertebrate fauna of three mountain streams in relation to the type of land use in each of the three stream catchments. Honours Thesis, Department of Zoology, Monash University, Clayton, Victoria.

Rounick J. S. & Winterbourn M. J. (1983) The formation, structure and utilization of stone surface organic layers in two New Zealand streams. *Freshwat. Biol.* **13,** 57-72.

Ryan P. (1987) Rum Jungle Mine Rehabilitation: Northern Territory. *J. Soil Conserv. N.S.W.* **43,** 17-27.

Shea S. R., Hatch A. B., Havel J. J. & Ritson P. (1975) The effect of changes in forest structure and composition on water quality and yield from the northern jarrah forest. *Proc. Ecol. Soc. Aust.* **9,** 58-73.

Silsbee D. G. & Larson G. L. (1983) A comparison of streams in logged and unlogged areas of Great Smoky Mountains National Park. *Hydrobiologia* **102,** 99-111.

Smith D. I. & Finlayson B. (1988) Water in Australia: its role in environmental degradation. In: *Land, Water and People. Geographical Essays in Australian Resource Management* (eds R. L. Heathcote & J. A. Mabbutt) pp. 7-48. Allen & Unwin, Sydney.

Standing Consultative Committee on River Improvement (1983) *The State of the Rivers.* Standing Consultative Committee on River Improvement, Melbourne.

Statzner B. & Higler B. (1986) Stream hydraulics as a major determinant of benthic invertebrate zonation patterns. *Freshwat. Biol.* **16,** 127-39.

Steering Committee for Research on Land Use and Water Supply (1987) The impact of logging on the water resources of the southern forests, Western Australia. Water Authority of Western Australia, Perth.

Tebo L. R. (1955) Effects of siltation, resulting from improper logging, on the bottom fauna of a small trout stream in the Southern Appalachians. *Progr. Fish Cult.* **12,** 64-70.

Tilzey R. D. J. (1976) Observations on interactions between indigenous Galaxiidae and introduced Salmonidae in the Lake Eucumbene catchment, New South Wales. *Aust. J. Mar. Freshwat. Res.* **27,** 551-64.

Tilzey R. D. J. (1977) Key factors in the establishment and success of trout in Australia. *Proc. Ecol. Soc. Aust.* **10,** 97-105.

Tilzey R. D. J. (1980) Introduced Fish. In: *An Ecological Basis for Water Resource Management* (ed. W. D. Williams) pp. 271-79. Australian National University Press, Canberra.

Towns D. R. (1985) Limnological characteristics of a South Australian intermittent stream. *Aust. J. Mar. Freshwat. Res.* **36,** 821-37.

Vannote R. L., Minshall G. W., Cummins K. W., Sedell J. R. & Cushing C. E. (1980) The River Continuum Concept. *Can. J. Fish. Aquat. Sci.* **37,** 130-7.

Warner R. F. (1984) Man's impacts on Australian drainage systems. *Aust. Geogr.* **16,** 133-41.

Weatherley A. H. & Dawson P. (1973) Zinc pollution in a freshwater ecosystem: analysis and proposed solution. *Search* **4,** 461-6.

Weatherley A. H., Beevers J. R. & Lake P. S. (1967) The ecology of a zinc polluted river. In: *Australian Inland Waters and Their Fauna: Eleven Studies* (ed. A. H. Weatherley) pp. 252-78. Australian National University Press, Canberra.

Weatherley A. H., Lake P. S. & Rogers S. C. (1980) Zinc pollution and the ecology of the freshwater environment. In: *Zinc in the Environment. Part I. Ecological Cycling* (ed. J. O. Nriagu) pp. 337-418. John Wiley & Sons, New York.

Webster J. R., Gurtz M. E., Hains J. J., Meyer J. L., Swank W. T., Waide J. B. & Wallace J. B. (1983) Stability of stream ecosystems. In: *Stream Ecology. Application and Testing of General Ecological Theory* (eds J. R. Barnes & G. W. Minshall) pp. 355-95. Plenum Press, New York.

Wharton J. C. (1969) Recreation and conservation in the aquatic environment. *Proc. Roy. Soc. Vic.* **83,** 47-54.

Wiederholm T. (1984) Responses of aquatic insects to environmental pollution. In: *The Ecology of Aquatic Insects* (eds V. H. Resh & D. M. Rosenberg) pp. 508-57. Praeger, New York.

Williams W. D. (1974) Freshwater Crustacea. In: *Biogeography and Ecology in Tasmania* (ed. W. D. Williams) pp. 63-112. W. Junk, The Hague.

Williams W. D. (1980) Catchment management. In: *An Ecological Basis for Water Resource Management* (ed. W. D. Williams) pp. 100-14. Australian National University Press, Canberra.

Williams W. D. (1983) *Life in Inland Waters.* Blackwell Scientific Publishers, Carlton, Victoria.

Williams W. D. (1987) Salinization of rivers and streams: an important environmental hazard. *Ambio* **16,** 180-6.

Williams W. D. & Campbell I. (1987) The inland aquatic environment and its fauna. In: *Fauna of Australia. Volume 1A. General Articles* (ed. G. R. Dyne) pp. 156-83. Australian Government Publishing Service, Canberra.

Wimbush D. J. & Costin A. B. (1983) Trends in drainage characteristics in the subalpine zone at Kosciusko. *Proc. Ecol. Soc. Aust.* **12,** 143-54.

Winterbourn M. J., Rounick J. S. & Cowie B. (1981) Are New Zealand stream ecosystems really different? *N.Z. J. Mar. Freshwat. Res.* **15,** 321-8.

Wood W. E. (1924) Increase in salt in soil and streams following the destruction of native vegetation. *J. Roy. Soc. West. Aust.* **10,** 35-47.

Woods L. E. (1984) *Land Degradation in Australia.* Australian Government Publishing Service, Canberra.

Proc. Ecol. Soc. Aust. 1990 — 16:93-114

From frontier to fragments: European impact on Australia's vegetation

R. J. HOBBS[1] and A. J. M. HOPKINS[2]

[1]*CSIRO, Division of Wildlife and Ecology, LMB , P.O. Midland, Western Australia 6056*

[2]*Department of Conservation and Land Management, Western Australian Wildlife Research Centre, P.O. Box 51, Wanneroo, Western Australia 6056*

Abstract

European settlement of Australia over the past 200 years has resulted in widespread and rapid modification of the vegetation. Four main categories of land-use can be recognised, depending on the degree of impact.

1. *Complete removal of the vegetation and disruption of ecological processes, such as occurs during urban development and mining.*
2. *Removal of vegetation and replacement with intensively managed systems such as agriculture or plantation forestry.*
3. *Utilization of existing vegetation with some consequent modification, as in pastoralism and timber harvesting in native forests.*
4. *Conservation with minimum deliberate modification.*

We suggest that the imposition of these new distubance regimes has commonly resulted in the alteration of the vegetation to new metastable states, often with accompanying land degradation. We identify how major ecosystem components are affected by human modification and present broad guidelines for rehabilitation and reconstruction of ecosystems based on treatment of these impacts. We also identify key directions for the future which will aid in developing methods for the maintenance, rehabilitation and reconstruction of Australian ecosystems. Current land degradation problems have their origins in the utilitarian attitude to vegetation which has prevailed since settlement. While this is changing slowly, there is an urgent need to promote a more sympathetic attitude towards the land and its associated biota and to develop more sustainable management regimes.

Introduction

In the 200 years since Europeans colonized Australia, the Australian landscape has changed dramatically. Aboriginal populations undoubtedly modified both vegetation and landscape for many thousands of years prior to 1788 (Bowler 1976; Hallam 1979; Singh *et al.* 1981; Singh 1982). However, the changes resulting from two centuries of European activity have undoubtedly had a much greater impact. The first colonists had cultural backgrounds and expectations that differed markedly from those of the Aborigines. They came to Australia expecting to live a European lifestyle in a new landscape, supported by crops of European origin and agricultural techniques developed in the Northern Hemispere, and bringing ungulates for food and fibre. The arrival of Europeans has been called an "apocalyptic event for Australian ecosystems" (Adamson & Fox 1982). To quote from Lowenthal (1976); "Nowhere else had technological man lately occupied so fragile an ecosystem; nowhere else had settlers so utterly failed to identify with their new landscapes; nowhere else was man's environmental impact so patently reprehensible".

In this chapter we examine the activities which have resulted in the widespread changes now seen in Australian vegetation. We adopt the approach of looking at how European land-use practices have affected basic population and ecosystem processes and how modifications to these have led to current management problems. We develop guidelines for ecosystem management based on a recognition of land-use impacts on ecosystem properties. Some insight into the present problems and future solutions is provided by an examination of attitudes towards vegetation and landscape. Finally, we suggest future directions for research and management.

General response to European settlement

The current pattern of land-use in Australia gives a good idea of the extent to which the continent has been modified over the past 200 years (Table 1). The 26% of the land surface that is not utilized in some way consists mainly of land in the arid zone. Much of the remainder of this zone, which in total accounts for 70% of the land surface, is utilized primarily for grazing. The remaining 30% of the continent contains all the major cities, 98% of the population, 75% of the livestock and all the agriculture and forestry. The area specifically devoted to nature conservation is 3.5%. Some areas are still largely unmodified. In others the degree of modification varies from complete removal of the native vegetation and replacement by urban or agricultural development to minor alteration of grazing, fire, nutrient or hydrologic regimes in otherwise natural plant communities (i.e., not substantially modified by technological intrusion, *cf* Taylor, this volume). Clearly land-use patterns are not static and have, in fact, undergone rapid change, as can be assessed from examining Wadham *et al.* (1964) and earlier editions of their book. Redistributions from one land-use category to another are still taking place. For instance, Hopkins & Saunders (1987) show a large increase in the area of national parks and nature reserves in Western Australia over the period 1969-84, and there have been recent proposals for reallocation of former State Forest areas to national parks or conservation reserves (Anon 1988).

TABLE 1. Percentage of total land area of Australia devoted to each major land-use (from Department of Arts, Heritage and Environment 1986), categorized according to degree of modification of the native vegetation.

1. Complete vegetation removal	
Mining	<0.1%
Urban	0.1%
Transport	1.2%
2. Replacement with agricultural systems	
Intensive cropping	0.3%
Extensive cropping	5.8%
3. Exploitation of native vegetation	
Forestry	2.0%
Non-arid grazing	17.4%
Arid grazing	43.7%
4. No deliberate modification	
Unused (mainly desert)	26.0%
Nature conservation	3.5%

Australia is not unique in possessing landscapes which have been significantly modified by humans (e.g., Thirgood 1981; Holzner *et al.* 1983). However, the rapidity of this modification is remarkable (Fig. 1; Aschmann 1973; Grubb & Hopkins 1986). For instance, Naveh and Dan (1973) have provided a concise history of land modification in Israel which spans over 11 000 years from the development of early agriculture through widespread land clearance, a period of agricultural decay and massive land degradation, and finally a period of reclamation and intensification of agriculture. This period essentially covers the entire development of western civilization through the Bronze and Iron Ages, followed by the rise and fall of empires and nations. During this period the Mediterranean landscape developed with technological humans as an integral part. In Australia, the equivalent historical series has been telescoped into only 200 years (or much less than this in most areas — e.g., Saunders & Curry, this volume) and involved an accelerated rate of change in a landscape which until then was affected only by the activities of sparse Aboriginal populations. Recognition of this provides the key to the understanding of both the attitudes of the original settlers towards the native vegetation and the effects which subsequent activities have had.

Because of the long association that has existed between intensive human utilization and landscape in Europe, the vegetation there as we see it now is largely made up of components which are resilient (*sensu* Westman 1978, 1986) to this type of human interference. European ecosystems may thus be viewed as mostly very resilient (i.e., they recover rapidly to their pre-disturbance state following disturbance) because they have already lost all their non-resilient components and/or gained aggressive and resilient species (e.g., Naveh 1967; Naveh & Dan 1973; Godron *et al.* 1981; Le Houérou 1981; Pignatti 1983). They also may be relatively inert, or resistant to change when stressed (see Westman 1978, 1986). We therefore hypothesise that continued human disturbance has led to the development of metastable systems which differ markedly from the original (Fig. 2). In Australia, on the other hand, the vegetation has not had this prolonged period of development in parallel with intensive human utilization and may thus be more vulnerable to European types of disturbance. The vegetation (and the entire ecosystem) may thus be likely to be much more rapidly degraded, especially when the rate of modification is high. Disturbances which may have relatively little impact in the Mediterranean basin can have large effects in Australia because Australian systems still

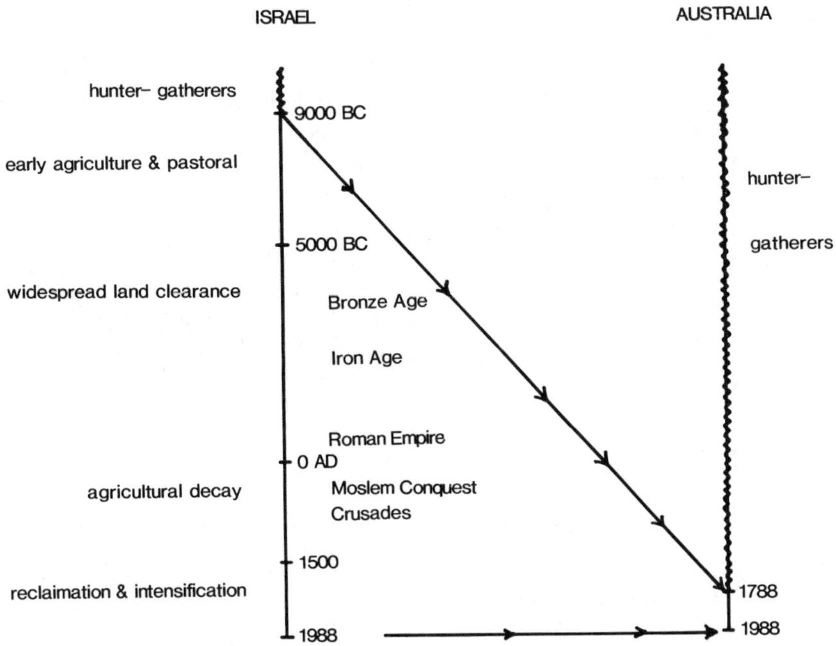

FIG. 1. Modification of natural vegetation has taken place much more rapidly in Australia than in Europe, where the present day patterns of vegetation have developed over thousands of years under the influence of European civilization and in parallel with a gradual intensification of land-use patterns.

contain many non-resilient components. Naveh (1967) has made this comparison in relation to the ecosystems of California and Israel. Effects of disturbance may take some time to manifest themselves because of lag effects and, while the slow rate of modification in Europe may have allowed for these, the rapidity of change in Australia possibly has not. Here we have continued to implement change without waiting to observe the effects of preceding steps. The disturbance regimes introduced with the newly imposed land-use practices differ markedly from those prevailing before European settlement. The capacity of the natural systems to recover (i.e., their resilience) could therefore be exceeded and the systems are now possibly progressing to new, usually simplified, metastable states. We suggest, therefore, that the degradation we now observe may be the combined effect of the immediate impact of disturbance plus the non-return of the system to the original state (Fig. 2).

Effects of different degrees of modification

Categories of land-use and degree of modification have been proposed previously for predominantly Northern Hemisphere systems by Godron and Forman (1983) and Forman and Godron (1986). We recognize four major land-use categories for Australia, based on the degree of modification of the natural vegetation.

1. REMOVAL. Complete removal of native vegetation and disruption of processes: urban settlement, transport, industrial development and extractive industries. Native vegetation and all natural processes are completely disrupted, except in small patches. Natural vegetation will not return without complete reconstruction of the ecosystem, including soil replacement.
2. REPLACEMENT. Removal of native vegetation and replacement with intensively managed systems: agriculture, horticulture and plantation forestry. This involves partial disruption of processes. Native vegetation (or components of it) can return after cessation of intensive management, although this may require management to restore natural processes.
3. UTILIZATION. Exploitation of native vegetation, with some consequent degree of modification: forestry (in native forests), pastoralism and recreation in natural areas. This corresponds to a partial removal of the natural vegetation and a partial disruption of processes.
4. CONSERVATION. Maintenance of natural vegetation for conservation and scientific purposes with minimum deliberate modification of natural processes. Lands designated as

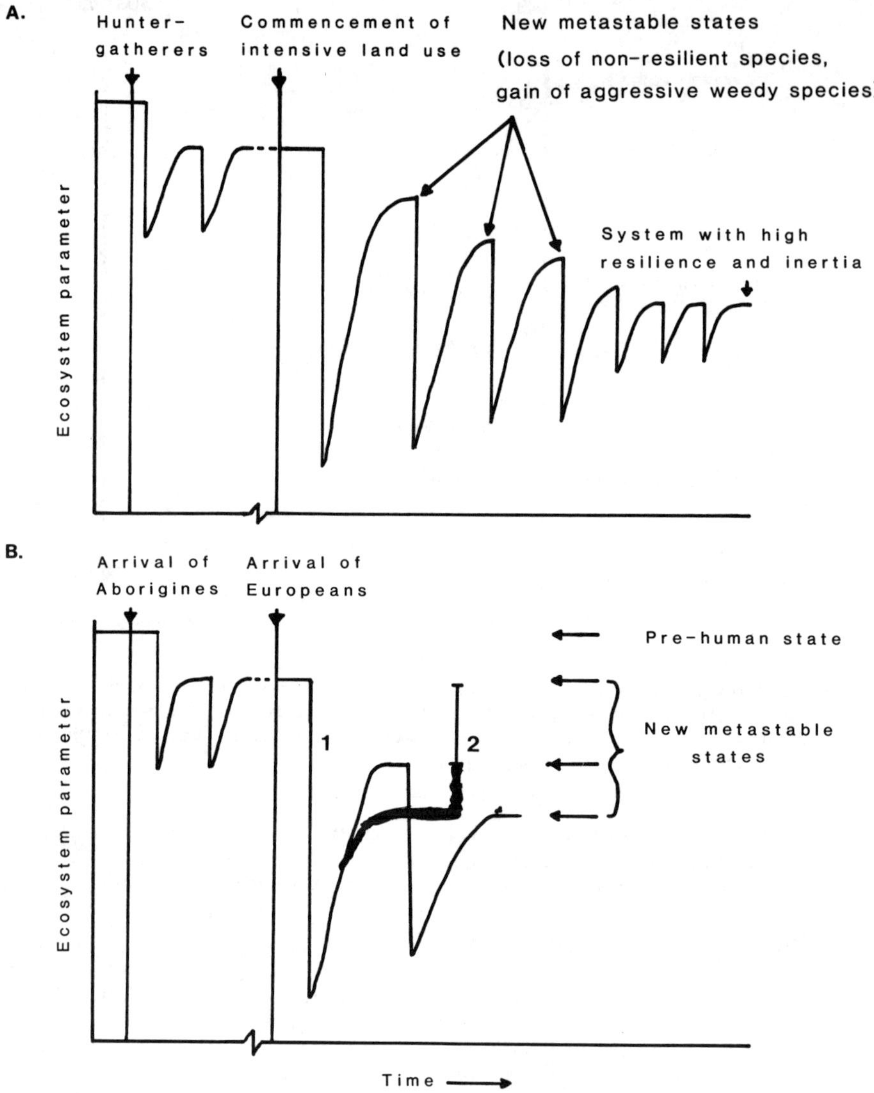

FIG. 2. Resilience and inertia of European ecosystems with a long history of intensive human use versus Australian ecosystems with relatively recent intensive disturbance. In European systems (A), some changes resulted from the development of hunter-gatherer societies, but development of agriculture and more intensive land-use led to greater change and the development of metastable, simplified communities which are now highly resilient and resistant to further change. Australian ecosystems (B) can be viewed in general as being at the early stages of this progression and large changes in vegetation now observed may be the result of response to European disturbance (1) coupled with a permanent change in ecosystem properties (2).

national parks and nature reserves fall into this category. In addition there are substantial areas of vacant or uncommitted government-owned land and Aboriginal lands. Designation for nature conservation does not preclude past or inadvertant modification. There are no direct or deliberate impacts on the native vegetation as a result of this land-use, but there may be indirect impacts from land-use activities on adjacent or nearby lands.

Each land-use category has a distinct set of impacts on ecosystem components and processes. We have recognized eight major sets of properties which are likely to be particularly affected by human disturbance (Table 2). We have selected these features on the basis of practical considerations as being the main ecosystem properties that can be manipulated in management and rehabilitation. For instance, energy or organic carbon

TABLE 2. Ecosystem components liable to be affected by human modification.

1. Vegetation	— above-ground biomass — below-ground organs
2. Propagule supply	— soil seed bank — on-plant store+vegetative propagules — dispersed seed — animal — wind, water
3. Other interacting organisms	— soil fungi, bacteria, mycorrhizae — soil and litter fauna — herbivores — pollinators — dispersal agents — other fauna
4. Soil properties	— structure/compaction — stability — pH
5. Hydrology	— water table — salinity
6. Nutrient cycling	— nutrient pools — cycling processes
7. Population processes	— supply of propagules (2 above) — germination and establishment — altered mortality — competition
8. Community properties	— species richness — stability/resilience — vegetation dynamics

cycles are more fundamental (e.g., Sprugel 1985), but are difficult to deal with at a practical level and are implicit in components such as "other interacting organisms" and "nutrient cycling". Impacts on these properties are particularly important in terms of ecosystem resilience and when rehabilitation is considered (see later). Other ecosystem properties may be of equal importance but have little practical relevance.

Using the four land-uses categories and the eight major ecosystem components, a matrix of impacts can be constructed. Table 3 indicates the major direct impacts, i.e., the impacts which arise as a direct consequence of human disturbance on a given piece of land. Direct impacts can be divided into two main types — i.e., deliberate modifications associated with the land-use and inadvertant modifications which may flow on from the land-use or arise independently of it. Thus, for instance, pastoralism (Category 3) primarily involves the deliberate introduction of grazing animals and harvesting of above-ground biomass, but this may also result in changed soil properties, redistribution of nutrients, altered vegetation dynamics and the spread of non-native species. Management for grazing may also entail changing fire regimes and alteration of surface and groundwater supplies. Each land-use thus entails a suite of deliberate and inadvertant impacts on the natural ecosystem.

In addition to the direct impacts there will also be indirect effects of particular land uses on adjacent or nearby native vegetation (Table 4). Landscape fragmentation in particular often means that actions in one part of the landscape often affect ecosystem properties elsewhere. In the following sections we examine the direct and indirect effects of each land use category in more detail.

TABLE 3. Impacts of major land use categories on ecosystem components and prescriptions for rehabilitation or reconstruction.

Land use category	Vegetation impacts	Vegetation reconstruction
1. Removal	Complete removal	Regenerate from seed or seedling
2. Replacement	Complete removal	Regenerate from seed or seedling
3. Utilization	Removal of all or part of above ground material. Below ground plant material little disturbed.	May regenerate from existing plant parts. This will require time with minimum disturbance. Episodic regeneration events may be important. May require some reintroductions.
4. Conservation	Affected by native and feral herbivores and fire.	Monitor and regulate as necessary.
Land use category	**Seed/Propagule supply: Impacts**	**Seed/Propagule Supply: Reconstruction**
1. Removal	Removal of all *in situ* seed (dispersed seed supply still available)	Replacement of seed store. On plant store can be replaced by introduction of brush. Soil store can be replaced in fresh topsoil. Both types of seed can be collected manually and sown.
2. Replacement	Removal of on plant seed store. Gradual decline of soil seed store. (Dispersed seed supply still available.	Replacement of seed store. On plant store can be replaced by introduction of brush. Soil store can be replaced in fresh topsoil. Both types of seed can be collected manually and sown.

TABLE 3 *continued*

Land use category	Seed/Propagule supply: Impacts	Seed/Propagule Supply: Reconstruction
3. Utilization	Store may be gradually depleted, possibly eliminated.	Recovery of seed store will require time free from further disturbance. Some introductions may be necessary.
4. Conservation	Seed store affected by indirect disturbances, fire and grazing.	Monitor and reintroduce seed as necessary.

Land use category	Other interacting organisms: Impacts	Other interacting organisms: Reconstruction
1. Removal	Local populations of organisms become extinct.	Organisms must be reintroduced. Soil micro-organisms can be reinoculated or reintroduced in fresh topsoil.
2. Replacement	Local populations of organisms decline.	Encourage regeneration of populations, may include reintroduction.
3. Utilization	Local populations of organisms decline.	Encourage regeneration of populations, may include reintroduction. Avoid additional disturbances.
4. Conservation	Pollinators and seed dispersal organisms may be reduced through competition, indirect disturbances, fire and grazing.	Monitor and reintroduce as necessary.

Land use category	Nutrient cycling processes: Impacts	Nutrient cycling processes: Reconstruction
1. Removal	Cycling processes disrupted. Nutrients may be lost or added.	Appropriate nutrient levels must be re-established. Nutrient fixing and cycling organisms must be reintroduced.
2. Replacement	Cycling processes disrupted. Nutrients may be lost or added.	Appropriate nutrient levels must be re-established. Nutrient fixing and cycling organisms may need to be reintroduced.
3. Utilization	Nutrients redistributed and removed as harvested biomass. Other nutrients may be converted to an inaccessible form. Cycling processes partially disrupted.	Nutrient removal rate must be adjusted to reflect rates of natural inputs. Adjust disturbance regimes to permit cycling processes to be sustained.
4. Conservation	May be input of nutrients from adjacent lands.	Manage causes off site as necessary.

Land use category	Soil properties: Impacts	Soil properties: Reconstruction
1. Removal	Soil profiles often completely disrupted. pH and nutrient levels altered substantially.	New soil profile can be reconstructed. Often requires addition of benign topsoil. Chemical properties must be ameliorated.
2. Replacement	Soil profiles often collapse, suffer compaction and erosion. pH and nutrient levels altered.	May require ripping, stabilization and other engineering works. Chemical properties must be ameliorated.
3. Utilization	Disturbance usually superficial, but includes fracture of soil crusts, compaction, wind and water erosion.	May require ripping, stabilization and other engineering works. Avoid further disturbance.
4. Conservation	Indirect effects only.	Manage causes off site as necessary.

Land use category	Hydrology/Salinization: Impacts	Hydrology/Salinization: Reconstruction
1. Removal	Often waterlogging results. The rising watertable may bring up soluble salts.	Plant tolerant species, may require mounding, bunding and other drainage works.
2. Replacement	Often waterlogging results. The rising watertable may bring up soluble salts.	Plant tolerant species, may require mounding, bunding and other drainage works. Re-establish deep-rooted vegetation.

TABLE 3 *continued*

Land use category	Hydrology/Salinization: Impacts	Hydrology/Salinization: Reconstruction
3. Utilization	May give rise to minor elevation in watertables and salinization. Reduced infiltration may cause increased run-off. Provision of watering points alters surface run-off.	Encourage regeneration of vegetation with particular attention to deep-rooting plant species.
4. Conservation	Indirect effects only.	Manage causes off site as necessary.

Land use category	Population processes: Impacts	Population processes: Reconstruction
1. Removal	Reduced to zero.	Re-establishment of processes requires time. Avoid additional disturbances and monitor.
2. Replacement	Generally reduced to very low levels.	Re-establishment of processes requires time. Avoid additional disturbances and monitor.
3. Utilization	Generally great increase in adult mortality through harvesting. Recruitment rates may be greatly increased or reduced. Interspecific competition affected.	Plant populations must be monitored to ensure sustainability. Fire and grazing regimes altered accordingly.
4. Conservation	Recruitment rates may be altered by changed fire regimes or changes in interacting organisms. Alien plants may compete with native species.	Native plant populations must be monitored. Control populations of interacting organisms including weeds, as necessary. Instigate appropriate fire regimes where necessary.

Land use category	Community properties: Impacts	Community properties: Reconstruction
1. Removal	Reduced to zero.	Re-establishment of processes requires time. Avoid additional disturbances and monitor. Re-establish successional processes.
2. Replacement	Generally reduced to monospecies systems. Few natural community processes.	Re-establishment of properties requires time. Avoid additional disturbances and monitor. Re-establish successional processes.
3. Utilization	Species richness altered at least in the short term. Vegetation may achieve a metastable state and lose non-resilient components.	Re-establishment of properties requires time. Avoid additional disturbances and monitor.
4. Conservation	All properties and processes at appropriate levels.	Maintain appropriate disturbance regimes, and prevent invasions.

Complete removal of vegetation

Complete removal of the vegetation results from urban development, the creation of transport networks and from extractive industries. Urban development has been rapid and the major cities which house most of Australia's population continue to expand. Mining has been an important industry since settlement began and gold was discovered, but the diversity of minerals extracted has increased greatly with extraction of things such as bauxite and uranium commencing within the last 40 years (e.g., Conybeare 1972; McCay 1972; Hore-Lacey 1982).

Under these land-uses, the land surface is covered over or excavated and subsequently returned in a new form. Mining differs from urban development in that it is transitional and the vegetation can potentially be reconstructed after mining ceases. The vegetation is replaced with mine spoil which may be toxic but is in any case often highly unsuited to vegetation growth. In the case of urban and transport development, the vegetation removal is usually permanent except for small remnant patches. Often urban parks retain no vestiges of native vegetation although some do. Road verges may also be left uncleared. Thus, for example, Kings Park in Perth still retains the *Banksia* woodland present originally on the site, although it is degraded and has many management problems (e.g., Wycherley 1984). Such small urban reserves really represent an extreme form of the fragmented systems discussed in the next section and are concomitantly more difficult to manage. Similarly, road reserves are equivalent to linear fragments with a maximum amount of edge and exposure to potential external disturbance.

TABLE 4. Indirect effects of human modification on adjacent native vegetation.

1. Fragmentation	
2. Introduction of alien organisms	— plant (weeds) — animals — herbivores — predators — competitors — pathogens
3. Alteration of fire regimes	
4. Burial (landslip, erosion)	
5. Input of additional nutrients (eutrophication)	
6. Alteration of groundwater	— rising water table — salinization
7. Toxic chemical inputs	— pollution — biocides
8. Regional or global climatic change	

Replacement with intensively managed systems

Replacement of the native vegetation by agriculture or plantation forestry is perhaps the most conspicuous European impact on native vegetation. Clearance for agriculture began slowly at first around the original settlements but then progressed rapidly as new areas were opened up and new farming techniques became available. Throughout the southern and north-eastern coastal parts of Australia, extensive areas have been cleared for agriculture resulting in the loss of large proportions of the original plant communities, together with the loss of many species and the potential to lose many more (Leigh *et al.* 1984). For instance, 93% of the Avon botanical district (roughly equivalent to the Western Australian wheatbelt, has been cleared. This has left very little natural vegetation except for what remains in a large number of small, scattered remnants (Beard & Sprenger 1984). In the case of the Avon district, development of land for agriculture began relatively slowly and increased in pace only during this century (e.g., Katahira 1985; Saunders *et al.* 1985; Saunders & Curry, this volume) resulting in the rapid isolation of patches of native vegetation. Government support for agricultural development hastened the process and there was often no incentive or apparent reason for retaining areas of native bush.

Clearing affected some vegetation types more than others, the types occupying better soils being utilized first. Thus woodlands are now severely under-represented in the remaining patches of native vegetation while some types which were unsuitable for agriculture, such as rock outcrops, are relatively unaffected. Similarly, forests now occupy approximately half of their original extent nationwide due to clearing for agriculture (Department of Arts Heritage and Environment 1986) and further large areas are modified significantly by timber extraction. Rainforests, in particular, have diminished greatly since settlement. Only about 25% of the original rainforests remain after clearing for agriculture and urban development, and much of the remainder has been logged. Virgin rainforest is thus now fragmentary and some types are almost extinct.

In forestry, the introduction of non-native tree species, especially conifers such as *Pinus radiata* and *P. pinaster,* has resulted in the development of extensive plantations which have quite different ecosystem characteristics to the original vegetation (e.g., Turner & Kelly 1977). The plantations have an effect analagous to agriculture, except that the crop rotations are longer. These forest plantations do, however, provide a valuable softwood resource which should reduce demand for native timbers if managed correctly (Turvey 1982; Department of Arts, Heritage and Environment 1985).

The direct, deliberate effect of land clearance is the removal of the native vegetation and replacement with an intensively managed artificial system. There are also potentially several unintentional impacts which can affect both the cleared land and adjacent areas. Fragmentation of the natural vegetation has led to many changes in overall landscape processes and, in particular, hydrologic and nutrient regimes. Removal of deep-rooted perennial vegetation and replacement with shallow-rooted pasture and crop species has in many areas reduced evapotranspiration, allowed water tables to rise and increased ground- and surface-water flows (Peck 1977; Burch 1986; Walker 1986). This has led to widespread erosion and salinity problems, both of which contribute significantly to land degradation and associated costs (O'Brien 1979; Conacher 1979, 1981; Greig & Devonshire 1981; Bettenay 1986). Wind erosion has also increased due to removal of perennial cover and this not only results in degradation but also alters the overall nutrient regime. A further important influence on the nutrient cycle is the addition of fertilizers to crop and pasture-land. Australian soils are generally rather poor in nutrients and hence have to be fertilized to obtain reasonable crop and pasture growth. Nutrients applied to crops can be transferred elsewhere, either in runoff or by transport of soil during wind erosion. Nutrients destined for crops may also end up elsewhere because of drift during application. Inputs to remnant patches of native vegetation may therefore be substantial and contribute to a slow nutrient enrichment of the natural system. As far as we know, such inputs into fragmented areas have not been quantified but appear to constitute a major problem for the long-term conservation of such areas, especially when coupled with invasion by non-native plant species (e.g., Fox 1982; see later).

Also affected by fragmentation are the population processes of the vegetation components. While the effects of landscape fragmentation may be obvious in the case of faunal movement and behaviour, there are also likely to be effects on the plant communities. Pollination and dispersal, in particular, are likely to be altered as the vegetation changes from a continuous system to one comprising isolated islands surrounded by agricultural land. Movement of pollen and propagules, and hence of genes, is likely to become more restricted, especially where animal vectors are also adversely affected by fragmentation. This may not be a problem for species which occurred in isolated small populations anyway before fragmentation (e.g., Moran & Hopper 1987), but will be a problem for more widespread species, especially obligate outcrossers. Such effects may take a long time to become apparent, particularly for organisms with long generation times.

Utilization of native vegetation

Pastoralism. The first Spanish merino sheep arrived in Sydney in 1797 and by the 1850s there were about 18 million sheep in Australia (Blainey 1980). There are now nearly ten times that number. Sheep and cattle spread rapidly into grassland areas across the continent and now the pastoral industry stretches the entire length of the continent from the relatively rich areas adjacent to the coast to the poorer arid areas of the interior (e.g., Adamson & Fox 1982). European stock differed from the native herbivores in many ways. They were ungulates, required more frequent access to water, and did not possess the marsupials' capacity to regulate reproduction in tune with drought conditions. Sheep and cattle therefore modified the native vegetation both by direct herbivory effects on community composition and structure and through trampling of the soil which led to compaction, alteration of soil structure and disruption of the surface soil crust (e.g., Graetz & Tongway 1986; Noble & Tongway 1986a,b; Williams, this volume; Wilson, this volume). Further modification of vegetation has resulted from human alteration of fire regimes in pastoral lands (e.g., Harrington *et al.* 1979, 1984a; Griffin & Friedel 1984a,b, 1985). The provision of watering points produced localized areas of very heavy stock use, leading to characteristic patterns of denudation and erosion (Lange 1969; Andrew & Lange 1986a,b; Stafford-Smith & Pickup, this volume). Severe erosion resulted from overgrazing and soil denudation by stock during drought periods such as those in the 1890s and early 20th century (Newman & Condon 1969; Roberts 1981; Buckley 1985; Clark, this volume). The main reason for this was that early pastoralists maintained excessively high stocking rates: stock numbers dropped dramatically after drought (see Wilson, this volume). There was, however, a tendency to maintain high stocking rates at the onset of drought in the hope of improvement of conditions (Friedel *et al.*, this volume). This "destocking lag" was a major cause of degradation. Early high stocking rates essentially consumed the capital and may have caused permanent changes that may now be difficult to perceive since in many areas there are few benchmarks left. In other words, new metastable states may have been formed, as indicated in Figure 2.

Problems still arise today where stocking rates are not low enough for drought conditions. Emphasis has been almost entirely on the enhancement of immediate productivity rather than maintenance of the vegetation and soil and hence protection of future production potential (Harrington *et al.* 1984b). More suitable management objectives have been proposed which aim to achieve efficient sustainable production while at the same time retaining future options and enhancing environmental and aesthetic qualities (Fig. 3; Young *et al.* 1984). These objectives could equally be applied to any form of land management in Australia. Management of grazing systems in rangelands is discussed further by Friedel *et al.* (this volume) and Wilson (this volume).

Timber harvesting. Although it does not have such a pronounced effect on the native vegetation as complete removal for agriculture, timber harvesting has nevertheless caused conspicuous changes in Australia's forests. Australia has a large array of unique forest communities, including many sclerophyll and rainforest types. Harvesting of these forests is perhaps one of the most contentious enviromental issues facing land managers today and the one which invokes the largest public reaction (e.g., Routley & Routley 1975). On the face of it, this should not be the case, since sound forestry practice should result in sustainable and ecologically sound production. However, the history of forestry in Australia has often been one of over-exploitation rather than sustainable yield.

Forestry can have varying impacts on the forest ecosystem. For instance, logging operations can significantly alter hydrologic and nutrient regimes (Likens *et al.* 1977; Bormann & Likens 1981; Mauger 1986; Borg *et al.* 1987; Hingston *et al.* 1988; Schofield *et al.* 1988) and can significantly alter long-term stand structure and fauna habitat (e.g., Smith 1985; Kavanagh *et al.* 1985). Recognition of the effect of deforestation on hydrologic regimes has led to the development of management for water catchments for the major cities. For example, Melbourne's water catchments are now fully protected from logging (Department of Arts, Heritage and Environment 1986).

Forestry practice has brought other incidental changes to forest ecosystems. In particular, the

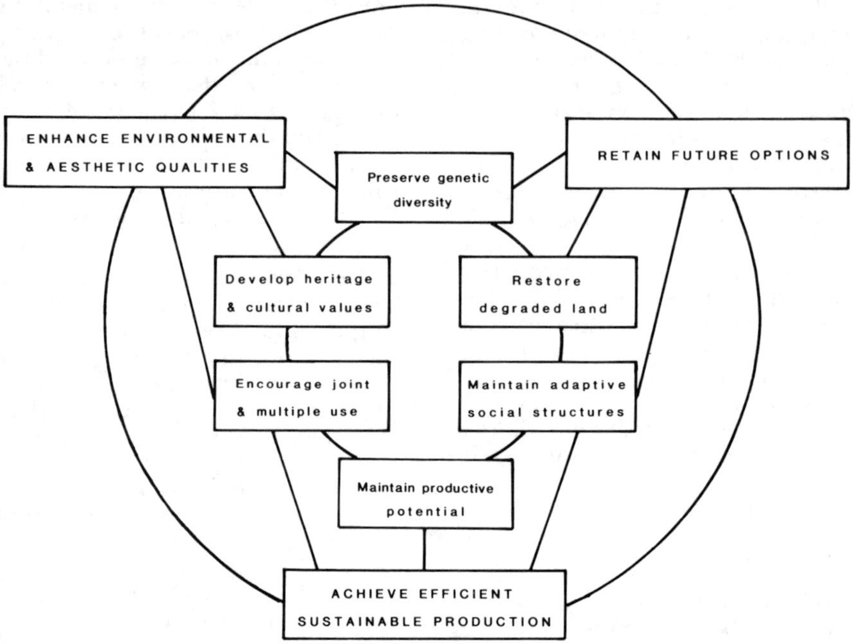

FIG. 3. Management aims for rangeland ecosystems (from Young *et al.* 1984, reproduced with permission of the publisher), which can be extended to all managed ecosystems. Broad aims in the outer circle depend on the more detailed aims in the inner circle being met.

instigation of fire protection policies has led to changes in the frequency and severity of fires (e.g., Clark 1981; Gill 1981a) which have in turn affected the overall dynamics of the forest system (e.g., Adamson & Fox 1982). Settlement often led to the burning of large areas during clearance or the removal of logging trash. Later the European response to fire in forests was to try to exclude it, but this resulted in massive fuel build-ups and the potential for extremely severe fires. Infrequent severe fires have now in many cases been replaced by frequent low-intensity fires (e.g., Underwood *et al.* 1985) reaching an extreme of almost annual buring in the north of the country (e.g., Lewis 1985; Ridpath & Corbett 1985). The imposed regime invariably involves burning at a different time of year for safety reasons. Although Australian vegetation is renowned for its ability to survive fire (e.g., Gill 1981b), there is plenty of evidence which suggests that such extreme alteration in fire regimes can lead to changes in species composition (e.g., Shea *et al.* 1979; Haynes 1985; Fox & Fox 1986a).

Further changes in forest systems have resulted from increases in phytophagous insects and disease organisms such as *Phytophthora cinnamomi*, a fungus which is causing extensive dieback in forest and non-forest systems in Western Australia and elsewhere (e.g., Old *et al.* 1980; Heatwole & Lowman 1986; Weste & Marks 1987). *P. cinnamomi* is spread mainly by current land-use activities, and there is evidence to suggest that management practices render the forest environment more conducive to the development of the fungus (Dell & Malajczuk 1989).

Recreation. In addition to the use of natural vegetation for primary production, a further important impact is that of people who visit natural areas during leisure time. While recreational activities may have less obvious direct effects on vegetation, their impacts may still be significant, and inclusion of recreation as a possible land-use can lead to important land-use conflicts. Increasingly, people seek outdoor recreational facilities, and outdoor activities such as camping, bushwalking, off-road vehicle use, skiing, boating and fishing are becoming more popular. Such activities can often lead to environmental degradation if not managed well, particularly in popular but sensitive regions such as rivers or coastal fringes (e.g., Hails 1982; Pen 1987). Increased recreational use also accentuates the problems of accidental bush fires, pollution and litter, and leads to conflicts with the traditional land uses, particularly forestry and potentially with conservation. Toyne (1974) has argued that "Since one of the main objectives of present-day society appears to be the acquisition of more time for leisure, the provision of adequate recreation and leisure facilities must surely take high priority in environmental planning policy". Pigram (1982) suggests that this has not yet been the case in Australia. Clearly the aim should be for a balanced approach which integrates recreation management with the other major land uses.

Conservation

Management for conservation implies that there should be little or no direct technological intervention by humans which would alter the natural processes occurring in the vegetation and landscape. Where large areas of native vegetation persist this may be possible, but often conservation areas are small and therefore open to the problems caused by landscape fragmentation (discussed above). Hopkins & Saunders (1987) suggest that any area less than 500 000 ha in any part of Australia will require active management. Even in large areas, managers have to make decisions concerning the disturbance regime, which may have been significantly altered since the arrival of Europeans (Hobbs 1987). This is particularly true of the fire regime, which is an important component of most ecosystems in Australia, with the exception of rainforest. Saxon (1984) provides an example of the development of a fire management regime for conservation purposes based on the pre-European (Aboriginal) land management practices.

In addition, grazing regimes may have altered considerably, both because of changes in the number of native herbivores and because of the introduction of non-native animals. Of particular note here are the rabbit, which has had a marked effect on northern Australia (e.g., Johnston 1968; Lange & Graham 1983; Leigh *et al.* 1984; Cooke 1987), and the water buffalo which has had a marked effect on northern Australia (e.g., Braithwaite *et al.* 1984). Goats, feral pigs, donkeys, horses and camels have also affected large areas.

Further problems are caused by the invasion of natural areas by non-native plant species which may alter stand and fire fuel structures, impede regeneration of the native species and lead to loss of conservation value (e.g., Amor & Piggin 1977; Williams 1985; Fox & Fox 1986b; Groves 1986; Hobbs & Atkins 1988).

Restoration and reconstruction of Australian ecosystems

In the context of this symposium it is appropriate to attempt to develop some general guidelines as to how Australian native vegetation might be better managed in the future. It is clear from earlier discussion that much of the landscape has been drastically altered by removal of all or parts of the original vegetation. Considerable areas of Australia can currently be considered as degraded or in danger of degradation in the absence of restorative measures (Department of Arts, Heritage and Environment 1986; Roberts 1987). Hence restoration and reconstruction (*sensu* Allen 1988) are major tasks for the future. In earlier sections we have attempted to identify the important impacts of human activities on the major vegetation components of ecosystems. Here we build on that by proposing practical measures that can be taken to manage sustainably and, where necessary, restore or reconstruct those ecosystems.

The relative ease or difficulty with which a system can be restored is largely a reflection of the degree of disturbance or modification it has experienced (Fig. 4). This is a further indication of the resilience of the system to various disturbances, although here we are looking at managed recovery rather than the system's inbuilt ability to recover unaided. Under this scheme, a system in which fire or grazing regimes have been modified should be relatively easier to restore than one in which erosion, salinity or nutrient enrichment has occurred. This is because, in the first case, only vegetation processes have been altered, but the basic ecosystem and landscape processes have remained intact. Where ecosystem or landscape processes have been disturbed much more effort must be expended in rehabilitating the physical environment before the vegetation processes can be restored. Thus, for example, it may prove relatively simple to regenerate salmon gum, *Eucalyptus salmonophloia,* woodland in the Western Australian wheatbelt by excluding stock and providing a suitable seedbed, but it would be very difficult to re-establish a similar woodland on an area of secondary salinity. Similarly, where ecosystems have been completely destroyed, there will be a spectrum of requirements for the reconstruction of these ecosystems. It should be noted that all human modifications and the requirements for subsequent rehabilitation may be affected by larger scale processes such as hypothesized global climatic change (e.g., Pearman 1988).

Before restoration or reconstruction projects are begun, several major problems have to be considered. Magnuson *et al.* (1980) and Cairns (1986) have outlined the major options available for management of surface-mined lands (Fig. 5), and this scheme can be applied more generally to restoration of other areas subject to other forms of disturbance. Firstly, we must decide on the objectives of restoration. Restoration for improved pastoral production will, for instance, be quite different from attempts to restore a natural ecosystem for conservation purposes. Secondly we must decide whether we want to return the system to a near-natural state or whether we want to produce a synthetic system (see Bridgewater, this volume). Although the synthetic system might not possess all the values of the pre-disturbance system, it may be easier to maintain and still restore many of the necessary processes. Many questions are imbedded

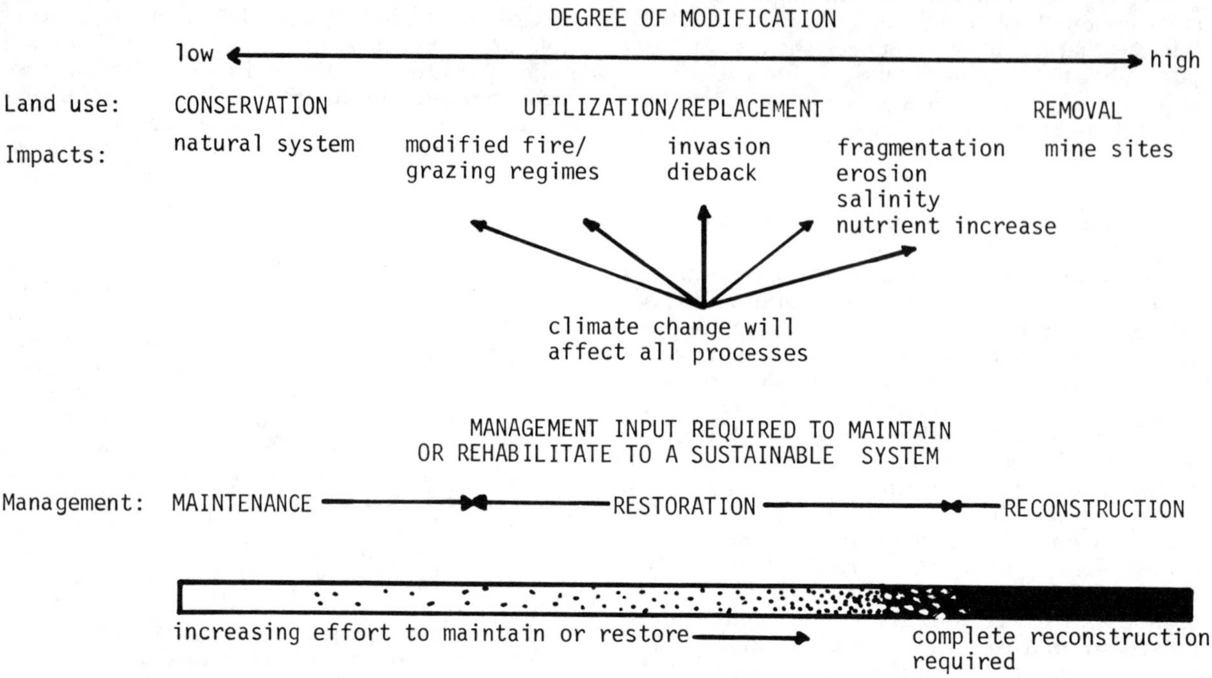

FIG. 4. Degree of modification of natural ecosystems resulting from various human impacts. Management options range from maintenance of relatively unmodified systems, through restoration of degraded systems to reconstruction of completely destroyed systems. Ease of restoration or reconstruction decreases from left to right.

in this problem — for instance, what was the natural system and is complete restoration to the natural state an achievable goal? Finally, there has to be some definable endpoint to the restoration — when can we say it is restored, and how do we go about measuring the degree of restoration? These questions have been examined in more detail by Cairns (1986) and in Jordan *et al.* 1987.

We already have considerable experience of ecosystem rehabilitation or reconstruction after mining operations. After cessation of mining vegetation will rarely establish unaided. Mine sites in Australia and elsewhere which have been abandoned for some time are usually only sparsely revegetated, often because of lack of propagule supply and toxic levels of metals in the soil, acidity or complete absence of nutrients (e.g., Bradshaw & Chadwick 1980; Hobbs & Streit 1986; Tasker & Chadwick 1978; Griffin & Hopkins 1981, 1985). Revegetation thus requires active rehabilitation including the reconstruction of the original environment, or at least its basic attributes (Table 3). This usually involves the replacement of topsoil removed during extraction and may be accompanied by active revegetation. This is particularly important in view of the poor dispersal ability of many components of Australian sclerophyll vegetation (e.g., Bell *et al.*, this volume). Replacement of some form of vegetation can be accomplished in most cases (e.g., Koch & Bell 1983; Nichols 1984; Ward *et al.*, this volume), but this may not bear any resemblance to the vegetation originally on the site. We also have some information on how systems can recover when, for instance grazing pressure is removed (e.g., Hall *et al.* 1964; Rolls 1984), but further information on active rehabilitation techniques is required.

Guidelines for management and/or rehabilitation have most often been specialized and case-oriented, or else theoretical and therefore not easily translated into action. Here we attempt to move from a systematic identification of impacts or "ailments" to a prescription for restorative treatments or "cures", so that the fundamental ecological basis for management action can be readily understood. This is not intended to be a final, definitive prescription for managers; rather it is a first iteration of an approach to be developed further.

In this approach we take the impacts on ecosystem components recognized earlier and identify the prescription for rehabilitation associated with each component (Table 3). The right-hand column of Table 3 thus indicates the management action required for maintenance, rehabilitation or reconstruction of the ecosystem. These are based on the eight ecosystem components listed in table 2. For this analysis we have distinguished two major vegetation types in Australia; rainforest and

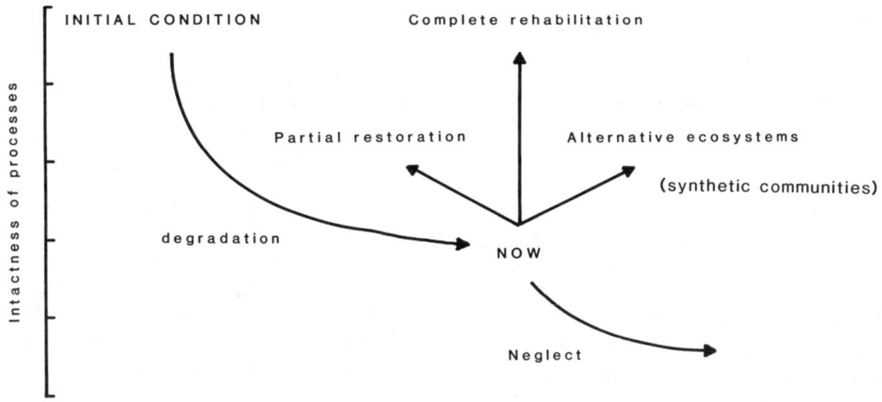

FIG. 5. Options available for the management of degraded or extensively altered ecosystems, indicating the degree of intactness of natural processes in the unaltered and treated systems (adapted from Cairns 1986, with permission from the publisher).

sclerophyllous vegetation. This distinction is made on the basis of characteristics that contribute to natural vegetation development. Rainforest tends to respond to disturbance in a way that corresponds to classical succession, i.e., the relay floristics model (Hopkins *et al.* 1976). In contrast, sclerophyllous vegetation types (and other minor types such as alpine grassland) regenerate in a way that more closely accords with the Initial Floristic Composition model of Egler (1954). We will deal predominantly with these latter types in the following discussion.

In the case of areas that have been subjected to land-uses 1 and 2 (removal or replacement of native vegetation) we consider two propositions. Either the land can be reconstructed to the pre-existing plant communities and ecosystems or else reconstruction can be directed towards the development of a self-sustaining synthetic ecosystem which has some natural-land attributes. As an example, it would be possible to rehabilitate an open-cut minesite by planting a carefully chosen mix of trees and shrubs that, while not native to the area, can still provide food and habitat for the local avifauna. In a like manner, selected areas of the rural landscape could be planted to enhance landscape values while at the same time providing a cash crop, land protection and some fauna habitat.

In the case of land in category 3 (Utilization) the option given is either to manage it better on a sustainable basis, or to restore it to the original native vegetation. In the case of land for conservation (land-use category 4) rehabilitation should not be necessary if it is properly managed, but effective process-based management is the key.

As an example of the approach we can consider land-use category 1, which according to Table 3 leads to complete removal of all living plant material including the soil and on-plant seed stores. Other interacting organisms such as soil mycorrhizae are also depleted. Soil properties, nutrient cycles and hydrology are drastically modified. There will be little or no recruitment of plants. Reconstruction will thus involve artificial reintroduction of propagules (or whole plants) and other organisms such as bacteria, mycorrhizae and soil fauna, once adverse soil properties have been ameliorated. Reintroduction may be a continuing need in order to maintain population processes and enhance community reconstruction.

In contrast, land-use 3 may lead to selective removal of above-ground biomass and *in situ* destruction of other parts. On-plant seed supply may be diminished and replenishment of the soil seed bank may decline. Interacting organisms may also decline and soil crusts may be fractured with resultant soil loss. Patterns of runoff and infiltration may also change. Nutrients will be removed in harvested biomass and be redistributed in dung. Plant recruitment may be affected by grazing and by competition from alien plants. The community will become metastable. Indirect impacts will include introduction of alien plants and animals (including pathogens), changed fire regimes and nutrient inputs. In this case, the management option most likely to be aimed at is sustainable utilization. This depends on maintaining viable populations of key species of plants and other organisms. Reintroductions of plants and other biota may be necessary. Vegetation cover and soil crusts need to be maintained to prevent erosion; this involves careful regulation of the fire and grazing regimes.

In addition to direct impacts, incidental and indirect impacts also have to be treated. It is becoming increasingly apparent that landscape elements are not independent of each other and that the major aim should be better management of the landscape as a whole. This includes the retention of existing vegetation, revegetation of important areas and sustainable production which minimizes soil loss. The possible importance of the effects of future global changes in climate on land-use patterns must also be recognized in this context.

Several points must be noted in the process of developing prescriptions for management and reconstruction.

1. Dominant features of most Australian sclerophyllous plant communities are the woody plant families Myrtaceae, Proteaceae and Mimosaceae/Fabaceae. A characteristic of these families is the very limited dispersal of seed. Many of the disturbance types that we have identified lead to depletion of the *in situ* seed reserves. Any proposals for reconstruction of disturbed plant communities must, therefore, depend heavily on the reintroduction of seed or other propagules. Related to this point is the role of animals in the long-distance dispersal of the Australian flora, a role which has been little studied to date.
2. A further characteristic of the Australian landscape is the very poor skeletal soil. The flora has in many instances developed synergistic relationships with soil biota to cope with this soil poverty (e.g., Lamont 1982, 1984). Successful rehabilitation may well require artificial reintroduction of the soil biota to facilitate plant growth. This can be effected by direct innoculation or by the reuse of top-soil.
3. In many of our native plant communities, recruitment events are often episodic and related to special combinations of climatic factors (e.g., Friedel *et al.*, this volume). Any management which is directed at sustainable utilization should take this into account. In particular, any modification to fire or grazing regimes could have major impacts through their long-term effects on recruitment. For instance grazing pressures may be increased in the semi-arid zone in response to high rainfall seasons which also promote plant recruitment. It also means that long-term management should be conservative because outcomes are contingent on largely unpredictable events.
4. Most forms of utilization of living resources are based on some concept of sustainability. Where resources are harvested, for example, the harvest rate is notionally related to the ability of the system to maintain the supply. Harvesting usually involves increasing the death rate, either through adult mortality (e.g., in timber harvesting) or in some cases through juvenile mortality where seedlings are grazed. Harvesting is thus based on population dynamics, but sustainability is rarely measured in these terms. Rather, surrogate measures are used. For instance, in a grazing system total cover is used while in forestry cellulose increment is used. These measures have not been calibrated adequately and may give misleading results concerning long-term sustainability.

Finally, it should be noted that rehabilitation and reconstruction can be costly and time consuming and are no substitutes for proper planning and land management. A large proportion of environmental problems arise as a consequence of inadequate land-use planning in the first instance. Good land-use planning is based on an adequate knowledge of the biophysical resource base and its capacity to support the various possible land-uses (land capability assessment). Inappropriate allocation of land to particular land uses leads to major costs associated with degradation and associated rehabilitation and reconstruction. What we require is a proactive approach associated with a better empathy for the land, a factor which we consider in the next section.

Attitudes towards vegetation

It is clear from the preceding discussions that the Australian landscape has undergone rapid and often deleterious modification following European settlement. It is instructive to consider how this has come about through an examination of the attitudes towards the land which prevailed during settlement and those which are current today.

Attitudes of the early settlers

The Europeans (initially British and Irish and later other nationalities) arriving in Australia had left landscapes consisting of agricultural systems, woods that were grazed and cut over for timber, and moors that were used for extensive grazing. The vegetation of Europe had developed with humans as a dominant influence, and most "natural" ecosystems were actually heavily altered by human activity. Natural forest had largely been removed from Britain and much of Northern Europe during the Bronze Age, to make way for open grazings and possibly to remove the danger of wild animals lurking in the forests (e.g., Pennington 1969; Gimingham 1972). Heathlands in Northern Europe, gentian-filled meadows in the Alps and pine and oak forests and olive groves in the Mediterranean countries were the types of "natural" vegetation that the European immigrants were used to. Coming from this background, the early settlers faced a foreign landscape, one that was not only unfamiliar but also apparently untamed and full of unfamiliar plants. Natural vegetation which had not been modified in some familiar way was probably rather terrifying, and the new settlers certainly had no great empathy with the bushland which confronted them. The initial task was to recreate a European landscape in this foreign land and to beat the natural vegetation into submission and make something useful out of it in order to survive; either clear it for crops, find areas suitable for grazing sheep, or find good timbers in the forests for building dwellings and for making ships.

With these European attitudes came the associated expectations, technology and knowhow. European crops and agricultural methods were transferred directly onto Australian soils. European livestock was put out to graze directly onto Australian grasslands. An entire way of life was thus transposed to a continent with a totally different biogeographic history which possessed dramatically different soils and climate and had never been trodden by an ungulate hoof. In retrospect it is possible to see that such activities would lead to large-scale problems, but this was not necessarily self-evident at the time. Indeed, perception of the new land was strongly coloured by conditions on first arrival. Settlement and agricultural expansion invariably coincided with years of good rains and hence good crop and pasture growth. Early settlers were not to know that droughts would later parch the land or that floods would wash away their soil. The people who introduced rabbits as game probably did not even stop to think about the possible implications and certainly could have had no idea of the degree of devastation their actions would ultimately cause.

Current attitudes

It is not helpful to condemn the people who first developed the land during the 19th century, but we can deplore the fact that the lessons of history have not been learned. Much current land development and utilization follows trends scarcely different from those of the past. Vegetation clearing and establishment of agriculture still progresses in good years, despite our knowledge that cyclical droughts are a fact of life and that most years are "below average" (Friedel *et al.*, this volume). Native bush is still cleared despite the knowledge that increased land degradatation is an almost inevitable result. For instance, 54% of the land cleared for agriculture in SW Western Australia was cleared between 1945 and 1982 (Saunders *et al.* 1985). Pastoral areas are still stocked to the carrying capacity of the vegetation in wet years, despite past experience of massive stock deaths and severe land degradation during drought (e.g., Friedel *et al.*, this volume). Non-native species are still becoming naturalized at an alarming rate (e.g., Specht 1981; Groves 1986).

To the first settlers Australia must have seemed like an endless frontier; there was little apparent need to conserve anything, be it timber, soil or pasture. We now know that Australia's natural resources are not inexhaustable, but we are still mining the soil and vegetation at an alarming rate.

It may seem, therefore, that attitudes towards vegetation and landscape have changed little in 200 years. The main change has been an intensification of land-use associated with the advent of increasingly sophisticated technology; this has not, however, been accompanied by an increase in sophistication of land-use. Part of the reason for this lies in the failure of the developing Australian nation to come to terms with the Australian environment. At the end of the last century, "Thousands of pioneers still hoped to plant England, its hedges and its fruit trees and green and pleasant fields, to the exclusion of almost everything Australian" (Blainey 1980). Blainey concluded that most Australians were still strangers in a new land. Seddon (1976) suggests that, even now, we are still largely aliens in an alien land. This has important consequences in a number of ways. Firstly, it has taken Australia a long time to develop a national identity and for Australians to take pride in things Australian. Historically, there was a small group of landscape artists (the Heidelberg school) and historians who recognized the unique character of the Australian landscape. However, this was largely an intellectual movement which scarcely affected the population as a whole. Only recently have Australians started looking at their land and considering it as a heritage which might be worth preserving in its own right. Only when people value what they've got is there a chance of keeping it. Nevertheless, even in more environmentally-aware times such as these, we have had difficulty preserving the last vestiges of Australian rainforest, and many additions to the conservation estate are still the result of a lengthy struggle.

Remaining as aliens in an alien land has also inhibited us from fully understanding the Australian environment. Only by gaining that understanding can we hope to maintain natural systems and to work within the constraints imposed by climate and soil. We are now seeing the consequences of early failure to work within these constraints; problems of degradation, loss of productive potential, soil erosion and salinity abound. As well as requiring new methods of land-use which will prevent future degradation (and repair past damage), we also require a shift in attitudes towards the land and vegetation. The development of attitudes towards vegetation can be viewed as a sequence of phases. The initial phase is an almost imperialistic attempt at complete dominance and subjugation (e.g., Passmore 1974; Taylor, this volume). Later, more tolerance is displayed and this finally leads to a symbiosis and a certain degree of interdependence. Australia has seen the initial phase of dominance, when the native vegetation was seen as an endless frontier to be tamed. This represents an entirely utilitarian view where the main consideration was maximization of short-term gain with no thought for longer-term effects (see also Roberts 1986, 1987). Now that much of the natural vegetation has gone from many areas, its value as a scarse resource is only just beginning to be realized. The shift to the idea of interdependence with the land and vegetation has, however, been slow to develop.

Such a shift is beginning to appear in some areas. As an example we can consider agriculture. Early farmers, and even those establishing farms as late as the 1960s, saw as the major requirement the extensive modification of the landscape to make it productive in European terms. Land has traditionally been regarded by European society as an indisputable object for ownership and use, and the landowner was free to use the land as he wished. Thus even into the last few decades many farmers continued to clear in the face of knowledge and advice as to its consequences. There is now, however, an encouraging trend of widespread acknowledgement of the problems and a willingness to take action. The Farm Tree Program in Victoria has been underway for some years now and in the past five years the Greening Australia programme has led to the revegetation of significant parts of the rural landscape. In addition, there has been widespread acceptance of the idea of community-based soil conservation advisory groups. This can be built on to begin reconstructing the landscape in a more comprehensive way with the promotion of revegetation, retention of remnant bush and better land management in general. There has thus been a gradual transition from the initial attitude of clearing everything except bits that were useless anyway to a positive desire to improve the landscape by planting trees and eventually, perhaps, by reconstructing natural communities. There is a gradual development of understanding of the limits to our use of the land and the emergence of a new conservation ethic. This ethic is centred around the idea of stewardship rather than dominance. We are thus slowly coming full circle to appreciate the concepts central to the relationship between the land and the original inhabitants of Australia, the Aborigines.

Wilderness

Further attitudinal shifts are apparent in the emerging concept of the Australian wilderness. Wilderness (literally "place of wild beasts") to early agriculturalists in Europe was a concept associated with untamed wasteland which harboured danger (Nash 1973). Wilderness was also, by definition, unoccupied and therefore unowned and available to the first takers (Wright 1980); hence the uncultivated wastelands of Australia were available to the settlers, despite their prior occupancy by the Aborigines. The concept of wilderness has now changed from meaning simply something which has not yet been brought under human dominance to something which has value in its own right. It now conveys ideas of wildness, inaccessibility and freedom from technological impacts. The subject of wilderness has been much discussed in the Northern Hemisphere (e.g. Dasmann 1968; Nash 1973; Stegner 1976). In the United States, there are many areas designated as "Wilderness" and there are strict guidelines for what can and cannot be considered as wilderness (Dasmann 1968). The United States has also seen the development of a wilderness philosophy, through the writings and activities of Thoreau, Emerson, Muir and Leopold (e.g., Leopold 1949; Nash 1973; Clarke 1979).

Such natural philosophers seem to have been missing in Australia and indeed it is only recently that the idea of preserving wilderness has caught on here (e.g., Wright 1980; Taylor, this volume). The reason for this may simply be that, for all of Australia's history since European settlement, wilderness has never been in short supply. In 1860 the printed map in the best guidebook to Australia portrayed the centre as completely blank. A vast oval region occupying over half of Australia became known colloquially as the Never Never, which is still being fully discovered (Blainey 1980). There was never any sense of scarcity of land, and there are still areas of Australia which are virtually inaccessible. In addition, Australia is still relatively sparsely settled for its size. The outback is in many ways the Australian version of wilderness. In the United States, however, settled for almost 200 years before Australia was colonized, population pressures have meant that wilderness areas have shrunk and are in danger of being over-used and of losing their wilderness charactieristics (e.g., Nash 1973).

The United States has already had many environmental battles such as those over Hetch Hetchy in California and the Glen Canyon and Grand Canyon in Arizona, and it now has a well established national park system with a high public profile. Australia has national parks, but the utility of these national parks and other wilderness areas is not greatly appreciated by politicians or the public as a whole. Although Australia may be seen as having wilderness to spare, this may not remain so, especially as remote areas are opened up for tourism. Kirkpatrick and Haney (1980) and Smith (1982) have discussed the loss of wilderness in eastern Australia, where population and industry pressures are highest. Taylor (this volume) presents proposals for a National Wilderness Inventory. The acceptance of the importance of wilderness requires a public sympathetic to the concept and to conservation in general. Taylor (this volume) has noted that there are still many inconsistences in the general notions of wilderness currently held in Australia. Nevertheless, the development of an appreciation of the Australian wilderness is symptomatic of the emergence of an Australian national identity and an empathy with the Australian landscape.

We are therefore seeing the awakening of interest in the land and how it operates amongst both the rural and urban communities. The further development of this is a challenge to the present ecological community. It goes beyond the mere development of a land or conservation ethic, but rather involves the need to develop an Australian national identity based on a mature appraisal of our history and an empathy for the land both in its near-natural state and in cultural landscapes. This national identity represents a "love of country" which includes the land, its people and an appreciation of the changes to the landscape over the past two centuries.

Future directions for research and management

In this chapter we have outlined the need to develop methods for maintenance, restoration and reconstruction of Australia's ecosystems which deal not only with vegetation assemblages but with the processes which support those assemblages. It is a matter of concern that many of the land-use practices which have in the past led to land degradation continue at present in much the same way. We can identify three major problem areas.

1. A major problem which we face is an incomplete inventory of basic resources and an incomplete understanding of the natural processes involved, far less an understanding of the human impacts on these. Ecological knowledge is lacking for many Australian systems. Without some basic information we can make only informed guesses about what may result from particular management actions, rather then being able to make useful predictions. We therefore need to establish priorities for research which target basic ecological processes and the potential human impacts on these. At the same time, we cannot wait for adequate information before making decisions or else we may be left with nothing to manage.
2. Land-use planning and management procedures are not well enough developed and are not sufficiently adaptive to cope with environmental uncertainties (e.g., Hopkins & Saunders 1987).
3. Attitudes towards land are still predominantly utilitarian; the changes we have discussed earlier are occurring, but not rapidly enough.

We suggest the following approaches to these problems:

1. The development of a comprehensive inventory of resources, involving resource mapping at the appropriate scale (1:1 million or better) and integration of data through the use of Geographic Information System (GIS) technology. This has to be supported by sound taxonomic research.
2. The development of a programme of long-term ecological studies. These are particularly important to the understanding of processes operating in an environment where unpredictable events are major driving forces (e.g., Graetz *et al.* 1988; Hobbs & Mooney 1989; see also Weatherhead 1986). Sound interpretation of short-term studies relies heavily on results from long-term studies.
3. Integration of research and management. Virtually all of Australia's vegetation is now managed in some way, whether this management is for production or for conservation. Even conservation management can rarely be completely passive. Some action is required to ensure that community and ecosystem processes are not disrupted by external influencess. Research and management should be able to proceed co-operatively since all management operations, if properly implemented, represent experimental manipulations. Much information can be gained from a synergistic relationship between research and management (Hopkins & Saunders 1987; Hobbs 1988).
4. There is great potential for learning from disturbed systems rather than focussing only on near-natural ones. The extensive human disturbance of much of Australia's vegetation which has already taken place can be used as huge experiments which provide information on its resilience and response to various levels of disturbance. Indeed, no researcher would be able to contemplate such large-scale and disruptive experiments for their own sake. We need to develop a "disturbance ecology" of Australian vegetation which allows us to predict the effects of each land-use category on population, ecosystem and landscape processes. We have presented an initial approach to this here, but further development is required.
5. A new approach to vegetation, which emphasizes function rather than composition. While basic taxonomy is essential for inventory, we cannot progress very fast in the development of a predictive ecology if we are constantly bogged down in species lists. Simplification could take the form of categorizing plants into functional types, based on their physiological characteristics and responses to various forms of disturbance. Such ideas have been partially explored by Gill (1981b), Noble and Slayter (1981) and Grubb and Hopkins (1986). The idea can be extended to consider functional vegetation types (Gillison 1981; Hopkins 1981). This is particularly important if, for example, we are to develop adequate methods for predicting the outcome of possible climatic changes superimposed on land-use changes.

6. In many ways, land-use planning and management planning are in their infancy in Australia. Although hampered by a lack of information, there are many instances where the planning process cannot handle existing information and make proper use of it. The whole management process is rather inflexible and inappropriate for dealing with natural systems where there is a high degree of uncertainty. There is therefore a need to incorporate adequate monitoring, appropriate feedback loops and review/re-evaluation procedures in the management process. This requirement runs parallel with the need to adopt an integrated management and research approach as outlined in point 2 (Hopkins & Saunders 1987).
7. There is an increasing need for ecologists to become actively involved in the decision-making process and in the promulgation of a more responsible attitude towards the land and in public debate on land management problems (e.g., McKeller 1987). This includes government scientists who should be able to comment publicly on land management issues.

Conclusions

The first part of this chapter outlined the changes brought about in Australian vegetation by the settlement and subsequent spread of Europeans, the management problems which have resulted and the approaches required for solving these problems. The second section dealt with the way in which Australian vegetation and landscapes have been viewed. The two are very closely linked; most of the current management problems with which we are now faced have their basis in the prevailing utilitarian view of vegetation and natural resources.

It is a major challenge to Australian ecologists and land managers to find management solutions to the problems involved in the maintenance of sustainable systems, whether for utilization or conservation. Beyond that is the further challenge to find ways to restore and reconstruct degraded ecosystems.

It may be too late in many instances to fight anything but a rearguard action to slow the rate of degradation. Solutions to the technical problems involved with this must, however, be matched by a change in attitudes of land managers, politicians and the public at large. Roberts (1984) has argued that the strongly developed materialism of the frontier society which developed in Australia is the greatest barrier to achieving sustainable land-use practices. We need a rapid transition from traditional utilitarian concepts of land management to a more symbiotic relationship with the land. This is not an idealistic viewpoint but one based on the reality that we depend directly on the continued functioning of all ecological processes at the population, ecosystem and landscape levels.

Some progress in this direction has been made. Areas previously designated as production forest are now being set aside for conservation (e.g., Anon 1988). Conservation movements have spread (e.g., Dunn 1979) and have developed considerable influence in some areas but the issues tackled by them are only a small subset of the land management problems which have to be faced. Conservation has to develop from being an issue to being a way of life before any large-scale success is likely. Conservationists standing in front of bulldozers may save the odd bit of forest from felling but will not stop salinity and erosion problems. Many of the problems to be faced are long-term; conservationists and politicians have a habit of responding to whatever is most urgent at the time and it is often difficult to tackle diffuse environmental problems by direct action. Hence some of the most serious problems are ignored. Ecologists cannot afford to treat these problems as purely academic, but instead have to take an active part in publicising them. Only once the population as a whole recognizes that there are problems will solutions be found. We have outlined some important developments in this direction and several other examples can be found throughout this volume.

Practical solutions have to be economically and socially viable as well as having a good biological and environmental basis. Land managers will be much more likely to take conservation or restorative measures if they see a definite return from their actions. Solutions will also be found only in a suitable political climate, where short-term expediency is replaced by longer-term strategic planning. This is true for all areas of land management from pure nature conservation through forestry, agriculture and mining. If Australian vegetation and landscapes are to have a future, we should be planning now for the next 200 years and beyond.

Acknowledgements

We thank A. A. Burbidge, M. Friedel, S. Hopper, D. Saunders, D. Spratt and two referees for their comments on the draft and Mike O'Brien for assistance with drafting the tables.

References

Adamson D. A. & Fox M. D. (1982) Change in Australasian vegetation since European settlement. In: *A History of Australasian Vegetation* (ed. J. M. B. Smith) pp. 109-46. McGraw-Hill, Sydney.

Allen E. B. (ed.) (1988) *The Reconstruction of Disturbed Arid Ecosystems – an Ecological Approach.* Westview, New York.

Amor R. L. & Piggin C. M. (1977) Factors influencing the establishment and success of exotic plants in Australia. *Proc. Ecol. Soc. Aust.* **10,** 15-26.

Anonymous (1988) *Forests for Tomorrow. A summary of the Management Plan for the Northern, Central and Southern Region and the Timber, Recreation and Conservation Strategies.* Department of Conservation and Land Management, Como, W.A.

Andrew M. H. & Lange R. T. (1986a) Development of a new piosphere in arid chenopod shrubland grazed by sheep. 1. Changes to the soil surface. *Aust. J. Ecol.* **11,** 395-410.

Andrew M. H. & Lange R. T. (1986b) Development of a new piosphere in arid chenopod shrubland grazed by sheep. 2. Changes to the vegetation. *Aust. J. Ecol.* **11,** 411-24.

Aschmann H. (1973) Man's impact on several regions with Mediterranean climates. In: *Mediterranean Type Ecosystems, Origin and Structure* (eds F. di Castri & H. A. Mooney) pp. 363-71. Springer, New York.

Beard J. S. & Sprenger B. S. (1984) Geographical data from the vegetation survey of Western Australia. Vegetation Survey of Western Australia, Occasional Paper 2. Vegmap Publications, Perth.

Bettenay E. (1986) Salt affected soil in Australia. *Reclam. Reveg. Res.* **5:** 167-79.

Blainey G. (1980) *A Land Half Won.* MacMillan, Melbourne.

Borg H., King P. D. & Loh I. C. (1987) Stream and groundwater response to logging and subsequent regeneration in the southern forest of Western Australia. Interim results from paired catchment studies. Surface Water Branch, Water Authority of Western Australia, Rep. WH 34.

Bormann F. H. & Likens G. E. (1981) *Pattern and Process in a Forested Ecosystem.* Springer, New York.

Bowler J. M. (1976) Recent developments in reconstructing late Quaternary environments in Australia. In: *The Origins of the Australians* (eds R. L. Kirk and A. G. Thorne) pp. 55-7. Australian Institute of Aboriginal Studies, Canberra.

Bradshaw A. D. & Chadwick M. J. (1980) *The Restoration of Land.* Blackwell, Oxford.

Braithwaite R. W., Dudzinski M. L., Ridpath M. G. & Parker B. S. (1984) The impact of water buffalo on the monsoon forest ecosystem in Kakadu National Park. *Aust. J. Ecol.* **9,** 309-22.

Buckley R. (1985) Development in the central Australian arid zone: current assessment. *J. Arid Env.* **8:** 173-89.

Burch G. J. (1986) Land clearance and vegetation disturbance. In: *Australian Soils: the Human Impact* (eds J. S. Russell & R. F. Isbell) pp. 159-84. University of Queensland Press, St. Lucia.

Cairns J. (1986) Restoration, reclamation, and regeneration of degraded or destroyed ecosystems. In: *Conservation Biology. The Science of Scarcity and Diversity* (ed. M. E. Soulé) pp. 465-84. Sinauer, Sunderland, Massachusetts.

Clark R. L. (1981) The prehistory of bushfires. In: *Bushfires, their Effect on the Australian Landscape* (ed. P. Stanbury) pp. 61-73. University of Sydney, Sydney.

Clarke J. M. (1979) *The Life and Adventures of John Muir.* The Word Shop, San Diego.

Conacher A. (1979) Salinity problems in south-western Australia. *Search* **19,** 162-4.

Conacher A. (1981) Dryland agriculture and secondary salinity. In: *Man and the Australian Environment* (eds W. Hanley & M. Cooper) pp. 112-25. McGraw-Hill, Sydney.

Conybeare C. E. B. (1972) Petroleum, coal and uranium: A case study in production, technology and consumption. In: *The Natural Resources of Australia: Prospects and Problems for Development* (ed. J. A. Sinden) pp. 143-62. Angus & Robertson, Sydney.

Cooke B. D. (1987) The effects of rabbit grazing on regeneration of sheoaks, *Allocasuarina verticellata* and saltwater ti-trees, *Melaleuca halmaturorum,* in the Coorong National Park, South Australia. *Aust. J. Ecol.* **13,** 11-20.

Dasmann R. F. (1968) *A Different Kind of Country.* Collier-MacMillan, London.

Dell B. & Malajczuk N. (1989) Jarrah dieback — a disease caused by *Phytophthora cinnamomi.* In: *The Jarrah forest. A complex Mediterranean Ecosystem* (eds B. Dell, J. J. Havel & N. K Malajczuk) pp. 67-87. Kluwer, Dordrecht.

Department of Arts, Heritage and Environment (1985) *Think Trees Grow Trees.* Australian Government Publishing Service, Canberra.

Department of Arts, Heritage and Environment (1986) *State of the Environment in Australia.* Australian Government Publishing Service, Canberra.

Dunn I. (1979) Ecology and environmental responsibility. In:. *A Natural Legacy: Ecology in Australia* (eds H. F. Recher, D. Lunney & I. Dunn) pp. 230-40. Pergamon, Sydney

Egler F. E. (1954) Vegetation science concepts: 1. Initial floristic composition — a factor in old-field vegetation develoment. *Vegetatio* **4,** 412-7.

Forman R. T. T. & Godron, M. (1986) *Landscape Ecology.* Wiley, New York.

Fox M. D. (1982) Vegetation changes in Mediterranean Australia since European settlement. In: *Dynamics and Management of Mediterranean-type Ecosystems* (eds C. E. Conrad & W. C. Oechel) pp. 122-7. Gen. Tech. Rep. PSW-58, Pacific Southwest Forest and Range Expt. Station, Berkeley.

Fox M. D. & Fox B. J. (1986a) The effect of fire frequency on the structure and floristic composition of a woodland understorey. *Aust. J. Ecol.* **11,** 77-85.

Fox M. D. & Fox B. J. (1986b) The susceptibility of natural communities to invasion. In: *Ecology of Biological Invasions: an Australian Perspective* (eds R. H. Groves & J. J. Burdon) pp. 57-66. Australian Academy of Science, Canberra.

Gill A. M. (1981a) Post-settlement fire history in Victorian landscapes. In: *Fire and the Australian Biota* (eds A. M. Gill, R. H. Groves & I. R. Noble) pp. 77-98. Australian Academy of Science, Canberra.

Gill A. M. (1981b) Adaptive responses of Australian vascular plant species to fires. In: *Fire and the Australian Biota* (eds A. M. Gill, R. H. Groves & I. R. Noble) pp. 243-72. Australian Academy of Science, Canberra.

Gillison A. N. (1981) Towards a functional vegetation classification. In: *Vegetation Classification in Australia* (eds A. N. Gillison & D. J. Anderson) pp. 30-41. Australian National University Press, Canberra.

Gimingham C. H. (1972) *Ecology of Heathlands.* Chapman and Hall, London.

Godron M. & Forman R. T. T. (1983) Landscape modification and changing ecological characteristics. In: *Disturbance and Ecosystems, Components of Response* (eds H. A. Mooney & M. Godron) pp. 12-28. Springer, New York.

Godron M., Guillerm, J. L., Poisonet J., Poissonet P., Thiault M. & Trabaud L. (1981) Dynamics and management of vegetation. In: *Mediterranean-type Shrublands* (eds F. di Castri, D. W. Goodall & R. L. Specht) pp. 317-44. Elsevier, Amsterdam.

Graetz R. D. & Tongway D. J. (1986) Influence of grazing management on vegetation, soil structure and nutrient distribution and the infiltration of applied rainfall in a semi-arid chenopod shrubland. *Aust. J. Ecol.* **11,** 347-60.

Graetz R. D., Walker B. H. & Walker P. A. (1988) The consequences of climatic change for seventy percent of Australia. In: *Greenhouse. Planning for Climatic Change* (ed. G. I. Pearman) pp. 399-420. CSIRO, Melbourne.

Greig P. J. & Devonshire P. G. (1981) Tree removals and saline seepage in Victorian catchments: some hydrologic and economic results. *Aust. J. Agric. Econ.* **25,** 134-48.

Griffin E. A. & Hopkins A. J. M. (1981) The short-term effects of brush harvesting on the kwongan vegetation south of Eneabba. *Dept. Fish Wildl. West. Aust. Rept.* **45,** 1-58.

Griffin E. A. & Hopkins A. J. M. (1985) The flora and vegetation of the Crown Land south of Eneabba. In: *Proceedings of a Seminar on the Plant Ecology of the Eneabba heathlands* (eds B. Lamont & B. Low) pp. 3-11. *WAIT School of Biology Bull.* **10,** West Australian Institute of Technology, Bentley.

Griffin G. F. & Friedel M. H. (1984a) Effects of fire on central Australian rangelands I. Fire and fuel characteristics and changes in herbage and nutrients. *Aust. J. Ecol.* **2:** 381-93.

Griffin G. F. & Friedel M. H. (1984b) Effects of fire on central Australian rangelands II. Changes in tree and shrub populations. *Aust J. Ecol.* **2,** 395-403.

Griffin G. F. & Friedel M. H. (1985) Discontinuous change in central Australia: some implications of major ecological events for land management. *J. Arid Env.* **2,** 63-80.

Groves R. H. (1986) Invasion of mediterranean ecosystems by weeds. In: *Resilience in Mediterranean-type Ecosystems* (eds B. Dell, A. J. M. Hopkins & B. B. Lamont) pp. 21-38. Junk, Dordrecht.

Grubb P. J. & Hopkins A. J. M. (1986) Resilience at the level of the plant community. In *Resilience in mediterranean-type Ecosystems* (eds B. Dell, A. J. M. Hopkins & B. B. Lamont) pp. 21-38. Junk, Dordrecht.

Hails J. (1982) The degradation of Australia's coastal environment: a review of competition, conflicts and compromises. In: *Man and the Australian Environment* (eds W. Hanley & M. Cooper) pp. 126- 54. McGraw-Hill, Sydney.

Hall E. A. A., Specht R. L. & Eardley C. M. (1964) Regeneration of the vegetation on Koonamore Vegetation Reserve, 1926-1962. *Aust. J. Bot.* **12,** 205-64.

Hallam S. J. (1979) *Fire and Hearth: a Study of Aboriginal Usage and European Usurpation in South-western Australia.* Australian Institute of Aboriginal Studies, Canberra.

Harrington G. N., Friedel M. H., Hodgkinson K. C. & Noble J. C. (1984a) Vegetation ecology and management. In: *Management of Australia's Rangelands* (eds G. N. Harrington, A. D. Wilson & M. D. Young) pp. 41-61. CSIRO, Melbourne.

Harrington G. N., Oxley R. E. & Tongway D. J. (1979) The effects of European settlement and domestic livestock on the biological system in poplar box *(Eucalyptus populnea)* lands. *Aust. Rangel. J.* **1,** 271-9.

Harrington G. N., Wilson A. D. & Young M. D. (1984b) Management of rangeland ecosystems. In: *Mangement of Australia's Rangelands* (eds G. N. Harrington, A. D. Wilson & M. D. Young) pp. 3-13. CSIRO, Melbourne.

Haynes C. D. (1985) The pattern and ecology of *munwag:* traditional Aboriginal fire regimes in north-central Arnhemland. *Proc. Ecol. Soc. Aust.* **13,** 203-14.

Heatwole H. & Lowman M. (1986) *Dieback: Death of an Australian Landscape.* Reed, Frenchs Forest, New South Wales.

Hingston F. J., O'Connell A. M. & Grove T. S. (1988) Nutrient cycling in jarrah forest. In: *The Jarrah Forest. A Complex Mediterranean Ecosystem.* (Eds B. Dell, J. J. Havel & N. Malajczuk) pp. 155-77. Kluwer, Dordrecht.

Hobbs R. J. (1987) Disturbance regimes in remnants of natural vegetation. In: *Nature Conservation: the Role of Remnants of Native Vegetation* (eds D. A. Saunders, G. W. Arnold, A. A. Burbidge & A. J. M. Hopkins) pp. 233-40. Surrey Beatty & Sons, Chipping Norton.

Hobbs R. J. (1988) What is ecological theory and is it of any use to managers? In: *Ecological Theory and Biological Management of Ecosystems* (eds D. A. Saunders & A. A. Burbidge) pp. 15-27. W.A. Dept. Conservation & Land Management Occ. Paper 1/88.

Hobbs R. J. & Atkins L. (1988) Effect of disturbance and nutrient addition on native and introduced annuals in plant communities in the Western Australian wheatbelt. *Aust. J. Ecol.* **13,** 171-9.

Hobbs R. J. & Mooney H. A. (1989) Effects of episodic events on Mediterranean-climate ecosystems. In: *Timescales of Water Stress Response of Mediterranean Biota* (eds F. di Castri, C. Floret, S. Rambal & J. Roy). In press.

Hobbs R. J. & Streit B. (1986) Heavy metal concentrations in plants growing on copper mine spoil in the Grand Canyon, Arizona. *Amer. Midl, Nat,* **115,** 277-81.

Holzner W., Werger M. J. A. & Ikusima I. (eds)(1983) *Man's Impact on Vegetation.* Junk, The Hague.

Hopkins A. J. M. & Saunders D. A. (1987) Ecological studies as the basis for management. In: *Nature Conservation: Role of Remnants of Native Vegetation* (eds D. A. Saunders, G. W. Arnold, A. A. Burbidge & A. J. M. Hopkins) pp. 15-28. Surrey Beatty & Sons, Chipping Norton.

Hopkins M. S. (1981) Disturbance and change in rainforests and the resulting problems of functional classification. In: *Vegetation Classification in Australia* (eds A. N. Gillison & D. J. Anderson) pp. 42-52. Australian National Univeristy Press, Canberra.

Hopkins M. S., Kikkawa J., Graham A. W., Tracey T. G. & Webb L. J. (1976) An ecological basis for the management of rainforest. In: *The Border Ranges. A Land-use Conflict in Regional Perspective* (eds R. Monroe & N. C. Stevens) pp. 58-66. Royal Society of Queensland, Brisbane.

Hore-Lacy I. (1982) Mining in Australia. In: *Man and the Australian Environment* (eds W. Hanley & M. Cooper) pp. 187-95. McGraw-Hill, Sydney.

Johnston T. N. (1968) The effect of sheep and rabbit grazing on regeneration of white cypress pine. *Aust. For. Res.* **4,** 3-12.

Jordan W. R. III, Gilpin M. E. & Aber J. D. (eds) (1987) *Restoration Ecology: A Synthetic Approach to Ecological Research.* Cambridge University Press, Cambridge.

Katahira H. (1985) Land-use history of part of Western Australian wheatbelt between Jerramungup and Newdegate. In: *Studies of Environmental Changes Due to Human Activities in the Semi-arid Regions of Australia* (eds H. Yoya, K. Takeuchi & H. Ohmori) pp. 160-8. Dept. of Geography, Tokyo Metropolitan University, Tokyo.

Kavanagh R. P., Shields J. M., Recher H. F. & Rohan-Jones W. G. (1985) Bird populations of a logged and unlogged forest mosaic at Eden, New South Wales. In: *Birds of Eucalypt Forests and Woodlands: Ecology, Conservation, Management* (eds A. Keast, H. F. Recher, H. Ford & D. Saunders) pp. 237-81. Surrey Beatty & Sons, Chipping Norton.

Kirkpatrick J. B. & Haney R. A. (1980) The quantification of developmental wilderness loss: the case of forestry in Tasmania. *Search* **11,** 331-5.

Koch J. M. & Bell D. T. (1983) Amelioration of acidic coal mine overburden from Collie, Western Australia. *Reclam. Reveg. Res.* **2,** 155-65.

Lamont B. B. (1982) Mechanisms for enhancing nutrient uptake in plants, with particular reference to mediterranean South Africa and Western Australia. *Bot. Rev.* **48,** 597-689.

Lamont B. B. (1984) Specialised modes of nutrition. In: *Kwongan. Plant Life of the Sandplain* (eds J. S. Pate & J. S. Beard) pp. 126-45. University of Western Australia Press, Nedlands.

Lange R. T. (1969) The piosphere: sheep track and dung patterns. *J. Range Manage.* **22,** 396-400.

Lange R. T. & Graham C. R. (1983) Rabbits and the failure of regeneration in Australian arid zone *Acacia. Aust J. Ecol* **8,** 377-81.

Le Houérou H. N. (1981) Impact of man and his animals on Mediterranean vegetation. In: *Mediterranean-type Shrublands* (eds F. di Castri, D. W. Goodall & R. L. Specht) pp. 479-521. Elsevier, Amsterdam.

Leigh J., Boden R. & Briggs J. (1984) *Extinct and Endangered Plants of Australia.* MacMillan, Sydney.

Leigh J. H., Wimbush D. J., Wood D. H., Holgate M. D., Slee A. V., Stanger M. G. & Forrester R. I. (1984) Effect of rabbit grazing and fire on a subalpine enviroment. 1. Herbaceous and shrubby vegetation. *Aust. J. Bot.* **35,** 433-64.

Leopold A. (1949) *A Sand County Almanac.* Oxford University Press, New York.

Lewis H. T. (1985) Burning the "Top End": kangaroos and cattle. In: *Fire Ecology and Mangement in Western Australian Ecosystems* (ed. J. R. Ford) pp. 21-31. Western Australian Institute of Technology, Bentley.

Likens G. E., Bormann F. H., Pierce R. S., Eaton J. S. & Reiners W. A. (1977) *Biogeochemistry of a Forested Ecosystem.* Springer, New York.

Lowenthal D. (1976) Perceiving the Australian environment: a summary and commentary. In: *Man and Landscape in Australia. Towards an Ecological Vision* (eds G. Seddon & M. Davis) pp. 357-65. Australian Government Publishing Service, Canberra.

McCay F. L. (1972) The natural resources of Australia. In: *The Natural Resources of Australia: Prospects and Problems for Development* (ed. J. A. Sinden) pp. 70-114. Angus & Robertson, Sydney.

McKeller R. (1987) Relevance, accountability and efficiency of research for management. In: *Nature Conservation: the Role of Remnants of Native Vegetation* (eds D. A. Saunders, G. W. Arnold, A. A. Burbidge & A. J. M. Hopkins) pp. 345-56. Surrey Beatty & Sons, Chipping Norton.

Magnuson J. J., Regier H. A., Cristie W. J. & Sonzongi W. C. (1980) To rehabilitate and restore Great Lakes ecosystems, In: *The Recovery Process in Damaged Ecosystems* (ed. J. Cairns Jr.) pp. 95-112. Ann Arbor Science Publishers, Ann Arbor, Michigan.

Mauger G. W. (1986) LANDSAT evaluation of the impact of logging, burning and dieback in the jarrah forest of south Western Australia. 1986 National Environmental Engineering Conference, Melbourne.

Moran G. F. & Hopper S. D. (1987) Geographic population structure of eucalypts and the conservation of their genetic resources. In: *Nature Conservation: the Role of Remnants of Native Vegetation* (eds D. A. Saunders, G. W. Arnold, A. A. Burbidge & A. J. M. Hopkins) pp. 151-62. Surrey Beatty & Sons, Chipping Norton.

Nash R. (1973) *Wilderness and the American Mind* (2nd Ed). Yale University Press, New Haven.

Naveh Z. (1967) Mediterranean ecosystems and vegetation types in California and Israel. *Ecology* **48,** 345-59.

Naveh Z. & Dan J. (1973) The human degradation of Mediterranean landscapes in Israel. In: *Mediterranean Type Ecosystems, Origin and Structure* (eds F. di Castri & H. A. Mooney) pp. 373-90. Springer, New York.

Newman J. C. & Condon R. W. (1969) Land-use and present condition. In: *Arid Lands of Australia* (eds R. O. Slatyer & R. A. Perry) pp. 105-32. Australian National University Press, Canberra.

Nichols O. (1984) Factors affecting understorey successional processes and the attainment of stability in revegetated bauxite minesites. In: *Proc. IV Conf. on Mediterranean Ecosystems* (ed. B. Dell) pp. 91-2. University of Western Australia, Nedlands.

Noble I. R. & Slatyer R. O. (1981) Concepts and models of succession in vascular plant communities subject to recurrent fire. In: *Fire and the Australian Biota* (eds A. M. Gill, R. H. Groves & I. R. Noble) pp. 311-35. Australian Academy of Science, Canberra.

Noble J. C. & Tongway D. J. (1986a) Pastoral settlement in arid and semi-arid rangelands. In: *Australian Soils: the Human Impact* (eds J. S. Russell & R. F. Isbell) pp. 217-42. University of Queensland Press, St. Lucia.

Noble J. C. & Tongway D. J. (1986b) Herbivores in arid and semi-arid rangelands. In: *Australian Soils: the Human Impact* (eds J. S. Russell & R. F. Isbell) pp. 243-70. University of Queensland Press, St. Lucia.

O'Brien B. J. (1979) Land-use: competitive and compatible. In: *Environment and Science* (ed. B. J. O'Brien) pp. 254-89. University of Western Australia Press, Nedlands.

Old K. M., Kile G. A. & Ohmart C. P. (1980) *Eucalypt Dieback in Forests and Woodlands.* CSIRO, Melbourne.

Passmore J. (1974) *Man's Responsibility for Nature.* Duckworth, London.

Pearman G. I. (ed.) (1988) *Greenhouse. Planning for Climatic Change.* CSIRO, Melbourne.

Peck A. J. (1977) Development and reclamation of secondary salinity. In: *Soil Factors in Crop Production in a Semi-arid Environment* (eds J. S. Russell & E. L. Greacen) pp. 301-19. University of Queensland Press, Brisbane.

Pen L. (1987) Peripheral vegetation of the Swan-Canning Estuary — past, present and future. In: *The Swan River Estuary Ecology and Management* (ed. J. John) pp. 221-31. Curtin University of Technology, Bentley.

Pennington W. P. (1969) *The History of British Vegetation.* English Universities Press, London.

Pignatti S. (1983) Human impact in the vegetation of the Mediterranean basin. In: *Man's Impact on Vegetation* (eds W. Holzner, M. J. A. Werger & I. Ikusima) pp. 151-61. Junk, The Hague.

Pigram J. (1982) Leisure and recreation in Australia. In: *Man and the Australian Environment* (eds W. Hanley & M. Cooper) pp. 266-75. McGraw-Hill, Sydney.

Ridpath M. G. & Corbett L. K. (1985) Fire in the wet-dry tropics: introduction. *Proc. Ecol. Soc. Aust.* **13,** 177-8.

Roberts B. R. (1981) Erosion in pastoral regions of Australia. *Aust. Rangel. Soc. Biennial Conf. Papers* **3,** 53-63.

Roberts B. R. (1984) Land ethics — a necessary addition to Australian values. In: *Proc. Land Degradation Conf.* Australian National University, Canberra.

Roberts B. R. (1986) *Soil Conservation — People, Religion and Land.* Soil Conservation Society of Australia, Toowoomba.

Roberts B. R. (1987) Australian land degradation and its control. *Ambio* **16,** 272-6.

Rolls E. (1984) *A Million Wild Acres.* Penguin, Ringwood, Victoria.

Routley R. & Routley V. (1975) *The Fight for the Forests: the Takeover of Australian Forests for Pine, Wood Chips and Intensive Forestry* (3rd edn.). Research School of Social Sciences, Australian National University, Canberra.

Saunders D. A., Rowley I. & Smith G. T. (1985) The effects of clearing for agriculture on the distribution of cockatoos in the southwest of Western Australia. In: *Birds of Eucalypt Forests and Woodlands: Ecology, Conservation, Management* (eds A. Keast, H. F. Recher, H. Ford & D. Saunders) pp. 309-21. Surrey Beatty & Sons, Chipping Norton.

Saxon E. C. (ed.) (1984) *Anticipating the Inevitable: a Patch-burn Strategy for Fire Management at Uluru (Ayers Rock-Mt Olga) National Park.* CSIRO, Melbourne.

Schofield N. J., Stoneman G. L. & Low I. C. (1988) Hydrology of the jarrah forest. In: *The Jarrah Forest. A Complex Mediterranean Ecosystem.* (eds B. Dell, J. J. Havel & N. Malajczuk) pp. 179-201. Kluwer, Dordrecht.

Seddon G. (1976) Introduction. In: *Man and Landscape in Australia. Towards an Ecological Vision* (eds G. Seddon & M. Davis) pp. 3-6. Australian Government Publishing Service, Canberra.

Shea S. R., McCormick J. & Portlock C. C. (1979) The effects of fires on the regeneration of leguminous species in the northern jarrah *(E. marginata)* forest of Western Australia. *Aust. J. Ecol.* **4,** 195-205.

Singh G., Kershaw A. P. & Clark R. (1981 Quaternary vegetation and fire history in Australia. In: *Fire and the Australian Biota* (eds A. M. Gill, R. A. Groves & I. R. Noble) pp. 23-54. Australian Academy of Science, Canberra.

Singh G. (1982) Environmental upheaval. The vegetation of Australasia during the Quaternary. In: *A History of Australasian Vegetation* (ed. J. M. B. Smith) pp. 90-108. McGraw-Hill, Sydney.

Smith J. (1982) The destruction of the Australian wilderness. In: *Man and the Australian Environment* (eds W. Hanley & M. Cooper) pp. 98-104. McGraw-Hill, Sydney.

Smith P. (1985) Woodchip logging and birds near Bega, New South Wales. In: *Birds of Eucalypt Forests and Woodlands: Ecology, Conservation, Management* (eds A. Keast, H. F. Recher, H. Ford & D. Saunders) pp. 259-71. Surrey Beatty & Sons, Chipping Norton.

Specht R. L. (1981) Major vegetation formations in Australia. In: *Ecological Biogeography of Australia* (ed. A. Keast) pp. 165-297. Junk, The Hague.

Sprugel D. G. (1985) Natural disturbance and ecosystem energetics. In: *The Ecology of Natural Disturbance and Patch Dynamics* (eds S. T. A. Pickett & P. S. White) pp. 335-52. Academic Press, New York.

Stegner W. (1976) The meaning of wilderness in American civilization. In: *The American Environment. Readings in the History of Conservation* (2nd Ed., ed. R. Nash) pp. 192-7. Adison-Wesley, Reading, Massachusetts.

Tasker A. & Chadwick M. J. (1978) The microdistribution of *Agrostis tenuis* on colliery spoil in relation to spoil chemical variability. *J. Appl. Ecol.* **15,** 551-63.

Thirgood J. V. (1981) *Man and the Mediterranean Forest: a History of Resource Depletion.* Academic Press, London.

Toyne P. (1974) *Recreation and Environment.* MacMillan, London.

Turner J. & Kelly J. (1977) Soil chemical properties under naturally regenerated Eucalyptus spp. and planted douglas fir. *Aust. For. Res.* **1,** 163-72.

Turvey N. (1982) Australia's forestry future. In: *Man and the Australian Environment* (eds W. Hanley & M. Cooper) pp. 37-46. McGraw-Hill, Sydney.

Underwood R. J., Sneeuwjagt R. J. & Styles H. G. (1985) The contribution of prescribed burning to forest fire control in Western Australia. In: *Fire Ecology and Management in Western Australian Ecosystems* (ed. J. R. Ford) pp. 153-70. Western Australian Institute of Technology, Bentley.

Wadham S., Wilson R. K. & Wood J. (1964) *Land Utilization in Australia* (4th edn.). Melbourne University Press, Parkville.

Walker P. H. (1986) The temperate southeast. In: *Australian Soils: the Human Impact* (eds J. S. Russell & R. F. Isbell) pp. 36-64. University of Queensland Press,St. Lucia.

Weatherhead P. J. (1986) How unusual are unusual events? *Am. Nat.* **128,** 150-4.

Weste G. M. & Marks G. C. (1987) The biology of *Phytophthora cinnamomi* in Australasian forests. *Ann. Rev. Phytopathol.* **25,** 207-29.

Westman W. E. (1978) Measuring the inertia and resilience of ecosystems. *BioScience* **28,** 705-10.

Westman W. E. (1986) Resilience: concepts and measures. In: *Resilience in Mediterranean-type Ecosystems* (eds B. Dell, A. J. M. Hopkins & B. B. Lamont) pp. 5-19. Junk, Dordrecht.

Williams O. B. (1985) Population dynamics of Australian plant communities, with special reference to the invasion of neophytes. In: *The Population Structure of Vegetation* (ed. J. White) pp. 623-35. Junk, Dordrecht.

Wright J. (1980) Wilderness, waste and society. *Habitat* **8,** 27-31.

Wycherley P. (1984) People, fire and weeds: can the vicious circle be broken? In: *The Management of Small Bush Areas in the Perth Metropolitan Area* (ed. S. A. Moore) pp. 11-7. Department of Fisheries & Wildlife, Perth.

Young M. D., Wilson A. D. & Harrington G. N. (1984) Management aims, objectives and responsiblities. In: *Management of Australia's Rangelands* (eds G. N. Harrington, A. D. Wilson & M. D. Young) pp. 15-22. CSIRO, Melbourne.

Proc. Ecol. Soc. Aust. 1990 — 16:115-127

Sydney's vegetation 1788-1988: utilization, degradation and rehabilitation

D. H. BENSON and J. HOWELL

National Herbarium of New South Wales, Royal Botanic Gardens, Mrs Macquarie's Road, Sydney, N.S.W. 2000, Australia

Abstract

Sydney, the site of the first European colonization in Australia, is fortunate in having so much infertile soil, supporting vegetation that is rich in species. Much of this has been preserved to provide an enviable urban setting of major national parks, and with bushland remnants conspicuous in many suburban areas. However, during 200 years, European settlement has removed most of the indigenous vegetation from the more fertile soils and destroyed or modified many wetlands and sand dune complexes.

The vegetation of 1788, inferred from geological, climatic, historical and present-day distributional information, has been described and grouped into eight major vegetation types. For each type, the impact of European settlement is outlined, and the remaining area estimated. Poorly conserved types have been identified. In bushland on soils of low fertility, the indigenous understorey is disappearing despite the preservation of canopy dominants, as weed invasion follows nutrient enrichment of drainage lines, and fire regimes change. Management strategies are being developed for individual sites, reflecting growing public concern to conserve remaining bushland in urban areas.

Introduction

Sydney (33°50′S; 151°00′E), the largest city on the eastern coast of Australia, was colonized by convicts, soldiers and administrators of the First Fleet in 1788, and has Australia's longest history of continuous European settlement. This chapter deals with the indigenous vegetation of the Sydney area, defined by the County of Cumberland, an area of almost 370 000 ha bounded to the north and west by the Nepean-Hawkesbury River and extending southwards to Bulli, north of Wollongong. This area represented the limit of the early colonial administration, and now contains most of Sydney's 3.5 million people. Its clearly-defined physiographic features represent the results of 280 million years of erosion, deposition, uplift and gentle subsidence associated with tectonic forces, occasional volcanic events and fluctuating sea levels. Its shape is that of a shallow saucer formed largely of two types of Triassic sedimentary rocks — shales and sandstones (Herbert & Helby 1980). The central part of the saucer is the Cumberland Plain where shales of the Wianamatta Group provide gently undulating to hilly country with an altitudinal range of 20-100 m a.s.l. This country extends westward from Parramatta to the Blue Mountains just beyond the Nepean-Hawkesbury River, and from Richmond in the north to Picton in the south. Between Windsor and Castlereagh are the main occurrences of Tertiary gravels, sands, silts and clays, deposited by an ancestral Nepean-Hawkesbury River system (Gobert 1978). Along the present Nepean-Hawkesbury River and its major tributaries, South, Rickabys and Eastern Creeks, are alluvial deposits of Quaternary age. The Georges River valley also has smaller areas of both Tertiary and Quarternary alluvials.

The eastern half of the saucer is formed by sandstones, principally of the Hawkesbury Sandstone. From an altitude of 20-50 m around Sydney Harbour, the Hornsby Plateau rises to over 250 m in the north before dropping to the Hawkesbury River. In the south the Woronora Plateau rises to over 200 m. Remnant Wianamatta Shale cappings persist on ridge-lines of the Hornsby Plateau, and over much of the flatter terrain between Sydney Harbour and the Georges River, but are less extensive south of the Georges River.

Quaternary sand deposits of mixed marine and aeolian origin are found around Botany Bay. These have been partly shaped by sea level fluctuations that are believed to have placed the sea 140 m below its present level as recently as 22 000 years ago (Chappell 1982). At this time the coastline would appear to have been up to 40 km east of Sydney. Occasional narrow basalt dykes and *ca* 30 small diatremes of volcanic breccia form only a minor part of Sydney's geology.

Climate in the Sydney area is subtropical (Gentilli 1972), with local conditions influenced by distance from the coast. Rainfall follows a decreasing gradient from the coast to the Cumberland Plain, temperature extremes become more pronounced and there is an increasing incidence of frost. Mean maximum temperatures increase from less than 26°C along the coast to over 29°C on the Cumberland Plain, while mean minima drop from 7-8°C at Sydney to 2-4° on the Cumberland Plain. Average annual rainfall decreases from 1 444 mm at Pymble, through 911 mm at Parramatta, generally less than 800 mm across the Cumberland Plain to a low of 683 mm at Menangle near Camden (Bureau of Meteorology 1979).

Major vegetation types

Native vegetation remains over about a third of the County of Cumberland. The vegetation existing at the time of European settlement has been deduced from present distributions, climatic and geological patterns and historical records. Some early writings need interpretation. For example, Watkin Tench sounded somewhat overwhelmed by new surroundings in 1789, when he described the woodlands of the Cumberland Plain as "the trackless immeasurable desert" (Tench 1979). However, Peter Cunningham's 1827 description relates vegetation to soils and gives a picture that is clearly recognizable.

"In Cumberland, the land immediately bordering upon the coast is of a light, barren, sandy nature, thinly besprinkled with stunted bushes; while from 10 to 15 miles interiorly, it consists of a poor clayey or ironstone soil, thickly covered with our usual evergreen forest timber and underwood. Beyond this commences a fine timbered country, perfectly clear of bush, through which you might, generally speaking, drive a gig in all directions, without any impediment in the shape of rocks, scrubs or close forest. This description of country commences immediately beyond Parramatta on one hand, and Liverpool on the other; stretching in length south easterly obliquely towards the sea, about 40 miles and varying in breadth near 20. The soil upon the immediate banks of the rivers is generally rich flooded alluvial, but in the forests partakes commonly of a poor clayey or ironstone nature, yet bearing tolerable crops, even without manure, at the outset." (Cunningham 1827).

Our field survey and mapping work, conducted since an earlier treatment of this topic (Burrell 1972), has identified at least 31 plant communities, differentiated by structural and floristic criteria. For convenience we have grouped these plant communities into eight major vegetation types, and given each a vernacular name. The postulated distribution of these types in 1788 is shown in Figure 1.

Blue Gum High Forest

Blue Gum High Forest is tall open-forest or wet sclerophyll forest dominated by *Eucalyptus saligna* and *E. pilularis*. Previously the dominants were probably over 40 m tall; there are trees 30 m tall in tiny remnants now. Other tree species included *Angophora costata, E. paniculata, E. globoidea, Syncarpia glomulifera* and *Allocasuarina torulosa*.

Blue Gum High Forest was found on the more easterly Wianamatta Shale where rainfall exceeds 1100 mm, along the central spine of the North Shore from Crows Nest to Hornsby and further west from Castle Hill to Eastwood. In well-drained situations the understorey had shrubs up to 2 m, together with grasses and herbs. In moister parts, along drainage lines, a ground cover of ferns was conspicuous beneath small trees of *Pittosporum undulatum* and *Glochidion ferdinandi*, and shrubs including *Breynia oblongifolia* and *Polyscias sambucifolia*. Trees that also occur in rainforest, *Ceratopetalum apetalum* and *Acmena smithii*, also grew along some creeks, but true rainforest appears to have occurred on the shale only around Brush Farm at Eastwood.

Turpentine-Ironbark Forest

Turpentine-Ironbark Forest grew on Wianamatta Shale country with a lower average rainfall (about 900-1100 mm) and/or shallower soil near boundaries with the underlying Hawkesbury Sandstone.

On the northern side of the Parramatta River, Turpentine-Ironbark Forest grew from Ryde westwards to Parramatta and from there discontinuously north to Castle Hill and beyond. On the south side of the harbour it probably began around Glebe-Newtown and extended westward to Parramatta and Fairfield.

This forest was not as tall as Blue Gum High Forest, its trees would probably have been 20-30 m tall, with *Syncarpia glomulifera, E. paniculata, E. globoidea, E. resinifera* and *Allocasuarina torulosa* the main species, growing above dense understorey which included *Pittosporum undulatum, Acacia* species and a variety of peas and climbers.

West of Auburn, rainfall decreases, the clay soils have increasing amounts of ironstone gravels, and drier country trees became components of this forest — *E. moluccana, E. longifolia* and *E. fibrosa*. The understorey here ranged from dense scrub up to 3 m high to open and grassy with scattered shrubs, with *Melaleuca* species in drainage lines.

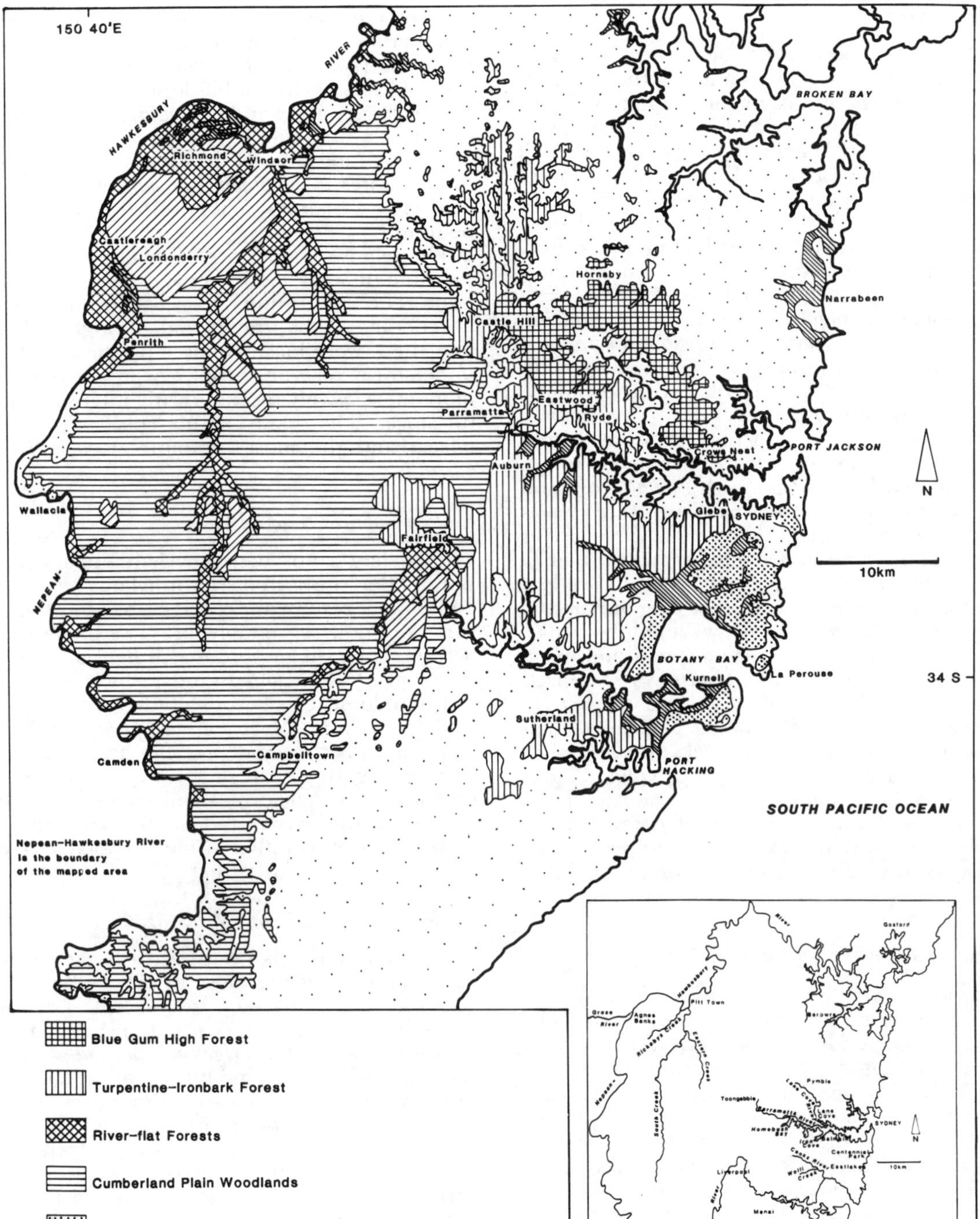

FIG. 1. Vegetation of the Sydney district east of the Nepean-Hawkesbury River at the time of European settlement. Inset: Localities mentioned in text.

River-flat Forests

River-flat Forests occupied some of the most fertile soil in the district — alluvium along the Nepean-Hawkesbury River and its major tributaries, South Creek and Eastern Creek, and smaller areas along the Georges River. Dominant species in these tall open-forests downstream from Penrith were *Eucalyptus tereticornis* and *Angophora subvelutina.* In the account of his 1791 expedition, John Hunter described River-flat Forest near Richmond as "fine, straight timber", growing above a dense understorey including shrubs, "long sedges" [*Gahnia* spp., *Lomandra longifolia*], "nettles" [*Urtica incisa*] and "a weed resembling ivy" [*Smilax australis* or *Dioscorea transversa*] (Hunter 1793). Species composition is difficult to describe with certainty, because it was rapidly cleared.

Upstream from the junction with the Grose River, the species dominants changed to *A. subvelutina* and *E. amplifolia,* with *E. elata* common as far downstream as Wallacia. The rare *E. benthamii* occurred occasionally down to the Grose River junction, and further upstream *E. viminalis* and *E. botryoides/saligna* intergrade grew around Camden. Along the river's length, *Casuarina cunninghamiana* and occasional *Tristaniopsis laurina* lined the banks, and limited numbers of *Toona australis* and *Melia azedarach* grew near Penrith. There were localized occurrences of *E. bauerana* and *E. bosistoana* on both the Georges River and the Nepean. Along smaller Cumberland Plain creeks *A. floribunda* appears to have replaced *A. subvelutina.* Paperbarks, *Melaleuca linariifolia, M. decora* and *M. styphelioides,* with a ground cover of *Juncus* spp. or sedges, were found in backswamps and periodically waterlogged depressions associated with the alluvial forests.

Cumberland Plain Woodlands

Woodlands originally occupied the bulk of the Cumberland Plain and about 30% of the Sydney district. They extended westward from Parramatta to the Nepean-Hawkesbury River, north to Richmond and Windsor, and south to Camden and Campbelltown. They occupied the driest part of Sydney, where rainfall is approximately 700-900 mm per annum, and soils are deep clays developed from shales of the Wianamatta Group — from the Bringelly Shale on the Plain itself and from the Ashfield Shale on the margins. In 1788 Governor Phillip described this country west of Parramatta as "singularly fine, level or rising in small hills of a very pleasing and picturesque appearance. The soil excellent, except in a few small spots where it was stony. The trees growing at a distance of from 20 to 40 feet [6-12 m] from each other, and in general entirely free from brushwood" (Phillip 1789).

Two tree species are most common across the Cumberland Plain, *E. moluccana* which tends to grow on rises, and *E. tereticornis* which is more often found on lower hillslopes and depressions. Ironbarks (*E. crebra* and *E. fibrosa*) accompany these on hilly country, and *E. eugenioides* and *E. longifolia* occur sporadically. Near creeks or on poorly-drained sites *E. amplifolia, E. bauerana, Angophora floribunda, Casuarina glauca* or *Melaleuca decora* may be found.

The relative abundance of shrubs and grasses at the time of settlement is now impossible to determine. The early writers describe "forest land" with a general lack of underwood, though indicating that there were locally denser patches. Today the understorey may be shrubby or grassy depending more on relatively recent grazing intensity and disturbance. The main shrub species is *Bursaria spinosa* and the most common grass is *Themeda australis.*

Eastern Suburbs Banksia Scrub

The aeolian sands between Centennial Park and Botany Bay supported sclerophyllous heath, scrub and low forest, with considerable local floristic variation, now regrettably lost. The name Banksia Scrub reflects the importance of *Banksia aemula,* an indicator of old leached sands, which occurs at its southern limit at La Perouse. Other common shrub species were *Monotoca elliptica, Eriostemon australasius, Ricinocarpos pinifolius* and *Xanthorrhoea resinosa,* together with *Angophora costata* generally growing as a mallee or small tree. Similar scrub and woodland was found on the sand dunes of Kurnell on the southern side of Botany Bay but *Banksia aemula* is absent here.

Freshwater and Estuarine Wetlands

Freshwater wetlands occurred in backswamps on the floodplain of the Nepean-Hawkesbury River. Expansive billabongs and lagoons with permanent water had emergent reeds and rushes, including *Eleocharis sphacelata, Triglochin procera, Philydrum lanuginosum* and, depending on conditions, *Phragmites australis* or *Typha* spp. Ephemeral swampy waterlogged sites were characterized by rushes, most commonly *Juncus usitatus,* together with some herbs (e.g., *Persicaria* spp.), bands of paperbark shrubland with *Melaleuca linariifolia* or *M. styphelioides,* and swamp woodland of *Eucalyptus robusta* on the drier edges, though still subject to flooding.

Freshwater wetlands were also associated with the Eastern Suburbs sand dunes; small swamps occurred in the swales, and larger expanses of open water along major drainage lines formed the Lachlan Swamps now in Centennial Park, and the Botany Swamps at Eastlakes. Here, tall emergent sedges, particularly *Eleocharis sphacelata,* and shorter sedges, including *Baumea articulata, B. rubiginosa* and *Juncus* spp., grew in and around standing water. Shrub species of surrounding waterlogged soil included *Melaleuca quinquenervia, Viminaria juncea, Callistemon citrinus* and *C. linearis.*

In estuarine wetlands, mangroves mainly of *Avicennia marina* with some *Aegiceras corniculatum* were conspicuous in 1788. Saltmarshes with *Sarcocornia quinqueflora* and *Suaeda australis* were quite extensive, meadows of *Juncus kraussii* with grasses *Sporobolus* and *Zoysia* occurred on rarely-flooded sites, and landward to these beyond normal tidal influence could be found swamp oak woodland with *Casuarina glauca* and occasionally *Melaleuca ericifolia* with a ground layer of *Baumea juncea* and *Juncus kraussii.*

Castlereagh Woodlands

Woodlands with distinctive shrubby understoreys grew on the Tertiary clay, silt, sand and gravel deposits centred on Castlereagh and Londonderry, with smaller areas along South Creek and Eastern Creek. These areas are mainly flat with shallow meandering drainage lines, and species composition appears to be determined by localized soil type. Soils are sands or clays, often containing ironstone gravels or large river stones, and poor in nutrients.

On the gravelly soils grow ironbarks, mainly *E. fibrosa* but also some *E. crebra* and *E. sideroxylon.* On sandy soil these are replaced by *E. sclerophylla* and *Angophora bakeri.* Changes in understorey shrubs and ground covers accompany changes in soil type and dominant species. Along poorly-drained watercourses and depressions grow *Melaleuca decora* and *E. parramattensis,* with grasses, herbs, and some now uncommon herbaceous species, e.g. *Isotoma fluviatilis.*

Woodlands and Heaths of the Sandstone

The rugged sandstone topography supports, as it did in 1788, a number of different plant communities determined by topography, drainage, and soils.

Ridge-tops and upper slopes support woodland with *Eucalyptus haemastoma, E. gummifera* and *Angophora costata* the most common co-dominants, and *E. racemosa, E. punctata, E. sieberi* and *E. oblonga* common in localized patches where there is some shale or ironstone influence. Where soil is particularly shallow there is scrubby heath with dense thickets of *Angophora hispida, Banksia ericifolia, Allocasuarina distyla* and *Hakea teretifolia.* Where drainage is impeded, the shrub layer diminishes and sedges predominate. On sandstone rock platforms low shrubs and moss mats grow in shallow pockets of soil. Mallees *(E. luehmanniana, E. multicaulis, E. obtusiflora)* are found sporadically on shallow sandy soils on or below ridge lines and may be associated with seepage zones.

On hillslopes, *E. piperita, Angophora costata, Allocasuarina littoralis* and some *E. gummifera* dominate open-forest in the exposed sites, while in sheltered situations *Syncarpia glomulifera* and the small trees *Elaeocarpus reticulatus, Ceratopetalum gummiferum* and *Pittosporum undulatum* become common. All ridge-top and hillslope communities include the well-known species-rich sclerophyllous shrub layer, except where very moist conditions favour mesic species and ferns.

Small trees and shrubs of *Callicoma serratifolia, Tristaniopsis laurina, Lomatia myricoides, Stenocarpus salignus* and *Austromyrtus tenuifolia,* together with ferns, are found along creeks. Where gully soil is enriched by fertile downwashed soil, trees and understorey shrubs such as *Syncarpia glomulifera, Eucalyptus pilularis, Polyscias sambucifolia* and vines normally associated with Turpentine-Ironbark Forest may grow. Further enrichment provides conditions suitable for rainforest-type species such as *Acmena smithii, Ceratopetalum apetalum, Acacia elata, Smilax australis* and ferns.

Coastal headland and dune vegetation, *Eucalyptus maculata* communities around Broken Bay and tall open-forest and rainforest south of Port Hacking on the Narrabeen Group shales and sandstones, and upland swamps of the Woronora Plateau, have been grouped with vegetation of the sandstone.

Aboriginal occupation and plant use

At the time of European settlement, Aborigines were concentrated in two parts of the Sydney district, the floodplain and the coast, and used a range of plants for food, shelter, canoes, tools, weapons, containers and medicines. For the Dharug people on the floodplain of the Nepean-Hawkesbury River, significant plant foods included vine tubers of the River-flat Forests, and bulbs and tubers of orchids, lilies and swamp plants of the Cumberland Plain Woodlands (Kohen 1984). Small game supplemented these plant food staples.

Along the coast, three peoples, Kuring-gai north of Port Jackson, coastal Dharug (Eora) between Port Jackson and Botany Bay, and Dharawal south of Botany Bay, lived on a diet rich in fish and shellfish. This diet was supplemented with small fleshy fruits and (mainly Proteaceous) flowers rich in nectar from the Eastern Suburbs Banksia Scrub and the sandstone heaths and woodlands, larger fleshy fruits of mesic plants in the moist sandstone gullies, and stems and rhizomes of freshwater wetlands reeds and rushes. From the forests, *Macrozamia* seeds and fern rhizomes, once detoxified, provided carbohydrate (Lampert & Sanders 1973; Poiner 1976; Ross 1976; Kohen & Lampert 1987).

Plants used in significant amounts for other purposes were species of *Xanthorrhoea,* for spear shafts and adhesive resin, *Livistona australis* for shelter and fibre, several shrubs and trees for fibre including *Brachychiton populneus, Hibiscus heterophyllus* and species of *Ficus* and *Pimelea,* and *Casuarina cunninghamiana* for bark canoes.

It is difficult to tell whether Aboriginal harvesting affected plant distributions. Current distributions seem to be explainable in terms of environmental factors, but the early European clearing of fertile river flats and Cumberland Plain soils would have obliterated any patterns there. Aboriginal use of fire however, would have affected the vegetation, the different types being subjected to differing burning patterns. For example, in the grassy Cumberland Plain woodlands, fires may have been used as often as every five years, to stimulate underground food resources and grass for game. On the less productive woodlands of the Hawkesbury Sandstone, areas may have remained unburnt for perhaps 20-30 years.

Plant resources used by Aborigines and known to increase after fire are *Macrozamia communis,* for which an eight-fold increase in seed production following a hot fire has been recorded (Beaton 1982), and *Xanthorrhoea australis* in which burning has caused earlier flowering, and doubled the incidence of inflorescence production (Gill & Ingwersen 1976).

Patterns of European Utilization

A few months after the First Fleet anchored in Sydney Cove in January 1788, Surgeon Worgan summarized the settlers' activities: "the principal Business has been the clearing of Land, cutting, Grubbing and burning down Trees, sawing up Timber & Plank for Building, ... " (Worgan 1978). Clearing the land was to become a national obsession for the next century and a typical achievement is shown in John Lewin's view of the Government Farm at Castle Hill around 1806 (Fig. 2). By 1822 Governor Brisbane had clearing operations well organized, boasting that "in order to accomplish the first process towards improvement I have a Thousand men employed in clearing the Country of its excess of Forest Timber and Brushwood" (Brisbane to Buchan 1822).

Although some clearing was necessary to provide agricultural and grazing lands, many of the early colonists had an over-zealous antagonism to the bush. Louisa Meredith complained in 1839: "The system of 'clearing' here, by the total destruction of every native tree and shrub, gives a more bare, raw, and ugly appearance to a new place. In

FIG. 2. View of the Government Farm at Castle Hill, about 1806, attributed to John Lewin. (Courtesy of the Mitchell Library, State Library of New South Wales).

England we plant groves and woods, and think our country residences unfinished and incomplete without them; but here the exact contrary is the case, and unless a settler can see an expanse of bare, naked, unvaried, shadeless, dry, dusty land spread all round him, he fancies his dwelling 'wild and uncivilized' " (Meredith 1844).

Clearing of vegetation took place at different rates and different times in the various vegetation types around Sydney in response to changing needs and population size.

In broad terms the vegetation gave way to agriculture for the first 100 years, and to suburban growth during the second. The decreasing area of Sydney's different vegetation types between 1788 and 1988 is shown schematically in Figure 3.

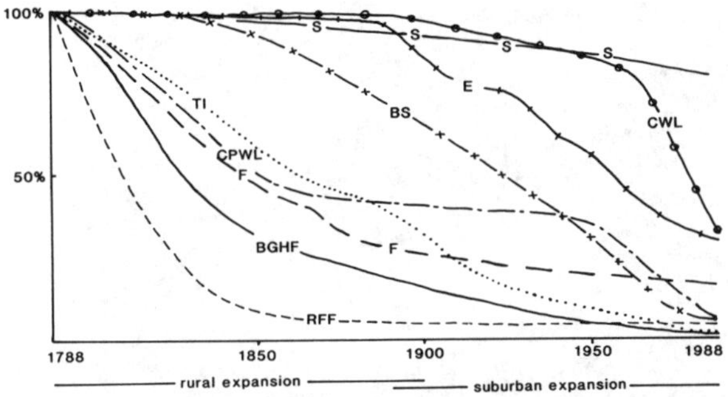

FIG. 3. Schematic diagram illustrating progressive loss of Sydney's vegetation types as a result of changing land use over 200 years. BGHF: Blue Gum High Forest, BS: Eastern Suburbs Banksia Scrub, CPWL: Cumberland Plain Woodlands, CWL: Castlereagh Woodlands, E: Estuarine Wetlands, F: Freshwater Wetlands, RFF: River-flat Forests, TI: Turpentine-Ironbark Forest, S: Woodlands and Heaths of the Sandstone.

River-flat Forests

River-flat Forests were on the first sites sought for agricultural activities, as they occupied some of the most fertile soil in the colony. Clearing the flats along the Nepean-Hawkesbury River commenced soon after 1789, and by 1826 James Atkinson was able to write "The greater part of the alluvial lands upon the Hawkesbury and Nepean have been cleared and are under cultivation" (Atkinson 1826).

The forests' demise is confirmed in early paintings; for example, John Lewin's "A View of the River Hawkesbury, N.S.Wales", of 1805, Figure 4, shows one lone, tall *Angophora subvelutina* and a few shrubs and *Casuarina* trees amongst cleared land. As early as 1803 Governor King issued a General Order against destroying trees on river banks (King to Hobart 1804) but in 1825 Judge Barron Field could still criticize the practice of clearing right to the river banks (Field 1825, reprinted in Mackaness 1965). River-flat Forests along the Hawkesbury's tributaries and on the Georges River were similarly cleared.

Cumberland Plain Woodlands

These woodlands, growing on deep clay soils, proved very attractive to the early settlers, and by 1821 the major portion of the Cumberland Plain had been allocated in land grants (Proudfoot 1987). Vegetation loss was not as rapid or as complete as for the River-flat Forests as much of the land was used for grazing. However, other areas were cleared for wheat-growing. The Government Farm at Toongabbie was established in 1791 originally for this purpose, and for a time Campbelltown was known as "the granary of the colony". Wheat-growing continued until the mid-ninteenth century, by which time wheat rusts had arrived and spread, and the new railway over the Blue Mountains opened up wheat-growing areas west of the Great Dividing Range.

The remaining Cumberland Plain Woodland has suffered gradual attrition. Completion of railway lines to Richmond and Penrith in the 1860s gave impetus to new sawmills, and extractive industries, which cleared vegetation to quarry sand, clay, gravel and blue metal to service the growing city. From the 1880s to 1960s, as land close to the city increased in value, traditional uses changed and large estates were fragmented. Some parts of previously-cleared land were used for more intensive agriculture, while on other parts there was intermittent regrowth of vegetation. From the 1970s, following Sydney's post World War II population boom, suburban expansion and hobby farming has spread across the Cumberland Plain and pressures on the few remaining areas of natural vegetation have escalated once more.

Floodplain Freshwater Wetlands

Freshwater wetlands of the Nepean-Hawkesbury floodplain were drained to provide land for grazing and agriculture. Their utilization and loss parallel the decline of the River-flat Forests.

Blue Gum High Forest

Blue Gum High Forest contained impressive trees up to 40 m tall, and the early settlers soon set about cutting these. Bush sawpits were operating on the North Shore in the early 1800s, and by 1810 Governor Macquarie reported timber "getting scarce" near the Government sawpit 10 km up the Lane Cove River. When the taller straighter trees had been removed for use in construction work (buildings, bridges, wharves, etc.), the smaller trees were cut to provide billets for navy vessels and firewood for Sydney's residents. She-oaks, species of *Casuarina* and *Allocasuarina,* were cut for shingles; in 1813, for example, Governor Macquarie instructed the Government sawpit at Lane Cove to supply 4 000 shingles each week per three-man gang, and 600 six-foot palings per week per three-man

FIG. 4. A view of the Hawkesbury River in about 1805, by John Lewin, showing the floodplain forest had been almost completely cleared by this date. (Courtesy of the Dixson Galleries, State Library of New South Wales).

gang. Each two-man team of sawyers was to saw 450 ft (150 m) of timber each week, in addition (Russell 1970). Hardwood paling fences were a very early Sydney suburban feature.

On the North Shore, the bulk of the Blue Gum High Forest had been cleared by the 1860s, and the land used for orchards. Suburbs had spread up the North Shore ridge by 1920.

Turpentine-Ironbark Forest

Turpentine-Ironbark Forest occupied country with lower rainfall than the Blue Gum High Forest. Its principal timbers were certainly useful in the early colonial days. *Syncarpia glomulifera* was invaluable for wharf and jetty pilings, *Eucalyptus paniculata* and other ironbarks were used extensively as fence posts, *E. resinifera* yielded timber prized for its strength and durability (Maiden 1889). Although cleared less purposefully than the Blue Gum High Forest, it has now been replaced by housing and little remains.

Eastern Suburbs Banksia Scrub

The Eastern Suburbs Banksia Scrub country was of little use for agricultural purposes, and its disappearance came about mainly because of its proximity to the city. The availability of fresh water in its swamps and aquifers led to the development of industries such as soap-making, tanning and wool-washing. Residential expansion in the late nineteenth century and early twentieth resulted in a steady decrease in the area of scrub. On the northern side of Botany Bay the few small areas now remaining are the subject of housing proposals.

On the southern side of Botany Bay, the Kurnell sand dune vegetation deteriorated through burning, clearing and grazing from the 1860s. Sand extraction on a major scale commenced in the 1940s, and in the 1950s a large area of dunes and swamps was flattened for an oil refinery. Large-scale sand extraction continues and use by off-road vehicles has become a problem.

Eastern Suburbs Freshwater Wetlands

The major freshwater wetlands associated with the Eastern Suburbs Banksia Scrub survived until the 1870s when the central swamps were dammed to supply Sydney's water. The system of freshwater lakes formed still exists today, in Centennial Park and Eastlakes Golf Course, but with permanent high water levels and grassy banks, supports relatively few of the original species. There are also heavy weed infestations, and alterations by major roadworks.

Estuarine Wetlands

Most Estuarine Wetlands survived into the twentieth century though small areas of shallow shoreline around Sydney Harbour were gradually filled and tidied up as Sydney developed. In Homebush Bay there was some modification of wetlands by salt evaporation works in the 1800s, and in the 1880s seawalls were built in preparation for filling, but it wasn't until after World War II that extensive wetlands of Sydney Harbour and Parramatta River were filled, particularly from Iron Cove to Homebush Bay and Parramatta. Filling has severely reduced the extent of both mangrove and saltmarsh communities, though some new opportunities for mangrove colonization have been provided by erosion and sedimentation along the Lane Cove River (McLoughlin 1985).

Wetlands of the Cooks River estuary were destroyed by the construction of Sydney Airport in the 1930s, and later the river was re-routed and turned into a canal, its neatened banks became ovals, golf courses and smaller parks. Much of the swampy creek line to the south was converted to market gardens, and transport corridors threaten tiny remnants along the tributary Wolli Creek.

In Botany Bay, mangroves were burnt to provide alkaline ash (barilla) for soap-making, from 1810 to 1850 (Bird 1981). *Casuarina glauca* was cut by oyster farmers in the late 1880s and early 1900s. The estuarine wetlands on the southern shore were conserved as Towra Point Nature Reserve in the 1970s, after being threatened with destruction for Sydney's second airport (Australian Littoral Society 1978). Its boundaries are inadequate, and in 1986 these wetlands narrowly escaped having a toxic chemical industry built on adjoining land.

Castlereagh Woodlands

The Castlereagh Woodlands survived destruction for a long period because their poor soils were not suitable for agriculture. In recent times however, woodland has been cleared during sand and clay extraction and the characteristic vegetation of the Agnes Banks sand deposits (Benson 1981) has almost been destroyed by mining. This vegetation is unusual because of its affinities with coastal plant communities 50-60 km away.

Woodlands and Heaths of the Sandstone

Finally, the vegetation of the sandstone areas has survived best in terms of the area remaining, because of low fertility soils and rugged terrain. Sydney's large national parks, Ku-ring-gai Chase, Marramarra, Heathcote and Royal, together with Muogamarra Nature Reserve and several large State Recreational Areas contain extensive sandstone communities. Smaller areas, generally along creek lines, remain within the suburbs. Though the sandstone around Sydney Harbour was occupied by housing last century, the stepped terrace housing of the period being suitable for the steeper slopes of Balmain and Paddington, it was not until after World War II that newer house design and construction capabilities led to the clearing and development of extensive sandstone areas. Provision of services through sandstone areas is still expensive, and engineers generally prefer softer bedrock for new housing development.

Conserving our remaining vegetation

Sydney's rich and diverse flora is spread between at least eight vegetation units but our conservation areas, particularly the major national parks, are almost all on the areas of Hawkesbury Sandstone. In earlier times such areas were seen as suitable for reserves for public recreation as they were scenically diverse yet could not be used for agriculture and were poor in minerals (except for coal). Conservation of the vegetation of the other units, most of which have now been almost totally cleared or degraded, is inadequate (Table 1), and deteriorating. Proposals for residential development, extractive industries and transport corridors threaten most unprotected vegetation remnants and it is inevitable that despite a great deal of effort, many will be lost.

A few examples

At Agnes Banks, near Richmond, about 450 ha of Pleistocene-Pliocene sand overlie Tertiary clays and silts. Although included in our broad Castlereagh Woodlands type, distinctive plant communities with coastal affinities grow on these sands, despite being 55 km inland in an area with only two-thirds of the average coastal rainfall. This vegetation is of great scientific interest because of information it contains about relationships between plant distributions and geomorphological events, including sea level fluctuations, in the Quaternary. Forster *et al.* (1977) included it in a list of areas of botanical conservation interest; detailed description and analysis of the five communities appears in Benson (1981). Despite initial proposals for reservation in 1968, about 70% of this vegetation has been removed by sand mining, and a nature reserve and Permanent Conservation Order protect about 9%.

Near Pitt Town, the Longneck Lagoon Wildlife Refuge and Field Studies Centre conserves one of the few remaining floodplain freshwater wetlands. Much of the catchment is Crown land with some of the best remaining areas of Cumberland Plain Woodland. This needs to be added to the conserved area. Until it is, it will remain vulnerable to proposals for residential or other development.

At Menangle in the south-west, the best remaining stand of the River-flat Forest which once flanked the Nepean occurs on the former Camden

TABLE 1. Areas of vegetation types within the County of Cumberland in 1788, and percentages remaining in 1980.

Vegetation types	1788 area (ha)	1988 % remaining
Blue Gum High Forest	11 000	0.9
Turpentine-Ironbark Forest	35 000	0.5
River-flat Forests	19 000	3
Cumberland Plain Woodlands	107 000	6
Eastern Suburbs Banksia Scrub	8 000	3.5*
Estuarine and	6 000	29
Freshwater Wetlands	3 000	18
	9 000	25
Castlereagh Woodlands	19 000	32
Woodlands and Heaths of the Sandstone	162 000	85
County of Cumberland	370 000	

* Of the original 7 000 ha of Eastern Suburbs Banksia Scrub between Bondi and Botany Bay, only 0.2% remains. Most remaining vegetation of this type is found on the Kurnell Peninsula. This is structurally similar but lacks *Banksia aemula*.

Park Estate, the historic property of the Macarthur family. The Department of Agriculture now manages this land and is trying to conserve the forest; however, pre-existing mining leases cover the vegetated area, and the sand and soil company's application to mine has been the subject of a Commission of Inquiry. Elsewhere remnants of this forest occur in narrow bands subject to attrition at the edges, often with only the canopy trees remaining from the former forest.

North of Botany Bay, about 10 ha of Eastern Suburbs Banksia Scrub remain, but even some of this is currently the subject of housing proposals.

Problems in managing conserved areas

Even where bushland is intended to be preserved, deterioration continues through the insidious polluting effects of urban run-off, weed invasion and changes to the fire regimes.

Soil nutrient enrichment and weed invasion

Weed invasion of natural communities follows disturbance that leads to alteration in the resource base (Fox & Fox 1986). In suburban bushland two situations are particularly susceptible to weed invasion; edges, where adjacent clearing increases the light and open space resource, usually next to sources of weed propagules; and drainage lines, where urban run-off increases soil nutrient and moisture levels, favouring growth of weeds and mesic native species over scleromorphic native species.

Increased soil phosphorus content has been shown to increase the weed component of the flora (Clements 1983) and weed growth is particularly conspicuous along creeks and drainage lines below disturbed areas. Urban stormwater run-off typically moves about 1 kg ha^{-1} per year of phosphorous (Harley Wright, pers. comm.), and in northern Sydney, soil phosphorous levels have been found to be increased twofold at suburban boundaries, threefold in creek lines, and eight times below stormwater outlets (Leishman 1986). Contributors are garden fertilizers, dumped refuse, sewer discharges, and, significantly, pet excrement. Pets may add 3-10 kg ha^{-1} per year of phosphorus to suburban areas (Wright, pers. comm.). Reduction of nutrient input requires improved design of stormwater drainage systems, as advocated by Bliss *et al.* (1983).

Weed propagules are dispersed into bushland in stormwater, dumped garden refuse, by fruit-eating birds and wind, and soon take advantage of any nutrient-enriched conditions. Weed-invaded bushland is readily apparent; many creeklines in and around Sydney have become overgrown tangles where *Ligustrum*, exotic *Rubus*, *Lonicera japonica*, *Cardiospermum grandiflorum* and *Ipomoea indica* have replaced the native creekside vegetation.

Changing fire regimes

Fire regimes have undoubtedly changed in the last 200 years. Currently in bushland there may be wildfires that are less frequent and more intense than in the past, or prescribed burns that are more frequent and less intense. Other areas may remain unburnt. Frequent fires can cause local disappearances of obligate seeders, changing floristic composition in favour of resprouters (Siddiqi *et al.* 1976; Bradstock & Myerscough 1981; Nieuwenhuis 1987). In woodland on the Hawkesbury Sandstone, intervals free of fire of at least 9-10 years are needed to prevent loss of obligate seeding species (Benson 1985). Even resprouters can lose vigour if burnt too often.

Fires of low intensity may fail to stimulate germination of hard-seeded species (e.g., Floyd 1976) or release of seeds stored in woody fruits (Bradstock & Myerscough 1981) even though adults are killed. Achieving periodic hot fires in small suburban reserves is, for obvious reasons, a major difficulty. Hot fires would be desirable to reduce the understorey dominance of *Pittosporum undulatum* in a number of reserves where fire has long been excluded. However, these reserves are generally too small and too close to suburban dwellings to make hot fires acceptable, and hand thinning needs to be used as a second best alternative where fire is impossible.

Fire exclusion may cause floristic change. The small tree *Pittosporum undulatum* is susceptible to fire, and in its absence forms a dense canopy in some sandstone and shale reserves (see Adamson

& Fox 1981; Clark & McLoughlin 1986). Growth and germination of other understorey species, and regeneration of canopy dominants are prevented.

Maintenance and rehabilitation

Although there is increasing awareness of the consequences of nutrient enrichment, weed invasion and altered fire regimes, bushland managers can find it hard to translate a bewildering collection of individual pieces of information, some seemingly contradictory, into an effective management plan for a specific location (Saunders *et al.* 1987). A generalized scheme of interactions between nutrients, weeds and fire for bushland on sandstone and shale in Sydney is shown in Figure 5. This gives a broad overview, designed to show the ways individual factors will interact, allowing prediction of outcomes from various combinations of conditions and actions. Soil fertility determines different sequences of events for vegetation on sandstone and shale.

To a certain extent, each site must be considered individually in the light of its past history of interacting factors. Some examples of burning and weeding programmes designed for specific areas are presented by Fox (1988).

In practice, there çan be conflicts; for example, between keeping conditions suitable for recruitment of rare native species by exposing soil, and preventing weed colonization by encouraging a ground cover of ferns. Such conflicts need to be recognized, and resolved either by obtaining more management resources, or by choosing the most desirable realistic goal. However, it is important that bushland managers do not feel daunted if there are gaps in our knowledge and events cannot be predicted with certainty. It is important in such cases to try various actions to find what works for a particular situation. Managers will almost always have to make decisions with incomplete information, but they should use their activities as experiments to build up the information available iteratively (Hopkins & Saunders 1987).

Conclusion

By accidents of geology and history, Sydney is fortunate to retain much indigenous vegetation close to the city. However, for the same reasons, this vegetation is not representative of the range of types previously found there. Most remaining vegetation is on the sandstone, and for five of Sydney's eight vegetation types little remains.

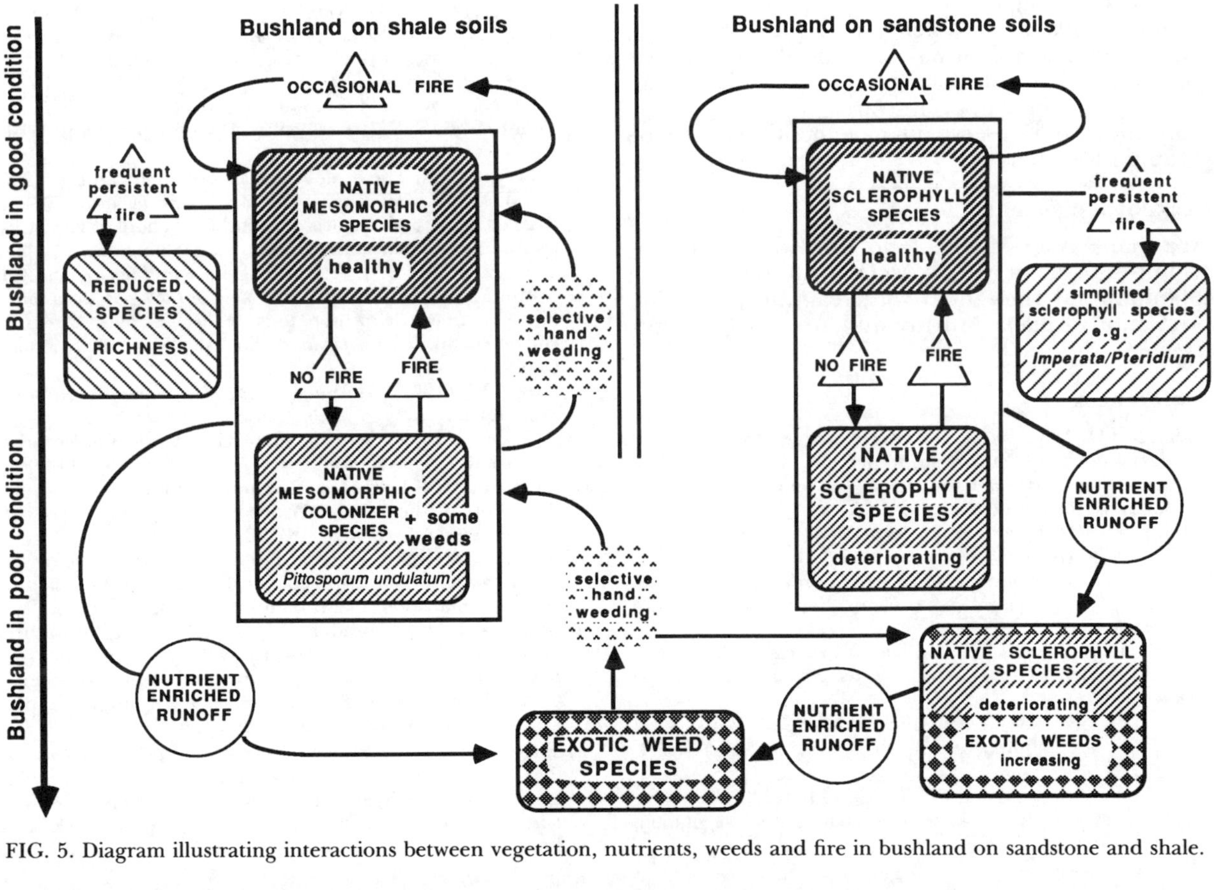

FIG. 5. Diagram illustrating interactions between vegetation, nutrients, weeds and fire in bushland on sandstone and shale.

Similarly, it is the species of the sandstone that are generally well conserved while species of other habitats survive as fragmented populations. Research is particularly needed on recruitment and maintenance of populations of uncommon species. Without adequate reservation and careful management, these fragmented populations are vulnerable to continuing attrition and possible extinction.

Research has provided guidelines for predicting management problems related to altered fire regimes and nutrient status in the large proportion of Sydney's remaining vegetation on low nutrient sandstone soils. Implementation of the results of this research by land managers, is now the main challenge in the retention of natural vegetation. For example, adequate protection of the catchments for reserves is the best long-term solution, to avoid weed invasion of gullies. However, new subdivisions are planned without regard to these factors, and bushland in suburbs such as Berowra and Menai will succumb to weed invasion in due course like the older suburbs that were developed before the consequences were evident. Planning may be more costly initially, but will allow our natural vegetation to survive and provide pleasure and value for many centuries. The deterioration of bush in new subdivisions cannot be blamed on our lack of knowledge.

For bushland in older areas, local solutions, perhaps simple or innovative, are needed to mitigate problems. The interface between bushland and housing is an important area where ecology, education and reserve management need to be brought together.

Acknowledgements

We thank Peter Mitchell and Jim Kohen for providing information, Lyn McDougall and Jeanette Stannard for their help with diagrams, and Sue McCahon, Angela Murray and Angela Benn for typing.

References

Adamson D. A. & Fox M. D. (1981) Change in Australasian vegetation since European settlement. In: *A History of Australasian Vegetation* (ed. J. M. B. Smith) pp. 109-46. McGraw-Hill, Sydney.

Atkinson J. (1826) *An Account of the State of Agriculture and Grazing in New South Wales.* Cross, London.

Australian Littoral Society (1978). *An Investigation of Management Options for Towra Point, Botany Bay.* Commonwealth of Australia.

Beaton J. M. (1982) Fire and water: aspects of Australian Aboriginal management of Cycads. *Archaeol. in Oceania* **17,** 51-8.

Benson D. H. (1981) Vegetation of the Agnes Banks sand deposit, Richmond, New South Wales. *Cunninghamia* **1,** 35-57.

Benson D. H. (1985) Maturation periods for fire-sensitive shrub species in Hawkesbury Sandstone vegetation. *Cunninghamia* **1,** 339-49.

Bird J. F. (1981) Barilla production in Australia. In: *Plants and Man in Australia* (eds D. J. Carr and S. G. M. Carr) pp. 274-80. Academic Press, Sydney.

Bliss P. J., Riley S. J. & Adamson D. (1983) Towards rational guidelines for urban stormwater disposal into flora preservation areas. *The Shire and Municipal Record* **76,** 181-5, 191.

Bradstock R. & Myerscough P. J. (1981) Fire effects on seed release and the emergence and establishment of seedlings of *Banksia ericifolia* L. f. *Aust. J. Bot.* **29,** 521-31.

Brisbane to Buchan (1822) *Historical Records of Australia, Series 1,* **10,** 723.

Bureau of Meteorology (1979) *Climatic Survey, Sydney, Region 5, New South Wales.* Department of Science and the Environment, Canberra.

Burrell J. P. (1972) Vegetation of the Sydney area: 1788 and 1961. In: *The City as a Life System?* (ed. H. A. Nix). *Proc. Ecol. Soc. Aust.* **7,** 71-8.

Chappell J. (1982) Sea levels and sediments: some features of the context of coastal archaeological sites in the tropics. *Archaeol. in Oceania* **17,** 69-78.

Clark S. S. & McLoughlin L. C. (1986) Historical and biological evidence for fire regimes in the Sydney region prior to the arrival of Europeans: implications for future bushland management. *Australian Geographer* **17,** 101-12.

Clements A. (1983) Suburban development and the resultant changes in the vegetation of the bushland of the northern Sydney region. *Aust. J. Ecol.* **8,** 307-19.

Cunningham P. (1827) *Two Years in New South Wales.* Colburn, London.

Field B. (1825) Journal of an Excursion Across the Blue Mountains of New South Wales, October, 1822. Appendix to *Geographical Memoirs on New South Wales,* John Murray.

Floyd A. G. (1976) Effect of burning on regeneration from seeds in wet sclerophyll forest. *Aust. For.* **39,** 210-20.

Forster G. R., Campbell D., Benson D. & Moore R. M. (1977) *Vegetation and soils of the western region of Sydney.* Technical Memorandum 77/10, CSIRO Division of Land Use Research, Canberra.

Fox M. D. (1988) The ecological status of alien plant species. In: *Weeds on public land — an action plan for today.* Weed Science Society of Victoria Inc. and the School of Environmental Science, Monash University, Melbourne.

Fox M. D. & Fox B. J. (1986) The susceptibility of natural communities to invasion. In: *Ecology of Biological Invasions: An Australian Perspective* (eds R. H. Groves & J. J. Burdon) pp. 57-66. Australian Academy of Science, Canberra.

Gentilli J. (1972) *Australian Climate Patterns.* Nelson, Melbourne.

Gill A. M. & Ingwersen F. (1976) Growth of *Xanthorrhoea australis* R. Br. in relation to fire. *J. Appl. Ecol.* **13,** 195-203.

Gobert V. (1978) Proposed nomenclature for the Cainozoic sediments of the Penrith-Windsor area. *Quarterly Notes of the Geological Survey of N.S.W.* **32,** 1-9.

Herbert C. & Helby R. (eds) (1980) *A Guide to the Sydney Basin.* Bulletin No. 26, Geological Survey of New South Wales. Department of Mineral Resources, Sydney.

Hopkins A. J. M. & Saunders D. A. (1987) Ecological studies as the basis for management. In: *Nature Conservation: The Role of Remnants of Native Vegetation* (eds D. A. Saunders, G. W. Arnold, A. A. Burbidge & A. J. M. Hopkins). pp. 15-28. Surrey Beatty, Chipping Norton.

Hunter J. (1793) *Historical Journal of the transactions at Port Jackson and Norfolk Island with the discoveries which have been made in New South Wales and in the Southern Ocean since the publications of Phillip's voyage.* John Stockdale, London.

King to Hobart (1804) *Historical Records of Australia, Series 1,* **5,** 67.

Kohen J. L. (1984) Computer analysis of Aboriginal-plant interactions on the Cumberland Plain. In: *Technology in the 80's* (ed. J. L. Kohen). Proceedings of a Regional Conference on Science Technology. ANZAAS-AIST, Macquarie University, Sydney.

Kohen J. L. & Lampert R. (1987) Hunters and Fishers in the Sydney Region. In: *Australians to 1788* (eds D. J. Mulvaney & P. J. White) pp. 342-65. Fairfax, Syme and Weldon Associates, Broadway.

Lampert R. J. & Sanders F. (1973) Plants and men on the Beecroft Peninsula, New South Wales. *Mankind* **9,** 96-108.

Leishman M. (1986) The Distribution of Soil Phosphorus within Urban Bushland in the area of Ku-ring-gai, Sydney. B.Sc. (Hons.) Report, School of Biological Sciences, Macquarie University.

Mackaness G. (ed) (1965) *Fourteen Journeys Over the Blue Mountains of New South Wales 1813-1841.* Horwitz-Grahame, Sydney.

McLoughlin L. (1985) *The middle Lane Cove River: a history and a future.* Monograph No. 1, Centre for Environmental and Urban Studies, Macquarie University.

Maiden J. H. (1889) *The Useful Native Plants of Australia.* Facsimile edition (1975), Compendium, Melbourne.

Meredith L. A. (1844) *Notes and Sketches of New South Wales during a residence in that colony from 1839 to 1844.* Facsimile edition (1973), Penguin Books, Harmondsworth, England.

Nieuwenhuis A. (1987). The effect of fire frequency on the sclerophyll vegetation of the West Head, New South Wales. *Aust. J. Ecol.* **12,** 373-85.

Phillip A. (1789). *Voyage of Governor Phillip to Botany Bay with an account of the establishment of the colonies of Port Jackson and Norfolk Island; compiled from authentic papers.* John Stockdale, London.

Poiner G. (1976). The process of the year among Aborigines of the central and south coast of New South Wales. *Arch. & Phys. Anthrop. in Oceania.* **11,** 186-206.

Proudfoot H. (1987). *Exploring Sydney's West.* Kangaroo Press, Kenthurst.

Ross A. (1976). Inter-tribal Contacts — What the First Fleet Saw. BA (Hons.) Thesis, Department of Anthropology, University of Sydney.

Russell E. (1970). *Lane Cove 1788, 1895, 1970 — A North Shore History.* The Council of the Municipality of Lane Cove, Lane Cove.

Saunders D. A., Arnold G. W., Burbidge A. A. & Hopkins A. J. M. (eds) (1987). *Nature Conservation: The Role of Remnants of Native Vegetation.* Surrey Beatty, Chipping Norton.

Siddiqi M. Y., Carolin R. G. & Myerscough P. J. (1976). Studies in the ecology of coastal heath in New South Wales, III, Regrowth of vegetation after fire. *Proc. Linn. Soc. NSW.* **101,** 53-63.

Tench W. (1979). *Sydney's First Four Years.* Library of Australian History in association with the Royal Australian Historical Society, Sydney.

Worgan G. B. (1978). *Journal of a First Fleet Surgeon.* Publication Number 16, The William Dixson Foundation. The Library Council of New South Wales in association with Library of Australian History, Sydney.

Proc. Ecol. Soc. Aust. 1990 — 16:129-134

The role of synthetic vegetation in present and future landscapes of Australia

P. B. BRIDGEWATER

Department of the Arts, Sport, the Environment, Tourism and Territories, G.P.O. Box 787 Canberra, A.C.T. 2601

Now at: Nature Conservancy Council, Northminster House, Petersborough, PE1 1UA, United Kingdom

Abstract

Synthetic vegetation is defined as a grouping of plant communities which are ad-mixtures of native, naturalized and exotic species. Such communities may be transitional phases in the degeneration of a native plant community, or they may be metastable communities capable of persistence through disturbance cycles. Genesis of synthetic vegetation is chiefly through breakdown of phytogeographic barriers by introduction of exotic species in the last 150-200 years. Maintenance of such vegetation depends on human intervention through varying management practices and little of such vegetation would persist in a stable form without "interventionist" management. Future landscapes in Australia will inevitably have large components of synthetic vegetation. Ecologists will have a major role in setting and developing appropriate management strategies and they will also have to communicate to the wider community the important contribution to biodiversity, landscape stability and maintenance that such vegetation makes.

Introduction

Grime (1984) in discussing the interaction between the British flora and landscape noted the current success in the British landscape of two plant families — Poaceae and Asteraceae. He related their success to ancestral characteristics of these families which predispose them to the prevailing ruderal environments of much of present-day British landscape. He further suggested that the future will show changes in vegetation and biodiversity as human-landscape interactions change.

Hildebrand (1983) documented a series of plant communities in southern Chile, in which native species have been complemented by species from other continents. In some cases the exotic species appear to have displaced some native species, whereas in other cases they appear to have had little effect. Benson and Howell (1988) writing about the bushland of Sydney, attribute vegetation changes to habitat fragmentation, nutrient enrichment, and alterations in fire frequency, which all allow the establishment of "invasive" or "uncontrolled weedy" species to the detriment of native species. For that reason it is not difficult to demonstrate that species such as *Cortaderia* spp., *Lantana camara* or *Cinnamonum camphora* cause significant ecosystem breakdown.

These works underline vegetation changes in both hemispheres resulting from a variety of effects, including the introduction of exotic species to native vegetation. This chapter suggests there is a new form of vegetation that can originate from a fusion of exotic/adventive/invasive species with native plant species. (I deliberately use the three terms because some authors have used them indiscriminately, and all three types of species are involved in synthetic vegetation.) Such vegetation I term synthetic, to imply the clear role that humans have in either actively or passively establishing these new mixtures.

Hobbs and Hopkins (this volume) discuss the establishment of new communities on highly disturbed or cleared ground — communities composed of species that may not have co-existed at the same site beforehand, but which may be self-sustaining. Such communities would be an extreme form of synthetic vegetation in my terms, but would be consistent with the general theme of human involvement in the genesis of the community.

The title of this chapter, as read to the symposium, used the term "anthropochorous" vegetation. This term is cumbersome compared with the simpler term synthetic, which I also used in Bridgewater (1988). Synthetic, however, does not translate well into some European languages, leading to a degree of confusion and uncertainty. Despite this, referees recommended that the term synthetic be retained, and I have done so in this published version.

There is a rapid increase world-wide, and certainly in Australia, of vegetation which includes synthetic plant communities. The general theme of this contribution is that, following European settlement, new communities of native and exotic plant species have arisen, some of which are metastable (i.e., given appropriate circumstances, capable of persistence through time). Naveh and Lieberman (1983) cite mediterranean shrub-, wood- and grasslands as examples of vegetation being in a metastable stage, and quote the work of Beddington *et al.* (1979) and Bernstein and Jung (1979) as showing both terrestrial and marine systems which are metastable — i.e., capable of persistence, despite being stressed and constantly perturbed.

Historical perspective

The present landscape in Australia has resulted from at least 40 000 years of human interaction (Singh *et al.* 1981) with increasingly technological sophistication, especially in the last 200 years. The focus of this present symposium is on the last 200 years — but we should not lose sight of the fact that prior to 1788 the landscape and its components have been modified by fire e.g., Nicholson (1981), Haynes (1985) and Press (1988). Apart from fire, which operated synergistically with a general trend of increasing aridity, the biggest change to our landscapes and their component ecosystems has been wrought by the introduction of species with an extra-territorial distribution.

Sukopp *et al.* (1979) and Sukopp and Werner (1983) recognize for Europe two groups of such species which they term archaeophytes (species introduced up to the end of the middle ages) and neophytes (species introduced after 1500 AD). From an Australian perspective exotic species are best regarded as neophytes — although there may well have been introductions of archeophytes along the northern coastline. Walsh (pers. comm.) also notes the propensity for Aboriginal manipulation of native plant distribution. Species such as *Microseris scapigera* and *Typha orientalis*, with tubers both as a food source and a means of vegetative reproduction, have certainly been spread by Aboriginal distribution (Mulvaney & White 1987).

The spread of native species since 1788 into new habitats by deliberate or accidental manipulation of populations, or by habitat change has also been documented e.g., *Leptospermum laevigatum* (Hazard & Parsons 1977) and *Pittosporum undulatum* (Benson & Howell 1988). In this discussion we should also not ignore lower plants such as fungi, where species such as *Phytophthora cinnamoni* have been spread or activated by human action, and have the potential to cause widespread vegetation changes.

Kornas (1983) points out that in the early stages of human-induced vegetation change there is a rise in species diversity; which in Europe has now become a decline. In Australia we have begun the down slope of diversity reduction, but only just.

Synthetic vegetation thus forms a part of the landscape matrix of plant communities created by management practices of European settlement. In some cases, however, such vegetation exists as changed forms of native communities, held in place by particular management practices. For example, Australia has no true equivalent to the European hedgerow. When hedgerow situations are created, the overstorey is typically formed from exotic species (e.g., *Ulex europaeus, Rubus* ssp., *Rosa rubiginosa*), with the ground flora often being a mixture of native species and exotics. In the absence of mowing or overburning such synthetic communities have the ability to persist.

There is a natural tendency to view all these possible changes with alarm and despondency. I think that is a mistaken attitude, and that rather we should regard this new vegetation as a major challenge for ecological management either to restore pre-existing communities, or to manage replacement synthetic communities.

Examples

1. In a number of sites along the eastern Victorian coast the native mangrove vegetation (*Avicennia marina,* with *Sarcocornia quinefloralSamolus repens* herb layer) has become invaded by *Spartina anglica.* While there is evidence that *Spartina* has the potential to cause degeneration of salt marsh communities and replace them with a *Spartina* community (Ranwell 1972) this synthetic vegetation mimics a natural plant community found along the shores of Florida and the northern Gulf of Mexico. There is thus a possibility that this species mixture may be potentially stable. An understanding of the environmental dynamics of the Floridian community should provide possibilities for management practices to hold the new community in balance.

2. Doing (1985) reported that southern Australian sea-shores have been invaded by introduced species and attributed that invasion to "empty niches". Certainly the tidemark regions of many sandy beaches in southern Australia are characterized by a two-species community of *Arctotheca populifolia* and *Cakile maritima.* While this community is most abundant along coasts which have been modified by European settlement, it is also found along remote coasts. Heyligers (1984, 1985) noted

that along with other species, these species have the potential to change dune physiography and induce further landscape change.

3. More complex than these few-species coastal communities are the wetlands of the entirely artificial Lake Burley Griffin in Canberra. Around the edge of the lake are communities dominated by the native *Typha orientalis.* A cosmopolitan species *(Lythrum salicaria)* grows as an admixture with *Typha,* and planted *Alnus glutinosa* now seeds into the *Typha* stands — much as it would into stands of *T. angustifolia* in Europe. This is not only an example of synthetic vegetation; it is also an example of seral change in such vegetation.

4. In the Adelaide Hills a number of sites occur where native shrubs have been replaced by *Ulex europaeus* and *Erica lusitanica.* These two species co-occur in communities in the coastal parts of Mediterranean Europe where they are native. In the Adelaide Hills they are mixed with a range of introduced and native forbs and grasses to produce a synthetic community which forms patches within the native eucalypt dominated communities. There is some evidence from other parts of southern Australia that continued disturbance of invaded vegetation will favour the introduced species (pers. obs.).

5. Norfolk Island, in the South Pacific, has had natural rainforest vegetation invaded by woody shrubs and trees, especially *Olea europaea* ssp. *africana* and *Psidium littorale* (Anon 1984). Such invasion has had dramatic effects on the edges of the rainforest. Nonetheless, the native species have strong ecological resilience [i.e., they have a high ability to regenerate following disturbance (Mooney & Hobbs 1986)], and can "invade the invaders", if an appropriate management regime, involving active removal of the woody weeds is pursued (Hallam, pers. comm.). Much of this vegetation is simply invasion replacement — but synthetic vegetation with *Citrus limon, Eupatorium riparium* and rainforest overstorey also occurs.

6. In New Zealand the Tongariro National Park has become colonized over large areas by the ericaceous species *Calluna vulgaris* and *Erica cinerea.* These species were introduced by the first Park Ranger, (Timmins, pers. comm.) and have now virtually replaced the native tussock grasses. Despite that replacement, native epacridaceous shrubs and the ground fern *Gleichenia microphylla* remain as part of the new synthetic community.

Ecological background

Neophytes and their success in integrating into native plant communities can be viewed in terms of the regeneration niche concept of Grubb (1977) and the three attribute model of plant performance suggested by Grime (1979). These two schemes emphasize the importance of different aspects of plant-environment interaction but both identify the existence and persistence of neophytes with some kind of environmental disturbance (see also Pickett & White 1985).

Johnstone (1986) has made a different contribution on this topic, using the "safe site" concept of Harper (1977) as the base for a universal explanation of invasions. Johnstone uses a matrix of four types of invasion barrier — botanical, non-botanical, species-selective, non species-selective to classify four invasion "windows". He defined these invasion windows as "time-dependent categories of safe site that are characterized by invasion barrier interaction categories. For the probability of an invasion window to be other than zero a suitable habitat must exist." He defined three types of window — stable, temporary and future.

Stable windows exist when a selective, non-botanical barrier to invasion is removed. Such barriers include fire prevention, herbivore control, nutrient enrichment, climate change. Note here the similarity with environmental factors canvassed by Benson and Howell (1988). Temporary windows exist when a non-selective botanical barrier is removed (e.g., forest canopy, grass sward). Future windows involve removal of selective botanical barriers to invasion — i.e., ingress of propagules is possible, but the invasive species remain dormant until a disturbance alters the balance of activity. Harper (1977) used safe site to suggest that presence and persistence of species depends not only on seed source, but on the availability of sites which provide the precise requirements of each species.

Disturbance to a plant community can result in the emergence of safe sites for exotic or invasive species. But, the implication of Johnstone's (1986) approach is that unless the disturbance is amplified or augmented, not all invasions are necessarily dramatic or will cause ecosystem breakdown. This point was made in a different way in Bridgewater and Backshall (1981) and Bridgewater (1988).

Grubb and Hopkins (1986) commented that "some communities subject to human disturbance have lost many species and, considered in their own right, they may be more resilient than the original communities".

Resilience of synthetic communities is also an important consideration. Fox and Fox (1986) defined damping, a property of resilience, as "the degree and manner by which the path of restoration is altered by any forces that change the normal restoring force". In the terms of Fox and Fox (1986) synthetic communities typically show critically damped responses to disturbance. The Norfolk Island rainforest example described above would exhibit a severely over damped response. The other examples, which are largely species-poor communities, exhibit damped or critically damped responses. Synthetic communities appear most commonly in species-poor vegetation, or in regions where critically damped responses are normal for natural vegetation (e.g., Mediterranean type climate). Vegetation not normally subject to natural disturbance is unlikely to contain synthetic communities, unless there are exceptional circumstances. Much more attention therefore needs to be given to the causes of synthetic vegetation — where invasion is better thought of as recruitment of new species, and management needs to focus on maintaining the synthetic vegetation in a balance.

Management

It is important to recognize that, good or bad, synthetic vegetation must be managed, and managed effectively. The aim of effective management must be to; maintain landscape heterogeneity in space and time, recognize the contribution of natural, semi-natural (Westhoff 1971) and synthetic vegetation to landscapes, and halt environmental degradation from poor land-use practice.

Forman (1986), in addressing remnant management commented that "... management require(s) maintaining the landscape structure (landscape elements sensitive to human activity), functioning (concentrations of flows within the landscape), and change (natural disturbance regimes). To accomplish this, remnants of natural landscapes also require management to combat isolation and human impacts from the matrix". The relevance of these comments lies in the fact that vegetation remnants often contain synthetic communities.

Main (1987) has given some useful frameworks for developing management standards, again for remnants of natural communities, but the principles are equally applicable to developing management strategies for synthetic communities in landscapes.

Micromanagement

There may be some synthetic communities which are unstable rather than metastable and are transitional to degraded communities dominated by invasive species. In such cases management to keep a balanced community intact is not possible and it may be best to attempt to recover the native vegetation and restore pre-existing communities. Techniques for this I term micromanagement, to distinguish them from the broader scale landscape-based management techniques needed to maintain vegetation. Micromanagement is usually labour or resource intensive.

It is possible to conceive of the *Opuntia* modified vegetation, which developed rapidly in western Queensland earlier this century, or the *Salvinia* monocultures which developed on open water in Mt. Isa (north-western Queensland), as extreme forms of synthetic vegetation. Effective management of these plant communities was possible only by bio-control — leading to extinction of the exotic species and thus the vegetation. Such techniques are "agricultural" in nature but there are more "ecological" methods also in use.

In New South Wales, the National Trust has had a fifteen year involvement with developing ecological strategies. These strategies have become known as the Bradley methods, after their proponents. As Bradley (1971) notes "Bush regeneration consists of weeding only, without re-planting. Its essence is that areas to be weeded are selected and their sizes graduated in such a way that each is promptly re-colonized, and quickly stabilized, by the spontaneous regeneration of native plants". Whilst the Bradley technique has had evident success in many areas, workers on other sites have opted for a different type of approach.

Robertson (pers. comm.) has used techniques similar to those of Bradley successfully in a nature reserve near Adelaide (South Australia). However, the deep rooted woody pest-plants she dealt with (*Olea, Fraxinus, Sarothamnus, Rosa*) were best removed by chemical hand swabbing after cutting. Other species (e.g., *Asparagus asparagoides* and *Rubus fruticosus*) could effectively be removed only by chemical spraying. Use of herbicides in any other way than swabbing or, rarely, careful targeted spraying, would destroy the native components, and lead to replacement by exotic ephemerals and creepers and further cycles of destabilization.

Macromanagement

Forman and Godron (1986) and Forman (1987) have defined a number of landscape ecological concepts which are useful for the present discussion. These concepts can be summarised as:

- a landscape is a cluster of interrelated and interacting ecosystems;
- ecosystem disturbance depends on neighbouring ecosystems in the landscape cluster;
- many animals require two or more ecosystems to survive and reproduce;
- ecosystems required by organisms for their life-cycle cannot exceed a certain distance apart to be energetically useful; and
- the greater the number of conduits (or higher the connectivity) the less likely landscape alterations can be isolated at one point.

Based on these concepts Forman (1987) gave many examples to substantiate what he termed the following principles:

- in general greater heterogeneity inhibits the spread of disturbance;
- the spread of disturbance decreases as boundary-crossing frequency increases; and
- where a corridor connecting patches (ecosystems) is narrow or has breaks (low connectivity) it acts as a barrier or filter to movement between patches. [For a discussion of connectivity see Forman and Godron (1986), Bridgewater (1987)].

From the earlier discussion that synthetic vegetation arises as a result of disturbances, these concepts and principles of landscape ecology give rise to some macromanagement strategies.

The first consideration is that synthetic communities may not be destructive to landscapes. In fact they may represent opportunities to provide an aesthetic and rugged landscape framework for the native communities to co-exist. They also represent an opportunity to study the processes of vegetation development and evolution. What is important is the need to understand clearly the vegetation dynamics on which management strategies can be based so that all landscape elements are able to be maintained in balance.

Using the ecological concepts outlined in the previous sections, it is clear that exotic species are exploiting aspects of the vegetation different from those of native species, producing metastable communities, often with a lower species richness than natural communities. The type of connectivity between landscape elements can help retard the spread and stability of synthetic vegetation.

Ultimately, however, the most profitable management option is to stabilize those new vegetation elements of landscape which can cope adequately with existing disturbance levels. Actual management processes to achieve this include: control of the fire regime, reduction in the level of nutrient enrichment, restricting public assess, monitoring and control of native and especially exotic animal species, and modification of vegetation dynamics. Note that fire control can be both increased frequency, or decreased frequency.

For example, in the New Zealand example quoted above, since fire is a natural part of *Calluna* management in Europe and the native epacridaceous shrubs are fire tolerant, burning should be considered as an option for the synthetic *Calluna* vegetation of New Zealand. At present the management strategy is to exclude fire, which achieves only a community of increasing senescence.

In the case of the Victorian mangrove community Ranwell (1972) also notes that grazing can induce changes to invasive *Spartina* communities in Europe and thereby suggests a management strategy.

Fauna and synthetic vegetation

Animal species richness typically declines with invasion of exotic plant species. However, there are examples where some native fauna species are able to persist in synthetic communities. An example is the eastern barred bandicoot *(Perameles gunnii)*, which survives in Victoria in restricted populations where protection is afforded by the introduced shrub *Ulex europaeus*. *P. gunnii* is an endangered species (Brown 1987). As *U. europaeus* is highly invasive, the particular management strategy followed has been to use a sterile strain, which provides the shelter effects without the possibility of unwanted invasion. Naturally, the long-term management of such communities must be intensive. In the New Zealand example quoted in the previous section the rare native fern bird *(Bowdleria punctata)* survives because the *Gleichenia* understorey has increased in biomass as a component of the synthetic vegetation.

On Norfolk Island, the green Parrot *(Cyanoramphus novaezelandiae cookii)*, with only 30 or so individuals left in the wild is being bred by the Australian National Parks and Wildlife Service. In captivity and in the wild it feeds from the introduced *Olea* species as well as native food sources (pers. obs.). While it is certain the total population structure of fauna in synthetic vegetation will change from pre-existing native communities the examples above show there may be short-term advantages for some species and there may well be opportunities to use synthetic vegetation in creative conservation programmes for endangered fauna.

The future

The management challenge is to ensure that synthetic communities in the landscape are self-maintaining and require the least expensive management input, while retaining the best quality landscapes. Grubb and Hopkins (1986) have pointed to this in the context of the pioneering work by Watt — "We must emphasize the issues raised by Watt: Plant communities are dynamic and the patterns we see now are a result of processes of the past; where these processes can be allowed to continue, then the general pattern will persist".

If the dynamic processes of the present are disrupted by the rise of synthetic communities, the inference is that patterns will change, but if we wish to maintain patterns of today, then management options must focus on ensuring the processes continue as before.

Traditional management approaches which do not recognize the importance of working with plant communities in a landscape context will fail the test of maintaining landscape quality, as defined by leaving stable, heterogeneous landscapes for the future.

References

Anon. (1984) *Plan of Management: Norfolk Island National Park.* Australian National Parks and Wildlife Service, Canberra.

Beddington J. R., Free C. A. & Lawton J. H. (1976) Concepts of stability and resilience in predator/prey models. *J. Anim. Ecol.* **45,** 791-816.

Benson D. H. & Howell J. (1988) *Sydney Bushland — Two Centuries of change.* Royal Botanic Gardens, Sydney.

Bernstein B. B. & Jung M. (1979) Selective pressures and coeolution in a kelp canopy community in southern California. *Ecol. Monogr.* **49,** 335-55.

Bradley J. (1971) *Bush Regeneration* Mosman Parkland and Ashton Park Association, Sydney.

Bridgewater P. B. (1987) Connectivity: an Australian perspective. In: *Nature Conservation: The Role of Remnants of Native Vegetation* (eds D. A. Saunders, G. W. Arnold, A. A. Burbidge & A. J. M. Hopkins), pp. 195-200. Surrey Beatty & Sons, Chipping Norton.

Bridgewater P. B. (1988) Synthetic plant communities: problems in definition and management. *Flora* **180,** 139-44.

Bridgewater P. B. & Backshall D. J. (1981) Dynamics of some Western Australian ligneous formations with special reference to the invasion of exotic species. *Vegetatio* **46,** 141-8.

Brown P. R. (1987) *Draft management plan for the conservation of the Eastern Barred Bandicoot* Perameles gunnii *in Victoria,* Arthur Rylah Institute for Environmental Research, Heidelberg.

Doing H. (1985) Coastal fore-dune zonation and succession in various parts of the World. *Vegetatio* **61,** 62-75.

Forman R. T. T. (1986) Emerging directions in landscape ecology and applications in natural resource management. In: *Proceedings of the conference on science in the National Parks. Volume One* (eds R. Herrmann & T. Bostedt-Craig) pp. 59-88. U.S. National Park Service, Colorado.

Forman R. T. T. (1987) The ethics of isolation, the spread of disturbance and landscape ecology. In: *Landscape Heterogeneity and Disturbance* (ed. M. G. Turner) pp. 213-29. Springer-Verlag, New York.

Forman R. T. T. & Godron M. (1986) *Landscape Ecology.* John Wiley, New York.

Fox B. J. & Fox M. D. (1986) Resilience of animal and plant communities to human disturbance. In *Resilience in Mediterranean-type Ecosystems* (eds B. Dell, A. J. M. Hopkins & B. B. Lamont) pp. 39-64. Dr W. Junk, Dordrecht.

Grime J. P. (1979) *Plant Strategies and Vegetation Processes.* John Wiley, New York.

Grime J. P. (1984) The ecology of species, families and communities of the contemporary British Flora. *New Phytol.* **98,** 15-33.

Grubb P. J. (1977) The maintenance of species richness in plant communities: the importance of the regeneration niche. *Biol. Rev.* **52,** 107-45.

Grubb P. J. & Hopkins A. J. M. (1986) Resilience at the level of the plant community. In: *Resilience in Mediterranean-type Ecosystems* (eds. B. Dell, A. J. M. Hopkins & B. B. Lamont) pp. 21-38. Dr W. Junk, Dordrecht.

Harper J. L. (1977) *Population Biology of Plants.* Academic Press, London.

Haynes C. D. (1985) The pattern of ecology of *munwag:* traditional Aboriginal fire regimes in north-central Arnhemland. *Proc. Ecol. Soc. Aust.* **13,** 203-14.

Hazard J. & Parsons R. F. (1977) Size-class analysis of coastal scrub and woodland, Western Port, southern Australia. *Aust. J. Ecol.* **2,** 187-98.

Heyligers P. C. (1984) Beach Invaders: sea rockets and beach daisies thrive. *Aust. Nat. Hist.* **21,** 212-4.

Heyligers P. C. (1985) The impact of introduced plants on foredune formation in south-eastern Australia. *Proc. Ecol. Soc. Aust.* **14,** 23-41.

Hildebrand R. (1983) Die Vegetation der Tieflandsgebusche des sudchilienischen Lorbeer Waldgebiets unter besondere Beruecksichtingung der Neophyten Problematick. *Phytocoenologia* **11,** 145-223.

Johnstone I. M. (1986) Plant invasion windows: a time-based classification of invasion potential. *Biol. Rev.* **61,** 369-94.

Kornas J. (1983) Man's impact upon the flora and vegetation in Central Europe. In: *Man's impact on vegetation* (eds M. J. A. Werger & I. Ikusma) pp. 277-286. Dr W. Junk, The Hague.

Main A. R. (1987) Management of remnants of native vegetation — a review of the problems and the development of an approach with reference to the wheatbelt of Western Australia. In: *Nature Conservation: The Role of Remnants of Native Vegetation* (eds D. A. Saunders, G. W. Arnold, A. A. Burbidge & A. J. M. Hopkins) pp. 1-13. Surrey Beatty & Sons, Chipping Norton.

Mooney H. A. & Hobbs R. J. (1986) Resilience and the individual plant level. In: *Resilience in Mediterranean-type Ecosystems* (eds B. Dell, A. J. M. Hopkins & B. B. Lamont) pp. 65-82. Dr W. Junk, Dordrecht.

Mulvaney D. J. & White J. P. (eds) (1987) *Australians to 1788.* Fairfax, Syme & Weldon, Sydney.

Naveh Z. & Lieberman A. S. (1983) *Landscape Ecology.* Springer-Verlag, Berlin.

Nicholson P. H. (1981) Fire and the Australian aborigine — an enigma. In: *Fire and the Australian Biota* (eds A. M. Gill, R. H. Groves & I. R. Noble) pp. 55-76. Australian Academy of Science, Canberra.

Pickett S. T. A. & White P. S. (eds) (1985) *The Ecology of Natural Disturbance and Patch Dynamics.* Academic Press, New York.

Press A. J. (1988) Comparisons of the extent of fire in different land management systems in the Top End of the Northern Territory. *Proc. Ecol. Soc. Aust.* **15,** 167-75.

Ranwell D. S. (1972) *Ecology of salt marshes and sand dunes.* Chapman & Hall, London.

Singh G., Kershaw A. P. & Clark R. (1981) Quarternary vegetation and fire. In: *Fire and the Australian Biota* (eds A. M. Gill, R. H. Groves & I. R. Noble) pp. 23-54. Australian Academy of Science, Canberra.

Sukopp H., Blume H. P. & Junick W. (1979) The soil, flora and vegetation of Berlin's waste lands. In: *Nature in Cities* (ed. I. C. Laurie) pp. 115-32. John Wiley, New York.

Sukopp H. & Werner P. (1983) Urban environments and vegetation. In: *Man's impact on vegetation* (eds M. J. A. Werger & I. Ikusma) pp. 247-60. Dr W. Junk, The Hague.

Westhoff V. (1971) The dynamic structure of plant communities in relation to the objectives of conservation. In: *The scientific management of Animal and Plant communities for conservation* (eds E. Duffy & A. S. Watt) pp. 3-14. Blackwell, Oxford.

Proc. Ecol. Soc. Aust. 1990 — 16:135-140

The revegetation of Victoria: the first stages. A description of the revegetation movement in Victoria

S. BURKE and R. YOUL

Land Protection Division, Department of Conservation, Forests and Lands, 240 Victoria Parade, East Melbourne 3002 Australia

Revegetation — a definition

In this chapter revegetation involves increasing the vegetative cover and the number of plant species in various altered rural and urban ecosystems by re-establishing vegetation components similar to those naturally occurring.

Background

Over the last decade, a major movement has developed, involving landholders, enthusiastic individuals, community groups, industry, private foundations, and government, to promote revegetation throughout rural Victoria. This movement, though dispersed and heterogeneous, nevertheless shares a common vision of productive, ecologically stable and aesthetically appealing landscapes resulting from the maintenance of indigenous vegetation, the adoption of revegetation and other conservation farming practices, and their integration with conventional agriculture.

Within cities and towns a parallel movement, called urban forestry, is underway for urban revegetation. This involves firstly the protection of remnants of indigenous vegetation, and secondly the creation of native forests which resemble natural stands, and line roads, easements and drainage channels and enclose factories and quarries. A third important element is the long-term planning of urban treescapes. Although progress has been less rapid in urban forestry than in the countryside, a number of promising projects are underway which involve community groups, municipal councils and private industry.

At present

Victoria, the smallest mainland state of Australia, still has over 33% of its area as public land, largely carrying vegetation in natural or semi-natural condition. A recent inventory (Woodgate & Black 1988) shows that 93% of public land is forested (i.e., woody vegetation with a top height greater than 2 m and a density in terms of projected foliar cover greater than 10%), and that this area remains relatively constant (reforestation is balanced by clearing). Whilst significant community interest exists regarding vegetation management on public land, we will not deal further with this, except where it closely affects revegetation.

About 64% of the total land area of Victoria is privately owned farmland, consisting of approximately 70 000 commercial holdings as well as hobby farms of various kinds. Farmland is generally in the western half of the State, the north-east and central Gippsland. Some regions in northern and western Victoria, having been alienated before even rudimentary conservation planning was undertaken, retain no substantial areas of public land suitable for conservation reserves. Nevertheless Victoria's agricultural country is criss-crossed by narrow lines of public land (railways, roads, unused roads, streamsides) and studded with small reserves for public purposes. Not all of this public land carries native vegetation or is well differentiated from adjacent freehold land.

The privately owned land in Victoria carrying native vegetation is not uniformly distributed. According to Woodgate and Black (1988), large (>15 ha) privately owned areas of forest comprise 4% of Victoria. The additional area of non-forest native vegetation associations on private land would be relatively small. The larger areas of freehold native vegetation are located mainly in the west, north-west and east of the State. In contrast, many farms in the agricultural regions of the State contain not a single remnant of indigenous flora. Between these extremes most properties carry small patches, strips, and isolated specimens of native vegetation in variable but generally poor condition.

Erosion, soil compaction, acidification, salinity, tree decline, water pollution, lack of native fauna and exposure also characterize many rural landscapes in Victoria.

The first comprehensive statewide inventory of forest cover (Woodgate & Black 1988) indicated that in the period 1972-87, 194 376 ha of freehold native forest were cleared, mostly in far western and north-western Victoria. Reforestation of freehold land in the same period amounted to 37 812 ha, being mainly pine plantations in central Gippsland and the south-west. Forest clearing on freehold land exceeded reforestation by 156 564 ha representing an average loss of 10 438 ha per year.

This ongoing, large-scale clearing, together with an insidious decline in both the total area and condition of the dispersed remnants, is most alarming.

Interspersed with privately owned forest and woodland is the second category of rural bushland; linear areas of vegetation on roadsides, unused roads, railway lines, streamsides and other small areas of publicly owned land. Management responsibility varies from State Government departments to municipalities. Nevertheless adjacent landholders and community groups can greatly influence and become directly involved in the management of these areas.

While a few suburbs and some country towns are desolate, there is also a surprising amount of remnant vegetation in urban areas that is particularly valuable for education and recreation. Townspeople are beginning to extend their deep concern for what grows within their own paling fences, to all urban vegetation.

The activities of the revegetation movement are just beginning to counteract the decline of native vegetation cover in Victoria. Evidence of this is the rapidly increasing sales of native trees and shrubs reported from State Government and private nurseries and the rising number of projects being funded from various government and corporate sources.

Why revegetate?

There are many reasons for the large and increasing revegetation effort on rural and urban land.

Pragmatic and economic motives drive many. Farmers in particular, being predominantly small businessmen, often revegetate land for pure financial reasons. Providing shade and shelter can improve profitability by increasing animal and plant growth (Lynch & Donnelly 1980). Woodlots and agroforests can raise and diversify farm incomes by producing timber for sale or use on the farm (Reid & Wilson 1985). Fodder trees may improve the carrying capacity of grazing properties. The simple presence of trees increases the capital value of rural and urban land (Campbell *et al.* 1988).

There is increasing awareness that re-establishment of deep-rooted perennial vegetation is essential for the long-term stability and productivity of large areas of agricultural land. Many land degradation problems (in particular dryland salinity and most forms of water erosion) are fundamentally due to changes in the hydrogeological balance caused by vegetation clearance. Revegetation is an important part of the strategy to correct such imbalances (Morris & Thomson 1983).

It appears that the reintroduction of trees and other forms of deep-rooted perennial vegetation, where part of the relatively stable natural climax communities, helps to restore a degree of ecological stability. The development of sustainable agricultural systems will depend on the incorporation of ecological principles into practical agriculture (Campbell 1989). We believe this will require agricultural systems with much greater biological diversity than those commonly found at present.

Non-economic factors certainly affect decision-making. Many in the revegetation movement are motivated by environmental or "quality of life" benefits of revegetation, including landscape improvement and flora and fauna conservation. Concern about the potential global effects of vegetation clearance stimulates further activity.

Participants

The foundations of the revegetation movement in Victoria are landholders and other enthusiastic citizens. Traditionally they have worked in relative isolation.

These individuals are now increasingly co-operating with their neighbours and local communities by organizing themselves into groups. Particularly significant for rural revegetation are the 48 Farm Tree and 53 LandCare groups with 1 900 and 2 200 landholder members respectively. In urban areas there has been a similar development of community group activity in revegetation.

On a broader scale still are regional community revegetation programmes. Two examples are the six-year-old Project Branchout in north-central Victoria and the Warrenbayne-Boho Land Protection Group outside Benalla. Both have employed full-time staff and sought funds from government, trusts, landowners and industry.

The Victorian Farmers Federation (VFF) has been involved for a decade. Assisted by Government funds, it sponsors the Farm Trees Group programme and co-sponsors the LandCare programme. Undoubtedly the VFF's support has been one of the most compelling features of rural revegetation in Victoria.

Another significant landholder organization dedicated to revegetation is the Natural Resources

Conservation League with 4 500 members. It is active in publicity, provides information, and operates two large nurseries.

The Victoria Conservation Trust promotes and organizes voluntary covenants on areas of freehold native vegetation. Registered on the land title, the covenant binds all future owners. The Trust also solicits tax-deductible donations for land purchase and other conservation projects.

The role of State Government is to co-ordinate activities through broad consultation with private individuals, groups and organizations; to conduct research; and to provide financial support, and extension and advisory services. Major agencies involved are the Departments of Conservation, Forests and Lands (CFL), and Agriculture and Rural Affairs (DARA).

Greening Australia and the Victorian Garden State Committee also play a valuable co-ordinating role having a broad membership from Government and community organizations involved in revegetation.

The State Conservation Strategy provides a broad framework for State Government action and strongly emphasizes the need for revegetation to meet its ecological, social and economic objectives.

A Land Protection Advisory System has been established under legislation to represent landholder interests in all aspects of land protection. This system allows landholders, and the general community, to comment upon and influence State Government action at both regional and central levels.

Local government plays an essential role in revegetation, firstly because it manages most of Victoria's roadsides. In many districts, these roadsides carry the last remnants of natural vegetation. Municipalities also have pre-eminent responsibility for managing urban vegetation. In addition some municipalities are drafting conservation strategies to complement the State document, and are employing conservation officers.

Local government can protect vegetation on private land using regulations and planning schemes that specify acceptable land use, though to date it has not been active in this area. Local government also directly affects rural vegetation through its fire protection requirements and activities.

Local government's flexibility, closeness to the community and tradition of leadership in other fields of welfare, suggest that its activities in revegetation will greatly increase. Another factor, of course, is that local government responds rapidly to community pressure, and, in the case of revegetation, pressure for action is growing.

There is an expanding role for consultants and private contractors in all aspects of revegetation. This is desirable as it involves more people, improves technical standards, provides absentee landowners with the means to revegetate, and leads, through competition, to innovations.

The Victorian revegetation movement is therefore a loose collection of individuals, groups, government and non-government organizations, generally pulling in the same direction through a shared vision and perhaps simplistic notion that more trees and other natural-looking vegetation in the landscape is a good thing. Its motives range from the pragmatic to the esoteric and altruistic.

The fact that the vision is simple helps generate public interest and support.

The tools

Funding

A useful principle is "beneficiary pays". Where individual landholders are the sole beneficiary of revegetation works, they should pay for them. Similarly adjoining landholders can share the costs of projects with mutual benefits. Landholder groups are an effective way to obtain both landholder co-operation and the redistribution of costs for land management issues which cross boundary fences. Where benefits extend to the local community, groups can solicit support, for example from local businesses and other community groups.

Government funding is also appropriate where there are broader community benefits. In Victoria approximately $1.1 million dollars is spent annually on revegetation grants to landholders and groups. Philanthropic trust and corporate sponsorship of projects is increasing. The Potter Farmland Plan (PFP) demonstrates this well. (The PFP is a major practical demonstration at Hamilton of Whole Farm Planning, with heavy emphasis on the value of revegetation to increase short- and long-term agricultural productivity and to facilitate good management). This source of funds has probably been underexploited in the past.

Giving interested parties an opportunity to contribute financially to revegetation projects through special trusts is a largely unexplored, but potentially powerful, funding strategy.

Expenditure on revegetation works for arresting land degradation can also be deducted from the taxable income of primary producers (under section 75D Income Tax Assessment Act).

Research

The basic technology required to revegetate Victoria exists. Research is required in some cases to refine techniques and improve efficiency.

Whilst planting is likely to remain the most common technique in revegetation, there are increasing roles for natural regeneration; direct seeding (mechanised techniques enable large areas to be covered cheaply and rapidly, but good work can be done with simple manual equipment); transplanting surplus natural regeneration; and fencing vegetation from domestic and pest animals.

To date there has been little State Government activity in revegetation research. Much of the useful information on revegetation has come from CFL research into land degradation and public land forestry. Current investigations include DARA and joint DARA-CFL research projects on agroforestry, fodder tree management and direct seeding for tree establishment.

Tertiary institutions are beginning to make a valuable contribution. Sociological research into land-manager attitudes to revegetation and the subsequent design of more effective extension programmes have been conducted at Melbourne University's School of Agriculture and Forestry. The School's recently opened Centre for Land Management and Farm Planning undertakes research into sustainable agricultural systems.

In practice, much of the practical research and development is done by innovative farmers. Notable examples are the renaissance of direct seeding for broadscale revegetation and the development of innumerable fencing and guarding techniques to protect vegetation. Most farmers generally confront problems with ingenuity. The ABC Countrywide Tree Care Award encourages this spirit of invention.

Labour — who does the work?

The task of revegetating large areas of rural and urban Victoria is a big enough job for everybody to be involved: certainly government and landholders cannot do the job on their own.

State Government must improve its management of linear reserves and small pockets of public land spread across the rural and urban landscapes. Local government must develop vision and take greater responsibility for revegetation, especially on roadsides and in urban areas.

Nevertheless, the greater part of the work will be done outside government: private landholders, landholder groups, community groups, regional revegetation programmes, employment programmes, contractors and consultants must do it.

Both urban and rural revegetation appear to be ideal activities to assist bridge the urban-rural social gap in Australia. Of particular value would be greater direct involvement of city people in rural revegetation activities. The Australian Trust for Conservation Volunteers is a growing organization which facilitates this sort of involvement for local and foreign volunteers.

Legislation

The Flora and Fauna Guarantee Act 1988 supports voluntary conservation and revegetation programmes. It provides for a period in which to negotiate a desirable solution where the existence of any species on private land (or any species or vegetation association on public land) is threatened. Compensation is payable if landholders are adversely affected. Complementary education and extension programmes will be very important for this Act to be effective.

Local government planning schemes and regulations, as stated above, could augment this State legislation.

Species

Indigenous species are used increasingly because they are generally long-lived and healthy. They regenerate well, particularly after fire, provide habitat for local wildlife, are aesthetically pleasing, conserve genetic resources and provide shelter and many useful economically important products such as timber and honey. As well, they have evolved in the local environment and so are capable of forming balanced and stable landscapes in the long term. Finally, by helping maintain our distinctive landscapes, they make a strong contribution to tourism, an increasingly important industry to Australia.

There has been a recent growth in the number of small local nurseries specializing in indigenous plants as well as nurseries willing to propagate plants for individuals from locally collected seed.

We have only started to comprehend the threat of environmental weeds (also called "bushland weeds") to indigenous ecosystems. Many well-meaning revegetators have employed such plants in the past, unaware, as the rest of us were also, of their capacity to spread.

Exotic plants and non-indigenous native plants can obviously still perform valuable functions in rural and urban areas, however, great care needs to be taken when making such introductions.

Education and extension

CFL and DARA conduct extension programmes in various aspects of revegetation. The quality of advice given varies greatly but is rapidly improving. Other non-government sources of information are equally important, or more so. The National Tree Program from Canberra also produces information, but perhaps its most valuable role is to engender broad, nation-wide awareness of the need to revegetate. Its closer links with Greening Australia, announced recently, are very welcome.

The rapid expansion of community and landholder groups active in revegetation has improved the flow of information. Groups also help accelerate the adoption of new ideas and techniques (O'Brien 1987). Group decisions are normally more stable, which is important for the long-term commitment to maintenance required for successful revegetation. Most groups organize meetings, seminars, field days and farm walks, as well as newsletters and local projects and publicity.

A variety of non-government organizations also publicise revegetation. The NRCL produces the excellent magazine "Trees and Natural Resources", and regularly attends rural and urban field days and shows. The VFF farm trees group co-ordinator produces and distributes the newsletter "Trunkline" to all members of farm tree and LandCare groups. The Australian Forest Development Institute often deals with agroforestry and management of small-scale forests in its magazine "Australian Forest Grower".

In Victoria the rural press and the major agricultural publications have given much space to revegetation.

Wider publicity and community education programmes therefore merit high priority to win the involvement and support of the Victorian people. Increased public awareness should lead to the whole community assuming a much greater responsibility for revegetation. Recent surveys show public awareness of the importance of revegetation is growing (Cary *et al.* 1986). We must sustain this trend to increase public support, ensure revegetation is more prominantly placed on local and State government agenda and encourage community groups and other organizations.

Future strategies

Groups

The significant development of landholder and other community groups active in revegetation must be sustained. These groups are accepting responsibility for broadscale land management issues previously considered the domain or the problem of government. This sense of ownership and empowerment will help provide longer term solutions to private land management problems than will government designed, sponsored and owned projects. State Government agencies can foster these groups by providing technical expertise and support as required. In return, groups can be useful to government by providing simpler and more effective means for disseminating information and funds.

Linear reserves and small blocks of government land

These areas of land are significant for revegetation but have been largely ignored by responsible government agencies. Management of these areas should have a higher priority.

Research co-ordination and initiatives

Revegetation research in Victoria is disparate and poorly co-ordinated. Better communication between the groups involved is necessary. There is also great unharnessed research potential in tertiary institutions. Enthusiastic and innovative landholders and groups should become more involved in formal research activities.

Research initiatives in a few key areas are required. Improving the reliability of broadscale direct seeding and natural regeneration techniques and the development of lower cost fencing systems would be particularly useful.

Whole farm planning

Whole farm planning is proving an extremely valuable tool for integrating ecological land management concepts into farming. Whole farm planning involves analysis of the land and identification of relatively homogeneous land management units. The plan then allocates preferred uses to land units according to their capability, and identifies and co-ordinates desirable improvements (including revegetation) that maintain and enhance sustainable productivity. A whole farm plan helps farmers place short-term management issues in the context of long-term land management objectives. Whole farm planning should be more widely promoted by government agencies.

Monitoring

Government must accept the responsibility for monitoring the status of native vegetation on private land. The existing data base is relatively poor overall, although good data are available for localized areas. Remote sensing by satellite now enables a broadscale assessment of the whole state. A continuous monitoring programme, with high priority for consulting community groups, interested individuals and other organizations is required.

Better information on the status of vegetation in Victoria, including identifying any prime and unique areas on private land and classifying the degree of degradation, will aid our planning of revegetation. It should also raise public and political awareness of the value of natural vegetation on private land.

Changing attitudes — sensitive use of public education, extension, incentives and legislation

Contrary attitudes are probably the greatest barrier to further expansion of the current revegetation effort.

Education and extension

Public education programmes must make landholders aware of their responsibility for good land management practices including revegetation and natural vegetation conservation. The wider community should also be targeted for such education. Citizens are becoming increasingly aware of their capacity to influence production systems through consumer preference and exerting public pressure.

Co-operative attitudes in both rural and urban areas are most likely to be fostered by extension, incentive and education. Education is needed to increase understanding of the value of revegetation and natural vegetation conservation. Recent research (Cary 1988) indicates that properly designed education and extension programmes can achieve desired attitude changes regarding revegetation.

Ultimately, and as soon as possible, a stewardship ethic amongst landholders must develop in Australia, based on managing land for sustainable production of goods and services, and underlining that we are temporary trustees of the land for future generations (Roberts 1986). This will also boost the revegetation movement.

Extension programmes must identify information that is likely to affect landholder decision-making. Extension research techniques are available to achieve this (Cary 1988).

Legislation

It is impossible to change attitudes through legislation, especially in an independent society like ours. We believe laws to enforce a particular land use, such as the maintenance of vegetative cover, are futile unless landholders and the wider community recognize the rationale behind those laws and actively support their objectives. General public support is required to achieve a revegetated Victoria. Unpopular laws will not achieve this.

Less severe forms of enforcement, such as cross-compliance regulations (for example, making salinity control assistance conditional upon maintenance of existing natural vegetation cover) may be more successful in encouraging desirable attitudes.

The major thrust should be to increase the number of co-operative landholders and improve community attitudes to revegetation by education, extension and incentive. We do believe however that a legislative "safety net" is needed to deal with landholders who persist in clearing large tracts of freehold native vegetation where it is clearly not in either the individual's or the community's interests to do so. This legislation should only be used as a last resort where other mechanisms fail. Compensation should be available to the landholder. In many situations it would be preferable if this type of land were purchased by the government.

Incentives

The use of incentives should be continued on a beneficiary pays basis. Government incentives should reflect community benefits from revegetation. Incentives however rarely change land management in the long term if unsupported by other approaches. Providing good technical information, encouraging groups that accelerate attitude change through social pressure, and generating publicity to raise awareness and reinforce desired community and landholder attitudes, all complement incentives.

Conclusion

The last decade in Victoria, though not without substantial losses of native vegetation on private land, has seen firm foundations laid for a revegetated Victorian landscape. The infrastructure is there and can be expanded, much of the technical knowledge is available, good demonstrations are common, community participation is a feature, many priorities have already been identified, and enthusiasm is rapidly developing. In the near future our activities must be increased by several degrees of magnitude and be accepted by the whole community.

References

Campbell C. A. (1989) Bridging the gap between conventional agriculture and sustainable agriculture — the role of whole farm planning. *Aust. J. Soil and Water Cons.* **2,** 43-52.

Campbell R., Chandler R. & Thomas G. (1988) *Victoria Felix: improving rural land with trees* (eds A. Lindsay & R. M. Youl) Monash University, Melbourne.

Cary J. W. (1988) Predicting the effect of communication strategies on attitudes and belief using galileo analysis. School of Agriculture and Forestry, University of Melbourne, Melbourne. (internal report).

Cary J. W., Beel A. J. & Hawkins H. S. (1986) Farmers Attitudes to Land Management for Conservation. Department of Conservation, Forests and Lands, Melbourne.

Lynch J. J. & Donnelly J. B. (1980) Changes in pasture and animal production resulting from the use of windbreaks. *Aust. J. Agric. Res.* **31,** 967-79.

Morris J. D. & Thomson L. A. J. (1983) The role of trees in dryland salinity control. *Proc. Roy. Soc. Vic.* **95,** 123-31.

O'Brien B. (1987) Old systems for new times — extension processes to suit the new agricultural scene. Australian Agricultural Extension Conference 1987. Working Papers Volume 1, 1-6.

Reid R. & Wilson G. (1985) *Agroforestry in Australia and New Zealand.* Goddard and Dobson, Box Hill.

Roberts B. R. (1984) Land Ethics — a necessary addition to Australian values. Proc. Soil Degradation Conference 1984. Australian National University, Canberra.

Woodgate P. & Black P. (1988) Forest cover changes in Victoria (1869-1987). Department of Conservation, Forests and Lands, Melbourne.

Proc. Ecol. Soc. Aust. 1990 — 16:141-147

Interactions of native and introduced species in new habitats

M. D. FOX

National Herbarium of New South Wales, Royal Botanic Gardens, Sydney, New South Wales 2000, Australia

Abstract

One outcome of the establishment of industrial and agricultural societies in Australia has been the creation of new habitats. These largely comprise highly disturbed urban habitats and modified rural areas. In both cases native and introduced organisms found new niches to occupy. This chapter reviews patterns of successful and unsuccessful transitions of native species becoming commensal and of introduced species becoming naturalized. In particular the urban environment, a highly modified environment that still retains some natural features, provides an abundance of resources. These are partitioned among a variety of native and introduced species. To a limited extent, native herbivores have switched to eat introduced plants and native predators to introduced prey. Similarly introduced herbivores consume some native plants and alien predators find prey among the native commensal animals. New habitats continue to be created and new communities established. Future trends will include more native weeds as well as continuing introductions.

Introduction

The movement of human populations from continent to continent has resulted in the mixing of previously isolated biotas. The relatively recent migration of populations of essentially European origins to Australia typifies this mixing. Perhaps more than on any other continent, the outcome involves combinations of organisms previously totally separated. The much earlier invasions by people from the near-north had introduced relatively few organisms. Added to this, the lifestyle of the first human occupants had not led to massive habitat alteration, although their widespread use of fire had doubtless led to fundamental changes in some vegetation types (Singh *et al.* 1981). However, the arrival of industrial and technological cultures two centuries ago accelerated changes already in place, and introduced entirely new disturbances as well as entirely new animals and plants.

This is one aspect of the theme "200 years of utilization, degradation and restoration", that has received little attention: the creation of new habitats, particularly of those highly modified areas in which people congregate in large numbers, our cities and towns. Had we arrived here unaccompanied and retained our isolation we could have created new habitats populated only by ourselves and by native animals and plants. That is clearly not the case. We have introduced, by accident and design, a large number of organisms from other parts of the world. The introduction of new species, although indirect, is as much a form of disturbance as stripping away topsoil or altering the fire regime.

There has been considerable ecological interest in the question of organisms introduced into Australia (Anderson 1977; Kitching & Jones 1981; Groves & Burdon 1986) and attention has been placed on the uniqueness of the Australian biota and the impact of the introductions (e.g., Adamson & Fox 1981; Fox & Adamson 1986). Because of the magnitude of the problems engendered by some of our recent arrivals, and the spectacular success of a few biological control agents, Australian examples feature in the world literature of biological invasions.

Weir (1977) considered past, present and future exotic invasions into Australia in the framework of a two-way table. The first distinction is based on the origin of the exotic species and whether that is a natural complex system or a modified habitat. The second is the system into which the species is introduced, again either natural or modified (or essentially undisturbed or disturbed, respectively).

It is not pertinent to this discussion to consider the situation where species migrate naturally from one natural complex system to another. Biological invasions occur in any combination of either the exotic species emanating from a modified system or being introduced to a modified system. In the first case the survival of the exotic species when its original habitat was modified preadapted it to survival in other habitats. Such species are commonly generalists (Elton 1958). Weir (1977) cites as examples lantana *Lantana camara* in natural forests, European carp *Salmo* spp. in rivers and water hyacinth *Eichhornia crassipes* in wetlands. In a review of the invasibility of natural ecosystems, Fox and Fox (1986) have suggested that invasions only occur in disturbed systems. Although some invaded systems may appear natural there has usually been some degree of modification. Of the three examples, lantana only occurs around the disturbed edges of forests or in gaps in the canopy, the persistence of carp in inland rivers is assisted by impoundments (Arthington & Mitchell 1986), and water hyacinth usually invades enriched waters (Sainty & Jacobs 1981).

Some native species make the transition from a natural system to a modified one and Weir (1977) suggests that brush-tailed possums *Trichosurus vulpecula* utilizing plantations, red-necked pademelons *Thylogale thetis* feeding on potato crops and red kangaroos *Megaleia rufa* surviving in semi-arid systems modified by artificial watering points, are typical of this form of invasion.

The most rapid invasion occurs when both the system yielding the exotic species and the system to which it is introduced are modified. All the agrestal weeds associated with crops the world over are typical of this category, as are the pests of agricultural and urban areas. Weir (1977) suggests that to date our knowledge of such invasions is relatively high whereas for the other three categories it is less so. In further considering introductions to urban and rural habitats, attention will focus on the cases in which the exotic species are more commonly introduced from modified systems elsewhere, while others may be from natural complex systems able to make the transition to commensalism.

In the Australian context relatively little attention has been paid to urban ecology (Recher 1972; Burrell 1972). The bulk of scientific enquiry into invasive organisms has concentrated on those species of economic significance which tend to be in agricultural systems. Little attention has focused on the highly modified urban areas and the interactions of introduced and native species there.

Cosmopolitan species

The urban environment is much the same the world over. Much of the land surface is sealed and covered with concrete or bitumen. Soils are altered by modified nutrient and watering regimes. The result is little difference between urban parklands of say Sydney, Hong Kong and Copenhagen as plant habitats except in the temperature-insolation regimes of each region. Accompanying this similarity there is a world-wide suite of urban weed species with only few regional variations. The pollen of common plantain *Plantago lanceolata* is associated with most neolithic sites in Europe (Corner 1964) and the plant is now a world-wide weed.

Similarly, there is only a small band of animals that have found niches in the modified urban environment. Some of them are found around the world. Domestic pigeons *Columba livia* sit atop Nelson in Trafalgar Square, perch on ledges of the Empire State Building in New York and befoul the Archibald Fountain in Sydney's Hyde Park. The high-rise buildings of the world's cities are structurally similar to the rocky seacliffs that are the pigeons' natural habitat (Goodwin 1978).

Widespread human commensals include the house mouse *Mus musculus,* black rat *Rattus rattus* and brown rat *R. norvegicus* as the main mammalian examples: cats and dogs require a greater level of coercion. The house sparrow *Passer domesticus,* domestic pigeon, Indian myna *Acridotheres tristis* and starling *Sturnus vulgaris* are the principal bird commensals. There are no good examples of cosmopolitan reptilian, fish or amphibian commensals. The list of invertebrates is similarly relatively short: including the house fly *Musca domestica,* cockroaches *Blatta* spp., Argentine ant *Iridomyrmex humilis,* flea *Pulex irritans,* garden snail *Helix aspersa* and earth worm *Lumbricus* sp.

The observation that successful establishments are few is confirmed by Recher's (1972) list of the numbers of vertebrate species recorded in Sydney since 1770. Comparative numbers for indigenous and introduced () species are: fish *ca* 300 (5), amphibians 35 (1), reptiles 50 (0), birds 386 (15) and mammals 25 (7).

Newsome and Noble (1986) used the introduced bird fauna of Australia to consider degrees of success in invasion. Of a recorded 107 attempted introductions of bird species (including translocating native species), only 59 (or 55%) were successful, of which 32 were native species. This leaves a continental total of 27 successfully established exotic bird species. This trend is not unique to Australia. Numbers of birds introduced to various regions of the world and the number and percentage of failures were published by Long (1981). The 54% of introductions which failed to establish in Australia (cf. 45% Newsome & Noble 1986) is comparable to values for North America, Hawaii, Fiji and Europe. South America is somewhat lower, at 40%, and Tahiti notably higher at 82%. Of the introduced birds the most abundant is the sparrow. In a recent bird census (Anon. 1988) it was the most numerous sighting in all states except Western Australia (which it has not colonized) and New South Wales. In New South Wales, the dominant bird is the Indian myna.

Of the mammal species, Myers (1986) lists 47 introduced to Australia of which 16 have died out and three persist only as small local populations. Fourteen are "widespread" species including three cosmopolitan human commensals (rats and mice), three domesticated (dingo, dog and cat), four herd animals (pig, sheep, goat, cattle), one domesticated draft animal (horse) and three sport animals (rabbits, hares, foxes). Few reptiles and amphibians are naturalized, they are: the house gecko *Hemidactylus frenatus* and arboreal gecko *Lepidodactylus lugubris* which occupy human dwellings in the coastal tropics and the marine toad *Bufo marinus*. Twenty-one species of introduced freshwater fish have self-maintaining populations (Arthington & Mitchell 1986).

Most of these vertebrate and invertebrate commensals co-occur with human populations, both urban and rural, all over the world. Added to these is the suite of other introductions that may be unique to one or a few regions. For example, in Sydney the red-whiskered bulbul *Pycnonotus jocosus* is particularly successful, and recently the white ibis *Threskiornis molucca* has become much more common in inner city parks after moving from Taronga Zoo (G. Phipps, pers. comm.).

Of the plant examples, there is a vast number of introduced species associated with any urban area. Of the 2000+ species of vascular plants in the Sydney region *ca.* 450 are introduced (Beadle *et al.* 1963). After only 78 years of settlement Woolls (1869) was able to list over 100 introduced plants occurring "in the neighbourhood of Sydney". Similarly, for South Australia, Schomburgk (1879) listed 99 naturalized weeds. Interestingly, this included 14 species which were introduced as garden plants and which had become acclimatized. This process is still occurring, for example, in Sydney the colourful, shade-tolerant balsams (*Impatiens* sp.) are spreading into the understorey of urban bushland. Such invasions only occur as a result of disturbance (Fox & Fox 1986), and often, in Sydney's oligotrophic soils, eutrophication is a major disturbance (Clements 1983).

An aspect of introductions in the urban environment which has received little attention is the interaction between the hardy native species which persist as development takes place, and the introduced species. Plant-herbivore relationships are examined first, followed by predator-prey relationships.

Native herbivores using introduced plants

The brush-tailed possum, whose natural diet comprises eucalypt leaves, fruits, buds and bark (How 1983), in urban settings will eat rosebuds and other valued horticultural plants. Though a generalized herbivore, nothing in the natural diet of the possum equates to succulent, perfumed rosebuds. The dietary adaptability of this marsupial can be gauged from its success as an introduction to New Zealand.

The preferred diet of the grey-headed fruit bat *Pteropus poliocephalus* is eucalypt blossum, native figs and rainforest flowers and fruits. In times of scarcity orchard fruit and bananas are eaten. Fruit bats accounted for significant losses from most fruit crops in Queensland with highest damage to lychees, paw paw, mango, stone fruits, apples and bananas at an estimated cost of $7 million per year (Jamieson 1988). That represented 3% of production whereas, by comparison, all combined bird damage amounted to just over 2%. Before the development of plantations fruit bats would have eaten native rainforest fruits including native bananas (Tidemann & Nelson 1987) and have opportunistically switched to the more readily available fruit. Fruit bats maintain large colonies in closely settled urban areas. The colony in the Sydney suburb of Gordon has been estimated at 25 000 (Parry-Jones 1988).

The introduction of native plants to new areas provides further examples of this kind. The Western Australian kangaroo paws *Anigozanthos* spp. introduced to gardens in eastern Australia are frequently visited by eastern honeyeaters such as the Eastern spinebill *Acanthorhynchus tenuirostris* that would not have encountered them previously. This bird also takes nectar from a variety of introduced garden plants, particularly fuchsias (Reader's Digest 1976). Other native birds have successfully switched to introduced garden plants. The eastern rosella *Platycercus eximius* and other small parrots eat berries of ornamental shrubs such as cotoneaster *Cotoneaster* spp.; Eastern silvereyes *Zosterops lateralis* eat the fruit of privet *Ligustrum* spp. and lantana. The white-headed pigeon *Columba leucomela* eats camphor laurel *Cinnamomum camphora* fruit in northern coastal New South Wales (Phipps, pers. comm.). The flowers of the introduced coral tree *Erythrina* sp. are eaten by many honeyeaters.

Some native invertebrates have reached pest proportions by switching to introduced crop plants as either food plants or oviposition sites. The light brown apple moth *Epiphyas postvittana* uses plants from 124 genera, of which 22 are natives (Geier & Briese 1981). Included in the crop plants are apples, and moth-injured fruit can be as high as 30% of the crop but this figure varies highly with both location and year.

The native cicada *Cicadetta quadricincta* has increased greatly in density in urban areas by using the roots of both native and introduced garden plants. Higher abundance of the cicada has led to greater densities of the common orb-weaving spider *Exiophora biapicata* (B. York Main, pers. comm.; Gwynne 1987).

Introduced herbivore-native plant

The introduced goldfinch *Carduelis carduelis* now common in south-eastern Australia, nests near settlements and eats a variety of weed seed but also eats the seed of native grasses. The spice finch *Lonchura punctulata* eats the seed of native grasses such as *Themeda* spp. and *Imperata cylindrica*. These are very adaptable birds, with a wide diet range, and in several areas are replacing native finches (Lreitzka in Wade, 1975).

The introduced garden snail *Helix aspersa* and slugs may eat some of the more mesomorphic native plants. Many species of introduced slugs have spread through relatively undisturbed forests of Tasmania and south-western Australia, presumably eating mesic native plants, however not much is known about their impact (Ponder, pers. comm.). Most of the scleromorphic native plants, seem only to suffer the depradations of their native foliovores such as leaf skeletonizers and are not affected by introduced species. One exception is the introduced honey bee *Apis mellifera* which takes nectar from a wide range of native plants (Wills *et al.* this volume). Taylor and Whelan (1988) have demonstrated that although the bees take nectar from *Grevillea* flowers they do not effect pollination.

The cabbage-white butterfly *Pieris rapae* is the only introduced butterfly that is a serious pest in Australia (Jones 1981). It oviposits on a wide range of crucifers, most of which are introduced plants such as the varieties of *Brassica oleracea* (cabbage, cauliflower, brussel sprouts, kale) but has also been reported using native species such as peppercress *Lepidium hyssopifolium* (C. Smithers, pers. comm.).

Native predators preying on introduced prey

The blue-tongued skink *Tiliqua scincoides* of urban backyards preys on introduced garden snails (H. Cogger, pers. comm.). *Amphibolurus* lizards also commonly take this prey. Goannas (*Varanus* spp.) are omnivorous, including in their diet a variety of native prey as well as *Mus, Rattus* and rabbits. The taipan *Oxyuranus scutellatus* of coastal tropical Australia and its arid counterpart, the small-scaled snake *O. micolepidotus,* now eat introduced house mice as well as native small mammals (Shine & Covacevich 1983).

Nicholls (1958) listed five native birds: the blue heron *Ardea novaehollandiae*, black duck *Anas superciliosa,* silver gull *Larus novaehollandiae* and bittern *Ixobrychus flavicollis,* as well as black cormorants *Phalacrocorax carbo* feeding on introduced trout mostly observed as captures from hatcheries. Water rats *Hydromys chrysogaster* also feed on young trout in hatcheries. Platypus *Ornithorhynchus anatinus* ingest trout eggs and emerging fry from spawning grounds in rivers. In outer urban areas and rural settings, the wedge-tailed eagle *Aquila audax,* brown goshawk *Accipiter fasciatus* and square-tailed kite *Lophoictinia isura* prey on rabbits, while the black-shouldered kite *Elanus notatus* and nankeen kestrel *Falco cenchroides* prey on house mice.

Many native spiders that can survive urbanization devour introduced insects captured in their webs. Good examples are the black house spider *Badumna robusta* and red-backed spider *Latrodectus hasseltii.* Around houses are many structures suitable for web construction and many suitable prey such as house flies; unnaturally high aggregations of spiders result (B. York Main, pers. comm.).

The long-nosed bandicoot *Perameles nasuta* and southern brown bandicoot *Isoodon obesulus* feed on introduced insects and larvae as do the native cats (*Dasyurus* spp.), however native predators are themselves the prey of cats, dogs and foxes.

Introduced predators — native prey

Many of the introduced sport fish are predatory unlike most native freshwater fish (Tilzey 1977). The introduced brown trout *Salmo trutta* eats native galaxias. Sometimes the interaction is not predation but competition and so the introduced mosquito fish *Gambusia affinis* competitively displaces small native fish such as hardyheads (Atherinidae) and blue eyes (*Pseudomugil* spp.) which are also surface feeders in fresh waters (J. Paxton, pers. comm.).

Introduced mammalian predators such as cats and foxes prey on small native marsupials as well as on native birds and reptiles. Recent reports by bird watchers of the decline in sightings of small native birds in the cities may be interpreted as predation of eggs and hatchlings by urban cats. Recher (1972) commented on letters in newspapers noting the decline of small native birds such as the superb fairy-wren *Malurus cyaneus* and the increase in pied currawongs *Strepera graculina* since the 1940s. He attributed the decline of wrens to the loss of shrubby cover and the increase of currawongs to the abundance of exotic berry plants. Recent reports continue this trend with concern now expressed over the general loss of small native passerines and the possibly unrelated increase in larger native birds (Anon. 1988). The ten most numerous sightings in Sydney gardens were: *Indian myna (* introduced), rainbow lorikeet *Trichoglossus haematodus,* *Indian dove *Streptopelia chinensis,* magpie *Gymnorhina tibicen,* pied currawong, *sparrow, *bulbul, *starling, kookaburra *Dacelo gigas* and "raven/crow" *Corvus* sp. The native species all have broad diets which now include introduced species of plants and/or animals (G. Phipps, pers. comm.). One trend to reverse this is the increased popularity of growing native shrubs, particularly those rich in nectar such as *Callistemon* spp., *Grevillea* spp. and *Banksia* spp., which attract native honeyeaters.

Except for the two geckos mentioned earlier no examples are known of introduced predatory reptiles.

The marine toad is the only introduced amphibian now supporting feral populations in urban and rural areas. This generalized predator feeds on a wide variety of vertebrate and invertebrate animals which would have previously constituted the diet of native predators (Covacevich & Archer 1975). Native snakes consume larvae and young toads (Stammer 1981) and there are reports of other native predators such as water rats learning to eat this poisonous animal (B. Fox, pers. comm.). There is inferential evidence that the arrival of the marine toad leads to the local extinction of some native frogs and snakes (Covacevich & Archer 1975).

The shovel-headed garden worm *Placocephalus kewensis* is a carnivorous terrestrial worm thought to be introduced from Britain to Australia (Anon. 1968). It feeds on earthworms, young snails, and insects and their larvae.

Extensions of ranges

Just as there have been native animals and plants disadvantaged by the development of urban areas, some native species have benefitted. The rare plant *Eucalyptus scoparia* is now commonly grown as a street tree throughout Sydney and extends well beyond its previous range of part of the northern Tablelands. Many birds of inland Australia, forced to the coast in times of drought have now established large breeding populations centred on urban parks. It is possible that some of these are individuals released by licensed trappers, they can persist in the metropolitan area but cannot breed due to a shortage of breeding hollows (Anon. 1988). Such extended ranges bring together not only native and introduced species, but also local native species and other native species previously isolated from each other leading to interactions previously precluded by geographic separation.

Rather than extensions of range, some native species now exhibit population concentrations previously not encountered. For example, the remaining fruit bat colonies in or near urban areas have problems with nutrient accumulation and propagule introduction by the animals. At Gordon, a northern suburb of Sydney, the native vegetation is dying and will be replaced in the near future by exotic species (Buchanan 1985).

Another interaction of native and introduced organisms is that of disease agents being carried by native vectors. Native mosquitos (*Anopheles annulipes* and *Culex annulirostris*) carry the *Myxoma* virus used in the biological control of rabbits (Fenner & Ratcliffe 1965). Heartworm *Dirofilaria immitis*, now widespread in Sydney, is carried by the native mosquitoes *C. annulirostris*, *Aedes vigila* and *Aedes notoscriptus* as well as by the introduced *C. fatigans* (Bemrick & Moorhouse 1968). A full list of the native and introduced mosquitoes and the diseases they carry is given in Kay *et al.* (1981). The introduced freshwater snail *Limnea columella* introduced to Australia two decades ago is host to the cattle liver fluke and is still spreading.

New niches and niche dimensions

In one respect there are no new urban niches here, just the same urban niches created elsewhere around the world reproduced here and filled by many of the same organisms. The way in which new niche dimensions have been created is by the interaction of native species and the introduced ones, in a variety of relationships. Native organisms that may have faced serious reductions in suitable habitat now have enlarged their food niches by eating introduced plants or prey. These food items may be available in seasons previously unavailable to the native species so that its seasonal movements are altered. Some native animals and a few plants are advantaged by human urban environments and have extended their ranges since European settlement.

Urban ecology warrants further investigation. Over the past 20 years efforts have concentrated on human impact, pollution, and sociological problems. Animal-plant interactions in an urban setting have been largely overlooked and would be a profitable avenue of research. More introductions continue, for example Swincer (1986) lists the following invertebrate introductions since 1960: five aphids, two Hymenoptera, one *Trogoderma* sp., two mite species and two species of snails. Among the plants an average of seven new species are naturalized each year (Groves 1986), compared to four species per year to urban Auckland (Esler & Astridge 1987). Given past patterns, many of these will fail to establish viable populations, but the future will see the establishment of new pest species, both animal and plant.

Acknowledgements

I am indebted to the following colleagues for examples in the named groups — Doug Benson (plants), Elizabeth Cameron (reptiles and amphibians), Hal Cogger (reptiles and amphibians), Barry Fox (mammals), Pat Hutchings (worms), Kerryn Parry-Jones (fruit bats), John Paxton (fish), Graeme Phipps (birds), Winston Ponder (snails and slugs), Harry Recher (birds), Courtney Smithers (butterflies), Peter Weston (plants), Barbara York

Main (spiders). I also thank Barry Fox for critically reading the manuscript.

References

Adamson D. A. & Fox M. D. (1981) Change in Australasian vegetation since European settlement. In: *A history of Australasian vegetation* (ed. J. M. B. Smith) pp. 109-47. McGraw-Hill Book Company, Sydney.

Anderson D. J. (ed.) (1977) Exotic species in Australia — their establishment and success. *Proc. Ecol. Soc. Aust.* **10,** 186 pp.

Anon. (1968) *The shovel-headed garden worm.* Leaflet No. 15, The Australian Museum, Sydney, 3 pp.

Anon. (1988) Bird count 1988. *Australian Wildlife,* September 1988.

Arthington A. H. & Mitchell D. S. (1986) Aquatic invading species. In: *Ecology of biological invasions: an Australian perspective* (eds R. H. Groves & J. J. Burdon) pp. 34-56. Australian Academy of Science, Canberra.

Beadle N. C. W., Evans O. D. & Carolin R. C. (1963) *Flora of the Sydney region.* A. H. & A. W. Reed, Sydney, 720 pp.

Bemrick W. J. & Moorhouse D. E. (1968) Potential vectors of *Dirofilaria immitis* in the Brisbane area of Queensland, Australia. *J. Med. Entomol.* **5,** 269-72.

Buchanan R. (1985) *Site assessment of the Gordon bat colony.* Ku-ring-gai Bat Colony Committee, 18 pp.

Burrell J. P. (1972) Vegetation of the Sydney area: 1788 & 1961. In: *The city as a life system?* (ed. H. A. Nix). *Proc. Ecol. Soc. Aust.* **7,** 71-8.

Clements A. (1983) Suburban development and resultant changes in the vegetation of the bushland of the northern Sydney region. *Aust. J. Ecol.* **8,** 307-19.

Corner E. J. H. (1964) *The life of plants.* The University of Chicago Press, Chicago, 315 pp.

Covacevich J. & Archer M. (1975) The distribution of the cane toad, *Bufo marinus,* in Australia and its effects on indigenous vertebrates. *Mem. Qld. Mus.* **17,** 305-10.

Elton C. S. (1958) *The ecology of invasions by animals and plants.* Methuen & Co. Ltd, London, 181 pp.

Esler A. E. & Astridge S. J. (1987) The naturalization of plants in urban Auckland, New Zealand. 2. Records of introduction and naturalization. *N.Z. J. Bot.* **25,** 523-37.

Fenner F. & Ratcliffe F. N. (1965) *Myxomatosis.* Cambridge University Press, Cambridge.

Fox M. D. & Adamson D. A. (1986) The ecology of invasions. In: *A natural legacy.* Second edition (eds H. F. Recher, D. Lunney and I. Dunn) pp. 100-21. Pergamon Press, Sydney.

Fox M. D. & Fox B. J. (1986) The susceptibility of natural communities to invasion. In: *Ecology of biological invasions: an Australian perspective* (eds R. H. Groves & J. J. Burdon) pp. 57-66. Australian Academy of Sciences, Canberra.

Geier P. W. & Briese D. T. (1981) The light-brown apple moth, *Epiphyas postvittana* (Walker): a native leafroller fostered by European settlement. In: *The ecology of pests* (eds R. L. Kitching & R. E. Jones) pp. 131-55. CSIRO, Australia.

Goodwin D. (1978) *Birds of man's world.* University of Queensland Press, St. Lucia, 286 pp.

Groves R. H. (1986) Plant invasions of Australia: an overview. In: *Ecology of biological invasions: an Australian perspective* (eds R. H. Groves & J. J. Burdon) pp. 137-49. Australian Academy of Sciences, Canberra.

Groves R. H. & Burdon J. J. (eds) (1986) *Ecology of biological invasions: an Australian perspective.* Australian Academy of Science, Canberra. 166 pp.

Gwynne D. T. (1987) Sex-biased predation and the risky mate locating behaviour of male tick-tock cicadas (Homoptera: Cicadidae) *Anim. Behav.* **35,** 571-6.

How R. A. (1983) Common brush-tail possum. In: *Complete book of Australian mammals* (ed. R. Strahan) pp. 147-8. Angus & Robertson Publishers, Sydney.

Jamieson G. I. (1988) Fruit losses from flying-foxes in Queensland. *Macroderma* **4,** 34-42.

Jones R. E. (1981) The cabbage butterfly, *Pieris rapae L.:* "a just sense of how not to fly". In: *The ecology of pests* (eds R. L. Kitching & R. E. Jones) pp. 217-28. CSIRO, Australia.

Kay B. H., Sinclair P. & Marks E. N. (1981) Mosquitoes: their interrelationships with man. In: *The Ecology of pests* (eds R. L. Kitching & R. E. Jones) pp. 157-74. CSIRO, Australia.

Kitching R. L. & Jones R. E. (eds) (1981) *The ecology of pests — some Australian case histories.* CSIRO, Australia. 254 pp.

Long J. L. (1981) *Introduced birds of the world.* A. N. & A. W. Reed, Sydney. 361 pp.

Myers K. (1986) Introduced vertebrates in Australia, with emphasis on the mammals. In: *Ecology of biological invasions: an Australian perspective* (eds R. H. Groves & J. J. Burdon) pp. 120-36. Australian Academy of Sciences, Canberra.

Newsome A. E. & Noble I. R. (1986) Ecological and physiological characters of invading species. In: *Ecology of biological invasions: an Australian perspective* (eds R. H. Groves & J. J. Burdon) pp. 1-20. Australian Academy of Sciences, Canberra.

Nicholls A. G. (1958) The population of a trout stream and the survival of released fish. *Aust. J. Mar. Freshw. Res.* **9,** 319-50.

Parry-Jones K. (1988) *Pteropus poliocephalus* (Chiroptera: Pteropodidae) in New South Wales. *Aust. Mammal.* **10,** 81-5.

Reader's Digest (1976) *Reader's Digest Complete Book of Australian Birds.* Reader's Digest Services, Surrey Hills. 615 pp.

Recher H. F. (1972) The vertebrate fauna of Sydney. In: *The city as a life system?* (ed. H. A. Nix). *Proc. Ecol. Soc. Aust.* **7,** 79-88.

Sainty G. R. & Jacobs S. W. L. (1981) *Waterplants of new South Wales.* Water Resources Commission NSW, Sydney. 550 pp.

Schomburgk R. (1879) *On the naturalized weeds and other plants in South Australia.* Government Printer, Adelaide, 13 pp.

Shine R. & Covacevich J. (1983) Ecology of highly venomous snakes: the Australian genus *Oxyuranus* (Elapidae). *J. Herp.* **17,** 60-9.

Singh G., Kershaw A. P. & Clark R., (1981) Quaternary vegetation and fire history in Australia. In: *Fire and the Australian biota* (eds A. M. Gill, R. H. Groves & I. R. Noble) pp. 23-54. Australian Academy of Science, Canberra.

Stammer D. (1981) Some notes on the cane toad *(Bufo marinus). Aust. J. Herp.* **1,** 61.

Swincer D. E. (1986) Physical characteristics of sites in relation to invasions. In: *Ecology of biological invasions: an Australian perspective* (eds R. H. Groves & J. J. Burdon) pp. 67-87. Australian Academy of Sciences, Canberra.

Taylor G. & Whelan R. J. (1988) Can honey bees pollinate *Grevillea? Aust. Zool.* **24,** 193-6.

Tidemann C. R. & Nelson J. E. (1987) Flying-foxes (Chiroptera:Pteropodidae) and bananas: some interactions. *Aust. Mammal.* **10,** 133-5.

Tilzey R. D. J. (1977) Key factors in the establishment and success of trout in Australia. In: *Exotic species in Australia* (ed. D. J. Anderson). *Proc. Ecol. Soc. Aust.* **10,** 97-105.

Wade P. (ed.) (1975) *Every Australian Bird Illustrated.* Rigby, Adelaide. 320 pp.

Weir J. S. (1977) Exotics: past, present and future. In: *Exotic species in Australia* (ed. D. J. Anderson). *Proc. Ecol. Soc. Aust.* **10,** 4-14.

Woolls W. (1869) Notes on introduced plants occurring in the neighbourhood of Sydney. *J. Linn. Soc. (London)* **X,** 35-42.

Proc. Ecol. Soc. Aust. 1990 — 16:149-165

Tree dieback and insect dynamics in remnants of native woodlands on farms

J. LANDSBERG, J. MORSE and P. KHANNA

CSIRO Division of Forestry and Forest Products, P.O. Box 4008, Queen Victoria Terrace, Canberra A.C.T. 2600, Australia

Abstract

In many of Australia's temperate regions, eucalypt woodland now occurs only as remnant stands on pastoral lands, and many of these remnants are being seriously degraded by rural tree dieback. From measurements made in contrasting matched stands of trees having understories dominated either by pasture (heavy use by livestock) or by native shrubs and grasses (light use by livestock), we concluded the following: (1) defoliation by insects, dieback and tree death were all more severe in the pasture stands; (2) despite site stress such as soil acidity, tree branches that were protected from herbivory produced prolific regrowth foliage; (3) insect abundance was not directly related to numbers of trees, since tree numbers were similar in both types of stands but insects were much more abundant in the pasture stands; (4) floristic and avian diversity were much reduced in pasture stands, but levels of predation of insects were similarly high in both types of stands during our study; and (5) insects feeding on trees with an understorey of pasture grew bigger and more quickly, probably because values of nitrogen in soils and foliage were higher in the pasture stands. The most likely cause of this nutrient enrichment was redistribution of nutrients by livestock. Thus nutrient enrichment may be a key factor contributing to the abundance of defoliating insects, and hence to rural dieback, in remnants of woodland used by livestock.

Introduction

The temperate eucalypt woodlands that once dominated large areas of the southern two-thirds of Australia are so poorly represented in conservation reserves that their continued survival as an ecosystem type is threatened (Pryor 1981). This has probably arisen, at least in part, because the main utilization of these woodlands by Europeans has been for pastoralism, and this has seldom involved complete removal of trees. For various reasons (e.g., shelter for livestock, timber for fencing, building and firewood, aesthetic appeal) large tracts of the original woodland were often left uncleared by the early pastoralists, and, until recently it was often assumed that continued use of these remnants of woodland for grazing by livestock posed little risk of further degradation. However, the current severity and extent of rural tree dieback in most of Australia's once-wooded pastoral regions (Fig. 1) indicate that many of these woodland remnants are seriously degrading under current management regimes, even in the absence of further clearing.

In the tablelands of New South Wales (Landsberg 1988; Mackay *et al.* 1984) and in south-eastern Queensland (Wylie 1986) rural dieback is often associated with chronically high levels of defoliation by native species of insects. There are at least five possible explanations for this chronic defoliation.

(1) Periods of high levels of defoliation may be natural in eucalypt woodlands, and rural dieback may therefore be the current manifestation of a naturally recurring phenomenon (White 1986).

(2) Trees in remnants of woodland may be chronically stressed by factors such as soil degradation, increased physical exposure or old age (Landsberg & Wylie 1988) and therefore they may be less able to replace foliage lost to defoliating insects.

(3) Selective removal of trees to promote growth of pasture may have resulted in increased localized abundance of insects on the remaining trees. This affect would be most pronounced for insects such as scarab beetles whose larvae feed on grass roots (Jacobs 1955).

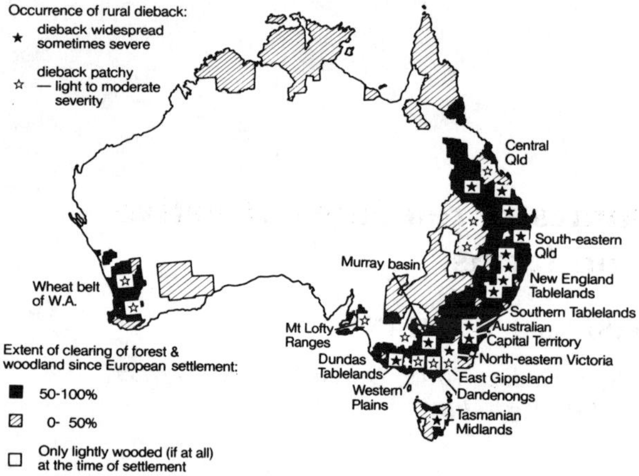

FIG. 1. The extent of modification of Australia's woody vegetation (from Wells *et al.* 1984) and reported occurrences of rural dieback (Landsberg & Wylie 1988).

(4) Floristic and structural simplification of remnant woodlands may, because of loss of habitat and alternative food sources, have reduced the diversity and abundance of the natural predators (parasitoids and insectivores) of defoliating insects, thereby allowing their numbers to increase (Davidson 1980).

(5) Pasture improvement, by increasing soil fertility, may also have improved the nutritional quality of tree foliage sufficiently to increase the abundance of defoliating insects (Landsberg 1986).

We investigated the extent to which each of these hypotheses is consistent with the pattern of dieback of stands of trees in remnants of native woodland in the southern tablelands of New South Wales. We selected stands of trees paired on the basis of their proximity and general similarity in all but understorey, which was dominated either by pasture or by native shrubs and grasses. The pasture stands were used for shelter and grazing by livestock, but the control stands were only lightly or infrequently grazed, if they had been grazed at all.

In these paired stands we measured tree health, the abundance of insects and the rates of defoliation of trees, soil chemistry, foliar nutritional status, floristic and structural diversity, diversity of bird species, and the rates of survival and growth of experimental insects.

Methods

Site selection

The control stands of trees were selected first. There were four selection criteria. (1) The stands must be relatively small (less than 50 ha) and isolated (surrounded by pasture and more than one kilometre from the nearest forest or woodland), so as to be comparable with most of the pasture stands in the study region. (2) They must include yellow box (*E. melliodora* A. Cunn. ex Shau.) and Blakelyi's red gum (*Eucalyptus blakelyi* Maiden) in their overstorey, since this alliance occurs extensively on the slopes and tablelands of New South Wales (Moore 1975) and appeared to us to be most affected by dieback in the study region. (3) They should contain an obvious shrub component in their understorey (so that we could test our fourth hypothesis). (4) They must be within two hours drive from our base in Canberra. We located only eight stands that met all four criteria.

The stands had all fortuitously escaped heavy grazing pressure from livestook. At the time of European settlement two had been gazetted as town commons and one as a cemetery reserve for townships that underwent little subsequent development. Two sites had been gazetted as reserves for travelling stock and have never been leased to adjacent property owners. One is a nature park near Canberra. One had never been used because it is small, and separated from the rest of the property by a busy highway. One had been retained by the descendents of the original settlers for aesthetic reasons. Only this last site appeared different from the pastoral country that it adjoined in any respect other than lack of use for pastoralism. Thus these control sites may well be reasonably representative of pre-European yellow box/red gum woodland ecosystems.

Each control stand was matched, as far as possible, with the nearest stand of trees with a pasture dominated understorey, but otherwise of comparable size, degree of isolation, geology, landscape position, and overstorey composition. If no such stand could be located within 10 km of a control stand, compromises were made in some of the matching criteria, but not in the similarity of overstorey, nor the contrast between understories.

Site management

The managers of each of the sites were asked to complete a detailed questionnaire about stocking rates, fertilizer application, usage of stands of trees by livestock, and other aspects of their management of the sites. From their responses, average annual stocking rates over the last 10 years were calculated. Sheep are the main livestock grazed, so stocking rates were expressed on the basis of sheep-equivalents, by assuming that one cow or bullock was equivalent to 10 sheep, one cow and calf to 13 sheep, one heifer to eight sheep and one horse to 11 sheep.

Tree health

We recorded species, diameter (DBHOB), and crown vitality of all trees in five randomly-located quadrats (10 m×50 m) at each site. Crown vitality was assessed visually, with scores being allocated for position, size and density of crown, and for absence of dead branches and epicormic shoots. Assessments were subjective, with the maximum score possible in each category (five points) being reserved for the best development we observed in each category for the woodland form of each tree species. The numbers of dead trees in the quadrats (excluding those that had been ring-barked or otherwise deliberately killed) were also recorded.

Abundance of insects

We confined our detailed insect studies to *E. blakelyi* trees, since, of the species common to all sites, it was the most heavily defoliated and the most affected by dieback.

The abundance of insects was assessed in two ways. Firstly, trays (50 cm×35 cm, with ends 15 cm and sides 7 cm high) were attached near the ends of branches on each of two trees at each site, and left in place from late January to mid-May 1988. The trays consisted of slippery, teflon-coated plastic ends fitted across a branch and joined by low teflon-coated sides with a terylene tray below to prevent the loss of dislodged insects. They were analogous to pitfall traps, in that insects that crawled or fell into them were unable to escape. During visits on alternate days from late January until early April (while the experiments were being conducted) and during weekly visits thereafter until mid-May, all insects found in the trays were collected into 70% ethanol. After the trays had been removed from the trees in May, the ethanol was filtered from the collected insects, which were allowed to dry for six hours before being weighed.

Flying and sedentary insects, which were poorly represented in the tray-trap samples, were sampled (along with crawling insects) by fogging trees with insecticide (Majer & Recher 1988; branch clipping was also attempted but yielded too few insects for analysis). Because of the danger of insecticidal drift harming experimental insects, fogging was delayed until early autumn when the experiments were complete and was carried out only once at each site, because the abundance of insects was very low by the time the first round of foggings was complete. No attempt was made to fog the sites in two districts (Bookham & Dalton, Appendix 1) in which the control site was very close to a township.

Two randomly-located trees with crown heights between about five and seven metres were fogged at each site. Calico cloths were pegged under the canopy of each tree and the trees were fogged with a solution of oil-based insecticide containing 2 gm.l^{-1} pyrethrins, dispensed by a "Pulsfog" motorized fog generator, until the fog enshrouded the complete canopy (usually within five minutes). All insects falling onto the cloths during the following two hours were collected into containers of 70% ethanol. The trees were fogged between 10 a.m. and 11 a.m. on calm, sunny days. The leaf area and biomass of each of the fogged trees was estimated by clipping a "branch unit" from the tree and then estimating the number of such units that comprised its crown. The branch unit was stored in a plastic bag until it could be taken to the laboratory where its leaf area and weight were measured using a leaf area meter and a balance. Unit areas ranged from about 900 to 5 000 cm^2. In a test of the method, values estimated in this way were compared, for three trees, with values estimated using a "demon", which calculates canopy leaf area from light transmittance (Lang & McMurtrie *in litt.*), and were found to be in reasonable agreement (unit estimated and demon estimated values were 51 and 60 m^2, 41 and 33 m^2 and 5.8 and 5.5 m^2 respectively). The weight of insects collected by fogging was measured after filtering and drying, as before, and expressed as a proportion of the leaf area and biomass fogged.

Defoliation rates

Leaf persistence was highly variable between sites. Unless disrupted by heavy defoliation, *E. blakelyi* has a two year cycle of leaf production and replacement (Clark & Dallwitz 1974). Many leaves at several of the control sites were more than one year old, because most of the trees at these sites had only minor leaf flushes during the summer of our study. In contrast, at the sites where defoliation levels were high, few leaves in any of the trees were more than two months old by the end of summer. We had intended to use an instantaneous measure of defoliation, by assessing the leaf area missing from samples of leaves collected at the end of the growing season (Fox & Morrow 1983; Landsberg 1988), but the big disparity in average age of leaves at different sites meant this technique was not appropriate. In early January we had placed metre-long insect-proof bags of terylene mesh over four branches on each of the two trees at each site to ensure that we had sufficient foliage available for the insect experiments (see later section). Two bags provided sufficient foliage for this purpose so we used the other two bags to assess defoliation rates.

We matched each protected branch with an adjacent, unprotected branch of similar length, diameter and number of leaf buds. There was little point in attempting a more precise matching, because this was not done until a month after the protective bags had been put out, by which time some leaf growth had already occurred inside the bags. The pairs of protected and unprotected bags were removed from the trees in mid-May. From the difference in their leaf areas, measured on a leaf area meter, we calculated relative defoliation rates by assuming that the greater leaf area development inside the protective bags was an indication of the leaf area lost from the unprotected branches because of herbivory.

Soil chemistry

Twelve cores (each 50 mm in diameter and 20 cm in length) were taken from each site by sampling at regular intervals along transects running through similar positions in the landscape at each pair of sites. The cores were sectioned in 5 cm depth increments, and bulked to give four replicates per site of each depth section. Exchangeable cations were extracted by leaching a 2.5 g field moist sample with 100 ml 0.1 M $BaCl_2$ for six hours; Ca^{+2}, Mg^{+2}, Na^+, K^+ and Mn^{+2} were determined by atomic absorption spectrophotometry, Al^{+3} was determined by colorimetry (Heffernan 1985), and H^+ was determined by calculation (Khana *et al.* 1986). Soil pH and the nitrogen measures were determined on filtered extracts of 10 g moist soil (fine earth faction) and 50 ml 1M KCl solution. Total mineral-N and NH_4^+-N were analyzed by automated distillation and colorimetric detection; NO_3^--N was determined from their difference (Raison *et al.* 1987). Mineralizable N values (total, NH_4^+ and NO_3^-) were determined by difference, following laboratory incubation at 20°C for 30 days for soils wetted to field capacity.

Foliar chemistry

Samples of 50 to 100 leaves were collected from the *E. blakelyi* trees on which insect experiments were conducted (two per site, see later section) on two occasions. Newly-flushed immature leaves were collected at the beginning of the experiment, and mature leaves were collected when the experimental insects on each tree had pupated. Total concentrations of N, P and K were determined in extracts obtained from the oven-dried, ground samples by automated colorimetry, following a modified Kjeldahl digestion (Heffernan 1985).

Floristic and avian diversity

Most of these data were collected between November 1987 and April 1988, as an adjunct to measurements of tree health and during the monitoring of the insect experiments and earlier trials. Specimens of plant species were collected as they flowered and these were used to build a field herbarium that was then used to construct species lists for each site. A score for abundance/cover for each species at each site was given when its presence was first noted, and this was checked toward the end of the study when greater familiarity with the sites and species had been gained. It was not always possible to identify all the species of grasses and forbs at the pasture sites because of a lack of identifiable material due to grazing.

Bird species were recorded both from opportunistic observations and from extra visits to each site specifically for bird observations, so that about 40 hours of observation were undertaken at each site. Morning activity of birds was observed following overnight camping at least once at all sites except the pair closest to Canberra.

Insect experiments

The aim of these experiments was to determine the relative contributions of predation (including parasitism) and food quality in controlling growth and survival of defoliating insects. As test animals we used the larvae of a chrysomelid beetle, *Paropsis atomaria* Oliver, which was a common defoliator at all the sites. Adult beetles reared in insectories on seedlings of *E. blakelyi* were used to provide a source of egg batches. Since larvae preferentially feed on newly-flushed foliage (Larsson & Ohmart 1988), experiments were delayed until sufficient suitable foliage became available at the sites, in late summer.

Experiments were conducted on each of two *E. blakelyi* trees at each site. Experimental trees were chosen that had newly-flushing foliage less than five metres from the ground. Where more than two such trees were present at a site, trees were selected randomly. To ensure that sufficient leaf material would be available for the experimental larvae, branches were protected from herbivory by mesh bags for a month prior to the commencement of the experiment.

Two egg batches were attached, just as they began to hatch, to each of the two branches on each tree. The batch on one of the branches was protected from predation by being enclosed inside a terylene mesh bag. The batch on the other branch was not protected but the hatching larvae were confined to that branch by a slippery-sided tray, as described in the section on insect abundance. On the day following the attachment of the egg batches, the numbers of newly emerged larvae were counted, and, if they were in excess of 15 per treatment branch they were thinned to that number. Thereafter until the larvae changed to pupae, the branches were visited on alternate days and the number and developmental stage of all surviving larvae were noted. When the larvae reached their fourth (and last) instar, peat was spread on the bottoms of the trays and bags to provide a pupating medium. Pupae were removed from the medium, placed in gelatin capsules, and removed to the laboratory, where their sex and weight were recorded. The pupae were then reared in insectories until adult beetles or parasitoids emerged.

The predation loss per tree, expressed as a proportion of the average number of newly-hatched larvae establishing on that tree (usually 15), was calculated from the relative difference in numbers of adult beetles that emerged following rearing in either protective bags or unprotected trays [i.e., predation per tree=(number of bag adults – number of tray adults) / 0.5 (total number of newly-hatched larvae in bag and tray)]. Some of the insects died from causes other than predation, indicated by the presence of their undamaged bodies in the bottom of the protective bags. The proportion of protected insects this represented was calculated from the difference between the number of larvae starting the experiment and the number of adults that successfully emerged, expressed as a proportion of the number of larvae at the start, for each bagged branch. Developmental time in the field was also calculated for the larvae reared in protective bags, from the time taken for the median number of larvae per treatment branch to develop from hatching to pupation.

Statistical analyses

The significance of any differences between data from control and pasture sites was determined from split-plot analyses of variance, in which differences between districts were treated as block effects and differences between paired sites as treatment effects (Alvey *et al.* 1982).

Results

Site management

Average annual stocking rates for the last 10 years confirmed our impression that the control sites were only very lightly grazed, if at all, and that stocking rates on the pasture sites were moderate to high (Table 1). Fertilizer had very seldom (or never) been spread deliberately under the trees at any of the sites, but superphosphate was applied routinely to adjacent paddocks, many of which had also been sown with clover. The managers of the pasture sites noted that livestock frequently camped under the trees.

Tree health

Trees at the control sites were consistently and significantly healthier than those at the pasture sites in the same districts (Table 1), although not all districts were equally affected by dieback. Trees were healthiest at the paired sites in the Gundaroo and Bookham districts (mean vitalities were 17.9 & 17.0 and 16.0 & 15.7 at the control and pasture sites respectively), and were least healthy at the sites at Murrumbateman and Tangmangaroo (vitalities were 14.6 & 10.2 and 15.2 & 12.5 at the control and pasture sites respectively). Tree health also varied with species. The species that was generally healthiest at the pasture sites was *E. melliodora* [mean vitality 16.2±3.0 (std. dev.) for 92 trees], followed by *E. mannifera* (15.7±2.4 for 46 trees), *E. rossii* (14.1±2.3 for 82 trees), *E. macrorhyncha* (13.4±2.5 for 59 trees), *E. blakelyi* (12.1±3.9 for 224 trees), and *E. goniocalyx* (12.1±2.9 for 44 trees).

There were also significantly more dead trees present at the pasture sites (Table 1), suggesting that the lower vitality of trees at the pasture sites might lead to more frequent tree death.

There was no significant difference between control and pasture sites in the numbers of trees per site, but the average basal area of trees was larger at the pasture sites (Table 1). There was little regeneration of any eucalypts at any of the pasture sites so this difference in basal areas probably reflects a paucity of younger trees and saplings at the pasture sites.

Abundance of insects

The biomass of insects killed by insecticide fogging or caught in the tray-traps was significantly greater at the pasture sites compared with their control counterparts; the weight of insects fogged was about 10 times greater, and the weight of insects trapped per day was double (Table 1).

Defoliation rates

The rates of defoliation were also much higher on the trees at the pasture sites (Table 1). The method by which defoliation was assessed was not very precise, given that the matching of protected and unprotected branches was only approximate, and that any effect of the protective bags on branch growth was not measured. The protective bags may have had a slight inhibitory effect on branch growth in the absence of herbivory, since negative defoliation rates were calculated for trees at several control sites where insect abundances were also very low. If this is the case, however, it means that the defoliation rates calculated are underestimates of the real impact of herbivores on branch growth, which may therefore have been very high at some sites. The highest defoliation rate measured was at the pasture site at Ginninderra, where the mean rate was 15.8 cm^2 day^{-1} (leaf areas on protected and unprotected branches were 1 280 and 180 cm^2 and 3 867 and 511 cm^2 respectively, on the two trees measured). This contrasts with the differences in leaf area on the branches at the control site in that district (2 375 and 2 486 cm^2 and 3 384 and 3 296 cm^2 on protected and unprotected branches respectively, giving a mean defoliation rate of 0.3 cm^2 day^{-1}). Any artifacts of the methodology were therefore probably insignificant by comparison with the impact of herbivory in determining differentials in branch growth with treatment.

Although the insects sampled have not yet been grouped according to diet, the preliminary data on insect abundance in combination with the data on defoliation rates provide strong support for the contention that tree dieback is associated with defoliation by insects at these sites.

Soil chemistry

Low pH values, the presence of exchangeable Al, and low base saturation at depth (Appendix 1), all indicate that soils at all the sites are moderately to

TABLE 1. Mean values of variables measured at the control and pasture sites.

Variable	Control sites	Pasture sites	Variance ratio[1]
Sheep stocking rate (10 yr mean.ha^{-1})	0.6	4.2	5.61 (1,5)*
Tree vitality (max=25)	15.9	13.7	189.55 (1,1259)***
Numbers of dead trees (ha^{-1})	12	54	3.73 (1,7)*
Numbers of trees (ha^{-1})	343	262	1.00 (1,7) ns
Basal area ($m^2 tree^{-1}$)	518	992	47.048 (1,1204)***
Insects fogged			
($g.m^{-2}$leaf area)	0.013	0.119	14.62 (1,17)***
($g.kg^{-2}$leaf weight)	0.031	0.323	12.03 (1,17)***
Insects trapped ($g.day^{-1}$)	0.017	0.037	4.71 (1,23)**
Defoliation rates ($cm^2 day^{-1}$)	1.24	7.09	14.66 (1,53)***
Soil values in top 5 cm (other depth sections followed the same trends)			
pH (in KCl)	4.26	4.17	0.50 (1,7) ns
C.E.C. ($meq.kg^{-1}$)	93	121	1.86 (1,7) ns
Exchangeable Al	4.70	3.92	0.17 (1,7) ns
Exchangeable Mn	1.60	1.89	0.43 (1,7) ns
Base saturation (%)	87.8	89.9	0.31 (1,7) ns
NH_4^+–N ($ug.g^{-1}$)	4.9	20.9	3.81 (1,7)*
NO_3^--N ($ug.g^{-1}$)	0.3	12.2	3.69 (1,7)*
Total mineralizeable–N	45.9	93.0	3.63 (1,7)*
Nutrients in young foliage:			
N (%)	1.44	1.88	7.57 (1,23)**
P (%)	0.16	0.21	6.06 (1,23)**
K (%)	1.79	1.77	0.02 (1,23) ns
Nutrients in mature foliage:			
N (%)	1.36	1.68	3.97 (1,23)*
P (%)	0.12	0.17	9.98 (1,23)***
K (%)	1.50	1.82	7.08 (1,23)**
Numbers of shrub species	18.4	4.0	58.48 (1,7)***
Numbers of bird species:			
All	29	23	7.23 (1,7)**
Mainly insectivorous spp.	24	19	9.32 (1,7)**
Insect experiments:			
Development time (days)	30.4	26.9	5.64 (1,23)**
% protected insects died	54.6	32.9	5.27 (1,23)**
Weight of female pupae (g)	0.114	0.141	91.62 (1,183)***
Weight of male pupae (g)	0.088	0.103	34.95 (1,110)***
% insects predated	29.2	32.8	0.12 (1,23) ns

[1]Variance ratio calculated from split-plot ANOVA; degrees of freedom in parenthesis; *** P<0.01, **P<0.05, *P<0.1, ns P>0.01.

highly acidic, but there was no clear difference in these parameters between soils from the pasture and control sites (Table 1). Toxicity to Mn does not appear to be a problem in any of the soils, but toxicity to Al may be, particularly at depths below 10 cm.

Nitrogen values were significantly higher in the soils from the pasture sites; these soils had, on average, about four times the concentration of NH_4^+-N, about 40 times the concentration of NO_3^--N and about twice the concentration of potentially mineralizable N (Table 1). The high concentrations of NO_3^--N in the pasture soils indicates that these soils are at risk of becoming very acid, because of the relative ease with which NO_3^- can be leached from soil profiles (Haynes 1983).

Foliar chemistry

There was significantly more nitrogen and phosphorus in young foliage, and significantly more nitrogen, phosphorus and potassium in mature foliage of the *E. blakelyi* trees growing on the pasture sites compared with the control sites (Table 1).

Floristic and avian diversity

The control sites not only supported a greater abundance of shrubs of all sizes (Appendix 2), but also supported significantly more shrub species (Table 1). Some pasture sites had a minor shrub component, but these shrubs were all very small and inconspicuous, and represented very few species (Appendix 2). Three plant families contributed to much of the shrub diversity: Mimosaceae (1 genus, 10 species), Fabaceae (9 genera, 15 species) and Epacridaceae (4 genera, 5 species). The data for forbs showed a similar pattern. Half of the 37 species recorded belonged to the families Asteraceae (8 genera, 11 species) and Liliaceae (5 genera, 7 species).

Significantly more species of birds were recorded at the control sites compared with the pasture sites (Table 1). The species observed are comparable with those recorded in the tablelands further south (Recher *et al.* 1985) and in similar woodland formations on the northern tablelands of New South Wales (Ford & Bell 1982). Of the 75 species recorded (Appendix 3), eight were particularly widespread (recorded at most sites) and abundant (observed on most visits). They were the grey fantail, willie wagtail, crimson rosella, white-throated treecreeper, Australian magpie, yellow-rumped thornbill, striated thornbill and scarlet robin. Thirty-five of the recorded bird species were rare, in that they were recorded at only one or two sites; three of the sites accounted for a large proportion of these rarer species. One site, the control site at Blakney Creek, was particularly rich in species, and accounted for nine species that were observed nowhere else, and a further six that were observed at only one other site. The efficacy of insectivorous birds in controlling the abundance of insects is likely to depend more on numbers of individual birds than on the numbers of bird species. Although no censuses of bird numbers were conducted, it was our impression that the abundance of birds, as well as their diversity, was higher on the control sites.

Insect experiments

The experimental larvae that were reared on the trees at the pasture sites developed significantly faster, and developed into significantly larger pupae (Table 1). The corollary was also true: relative nutrient impoverishment at the control sites appeared to significantly depress growth of experimental larvae, as evidenced by longer development times and lower pupal weights. In addition, significantly more of the protected larvae reared at the control sites died before reaching adulthood (Table 1), and at least some of these deaths may have been the result of poor nutrition.

There was no significant difference between control and pasture sites in the number of experimental insects lost to predation, although predation accounted for about 30% of the insects in the study (Table 1). Ants may have been important predators, since we saw ants removing larvae on a number of occasions, but did not actually see any other acts of predation. Parasitism was a relatively minor component of total predation: only 9 adult parasitoids emerged, all from larvae reared at two of the pasture sites.

Discussion

Our results indicate the extent to which remnants of woodland used for pastoralism are degraded, by comparison with otherwise similar woodland remnants. The floristic and structural diversity of their understorey is greatly impoverished, and, to a lesser degree, so too is the diversity of avian fauna that uses them. Tree health is not always as severely affeced (both pairs of sites at Gundaroo and Bookham were still relatively healthy), but there were consistent differences in the health of woodland remnants in the districts. In sites that are heavily used for pastoralism greater numbers of trees are unhealthy or dead. Thus our first hypothesis, that dieback of rural trees may be a natural phenomenon unrelated to pastoral use, is not supported.

Our second hypothesis also lacks support because abundant leaf growth occurred on all branches that were protected from insects, regardless of site use. All the sites may be under chronic chemical stress because of the acid nature of their soils, and high nitrate concentrations in soils at some sites pose further risk of their becoming increasingly acid.

Data about soil acidity related to tree vitality are not available for regions where rural dieback is not severe, but it is interesting to speculate about the extent to which soil acidity in the southern tablelands may predispose trees in this region to dieback. Old age may be an additional stress experienced by more trees at pasture sites than at control sites. None of our study trees, however, was sufficiently stressed to hinder abundant leaf growth when herbivory was excluded.

The relatively poor condition of tree crowns at the pasture sites was associated with significantly higher rates of defoliation by insects, and insects were generally more abundant at these sites. Thus there is also little support for our third hypothesis. If increased abundance of insects were simply a function of decreased abundance of trees we would expect little difference between control and pasture sites in insect abundance, since the numbers of trees at both sites were similar.

The evidence relating to the fourth hypothesis was less straightforward. Although the diversity and abundance of birds appeared to be lower at the pasture sites, this did not markedly influence relative predation on the experimental larvae at these sites. Since many of the common parasitoids of *P. atomaria* oviposit into eggs rather than larvae (Tanton & Epila 1984), our results probably underestimate the level of parasitism likely to be encountered in natural populations. Likewise, our results probably underestimate the possible extent of avian control of natural populations, particularly during spring when birds are feeding nestlings. Notwithstanding this possible underestimation of potential predation, it is noteworthy that parasitoids were present at at least two of the pasture sites, despite their floristic and structural impoverishment, and that other predators were apparently similarly active at both pasture and control sites. Thus, at these sites, any influence of ecosystem simplification in controlling the abundance of defoliating insects was probably not as dominant as was the influence of nutrient enrichment.

The evidence in support of the hypothesis relating to a major role for nutrient enrichment is very strong. Nitrogen levels indicate that the soils at the pasture sites tended to be much more fertile than the soils at the control sites, and the nutritional quality of their tree foliage is also significantly enhanced. The growth of experimental larvae on these enriched trees indicates that nutritional enrichment may have a strong influence on insect abundance at these sites. Since *P. atomaria* can develop through several generations during favourable growing seasons (Carne 1966), increased rates of growth of larvae could contribute to increased populations of beetles. Also, since the fecundity of female *P. atomaria* beetles depends on both the nitrogen content of their diet and on their weight (Ohmart *et al.* 1985), it is probable that the female pupae that developed on the pasture sites would also develop into more fecund adult beetles, further contributing to increases in the total populations of this defoliator.

What causes the nutrient enrichment? Although we did not investigate this explicitly, some inferences can be drawn from the data we collected. Most managers avoid deliberately spreading fertilizer under stands of trees, but adjacent paddocks are routinely fertilized. Thus inadvertent drift of fertilizer probably does occur. However, the extent of drift is likely to be similar at both pasture and control sites, since both are adjacent to fertilized paddocks. Similarly, movement of nutrients through soils or lateral growth of tree roots into paddocks, if they do occur, are likely to be similar at both pasture and control sites. The sites differ, however, in the extent to which livestock may contribute to nutrient movement. Robinson *et al.* (1983) and Russell (1986), for example, have noted that the general improvement in nutrient status of soils under improved pasture is often associated with a very uneven pattern of concentration of nutrients on topsoils because of their redistribution in animal excreta. In particular, nutrients often become concentrated in areas where stock tend to congregate. Thus the tendency of livestock to camp under stands of trees may be a cause of substantial nutrient enrichment of the stands.

Conclusions

In the southern tablelands of New South Wales, remnants of red gum/yellow box woodlands that are used for pastoralism are severely degraded. Impoverishment of floristic and avian diversity may have serious consequences for their conservation, since, after 200 years of use, woodland remnants are all that remain of this once widespread ecosystem. Dieback of rural trees, which is currently degrading the only more or less intact component of the ecosystem, has serious consequences in addition to those affecting conservation of flora and fauna, since rural trees play a key role in protecting soil and water resources.

Tree dieback in the woodland remnants is associated with increases in the abundance of defoliating insects. Nutrient enrichment is probably a major cause of the increased abundance of insects, and inputs from livestock may be the main cause of such enrichment. Thus rehabilitating stands of trees suffering from this form of dieback will probably require that nutrient inputs be reduced, and this may entail reducing the intensity of livestock usage of the woodland.

Acknowledgements

For allowing us free access, and for their assistance and co-operation, we thank the managers of the properties on which we worked: Mr A. Asquith, Mr R. Bennett and Mr J. McKinnon, of Canberra,

Mr K. Chown and Mr D. Dwyer of Tangmangaroo via Yass, Mr J. P. Foster-Spink of Gundaroo, Mr K. R. Friend and Mr T. J. Thorn of Dalton, Mr. R. Julian of Bookham, Mr S. Merriman and Mr F. L. Hodgkinson of Murrumbateman, Mr R. Southwell of Blakney Creek, the Yass Pastures Protection Board, the Yass Shire Council and the Canberra Nature Park. We also thank Ms J. A. Harragan and Mr S. J. Smith of the Division of Forestry and Forest Products, CSIRO, Canberra, for the analysis of soil and foliar samples, and for assistance with collection of soils, and Dr A. R. G. Lang of the Division of Environmental Mechanics, CSIRO, Canberra, for advice and assistance in measuring leaf areas of trees.

References

Alvey, N., Galwey N. & Lane P. (1982) *An Introduction to Genstat.* Academic Press, London.

Carne P. B. (1966) Ecological characteristics of the eucalypt-defoliating chrysomelid *Paropsis atomaria* Ol. *Aust. J. Zool.* **14,** 647-72.

Clark L. R. & Dallwitz M. J. (1974) On the relative abundance of some Australian Psyllidae that coexist on *Eucalyptus blakelyi. Aust. J. Zool.* **22,** 387-415.

Davidson R. L. (1980) Local management. In: *Focus on Farm Trees* (eds. N. M. Oates, P. J. Greig, D. G. Hill, P. A. Langley & A. J. Reid) pp. 89-98. Capitol Press, Box Hill, Victoria.

Ford H. A. & Bell H. (1982) Density of birds in eucalypt woodland affected to varying degrees by dieback. *Emu* **81,** 202-8.

Fox L. R. & Morrow P. A. (1983) Estimates of damage by insect grazing on *Eucalyptus* trees. *Aust. J. Ecol.* **8,** 139-47.

Haynes R. J. (1983) Soil acidification induced by leguminous crops. *Grass and Forage Science* **38,** 1-11.

Heffernan B. J. (1985) A handbook of methods of inorganic chemical analysis for forest soils, foliage and water. CSIRO Australia, Division of Forest Research, Canberra, Internal Report.

Jacobs M. R. (1955) *Growth habits of the Eucalypts.* Commonwealth of Australia, Forestry and Timber Bureau, Canberra.

Khanna P. K., Raison R. J. & Falkiner R. A. (1986) Exchange characteristics of some acid organic-rich forest soils. *Aust. J. Soil Res.* **24,** 67-80.

Landsberg J. (1986) Dieback of rural eucalypts: Dietary quality of foliage and insect herbivory. Ph.D. Thesis, Australian National University, Canberra.

Landsberg J. (1988) Dieback of rural eucalypts: Tree phenology and damage caused by leaf-feeding insects. *Aust. J. Ecol.* **13,** 251-68.

Landsberg J. & Wylie F. R. (1988) Dieback of rural trees in Australia. *Geojournal* **17,** 231-7.

Larrson S. & Ohmart C. P. (1988) Leaf age and larval performance of the leaf beetle *Paropsis atomaria. Ecol. Entomol.* **13,** 19-24.

Mackay S. M., Humphreys F. R., Clark R. V., Nicholson D. W. & Lind P. R. (1984) Native tree dieback and mortality on the new England Tablelands of New South Wales. Forestry Commission of NSW Research Paper No. 3.

Majer J. D. & Recher H. F. (1988) Invertebrate communities on Western Australian eucalypts: A comparison of branch clipping and chemical knockdown procedures. *Aust. J. Ecol.* **13,** 269-78.

Moore R. Milton (1975) South-eastern temperate woodlands and grasslands. In: *Australian Grasslands* (ed. R. Milton Moore), pp. 169-90. Australian National University Press, Melbourne.

Ohmart C. P., Stewart L. G. & Thomas J. R. (1985) Effects of nitrogen concentrations of *Eucalyptus blakelyi* foliage on the fecundity of *Paropsis atomaria* (Coleoptera: Chrysomelidae). *Oecologia* (Berl.) **68,** 41-4.

Pryor L. D. (1981) Australian Endangered Species: Eucalypts. Australian National Parks & Wildlife Service Special Publication No. 5.

Raison R. J., Connell M. J. & Khanna P. K. (1987) Methodology for studying fluxes of soil mineral-N in situ. *Soil Biol. Biochem.* **19,** 521-30.

Recher H. F., Holmes R. T., Schultz M., Shields J. & Kavanagh R. (1985) Foraging patterns of breeding birds in eucalypt forest and woodland of southeastern Australia. *Aust. J. Ecol.* **4,** 399-420.

Robinson G. G., Whalley R. D. B. & Taylor J. A. (1983) The effect of prior history of superphosphate application and stocking rate on faecal and nutrient distribution on grazed natural pastures. *Aust. Rangel. J.* **5,** 79-82.

Russell J. S. (1986) Improved pastures. In: *Australian Soils The Human Impact* (eds J. S. Russell & R. F. Isbell), pp. 374-96. University of Queensland Press, St Lucia, Queensland.

Tanton M. T. & Epila J. S. O. (1984) Parasitization of larvae of *Paropsis atomaria* Ol. (Coleoptera: Chrysomelidae) in the Australian Capital Territory. *Aust. J. Zool.* **32,** 251-9.

Wells K. F., Wood N. H. & Laut P. (1984) Loss of forests and woodlands in Australia: A summary by state, based on rural local government areas. CSIRO Division of Water and Land Research, Technical Memorandum 84/4.

White T. C. R. (1986) Weather, *Eucalyptus* dieback in New England, and a general hypothesis of the cause of dieback. *Pac. Sci.* **40,** 69-89.

Wylie F. R. (1986) Native tree dieback in southern Queensland: Its occurrence, severity and aetiology. Ph.D. Thesis, University of Queensland, St Lucia, Queensland.

APPENDIX 1. Some characteristics of the fine earth fraction of soils from the sites.

	Depth (cm)	District[1] and site[2]: Gin C	Gin P	Mur C	Mur P	Book C	Book P	Tang C	Tang P	Blak C	Blak P	Dalt 1 C	Dalt 1 P	Dalt 2 C	Dalt 2 P	Gund C	Gund P
pH	0-5	4.30	4.41	4.21	4.30	4.27	4.30	4.63	4.04	4.22	4.01	3.90	4.50	4.12	3.83	4.46	3.95
	5-10	4.10	4.27	4.07	4.02	4.04	4.10	4.41	4.07	4.21	4.01	3.81	4.13	3.89	3.69	4.03	3.74
	10-15	4.02	4.25	4.06	4.01	4.08	4.10	4.39	4.13	4.20	4.22	3.84	3.97	3.88	3.74	3.85	3.80
	15-20	3.98	4.16	4.04	4.03	4.08	4.12	4.38	4.17	4.18	4.27	3.86	3.94	3.93	3.74	3.74	3.82
Total mineral N	0-5	6.1	86.1	5.6	53.5	6.1	5.6	5.4	10.9	4.8	90.5	2.9	3.7	2.5	3.8	7.8	10.8
(mg kg^{-1};	5-10	2.9	23.5	2.1	13.5	1.7	1.9	3.1	3.8	3.4	12.7	2.3	2.4	1.9	1.5	2.1	2.5
not incubated)	10-15	1.5	8.4	2.6	5.5	1.2	1.3	2.7	2.7	3.3	6.9	1.5	1.5	1.0	2.4	1.7	1.5
	15-20	1.0	7.7	1.0	4.2	1.0	0.9	2.4	2.5	3.1	4.7	1.4	1.1	1.2	1.1	1.4	1.5
Total	0-5	34.5	130.4	24.3	58.7	31.4	28.1	107.3	181.5	80.9	270.6	23.3	22.2	25.6	17.4	40.0	35.3
mineralizeable N	5-10	6.8	36.2	7.2	15.4	9.9	9.7	32.6	21.0	4.9	30.7	12.3	8.4	7.3	3.3	11.0	11.9
(mg kg–1)	10-15	3.5	16.7	11.5	8.0	14.6	4.9	11.7	20.1	5.5	12.0	7.2	5.9	2.0	0.6	6.2	4.8
	15-20	0.8	12.2	2.3	4.5	1.8	3.3	4.2	11.1	1.4	7.9	1.6	0.5	1.9	0.0	5.4	2.9
Total exchangeable	0-5	90.5	210.5	83.4	132.4	79.8	67.5	115.7	102.7	100.2	170.8	50.0	110.9	71.6	98.6	149.1	77.5
cations	5-10	44.4	93.3	38.6	42.5	35.3	45.7	48.3	42.7	56.9	75.8	30.5	41.6	36.5	54.5	72.6	34.1
(CEC; meq.	10-15	40.8	56.6	27.6	30.6	28.4	39.0	28.0	32.3	47.5	49.7	26.3	31.3	33.2	52.2	59.5	32.0
kg^{-1})	15-20	41.6	46.3	26.3	29.6	21.8	37.6	22.1	25.6	46.2	41.4	27.5	31.6	32.9	50.5	49.7	31.8
% base	0-5	89.2	97.6	85.7	93.3	88.3	92.7	96.2	89.6	81.4	86.6	76.8	95.6	86.4	80.0	98.7	84.0
saturation	5-10	75.5	95.7	62.8	78.7	69.1	80.8	76.5	57.5	46.6	58.4	55.2	81.2	66.0	48.6	88.5	59.3
	10-15	72.9	94.9	54.9	74.0	63.8	80.9	67.3	55.5	42.6	45.1	55.5	68.3	65.3	47.9	82.8	51.8
	15-20	64.3	93.9	52.8	73.9	63.7	83.7	68.5	54.2	42.5	42.7	57.1	67.0	66.5	49.6	74.7	54.8
Exchangeable	0-5	3.5	0.0	4.2	0.8	3.2	1.6	1.2	4.4	13.0	6.0	7.4	1.2	4.9	12.0	0.3	5.4
Al (meq. kg^{-1})	5-10	8.4	0.2	10.6	4.9	7.0	5.4	6.2	14.2	27.0	19.3	11.2	4.5	9.3	23.7	5.7	11.0
	10-15	9.8	0.8	9.8	5.6	7.4	4.9	6.5	12.0	25.3	22.4	9.4	7.3	9.4	24.0	6.9	13.3
	15-20	13.0	1.1	9.2	5.9	6.0	4.0	5.4	9.9	24.5	20.6	8.6	7.9	9.4	22.9	9.9	12.4
Exchangeable	0-5	2.6	2.8	3.4	4.2	3.2	1.0	0.8	2.0	0.8	2.8	0.6	0.5	0.9	1.0	0.6	0.8
Mn (meq. kg^{-1})	5-10	0.4	0.7	0.5	0.6	0.7	0.3	0.3	0.3	0.1	0.3	0.1	0.1	0.2	0.2	0.3	0.1
	10-15	0.2	0.2	0.3	0.2	0.3	0.1	0.1	0.1	0.0	0.1	0.0	0.1	0.1	0.0	0.2	0.1
	15-20	0.2	0.2	0.3	0.1	0.2	0.4	0.1	0.1	0.0	0.0	0.0	0.0	0.0	0.0	0.1	0.0

1. Districts are: Ginninderra (Gin);Murrumbateman (Mur); Bookham (Book); Tangmangaroo (Tang); Blakney Creek (Blak); Dalton district (Dalt 1); Dalton township (Dalt 2); Gundaroo (Gund).
2. Control (C) or Pasture (P) sites.

APPENDIX 2. Relative abundance of shrub species at the sites.

		District and site[3]:															
		Gin		Mur		Book		Tang		Blak		Dalt 1		Dalt 2		Gund	
Family[1]	Species[2]	C	P	C	P	C	P	C	P	C	P	C	P	C	P	C	P
		3	.	3		.		2		3	1	2	1	1			
		2	.	2		.		3		3	.	.	.	1			
Mim	*Acacia dealbata*	1	1	3		1		2		3	.	1	1	1			
		1		1						3		2		2		3	
		.		.						2		.		2		.	
	A. mearnsii	.		.						.		.		.		.	
		.				1						.		.	.		
		2				.						3		1	1		
	A. brownii	.				.						.		.	.		
						1						1		1			
						.						.		.			
	A. lanigera					.						.		.			
						2		1				1					
						2		2				.					
	A. implexa					2		.				.					
						1											
						.											
	A. decora					.											
						2											
						.											
	A. buxifolia					3											
																2	
																3	
	A. rubida															.	
		.								.		.		.	.	.	
		1								1		.		1	.	.	
	A. gunnii	.								.		1		1	1	1	
																.	
																.	
	A. cultriformis															1	
								3									
								2									
Fb	*Daviesia latifolia*							.									
										.				.			
										1				1			
	D. genistifolia									.				.			
		.		.		.				.		.				.	
		4		3		1				1		3				3	
	D. virgata	.		.		.				.		.				.	
																.	
																.	
	D. sp.															1	
		.	.	.		.				.				.		.	
		2	.	2		2				1				.		2	
	Indigofera australis	1	1	2		1				1				1		.	
						.											
						1											
	Pultenaea foliosa					.											
		.						.		.							
		.						.		.							
	P. procumbens	2						1		3							
		.	.							.						.	
		.	.							.						.	
	P. subspicata	2	1							2						2	
										.							
										.							
	Hovea linearis									1							
		.	.														
		.	.														
	H. heterophylla	1	1														

Appendix 2 (continued)

		District and site[3]: Gin		Mur		Book		Tang		Blak		Dalt 1		Dalt 2		Gund	
Family[1]	Species[2]	C	P	C	P	C	P	C	P	C	P	C	P	C	P	C	P
	Kennedia sp.																
													1				
	Gompholobium hueglii																
				1		1		1		3						1	
	Dillwynia sericea																
		3	1	1		3		1		3		2	1	1	1	1	1
	Hardenbergia violacea																
		2	1	2		2				2		2	1	1	1	2	
	Glycine clandestina																
		1	1														
Ep	*Brachyloma*			1													
	daphnoides	1		2		1		2		3		2		3		2	
	Leucopogon aff. fletcheri									2							
														2			
	L. virgatus																
										2		2		2		1	
	Lissanthe strigosa	1		1						3				3			
		2	1	2						3		2	1	3	2		
	Melichrus urceolatus																
		2	1	2		1				3		2	1	2	1	1	
Myr	*Leptospermum multicaule*	1								4							
Dil	*Hibbertia obtusifolia*																
		2	2	2		1		1		3		2	1	1	1	2	
	H. stricta							1		2						1	
As	*Cassinia*	2								2		1					
	aculeata	2								2							
	Helichrysum sp.a	2												1	1		
	H. thyrsoideum												1	2	1		
	H. sp.b	2															
Ro										1			1			1	
	Rosa rubignosa	2		1													
				1									1				
			2														
	Cretageous sp.	2	1	1													
Pit	*Bursaria spinosa*													1			
														1			
	Cheiranthera linearis																
				1		1		1		2		1	1			1	
	Billardiera scandens																
		1						1		1							

Appendix 2 (continued)

		District and site[3]: Gin		Mur		Book		Tang		Blak		Dalt 1		Dalt 2		Gund	
Family[1]	Species[2]	C	P	C	P	C	P	C	P	C	P	C	P	C	P	C	P
Thy	*Pimelia glauca*															· · 1	
Sap	*Dodonea viscosa*					· 3 ·											
San	*Exocarpus strictus*	1 · ·												1 · ·			
	E. cupressiformis			1 1 ·				1 · ·									
Rha	*Pomaderris sp.*					· 1 ·											
Total number of species		24	11	15	0	18	0	12	0	24	1	16	10	19	9	19	1

[1] Mim=Mimosaceae; Fb=Fabaceae; Ep=Epacridaceae; Myr=Myrtaceae; Dil=Dilleniaceae; As=Asteraceae; Ro=Rosaceae; Pit=Pittosporaceae; Thy=Thymelaeaceae; Sap=Sapindaceae; San=Santalaceae; Rha=Rhamnaceae.

[2] Abundance in each of three strata is indicated: upper (1.5-3 m); middle (0.5-1.5 m) and lower (<0.5 m). Abundance ratings are: 5=dominant; 4=co-dominant; 3=obvious; 2=apparent; 1=inconspicuous; ·=not observed in that stratum.

[3] Districts and sites as in Appendix 1.

APPENDIX 3. Bird species recorded at sites between November 1987 and April 1988.

Species	District and site															
	Gin		Mur		Book		Tang		Blak		Dalt 1		Dalt 2		Gund	
	C	P	C	P	C	P	C	P	C	P	C	P	C	P	C	P
Quail[S,i] *(Turnix spp.)*									+						+	
Common bronzewing[S] *(Phaps chalcoptera)*										+						+
Galah[S] *(Cacatua roseicapilla)*		+	+				+	+	+	+	+	+	+			+
Sulphur-crested cockatoo[S,i] *(C. galerita)*				+		+			+	+	+				+	+
Superb parrot[S] *(Polytelis swainsonii)*				+												
Crimson rosella[S,i] *(Platycercus elegans)*	+	+	+	+	+	+	+	+	+	+	+	+		+	+	+
Eastern rosella[S,i] *(P. eximius)*	+		+			+		+	+	+		+		+	+	+
Red-rumped parrot[S] *(Psephotus haematonotus)*												+				
Southern whiteface[S,I] *(Aphelocephala leucopsis)*								+								
Mistletoe bird[S,i] *(Dicaeum hirundinaceum)*					+											
Diamond firetail[S,i] *(Emblema guttata)*									+		+					
Double-barred finch[S,I] *(Poephila bichenovii)*											+					
White-winged chough[S,I] *(Corcorax melanorhamphos)*			+		+	+			+		+		+	+		+
Pallid cuckoo[I] *(Cuculus pallidus)*									+							
Owlet nightjar[I] *(Aegotheles cristatus)*															+	
Spine-tailed swift[I] *(Hirundapus caudacutus)*									+							
Rainbow bee-eater[I] *(Merops ornatus)*			+					+						+		
Dollarbird[I] *(Eurystomus orientalis)*												+				
Welcome swallow[I] *(Hirundo neoxena)*		+							+	+						
Tree martin[I] *(Cecropis nigricans)*	+							+								
Black-faced cuckoo shrike[I] *(Coracina novaehollandiae)*	+	+	+		+	+		+	+	+		+	+		+	
White-winged triller[I] *(Lalage sueurii)*									+							

APPENDIX 3 (continued)

Species	District and site: Gin C	Gin P	Mur C	Mur P	Book C	Book P	Tang C	Tang P	Blak C	Blak P	Dalt 1 C	Dalt 1 P	Dalt 2 C	Dalt 2 P	Gund C	Gund P
Flame robin[I] (*Petroica phoenicea*)						+									+	
Scarlet robin[I] (*P. multicolor*)	+	+	+	+	+	+	+		+	+	+	+	+	+	+	+
Hooded robin[I] (*Melanodryas cucullata*)									+							
Eastern yellow robin[I] (*Eopsaltria australis*)									+							
Jacky winter[I] (*Microeca leucophaea*)			+						+							
Crested shrike-tit[I] (*Falcunculus frontatus*)					+				+	+		+				
Golden whistler[I] (*Pachycephala pectoralis*)			+	+	+		+		+			+			+	+
Rufous whistler[I] (*P. rufiventris*)			+		+		+	+	+	+	+	+	+	+	+	+
Grey shrike-thrush[I] (*Colluricincla harmonica*)			+						+	+	+			+	+	+
Leaden flycatcher[I] (*Myiagra rubecula*)				+							+					
Satin flycatcher[I] (*M. cyanoleuca*)									+							
Restless flycatcher[I] (*M. inquieta*)			+			+		+				+	+			+
Grey fantail[I] (*Rhipidura fuliginosa*)	+	+	+	+	+	+	+	+	+	+	+	+	+	+	+	+
Willie wagtail[I] (*R. leucophrys*)	+	+	+	+	+	+	+	+	+	+	+	+	+	+	+	+
White-browed babbler[I] (*Pomatostomus superciliosus*)									+							
Superb blue wren[I] (*Malurus cyaneus*)	+	+					+	+	+			+		+	+	+
White-browed scrubwren[I] (*Sericornis frontalis*)									+							
Speckled warbler[I] (*Chthonicola sagittatus*)									+							+
Weebill[I] (*Smicornis bevirostris*)		+	+	+	+		+			+	+	+	+			+
White-throated warbler[I] (*Gerygone olivacea*)	+				+		+				+					
Brown thornbill[I] (*Acanthiza pusilla*)			+	+					+						+	
Buff-rumped thornbill[I] (*A. reguloides*)	+	+	+	+	+	+	+		+	+	+		+	+	+	+

APPENDIX 3 (continued)

Species	District and site Gin C	Gin P	Mur C	Mur P	Book C	Book P	Tang C	Tang P	Blak C	Blak P	Dalt 1 C	Dalt 1 P	Dalt 2 C	Dalt 2 P	Gund C	Gund P
Yellow-rumped thornbill[I] *(A. chrysorrhoea)*	+	+	+		+	+	+	+	+		+	+	+	+	+	+
Yellow thornbill[I] *(A. nana)*	+		+													
Striated thornbill[I] *(A. lineata)*			+	+	+	+	+	+	+	+	+	+	+	+	+	+
Varied sitella[I] *(Daphoenositta chrysoptera)*	+													+		
White-throated treecreeper[I] *(Climacteris leucophaea)*	+		+	+	+	+	+	+	+	+	+	+	+	+	+	+
Brown treecreeper[I] *(C. picumnus)*									+	+	+					+
Noisy miner[I] *(Manorina melanocephala)*												+				
White-eared honeyeater[I] *(Lichenostomus leucotis)*			+	+					+		+		+		+	
Brown-headed honeyeater[I] *(Melithreptus brevirostris)*	+		+					+		+				+	+	
Spotted pardalote[I] *(Pardalotus punctatus)*	+		+	+	+	+	+		+	+		+	+		+	+
Striated pardalote[I] *(P. striatus)*		+	+	+	+	+	+	+		+	+	+			+	+
Starling[I] *(Sturnus vulgaris)*	+	+	+					+		+		+				
Olive-backed oriole[I] *(Oriolus sagittatus)*	+		+												+	
White-browed woodswallow[I] *(Artamus superciliosus)*									+							
Dusky woodswallow[I] *(A. cyanopterus)*									+							
Australian magpie lark[I] *(Grallina cyanoleuca)*	+	+		+	+		+		+	+	+	+	+		+	
Australian magpie[I] *(Gymnorhina tibicen)*	+	+	+	+	+	+	+	+	+	+	+	+	+	+	+	+
Red wattlebird[N,i] *(Anthochaera carunculata)*	+				+								+			
Noisy friarbird[N,I] *(Philemon corniculatus)*	+		+			+					+			+	+	+
Yellow-faced honeyeater[N,I] *(Lichenostomus chrysops)*			+	+						+		+	+	+	+	+
White-plumed honeyeater[N,I] *(L. penicillatus)*		+	+	+	+	+			+	+	+		+		+	+
White-naped honeyeater[N,I] *(Melithreptus lunatus)*			+													

APPENDIX 3 (continued)

Species	District and site Gin C	Gin P	Mur C	Mur P	Book C	Book P	Tang C	Tang P	Blak C	Blak P	Dalt 1 C	Dalt 1 P	Dalt 2 C	Dalt 2 P	Gund C	Gund P
Eastern spinebill^{N,i} *(Acanthorhynchus tenuirostris)*									+				+		+	
Black-shouldered kite^{V,i} *(Elanus notatus)*				+												
Wedge-tailed eagle^{V} *(Aquila audax)*																+
Australian kestrel^{V,i} *(Falco cenchroides)*									+							
Boobook owl^{V,I} *(Ninox novaeseelandiae)*	+										+					
Kookaburra^{V,i} *(Dacelo gigas)*	+		+	+	+	+	+		+		+	+			+	+
Sacred kingfisher^{V,i} *(Halcyon sancta)*			+						+							
Pied currawong^{V,I} *(Strepera graculina)*	+	+	+	+							+	+			+	+
Australian raven^{V,I} *(Corvus coronoides)*	+			+	+			+			+		+			
Total number of species	26	17	35	24	24	20	19	20	44	26	29	27	22	19	32	30
Total number of mainly insectivorous species (includes all ^{I})	22	15	30	9	20	16	16	17	34	21	24	22	19	17	26	23

Superscripts refer to typical dietary items: ^{Ss} seeds and fruits, ^{I,i} insects, ^{N,n} nectar and ^{V,v} to various prey and carrion. Upper case superscripts indicate major, and lower case superscripts indicate minor, dietary components.

Species are arranged primarily by diet; within each diet type species are arranged according to the order in which they appear in Pizzey, G. (1980). A Field Guide to the Birds of Australia. Collins Australia. Nomenclature also follows Pizzey.

Proc. Ecol. Soc. Aust. 1990 — 16:167-176

The European honey bee in Western Australian kwongan: foraging preferences and some implications for management

R. T. WILLS[1], M. N. LYONS[2] and D. T. BELL

Department of Botany, The University of Western Australia, Nedlands, W.A. 6009, Australia.

[1]Present address: Department of Conservation and Land Management, Research Division, Brain Street, Manjimup, W.A. 6258

[2]Present address: Department of Conservation and Land Management, Research Division, Woodvale, W.A. 6026

Abstract

Utilization of floral resources in native shrublands in the Northern Sandplain of Western Australia by the introduced European honey bee (Apis mellifera) *was examined over 30 months. Honey bees used 30% of the 413 species recorded as a source of pollen and/or nectar, preferentially foraging on the most common species. A second survey of seven months revealed that native bees also favoured species with high relative cover values and foraged on 70% of the 51 plant species visited by honey bees in the same period. Although only 30% of the regional flora are post-fire seed regenerators, these species were preferentially used by both the honey bee and native bee foragers. Fire, especially close-interval recurrent fires, can have a deleterious effect on the major melliferous species. Perennial reseeding species had a higher projective foliage cover than perennial sprouting species, but had lower cover at sites with shorter intervals between fires. Some possible effects of honey bee foraging preferences on plant seed set are examined. A short interval between fires which decreases the local abundance of reseeding species may reduce the ability of a population to attract pollinators, and, combined with the effect of a short interval between fires, may lead to the decline and eventual extinction of that population. The management requirements for areas vested for apicultural production are explored based on the attributes of melliferous species identified in the study, and the need for an ecological approach to management is discussed.*

Introduction

Since the settlement of Europeans in Australia 200 years ago, many species of animal alien to Australian ecosystems have been introduced and have readily established in this environment. The European honey bee, *Apis mellifera,* was first introduced into Western Australia in 1846 (Smith 1963; Chambers 1979). However, it was not until the 1880s that commercial beekeeping commenced. By the time first apicultural records were made in 1896 there were 2 267 registered hives in the state (Smith 1963). In its early years, the apicultural industry relied mainly on local stands of jarrah *(Eucalyptus marginata),* marri *(E. calophylla)* and wandoo *(E. wandoo)* for production (Cook 1987). Today, most commercial apiarists in Western Australia practice migratory beekeeping and are dependant on native plant communities as the major source of pollen and nectar. Apiaries are moved with the changing season to utilize the peak flowering of *Eucalyptus* species in forests in the south-west of Western Australia in late spring, summer and autumn, and relocated in the kwongan (native shrubland) of the Northern Sandplain for the winter flowering.

The study region, centred in the Northern Sandplain approximately 300 km north of Perth, covers about 1 300 km^2, more than half of which is included in the Beekeepers' Reserve (Fig. 1). This Reserve, unvested but set aside for the purposes of apiculture and flora conservation, provides an important resource to the beekeeping

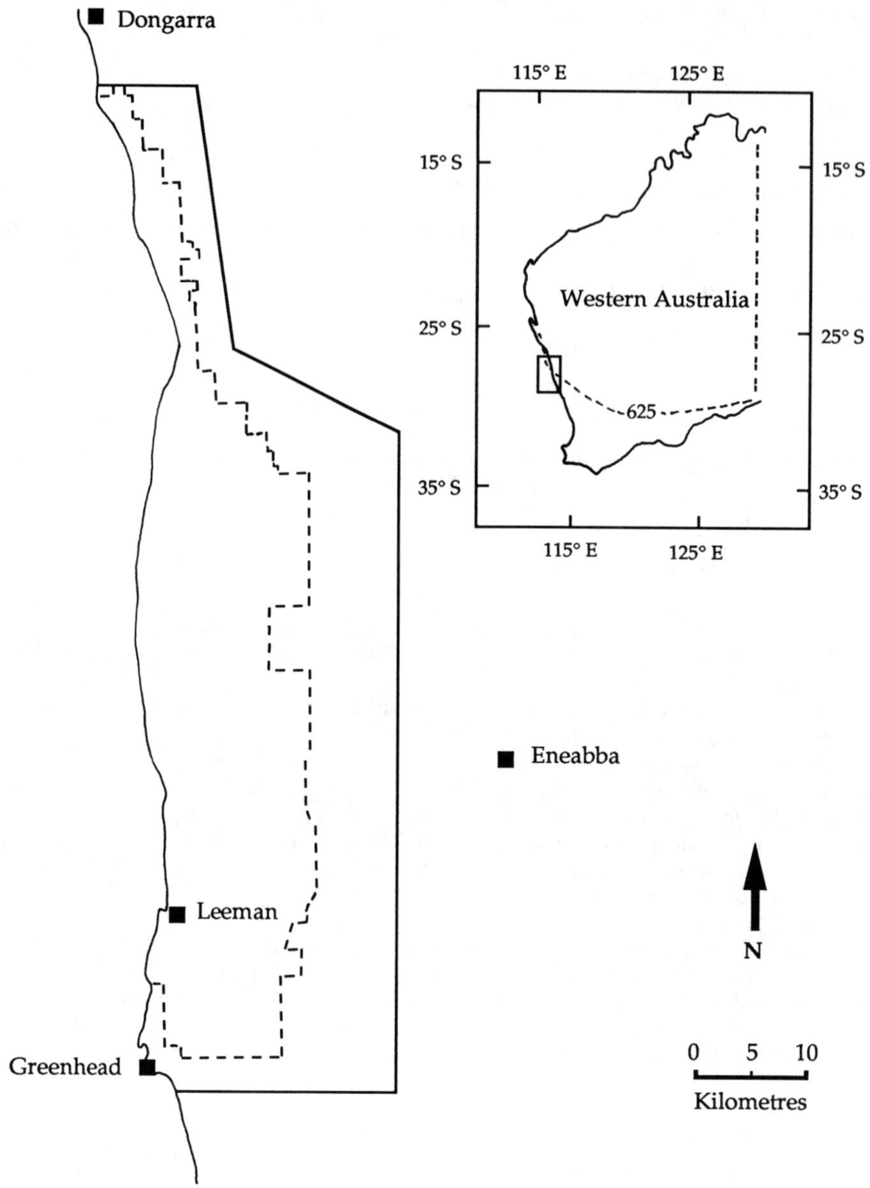

FIG. 1. Location of the Beekeepers' Reserve (dashed line) and the area of the Northern Sandplain region (solid line) included in this survey. The inset includes the 625 mm rainfall isohyet.

industry. The apicultural potential of the Beekeepers' Reserve has in recent times been degraded by frequent fires which have reduced the frequency of species considered important by apiarists (Burking 1986). At a time when land clearing is decreasing the resources available to beekeepers, considerable concern has been voiced by land managers over the possible impacts of honey bees on native plant and animal species (see Anon. 1983; Bell *et al.* 1984; Burking 1986; Blyth 1987). While introductions of alien species have often been to the detriment of the native flora and fauna, the honey bee was believed until recently to be one of the least harmful of European introductions. Now, however, it has become a contentious issue between conservationists and the apicultural industry. Management of the Beekeepers' Reserve and other areas to maintain apicultural values in harmony with flora conservation is of great importance. Since 1984, we have been studying the environmental and biological factors of importance in determining the floristic resource for the foraging honey bee. These data will assist in the development of an ecologically-based management programme for the Beekeepers' Reserve to enhance apicultural production without deleteriously affecting the conservation values of the Reserve area. In this chapter, we survey the floral resources used by the European honey bee in the study region and determine the species utilized by native bee species. Emphasizing the species of plants utilized by honey bees, we then explore some of the implications of fire management on the potential of the Reserve to continue to serve the apiculture industry in the future. Finally, we discuss the importance of considering plant pollination biology in forming management strategies for areas of native vegetation utilized for apiculture.

Methods

A total of 90 permanent monitoring sites representative of the diversity of vegetation types in the study region were selected. These included a range of sites differing in climatic, edaphic and fire history. Site fire history was determined by field estimation of site age, aerial photography and satellite remote sensing data. The relative abundance of each species within a 10 m × 10 m quadrat at each site was determined by visual estimation of percentage projective foliage cover. Assessment of species projective cover was carried out in March and April of 1985 and 1986. The growth form of each species was noted and the primary mode of plant regeneration after fire (either reseeding or sprouting) was recorded. Observations on plant flowering and honey bee foraging were made at all sites on 31 field trips held at intervals of three to eight weeks between April 1984 and December 1986. Where possible, flowering specimens of plant species were obtained and preserved in a herbarium collection. It is intended that voucher specimens of these species will be lodged with the Western Australian Herbarium. Plant species nomenclature follows that of Green (1985).

Honey bee foraging observations were carried out on each field visit and, therefore, covered a wide range of time of day, time of year and environmental conditions. Observation time at each site was dependent on the time required to assess the flowering phenology of each species and, therefore, varied from a minimum of five minutes in some sites to nearly an hour in full flowering, species rich sites. When discovered on flowers, the foraging behaviour of honey bees was observed to determine whether individuals were collecting nectar or pollen. Pollen-gathering honey bees can be differentiated from those foraging for nectar by the characteristic leg movements associated with the transfer of pollen to the pollen baskets on the hind tibiae and by the presence of pollen loads. On large flowers, the position of a honey bee will also indicate if it is obtaining pollen or nectar. The primary native pollinator of all species was determined by opportunistic observation of pollinator visitations in the field and inference from floral morphology, and confirmed by information from the literature when available.

In a second survey from March to September in 1987 during six field trips at intervals of four to six weeks, a further 12 sites were monitored more intensively to record visits to plants by both native bees and honey bees. At the more intensive sites, the type and number of bees encountered along a permanently marked 2 m wide 50 m transect was scored as the transect was traversed over a five minute period. A second traverse was made approximately 15 minutes after the first. All counts were made under sunny conditions between 1100 and 1500 hours. Further documentation of pollen foraging by native bees was obtained by the identification of pollen found on individuals obtained in the field.

Results and Discussion

Floral resources of honey bees

A total of 413 vascular plant species from 192 genera in 66 families were identified from the study region. Over one half of these species were from just eight families, with the largest numbers from the Myrtaceae and Proteaceae (Table 1). *Apis mellifera* visited 30% of the 413 species as a source of pollen and/or nectar. More than 70% of species visited by foraging honey bees were from the eight largest families.

Autumnal measurement of relative projective foliage cover precluded annuals and geophytes from assessment. A total of 228 native perennial species were recorded in 100 m^2 quadrats at the 90 permanent monitoring sites. Of these, 61 species were widespread (occurring in 10 or more quadrats) while 59 species were locally abundant

TABLE 1. Number of genera and species of the most common families recorded in the study region with number of species visited by honey bees.

Family	# Genera	# Species	# Visited by Honey bee
All families	192	413	125
Mytraceae	15	56	23
Proteaceae	11	42	23
Asteraceae	18	28	2
Goodeniaceae	5	21	6
Papilionaceae	14	21	13
Mimosaceae	1	18	11
Orchidaceae	7	17	0
Epacridaceae	6	16	11

TABLE 2. Frequency, projective foliage cover, fire response and primary pollinators of the 18 perennial species which were both widespread and locally abundant.

Family	Species	# Sites[1]	Mean % cover[2]	Post-fire	Pollinated by:
Cyperaceae	*Mesomelaena stygia*	36	5.4	Sprout	Wind
Dilleniaceae	*Hibbertia glabberima*	16	8.0	Sprout	Insects
Goodeniaceae	*Scaevola thesioides*	19	5.2	Reseed	Insects
Mimosaceae	*Acacia lasiocarpa*	23	7.0	Reseed	Insects
	A. rostellifera	20	13.4	Reseed	Insects
	A. spathulifolia	33	14.0	Reseed	Insects
Myrtaceae	*Beaufortia squarrosa*	13	13.0	Reseed	Birds
	Calothamnus quadrifidus	31	6.5	Sprout	Birds[3]
	Eucalyptus erythrocorys	16	9.4	Sprout/Reseed	Birds
	Melaleuca acerosa	78	8.5	Sprout/Reseed	Insects
	M. huegelii	13	14.9	Sprout	Insects
	Scholtzia umbellifera	31	6.3	Sprout	Insects
Papilionaceae	*Jacksonia spinosa*	25	12.1	Reseed	Insects
Proteaceae	*Banksia attenuata*	14	5.2	Sprout	Birds[3]
	B. leptophylla	42	15.6	Reseed	Mammals[4]
	B. prionotes	14	12.6	Reseed	Birds
	Dryandra sessilis	34	8.8	Reseed	Birds[3]
	Hakea trifurcata	33	5.6	Reseed	Insects

[1]Site refers to 10 m × 10 m quadrat; [2]Mean species projective foliage cover per site; [3]Insects may also be important in these species; [4]Birds may also be important in these species.

(having an average projective cover of 5% or more per quadrat). Most of the widespread species were generally not the most locally abundant, while conversely the more locally abundant species were often restricted in distribution so that only 18 species shared both attributes (Table 2). These 18 species make up only 8% of the total number of species recorded within quadrats, yet account for 51% of the total projective cover recorded for all species at all sites. These results compare favourably with those from other studies of kwongan vegetation which report that most species account for only a small proportion of foliage cover at a site (Griffin *et al.* 1983; Hopkins & Griffin 1984; Froend 1987).

While only 55% of all species collected from the study region occurred within the quadrats, about 81% of melliferous species were found within the quadrats. The species selected by honey bees within the quadrat area had an average projective cover per site of 6.1% compared with an average cover of 2.9% for those species which were not visited. In addition, melliferous species in the quadrats occurred, on average, at 10.5 sites compared with 7.0 sites for the remainder. Thus, honey bees tended to favour species which were either widespread and/or locally abundant. With the exception of *Scholtzia umbellifera*, all the species which were both widespread and locally abundant were used by honey bees (Table 2). Other studies of honey bee foraging on the Northern Sandplain have noted that species flowering locally in abundance were preferred, but, as observed in this study, some abundant species were never visited (van der Moezel *et al.* 1987a).

Floral resources of native bees

The collection of native bees and the identification of plant species used by native bees revealed that native bee foraging preferences were very similar to those observed for honey bees. A total of 72 taxa of native bees were identified during the 1987 survey period (Table 3). Species of *Leioproctus* and *Hylaeus* (Colletidae) were the most prominent of the native bee fauna. The various native bee taxa of the region were observed to forage on 42 native plant species. Almost 90% of plant species visited by native bees as sources of pollen and/or nectar were from the eight families listed earlier (Table 1), with the Myrtaceae, Proteaceae, Papilionaceae and Goodeniaceae the most important of these. Significant overlap in the resources used by honey bees and native bees was observed with almost 70% of the 51 species used by honey bees in the seven month survey period also visited by native bees (Table 4). The sharing of resources by native bees and honey bees indicates the potential for inter-species competition. Notably, studies in eastern Australia by Pyke and Balzer (1985) observed significant declines in the densities of native bees near experimental honey bee hives. Although the current study indicates some species of native bee and the European honey bee do presently co-exist, other species of native bee which may have formerly inhabited this region may have already been displaced due to the effects of either direct or indirect competition with *Apis mellifera*.

Post-fire regeneration

The determination of post-fire regenerative strategy for all 413 plant species revealed that 70% of species recovered post-fire primarily by sprouting, while 30% recovered primarily by reseeding (Table 5). If only the 360 native perennial species

TABLE 3. Taxonomic list of native bees collected during the winter 1987 study of the central Beekeepers' Reserve. () Indicates sub-genus. *Indicates unconfirmed species.

Family	Sub-family	Species
Colletidae	Colletinae	*Leioproctus * cyanescens*
		L. (Leioproctus) spp. 1-18
		*L. (*Nodocolletes)* sp.
		L. (Euryglossidia) spp. 1-11
		L. (Colletellus) spp. 1-6
		Trichocoletes (Trichocoletes) spp. 1-8
	Euryglossinae	*Euryglossa (Euhesma)* sp.
		**Mellitosmitha* sp.
	Hyalaeinae	*Hyaleus (Euprosopsis) violaceus*
		H. (Macrohyaleus) alcyoneus
		H. (Prosopisteron) spp. 1-2
		H. (Prosopisteron) sp. A174B200
		H. (Prosopisteron) distractus
		*H. (Prosopisteron) *distractus*
		H. (Prosopisteron) sanguinipictus
		H. (Prosopisteron) vittatifrons
		H. (Sphaerhyaleus) globuliferus
Halictidae	Nomiinae	*Nomia *flavoviridis*
	Halictinae	*Lasioglossum (Chilalictus)* spp. 1-4
		Homalictus sp.
Megachilidae	Megachilinae	*Megachile *chrysopyga*
		Chalicodoma (Hackeriapsis) spp. 1-2
		C. (Austrochile) sp.
Anthophoridae	Anthophorinae	*Amegilla *pulchra*

are considered, reseeders make up 20% of the total. Since the assessment of projective cover did not include annuals and geophytes, it is not surprising to find that only 22% of the 228 species recorded within quadrats were post-fire reseeders. Notably, honey bees prefer perennial species so that only three species of annuals were visited by honey bees. An examination of honey bee foraging according to post-fire regenerative strategies of native woody perennials reveals that sprouters form the major fraction of species used by honey bees. However, whereas 74% of all native woody perennials were post-fire sprouting species, only 61% of species visited by honey bees were sprouters. Indeed, honey bees foraged on only 36% of available sprouters compared with 63% of available reseeding species. Thus, these data indicate that the post-fire reseeding species are of major importance to the apiculture industry.

In addition to species preferences, an analysis of the individual plant cover data indicates that the relative importance of reseeders was substantially greater than might have been expected from reseeder frequency within quadrats, with reseeders accounting for 42% of the projective cover recorded for all sites. Reseeding species had a mean relative projective cover per species of 7.3%, twice the amount of cover per species contributed by sprouters. Notably, the group of melliferous reseeders had an average projective foliage cover of 7.7% compared with 4.8% for those reseeders which honey bees did not visit. Similarly, melliferous sprouters had an average cover per site of 4.9%, almost twice the average cover of sprouters which were not used by honey bees (2.7%). However, while honey bees used all but two of the locally abundant reseeders, only one half of the sprouters which were locally abundant were visited. This suggests that locally abundant reseeders may generally be a more profitable resource for honey bees than sprouting species with similar cover.

As was observed with honey bees, native bees tended to preferentially forage on reseeding species. In particular, 11 species of native bees visited *Dryandra sessilis* and 12 species of native bee visited *Hakea trifurcata* for either pollen or nectar. Several of the bee taxa were only found in the study region during the flowering of these two species. These plant species are of primary concern in this study as they are important sources of nectar and pollen for the apicultural industry (Bell *et al.* 1984; Bell 1985; Burking 1986). Both species are post-fire reseeders and were recorded as widespread and locally abundant (Table 2).

Fire regimes

Floristic composition of plant communities adapted to a particular fire regime should remain much the same following a fire unless the fire regime has been modified. The Northern Sandplain has probably been subjected to recurrent fire since the start of the current climatic period began

TABLE 4. Plant species of the central Beekeepers' Reserve visited by honey bees during the winter 1987 study period with records for species also visited by native bees (Apoidea).

Family	Species	Honey bee Resource	Native bee[1]
Casuarinaceae	*Allocasuarina humilis*	Pollen	
Chloanthaceae	*Pityrodia bartlingii*	Nectar	
Dilleniaceae	*Hibbertia crassifolia*	Pollen	
	H. polystachya	Pollen	
Epacridaceae	*Andersonia heterophylla*	Pollen	
	A. lehmanniana	Pollen	
	Leucopogon allittii	Pollen	
	L. conostephioides	Pollen	++
	L. crassifolia	Pollen/Nectar	++
	L. polymorphus	Pollen	
Goodeniaceae	*Scaevola canescens*	Nectar	+?
	S. globulifera	Pollen/Nectar	++
	S. paludosa	Nectar	+?
	S. thesioides	Pollen/Nectar	++
Mimosaceae	*Acacia pulchella*	Pollen	
	A. rostelifera	Pollen	
	A. spathulifolia	Pollen	++
Myrtaceae	*Beaufortia squarrosa*	Nectar	
	Colothamnus quadrifidus	Nectar	++
	C. sanguineus	Nectar	++
	Darwinia pauciflora	Nectar	
	Eremaea beaufortioides	Pollen	
	Eucaluptus calophylla	Pollen/Nectar	++
	E. erythrocorys	Pollen/Nectar	++
	Hypocalymma xanthopetalum	Nectar	
	Melaleuca scabra	Pollen	
	Scholtzia laxiflora	Nectar	+?
Papilionaceae	*Bossiaea eriocarpa*	Pollen	
	Daviesia aff. *polyphylla*	Pollen	++
	D. divaricata	Pollen	+
	D. nudiflora	Pollen	++
	D. quadrilatera	Pollen/Nectar	++
	D. triflora	Nectar	++
	Jacksonia floribunda	Pollen	+?
	J. furcellata	Pollen	++
	J. spinosa	Pollen	++
	J. ulicina	Pollen	++
	Oxylobium capitatum	Pollen	
Proteaceae	*Banksia attenuata*	Nectar	+
	B. leptophylla	Nectar	+
	B. menziesii	Pollen/Nectar	
	B. prionotes	Pollen/Nectar	++
	Dryandra sessilis	Pollen/Nectar	++
	Grevillea thelemanniana	Nectar	++
	Hakea costata	Pollen/Nectar	++
	H. lissocarpha	Pollen/Nectar	
	H. trifurcata	Pollen/Nectar	++
	Isopogon dubius	Pollen	++
Solonaceae	*Anthocercis littorea*	Nectar	
Stylidicaeae	*Stylidium repens*	Pollen	
Xanthorrhoeaceae	*Xanthorrhoea drummondii*	Nectar	

1. ? Native bee foraging for unknown resource.
\+ Plants visited by native bees.
++ Plants visited by native bees for same resource.

about 5 000 to 6 000 years ago (Wyrwoll 1979; Hopkins *et al.* 1983). Fire frequency in this time has been estimated to be in the order of once every 2-3 decades (Bell *et al.* 1984; Bell 1985), although more recent research suggests a fire cycle of 15 years or less could have occurred in kwongan near Badgingarra (van der Moezel *et al.* 1987b). However, since the rapid expansion of farming land

TABLE 5. Post-fire regeneration strategies of perennial plant species compared with species found in quadrats and species visited by honey bees.

Group	# Species	Reseeder	Sprout
All species	413	125	288
Native perennial species	360	72	288
# perennials in quadrats	228	50	178
# perennials visited by honey bee	120	45	75

and mining leases in the 1970s, the interval between fires in the Northern Sandplain in recent times has been substantially shorter than this (R. Place, W. A. Bush Fires Board, pers. comm.).

At the time of the survey, the average age since fire of the 90 permanent sampling sites was 7.9 years, and the mean interval between the last fire and the immediately preceding fire at site was 9.4 years. Therefore, the sites have on average experienced two fires in less than 18 years. Indeed, some sites have been burnt as many as four times in that period. If the relative abundance of reseeding species is compared with the interval between the last two fires at a site, it is apparent that, as the between-fire interval decreases, the contribution made by reseeders to relative cover also decreases (Fig. 2). Other workers report substantial reductions in the projective cover of reseeding species in more frequently burnt sites (e.g., Nieuwenhuis 1987), although this effect may in part reflect the rapid post-fire regeneration of foliage projective cover by sprouters relative to reseeders (Keeley 1986; Delfs *et al.* 1987; Nieuwenhuis 1987). For these reasons, communities dominated by a high number of sprouters are more resilient to a short interval between fires (Keeley 1986; Westman & O'Leary 1986). Progressive reduction in reseeding species cover, however, means reduced resource availability for native bees and honey bees on the region.

General discussion

When extrapolating the information on pollination ecology recorded here to aspects of community management in the Northern Sandplain region of Western Australia, a range of biological and environmental data needs to be considered. The kwongan flora generally has low dispersive powers (Hopkins & Griffin 1984; Pate *et al.* 1984; Cowling 1987). Low dispersability of seed may be a consequence of large seed weight, reflecting a physical consequence of the level of resource allocation to seeds in either nutrient-poor habitats (Howe & Westley 1986) or in relation to the importance of seeds in plant post-fire regeneration (Carpenter & Recher 1979). Alternatively, low dispersability of propagules may be adaptive since it would promote the close grouping of individuals and maintain microgeographic genetic differentiation to local edaphic and biotic conditions (Levin & Kerster 1974; Price & Waser 1979; Slatkin 1985; Waser 1987). Some flower visitors preferentially forage on individuals or clumps of individuals with the largest floral displays (Stephenson 1982; Grey 1985), and so plants which are growing in patches may act collectively to attract pollinators. Furthermore, pollinator foraging has been found to be lower on the edge of a clump of plants than in its centre, resulting in lower seed set on the edge of a population (Augspurger 1980, 1981). Similarly, if plants which usually grow in dense stands were to become scattered or rare, they may receive fewer visits from pollinators, and so seed set may be reduced. Notably, if seed weight is greater in reseeders for reasons of resource allocation, dispersability should therefore be lower and so clumping should be promoted.

If density-dependent gene-flow is responsible, even in part, for the patterns of frequency and abundance reported for reseeders, changes in stand density associated with altered fire regimes may cause serious perturbations of plant-pollinator interactions and so have serious consequences for the management of melliferous species. It was shown earlier that reseeder abundance is reduced as the interval between fires is shortened. Apart from directly affecting the availability of resources for both honey bees and native flower visitors, a reduced abundance of reseeding species may cause the loss of genetic diversity and so founder effects could influence regeneration of reseeders. Similarly, the effect of incompatibility and other mating system barriers could also result in lower than usual seed set restricting population expansion or even lead to exinction in small populations of founding species (McClanahan 1986). In addition, if reduction in stand densities caused plants to become less attractive to pollinators, seed set may be reduced further.

In areas that are now subject to a shorter interval between fires, further interruption to pollination in reseeding species may decrease seed banks and so have an impact on the continued stability of plant communities. It has been suggested that honey bees may reduce seed set in native plant species because they compete with native pollinators for food and so lower the carrying capacity of an area for native flower visitors (Bond & Brown 1979; Paton 1979, 1986; Hopper 1987; Keighery 1985; Pyke & Balzer 1985). Furthermore, it has been suggested that honey bees may reduce seed set in native plant species because of: the competitive displacement of natural pollinators; the reduction of pollen loads available for legitimate pollinators; or due to ineffective

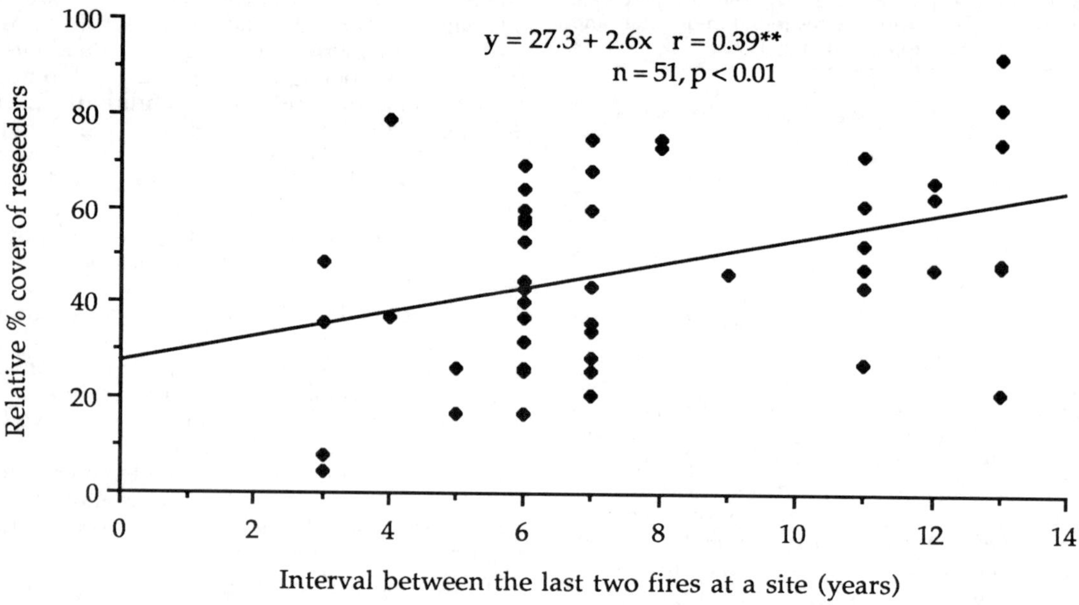

FIG. 2. Relative projective foliage cover of reseeders for 51 sites occurring exclusively on limestone plotted against the interval between the last two fires at a site. N.B. 39 sites excluded from the analysis on the basis of floristic dissimilarity.

pollination of native plants by honey bees (Keighery 1985; Paton 1986; Hopper 1987). Although the available evidence is not conclusive, if honey bee foraging does result in loss of seed set, preferential foraging by honey bees on reseeders may aggravate the effects of altered fire regimes, and may require special attention from land managers.

In the light of this discussion, we may consider as an example the implications for the key honey producing species, *Dryandra sessilis*. The plant grows extensively throughout Western Australia in dense stands or clumps (Grey 1985; Collins & Rebelo 1987; this study), is obligately xenogamous (Grey 1985; Collins & Rebelo 1987), and is a post-fire reseeder (Table 2). Studies have demonstrated that honeyeaters are the primary pollinator and that birds preferentially forage on clumped plants, avoiding solitary plants (Grey 1985). Frequent fires which produce scattered individuals will reduce the probability of attracting flower visitors necessary for pollination and may result in the loss of populations of pollinating honeyeaters.

Clearly, many of the attributes of *Dryandra sessilis* make it predisposed to the possible deleterious effects of foraging honey bees outlined above. It is not known if honey bees do have any deleterious or beneficial effects on the pollination of this species. Information on the effects of honey bee foraging on this and other native plant species is lacking, but must be gained to allow ecologically-based management decisions. Any management strategy must consider the phenology and the pollination ecology of the species involved to ensure the diversity of plant and animal species and the continued viability of ecosystems.

While honey bees may have a number of negative effects upon Australian ecosystems, the value of the honey industry as a vehicle for conservation of substantial areas of native vegetation all around Australia should not be overlooked. The contribution that the honey industry can make to conservation is great provided the industry approaches any apparent conflicts in a responsible and enlightened manner. Where most land managers today are being increasingly pressured by politicians and lobby groups to put a dollar value on conservation, or to justify the cost of maintaining reserve areas in deference to other land-uses such as agriculture, it is a powerful argument to point out that these areas may produce a cash crop of honey and pollen without the need for clearing. However, the honey industry needs to consider carefully the ecological implications of utilizing native floral resources for honey production. Intelligent utilization and management of native vegetation should ensure the maintenance of productive, self-maintaining pastures, and could provide the vehicle for the retention of many remnants of vegetation.

Acknowledgements

Funds for this study were contributed by the Honey Research Council of the Department of Primary Industry to Dr D. T. Bell. Research formed part of the Ph.D. dissertation research of Mr R. T. Wills and Honours research of Mr M. N. Lyons. Dr T. F. Houston of the Western Australian Museum provided assistance with the identification of native bee species. We gratefully acknowledge Dr W. A. Loneragan and Dr R. H. Froend for helpful comments on early drafts of this paper, and thank Dr R. H. Froend and Ms C. P. Watson for their support throughout this project.

References

Anon. (1983) *Apiculture Workshop Papers. Workshop papers and documents for the Standing Committee on Agriculture. Hawkesbury College, Richmond, New South Wales. 18-22 July, 1983.* N.S.W. Dept. Agric. Anim. Prod. Div., Richmond, New South Wales.

Augspurger C. K. (1980) Mass flowering of a tropical shrub *(Hybanthus prunifolius):* Influence on pollinator attraction and movement. *Evolution* **34,** 475-88.

Augspurger C. K. (1981) Reproductive synchrony of a tropical shrub; experimental studies on effects of pollinators and seed predators on *Hybanthus prunifolius* (Violaceae). *Ecology* **62,** 775-88.

Bell D. T. (1985) Aspects of response to fire in the Northern Sandplain heathlands. In: *Fire Ecology and Management in Western Australian Ecosystems* (ed. J. Ford) pp. 33-40. Western Australian Institute of Technology Environmental Studies Group Rep. No. 14, Western Australian Institute of Technology, Bentley, Western Australia.

Bell D. T., Hopkins A. J. M. & Pate J. S. (1984) Fire in the kwongan. In: *Kwongan: Plant Life of the Sandplain* (eds J. S. Pate & J. S. Beard) pp. 178-204. University of Western Australia Press, Nedlands, Western Australia.

Blyth J. D. (ed.) (1987) *Beekeeping and Land Management. Proceedings of a workshop held at the Department of Conservation and Land Management (CALM) Hayman Rd, Como, Western Australia. November 4, 1985.* Department of Conservation and Land Management, Perth, Western Australia.

Bond H. W. & Brown W. L. (1979) The exploitation of floral nectar in *Eucalyptus incrassata* by honeyeaters and honey bees. *Oecologia* **44,** 105-11.

Burking R. C. (1986) The effects of wildfires on the Western Australian Beekeeping Industry. *Australas. Beekeeper* **87,** 255-8.

Carpenter F. L. & Recher H. F. (1979) Pollination, reproduction and fire. *Amer. Nat.* **113,** 871-9.

Chambers S. R. (1979) The Honey Bee. In: *Agriculture in Western Australia: 150 years of development and achievement 1829-1979* (ed. G. H. Burvill) p. 248. University of Western Australia Press, Nedlands, Western Australia.

Collins B. G. & Rebelo T. (1987) Pollination biology of the Proteaceae in Australia and southern Africa. *Aust. J. Ecol.* **12,** 387-421.

Cook S. C. (1987) Resource utilization — past. In: *Beekeeping and Land Management. Proceedings of a workshop held at the Department of Conservation and Land Management (CALM) Hayman Rd, Como, Western Australia. November 4, 1985.* (ed. J. D. Blyth) pp. 29-33. Department of Conservation and Land Management, Perth, Western Australia.

Cowling R. M. (1987) Fire and its role in co-existence and speciation in Gondwanan shrublands. *S. Afr. J. Sci.* **83,** 106-11.

Delfs J. C., Pate J. S. & Bell D. T. (1987) Northern Sandplain Kwongan: community biomass and selected species response to fire. *J. Roy. Soc. West. Aust.* **69,** 133-8.

Froend R. H. (1987) *Investigations into species richness patterns in the Northern Sandplain region of Western Australia.* Ph.D. Thesis, Department of Botany, University of Western Australia, Perth, Western Australia.

Green J. W. (1985) *Census of the Vascular Plants of Western Australia* (2nd edition). Western Australian Herbarium, Perth, Western Australia.

Grey J. (1985) *Behavioural energetics of honeyeaters in relation to pollination and resource selection.* M. Appl. Sc., Curtin University of Technology, Bentley, Western Australia.

Griffin E. A., Hopkins A. J. M. & Hnatiuk R. J. (1983) Regional variation in mediterranean-type vegetation at Eneabba, south Western Australia. *Vegetatio* **52,** 103-27.

Hopkins A. J. M. & Griffin E. A. (1984) Floristic patterns. In: *Kwongan: Plant Life of the Sandplain* (eds J. S. Pate & J. S. Beard) pp. 69-83. University of Western Australia Press, Nedlands, Western Australia.

Hopkins A. J. M., Keighery G. J. & Marchant N. G. (1983) Species-rich uplands of south-western Australia. *Proc. Ecol. Soc. Aust.* **12,** 15-26.

Hopper S. D. (1987) Impact of honeybees on Western Australia's nectarvorous fauna. In: *Beekeeping and Land Management. Proceedings of a workshop held at the Department of Conservation and Land Management (CALM) Hayman Rd, Como, Western Australia. November 4, 1985.* (ed. J. D. Blyth) pp. 59-72. Department of Conservation and Land Management, Perth, Western Australia.

Howe H. F. & Westley L. C. (1986) Ecology of pollination and seed dispersal. In: *Plant Ecology* (ed. M. J. Crawley) pp. 185-215. Blackwell Scientific Publications, Oxford, England.

Keeley J. E. (1986) Resilience of mediterranean shrub communities to fires. In: *Resilience in Mediterranean-type Ecosystems* (eds B. Dell, A. J. M. Hopkins & B. B. Lamont) pp. 95-112. Dr W. Junk Publishers, Dordrecht, The Netherlands.

Keighery G. J. (1985) Of birds, bees and others. *West. Aust. Wildfl. Soc. Newsletter* **23(3),** 27-31.

Levin D. A. & Kerster H. W. (1974) Gene flow in seed plants. *Evol. Biol.* **7,** 139-220.

McClanahan T. R. (1986) Pollen dispersal and intensity as criteria for the minimum viable population and species reserves. *Environ. Manage.* **10,** 381-3.

Nieuwenhuis A. (1987) The effect of fire frequency on the sclerophyll vegetation of the West Head, New South Wales. *Aust. J. Ecol.* **12,** 373-85.

Pate J. S., Dixon K. W. & Orshan G. (1984) Growth and life form characteristics of kwongan species. In: *Kwongan: Plant Life of the Sandplain* (eds J. S. Pate & J. S. Beard) pp. 84-100. University of Western Australia Press, Nedlands, Western Australia.

Paton D. C. (1979) *The behaviour and feeding ecology of the New Holland honeyeater* Phylidonyris novaehollandiae *in Victoria.* Ph.D. Thesis, Department of Zoology, Monash University, Melbourne, Victoria.

Paton D. C. (1986) Honeyeaters and their plants in south-eastern Australia. In: *The Dynamic Partnership: Birds and Plants in Southern Australia* (eds H. A. Ford & D. C. Paton) pp. 9-19. The Flora and Fauna of South Australia Handbooks Committee, Adelaide, South Australia.

Price M. V. & Waser N. M. (1979) Pollen dispersal and optimal outcrossing in *Delphinium nelsonii. Nature* **277,** 294-7.

Pyke G. H. & Balzer L. (1985) *The effects of the introduced honeybee* (Apis mellifera) *on Australian bees.* New South Wales National Parks and Wildlife Service Occasional Paper No. 7, New South Wales Govt. Printer, Sydney, New South Wales.

Slatkin M. (1985) Gene flow in natural populations. *Ann. Rev. Ecol. Syst.* **16,** 393-430.

Smith J. N. (1963) *The development of apiculture in Western Australia with special reference to migratory beekeeping and present day problems.* Teachers' Higher Certificate Optional Thesis, Education Department, East Perth, Western Australia.

Stephenson A. G. (1982) When does outcrossing occcur in a mass-flowering plant? *Evolution* **36,** 762-7.

van der Moezel P. G., Delfs J. C., Pate J. S., Loneragan W. A. & Bell D. T. (1987a) Pollen selection by *Apis mellifera* in shrublands of the Northern Sandplains of Western Australia. *J. Apicult. Res.* **26,** 224-32.

van der Moezel P. G., Loneragan W. A. & Bell D. T. (1987b) Northern Sandplain Kwongan: regeneration following fire, juvenile period and flowering phenology. *J. Roy. Soc. West. Aust.* **69,** 123-32.

Waser N. M. (1987) Spatial genetic heterogeneity in a population of the montane perennial plant *Delphinium nelsonii* **58,** 249-56.

Westman W. E. & O'Leary J. F. (1986) Measures of resilience: the response of coastal sage scrub to fire. *Vegetatio* **65,** 179-89.

Wyrwoll K. -H. (1979) Late Quarternary climates of Western Australia: evidence and mechanisms. *J. Roy. Soc. West. Aust.* **62,** 129-42.

Proc. Ecol. Soc. Aust. 1990 — 16:177-183

Fire and management of south-western Australian banksias

R. M. COWLING[1], B. B. LAMONT[2] and N. J. ENRIGHT[3]

[1]*Botany Department, University of Cape Town, Rondebosch 7700, South Africa*

[2]*School of Biology, Curtin University of Technology, G.P.O. Box U1987, Perth 6001, Western Australia*

[3]*Geography Department, University of Auckland, Auckland, New Zealand*

Abstract

Sixty of the 75 described species of Banksia *are endemic to mediterranean south-western Australia where they are often the dominant plants in fire-prone scrub-heath and woodland on nutrient-poor soils. Data on the fire-related reproductive and population biology of 11 banksias, largely confined to kwongan vegetation north of Perth, are now available. These data are reviewed here in order to provide guidelines for the maintenance of banksia populations through the judicious use of fire.*

In kwongan, populations of non-sprouting banksias are largely even-aged: recruitment is confined to the immediate post-fire period. Many banksias, particularly non-sprouting species, are strongly serotinous and maintain large viable seed banks. These traits maximize the amount of seed available for post-fire recruitment. Patterns of annual seed production indicate that many species take at least 10 years to accumulate seed reserves sufficient for adequate post-fire recruitment. Post-fire seed mortality, resulting from predation and the effects of high soil temperatures, is higher after spring than autumn burns, resulting in lower recruitment. However, delayed seed release and the occurrence of safe sites after spring burns ensures that recruitment is well above replacement levels after fires in this season. Differential post-fire recruitment of up to six banksias that may co-occur in a community, usually in response to variable fire regimes and post-fire conditions, ensures their long-term co-existence. Elimination of non-sprouting species from a site is most likely following small fires at short intervals in climates of unreliable rainfall.

Introduction

Sixty of the 75 described species of *Banksia* (Proteaceae) are endemic to mediterranean south-western Australia (Taylor & Hopper 1988). Many of these species dominate the fire-prone kwongan (scrub-heath) on infertile soils which are widespread in the region (Beard 1984). Many banksias have narrow distribution ranges and seven taxa are gazetted as rare and endangered (Anon. 1988). Up to six species may co-exist in a community and species turnover along geographical gradients is extremely high (Cody 1986).

Over the past five years we have studied the reproductive and population biology of 11 banksias occurring in the kwongan and woodlands north and east of Perth. The stimulus for this work arose from both practical and theoretical considerations. Some of the species are rare, while others are harvested for the wildflower trade (Burgman & Hopper 1982). Therefore, we wished to provide guidelines for fire regimes which would optimize the post-fire recruitment of banksias. All banksias store seed in the canopy (serotiny) to various degrees, and the size and fate of each annual seed crop can be determined readily (e.g., Cowling *et al.* 1987; Lamont & Barker 1988). As a result, we were able to collect demographic data necessary for studying the effects of variation in the different components of the fire regime on post-fire reproductive success (Lamont 1985; Cowling and Lamont 1987). Secondly, banksias provide excellent opportunities for exploring the evolutionary significance of fire-adapted reproductive traits and the mechanisms that promote co-existence of species.

In this chapter, we concentrate on post-fire determinants of population size, since it is possible to manipulate the fire regime as a management tool. Pre-fire factors which influence reproductive output include breeding systems, pollinator efficiency, resource-limited ovule and seed abortion, flower and seed predation, seed release and senescence (Lamont *et al.* 1985; Collins & Rebelo 1987; Lamont & Barrett 1988). Their year-by-year interaction between fires determines the size of the pre-fire seed banks available for post-fire recruitment.

Population age structure

In kwongan, populations of non-sprouting banksias are usually single-aged (Cowling *et al.* 1987); some interfire recruitment can occur after drought-induced mortality of parent plants (Hnatiuk & Hopkins 1980). Seedlings which establish in mature vegetation during the wet winter period, invariably perish in summer (Lamont & Barker 1988; Lamont & van Leeuwen 1988), probably as a result of competition with mature plants for moisture (Cowling & Lamont 1987). Thus, population growth is largely episodic, as it is confined to the immediate post-fire period.

Seed banks

Banksias store their seeds in woody fruits (follicles) embedded in massive infructescences (cones) (George 1981). The degree of serotiny varies both between and within species. Cowling and Lamont (1985a) showed that for three banksias, the degree of serotiny was negatively correlated with mean annual rainfall and length of the growing season. Only at the mesic (woodland) end of the climatic gradient were conditions suitable for establishment in the inter-fire period and intraspecific serotiny was minimal. Other factors responsible for the evolution of serotiny in banksias are discussed by Cowling *et al.* (1987).

Patterns in the post-fire accumulation of seed banks, as estimated from data on age-specific follicle production, provide important management insights. For example, it is clear from Figure 1 that a fire frequency of less than six years could result in the local disappearance of non-sprouting species. In most cases it takes more than 10 years for substantial follicle production to occur (Cowling *et al.* 1987; Lamont & Barker 1988). Also, of interest in Figure 1 are the year-to-year fluctuations in reproductive output which are not synchronous among species, although they are regional congeners. Non-sprouting, weakly serotinous species such as *B. prionotes* would be expected to show relatively low recruitment when a fire coincided with a period of low reproductive output (Cowling *et al.* 1987). Upper limits to fire intervals need to be considered also, as reproductive output declines in senescent stands, particularly when populations are dense. Although this has been documented for South African *Protea*

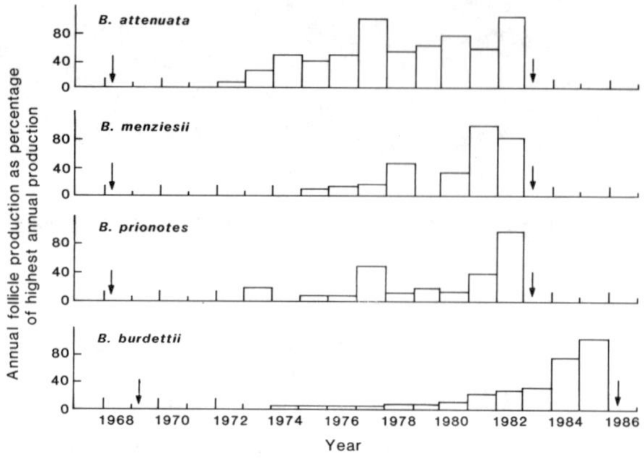

FIG. 1. Annual follicle production by four banksias occurring in the northern kwongan of south-western Australia. *B. attenuata* and *B. menziesii* are sprouters, while *B. prionotes* and *B. burdettii* are non-sprouters. The first arrow pinpoints the date of the fire, and the second arrow indicates when the site was burnt again (adapted from Cowling 1987).

spp. (Bond *et al.* 1984), no data are available for banksias. However, it appears that mortality of *B. coccinea,* a non-sprouting species confined to the southern coastal shrublands, increases dramatically in 15-20 year-old stands (S. Connell, unpub. data). Since seeds on dead plants are soon released, unlike more serotinous species (Lamont & Barker 1988), fires at greater than 20 year intervals could be disastrous for *B. coccinea.*

Generally, non-sprouting species have the largest viable canopy-stored seed crops (Table 1). The upper limit is represented by 24-year-old individuals of the strongly serotinous, non-sprouting *B. cuneata* which stores 16 900 viable seeds per plant (S. Connell, B. Lamont & S. Bergl, unpubl. data). The lower limit is provided by *B. elegans,* a strongly serotinous sprouter which stores less than one seed per plant (Lamont & Barrett 1988). This species is unique in that it regenerates after fire by both sprouting and extensive suckering from its surface roots; no seedlings have been observed (Lamont 1989).

Seed release

In banksias, both heat and drying promote follicle opening for the subsequent release of seed (Gill 1976). In strongly serotinous species fire is essential for synchronized follicle opening. Summer dry conditions are sufficient to open many follicles in weakly serotinous banksias (Cowling & Lamont 1987). Lamont and Cowling (1984) showed that when the cones of serotinous species were exposed to a brief flame, the heat released after the ignition of persistent florets was sufficient to open most follicles. Cones with deciduous florets failed to ignite and follicles remained closed in the absence of any significant temperature rise.

TABLE 1. Size of viable canopy-stored seed bank in 14-16 year-old post-fire populations of banksias in south-western Australia. a=Mt Adams; b=Eneabba. Data from Cowling *et al.* (1987), Lamont & Barker (1988), Lamont & Barrett (1988), S. Connell, N. Enright & B. Lamont (unpub. data), Lamont & van Leeuwen (1988).

	Number of viable seeds per plant
Weakly serotinous, non-sprouter	
B. prionotes	124
Strongly serotinous, non-sprouter	
B. burdettii	832
B. cuneata	785
B. hookeriana	374
B. lanata	135
B. leptophylla	1 344
Weakly serotinous, sprouter	
B. menziesii (a)	<1
B. menziesii (b)	1
B. tricuspis	46
Strongly serotinous, sprouter	
B. attenuata (a)	15
B. attenuata (b)	13
B. candolleana	10
B. elegans	<1

Enright and Lamont (1989) investigated temperature requirements for follicle rupture for 10 banksias and corroborated these findings. However, there was great variation in temperature requirements for follicle opening (e.g., 50% of *B. tricuspis* follicles opened at 145°C while *B. hookeriana* opened at 390°C). They found no overall correlation of temperature requirements for follicle opening with degree of serotiny, cone height, follicle volume, dependence on seed for post-fire recruitment or follicle number per fertile cone.

Many kwongan banksias show delayed seed release after follicle rupture (Cowling & Lamont 1987; Lamont & Barker 1988; Lamont 1989). Cowling and Lamont (1985b) demonstrated that seed release, particularly in serotinous species, is promoted by a series of wet-dry cycles. The seeds are held in a two-winged plate termed a separator (George 1981). The wings of the separator are hygroscopic and act like a lever, drawing the adpressed seeds out of the follicle with each wet-dry cycle. More seeds are released when cones are subject to wet-dry cycles under cool (15-20°C) rather than warm (<20°C) temperatures. Thus, more seeds are released at a faster rate after autumn (cool, wet) rather than spring, or especially summer (warm, dry) burns (Cowling & Lamont 1987; Enright & Lamont 1989).

The rapid release of seeds with the onset of late autumn/early winter rains ensures optimum germination and establishment. It also means that, in the event of follicle opening after spring or summer burns, a portion of the seed bank is retained in the cones until autumn and thus escapes post-dispersal seed predation and exposure to high summer soil temperatures (Cowling & Lamont 1987). However, *B. hookeriana* releases most of its seeds within 40 days of both spring and autumn fires (Enright & Lamont 1989). Its small seeds readily disperse into microsites where they remain protected from heat and granivores.

Seed germination and mortality

Seeds of the kwongan banksias we have studied show no innate dormancy and germinate optimally at 15-20°C. Viability is highest for non-sprouting species (*c* 80-100%) and lowest for sprouters (13-50%) (Cowling & Lamont 1987; Lamont & Barker 1988; Lamont 1989). Seeds left exposed on the soil during summer show a marked loss of viability relative to untreated controls (Cowling & Lamont 1987) and buried seed (Enright & Lamont 1989).

Field experiments indicate negligible seed predation after an autumn burn but high levels after a spring burn (Cowling & Lamont 1987). This may be due to the longer period that seed is exposed to post-dispersal granivores after spring than after autumn burns (cf. Bond 1984). Enright and Lamont (1989) noted that seeds located in "safe sites" (see below) after dispersal may escape predation.

Seedling establishment and mortality

Seedlings of kwongan banksias emerge during the cool, wet, winter months. Mortality during the first winter is very low (*c* 1%) but increases sharply in late summer (Cowling & Lamont 1987; Enright & Lamont 1989). Although there were no differences in germination patterns from seed planted in burnt and mature vegetation, mortality in the latter site was 100% (Cowling & Lamont 1987). In both burnt and unburnt vegetation, mortality appears to be drought-induced (Cowling & Lamont 1987; Lamont *et al.* 1989), exacerbated by competition with the parent plants when unburnt. Grazing also plays an important role after small or patchy burns (Whelan & Main 1979; Cowling & Lamont 1987).

We would predict poor seedling establishment after spring burns due to the enormous post-dispersal attrition (predation, loss of viability) of the seed bank. Seedling to parent ratios, measured at the end of the establishment phase (late spring), were more than twice as high after an autumn burn than a spring burn for four co-occurring banksias at Mt Adams (Cowling & Lamont 1987) and five co-occurring banksias at Eneabba (Enright & Lamont 1989), both about 300 km north of Perth. However, by the end of the first summer after establishment, seedling to parent ratios were only marginally higher for the autumn than the spring burns. For all non-sprouting species seedling numbers were well above replacement levels, irrespective of fire season.

What is surprising is that any seedlings establish at all after spring burns. Delayed seed release could reduce the time that seeds are exposed to predation and high soil temperatures after the spring burn. Furthermore, after the cooler spring burns, there is a large amount of scorched litter and other plant debris which accumulates in depressions on the soil surface. These areas are safe sites which offer seed protection from predation by vertebrates and high summer soil temperatures (Cowling & Lamont 1987; N. Enright, unpubl. data). Enright and Lamont (1989) observed that although safe sites comprised 14% and 30% of the ground area in autumn and spring burn sites respectively, they supported 37% (autumn) and 78% (spring) of the total seedlings. Clearly, safe sites would be unimportant after autumn burns when seed are exposed to cool conditions for a very short period prior to germination (Cowling & Lamont 1987). Furthermore the concept of safe site applies only to seed survival as seedling growth rates in bare areas and safe sites were not significantly different (N. Enright, unpubl. data).

The very high levels of seedling mortality recorded over the first summer after establishment are both density-dependent and species- and habitat-specific. Summertime mortality rates were higher after autumn (65%) than spring burns (32%), where seedling density, and hence competitive effects, were much lower (Enright & Lamont 1989). Mortality rates of *B. leptophylla* were consistently lower than those of co-occurring congeners (Cowling & Lamont 1987; Enright & Lamont 1989). Its linear ericoid leaves may confer a drought tolerance not shared with the other species. Lamont *et al.* (1989) have explained the distribution of *B. hookeriana* along a soil moisture gradient in terms of interspecific competition between seedlings. *B. hookeriana* is restricted to the midrange of a gradient from a sand dune swale to a sand-over-limestone ridge. As a monoculture, *B. hookeriana* seedlings established at three sites along the gradient, but when mixed with seedlings of four other banksias, co-dominant along the gradient, *B. hookeriana* was eliminated from all sites except where it occurs naturally.

Ramet recruitment

Forty-nine percent of south-western Australian banksias resprout after fire, either from lignotubers or epicormic buds (George 1981). *B. elegans* is unique in the genus in that it recruits post-fire ramets by both sprouting as above and extensive suckering from surface roots (Lamont 1988). Generally, less attention has been given to ramet than seedling recruitment, possibly because sprouting banksias are considered less prone to local extinction.

Cowling and Lamont (1987) observed that almost all pre-fire individuals of *B. attenuata* resprouted from lignotubers after both autumn and spring burns. Corresponding values for recovery of *B. menziesii* individuals via epicormic buds were 93% and 70%. The minor mortality of *B. attenuata* adults was more than compensated for by seedling recruitment. However, recruitment of *B. menziesii* was below replacement levels, which suggests a possible long-term population decline in this species.

There is a general impression that sprouting species are resilient to short interval (4-6 year) burns (Bell *et al.* 1984). However, this does not take into account the extended primary juvenile period in this group. *B. tricuspis,* a highly localized species, takes over 20 years to become fire tolerant but thereafter survives fires at short intervals (Lamont & van Leeuwen 1988). Post-fire population size of a 15-year-old stand of *B. elegans* increased by more than threefold, entirely due to suckering from parent lateral roots over the next three years, in a way analagous to seedling recruitment in non-sprouters (Lamont 1988).

Co-existence

Any management strategy for banksia-dominated kwongan which has as its goal the maintenance of populations of all or selected species, must consider the mechanisms that mediate co-existence and hence maintain species richness. The species-rich banksia shrublands in particular pose challenging problems for understanding the processes that mediate co-existence (Grubb 1986; Cowling 1987).

Much attention has been given to the development of theoretical models to explain and predict the maintenance of community (alpha) richness. Ecologists usually distinguish between equilibrium and non-equilibrium models (Chesson & Case 1986). The former attribute co-existence to niche differentiation; species co-exist if differences in resource utilization are large enough to prevent competitive exclusion (e.g., Cody 1986). Non-equilibrium models invoke a wide array of plant processes or disturbances which prevent competitive exclusion (e.g., Huston 1979). Elements of both models usually apply to real situations (Silvertown & Law 1987), as we show below for banksia co-existence.

In terms of seasonal patterns of water utilization and growth, there was no evidence for niche differentiation of three co-occurring banksias in a 14-year-old shrubland, probably because of broadly similar above- and below-ground architecture (Bergl & Lamont 1988; B. Lamont & S. Bergl, unpubl. data). Cody (1986) has argued that the non-overlapping leaf areas of co-occurring fynbos *Protea* spp. reflects structural niche differentiation which could explain the co-existence of these species. However, two of the species (a resprouter and non-sprouter) have almost identical leaf morphology so that this hypothesis can be dismissed.

The key to understanding the co-existence of trophically equivalent banksias rests with post-fire regeneration processes (Cowling 1987). Although many components of the regeneration niche (Grubb 1977) of banksias are similar (e.g., germination requirements), post-fire population sizes may vary for each species, as a result of variation in other regeneration niche components, such as size of the pre-fire viable seed bank, moisture conditions, and seed and seedling predation. Many of these factors vary stochastically so that each fire is unique in its effect on the recruitment potential of any given co-occurring species. This results in stochastic fluctuations in population sizes, and hence shifting competitive hierarchies, after each fire: under this scenario no single species necessarily dominates for any length of time and co-existence is maintained (Cowling 1987). This model incorporates elements of Chesson & Warner's (1981) lottery model with a storage effect (Warner & Chesson 1985) associated with conditions (e.g., seed bank size) favouring occasional strong recruitment of all co-existing species.

There are some empirical data to evaluate this model. The low winter rainfall and exceptionally dry summer after an autumn burn resulted in the local disappearance of *B. cuneata* but some *B. prionotes* seedlings survived (B. Lamont, unpubl. data). Unusually dry post-establishment conditions would tend to favour the seedlings of more drought resistant species (e.g., *B. leptophylla*, Enright & Lamont 1989). Fires at intervals of greater than 20 years might result in very poor recruitment of *B. coccinea* at some sites, due to senescence and death, but good recruitment of the other co-occurring non-sprouting banksias (*B. speciosa*, *B. baxteri*) which are longer lived (H. King, S. Connell & B. Lamont, unpubl. data).

As a result of temporal differences in the sizes of seed banks, short fire intervals could result in differential recruitment of species with different juvenile periods. Density-dependent interspecific interactions among post-fire seedlings also have a profound effect on recruitment levels (Lamont *et al.* 1989) and could result in population shifts along habitat gradients. The net result, at a landscape level, could be stochastically "drifting clouds of species abundance" (Grubb 1986).

Management

What are the implications of our studies for the management of banksia-dominated vegetation? We need to examine all components of the fire regime, namely frequency, season, size and intensity. However, these components also interact in important ways.

Our data indicate that for the maintenance of the populations of most of the banksias we have studied, fires should occur at intervals not less than ten years. Non-sprouting species take three to five years to first set seed and production then increases exponentially. Seedlings of *B. tricuspis*, a resprouter, take >20 years to develop fire resistance (Lamont & van Leeuwen 1988). Since very few mature individuals of this species are killed by fire, shorter intervals may not be so critical in an established stand. *B. attenuata* becomes fire-tolerant after 2-3 years, but takes much longer to begin flowering (Enright, unpubl. data).

Upper limits to fire intervals are more difficult to determine, since we have rarely located stands older than 20 years in the northern kwongan. However, A. Hopkins & T. Griffin (pers. comm.) have recorded senescence and interfire seedling recruitment by *B. prionotes* stands >30 years since the last fire. The fact that marginal populations of *B. coccinea* (a southern kwongan species) begin to senesce at about 15 years suggests that the maintenance of this species would require fire intervals less than this.

As a result of potential wildfire hazards, fire season options for prescribed burning are generally restricted to spring and autumn (Bell *et al.* 1984). Although there is evidence that initial recruitment is higher after autumn than spring burns, seedling to parent ratios converge at the end of the first summer after establishment (Cowling & Lamont 1987; Enright & Lamont 1989). Nonetheless, we would still recommend autumn fires, as this minimizes the vulnerable period between follicle opening and germination (Bond 1984). For example, a hot spring burn could consume all litter and hence reduce the area of safe sites available for seeds. Post-dispersal seed predation is evidently patchy but could be significant after spring burns (Cowling & Lamont 1987). The effect of granivory after autumn burns is insignificant. However, we accept that fire season will have less impact when pre-fire seedbanks are large because of a long interval since the last fire (Enright & Lamont 1989).

Small fires are problematic in that predators of both seeds and seedlings may converge on burnt, resource-rich patches and dramatically reduce recruitment (Cowling & Lamont 1987). In the case of firebreaks, where it may be desirable to eliminate the fast-growing, non-sprouting banksias, spring burns would be more effective than autumn burns. However, these should be intense so that as much litter as possible is consumed to reduce the accumulation of canopy-stored seed (there might be a reverse effect on soil-stored seed). Elsewhere, the use of a particular fire regime over repeated cycles could be hazardous in that it may promote the recruitment of some species at the expense of others. This would be offset by seasonal effects on sizes and post-fire conditions (e.g., severity of summer drought). However, monitoring of actual recruitment patterns should be undertaken and the future fire regime to be imposed modified according to the principles outlined here.

Future research directions

Our research programme on south-western Australian banksias has provided important insights for the management of banksia populations. We have also made some contribution to understanding the fire-adapted traits of banksias as well as the mechanisms that mediate their co-existence. We are now in a position to refine and formalize these models. Below we suggest two lines for future research, both of which have applied and theoretical implications.

The first topic addresses the question of minimum viable population size. This is particularly pertinent for south-western Australia where populations are becoming increasingly fragmented as a consequence of habitat destruction. Using data on seedling to parent ratios (which are equivalent to estimates of R, the per capita population growth) collected for non-sprouting fynbos Proteaceae after a wide range of fires, Bond (1989) developed a model to calculate the probability of local extinction at different values of R and for populations of different sizes. The model predicts that where $R < 4$, the probability of extinction is high, even where patches could theoretically support populations in excess of 100 000 individuals. Immersed in this model are both the effects of demographic and environmental stochasticity. The value of the model is its simplicity and the fact that the data (seedling to parent ratios) are easy to collect.

A second, but equally important, research field is the phylogenetic systematics of *Banksia*. In order to interpret the traits of banksias in an evolutionary sense, the traits must be placed in a phylogenetic perspective (Wanntorp 1983; Midgley 1987). This could enable the prediction of environmental conditions and disturbance regimes under which various traits and lineages diversified, and identify the phylogenetic constraints operating on these traits.

Acknowledgements

Aspects of this research were funded by Associated Minerals Consolidated (N.J.E.), Australian Research Council (B.B.L.), C.R.A. Exploration Pty Ltd (B.B.L.), Curtin University of Technology (N.J.E.), Foundation for Research Development (R.M.C.), University of Auckland (N.J.E.) and the University of Cape Town (R.M.C.). We thank the Western Australian Bushfires Board and Eneabba Volunteer Fire Brigade for essential support.

References

Anon. (1988) Declared rare flora. Western Australian Government Gazette. Pp. 2455.

Beard J. S. (1984) Biogeography of the kwongan. In: *Kwongan: Plant Life of the Sandplain* (eds J. S. Pate & J. S. Beard) pp. 1-26. University of Western Australia Press, Nedlands.

Bell D. T., Hopkins A. J. M. & Pate J. S. (1984) Fire in kwongan. In: *Kwongan: Plant Life of the Sandplain* (eds J. S. Pate & J. S. Beard) pp. 178-204. University of Western Australia Press, Nedlands.

Bergl S. M. & Lamont B. B. (1988) The water relations, rooting patterns and phenologies of two sclerophyllous shrubs. In: *Proceedings of Fifth International Conference on Mediterranean Ecosystems.* pp. 569-73. International Union Bio. Sciences, Paris.

Bond W. J. (1984) Fire survival of Cape Proteaceae: influence of fire season and seed predators. *Vegetatio* **56,** 65-74.

Bond W. J. (1989) Describing and conserving biotic diversity. In : *Biotic Diversity in Southern Africa. Concepts and Conservation* (ed. B. J. Huntley) pp. 1-18. Springer, Berlin.

Bond W. J., Vlok J. & Viviers M. (1984) Variation in seedling recruitment of Cape Proteaceae after fire. *J. Ecol.* **72,** 209-21.

Burgman M. A. & Hopper S. D. (1982) The Western Australian Wildflower Industry. Department of Fisheries and Wildlife of Western Australia Report 53.

Chesson P. L. & Case T. J. (1986) Overview: nonequilibrium community theories: chance, variability, history and coexistence. In: *Community Ecology* (eds J. Diamond & T. J. Case) pp. 229-39. Harper and Row, New York.

Chesson P. L. & Warner R. R. (1981) Environmental variability promotes co-existence in lottery competitive systems. *Am. Nat.* **117,** 923-43.

Cody M. L. (1986) Diversity, rarity and conservation in mediterranean-climate regions. In: *Conservation Biology. The Science of Scarcity and Diversity* (ed. M. D. Soulé) pp. 122-52. Sinauer, Sunderland.

Collins B. G. & Rebelo T. (1987) Pollination biology of the Proteaceae in Australia and southern Africa. *Aust. J. Ecol.* **12,** 387-422.

Cowling R. M. (1987) Fire and its role in co-existence and speciation in Gondwanan shrublands. *S. Afr. J. Sci.* **83,** 106-12.

Cowling R. M. & Lamont B. B. (1985a) Variation in serotiny of three Western Australian *Banksia* species along a climatic gradient. *Aust. J. Ecol.* **10,** 345-50.

Cowling R. M. & Lamont B. B. (1985b) Seed release in *Banksia:* the role of wet-dry cycles. *Aust. J. Ecol.* **10,** 169-71.

Cowling R. M. & Lamont B. B. (1987) Post-fire recruitment of four co-occurring *Banksia* species. *J. Appl. Ecol.* **24,** 645-58.

Cowling R. M., Lamont B. B. & Pierce S. M. (1987) Seed bank dynamics of four co-occurring *Banksia* species. *J. Ecol.* **75,** 289-302.

Enright N. J. & Lamont B. B. (1989) Fire temperatures and follicle opening requirements in ten *Banksia* species. *Aust. J. Ecol.* **14,** 107-13.

Enright N. J. & Lamont B. B. (1989) Seed banks, fire season, safe sites and seedling recruitment in five co-occurring *Banksia* species. *J. Ecol.* **77,** 1111-22.

Gill A. M. (1976) Fire and the opening of *Banksia ornata* F. Muell. follicles. *Aust. J. Bot.* **24,** 329-35.

George A. S. (1981) The genus *Banksia. Nuytsia* **3,** 239-473.

Grubb P. J. (1977) The maintenance of species richness in plant communities: the importance of the regeneration niche. *Biol. Rev.* **52,** 107-45.

Grubb P. J. (1986) Problems posed by sparse and patchily distributed species in species-rich plant communities. In: *Community Ecology* (eds J. Diamond & T. J. Case) pp. 207-26. Harper and Row, New York.

Hnatiuk R. J. & Hopkins A. J. (1980) Western Australian species-rich kwongan (sclerophyllous shrubland) affected by drought. *Aust. J. Bot.* **28,** 573-85.

Huston M. (1979) A general hypothesis of species diversity. *Am. Nat.* **113,** 81-101.

Lamont B. B. (1985) Fire responses of sclerophyll shrublands — a population ecology approach with particular reference to the genus *Banksia*. In: *Ecology and Management of Fire in Natural Ecosystems of Western Australia* (ed. J. Ford) pp. 41-6. Western Australian Institute of Technology. Environmental Studies Group Report 10.

Lamont B. B. (1988) Sexual versus vegetative reproduction in *Banksia elegans*. *Bot. Gaz.* **149,** 370-5.

Lamont B. B. & Barker M. J. (1988) Seed bank dynamics of a serotinous, fire-sensitive *Banksia* species. *Aust. J. Bot.* **36,** 193-203.

Lamont B. B. & Barrett G. J. (1988) Constraints on seed production and storage in a root-suckering *Banksia*. *J. Ecol.* **76,** 1069-82.

Lamont B. B., Collins B. G. & Cowling R. M. (1985) Reproductive biology of the Proteaceae in Australia and South Africa. *Proc. Ecol. Soc. Aust.* **14,** 213-24.

Lamont B. B. & Cowling R. M. (1984) Flammable infructescences in *Banksia:* a fruit-opening mechanism. *Aust. J. Ecol.* **9,** 295-6.

Lamont B. B., Enright N. J. & Bergl S. M. (1989) Co-existence and competitive exclusion of *Banksia hookeriana* in the presence of congeneric seedlings along a topographic gradient. *Oikos* **56,** 39-42.

Lamont B. B. & van Leeuwen S. J. (1988) Seed production and mortality in a rare *Banksia* species. *J. Appl. Ecol.* **25,** 551-9.

Midgley J. J. (1987) Aspects of the evolution and ecology of the Proteaceae with emphasis on the genus *Leucadendron* and its phylogeny. Ph.D. Thesis, University of Cape Town.

Silvertown J. & Law R. (1987) Do plants need niches? Some recent developments in plant community ecology. *Trends Ecol. Evol.* **2,** 24-6.

Taylor A. & Hopper S. (1988) *The Banksia Atlas.* Australian Government Printing Service, Canberra.

Wanntorp H. E. (1983) Historical constraints in adaptation theory: traits and non traits. *Oikos* **41,** 157-60.

Warner R. R. & Chesson P. L. (1985) Co-existence mediated by recuitment fluctuations: a field guide to the storage effect. *Am. Nat.* **125,** 769-87.

Whelan R. J. & Main A. R. (1979) Insect grazing and post-fire plant succession in a south-west Australian woodland. *Aust. J. Ecol.* **4,** 387-98.

Proc. Ecol. Soc. Aust. 1990 — 16:185-194

Where the creeks run dry or ten feet high: pastoral management in arid Australia

M. H. FRIEDEL, B. D. FORAN and D. M. STAFFORD SMITH

CSIRO Division of Wildlife and Ecology, P.O. Box 2111, Alice Springs, N.T. 0871, Australia

Abstract

Sustained production in our arid lands is possible, provided the ecological limits are understood.

Over the last 150 years, introduced animals, changed fire regimes and inappropriate government requirements have led to lowered productivity and soil erosion in the pastoral lands.

Now, a number of pastoral land managers are showing that practical experience, coupled with scientific research, points to management systems which can prevent further degradation and, in some cases, repair past damage without loss of viability of their enterprise.

We outline some key ecological concepts and some bureaucratic constraints, and we present examples of successful arid zone pastoralists who have maintained or repaired their environment.

Introduction

"... what does the opening up and stocking of the immense tract (from the South Australia-Northern Territory border to Barrow Creek) mean to us? At the moderate estimate given for a portion of it, the amount of wool would be 80 000 000 lbs., having a money value of £2 000 000, nearly double the quantity we now produce, to say nothing of the annual surplus from 16 000 000 sheep, which may be roughly reckoned at another £1 000 000." such was the enthusiasm of Mr. Simpson Newland in 1887, quoted by Day (1916), for an area forming only two-thirds of the present Alice Springs District. The entire District now supports about 400 000 beef cattle in a fair season, or about one quarter of the equivalent in the forecast number of sheep.

Mr. Newland's optimistic predictions came only 20 years after the publication of the Report of (the) Commission to Inquire into the State of the Runs Suffering from Drought (Anon. 1867), by the South Australian Government. The Commission heard of runs where, after a few years of occupation, the saltbush was all dead, the herbage gone and 3-6 inches of soil blown away. Massive stock losses had been incurred and the Commissioners concluded that "... the carrying capabilities of the runs, within the country indicated, have been undoubtedly over-estimated, both by the lessees and by Government officers. . .". Overstocking, combined with a severe drought and an inadequate system of tenure, had brought about disastrous losses to lessees and financial depression to the entire Colony.

Little was gained from this lesson as region after region of inland Australia took the same path of boom and bust over the next 60 years (Newman & Condon 1969). Drought inevitably accompanied the collapse. Sometimes an invasion by rabbits hastened the process. The legacy was a damaged resource and a lower carrying capacity when the rains returned (Fig. 1). As well, there was a dramatic decline in the medium-sized native mammals occupying inland Australia (Morton, this volume).

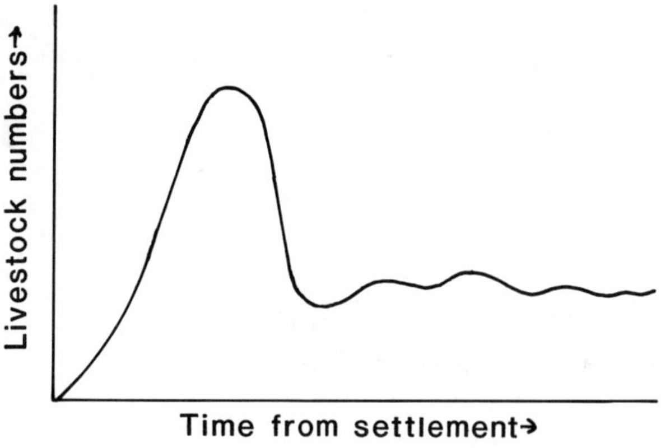

FIG. 1. General trend in livestock numbers on Australian rangelands (from Perry 1977). Time scale on the x-axis is in the order of decades.

What went wrong? Are we now able to get it right? We argue that there are a number of key perceptions that early settlers and governments lacked, which led to misuse of the arid lands. As experience increased, understanding and better management in some cases followed, although we have yet to convert some of our scientific concepts into management action. In the discussion which follows, we focus on pastoralism in Australia's arid lands. The potential for other land uses is an issue beyond the scope of our paper and so we do not consider whether pastoralism is the optimal use for the country. However, we believe that vast areas will remain under pastoral management for the foreseeable future and so it is vital that we get it right.

Episodic events

Rainfall outside the winter-rain regions of southern Australia is very erratic. It becomes increasingly variable inland and northwards and depends on unpredictable storms rather than regular showers (Leeper 1970). The early pastoralists had no experience of similar climates to call on. The rainfall in most of pastoral Australia is 10-20% more variable than the world mean for comparable areas (Leeper 1970).

Even the mean annual rainfall is misleading, because it is strongly influenced by a few extremely high rainfalls. As a consequence, the annual rainfall in an area is below average for considerably more than 50% of years. At Nappa Merrie on the edge of Sturt's Stony Desert, the mean and median rainfall are 160 and 117 mm respectively (Williams & Calaby 1985). In south-west Queensland, only 35 of the 100 "growing seasons" (October-March) to 1981 were above average in rainfall, and drought was officially declared in a local shire for almost 50% of the preceding 17 years (Silcock 1981). There was an assumption in that shire that dry seasons were not normal, although the rainfall data clearly show that they were.

The early pastoralists did not recognize that the good seasons were relatively rare events which were crucial to the regeneration of both pastures and woody species. The inland characteristically experiences occasional sequences of above average rainfall (Fig. 2) which generate in turn high plant production, and fuel for fires. With extended wet periods, soil moisture is sufficient for slower-growing perennials as well as annuals to establish and grow.

Plant species have life histories which are favoured by different combinations of soil moisture and type of substrate, as well as fire, grazing and plant competition (Noy-Meir 1973, 1985; Westoby 1980). In general, plants that allocate more effort to large root structures are more tolerant of drought and of low levels of nutrients. Since they remain active when soil moisture is depleted, they can make use of rainfalls that are unavailable to plants that must first grow from seed. These advantages are offset by slower growth rates and lower competitiveness as a seedling, due to the resources which are invested in non-photosynthetic structures. By comparison, short-lived plants tend to invest less in roots, grow faster, use nutrients more quickly, but be less drought-tolerant. Plants which are drought-tolerant may thus need unusual conditions of soil moisture for establishment.

So it is that perennial grasses and woody species require extensive periods of rainfall to regenerate on a large scale. The first recorded episode of major regeneration of trees and shrubs in western New South Wales ocurred in the 1890s, 20 years after settlement, and followed prolonged high rainfall. The regrowth caused a marked reduction in carrying capacity because it suppressed the pasture layer. Within a few years came drought, rabbits and depression and, in 1921, wildfire (Anon. 1969). Further massive regrowth in the 1950s, after an extended period of above-average rain, was again followed by drought and lower carrying capacity, but this time rabbits had been drastically reduced by myxomatosis (Anon. 1969; Harrington *et al.* 1979) and nothing held back the development of excessively dense trees and shrubs or "woody weeds". There were two more such periods in the 1960s and 70s and major wildfires in 1974/5 (Condon 1979; Wilson & Mulham 1979). Climatic peaks followed by increases in livestock, rabbits and fire can also be identified for central Australia (Fig. 2; Griffin & Friedel 1985). Anecdotal evidence links major shrub increases to the high rainfalls of the 1970s and very likely with earlier wet periods in central Australia as well (Friedel 1981) particularly in woodlands and mulga shrublands.

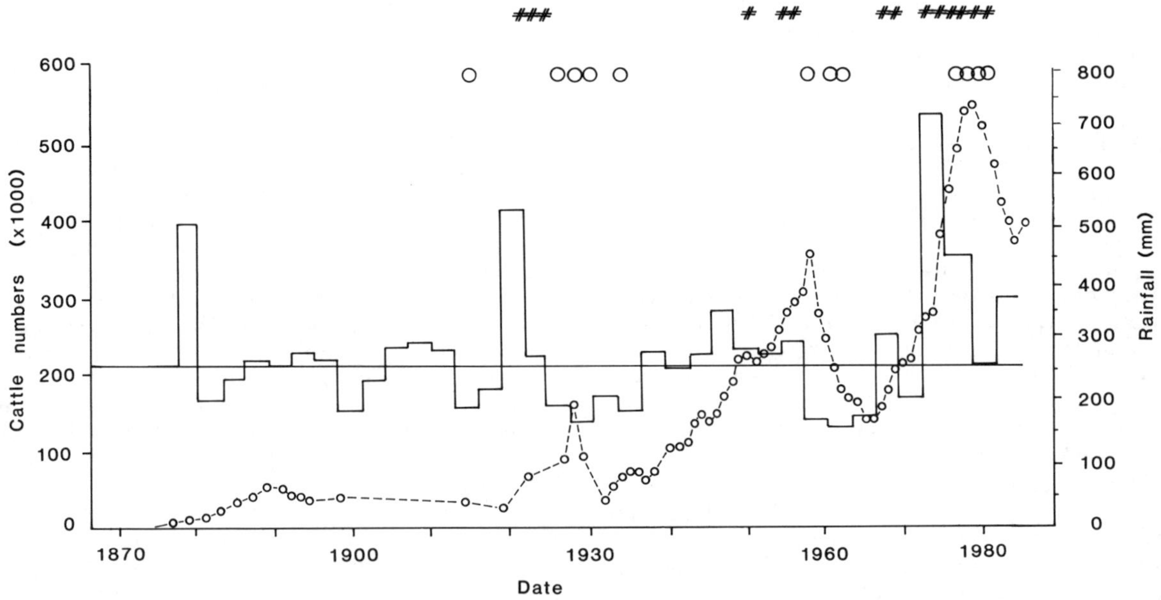

FIG. 2. Episodic nature of rainfall and subsequent events. Mean 3 year rainfalls for Alice Springs; cattle numbers (=-o-) and occurence of extensive wildfires (=#) and rabbit plagues (=○) in the Alice Springs pastoral district.

Although it is normal that regeneration should occur episodically, the checks and balances that were in place before Europeans arrived with their grazing animals were no longer effective. Excessive numbers of woody plants established because the grass layer was depleted by grazing and offered less competition to shrub seedlings. Thereafter the shrub layer with its perennial root systems out-competed the grass layer for moisture and became increasingly dominant.

Historical and experimental evidence also links shrub increases to the failure of pastoralists and their advisers to recognize the importance of fire, another episodic event. They did not burn their land as the Aborigines had done, when fuel was abundant and the seedling shrubs were susceptible to fire (Griffin & Friedel 1984, 1985; Hodgkinson & Harrington 1985). When wildfires were later ignited by lightning, huge areas were burnt but, by then, the young shrubs were too well established and many survived. Moreover, opportunities to burn were reduced because potential fuel was eaten by stock.

The chenopod shrublands fared differently from the woodlands described above. Overstocking with sheep and ensuing drought very quickly led to widespread death of saltbush and soil erosion in South Australia (Anon. 1867). In New South Wales, the peak of sheep numbers was reached between 1892 and 1894. The inevitable decline was accompanied by lower rainfall, plagues of rabbits, and loss of the cottonbush, saltbush and grass (Williams & Oxley 1979). A similar collapse in the arid north of South Australia came after a run of dry years which began in 1921. Ratcliffe (1936) put the blame squarely on the stock, although rabbits were present, and noted the destruction of seedlings by soil drift. Despite germination events, the chenopod shrubs did not recover and there has been a progressive decline in shrub density in South Australia since the inception of pastoralism (Lange *et al.* 1984).

Landscape degradation is often equated with soil erosion, and there is clearly a relationship between the advent of European settlement and increased erosion (Wasson & Galloway 1986). The fore-going historical summary shows that loss of pastoral productivity is also caused by excessive regeneration of trees and shrubs in some landscapes and lack of regeneration in others. In addition, selective grazing can reduce productivity by increasing the proportion of unpalatable species in the pasture layer.

The important conclusion is that change is episodic. A pulse of high rainfall initiates a sequence of events somewhat out of phase with the rainfall, with a consequence that peak animal numbers (domestic and feral) usually occur when the pasture reponse is fading. Excessive numbers of woody plants will establish in some landscapes unless controlled by strategic fires lit soon after major rainfall periods. Wildfire will inevitably follow, unless management burns are initiated within a critical timespan. Important perennial plants may have only infrequent opportunities, sometimes decades apart, to regenerate. Large rainfall events can also initiate erosion of susceptible soils and alter the productivity of entire landscapes by stripping and depositing thin veneers on the surface (Pickup 1985).

Spatial heterogeneity

Plant growth in arid zones of the world is closely correlated with rainfall, in space as well as time. At very low rainfalls, soil moisture is lost by evaporation before much growth can occur but its effectiveness can be increased if water is redistributed to run-on areas by uneven topography (Noy-Meir 1985). In addition, water which moves across the landscape can transport particles of soil and organic matter and dissolved nutrients, causing run-on areas to accumulate nutrients and seeds (Pickup 1985).

The process operates at a variety of scales, including gilgais where the soil surface has developed into mounds and depressions of a few metres across, mulga grove/intergrove bands in which the mulga groves can be in the order of 50 m wide and 500 m long, and floodplains of several kilometres length. Run-on areas are relatively stable and soil moisture conditions favourable, so that the establishment of perennial plants tend to follow. In central Australia, a veneer of deposited material only 20 cm deep was sufficient for the development of a woodland on what had been eroded and probably bare ground (Pickup 1985). This change in spatial distribution of resources is not desirable for pastoral management if the thinly-spread productive soil from the source area is concentrated in deep deposits in the sink area. Much of the productive soil may be effectively buried, and the establishment of woody plants can become excessive, suppressing pasture growth.

Australia's arid zone soils are generally poor in nitrogen and phosphorus (Charley & Cowling 1968) but nutrients are not evenly spread across the soil surface. In addition to the redistribution of nutrients by water, shrubs in arid shrublands have an important role in creating micro-environments which enhance soil fertility. These "fertile islands", first named by Garcia-Moya and McKell in 1970, are a widely recognized phenomenon of arid lands (West & Skujins 1978; Noy-Meir 1985) and can be as fertile as the entire area of more mesic ecosystems. They persist for decades and, in some cases, centuries. The micro-climate and soil surface conditions created by shrubs encourage biological activity. Litter fall and trapped wind-blown particles improve the nutrient and organic matter content of the soil, and seedling plants find a mild environment for establishment. Animals are attracted to the fertile patches and in turn enrich them further. Once the shrub vegetation is destroyed and the soil dispersed over the entire area, as it was by early pastoralism, nutrients and organic matter are often inadequate to sustain seedling establishment and plant production (Romney *et al.* 1978). Restoration of damaged shrublands will not succeed unless fertile patches are re-created.

Environments other than shrublands have their own fertile islands which persist for decades. In mulga and other woodlands, fallen timber creates log mounds by trapping wind blown and water washed material (D. J. Tongway, J. A. Ludwig & W. G. Whitford, pers. comm.). Infiltration rates, soil nutrients and plant growth are all greater on the mounds than adjacent to them, and wood-eating termites are active. In hummock grasslands and elsewhere, termite mounds themselves are probably fertile islands. Work in the arid tropics (Beckman 1988) suggests that microbial activity, nutrients and organic matter are all greater within mounds than away from them. If the termite colony survives 20-40 years and then weathers away over a similar time, as educated guesswork indicates (Holt, in Beckman 1988), an island of fast-growing, nutrient-demanding species is created (Spain and McIvor 1988) on a nutrient-enriched patch. Indeed, perennial grasses themselves may create fertile patches on a smaller spatial- and time-scale.

Spatial heterogeneity is not restricted to the horizontal plane. Strongly truncated distributions of nutrients and organic matter are common in arid zone soil profiles (Charley & Cowling 1968; West & Klemmedson 1978; Noy-Meir 1985), so that loss of just a few centimetres of topsoil can be critical.

Spatial heterogeneity was, and is, important to the pastoralists, although it may not always have been recognized (e.g., Fitzgerald & Burnside 1988). The potential productivity of the arid zone was rapidly reduced by the loss of fertile patches, topsoil and perennial plants. At a different scale, the landscape level, the differential productivity of landscape units had implications for management too. Initial stocking of new country focussed on the reliable waters (e.g., Wilcox 1963), and the fertile floodplains of the rivers were quickly degraded. Similarly, runs were stocked as if the whole area had the potential of the best country (e.g., Purvis 1986) and, with minimal fencing to control access, the livestock grazed the best pasture until it was almost worthless, and left adjacent poor country alone.

Differences in the productivity of landscape elements have fundamental implications for the distribution and density of many organisms, including people. The riverine plains, the hills and footslopes where water is more reliable (and the soils presumably richer nutritionally) supported many more Aboriginal people than did the sandplains (Meggitt 1965; Tindale 1972). Native grazing mammals also preferred the better watered drainage lines and flanking plains of central Australian ranges (Newsome 1980). Similarly, when livestock are allowed to graze freely in a mix of pasture types, they concentrate on the productive foothills and associated woodlands and floodplains (Low *et al.* 1981).

Recognizing the significance of spatial pattern enables us to identify priorities for management. For instance, one quarter of a central Australian paddock supported half the grazing (Low *et al.* 1981). The favoured areas were "indicator landscapes", requiring particular attention to maintain their productivity. If they were in good order, less preferred areas would be too. Moreover careful pastoral management of favoured areas should serve broader conservation objectives for native flora and fauna, by minimizing damage to resource-rich environments. Stafford Smith and Pickup (this volume) have developed techniques, based on spatial processes, which should provide better indices of landscape response to management than point-based methods which are confounded by spatial pattern.

Impact of non-biological influences

Unfamiliarity with the environment was not the only problem for the pastoralists. Government policy and national or international economic forces put pressure on management that was not in keeping with the variable income which the land generated. Today, market requirements and sales returns are still the major forces shaping land management decisions in pastoral Australia. It may be too obvious to state that, in years of high income, good land management policies are easier to implement, while the reverse is true in poor income years. Cash flow is driven directly by rainfall and is, therefore, extremely variable.

Of the many options available to a manager, there are two approaches which illustrate the extremes of the management continuum. The first, *drought resistance,* relies on lower animal numbers and high *per animal* production, often for speciality markets, to give a relatively constant income whether the seasons are good or bad (Heussler 1986; Purvis 1986; J. G. Morrissey & R. E. Y. O'Connor, pers. comm.). The opposite of this is a *high utilization/quick turnoff* approach, which aims to reap the windfall production from good years but destock quickly at the start of drier seasons. Both require high management skills, but the former is characterized by less risk and perhaps lower returns in good years. High utilization forms of management can be extremely successful economically, when the manager is "risk aware" and when pastoral landscapes are resilient. However, if the manager retains stock in a poor season because prices are depressed, and the enterprise has no pastures in reserve, susceptible landscapes are likely to degrade.

The need to generate a steady cash income for basic living from a variable resource is not aided by the requirements of finance managers from either the public or private sectors. Taxation policies now include the concept of averaging farm income (and resulting taxation payments) over a five year period and, from 1989, the reintroduction of an Income Equalization Deposit scheme which will allow occasional high income to be used in years of poor income. These provisions should help pastoral managers ride out the income fluctuations caused primarily by season, in a manner of their own choosing. However, the introduction of these income buffering mechanisms must be balanced by the phasing out of Drought Relief schemes which, on many occasions, have seen most of the financial assistance go to propping up high utilization managers caught short by a dry time, so that intemperate management is rewarded by government subsidy.

Banks, finance houses and agri-business firms also put pressure on management for yearly or even monthly cash flows, e.g., interest payments, that are not in keeping with either the seasonal forage situation or the biological requirements of the herds/flocks and the pastoral landscapes on which they graze.

Dominating all these considerations is the failure to include landscape condition in the valuation of pastoral properties for sale, or to include change in condition in subsequent re-evaluations. Too often, sales are transacted on the basis of herd numbers in a good season, leaving a new owner to pursue management goals based on numbers of stock, rather than the quality of the product and the efficiency of production.

The failure of bureaucrats, researchers and extension officers to combine the biological and financial streams of knowledge into meaningful management advice has contributed to the present state of our rangeland ecosystems. In the past and right now, unrealistically high stocking rates have been required or condoned by government (Fig. 3; Purvis 1986). Restrictions on property size and adjustment, particularly in New South Wales and Queensland, have produced sub-economic living areas which tend to be in poorer condition than larger properties (Young *et al.* 1984). Pastoral managers seeking advice are obliged to search out "experts" in a number of different government departments, who may give conflicting advice or be unaware of research done long ago.

Researchers have not always seen communication of their results outside scientific journals as a priority. For research to reach its target and be implemented, information has to be made readily available and easily understood. Computer-based decision support systems can provide access to

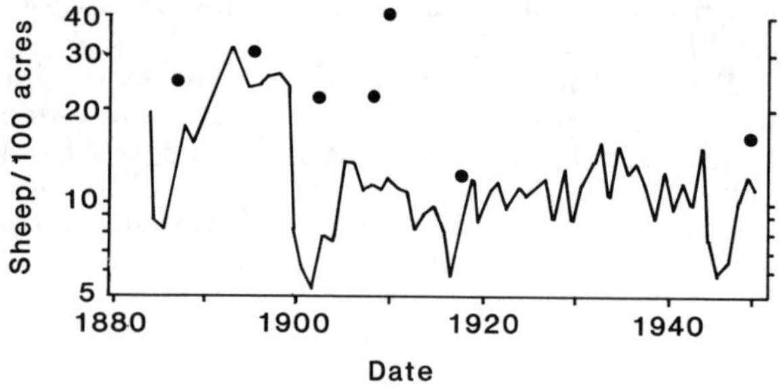

FIG. 3. Sheep densities 1883-1949 for Thurrulgoonia Qld (from Heathcote, in Williams & Oxley 1979). Official estimates of carrying capacity = ●.

increasingly complex information and researchers will have to develop them and promote their use to government and pastoralists alike.

All States with rangeland areas have departments dealing with agriculture, conservation and lands, but there are few effective mechanisms which ensure that they work together to produce mutually compatible policies which are best for the land resource. Experience in the Northern Territory is almost to the contrary. During 1986 and 1987 an attempt was made to discuss land management issues, and to interweave departmental land management policies through workshops, discussions and briefings for key decision makers. This procedure caused each bureaucracy to become even more isolated, and to develop tactics designed primarily to protect their own job definitions. Thus Lands Inspectors deal only with fences and windmills, Soil Conservation Officers with soil, Range Management Officers with pastures, Animal Production Officers with livestock, and Bushfires Control Officers with bushfires. Yet it is only by the combined efforts of these different people to give integrated advice at the whole-property level that buoyant economics, productive herds and flocks, and healthy ecosystems will result.

Managing the land

In an industry where costs are increasing without a corresponding increase in returns, survival depends on improved efficiency. Fewer top quality stock will bring good prices at lower cost, and put less pressure on the land. Pastoralists who manage for the good health of their land as well as their stock are in no way financially disadvantaged in the long term, although short-term profits may be lost when income is put back into the property for development purposes (Lange *et al.* 1984; Fitzgerald & Burnside 1988). Of 12 properties surveyed in the Meekatharra area of Western Australia, the enterprise with the lowest stocking rate had the highest net income per hectare (Fig. 4). As well as economic advantage, there is a long term benefit to forage production of investing money in development or reclamation on-property (Lange *et al.* 1984; Purvis 1986; Fitzgerald & Burnside 1988; J. G. Morrissey & R. E. Y. O'Connor, pers. comm.).

Good management systems integrate land care with efficient animal husbandry, carefully planned infrastructure and sound economic management. Conservative stocking is a common feature, combined with subdivision of paddocks and sometimes provision of additional watering points to spread grazing pressure more evenly. Small flocks or herds per watering point are the norm and some form of pasture resting is practised. In central Australia, paddocks on "Atartinga" (Purvis 1986) are destocked for six months per year and sometimes longer, ensuring rest in both summer and winter seasons. The chenopod shrub pastures on "Annean", near Meekatharra, are grazed in winter while the wanderrie grass and short grass chenopod pastures are grazed in summer (J. G. Morrissey & R. E. Y. O'Connor, pers. comm.). On the Nicolson properties near Whyalla (Lange *et al.* 1984), no deliberate destocking of paddocks is practised; instead, portions of paddocks are rested for several months by shutting off watering points. The consequence is recovery or maintenance of the perennial component of the pasture.

These pastoralists know their country well. Their management systems allow for episodic events. Despite more reliable pasture growth in winter at Meekatharra " . . . the years when good summer rain falls are considered to provide the best opportunities for the recruitment of perennial plant species and for range rehabilitation . . . Summer spelling is designed to facilitate recruitment of shrubs in the chenopod pastures" (J. G. Morrissey & R. E. Y. O'Connor, pers. comm.). At the same time recruitment of undesirable woody species has been general throughout the wanderrie pastures on "Annean". Prescribed burning is not common practice in Western Australia's arid zone at the moment, but its potential value is recognized and will be tested at the earliest opportunity (J. G. Morrissey, pers. comm.).

Fire is a familiar management tool to Purvis (1986). "The first major bushfire I fought was also my last! . . . The money wasted should have been spent in the previous year using fire for woody weed control and fuel reduction." The watercourses and some of the best woodland country on the property were thick with woody shrubs. "Fire was the only tool which could cope with this problem." From careful beginnnings with cool season fires, he will now burn in summer wildfire conditions to win back some of the most densely shrub-infested areas.

Recognition of spatial pattern at the scale of pasture types was a critical factor in improving pasture management. "My father saw only the

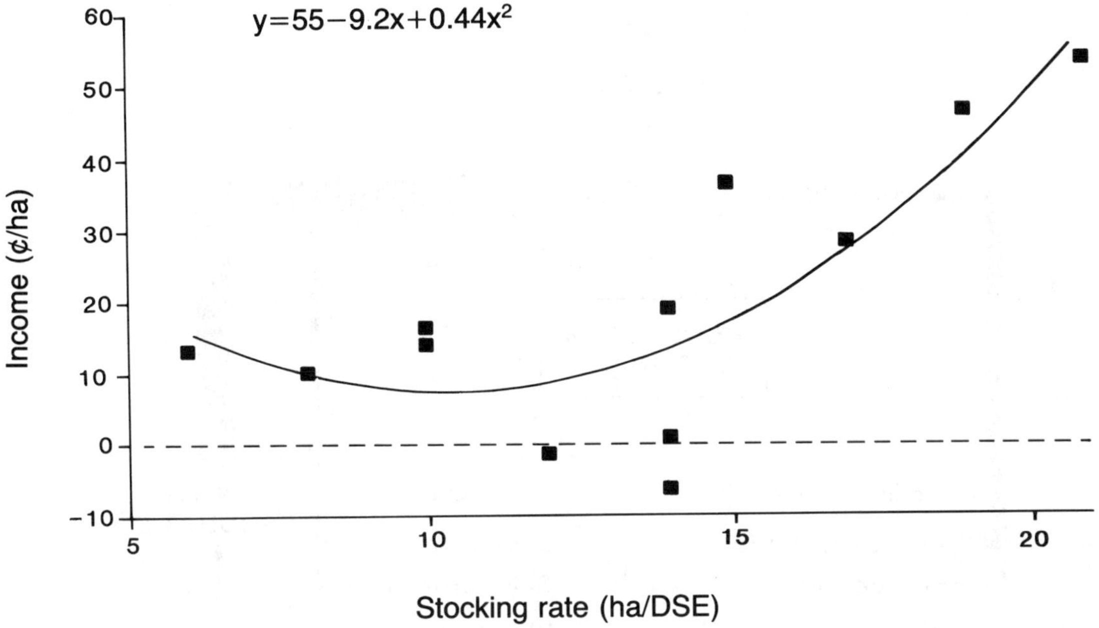

FIG. 4. Results of an economic survey of 12 station properties in the winter rainfall zone of Western Australia (from J. G. Morrissey & R. E. Y. O'Connor, pers. comm.). Stocking rate, in ha/DSE (Dry Sheep Equivalents), *decreases* along the x-axis; the lowest rate (21 ha/DSE) occurs on "Annean". For the regression $y = 55\text{-}9.2x + 0.44x^2$, $r^2 = 0.68$; $p<0.002$; df = 11.

pastoral productivity of the small areas of good country and stocked the total area around his permanent waters up to that potential. Wholesale pasture and land degradation resulted." (Purvis 1986). The son fenced the property into a number of smaller paddocks and learned by careful observation how many stock each type of country could carry. Similarly, on "Nambi" near Leonora, "Early development for sheep grazing did not consider the diverse nature of the pastures or the complex mosaic of the land systems" (Fitzgerald & Burnside 1988). Highly preferred lakeside country has now been fenced off from wanderrie/watercourse areas, and overall carrying capacity is predicted to increase once the lakeside chenopods have recovered. Fencing at "Annean" in on land system boundaries (J. G. Morrissey & R. E. Y. O'Connor, pers. comm.).

When damage to country has been great, resting it from grazing will not be a realistic means of returning it to production; some form of reclamation will be necessary if the soil is significantly damaged or the seed pool is deficient. Seeds of exotic or native perennials may need to be introduced (Purvis 1986; Fitzgerald & Burnside 1988), but success is unlikely unless resources are optimized by re-creating fertile patches. Ponding banks, pitting and tyning fulfil this function by concentrating water and nutrients that would otherwise run off scalded soils. Another option currently under investigation is the placement of felled mulga trees along the contour of degraded mulga lands to re-create banding patterns (Tongway & Ludwig 1988).

Pastoralists with a comprehensive understanding of their country, as in our examples here, probably have little need of professional ecologists' expertise, as they are already ecologists in their own right. There are many others who make management decisions largely on the basis of the condition of their stock (Fig. 5), not recognizing that damage to pasture can occur well before stock condition declines. They take the view that "the country always comes back when it rains" simply because rain initiates a green flush of growth. Some may be influenced by the state of the vegetation to the extent that they move stock to new pastures when the amount of forage in the current paddock gets low, but they do not detect long-term indicators of degradation such as shifts in pasture composition or mass germination of unpalatable shrubs (Fig. 5). The delayed feedbacks from damaged soil and declining pasture production to animal production are rarely recognized.

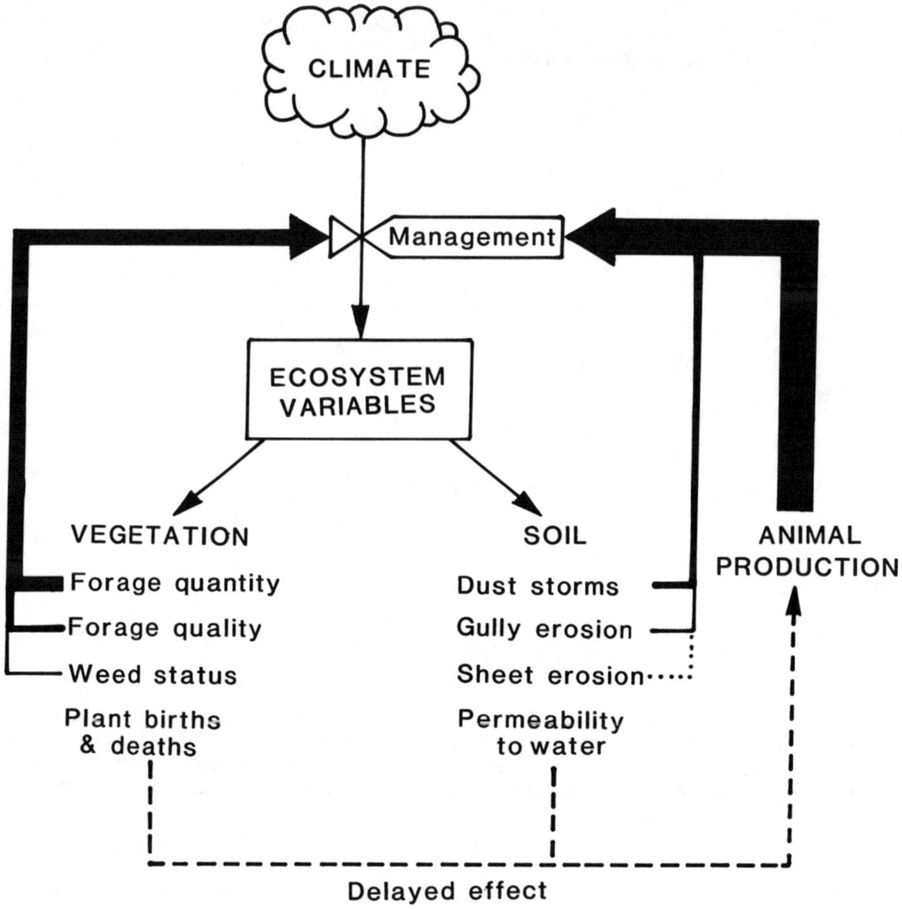

FIG. 5. The relative importance of factors which influence decision-making by pastoral managers (from Harrington *et al.* 1984). Thickness of lines indicates degree of importance e.g., managers place most emphasis on immediate animal production.

Most managers are not responsive to ecologically-based advice from extension services, and they are unlikely to change unless the advice is couched in economic terms. We have shown that ecologically sound management can be economically rewarding, when it has the following features and is part of an integrated management system.

1. Conservative upper limit for stocking rate.
2. Small flocks or herds per watering point.
3. Small paddocks.
4. Pasture resting.
5. Recognition of importance of key seasonal events for vegetation recovery.
6. Placement of fences and watering points according to pasture type and stock distribution.
7. Burning for shrub control.
8. Re-establishment of "fertile islands".

Research and extension advice clearly must be provided within the same integrated framework, which links good land care with comprehensive infrastructure planning and sound financial and livestock management.

Conclusions

As Australia's pastoral industries move towards the year 2000, rapid change is being forced upon them. Unless this change is facilitated, pastoral landscapes, as ever, will bear the brunt. If, as ecologists, we are able to help pastoral mangers, we must bring together streams of knowledge at an ecosystem level which will help a manager make the best decision for the enterprise, and for the land resource. The most important elements are listed below.

1. Financial management policies at government, corporate and property levels must not constrain the flexibility required by the inherent biological variability of arid ecosystems.

2. Ecologists must synthesize information so that it matches pastoralists' needs, and also present it in an easily available manner using new information technologies.
3. Pastoral managers must target quality and efficiency of livestock production rather than high stock numbers.
4. Co-operation between government land management agencies should be promoted so that land care and sustained production form the basis of a suite of compatible land management policies at all levels.

References

Anon. (1867) *Report of Commission Appointed by the Governor-in-Chief to Inquire into the State of the Runs Suffering from Drought; Together with Minutes of Evidence and Appendix.* Government Printer, Adelaide.

Anon. (1969) *Report of the Inter-Departmental Committee on Scrub and Timber Regrowth in the Cobar-Byrock District and Other Areas of the Western Division of N.S.W.* Government Printer, Sydney.

Beckman R. (1988) Termites and nutrient cycles in the north. *Ecos* **54,** 15-8.

Charley J. L. & Cowling S. W. (1968) Changes in soil nutrient status resulting from overgrazing and their consequences in plant communities of semi-arid areas. *Proc. Ecol. Soc. Aust.* **3,** 28-38.

Condon R. W. (1979) Pastoral management, economics and land use options in the poplar box *(Eucalyptus populnea)* lands of New South Wales. *Aust. Rangeld J.* **1,** 260-6.

Day T. E. (1916) Report and plans of explorations in central Australia. *Bull. N.T.,* No. 20. Home and Territories Department (Australia), Melbourne.

Fitzgerald D. R. & Burnside D. G. (1988) Good land management is profitable. *Working Papers 5th Biennial Conf. Australian Rangeld Soc., Longreach Qld.* pp. 78-81. (Mimeo).

Friedel M. H. (1981) Potential problems with tree and shrub regeneration in central Australia. *Working Papers 3rd Biennial Conf. Australian Rangeld Soc., Alice Springs N.T.* pp. 75-85. (Mimeo).

Garcia-Moya E. & McKell C. M. (1970) Contribution of shrubs to the nitrogen economy of a desert-wash plant community. *Ecology* **51,** 81-8.

Griffin G. F. & Friedel M. H. (1984) Effects of fire on central Australian rangelands. II. Changes in tree and shrub populations. *Aust. J. Ecol.* **9,** 395-403.

Griffin G. F. & Friedel M. H. (1985) Discontinuous change in central Australia: some implications of major ecological events for land management. *J. Arid Environments* **9,** 63-80.

Harrington G. N., Oxley R. E. and Tongway D. J. (1979) The effects of European settlement and domestic livestock on the biological system in poplar box *(Eucalyptus populnea)* lands. *Aust. Rangeld J.* **1,** 271-9.

Harrington G. N., Wilson A. D. & Young M. D. (1984) Management of rangeland ecosystems. In: *Management of Australia's Rangelands* (eds G. N. Harrington, A. D. Wilson & M. D. Young) pp. 3-13. CSIRO (Australia), Melbourne.

Heussler J. H. S. (1986) Australian rangelands — keys to development. In: *Rangelands: A Resource Under Siege* (eds P. J. Joss, P. W. Lynch & O. W. Williams) pp. 183-4. Australian Academy of Science, Canberra.

Hodgkinson K. C. & Harrington G. N. (1985) The case for prescribed burning to control shrubs in eastern semi-arid woodlands. *Aust. Rangeld J.* **7,** 64-74.

Lange R. T., Nicolson A. D. & Nicolson D. A. (1984) Vegetation management of chenopod rangelands in South Australia. *Aust. Rangeld J.***6,** 46-54.

Leeper G. W. (1970) Climates. In: *The Australian Enviroment* (ed G. W. Leeper) pp. 12-20. CSIRO (Australia) and Melbourne University Press, Melbourne. 4th edn.

Low W. A., Dudzinski M. L. & Mueller W. J. (1981) The influence of forage and climatic conditions on range community preference of Shorthorn cattle in central Australia. *J. Appl. Ecol.* **18,** 11-26.

Meggitt M. J. (1965) *Desert People: A Study of the Walbiri Aborigines of Central Australia.* University Chicago Press, Chicago.

Newman J. C. & Condon R. W. (1969) Land use and present condition. In: *Arid Lands of Australia* (eds R. O. Slatyer & R. A. Perry) pp. 105-32. Australian National University Press, Canberra.

Newsome A. E. (1980) The eco-mythology of the red kangaroo in central Australia. *Mankind* **12,** 327-33.

Noy-Meir I. (1973) Desert ecosystems: environment and producers. *Ann. Rev. Ecol. Systematics* **4,** 25-51.

Noy-Meir I. (1985) Desert ecosystem structure and function. In: *Ecosystems of the World 12A: Hot Deserts and Arid Shrublands* (eds M. Evenari, I. Noy-Meir & D. W. Goodall) pp. 93-103. Elsevier, Amsterdam.

Perry R. A. (1977) Rangeland management for livestock production in semi-arid and arid Australia. In: *The Impact of Herbivores on Arid and Semi-arid Rangelands. Proc. 2nd US/Aust. Rangeld Panel, Adelaide 1972* pp. 311-6. Aust. Rangeld Soc., Perth.

Pickup G. (1985) The erosion cell — a geomorphic approach to landscape classification in range assessment. *Aust. Rangeld J.* **7,** 114-21.

Purvis J. R. (1986) Nurture the land: my philosophies of pastoral management in central Australia. *Aust. Rangeld J.* **8,** 110-7.

Ratcliffe F. N. (1936) *Soil Drift in the Arid Pastoral Areas of South Australia.* CSIR (Australia) Pamphlet No. 64, Melbourne.

Romney E. M., Wallace A. & Hunter R. B. (1978) Plant response to nitrogen fertilization in the northern Mojave Desert and its relationship to water manipulation. In: *Nitrogen in Desert Ecosystems* (eds N. E. West & J. J. Skujins) pp. 232-43. (US/IBP Synthesis Series 9) Dowden, Hutchinson & Ross, Stroudsberg.

Silcock, R. G. (1981) Are drought relief schemes desirable? *Working Papers 3rd Biennial Conf. Australian Rangeld Soc., Alice Springs N.T.* pp. 163-7. (Mimeo).

Spain A. V. & McIvor J. G. (1988) The nature of herbaceous vegetation associated with termitaria in north-eastern Australia. *J. Ecol.* **76,** 181-91.

Tindale N. B. (1972) The Pitjandjara. In: *Hunters and Gatherers Today* (ed M. G. Bicchieri) pp. 217-68. Holt, Rinehart & Winston, New York.

Tongway D. J. & Ludwig J. A. (1988) Mulga woodlands: restoration of degraded landscapes. *Working Papers 5th Biennial Conf. Australian Rangeld Soc., Longreach Qld.* pp. 179-80. (Mimeo).

Wasson R. J. & Galloway R. W. (1986) Sediment yield in the Barrier Range before and after European settlement. *Aust. Rangeld J.* **8,** 79-90.

West N. E. & Klemmedson J. O. (1978) Structural distribution of nitrogen in desert ecosystems. In: *Nitrogen in Desert Ecosystems* (eds N. E. West & J. J. Skujins) pp. 1-16. (US/IBP Synthesis Series 9) Dowden, Hutchinson & Ross, Stroudsberg.

West N. E. & Skujins J. J. (1978) *Nitrogen in Desert Ecosystems* (US/IBP Synthesis Series 9) Dowden, Hutchinson & Ross, Stroudsberg.

Westoby, M. (1980) Elements of a theory of vegetation dynamics in arid rangelands. *Israel J. Bot.* **28,** 169-94.

Wilcox D. G. (1963) The pastoral industry of the Wiluna-Meekatharra area. In: *Lands of the Wiluna-Meekatharra Area, Western Australia, 1958* pp. 194-212. CSIRO (Australia) Land Res. Series No. 7, Melbourne.

Williams O. B. & Calaby J. H. (1985) The hot deserts of Australia. In: *Ecosystems of the World 12A: Hot Deserts and Arid Shrublands* (eds M. Evenari, I. Noy-Meir & D. W. Goodall) pp. 269-312. Elsevier, Amsterdam.

Williams O. B. & Oxley R. E. (1979) Historical aspects of the use of chenopod shrublands. In: *Studies of the Australian Arid Zone IV. Chenopod Shrublands* (eds R. D. Graetz & K. M. W. Howes) pp. 5-16. CSIRO (Australia), Melbourne.

Wilson A. D. & Mulham W. E. (1979) A survey of the regeneration of some problem shrubs and trees after wildfire in western New South Wales. *Aust. Rangeld J.* **1,** 363-8.

Young M. D., Gibbs M., Holmes W. E. & Mills D. M. D. (1984) Socio-economic influences on pastoral management. In: *Management of Australia's Rangelands* (eds G. N. Harrington, A. D. Wilson & M. D. Young) pp. 79-93. CSIRO (Australia), Melbourne.

Proc. Ecol. Soc. Aust. 1990 — 16:195-200

Pattern and production in arid lands

D. M. STAFFORD SMITH and G. PICKUP

CSIRO Division of Wildlife and Ecology, P.O. Box 2111, Alice Springs, N.T. 0871

Abstract

Attempts to detect biological change in Australia's ancient arid landscapes are often confounded by complex spatial patterns in soils, and physical processes such as runoff. These patterns are exploited and exaggerated by plants and animals, many of which survive because of local concentrations of resources. They are also crucial to land managers, who must deal with spatial heterogeneity within, rather than between, their management units such as paddocks or parks.

We show how a few simple assumptions about the effects of erosion and grazing in an hypothetical paddock can result in complex ecological patterns involving changes in variances as well as means across the landscape. The apparent noise that would be obtained by sampling these patterns at a few individual points can thus be rendered meaningful in terms of the spatial relationships between points. We discuss the implications of this for data collection and resource management.

Introduction

Australia's arid landscapes are characterized by high degrees of both spatial and temporal variability. The spatial variability originates from diverse geologic source material, but has been enhanced by a long and complex climatic history. This has produced highly sorted soils which, although generally impoverished, are expressed as a mosaic of rich and poor patches. The mosaic structure is enhanced by contemporary processes of runoff and sediment transport, which concentrate both water and eroded soil (Pickup 1985). Temporal variability arises from short-term climatic fluctuations; these add to the effect of spatial patterns on the vegetation because they contribute to the possible diversity of growth strategies and opportunities.

The problems caused by heterogeneity have been widely recognized (Noy-Meir 1981; Ludwig 1987; Pickup 1988), but few attempts have been made to understand landscape behaviour in terms of spatial processes. The result has been that supposedly independent point-based measurements of landscape status, such as range condition (Lendon & Lamacraft 1976), have been confounded by "noise" arising from spatial variability. However, because points are actually not independent from their neighbours, this so-called noise contains important information about the landscape, once the data are examined in terms of factors such as lateral movement of soil, water, or grazing animals.

In this chapter we explore how to replace or improve point measurements of high intrinsic variability and low statistical discrimination with pattern-based sampling methods, which integrate the relationships between points and consequently may provide more sensitive indices of landscape change. These methods are based on spatial processes, and incorporate regional trends in mean, variance and the spatial relationship between points, measured as spatial autocorrelation. To develop such approaches, we first consider the interactions of grazing and erosion on a simple landscape, then discuss some examples of their use in monitoring and management.

Grazing animals on a simple landscape

To illustrate how simple spatial processes can result in relatively complex patterns, consider a square paddock, sloping uniformly to the east, with a waterpoint north-west of centre (Fig. 1). It contains a uniform vegetation type, and is being grazed by sheep. Similar results would be obtained, at a different scale, if the herbivores were cattle.

Although various detailed models exist of how animals graze out from a waterpoint (Stafford Smith 1988), all result in some area around the water being eaten out, so that patterns of grazing use and trampling density develop with a broadly sigmoidal form. This pattern has been confirmed by models of sheep and cattle distributions (Stafford Smith 1988), and in the development of piospheres

FIG. 1. Plan view of a simple, uniform paddock with one waterpoint (W), a tributary drainage system, and a general slope to the right. Areas of deposition are shown by cross-hatching, stable areas by stippling, and eroding areas are bare. A-B is a transect used in subsequent figures.

(Lange 1969; Graetz & Ludwig 1979; Andrew & Lange 1986a,b). Grazing pressure will therefore vary along transect A-B according to Figure 2(a), with the magnitude of the impact on soil and vegetation being determined by the absolute stocking rate. The resulting spatial pattern is concentric (Fig. 2(b)) in the absence of wind effects and other environmental heterogeneity.

Erosion patterns on a simple landscape

Erosion of soil does not occur uniformly in space, even on simple arid landscapes. Instead, a series of scour-transport-fill (STF) sequences develop (Pickup 1988). These sequences are inherent to the sediment transport process and consist of areas which shed sediment (source zones), areas which gain sediment (sinks), and, in between, areas of alternating scour and fill (transfer zones) where the sediment is intermittently in transit. There will also be areas of stability. The two-dimensional equivalent of the STF sequence is the erosion cell, and many arid landscapes are made up of interlocking and overlaid cells which vary in size and intensity to form a mosaic. Erosion cell structures may develop on distributary drainage systems (Pickup 1985) or tributary systems, as shown in Figure 1.

Erosion status may be expressed quantitatively with erosion at one end of the scale, deposition at the other, and stability in the middle. Complex erosion patterns may then be described statistically by the mean erosion status of an area, its variance and its spatial auto-correlation (Pickup & Chewings 1988). Variance is a measure of the intensity of erosion and deposition, or erosion cell amplitude. Spatial auto-correlation describes the level of spatial uniformity, and is closely related to average erosion cell size. These three parameters are initially constant across the paddock for our simple landscape (Fig. 3(a)).

Erosion and grazing on a simple landscape

We now explore the result of adding the grazing patterns to the erosion patterns. An increase in grazing can result in a reduction in cover, and hence an increase in runoff and erosion cell intensity (Fig. 4(a)). An increase in trampling can result in greater soil erodibility and reduced infiltration, and hence in increased runoff, erosion cell intensity and loss of vegetation cover (Fig. 4(b)). There are additional feedbacks to grazing via vegetation changes which may modify the grazing and trampling patterns, but we ignore these for now.

Thus, if grazing destabilizes a landscape, the intensity and amount of erosion and deposition increase. The larger erosion cells expand, often at the expense of smaller ones. Associated changes in

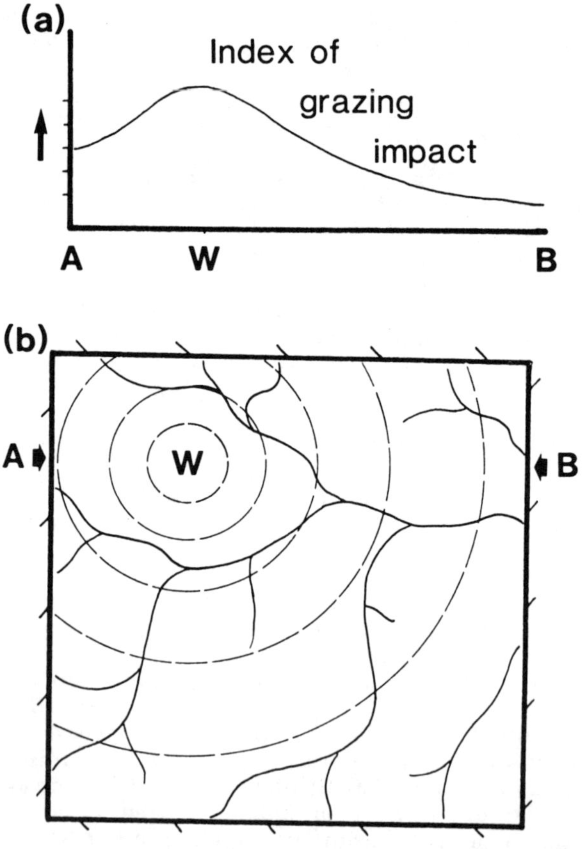

FIG. 2. Distribution of grazing animal activity: (a) grazing pressure pattern expected across transect A-B, and (b) derived pattern of isopleths of grazing impact across the whole paddock.

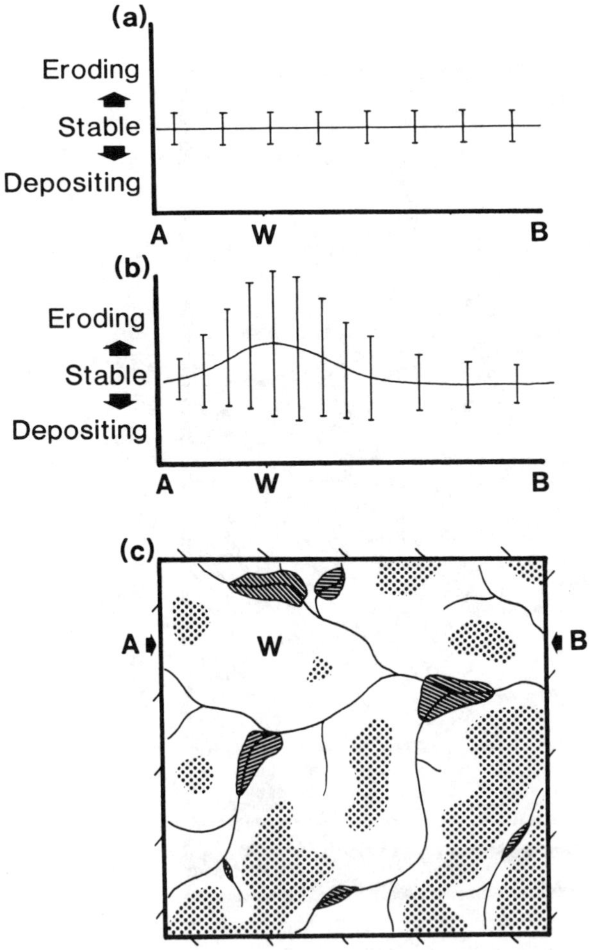

FIG. 3. Distribution of erosion cells: (a) erosion status of locations along transect A-B, schematically showing the mean and variance (1 standard deviation) at each point (imagine multiple measurements in strips parallel to A-B); (b) erosion status when grazing impact is added to the system — note the increased variance near and downslope of the water-point W; (c) potential changes in erosion patterns resulting over the whole paddock (cf. Fig. 1: note that many small sinks tend to be replaced by fewer, bigger ones).

mean erosion status may be small, but there will be substantial increases in variance, auto-correlation and cell size (Fig. 4(c); Pickup & Chewings 1988).

The result of grazing on the average erosion status of the paddock is fairly simple (Fig. 3(b)). There is a tendency for a greater proportion of the area close to water to be in an eroding state. There are also implications for spatial patterns in variance, such that there is an increase in both erosion and deposition close to water. However, runoff upslope of the waterpoint will contribute to the grazing impact downslope, so that greater variance in the erosion patterns can be expected downslope than upslope. The effect as simulated in space therefore appears far more messy (Fig. 3(c)). Cells are affected by their spatial relationship to those upslope, so that the impact is exaggerated downslope of the waterpoint. The critical point is that the variability close to the water is now at a different level to that found further away.

To illustrate this on a landscape with a realistic level of complexity, we have simulated the effect of erosion without and with the addition of a grazing pattern on an area selected from a LANDSAT scene of central Australia (Fig. 5). In this example,

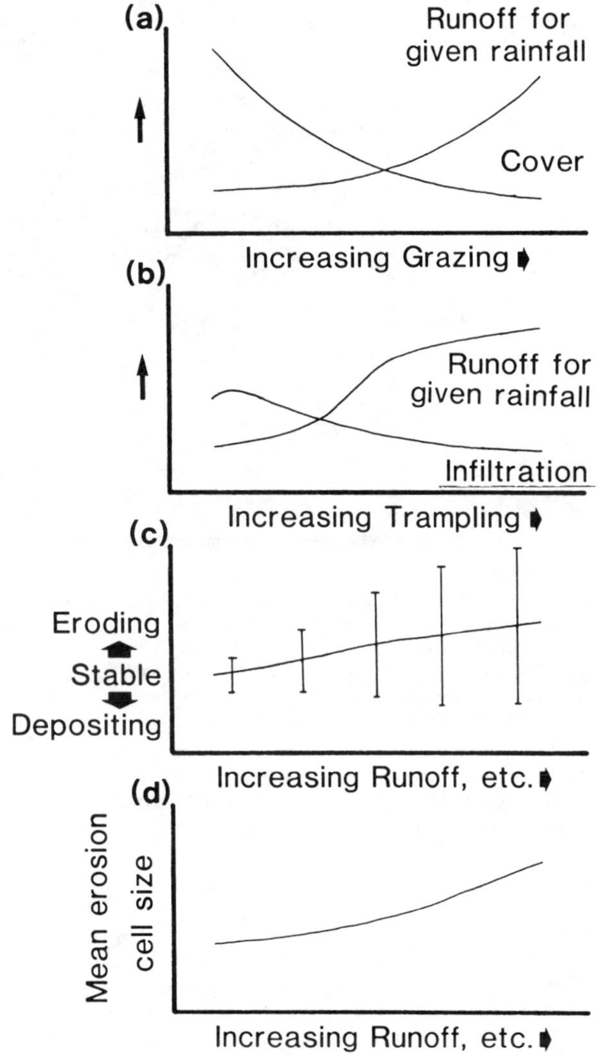

FIG. 4. Posited functional relationships between the impact of grazing, and vegetation and soil characteristics: (a) effects of grazing on cover and runoff, (b) effect of trampling on infiltration and runoff (for a rainfall of a given intensity and duration); effect of changes in runoff and erosiveness on (c) erosion status (mean and variance), and (d) on mean erosion cell size. The ordinate axis scales are arbitrary.

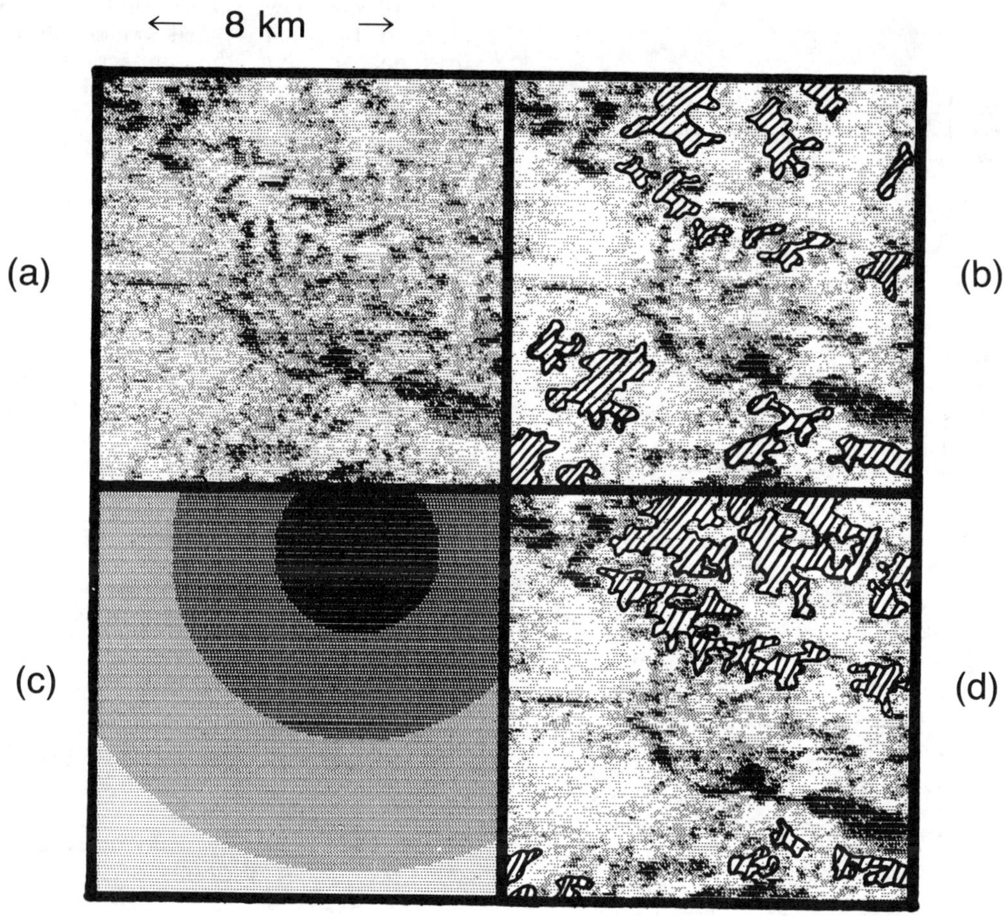

FIG. 5. Effect of adding grazing to the natural erosion cell pattern on a real landscape: (a) LANDSAT image of cover values on an 8 km×8 km area of central Australia; (b) same area with a severe erosion forecast applied evenly to the landscape; (c) assumed pattern of grazing pressure around a waterpoint in the north-east of the paddock; (d) the same severe erosion forecast as in (b) but modified to take account of the grazing pattern in (c). (b) and (d) have areas of severe erosion hatched: note the increase of these in the north-east and reduction in the south-west in (d), but also the patchy and, at first sight, confusing nature of the landscape in both.

the true patchiness and complexity of the environment becomes clear (Fig. 5(b & d)), as does the importance of a spatial approach.

Implications on a simple landscape

The functional interrelationships between factors on our simple landscape are summarized in Figure 6, although it must be emphasized that this type of representation removes precisely the spatial element that is so important. What Figure 6 fails to indicate is that factors such as grazing can cause the level of variability to change across the landscape (Fig. 3(b)), so that different sampling intensities are needed in different areas. In the absence of an underlying model which takes account of these changes, traditional sampling methods are likely to be deeply confounded, and unable to pick up any but the most obvious trends.

The model that we provide here is a start towards interpreting landscapes such as these. For example, using the underlying concepts, it becomes easier to determine where sampling sites should be located. Clearly stratification must provide replication to assess variance at different distances from water across the simple gradient. The model also helps in the analysis of the data by indicating how points in the landscape are likely to be related to each other. Its initial validation may depend on numerous measurements such as those which might be collected from satellite imagery, but, once validated, the model will permit sense to be made of smaller numbers of correctly located ground measurements. Clearly the purpose is to reduce spatial complexity comparable to that found in Figure 5 to simpler relationships by recognizing the spatial interactions.

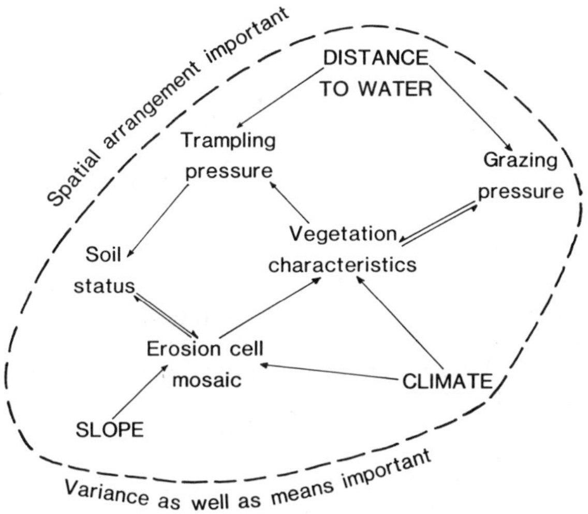

FIG. 6. Key interrelationships in the simple paddock system.

Vegetation on the simple landscape

Ecologically-based management procedures require the measurement and interpretation of vegetation patterns. We therefore briefly explore how vegetation may respond to patterns generated by spatial processes. These responses are less well understood. However, we assume (Fig. 7(a)) that increasing erosion cell intensity results in a reduced chance of germination on source zones (due to increasing runoff and removal of topsoil) and an increased chance on deposition areas (due to run-on). We also assume that increasing grazing pressure will result independently in the patterns of responses shown in Figure 7(b) for ephemeral plants, and palatable and unpalatable perennials (e.g., Foran 1986; Friedel 1987). We expect palatable perennials to decline with increasing grazing pressure. Unpalatable perennials should initially increase as they are released from competition with palatable perennials but then decline at the highest grazing pressure where even they are nibbled or trampled. Ephemerals should increase with loss of perennials but decline at the highest grazing pressure. Note that this is a result of grazing selection as well as changed soil conditions in the erosion cells.

These relationships would cause a pattern along the transect A-B similar to Figure 8(a) independently of the erosion cells. The distance from water at which the responses curve will be highly dependent on stocking rate and vegetation type, of course. These responses add a further dimension of complexity to the patterns caused by the erosion cell mosaic. One element of the responses is that some areas might be expected to have increased problems with woody weeds (Fig. 8(b)). Needless to say, the comments on measurement of the previous section apply all the more strongly here. Furthermore, the new pattern will have further feedbacks on the grazing patterns, due, for example, to reduced access in heavily wooded areas, and increased grazing concentration on the remaining preferred vegetation types.

Realistic landscapes

Real landscapes are affected by many other sources of variability. Slope may vary, animals are affected by barriers of slope or vegetation, sheep move into the prevailing winds, soil types have different susceptibilities to erosion, as well as runoff and infiltration rates, vegetation responses vary with soil type, animals change their preferences for different vegetation types, more variability is introduced by local weather patterns, and so on. For instance, Figure 2(c) can be modified to take account of the effect of wind on sheep grazing. This biases the grazing pattern into the wind, and results in greater changes to the erosion patterns in that direction than would be predicted by a radially symmetrical model.

As another example, different vegetation types (even those likely to grow on erosion cell sinks as compared to source zones) may have a different preference to grazers. The degree to which this

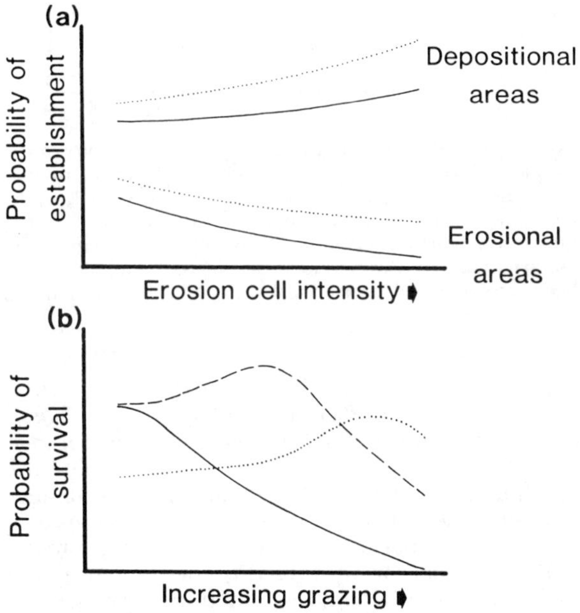

FIG. 7. Posited indirect and direct effects of grazing on key elements of the vegetation: (a) changes in the probability of ephemeral (dotted) and perennial (solid) plant establishment on source and sink areas of erosion cells with increasing cell "intensity" due to grazing (cf. Fig. 4), and (b) changes in the probability of survival of ephemeral (dotted), unpalatable (dashed) and palatable (solid) plants with increasing grazing intensity. (Note that the arbitrary ordinate axis would be scaled differently for different soils and species).

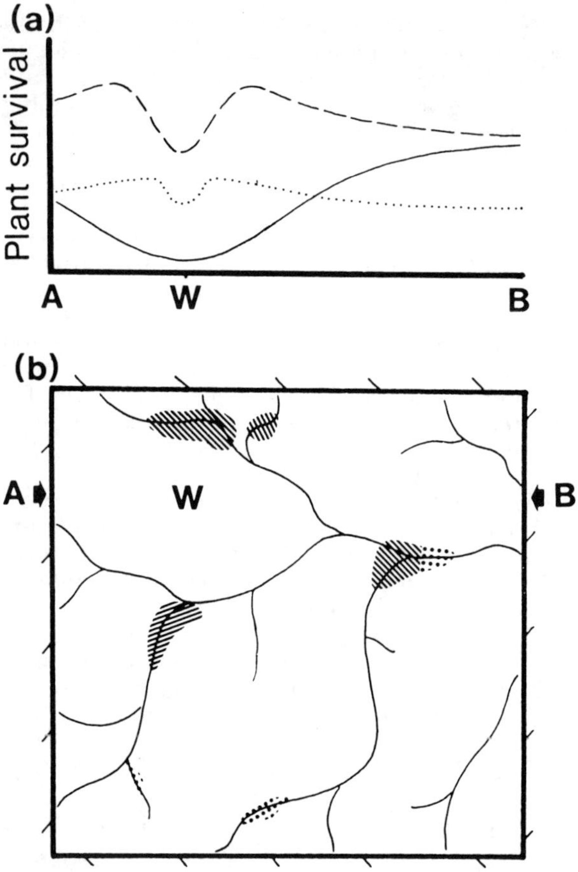

FIG. 8. Posited effects of grazing (cf. Fig. 7) on plants in the simple paddock: (a) survival of ephemeral (dotted), unpalatable (dashed) and palatable (solid) perennials on transect A-B, and (b) the areas which would consequently have a high (cross-hatching) or lower (stippling) risk of woody weed encroachment in the paddock due to the combination of heavy grazing and erosion patterns. This result is not a simple pattern.

may affect animal distributions depends on what preference other parts of the landscape may have to the grazing animal. A small patch of preferred forage in a sea of poor feed will be eaten, whilst the same patch in a large palatable area may not be touched. It is not enough to know what is happening at a point, unless that point is in a landscape context.

Despite the complexity, this sort of variability must be accommodated in our thinking if we are to understand the functioning of our arid landscapes. Many of the erosion, and animal, related sources can now be modelled reasonably, for example. Our concepts lag with respect to the effects of climate, however, and the responses of vegetation (but see Pickup & Foran 1987). The greatest challenge is to recognize the implications of changing patterns in both means and variances.

Conclusions

This volume reviews the past and is looking to the future of Australian ecology. It is fair to say that the discipline has been dogged by problems of spatial heterogeneity. Our arid lands are spatially diverse, and management is on such a scale that the heterogeneity is usually reflected within, rather than between, management units such as paddocks or parks.

Furthermore, our ecological thinking has been constrained by work from areas of the world where spatial patterns can be more safely ignored. The development of techniques which come to grips with spatial heterogeneity and the inter-dependence of points in the landscape is one of the several great advances which Australian ecology can offer the world. Past work has taught us that we need them. Future management will rest upon them.

References

Andrew M. H. & Lange R. T. (1986a) Development of a new piosphere in chenopod shrubland grazed by sheep. 1. Changes to the soil surface. *Aust. J. Ecol.* **11,** 395-409.

Andrew M. H. & Lange R. T. (1986b) Development of a new piosphere in chenopod shrubland grazed by sheep. 2. Changes to the vegetation. *Aust. J. Ecol.* **11,** 411-24.

Foran B. D. (1986) Thc impact of rabbits and cattlc on an arid calcareous shrubby grassland in central Australia. *Vegetatio* **66,** 49-59.

Friedel M. H. (1987) A preliminary investigation of woody plant increase in the western Transvaal and implications for veld assessment. *J. Grassld. Soc. South Afr.* **4,** 25-30.

Graetz R. D. & Ludwig J. A. (1979) A method for the analysis of piosphere data applicable to range assessment. *Aust. Rangel. J.* **1,** 126-36.

Lange R. T. (1969) The piosphere: sheep track and dung patterns. *J. Range Manag.* **22,** 396-400.

Lendon C. & Lamacraft R. R. (1976) Standards for testing and assessing range condition in central Australia. *Aust. Rangel. J.* **1,** 40-8.

Ludwig J. A. (1987) Primary productivity in arid lands: myths and realities. *J. Arid Environ.* **13,** 11-26.

Noy-Meir I. (1981) Spatial effects in modelling arid ecosystems. In: *Arid lands ecosystems: structure, functioning and management* (eds D. W. Goodall & R. A. Perry) Vol. 2, pp. 411-32. Cambridge University Press, Cambridge.

Pickup G. (1985) The erosion cell — a geomorphic approach to landscape classification in range assessment. *Aust. Rangel. J.* **7,** 114-21.

Pickup G. (1988) Modelling arid zone soil erosion at the regional scale. In: *Fluvial Geomorphology of Australia* (ed. R. F. Warner) pp. 105-27. Academic Press, Sydney.

Pickup G. & Chewings V. H. (1988) Forecasting patterns of soil erosion in arid lands from Landsat MSS data. *Int. J. Remote Sensing* **9,** 69-84.

Pickup G. & Foran B. D. (1987) The use of spectral and spatial variability to monitor cover change on inert landscapes. *Remote Sens. Environ.* **23,** 351-63.

Stafford Smith D. M. (1988) Modeling: three approaches to predicting how herbivore impact is distributed in rangelands. *New Mex. Agr. Sta. Reg. Res. Rep.* **628.** 56 pp.

Proc. Ecol. Soc. Aust. 1990 — 16:201-213

The impact of European settlement on the vertebrate animals of arid Australia: a conceptual model

S. R. MORTON

CSIRO Division of Wildlife and Ecology,
P.O. Box 2111, Alice Springs, N.T. 0871, Australia

Abstract

The mammals of inland Australia were seriously affected by European settlement, and a high proportion declined dramatically in range or became extinct. Native birds fared much better, and no species are certainly extinct. Reptiles have suffered no extinctions and no apparent reductions of range. Among the mammals there are clear biases in the species which suffered most; medium-sized species, most of them herbivorous or omnivorous, were devastated. A conceptual model to account for these patterns is developed. It proposes that the environment of inland Australia was originally difficult for herbivorous and omnivorous mammals to inhabit because of the infertility of the soils, and the consequent restriction of digestible production to small fertile areas. This problem was exacerbated by the uncertain climate with its droughts of irregular length. The consequent restriction of these types of mammals to scattered pockets of suitable habitat during droughts exposed them to high probabilities of local disappearance.

The arrival of grazing stock and the rabbit upset the balance between local disappearance and reinvasion after drought. The introduced herbivores maintained extraordinarily high populations for a short time and, in doing so, altered the vegetational composition of the very habitats upon which the native species had been dependent for refuge during droughts. This degradation, together with the pressure of introduced predators and, in some places, altered patterns of fire, caused increased probabilities of local disappearance. Several droughts later, extinctions resulted.

A variety of apparently conflicting evidence is weighed up and resolved. I conclude that the model not only has potential explanatory validity but also has substantial management implications.

Introduction

The mammals of inland Australia have suffered badly in the 200 years since European settlement. They have experienced numerous extinctions and contractions of range, in some places to such an extent that mere fragments of the original fauna now remain (Finlayson 1961; Newsome 1971; Newsome & Corbett 1977; Smith & Medlin 1982; Low 1984, 1986; Morton & Baynes 1985; Caughley 1986; Burbidge *et al.* 1988; Johnson *et al.* 1989). Such a catastrophic impact demands explanation, especially in a nation increasingly interested in its unique biological heritage. Despite considerable attention in the literature to possible causes of this impact, however, there is no general agreement about the fundamental forces that are responsible for it.

In this chapter, I put forward a conceptual model which attempts to account for the contraction and extinction of certain types of vertebrate animals. I begin by outlining a series of facts about the extinctions that I believe are important to an understanding of the causal factors and then proceed to a review of the previously suggested causes. These facts and previous interpretations are then incorporated into an explanatory model. The model is briefly tested but requires further rigorous examination in the light of available ecological and historical data.

Attributes of the extinctions

Extinctions among vertebrates biased towards mammals

The most striking fact is that mammals suffered extinction and contractions of range to a far greater degree than any other vertebrate group in the arid zone. No desert-dwelling reptile or amphibian has become extinct despite the fact that some species are highly restricted geographically (Ehmann & Cogger 1985). Even the bronzeback, *Ophidiocephalus taeniatus*, a pygopodid apparently confined to tiny patches of riparian habitat that are subject to considerable disturbance by cattle, has persisted through the past 200 years (Ehmann 1981).

No birds have become extinct from the arid zone for certain, although the night parrot, *Geopsittacus occidentalis*, is extremely rare (Blakers *et al.* 1984), the malleefowl, *Leipoa ocellata*, has disappeared from much of the arid zone (Blakers *et al.* 1984; Kimber 1985), and the flock bronzewing, *Phaps histrionica*, apparently has declined from "countless multitudes" to its present patchy occurrence (Blakers *et al.* 1984). Several other species are uncommon or rare, but it is difficult to show conclusively that they have declined since European settlement. Among them are the thick-billed grass wren, *Amytornis textilis*, which has probably declined in some areas but which seems to have adjusted to changes brought about by pastoralism in others (Blakers *et al.* 1984; Curry 1986); the chestnut-breasted whiteface, *Aphelocephala pectoralis*, which may always have been rare (Blakers *et al.* 1984); the plains-wanderer, *Pedionomus torquatus*, which has declined in the more productive and agriculturally-developed parts of its range (Bennett 1983); and Alexandra's parrot, *Polytelis alexandrae*, which has been considered rare and declining (Smith 1979) but may simply be uncommon, nomadic and little changed in abundance (M. R. Fleming, pers. comm.). Considering the size of the arid Australian avifauna, the number of species currently at risk as a result of European settlement appears small.

There is a striking contrast between the birds and reptiles of arid Australia and the mammals, among which there has been devastation. Of the 72 species of terrestrial mammals known originally from the arid zone, 11 have become extinct, five have disappeared from the mainland and are now confined to offshore islands, and 15 more have declined dramatically in range, often becoming absent from the arid zone and persisting only in the semi-arid fringes (Table 1; see also Strahan 1983). The disparity between the responses of arid Australian mammals to European settlement and those of the other vertebrates has never been examined before but it needs to be incorporated into any explanatory framework.

Extinctions among Australian mammals biased towards the arid zone

Of the 38 terrestrial Australian mammals considered endangered or extinct, 23 (60%) are or were present in the arid zone (CONCOM 1988). Only 38% of the continental fauna of terrestrial mammals originally occurred in the arid zone but 32% of those species are listed by CONCOM. In contrast, only 13% of species of mesic environments of Australia are listed, even though these species make up 62% of the terrestrial fauna. Thus, species of the arid zone have been considerably more affected by European settlement than those of mesic environments and this fact must be accounted for by any theory.

Extinctions among desert-dwelling mammals biased towards medium-sized species

Several authors have shown that species disadvantaged following European settlement tend to be of medium size (Newsome 1975; Burbidge & Fuller 1979; Burbidge 1985; Morton & Baynes 1985; Low 1986; Johnson *et al.* 1989) and quantitative analysis confirms that species within the range 35 to 5 500 g were by far the most susceptible (Burbidge & McKenzie 1989). With few exceptions, most of these medium-sized species were herbivorous or omnivorous (see Table 1). Medium-sized herbivorous mammals are predominant among extinct

TABLE 1. Mammals of the Australian arid zone severely affected by European settlement, based on CONCOM (1988). Status is defined by three letters: E=Extinct; I=now found only on islands; D=declined dramatically to small proportions of the original range. Size class is defined by two letters: S=less than 0.1 kg; M=between 0.1 and 5.5 kg. Trophic class is defined by four letters: H=herbivore; G=granivore; C=carnivore or insectivore; O=omnivore.

Species	Status	Size class	Trophic class
Bettongia lesueur	I	M	H
B. penicillata	D	M	H
Caloprymnus campestris	E	M	H
Lagorchestes asomatus	E	M	H
L. hirsutus	D	M	H
L. leporides	E	M	H
Onychogalea lunata	E	M	H
Chaeropus ecaudatus	E	M	O
Isoodon auratus	D	M	O
Perameles bougainville	I	M	O
P. eremiana	E	M	O
Macrotis lagotis	D	M	O
M. leucura	E	M	O
Trichosurus vulpecula	D	M	H
Myrmecobius fasciatus	D	M	C
Dasyurus geoffroii	D	M	C
*Sminthopsis longicaudata**	D?	S	C
*S. psammophila**	D?	S	C
Leporillus apicalis	E	M	H
L. conditor	I	M	H
Notomys amplus	E	S	G/O
N. fuscus	D	S	G/O
Pseudomys fieldi	I	S	G/O
P. gouldii	E	S	G/O
Zyzomys pedunculatus	D	S	G/O

*Detailed studies may well show that these two species have not declined dramatically.

and endangered species in mesic Australian environments; of 20 such species on the CONCOM (1988) list, 17 are herbivorous or omnivorous and 13 are of medium size. Why were these species particularly vulnerable?

Causes of the extinctions: previous hypotheses

There are seven commonly cited reasons for the decline of arid Australian mammals since European settlement (e.g., Low 1986; Burbidge *et al.* 1988; Johnson *et al.* 1989). Here I review them briefly, concentrating not only on the characteristics that make them unsatisfactory as complete explanations for the observed patterns of extinctions but also on those aspects that seem to me to be of value.

Droughts and climatic change

Interpretation of the historical record suggests that mammalian species frequently disappeared during or just after prolonged droughts (Low 1984, 1986; Johnson *et al.* 1989). However, populations of these animals must have experienced droughts throughout the millenia before the arrival of Europeans; hence, this explanation cannot be the ultimate reason for extinctions. Despite this conclusion, I believe that the correlation of extinctions with drought is of significance and will return to this topic below.

A related suggestion, made especially by earlier writers on the issue of extinctions, is that some species had been adversely affected by long term climatic change and were declining long before the impact by Europeans was felt (Finlayson 1961; Calaby 1963). However, we know now that the climate of inland Australia has been ameliorating over the past 2 000 years (Singh 1981) and that this suggestion is unlikely to be of much importance. In any case, it fails to explain why the impact should be most severe upon medium-sized mammals.

Introduced predators

Cats, *Felis catus,* and foxes, *Vulpes vulpes,* became widespread in inland and northern Australia in the late 19th and early 20th centuries and have been widely blamed for the disappearance of numerous native animals (e.g., Jones 1925; Finlayson 1961; Rolls 1969; Burbidge & Fuller 1979; King *et al.* 1981; Strahan 1983; Johnson *et al.* 1989). There is no doubt that these two predators eat a wide range of native vertebrates in the arid zone (Martensz 1971; Coman & Brunner 1972; Ryan & Croft 1974; Bayly 1976, 1978; Croft & Hone 1978; Jones & Coman 1981; Catling 1988), but three anomalies suggest that the presence of introduced predators is unlikely to be the sole and ultimate cause of the extinction of native animals. First, why would introduced predators cause extinctions only in the arid zone? As was noted above, extinctions have not been dramatic in the moister parts of Australia where the predators are widespread and prominent. There is clearly some extra factor which made desert-dwelling animals more susceptible to predation than their mesic relatives.

Secondly, why should introduced predators wreak havoc on medium-sized species rather than small granivorous or insectivorous mammals? Cats especially take numerous small mammals (Jones & Coman 1981) and yet few small mammals seem to have suffered. Again, if predation is important then it is highly dependent upon the characteristics of the prey. An associated issue is that predation does not appear to have driven birds or reptiles to extinction. Cats, for example, eat considerable numbers of both these groups (Bayly 1976, 1978; Jones & Coman 1981), although it remains possible that ground-dwelling predators have less impact on birds and reptiles than they do on mammals. Thus, the apparent lack of impact on birds and reptiles casts some further doubt upon this hypothesis.

Thirdly, if the introduction of a eutherian predator were to result in rapid and widespread extinctions, then one would predict such an effect following the arrival of the dingo, *Canis familiaris,* some 5 000 years ago (Corbett 1985). There have been substantial changes in the distributions and relative abundances of mammals during the past 10 000 years (e.g., Lundelius 1983), but there was no wholesale extinction among small or medium-sized species and, as far as I am aware, no-one has suggested that the dingo was responsible for anything except the competitive displacement of the thylacine, *Thylacinus cynocephalus,* and the Tasmanian devil, *Sarcophilus harrisii* (Calaby 1971; Archer 1974). Dingoes eat a wide range of small and medium-sized mammals (Whitehouse 1977; Newsome *et al.* 1983; Robertshaw & Harden 1985; Corbett & Newsome 1987) and so why should the cat and the fox have had a dramatic impact but not the dingo? These considerations lead me to conclude that although the introduction of predators had significant consequences, it is not a sufficient explanation for widespread extinction.

Introduced mice

Newsome and Corbett (1975) suggested that introduced house mice, *Mus musculus,* displaced some native rodents in arid Australia. Morton and Baynes (1985) disagreed and postulated instead that house mice had merely occupied the remnants of niche spaces made vacant by the extinction of the native species. I still believe that the latter explanation is more adequate, as the house mouse is an opportunistic "weedy" species with no apparent competitive superiority over native rodents (Fox & Pople 1984). Furthermore, the introduction of house mice would not seem to provide an explanation for the marked disappearance of the medium-sized mammals.

Disease

The possibility that diseases introduced with domestic and feral animals contributed towards extinctions has been mentioned by several authors (e.g., Finlayson 1961; Low 1986; Johnson *et al.* 1989). Epidemics may certainly have drastic yet poorly appreciated effects on mammalian populations (e.g., Prins & Weyerhaeuser 1987) although Australian examples are inconclusive (Johnson *et al.* 1989). There is no reason to believe that such diseases should have been selective for medium-sized mammals (Low 1986; Johnson *et al.* 1989), nor is there any apparent reason why disease should have had a more pronounced effect on mammals of the arid zone than in the moister parts of the continent.

Pastoralism

The arrival of sheep, cattle, horses and goats had a major impact on most of inland Australia except for the sandy spinifex country. Numerous studies attest to the environmental degradation arising from the initial over-optimistic assessment of carrying capacity, the subsequent heavy overstocking, the futile attempts to keep stock on country gripped by what we now know are inevitable droughts, the eventual catastrophic losses, and the failure of the country ever to carry the original numbers of stock again (e.g., Heathcote 1965; Harrington *et al.* 1984; Griffin & Friedel 1985; Payne *et al.* 1987; Friedel *et al.*, this volume).

All those who have written about extinctions express the opinion that the resulting wholesale alteration of habitats, the competition for forage, and the removal of cover was a primary cause of the decline of many native animals (Finlayson 1958, 1961; Newsome 1971; Newsome & Corbett 1977; Low 1984, 1986; Caughley 1986; Burbidge & McKenzie 1989; Johnson *et al.* 1989). I agree, and note that this issue will recur in my attempts to present a comprehensive argument below. Regardless of how significant the effect was, though, it could not account for the widespread extinctions in the non-pastoral country of the western arid zone (Finlayson 1961; Burbidge & Fuller 1979; Low 1984, 1986; Burbidge 1985; Morton & Baynes 1985; Johnson *et al.* 1989). Furthermore, it is not completely clear why the medium-sized mammals were so badly disadvantaged by pastoral occupation in comparison to small species, although Newsome (1971) and Newsome & Corbett (1977) may have gone part way towards accounting for this trend. To explain all these aspects satisfactorily, we have to look for other causes.

The rabbit

Rabbits, *Oryctolagus cuniculus*, poured through the inland from the early 1870s until they reached their limits in northern Australia sometime during the first two decades of the 20th century (Rolls 1969; Myers 1971; Cooke 1977; Wood 1984). Like domestic stock, rabbits had a dramatically negative effect on the vegetation of those habitats which suited them (e.g., Wood 1936; Hall *et al.* 1964; Crisp & Lange 1976; Lange & Purdie 1976; Crisp 1978; Lange & Graham 1983; Foran *et al.* 1985; Foran 1986), and no author has disputed that the resulting competition for food and habitat alteration were major factors in the subsequent declines among native animals.

Nevertheless, doubt has been expressed that the rabbit could have been the sole cause of the disappearance of many mammals. First, rabbits were apparently never common in the northern parts of their distribution, particularly in the Great Sandy and Tanami Deserts, and yet extinctions have been evident there (Burbidge 1985; Johnson *et al.* 1989). Secondly, rabbits were rarely common in much of the sandhill and sandplain country or the mulga shrublands of the central arid zone, and yet such habitats were apparently used by certain native animals that have become extinct (Johnson *et al.* 1989). Thirdly, how could rabbits have been responsible for the demise of animals with which they would seemingly never compete for food, such as the numbat, *Myrmecobius fasciatus*, the western quoll, *Dasyurus geoffroii*, or the brushtail possum, *Trichosurus vulpecula*? Fourthly, Finlayson (1958) described burrowing bettongs, *Bettongia lesueur*, living together with rabbits in warrens, and suggested that the two had coexisted for 60 years. If competition between rabbits and herbivores was a dominant force, how was it possible for these two species to coexist for such long periods? I think that these objections can be answered (see below), but they emphasize the fact that a coherent picture of the extinctions is lacking.

Altered patterns of fire

Aboriginal people used fire in the arid zone for a variety of purposes, especially in the spinifex grasslands (Kimber 1983; Latz & Griffin 1978; Griffin 1984). It has been suggested that the departure of these people from their traditional lands resulted in a change from the original mosaic of relatively small-scale burns to less frequent but much larger wildfires begun by lightning. The consequence of larger wildfires may have been a reduction in the fine-scale mosaic favoured by some animals because of its apposition of dense vegetation providing shelter and regenerating areas providing a greater range of plant foods (Bolton & Latz 1978; Burbidge & Fuller 1979; Friend *et al.* 1982; Griffin 1984; Burbidge 1985; Johnson *et al.* 1989).

But why should the impact of fire be so marked for mammals, and particularly why should the effect be concentrated on medium-sized species? If patterns of burning are important for such mammals, then an extra factor seems necessary to account for it. In addition, it is not immediately clear how the hypothesis can explain why other regions in which Aboriginal patterns of burning have been similarly disrupted, for example the savannahs of northern Australia (Braithwaite & Estbergs 1985; Haynes 1985), have not experienced similar problems. Of all regions of Australia, these woodlands and forests have suffered least in terms of contractions of range (Braithwaite *et al.* 1985; Morton & Baynes 1985; Braithwaite & Werner 1987). Thus, it is not yet clear what part altered patterns of burning have played in causing extinctions.

A model to account for extinctions

The attempt to deduce a model to account for the extinctions begins at the physical environment. Following this, I proceed to other climatic and biological elements that seem of importance. Finally, the various elements are brought together in a conceptual flow-chart.

Element 1: the abiotic environment

One fact underpins the entire model and it is that the ancient, weathered and highly-sorted soils of arid Australia are unusually infertile. Almost all the soil types show extreme poverty of phosphorus and nitrogen (Jackson 1957; Wild 1958; Charley & Cowling 1968; Williams & Raupach 1983) and they cover a large proportion of the arid zone; an example is the vast sandplains supporting spinifex grasslands of *Triodia* and *Plectrachne*. Nevertheless, regions of greater topographic diversity almost always contain run-on areas where the flow of water has produced patches with enhanced fertility and moisture supplies (e.g., floodouts), and in some regions there are soils with higher nutrient levels (e.g., calcareous outcrops). In general, such fertile areas exist as islands in a sea of infertility. These relationships appear most pronounced in the northern half of the arid zone and less so in the southern half.

The patterns outlined here have substantial consequences for plant production, as perennial plants growing on infertile soils tend to be poorly digestible because they have developed either chemical or structural defences (Mattson 1980; Coley *et al.* 1985; Campbell 1986; Owen-Smith & Cooper 1987). Thus, much of the plant production in the large expanses of infertile country is almost certainly low in nutrients and high in inhibitory compounds. From the herbivorous or omnivorous mammal's point of view, therefore, much of inland Australia is nutritionally hostile (Kinnear & Main 1979).

A second characteristic, uncertainty of rainfall, modifies and accentuates the patterns of infertility. Arid Australia is not especially dry on a global scale, but the temporal variability of its rainfall is comparatively high (Leeper 1970; Gentilli 1971; Bell 1979). Virtually all of the arid zone experiences annual rainfalls that are skewed towards lower totals than the mean would indicate, and occasional cyclonic or monsoonal depressions that dump several times the median annual rainfall, recharge water tables, and cause sub-surface flow through extensive drainage systems. In summary, periods of substantial plant production are separated by very dry spells of irregular length, and the unpredictable timing and severity of the ensuing droughts have severe consequences for the animals inhabiting arid Australia.

Together, these two features describe an environment with highly patterned productivity. Those areas of the landscape into which water is channeled not only have better moisture supplies for plant growth, but they also tend to be more fertile because the big flows of water carry the nutrients with them. In addition, the fertile areas frequently display more dependable production because water is being funnelled into them more regularly. This image of arid Australia, therefore, is of a series of patches of fertile, dependable and digestible production in all shapes and sizes dotted throughout a vast infertile landscape with relatively unpredictable and unusable plant growth. In the occasional wet years the patches tend to expand and coalesce. In the long dry times which intervene they shrink and break up into smaller and smaller units, and as the drought stretches out, more and more of them disappear altogether.

Element 2: physiological and morphological constraints acting upon herbivorous and omnivorous mammals

All animals must secure nutritionally adequate food from an environment chronically poor in nitrogen (White 1978; Mattson 1980). The difficulties of doing so are exaggerated among the homeotherms, which have to obtain high maintenance energy requirements every single day. Most homeothermic grazers are mammals and studies suggest that small herbivorous mammals face greater problems relative to larger species in meeting their metabolic requirements (Jarman 1974; Parra 1978; Bell 1982; Batzli 1985). This problem arises because smaller mammals have similar relative gut sizes but larger relative metabolic rates than do larger animals and so intake and digestion of poor quality diets will provide less energy relative to metabolic requirements for smaller than larger species. Consequently, smaller herbivores and omnivores must eat more digestible and nutritious food. In fact, this trend is inescapable to such an extent that below a certain body size (about 50 g) there are virtually no herbivorous mammals because a diet of green vegetation cannot supply sufficient energy; mammals of this size must be insectivorous, granivorous or omnivorous upon highly nutritious food. As a result, the medium-sized species, most of which are herbivores, are confined to much smaller patches of suitable habitat than the larger species.

A second problem facing the mammals under consideration is that they cannot fly. This is important because considerable mobility is necessary if the animals are to recolonize the expanded habitat made available after a major drought-breaking rain, and in particular if they are to re-occupy the patches that failed during the preceding drought. Smaller mammals have less mobility than larger species simply because of their smaller sizes and therefore must of necessity recolonize vacant habitat more slowly.

However, body size affects the ability of populations to persist in patches of varying sizes in yet another way. All else being equal, smaller animals need less energy per individual than larger species and so populations of smaller species can persist on refuge patches of smaller sizes than is possible for larger species. As a result, there are more suitable patches available to small species in any particular region. Consequently, small mammals experience patch failure less frequently than larger species, and they do not have as far to disperse to locate another suitable patch. The contrasting trends between mobility and changing availability of patches for differing body sizes are depicted schematically in Figure 1, which suggests that both large and small mammals have advantages accruing to them. In the middle are the medium-sized species, which appear to be getting the worst of both trends — they still require the relatively rare large patches, but their mobility lags behind that of large mammals.

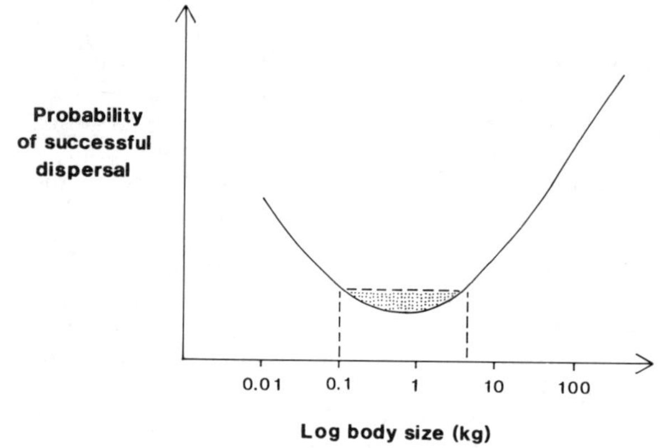

FIG. 1. The probability of successful dispersal between suitable patches of habitat in arid Australia by mammals of different body sizes. The graph represents the fact that at least two forces govern successful dispersal and, therefore, persistence. One of these is mobility, which is greater in larger mammals. The other is patch suitability as perceived by mammals of different sizes; small mammals can persist as viable populations on much smaller patches than large mammals, and so there are many more patches suitable for the small species. Consequently, the probability of successful dispersal and persistence is lowest for medium-sized species, and the hatched area outlines the size range of arid Australian mammals most affected by European settlement.

It is probable that the occupation by medium-sized mammals of all suitable habitat took place only during unusually extended wet periods when the infertile country was temporarily productive. Hence, recolonization of certain patches of habitat after a long drought may well have been a lengthy process. It is also worth noting that Kitchener *et al.* (1980) suggested that medium-sized mammals reacted to environmental patchiness in the western Australian wheatbelt in a similar manner to that outlined here.

Synthesis of elements 1 and 2

It is impossible to prove conclusively the two propositions about medium-sized mammals because so few of these animals still exist. I will mention circumstantial evidence concerning these animals later in the chapter, but want to look briefly at some corroborative data from outside arid Australia.

First, there is a considerable body of literature which examines in a theoretical way the survival of populations in variable environments. This literature largely confirms common sense: the persistence of species in patchy habitats depends to a marked extent on the sizes of patches, their degree of isolation, the disturbance regime and the dispersal ability and demographic potential of the animals (e.g., den Boer 1981; Seagle & Shugart 1985; Lande 1987; Stamps *et al.* 1987). In particular, it is becoming clear that the fundamental driving force behind habitat patchiness is edaphic (Wiens *et al.* 1985) and that the energetic constraints faced by particular trophic groups have a considerable effect on their vulnerability to local disappearance (Brown & Maurer 1987). Further, because even a small number of large local populations can be crucial in regional population dynamics, then the collapse of regional populations may be unexpectedly rapid and catastrophic (Hanski 1985). Finally, the importance of immigration in reducing the probability of local disappearance is being increasingly recognized, and has been termed, appropriately, the "rescue effect" (Brown & Kodric-Brown 1977). Indeed, Pulliam (1988) has suggested that, in some populations, large fractions of the individuals may occur in "sink" habitats, where within-habitat reproduction is insufficient to balance local mortality. He argues that populations can still persist in such habitats because they are maintained by continued immigration from richer "source" areas nearby. The emphasis on soil, energetic problems and vulnerability resulting from extreme patchiness are all consistent with the two elements outlined above for arid Australia.

Secondly, there are several pertinent examples of these issues. On a regional scale, the isolated distributions of montane mammals in the south-western United States are explicable in terms of vulnerability to extinction and relative inability to recolonize, and contrast dramatically with the more vagile birds of the area (Brown 1978). Again, at the regional level, the distribution and abundance of field mice (*Apodemus* spp.) in Sweden is governed by geomorphology, because the mice are confined to richer soils; their presence in fragmented areas is sporadic and is dependent on a balance between immigration and local disappearance (Angelstam *et al.* 1987). At a local scale, Smith's (1974) studies of the North American pika, *Ochotona princeps*, demonstrate in detail how the distribution of a medium-sized herbivorous mammal may be governed intimately by the availability of habitat and the balance between disappearance and re-colonization. Such pragmatic results are also under increasing study in Australia (Saunders *et al.* 1987).

In summary, I believe that there is considerable theoretical support for the proposition that inland Australia represented an unusually difficult environment for medium-sized mammals, especially the herbivorous and omnivorous species. The nature of the soils and of the plants which they support, together with the unusually uncertain and drought-prone climate, probably exposed these animals to frequent local disappearance. These two elements provide us with a conceptual model (Figure 2) of how mammalian herbivores were living when Europeans arrived with their Pandora's box of foreign animals.

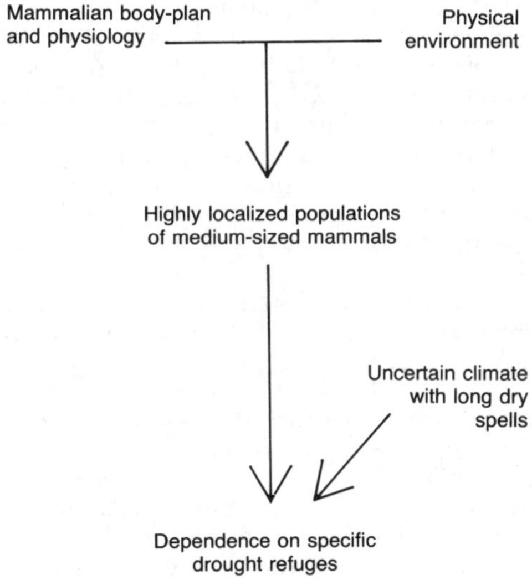

FIG. 2. A flow-chart showing the principal forces postulated as acting upon medium-sized mammals in arid Australia before the arrival of Europeans.

Element 3: the introduction of European herbivores

I believe that the causal chain shown in Figure 2 resulted in the medium-sized mammals being inherently prone to disturbance of their precious patches of habitat. I argue that the primary cause of extinctions was habitat modification brought about by waves of introduced herbivores, the cattle, horses, sheep, goats and, especially, rabbits. My conclusion is that in most cases the other factors were secondary and would not have had a dramatic impact if it were not for the devastating changes wrought by the introduced herbivores. The argument is summarized in Figure 3.

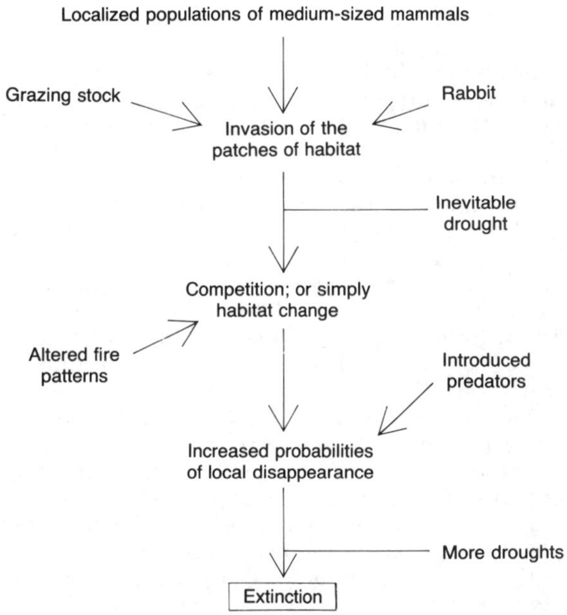

FIG. 3. A flow-chart summarizing the forces which are suggested to have caused the extinctions of medium-sized mammals after the arrival of Europeans in arid Australia.

The basic relationship between the herbivore and the inland Australian environment is the same for introduced animals as it is for native species. Introduced animals are distributed in a highly localized and patchy manner, less so in the southern arid zone and markedly so in the north (e.g., Myers 1971; Myers & Parker 1975; Low *et al.* 1980; Harrington *et al.* 1984; Foran *et al.* 1985; Griffin & Friedel 1985; Parer & Libke 1985). In essence, these herbivores are forced to select the same richer, more dependable portions of the landscape as I have postulated for the native species, and this relationship becomes more pronounced as a drought lengthens.

The important point about both grazing stock and rabbits is that their numbers shot well beyond anything that might be considered as long-term carrying capacity. The stock were held on country long after the vegetation was able to support them; the rabbit had escaped its normal suite of pathogens, predators and competitors, and so exploded into enormous populations before suffering extreme mortality in the next inevitable droughts. I suggest that the immediate outcome was removal of the sources of dependable and nutritious growth which in prior times had supported medium-sized native mammals, especially the herbivores and omnivores. In the case of the rabbits, it seems likely to me that they have altered the vegetation to such an extent that they can no longer permanently occupy many places themselves.

In pre-European times, recolonization of patches from which the native mammals had been eliminated would gradually have followed a severe drought. However, in the new situation there were two barriers to this process. First, much of the previously suitable habitat may simply have disappeared through degradation and destruction by stock and rabbits. Secondly, when recolonization was possible, stock were being pushed back onto the land or rabbits were pouring back in from the refuges that they had occupied at the expense of the native species. In short, each drought saw fewer and fewer refuges containing native species, because the probabilities of local disappearance were impossibly high under such conditions. Thus, astonishingly rapid regional collapses often took place as predicted on theoretical grounds by Hanski (1985).

Two other potential forces were at work — altered fire patterns and introduced predators (Fig. 3). Before looking at their role in this story, however, I want to examine the objections to the notion that blame for the extinctions can be laid completely at the feet of rabbits and grazing stock. This notion is not a new one by any means (Finlayson 1958, 1961; Newsome 1971; Newsome & Corbett 1977; Low 1984, 1986; Caughley 1986; Johnson *et al.* 1989) although I have tried to show in more detail than my predecessors why such a conclusion seems inevitable. So. why hasn't the conclusion been universally accepted?

One objection is that several extinctions took place outside the range of grazing stock and the rabbit, and especially in parts of the rabbit's range, particularly in the north-west, where rabbits were very sparsely distributed. All recent maps of the range indeed show that the rabbit occurs sparsely in the north-western arid zone (Fig. 4). However, I don't believe this fact to be inconsistent with the proposition that the rabbit was the major cause of extinctions beyond the pastoral country. This part of the arid zone would be expected to display the distributional pattern described earlier most dramatically, for in the sandy deserts that constitute this region the soils are universally poor, and fertile pockets appear to be sparser and smaller. Thus, according to the model developed in Figures 2 and 3, the mammals of this region were more vulnerable than almost anywhere else on the continent.

Further, it seems likely to me that the rabbit behaved in this region in exactly the same way as it did elsewhere in the arid zone. That is, it arrived in areas of pristine habitat, exploded in numbers, depleted the resource, and then occupied those areas either sporadically or at much lower densities than before. Few non-Aboriginal observers are known to have been present in these deserts at the time of the arrival of the rabbit, and so in my opinion we may have under-estimated the impact of rabbits.

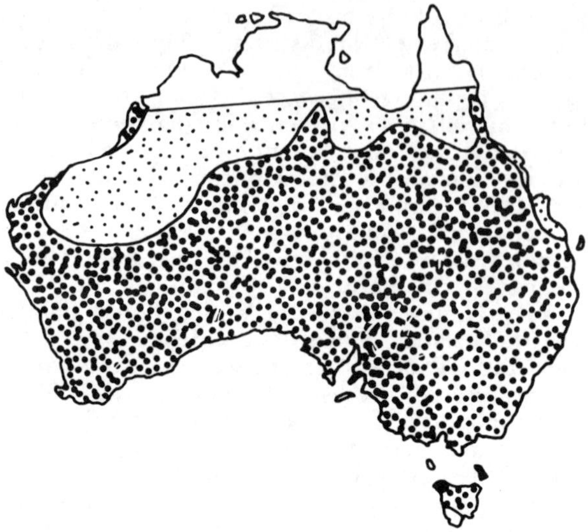

FIG. 4. The distribution of the rabbit (after Strahan 1983). The shaded area may be occupied in years of favourable rainfall.

In short, I think it likely that the degree of subtle habitat change in these northern regions has so far been unappreciated and that the rabbit has damaged the relatively tiny pockets of dependable and nutritious vegetation precisely as it did further south. Of course, it is also possible that other forces were more important in those northern areas (see below).

A second objection is that the rabbit never really occupied certain habitats from which some native species have become extinct. This objection does not appear to be fatal to the model shown in Figures 3 and 4 because during wetter years the species dependent upon particularly nutritious food supplies would be able to inhabit a much broader swathe of country. The critical point is that the populations would retreat from such areas during droughts into the secure and dependable habitats, just as rabbits themselves do today. Thus, records of extinct native animals from areas not permanently occupied by rabbits are to be expected.

Thirdly, if the impact of rabbits and stock was so severe, how did they manage to coexist with certain native species for such extended periods of time in some places? My interpretation is that coexistence occurred within the largest and most dependable patches of habitat in any particular region — in the very cream of the country. In these places, the resource was of such quality that several droughts may have passed by before the native animals disappeared. Consequently, native species may have persisted for considerably longer than in the less favoured parts of their range.

These interpretations are supported by the observation that the range of the rabbit matches almost perfectly the composite geographic distribution of those mammals that have become extinct or declined dramatically (Fig. 4). In addition, I am impressed by the fact that the spectacled hare-wallaby, *Lagorchestes conspicillatus,* and the northern nailtail wallaby, *Onychogalea unguifera,* whose relatives have suffered so badly in the arid zone, have persisted reasonably well since European settlement (see Strahan 1983). They have done so despite being subject to all the pressures that have been invoked to explain extinctions elsewhere — all, that is, except one. I suggest that they have survived because the heart of their distributions lies to the north of the country occupied by the rabbit.

Element 4: introduced predators and altered patterns of fire

I concluded earlier that it seemed unlikely on a variety of grounds that the cat and fox could be held responsible *by themselves* for the extinctions. It seems highly likely to me that the waves of rabbits carried with them unusually high densities of predators which almost certainly contributed to local disappearances, or even actually caused them in some places. Corbett and Newsome (1987) showed how dingoes switch from rabbits to other prey after the collapse of a plague, and it is easy to imagine how such focussing of predation could have placed intolerable pressure upon small colonies of native species. However, I believe that the impact of predation has been over-emphasized simply because it is so obvious. Habitat destruction is by far the most common cause of rarity in the world today and, given that we have here an obvious and widespread agent of such change in the introduced herbivores, it seems most likely that the predators were a subsidiary although important contributor to the extinctions. Thus, Figure 3 includes these animals as a secondary force.

Similarly, the final few paragraphs above which completed the section on Element 3 suggest that it is unnecessary to invoke altered patterns of fire as a *sole* explanation for the extinctions throughout the spinifex grasslands. This does not mean, however, that altered patterns of fire have not had important synergistic effects with the impact of rabbits, as is the case with introduced predators and rabbits. My model is based on the notion that the major problem facing herbivores is the location of dependable supplies of nutritious plant growth. An immediate impact of burning in these infertile landscapes is to release nutrients from persistent tissues of perennial plants and thereby to produce a flush of relatively nutritious and digestible growth. Thus, it is conceivable that Aboriginal mosaic-burning increased the dependability of certain types of habitats, expanded the sizes of suitable patches, and provided pathways for recolonization of dependable patches. If this were so, then the departure of Aboriginal people could well have resulted in a decline in the area of suitable habitat and increased probabilities of local disappearance. In the most northerly regions where rabbits are especially sparse, e.g., the Great Sandy Desert, such effects may well have been sufficient to cause the disappearance of medium-sized species. However, the underlying force remains the restriction of these native mammals to refuge habitats and their consequent exposure to the disturbances set in train by Europeans. My conclusions are, in general, similar to those reached by Burbidge and McKenzie (1989).

Historical and ecological evidence

Evidence from the reptiles and birds

The model described here is consistent with the lack of impact of European settlement on the reptiles and birds. Reptiles are metabolically worlds apart from mammals as they can remain inactive for extended periods and reduce energy expenditure dramatically. The energetic costs of reptiles are tiny compared to endothermic vertebrates (Nagy 1987) and it seems likely that these reduced costs allow them to be universally successful in deserts because it enables them to persist through dry times (Pianka 1986). In addition, very few reptiles are herbivorous, and so the removal of certain types of plants by rabbits and stock would have less immediate effects upon them. In short, reptiles are not limited to the same degree as mammals by infertility and unpredictable drought, and so their populations are less liable to extinction.

Birds experience the same metabolic problems as mammals, but escape two of the difficulties that made mammals so vulnerable. No birds are herbivorous in the sense that they eat green vegetation and so there was no direct conflict with rabbits and stock. Further, their dispersal abilities are in general far greater than those of mammals, and their ability to recolonize after a drought is consequently better. Thus, birds probably experienced less frequent local disappearance following the arrival of rabbits and stock, and were able to repopulate any areas from which they were excluded much more readily than mammals.

Evidence from the mammals

Complete testing of the model is required, but there is insufficient space here to collate all the available information. My purpose is merely to note some evidence that shows the model to be at least plausible.

First, the distribution of red kangaroos, *Macropus rufus,* in central Australia is highly patchy, and the animals congregate on well-watered fertile country where forage is produced with above-average regularity (Newsome 1975). The red kangaroo is the largest macropodid and, other factors being equal, it would be expected to suffer least from the limitations imposed by weight-specific nutritional and metabolic needs. Despite this, Newsome (1980) has documented the fact that only a tiny proportion of the country is sufficiently productive to support populations during extended droughts. Thus, this example seems completely consistent with the model.

Secondly, analysis of the distributions of the herbivorous stick-nest rats, *Leporillus conditor* and *L. apicalis,* reveals that they were dependent upon stony or calcareous outcrops and other types of country where nutritional quality and dependability of plant foods were enhanced (Copley 1988). The first species now survives only on one island off the southern coast and the second is gone forever. This case also seems consistent with the model.

Thirdly, the plains rat kangaroo, *Caloprymnus campestris,* appears to have been confined to the fringes of claypans in the Lake Eyre basin (Finlayson 1932), where I would predict that the supply of nutritious vegetation was originally surprisingly dependable even though thc rainfall is exceptionally low and erratic. In this case, the supply of food may have been made more dependable by the input of moisture from the huge river systems; nevertheless, the outcome is in line with the model.

Fourthly, Finlayson's (1958) remarkable account of the burrowing bettong is perhaps the best test of the model that is available. He noted that the bettong's "numbers fluctuate greatly and its occurrence is local and discontinuous and not uniform. Warrens housing a big population during one season may be found quite deserted the next" (p. 242). His descriptions of the habitats particularly favoured by this species are generally in line with the theory advanced here.

Finally, Spencer's (1896) narrative of the Horn Expedition through central Australia is characterized by a marked paucity of records of mammals. Indeed, at one point Spencer exclaims "Mammals both here (in the MacDonnell Ranges) and elsewhere were very difficult indeed to obtain" (p. 109). The narrative mentions only two or three colonies of burrowing bettongs, although one such colony within the giant drainage basin of Lake Amadeus was apparently quite populous. The overall paucity of these types of mammals and their confinement to certain places in the landscape are as predicted on the basis of the model.

Conclusion

The model developed here has the potential to explain major patterns in the response of the Australian vertebrate fauna to European settlement. It seems to account for the fact that the reptiles have not suffered any extinctions and that the birds have been affected to a much lesser extent than the mammals. It apparently is able to tell us why the mammals of the arid zone bore the brunt of the extinctions and what factors predisposed certain types of mammals to regional decline. It focuses attention back onto habitat alteration, often subtle, and lays the blame principally (although not entirely) with the rabbit. In my opinion, we still do not fully comprehend the extent of the changes wrought in inland Australia by this most unfortunate of all the European introductions.

However, this chapter is not just meant to be an exercise in historical ecology. There is currently a considerable amount of interest in the reintroduction of several of the native mammals discussed in this chapter to parts of their former ranges (e.g., Delroy *et al.* 1986). This is an admirable objective from a variety of perspectives, not the least being a desire to increase the number of surviving populations so as to reduce the chances of yet another extinction. However, if the ecological viewpoint advanced here is correct, then reintroductions will only be successful if the habitats into which the animals are released are of exceptionally high quality in terms of the dependability and nutritional character of their plant growth. In the short term, it is obviously of great importance to control the introduced predators that can destroy small populations. Nevertheless, such populations will only be viable in the long term if they are situated in the uncommon pockets of relatively fertile, moist country that are scattered across the arid zone. The identification and management of these refuges are among the greatest challenges faced by ecologists in inland Australia today.

Acknowledgements

This chapter is dedicated to H. H. Finlayson. Without his extraordinary voluntary efforts earlier this century we would know little today of what life used to be like for the mammals of inland Australia.

I have substantial intellectual debts to those people who have helped in the search for understanding, although not all of them necessarily agree with my opinions. I acknowledge Bert Main and Alan Newsome for their important writings, and am indebted to Alex Baynes for stimulating me to think more deeply about these issues. The background to the model presented here was developed in collaboration with Mark Stafford Smith and Graham Griffin, and the paper springs directly from their ideas. I am indebted to a local community of people who care about what has happened and who have helped form the concepts expressed here; they are Dick Kimber, Mike Fleming, Peter Latz, Bill Low, Ken Johnson, Anne Kerle, Penny van Oosterzee, Rick Southgate and Dave Gibson. Finally, thanks to Norm McKenzie,

Darrell Kitchener, Alan Newsome, Steve Cork, Denis Saunders and Barry Fox for their helpful comments on this chapter.

References

Angelstam P., Hansson L. & Pehrsson S. (1987) Distribution borders of field mice *Apodemus:* the importance of seed abundance and landscape composition. *Oikos* **50,** 123-30.

Archer M. (1974) New information about the quaternary distribution of the thylacine (Marsupialia, Thylacinidae) in Australia. *J. Roy. Soc. W. Aust.* **57,** 43-50.

Batzli G. O. (1985) Nutrition. In: *Biology of New World Microtus* (ed. R. H. Tamarin) pp. 779-811. Special publication of the American Society of Mammalogists, 2.

Bayly C. P. (1976) Observations on the food of the feral cat *(Felis catus)* in an arid environment. *S. Aust. Nat.* **51,** 22-4.

Bayly C. P. (1978) A comparison of the diets of the red fox and the feral cat in an arid environment. *S. Aust. Nat.* **53,** 20-8.

Bell F. C. (1979) Precipitation. In: *Arid-land Ecosystems: Structure, Functioning and Management* (eds D. W. Goodall and R. A. Perry) pp. 51-72. Cambridge University Press, Cambridge.

Bell R. H. V. (1982) The effect of soil nutrient availability on the community structure in African ecosystems. In: *Ecology of Tropical Savannas* (eds B. J. Huntley & B. H. Walker) pp. 193-216. Springer-Verlag, Berlin.

Bennett S. (1983) A review of the distribution, status and biology of the plains-wanderer *Pedionomus torquatus* Gould. *Emu* **83,** 1-11.

Blakers M., Davies S. J. J. F. & Reilly P. N. (1984) *The Atlas of Australian Birds.* Melbourne University Press, Melbourne.

Bolton B. L. & Latz P. K. (1978) The western hare-wallaby, *Lagorchestes hirsutus* (Gould) (Macropodidae), in the Tanami Desert. *Aust. Wildl. Res.* **5,** 285-93.

Braithwaite R. W. & Estbergs J. A. (1985) Fire patterns and woody vegetation trends in the Alligator Rivers Region of northern Australia. In: *Ecology and Management of the World's Savannas* (eds J. C. Tothill & J. J. Mott) pp. 359-64. Australian Academy of Science, Canberra.

Braithwaite R. W. & Werner P. A. (1987) The biological value of Kakadu National Park. *Search* **18,** 296-301.

Braithwaite R. W., Winter J. W., Taylor J. A. & Parker B. S. (1985) Patterns of diversity and structure of mammalian assemblages in the Australian tropics. *Aust. Mammal.* **8,** 171-86.

Brown J. H. (1978) The theory of insular biogeography and the distribution of boreal birds and mammals. *Great Basin Nat. Memoirs.* **2,** 209-27.

Brown J. H. & Kodric-Brown A. (1977) Turnover rates in insular biogeography: effect of immigration on extinction. *Ecology* **58,** 445-9.

Brown J. H. & Maurer B. A. (1987) Evolution of species assemblages: effects of energetic constraints and species dynamics on the diversification of the North American avifauna. *Amer. Nat.* **130,** 1-17.

Burbidge A. A. (1985) Fire and mammals in hummock grasslands of the arid zone. In: *Fire Ecology and Management of Western Australian Ecosystems* (ed. J. R. Ford) pp. 91-4. Environmental Studies Group Report No. 14, Western Australian Institute of Technology, Perth.

Burbidge A. A. & Fuller P. J. (1979) Mammals of the Warburton Region, Western Australia. *Rec. West. Aust. Mus.* **8,** 57-73.

Burbidge A. A., Johnson K. A., Fuller P. J. & Southgate R. I. (1988) Aboriginal knowledge of the mammals of the central deserts of Australia. *Aust. Wildl. Res.* **15,** 9-39.

Burbidge A. A. & McKenzie N. L. (1989) Patterns in the modern decline of Western Australia's vertebrate fauna: causes and conservation implications. *Biol. Conserv.* **50,** 143-98.

Calaby J. H. (1963) Australia's threatened mammals. *Wildlife* **1,** 15-8.

Calaby J. H. (1971) Man, fauna and climate in Aboriginal Australia. In: *Aboriginal Man and Environment in Australia* (eds D. J. Mulvaney & J. Golson) pp. 80-93. Australian National University Press, Canberra.

Campbell B. M. (1986) Plant spinescence and herbivory in a nutrient poor ecosystem. *Oikos* **47,** 168-72.

Catling P. C. (1988) Similarities and contrasts in the diets of foxes, *Vulpes vulpes,* and cats, *Felis catus,* relative to fluctuating prey populations and drought. *Aust. Wildl. Res.* **15,** 307-17.

Caughley G. (1986) Rangelands, livestock and wildlife, the ecological equivalent of sulphur, saltpetre and charcoal. In: *Rangelands: A Resource Under Siege* (eds P. J. Joss, P. W. Lynch & O. B. Williams) p. 545. Australian Academy of Science, Canberra.

Charley J. L. & Cowling S. W. (1968) Changes in soil nutrient status resulting from overgrazing and their consequences in plant communities of semi-arid areas. *Proc. Ecol. Soc. Aust.* **3,** 28-38.

Coley P. D., Bryant J. P. & Chapin F. S. (1985) Resource availability and plant antiherbivore defense. *Science* **230,** 895-9.

Coman B. J. & Brunner H. (1972) Food habits of the feral house cat in Victoria. *J. Wildl. Mgmt.* **36,** 848-53.

CONCOM (1988) *List of Australian endangered vertebrate fauna.* Mimeo, Council of Conservation Ministers, Canberra.

Cooke B. D. (1977) Factors limiting the distribution of the wild rabbit in Australia. *Proc. Ecol. Soc. Aust.* **10,** 113-20.

Copley P. (1988) *The Stick-nest Rats of Australia.* South Australian Department of Environment and Planning, Adelaide.

Corbett L. K. (1985) Morphological comparisons of Australian and Thai dingoes: a reappraisal of dingo status, distribution and ancestry. *Proc. Ecol. Soc. Aust.* **13,** 277-91.

Corbett L. K. & Newsome A. E. (1987) The feeding ecology of the dingo. III. Dietary relationships with widely fluctuating prey populations in arid Australia: an hypothesis of alternation of predation. *Oecologia* **74,** 215-27.

Crisp M. D. (1978) Demography and survival under grazing of three Australian semi-desert shrubs. *Oikos* **30,** 520-8.

Crisp M. D. & Lange R. T. (1976) Age structure, distribution and survival under grazing of the arid-zone shrub *Acacia burkittii. Oikos* **27,** 86-92.

Croft J. D. & Hone L. J. (1978) The stomach contents of foxes, *Vulpes vulpes,* collected in New South Wales. *Aust. Wildl. Res.* **5,** 85-92.

Curry P. J. (1986) Habitat characteristics of the thick-billed grasswren *Amytornis textilis* in grazed shrublands in Western Australia. In: *Rangelands: A Resource Under Siege* (eds P. J. Joss, P. W. Lynch & O. B. Williams) p. 566. Australian Academy of Science, Canberra.

Delroy L. B., Earl J., Radbone I., Robinson A. C. & Hewett M. (1986) The breeding and re-establishment of the brush-tailed bettong, *Bettongia penicillata,* in South Australia. *Aust. Wildl. Res.* **13,** 387-96.

den Boer P. J. (1981) On the survival of populations in a heterogeneous and variable environment. *Oecologia* **50,** 39-53.

Ehmann H. (1981) The natural history and conservation of the bronzeback (*Ophidiocephalus taeniatus* Lucas and Frost) (Lacertilia, Pygopodidae). In: *Proceedings of the Melbourne Herpetological Symposium* (eds C. B. Banks & A. A. Martin) pp. 7-13. Zoological Board of Victoria, Melbourne.

Ehmann H. & Cogger H. (1985) Australia's endangered herpetofauna: a review of criteria and policies. In: *Biology of Australasian Frogs and Reptiles* (eds G. Grigg, R. Shine & H. Ehmann) pp. 435-47. Royal Zoological Society of N.S.W., Sydney.

Finlayson H. H. (1932) *Caloprymnus campestris*. Its recurrence and characters. *Trans. R. Soc. S. Aust.* **50,** 148-67.

Finlayson H. H. (1958) On central Australian mammals (with notice of related species from adjacent tracts). Part III — the Potoroinae. *Rec. S. Aust. Mus.* **13,** 235-302.

Finlayson H. H. (1961) On central Australian mammals. Part IV — the distribution and status of central Australian species. *Rec. S. Aust. Mus.* **14,** 141-91.

Foran B. D. (1986) The impact of rabbits and cattle on an arid calcareous shrubby grassland in central Australia. *Vegetatio* **66,** 49-59.

Foran B. D., Low W. A. & Strong B. W. (1985) The response of rabbit populations and vegetation to rabbit control on a calcareous shrubby grassland in central Australia. *Aust. Wildl. Res.* **12,** 237-47.

Fox B. J. & Pople A. R. (1984) Experimental confirmation of interspecific competition between native and introduced mice. *Aust. J. Ecol.* **9,** 323-34.

Friend J. A., Fuller P. J. & Davis J. A. (1982) The numbat in central Australia. *SWANS* **12(3),** 21-6.

Gentilli J. (1971) The main climatological elements. In: *Climates of Australia and New Zealand* (ed. J. Gentilli) pp. 119-88. Elsevier, Amsterdam.

Griffin G. F. (1984) Climate, vegetation, man and fire. In: *Anticipating the inevitable: a patch-burn strategy for fire management at Uluru (Ayers Rock–Mt. Olga) National Park* (ed. E. C. Saxon) pp. 7-12. CSIRO, Melbourne.

Griffin G. F. & Friedel M. H. (1985) Discontinuous change in central Australia: some implications of major ecological events for land management. *J. Arid Environ.* **9,** 63-80.

Hall E. A. A., Specht R. L. & Eardley C. M. (1964) Regeneration of the vegetation on Koonamore Vegetation Reserve, 1926-1962. *Aust. J. Bot.* **12,** 205-64.

Hanski I. (1985) Single-species spatial dynamics may contribute to long-term rarity and commonness. *Ecology* **66,** 335-43.

Harrington G. N., Wilson A. D. & Young M. D. (eds) (1984) *Management of Australia's Rangelands.* CSIRO, Melbourne.

Haynes C. D. (1985) The pattern and ecology of *munwag:* traditional Aboriginal fire regimes in north-central Arnhemland. *Proc. Ecol. Soc. Aust.* **13,** 203-14.

Heathcote R. L. (1965) *Back of Bourke.* Melbourne University Press, Melbourne.

Jackson E. A. (1957) Soil features in arid regions with particular reference to Australia. *J. Aust. Inst. Agric. Sci.* **23,** 196-208.

Jarman P. J. (1974) The social organization of antelope in relation to their ecology. *Behaviour* **48,** 215-68.

Johnson K. A., Burbidge A. A. & McKenzie N. L. (1989) Australian macropods: status, causes of decline, and future research and management. In: *Kangaroos, Wallabies and Rat-kangaroos* (eds I Hume, G. Grigg & P. Jarman) pp. 641-57. Surrey Beatty & Sons, Chipping Norton.

Jones E. & Coman B. J. (1981) Ecology of the feral cat *Felis catus* (L.) in south-eastern Australia. I. Diet. *Aust. Wildl. Res.* **8,** 537-47.

Jones F. W. (1925). *The Mammals of South Australia. Part III. The Monodelphia.* Government Printer, Adelaide.

Kimber R. (1983) Black lightning: Aborigines and fire in central Australia and the Western Desert. *Archaeology in Oceania* **18,** 38-45.

Kimber R. G. (1985) The history of the malleefowl in central Australia. *Royal Australasian Ornithologists Union, Newsletter* **64,** 6-8.

King D. R., Oliver A. J. & Mead R. J. (1981) *Bettongia* and fluoroacetate: a role for 1080 in fauna management. *Aust. Wildl. Res.* **8,** 529-36.

Kinnear J. E. & Main A. R. (1979) Niche theory and macropodid nutrition. *J. Roy Soc. W. Aust.* **62,** 65-74.

Kitchener D. J., Chapman A., Muir B. G. & Palmer M. (1980) The conservation value for mammals of reserves in the Western Australian wheatbelt. *Biol. Conserv.* **18,** 179-207.

Lande R. (1987) Extinction thresholds in demographic models of territorial populations. *Amer. Nat.* **130,** 624-35.

Lange R. T. & Graham C. R. (1983) Rabbits and the failure of regeneration in Australian arid zone *Acacia. Aust. J. Ecol.* **8,** 377-81.

Lange R. T. & Purdie R. W. (1976) Western myall *(Acacia papyrocarpa),* its survival prospects and management needs. *Aust. Rangel. J.* **1,** 64-9.

Latz P. K. & Griffin G. F. (1978) Changes in Aboriginal land management in relation to fire and to food plants in central Australia. In: *The Nutrition of Aborigines in Relation to the Ecosystem of Central Australia* (eds B. S. Hetzel & H. J. Frith) pp. 77-85. CSIRO, Melbourne.

Leeper G. W. (1970) Climates. In: *The Australian Environment* (ed. G. W. Leeper) pp. 12-20. CSIRO, Melbourne.

Low W. A. (1984) Interactions of introduced and native mammals in the arid regions of Australia. In: *Arid Australia* (eds H. G. Cogger & E. E. Cameron) pp. 301-13. Australian Museum, Sydney.

Low W. A. (1986) The disappearance of mammal species from pastoral areas of central Australia: a land management problem? In: *Rangelands: A Resource Under Siege* (eds P. J. Joss, P. W. Lynch & O. B. Williams) pp. 551-2. Australian Academy of Science, Canberra.

Low W. A., Muller W. J. & Dudzinski M. L. (1980) Grazing intensity of cattle on a complex of rangeland communities in central Australia. *Aust. Rangel. J.* **2,** 76-82.

Lundelius E. L. Jr. (1983) Climatic implications of late Pleistocene and Holocene faunal associations in Australia. *Alcheringa* **7,** 125-49.

Martensz P. N. (1971) Observations on the food of the fox, *Vulpes vulpes* (L.), in an arid environment. *CSIRO Wildl. Res.* **16,** 73-5.

Mattson W. J. Jr. (1980) Herbivory in relation to plant nitrogen content. *Ann. Rev. Ecol. Sys.* **11,** 119-61.

Morton S. R. (1979) Diversity of desert-dwelling mammals: a comparison of Australia and North America. *J. Mammal.* **60,** 253-64.

Morton S. R. & Baynes A. (1985) Small mammal assemblages in arid Australia: a reappraisal. *Aust. Mammal.* **8,** 159-69.

Myers K. (1971) The rabbit in Australia. In: *Dynamics of Populations* (eds P. J. den Boer & G. R. Gradwell) pp. 478-503. Centre for Agricultural Publishing and Documentation, Wageningen, Netherlands.

Myers K. & Parker B. S. (1975) A study of the biology of the wild rabbit in climatically different regions in eastern Australia. VI. Changes in numbers and distribution related to climate and land systems in semiarid north-western New South Wales. *Aust. Wildl. Res.* **2,** 11-32.

Nagy K. A. (1987) Field metabolic rate and food requirement scaling in mammals and birds. *Ecol. Monogr.* **57,** 111-28.

Newsome A. E. (1971) Competition between wildlife and domestic livestock. *Aust. Vet. J.* **47,** 577-86.

Newsome A. E. (1975) An ecological comparison of the two arid-zone kangaroos of Australia, and their anomalous prosperity since the introduction of ruminant stock to their environment. *Quart. Rev. Biol.* **50,** 389-424.

Newsome A. E. (1980) The eco-mythology of the red kangaroo in central Australia. *Mankind* **12,** 327-33.

Newsome A. E. & Corbett L. K. (1975) Outbreaks of rodents in semi-arid and arid Australia: causes, preventions, and evolutionary considerations. In: *Rodents in Desert Environments* (eds I. Prakash & P. K. Ghosh) pp. 117-53. Junk, The Hague.

Newsome A. E. & Corbett L. K. (1977) The effects of native, feral and domestic animals on the productivity of the Australian rangelands. In: *The Impact of Herbivores on Arid and Semi-Arid Rangelands* (ed. Australian Rangeland Society) pp. 331-56. Australian Rangeland Society, Perth.

Newsome A. E., Corbett L. K., Catling P. C. & Burt R. J. (1983) The feeding ecology of the dingo. I. Stomach contents from trapping in south-eastern Australia, and the non-target wildlife also caught in dingo traps. *Aust. Wildl. Res.* **10,** 477-86.

Owen-Smith N. & Cooper S. M. (1987) Palatability of woody plants to grazing ruminants in a South African savanna. *Ecology* **68,** 319-31.

Parer I. & Libke J. A. (1985) Distribution of rabbit, *Oryctolagus cuniculus,* warrens in relation to soil type. *Aust. Wildl. Res.* **12,** 387-405.

Parra R. (1978) Comparison of foregut and hindgut fermentation in herbivores. In: *The Ecology of Arboreal Folivores* (ed. G. G. Montgomery) pp. 205-29. Smithsonian Institute Press, Washington, D. C.

Payne A. L., Curry P. J. & Spencer G. F. (1987) An inventory and condition survey of rangelands in the Carnarvon Basin, Western Australia. *Technical Bulletin, Western Australian Department of Agriculture* **73,** 1-478.

Pianka E. R. (1986) *Ecology and Natural History of Desert Lizards.* Princeton University Press, Princeton, New Jersey.

Prins H. H. T. & Weyerhaeuser F. J. (1987) Epidemics in populations of wild ruminants: anthrax and impala, rinderpest and buffalo in Lake Manyara National Park, Tanzania. *Oikos* **49,** 28-38.

Pulliam H. R. (1988) Sources, sinks, and population regulation. *Amer. Nat.* **132,** 652-61.

Robertshaw J. D. & Harden R. H. (1985) The ecology of the dingo in north-eastern New South Wales. II. Diet. *Aust. Wildl. Res.* **12,** 39-50.

Rolls E. C. (1969) *They All Ran Wild.* Angus & Robertson, Sydney.

Ryan G. E. & Croft J. D. (1974) Observations on the food of the fox, *Vulpes vulpes* (L.), in Kinchega National Park, Menindee, N.S.W. *Aust. Wildl. Res.* **1,** 89-94.

Saunders D. A., Arnold G. W., Burbidge A. A. & Hopkins A. J. M. (eds) (1987) *Nature Conservation: The Role of Remnants of Native Vegetation.* Surrey Beatty & Sons, Chipping Norton.

Seagle S. W. & Shugart H. H. (1985) Faunal richness and turnover on dynamic landscapes: a simulation study. *J. Biogeog.* **12,** 499-508.

Singh G. (1981) Late Quaternary pollen records and seasonal palaeoclimates of Lake Frome, South Australia. *Hydrobiologia* **82,** 419-30.

Smith A. T. (1974) The distribution and dispersal of pikas: consequences of insular population structure. *Ecology* **55,** 1112-9.

Smith G. T. (1979) The status of Australian parrots. In: *The Status of Endangered Australasian Wildlife* (ed. M. J. Tyler) pp. 101-8. Royal Zoological Society of South Australia, Adelaide.

Smith M. J. & Medlin G. C. (1982) Dasyurids of the northern Flinders Ranges before pastoral development. In: *Carnivorous Marsupials* (ed. M. Archer) pp. 563-72. Royal Zoological Society of New South Wales, Sydney.

Spencer W. B. (1896) Through Larapinta Land: a Narrative of the Horn Expedition to Central Australia. In: *Report of the Work of the Horn Scientific Expedition to Central Australia. Part 1.* (ed. W. B. Spencer) pp. 1-136. Melville, Mullen and Slade, Melbourne.

Stamps J. A., Buechner M. & Krishnan V. V. (1987) The effects of edge permeability and habitat geometry on emigration from patches of habitat. *Amer. Nat.* **129,** 533-52.

Strahan R. (ed.) (1983) *Complete Book of Australian Mammals.* Angus & Robertson Publishers, Sydney.

White T. C. R. (1978) The importance of a relative shortage of food in animal ecology. *Oecologia* **33,** 71-86.

Whitehouse S. J. O. (1977) The diet of the dingo in Western Australia. *Aust. Wildl. Res.* **4,** 145-50.

Wiens J. A., Crawford C. S. & Gosz J. R. (1985) Boundary dynamics: a conceptual framework for studying landscape ecosystems. *Oikos* **45,** 421-7.

Wild A. (1958) The phosphate content of Australian soils. *Aust. J. Agric. Res.* **9,** 193-204.

Williams C. M. & Raupach M. (1983) Plant nutrients in Australian soils. In: *Soils: An Australian Viewpoint* (ed. CSIRO Division of Soils) pp. 777-94. CSIRO Melbourne and Academic Press, London.

Wood D. H. (1984) The rabbit (*Oryctolagus cuniculus* L.) as an element in the arid biome of Australia. In: *Arid Australia* (eds H. G. Cogger & E. E. Cameron) pp. 273-87. Australian Museum, Sydney.

Wood J. G. (1936) Regeneration of the vegetation on the Koonamore Vegetation Reserve, 1926 to 1936. *Trans. R. Soc. S. Aust.* **60,** 96-111.

Proc. Ecol. Soc. Aust. 1990 — 16:215-220

Uluru (Ayers Rock-Mt Olga) National Park: an assessment of a fire management programme

G. ALLAN[1] and L. BAKER[2]

[1]*CSIRO Division of Wildlife and Ecology, P.O. Box 2111, Alice Springs, N.T. 0871*
Current address: CCNT, P.O. Box 1046, Alice Springs, N.T. 0871

[2]*ANPWS, Uluru National Park, P.O. Box 119, Yulara, N.T. 0872*

Abstract

A fire management strategy for Uluru (Ayers Rock-Mt Olga) National Park was developed by CSIRO in 1982. The strategy aims to re-introduce a mosaic of patch burns, similar to those imposed by traditional Aboriginal people, that will disrupt the flow of the inevitable wildfires in spinifex grasslands. After six years of fire management we have assessed the ability of the patch burns to inhibit the spread of wildfires and to promote habitat diversity. An effective mosaic of patch burns within the Park for wildfire protection has not yet been created. The management burns have been too few and too small. However, the patches are effectively increasing localized species richness and Park biotic diversity by providing a variety of post-fire successional states in the vegetation. The Park managers now have the opportunity to improve the use of fire as an effective management tool.

Introduction

Uluru (Ayers Rock-Mt Olga) National Park is situated in central Australia, 350 km south-west of Alice Springs. The Park is internationally recognized as an important conservation area with designations as a World Heritage property and as a Biosphere Reserve representing arid lands. It covers an area of 132 566 hectares and is located on and surrounded by Aboriginal-owned land. Uluru National Park has been co-operatively managed by traditional Aboriginal owners and the Australian National Parks and Wildlife Service (ANPWS) through the Uluru-Kata Tjuta Board of Management since 1985. Throughout this chapter the Pitjantjatjara orthography as adopted within the Uluru (Ayers Rock-Mt Olga) National Park Plan of Management (ANPWS 1986) has been used.

The co-operative nature of the Park's management is reflected in the fire management programme being carried out by a team of Park staff and Aboriginal traditional owners. The programme is based on the fire management strategy developed by the Commonwealth Scientific and Industrial Research Organization (CSIRO) for Uluru National Park (Saxon 1984). It was designed to re-establish a fire regime within the Park similar to that practised by Aboriginal people prior to European settlement. Historically, the region was actively managed by Aboriginal people through the use of fire. The displacement of the Aboriginal people from their homelands after the arrival of European people in central Australia restricted the opportunities for traditional burning practices and for the last 50 years fire has not been used as a management tool in the Uluru area. As a result the new fire regime in the region is dominated by infrequent large scale wildfires (Griffin *et al.* 1983). In 1976, lightning started a wildfire near Uluru which burnt an area of 20 000 km^2, including 76% of the Park (Allan 1984). The wildfire followed three years of above average rainfall which contributed to a build up of fuel throughout the Park. In an effort to prevent future recurrences of the 1976 wildfire, CSIRO began the development of the fire management strategy in 1982 on behalf of the Conservation Commission of the Northern Territory (CCNT).

The strategy focused on the fire-prone spinifex grasslands that are the dominant vegetation type on the sandplain and dunefield land systems within and surrounding Uluru National Park. The sandplain and dunefield areas have been divided between those areas dominated by hard spinifex *(Triodia basedowii)* and the remainder which is dominated by soft spinifex *(T. pungens)*. Locally important non-spinifex areas of the Park surround the geologic features of Uluṟu and Kata Tjuṯa and are composed of communities of open woodlands and *Acacia* shrublands (Fig. 1). The aim of the strategy is to reinstitute a mosaic of small-area post-fire aged vegetation communities, to protect the non-spinifex communities which may suffer undesirable compositional change from too frequent burning and to promote the diversity of flora and fauna habitats throughout the Park. The main basis of the strategy is the strategic burning of patches in the spinifex grasslands to create areas of low fuel which will interrupt and attenuate the flow of large wildfires.

After six years of conducting management burns within Uluṟu National Park an assessment of the programme has been undertaken. This chapter will examine the process of using fire as a management tool to inhibit the spread of wildfires and promote habitat diversity.

Methods

Three data sets were used for our assessment of the fire management programme at Uluṟu National Park. These were:

a) an update of the fire history map,
b) sampling the vegetation species in the range of fire ages in both spinifex communities to measure changes in species richness and frequency through time since fire, and
c) a regional assessment of the fire mosaic in the spinifex communities within the Park and immediately adjacent to the Park using satellite imagery.

Fire history update

The fire history update of Uluṟu National Park was prepared in December 1987. The update shows the size and location of all fires within the National Park between 1982 and 1987, including management burns and wildfires. The fires were mapped from two sets of aerial photographs, 1986 verticals and supplementary 70 mm obliques taken in November 1987. The fire history update was prepared at 1:100 000 to match the fire history map prepared as part of the development of the fire management strategy (Allan 1984).

Vegetation sampling

Twelve sample sites were selected to sample the range of post-fire ages in both the *Triodia pungens* and *T. basedowii* spinifex communities within the Park. Six sites were located in each community (Fig. 1). The fire ages sampled in the *T. pungens* sites were pre 1950, 1976, 1982, 1985, 1986, and 1987 whereas the ages in the *T. basedowii* sites were 1950, 1976 (two sites), 1982, 1984 and 1986. The pre 1950 fire age is estimated to be 63 years, which corresponds to a period of numerous wildfires in the early 1920s (Griffin, pers. comm.). The ages chosen depended on availability within the two spinifex types. Further sampling of other fire ages was not possible due to the absence of wildfires and the lack of a fire management programme prior to 1982.

The vegetation was sampled using the technique of previous surveys at Uluṟu National Park (Griffin & Nelson, pers. comm.). For each site, all species in an area of 20 m radius were recorded for a series of sample points at 50 m intervals along a 500 m transect. For each spinifex community the data collected were analysed separately because *Triodia pungens* and *T. basedowii* communities react differently after burning (Griffin, pers. comm.).

The data were compared for species richness and species frequency between sites. Site versus species frequency was transformed (1+log (freq)) and a measure of dissimilarity between each of the sites was derived with the PATN analysis package (Belbin 1987) using the Bray-Curtis metric. The measure of dissimilarity was then plotted against the age difference in years for each pair of sites.

Regional assessment of the fire mosaic

The regional assessment of the fire mosaic was designed to compare the mosaic within the Park with a similar area immediately north of the Park. The mosaic was assessed using two dates of Landsat images (30 August 1985 and 24 November 1987) to determine changes through time associated with the fire management strategy within the Park.

Using the microBRIAN image processing system two Landsat 5 MSS satellite images (WRS 104-77E) were rectified to the Australian Map Grid at 100 m pixel resolution. The 7 000 km^2 area was divided into the Park and non-park components for separate image processing (Fig. 1). Principal components analysis (PCA) was used to reduce the dimensionality of the data (Kauth & Thomas 1976) and to separate the areas of change between the two image dates (Byrne *et al.* 1980; Richards & Milne 1984). A PCA image was created for both 1985 and 1987, and a third PCA image was created from a combined 1985-1987 image of channels 2 and 4 from each date. The four channel processing limitation of the microBRIAN system precluded an 8 channel PCA of the multitemporal images (Ingebritsen & Lyon 1985).

Results

Fire history update

The fire history update (Fig. 1) provided an assessment of the progress made in implementing the fire management strategy. It has also provided a perspective on the size and nature of the few wildfires which occurred within the Park. Within the Park the management burns included the

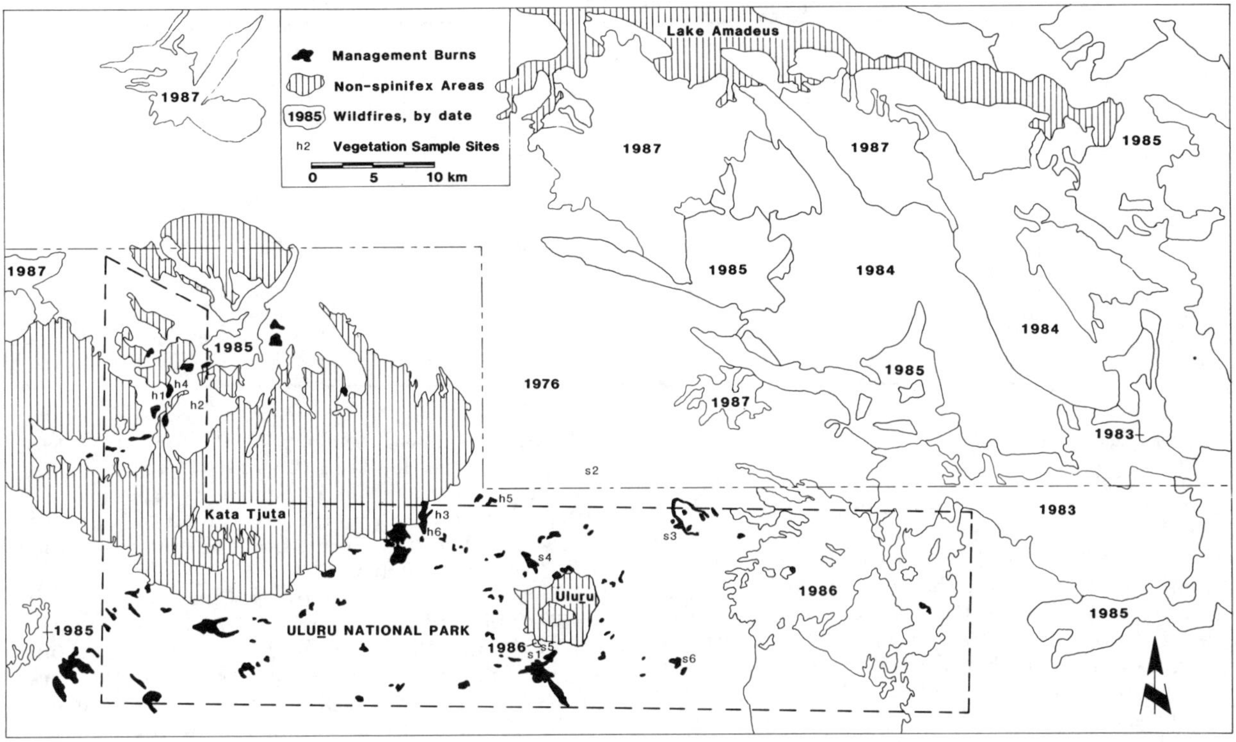

FIG. 1. Fire mosaic map.

series of experimental fires lit by CSIRO in 1982 and 1983 as part of the development of the fire management strategy, the management burns conducted by the CCNT from 1983 to 1985, and the most recent series of management burns by ANPWS since 1985.

Table 1 summarizes the fires in the spinifex communities within Uluṟu National Park during the past six years. The patch size groups were based on the fire management strategy which prescribed the positioning of a large number of patches in the 0.5 to 5 km² patch size range (Saxon & Allan 1984). It is our contention that fires smaller than 0.5 km², even with a linear to elliptical shape to maximize their areal effect, would be too small to modify the progress of a wildfire. We also believe that patches which exceed 5 km² may reduce the probabilities of some components of the fauna moving between the patches. Only one-third of the management burns have created effective patches for wildfire attenuation within the Park.

Vegetation sampling

Figure 2 shows plant species richness plotted against age (years since burnt) for *Triodia pungens* and *T. basedowii.* The trend from higher species richness in younger communities to lower species richness in older communities is reflected in both communities. The highest species diversity occurs about six years after fire, and can be attributed to the presence of annual and short-lived forbs and grasses combined with resprouting perennial shrubs and trees. After six years the species richness rises again in the pre 1950 *T. pungens* community where spinifex becomes senescent and the presence of perennial shrubs plus some grasses and forbs increases.

Figure 3 shows the age difference in years for each pair of *T. pungens* and *T. basedowii* sites plotted against a measure of dissimilarity for the same pairs. The measure indicates the level of association between sites, the higher the value (with a range between 0 and 1) the more dissimilar the sites. Despite the limited number of sites the trend is the same for both communities. The pairs of sites that are recently burnt and have a small age difference are most variable in their dissimilarity. This reflects the importance of time and rainfall as the driving forces in the early stages of recovery after a fire. The pairs of sites with a large age difference, representing a comparison between recently burnt and long unburnt sites, are less variable in their measure of dissimilarity. The oldest sites become dominated by spinfex and a few longer lived shrubs, and are most dissimilar to the more recently burnt sites. This is consistent with the separation of the sites on the basis of species richness in Figure 2.

TABLE 1. Spinifex Fires within Uluru National Park 1982-1987.

Year	Number of fires: Management Burns	Number of fires: Wildfires	Patch size (km^2): <.5	Patch size (km^2): .5-1	Patch size (km^2): 1-5	Patch size (km^2): >5
1982	27	—	15	11	1	—
1983	4	1	4	—	—	1
1984	12	—	12	—	—	—
1985	28	2	23	3	3	1
1986	13	2	10	2	2	1
1987	33	—	26	6	1	—
	117	5	90	22	7	3

Regional assessment of the fire moasic

The regional fire mosaic for both the Park and non-park areas (Fig. 1) is dominated by the 1976 wildfire. The 1985 Landsat image represented the mosaic just before the Park managers increased their activity levels in the fire management programme. Within the Park there had been only a small number of management burns, and most were associated with the development of the fire behaviour model by CSIRO. Beyond the Park boundaries the 1983 to 1985 wildfires were associated with an area unburnt in the 1976 wildfire.

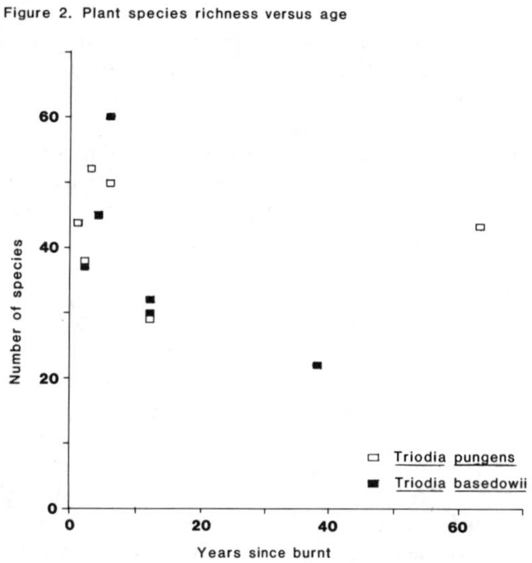

FIG. 2. Plant species richness versus age.

The 1987 Landsat image represents the current fire mosaic. In the area outside the Park, four wildfires in 1987 were associated with areas unburnt in 1976. The four fires burnt an area of 330 km^2, with an average size of 82 km^2. From the pattern of wildfires between 1983 and 1987 we feel that lightning will not yet be effective in starting fires in the fuel loads of the *Triodia basedowii* communities north of the Park which were burnt in 1976. Further work on the assessment of fuel loads in the spinifex grasslands using Landsat reflectance data is underway at CSIRO.

Within the Park area two wildfires occurred in 1986. The first burnt a small area of pre 1950 *T. pungens* immediately south of Uluru but the second burnt a large area in the eastern portion of the Park. It has been the only wildfire within the area burnt in 1976. This exception was a function of its location within the floodout of Britten-Jones Creek and its spinifex type. The fuel loads in the *T. pungens* community had built up to a higher level than the surrounding area because of extra soil moisture in the creek system following a large rainfall event in 1983.

The mosaic of wildfires in the region dwarfs the scale of the management burns within the Park. The small size of the management burns makes them difficult to detect accurately with image processing routines. The principal component analysis discriminated the wildfires and larger management burns in excess of 1 km^2 but failed to accurately detect the smaller fires.

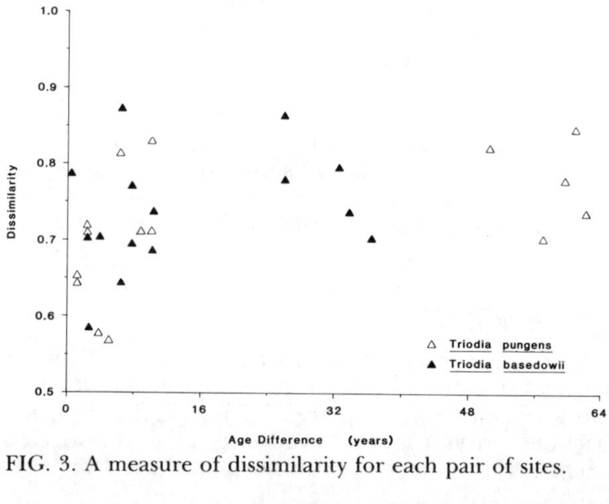

FIG. 3. A measure of dissimilarity for each pair of sites.

Discussion

The use of fire as a management tool has progressively increased over the past six years but its effects have not yet met the aims of the fire management strategy. An effective mosaic of small patches in a range of post-fire ages and successional states has not been created but significant advances have been made.

The patches burnt by the Park managers from 1984 to 1987 have been small due to low fuel loads in the 1976 spinifex communities and burning only when the weather factor of the fire behaviour model (Griffin 1984) is low and the spread of fires is limited. Despite their size many of the patches are aligned in a NE to SW direction to maximize their influence on wildfires from the NW. Fires have also been lit adjacent to roads and tracks to increase the impeding effect of these features to wildfires. The burning programme has shown an annual increase in the number of fires as the Park managers are becoming more confident with the use of fire and the accuracy of the fire behaviour model.

Although it is not the intention of the fire management strategy to burn large areas, the fire history update and the regional assessment of the fire mosaic has shown that the number of management burns and the total area burnt must increase dramatically if the strategy to disrupt the flow of wildfires is to be successful. We feel that the current mosaic of management burns and wildfires would be ineffective in modifying the progress of a large wildfire because there are too few patches, most patches are too small, and the relative position of each patch is not complementary. It is also important that a large number of management burns are positioned within the area burnt by the 1976 wildfire before that area begins to carry wildfires.

The small number of management burns has also limited the opportunity for the diversity of habitats within the Park to be enhanced. The vegetation sampling indicated that habitat diversity within the spinifex grasslands can be increased by the creation of a variety of post-fire aged communities. The vegetation data supports the notion that sites in different successional states are botanically different and that the local diversity is enhanced where a range of patch states exist. The trend for plant species richness and diversity to increase after fire and then decline as the community matures, is also shown by the results of work done over a larger area of central Australia (Griffin, pers. comm.). Similar trends have been noted for various vegetation communities throughout the world (Fox & Fox 1986). To complement the vegetation diversity, Griffin and his fellow workers (pers. comm.) and Morton *et al.* (1988) found that each group of reptiles and birds react in different ways to different post-fire successional states in the vegetation. Some species display a preference for early post-fire states and others for later post-fire states, therefore more species will be abundant in an area that contains a greater diversity of patches in different post-fire successional states.

This assessment of fire management at Uluṟu National Park has been of great value to the Park managers. It has been recognized that the use of fire must increase dramatically in order to meet the aims of the fire management strategy. Managers have gained confidence in the implementation of the strategy and realize the necessity to use appropriate weather conditions more effectively to burn larger patches. With several recent seasons of good rainfall the fuel load throughout the region has increased and a major effort will be put into the fire management programme in the near future. It will also be combined with other work focussed on gaining more insight into the traditional fire practices in the region before the arrival of European settlers. The traditional owners have been consistently involved in the fire management programme at Uluṟu National Park and continue to provide valuable insight into traditional practices of land management. A continuing assessment of the fire programme will help improve the use of fire, provide a stimulus for further research and improve land management practices.

Acknowledgements

The authors would like to thank A. Gee and B. Nesbitt (ANPWS), D. Nelson (CSIRO) and G. Leach (N.T. Herbarium), for their assistance with various stages of the vegetation survey. We would particularly like to thank G. Griffin (CSIRO), for help with data analysis and review. Thanks also to B. Nesbitt and M. Chuk (ANPWS), for providing editorial comments on the manuscript. Final thanks are due to the CCNT for their support of fire research and image processing and to ANPWS for permission to use the fire history update.

References

Allan, G. E. (1984) Fire History and Land Units of Uluṟu. In: *Anticipating the Inevitable: A Patch-burn Strategy for Fire Management at Ulurṟu (Ayers Rock-Mt Olga) National Park* (ed. E. C. Saxon) pp. 13-24. CSIRO Australia, Melbourne.

ANPWS (1986) *Uluṟu (Ayers Rock-Mt Olga) National Park Plan of Management.* Uluṟu Kata Tjuṯa Board of Management and Australian National Parks and Wildlife Service, Canberra.

Belbin, L. (1987) *PATN: Pattern Analysis Package, Reference Manual.* CSIRO Division of Wildlife and Rangelands Research, Canberra.

Byrne G. F., Crapper P. F. & Mayo K. K. (1980) Monitoring land-cover change by principal component analysis of multitemporal Landsat data. *Remote Sensing of Env.* **10,** 75-184.

Fox B. J. & Fox M. D. (1986) Resilience of animal and plant communities to human disturbance. In: *Resilience in Mediterranean-type Ecosystems* (eds B. Dell, A. J. M. Hopkins & B. B. Lamont), pp. 39-64. Junk, Dordrecht.

Griffin G. D. (1984) Patch-burn Management: Timing. In: *Anticipating the Inevitable: A Patch-burn Strategy for Fire Management at Uluru (Ayers Rock-Mt Olga) National Park* (ed. E. C. Saxon), pp. 79-87. CSIRO Australia, Melbourne.

Griffin G. F., Price N. F. & Portlock H. F. (1983) Wildfires in the central Australian rangelands, 1970-1980. *J. Environ. Manage.* **17,** 311-23.

Ingebritsen S. E. & Lyon R. J. P. (1985) Principal components analysis of multitemporal image pairs. *Int. J. Remote Sensing* **6,** 687-96.

Kauth R. J. & Thomas G. S. (1976) The tasselled cap — a graphic description of the spectral-temporal development of agricultural crops as seen by Landsat. In: *Proceedings of Symposium on Machine Processing of Remotely Sensed Data*, pp. 45-51. LARS, Purdue.

Morton S. R., Gillam M. W., Jones K. R. & Fleming M. R. (1988) Relative efficiency of different pit-trap designs for sampling reptiles in spinifex grasslands. *Aust. Wildl. Res.* **15,** 571-7.

Richards J. A. & Milne A. K. (1984) Mapping Fire Burns and vegetation regeneration by classification of multitemporal Landsat MSS image data. In: *LANDSAT 84 Proc. 3rd Australasian Remote Sensing Conf.* (ed. E. Walker), pp. 304-14. Brisbane.

Saxon E. C. (ed.) (1984) *Anticipating the Inevitable: A Patch-burn Strategy for Fire Management at Uluru (Ayers Rock-Mt Olga) National Park.* CSIRO Australia, Melbourne.

Saxon E. C. & Allan G. E. (1984) Patch-burn Management: Site Selection. In: *Anticipating the Inevitable: A Patch-burn Strategy for Fire Management at Uluru (Ayers Rock — Mt Olga) National Park.* (ed. E. C. Saxon) pp. 69-78. CSIRO Australia, Melbourne.

Proc. Ecol. Soc. Aust. 1990 — 16:221-233

Application of fire ecology to fire management: a manager's perspective

N. PREECE

Conservation Commission of the Northern Territory, P.O. Box 1046, Alice Springs, Northern Territory 0871

Abstract

Successful natural ecosystem management depends on a good understanding of the natural resources of the system as well as good management application skills. Most management to date has been implemented in the absence of adequate ecological data by managers who generally lack ecological training, while few ecologists are directly involved in management. Ecological theory therefore is rarely being applied to management. Managers need operational guidelines and rules for the use of fire in managing ecosystems as well as the best ecological data available to develop management prescriptions.

Over the past four years a fire management manual for central Australian parks and reserves has been developed in order to fulfil this need. The manual provides a collation and synthesis of material on fire behaviour, safety, and procedures and on arid zone fire ecology. While technical information on fire behaviour and safety is generally adequate, information on fire ecology for the central Australian region is inadequate for making informed decisions on management. The guidelines in the manual for planning and managing fire in ecosystems have been developed using the ecological information which was considered by the majority of ecologists working in central Australia to be appropriate. The guidelines are therefore conservative.

The manual is apparently the only one of its kind for the arid and semi-arid zones of Australia. As such it is a starting point for professional management of ecosystems using fire in the arid zone.

Introduction

Need to manage natural ecosystems on the basis of ecological principles

Ecosystem management based on ecological principles is a relatively recent development, despite decades of "control burns" and other management practices on parks and state forests in Australia. Fire is one essential component of the dynamics of many of these ecosystems. Land management authorities now recognize that well considered prescriptions with specific objectives for the application of fire must be introduced to adequately manage natural reserved lands. This has been demonstrated particularly for a number of hot arid and semi-arid lands throughout the world, including Australia (Dobrowolski *et al.* 1986; Edroma 1986; Trabaud 1986; Trollope 1986; Wright 1986; Xaba and Nxumalo 1986).

This chapter discusses some trends in management of fire in Australia, and outlines the scope of a manual which was produced for central Australian parks and reserves. It then discusses some of the problems encountered during the development of the manual, the perceived response to the manual, some comments on translating sometimes complex scientific concepts and data into forms usable by managers, and some suggestions for further work.

Fire — the available data

Despite three decades of scientific study of behaviour of fire in Australia (see e.g., McArthur 1968), the role of fire in natural ecosystem management has lagged behind. While important works such as Luke and MacArthur's (1978) *Bushfires in Australia* improved the situation, they were mostly based on fire control and forestry practices, with scant regard to natural ecosystem ecology. A significant step in rectifying this situation was taken in 1981 when a comprehensive review of biological and ecological aspects of fire

(Gill *et al.* 1981) synthesized much recent knowledge and thinking about fire and its effects on Australian ecosystems.

More recently, fire has been identified as a critical factor in Australian ecosystem dynamics and species survival in such volumes as Barker and Greenslade (1982), Ford (1985), Sattler (1986), and Saunders and Burbidge (1988), and in some of the papers presented in this symposium (Morton; Allan & Baker). Others have recognized the uncertain nature of our understanding of species and ecosystem management, going so far as to say that every management activity is an experiment and should be treated as such (e.g., Recher, Hobbs & Hopkins, this volume).

These considerations are particularly important in the Australian arid zone where some 60% of the original complement of small mammal species and 40% of the total mammal complement are now extinct or under threat (Morton & Baynes 1985; Morton, this volume). A comparable rate of extinction and decline of species has occurred in Western Australia, although the percentage decline appears to be lower (Burbidge 1988). Changed fire regimes, along with other causal factors have been implicated in these extinctions (Griffin *et al.* 1987; Burbidge & McKenzie 1989), and may still be contributing to the decline of species such as mala *(Lagorchestes hirsutus)* (Johnson 1988) and bilby *(Macrotis lagotis)* (Southgate 1987), although the relative importance of changed fire regimes is still a matter of scientific debate (see, for example, Morton, this volume; Burbidge & McKenzie 1989).

Data on the ecological effects of fire in central Australian ecosystems are scarce. The only detailed study of fire in central Australia is of spinifex dunefields and sandplains within Uluṟu National Park, in Saxon (1984).

Less comprehensive work includes documentation of fire histories for the southern Northern Territory (Bolton & Latz 1978; Griffin *et al.* 1988; Allan & Griffin 1986), some important work on the Burt Plain north of Alice Springs (Griffin & Friedel 1984a,b), preliminary work on the Tanami Desert sandplains (Griffin *et al.* 1987), summary work on responses of hummock grasslands of northwestern Australia to fire (Suijdendorp 1981), and reports on fire history and Aboriginal use of fire in the Tanami Desert from Gibson (1986), and from Southgate (1987).

Only twenty or so references specifically on the ecological effects of fire in the Australian arid zone which are relevant to land management were found for the manual introduced in this chapter. Many of the references only peripherally discuss fire and ecology in a way that is useful for the purpose of prescribing fire for ecosystem management. For example, several papers presented to the *Mulga Lands* symposium (Sattler 1986) argued for the benefits of prescribed burning in mulga lands to reduce the incidence of shrub infestation and increase productivity, and some discuss the effects fire has on certain species.

None of the published material available provides detailed guidance on how to actually apply a fire except in general terms with mostly vague notions of when to burn and how frequently (Mills 1986; Booth 1986; Harrington 1986; Start 1986 and Griffin & Hodgkinson 1986). The same observation applies to papers in the proceedings of the International Rangelands Conference (Joss *et al.* 1986) such as those by Fox (1986) and Pressland *et al.* (1986). Certainly none are specific to the central Australian situation.

Changing emphasis on fire management

It is useful in this discussion to review briefly some trends in fire management, particularly as they relate to parks in Australia.

To ensure that management of parks is professional and effective, rules and procedures are needed. They have most commonly been developed for day-to-day functions such as administration, purchasing, law enforcement, vehicle maintenance, and search and rescue, among others.

Ecosystem management, being a more difficult area to deal with, has usually missed the benefits of procedures or instruction manuals. A quick review of a dozen plans of management produced for parks around Australia showed that they have not fulfilled the need for such guidelines, but have either only addressed management issues such as picnic area proposals and rubbish collection, or identified the need for Fire Management Plans or Weed Management Plans or the like.

The basic problem for managers is that general rules cannot apply in all situations, and that prescriptions must be area-specific, but this knowledge does not resolve the problem of not being able to obtain the information necessary to make informed decisions or prepare reliable prescriptions.

The trend around Australia towards developing fire management guidelines, rules or procedures has tended to be technically orientated in most cases. Some examples include the New South Wales National Parks and Wildlife Service, North Metropolitan District Action Fire Plan of 1984, the Nowra/Goulburn District Fire Management Plan of 1984/85, the Victorian Department of Conservation, Forests and Lands Fire Management Techniques Notes, and Fire Control Notes of 1987.

In most cases the manuals are "in-house" documents which almost always address fire control techniques, contact lists for the districts, equipment lists, human resources, and locations of fire trails and water supplies for fire fighting, but rarely ecosystem management. I agree that these issues are important, especially considering the adverse publicity of and coroners' inquests into the alleged incapacity of authorities to protect people and property from wildfire.

Over the last decade or so, however, ecologically based fire management has gained prominence. Examples include the PREPLAN computer modelling system developed for Kosciusko National Park in New South Wales, the first of its kind in Australia (Kessell *et al.* 1982), and modifications of PREPLAN for the Blue Mountains and Morton-Budawang National Parks of New South Wales (Gellie 1987), and the Patch-burn strategy for Uluru National Park (Saxon 1984), among a few others.

Workshops such as the "Bushfire Control or Management by Fire" workshop at the Mt Keira Scout Camp in Wollongong in 1981, the "Third Queensland Fire Research Workshop" held at Queensland Agricultural College, Gatton in 1986 (Roberts 1986), the "Bushfire Management in Natural Areas" workshop held around the Central Coast of New South Wales in 1987 (Conroy 1987), and the workshop on "Ecological Impact and Management of Fire in Northern Australia" held in Alice Springs in 1987 (Morton & Andrew 1987), have addressed management of fire and ecosystems in particular.

The situation in central Australia

At a recent workshop on fire ecology in Northern Australia, researchers and managers recognized and discussed the problems of insufficient data on the importance of fire and the impacts of fire on ecosystems, and the diverse priorities of managers with insufficient resources to undertake all the important tasks required of them. The workshop conclusions were that managers needed to implement management practices on the basis of the data available, and be prepared to modify their thinking as more data surface and knowledge develops:

> *"intervention through some form of prescribed burning (i.e., judicious patch-burning) is the best means of ensuring the long-term persistence of biotic resources in fire-prone environments" (Morton & Andrew 1987).*

The use of the term *judicious* with patch-burning needs some clarification or definition. Prescribed burning can be prescribed for reasons ranging from fuel or hazard reduction for protection of people or buildings and other fixed assets, to manipulation of basic ecological processes by applying fire to manage organisms or communities (Hobbs 1988). Judicious patch-burning will then be determined by the purpose of burning.

Ecologists cited in this chapter have repeatedly emphasized that natural ecosystems need active and professional management if the rate of species extinctions and undesirable, irreversible changes to natural communities are to be slowed or halted. Fire management programmes in central Australia have rarely considered the ecological needs of species or ecosystems. On the contrary, prescribed burns in central Australian parks have been, until five or six years ago, spasmodic at best, sometimes poorly planned, and prescribed mostly for protection of people and property. Where more standardized procedures have been developed to plan for fire on parks they have all lapsed after the instigators departed.

It is widely recognized that much of central Australia was burned regularly by the Aboriginal people of the area for possibly tens of thousands of years (Latz & Griffin 1978; Allan & Griffin 1986; Griffin & Allan 1986; Allan & Baker, this volume). Prescribed patch burns, possibly similar to the patch burns lit by Aboriginal people, appear to be essential to maintain the structure, composition and dynamics of these semi-arid and arid natural ecosystems (Morton & Andrew 1987).

While some data, theories, and expert advice on aspects of central Australian arid and semi-arid zone ecosystem management using fire have been available for about a decade (for example, Latz & Griffin 1978), application of the data and theories, and recourse to the expert advice have not taken effect due to, I suggest, a number of reasons.

1. While some managers have scientific training they lack time to interpret the available data and translate them into practice.
2. Most researchers have limited understanding of the complications of applying theory to practice and extracting the parts of the data important for management.
3. Priorities and interests of managers have mostly been day-to-day essential park maintenance functions.
4. The ecological theories are poorly developed, are too general to be of use, and are subject to significant debate between ecologists.
5. Expertise in prescribed burning has resided in very few people.
6. There are few reliable data on which confident ecosystem management decisions can rely.

Parks and reserves have existed in central Australia since 1924 when Palm Valley was set aside as a reserve. Now major conservation lands in central Australia include Finke Gorge (expanded from Palm Valley Reserve), Watarrka, Simpsons Gap, and Ormiston Gorge National Parks managed by the Conservation Commission of the Northern Territory (CCNT), and Uluru National Park, managed now by the Australian National Parks and Wildlife Service (ANPWS). Most of the parks lie within the MacDonnell, Krichauff and James Ranges (the Central Ranges).

Since the inception of the reserves and parks, management practices have mostly been dictated by the managers' personal endeavour, skills, experience and interests, and have often lacked continuity and consistency.

Fire management in particular has been left mostly to the individual whims of managers. Historically, this management has emphasized protection from wildfire. Where prescribed burning has been tried, it has often been based on experience in temperate and tropical forests and woodlands as modified by trial and error. This experience however is not easily translated because of the great differences in climate, vegetation and other ecological factors. A major difference between the arid and the temperate and tropical areas is the unpredictability of rainfall and thus fuel production.

Coupled with this unsatisfactory management situation in relation to fire and ecosystems is a relatively high turnover of staff, and insufficient transfer of information and expertise, neither of which is confined to central Australia.

Documentation of the natural resources of the parks is of course critical to ecosystem management. It is surprising therefore that there is almost no existing documentation for any of the parks, at least in a form useful for prescribed burning or other ecosystem management.

Managers need also to document carefully considered objectives for a prescribed burning strategy. The success or otherwise of the prescribed burning programme is not otherwise measurable, and no improvements can be made. Objectives of prescribed burning in the past, at least from 1977 to 1984 have included such as those listed below for Devils Marbles Conservation Reserve, Northern Territory.

1. To utilize fire to the greatest advantage in the management of the Devils Marbles Reserve.
2. To protect neighbouring properties and/or the Reserve from wild fires by the installation of fire breaks.
3. To simulate the timing and effect of natural fires deemed beneficial to the Reserve, in the event that such natural fires be hindered by various human factors.

These objectives are valuable as they were defined by the manager of the day and thus provide some insight into the attitudes of the time, even though they may now seem inadequate with hindsight and more recent ecological understanding. Stating management objectives is important if management is to improve.

New developments in central Australia

Applicability of the Uluru patch burn strategy

The patch burn strategy for Uluru National Park is a valuable contribution to management of parks in the central Australian region. It has provided a model for management of spinifex grasslands and will undoubtedly assist in management of other parks in the region. It cannot be directly applied, however, to the other parks. The majority of parks and reserves in central Australia which are managed by the Conservation Commission of the Northern Territory lie within the Central Ranges. While hummock grasslands of spinifex *(Triodia clelandii, T. hubbardii, T. longiceps)* cover a large proportion of these ranges, the species are different from the spinifex at Uluru National Park *(T. pungens, T. basedowii, Plectrachne schinzii)* where detailed studies have been undertaken, and probably exhibit different burning properties and community responses to fire (G. Griffin, pers. comm.).

Vegetation formations found in the Central Ranges include, for example, 18 land units in Simpsons Gap National Park, each with characteristic vegetation (Wood *et al.* 1982), and 81 plant communities identified in the George Gill Range, now Watarrka National Park (Latz *et al.* 1981). Considering that these are the only two vegetation studies done for the Central Ranges [other than the broad land systems of the Alice Springs area undertaken by Perry and others in the 1960s (Perry 1960, 1962)], that there is only limited ecological similarity between the two areas, and with Uluru, and that the other parks in the ranges are on different land systems and have different suites of species, it is perhaps not surprising that prescribing fire with ecologically sound objectives is a difficult task in the ranges.

The other factors which do not allow direct transfer of the research findings and prescriptions for fire from Uluru are the topographic differences, a significant factor in influencing fire behaviour and vegetation responses.

Other developments in central Australia

In addition to the Uluru strategy, two recent developments relevant to central Australian fire and ecology are worth mentioning here. A collaborative fire ecology research project between the CCNT and the Commonwealth Scientific and Industrial Research Organization (CSIRO) in Alice Springs has been underway since 1984 and is nearing completion. The study includes investigation into broad scale fire history, vegetation measurement, fuel load assessment, fire behaviour, satellite imagery analysis, techniques for monitoring from aircraft, and vegetation responses to fire for the arid zone of the Northern Territory (Allan 1984; Griffin & Allan 1984, and others). More detailed work at Uluru and in the Tanami Desert is being carried out by CCNT, ANPWS and CSIRO researchers and is providing valuable data on fire ecology in central Australia.

Second, the MacDonnell Ranges are the focus of a comprehensive collaborative study commenced in 1989 by CCNT and CSIRO which will produce a Geographic Information System (GIS) and predictive modelling modules for a range of management needs, including fire, over the next three years. The study should provide a more precise database for managers and identify the priorities and opportunities for experimental management and monitoring.

The manual

Preparing a fire management manual for central Australia

Recognition of the need for appropriate fire management strategies and practices for central Australian parks and reserves stimulated some managers and researchers to begin preparation of a fire manual in 1984. The objectives were to collate methods of fire management and control suitable to the arid zone, develop planning procedures for management and application of fire, provide safety information and rules, and collate and synthesize the diffuse ecological and technical knowledge on fire applicable to the central Australian region, in order to set up a soundly-based fire management strategy for our parks and reserves.

In preparing the manual, all available published material on central Australian ecology was considered. As the data are limited, material from environments as diverse as Victorian mallee, Kosciusko ash, and Western Australlian jarrah were considered and included in the manual where the authors and external advisors considered them appropriate. Data which were found to be anecdotal, or unquantified for central Australia, were of course rejected. Utilization of data from other environments mainly applied to the sections orientated towards the practical application of fire and fire behaviour. It was mostly impossible to transfer ecological data from elsewhere to the central Australian region.

An important problem for the writers of the manual was that it was difficult to develop definitive prescriptions for ecologically-based fire management from the available literature, a key element of the purposes of the manual. The reviewed data ranged from complex multivariate analyses to broad statements on vegetation responses, few of which were directly useful for managers of the ecosystems under consideration. Many discussions about the data with ecologists from the CSIRO Centre for Arid Zone Research at Alice Springs, and from CCNT resolved only sufficient differences of opinion to formulate what I consider excessively broad guidelines for managers, leaving managers who may or may not be expert in fire management to determine prescriptions for fire in the absence of adequate knowledge.

While I appreciate the difficulties of making definitive statements and producing precise prescriptions, as a manager I am also aware that management practices are being implemented regardless of the level of ecological knowledge of the systems being managed. This is not a new observation and is at least one of the stimuli for the recent conferences and symposia which have addressed these issues (Roberts 1986; Sattler 1986; Conroy 1987; Saunders *et al.* 1987; Saunders & Burbidge 1988; and this conference).

Managers need at least a secure position from which to manage, a broad-brush set of guidelines which establish what is allowable, desirable and essential in ecosystem management. A "critical data set" (Morton & Andrew 1987) appears to be the best approach, so that while we as managers may not be making the absolutely best decisions on management practices, we will not at the same time be making decisions based on false assumptions with consequences which are irreversibly undesirable. The manual has attempted to provide this critical data set.

The manual

The manual which we have produced provides a practical tool and comprehensive resource on fire in central Australia. It is also proving to be a valuable resource for training.

To ensure that it was a practical solution, managers were involved in the preparation and review stages. Two draft editions were distributed for use and testing by managers, to obtain constructive criticism from their practical observation and experience. Many comments were received and consequent changes made. The manual has now undergone external review and is due for publication shortly.

The contents of the manual are detailed in Table 1.

From an ecosystem management point of view, some parts of the manual are of more immediate interest than others. I will discuss aspects of some briefly.

Fire management plans

The plans required of managers include the natural and man-made resources of the park in the form of maps and overlays, or documented where not mappable. They also include fire histories and maps of last fires. Objectives of prescribed burning on the parks are to be documented and a broad strategy for prescribed burns over a series of two to three years outlined. In addition, details on staff and other resources are required.

TABLE 1. Contents of the Fire Management Manual for Central Australian Parks and Reserves.

Fire Management Manual

Central Australian Parks and Reserves

Conservation Commission of the Northern Territory
1989

CONTENTS PAGE

PREFACE ..... (i)
1.0 INTRODUCTION .....
1.1 PURPOSE OF THE MANUAL 1
1.2 OBJECTIVES 1
1.3 DEFINITIONS 2
1.4 REASONS FOR PRESCRIBED BURNING 2
1.5 FIRE & PEOPLE 4
1.5.1 Factors affecting people 4
1.5.2 Community attitudes 4
1.6 SUMMARY 5
2.0 ADMINISTRATIVE & LEGAL REQUIREMENTS. 6
2.1 LEGISLATIVE & POLICY BASE 6
2.1.1 Bushfires Act (1980) 6
2.1.2 Fire Service Act 6
2.1.3 Territory Parks & Wildlife Conservation Act 7
2.1.4 By-laws under T.P & W.C. Act 7
2.1.5 Policy 8
2.2 FIRE MANAGEMENT COMMITTEE FUNCTION 8
2.3 PRESCRIBED BURNING APPROVALS 8
3.0 PLANNING & PRE-BURNING PREPARATIONS . 9
3.1 INTRODUCTION 9
3.1.1 Calendar of activities 9
3.2 FIRE MANAGEMENT PLANS 9
3.2.1 Contents .....
3.2.2 Map Preparation .....
3.3 AERIAL SURVEY .....
3.4 GROUND CHECKING OF PROPOSALS .....
3.4.1 Weather .....
3.4.2 Slope .....
3.4.3 Fuel cover and patchiness .....
3.4.4 Control line planning .....
3.4.5 Burning technique assessment .....
3.5 MONITORING PREPARATION AND IMPLEMENTATION .....
3.6 CONTROL LINE PREPARATION .....
3.7 EQUIPMENT CHECK .....
3.7.1 Special equipment note .....
3.8 PREPARATION & SUBMISSION OF PROPOSALS .....
3.8.1 Submissions .....
3.8.2 Approvals .....
3.8.3 Implementation .....
3.9 LANDHOLDER LIAISON .....
4.0 IMPLEMENTATION .....
4.1 PERMITS .....
4.1.1 Fire Protection Zones .....
4.1.2 Fire Danger Period; Fire Ban Area .....
4.2 OTHER NOTIFICATIONS .....
4.3 VISITOR SAFETY .....
4.4 MINIMUM CREW SIZE .....
4.5 WEATHER .....
4.5.1 To burn or not to burn .....
4.5.2 Fuel moisture .....
4.5.3 Weather reading & recording .....

CONTENTS PAGE

4.0 IMPLEMENTATION — *continued*
4.6 RECORDING FIRE BEHAVIOUR .....
4.7 BURNING TECHNIQUES .....
4.7.1 Ground ignition .....
4.7.2 Aerial ignition .....
4.8 MOPPING UP .....
4.9 FIELD ASSESSMENT OF RESULTS .....
5.0 POST-BURNING ACTION .....
5.1 MAPPING .....
5.2 REPORTS .....
5.3 PLOT MONITORING .....
5.4 EQUIPMENT CHECK & REPAIR .....
5.5 POST-FIRE DISCUSSIONS .....
6.0 SAFETY .....
6.1 INTRODUCTION .....
6.2 EQUIPMENT AND YOUR PERSONAL SAFETY
6.2.1 Dress .....
6.2.2 Water .....
6.2.3 Tools .....
6.2.4 Fire Line Safety .....
6.2.5 Vehicles .....
6.2.6 Aircraft .....
6.2.7 Survival .....
6.2.8 Threats .....
6.3 PUBLIC SAFETY .....
6.4 BASIC FIRST AID FOR FIRE MANAGERS .....
6.4.1 Basics .....
6.4.2 Essential rules of First Aid .....
6.4.3 Recovery position .....
6.4.4 Bleeding .....
6.4.5 Fractures .....
6.4.6 Burns .....
6.4.7 Shock .....
6.5 CALLING FOR HELP .....
6.5.1 Situations .....
6.5.2 Common Sense .....
6.5.3 Radios .....
6.5.4 Time for response .....
6.5.5 Maintenance of casualty .....
APPENDICES
APPENDIX A FORMS
A.1 Fire histories A.1
A.2 Prescribed Burning Proposal A.2
A.3 Permit to Burn (Bushfires Act) A.3
A.4 Aerial incendiary burn indemnity (Pastoral Stations) A.4
A.5 Park Fire Report A.5
A.6 Site monitoring and Vegetation Record sheets A.6
APPENDIX B WEATHER TABLES
B.1 Relative humidity B.1
B.2 Dew-point table B.2
B.3 Beaufort wind velocity scale B.3

Table 1 — *continued*

CONTENTS PAGE

APPENDIX C

TECHNICAL INFORMATION

C.1 MAPPING & VEGETATION ASSESSMENT
- C.1.1 Vegetation Record Sheet — Def's and method C.1.1
- C.1.2 Crown separation and % crown cover ratio etc. C.1.2
- C.1.3 Percentage cover estimation chart C.1.3
- C.1.4 Mapping techniques C.1.4
  - Map 1: Soils, Geology, and Land Systems
  - Map 1: Vegetation
  - Map 2: Features and facilities
  - Map 3: Fire histories
  - Map 4: Last fire

C.2 EQUIPMENT
- C.2.1 Kit contents — vehicle, first aid, chainsaw C.2.1
- C.2.2 Drop-on unit C.2.2

C.3 AIRCRAFT STATISTICS
- C.3.1 Aviation spirit and jet fuel guidelines . C.3.1
- C.3.2 Pilot Limitations C.3.2
- C.3.3 Aircraft Statistics C.3.3
- C.3.4 Airspeed conversion chart C.3.4
- C.3.5 Helicopter safety and use C.3.5
- C.3.6 Ground to air signals and v.v C.3.6

C.4 PLANT (HEAVY EQUIPMENT) STATISTICS
- C.4.1 Fuel and oil per hour operation C.4.1
- C.4.2 Weight of typical plant items C.4.2

C.5 PHONETIC ALPHABET C.5.0

C.6 FIRE CONTROL TECHNIQUES
- C.6.1 Control lines C.6.1
- C.6.2 Fireline construction times C.6.2
  - — by hand C.6.2
  - — by machinery C.6.2
- C.6.3 Backburning C.6.3
  - — Backburning worksheet C.6.3
- C.6.4 Rate of spread/area/perimeter table .. C.6.4

APPENDIX D

FIRE ECOLOGY AND TRENDS

- D.1 Fire and ecological effects in Central Australia D.1
- D.2 Trends in fire management D.4
  - — remote sensing, computer and fire behaviour . modelling D.4

CONTENTS PAGE

APPENDIX E

FIRE BEHAVIOUR

E.1 WEATHER FACTORS E.2
- E.1.1 Relative humidity and dew point E.2
- E.1.2 Temperature E.3
- E.1.3 Wind velocity E.4
- E.1.4 Rainfall E.5
- E.1.5 Lightning E.5

E.2 FUELS E.6
- E.2.1 Size, arrangement, degree of curing E.6
- E.2.2 Measurement E.8

E.3 SLOPE, ASPECT E.11

E.4 FLAME & FIRE BEHAVIOUR E.12
- E.4.1 Flame behaviour E.12
- E.4.2 Fire front characteristics E.12
- E.4.3 Spotting, fire spread E.14
- E.4.4 Residence and burn-out time E.14
- E.4.5 Fuel consumption E.14

E.5 RATE OF SPREAD E.15

E.6 INTENSITY E.17

APPENDIX F

REFERENCES, FURTHER READING

LIST OF FIGURES
- Fig. 4.1 Fire Protection Zone, Alice Springs .....
- Fig. 4.2 Common Ignition patterns .....
- Fig. 6.1 Approach to and departure from helicopters .
- Fig. 6.2 Recovery position for first aid .....
- Fig. 6.3 Shock management .....
- Fig. C.1.4 Last fire map C.1.4
- Fig. E.1 Combustion Triangle E.2
- Fig. E.2 R.H. & Dew Pt. relationships E.3
- Fig. E.3 Rainfall, Thunderdays, fire numbers E.6
- Fig. E.4 Fuel sampling block E.9
- Fig. E.5 Wheel point E.10
- Fig. E.6 Flame behaviour E.12
- Fig. E.7 Potential scorch heights E.13

List of Tables not otherwise listed in contents
- Table D.1 Characteristics of plant survival D.2
- Table E.1 Flame heights, pasture condition and rates of spread E.13
- Table E.2 Derived shape factors of fires in grasslands E.17

Objectives of fire management as stated in the manual are:

1. To protect human life and capital assets;
2. To maintain the full range of natural communities and species in a vigorous condition by providing diversity in fire patch sizes, ages and distribution; and,
3. To reduce the chance of major portions of parks and reserves being burnt in wildfires.

There are of course potential conflicts in attempting to meet all these objectives. The purpose of planning is to resolve those conflicts.

Prescribed burning proposals

Managers are required to submit proposals for all prescribed burning to the Fire Management Committee and Regional Manager on the form shown in Table 2. The form was designed to cater for all types of prescribed burning and includes a section on purposes (or objectives) of the proposed burns. It is also designed to be simple and quick to complete, while obtaining essential information for assessing a proposal.

Reporting

An example of the Park Fire Report is shown in Table 3. The format is similar to many forms I

TABLE 2.

PRESCRIBED BURNING PROPOSAL APPENDIX A2

(To be completed for all proposals, in as much detail as is reasonably possible, for submission to F.M.C.).

PARK or RESERVE: .... PROPOSAL NO.: P/..................

REPORTING OFFICER: .... DATE:................

PROPOSED PRESCRIBED BURN LOCATION

AIR PHOTO NAME:...................... MAP NAME: ....

DATE:.............. RUN:.............. GRID REF:......—...... (E-N) SCALE: ....

PHOTO NO.: .... OR LAT:............ LONG:............

EAST (mm): .... NORTH (mm): .... (Map at 1:50,000 or 1:100,000)

ESTIMATED AREA: ....(ha)

ASPECT:

SLOPE: From:............. degrees in .... portion (NE, SW, etc.)
To:............. degrees in .... portion

CONTROL LINE: — TYPE ....

WIDTH: .... PREPARATION REQUIRED ALL/PART (WHERE?) .. /NIL

POTENTIAL PROBLEM AREAS: ....

....

DOMINANT FUEL TYPES:

SPECIES (if known)	% COVER FUEL SAMPLE NO.:
SPINIFEX	
BUNCH GRASSES	
OTHER (SPECIFY)	

LIST DOMINANT PLANT SPECIES

....

....

....

....

FIRE STRATEGY

PROPOSED IGNITION POINTS — SKETCH BELOW

DESIRED WEATHER:

WIND DIR.:.................... VEL.:.................... km/hr

MAXIMUM TEMPERATURE:.........................

TIME OF IGNITION: .... **PROPOSED WEEK:** ....../....../......

CREW SIZE (MINIMUM 3): .... **TIME TO COMPLETE:** ....

TIME TO REACH FIRE: .... HOURS — HELICOPTER/WALK/DRIVE?

NOTES: (Special features, rare species etc.)

....

....

....

PURPOSE OF BURN: ....

....

....

SCALE (metres)

Checked: .... __/__/8_ Approved:............................ __/__/8_ have seen and used, and is almost the same as the one used in central Australia until 1984. The important difference between this one and the others is that the background to the information required on the form is explained in some detail in the manual. In other words, users can gain understanding of the purpose of filling in the forms.

Other differences include more space for weather data, and a simplified layout for recording data.

Monitoring preparation

I have observed that enthusiasm for monitoring waxes and wanes both within individuals according to their work priorities, and between individuals who differ in the skills required to undertake monitoring. Much effort has been wasted on monitoring ecosystems where the one setting up the monitoring programme has either lacked all the necessary knowledge to monitor adequately and has not sought the knowledge, has failed to set up an adequate records system so that the data are lost, or has moved to another area without providing the new managers with the information to continue the monitoring programme.

The manual provides monitoring forms and methods, plus information on the value and pitfalls of monitoring, and guidelines on how to set up monitoring programmes. The Site Monitoring Sheet provided in the manual (Table 4) and the Vegetation Record Sheet (also provided in the manual) were designed considering the principles later outlined by Hopkins (1988) and utilizing data sheets in McDonald *et al.* (1984).

Fire ecology and trends

The section on ecological aspects of fire is relatively small and general. This leaves a gap in the information base available to managers to determine appropriate fire management strategies for their parks and reserves. There are a number of reasons for this inadequacy which I have discussed above. Recommended priorities for further work are provided below.

Conclusions and recommendations

Ecosystem management requires balancing a diverse range of day-to-day and long-term priorities against decreasing resources available to respond to those priorities. The intention of the writers of the fire management manual was to assist managers in planning and applying fire to the ecosystems they manage by simplifying procedures while ensuring that the procedures were determined on the basis of reliable facts. These procedures along with sufficient information on the reasons for applying fire to manage ecosystems, and current information on a wide range of matters of practical importance to managers are provided in the manual.

The Fire Management Manual addresses the broad range of prescribed burning purposes, while focussing on the need for ecologically based prescribed burning for ecosystem management. Managers are asked specifically to consider their objectives in prescribing fire.

It is clear from the experience gained from development of the manual, from the review of the literature and from lengthy and detailed discussions with researchers and managers alike that much more work is needed on fire ecology in the central Australian region.

Management of ecosystems in central Australia requires urgent research in critical areas including responses of particular species and communities to various fire regimes, the desirable frequencies and intensities of prescribed fires in a number of ecosystems, species and communities requiring urgent attention due to their vulnerability, and methods of applying fire to manipulate ecosystems to a desired state.

Management of natural areas is continuous and some of the decisions on management may be detrimental to the conservation of species or ecosystems. Without accurate data, it is impossible to know whether or not those decisions are right.

Coupled with this is the need to develop fire models such as those developed for Uluṟu National Park (Saxon 1984) and Kosciusko National Park (Kessell *et al.* 1982). Natural resource data also need to be produced for each management area, especially vegetation maps. Some of this could be done by managers, including testing the models developed for Uluṟu for their applicability in the ranges.

Apart from some minor aspects, particularly the effects on fire behaviour of slope, fire behaviour is reasonably well covered in the literature and in the manual, as are the practical operational aspects.

Managers of the parks are now using the manual and relying on the information presented in it. It has been well-received both for its practical application and its ecological and planning content.

I feel the manual is a worthwhile product for several reasons. It provides a practical example of how research and management can be integrated, even with limited data available. It provides a starting point and framework for professional fire management in central Australia, and it collates a broad range of data of relevance to managers and ecologists alike.

TABLE 3.

PARK FIRE REPORT APPENDIX A5

Note: Complete in duplicate immediately after any fire on or near a park or reserve. Keep a copy in the district and forward one to the Regional Office, with map attached.

PARK NAME: ....

PROPOSED PRESCRIBED BURN NO.: .... FIRE NUMBER: ....

OFFICER-IN-CHARGE of burn: ....

Reporting Officer: ....

A. FIRE DETAILS .... Date of Burn:........./........./.........

Time of Fire: From:................... hours to .... hours.

Cause of Fire: ....

Location of Fire:

MAP NAME:................................ AIR PHOTO NAME: ....

Grid Ref: .....–..... (E-N) Scale: .... DATE: ....RUN: ....

or: Lat.: .... Long.: .... PHOTO NO.: ....
EAST (mm): ...... NORTH (mm): ......

MAP on 1:50,000 or 1:100,000 map and show final edge, areas unburnt within perimeter and ignition points. For planning and if no map available, use back of form to sketch approximate location of fire.

B. EFFECTS ON PLANT COMMUNITIES (complete table):

	WOODLAND	SCRUB	SPINIFEX	OTHER (specify)
DOMINANT SPECIES (most frequent or obvious)				
TOTAL AREA BURNT (ha.) (incl. unburnt patches)				
% OF TOTAL AREA BURNT				

C. CONTROL EFFORT

No. of people actively involved: .... Total staff-hours: ....
Aircraft used (type): .... Total flying time: .... hrs
Equipment used: ....

D. WEATHER

Rainfall: Last Rain: Date....../....../...... Amount: .... (mm)

Total rain last .... months .... (mm)

Time (24 hr)	Cloud %	Temp. C Dry	Temp. C Wet	R.H. %	WIND Ave. (km/hr)	WIND Max. (km/hr)	WIND Dir. (N.E. etc.)	FLAME HT. (m)	FLAME ANG deg.	FLAME ROS m/s km/h
PRE-FIRE a.m. .... hrs.										
p.m .... hrs.										
FIRE IGNITION TIME: .... hrs.				#	#	#	#	#	#	#
DURING BURN										
.... hrs.										
.... hrs.										
.... hrs.										
.... hrs.										
.... hrs.										
POST BURN .... hrs.										

E. NOTES FIRE NUMBER: ....

(Any other relevant information, e.g., success, problems, accidents, equipment failure, unexpected behaviour).

....

....

....

....

....

....

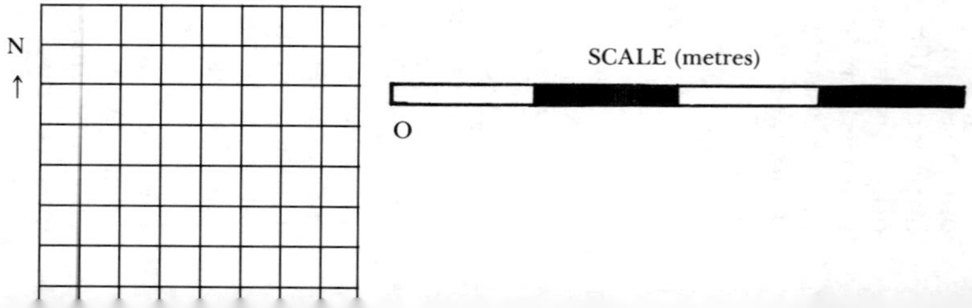

TABLE 4.

SITE MONITORING SHEET APPENDIX A.6

PARK NAME: .... SITE NO.:..........................
(std. N.T. No., see VEG. SHEET)

Vegetation information should be recorded on the attached VEGETATION RECORD SHEET. This sheet is to be used to record details of the fire history and monitoring history relevant to the site. Complete details should be included to keep a running history of the site. Copies should be kept in the district or park, in the regional office, and on file to ensure the information is not lost.

MONITORING HISTORY

DATE SURVEYED	SURVEYOR	COPY of record sheets: DISTRICT	REGION	FILE

SITE PLOTTED ON:

MAP NAME: .... AIR PHOTO NAME:........................

Grid Ref.:......–...... (E-N) Scale: .... DATE:............ RUN:...........

or: Lat.: .... Long.:.............. PHOTO NO.: ....

EAST (mm) .... NORTH (mm) ....

COPY in DISTRICT (Y/N); REGION (Y/N); FILE (Y/N)?

FIRE HISTORY

DATE (or season & year) OF FIRE	FIRE NUMBER (Code: see Ch.3)	PRESCRIBED BURN OR WILDFIRE

COMMENTS ....

....

PHOTOGRAPHIC RECORDS (from fixed point)

DATE OF PHOTO	AZIMUTH	LENS	STORED AT: (file a print copy)

**ATTACH VEGETATION RECORD SHEET

Acknowledgements

I thank Grant Allan, Penny van Oosterzee and an anonymous reviewer for their constructive criticisms of this paper. The work would not have been possible without the support of the Conservation Commission of the Northern Territory and the many people who worked in some way on the manual; for this I am grateful.

References

Allan G. E. (1984) Fire history and land units of Uluru. In: *Anticipating the inevitable: a patch-burn strategy for fire management at Uluru (Ayers Rock-Mt Olga) National Park* (ed. E. C. Saxon) pp. 13-24. CSIRO Australia, Melbourne.

Allan G. E. & Griffin G. F. (1986) Fire ecology of the Hummock Grasslands of Central Australia. *Proc. 4th Aust. Rangel. Soc.* Armidale, New South Wales.

Barker W. R. & Greenslade P. J. M. (eds) (1982) *Evolution of the Flora and Fauna of Arid Australia.* Peacock Publications, Frewville.

Bolton B. L. & Latz P. K. (1978) The Western Hare-wallaby, *Lagorchestes hirsutus* (Gould) (Macropodidae), in the Tanami Desert. *Aust. Wildl. Res.* **5,** 285-93.

Booth C. A. (1986) Management of Woody Weeds for Sustainability of the Ecosystem. In: *The Mulga Lands* (ed. P. S. Sattler) pp. 84-9. Proc. Roy. Soc. Qld., Brisbane.

Burbidge A. A. (1988) The why and how of managing biological resources. In: *Ecological Theory and Biological Management of Ecosystems* (eds D. A. Saunders & A. A. Burbidge) pp. 9-14. Occasional Paper No. 1/88. Dept. Cons. and Land Management, Western Australia.

Burbidge A. A. & McKenzie N. L. (1989) Patterns in the Modern Decline of Western Australia's Vertebrate Fauna: Causes and Conservation Implications. *Biol. Conserv.* **50,** 143-98.

Conroy B. (1987) Bushfire management planning in natural areas. In: *Bushfire Management in Natural Areas* (ed. B. Conroy) pp. 1-9. N.S.W. National Parks and Wildlife Service, Turramurra, Sydney.

Dobrowolski J. P., Blackburn W. H. & Grelen H. E. (1986) Effects of 20 years of prescribed burning on infiltration rates, sediment production, and water quality of southern rangeland soils. In: *Rangelands: A Resource under Siege* (eds P. J. Moss, P. W. Lynch & O. B. Williams) pp. 581-2. Proc. 2nd Int. Rangeland Cong. Australian Academy of Science, Canberra.

Edroma E. L. (1986) Effects of fire on the primary productivity of grasslands in Queen Elizabeth National Park, Uganda. In: *Rangelands: A Resource under Siege* (eds P. J. Joss, P. W. Lynch & O. B. Williams) pp. 599-600. Proc. 2nd Int. Rangeland Cong. Australian Academy of Science, Canberra.

Fire Protection Training Branch (1987) *Fire Management (Techniques) Notes, and Basic Fire Control; Fire Control Notes.* Conservation, Forests and Lands, Victoria.

Ford J. (ed) (1985) *Fire Ecology & Management in Western Australian Ecosystems.* WAIT Environmental Studies Group Rept 14. Western Australian Institute of Technology, Perth.

Fox J. E. D. (1986) Fire and its effects on mulga *(Acacia aneura)* in Western Australia. In: *Rangelands: A Resource under Siege* (eds P. J. Joss, P. W. Lynch & O. B. Williams) pp. 601-2. Proc. 2nd Int. Rangeland Cong. Australian Academy of Science, Canberra.

Gellie N. (1987) Practical Problems — Fire Modelling. In: *Bushfire Management in Natural Areas* (ed. B. Conroy) pp. 69-82. N.S.W. National Parks and Wildlife Service, Turramurra, Sydney.

Gibson D. F. (1986) *A Biological Survey of the Tanami Desert in the Northern Territory.* Technical Report No. 30. Conservation Commission of the Northern Territory, Alice Springs, Northern Territory.

Gill A. M., Groves R. H. & Noble I. R. (eds) (1981) *Fire and the Australian Biota.* Australian Academy of Science, Canberra.

Griffin G. F. & Allan G. E. (1984) Fire behaviour. In: *Anticipating the inevitable: a patch-burn strategy for fire management at Uluru (Ayers Rock-Mt Olga) National Park* (ed. E. C. Saxon) pp. 55-68. CSIRO, Australia, Melbourne.

Griffin G. F. & Allan G. E. (1986) Fire and the Management of Aboriginal owned lands in central Australia. In: *Science and Technology for Aboriginal Development* (eds B. Foran & B. Walker) sn.2.6. Project Report 3. CSIRO Division Wildlife and Rangelands Research, Alice Springs.

Griffin G. F. & Friedel M. H. (1984a) Effects of fire on central Australian rangelands. I Fire and Fuel characteristics and changes in herbage and nutrients. *Aust. J. Ecol.* **9,** 381-93.

Griffin G. F. & Friedel M. H. (1984b) Effects of fire on central Australian rangelands. II Changes in tree and shrub populations. *Aust. J. Ecol.* **9,** 395-403.

Griffin G. F. & Hodgkinson K. C. (1986) The Use of Fire for the Management of the Mulga Land Vegetation in Australia. In: *The Mulga Lands* (ed. P. S. Sattler) pp. 93-7. Proc. Roy. Soc. Qld., Brisbane.

Griffin G. F., Morton S. R., Allan G. E. & Fleming M. R. (1987) Fire created patch — dynamics for conservation management in the hummock grasslands of central Australia. *Proc. Int. Grasslands Symposium,* Huhhot, China.

Griffin G. F., Price N. F., & Portlock H. F. (1983) Wildfires in the Central Australian Rangelands, 1970-1980. *J. Envion. Manage.* **17,** 311-23.

Harrington G. N. (1986) Critical Factors in Shrub Dynamics in Eastern Mulga Lands. In: *The Mulga Lands* (ed. P. S. Sattler) pp. 90-2. Proc. Roy. Soc. Qld., Brisbane.

Hobbs R. J. (1988) What is ecological theory and is it of any use to managers? In: *Ecological Theory and Biological Management of Ecosystems* (eds D. A. Saunders & A. A. Burbidge) pp. 15-27. Occas. Paper 1/88. Department of Conservation and Land Management, Western Australia.

Hopkins A. (1988) Monitoring: Is it of use in integration of research and management? In: *Ecological Theory and Biological Management of Ecosystems* (eds D. A. Saunders & A. A. Burbidge) pp. 55-7. Occas. Paper 1/88. Department of Conservation and Land Management, W. A.

Johnson K. A. (1988) Rufous Hare-wallaby. *Aust. Nat. Hist.* **22,** 406-7.

Joss P. J., Lynch P. W., & Williams O. B. (eds) (1986) *Rangelands: A Resource under Siege.* Proc. 2nd Int. Rangeland Congress, Australian Academy of Science, Canberra.

Kessell S. R., Good R. B. & Potter M. W. (1982) *Computer Modelling in Natural Area Management,* ANPWS Spec. Pub. 9, Canberra, Australia.

Latz P. K. & Griffin G. F. (1978) Changes in Aboriginal land management in relation to fire and to food plants in central Australia. In: *The nutrition of Aborigines in relation to the ecosystem of central Australia* (eds B. S. Hetzel & H. J. Frith). CSIRO, Canberra.

Latz P. K., Johnson K. A. & Gillam M. W. (1981) *Preliminary Report on the Fauna and Flora of the Kings Canyon Area of the George Gill Range.* Internal Report to the Conservation Commission of the Northern Territory.

Luke R. H. & McArthur A. G. (1978) *Bushfires in Australia.* Australian Government Publishing Service, Canberra.

McArthur A. G. (1958) The preparation and use of fire danger tables. *Proc. Fire Weather Conf.* Paper 11, Commonwealth of Australia, Bureau of Meteorology, Melbourne.

McDonald R. C., Isbell R. F., Speight J. G., Walker J. & Hopkins M. S. (1984) *Australian Soil and Land Survey Field Handbook.* Inkata Press, Melbourne.

Mills J. R. (1986) Degradation and Rehabilitation of the Mulga Ecosystem. In: *The Mulga Lands* (ed. P. S. Sattler) pp. 79-83. Proc. Roy. Soc. Qld., Brisbane.

Morton S. R. & Andrew M. (1987) Ecological Impact and Management of Fire in Northern Australia. *Search* **18,** 77-82.

Morton S. R. & Baynes A. (1985) Small mammal assemblages in arid Australia: a reappraisal. *Aust. Mammal.* **8,** 159-69.

Perry R. A. (1960) *Pasture Lands of the Northern Territory, Australia.* Land Research Series No. 5. Commonwealth Scientific and Industrial Research Organization, Australia, Canberra.

Perry R. A. (Compiler) (1962) *Lands of the Alice Springs Area, Northern Territory, 1956-57.* Land Research Series No. 6. Commonwealth Scientific and industrial Research Organization, Australia, Canberra.

Pressland A. J., Cowan D. C., Evenson C. J. & Bowly P. S. (1986) Benefits of infrequent fire in the Mulga *(Acacia aneura)* rangelands of Queensland. In: *Rangelands: A Resource under Siege* (eds P. J. Joss, P. W. Lynch & O. B. Williams) pp. 608-9. Proc. 2nd Int. Rangeland Cong. Australian Academy of Science, Canberra.

Roberts B. (ed.) (1986) *Third Queensland Fire Research Workshop.* Working Papers. Darling Downs Institute of Advanced Education, Toowoomba, Qld.

Sattler P. S. (ed.) (1986) *The Mulga Lands.* Roy. Soc. Qld., Nth. Quay, Brisbane.

Saunders D. A., Arnold G. W., Burbidge A. A. & Hopkins A. J. M. (eds) (1987) *Nature Conservation: The Role of Remnants of Native Vegetation.* Surrey Beatty & Sons, Chipping Norton.

Saunders D. A. & Burbidge A. A. (eds) (1988) *Ecological Theory and Biological Management of Ecosystems.* Occas. Paper 1/88. Department of Conservation and Land Management, Perth, Western Australia.

Saxon E. C. (ed) (1984) *Anticipating the inevitable: a patch-burn strategy for fire management at Uluru (Ayers Rock-Mt Olga) National Park.* CSIRO Australia, Melbourne.

Southgate R. (1987) *Conservation of the Bilby* Macrotis lagotis *(Reid 1937).* World Wildlife Fund (Aust.) Proj. 6. Conservation Commission of the Northern Territory, Alice Springs.

Start A. N. (1986) Status and Management of Mulga in the Pilbara Region of Western Australia. In: *The Mulga Lands* (ed. P. S. Sattler) pp. 136-8. Proc. Roy. Soc. Qld., Brisbane.

Suijdendorp H. (1981) Responses of the Hummock Grasslands of Northwestern Australia to Fire. In: *Fire and the Australian biota* (eds A. M. Gill, R. H. Groves & I. R. Noble) pp. 417-24. Australian Academy of Science, Canberra.

Trabaud L. (1986) Fire effects on soils of the Mediterranean Basin region. In: *Rangelands: A Resource under Siege* (eds P. J. Joss, P. W. Lynch & O. B. Williams) pp. 582-5. Proc. 2nd Int. Rangeland Cong. Australian Academy of Science, Canberra.

Trollope W. S. W. (1986) The integration of episodic fire into arid and semi-arid rangeland management systems in southern and eastern Africa. In: *Rangelands: A Resource under Siege* (eds P. J. Joss, P. W. Lynch & O. B. Williams) pp. 586-7. Proc. 2nd Int. Rangeland Cong. Australian Academy of Science, Canberra.

Wood B., Latz P. & Lucas S. (1982) *The Land Resources of Simpsons Gap National Park.* Tech. Report. Land Conservation Unit, Conservation Commission of the Northern Territory, Darwin.

Wright H. A. (1986) Effects of fire on arid and semi-arid ecosystems — North American continent. In: *Rangelands: A Resource under Siege* (eds P. J. Joss, P. W. Lynch & O. B. Williams) pp. 575-6. Proc. 2nd Int. Rangeland Cong. Australian Academy of Science, Canberra.

Xaba B. B. & Nxumalo S. M. (1986) The role of fire in the Swaziland bushveld. In: *Rangelands: A Resource under Siege* (eds P. J. Joss, P. W. Lynch & O. B. Williams) pp. 581-2. Proc. 2nd Int. Rangeland Cong. Australian Academy of Science, Canberra.

Proc. Ecol. Soc. Aust. 1990 — 16:235-244

The effect of grazing on Australian ecosystems

A. D. WILSON

CSIRO Division of Wildlife and Ecology, Deniliquin 2710

Abstract

This chapter reviews the effect of grazing on the extensive grazing lands of semi-arid, tropical, subtropical and highland Australia that still remain as semi-natural ecosystems.

It outlines the plant, animal and fire factors that have contributed to changes in the biota, with specific examples taken from the chenopod shrublands, semi-arid woodlands and subalpine grasslands. Overall there has been a decrease in perennial herbs and an increase in annual herbaceous, inedible woody plants and soil erosion. Few plant species have been lost entirely, but there has been widespread local disappearance of mammals.

In southern Australia the rate of change is now slow. Rabbit populations have been greatly reduced and livestock grazing is now more controlled and responsive to forage supply. Restoration has occurred in some regions, although there is a continuing trend to shrub increase in others. New threats arise in northern Australia from exotic invasions and from increased utilization by cattle.

A continuation of restoration is anticipated over the next 200 years. Pastoral activities will remain dominant and become established as a sustainable land use, with the further development of the conservation ethic among landholders, improved monitoring of vegetation change and further research into management. The development of joint management plans for sensitive areas and the completion of the reserves network are other avenues for improvement.

Introduction

Grazing by sheep and/or cattle on natural or semi-natural pastures is the major land use over 4.5 million km^2 or 60% of the land surface of Australia. The grazing lands are predominantly semi-arid or subtropical, but vary widely in accordance with soil type and season of rainfall, as well as amount of rainfall. There are also smaller areas of tablelands and subalpine grazing lands. They are found in the extremes of climate and terrain, where intensive agriculture or sown pastures cannot be sustained. Still other lands, principally the hummock grasslands of the north and the interior are neither grazed or farmed. The grazing lands are home to approximately 50 million sheep and 15 million cattle. They are also grazed by approximately 20 million kangaroos, 0.5 million feral goats and perhaps 100 million wild rabbits. This is a major increase in the intensity of defoliation compared with that of 200 years ago. Herbivore numbers were then comparatively low (although more diverse), and controlled more by scarcity of water and predators than by forage supply. There has also been a significant change in fire frequency on these lands. Fires have been reduced by the removal of grasses which are the main fuel, and by a change in management from fire encouragement (Aboriginal) to fire suppression (European).

The effect has been profound. A palaeoecologist of a future civilization will note the sharp change in floristic composition and mammal population circa 1900. Whilst some communities have proved resilient to grazing (e.g., Mitchell grasslands, Orr 1980a,b), most others show major changes in composition and loss of productive value (e.g., West Kimberley, Payne *et al.* 1972).

Nevertheless, the degree of change must be kept in perspective. For the semi-arid woodlands of eastern Australia, Harrington *et al.* (1979) and Beale *et al.* (1986) paint a gloomy picture of shrub encroachment, and loss of perennial herbs. Yet Purdie (1986) finds at least some joy in the same communities because, they are still "composed predominantly of native shrubs and herbs". Furthermore, few if any plant species have been made extinct by grazing (~0.2%; Leigh *et al.* 1984). This is in contrast to the arable regions which have been cleared, ploughed, fertilized and sown to introduced grasses. The grazing lands are still semi-natural. They are used for agriculture, but that agriculture is based on the harvesting of the native flora by domestic animals.

In this review I will begin by outlining the nature of grazing and the type of effects that it has on ecosystems in general. I will then survey the extent of changes in our biota in some representative ecosystems. This will lead to conclusions about the future impact of grazing, and positive action to restore our grazed lands to a better condition.

The nature of grazing

The primary impact of a herbivore is to remove herbaceous biomass from a plant community. This may in itself have a direct effect on other animals that use the herbaceous layer for forage or shelter. The major effect, however, is the long-term response of the plant community. This will be dependent on the intensity of defoliation arising from the abundance and forage preference of each herbivore. It also depends on the sensitivity of each forage species to defoliation. There are also interactions with climate, fire and soil.

Defoliation tolerance

Plant species differ widely in their ability to withstand defoliation either directly or as an interruption to flowering and recruitment. The arid zone shrub *Atriplex vesicaria* is an example of a sensitive species, which may not regrow after a single complete defoliation (Leigh & Mulham 1971). Grasses are generally more tolerant of defoliation, but some may still be classified as sensitive. *Themeda australis*, for instance, was once widespread in the eastern temperate grazing lands, but is now quite rare. On the other hand, the short-grass *Danthonia caespitosa* is encouraged by grazing (Williams 1969). Other species are sensitive to defoliation in the recruitment phase (e.g., *Thyridolepis mitchelliana, Enteropogon acicularis*) and either fail to set seed or succumb to defoliation when young. These species are still widespread, but with reduced density. Still others are resistant to defoliation and are prolific seeders (e.g., *Aristida armata*, see Brown 1987) leading to enhanced abundance in grazed lands.

Herbivore preference and uneven distribution of grazing modify the outcome of grazing at a landscape level. Generalist herbivores, such as sheep and cattle, prefer those plants that are soft and green, choosing ephemerals while they are plentiful and the softer perennial grasses or forbs that remain green during the dry season. Browse shrubs are eaten mainly when other herbage has been consumed. As a consequence the grazing pressure on browse shrubs is intermittent and only heavy during droughts. The grazing pressure on ephemerals is also less than on perennials because of brief periods of rapid growth which exceed consumption and avoidance of grazing during the dry season. In contrast, the defoliation on perennial herbs, which provide nutritious forage into the dry season, is prolonged and intense. These species have therefore suffered a widespread decline in abundance in grazed communities (Pressland 1984). Unpalatable shrubs, on the other hand, have been encouraged by the release from competition. The reduction in perennial grasses increases the survival of shrub seedlings in semi-arid regions (Harrington 1986). In subalpine environments, disturbance of the perennial grass cover also encourages shrub establishment (Williams, this volume). The consequence of grazing by domestic herbivores, in all but the more arid ecosystems, is towards an increase in woody species.

Herbivore species

Each species of herbivore has a different effect on vegetation. Cattle are unable to graze as closely as sheep, so that the survival of herbaceous species is always higher with cattle grazing than with sheep (e.g., Graetz 1980). Furthermore, body growth is much more sensitive to overstocking than wool growth (wool continues to grow at bodyweight maintenance, although at a slower rate), so that optimal stocking rates for cattle are relatively lower than for sheep.

Other herbivores are less abundant, but deserve mention. The large kangaroos (*Macropus* spp.) represent a significant additional grazing impact in semi-arid areas where numbers are now greater than before settlement (Caughley *et al.* 1980). They have the ability to reduce herbage to the same low residues as sheep (Short 1985) and they directly affect perennial grasses, which are their preferred species (e.g., Gardiner 1986). Their impact is greatest in semi-arid woodlands, where kangaroos are abundant. Their grazing is not benign simply because they are native; the replacement of sheep farming by a similar intensity of kangaroo farming would not eliminate land degradation as surmised by Grigg (1988). Nevertheless their effect is different, with impact on a narrower range of species (Barker 1987) and an absence of the soil degradation arising from trampling around water points that is a characteristic of sheep or cattle.

The introduced rabbit *(Oryctolagus cuniculus)* is also credited with a major impact. It is still abundant in subalpine and temperate semi-arid regions. It was present in large numbers in southern Australia from 1880 to 1950 with density controlled only by the limits of the feed supply. It is estimated that numbers reached 750 million at their peak and now number 200 million for the whole of Australia (Flavel 1988). In contrast to sheep, rabbit grazing is severe on perennial forbs in the subalpine (Leigh *et al.* 1987) and on tree seedlings in those parts of the arid areas where they are common (Lange & Graham 1983; Friedel 1985). However, perhaps their greatest impact has been on the fauna. They competed directly with smaller marsupials, such as the bilby *(Macrotis lagotis)*, but more importantly are thought to have sustained an abundant population of predators, thus indirectly causing the demise of many species of small mammals (Jarman & Johnston 1977).

Apart from the rabbit, sheep have been by far the greatest agents for change in the Australian biota. They have been the most abundant herbivore, have an ability to graze close to the ground and are grazed more intensely than other herbivores.

Grazing distribution

The distribution of grazing across the landscape is uneven, being influenced by terrain, wind direction, the distribution of preferred plants (Low *et al.* 1973), proximity to cover (Hill 1981), suitable soils for burrowing (Parer & Libke 1985) and above all, water supply. The influence of water is particularly noticeable in the chenopod shrublands. The shrubs are high in minerals, so that animals grazing on them have high water requirements (Wilson & Graetz 1980). Thus livestock are closely tied to centres of water supply and this leads to a gradient of grazing pressure out from water. The result is a mosaic of denuded patches near water, vegetated only by unpalatable species, in sharp contrast to the lightly grazed areas far from water. These are usually accepted as representative of the original vegetation and are used as reference areas in determining the degree of vegetation change (e.g., Hacker 1987).

Interaction with rainfall and fire

Forage supply provides the effective limit on herbivore populations in the grazing lands, so grazing is usually intense when rainfall is low. The corollary is that grazing pressure is relatively light when rainfall is above average. Vegetation change is therefore a stepwise rather than a continuous process. Rainfall variation also introduces a problem for our perception of vegetation change. The extremes of high and low rainfall, and the co-occurrence of two rare events (e.g., very wet and fire), lead to occasional major shifts in species density that persist for decades (Noble 1986). Such changes mean that not all vegetation change since settlement can be attributed to grazing.

Fire is also a significant factor in grazed lands. In the semi-arid woodlands, the extent and frequency of fires have been sharply reduced under grazing land-use, partly because fuel has been consumed as forage and partly because natural fires have been suppressed. The result has been a widespread increase in woody species and a decline in pastoral value. In the subtropical woodlands and subalpine grazing lands, fire frequency has increased because it is used as a management tool to produce green foraged of improve nutritive value. This does not seem to have been detrimental to the subtropical woodlands (Mott & Tothill 1984), but causes considerable degradation in the subalpine where the cover of herbaceous plants and shrubs is reduced, the number of rabbits is increased, erosion rate is increased and the invasion of exotic species is enhanced (Leigh *et al.* 1987).

Plant communities

The total community response to these various influences of grazing is seen as a decrease in those species that are both sensitive to defoliation and palatable (referred to as decreasers), and an increase in those species that are resistant to grazing or are unpalatable (increasers). A further useful classification is to subdivide the increasers into Type I (showing a bell-shaped response to increasing grazing intensity) and Type II (increasing monotonically with grazing intensity) species (Hacker 1987). Species diversity usually shows a bell-shaped response, with increases at moderate grazing intensity because of a release from competitive exclusion by the dominant shrubs or tussock grasses (Grime 1973). Sustained high grazing pressure leads to a subsequent decline in diversity, because of the combined effect of heavy defoliation pressure and the loss of landscape variability through soil erosion. Large areas of the rangelands, particularly those with plant communities that are not favoured by grazing animals, or are far from water, have not been changed to a demonstrable degree. Other large areas now carry a different, but still diverse assemblage of native species. Smaller, but significant areas of favoured plant communities have lost both cover and diversity, and must be classed as degraded.

Fauna

Changes in the fauna have been much more pronounced than changes in the flora. Aslin's (1983) survey of the marsupials of the arid zone shows 46% have become rare or locally extinct since settlement. A similar, but less severe decline (34%) has occurred in the Nullarbor (Boscacci *et al.* 1987) where grazing by domestic stock is sparse or absent. Denny (1983) also lists 26 arid zone species of birds that are of diminished or endangered status although none are listed as extinct. We know little about the reasons for these changes, but introduced herbivores have had a direct effect by removing shelter and specialized food plants (e.g., for ground dwelling birds) and by direct competition with native herbivores. However, the introduction of new predators, competition with rabbits and changes in fire regimes are also major factors (Aslin 1983). The changes are catastrophic. Unfortunately they seem to be unrecognized by the general population. An animal that has never been seen in a lifetime is not perceived to have changed and the abundance of some species [e.g., large kangaroos, galahs *(Cacatua roseicapilla)*] that have taken advantage of improved water distribution is often erroneously used by landholders as evidence that the wildlife of the grazing lands is thriving.

Changes in the composition and cover of the vegetation has also influenced populations of ants (Briese & Macauley 1977), termites (Watson *et al.* 1973) and topsoil micro-arthropods (King & Hutchinson 1983). Some of these changes may have more ramifications than we may realize at first, because of their long-term effect in maintaining soil structure (e.g., macropores that are the avenue for deep water infiltration) (Tongway *et al.* 1989).

Representative ecosystems

Some of the major changes since settlement in Australian ecosystems subject to grazing are noted in Table 1. All communities have probably changed in botanical composition to a greater (e.g., semi-arid woodlands) or lesser (e.g., Mitchell grasslands) degree. Changes in the balance between herbaceous and shrub layers are common. Increased erosion is widespread in the semi-arid woodlands, and in some of the Western Australian ecosystems, but only localized in others (Woods 1983). Other features are the low numbers of invading exotic species and the minor extinction of plant species. I will present three examples in more detail.

Chenopod shrublands

The chenopod shrublands, covering approximately 500 000 km^2, are found in the winter rainfall arid and semi-arid areas of the continent. They are dominated by chenopod shrubs (*Atriplex* and *Maireana* spp.) that have a canopy cover of approximately 10%. The *Maireana* spp. are long-lived perennials (>100 years) and the *Atriplex* short-lived perennials (~10 years). Ephemeral and perennial herbaceous species grow between the shrubs, but vary widely in abundance according to rainfall (Graetz & Wilson 1984).

The principal impact of grazing has been the elimination of the dominant *Atriplex* or *Maireana* shrub and its replacement by a less palatable or

TABLE 1. The principal effects of grazing on natural ecosystems.

		Examples of species changes		
Vegetation Type/Region	Major effects or features	Decreasers	Increasers	Reference
Chenopod shrublands	Loss of palatable shrubs Concentration of grazing near water supply	*Atriplex vesicaria* *Maireana* spp.	*Sclerolaena* spp.	Barker & Lange 1969 Hacker 1987 Mitchell *et al.* 1979
Astrebla grasslands	Resilience of herbaceous layer. Invasion by exotic tree	*Astrebla* spp. (in some areas) *Aristida latifolia*	*Solanum* spp. *Sclerolaena* spp. *Acacia nilotica*	Orr 1980a, 1980b Pressland 1984
Semi-arid shrub woodlands	Loss of perennial grasses Widespread soil erosion Increase in shrub density associated with fire exclusion	*Thyridolepis mitchelliana* *Themeda australis*	*Acacia aneura* *Eremophila spp.* *Aristida* spp.	Harrington *et al.* 1979 Pressland 1984
Arid Acacia (W.A.)	Decrease in palatable shrubs and perennial grasses	*Maireana convexa* *Eremophila latrobei*	Annual forbs *Cassia sturtii* *Eremophila fraseri*	Hacker 1984a, 1984b Gardiner 1986 Payne *et al.* 1982, 1987
Alpine grass and heathland	Increase in bare ground and shrub cover Fire has major effect on ground cover and shrub abundance	*Poa* spp.	*Prostanthera cuneata* *Phebalium squamulosum* *Bossiaea foliosa*	Williams & Ashton 1987b Leigh *et al.* 1987
Temperate grassy woodlands	Change from tall grass to short grass	*Themeda australis*	*Aristida ramosa* Annuals	Whalley *et al.* 1978
Tropical savannah	Poor quality limits grazing intensity. Overgrazing occurs when cattle given supplement Annual burning Legume introduction	*Themeda australis*	*Heteropogon contorta*	Mott & Tothill 1984

less perennial chenopod, or a forb-grassland. This invariably occurs on areas that are close to supplies of stock water, but may extend to whole paddocks and stations. The boundary between lost and retained shrub is often sharp (Graetz & Ludwig 1978). Surprisingly there is usually little change in pastoral value in the New South Wales and South Australian rangelands because the chenopods are replaced by grasses or sub-shrubs of similar value (Graetz 1986). This is not so in Western Australia where sheep carrying capacity has been sharply reduced. In all regions, shrub loss leads to an increased exposure of the soil to erosion and the landscape is perhaps less appealing aesthetically because of the loss of vertical elements.

On the basis of *Atriplex* cover, Stanley (1983) estimated that 43% of the land in the Barrier Range area remained essentially unchanged, 35% had been denuded of saltbush and 22% was in an intermediate state of either degeneration or regeneration. Despite the large visual difference, the change is confined largely to a few dominant species. On a group of nine ha sites on the footslopes of the Barrier Range, chosen for similarity in position in the landscape and difference in apparent grazing pressure (i.e., *Atriplex* cover, see Wilson *et al.* 1982) there was no significant change in the number of plant species present (mean 55). There were six exotic species within a total count of 94. Several plants from these communities are listed as being extinct or rare by Leigh *et al.* (1984) but they are from specialized habitats, such as saline areas or springs and were never common.

The most dramatic and widespread changes occurred between settlement and 1945. Subsequently, through better management, better rainfall and less rabbits, there has been considerable regeneration. Nevertheless, further slow changes may be expected in future through the gradual demise of existing populations of some long-lived species. Two such species (*Acacia papyrocarpa* and *Maireana sedifolia*) have been identified in the chenopod shrublands. They are extremely long-lived (>300 years), but will eventually become rare because of absence of recruitment in grazed areas (Lange & Purdie 1976; Crisp 1978; Lange & Graham 1983). There may also be secondary invasion by unpalatable shrubs, such as *Cassia* and *Eremophila* spp.

Semi-arid shrub woodlands

The semi-arid woodlands are located in the eastern inland region, where rainfall is 200-500 mm, with no marked seasonal incidence. They cover an area of approximately 500 000 km^2. They have a tree and shrub layer of approximately 20 species at any one site, and a total canopy cover of between five and 30%. *Acacia aneura* and *Eucalyptus populnea* are common dominants (Beeston *et al.* 1980). The herbaceous layer contains a range of perennial grasses, and of ephemeral and perennial forbs (Harrington *et al.* 1984). The region is grazed by both sheep and cattle, and kangaroos (*Macropus* spp.) may constitute 20-40% of the total grazing pressure (Short & Bayliss 1985).

The principal structural change in these communities since settlement has been an increase in the density of the shrub layer, dominated by species of *Eremophila, Dodonea* and *Acacia,* as a consequence of both grazing and fire exclusion (Harrington *et al.* 1979). In addition there has been widespread loss of perennial grasses.

Probably all of these woodlands have been changed substantially, either by shrub increase or loss of grasses. Newman & Condon (1969) estimate a major degeneration of pastoral value over 30% of the woodlands (including Western Australia) and a further 30% with minor degeneration.

On a group of nine ha sites selected for similarity of position in landscape and difference in shrub cover and erosion (see Wilson *et al.* 1988 for details), an average of 49 plant species was recorded on the "good condition" site, 48 species on "average condition" sites and only 28 on sites in poor condition (i.e., eroded). The difference was accounted for mainly by an absence of herbaceous perennials. This suggests that the degree of change is more severe on poor sites than on any of the chenopod shrubland sites mentioned earlier. The axis of environmental stress, outlined by Grime (1973), has been pushed to a further extreme. Eleven exotic species (all herbaceous) were recorded out of a total of 97 species. Several rare plant species are listed as endangered by grazing (Leigh *et al.* 1984).

A major feature of these lands is the widespread erosion of the soil on the slopes, even though the actual slope may be no more than 1%. This represents a more serious change than a change in botanical composition, because of the greater time required for recovery. Nevertheless, substantial restoration can be achieved following grazing reduction or exclusion. An example is the regeneration area around the town of Cobar, which is now well vegetated after 25 years of exclosure from grazing and several fires (see Walker 1976).

Alpine grass and heathlands

The subalpine grazed lands are tiny in comparison with the semi-arid, (approximately 5 000 km^2) but are of special interest because of their proximity to urban centres and value for water yield. The grasslands and heathlands of these mountain regions of New South Wales and Victoria have been grazed seasonally (Dec-Apr) by livestock since 1850. The grasslands (dominated by *Poa* spp. — snowgrasses) occur on the flatter areas that are subject to frost and wind. There is a progression to open grassy heath and closed heath as the terrain becomes steeper and more rocky (Williams & Ashton 1987a). Grazing has led to an expansion of the heathland, due to an invasion of the *Poa* grassland by shrubs. This arises from an increase in bare ground, which is the prerequisite for the establishment of shrubs (Williams & Ashton 1987b; Williams, this volume). The type of shrub that predominates is influenced by the grazing regime. The shrubs that are of low palatability and reproduce vegetatively become dominant with continued grazing, while the shrubs that are palatable and reproduce from seed become dominant with grazing exclusion. Grazing leads to a selective reduction in palatable herbs (Wimbush & Costin 1979a, 1979b) which are regarded as the more nutritious component of cattle diets (van Rees 1982). The regeneration of snowgum *(Eucalyptus pauciflora)* is also reduced. Cattle also damage the moss beds by trampling. These areas are unattractive for grazing but cattle enter them to obtain drinking water (van Rees & Hutson 1983). The ensuing damage poses a threat to the hydrological efficiency of alpine areas. A similar area of highland grasslands in Tasmania is also grazed and has similar problems (Fitzpatrick & Duncan 1987).

Other factors are fire and rabbit grazing. Fire reduces the cover of both shrubs and grasses and encourages an invasion of alien species. The reduction in shrub cover from even a single fire is substantial, so the effect is significant even though fire is infrequent (Leigh *et al.* 1987).

Rabbits graze mostly on the perennial forbs, reducing both cover and diversity, and the display of wildflowers. Soil disturbance arises from scratching and burrowing. Their effects are magnified after fire when grazing is concentrated on resprouting plants and rabbit numbers are encouraged by the more nutritious short grass. It is thought that rabbits are found only in areas with a history of sheep grazing, because of a requirement for short immature herbage (Leigh *et al.* 1987). Twenty exotic naturalized species were recorded at Kiandra by Leigh.

The areas grazed and the numbers of livestock have been gradually reduced over the last 40 years. Also, the practice of burning the *Poa* to improve palatability has ceased. It is nevertheless predicted that the trend to increased shrub density will continue in the short-term, even if grazing is excluded, because of the large number of shrub seedlings already present in disturbed areas (Williams, this volume). The interaction of fire and grazing is significant, but not yet well understood. It is difficult to study because of the low fire frequency and its interaction with grazing intensity.

In botanical terms the extent of change has been far less in the subalpine than in the semi-arid, but it evokes a greater concern. This is because there are alternative uses for this land, including recreation and water catchment, that are potentially of greater value than cattle grazing. No such alternatives are apparent for the greater part of the flat arid and semi-arid landscapes.

Overall impact: past and present

Despite the differences in species composition and the degree of change between communities, some generalizations about the impact of grazing by introduced herbivores can be made. The trend has always been towards a reduction in the abundance of perennial forbs and grasses, and an increase in annuals and ephemerals. The remaining perennials are usually of reduced palatability. There has also been an increase in shrub density because of reduced competition from perennial grasses and changes in fire regime. The exception is the more arid areas where shrubs and trees have been reduced, primarily by the predation of their slow-growing seedlings by either sheep or rabbits. A few of the more palatable trees are no longer regenerating and are expected to decline in density over hundreds of years. An increase in bare ground and erosion rate is common, particularly where animals congregate around water or along fences.

The rate of change is now slow, with few new areas of land being changed botanically or eroded (Condon 1983). This arises from a number of factors. One is the partial control of rabbits by myxomatosis over the last 35 years. Rabbits were once an enormous uncontrolled force of grazers and burrowers. A second factor is the gradual change in the structure of the grazing industries. There is now more return for quality (e.g., well grown animals at a young age), which can only be produced at moderate stocking rates. There is also a ready availability of road transport, so that animals in drought stricken areas can be removed to distant markets. The modern commercial world now has a system which combines the advantages of nomadic grazing (varying livestock number with rainfall) and private land control (where the individual is rewarded in the long term for conservative grazing). The major continuing change is a trend towards higher shrub density in semi-arid and subalpine areas. This cannot be controlled by grazing reductions alone since large populations of juvenile shrubs are already present. If these areas are to remain in grazing use, management systems based on the strategic control of grazing pressure and the controlled use of fire must be introduced.

In other areas, notably the chenopod shrublands, there has been a gradual regeneration of soil and shrub cover over the last 40 years. The steady repair of large bare areas on the Riverine plain has been measured by Jones (1966) and N. A. Abraham (pers. comm.). A significant increase in the cover of chenopod shrubs and a reduction in soil movement in the Western Division of New South Wales has been documented by Condon (1983) and Bolton-Smith (1987). The improvement in stock management and a reduced rabbit population are certainly factors in this improvement, but climatic change has also played a part. As an example, the mean rainfall at Broken Hill for the period 1946-1986 was 271 mm, compared with 223 mm for the period 1888-1945. The projection of Pittock (1983), based on modelling world climatic circulation under an increasing atmospheric CO_2 concentration, is that this trend will continue. This can only favour the continued regeneration of vegetation over most of the grazing lands.

Whilst the southern grazing lands have now stabilized, new threats arise in the north. The genetic and nutritional problems that previously limited cattle numbers in the tropical savannahs have now been overcome, giving rise to new areas of overgrazing (see Gardener & McIvor, this volume). Also, the northern Mitchell grass, which have been little changed by grazing, are now threatened by an invasion of the introduced tree *Acacia nilotica* (Burrows *et al.* 1986).

The future

The changes are presented above. What judgement do we place upon them? As mentioned in the introduction, the value we place on these changes depends greatly on our perspective. If conservation and aesthetic values are uppermost, then any change in composition brought about by grazing will be seen as detrimental, and grazing pressure must be very light indeed to ensure that these values do not deteriorate. This is the situation with alpine and subalpine areas. If on the other hand, the land is primarily dedicated to pastoral production, albeit one based on the utilization of natural vegetation, then change in the composition of the vegetation is not seen to be deleterious unless that change affects long-term productivity or soil stability. Changes between one palatable plant and another often do not affect production, as has been observed in the chenopod shrublands (Graetz 1986) and should not be classified as degradation.

Since the biota is a dynamic entity, changing with climatic fluctuation and differing fire regimes, it is difficult to measure long-term change that may be attributed to land use. Original vegetation is often not a viable concept, even in reserves, because of permanent changes in predator populations, fire frequency and the introduction of exotic species. The setting of goals for the vegetation and wildlife on various landholdings, both public and private, is therefore essential. Without such goals there is unlikely to be an easy resolution of controversial issues, (e.g., the control of kangaroo numbers, fire policy) on both reserves and grazing land.

One appropriate goal is to preserve examples of all our biota within reserves, requiring the exclusion of domestic stock and artificial watering points, and appropriate control of native grazers where required. Reserves have an important role to play in maintaining examples of our ecosystems (e.g., Purdie 1986) and for preserving threatened species, such as the nail-tailed wallaby *(Onychogalea fraenata)* (Gordon & Lawrie 1980). The reserve network throughout the grazing lands should be completed as soon as practicable. These reserves require appropriate management policies since reservation alone does not guarantee naturalness with respect to fire, exotic species and extinct species. The management of grazing by kangaroos on these reserves is dependant on the objectives of each reserve (Shepherd & Caughley 1987). The current policy in New South Wales of maintaining man-made waters is inconsistent with the proposition that these waters have upset the competitive balance between species, encouraging the expansion of one bird at the expense of another (Saunders & Curry, this volume). The control of rabbits, and possibly foxes and cats, is also desirable, although difficult to achieve. Such reserves are not useful as reference areas for measuring vegetation change on grasslands primarily because the ungrazed composition is not necessarily a useful goal for grazed lands and other factors such as fire regime and rabbit grazing are as important as livestock density in determining composition.

Another goal is to maintain the natural vegetation of the remaining grazed lands, with sufficient cover to prevent erosion and in a state that will optimize production. This will also ensure the conservation of the greater part of the flora throughout the natural range of each species; a task beyond the reserve system by itself. Conservative grazing is an appropriate management aim for conservationists and pastoralists alike and is one of the important aims for the future.

The primary means of achieving this aim is to foster *a conservation ethic* amongst landholders. They already accept the desirability of conserving endangered fauna and of preventing erosion, but they have less appreciation of the flora, its changes in relation to climate, fire and grazing, or the importance of these changes. Regional soil conservation committees are ideal venues for discussion and interaction on conservation ethics. An important avenue for education is the development of monitoring sites on their holdings, that are both measured (for interpretation) and photographed (for visual impact). The Department of Agriculture in Western Australia (Holm 1986) and the Pastoral Board in South Australia are active in this area, and it is hoped that New South Wales and Queensland will follow. In 10 years they will start to be of enormous value in promoting improved management on pastoral holdings. In 20 years they will also serve as an information source to the public and a data bank for research on the long-term path of degradation or restoration on our pastoral lands (Curry & Hacker 1986; Wilcox 1988). Where management is good, it will serve to reinforce the progress that is being made and assure the public that pastoralism is a sustainable land use. Where management is poor, it will serve to focus the energies of research and advisory services on areas of need. To be effective they must be promoted as an avenue for education and not as a means of exerting control.

Monitoring sites should also be obligatory in parks and reserves. Vegetation change will also occur on these and be otherwise unnoticed by the passing parade of park employees. Some locations, on both pastoral and reserve land, would become research sites with appropriate grazing exclosures and more detailed measurement of the vegetation.

An additional step is the development of joint management plans for areas of public value or special interest. Riparian zones provide an example. Grazing requires access to water, but in the process does considerable damage to the unique vegetation along the waters edge. The watered areas may also be required for wildlife or public use. A simple solution is to fence off the spring, moss-bed or river edge and pipe the water to the stock. This allows all uses and values to be retained at minimal cost.

The Victorian Flora and Fauna Guarantee Act (Oates 1988) provides another model appropriate for endangered species. It seeks to identify and protect endangered species on crown and private land by engaging in co-operative management schemes with landholders. The landholders are compensated for extra costs or loss of production, without adding any rights of public access. In other states, such as Queensland, there is provision for the establishment of wildlife refuges and habitat conservation contracts (A. Don, Queensland National Parks and Wildlife Service, pers. comm.), which may achieve the same objective. We should be aware that grazing exclusion is not always necessary, or even desirable. The plains wanderer *(Pedionomus torquatus)* is an example. This bird is now largely confined to the plains of the Riverina, where it occurs on lightly grazed grasslands. Observations by Harrington *et al.* (1988) show that it is not found where the grass has been heavily grazed and is uncommon in the nearby chenopod shrublands. It prefers the areas with spaced tussocks of perennial grass, which arise where grazing has been heavy in the past. This would seem to be an opportunity for a co-operative management programme involving both government and landholders.

The education and conservation goals mentioned above must be underpinned by continuing research. With rare species, protection is not simply a matter of grazing exclusion. Some species, both plant and animal, require fire disturbance or a mosaic of stages of post-fire succession. The malleefowl *(Leipoa ocellata)* is an example. It is becoming rare in long unburnt mallee partly because of declining populations of Acacia spp. that are an important seed resource for this bird during the autumn (Noble 1984).

For grazed lands there is still much to be learnt about the relationships between land use, the management of that use (e.g., intensity and timing of grazing, fire regime), vegetation change, soil stability and productivity. There is a particular need to know more about the circumstances of grazing damage. Is it through the overgrazing of established plants or the loss of seed stores? Does it occur in droughts or during drought recovery? If such thresholds were known it might be possible to achieve some conservation goals by changing the distribution of grazing without reducing livestock numbers. In scientific terms this is disturbance ecology, with emphasis on intensity, rates and timing, rather than simply presence or absence.

Another important avenue involves restoration. We need to know more about the natural restoration process and how this may be aided by grazing control or other management. It also needs to be related to the intensity and timing of use. The major part of restoration will occur, with appropriate management, in conjunction with continued use, rather than by grazing exclusion. Reserves also require restoration, which might entail the limitation of marsupial herbivores, the removal of artifical waters and where appropriate the re-introduction of locally extinct fauna.

Conclusion

Grazing by domestic stock on the extensive marginal lands is an accepted and acceptable land use that will continue for the foreseeable future. A few holdings may be withdrawn from grazing because they are too marginal to sustain economic enterprises or because of land requirements for other purposes. However, the majority will remain under pastoral use in one form or another. We should not now view grazing as either alien or temporary. Much progress has been made over the past 40 years in reducing overgrazing and erosion in some regions, whilst in others we still have much to learn about control and management of grazing. I am optimistic that the next 200 years will see a further improvement in all the production and conservation values of the grazed lands.

Acknowledgements

I wish to thank Drs J. C. Noble, J. A. Ludwig and D. A Saunders for their assistance in the revision of this paper.

References

Aslin H. J. (1983) Marsupials in the arid zone. In: *What Future for Australia's Arid Lands* (eds J. Messer & G. Mosley) pp. 40-4. Australian Conservation Foundation, Melbourne.

Barker R. D. (1987) The diet of herbivores in the sheep rangelands. In: *Kangaroos: their ecology and management in the sheep rangelands of Australia* (eds G. Caughley, N. Shepherd & J. Short). Cambridge University Press, Cambridge.

Barker S. & Lange R. T. (1969) Effects of moderate sheep stocking on plant population of a black oak-bluebush association. *Aust. J. Bot.* **17,** 527-37.

Beale I. F., Orr D. M. & Mills J. R. (1986) Pastoral impacts on the mulga ecosystem. In: *The Mulga Lands* (ed. P. S. Sattler) pp. 50-3. Royal Society of Queensland, Brisbane.

Beeston G. R., Walker P. J., Purdie R. & Pickard J. (1980) Plant communities of the poplar box *(Eucalyptus populnea)* lands of eastern Australia. *Aust. Rangel. J.* **2,** 1-16.

Bolton-Smith B. (1987) Anecdotes from a past era. Range Management Newsletter 87/1 (The Australian Rangeland Society).

Boscacci L. J., McKenzie N. L. & Kemper C. M. (1987) Mammals. In: *A Biological Survey of the Nullabor Region of South and Western Australia in 1984* (eds N. L. McKenzie & A. C. Robinson) pp. 103-37. Government Printer, Adelaide.

Briese D. T. & Macauley B. M. (1977) Physical structure of an ant community in semi-arid Australia. *Aust. J. Ecol.* **2,** 107-20.

Brown R. F. (1987) Defoliation regrowth and dry matter partitioning in the two arid zone grasses, *Aristida armata* and *Thyridolepis mitchelliana. Aust. J. Agric. Res.* **38,** 881-93.

Burrows W. H., Carter J. O., Anderson E. R. & Bolton M. P. (1986) Prickly acacia *(Acacia nilotica)* invasion of Mitchell grass (*Astrebla* spp.) plains in central and north-west Queensland. Working papers, 4th Biennial Conference, Australian Rangeland Society, Armidale pp. 104-6.

Caughley G., Grigg G. C., Caughley J. & Hill G. J. E. (1980) Does dingo predation control the densities of kangaroos and emus. *Aust. Wild. Res.* **7,** 1-12.

Crisp M. D. (1978) Demography and survival under grazing of three Australian semi-desert shrubs. *Oikos* **30,** 520-8.

Condon R. W. (1983) Pastoralism In: *What Future for Australia's Arid Lands* (eds J. Messer & G. Mosley) pp. 54-60. Australian Conservation Foundation, Melbourne.

Curry P. J. & Hacker R. B. (1986) The potential of rangeland monitoring to facilitate production and conservation objectives in Western Australian rangelands. Proceedings 4th Biennial Conference of the Australian Rangeland Society, August 1986. University of New England, Armidale, New South Wales.

Denny M. (1983) Animals — native and feral. In: *What Future for Australia's Arid Lands* (eds J. Messer & G. Mosley) pp. 19-25. Australian Conservation Foundation, Melbourne.

Fitzpatrick J. B. & Duncan F. (1987) Tasmanian high altitude grassy vegetation: its distribution, community composition and conservation status. *Aust. J. Ecol.* **12,** 73-86.

Flavel N. (1988) Rabbit population is 200 million and growing by 200 a minute. The Age, 7th July p.3.

Friedel M. H. (1985) The population structure and density of central Australian trees and shrubs, and relationships to range condition, rabbit abundance and soil. *Aust. Rangel. J.* **7,** 130-9.

Gardiner H. G. (1986) Dynamics of perennial plants in the mulga (*Acacia aneura* F. Muell) zone of Western Australia. *Aust. Rangel. J.* **8,** 18-27.

Gordon G. & Lawrie B. C. (1980) The rediscovery of the bridled nail-tailed wallaby, *Onychogalea fraenata* (Gould) (Marsupialia:Macropodidae) in Queensland. *Aust. Wildl. Res.* **7,** 339-45.

Graetz R. D. (1980) The influence of grazing by cattle and sheep on the population dynamics of bladder saltbush *(Atriplex vesicaria)* and white top *(Danthonia caespitosa) Aust. Rangel. J.* **2,** 208-13.

Graetz R. D. (1986) A comparative study of sheep grazing a semi-arid saltbush pasture in two condition classes. *Aust. Rangel. J.* **8,** 46-56.

Graetz R. D. & Ludwig J. A. (1978) A method for the analysis of piosphere data applicable to range assessment. *Aust. Rangel. J.* **1(2),** 126-36.

Graetz R. D. & Wilson A. D. (1984) Saltbush and bluebush. In: *Management of Australia's Rangelands* (eds. G. N. Harrington, A. D. Wilson & M. D. Young) pp. 209-22. CSIRO, Melbourne.

Grigg G. (1988) Harvest roos to help halt soil erosion — Professor. Sun Herald, 17th July p. 117.

Grime J. P. (1973) Competitive exclusion in herbaceous vegetation. *Nature* **242,** 344-7.

Hacker R. B. (1984a) Vegetation dynamics in a grazed mulga shrubland community. I. The mid-storey shrubs. *Aust. J. Bot.* **32,** 239-49.

Hacker R. B. (1984b) Vegetation dynamics in a grazed mulga shrubland community. II. The ground storey. *Aust. J. Bot.* **32,** 251-61.

Hacker R. B. (1987) Species response to grazing and environmental factors in an arid halophytic shrubland community. *Aust. J. Bot.* **35,** 135-50.

Harrington G. N. (1986) Critical factors in shrub dynamics in eastern mulga lands. In: *The Mulga Lands* (ed. P. S. Sattler) pp. 90-2. Royal Society of Queensland, Brisbane.

Harrington G. N., Oxley R. E. & Tongway D. J. (1979) The effects of European settlement and domestic livestock on the biological system in poplar box *(Eucalyptus populnea)* lands. *Aust. Rangel. J.* **1,** 271-9.

Harrington G. N., Mills D. M. D., Pressland A. J. & Hodgkinson K. C. (1984) Semi-arid woodlands. In: *Management of Australia's Rangelands* (eds G. N. Harrington, A. D. Wilson & M. D. Young) pp. 189-207. CSIRO, Melbourne.

Harrington G. N., Maher P. N. & Baker-Gabb D. J. (1988) The biology of the Plains Wanderer *Pedionomus torquatus* on the Riverine plain of New South Wales during and after drought. *Corella* **12,** 7-13.

Hill G. J. E. (1981) Distribution of grey kangaroos in south inland Queensland. *Aust. Rangel. J.* **3,** 58-66.

Holm A. McR. (1986) The assessment of range trend in Western Australian pastoral shrublands. In: *Rangelands: A Resource Under Siege, Proc. 2nd Inter. Range. Congr.*, Adelaide.

Jarman P. J. & Johnson K. A. (1977) Exotic mammals, indigenous mammals and land use. *Proc. Ecol. Soc. Aust.* **10,** 146-66.

Jones R. M. (1966) Scald reclamation studies in the Hay district. Part I. Natural reclamation of scalds. *J. Soil Conserv. Serv. N.S.W.* **22,** 147-60.

King K. L. & Hutchinson K. J. (1983) The effects of sheep grazing on invertebrate numbers and biomass in unfertilized natural pastures of the New England Tableland. *Aust. J. Ecol.* **8,** 245-55.

Lange R. T. & Graham C. R. (1983) Rabbits and the failure of regeneration in Australian arid zone Acacia. *Aust. J. Ecol.* **8,** 377-82.

Lange R. T. & Purdie R. (1976) Western myall *(Acacia sowdenii),* its survival prospects and management needs. *Aust. Rangel. J.* **1,** 64-9.

Leigh J. H. & Mulham W. E. (1971) The effects of defoliation on the persistence of *Atriplex vesicaria. Aust. J. Agric. Res.* **22,** 239-44.

Leigh J. H., Boden R. & Briggs J. (1984) *Extinct and Endangered Plants of Australia.* MacMillan, Melbourne.

Leigh J. H., Wimbush D. J., Wood D. H., Holgate M. D., Slee A. V., Stanger M. G. & Forrester R. I. (1987) Effects of rabbit grazing and fire on a subalpine environment. I. Herbaceous and shrubby vegetation. *Aust. J. Bot.* **35,** 433-64.

Low B. S., Birk E., Lendon C. & Low W. A. (1973) Community utilization by cattle and kangaroos in Mulga near Alice Springs, N.T. *Trop. Grassld.* **7,** 149-56.

Mitchell A. A., McCarthy R. & Hacker R. B. (1979) A range inventory and condition survey of part of the Western Australia Nullabor Plain, 1974. Technical Bulletin 47. Department of Agriculture, Western Australia.

Mott J. J. & Tothill J. C. (1984) Tropical and subtropical woodlands. In: *Management of Australia's Rangelands* (eds G. N. Harrington, A. D. Wilson & M. D. Young) pp. 255-69. CSIRO, Melbourne.

Newman J. C. & Condon R. W. (1969) Land use and present condition. In: *Arid Lands of Australia* (eds R. O. Slatyer & R. A. Perry) pp. 105-32. Australian National University Press, Canberra.

Noble I. R. (1986) The dynamics of range ecosystems. In: *Rangelands: A Resource Under Siege, Proc. 2nd Inter. Rangel. Congr.*, Adelaide. pp. 3-5.

Noble J. C. (1984) Mallee. In: *Management of Australia's Rangelands* (eds G. N. Harrington, A. D. Wilson & M. D. Young) pp. 223-40. CSIRO, Melbourne.

Oates N. (1988) Flora and fauna guarantee act. *Trees and Natural Resources* **30,** 11-3.

Orr D. M. (1980a) Effects of sheep grazing on *Astrebla* grasslands in central western Queensland. I. Effect of grazing pressure and livestock distribution. *Aust. J. Agric. Res.* **31,** 797-806.

Orr D. M. (1980b) Effects of sheep grazing on *Astrebla* grasslands in central western Queensland. II. Effects of seasonal rainfall. *Aust. J. Agric. Res.* **31,** 807-20.

Parer I. & Libke J. (1985) Distribution of rabbit, *Oryctolagus cuniculus,* warrens in relation to soil type. *Aust. Wildl. Res.* **12,** 387-405.

Payne A. L., Kubicki A., Wilcox D. G. & Short L. C. (1972) A report on erosion and range condition in the west Kimberley area of Western Australia. Technical Bulletin 42. Department of Agriculture, Western Australia.

Pittock A. B. (1983) Recent climatic change in Australia: Implications for a CO_2-warmed earth. *Climatic Change* **5,** 321-40.

Pressland A. J. (1984) Productivity and management of Western Queensland's rangelands. *Aust. Rangel. J.* **6,** 26-45.

Purdie R. W. (1986) Development of a National Park system for Queensland's mulga region. In: *The Mulga Lands* (ed. P. S. Sattler) pp. 122-7. Royal Society of Queensland, Brisbane.

Rees van H. (1982) The diet of free-ranging cattle on the Bogong High Plains, Victoria. *Aust. Rangel. J.* **4,** 29-33.

Rees van H. & Hutson G. D. (1983) The behaviour of free-ranging cattle on an alpine range in Australia. *J. Range. Manage.* **36,** 740-3.

Shepherd N. & Caughley G. (1987) Options for management of kangaroos. In: *Kangaroos: their Ecology and Management in the Sheep Rangelands* (eds G. Caughley, N. Shepherd & J. Short) pp. 188-219. Cambridge University Press, Cambridge.

Short J. (1985) The functional response of kangaroos, sheep and rabbits in an arid grazing system. *J. Appl. Ecol.* **22,** 435-47.

Short J. & Bayliss P. (1985) Bias in aerial survey estimates of kangaroo density. *J. Appl. Ecol.* **22,** 415-22.

Stanley R. J. (1983) Soils and vegetation: an assessment of current status. In: *What Future for Australia's Arid Lands* (eds J. Messer & G. Mosley) pp. 8-18. Australian Conservation Foundation, Melbourne.

Tongway D. J., Ludwig J. A. & Whitford W. G. (1989) Mulga log soil mounds: fertile patches in the semi-arid woodlands of eastern Australia. *Aust. J. Ecol.* (in press).

Walker P. J. (1976) Cobar regeneration area — the first sixteen years. *Soil Convserv. Serv. N.S.W.* **32,** 119-30.

Watson J. A. L., Lendon C. & Low B. S. (1973) Termites in mulga lands. *Trop. Grassl.* **7,** 121-6.

Whalley R. D. B., Robinson G. G. & Taylor J. A. (1978) General effects of management and grazing by domestic livestock on the rangelands of the Northern Tablelands of New South Wales. *Aust. Rangel. J.* **1,** 174-90.

Wilcox D. G. (1988) Fair use and a fair go. *Aust. Rangel. J.* **10,** 76-81.

Williams O. B. (1969) Studies in the ecology of the Riverine plain. V. Plant density response of species in a Danthonia caespitosa grassland to 16 years of grazing by Merino sheep. *Aust. J. Bot.* **17,** 255-68.

Williams R. J. & Ashton D. H. (1987a) The composition, structure and distribution of heathland and grassland communities in the subalpine tract of the Bogong High Plains, Victoria. *Aust. J. Ecol.* **12,** 57-71.

Williams R. J. & Ashton D. H. (1987b) Effect of disturbance and grazing by cattle on the dynamics of heathland and grassland communities on the Bogong High Plains, Victoria. *Aust. J. Bot.* **35,** 413-31.

Wimbush D. J. & Costin A. B. (1979a) Trend in vegetation at Kosciusko. I. Grazing trials in the subalpine zone 1957-1971. *Aust. J. Bot.* **27,** 741-87.

Wimbush D. J. & Costin A. B. (1979b) Trends in vegetation at Kosciusko. II. Subalpine range transects 1959-1978. *Aust. J. Bot.* **27,** 789-831.

Wilson A. D. & Graetz R. D. (1980) Cattle and sheep production on an *Atriplex vesicaria* (saltbush) community. *Aust. J. Agric. Res.* **31,** 369-78.

Wilson A. D., Tupper G. J. & Tongway D. J. (1982) Range condition assessment in bladder saltbush *(Atriplex vesicaria)* communities. *Aust. Rangel. J.* **4,** 41-51.

Wilson A. D., Tongway D. J. & Tupper G. J. (1988) Factors contributing to differences in forage yield in the semi-arid woodlands. *Aust. Rangel. J.* **10,** 13-7.

Woods L. E. (1983) *Land degradation in Australia.* Australian Government Printing Service, Canberra.

Proc. Ecol. Soc. Aust. 1990 — 16:245-253

Analysis of stocking records from 1884 to 1988 during the subdivision of *Momba*, the largest property in semi-arid New South Wales

JOHN PICKARD

Western Lands Commission, G.P.O. Box 4351, Sydney, N.S.W. 2001

Present address: Graduate School of the Environment, Macquarie University, N.S.W. 2109

Abstract

Over 100 years of grazing by domestic animals have wrought substantial changes to the ecosystems of the semi-arid zone of New South Wales. There is not yet sufficient data to determine which combinations of weather, fire and pastoral management are primarily responsible for the changes. However, conventional wisdom suggests that high stocking rates bear the brunt of the blame. This chapter briefly reviews the changes and examines the subdivision of Momba, *once the largest property in New South Wales. Historical data on stocking rates, and estimates of carrying capacity are analysed. The high stocking rates of the 1880s (1 dry sheep equivalent (dse) to 3.1 ha) were followed by a decline to 1 dse:5.7 ha in the 1940s and then increased to 1 dse:4.9 ha in the 1980s. Estimates of carrying capacity are generally higher than the stock actually recorded. Closer settlement or breaking up large properties into smaller units led to social problems and environmental degradation. Because of economic pressure, properties are now increasing in size through amalgamation. Stocking rates have increased since the 1940s and this is attributed to pastoralists responding to a combination of better rainfall, improved technology and economic changes.*

Introduction

Two hundred years ago the convict settlement at Sydney was founded and Europeans commenced the political, social, economic and biological conquest of Australia. In 1788 the biological environment was not free from the impact of man. Aborigines had been in Australia for over 40 000 years. They had probably wrought substantial changes in the ecosystems by using fire to aid their hunting and gathering. However, it seems likely that the Aborigines had a kinship with the land that the Europeans lacked completely. As a consequence, the Australian environment now bears what may be indelible scars of the European conquest: destroyed or disrupted ecosystems, new ecosystems, extinct species, feral animals and naturalized plants, agriculture, pastoralism, mining, industry, urban sprawls and transport networks.

Some economists (e.g., Davidson 1987) argue that these environmental costs are minimal and have been necessary for the development of Australia. In the arid zone, Heathcote (1965) suggested that "exploitation of the evanescent resources" or "opportune use" may be the best approach. He challenged the alternative of conservative use in these environments. Whereas the first European explorers dismissed the semi-arid zone as totally without value, subsequent entrepreneurs found that grazing the rangelands with domestic sheep and cattle could be extremely profitable. In New South Wales, the semi-arid zone was almost fully occupied by the end of the first century of European settlement.

Davidson (1981) documented the economic background to the settlement of Australia, and King (1957) summarized the complex history of land legislation in New South Wales. There have been several studies of both the pioneer and the consolidation phases of settlement in the semi-arid zone (e.g., Heathcote 1965; Noble & Tongway 1986a; Oxley 1987a,b; Williams & Oxley 1979); and reviews of the impact of European stock on the rangelands (e.g., Harrington *et al.* 1979; Noble & Tongway 1986b). However, there are few studies of the history of stocking on an individual property. As the present vegetation is primarily a consequence of the imposition of European domestic stock on the original vegetation, then stocking is a critical component of the overall picture.

In this chapter, I examine changes in stocking rates during the progressive subdivision of *Momba*, once the largest property in New South Wales (Fig. 1). The present paper is part of a larger overall programme to document the nature and extent of vegetation changes in the Western Division of New South Wales. The Division is administered by the Western Lands Commission under the Western Lands Act and includes most of the semi-arid zone of the State (Fig. 1). I am seeking to integrate the individual and collective impacts of changes in land use, the environment, legislation and economic conditions to explain the observed changes in the vegetation. From these explanations will come understanding of the present and lessons for the future.

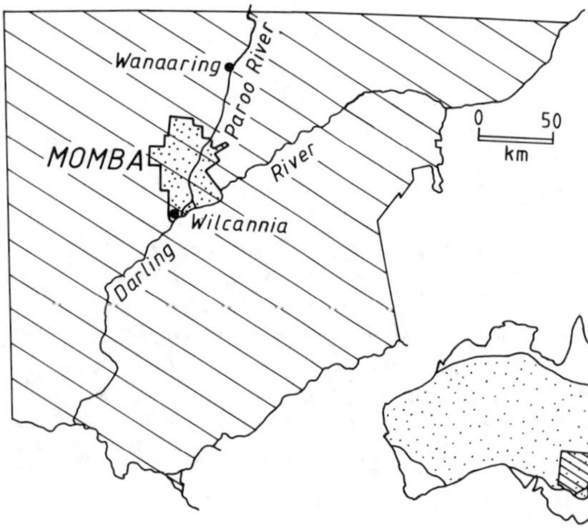

FIG. 1. Map of the Western Division of New South Wales showing the location of *Momba* and other place names mentioned in text. Inset map of Australia shows the location of the Western Division (hatched) in the arid zone (stippled) of Australia.

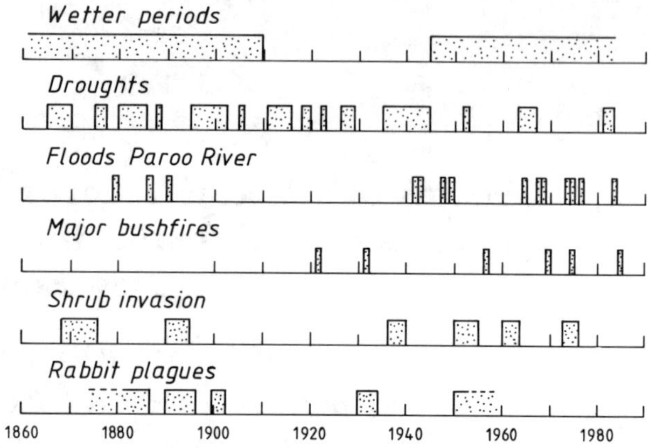

FIG. 2. Occurrences of rainfall cycles and various natural disasters that have befallen graziers in the Western Division since 1860. Data are generalized and do not necessarily reflect conditions at any one locality. Absence of an event (especially floods on the Paroo River before 1926, and bushfires) may be due to lack of data. Sources: rainfall (Gentilli 1971; Pittock 1981), droughts (Heathcote 1965), major floods of Paroo River at Wanaaring (Hydro-meteorological Section, Bureau of Meteorology), bushfires (Western Lands Commission), shrub invasion and rabbit plagues (Denny 1987).

Environmental changes in the Western Division

There is ample evidence of changes in the environment of the Western Division over the last century (Fig. 2). Rainfall records form three distinct groups with breaks at 1910 and 1945. The first period (pre-1910) was generally wetter than the second (1911-1945) (Gentilli 1971). The third period (post-1947) has also been substantially wetter and there has been a marked increase in the amount of summer rainfall (Pittock 1981).

Regardless of how they are defined, droughts are a feature of the Western Division. The succinct opinion of the 1901 Royal Commission to Inquire into the Conditions of the Crown Tenants has not been bettered. "... drought is the predominant characteristic of the west, and not merely an enemy to be occasionally encountered" (Anon. 1901). Paradoxically, series of wet years may have more profound long-term effects that the more dramatic droughts. After a sequence of years with above average rainfall, the rangelands are invaded by unpalatable native shrubs which are able to persist because of marked changes in the natural fire regime (Booth undated).

The Paroo River may flood even when the area is gripped by severe drought, e.g., the early 1940s. Such floods, caused by heavy rainfall many kilometres upstream, are a mixed blessing for pastoralists. Low-lying country is submerged thus increasing the grazing pressure on any remaining dry land. If the property is also gripped by a drought, then there may be little feed available.

Fire is a major force in semi-arid ecosystems (e.g., Noble *et al.* 1980; Walker *et al.* 1980). Indeed, the lack of fires has been blamed for survival of major waves of establishment of unpalatable trees and shrubs during years of heavy rainfall (Booth undated).

Despite an abundant literature summarizing the invasion of rabbits (e.g., Rolls 1984), there are surprisingly few readily available data which go beyond the level of anecdotes. However, there seems no question that the numbers were huge and the impact devastating.

Given the amount and variety of environmental change in the Western Division, it is well-nigh impossible to apportion blame for vegetation change. The various forces acting on rangelands affect different processes in the ecosystem but the visual manifestations may be very similar. A related problem is the perception of "naturalness" in the semi-arid rangelands. Because they have not obviously been cleared, they are often judged to be natural communities. However, this may be a false impression. At least some of the extensive forests of white cypress pine, *Callitris glaucophylla* in the Western Division are a consequence of European interference (Wells 1974).

It is against this background of environmental change and uncertainty about what is natural and what is induced by European settlement that I examine changes in stock numbers on *Momba* following subdivision for closer settlement. There have also been major technological (Duncan 1972) and economic changes during this period. I return to these later in the chapter.

Data

Before examining the history of *Momba* and the changes in stocking rates, it is necessary to examine problems in the data. As in all historical studies, there is no single complete source of data. Rather, there are numerous sources of incomplete and imperfect data which must be merged into a coherent body. I have used data recorded by many inspectors employed by various government agencies, data in returns from property holders to Pastures Protection Boards, and data submitted by lessees appealing against increases in rents.

The most complete sets of stock numbers are in returns to the Wilcannia Pastures Protection Board. In a sample of properties in the *Momba* area, I found that the returns underestimate the actual stock numbers by 20-70% compared with shearing tallies. While this is higher than the average 9% underestimate found by Young and Miles (1982), these data are the most extensive and they accurately show trends. The underestimate is primarily caused by the timing of the returns on 31 December of each year. Stock numbers are usually at their lowest on most properties in summer.

A problem with some other records is whether sheep numbers are summarized as actual counts (flock sheep) or as dry sheep equivalents (dse). Related to this are the conversions used to estimate dry sheep equivalents. These have varied from one beast (cattle, horse) as equivalent to 10 dse, to 1:6. In this chapter, I have used 1:10. Rabbits are converted to dse at the rate of 16 rabbits to 1 dse.

Stocking rates and carrying capacities are quoted in this chapter as hectares per dse. This is reciprocal of the more usual measure of density (sheep ha^{-1}), but is standard practice in the semi-arid zone. Thus a stocking rate of 5 ha per dse is equivalent to a density of 0.2 sheep ha^{-1}.

Subdivision of *Momba*

The history of land use in the Western Division is also the history of legislation attempting to control the distribution of land and later to subdivide it for closer settlement (King 1957). Many details of the history of *Momba* (Fig. 3) are not yet clarified as my research is still incomplete. The broad outline below is accurate and will suffice, although some of the details are missing. The following chronology also lists the major legislation.

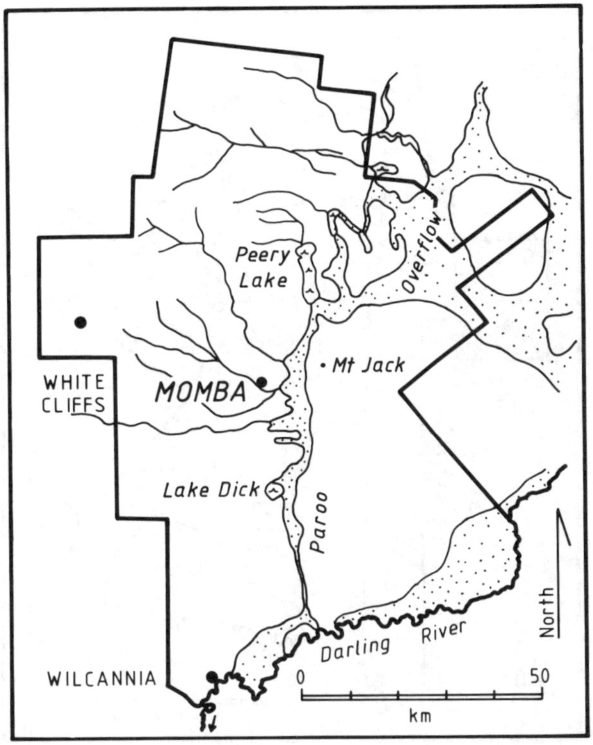

FIG. 3. Map showing *Momba* at its maximum, and generalized topography. Lakes and creeks are ephemeral. Stipple shows major flood plains of Paroo and Darling Rivers.

1846 *Waste Lands Occupation Act* and 1847 *"Orders in Council"* divided western New South Wales into rectangular "runs" of 50 sq. miles (12 950 ha) available for settlement.

1861(?) Thirty-eight runs taken up as *Momba,* total area approximately 2.09 million acres (848 000 ha) (Fig. 4a).

1884 *Crown Lands Act* divided leases into two parts (Fig. 4b). Pastoral Leases for 15 years granted over the Leasehold Area (439 160 ha), Occupation Licences granted to original occupier over the Resumed Area (408 834 ha). The Resumed Area was made available for selection as 10 240 acre (4 143 ha) Homestead Leases for 15 years.

1899-1901 *Royal Commission to inquire into the conditions of the Crown tenants* (Anon. 1901).

1901 *Western Lands Act* Western Lands Leases of 42 years, expired on 30 June, 1943.

1902 Parts of *Momba* Resumed Area selected as Homestead Leases, alienated as freehold, or held in reserves. Occupation Licence granted over 332 687 ha of the remainder (Fig. 4c).

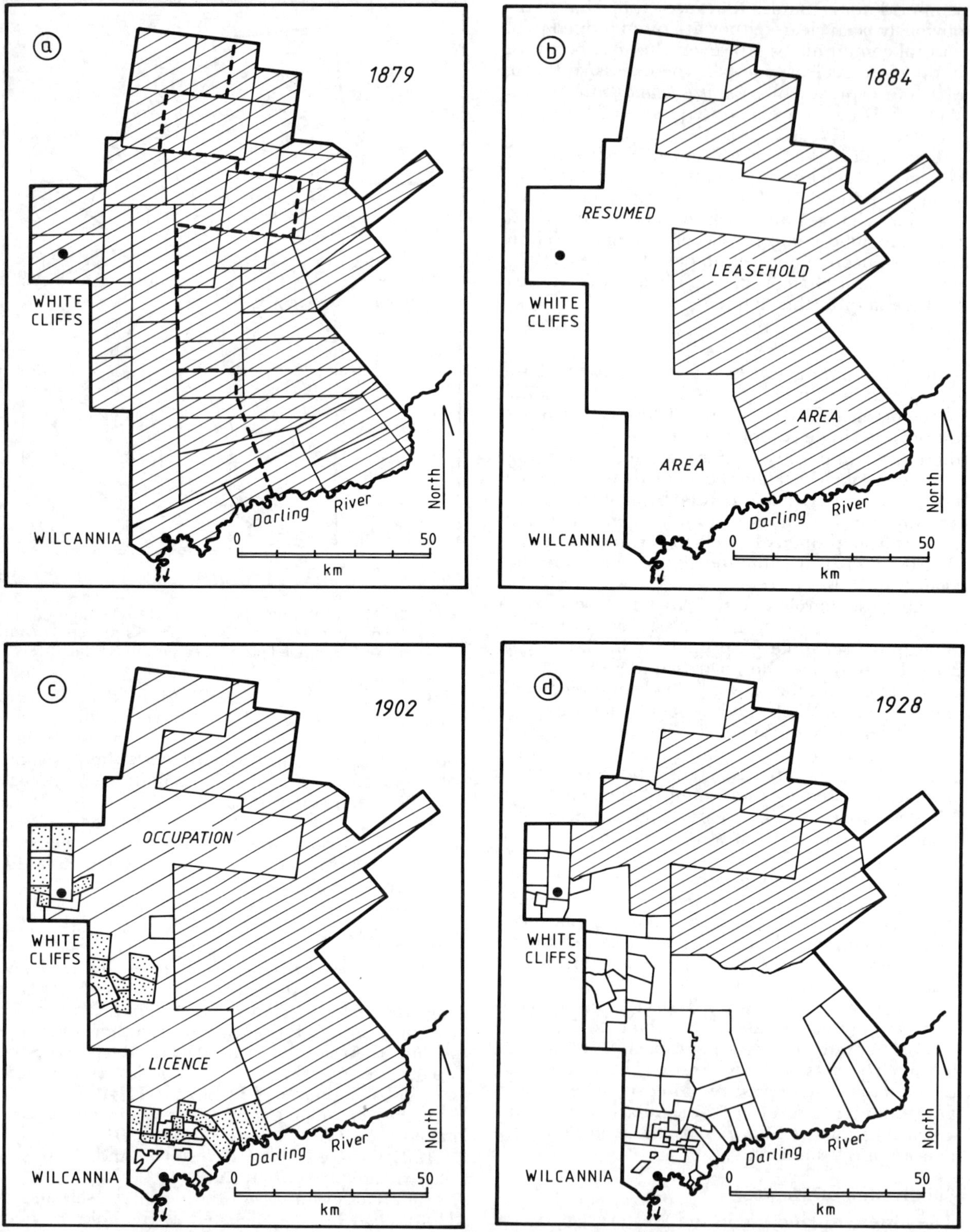
a
1879
WHITE CLIFFS
WILCANNIA
Darling River
North
0
50
km
b
1884
RESUMED
AREA
LEASEHOLD
AREA
WHITE CLIFFS
WILCANNIA
Darling River
North
0
50
km
c
1902
OCCUPATION
LICENCE
WHITE CLIFFS
WILCANNIA
Darling River
North
0
50
km
d
1928
WHITE CLIFFS
WILCANNIA
Darling River
North
0
50
km

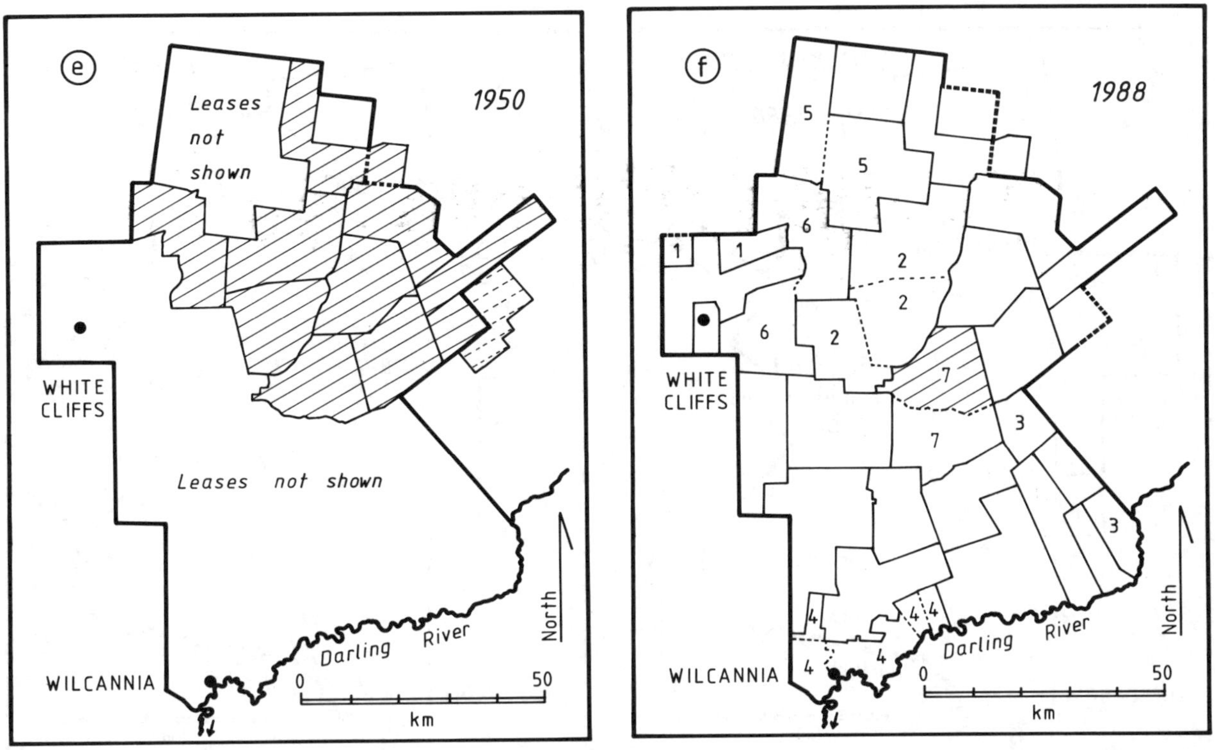

FIG. 4 a-f. Progressive subdivision of *Momba* from 1879 to 1988. The boundaries of the component runs are shown in 1879 along with the future subdivision of 1884 (dotted). In 1884, the property was divided into the Leasehold Area (hatched) and the Resumed Area. By 1902, many Homestead Leases (stippled) had been granted in the Resumed Area, but the bulk was held under an Occupation Licence (coarse hatch) and run in conjunction with the Leasehold Area. The Homestead Leases were converted into Western Lands Leases by 1928. In 1950, the residue of *Momba* was subdivided into 10 properties. The 1988 map shows the remnant of *Momba* (hatched) and all other properties. Each is a single management unit except those with a common number. Dotted lines on original *Momba* boundary show where leases form part of an adjoining property. Sources: Plans and maps held by the Western Lands Commission, and State Archives Office.

1928 Further subdivision of *Momba,* some amalgamation of leases (Fig. 4d).

1930 *Western Lands (Amendment) Act* Leases extended 25 years, small withdrawals for closer settlement.

1932 *Western Lands (Amendment) Act* Leases extended to perpetuity, large leasehold such as *Momba* not included.

1934 *Western Lands (Amendment) Act* Lessees of large leases given option of continuing until expiry in 1946, or surrendering 1/4 immediately, 1/8 in 1943, and 1/8 in 1948, in which case the lease would be extended a further 25 years.

1949 *Western Lands (Amendment) Act* All properties greater than two home maintenance areas resumed and subdivided into "living areas".

1950 Expired *Momba* leasehold land (approximately 309 400 ha) divided into 10 blocks for disposal as new holdings (Fig. 4e).

The subdivision of *Momba* was essentially completed in 1950 following the 1949 legislation.

However, existing policy of the Western Lands Commission limited the area of land available to a family to one "home maintenance area": the area which in average seasons could generate sufficient income to support one family. The legislators and administrators forgot that economies are not static. Prices increased more rapidly than returns and many lessees on small properties quickly found that what had been an adequate property in the 1940s was too small to generate a liveable income. This policy locked many smaller lessees into an invidious position. As they struggled to maintain their income, they ran increasing numbers of stock. Thus, smaller properties were generally more prone to deterioration from overstocking than larger properties. The policy was effectively abolished in 1987 to encourage amalgamation of properties. Today, many of the pastoral management units in *Momba* consist of more than one property (Fig. 4f). It is doubtful if the process will return full circle to the original size of *Momba,* but further amalgamations will certainly occur.

Stocking rates on *Momba*

In 1884 there were 220 000 sheep, 2 500 cattle and 350 horses grazing on *Momba* (Anon. 1901: 831) giving a total stocking rate of 1 dse:3.1 ha (Fig. 5a).

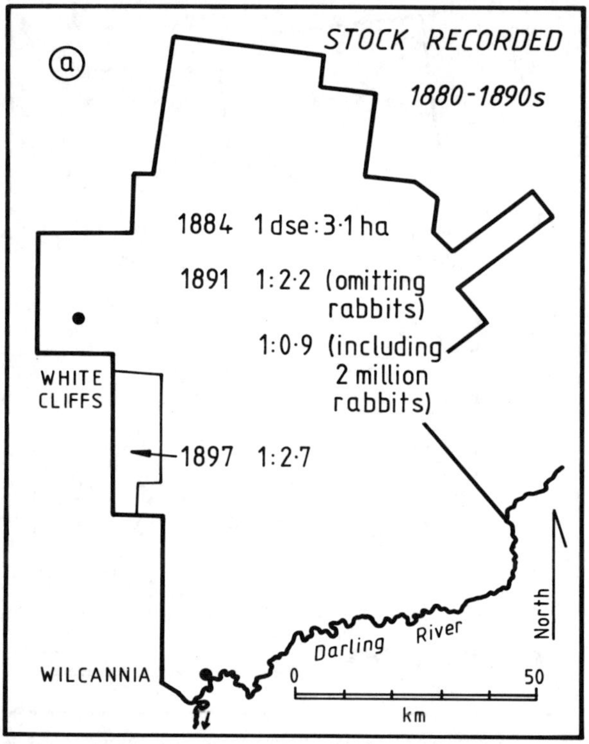

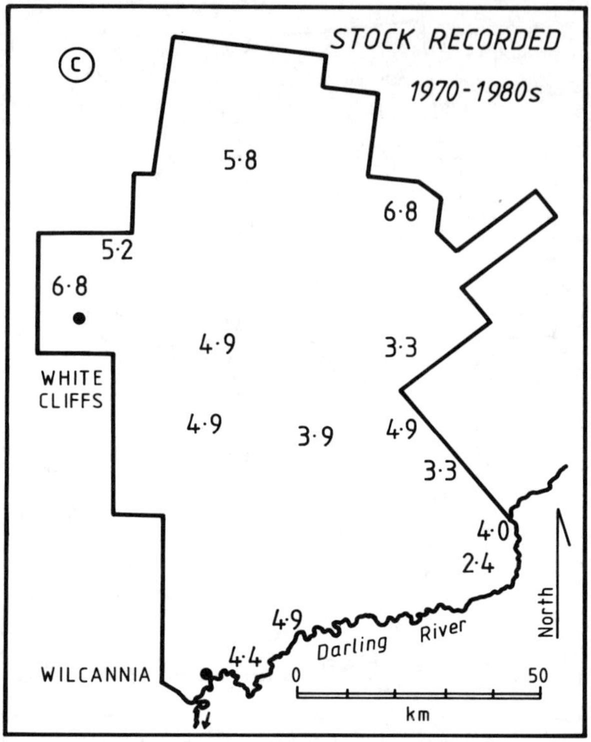

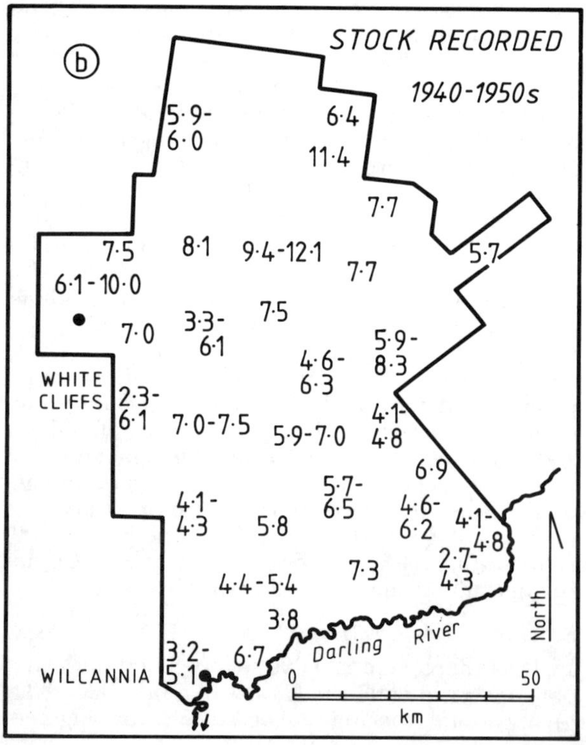

FIG. 5 a-c. Stock numbers recorded on *Momba* from the 1880s to the 1980s. Numbers represent the number of hectares required to support one sheep or dry sheep equivalent (dse). In the 1940-1950s and 1970-1980s, numbers show the approximate locations of the properties or leases carrying the stock. Gaps represent missing data. Rabbits converted to dse at the rate of 16 rabbits:1 dse. Sources: Western Lands Commission files, Hamilton (1892).

By 1891, the stock numbers had increased substantially and the stocking rate was 1 dse:2.2 ha.

However, both these figures omit rabbits which had reached *Momba* by the early 1880s. Hamilton (1892) quotes a contemporary newspaper report that two million rabbits were killed on *Momba* in 1891. While it is easy to accept a total population of two million rabbits, to kill this number involves killing four rabbits every minute of the year! However, as no other estimates are available, I use the two million as a conservative indicator of the total population. Including two million rabbits in the total grazing pressure increases the stocking rate to 1 dse:0.9 ha, a prodigious rate for the semi-arid zone.

By the 1940s, the stocking rate had fallen to a median of 1 dse:5.7 ha (Figs 5b, 6). There is considerable variation from property to property which is not simply a consequence of distance from the two major rivers. Stocking rates have subsequently increased in the 1970s and 80s (Figs 5c, 6).

The overall changes from the 1880s show a decline till the 1940s and a subsequent increase (Fig. 6). This pattern virtually mirrors the broad changes in rainfall over the Western Division (Fig. 2). One could speculate that prior to the 1880s, stocking rates were increasing rapidly as the pastoralists exploited what appeared to be an infinite resource.

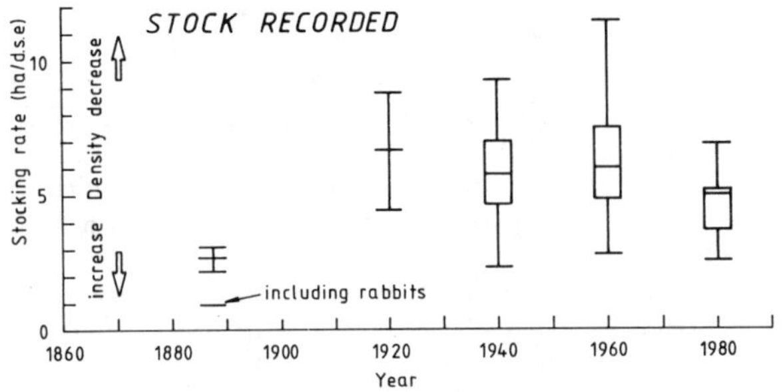

FIG. 6. Boxplots summarizing changes in numbers of stock carried on *Momba* from 1880s to 1980s. (Outer ticks represent extreme values, boundaries of boxes represent fourths and mid line is the median. For further details see Hoaglin *et al.* 1983). Source: compiled from data in Figure 5 and supplementary data in files of Western Lands Commission.

Using similar data, the pattern of change in the *estimated* carrying capacity of *Momba* can be examined (Fig. 7). Rents on leasehold grazing properties in the Western Division are charged on the estimated number of sheep that the property can carry. Data are available from Western Lands Commission files on every lease and each property. Not surprisingly, estimates by government inspectors are usually higher than those of property owners. Careful examination of the records reveals that the normal method of estimating carrying capacity in the past was to obtain data from the Pastures Protection Board for the previous decade and use the average. Thus, previous stock management of the landholder determined the estimates of the inspectors.

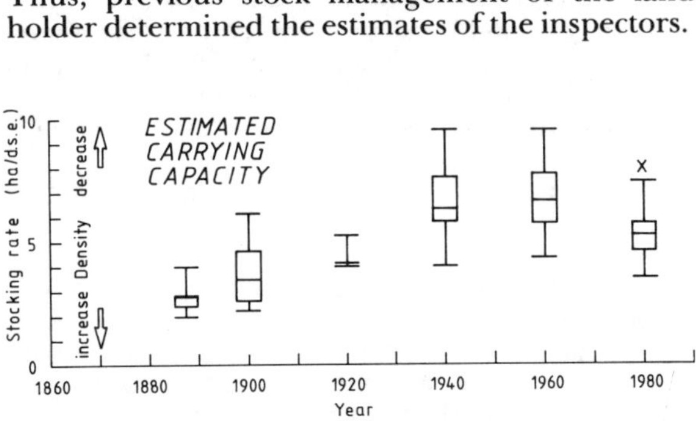

FIG. 7. Boxplots summarizing changes in carrying capacity of *Momba* from 1880s to 1980s. (See Figure 6 for explanation. Cross indicates an outlying extreme value). Source: Western Lands Commission files.

Consequently, the trend in estimated carrying capacity (Fig. 7) is similar to that of stock recorded (Fig. 6) but with the decline lagging until the 1960s. While the variability or range is smaller than in the actual stocking rates, the median estimated carrying capacity is always greater than the actual stocking rates. There are at least three possible explanations. The inspectors estimating stocking rates were overly optimistic, or the pastoralists are cautious, both of which are unlikely. Alternatively, the decadal data used for the estimates carried with it a component from previous higher rainfall periods.

Even so, any apparent correlation between rainfall and stocking rates is probably due more to coincidence than causation. Pastoralists respond to economic, technological and legislative changes and not just rainfall events as Duncan (1972) concluded. Thus, any complete analysis must include an assessment of changes in prices for wool, sheep and cattle over the last century. Even without an economic analysis we can consider the environmental implications of the subdivision legislation and the consequent changes in stocking rates.

Technology has not remained constant over the last century. Pastoralists of today do not use the same management techniques as their forebears. In the 1880s, there was a median of three artificial watering points per block (Fig. 8). By the 1980s, the range was 3 to 21 with a median of 11.5 per lease. Thus there are now many fewer stock per watering point than in the 1880s. As a consequence, the original "sacrifice areas" of 5 km diameter around each watering point have shrunk to less than 1 km. A similar change has occurred with stock yards. In 1887, there were only 13 yards on the *Momba* Leasehold Area. These were assessed as holding from 7 000 to 18 000 sheep each. Today, there are many more yards and none would hold such large numbers of sheep. These two technological and management changes have reduced the impact of stock on the land. They have allowed

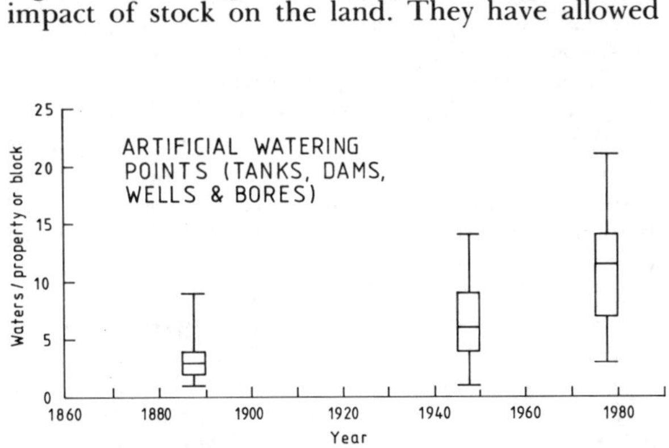

FIG. 8. Boxplots summarizing changes in numbers of artificial watering points on *Momba* from 1880s to 1980s. (See Figure 6 for explanation). Source: Western Lands Commission files.

more stock to be carried, but spread more evenly across the landscape at lower stocking rates per unit area.

Finally, there have been substantial reductions in the labour force in the semi-arid zone. Most properties now operate as a family unit with few if any permanent hired staff. Thus, when amalgamations occur, stock numbers do not increase proportionally. The basic family unit cannot manage more than a certain number of stock. Once this limit is reached, any further land available means a decrease in the stocking rates as there is more land available for the same maximum number of stock.

Discussion

There is considerable debate about the "condition" of the semi-arid lands of the Western Division (good or bad), and the trends in condition (improving, static or deteriorating). Condition is often vaguely defined as the state of the land resource and is estimated by measuring or observing various indicators. While most pastoralists are generally more interested in the condition of their stock than the land, a pastoralist is generally likely to define condition as the ability of the land to make the maximum profit over the longest period of time. Pastoralists generally assume that this means a higher stocking rate. Interestingly, this view is now being questioned by pastoralists themselves (e.g., Purvis 1986; Morrissey & O'Connor 1988) who have found significant increases in income from lowering stocking rates by up to 50%.

A high biomass of grazing animals should indicate a high biomass of feed. Thus, the increased stock numbers since the late 1940 appears to indicate either a general improvement in condition of *Momba,* or lessees seeking more money regardless of the condition of the land. As there is little evidence to indicate deterioration that the latter would cause, I suggest that there has been a general increase in plant biomass (feed) since the 1940s as a consequence of generally higher rainfall.

These trends have been recognized by others. For example Condon (1983) listed eight factors contributing to regeneration of the rangelands since 1945: increased rainfall, rabbit control, closer settlement, investment in watering points, road transport, drought relief measures and security of tenure. I see little evidence at *Momba* to suggest that closer settlement has contributed to this trend. Indeed, Condon admits that the impact of closer settlement is through improved management. This could have occurred regardless of whether the large properties were broken up. The other factors are technological, legislative and economic which have all influenced the stock numbers and the management of stock on the properties and it is through the stock that they influence the condition of the range.

The future

In this chapter I have evaluated only one causative agent of changes in vegetation in the semi-arid zone. While stock may well be the predominant force, they are not the only force. Thus we need rules to evaluate hypotheses about vegetation change. In their seminal work, Hastings & Turner (1965) evaluated alternative explanations (cattle grazing, effect of rodents, effect of fires, climatic changes) of vegetation changes in arid areas of south-west USA. They found many of the arguments unconvincing. A similar critical approach is still needed in the semi-arid zone of Australia. We need to carefully examine historical data to evaluate some of the widely cherished but seldom examined beliefs that seem to dominate much of our thinking of the arid zone.

The major problem is how to interpret the changes and to apportion cause. For it is only when the cause is known with reasonable certainty that we can attempt to reconstruct if needed. This of course begs the question of "reconstruct to what?". Current views on the "naturalness" of the vegetation do not help because they generally ignore substantial but subtle changes that have occurred e.g., the spread of white cypress pine.

This study provides only an initial glimpse of the history of *Momba.* Two points are worth highlighting. Firstly, stocking rates have been increasing over the last 40 years at the same time as the condition of the landscape has been improving. This may be peculiar to *Momba* or it may apply widely across the Western Division and other parts of the semi-arid zone of Australia. It is certainly worth further investigation.

Secondly, subdivision is finished and has been superseded by amalgamations of properties into single management units. Closer settlement through subdivision of large properties has permeated land legislation in New South Wales for well over 100 years. Successive governments of all political persuasions had a clear but unrealistic image of small family farms across the entire State. Unfortunately the dreams of the legislators often became the nightmares of the settlers. The subdivisions were simply too small; closer settlement in the Western Division was generally a failure. One consequence has been environmental damage by overstocking.

The present amalgamations driven by (more or less) free market forces are leading back to larger, but not enormous properties. We can only trust that the policy makers carefully examine the social and environmental consequences of past closer settlement legislation. Unfortunately even a cursory glance at history indicates that legislators rarely learn by the errors of their elected predecessors. Thus another couple of decades could see a resurrection of the rather fuzzy socialism that is the usual justification for subdivision (i.e., spreading the land among the people, rather than the chosen wealthy few).

Ecologists have a vital role in preventing any further attempts at closer settlement. Historical-ecological research is the ideal approach to integrate disparate strands of evidence from anecdotes, journals, archives, station diaries, etc. into a coherent explanation (e.g., Denny 1987; Oxley 1987a,b). The necessity for such long time scales can be seen by considering the ages of many of the trees that make up the Australian landscape. Many of the mature individuals of most tree species probably germinated before the European settlement in 1788.

With this sobering thought, we should consider the reality of living in the third century of European occupation of Australia. Australian vegetation and ecosystems have irrevocably changed. We could not reconstruct them even if we wanted to simply because we have insufficient information. Most, if not all of this change, has been driven by economics and legislation, two factors generally conspicuous by their absence in most ecological studies. The early owners and managers of *Momba* almost certainly had no idea of the damage that their management would cause. Some economists argue that the level of damage is acceptable. The challenge to all Australians for the next century is to accept our responsibility as stewards of the lands and to apply our ecological and land management knowledge to minimize further damage.

Acknowledgements

I gratefully acknowledge the support of the staff of the Administration and Operations Branches of the Western Lands Commission during preparation of this paper. They located many files and maps during my search for historical data. Naturally, the views expressed in this paper are my own and must not be taken as a statement of official policy.

References

Anon. (1901) *Royal Commission to inquire into the condition of the crown tenants. Part I Report and summary of evidence.* Legislative Assembly of New South Wales, Sydney.

Booth C. A. (no date) Woody weeds. Soil Conservation Service of New South Wales, Sydney.

Condon R. W. (1983) Pastoralism. In: *What future for Australia's arid lands?* (eds J. Messer & G. Mosely) pp. 54-60. Australian Conservation Foundation, Hawthorn.

Davidson B. R. (1981) *European farming in Australia. An economic history of Australian farming.* Elsevier Scientific Publishing Company, Amsterdam.

Davidson B. R. (1987) Comments. Appendix B. In: *Land degradation. Problems and policies.* (eds A. Chisholm & R. Dumsday) pp. 357-62. Cambridge University Press, Cambridge.

Denny M. (1987) Historical and ecological study of the effect of European settlement in inland N.S.W. Report by the Nature Conservation Council of New South Wales to the Heritage Council of New South Wales.

Duncan R. C. (1972) Technological change in the arid zone of New South Wales. *Australian Journal of Agricultural Economics* **16,** 22-33.

Gentilli J. (ed) (1971) *Climates of Australia and New Zealand. World Survey of Climatology Volume 13.* Elsevier Publishing Company, Amsterdam.

Hamilton A. G. (1892) On the effect which settlement in Australia has produced upon indigenous vegetation. *Journal and Proceedings of the Royal Society of New South Wales* **26,** 178-239.

Harrington G. N., Oxley R. E. & Tongway D. J. (1979) The effects of European settlement and domestic livestock on the biological system in Poplar Box *(Eucalyptus populnea)* lands. *Australian Rangelands Journal* **1,** 271-9.

Hastings J. R. & Turner R. M. (1965) *The changing mile.* University of Arizona Press, Tucson.

Heathcote R. L. (1965) *Back of Bourke.* Melbourne University Press, Melbourne.

Hoaglin D. C., Mosteller F. & Tukey J. W. (eds) (1983) *Understanding robust and exploratory data analysis.* John Wiley and Sons, New York.

King C. J. (1957) An outline of closer settlement in New South Wales. Part. I. The sequence of land laws 1788-1956. *Review of Marketing and Agricultural Economics* **25,** 1-289.

Morrissey J. G. & O'Connor R. E. Y. (1988) Twenty-eight years of station management: "fair use and a fair go". Paper presented at 5th Biennial Conference of the Australian Rangelands Society, Longreach Queensland, August 1988.

Noble J. C., Smith A. W. & Leslie H. W. (1980) Fire in the mallee shrublands of western New South Wales. *Australian Rangelands Journal* **2,** 104-14.

Noble J. C. & Tongway D. J. (1986a) Pastoral settlement in arid and semi-arid rangelands. In: *Australian soils: the human impact.* (eds J. S. Russell & R. F. Isbell) pp. 217-42. University of Queensland Press, St Lucia.

Noble J. C. & Tongway D. J. (1982b) Herbivores in arid and semi-arid rangelands. In: *Australian soils: the human impact.* (eds J. S. Russell & R. F. Isbell) pp. 243-70. University of Queensland Press, St Lucia.

Oxley R. E. (1987a) Analysis of historical records of a grazing property in south-western Queensland. 1. Aspects of the patterns of development and productivity. *Australian Rangelands Journal* **9,** 21-9.

Oxley R. E. (1987b) Analysis of historical records of a grazing property in south-western Queensland. 2. Vegetation changes. *Australian Rangelands Journal* **9,** 30-8.

Pittock A. B. (1981) Long term climatic trends in eastern Australia. In: *Cropping at the margin: potential for overuse of semi-arid lands* (eds D. R. de Krantzow & B. G. Sutton) pp. 23-39. Australian Institute of Agricultural Science Symposium Series No. 1.

Purvis J. R. (1986) Nurture the land: my philosophies of pastoral management in central Australia. *Australian Rangelands Journal* **8,** 110-7.

Rolls E. C. (1984) *They all ran wild.* Angus & Robertson, Sydney.

Walker J., Condon R. W., Hodgkinson K. C. & Harrington G. N. (1980) Fire in pastoral areas of Poplar Box *(Eucalyptus populnea)* lands. *Australian Rangelands Journal* **3,** 12-23.

Wells K. F. (1974) Pasture and forest: the situation in north-western New South Wales. *CSIRO Division of Land Research, Technical Memorandum.* 74/6.

Williams O. B. & Oxley R. E. (1979) Historical aspects of the use of chenopod shrublands. In: *Studies in the Australian Arid Zone IV Chenopod shrublands* (eds R. D. Graetz & K. M. W. Howes) pp. 5-16. CSIRO Division of Land Resources Management, Deniliquin.

Young M. D. & Miles G. E. (1982) The efficiency and equity of livestock rating systems and the accuracy of data obtained from them. *Australian Rangelands Journal* **4,** 21-8.

Proc. Ecol. Soc. Aust. 1990 — 16:255-265

Cattle grazing within subalpine heathland and grassland communities on the Bogong High Plains: disturbance, regeneration and the shrub-grass balance

R. J. WILLIAMS

School of Botany, University of Melbourne, Parkville, Victoria 3052 Australia

Present Address: Department of Botany and Zoology, Monash University, Clayton, Victoria 3168 Australia

Abstract

The Bogong High Plains have been utilized for summer grazing of domestic livestock since the 1850s. This chapter reviews the factors affecting the dynamics of heathland and grassland communities, with special reference to the effects of cattle on disturbance regimes, and the shrub-grass balance. The dynamics of these subalpine communities depend upon interactions between site, disturbance and the life history of the shrubs. The distribution of heathland and grassland is correlated closely with exposure and topography, with heathland dominant on the more sheltered, steeper slopes. Establishment of shrubs depends largely upon disturbance, and both past and present grazing by cattle is a primary agent of disturbance, causing bare ground patches on which shrubs may establish. In grassland, such grazing-related disturbances may facilitate the invasion of disturbed inter-tussock spaces by shrubs. However, the subsequent development and persistence of shrubs depends upon both their palatability and life history characteristics. The expansion of palatable shrubs may be inhibited by cattle activity. Obligate seeders are generally replaced at senescence by grasses, whereas facultative seed reproducers may persist long after establishment. Both the direction and rate of such vegetation changes are affected by the co-incidence of infrequent biotic and climatic events. Discontinuation of grazing is likely to be followed by an expansion of palatable, obligate seeders, but the frequency of disturbances necessary for the regeneration and persistence of these shrubs is likely to diminish substantially. Continuation of grazing may inhibit the growth of palatable shrubs in some sites, but is likely to facilitate the persistence of non-palatable shrubs in other more sheltered sites.

Introduction

Alpine and subalpine ecosystems in continental Australia occupy less than 1% of the area of the land, yet they are extremely important regions for water catchment, nature conservation and recreational activities. They have also been utilized for more than a century as summer grazing areas for free-ranging domestic livestock. As in other Australian ecosystems used for grazing, the impact of grazing (and associated practices such as regular burning) on the native plant communities in this fragile environment has been substantial, with severe degradation reported from numerous areas (Costin 1983).

Research on the Bogong High Plains has been active since the 1940s, following the establishment of trials by Carr and Turner (1959a,b) to monitor the effects of grazing on a number of plant communities. Monitoring of these trials has continued (van Rees *et al.* 1985; Ashton & Williams 1989), and there have been detailed studies of the diet and behaviour of cattle (van Rees 1984), and the ecology of the heathland and grassland communities (Williams 1987; Williams & Ashton 1987a,b, 1988). The locations of some of the key study sites on the High Plains are indicated in Figure 1, and the general vegetation at one of the long-term monitoring sites, the Pretty Valley experimental plots, is shown in Figure 2.

This chapter reviews the dynamics of heathland and grassland communities within the subalpine tract of the Bogong High Plains in north-eastern Victoria, with special reference to the effects of grazing by cattle. As in many Australian ecosystems subject to grazing by livestock, there has been an increase in the cover of shrubs on the Bogong High Plains over the last 40 years, and many areas

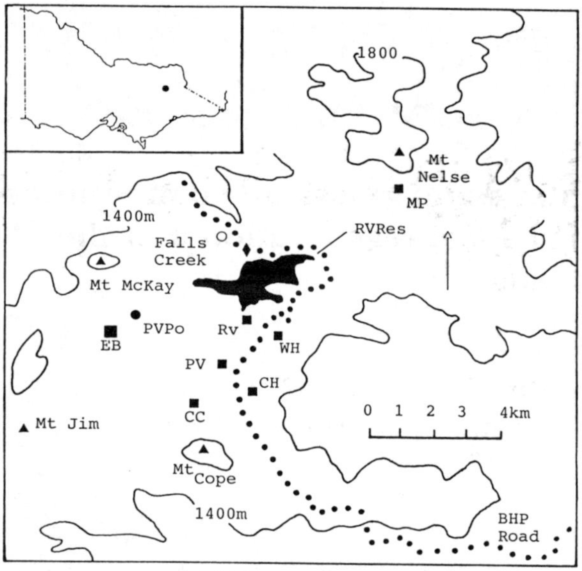

FIG. 1. Map of Bogong High Plains area. Falls Creek village (1 500 m); Rocky Valley experimental plots (RV, 1 670 m); Pretty Valley experimental plots (PV, 1 700 m); Cope Hut (CH, 1 680 m); Wallace's Hut (WH, 1 640 m); Cope Creek Valley (CC, 1 700 m); East Base, on the floor of Pretty Valley (EB, 1 650 m); Mt Cope (1 837 m); Mt Jim (1 800 m); Mt McKay (1 843 m); Mt Nelse (1 880 m); Marm Point-Hollonds Knob (MP, 1 750 m); Rocky Valley Reservoir (RV Res, 1 600 m); Pretty Valley Pondage (PVPo, 1 640 m); Bogong High Plains Road (BHP Road).

of grassland have been invaded by shrubs (Carr 1977; Williams & Ashton 1987b). The relationship between grazing and such changes in the shrub-grass balance within heathland and grassland has long been the subject of debate. There is much evidence linking such changes with disturbances associated with cattle grazing. The aims of this chapter, therefore, are listed below.

1. Describe the composition, structure, distribution of the major heathland and grassland communities on well-drained, alpine humus soils.
2. Interpret the dynamics of these communities in relation to disturbance regimes, especially those associated with cattle activity.
3. Discuss the likely effects of various management options on the dynamics of these plant communities.

The environment of the Bogong High Plains

The Bogong High Plains are a series of alpine and subalpine plateaux, the terrain of which is relatively gentle. There is a well-developed mantle of soil, the dominant type being alpine humus soil (Costin 1962a,b), which develops on sedimentary, igneous and metamorphic substrates. Precipitation is high (2 000-2 500 mm p.a.), much of which falls as snow during May-October. Maximum air temperatures rarely exceed 30°C; whilst minima rarely fall below -15°C; frosts may occur 100-200 times per year (Land Conservation Council 1977; Williams 1987). Winds are predominately north-westerly to south-westerly, hence easterly aspects tend to be more sheltered.

The vegetation of the subalpine tract (1 400-1 800 m) consists of woodland and low open forest of *Eucalyptus pauciflora* (snowgum), which, at altitudes between 1 600-1 800 m, may occur in mosaic with treeless communities such as heaths, grasslands, herbfields and wetlands. The predominant treeless communities on well-drained alpine humus soils are heaths and grasslands; these grade to bogs and fens on peats in sites of impeded drainage, and to short herbfield on leeward, sheltered slopes where snow accumulates and persists well into the growing season. Many treeless communities below the treeline are floristically and structurally similar to those occurring above the treeline (McDougall 1982; Williams & Ashton 1987a).

Structure, composition and distribution of plant communities of the heath-grass complex

There are four primary plant communities in the heathland-grassland complex on well-drained alpine humus soils (McDougall 1982; Williams & Ashton 1987a).

1. Closed heath, 1-1.5 m tall, with a shrub cover of 70-100%, variously dominated by *Prostanthera cuneata, Phebalium squamulosum, Orites lanciflolia* and *Bossiaea foliosa*. Grass cover is usually less than 10%, the predominant species being the rhizomatous *Poa hothamensis,* which occurs primarily beneath the shrub stratum. Small grassy enclaves, 2-50 m^2 in area, dominated by the tussock-forming *Poa hiemata* may occur between patches of shrub canopy but this species is rare beneath the shrub stratum.

2. Closed heath, 1-1.5 m tall with a shrub cover of 50-80%, dominated by *Phebalium squamulosum* and *Grevillea australis.* Grass cover varies from 5-50% and both *Poa hothamensis* and *Poa hiemata* may occur beneath the shrub stratum, and in enclaves between areas of canopy.

3. Open grassy-heath, 0.3-0.5 m tall, with a shrub cover of 20-50%, the predominant shrubs being *Grevillea australis, Hovea longifolia, Kunzea muelleri* and *Asterolasia trymalioides.* Inter-shrub areas are occupied by *Poa hiemata.*

4. Tussock grassland, variously dominated by *Poa hiemata, P. costiniana* or *P. fawcetteae,* in association with major herbs such as *Celmisia* sp., *Craspedia* spp., *Podolepis robusta, Brachycome* spp., and *Asperula gunnii,* and scatered shrubs such as *Grevillea* and *Asterolasia.*

The distribution of these four vegetation units is affected by slope, soil depth and rockiness, and exposure to wind and frost. Closed *Prosthanthera*

FIG. 2. (a) Experimental exclosure (*c.* 10 m×40 m) established in 1946 by the late Mrs S. G. M. Carr and Prof. J. S. Turner at Pretty Valley experimental site. Grazed plot on left. (b) Detail of vegetation within ungrazed plot. Species more common within ungrazed plot include *Celmisia* sp. (silver snow daisy), *Craspedia* spp. (billy buttons) and *Asterolasia trymalioides* (alpine star bush). Cover of *Poa* spp. (snow grass) approximately equal in both plots.

heath predominates on steeper, more sheltered slopes where soils are relatively shallow and rocky. *Poa* grassland predominates on more gentle slopes where soils are deeper, especially in hollows subject to cold air accumulation and strong winds. Both *Phebalium-Grevillea* closed heath and *Grevillea-Poa* open grassy heath are ecotonal communities and occupy sites of intermediate steepness and exposure. There are definite limits to the distribution of the major heath species. The absence of both adult and seedling shrubs of *Prostanthera* and *Phebalium* from the vast majority of frost hollows and windy saddles on the High Plains indicates that these two shrubs in particular are intolerant of exposed conditions (Williams 1987; Williams & Ashton 1987a).

Dynamics of the heath-grass complex

Establishment and regeneration of the major shrubs

All shrubs which occur on the better drained alpine humus soils establish primarily by seed. A few shrubs, such as *Asterolasia* and *Grevillea* are obligate seed reproducers (Williams & Ashton 1987b, 1988). Others, such as *Hovea, Kunzea, Phebalium, Prosthanthera* and *Orites,* are facultative seeders, capable of vegetative resprouting via stem layering or root-stock, during the mid to later phases of development. *Hovea* in particular appears capable of persisting in a suppressed form in grasslands, via leafless stems up to 50 cm long.

Shrubs exhibit developmental phases similar to those described by Watt (1947, 1964) for *Calluna vulgaris.* Pioneer shrubs occur within gaps in both grassy swards and the canopies of shrubs. Seedling shrubs will establish only on microsites where bare ground has been exposed following damage to either the local grass, herb or shrub cover. Seedling shrubs are incapable of establishing on those microsites where there is a complete vegetative cover, either live or due to fixed *Poa* litter (Williams & Ashton 1987b). Establishment of seedling shrubs is episodic, with major pulses of establishment occurring only during those growing seasons which are relatively warm, moist and with a minimum of frosts (Williams 1985; Williams & Ashton 1987b). Such seasons may occur only every 10-20 years (Williams 1985), hence successful shrub establishment is likely to be a relatively rare event within intact grassy swards.

Growth of shrubs is greatest during the building and mature phases, but is influenced greatly by exposure, with greater growth occurring within closed heath on relatively sheltered sites than within open heath on more exposed sites (Williams 1985). Canopy gaps are a feature of senescent shrubs in both open and closed heaths, although the causes of shrub senescence are as yet unknown. However, the weevil *Merimnetes viridis* Lea has been associated with localized death of *Kunzea ambigua,* and the geometrid *Chloroclystis laticostata* Walker and a tortricid species of the Hermenias group of genera have been isolated from juvenile, dying individuals of *Baeckea gunniana* (Ashton & Williams 1989). The obligate seeders, such as *Grevillea* and *Asterolasia,* appear to senesce and die in 30-50 years, whereas the facultative seeders such as *Phebalium, Prostanthera* and *Orites* may resprout vigorously following initial senescence, and appear to be relatively long-lived (William & Ashton 1988).

The gaps beneath senescent shrubs may be recolonized variously by both shrubs and grasses. The recolonization of gaps in the canopy of *Grevillea* in the *Grevillea-Poa* open grassy heath is dominated by snow grass. Recolonization of gaps in the canopy of *Prostanthera* and other shrubs in the *Prostanthera* closed heath is dominated by various shrubs. In the *Phebalium-Grevillea* closed heath, the recolonization of canopy gaps in both *Phebalium* and *Grevillea* closed heath may be dominated locally by either snow grass or shrubs (Williams & Ashton 1988). The life cycles of these three shrubs (*Grevillea australis* in the *Grevillea-Poa* open grassy heath, *Phebalium squamulosum* in the *Phebalium-Grevillea* closed heath and *Prostanthera cuneata* in the *Prostanthera* closed heath) are summarized in Figure 3.

The role of disturbance in shrub-grass dynamics

Of paramount importance to the establishment of juvenile shrubs, especially seedlings, is the occurrence of bare ground, which results from disturbances to both grass and shrub cover. Recent ecological work has emphasized the role of disturbances, both natural and man-induced, as being integral to the development of organization of both terrestrial and aquatic plant communities (Connell 1978; White 1979; Bazzaz 1983; Pickett & White 1985; Hobbs 1987). Disturbances within alpine and subalpine ecosystems may be relatively frequent (Churchill & Hanson 1958), and may be both abiotic (due primarily to the extreme and variable nature of the climate) and biotic (due to both native and introduced biota). In sites subject to disturbance, it is the various life history characteristics of species which may primarily determine the course of vegetation change following disturbance (Noble & Slatyer 1980; Sousa 1984; Leps *et al.* 1983). Thus, the understanding of vegetation change depends intimately on an appreciation of disturbance regimes and patch dynamics, and may require the application of a number of dynamic models. This is especially so in ecosystems where the prevailing climate is both extreme and highly variable, such as alpine and arid regions, and where the vegetation is grazed by large herbivores (Miles 1979; Noble 1984).

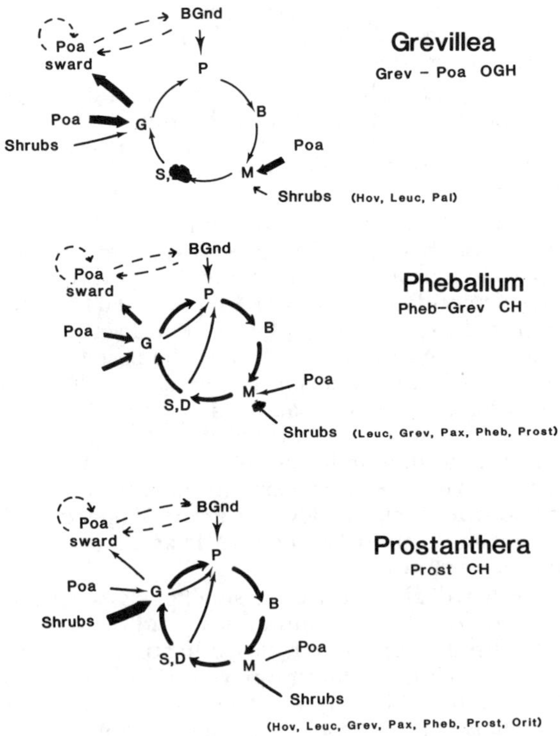

FIG. 3. Diagrammatic models of the life cycles of the three major shrub species in the subalpine tract on the Bogong High Plains. *Grevillea australis* in the *Grevillea-Poa* open grassy heath; *Phebalium squamulosum* in the *Phebalium-Grevillea* closed heath and *Prostanthera cuneata* in the *Prostanthera* closed heath. Each cycle begins on a disturbed bare ground patch (BG) and proceeds through the various growth phases (pioneer, P; building, B; mature, M; senescent and degenerate, S, D; and the formation of a canopy gap, G). During the latter stages, the area occupied by individual shrubs is subject to colonization by grasses (and forbs) and renascent and seedling shrubs. The width of the arrow indicating the direction of each progression in each cycle is a relative measure of the rate at which a given patch is affected by that particular progression. Broken arrow =rate not estimated. Abbreviations for shrubs are: Hov, *Hovea longifolia;* Leuc, *Leucopogan hookeri;* Pal, *Pimelea alpina;* Grev, *Grevillea australis;* Pax, *Pimelea axiflora;* Pheb, *Phebalium squamulosum;* Prost, *Prostanthera cuneata;* Orit, *Orites lancifolia.*

Natural disturbance regimes

Alpine and subalpine ecosystems in Australia are subject to a number of natural disturbance regimes (*sensu* Hobbs 1987), all of which may alter community/population structure, substrate availability or local physical conditions (*sensu* Pickett & White 1985). Wildfires, wind, frosts, heavy rainfall, unusual co-occurrences of climatic events (e.g., unseasonally severe frosts during winters with meagre cover of insulating snow) and insect attacks are some of these natural disturbances (Costin 1954; Carr & Turner 1959a,b; Costin & Wimbush 1973; Wimbush & Costin 1979a,b,c; Banks 1982; Kirkpatrick & Dickinson 1984; Leigh *et al.* 1987; Williams 1987).

With respect to shrub-grass dynamics on the Bogong High Plains, an important natural agent of disturbance, and hence potential cause of bare ground, within grassland appears to be attack by swift moths *(Oncopera, Hepilalidae)* and case moths (*Plutorectis caespitosa* Oke; Costin 1954; Carr 1977; Carr & Turner 1959a). The frequency of occurrence of the outbreaks is not known precisely, but they appear to be highly episodic and relatively infrequent, occurring once every 3-15 years (S. G. M. Carr, pers. comm.). The swift moth appears to be the more widespread and damaging insect (Carr & Turner 1959a). Outbreaks may occur in grassland, open grassy-heath and within the small enclaves of *Poa* which occur between patches of closed heath. These disturbances may be small-scale, affecting only individual grass tussocks, or large-scale, affecting thousands of square metres of grassland.

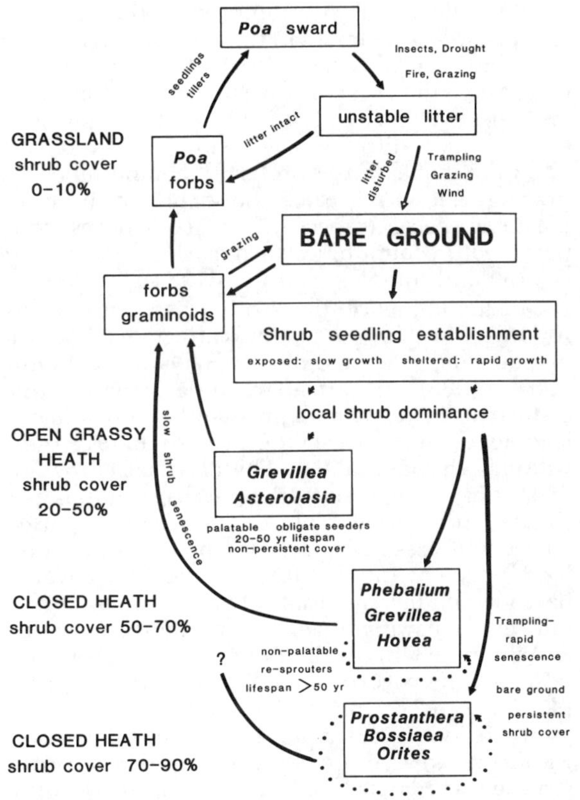

FIG. 4. Diagramatic summary of the major interactions involving disturbance, regeneration and senescence of dominants within grassland, open grassy heath and closed heath on the Bogong High Plains.

Bare ground may result from such attacks if the litter is displaced by subsequent frost heave, wind, and/or rainfall. However, the litter is generally persistent if protected from livestock. At a small 0.25 ha ungrazed sward of *Poa hiemata* at the Rocky Valley experimental exclosure, litter patches, 10-40 cm in diameter, resulting from case moth attack in 1982 were still fixed to the ground in 1985 (Williams 1985), which indicates that following this scale of attack, any subsequent formation of bare ground is not immediate. At a larger scale, litter resulting from wide-spread death of *Poa* tussocks following drought damage in the Kosciusko region was also persistent for a number of years if protected from grazing by domestic livestock, allowing recolonization by grasses and other herbs (Wimbush & Costin 1979c). Shrub seedlings are incapable of establishing on such fixed litter microsites (Williams & Ashton 1987b).

The effects of cattle on disturbance regimes

At present, approximately 4 000 cattle graze the vegetation of the Bogong High Plains each summer between December and March, although in the past stocking rates have been much higher (7 000 cattle in the early 1950s; more than 100 000 sheep, cattle and horses in the 1901 drought; Carr & Turner 1959a; Cabena 1980). The major effects of cattle on the vegetation occur via both trampling and selective grazing. Cattle utilize the major plant communities differentially, with grassland and open grassy heath the preferred communities for grazing, due to a greater abundance of the more palatable species (especially the major herbs) compared with communities such as closed heath and the wetland complex (van Rees 1984).

Grazing influences the shrub-grass complex profoundly. Shrub development is affected at both the juvenile and senescent phases (Williams & Ashton 1987b, 1988). In particular, cattle activity causes disturbances within both grassland and heathland, leading to increases in the area of bare ground within both communities. This occurs via browsing of shrubs, grasses and other herbs, trampling of shrubs, and trampling of *Poa* litter following insect attack. On experimental 1 m^2 plots within grassland grazed by cattle at Rocky Valley, the cover of bare ground increased sharply from two to 16% in one year following attack by case moths (Williams & Ashton 1987b). At the Pretty Valley experimental plots, the cover of bare ground on the grazed plot has ranged from seven to 15% over the past 40 years (van Rees *et al.* 1985). In comparison, within the adjacent ungrazed plot, protection from grazing caused a significant and persistent decrease in the cover of bare ground within 20 years; it is presently <1% and has been so for 20 years (Carr 1977; Papst & Williams, unpubl. data). If grassland is grazed by cattle, then the formation of bare inter-tussock spaces occurs very frequently, and hence the probability of establishment of shrubs is increased greatly.

Once established, however, the subsequent growth of juvenile shrubs may be inhibited by cattle activities such as trampling and browsing (Williams & Ashton 1987b). Shrubs such as *Asterolasia* and *Grevillea* are palatable to cattle (van Rees 1984), and are damaged easily by trampling. Growth of these two shrubs has been relatively slow on the grazed experimental plot at Pretty Valley, but they have expanded (*Asterolasia* in particular) on the ungrazed experimental plots at both Pretty Valley (Figure 2) and Rocky Valley over the past 40 years. Whilst such an expansion is partly a result of the cessation of browsing and trampling, the initial presence of seedlings of these shrubs within grassland was undoubtedly facilitated greatly by prior cattle activity. Hence the expansion of these shrubs at both the Rocky Valley and Pretty Valley experimental sites cannot be seen solely as a response to the removal of cattle.

Cattle will also browse the small grassy swards of the closed heath communities, and within the heath itself, even though closed heath is not a preferred grazing community (van Rees 1984). In these more sheltered sites, shrubs such as *Phebalium* and *Prostanthera* may colonize bare ground which result from disturbance to either the grass tussocks or the shrub canopy. These two shrubs are non-palatable to cattle (van Rees 1984), and have higher growth rates within closed heath sites than either *Asterolasia* or *Grevillea* within open grassy heath (Williams 1985). Thus, once established, the growth of the former two shrubs is less likely to be inhibited by cattle activity than that of the latter two. The rate of expansion of *Phebalium* and *Prostanthera* on experimental grassland monitoring plots at the Rocky Valley exclosure since 1947 has been similar to that on plots subject to continued grazing (Carr 1977; Williams & Ashton 1987b).

Cattle activity may also affect the patterns of gap phase regeneration in the major shrub species. Whilst the canopy gaps of senescent individuals of *Grevillea* appear to be recolonized primarily by *Poa* under both grazed and ungrazed cattle conditions, cattle may inhibit the extent to which *Poa* and other herbs colonize canopy gaps within both *Phebalium* and *Prostanthera,* due to selective grazing and trampling, thereby increasing the area of bare ground and making shrub replacement at the gap phase more likely (Williams & Ashton 1988).

Cattle grazing, therefore, continually creates and maintains bare inter-tussock spaces. Such a modification of the frequency of disturbance therefore vastly increases the chances of successful establishment of shrubs. Under ungrazed conditions, in comparison, the coincidences of insect attack, subsequent exposure of bare ground at a site within the ecological range of shrubs, followed by climatic conditions suitable for the large-scale establishment of shrub seedlings, are relatively rare.

Grazing therefore facilitates the persistence of shrubs in both space and time, because cattle activity increases the likelihood of establishment of shrubs, through increasing the cover of bare ground, and also promotes shrub replacement in the gaps of senescent shrubs. This is particularly so in both the *Phebalium- Grevillea* closed heath and *Grevillea-Poa* open grassy heath. Disturbances due to cattle facilitate the establishment of seedlings of both *Phebalium* and *Grevillea.* Moreover, in the *Phebalium-Gevillea* closed heath, gaps within senescent *Phebalium* shrubs are more likely to be recolonized by shrubs than by *Poa hiemata* if the local patch of heath is subject to continued grazing by cattle. Ungrazed, gaps are more likely to be replaced by *Poa.* In the open grassy heath, whilst grazing may inhibit the development of the palatable obligate seeders, and does not inhibit drastically the recolonization of *Grevillea* gaps by *Poa,* it does continually provide potential microsites within the community for the establishment of subsequent generations of seedling shrubs (Williams & Ashton 1988). In the absence of continual disturbance, the potential for the development and persistence of heath species such as *Grevillea* and *Phebalium* is lessened. Similar conclusions were reached by Costin *et al.* (1959) concerning similar heath communities in areas subject to grazing and burning in the Kosciusko region. Thus, whilst the present distribution of closed heath, open heath and grassland may be correlated with environmental gradients (Williams & Ashton 1987a) the heath communities, especially the *Grevillea-Poa* open grassy heath and the *Phebalium-Grevillea* closed heath, are also a legacy of past interactions between disturbance and climate on sites which, under relatively undisturbed conditions, may potentially support grassland, with few or no shrubs. Figure 4 summarizes the interactions between disturbance and species life history for the heathland and grassland communities.

Implications for management

There were few constraints to the grazing of domestic livestock in the Australian high country prior to the late 1940s. In addition to grazing, the vegetation was burnt regularly, particularly in the Kosciusko region (Costin 1983). Controls upon grazing and burning of the high country in both Victoria and New South Wales were first imposed in the 1940s, and since then the Soil Conservation Authority in Victoria (now the Land Protection Division of the Department of Conservation, Forests and Lands) has regulated lease areas, stock numbers and the duration of the grazing season. Grazing has ceased on the highest peaks of Mts Bogong, Feathertop and Hotham, and burning off is prohibited.

Recently, the Land Conservation Council of Victoria (1983) recommended that cattle grazing be discontinued by 1991 in several extra areas of the Bogong High Plains currently subject to grazing. These areas, to the north and east of the Rocky Valley Reservoir, include Hollonds Knob/Marm Point, Mt Nelse, Spion Kopje and areas around Ropers Hut. However, the Council also recommended that utilization of other parts of the Bogong High Plains for grazing purposes be permitted to continue. These latter areas, to the south and west of Rocky Valley, the Cope region, the floor of Pretty Valley, Mt Jim and the Fainter region are, like the former, all within the Bogong National Park. There is ample evidence indicating that the grazing of domestic livestock within the Australian high country is incompatible with nature conservation values (Wimbush & Costin 1979a,b,c; Costin 1983; Ashton & Williams 1989), although the argument is put frequently by members of the pro-grazing lobby that continued grazing will reduce the fuel load and thereby reduce the risk of fire on the Bogong High Plains (Commins 1988). The likely consequences of discontinued grazing and continued grazing, with respect to heathland and grassland communities on well-drained alpine humus soils, are considered below.

Discontinued grazing

The likely consequence of the discontinuation of grazing on the Bogong High Plains is an increase in the cover of heath species at those sites where shrubs, especially palatable species such as *Grevillea* and *Asterolasia,* have established on disturbed micro-sites. This fundamental conclusion was reached by Carr (1977). However, the overall direction, specific composition and rate of such an increase (especially within grassland) will vary in both space and time, according to the nature of the environmental gradients, and both the composition and density of the populations of established shrub seedlings at various sites.

The good correlation between the distribution and structure of the communities of the heath-grass complex and topographic-exposure gradients suggests that the taller shrubs such as *Prostanthera* and *Phebalium,* are intolerant of the most exposed conditions. Hence, on the most exposed grassland sites, extensive establishment and expansion of the taller shrubs following disturbance is unlikely. In comparison, seedlings of *Asterolasia* are distributed more widely within grassland areas, although there is much site to site variation in the density of shrub seedlings (Williams 1985). Where relatively high densities of *Asterolasia* seedlings have established within grassland (e.g., at Hollonds Knob near Mt Nelse, in the valley of Cope Creek and around Wallace's Gap), then, if grazing is discontinued, a low, open heath, rich in herbaceous species, similar to the present ungrazed plot at Pretty Valley (Figure 2) may be expected to develop. Shrub expansion is unlikely to occur, however, on those grassland sites where the density of both adult and juvenile shrubs is currently low [e.g., <0.05 m^{-2} at East Base, on the floor of Pretty Valley (Williams 1985)].

Where *Grevillea* seedlings have established in inter-tussock areas, an expansion of this species can also be expected if grazing is discontinued, and grassland may change to open grassy heath as these shrubs mature. Such changes have been demonstrated for a number of sites, both grazed and ungrazed, in the vicinity of Rocky Valley (Carr 1977; Williams & Ashton 1987b; K. McDougall, unpubl. data). However, the potential expansion of *Grevillea* is difficult to predict, as the establishment of *Grevillea* seedlings appears to be highly episodic, even in areas in which *Grevillea* is presently the dominant shrub (Williams 1985).

Shrub expansion will be most rapid on the relatively sheltered grassy enclaves within closed heath, where seedlings of *Phebalium* and *Prostanthera* have established, as has happened at a number of the experimental plots at the Rocky Valley exclosure (Carr 1977). These shrubs establish by both seed and vegetatively, are relatively long-lived, and under sheltered conditions are relatively fast-growing (Williams 1985; Williams & Ashton 1988).

Whilst local increases in the shrub cover at some sites can be expected in the short term (5-50 years), in the longer term the shrub component is not necessarily persistent, as much of the expanded heath can be expected to be replaced gradually by snow grass as the adult shrubs senesce and die. This is particularly so in open grassy heath where *Grevillea* and/or *Asterolasia* are the predominant shrubs, and the *Phebalium-Grevillea* closed heath. With cessation of grazing, and the consequent decrease in the cover of bare ground, the chances of subsequent generations of shrubs in these two communities will become less and less, and ultimately the overall cover of shrubs will decrease. In comparison, the *Prostanthera-Phebalium* closed heath is likely to persist on those steeper, rocky slopes where it is currently well developed (Williams & Ashton 1988).

Continued grazing

Continued grazing will maintain the potential for shrub regeneration in both grassland and heathland, due to continual disturbances which favour shrub establishment (Carr 1977; Costin 1983; Williams & Ashton 1987a, 1988). Whilst the growth of established, palatable shrubs will be inhibited by cattle activity in some sites, where non-palatable and fast growing shrubs establish, their growth is not likely to be inhibited by cattle activity. Within grassland and open grassy-heath, grazing of palatable herbs, especially *Celmisia,* which compete with shrubs on inter-tussock bare ground patches, may also increase the chances of the growth of established shrubs (Carr 1977). Within *Grevillea-Poa* open grassy heath, disturbances associated with continued grazing will maintain the potential for the regeneration of subsequent generations of *Grevillea* seedlings, once the current generation of dominants senesces (Williams & Ashton 1988). Within closed heathland, especially within the *Phebalium-Grevillea* closed heath, selective grazing of palatable herbs will decrease the chance of establishment and growth of grassy and other herbaceous species within the gaps of senescent shrubs. Trampling will cause bare ground, and will initiate both seedling and vegetative regeneration of the major heath species. Thus, in both open heath and closed heath, continued grazing will increase the chance of persistence of the shrub component, in both space and time.

There is no doubt that continued grazing by cattle will inhibit the growth of some shrub species in some open heath sites on the Bogong High Plains due to trampling and selective grazing. That grazing by domestic and feral animals may inhibit the development of woody species within woody-grass ecosystems has been demonstrated by a number of studies. Fitzgerald and Bailey (1983) discussed the control of aspen *(Populus tremula)* regrowth by cattle grazing in the USA. In Europe and Britain, sheep and rabbit grazing may inhibit the development of *Calluna vulgaris* heathland and maintain herb-graminoid communities. The processes which inhibit the growth of shrubs are complex, but include the intensity and frequency of trampling and grazing, increases the non-palatable grasses, and local increases in grass cover following nutrient enrichment as a result of dung concentration (Bakker *et al.* 1983; Gimmingham 1972; Heil & Diemont 1983; Miles 1979). In arctic and alpine areas in the northern hemisphere, both native and domestic ungulates may graze woody plants selectively (Bliss 1975), as many tundra shrubs have relatively high concentrations of nutrients in the foliage (Chapin 1980).

However, as a generalization, the assertion that cattle grazing on the Bogong High Plains necessarily controls the spread of shrubs and encourages grassland (e.g., Parry 1978; Holth 1980; Jamieson 1986; Commins 1988), ignores a number of processes related to cattle activity which, quite clearly, achieve the opposite. Whilst cattle may inhibit the growth of some palatable shrubs in some sites, I would argue that the contention that grazing is a primary (or even *the* primary) factor preventing the spread of all shrubs on the High Plains is a naive and inappropriate application of the ecological evidence. There are a number of fundamental objections to the above hypothesis, the most important of which are discussed below.

DISTURBANCE

The role of disturbance is of paramount importance, as has been discussed at length. Much of the heath on the High Plains at present is a legacy of past disturbances. Thus, on alpine humus soils where grassland has changed to heathland, a clear distinction must be made in the invasion process between establishment and the subsequent growth of shrubs. Whilst cattle may inhibit the growth of some shrubs, there is no question that they encourage the establishment of all shrubs.

CYCLICAL PROCESSES

The hypothesis implies that without grazing, heath communities will persist on the High Plains. In the *Grevillea-Poa* open heath that is not so, as the dominant shrub, *Grevillea australis,* has a definite life span of about 50 years, with little or no capacity for vegetative resprouting. As it dies, it is subject to replacement by snow grass. Whilst this process occurs on grazed sites, grazing maintains the potential for subsequent generations of shrubs to establish on inter-tussock spaces. In *Phebalium-Grevillea* heath, in the absence of grazing, the chances of the persistence of the heath are also ultimately decreased.

SPECIES AND SITE EFFECTS

Whilst cattle activity may have inhibited the growth of palatable shrubs such as *Asterolasia* at the Pretty Valley experimental site, this is not a general effect for all shrubs at all sites. Unpalatable, re-sprouting shrubs such as *Hovea, Kunzea, Phebalium, Orites* and *Prostanthera* are less likely to be inhibited by cattle activity than are the palatable shrubs. Indeed, the spread of unpalatable shrubs into disturbed grassland areas is likely to be encouraged by continued grazing. The air-photo analysis of vegetation change between 1936 and 1980 by McDougall (unpublished) and the long-term monitoring studies of Carr (1977) show quite unequivocally that, on a scale of hectares, grassland has changed to open heath, and open heath to closed heath, at a number of sites subject to continued grazing pressure.

ENVIRONMENTAL FACTORS

Within the subalpine tract, the distribution of grassland and heathland is controlled primarily by topographic/exposure gradients. Most importantly, there are definite climatic limits to the distribution of the shrub communities. Thus, grassland sites such as the floor of Pretty Valley, which are beyond the ecological range of the taller shrubs such as *Phebalium* and *Prostanthera* and even *Grevillea,* owe their grassland structure to their position along an environmental gradient, and not to the inhibition of shrub growth by cattle activity. By the same principle, on steeper, rocky, more sheltered slopes, taller shrubs such as *Prostanthera* are the natural potential dominants (*sensu* Küchler 1964). These closed heaths consist of relatively long-lived species, which are capable of regeneration and persistence at the site of establishment. Such sites are likely to have been dominated by shrubs well before the advent of grazing. The heath in such sites is unlikely to be inhibited by grazing, as the community is not a preferred grazing community, cattle movement through the community is restricted to narrow pathways hence damage to the shrub canopy is localized (Williams & Ashton 1987b), the dominant shrubs are non-palatable, with the capacity to resprout vigorously following trampling damage.

On sites between the above two extremes, there are ecotonal areas where both grassland and heathland are capable of developing, in varying proportions. At such sites, disturbance will favour the development of shrubs. It is in these ecotonal sites in particular that disturbance has had the most noticeable effect on the increases and decreases of the cover of heath, both following the removal of grazing, and in response to its continuation. At present, some 30% of the High Plains massif (85 km^2 excluding the summits of Mts Feathertop and Bogong) is occupied by open grassy heath (McDougall 1982); hence a sizeable proportion of the High Plains can be expected to show vegetation changes in response to these management options.

Thus, statements concerning the role of cattle in controlling heath must be made in the context of environmental gradients, disturbance, and on a time scale much longer than one generation of heath species. The inhibitory effects which cattle activity may have on shrubs vary considerably from site to site (Williams 1985) and, where inhibition has been demonstrated, the results do not necessarily apply to all shrubs over the High Plains as a whole. In over 100 years of utilization, cattle have undoubtedly affected both the composition and structure of every major treeless subalpine vegetation type on the Bogong High Plains (Ashton & Williams 1989). With respect to the communities of the heath-grass complex, cattle have contributed undoubtedly to the spread and persistence of shrubs during this time. The continuation of grazing as a means of controlling the cover of shrubs on the Bogong High Plains, therefore cannot be recommended as a management option, given the weight of the experimental evidence against the practice collected over four decades, both on the Bogong High Plains and in the Kosciusko region.

Acknowledgements

This work was supported by the Victorian Department of Conservation, Forests and Lands. D. H. Ashton, P. S. Lake, R. J. Hobbs and P. van der Haar made helpful comments on the manuscript.

References

Ashton D. H. & Williams R. J. (1989) The dynamics of subalpine vegetation in the Victorian Region. In: *The Scientific Significance of the Australian Alps* (ed. R. B. Good). Australian Alps Liaison Committee, PCN Publications, Canberra.

Bakker J. P., Bie S. de, Dallinga P. T. & Vries Y. de (1983) Sheep grazing as a management tool for heathland conservation and regeneration in the Netherlands. *J. App. Ecol.* **20,** 541-60.

Banks J. (1982) The use of dendrochronology in the interpretation of the dynamics of the snowgum forest. Ph.D. Thesis, Australian National University, Canberra.

Bazzaz F. W. (1983) Characteristics of populations in relation to disturbance in natural and man-modified ecosystems. In: *Disturbance and ecosystems* (eds H. A. Mooney & M. Gordon). Springer-Verlag, Berlin.

Bliss L. C. (1975) Tundra grasslands, herblands and shrublands and the role of the herbivores. *Geoscience and Man.* **10,** 51-79.

Cabena P. (1980) Grazing the High Country: An historical and political geography of high country grazing in Victoria, 1835-1935. MA Thesis, University of Melbourne, Melbourne.

Carr S. G. M. (1977) Report on inspection of the Bogong High Plains. Land Conservation Council, Victoria.

Carr S. G. M. & Turner J. S. (1959a) The ecology of the Bogong High Plains. I. The environmental factors and the grassland communities. *Aust. J. Bot.* **7,** 12-33.

Carr S. G. M. & Turner J. S. (1959b) The ecology of the Bogong High Plains. II. Fencing experiments in grassland C. *Aust. J. Bot.* **7,** 34-63.

Chapin F. S. (1980) Nutrient allocation and responses to defoliation in tundra plants. *Arct. Alp. Res.* **12,** 553-63.

Churchill E. D. & Hanson H. C. (1958) The concept of climax in arctic and alpine vegetation. *Bot. Rev.* **24,** 128-91.

Commins J. (1988) Caring for the High Country: Thoughts of a Victoria mountain cattleman. *Snowy River Country.* **8,** 13-5.

Connell J. H. (1978) Diversity in tropical rainforests and coral reefs. *Science* **199,** 1302-10.

Costin A. B. (1954) *A Study of the Ecosystems of the Monaro Region of New South Wales.* Government Printer, Sydney.

Costin A. B. (1962a) Ecology of the High Plains I. *Proc. Roy. Soc. Vict.* **75,** 327-37.

Costin A. B. (1962b) Soils of the High Plains. *Proc. Roy. Soc. Vict.* **75,** 291-9.

Costin A. B. (1983) Mountain lands in the Australasian region. *Proc. Ecol. Soc. Aust.* **12,** 1-13.

Costin A. B. & Wimbush D. J. (1973) Frost cracks and earth hummocks at Kosciusko, Snowy Mountains, Australia. *Arct. Alp. Res.* **5,** 111-20.

Costin A. B., Wimbush D. J., Kerr D. & Gay L. W. (1959) Studies in catchment hydrology in the Australian Alps. I. Trends in soils and vegetation. CSIRO Division of Plant Industry Technical Paper No. 13, Melbourne.

Fitzgerald R. D. & Bailey A. W. (1983) Control of Aspen regrowth by grazing with cattle. *J. Range Manage.* **37,** 156-8.

Gimingham C. H. (1972) *Ecology of heathlands.* Chapman and Hall, London.

Heil G. W. & Diemont W. H. (1983) Raised nutrient levels change heathland into grassland. *Vegetatio* **53,** 113-20.

Hobbs R. J. (1987) Disturbance regimes in remnants of natural vegetation. In: *Nature Conservation: The Role of Remnants of Vegetation* (eds D. A. Saunders, G. W. Arnold, A. A. Burbidge & A. J. M. Hopkins). Surrey Beatty & Sons, Chipping Norton.

Holth T. (1980) *Cattlemen of the High Country: The Story of the Mountain Cattlemen of the Bogongs.* Rigby, Adelaide.

Jamieson B. (1986) *The Revolt of the Mountain Cattlemen.* Rigby, Melbourne.

Kirkpatrick J. B. & Dickinson K. J. M. (1984) The impact of fire on Tasmanian alpine vegetation and soils. *Aust. J. Bot.* **32,** 613-29.

Küchler A. W. (1964) Potential natural vegetation of the conterminous United States. Amer. Geogr. Soc. Spec. Publ. No. 36.

Land Conservation Council of Victoria (1977) Report on the Alpine Study Area. Melbourne.

Land Conservation Council of Victoria (1983) Alpine Study Area. Final Recommendations. Melbourne.

Leigh J. H., Wimbush D. J., Wood D. H., Holgate M. D., Slee A. V., Stanger A. G. & Forrester R. I. (1987) Effects of rabbit grazing and fire on a subalpine environment. I. Herbaceous and shrubby vegetation. *Aust. J. Bot.* **35,** 433-64.

Leps J., Osbornova-Kosinova J. & Rejmanek M. (1982) Community stability, complexity and species life history strategies. *Vegetatio* **50,** 53-63.

McDougall K. N. (1982) The alpine vegetation of the Bogong High Plains. Environmental Studies Publication No. 357. Ministry for Conservation, Victoria.

Miles J. (1979) *Vegetation Dynamics.* Chapman and Hall, London.

Noble I. R. (1984) The dynamics of range ecosystems. *Proc. 2nd Int. Rangel. Congr.* Adelaide.

Noble I. R. & Slatyer R. O. (1980) The use of vital attributes to predict successional changes in plant communities subject to recurrent disturbances. *Vegetatio* **43,** 5-21.

Parry J. (1978) Cattlemen shed light on grazing in the alps. *Weekly Times,* August 23, Melbourne.

Pickett S. T. A. & White P. S. (1985) *The Ecology of Natural Disturbance and Patch Dynamics.* Academic Press, New York.

Rees H. van (1984) Behaviour and Diet of Free-ranging Cattle on the Bogong High Plains, Victoria. Environmental Studies Publication No. 409. Department of Conservation, Forests and Lands, Victoria.

Rees H. van, Papst W. A., McDougall K. & Boston R. C. (1985) Trends in vegetation cover in the grassland community on the Bogong High Plains, Victoria. *Aust. Rangel. J.* **7,** 93-8.

Sousa W. P. (1984) The responses of a community to disturbance: the importance of successional age and species' life histories. *Oecologia* **45,** 72-81.

Watt A. S. (1947) Pattern and process in the plant community. *J. Ecol.* **36,** 1-22.

Watt A. S. (1964) The community and the individual. *J. Ecol.* **52,** 203-11.

White P. S. (1979) Pattern, process and natural disturbance in vegetation. *Bot. Rev.* **45,** 229-99.

Williams R. J. (1985) Aspects of shrub-grass dynamics on the Bogong High Plains, Victoria. Ph.D. Thesis, University of Melbourne, Melbourne.

Williams R. J. (1987) Patterns of air temperature and accumulation of snow in subalpine heathland and grassland communities on the Bogong High Plains, Victoria. *Aust. J. Ecol.* **12,** 153-63.

Williams R. J. & Ashton D. H. (1987a) The composition, structure and distribution of heathland and grassland communities in the subalpine tract of the Bogong High Plains, Victoria. *Aust. J. Ecol.* **12,** 57-71.

Williams R. J. & Ashton D. H. (1987b) The effects of disturbance and grazing by cattle on the dynamics of heathland and grassland communities on the Bogong High Plains, Victoria. *Aust. J. Bot.* **35,** 413-31.

Williams R. J. & Ashton D. H. (1988) Cyclical patterns of regeneration in heathland communities on the Bogong High Plains, Victoria. *Aust. J. Bot.* **36,** 605-19.

Wimbush D. J. & Costin A. B. (1979a) Trends in vegetation at Kosciusko. I. Grazing trials in the subalpine zone, 1957-1971. *Aust. J. Bot.* **27,** 741-87.

Wimbush D. J. & Costin A. B. (1979b) Trends in vegetation at Kosciusko. II. Subalpine range transects, 1959-1978. *Aust. J. Bot.* **27,** 789-831.

Wimbush D. J. & Costin A. B. (1979c) Trends in vegetation at Kosciusko. III. Alpine range transects, 1959-1978. *Aust. J. Bot.* **27,** 833-71.

Proc. Ecol. Soc. Aust. 1990 — 16:267-272

Regeneration in populations of the arid zone plants *Acacia carnei* and *A. oswaldii*

T. D. AULD

National Parks and Wildlife Service, N.S.W. P.O. Box 1967, Hurstville N.S.W. 2220

Abstract

The current conservation status of populations of two arid zone perennial plants, Acacia carnei *and* A. oswaldii, *was examined at Kinchega National Park in western New South Wales by quantifying size structure, controls on germination and survival of seedlings or suckers. In* A. carnei, *regeneration has been limited by rabbit grazing. Some form of rabbit control would be necessary to maintain populations of this species in the long-term. In* A. oswaldii, *some recruitment of juveniles has occurred, although grazing by rabbits and kangaroos may have influenced the magnitude of this recruitment. Germination is a relatively common event in this species with moisture availability regulating seedling recruitment.*

Introduction

Long-lived plant species are a predominant feature of arid and semi-arid landscapes of Australia. The advent of European settlement has resulted in a change to the grazing regime experienced by these plant populations. This change followed the introduction of a new suite of herbivores, i.e., domestic stock and other feral animals, most notably rabbits and goats, and changed densities of existing native herbivores, i.e., the loss of a number of small native mammal species and a suspected increase in the large kangaroo species via the provision of abundant stock watering points (Caughley *et al.* 1987).

To achieve the long-term conservation of any plant species, rapid or steady declines in plant numbers must be prevented. Predictions about whether a species is likely to decline, remain stable or increase in numbers need to be made. This ability to predict will depend on a sound understanding of the life-cycle stages that play a crucial role in the maintenance of populations and the mechanisms that control the movement of individuals through the life-cycle (Harper 1977). In any plant population, some recruitment is essential to replace those individuals that die in order to maintain stable populations. An understanding of the process of recruitment is vital if an adequate assessment of the long-term status of a population is to be made.

The effect of current grazing regimes on arid zone plant populations have been examined in several studies. As the accurate ageing of long-lived plants is rarely possible (Crisp 1978; Silander 1983), an initial assessment of the age structure of populations has largely depended on inferring a correlation between the size of plants and their age. An examination of age/size structures may indicate periods where recruitment has been poor or non-existent in the past (e.g., Crisp & Lange 1976; Chesterfield & Parsons 1985; Cooke 1988). This information can then be assessed in relation to how rainfall affects the timing and magnitude of fecundity and germination events, as documented for *Acacia aneura* (Preece 1971a,b). Should periods of poor recruitment coincide with a series of germination events, the survival of seedlings should be examined to identify the causes behind the poor level of recruitment. For those plants that can produce vegetative suckers as well as seedlings, the survival of such suckers can also be followed through time. While the survival of vegetative suckers does not lead to the production of new genetic individuals, it can act to either maintain existing populations or increase the area occupied by genetic individuals. This, in turn may increase the opportunity for future recruitment from seedlings. Many arid and semi-arid plant populations have their recruits eliminated or severely reduced under the current grazing regime, whilst others do not (Table 1). This has the potential to produce population declines or even the loss of species in the long-term.

This study examines the biology of two species of *Acacia* at Kinchega National Park in New South Wales. The current conservation status of these

TABLE 1. Summary of studies examining the survival of recruits under grazing.

Plant species	Herbivores present[1]	Recruitment	Reference[2]
Acacia aneura	R S	nil	2,4
A. burkittii	R S	nil; slight	3,4,5
A. kempeana	R	nil	5
A. oswaldii	R	nil	5
A. papyrocarpa	R	nil	5,6
Atriplex stipitata	R	excellent	4
A. vesicaria	R	excellent	4
Cassia nemophila	R S	nil	7
C. sturtii	R	excellent	4
Casuarina cristata	K R S	nil; slight	1,4
Heterodendrum oleifolium	K R S	nil; slight	1,4
Maireana sedifolia	R	nil	2,4
Myoporum platycarpum	K R S	nil; slight	1,4,8

[1]kangaroo (K); rabbit (R); sheep (S)
[2]1 — Chesterfield & Parsons 1985; 2 — Crisp 1978; 3— Crisp & Lange 1976; 4 — Hall *et al.* 1964; 5 — Lange & Graham 1983; 6 — Lange & Purdie 1976; 7 — Silander 1983; 8 — Wood 1936.

species is assessed by quantifying the size structure of populations, fecundity and germination characteristics, and the survival of seedlings or suckers.

Study site

Kinchega National Park, 32°28′S. 142°20′E., lies adjacent to the Darling River in western New South Wales, 110 km south-east of Broken Hill. The park covers 44 000 ha and has been fenced to exclude domestic stock since 1967. The climate is one of low erratic rainfall (average 236 mm per annum) combined with high summer temperatures in excess of 40°C (Caughley *et al.* 1987). The vegetation includes *Eucalyptus largiflorens* woodland on the floodplain near the Darling River; low shrublands, usually dominated by *Maireana pyramidata* but also with *Eremophila sturtii, Dodonaea attenuata* and *Cassia nemophila;* open woodlands of *Casuarina cristata* and *Heterodendrum oleifolium;* and tall shrublands, mainly *Acacia victoriae, A. loderi* and *A. carnei,* on sandplain and dunefield land systems in the western half of the park. The main mammal grazers on the park are red (*Macropus rufus*) and western grey (*M. fuliginosus*) kangaroos and rabbits (*Oryctolagus cuniculus*). Small populations of eastern grey kangaroos (*M. giganteus*), euros (*M. robustus*) and goats (*Capra hircus*) also occur.

Study species

Acacia oswaldii F. Muell. (miljee) is a large shrub to small tree growing up to 8 m in height. This species is commonly found in semi-arid and arid areas of New South Wales, Queensland, Victoria, South Australia and to a lesser extent Northern Territory and Western Australia (Maslin & Pedley 1982). At Kinchega, this species occurs in all main vegetation types as scattered individuals. *A. oswaldii* does not produce suckers and relies totally on seedling recruitment to maintain populations.

Acacia carnei Maiden (purple-wood wattie) is a rounded shrub to 4 m and is listed as a rare and threatened plant (code 3R) by Leigh *et al.* (1981). Its distribution is limited to a few localities in arid western New South Wales and eastern South Australia (Maslin & Pedley 1982). At Kinchega, it is largely confined to the red earth dune soils as a dominant (tall shrubland) or occasionally as a co-dominant with *Heterodendrum oleifolium* or *Casuarina cristata. A. carnei* occurs as small to large clusters of aerial stems usually on dune crests or slopes, over a broad area in the south-western part of the park. *A. carnei* rarely sets seed (Whibley 1980; T. Auld, pers. obs.) but new suckers are produced, independent of root disturbance, either annually or biannually in reponse to two seasonal growth pulses (autumn, spring), at least in average or above average rainfall years. While some seed production and subsequent seedling establishment is necessary to initiate new populations on dunes, existing populations may be maintained or increase the area they occupy by the production of suckers.

Both species have large seeds (seed weights range from an average of 60-130 mg in *A. oswaldii* individuals) with a large orange aril which promotes the dispersal of some seeds by birds (T. Auld, pers. obs.; *cf.* Davidson & Morton 1984).

Methods

Several size measurements were made on all sampled plants: i) plant height (m); ii) two orthogonal canopy diameters (m); and iii) trunk diameter at 0.5 m above ground (cm). Where trunks were fallen or at a large angle away from vertical, the 0.5 m distance was measured along the trunk. For *A. carnei,* two discrete populations of 100 and 92 plants, respectively, were sampled and the data pooled. As this species produces suckers, a plant was classified as an isolated aerial stem. For *A. oswaldii,* 300 plants were sampled across the western half of the park along east/west transect lines. Seedlings or newly emerged suckers were not sampled.

Seed production was estimated in 1987 and 1988 in *A. oswaldii* by counting the number of fruits produced on 300 individuals. Virtualy no seeds were produced by any *A. carnei* plant in the park in either year. For *A. oswaldii,* viability and seed dormancy were estimated from a pooled seed lot. Six replicates of 50 seeds were placed on moistened filter paper in petri dishes on a laboratory bench at room temperature. Germination was followed for four weeks after which any seeds that had not imbibed were scarified and allowed to germinate to estimate seed viability.

Cohorts of newly emerged seedlings or suckers were tagged and their survival followed through time under the following caging treatments:

i) no cage — all grazers present;
ii) rabbit entry cage — rabbits in, other mammals out;
iii) rabbit exclusion cage — rabbits out, other mammals in;
iv) total exclosure — all mammals out.

Insect and avian grazers were potentially present in all treatments. Insufficient time and cages were available to replicate each treatment evenly. In *A. carnei,* 100 suckers were sampled for each site across two seasons with treatments being applied in the following ratio i) 40; ii) 20; iii) 20; and iv) 20. For *A. oswaldii,* some 900 seedlings were sampled, with the proportions allocated to each treatment given in Table 2. The total exclusion treatment was effective in preventing mammal grazers from reaching suckers/seedlings. Observations in the field indicated that the rabbit entry cage, which was open in the lower 20 cm, effectively allowed the entry of rabbits while excluding kangaroos. The rabbit exclusion cages, mesh to a height of 25 cm with an overhang, probably kept most rabbits out, however, they may also have deterred kangaroos from grazing suckers/seedings.

Several hundred (341) small, established *A. oswaldii* plants (juveniles — up to 30 cm high) were also tagged, and their survival, along with seedlings and *A. carnei* suckers, were followed at roughly one-monthly intervals for up to 18 months. For each *Acacia* species, the proportion of plants surviving in each treatment at the end of 18 months was compared using the test statistic:

$$z=\frac{T_1-T_2}{\sqrt{se_1^2+se_2^2}}$$

where T_1 and se_1 are treatment means and standard errors, repectively. Standard errors were calculated by assuming the data were binomial (i.e., plants were alive or dead).

Results

The two *Acacia* spp. showed markedly different size distributions. *A. carnei* was characterized by a lack of individuals in the small size classes in both populations (Fig. 1a). The population of *A. oswaldii* was skewed towards small to medium size classes with a steady decline in the larger size classes (Fig. 1b).

A. oswaldii plants produced 154(±47) and 71(±14) fruits/plant in 1987 and 1988, respectively. These were both average rainfall years (Fig. 2). The seeds produced were mostly viable (96±2%) and had a high level of non dormancy (58±3%) compared to other *Acacia* spp. (Auld 1986). Most seeds simply fall to the ground below the parent while a few are moved by birds. Seeds remain on the soil surface or are incorporated into the soil by the movement of sand. Of the 42% of seeds released as dormant most become non dormant in the first year after release. Consequently, germination occurred in response to significant rainfall events (Fig. 2), independent of the time of year.

Newly emergent *A. carnei* suckers not protected from grazing by rabbits were mostly eaten (stem nipped off at base) or had disappeared, while those that were protected showed significantly higher survival (Fig. 3a, Table 2). Suckers which disappeared were mostly eaten, although some may have been buried and others withered *in situ* between sampling dates. The rabbit entry treatment yielded results matching the uncaged treatment, while the rabbit exclusion treatment gave a level of sucker survival intermediate between total exclusion and uncaged treatments.

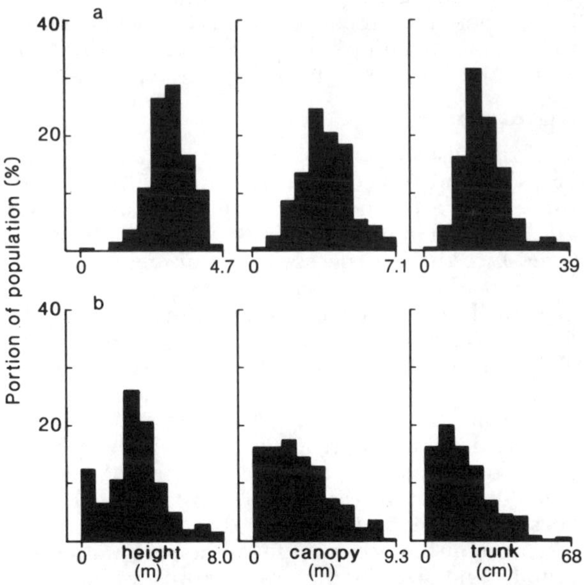

FIG. 1. Size class frequency distributions for (a) *Acacia carnei* and (b) *A. oswaldii.* Size class intervals are one tenth the maximum size value.

TABLE 2. Causes of mortality in seedlings, juveniles and suckers of two *Acacia* species.

Plant species	Treatment[1]	Eaten	Disappeared	Mortality (%) Withered	Buried	Trampled	Number of plants sampled[2]	Percentage of plant dead after 18 months (±s.e.)
A. carnei suckers	NC	9.4	90.6	0.0	0.0	0.0	151	99(1)
	R−	7.5	90.6	1.9	0.0	0.0	81	65(5)
	R+	8.9	91.1	0.0	0.0	0.0	80	99(1)
	T	0.0	71.4	28.6	0.0	0.0	80	18(4)
A. oswaldii seedlings	NC	7.7	53.7	35.2	1.5	2.0	636	86(1)
	R−	0.0	43.3	56.7	0.0	0.0	51	59(2)
	R+	8.6	62.9	27.1	1.4	0.0	96	73(5)
	T	2.2	26.7	74.4	0.0	0.0	121	77(4)
A. oswaldii juveniles	NC	15.8	60.5	23.7	0.0	0.0	341	33(3)

[1]NC no cage; R− rabbits excluded; R+ only rabbits in; T total exclosure.
[2]some plants were lost when tags removed.

Survival of *A. oswaldii* seedlings showed a rapid decline in the initial establishment phase and in the following summers in all treatments (Fig. 3b). This was mainly attributable to desiccation of seedlings. While a limited amount of grazing occurred, including insect grazing in the total exclusion treatment, this was overshadowed by seedling losses to desiccation. No clear trend of survival across treatments after 18 months emerged. Those seedlings which disappeared were probably desiccated, although a component may also have been eaten.

Uncaged juveniles of *A. oswaldii* survived significantly better than seedlings in any caging treatment (Fig. 3b, $P<0.001$). Most deaths occurred in summer months. More losses were directly identified as eaten (i.e., stems snipped off at the base) than in the seedling treatments, while a mixture of predation and desiccation probably accounted for those which disappeared.

Discussion

The size distributions of the two *Acacia* spp. (Fig. 1a,b) indicate two contrasting patterns. In *A. carnei*, successful recruitment of seedlings or suckers has not occurred in the recent past. An estimate of the time since the last successful recruitment could be made if data were available on rates of plant growth. In the absence of such information, anecdotal measurements may provide an indication. For example, few plants less than about 1.5 m in height were present in the populations sampled at Kinchega. A tagged and fully caged sucker, some four years old, had reached 30 cm in height. At that growth rate it would take some 16-20 years to reach 1.5 m in height. When grazed, growth may be slower. Thus, it is likely that little successful recruitment has occurred since the park was established. The size distribution for this species also shows that most plants are in the middle to upper size classes, with a few very large plants. This would imply that recruitment has been limited for a period well in excesss of 20 years. As the population ages, if no recruitment occurs, there will be a decline in density as no plants will be available to replace those that die of old age.

This failure to establish new plants, i.e., ramets, in the population is a result of grazing (Fig. 3a) and not a lack of sucker production (Fig. 2). When grazers are excluded survival of suckers is high. While kangaroos may be responsible for a decline in sucker densities, in this study, they did not eliminate suckers, even though the density of kangaroos is high in the park (Caughley *et al.* 1987). Rabbits, however, build burrows in dunes, underneath existing *A. carnei* populations and even in low densities will eliminate suckers (Fig. 3a). Similarly, Correll and Lange (1966), Crisp and Lange (1976) and Chesterfield and Parsons (1985) have all found gaps in the smallest size, or youngest age class of some populations of perennial arid zone plants (*Acacia burkittii*, *A. papyrocarpa*, *Casuarina cristata*). In each case rabbits were identified as the prime agent of mortality.

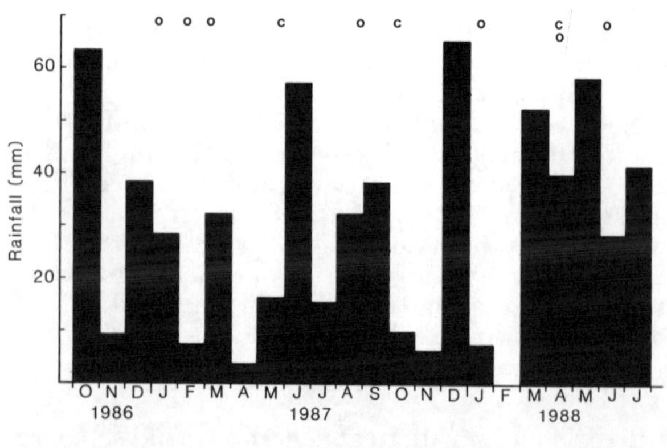

FIG. 2. Rainfall data for Menindee, adjacent to Kinchega National Park. Seedling/sucker emergence dates are indicated: c — *A. carnei* suckers; o — *A. oswaldii* seedlings.

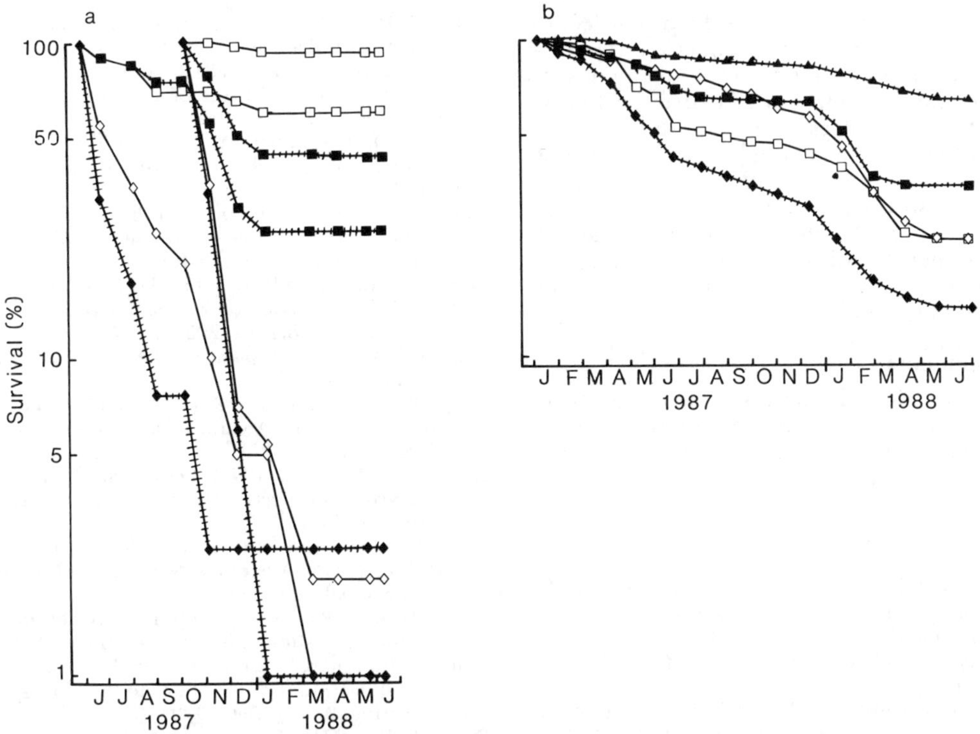

FIG. 3. Survival of seedlings/juveniles/suckers under various caging treatments. a) *A. carnei* suckers; b) *A. oswaldii* seedlings and juveniles. (◆) uncaged; (◇) rabbit entry cage; (■) rabbit exclusion cage; (□) total exclosure cage; and (▲) uncaged juveniles of *A. oswaldii*.

In *A. oswaldii,* there was a greater spread of plant sizes, including a greater proportion of small plants (Fig. 1b), indicating that some successful seedling recruitment has occurred in the recent past. Seed production is likely to be regular in this species, although it is probably influenced by the magnitude of rainfall events (*cf.* Davies 1976), and many seeds were produced in the average rainfall years of 1987 and 1988. Given that many seeds are non dormant when released and the remaining seeds become non dormant in a short space of time, germination should occur on a regular basis in relation to available rainfall, as was shown by field observations (Fig. 2). However, many emergent seedlings failed to establish because of moisture stress. There was little difference in the survival of seedlings between the uncaged and total exclosure treatments (Table 2, Fig. 3b). Hence, any difference between these and the other caged treatments was probably not related to grazing but to variation in the microhabitat of individual seedlings.

Once seedlings become established as juveniles, survival is much higher (Fig. 3b) though both desiccation and grazing cause mortality. Several juveniles were nibbled back to the stem and some managed to resprout. Presumably, as plants grow, the potential to resprout after grazing by herbivores may increase. Additionally, many uncaged juveniles and seedlings avoided being grazed because they were located inside shrubs such as *Maireana pyramidata, Enchylaena tomentosa* and *Einadia nutans.* These shrubs prevented grazers from reaching seedlings and eventually the seedlings may grow to overtop the shrubs. Many, but not all, of the 300 *A. oswaldii* plants sampled for size measurements showed evidence of having grown up and over-topped another species of shrub. This pattern of distribution of seedlings reflects the dispersal of some seeds by birds. For example, *A. oswaldii* plants were located under the branches of *Eucalyptus largiflorens, Acacia aneura* and *Hakea leucoptera* trees and under windmills.

Consequently, successful recruitment in *A. oswaldii* has occurred under the existing grazing regime at Kinchega. Whether this is sufficient to maintain the population at current levels is unknown. More work is needed on the survival of different seedling and adult cohorts.

Of the two species studied, *A. carnei* is the more sensitive to grazing pressure, with sucker recruits being eliminated under the current grazing regime. Given the aim of long-term conservation of species, some form of rabbit control/exclusion is needed to allow ramet and presumably genet recruits to survive in *A. carnei.* The alternative is a population decline. For *A. oswaldii,* some recruitment is occurring and while it is too early to decide whether this is sufficient to maintain current population density, the prospects for long-term conservation of the species seem more secure than in *A. carnei.* Control measures which reduce rabbit numbers should have a favourable effect on not only the recruitment of a sensitive species such as *A. carnei,* but also on less sensitive species in the park, e.g., *A. oswaldii* and also other species at Kinchega in which recruitment may also be a problem (See Table 1) e.g., *Casuarina cristata* and *Heterodendrum oleifolium.*

Acknowledgements

Many thanks to M. Ellis for his help in the field and for the preparation of figures. Thanks also to numerous volunteers for help with field work. Ross Bradstock made helpful comments on the manuscript.

References

Auld T. D. (1986) Dormancy and viability in *Acacia suaveolens* (Sm.) Willd. *Aust. J. Bot.* **34,** 463-72.

Caughley G., Shepherd N. & Short J. (1987) *Kangaroos: their ecology and management in the sheep rangelands of Australia.* Cambridge University Press, Cambridge.

Chesterfield C. J. & Parsons R. F. (1985) Regeneration of three tree species in arid south-eastern Australia. *Aust. J. Bot.* **33,** 715-32.

Cooke B. D. (1988) The effects of rabbit grazing on regeneration of sheoaks, *Allocasuarina verticilliata* and saltwater ti-trees, *Melaleuca halmaturorum,* in the Coorong National Park, South Australia. *Aust. J. Ecol.* **13,** 11-20.

Correll R. L. & Lange R. T. (1966) Some aspects of the dynamics of vegetation in the Port Augusta-Iron Knob area, South Australia. *Trans. R. Soc. S. Aust.* **90,** 41-3.

Crisp M. D. (1978) Demography and survival under grazing of three Australian semi-desert shrubs. *Oikos* **30,** 520-8.

Crisp M. D. & Lange R. T. (1976) Age structure, distribution and survival under grazing of the arid-zone shrub *Acacia burkittii. Oikos* **27,** 86-92.

Davidson D. W. & Morton S. R. (1984) Dispersal adaptations of some *Acacia* species in the Australian arid zone. *Ecology* **65,** 1038-51.

Davies S. J. J. F. (1976) Studies of the flowering season and fruit production of some arid zone shrubs and trees in Western Australia. *J. Ecol.* **64,** 665-87.

Hall E. A. A., Specht R. L. & Eardley C. M. (1964) Regeneration of the vegetation on Koonamore vegetation reserve, 1926-62. *Aust. J. Bot.* **12,** 205-64.

Harper J. L. (1977) *Population Biology of Plants.* Academic Press, London.

Lange R. T. & Graham C. R. (1983) Rabbits and the failure of regeneration in Australian arid zone *Acacia. Aust. J. Ecol.* **8,** 377-82.

Lange R. & Purdie R. (1976) Western myall (*Acacia sowdenii*) its survival prospects and management needs. *Aust. Rangel. J.* **1,** 64-9.

Leigh J., Briggs J. & Hartley W. (1981) *Rare or Threatened Australian Plants.* Australian National Parks and Wildlife Service Special Publication No. 7.

Maslin B. R. & Pedley L. (1982) The distribution of *Acacia* (Leguminosae: Mimosoideae) in Australia. Part 1. Species distribution maps. *West. Aust. Herb. Res. Notes.* **6,** 1-128.

Preece P. B. (1971a) Contributions to the biology of mulga. I. Flowering. *Aust. J. Bot.* **19,** 21-38.

Preece P. B. (1971b) Contributions to the biology of mulga. II. Germination. *Aust. J. Bot.* **19,** 39-49.

Silander J. A. (1983) Demographic variation in the Australian desert cassia under grazing pressure. *Oecologia* **60,** 227-33.

Whibley (1980) *Acacias of South Australia.* Government Printer, Adelaide.

Wood J. G. (1936) Regeneration of the vegetation on the Koonamore vegetation reserve, 1926 to 1936. *Trans. R. Soc. S. Aust.* **60,** 96-111.

Proc. Ecol. Soc. Aust. 1990 — 16:273-277

Soil and vegetation characteristics of dry tropical rangelands for predicting pasture regeneration in exclosures

J. G. McIVOR and C. J. GARDENER

CSIRO Division of Tropical Crops and Pastures, Davies Laboratory, Private Mail Bag, Post Office, Aitkenvale, Qld. 4814

Abstract

Effects of high grazing pressure (less available herbage, changed species composition, reduced plant cover) are apparent in some semi-arid woodlands in north-eastern Australia. Preventing grazing or reducing grazing pressure is frequently suggested to reverse the changes and allow the pastures to regenerate. If reduced grazing pressure is to be recommended, we need to be able to predict how a pasture will respond. Relationships between soil and vegetation characteristics of pastures grazed to different degrees, and the growth of pastures when grazing was prevented, were studied for 20 pastures near Collinsville, north Queensland. Pasture growth and botanical composition were combined into a single index (site condition value) to provide an estimate of the values of the different pastures. The soil and pasture characteristics were then regressed against the site condition values to determine which characteristics were the best predictors of pasture growth and botanical composition the following year. Basal area of the perennial grasses was the most useful characteristic and explained 84% of the variation in site condition values.

Introduction

Cattle grazing is the major form of land use in the semi-arid woodlands of north-eastern Australia. Grazing with domestic animals commenced approximately 120 years ago. During the first century of domestic grazing, the herbaceous layer remained dominated by tall, tussock grasses although there were major changes to the species composition (e.g., *Themeda triandra* to *Heteropogon contortus*). Apart from small areas along river frontages, the vegetative cover was maintained. However, during the last 25 years, there has been a marked increase in grazing pressure (Gardener *et al.*, this volume) and the effects of this increase have become obvious in some areas with less available herbage, changed species composition and reduced plant cover. Since these changes are likely to be associated with increased soil erosion and reduced productivity, prevention of grazing or a reduction in grazing pressure is frequently suggested to allow the pastures to regenerate, i.e., produce a productive, nutritious pasture with good ground cover.

It is not always clear if an overgrazed pasture will regenerate and before a reduction in grazing pressure is recommended, we need to be able to predict how pastures will respond. We need to establish whether a pasture can recover or whether it will need inputs of seed (and possibly some cultivation or other soil disturbance). This chapter reports the results of a study of the relationships between soil and vegetation characteristics of pastures grazed to different degrees, and the growth and botanical composition of these pastures when grazing was prevented. The objectives were to relate pasture growth following the cessation of grazing to soil and vegetation characteristics, and to develop criteria that could be used to predict the capacity of overgrazed pasture lands to regenerate.

Materials and Methods

Location

The study was conducted at Kerale Station, near Collinsville (20°34′S, 147°51′E), approximately 180 km south-east of Townsville in north-east Queensland. The average annual rainfall at Collinsville is 766 mm with 77% falling during December to April. Temperatures vary from 34/21°C (mean maximum/minimum) in December to 25/8°C in July.

Twenty native pasture sites were selected to cover a range of vegetation classes and conditions on different soil types. At each site, two adjacent areas (each 25 m x 25 m) were selected. One (chosen at random) was fenced to exclude large herbivores and the other was grazed in conjunction with the surrounding paddock.

Weather records

Daily records were made of rainfall, evaporation, and maximum and minimum temperatures at the Kerale homestead.

Initial measurements of soil and vegetation

Twenty soil samples (each 10 cm x 10 cm x 10 cm) were taken at random from each site. Measurements were made of particle size (Coventry & Fett 1979), pH, electrical conductivity, total carbon, extractable phosphorus, and exchangeable cations (Murtha 1986). Soil seed banks were measured using the techniques described by McIvor (1987) except that a total of five germination runs was used. The soil was allowed to dry and was then mixed thoroughly between runs. Six cores (7.5 cm diameter x 5 cm deep) were taken at each site for measurements of bulk density, saturated conductivity, and soil moisture characteristics (Forrest *et al.* 1985).

At each site, trees and shrubs were counted. Herbage yield and botanical composition were estimated during the late dry season (November) using the BOTANAL technique (Tothill *et al.* 1978), and a visual estimate was made of the basal area of perennial grasses.

Pasture growth measurements

Estimates of the yield and botanical composition on the grazed and ungrazed areas were made during April, one wet season after the plots had been fenced.

Assessment of site condition values

A site condition value based on pasture yield and botanical composition after exclosure for one year was calculated to provide a simple measure of the pasture value at a site. These values were related to the soil and vegetation characteristics in order to determine which characteristics were most closely related to pasture value and thus of most use in predicting the capacity of overgrazed lands to regenerate following the cessation of grazing.

The contribution of pasture yield to the site value was calculated by expressing the yield recorded in the ungrazed area after exclosure for one growing season as a proportion (0-1.0) of the estimated climatic potential yield. Climatic potential yield (Mott *et al.* 1985) was calculated from rainfall, evaporation and temperature records using the simple water balance model, WATBAL (Keig & McAlpine 1974) to estimate the number of growth weeks (McCown 1981). Maximum short term growth rates in pastures such as these are approximately 3 g/m^2/day (Mott *et al.* 1985) or 210 kg/ha/wk. Climatic potential yield was estimated by multiplying the number of growth weeks by this weekly growth rate. Although this is a simplistic model of pasture growth, such models have given useful results elsewhere (Williams & Gardener 1984; Mott *et al.* 1985) and can be readily calculated from available information.

The contribution of botanical composition to the site value was assessed on the basis of the desirability of species at each site. Species were divided into three groups: legumes and all perennial grasses (except *Aristida* spp.) were considered desirable, annual grasses intermediate, and other species (mainly *Alternanthera, Aristida, Portulaca, Sida, Stachytarpheta*) undesirable. The proportion of the herbage contributed by a species was multiplied by 1.0 for desirable species, by 0.5 for the intermediate species and by 0 for the undesirable species, and the products were summed to give a botanical composition index with values between 0 and 1.0.

The site condition value (0-1.0) is the product of the yield and botanical composition values.

Results

Initial conditions

There were large differences between sites for all characteristics measured (Table 1). Soils varied from sands to heavy clays, from infertile to fertile and from acid to alkaline. Some sites were almost bare of vegetation and others carried a dense stand of perennial grasses. Although most sites had total seed numbers of more than 1 000/m^2 in the soil there were few seeds of perennial grasses; *Bothriochloa pertusa* had 620 seeds per square metre but the mean value for all other perennial grasses combined was only 150 seeds per square metre.

Regeneration of ungrazed plots

The pasture yields on the grazed and ungrazed areas after one growing season are shown in Table 2. At nine of the 20 sites, there were no yield differences between the grazed and ungrazed plots, but at the other 11 sites the ungrazed plots had higher yields. There were no consistent effects of exclosure on botanical composition.

For the nine sites with similar yields on the grazed and ungrazed plots, five had yields below 600 kg/ha. These five sites (2, 8, 11, 14 and 19) had a number of common features.

(i) All 5 sites had below average basal area — site 11 had the highest values (0.93%) but this site consisted of dense patches of *Bothriochla*

TABLE 1. Soil and vegetation characteristics during the late dry season of native pasture sites at Kerale, near Collinsville, north Queensland.

Characteristic	Minimum	Maximum	Mean	Median
Sand (%)	16	90	61	69
Clay (%)	6	70	24	18
Bulk density (g/cc)	0.74	1.34	1.14	1.20
Water holding capacity (cc/cc)	0.09	0.21	0.15	0.14
Saturated conductivity (cm/day)	4	7812	1184	365
Electrical conductivity (mS/cm)	0.02	0.14	0.08	0.09
pH	4.9	7.7	6.7	6.9
Total carbon (%)	0.5	2.1	1.2	1.1
Extractable phosphorus (ppm)	5	480	73	19
Cation exchange capacity (me/100 g)	2.5	56.6	14.8	10.7
Pasture yield (kg/ha)	20	3240	840	460
Perennial grass yield (kg/ha)	0	3180	550	240
Perennial grass (%)	0	98	60	69
Total soil seeds (no/m²)	210	9770	3990	2740
Grass seeds (no/m²)	0	5640	1040	700
Perennial grass seeds (no/m²)	0	5510	780	200
Trees and shrubs (no/ha)	0	860	150	70
Basal area (%)	0	5.4	1.1	0.7

pertusa with large bare areas between the patches. The other four sites had basal areas below 0.2%.

(ii) All sites had below average levels of extractable soil phosphorus. Since the sites also had average or below average cation exchange capacities, these values suggest the sites were the lower fertility sites.

(iii) Site 11 had an average bulk density but the other four sites had above average values.

(iv) Four of the five sites had a below average proportion of perennial grasses during the late dry season.

(v) Four of the five sites had exposed sub-soil.

Relationship between site condition value and soil and vegetation characteristics

When site condition values for the 20 sites (Table 2) were related to the soil and vegetation characteristics (Table 1), the relationships were poor for all except the basal area of perennial grasses (Figure 1). The equation:

$$SCV = 1.0 - e^{-0.5 \times BA}$$

explained 84% of the variation in site condition value.

TABLE 2. Pasture yields of grazed and ungrazed plots, after one wet season of exclosure, proportion of perennial grasses and site condition values for twenty sites at Kerale, near Collinsville, north Queensland.

Site	Pasture yield Ungrazed (kg/ha)	Pasture yield Grazed (kg/ha)	Proportion of perennial grass (%)	Site condition value
1	780	600	59	0.16
2	240	280	39	0.05
3	1 860***[1]	1 140	91	0.76
4	2 400*	1 980	89	0.91
5	1 060	1 140	94	0.44
6	1 200*	940	85	0.30
7	1 000***	480	24	0.12
8	460	500	70	0.17
9	3 160***	2 140	88	0.96
10	780	800	73	0.25
11	580	590	75	0.21
12	1 130***	520	87	0.43
13	1 160***	960	95	0.53
14	290	180	41	0.04
15	2 190***	1 050	97	0.95
16	1 590	1 370	0	0
17	620*	400	65	0.03
18	930***	410	24	0.10
19	<10	60	40	0
20	1 070***	630	37	0.19

[1]Yields of grazed and ungrazed plots are significantly different (*$P<0.05$; **$P<0.01$; ***$P<0.001$) by t-test (Sokal & Rohlf 1969).

Discussion

Many soil and plant characteristics were measured in this study but basal area of perennial grasses was the only useful predictor of future pasture performance.

Since the perennial grasses are important for both ground cover and as a source of feed for animals, the value of a pasture is related to the growth of the perennial grasses. The growth of a grass sward depends on the density of growing points. These growing points may be either from germinating seeds or buds on perennial plants. Apart from *Bothriochla pertusa,* there are few seeds of the perennial grasses in the soil under these pastures or similar pastures elsewhere in north-east

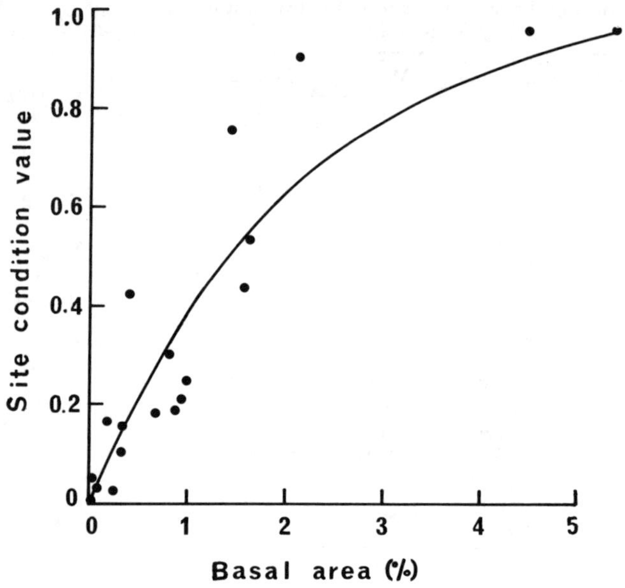

FIG. 1. The relationship between the site condition value and basal area of perennial grasses for native pasture sites at Kerale.

Queensland (McIvor 1987; McIvor and Gardener, unpubl. data) so the survival of buds as potential growing points is critical for the future productivity. The number of buds is related to basal area and hence basal area is a good predictor of future growth.

This study was only concerned with regrowth in the first year after grazing ceased. Regeneration (i.e., return to a productive pasture with good ground cover) may take many years and some sites which did not recover during one year may respond after a longer period without grazing. However, if reduced grazing is to be an acceptable method for regenerating degraded areas, it will be important that regeneration is rapid since income is foregone during the period of reduced grazing. For areas where regeneration is slow, it may be more economic to seed such areas with introduced species than to wait for them to regenerate naturally. Where the perennial grasses are lost from a pasture (i.e., the basal area declines to zero), recovery will be slow as the new plants will have to come from seeds from other areas. In the Collinsville area and other parts of the Burdekin River catchment *Bothriochloa pertusa*, a prostrate, introduced grass, is likely to be the most prominent perennial grass during any recovery phase as it produces more seed than the native grasses and also spreads by strong stolons (Howden, pers. comm.). Three sites (7, 17 and 20) produced reasonable yields (more than 1 000 kg/ha when grazing was prevented) but were dominated by unpalatable species (*Argemone mexicana* and *Stachytarpheta* spp.). These species are of little or no grazing value and are likely to slow the rate of regeneration by reducing the establishment of grass seedlings.

Basal area has also been correlated with the ability of Mitchell grasslands (*Astrebla* spp.) to recover following grazing (Orr 1986). Although basal area can be measured by a number of methods it has disadvantages for use as a routine management tool; it is time consuming and different observers may produce different values for a pasture. For these reasons we are currently studying other characteristics which are easier to estimate (especially ones that can be done visually) but which are related to pasture performance. A combination of foliage cover and the proportion of perennial grasses is showing promise.

Acknowledgements

We thank Messrs. P. E. J. Allen, W. A. Beyer, M. G. Fulloon and L. V. Whiteman for technical assistance, and Colinta Holdings Pty Ltd for provision of land and financial support.

References

Coventry R. J. & Fett D. E. R. (1979) A pipette and sieve method of particle-size analysis and some observations on its efficacy. CSIRO Division of Soils Divisional Report No. 38.

Forrest J. A., Beatty J., Hignett C. T., Pickering J. & Williams R. G. P. (1985) A survey of physical properties of wheatland soils in eastern Australia. CSIRO Division of Soils Divisional Report No. 78.

Keig G. & McAlpine J. R. (1974) WATBAL: A computer system for the estimation and analysis of soil moisture regimes from simple climatic data. CSIRO Division of Land Research Technical Memorandum No. 29.

McCown R. L. (1981) The climatic potential for beef cattle production in tropical Australia: Part III — Variation in the commencement, cessation and duration of the green season. *Agric. Syst.* **7,** 163-78.

McIvor J. G. (1987) Changes in germinable seed levels in soil beneath pastures near Townsville, north Queensland. *Aust. J. Exp. Agric.* **27,** 283-9.

Mott J. J., Williams J., Andrew M. H. & Gillison A. N. (1985) Australian savanna ecosystems. In: *Ecology and Management of the World's Savannas* (eds J. C. Tothill & J. J. Mott) pp. 56-82. Australian Academy of Science, Canberra.

Murtha G. G. (1986) Soils of the Tully-Innisfail Area, north Queensland. CSIRO Division of Soils Divisional Report No. 82.

Orr D. M. (1986) Factors affecting the vegetation dynamics of *Astrebla* grasslands. Ph.D. Thesis, University of Queensland.

Sokal R. R. & Rohlf F. J. (1969) *Biometry.* W. H. Freeman and Co., San Francisco.

Tothill J. C., Hargreaves J. N. G. & Jones R. M. (1978) BOTANAL — A comprehensive sampling and computing procedure for estimating pasture yield and composition. 1. Field Sampling. CSIRO Division of Tropical Crops and Pastures Tropical Agronomy Technical Memorandum No. 8.

Williams J. & Gardener C. J. (1984) Environmental constraints to growth and survival of *Stylosanthes*. In: *The Biology and Agronomy of Stylosanthes* (eds H. M. Stace & L. A. Edye) pp. 181-201. Academic Press, Sydney.

Proc. Ecol. Soc. Aust. 1990 — 16:279-286

Dry tropical rangelands: solving one problem and creating another

C. J. GARDENER, J. G. McIVOR and J. WILLIAMS

CSIRO, Davies Laboratory, Private Mail Bag, Post Office, Aitkenvale, Qld. 4814

Abstract

The open woodlands of northern Queensland have been grazed by domesticated livestock for approximately 120 years. Botanic shifts occurred, but there was relatively little degradation during the first 100 years. The main limitation to cattle production was poor quality herbage. Stocking rates were low to enable maximum selection of the most nutritious plants. Mortalities in the British breeds were high during droughts and subsequent herd build up was limited by poor breeding performance.

To improve cattle nutrition, graziers adopted feed supplements and better-adapted zebu cattle. This resulted in greatly improved survival and growth rates, with an accompanying increase in herbage consumption and grazing pressure. Some pastures, particularly in the "better" areas (e.g., Burdekin catchment) are now suffering from severe overgrazing with loss of herbaceous cover, shrub invasion and soil erosion. Only a reduction in grazing pressure can prevent rapid expansion of this problem. Pasture improvement can only help solve the problem if stocking rates are not again increased to recoup the investment. In the Burdekin catchment, silt and suspended colloids in the dam, irrigation aquifers and water filtration plants will lead to pressures to moderate land use.

Introduction

This chapter is concerned with the utilization of the dry tropical, open eucalypt woodland that lies in a belt extending inland from the coast of most of northern Australia. The examples given come from the Burdekin catchment in north-eastern Queensland but the principles are relevant to the whole area.

The land was initially settled by Europeans approximately 120 years ago and cattle raising rapidly became the dominant industry. Initially, overgrazing of the grasses was not a problem as pasture quality was extremely low in the dry season and the cattle tended to die at grazing pressures still safe for the majority of pasture species. However, recently, highly innovative technology has enabled graziers to increase stocking rates greatly without the previous accompanying mortality, and some areas such as the Burdekin basin are becoming badly degraded. We believe that unless different management practices based on sound ecological principles are adopted to go with the improved technology, land degradation will become more widespread in areas such as Cape York and will eventually impinge on other sectors of the community.

Geography, climate, soils and vegetation

The area of Australia considered in this chapter covers about 1 500 000 km^2 (Fig.1). It stretches in an arc from Broome in northern Western Australia to Rockhampton in central Queensland and lies between the coast and the 500 mm isohyet. It does not include the narrow strip of rainforest and sclerophyll forest adjacent to the coast, the brigalow, *Acacia harpophylla,* area or the *Astrebla-Iseilema* and *Dichanthium* grasslands growing on clay soil plains.

The dry tropical climate in the region is characterized by hot wet summers and mild dry winters. The rainfall is strongly seasonal with 80-95% falling in the summer period (mid November-mid April). Rainfall variability is greatest in the south and inland, e.g., there is an almost eleven-fold variation in annual rainfall at Hillgrove 130 km inland from Townsville. This rainfall variability leads to highly variable growing seasons, e.g., 5-28 weeks at Woodstock near Townsville (McCown 1982). The high intensity of the rainfall provides sufficient energy for the detachment and transport of insufficiently protected surface soil.

Shallow, stoney or leached soils are widespread with lithosols, red, grey and yellow earths, neutral red duplexes and solodics among the major soil

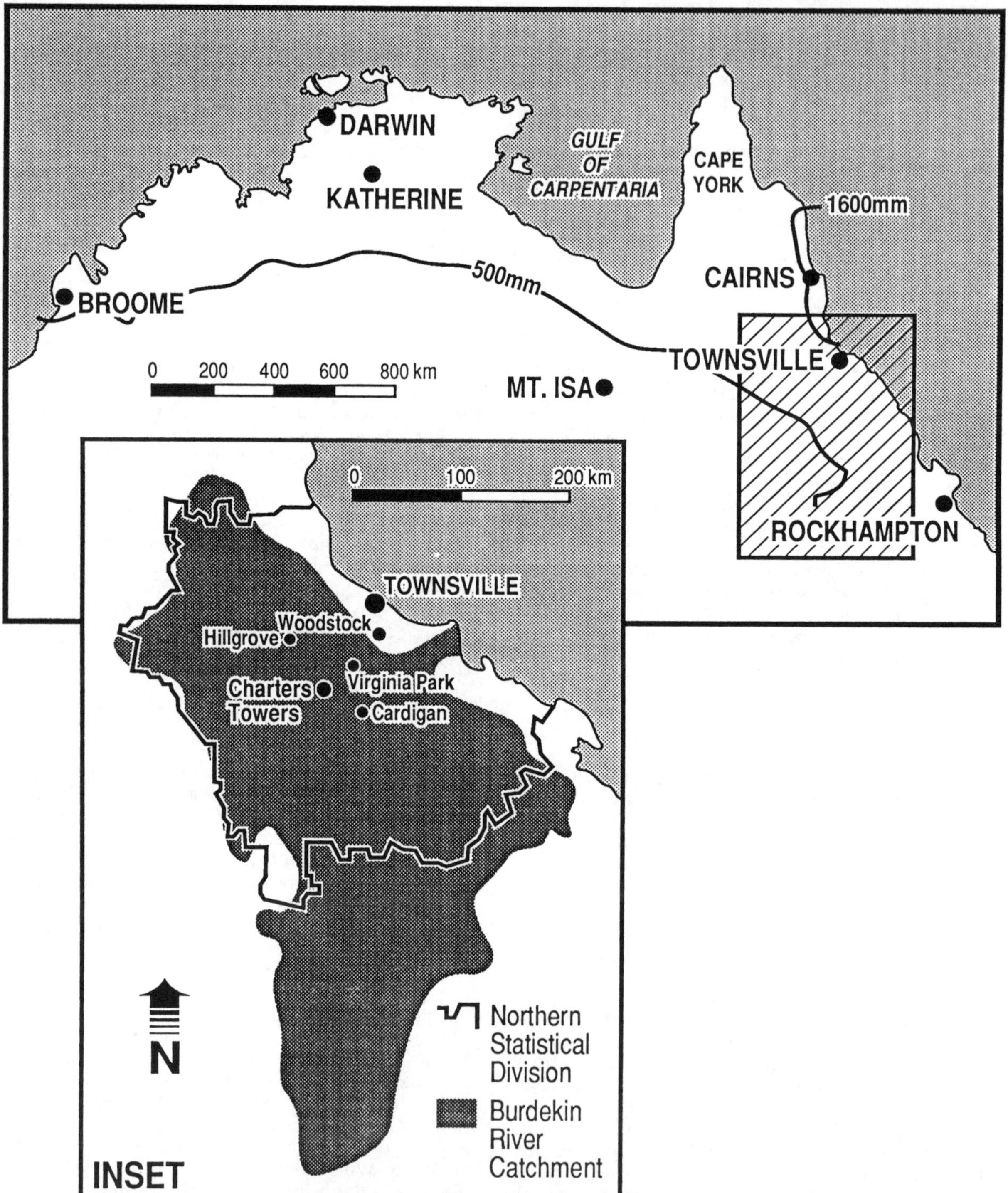

FIG. 1. Dry tropical rangelands of northern Australia. The zone extends inland from the coast to the 500 mm annual rainfall isohyet excluding the area on the east Queensland coast where rainfall exceeds 1 600 mm.

groups (Isbell 1983; Hubble & Isbell 1983). Many soils have a low water holding capacity and most are low in organic matter. Nitrogen and phosphorus are commonly deficient with local deficiencies of sulphur, copper and zinc.

The vegetation is usually an open eucalypt woodland from 6 to 18 m high with an understorey of perennial, bunch grasses up to 1.8 m tall. The most important grass species are *Themeda triandra, Sorghum plumosum, Chrysopogon fallax, Sehima nervosum, Heteropogon contortus, Aristida* spp. and *Bothriochloa* spp. (Moore 1970). Forbs, annual grasses and legumes comprise a consistent but minor part of the sward.

Industry 1860-1960

The first explorers saw . . . "Feed everywhere most luxuriant, averaging from four to five feet in height" (Cadell 1866). The overlanders brought large flocks of sheep and cattle to graze this resource. The sheep industry was short-lived due to a combination of factors: flesh-piercing grass seeds, grass fires, dingoes, low wool prices, footrot and loss of shepherds to the goldfields. Dairy farming and many types of crops were also tried without long-term success because of the infertility of the soil and the variability and severity of the climate.

Cattle raising therefore became the dominant agricultural industry. Management initially consisted largely of mustering and droving, and later included some property development such as providing more watering points, fencing and cattle dips. British breeds of cattle were used exclusively, and little clearing or pasture management apart from fire took place. Stocking rates were low to allow each beast the maximum chance to select the most nutritious component of the sward from the large amount of forage available, and thus obtain the highest quality diet possible. Stocking rates were also kept low to provide sufficient herbage in dry years. Excess herbage was regularly burnt in the dry season to provide access to any higher quality new growth that occurred.

The tall tussock grasses continued to dominate the pastures during this period although there were some botanical changes. Native legumes declined and *Themeda triandra* was replaced by *Heteropogon contortus* in some areas due to the higher incidence of burning and grazing with sheep (Tothill 1969). The annual legume *Stylosanthes humilis* was introduced in the 1880s and spread along stockroutes (Humphreys 1967), while later the creeping and prostrate perennial grass *Bothriochloa pertusa* became established in heavily grazed areas.

The primary problem

Many problems faced the early graziers in the dry tropics; isolation, distance to markets, cattle pests and diseases, lack of feed in droughts and climatic stress, but the primary problem was the low quality of the herbage in the dry season.

At the start of the wet season the young green grass had adequate nutrients for cattle growth (Table 1), but the quality fell rapidly once stem elongation occurred and the plants matured. Herbage quality fell further as the season advanced and the most nutritious fractions were eaten first. There was relatively little the graziers could do to stem this seasonal decline in quality. Typical values of <0.4 %N and <0.03 %P were recorded for forage in the dry season at Katherine, Northern Territory (Norman 1963) and Woodstock, Queensland (McIvor 1981). Since a value of 0.9%N has been suggested as necessary for maintenance (Hunter *et al.* 1976), and 0.12%P for maintenance + 0.5 kg day^{-1} gain (Little 1980), the herbage was grossly deficient in nitrogen and phosphorus.

TABLE 1. Seasonal changes in the proportion and quality of plant components of *Heteropogon contortus* growing in an ungrazed pasture at Woodstock, near Townsville (from McIvor 1981).

Component	Measurement		Early wet (November)	Late wet (April)	Late dry (October)
Green leaf	Proportion	(%)	78	18	1
	N conc.	(%)	2.08	0.79	0.63
	P conc.	(%)	0.20	0.08	0.05
	Digestibility	(%)	64	40	39
Dry leaf	Proportion	(%)	0	9	26
	N conc.	(%)		0.22	0.33
	P conc.	(%)		0.02	0.02
	Digestibility	(%)		29	34
Stem	Proportion	(%)	22	73	73
	N conc.	(%)	1.35	0.23	0.17
	P conc.	(%)	0.15	0.04	0.01
	Digestibility	(%)	66	23	19

Cattle growth rates were high at the start of each wet season, but fell progressively as the season advanced and both the amount of the highest quality component (green leaf) and its quality fell. Cattle took 5-7 years to reach a turn-off weight of 500 kg (Mott *et al.* 1981), with part of each year's gains in the wet season being lost in the dry season.

Herd fertility was low with calving percentages of 45-55% (Mott *et al.* 1981). First calf heifers, thin and with a calf suckling, in particular had problems conceiving. These problems of growth and reproduction were greatly accentuated in drought years, and the number of mortalities also increased. The cattle numbers in Queensland after a run of dry years culminating in the 1902 drought fell to 35% of the number before the drought. Cattle numbers took 30 years to recover. Keeping as many female cattle as possible alive during a drought was of prime importance to graziers due to the long time needed to rebuild herd numbers once the drought finished.

Graziers solution to the problem of low herbage quality

The graziers solution to the problem was twofold: introduce better-adapted cattle and feed the cattle additional nutrients to provide those in short supply in the forage. The first objective was accomplished by crossing tropical zebu, *Bos indicus* cattle with the British, *Bos taurus* breeds. *Bos indicus* cattle have enhanced animal performance, greater resistance to heat stress, stronger resistance to cattle ticks, increased foraging ability and a higher forage intake (Frisch & Vercoe 1977; Siebert 1982). Acceptance among the graziers was rapid and, by 1983, over 90% of the cattle in the northern statistical division of Queensland had some zebu genes.

The use of supplements also increased markedly in the late 1960s and the 1970s. A wide range of supplements is used; grain, salt, animal or plant protein (meat, fish, cottonseed meal), molasses, non-protein nitrogen (urea), superphosphate, phosphoric acid or mono ammonium phosphate. The supplements increase the intake of low quality forage and reduce or prevent loss of liveweight and cattle death (e.g., Norman 1960; Siebert *et al.* 1976). The use of supplements has been adopted most readily in the more intensively grazed areas such as the Burdekin basin which have easy access on properties, lower freight costs and fewer other constraints to production.

The induced problem

The result of the use of zebu cattle and supplements was reduced mortality, increased calving percentages and more rapid growth rates. These, coupled with low beef prices during the 1970s, resulted in the doubling of cattle numbers in the northern statistical division of Queensland (essentially the northern part of the Burdekin basin) between 1961 and 1978 (Figure 2). The larger cattle and their greater growth rates resulted in a greater intake of forage per beast. The greater consumption of herbage was largely masked by the high production of biomass during the years of above average rainfall in the 1970s and overgrazing did not become commonly apparent until the drier 1980s.

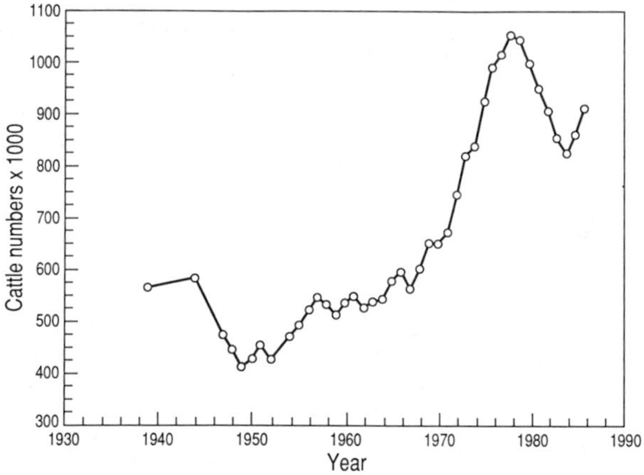

FIG. 2. Total cattle numbers in the northern statistical division of Queensland between 1939 and 1986.

Drought declarations by the State Government have exacerbated the overgrazing problem. A drought declaration is based on the condition and mortality of stock rather than the condition of the pasture and so encourages graziers to keep stock on the land until relief is provided. Drought assistance in the form of freight subsidies also enables cattle to remain on the overgrazed pasture, while destocking transport subsidies enables their removal once all the grass is eaten and return at the first sign of new growth. The policy of drought assistance therefore provides financial assistance for those who overgraze, while graziers who stock conservatively, allowing for drought, do not benefit from the scheme.

Effect of overgrazing on pastures and soil

Overgrazing is not new in the Australian dry tropics (e.g., the Ord River) but was often restricted to river frontages, permanent water holes and artificial watering points where cattle congregated in the dry season. The combination of additional watering points, zebu cattle, supplements and a policy of high stocking rates enable whole properties, shires and catchments to be degraded. Land degradation in the Burdekin catchment shows up as five major effects. These are: change in species composition; reduction in vigour of grasses; reduction of vegetative cover resulting in increased run-off, and loss of soil and nutrients; gully erosion and increased sediment transport along creeks and rivers; and invasion of shrubs. Examples of these effects are given below.

Between 1979 and 1988, a heavily grazed pasture at Virginia Park changed from one dominated by tall perennial grass species with some legumes and dicots, to one comprising much bare soil, a prostrate creeping grass *(B. pertusa)*, annual grasses and low growing xerophytic species such as *Portulaca* sp. (Table 2). Although neither its effect on plant or animal production is known, the spread of the introduced *B. pertusa* in overstocked areas in the Burdekin basin is fortuitous. Without it, there would be even worse erosion, and less chance of re-vegetation following the reduction of grazing pressure.

TABLE 2. Botanical changes in a grassland at Virginia Park in the Burdekin basin between 1979 and 1988 (from McIvor, unpubl. data).

	1979 % DM	1988 % DM
Tall perennial grasses (e.g., *Heteropogon, Dichanthium*)	78	5
Prostrate perennial grasses (e.g., *Bothriochloa pertusa*)	0	56
Annual grasses (e.g., *Sporobolus*)	1	10
Non-legume dicots (e.g., *Portulaca*)	1	16
Legumes (e.g., *Indigofera, Stylosanthes*)	20	13

Heavy grazing of pastures at Hillgrove for one year had little effect on vigour or productivity, but after three years of heavy grazing the pasture produced only 40% of the dry matter present at the low stocking rate (Fig. 3).

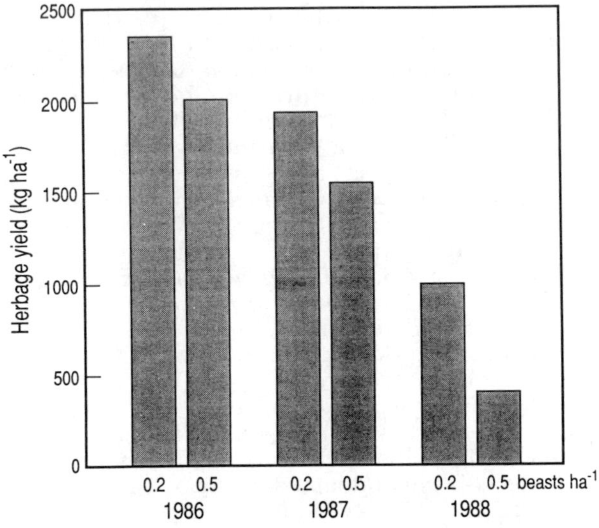

FIG. 3. Yield of herbage produced in exclosures during the wet season following one (1986), two (1987) and three (1988) years of grazing at a high (0.5 beast ha^{-1}) stocking rate, compared with yield on pastures grazed at 0.2 beast ha^{-1}. Data from euchrozem soil at Hillgrove in Burdekin basin (Gardener & McIvor, unpubl. data).

Foliage cover and basal area of the perennial grasses also fall under heavy stocking. After three years of grazing, the basal area of pastures grazed at 1 beast ha^{-1} was only 28% of those grazed at 0.2 beast ha^{-1} (Fig. 4). The amount of cover, in turn, was negatively correlated with both water runoff (Fig. 5) and soil loss (Fig. 6). A cover of approximately 50% appears necessary to minimize runoff and 40% to minimize soil loss. It is hard to put a monetary value on the lost soil and water. Annual rates of soil loss from the Burdekin catchment have been estimated to exceed 20 million tonnes (Fleming *et al.* 1981). Fertilizers containing similar amounts of nutrients to those in this amount of "typical" top soil from the Burdekin catchment could cost $90 million.

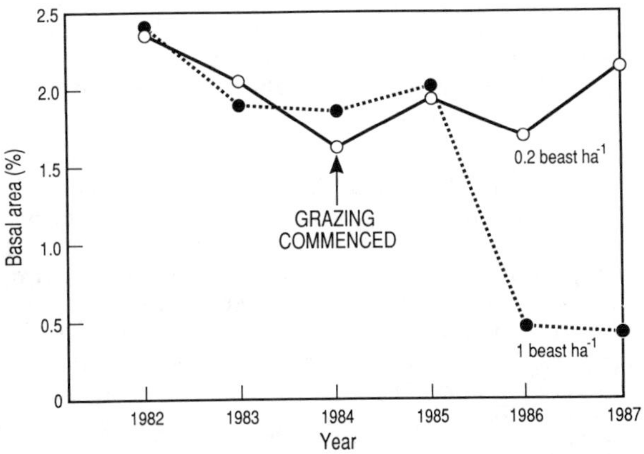

FIG. 4. Basal area of pastures at Hillgrove in the Burdekin basin. The pastures were not grazed in 1982 and 1983 and were subsequently grazed at high (1 beast ha^{-1}) and low (0.2 beast ha^{-1}) stocking rates. Values for native grasses oversown with *Stylosanthes* spp. under live trees (Gardener & McIvor, unpubl. data).

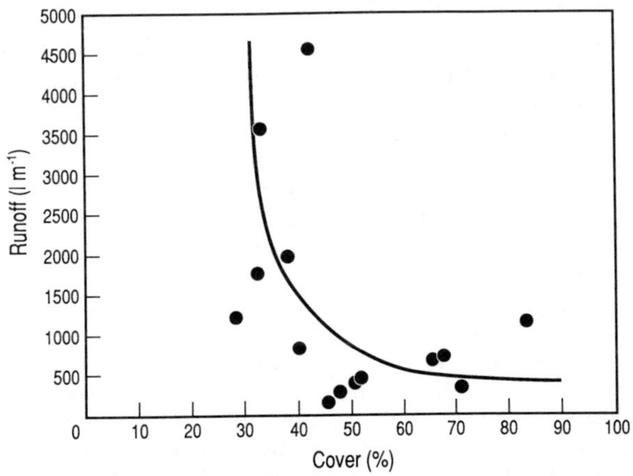

FIG. 5. Relationship between water runoff and percentage cover at Cardigan in the Burdekin basin (Williams, Gardener & McIvor, unpubl. data).

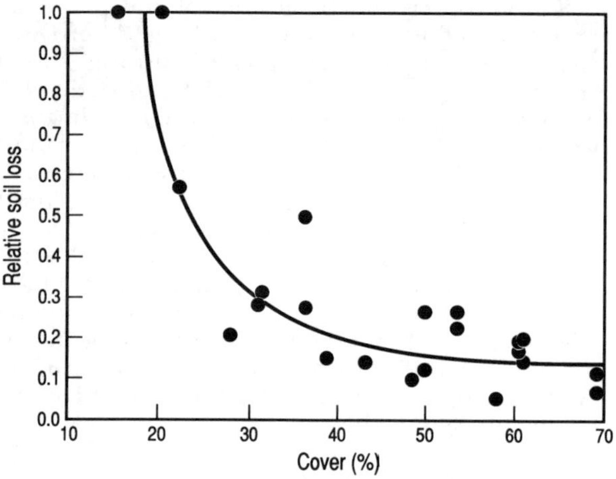

FIG. 6. Relationship between relative soil loss and percentage cover at Cardigan in the Burdekin basin (Williams Gardener & McIvor, unpubl. data).

The invasion of inedible shrubs onto overgrazed land is also becoming widespread. Without either regular burning because of lack of fuel or strong competition from perennial grasses to prevent their establishment, there has been a huge increase of both introduced shrubs such as chinee apple, *Ziziphus mauritiana,* and rubber vine, *Cryptostegia grandiflora,* and native shrubs like currant bush, *Carissa ovata.*

Solutions to the problem of overgrazing

Clearly there are only two ways to overcome or prevent the problem of overstocking; either stock numbers have to be reduced or the supply of forage has to be increased.

Reduced stock numbers

Reducing the stocking rate requires a fundamental change of attitude by graziers. Graziers overgraze land for a number of reasons. Economic pressures force some to overgraze their land to maximize current income to survive, some overgraze out of ignorance of effects on current and future productivity, while some are greedy.

The most convincing argument for lower stocking rates is likely to be one showing that a reduction in stocking rate results in increased profits. Little information is available on the relationship between animal production and stocking rate as pasture condition deteriorates, but the situation is likely to resemble that shown in Figure 7. Animal production is greatest at A for a short period, when the vigorous perennial grassland is first grazed heavily, but as heavy grazing continues over time and the amount of pasture available declines, both the potential for animal production and the stocking rate for maximum animal production fall (from A to B then to C). Continuing to stock at A will be less profitable than at B or C.

Restoring the grassland, where vestigial amounts of the native perennial grasses or *B. pertusa* remain, could require a period at an even lower stocking rate (D) in order to let plants grow, produce new seed and re-establish. Little seed of the tall perennial grass species is present in the seed bank (McIvor 1987), and when all plants are lost, regeneration of these species may depend on the dissemination of seed from elsewhere.

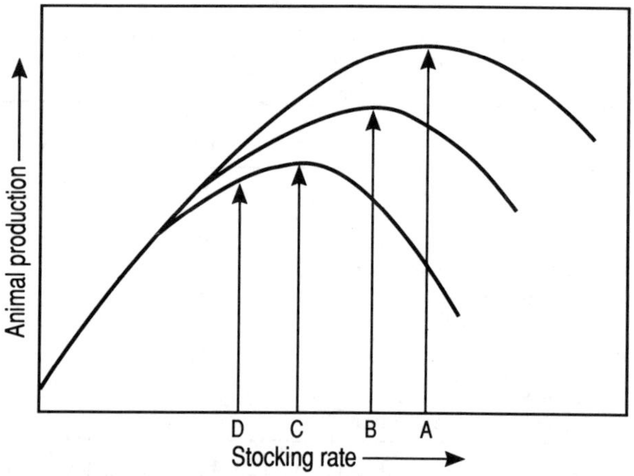

FIG. 7. Theoretical relationships between animal production per unit area and stocking rate at three stages of increasing degradation over time.

Most of the land in Queensland is leasehold so the State Government has the right to set stocking rates.

Subsection 233 of the Land Act 1962-1984 states:

> If the Minister is of opinion that the lessee of any holding is depasturing on the land included in his holding such a number of stock that the land is likely to be permanently injured thereby, the Minister may, by notice in writing, require the lessee, within the time specified in the notice, to reduce the number of stock so depastured to the number specified in the notice as to the stocking of such land, and if the lessee fails to comply with the terms of the notice the holding shall be liable to be forfeited.

To our knowledge, subsection 233 has never been applied.

There is an increased awareness of the problem of land degradation in the Burdekin basin among some graziers and a land care committee has been set up. This will undoubtedly influence the graziers most receptive to new information, but is less likely to convince those grazing most heavily.

Increased forage supply

Increasing the forage supply can be achieved by sowing a pasture legume to increase the nitrogen supply, by killing the trees to remove competition with the herbaceous layer, and by applying superphosphate to correct deficiencies of phosphorus and sulphur. Cultivars of *Stylosanthes* are well adapted to the region and have given large increases in forage and animal production (Gillard and Winter 1984).

An example of the large effect of killing trees (273% increase) and sowing legumes (*Stylosanthes* spp.) (52% increase) on the amount of dry matter in a year of below average rainfall is shown in Figure 8. The increased herbage cover greatly reduced soil losses in the treatments with sown legumes and trees killed (Fig. 9). Cultivating the soil surface, however, to establish sown grasses increased soil loss.

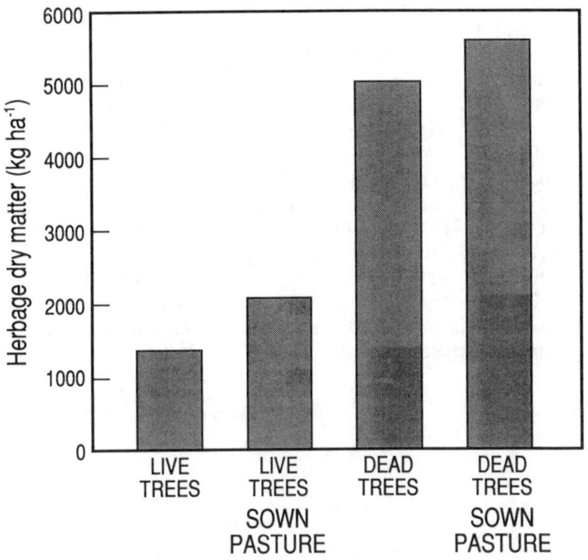

FIG. 8. Yield of herbage produced at Hillgrove in the Burdekin basin with dead or live trees and oversown or native pasture. Means of paddocks stocked at 0.2, 0.33 and 0.5 beast ha^{-1} and two levels of superphosphate (0 and 100 kg $ha^{-1}yr^{-1}$). (Gardener & McIvor, unpubl. data).

Killing trees and sowing pasture species can therefore be an effective method of halting or overcoming grassland degradation provided certain conditions are met. Firstly, landscapes likely to suffer from salinization should not be cleared, although such landscapes may not be easy to recognize. Secondly, land with steep slopes, unstable soils or economically uncontrollable regrowth should also not be deforested. Thirdly, the graziers' philosophy on overgrazing must also change. If a grazier decided to recoup his additional expenditure on pasture development as quickly as possible, and increased his previous stocking rate in proportion to the feed available, land degradation and particularly the invasion of inedible shrubs without any tree competition, could be far greater. Sowing pasture legumes under live trees therefore might

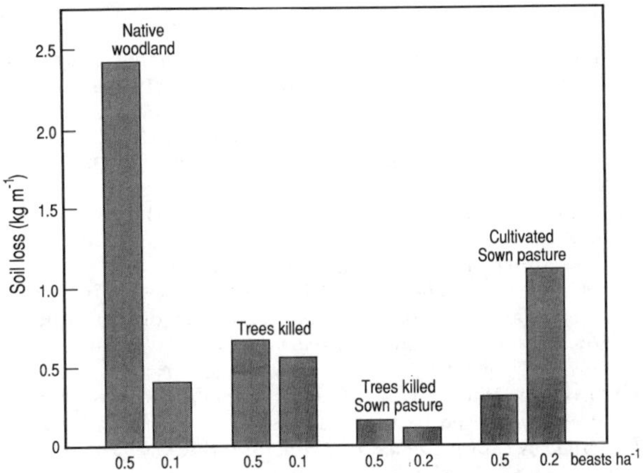

FIG. 9. The effect of killing trees and sowing pasture legumes and grasses on the loss of soil through increased soil cover at Cardigan in the central Burdekin basin. Soil loss occurred after a fall of 62 mm in March, 1987 (Williams, Gardener & McIvor, unpubl. data).

be a less risky management strategy than killing the trees, especially in situations where overgrazing could periodically occur at times of drought.

Future of grazing in the dry tropics

In the Burdekin catchment there will be little net change in the extent of land degradation in the short term. Some graziers will either stock leniently or sow improved pastures while others will continue to overutilize their land. Of most concern is that overgrazing will spread to other regions of the dry tropics. Widespread overgrazing occurred in the Burdekin catchment first because the properties were more intensively managed than elsewhere — the infrastructure was good, access to markets easy, properties relatively small and developed, and soils relatively fertile. Access and infrastructure could improve remarkably in Cape York and the Eastern Gulf if a space port is built towards northern Cape York, allowing greater intensification there on far more infertile and fragile soils.

Changes in the grazing industry are likely to come about in the longer term as a result of outside pressure. Future floods could carry an unacceptable amount of sediment into the Burdekin dam as well as out to the Great Barrier Reef and associated tourist areas.

The permanently suspended colloids, formed by the mixture of saline water and clay colloids in the dam, are already causing problems with recharge of the underground aquifers of the Burdekin irrigation area, and will probably increase the costs of the water filtration plants once Townsville starts to use Burdekin water. In the long term, we see considerable pressure being applied to the grazing industry to change management, and to the State Government to use its hitherto latent powers over the grazing industry.

References

Cadell F. (1866) South Australian Parliamentary Papers 24/1868-9, p.1. Quoted by Bauer, F. H. Historical geography of white settlement in part of northern Australia Part 2. The Katherine-Darwin Region. CSIRO Australia Division Land Research and Regional Survey Div. Rep. No. 64-1. 1964. p.59.

Fleming P. M., Gunn R. H., Reece A. M. & McAlpine J. R. (eds) (1981) *Burdekin Project Ecological Study.* Australian Government Publishing Service, Canberra.

Frisch J. E. & Vercoe J. E. (1977) Food intake, eating rate, weight gains, metabolic rate and efficiency of feed utilization in *Bos taurus* and *Bos indicus* crossbred cattle. *Anim. Prodn.* **25,** 343-58.

Gillard P. & Winter W. H. (1984) Animal production from *Stylosanthes* based pastures in Australia. In: *The Biology and Agronomy of Stylosanthes* (eds Helen M. Stace & L. A. Edye) pp. 405-32. Academic Press, Sydney.

Hubble G. D. & Isbell R. F. (1983) Eastern Highlands (VI). In: *Soils: an Australian Viewpoint.* CSIRO Division of Soils. CSIRO, Melbourne: Academic Press, London.

Humphreys L. R. (1967) Townsville lucerne: history and prospect. *J. Aust. Inst. Agric. Sci.* **33,** 3-13.

Hunter R. A., Siebert B. D. & Breen M. J. (1976) The botanical and chemical composition of the diet selected by steers grazing Townsville stylo-grass pastures during a period of liveweight gain. *Proc. Aust. Soc. Anim. Prodn.* **11,** 457-60.

Isbell R. F. (1983) Kimberley-Arnhem-Cape York. (III). In: *Soils: an Australian Viewpoint.* CSIRO Division of Soils CSIRO, Melbourne. Academic Press, London.

Little D. A. (1980) Observations on the phosphorus requirement of cattle for growth. *Res. Vet. Sci.* **28,** 358-60.

McCown R. L. (1982) The climatic potential for beef cattle production in tropical Australia: part IV — Variation in seasonal and annual productivity. *Agric. Syst.* **8,** 3-15.

McIvor J. G. (1981) Seasonal changes in the growth, dry matter distribution and herbage quality of three native grasses in northern Queensland. *Aust. J. Exp. Agric. Anim. Husb.* **21,** 600-9.

McIvor J. G. (1987) Changes in germinable seed levels in soil beneath pastures near Townsville, North Queensland. *Aust. J. Exp. Agric.* **27,** 283-9.

Moore R. M. (1970) Australian Grasslands. In: *Australian Grasslands* (ed. R. M. Moore). Australian National University Press, Canberra.

Mott J. J., Tothill J. C. & Weston E. J. (1981) Animal production from the native woodlands and grasslands of northern Australia. *J. Aust. Inst. Agric. Sci.* **47,** 132-41.

Norman M. J. T. (1960) Grazing and feeding trials with beef cattle at Katherine, N.T. CSIRO Division of Land Research and Regional Survey, Tech. Pap. No. 12.

Norman M. J. T. (1963) The pattern of dry matter and nutrient content changes in native pastures at Katherine, N.T. Aust. *J. Exp. Agric. Anim. Husb.* **3,** 119-24.

Siebert B. D. (1982) Research findings in relation to future needs. *Proc. Aust. Soc. Anim. Prodn.* **14,** 191-6.

Siebert B. D., Playne M. J. & Edye L. A. (1976) The effects of climate and nutrient supplementation on the fertility of heifers in north Queensland. *Proc. Aust. Soc. Anim. Prodn.* **11,** 249-52.

Tothill J. C. (1969) Soil temperatures and seed burial in relation to the performance of *Heteropogon contortus* and *Themeda australis* in burnt native woodland pastures in eastern Queensland. *Aust. J. Bot.* **17,** 269-75.

Proc. Ecol. Soc. Aust. 1990 — 16:287-301

A review of current ideas of the extinction, conservation and management of Australia's terrestrial vertebrate fauna

H. F. RECHER[1] and L. LIM[2]

[1]*Department of Ecosystem Management, University of New England, Armidale, N.S.W. 2351*

[2]*Taronga Zoo, P.O. Box 20, Mosman, N.S.W. 2088*

Abstract

Of Australia's 1 600 species of terrestrial vertebrates, nearly 300 are considered endangered. Seventeen mammals, three birds, and one species of lizard are known to have become extinct since European settlement in 1788. Extinction rates and the decline in the distribution and abundance of species has been greatest in the drier regions of the continent and where European settlement first occurred. East-west and north-south differences in the decline of native fauna can be attributed to settlement history related to land use and the spread of exotic predators and herbivores. The loss of species cannot be attributed to a single cause. Declines in the terrestrial vertebrate fauna have resulted from a combination of factors including habitat modification and fragmentation, over-exploitation, exotic diseases, the effects of pesticides, competition with introduced herbivores, and predation by foreign predators.

An important observation is the rapidity of change in species abundances as each part of the continent was settled. The time from first settlement to decline and extinction is measured in decades, not centuries. Despite this, over most of the continent the native fauna has not reached equilibrium with introduced species and post-settlement habitat changes.

As well as continued losses of mammal species, it is predicted that there will be an accelerating loss of avian species paralleling the losses in the mammal fauna which occurred in Australia's pastoral and agricultural areas from 1900 onwards. Declines in the abundance of reptiles and amphibians are expected to follow a similar pattern to birds and mammals. While the dedication of reserves which meet the minimum population size requirements of species and the control of feral animals are important in reducing the rate of loss of critical ecosystems and their associated fauna, the conservation of Australia's terrestrial vertebrates depends on significant changes in land management practices. Clearing of native vegetation, prescription or hazard reduction burning, and grazing by domestic stock are among the practices which should cease or be better managed, if we are to ensure the survival of Australia's terrestrial vertebrates into the twenty-second century.

Introduction

The Bicentennial was an opportunity to reflect on our past and to consider where Australia will be in another 200 years. When the history of European settlement in Australia is reviewed, the changes found are disturbing. Given Australia's relatively small population and the short time that Europeans have occupied the continent, it is doubtful that any group of people has so profoundly altered their environment so rapidly as Australians have done in 200 years of settlement.

Recognition of the impact of European settlement on the Australian landscape and its flora and fauna has been slow to emerge. Most Australians view the continent as vast, unspoiled and underpopulated. Yet in historical times, no other continent has experienced the massive changes in the distribution and abundance of such a large proportion of its vertebrate fauna as has Australia. On no other continent has such an array of exotic animals been established and with such a devastating impact on native wildlife (see Kitching 1986; Myers 1986). In 200 years Europeans have recast the evolutionary biogeography of the continent.

In reviewing the state of Australia's environment, it is easy to become lost in emotion. This chapter reviews the status of Australia's terrestrial vertebrate fauna. We are not the first to take on this task and the efforts of others has made our work easier. What we seek to do, however, is different. We seek the future. Part of our task is to review the status of Australia's terrestrial vertebrates, to analyse the reasons for changes in patterns of abundance, and to use our understanding of ecological principles to predict the future. The remainder of our task is to suggest ways in which the tide of events can be changed. We are not optimistic, nor are we totally pessimistic. Australia has the resources, the knowledge, the skills and the people to reconstruct degraded environments and restore endangered species. Australians may not feel morally responsible nor have the will to act, but these are political and social challenges, not ecological or technical problems.

The conference at which this chapter was presented may join others that preceeded it into obscurity or it may capture the minds of the nation. We can identify the different paths to the future that Australia may travel, but which path Australians finally adopt will depend on how forcefully the words written here and spoken at the conference are followed by action. Action must first come from Australia's ecologists.

Australia's terrestrial vertebrates

Excluding fish, *ca.* 1 630 species of terrestrial and freshwater vertebrates have been described for the Australian continent. Reptiles and birds are the most diverse taxa with *ca.* 530 species of birds (Slater *et al.* (1986) and *ca.* 680 reptiles (Cogger 1986). Of the reptiles, approximately 495 are lizards, 165 are snakes, 15 are turtles, and there are two crocodiles. There are *ca.* 245 species of mammals, half of which are marsupials (Strahan 1983). There are *ca.* 55 species of rodents and 55 species of bats, although the taxonomy of these groups is currently being reviewed. Two species of monotremes, the platypus, *Ornithorhynchus anatinus,* and the echidna, *Tachyglossus aculeatus,* complete the mammal fauna. There are one hundred and seventy-eight species of frogs (Cogger 1986), but this underestimates the diversity of the fauna as new species are still being discovered (Tyler 1978).

Excluding New Guinea, which has its own rich fauna and shares many species with Australia, more than 90% of Australian vertebrates are endemic. In addition to the marsupials, the major families of Australian rodents, birds, frogs and reptiles radiated on the Australian continent independent of the faunas of other continents (e.g., see Sibley & Ahlquist 1985; Cogger & Heatwole 1981; Archer 1981).

Status of Australia's terrestrial vertebrates

What is the current status of Australia's terrestrial vertebrate fauna? How many species are extinct? How many others are rare, endangered or threatened in some way that requires action to prevent their loss? Unfortunately, there are no formal criteria for determining the status of a species. Generally, species are considered endangered or threatened, if they are rare or have a restricted distribution. Rarity itself has different meanings and is open to widely varying interpretations (see Main 1982). For these reasons, different organizations have very different interpretations of the status of Australia's terrestrial vertebrate fauna.

Seventy-eight species of birds, mammals, reptiles and frogs are listed officially by the Australian Council of Nature Ministers (CONCOM) as rare, endangered or threatened (Appendix 1). This includes five species of lizards, two snakes, the western swamp turtle, *Pseudemydura umbrina,* six frogs, 27 birds, and 37 mammals (13 rodents and 24 marsupials). Internationally, the Red Data Book (ICBP 1981) lists 14 taxa (species and subspecies) of Australian birds as endangered. Eight of these are parrots.

Using different criteria, the Total Environment Center (TEC 1983) identified 62 marsupials, 20 rodents, 18 bats, 69 birds, two crocodiles, two turtles, six snakes, 22 lizards and 14 frogs as vulnerable. Only 23 of the marsupials, 10 of the rodents, 17 of the birds, one turtle, two of the snakes and five of the frogs listed by TEC appear on the CONCOM list. In addition the TEC (1983) identified one turtle, 75 lizards, 18 snakes and 20 frogs which had restricted distributions and could prove to be endangered after further investigation. The TEC report emphasized the threat to any species with a restricted distribution whether or not it was common where found and whether or not its restricted distribution was the result of human activities. Although rarity does not necessarily mean species are at risk of extinction (Ride & Wilson 1982), their small populations make them vulnerable to the broad area habitat changes associated with modern agriculture, forestry and land development. The CONCOM, ICBP and TEC lists include species already extinct (e.g., thylacine, *Thylacinus cynocephalus,* and paradise parrot, *Psephotus pulcherrimus).*

Each state has its own lists of endangered and threatened fauna. For example, the New South Wales National Parks and Wildlife Act (1974) lists 182 species of terrestrial vertebrates as fauna of special concern. This includes 69 mammals (45 marsupials, 14 rodents, nine bats and the platypus, 21 reptiles (two crocodiles, one turtle, 11 lizards and seven snakes). Ninety-two birds, including 38 Palaearctic waders protected by an international agreement, are also listed. Under the Act, "fauna of special concern" includes species which occur in Australia, but not necessarily in New South Wales. Among the vertebrates found in New South Wales, 51 species (11 mammals, four lizards and 36 birds) are considered vulnerable and/or rare. Nineteen species (four mammals, five reptiles and 10 birds) are threatened and 30 species (25 mammals and 5 birds) are in imminent danger of extinction or are already extinct in New South Wales. The mammals listed in these three categories include 23 marsupials, two bats and 15 rodents. Four lizards and five snakes are listed.

All lists of threatened fauna suffer from a lack of precise data on the distribution and abundance of species. Rediscovery of animals thought to be extinct or extremely rare has occurred frequently during the past 20 years as the level of field research and biological survey throughout Australia has increased. For example, in 1987, a population of the newly described Pilliga mouse, *Pseudomys pilligaensis* (Fox & Briscoe 1980) was located in the Pilliga Scrub (Lim, unpubl.). It was previously known only from seven specimens collected between 1975 and 1979. Our knowledge of the status of several genera of small mammals (e.g., *Mastacomys, Planigale, Ningaui*) has changed dramatically in the past decade with more intensive surveys and the use of appropriate trapping procedures. Surveys of rainforest reptiles and amphibians in north-eastern New South Wales during 1988 have demonstrated that some species which had been considered rare are abundant when the correct survey procedures are used (A. Smith, pers. comm.). While these examples illustrate the potential for misclassifying the status of common but cryptic species, it is unlikely that a significant number of species classed as endangered will be found to be abundant.

Often rare fauna are not listed. For example, the Julia Creek dunnart, *Sminthopsis douglasi,* that was described from only two specimens collected in 1931 (Archer 1979), was not listed by CONCOM. The TEC listed 114 species for which additional data were required before a decision could be reached on their status. Yet neither the TEC nor the New South Wales National Parks and Wildlife Service listed the regent honeyeater, *Xanthomyza phrygia,* a species that has declined significantly in abundance (Blakers *et al.* 1984) and should be considered endangered.

Despite differences in selection criteria and the problems associated with assessing status, the numbers of species involved vividly illustrate the magnitude of the problems associated with the conservation and management of Australia's terrestrial vertebrates. About 280 species or nearly 17% of Australia's terrestrial vertebrate species are listed by one organization or another as endangered, rare or threatened. As illustrated by the Julia Creek dunnart and the regent honeyeater, these lists are incomplete.

Decline and extinction

Not all the species for which there is concern are in imminent danger of extinction. Nor are all the species listed, rare or threatened as a result of European settlement. Many may be rare or restricted in distribution by climate, habitat or chance. For example, the long-footed potoroo, *Potorous longipes,* occupies a narrow range of forest habitats and occurs only in isolated areas in East Gippsland and south-eastern New South Wales (see Strahan 1983). There is no evidence it was more widely distributed prior to European settlement. However, the number of species that are considered to be threatened or endangered is indicative of the deteriorating quality of the Australian landscape following the arrival of Europeans on the Australian continent.

The impact of European settlement has been greatest on mammals and birds. In the 200 years since the colony was established in 1788 at Port Jackson, 17 species of terrestrial mammals, three taxa of birds (i.e., the paradise parrot and two island forms of the emu, *Dromaius novaehollandiae, D. n. diemenianus* and *D. n. ater,* and the Adelaide blue-tongue lizard, *Tiliqua adelaidensis,* are known to have become extinct.

Eight species of mammals formerly occurring on the mainland survive only on isolated, small offshore islands (Burbidge & McKenzie 1989). These include the western barred bandicoot, *Perameles bougainville,* the burrowing bettong, *Bettongia lesueur,* the banded hare-wallaby, *Lagorchestes fasciatus,* and the greater stick-nest rat, *Leporillus conditor.* Nineteen others have declined significantly in abundance with many persisting as relict populations (Burbidge & McKenzie 1989). The eastern quoll, *Dasyurus viverrinus,* and the eastern bettong, *B. gaimardii,* which were formerly widespread on the mainland now survive only in Tasmania.

Until recently, the impact of European settlement on the mammal fauna was less severe in south-western Australia than in the east or in the more arid interior. In the humid forests and heaths of the south-west, most mammals remained abundant, if restricted in distribution. However, in the 1970s a number of mammals declined rapidly. These included the western form of the ringtail possum, *Pseudocheirus peregrinus occidentalis,* and the numbat, *Myrmecobius fasciatus.* Although a number of species are less abundant than formerly, the mammal fauna of northern Australia has been less severely affected than that of southern Australia. Probably this is the result of fewer people, less intensive agriculture, more recent settlement, and the absence of the fox, *Vulpes vulpes,* and rabbit, *Oryctolagus cuniculus,* but reports of species declining in abundance in north-western Australia (Burbidge & McKenzie 1989) suggest the pattern of the southern States will be repeated.

Ground-dwelling mammals in the size range of 35-5 000 g have been most affected by European settlement (Burbidge & McKenzie 1989; Giles & Lim 1987; Morton & Baynes 1985; Morton, this volume; Newsome 1971, 1975; Rosel & Smith 1985). In the arid and semi-arid regions of the continent, few mammals in this critical size range survived. This corresponds to Australia's main pastoral and agricultural areas and encompasses most of the continent's woodland, savannah, grassland, and shrub communities and deserts. Native rodents, the peramelids and the potoroids were the most affected. Animals that burrowed or used nests, carnivores as well as herbivores, colonial and solitary species, were affected equally. Some larger species, such as the northern hairy-nosed wombat, *Lasiorhinus krefftii,* were also affected adversely. Species which occupy rugged terrain or rock piles (e.g., rock wallabies *Petrogale*), very mobile species such as bats, and most large species (e.g., kangaroos *Macropus*) were less seriously affected.

Apart from two large dasyurid marsupials, the thylacine and the eastern quoll, no mammals have been lost from forest habitats. However, many forest mammals have declined in abundance and regional extinctions are common (see Lunney & Leary 1988). About 50% of Australia's forest have been cleared (Anon. 1986) and populations of most forest species are now fragmented. As such they may be adversely affected by the more intensive harvesting and management of forests associated with modern forestry (e.g., integrated logging) and broad area hazard reduction burning.

There is less information on the status of Australian birds than mammals. One of the disturbing results of preparing this review was finding how few data are available on the distribution and abundance of Australian birds. Despite the lack of quantitative data, the impact of Europeans on the Australian avifauna has been significant, but attracts less public and official concern than the decline in native mammals. There are reasons for this. Many species of birds, including terrestrial and freshwater species, have declined in abundance and have more restricted distributions than formerly (see Blakers *et al.* 1984). However, there have been few extinctions and, combined with the abundance of many common species, the ease of finding and seeing birds creates an illusion of well-being.

As with mammals, the greatest impact of settlement has been on birds of the pastoral and agricultural regions. Although ground-dwelling and ground-foraging birds have been most affected, there is no simple pattern or set of characteristics, such as bodyweight, which describe the birds that have declined in abundance following European settlement and which may be endangered. Some, such as the scrub-birds, *Atrichornis,* and the grasswrens, *Amytornis,* were already restricted by specialized habitat requirements and have been affected by changes to their habitats from altered fire regimes and from grazing by introduced herbivores. Others, such as the red-lored whistler, *Pachycephala rufogularis,* and regent honeyeater, have been affected by habitat clearing and fragmentation.

Parrots are among the best studied Australian birds; changes in the status of parrot species are typical of the terrestrial avifauna. Half of Australia's 53 species of parrots have declined in abundance since the 19th century (Blakers *et al.* 1984). One species, the paradise parrot, is extinct and at least seven others are endangered (ICBP 1981). The New South Wales National Parks and Wildlife Service lists 16 species of parrots (including the paradise parrot) as endangered. No other group of birds is so well represented. Parrots are conspicuous on endangered lists because they are important avicultural species, they are large and attractive, and they command high prices on illegal markets. However, many small birds are equally threatened. Small brown birds may become extinct before official notice is paid to their plight.

Although the decline in species numbers is less noticeable than for parrots or for birds in pastoral and arid environments, forest birds have been affected by fragmentation of forests through clearing (Howe 1984; Howe *et al.* 1981) and the loss of forest old growth through logging (Recher *et al.* 1980; Loyn 1985). Declines in the abundance of such birds as the glossy black cockatoo, *Calyptorhynchus lathami,* and the shrike-tit, *Falcunclus frontatus,* are typical. Both are specialized foragers, require extensive areas of habitat in which to forage, and are among the first birds to disappear from forest remnants (Saunders & Curry, this volume; Howe 1984). Shrike-tits feed on invertebrates located within strips and coils of decorticating bark on eucalypts and probably require large, relatively mature trees with a complex invertebrate fauna linking bark inhabitants to the soil and litter organisms at the base of the tree. If this hypothesis is correct, then shrike-tits may be adversely affected by changes to the soil litter layer, although they are not ground-foragers.

Ground-foraging or ground-dwelling birds, like ground-dwelling mammals, are particularly sensitive to the changes associated with European settlement. Degradation of ground vegetation, diversion of resources to domestic stock and introduced herbivores, trampling and destruction of the litter layer, and soil erosion reduce the food resources available to ground-foraging species. Removal of low vegetation removes cover and increases the risk of predation. It also diminishes the availability of nest sites. Of 13 species of birds which declined in abundance between 1927 and 1988 in Kings Park, Perth, 10 were ground-foragers. This is attributed primarily to changes in ground vegetation and litter layers brought about by the spread of exotic weeds and changed fire regimes (Recher & Serventy, unpubl. data).

Similar habitat changes occur in forest areas where prescription burning is conducted on a regular basis. These regular low intensity fires remove the low shrub and litter layer depriving ground-dwelling organisms of food resources and cover. Of particular concern is the continued loss of energy released by the fire from litter. The slow decomposition of litter is a rate regulating mechanism (*sensu* Springett 1978) through which a large part of the primary production of forest is cycled. The soil and litter organisms sustained by the energy incorporated into the litter layer with branch and leaf fall are the principal food source for ground-foraging birds. They are also a significant source of food for bark and foliage foragers as well a aerial feeders such as fantails *Rhipidura*, flycatchers *Myriagra*, treecreepers *Climacteris*, and warblers *Gerygone*.

There is less evidence of change in the distribution and abundance of frogs and reptiles than of birds and mammals. The small size of most reptiles and frogs, their relatively low energy requirements and ability to avoid adverse conditions through torpor and other physiological adaptations (Heatwole & Taylor 1988) enables them to persist in regions where other vertebrates have been excluded by European modifications to the environment. Although clearing of native vegetation restricts the number and kinds of reptiles and amphibians found in agricultural and urban environments, even quite large reptiles will persist in remnant patches and corridors of native vegetation. Kitchener *et al.* (1980) found that the number of species of lizard in Western Australian reserves was correlated with the number of vegetation types. Reserve size was relatively unimportant. Most reptiles also persist in narrow corridors of vegetation retained in pine plantations (Recher *et al.* 1987). In the longer term, residual populations will probably decline in a pattern similar to the more sensitive mammals and birds; large species and the more predatory reptiles (e.g., varanid lizards) will decline first. Ultimately those species most dependent on soil and litter ecosystems will decline to extinction.

Increases in abundance

Not all species have been adversely affected. Many birds and mammals have benefited from the changes associated with European settlement and have increased in numbers and/or extended their distribution. Generally the greatest increases in abundance or distribution of native fauna have occurred among animals originating in the arid and semi-arid habitats where the greatest losses of species have also occurred. Among the marsupials, populations of the large kangaroos (*Macropus rufus*, *M. giganteus* and *M. fuliginosus*) have increased to unprecedented numbers. This may be due to the increase in the availability of watering points on farms and pastoral leases (Frith 1964; Newsome 1975), but other factors may also be important. An understanding of why some species increase in abundance, while others decrease, assists in identifying important environmental changes and in developing plans of management for the conservation of native fauna.

Several species of parrots including the galah, *Cacatua roseicapilla*, little corella, *C. pastinator*, and sulphur-crested cockatoo, *C. galerita*, the crested pigeon, *Ocyphaps lophotes*, pied currawong, *Strepera graculina*, and Australian raven, *Corvus coronoides*, are among the birds which have increased in abundance (e.g., Heron 1973; Serventy 1977; Blakers *et al.* 1984; Jack 1973; Saunders *et al.* 1985; Saunders 1989). Birds, such as the striped honeyeater, *Plectorhynchus lanceolata*, galah, crested pigeon, and red-rumped parrot, *Psephotus haematonotus*, have extended their distribution from semi-arid habitats to the coast. Most of the birds that have increased in abundance and expanded their distribution responded to an increase in suitable habitat resulting from clearing of forest for agriculture and/or increased food resources. Not only do some birds benefit from crops as a food source, but weed seeds and ornamental fruits provide abundant year round food resources. Birds probably also benefit from greater access to water.

The pattern of change

As Europeans moved with their livestock across the continent, they were accompanied by massive changes in the abundance of native fauna. Associated with these changes was a wave of extinctions (Marshall 1966; Serventy 1977). The earliest losses were in eastern Australia. Many arid zone animals remained relatively abundant through the 1920s, or later in more remote regions, and then declined rapidly with the advent of modern agriculture and the spread of the rabbit and fox. Such species included the eastern hare-wallaby, *Lagorchestes leporides*, the lesser stick-nest rat, *Leporillus apicalis*, the desert rat-kangaroo, *Caloprymnus campestris*, the white-footed rabbit-rat, *Conilurus albipes*, and the pig-footed bandicoot, *Chaeropus ecaudatus*.

The precise reasons for these extinctions are unknown (see below), but the large number of species involved and their rapid decline in abundance testifies to the artificial nature of events. In describing the decline and extinction of the Australian terrestrial vertebrate fauna, we are not describing changes over 200 years of settlement. Instead, most events occurred within 20 or 30 years of an area being settled. For most of the continent the changes described occurred within this century. The impact of Europeans on the Australian environment is not measured in centuries, but in decades. This pattern is now being repeated in the south-west and north-west of the continent as settlement of these regions intensifies (Burbidge & McKenzie 1989).

Jarman and Johnson (1977) described the pattern of change associated with settlement of the tablelands and slopes of eastern New South Wales.

Within 20 years of a district being settled by graziers, numbers of indigenous herbivores increased to pest proportions followed by a rapid decline. The pattern reported by Jarman and Johnson (1977) was first an increase in the numbers of kangaroos, *Macropus,* followed by wallabies, *Macropus* and *Wallabia,* and then rat-kangaroos *(Potorous, Aepyprymnus, Bettongia).* There was then a decrease, often to local extinction, in the same sequence of species. Other native mammals, including bandicoots *(Perameles, Isoodon),* brush-tailed possum, *Trichosurus vulpecula,* and the introduced hare, *Lepus capensis,* had similar patterns of increase and decline in some districts. The pattern of increase and decline for each group or species occurred within a 10 to 20 year period. Declines were associated with the invasion of each district by the fox and rabbit.

Lunney and Leary (1988) found that similar patterns of increase and decline occurred among native mammals in the moist coastal district of Bega in south-eastern New South Wales. As on the tablelands, numbers of hares, bandicoots, kangaroos, wallabies and rat-kangaroos increased between the 1860s and 1900 only to decline precipitously with the invasion of the region by rabbits in 1900 and foxes a few years later. They also showed that arboreal possums and gliders *(Petaurus, Petauroides)* and the koala, *Phascolarctos cinereus,* increased in abundance between 1880 and 1900 only to decline in numbers between 1900 and 1910. At least six species of native mammal disappeared from the Bega district, four are endangered and many others are rare or uncommon. This resulted largely from the clearing of native vegetation from areas of richer soil. As Lunney and Leary (1988) suggest, the forest fauna at Bega is now restricted to habitats on the poorest soil in what may be marginal situations.

The changes in abundance described by Jarman and Johnson (1977) and Lunney and Leary (1988) were associated with the development of pastoral industries. Although forests and woodlands were cleared and opened up to encourage the growth of grass, considerable natural cover was initially retained. This differs from the clearing of native vegetation for crops where native mammals are simply displaced.

Among birds, there is no evidence of a similar pattern of increase and decline by native species. Instead the pattern is the gradual decline, often to local extinction, of most species accompanied by the massive increase in abundance of others. For example, Kitchener *et al.* (1982), in their survey of West Australian reserves, suggested that no species of bird had been lost from the wheatbelt. However, many species had declined in abundance and were restricted to remnants of native vegetation. Smith (1987) also concluded that no species of birds had been lost from the wheatbelt. Morton (this volume) points out that, in the arid zone, birds have been less affected by European settlement than mammals. No species of bird in the arid zone is known to be extinct compared to the loss of 11 species of mammal and the restriction of five other species to islands.

In a recent analysis, Saunders (1989) found a rapid loss of species from wheatbelt reserves since the original clearing of the native vegetation 30-50 years ago. Most of these remnants are small (>95% are less than 100 ha in area) and even the largest are unable to retain the full complement of the native avifauna. Two species, thick-billed grasswren, *Amytornis textilis,* and Gilbert whistler, *Pachycephala inornata,* are extinct in the wheatbelt. East Yorkrakine Reserve near Kellerberrin, with an area of 81 ha, has lost three species in the last decade. The larger (1 100 ha) Durokoppin Reserve has lost 10 species since the surrounding vegetation was cleared. Species which have declined in the wheatbelt remain abundant further east and along the wheatbelt's western periphery where clearing has been less extensive.

These changes in bird species abundances have been repeated throughout Australia. Ford and Howe (1980) described the status of the avifauna in the Mt Lofty Ranges of South Australia. Of 116 species originally present, only 90 were recorded in a survey during 1977. Ford and Howe (1980) attributed these losses to the clearing and fragmentation of the original forest. In 120 years of European settlement of South Australia, 90% of the forest of the Mt Lofty Ranges was cleared. The remainder was fragmented with the largest patches a few thousand hectares in extent compared with an original forest area of 500 000 ha.

Despite the loss of 26 species, bird species richness in the Mt Lofty Ranges may not yet be at equilibrium following the fragmentation of the forests. Using the relationship between the number of bird species and area of individual forest patches, Ford and Howe (1980) calculated that the patches of forest remaining in the Mt Lofty Ranges were capable of supporting between 65 and 80 species of birds. Thus, they predicted the loss of a further 35 to 50 species. Ford and Howe (1980) pointed out that the predicted losses may occur over a very long time and the result may not be as extreme as the simple area-relationships suggest. The ability of some birds to use habitats other than forests and the possibility that a number of small patches may be able to support a greater number of species than a single patch of the same area may result in fewer extinctions than expected.

Loyn (1987) studied the avifauna of forest patches in the La Trobe Valley of Victoria. About 70% of the forest in the La Trobe Valley had been cleared and forest birds were less abundant than formerly, but all species expected were present

(Loyn 1987). Loyn (1987) predicted that only nine species would be lost as patches went to equilibrium. The smaller area of forest cleared (70% vs 90%), the presence of the different forest types originally occurring in the valley, and the proximity of large areas of uncleared forest probably contributed to the greater retention of bird species in the La Trobe Valley than in the Mt Lofty Ranges. Porter (1982), working in an area of south-eastern Queensland where 30% of the forest had been cleared, could find no loss of forest bird species.

In the examples described above, there is evidence that patches of remnant vegetation act as islands. The number of species which occur in a forest patch is determined by the size of the patch (area) (Ford & Howe 1980; Howe 1984; Howe *et al.* 1981; Loyn 1987), its degree of isolation (Howe *et al.* 1981), shape (Recher *et al.* 1987), history of disturbance (Howe *et al.* 1981; Recher *et al.* 1987; Loyn 1987) and habitat heterogeneity (Loyn 1987). Other environmental factors may be regionally significant. For example, the number and kinds of birds surviving in forest patches is inversely related to tree health (Ford & Bell 1981; Loyn 1987) and the presence or absence of particularly aggressive species, such as noisy miner, *Manorina melanoleuca* (Loyn 1987). Although several studies have shown that forest fragments support similar numbers of birds to similar sized patches within continuous forest blocks (e.g., Howe 1984; Loyn 1987), the proportion of edge and open-country species increases and the number of forest species decreases (Howe 1984; Loyn 1987; Recher *et al.* 1987). Species with narrow habitat requirements and those with low dispersal abilities are the first to disappear from forest patches.

Non-quantitative accounts of changes in the abundance of species and the composition of regional avifaunas have been provided for New South Wales by Heron (1973) for the Orange area on the central tablelands, for the Sydney region (County of Cumberland) by Hindwood & McGill (1958), and for the Inverell district on the north-western slopes by Baldwin (1975). Jack (1973) described changes in avian numbers for the Brisbane area in south-eastern Queensland. In all instances there have been significant losses of bird species associated with the clearing of native vegetation, although these did not always occur immediately.

The decline of species in the Inverell and Orange districts are typical. Six formerly common species including jacky winter, *Microeca leucophaea,* crested bellbird, *Oreoica gutturalis,* bush stone curlew, *Burhinus magnirostris,* and hooded robin, *Melanodryas cucullata,* are now uncommon or rare in the Inverell district (Baldwin 1975). The Australian crow, was displaced by the Australian raven, *C. coronoides.* At Orange the emu is locally extinct and five other species are described by Heron (1973) as formerly abundant and now rare. These include hooded robin, spotted quail-thrush, *Cinclosoma punatatum* and grey-crowned babbler, *Pomatostomus temporalis.* All the species identified by Heron (1973) and Baldwin (1975) as formerly common and now uncommon are ground-dwellers or ground foragers. Nine species in the Brisbane area and 14 near Sydney which declined in abundance are also ground or low shrub foragers. In both areas the black-eared cuckoo, *Chrysococcyx osculans,* declined in abundance along with its preferred host species, the speckled warbler, *Chthonicola sagittata,* a ground-dwelling species.

Although individual species, such as the purple-crowned wren, *Malurus coronatus,* and groups of species, such as finches, have declined in abundance (e.g., Blakers *et al.* 1984; Rowley 1988), there is no indication of the massive dislocations in the north or north-west of the continent that occurred throughout eastern and southern Australia. As for mammals, changes in the avifauna have been greatest where European settlement has been the longest.

The lists referred to in the preceding paragraphs are incomplete. The authors only highlighted the most obvious (to them) changes in avian abundance. There is a risk when considering the numbers of birds, or for that matter of any of the smaller vertebrates, to assume that, because it is easy to find a species by visiting the right place or habitat, it is common. Some birds, for example, galah, Australian magpie, *Gymnorhina tibicen,* welcome swallow, *Hirundo neoxena,* and black-faced cuckoo-shrike, *Coracina novaehollandiae,* are abundant and nearly ubiquitous in distribution. Hidden by the conspicuousness of these birds are a large number of species which have declined in abundance over the past 20-30 years and are now patchily distributed.

Explanations of change

Newsome (1971, 1975) suggested that in central Australia the stock of the first settlers lived on the climax vegetation with all its reserves of fodder. Overstocking resulted in the depletion of this vegetation that the native herbivores depended on for drought fodder. As well as losing their food source, the small and medium size marsupials that depended on the tall grasses for shelter were placed at greater risk of predation.

There is probably no single cause for the changes in the distribution and abundance of Australian terrestrial vertebrates which have taken place since European settlement. Predation by foxes (Kinnear *et al.* 1988), competition from introduced herbivores (Lim 1987), clearing for agriculture and fragmentation of habitats (Lim 1987; Saunders & Ingram 1987; Howe 1984; Ford & Howe 1980), changes in vegetative structure and/or plant species composition as a result of logging (Recher *et al.* 1980; Loyn 1985; Smith

1985), disease, changed fire regimes (Burbidge & McKenzie 1989; Morton, this volume), hunting (Frith 1973), and predation or competition from native fauna which have increased in abundance have been presented as possible causes. Explanations for the increased abundance of species include the increased availability of water following the provision of stock watering points (Newsome 1975, Denny 1981), increased or novel food sources (e.g., seeds of exotic weeds, rabbits), creation of more open habitats and the increased area of edge habitats.

Some species only survived in the higher rainfall areas or rocky parts of their distribution. However, not all species that have suffered a reduction in range follow that pattern. For example, the northern hairy-nosed wombat and the bridled nail-tail wallaby, *Onychogalea fraenata,* have declined in abundance, but have not survived in the wetter or rockier parts of their original distribution. The wombat has survived in areas providing suitable soils in which to burrow and that have sufficient rainfall during *both* winter and summer to sustain domestic livestock and native grazers (pers. comm. D. Crossman QNPWS). In addition, stocking rates are low. Thus the survival of the hairy-nosed wombat involves a complex set of interactions between local patterns of rainfall, substrate and stocking rates. The reason for the wallaby's survival may be more complicated, but involves an interaction between the structure of the available vegetation and the interaction between dingoes, foxes and alternative prey species may be involved (Lavery & Tierney 1985; Tierney 1988; Tierney, pers. comm.).

At least one omnivore, the bilby, *Macrotis lagostis,* was displaced from most of its range as its habitat was altered. Bilbies persist in south-western Queensland where suitable insect prey remains relatively abundant in the channel country of the Diamantina River. In this region dingoes are abundant and fox and rabbit numbers are low. In addition the area is used primarily to run cattle and stocking rates are lower than in other pastoral regions (Lim, pers. obs.). Bilbies also survive in the Tanami Desert where they feed mainly on bulbs and seeds of native plants. Like the Diamantina in south-western Queensland, the Tanami Desert represents the more arid regions of the bilbies' original distribution. A fire regime which promotes a particular mosaic of vegetation, and low rabbit numbers and few foxes are thought to be significant factors in the survival of bilbies.

None of the changes to the Australian environment given above occurred in isolation. Clearing and fragmentation of habitats, the spread of exotic plants and animals, increases and decreases in the abundance of native fauna and exploitation by humans occurred more or less in synchrony. Separating their effects on native species is difficult. Serventy (1977), for example, attributed most of the changes in species abundances since European settlement to changed climatic conditions. He argued that rainfall was particularly plentiful towards the end of the 19th century (1860-1890). This was the time when large areas of Australia's pastoral areas were settled. It then became drier through to the 1940s. Increased numbers of native animals were associated with the high rainfall, but then declined in abundance as conditions became drier. Changes associated with European settlement probably enhanced the effects of the declining rainfall. He argued that the spread of inland birds, such as galah and crested pigeon, to the coast was the result of drier conditions and not solely the result of agricultural clearing (Serventy 1977). Burbidge and McKenzie (1989) suggest that the extreme sensitivity shown by the Australian fauna to change and the inability of native mammals to adapt to altered conditions is related in part to the low fertility of the continent's soils and the erratic pattern of rainfall particularly in semi-arid and arid regions. As a consequence of low nutrients and rainfall, the productivity of these regions is low (see Nix 1976). In most years the numbers of native fauna would be low and during droughts reproduction would be depressed and populations restricted to the most favourable habitats (Morton, this volume; Lim 1988). The introduction of exotic herbivores and possibly changed fire regimes reduced the amount of food for native species. Morton (this volume) points out that during drought native fauna are restricted to patches of highly favourable (suitable vegetation, water) habitat. These refuges are critical for the survival of species during dry periods. They are also favoured by introduced herbivores for the same reasons. Competition with introduced herbivores during dry periods therefore has a particularly severe impact on native fauna. Under these conditions, they are also highly vulnerable to predation. The greatest impact of introduced predators may be related primarily to dry periods when populations are minimal and restricted to patches of vegetation.

Fragmented and isolated populations of native animals, already reduced in abundance through reduced food resources, would be particularly vulnerable. If the numbers of predators (introduced or native) were maintained at high levels by feeding on abundant domestic stock or rabbits, the impact on native herbivores would be intensified. Maintaining domestic stock at high levels during drought also has a significant impact on food resources available to native animals and may be particularly critical when fauna is forced to rely on limited areas of particularly favourable habitat during times of stress. Although these interpretations are speculative, the places where diverse faunas have survived, for example in the Kimberley, in Kakadu, and between the Simpson Desert and Diamantina River in Queensland, are characterized by the absence or low numbers of domestic stock, foxes and rabbits. The same is true for many offshore islands where native fauna survives.

Davey and Norton (this volume) demonstrated that similar patches of particularly favourable habitat also occur in forest environments. Such patches are characterized by rich soils and abundant water and support a diverse and abundant native fauna. Management of these forest patches to protect these values is therefore necessary for the long-term survival of forest wildlife in forests exploited for wood production. This becomes critical if fauna is already restricted to marginal areas as suggested by Lunney and Leary (1988).

Future events

In reviewing the status of Australia's terrestrial vertebrates and the reasons why the Australian fauna may be particularly sensitive to the changes associated with European settlement, an important observation to emerge is the short time from when the fauna is first affected to when it is lost. In a pattern repeated across Australia, twenty or thirty years sees species decline from abundance to extinction. Can this observation provide a mirror for future events? There are suggestions that the decline of vertebrates first observed in southern Australia has extended to the northern parts of the continent and that fauna in the wetter coastal regions may also be affected. Similarly, there is evidence that birds are declining in abundance with an increasing number of local extinctions. If these suggestions are correct, and we see no reason to doubt their accuracy, then we do not have the time to wait for the trends to be quantified. If the observations are correct, then we will be able to observe the rapid extinction of species throughout Australia in the next few decades.

We predict that there will be further significant declines in the distribution and abundance of birds throughout the same regions where the native mammal fauna has already disappeared. These will follow the same pattern as the mammals with the rapid extinction of species or a reduction in abundance to small, tenuous relict populations persisting where conditions remain marginally favourable. There will be a major loss of bird species and possibly some reptiles from the more intensive agricultural and pastoral regions. We suggest that the presence of many species in these regions is illusory and that what we have already are isolated populations restricted to patches and strips of remnant vegetation. As Saunders (1989) has shown for the wheatbelt reserves in Western Australia, species are lost rapidly from remnants. As each population becomes extinct, the chance of recolonization is diminished. With continued clearing of native vegetation, use of chemicals for weed and pest control, and the increasing deterioration of soils in these regions, the process of decline and extinction of the avifauna will accelerate.

In wetter forest and woodland habitats along the coast the prospects for the future are only marginally better. The continued clearing and fragmentation of forest for agriculture and the establishment of plantations, the greater use of prescription burning by land management authorities, loss of important wintering habitats along the coast to residential development and tourism, the loss of mature forest from logging, and the spread of rural dieback place intense pressure on forest animals which may already be restricted to marginal habitats (see Lunney & Leary 1988) and reduced in abundance. Ground-dwelling and ground-foraging animals, species which require tree hollows as den or nest sites, fauna with specialized foraging substrate requirements (e.g., decorticating bark), and migratory species which require continuous habitat for foraging as they migrate (see Recher *et al.* 1983) will be particularly affected.

What is the future of Australia's terrestrial vertebrate fauna? We said at the outset of this chapter that it is easy to become emotional when reviewing the status of the continent's fauna. The future is grim. It is grim largely because so few people can read the past and see the present. The loss of Australia's wildlife is a trivial event to the majority of Australians. Were it otherwise, it would not be necessary to beg funds for research on the conservation of endangered species; to fight to protect the last scattered remnants of native vegetation; to protest over the logging of the last stands of undisturbed forest or to yield to the development of tourist facilities in national parks as a better alternative than mining or residential subdivision.

Although we talk of species, the patterns described are of entire ecosystems disappearing. The scale of change is not of individuals or populations, but of functional units of interacting organisms. Once destroyed it is no more likely that an ecosystem can be restored to its original state any more than an extinct species can be ressurected from its bones recovered from a museum drawer. Half of Australia's soils are affected by erosion and/or salinization (Anon. 1986; SPCC 1987). Over four million hectares of the pastoral inland are on the verge of becoming perpetual desert. Australia's soils have been mined of their nutrients and organic matter. The scale of this disaster is unprecedented. Viewed in this context the decline of the continent's vertebrate fauna is almost meaningless; the extinction of species becomes a symptom of far more pervasive changes to the land. As ecologists we should view the decline of Australia's fauna as a symptom of more serious events and seek to remedy the problem by correcting the underlying causes and not by treating the symptoms.

Australia, no differently from other nations, has approached the conservation of native wildlife species by species. It has responded to events and public outcry with rules and regulations and the creation of a parks and reserves system designed more to mollify environmentalists than conserve wildlife. If Australia is to conserve its terrestrial fauna, it must take a radically different approach. Emphasis needs to shift from species preservation to the management of ecosystems. The landscape must be managed as a whole and not parcelled into separate units each with a different set of goals and managed by people who lack common understanding. Australia is not a set of political or economic land units, but is a whole functional ecosystem. Management by units, each in isolation, will only perpetuate the decline of the nation's land systems and accelerate the extinction of its wildlife. Land management and the conservation of wildlife must be extended to include and to involve all lands irrespective of tenure. There must be national priorities which transcend political boundaries and the limits imposed by the Constitution on National and State responsibilities. By themselves national parks and nature reserves will not preserve Australia's biota. This can only be achieved by the full co-operation of land managers, farmers and public servants alike, working toward specified national objectives. The National Conservation Strategy was a start; it is time to implement its objectives.

While State and Federal Wildlife Acts extend general protection to almost all native species of Australia and its territories, prohibitive provisions dominate existing legislation. That is, management authorities rely on the imposition of maximum fines and other penalties for violation of fauna regulations involving species listed in gazetted schedules. They do not compel those authorities to undertake applied research to establish the minimal ecological and critical habitat requirements of the species listed. Nor do they require the protection and management of critical habitats. This is in marked contrast to the Endangered Species Act of the United States (1983) which includes an array of positive provisions to compel wildlife authorities to determine critical wildlife habitat for designated endangered species and then to manage those areas to ensure the survival of the designated species.

On the user-pays principle, legislation could be put in place in Australia with positive provisions (see U.S. Endangered Species Act 1983) to encourage governments and government departments to commit a set proportion of their allocated funds and part or the whole of their revenue from licence fees and park entrance fees to support wildlife conservation. These funds could be used for research to manage ecosystems and harvested species, to undertake adequate surveys, to monitor changes of species and habitats at risk, to control pest and weed species, to study the effects of proposed and introduced biological agents on Australian ecosystems, to document the effects of agricultural chemicals on wildlife, and to rehabilitate rare and endangered species.

It is critical for the survival of Australia as a nation, as a functional ecosystem, to reverse the rundown of the nation's research capabilities. Rather than creating an environment of educational mediocrity, where researchers have inadequate funding, there should be an emphasis on excellence and a commitment to research, regardless of what may or may not be immediate returns on the funds invested. It is necessary to recognize that not all research has immediate application and that there are programmes of investigation which may require decades for completion. An understanding of the effects of fire on natural ecosystems, for example, requires a succession of fire events. Even with intervals as short as five years it may be decades and a series of fires before an effect appears. The establishment of long-term monitoring programs of environmental change must become a national research priority. Such studies are not the province of short-lived individuals, but are the responsibility of institutions which can ensure that each generation of students or researchers makes their collective contribution to a common goal of understanding European impact on the Australian environment.

Epilogue

Writing about the settlement of the pastoral regions of South Australia during the 1920s, Ratcliffe (1938) described the sequence of rain, drought, over-grazing and soil erosion which typified most of the interior. At first conditions were good. It rained, the flocks grew, and money was made. Then the rabbits came and the rain failed. Between them, the rabbits and sheep denuded the landscape. As put by Ratcliffe, the rabbits drove out the bilbies. Along with the vegetation and the bilbies, the other native animals disappeared. With the plants and animals, the soil disappeared, but the rabbits and sheep remained. It took only one such sequence to bring about the demise of native mammals over the vast pastoral areas of New South Wales, Victoria and South Australia.

Ratcliffe could have been writing about anywhere in Australia. The image he presented was of a fragile land ill-suited to pastoral and agricultural activities developed on other, more resilient continents. His is the same story that Saunders (1989) and Saunders and Curry (this volume) tell of the West Australian wheatbelt, that Ford and Howe (1980) present of the Mt Lofty Ranges, that Lunney and Leary (1988) describe for south-eastern New South Wales, and that Burbidge and McKenzie (1989) offer for the arid interior. It is the same story that is being repeated in the north of Australia, on Kangaroo Island, in Tasmania and in northern Queensland. It is not a scientific story nor one that lends itself to statistical analysis. There is no experimental design nor can the experiment be repeated.

Acknowledgements

Hugh Ford, Peter Jarman, Norm McKenzie and Denis Saunders constructively criticized an early draft of this paper. Their assistance was most helpful and gratefully appreciated. Preparation of this chapter was assisted by a grant from the University of New England.

References

Anon. (1986) *State of the Environment in Australia.* Australian Government Publishing Service, Canberra.

Archer M. (1979) Two new species of *Sminthopsis* Thomas (Dasyuridae: Marsupialia) from northern Australia, *S. butleri* and *S. douglasi. Aust. Zool.* **20,** 327-46.

Archer M. (1981) A review of the origins and rediations of Australian mammals. In: *Ecological Biogeography of Australia.* (ed. A. Keast) pp. 1435-88. Dr W. Junk, The Hague.

Baldwin M. (1975) Birds of the Inverell district, New South Wales. *Emu* **75,** 113-20.

Blakers M., Davies S. J. J. F. and Reilly P. N. (1984) *The Atlas of Australian Birds.* Melbourne University Press, Melbourne.

Burbidge A. A. & McKenzie N. L. (1989) Patterns in the decline of the mammal fauna of Western Australia: causes and implications for future conservation. *Biol. Conserv.* **50,** 143-98.

Cogger H. G. (1986) *Reptiles and Amphibians of Australia.* Reed Books Pty. Ltd., Sydney.

Cogger H. G. & Heatwole H. (1981) The Australian reptiles: origins, biogeography, distribution patterns and island revolution. In: *Ecological Biogeography of Australia* (ed. A. Keast) pp. 1331-74. Dr W. Junk, The Hague.

Denny M. (1981) Kangaroos: A historic perspective. In: *Kangaroos and other macropods of New South Wales* (ed. C. Haigh). National Parks and Wildlife Service (N.S.W.), Sydney.

Ford H. A. & Bell H. (1981) Density of birds in eucalypt woodland affected to varying degrees by dieback. *Emu* **81,** 202-8.

Ford H. & Howe R. (1980) The future of birds in the Mt Lofty Ranges. *South Aust. Ornith.* **28,** 85-9.

Fox B. J. & Briscoe D. R. (1980) *Pseudomys pilligaensis.* A new species of murid rodent from the Pilliga Scrub, Northern New South Wales. *Aust. Wildl. Res.* **9,** 1-19.

Frith H. (1964) *Mobility of the Red Kangaroo.* CSIRO, Canberra.

Frith H. (1973) *Wildlife Conservation.* Angus & Robertson, Sydney.

Giles J. R. & Lim L. (1987) Conservation of the Endangered Species of New South Wales: An Assessment of Current Status, Threats and Requirements for Maintenance of Wildlife Populations. Proc. Nat. Conf. Conserv. Threat. Spp. and their Habitat (ACIUCN, 3-6 March 1987, Sydney).

Heatwole H. F. & Taylor J. (1988) *Ecology of Reptiles.* Surrey Beatty & Sons, Chipping Norton.

Heron S. J. (1973) Birds of the Orange district, NSW. *Emu* **73,** 1-8.

Hindwood K. A. & McGill A. R. (1958) *The Birds of Sydney.* Royal Zoological Society of N.S.W., Sydney.

Howe R. H. (1984) Local dynamics of bird assemblages in small forest habitat islands in Australia and North America. *Ecology* **65,** 1585-601.

Howe R. W., Howe T. D. & Ford H. A. (1981) Bird distributions on small rainforest remnants in New South Wales. *Aust. Wildl. Res.* **8,** 637-51.

ICBP (1981) *Endangered Birds of the World: The ICBP Bird Red Data Book* Smithsonian Institute Press, Washington, D.C.

Jack N. (1973) November general meeting. *Queensland Ornith. Soc. Newsletter,* November 1973, pp. 1-3.

Jarmen P. J. & Johnson K. A. (1977) Exotic mammals, indigenous mammals and land-use. *Proc. Ecol. Soc. Aust.* **10,** 146-65.

Kinnear J., Onus M. & Bromilow B. (1988) Fox control and rock-wallaby population dynamics. *Aust. Wildl. Res.* **15:** 435-50.

Kitchener D. J., Chapman A., Dell J. & Muir B. G. (1980) Lizard assemblage and reserve size and structure in the Western Australian wheatbelt — some implications for conservation. *Biol. Conserv.* **17,** 25-62.

Kitchener D. J., Dell J., Muir B. G. & Palmer M. (1982) Birds in Western Australian wheatbelt reserves — implications for conservation. *Biol. Conserv.* **22,** 127-63.

Kitching R. L. (ed.) (1986) *The Ecology of Exotic Animals and Plants: Some Australian Case Histories.* John Wiley & Sons, Brisbane.

Lavery H. J. and Tierney P. J. (1985) Scarcity and Extinction. In: *The Kangaroo Keepers* (ed. H. J. Lavery) pp. 25-54, University of Queensland Press, St Lucia.

Lim L. (1987) Ecology and Management of the Rare Yellow-footed Rock Wallaby, *Petrogale xanthopus* Gray 1854 (Macropodidae) Ph.D. Thesis. Macquarie University, Sydney.

Lim L. (1988) The conservation of fauna in Western New South Wales. *J. Nat. Parks Assoc. (N.S.W.)* **32,** 5-10.

Loyn R. H. (1985) Bird populations in successional forests of Mountain Ash *Eucalyptus regnans* in central Victoria. *Emu* **85:** 213-30.

Loyn R. H. (1987) Effects of patch area and habitat on bird abundances, species numbers and tree health in fragmented Victorian forests. In: *Nature Conservation: The Role of Remnants of Native Vegetation* (eds D. A. Saunders, G. W. Arnold, A. A. Burbidge & A. J. M. Hopkins) pp. 65-77. Surrey Beatty & Sons, Chipping Norton.

Lunney D. & Leary T. (1988) The impact on native mammals of land-use changes and exotic species in the Bega district, New South Wales, since settlement. *Aust. J. Ecol.* **13,** 67-92.

Main A. R. (1982) Rare species: precious or dross? In: *Species at Risk: Research in Australia* (eds R. H. Groves and W. D. L. Ride) pp. 163-74. Australian Academy of Science, Canberra.

Marshall A. J. (ed) (1966) *The Great Extermination.* The Griffin Press, Adelaide.

Morton S. R. & Baynes A. (1985) Small mammal assemblages in arid Australia: A reappraisal. *Aust. Mamm.* **8,** 159-69.

Myers K. (1986) Introduced vertebrates in Australia, with emphasis on mammals. In: *Ecology of Biological Invasions: An Australian Perspective* (eds R. H. Groves & J. J. Burdon) pp. 120-36.

Newsome A. E. (1971) Competition between Wildlife and Domestic Livestock. *Aust. Vet. J.* **47,** 577-86.

Newsome A. E. (1975) An ecological comparison of the two arid-zone kangaroos of Australia, and their anomalous prosperity since the introduction of ruminant stock to their environment. *Quart. Rev. Biol.* **50,** 389-424.

Nix H. A. (1976) Environmental control of breeding, post-breeding dispersal and migration of birds in the Australian region. *Proc. Int. Ornith. Cong.* **16,** 272-306.

Porter J. W. (1982) Terrestrial birds of the coastal lowlands of south-east Queensland. *Tech. Paper* No. 30. Dept. of Forestry, Queensland.

Ratcliffe F. (1938) *Flying Fox and Drifting Sand.* Angus & Robertson, Sydney.

Recher H. F., Gowing G., Kavanagh R., Shields J. & Rohan-Jones W. (1983) Birds, resources and time in a tablelands forest. *Proc. Ecol. Soc. Aust.* **12,** 101-24.

Recher H. F., Rohan-Jones W. & Smith P. (1980) Effects of the Eden woodchip industry on terrestrial vertebrates with recommendations for management. Research Note No. 42, Forestry Commission of N.S.W.

Recher H. F., Shields J., Kavanagh R. & Webb G. (1987) Retaining remnant mature forest for nature conservation at Eden, New South Wales: a review of theory and practice. In: *Nature Conservation: The Role of Remnants of Native Vegetation* (eds D. A. Saunders, G. W. Arnold, A. A. Burbidge & A. J. M. Hopkins) pp. 177-94. Surrey Beatty & Sons, Chipping Norton.

Ride W. D. L. & Wilson G. R. (1982) The conservation status of Australian mammals. In: *Species at Risk: Research in Australia* (eds W. D. L. Ride & R. H. Groves) pp. 27-45. Australian Academy of Science, Canberra.

Rosel M. & Smith R. (1985) Australia backs conservation of flora and fauna. *Aust. Now* **11,** 22-7.

Rowley I. (1988) *The Purple-crowned Fairy-wren.* RAOU, Melbourne.

Saunders D. A. (1989) Changes in the avifauna of a region, district and remnant as a result of fragmentation of native vegetation: the wheatbelt of Western Australia. A case study. *Biol. Conserv.* **50,** 99-135.

Saunders D. A. & Ingram J. A. (1987) Factors affecting survival of breeding populations of Carnaby's cockatoo *Calyptorhynchus funereus latirostris* in remnants of native vegetation. In: *Nature Conservation: The Role of Remnants of Native Vegetation* (eds D. A. Saunders, G. W. Arnold, A. A. Burbidge & A. J. M. Hopkins) pp. 249-58. Surrey Beatty & Sons, Chipping Norton.

Saunders D. A., Rowley I. & Smith G. T. (1985) The effects of clearing for agriculture on the distribution of cockatoos in the south-west of Western Australia. In: *Birds of Eucalypt Forests and Woodlands: Ecology, Conservation, Management* (eds A. Keast, H. F. Recher, H. Ford & D. Saunders) pp. 309-21. Surrey Beatty & Sons, Chipping Norton.

Serventy D. L. (1977) The use of data on the distribution of birds to monitor climatic changes. *Emu* **77,** 162-6.

Sibley C. O. & Ahlquist J. E. (1985) The phylogeny and classification of the Australo-Papuan passerines. *Emu* **85,** 1-14.

Slater P., Slater P. & Slater R. (1986) *The Slater Field Guide to Australian Birds.* Rigby, Dee Why West, N.S.W.

Smith P. (1985) Effects of intensive logging on birds in eucalypt forest near Bega, New South Wales. *Emu* **85,** 200-10.

Smith G. T. (1987) The changing environment for birds in the south-west of Western Australia. In: *Nature Conservation: The Role of Remnants of Native Vegetation* (eds D. A. Saunders, G. W. Arnold, A. A. Burbidge & A. J. M. Hopkins) pp. 269-77. Surrey Beatty & Sons, Chipping Norton.

SPCC (1987) Murray-Darling Basin Environmental Resources Study. Murray-Darling Basin Ministerial Council, Canberra.

Springett B. P. (1978) On the ecological role of insects in Australian eucalypt forests. *Aust. J. Ecol.* **3,** 129-40.

Strahan R. (ed) (1983) *Complete Book of Australian Mammals.* Angus & Robertson, Sydney.

TEC (1983) *Our Wildlife in Peril.* A. H. & A. W. Reed, Sydney.

Tierney P. (1989) Habitat preferences of the Bridled Nail-tail Wallaby *(Onychogalea fraenata).* In: *Kangaroos, Wallabies and Rat-kangaroos* (eds G. Grigg, I. Hume & P. Jarman). Surrey Beatty & Sons, Chipping Norton (in press).

Tyler M. J. (ed) (1978) *The Status of Endangered Australasian Wildlife.* Royal Zoological Society of South Australia, Adelaide.

APPENDIX 1

CONCOM

LIST OF AUSTRALIAN ENDANGERED VERTEBRATE FAUNA[1]

(last amended in April 1988)

	Scientific Name	Common Name
MAMMALIA		
Marsupialla		
Macropodidae	*Bettongia lesueur*	Burrowing bettong (boodie)
	B. penicillata	Brush-tailed bettong (woylle)
	B. tropica	Northern bettong
	Caloprymnus campestris (1935)	Desert rat-kangaroo
	Lagorchestes asomatus (1931)	Central hare-wallaby
	L. hirsutus	Rufous hare-wallaby
	L. leporides (1980)	Eastern hare-wallaby
	L. fasciatus	Banded hare-wallaby
	Macropus grevi (1927)	Toolache wallaby
	Onychogalea fraenata	Bridled nail-tail wallaby
	O. tunata (1930)	Crescent nail-tail wallaby
	Petrogale persephone	Proserpine rock-wallaby
	Potorous platyops (ca. 1875)	Broad-faced potoroo
	P. longipes	Long-footed potoroo
Petauridae	*Gymnobelideus leadbeateri*	Leadbeater's possum
Vombatidae	*Lasiorhinus krefftii*	Northern hairy-nosed wombat
Peramelidae	*Chaeropus ecaudatus* (1926)	Pig-footed bandicoot
	Perameles bougainville	Western barred bandicoot
	P. eremiana (1931)	Desert bandicoot
Thylacomyidae	*Macrotis lagotis*	Greater bilby (dalgyte)
	M. leucura	Lesser bilby
Thylacinidae	*Thylacinus cynocephalus* (1933)	Thylacine (Tasmanian tiger)
Myrmecobiidae	*Myrmecobius fasciatus*	Numbat
Dasyuridae	*Sminthopsis longicaudata*	Long-tailed dunnart
	S. psammophila	Sandhill dunnart
RODENTIA		
Muridae	*Conilurus albipes* (1875)	White-footed rabbit-rat
	Leporillus apicalis (ca. 1840)	Lesser stick-nest rat
	L. conditor	Greater stick-nest rat
	Notomys amplus (1896)	Short-tailed hopping-mouse
	N. aquilo	Northern hopping-mouse
	N. fuscus	Dusky hopping-mouse
	N. longicaudatus (1906)	Long-tailed hopping-mouse
	N. macrotis (ca. 1840)	Big-eared hopping-mouse
	N. mordax (1840)	Darling Downs hopping-mouse
	Pseudomys fieldi (1895)	Alice Springs mouse
	P. praeconis	Shark Bay mouse
	Xeromys myoldea	False water-rat
	Zyzomys pedunculatus	Central rock-rat
CETACEA		
Balaenopteridae	*Balaenoptera musculus*	Blue whale
	Megaptera novaeangliae	Humpback whale
Balaenidae	*Eubalaena australis*	Southern right whale
AVES		
Procellariiformes		
Procellariidae	*Pterodroma leucoptera leucoptera*	Gould's petrel
	P. solandri	Providence petrel
Pelecaniformes		
Sulidae	*Sula abbotti*	Abbott's booby
Fregatidae	*Fregata andrewsi*	Christmas Island frigatebird

APPENDIX 1 — *continued*

	Scientific Name	Common Name
FALCONIFORMES		
	Accipter radiatus	Red goshawk
GRULFORMES		
Pedionomidae	*Pedionomus torquatus*	Plains-wanderer
Palidae	*Tricholimnas sylvestris*	Lord Howe Island woodhen
	Rallus philloppensis andrewsi	Cocos buff-banded rail
CHARADRIIFORMES		
	Anous tenuirostris	Lesser noddy
	Sterna albifrons	Little tern
PSITTACIFORMES		
Psittacidae	*Cyanoramphus novaezelandiae cookii*	Norfolk Island parrot
	Psittaculirostris diophthalma coxeni	Coxen's fig-parrot
	Geopsittacus occidentalis	Night parrot
	Neophema chrysogaster	Orange-bellied parrot
	Pezoporus wallicus	Ground parrot
	Polytelis alexandrae	Alexandra's parrot
	Psephotus chrysopterygius	Golden-shouldered parrot
	P. dissimills	Hooded parrot
	P. pulcherrimus (1922)	Paradise parrot
STRIGIFORMES		
Strigidae	*Ninox undulata*	Norfolk Island boobook owl
	N. squamipila natalis	Christmas Island hawk-owl
CAPRIMULGIFORMES		
Caprimulgidae	*Podargus ocellatus plumiferus*	Plumed frogmouth
PASSERIFORMES		
Atrichornithidae	*Atrichornis clamosus*	Noisy scrub bird
Maluridae	*Malurus coronatus*	Purple-crowned fairy-wren
	Amytornis dorotheae	Carpentarian grasswren
	A. textilis textilis	Thick-billed grasswren
Acanthizidae	*Dasyornis longirostris*	Western bristlebird
	D. brachypterus	Eastern bristlebird
	D. broadbenti litoralis (1940)	Rufous bristlebird
Cracticidae	*Strepera graculina crissalis*	Lord Howe Island currawong
Timallidae	*Psophodes nigrogularis*	Western whipbird
Pardalotidae	*Pardalotus quadragintus*	Forty-spotted pardalote
Zosteropidae	*Zosterops alboqularis*	Norfolk Island silvereye
Meliphagidae	*Lichenostomus melanops cassidix*	Helmeted honeyeater
	Manorina melanotis	Black-eared miner
Muscicapidae	*Drymodes supercilaris colcloughi* (1915)	Northern scrub-robin
	Turdus poliocephalus poliocephalus	Norfolk Island thrush
AMPHIBIA		
Anura		
Leptodactylidae	*Arenophryne rotunda*	Round frog
	Philoria frosti	Mt Baw Baw frog
	Rheobatrachus silus	Platypus frog
Hylidae	*Cophixalus concinnus*	Elegant microhylid
	C. saxatillis	Rock-dwelling microhylid
	Litoria longirostris	Long-nosed tree frog
REPTILIA		
Chelonia		
Dermochelyidae	*Dermochelys coriacea*	Leathery turtle
Chelidae	*Pseudemydura umbrina*	Western swamp tortoise

APPENDIX 1 — *continued*

	Scientific Name	Common Name
SQUAMATA		
Scincidae	*Lerista lineata*	Lined burrowing skink
	Ctenotus lancelini	Lancelin Island striped skink
	Pseudomoia palfreymani	Pedra Branca skink
Pygopodidae	*Ophidiocephalus taeniatus*	Bronzebacked legless lizard
	Aprasia parapulchella	Pinktailed legless lizard
Elapidae	*Vermicella calonota*	Black-striped snake
	Hoplocephalus bungaroides	Broad-headed snake
OSTEICHTHYES		
Salmoniformes		
Galaxiidae	*Galaxias fontanus*	Swan galaxias
	G. johnstoni	Clarence galaxias
PERCIFORMES		
Percichthyidae	*Maccullochella ikei*	Eastern freshwater cod
	M. macquariensis	Trout cod

[1]Year in parenthesis is year of last confirmed sighting or specimen of species presumed to be extinct.

Proc. Ecol. Soc. Aust. 1990 — 16:303-321

The impact of agricultural and pastoral industries on birds in the southern half of Western Australia: past, present and future

D. A. SAUNDERS[1] and P. J. CURRY[2]

[1]*CSIRO Division of Wildlife and Ecology, LMB No. 4, P.O. Midland, W.A. 6056*

[2]*Department of Agriculture, Division of Resource Management, Baron-Hay Court, South Perth, W.A. 6151*

Abstract

The agricultural industry in Western Australia has resulted in extensive clearing of native vegetation for cereal cropping and pasture, leaving scattered remnants across the landscape and providing water points every 2.5 km². In contrast, the pastoral industry of the southern arid zone grazes stock on native vegetation and, on average, provides water points every 50 km² across the landscape.

The avifauna of the central wheatbelt (agricultural) has changed markedly over the last 80 years. Species have become reduced in numbers or disappeared from districts. Other species have invaded the central wheatbelt from the adjacent pastoral country, with only a few species not undergoing some change in status. The present-day avifauna of the Murchison River catchment (pastoral) is composed of a majority of species for which there are no clear indications of any change in status. A small group has declined or disappeared from the catchment and a larger group of species has increased in range, abundance or both.

Facts underlying these changes are discussed, as are possible scenarios for future changes in these environments and their management.

Introduction

Since Australia was settled by Europeans in 1788 there have been changes in land use which, in most cases, have resulted in marked alterations to the distribution and abundance of the original vegetation.

This chapter looks at impacts of European development on the avifaunas of the Kellerberrin district and the Murchison River catchment, two study areas representative of the semi-arid and arid zones of south-western Australia and now used for extensive primary production industries. We examine some of the conservation issues raised. Some changes in management actions are required to counter the trend of continuing loss of species from the two study areas and more widely in their respective climatic zones.

Study areas

The agricultural area or wheatbelt of Western Australia

European settlement of south-eastern Australia began in 1788, but the south-west was not settled until 1829. In the early stages of the Swan River Colony the changes brought about were small and by 1890 only 500 km² of land in the south-west had been cleared, mainly around farmsteads and towns. The pace of clearing increased and by 1968 nearly 130 000 km² of land in the south-west had been cleared; this amounts to nearly 93% of the 140 000 km² area known as the wheatbelt of Western Australia (Fig. 1). Fifty-four percent of cleared land was stripped of its native vegetation in the short period between 1945 and 1982 (Saunders *et al.* 1985).

The wheatbelt occurs in that area which receives an average annual rainfall of between 300 and 650 mm (Fig. 1). Substantial variation in climate, pedology and vegetation throughout the wheatbelt means that it does not constitute a well-defined natural region (Gentilli 1958) but it had a rich and diverse flora with a high degree of heterogeniety associated with the mosaic of different soil types and possibly fire history (Beard 1981). Nevertheless, extensive clearing of native vegetation has resulted in highly modified landscapes which have had their appearance reduced towards monotony over 14 million ha. The wheatbelt now consists

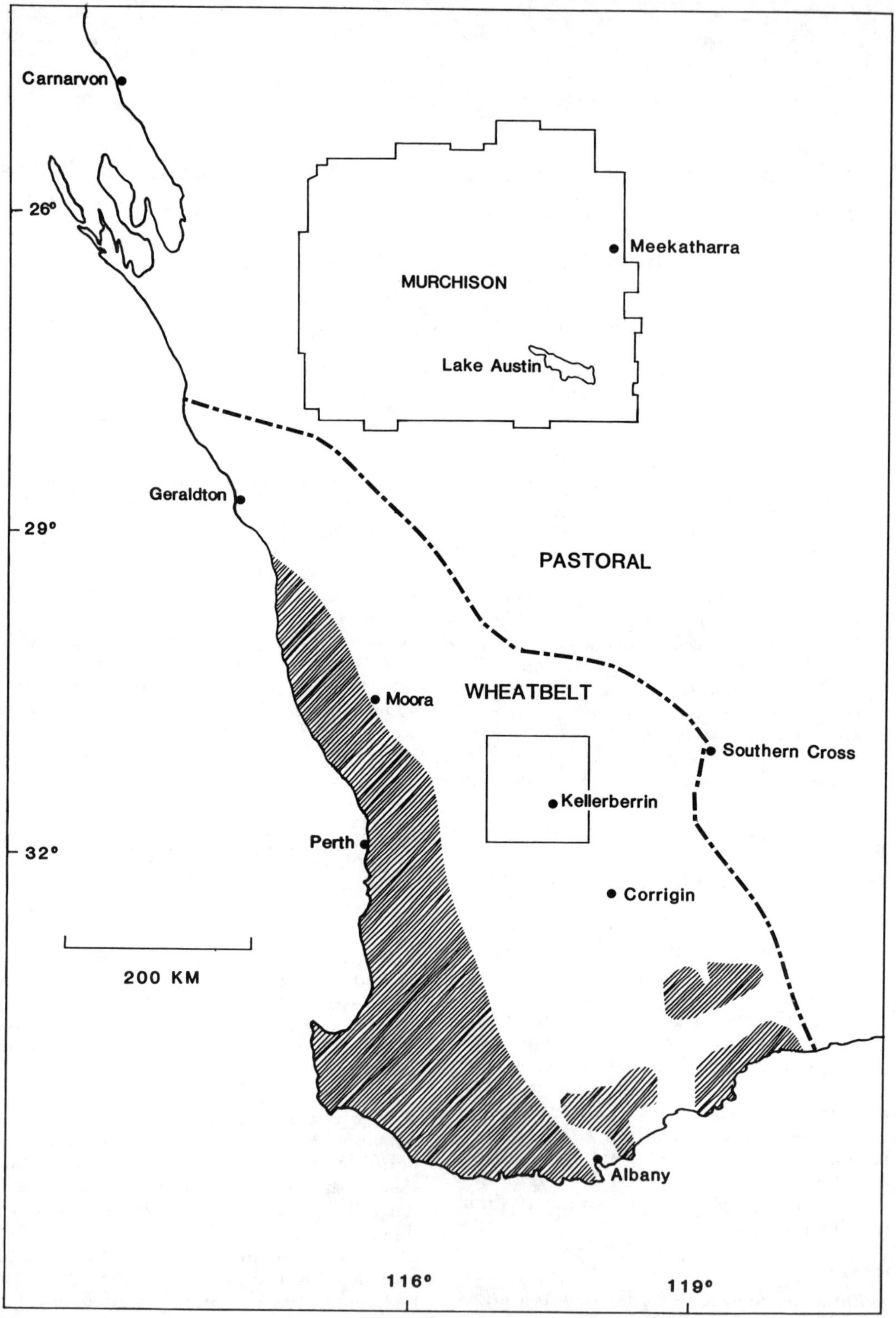

FIG. 1. The south-west of Western Australia showing the Kellerberrin district of the wheatbelt and the Murchison River catchment of the pastoral region. The hatched areas are not cleared for wheat production.

mainly of extensively cleared areas dominated by winter-growing annual grasses (mainly wheat, oats, lupins, wild oats, pasture grasses and associated exotic species) but with thousands of small remnants of native vegetation scattered over the landscape. These vary in size, shape, vegetation, degree of isolation and intensity of human disturbance (see Saunders 1989; Saunders *et al.* 1987). Sheep are the main livestock. Stock troughs and farm dams are now abundant across a region that formerly had very limited available surface water during the summer dry season.

During development of the wheatbelt for agriculture, few areas of native vegetation were set aside for the conservation of the biota. As a result this region is poorly served by conservation reserves. The existing nature reserves in the wheatbelt range in size from 0.4 to 309 000 ha, but have a median size of only 114 ha. They were acquired mainly in the 1960s and 1970s after active agricultural development had ceased and consist of parcels of land originally set aside for purposes unrelated to conservation (e.g., water catchment, town or other amenity sites, camping areas, road reserves). Together these reserves contain 6.7% of the area of the wheatbelt, but if the three largest are excluded this drops to 2.4% (Wallace & Moore 1987).

The Kellerberrin district lies between 31-32°S. and 117-118°E. and the town of Kellerberrin has an annual average rainfall of 336 mm and an average of 77 rain days per year (Anon. 1979). It is in the area known as the central wheatbelt and the district is representative both naturally and sociologically of most of the wheatbelt region (W. M. McArthur, pers. comm.). The district was settled in the early 1860s (Leake 1962) with pastoral activities predominating initially. The opening of the railway line from Perth in the early 1890s allowed easier access and closer settlement followed. By 1930/31 nearly 70% of the areas of native vegetation had been removed. The Depression and World War temporarily halted development, but by 1959/60 90% of the area of original vegetation had disappeared and this figure had risen to 94% by 1967/68. There has been little further clearing since then (Australian Bureau of Census and Statistics).

The pastoral area

Grazing of sheep and cattle on the unimproved native vegetation of extensive pastoral leases is the only widespread form of agriculural production in the arid zone of Western Australia. Within 1.7×10^6 km^2 lying between the 250 mm mean annual rainfall isohyet in the south and the 350 mm isohyet in the north, pastoral leases occupy about 45% of the area. In the winter-moisture-dominated southern and western areas of the arid zone (Nix 1982) both leases and stock occur at a higher density than the summer-moist areas, which have a more severe climate and are largely dominated by spinifex, *Triodia* and *Plectrachne,* associations. This chapter focuses on a part of the winter-moist sector, which is mostly dominated by shrublands, rather than hummock grasslands.

Pastoral use and developmment of the area began in the 1870s when sheep were brought up the Greenough and Murchison Rivers. A fairly rapid spread of sheep flocks throughout most of the river frontage areas occurred during the next 30 years. Wells were dug to obtain permanent ground-water for stockmen and their flocks which were habitually shepherded and allowed to graze at high densities on the much favoured river country. Paddock subdivisions, by plain wire fencing, were not in general use for the unshepherded control of stock until after 1920. It was not until this latter period that the broader landscapes of the river catchment areas were brought into general use for grazing in this way. Accounts of the pastoral industry and its land use in parts of the arid southern shrublands are given by Mabbutt *et al.* (1963), Williams *et al.* (1980) and Nixon and Lefroy (1988).

The Murchison River catchment study area lies in southern central Western Australia and extends over approximately 80 000 km^2 to the south and west of Meekatharra (26°35′S., 118°30′E.) (Fig. 1). It includes almost all of the catchment of the Murchison River and its tributaries, plus parts of the catchments of the Wooramel and Greenough Rivers and other marginal areas that drain into lake systems (e.g., Lakes Annean, Austin, Nerramyne and Muggon). As described by Beard (1976), mean annual rainfall is below 250 mm everywhere except for the extreme south-west corner and is close to 200 mm over most of the area. The seasonal moisture index is winter-dominated throughout and most rain falls between January and July. Vegetation in the study area is dominated by shrublands of the Eremaean Botanical Province (Beard 1976).

By the 1950s most Murchison stations had developed an infrastructure of fenced paddocks and stock watering points similar to that maintained at present. Stocking rates are generally one sheep to between 10 and 40 ha. Besides sheep, some cattle and small numbers of horses are pastured on some leases.

In the hundred or so years since the advent of pastoralism and the simultaneous loss of Aboriginal land use practices, the landscapes and their vegetation have undergone extensive but variable degrees of alteration under grazing. Appreciable populations of feral mammals, mainly goats and to a lesser extent rabbits, contribute to grazing pressures as do native mammals (red kangaroos and euros). Severely overgrazed areas have become denuded and eroded in some places, with secondary invasions of unpalatable shrubs. The presence of kite leaf poison, *Gastrolobium laytonii,* which is highly toxic to stock, in the granite outcrops and stream-lines of parts of the southern half of the catchment has resulted in areas of up to 25 000 ha remaining rarely grazed or ungrazed by stock.

There are no national parks in the study area, nor are nature reserves or other areas set aside specifically for conservation purposes.

The avifauna

The wheatbelt

Methods. The avifauna of the Kellerberrin degree block (Fig. 1). was compiled from records published by Storr (1953), Ford and Stone (1957), Leake (1962), Dell (1978, 1980, 1981), Blakers *et al.* (1984) and from the personal observations during 1987 of 16 people who lived (N=12) in the district or visited it during field work (N=4) and noted every species they saw in any recording week [See Saunders (1989) for details].

From these records, past and present status of each species was assessed, and where some measure of relative abundance was noted, this was adjusted subjectively to a scheme ranging from very common to extinct. Some authors mentioned species but did not provide enough information to gauge abundance and so no classification of abundance was noted for these species (* on Table 1).

Results. Abundance and habitat preferences for the 166 species recorded in the Kellerberrin district are shown in Table 1. Thirty of these are waterbirds (recorded only in aquatic habitats including farm dams) and five are vagrants which have been recorded rarely in this area. This latter group include the letter-winged kite, square-tailed kite, black kite, fork-tailed swift and orange chat. These thirty-five species have been excluded from subsequent analysis.

Possible changes in status of the other 131 species were assessed from Table 1 and from the interpretation of comments by the various sources cited.

Fourteen (23%) species of non-passerine have decreased in range and/or abundance over the last 80-90 years (Table 2). The brown falcon, nankeen kestrel, little button quail, bush stone curlew, purple-crowned lorikeet, Horsfield's bronze cuckoo, tawny frogmouth and owlet-nightjar have all declined, in most cases going from common to uncommon or rare (Table 1). The emu, mallee fowl, Carnaby's cockatoo, regent parrot, red-capped parrot and Western rosella have all disappeared from the district. Of the latter six species, only the regent parrot was formerly common. Fifteen (25%) species of non-passerine have increased in range and/or abundance (Table 2). Only four of these (wood duck, black-shouldered kite, wedge-tailed eagle and stubble quail) were probably present in the district at European settlement. Three species (domestic pigeon, laughing turtledove and laughing kookaburra) are exotics which were released in Perth and expanded from there. In the case of the pigeon and dove the expansion mainly followed the railway lines. None of the three species is very plentiful today. The remaining eight species (mountain duck, banded plover, crested pigeon, red-tailed black cockatoo, galah, long-billed corella, mulga parrot and elegant parrot) have all extended their ranges into the wheatbelt. In fact, two of the most widespread and abundant species of non-passerine occurring in the district are the galah (seen in 100% of the recording weeks by observers in 1987) and the crested pigeon (97%); the former reached Kellerberrin in the early 1920s and the latter in the mid 1940s (Leake 1962). The remaining 32 species of non-passerine showed no apparent change in abundance or range over the period in question.

Passerines have apparently fared much worse than the non-passerines. Twenty-four (34%) of 70 passerine species have decreased in range and/or abundance this century. Sixteen species (white-backed swallow, fairy martin, white-winged triller, southern scrub robin, jacky winter, restless flycatcher, western warbler, varied sittella, red wattlebird, spiny-cheeked honeyeater, white-eared honeyeater, brown-headed honeyeater, tawny-crowned honeyeater, silvereye, masked woodswallow and grey currawong) have declined but still occur in the area, even if infrequently in some cases. For example the restless flycatcher was noted by Leake (1962) as undergoing a tremendous reduction in numbers and he said it had practically disappeared. One restless flycatcher was seen during 273 recording weeks during 1987 and it is possibly extinct as a breeding species in the Kellerberrin district.

The ground cuckoo-shrike, hooded robin, crested shrike-tit, Gilbert's whistler, chestnut quail-thrush, rufous treecreeper, yellow-plumed honeyeater and white-naped honeyeater all have disappeared from the district (Tables 1 and 2). Only three (4%) species of passerine (black-faced cuckoo-shrike, striated pardalote and pied butcherbird) have apparently increased in abundance over the last 80 or so years. Of these, only the pied butcherbird has invaded the area, coming from the semi-arid areas to the north and east and arriving in Kellerberrin at the turn of the century (Leake 1962).

The 75 (32 species of non-passerine and 43 species of passerine) species of bird not apparently changed in status may be deceptive. Those species which are restricted to native vegetation remnants but which still occur in those remnants will have undergone major decreases in overall population size simply because there is considerably less native vegetation (habitat) available to them. The golden whistler and grey shrike-thrush are examples of this pattern.

TABLE 1. Habitat, past and present status of birds of the Kellerberrin district of the wheatbelt and the Murchison River catchment of the pastoral region of the winter rainfall district of Western Australia. Data for Kellerberrin obtained from Storr (1953), Ford and Stone (1957), Leake (1962), Dell (1978, 1980, 1981), Blakers *et al.* (1984) and personal observations during 1987 of J. Brown, P. Cale, A. Carter, J. Dixon, P. Godfrey, L. Harris, J. Lynn, J. Metcalfe, J. Riches, D. Saunders, G. Smith, B. Stokes, D. Tyler, D. Whisson, K. York and D. Zis. Figures given in the "1987 observers" column are the percentage of total recorder weeks (N=273) species was seen (less than 5%=Ø). Data for the Murchison River catchment are from Austin (1855), North (1897), Whitlock (1910 and diaries), C. P. de Rebeira and P. J. Curry.

	Wheatbelt Kellerberrin					Pastoral Murchison River Catchment			
Name	Habitat	Past status	Present status	1987 observers	Seasonality	Habitat	Past status	Present status	Seasonality
Emu *Dromaius novaehollandiae*	W,M,S,H,F	U	E		N	P,S,Mu, Dr,E,C	U	C	N
Hoary-headed grebe *Poliocephalus poliocephalus*	Aq,D	*	*		V	Aq	*	U	N
Australasian little grebe *Tachybaptus novaehollandiae*	Aq,D		*		V	Aq	*	U	N
Australian pelican *Pelecanus conspicillatus*	Aq		*	Ø	V				—
Little black cormorant *Phalacrocorax sulcirostris*	Aq,D	*	*		V	Aq		R	V
Black cormorant *P. carbo*	Aq,D				—	Aq		R	V
Little pied cormorant *P. melanoleucos*	Aq,D	*	*		V	Aq		U	N
Pied cormorant *P. varius*					—	Aq		R	V
White-necked heron *Ardea pacifica*	Aq,F	*	*	Ø	V	Aq	U	U	N
White-faced heron *A. novaehollandiae*	Aq,F,D	*	U	Ø	V	Aq	U	U	N
Large egret *Egretta alba*					—	Aq		U	N
Glossy ibis *Plegadis falcinellus*					—	Aq		R	V
Straw-necked ibis *Threskiornis spinicollis*	Aq,F	R	U		V	Aq	U	U	?
Royal spoonbill *Platalea regia*					—	Aq		R	V
Yellow-billed spoonbill *P. flavipes*					—	Aq		U	N
Grass whistling duck *Dendrocygna eytoni*					—	Aq		R	V
Black swan *Cygnus atratus*	Aq,D	*	*	Ø	V	Aq	U	U	N
Freckled duck *Strictonetta naevosa*	Aq	*			V	Aq		R	N
Mountain duck *Tadorna tadornoides*	W,Aq,F,D		C	49	M	Aq	C	C	N
Black duck *Anas superciliosa*	Aq	C	U	12	V	Aq	?	U	N
Grey teal *A. gibberifrons*	Aq,D	*	U	16	V	Aq	C	C	N

Table 1 — *continued.*

	Wheatbelt Kellerberrin					Pastoral Murchison River Catchment			
Name	Habitat	Past status	Present status	1987 observers	Season-ality	Habitat	Past status	Present status	Season-ality
Chestnut teal *A. castanea*	Aq		R	Ø	V	Aq		R	V
Shoveller *A. rhynchotis*	Aq		*		V	Aq		R	V
Pink-eared duck *Malacorhynchus membranaceus*	Aq	*	*	Ø	V	Aq	?	U	N
White-eyed duck *Aythya australis*	Aq		*		V	Aq	C	C	N
Wood duck *Chenonetta jubata*	W,Aq,F,D	*	C	29	M	Aq	U?	C	N
Black-shouldered kite *Elanus notatus*	W,F		U	26	R			R	V
Letter-winged kite *E. scriptus*	W,F	*			V				—
Square-tailed kite *Lophoictinia isura*	W,H,F	*			V	E	R	R	?
Black kite *Milvus migrans*	W,H,F	*			V		R	R	Nb
Black-breasted kite *Hamirostra melanosternon*					—	P,Mu,E,C		U	R
Whistling kite *Haliastur sphenurus*	W,H,Aq,F		R		R	E	U	U	R
Brown goshawk *Accipiter fasciatus*	W,M,S,F	*	R	8	R	Mu,Dr, E,C		U	Nb
Collared sparrowhawk *A. cirrhocephalus*	W,M,S,H	U	U	6	R	S,Mu,D,E	C	C	R
Wedge-tailed eagle *Aquila audax*	W,M,S,H,F	*	U	28	R	Hi,P,S, Mu,E,C	U	C	R
Little eagle *Hieraaetus morphnoides*	W,M,S,H,F	*	R	8	R	S,Mu,E,C	R	R	R
Spotted harrier *Circus assimilis*	S,H,F	U	U	13	R	S,Mu,C	U	U	N
Marsh harrier *C. aeruginosus*					—	C		R	V
Peregrine falcon *Falco peregrinus*	W,M,S,H	R	R	6	R	Hi,B		R	R
Little falcon *F. longipennis*	W,M,S,H,F	U	R	8	R	P,S,Mu, Dr,E	U	U	R
Grey falcon *F. hypoleucos*					—			R	V
Brown falcon *F. berigora*	W,M,S,H,F	C	U	18	R	P,S,Mu, E,C	C	C	R
Nankeen kestrel *F. cenchroides*	W,M,S,H,F	C	U	46	R	Hi,P,S,B, Mu,E,C	C	C	R
Mallee fowl *Leipoa ocellata*	W,M,S	R	E		R	S	U	R	R
Stubble quail *Coturnix novaezealandiae*	S,H,F		U	25	R	Dr,E,C	U	U	N

Table 1 — *continued.*

	Wheatbelt Kellerberrin					Pastoral Murchison River Catchment			
Name	Habitat	Past status	Present status	1987 observers	Season-ality	Habitat	Past status	Present status	Season-ality
Brown quail *C. australis*	H	R	*		V				—
Painted button quail *Turnix varia*	W,M,S,H		R	Ø	V				—
Little button quail *T. velox*	W,H	C	*	Ø	V	Mu,Dr, E,C	C	U	N
Marsh crake *Porzana pusilla*	Aq		*		V			R	V
Black-tailed native hen *Gallinula ventralis*	Aq,D,F	R	R	Ø	V		C	C	N
Dusky moorhen *G. tenebrosa*	Aq	R			V				—
Coot *Fulica atra*	Aq	U	*	Ø	V			U	N
Bustard *Ardeotis australis*	W,H,F	*	*	Ø	N	P,S,Mu, E,C	U	U	N
Bush stone curlew *Burhinus magnirostris*	W	C	R	15	R	P,S,Mu	U	U	R
Banded plover *Vanellus tricolor*	F		U	41	R	O	—	C	N
Red-kneed dotterel *Erythrogonys cinctus*	Aq	U	*	Ø	V	Aq		U	N
Hooded dotterel *Charadrius rubicollis*	Aq	R			V				—
Red-capped dotterel *C. ruficapillus*	Aq	U	*	Ø	V	Aq	C	C	N
Black-fronted dotterel *C. melanops*	Aq	*	*	Ø	V	Aq	C	C	N
Australian dotterel *Peltohyas australis*	F	R	*		V	P,C,O	U	U	N
Pied stilt *Himantopus himantopus*	Aq	C	*	5	V	Aq	C	C	N
Banded stilt *Cladorhynchus leucocephalus*	Aq	R	*		V	Aq	*	U	N
Red-necked avocet *Recurvirostra novaehollandiae*	Aq	C	*	5	V	Aq		U	N
Wood sandpiper *Tringa glareola*					—	Aq		R	Nb
Greenshank *T. nebularia*	Aq,D		R		V	Aq		U	Nb
Marsh sandpiper *T. stagnatilis*	Aq,D		R		V				—
Common sandpiper *T. hypoleucos*	Aq				—	Aq		U	Nb
Sharp-tailed sandpiper *Calidris acuminata*	Aq	U			V	Aq		U	Nb
Red-necked stint *C. ruficollis*	Aq	C	*		V	Aq	U	U	Nb

Table 1 — *continued.*

	Wheatbelt Kellerberrin					Pastoral Murchison River Catchment			
Name	Habitat	Past status	Present status	1987 observers	Seasonality	Habitat	Past status	Present status	Seasonality
Long-toed sandpiper *C. subminuta*					—	Aq		R	Nb
Curlew sandpiper *C. ferruginea*	Aq	U			V	Aq	R	R	Nb
Oriental pratincole *Glareola maldivarum*					—	Aq		R	V
Australian pratincole *Stiltia isabella*					—	Aq		R	V
Silver gull *Larus novaehollandiae*	Aq		*		V	Aq		R	V
Whiskered tern *Chlidonias hybrida*	Aq	R			V	Aq		U	N
Gull-billed tern *Gelochelidon nilotica*					—	Aq		R	N
White-winged black tern *C. leucoptera*					—	Aq		R	V
Domestic pigeon *Columba livia*	W,M,T,F		U	45	R		—	R	—
Laughing turtledove *Streptopelia senegalensis*	W,T,F		U	12	R	T		R	R
Peaceful dove *Geopelia placida*					—	E		R	R
Diamond dove *G. cuneata*	W	R	R		V	P,Mu, Dr,E	U	U	R
Common bronzewing *Phaps chalcoptera*	W,M,S,H	U	U	38	R	P,S,Dr,E	C	C	R
Crested pigeon *Ocyphaps lophotes*	W,M,S,H,F		VC	97	R	P,S,B,Mu, Dr,E,C	U	VC	R
Spinifex pigeon *Petrophassa plumifera*					—			R	R
Red-tailed black cockatoo *Calyptorhynchus magnificus*	W,F		U	Ø	R	Mu,E		U	N
Carnaby's cockatoo *C. funereus latirostris*	W,S,H	U	E		M				—
Galah *Cacatua roseicapilla*	W,M,F		VC	100	R	P,S,B,Mu, Dr,E,C	U	VC	R
Little corella *C. sanguinea*					—	E	U	U	?
Long-billed corella *C. pastinator*	W,F		U	36	R				—
Pink cockatoo *C. leadbeateri*	W,M,F	*	*		V	E,Mu		R	R
Purple-crowned lorikeet *Glossopsitta porphyrocephala*	W,M	C	*	6	V				—
Regent parrot *Polytelis anthopeplus*	W,M,S	C	E		M				—
Cockatiel *Nymphicus hollandicus*	W,F	U	U	Ø	V	Hi,P,S,Mu Dr,E,C	C	C	N

Table 1 — *continued.*

Name	Wheatbelt Kellerberrin Habitat	Past status	Present status	1987 observers	Season-ality	Pastoral Murchison River Catchment Habitat	Past status	Present status	Season-ality
Night parrot *Geopsittacus occidentalis*					—	B		R	R?
Budgerigar *Melopsittacus undulatus*	W,S,F	U	U	Ø	V	P,S,Mu, Dr,E	C	C	N
Red-capped parrot *Purpureicephalus spurius*	W	*	E		R				—
Western rosella *Platycercus icterotis*	W	U	E		R				—
Port Lincoln ringneck *Barnardius zonarius*	W,M,S,F	VC	VC	99	R	Hi,P,S,B, Mu,Dr,E	C	C?	R
Mulga parrot *Psephotus varius*	W,S,F		U	32	R	P,S,Mu,E	C	C	R
Bourke's parrot *Neophema bourkii*					—	P,S,B, Mu,E,C	R	C	R
Scarlet-breasted parrot *N. splendida*					—	E	R	E?	?
Elegant parrot *N. elegans*	W,S,F		U	21	R				—
Pallid cuckoo *Cuculus pallidus*	W,M,F	*	U	26	M	P,S,B,Mu, Dr,E,C	U	U	M
Fan-tailed cuckoo *C. pyrrhophanus*	W,M,F		R	Ø	M				—
Black-eared cuckoo *Chrysococcyx osculans*	W,M,H,F	U	*	Ø	N	Hi,P,S, Mu,Dr,E	C	C	N
Horsfield's bronze cuckoo *C. basalis*	W,M,S	C	R	9	M	E,C	C	C	N
Shining bronze cuckoo *C. lucidus*		R	R	Ø	M				—
Boobook owl *Ninox novaeseelandiae*	W,M,S	*	U	9	R	P,Dr,E	U	U	R
Barking owl *N. connivens*	W,M		*	Ø	R?				—
Barn owl *Tyto alba*	W,F	U	U	Ø	R			R	R?
Tawny frogmouth *Podargus strigoides*	W,M,S	C	U	15	R	P,S,B, Mu,Dr,E	U	U	R
Owlet-nightjar *Aegotheles cristatus*	W,M	C	U	5	R	Hi,P,S,B, Mu,Dr, E,O	C	C	R
Spotted nightjar *Caprimulgus guttatus*	W,M	R	*	Ø		Hi,P,S, B,Mu,E	U	U	R
Fork-tailed swift *Apus pacificus*	Sky	R	R		V	Sky		R	Nb
Laughing kookaburra *Dacelo gigas*	W		*		V				—
Blue-winged kookaburra *D. leachii*					—	E		R	R
Red-backed kingfisher *Halcyon pyrrhopygia*	W,M	R	R	Ø	M	P,Mu	R	R	N?

Table 1 — *continued.*

	Wheatbelt Kellerberrin					Pastoral Murchison River Catchment			
Name	Habitat	Past status	Present status	1987 observers	Season-ality	Habitat	Past status	Present status	Season-ality
Sacred kingfisher *H. sancta*	W,H		R	Ø	M	E		U	Nb
Rainbow bee-eater *Merops ornatus*	W,M,H,F	C	C	10	M	E,C	U	U	M
White-backed swallow *Cheramoeca leucosternum*	H,F	C	R	7	R	C,Sky	U	U	R
Welcome swallow *Hirundo neoxena*	W,T,F	C	C	63	R	P,B,Mu,E	U	C	R
Tree martin *Cecropis nigricans*	W,M,T,F	C	C	34	R	E,Sky		U	N
Fairy martin *C. ariel*	F	U	*	Ø	?	B,Sky	U	U	N
Richard's pipit *Anthus novaeseelandiae*	F	C	C	58	R	Hi,P,B, Mu,C,O	C	C	R
Black-faced cuckoo-shrike *Coracina novaehollandiae*	W,M,H,T,F	U	C	49	R	Hi,P,S,B, Mu,Dr,E	C	C	N
Ground cuckoo-shrike *C. maxima*	W	*	E		?	Hi,P, Mu,Dr	U	U	R
White-winged triller *Lalage sueurii*	W,M,S,H	C	U	8	N	P,S,Mu, Dr,E	C	C	N
Southern scrub robin *Drymodes brunneopygia*	M,S,H	U	R	Ø	R				—
Scarlet robin *Petroica multicolor*	W	*	*	Ø	?				—
Red-capped robin *P. goodenovii*	W,M,S,H	C	C	21	R	Hi,P,S,B, Mu,C	C	C	R
Hooded robin *Melanodryas cucullata*	W,M,S	R	E		?	Hi,P, S,B,Mu	U	U	R
Western yellow robin *Eopsaltria griseogularis*	W,M,S	U	U	10	R				—
Jacky winter *Microeca leucophaea*	W,M	C	U	25	R				—
Crested shrike-tit *Falcunculus frontatus*	W,M	R	E		R				—
Gilbert's whistler *Pachycephala inornata*	W,M	R	E		R				—
Golden whistler *P. pectoralis*	W,M,S,H		R	Ø	R				—
Rufous whistler *P. rufiventris*	W,M,S,H	C	C	16	R	Hi,P,S,B, Mu,Dr,E	C	C	R
Grey shrike-thrush *Colluricincla harmonica*	W,M,S,H	C	C	20	R	Hi,P,S,B, Mu,Dr,E	C	C	R
Crested bellbird *Oreoica gutturalis*	W,M,S,H	C	C	15	R	P,S,B, Mu,Dr	C	C	R
Restless flycatcher *Myiagra inquieta*	W,M	R	E	Ø	R				—
Grey fantail *Rhipidura fuliginosa*	W,M,S	*	U	15	M	B,Dr,E	R	R	N/R

Table 1 — *continued.*

Name	Wheatbelt Kellerberrin Habitat	Wheatbelt Kellerberrin Past status	Wheatbelt Kellerberrin Present status	Wheatbelt Kellerberrin 1987 observers	Wheatbelt Kellerberrin Seasonality	Pastoral Murchison River Catchment Habitat	Pastoral Murchison River Catchment Past status	Pastoral Murchison River Catchment Present status	Pastoral Murchison River Catchment Seasonality
Willie wagtail *R. leucophrys*	W,M,S,H,F	C	C	90	R	Hi,P,S, Mu,E	C	C	R
Chiming wedgebill *Psophodes occidentalis*					—	P,Mu,C	U	U	R
Chestnut quail-thrush *Cinclosoma castanotum*	W,M,S,H	*	E		?				—
Cinnamon quail-thrush *C. cinnamomeum*					—	Hi,P,S, B,Mu,Dr	C	C	R
Grey-crowned babbler *Pomatostomus temporalis*					—	Hi,P,B, Mu,Dr	U	C	R
White-browed babbler *P. superciliosus*	W,M,S,H	C	C	16	R	Hi,P,S,B, Mu,Dr,E	C	C	R
Rufous song lark *Cinclorhamphus mathewsi*	F	U	U	Ø	N	P,Dr,E	U	U	N
Brown song lark *C. cruralis*	F	C	C	26	M	C	U	U	N
Splendid wren *Malurus splendens*					—	S,Dr	U	U	R
Variegated wren *M. lamberti*					—	Hi,P,S, Mu,Dr, E,C	C	C	R
Blue-breasted wren *M. pulcherrimus*	W,M,S,H	*	R	7	R				—
White-winged wren *M. leucopterus*	M,H		*	Ø	?	C	C	C	R
Thick-billed grass wren *Amytornis textilis*					—	C	U	E?	R
Shy hylacola *Sericornis cautus*	S,H		R	Ø	R				—
Redthroat *S. brunneus*	W,M	*	R	Ø	R	Hi,P,S, Mu,Dr,C	C	C	R
Field wren *S. fuliginosus*	S,H	*	R	Ø	R	C	C	U	R
Southern whiteface *Aphelocephala leucopsis*	W	R	*	Ø	V	Hi,P,S,B, Mu,Dr, E,C	VC	VC	R
Banded whiteface *A. nigricincta*					—	P,Mu,C	*	U	N
Weebill *Smicrornis brevirostris*	W,M,S	C	C	25	R	S,E	U	U	R
Western warbler *Gerygone fusca*	W,M,S	C	U	8	Nb	B,Mu, Dr,E	U	U	?
Broad-tailed thornbill *Acanthiza apicalis*	W,M,S,H	*	C	15	R	Dr,E,C	U	U	R

Table 1 — *continued.*

Name	Wheatbelt Kellerberrin Habitat	Past status	Present status	1987 observers	Season-ality	Pastoral Murchison River Catchment Habitat	Past status	Present status	Season-ality
Slaty-backed thornbill *A. robustirostris*					—	Hi,P,S, Mu,Dr	U	U	R
Samphire thornbill *A. iredalei*					—	C	U	U	R
Chestnut-rumped thornbill *A. uropygialis*	W,M,S	C	C	13	R	Hi,P,S,B, Mu,Dr E,C	VC	VC	R
Yellow-rumped thornbill *A. chrysorrhoa*	W,M,S	C	C	27	R	Hi,P,S,B, Mu,Dr, E,C		C	R
Varied sittella *Daphoenositta chrysoptera*	W	C	U	7	R	Hi,P,S, Mu,Dr,E	C	C	R
White-browed treecreeper *Climacteris affinis*					—	Hi,P, Mu,E	U	U	R
Rufous treecreeper *C. rufa*	W	*	E		R				—
Red wattlebird *Anthochaera carunculata*	W,M,S,H	C	U	14	R				—
Little wattlebird *A. chrysoptera*	W,S,H	*	*		?				—
Spiny-cheeked honeyeater *Acanthagenys rufogularis*	W,M,S,H	C	U	6	R	Hi,P,S,B, Mu,Dr,E	C	VC	R
Yellow-throated miner *Manorina flavigula*	W,M,S,F,T	C	C	82	R	P,S,B, Mu,Dr,E	C	C	R
Singing honeyeater *Lichenostomus virescens*	W,M,S,H	C	C	36	R	Hi,P,S,B, Mu,Dr,E	C	C	R
White-eared honeyeater *L. leucotis*	W,M,S,H	C	U	16	R				—
Yellow-plumed honeyeater *L. ornatus*	W,M	VC	E		R				—
Grey-fronted honeyeater *L. plumulus*					—	Dr	*	U	?
White-plumed honeyeater *L. penicillatus*					—	Dr,E	C	C	R
Brown-headed honeyeater *Melithreptus brevirostris*	W,M,S	C	U	15	R				—
White-naped honeyeater *M. lunatus*	W	*	E		?				—
Brown honeyeater *Lichmera indistincta*	W,M,S,H	C	C	26	R	E	U	U	R
White-fronted honeyeater *Phylidonyris albifrons*	W,S,H	R	R	Ø	R	Dr,E		U	R
Tawny-crowned honeyeater *P. melanops*	M,S,H	C	R	Ø	V				—
Grey honeyeater *Conopophila whitei*					—	Dr		R	N?
Black honeyeater *Certhionyx niger*	M,S	R	R		V	Hi,P,B, Mu,Dr,C	C	C	N
Pied honeyeater *C. variegatus*					—	P,B,Mu, Dr,C	C	C	N

Table 1 — *continued.*

Name	Wheatbelt Kellerberrin					Pastoral Murchison River Catchment			
	Habitat	Past Status	Present Status	1987 Observers	Seasonality	Habitat	Past Status	Present Status	Seasonality
Crimson chat *Ephthianura tricolor*	H	R	*		V	Hi,P,S, B,Mu,C	C	C	N
Orange chat *E. aurifrons*	S		*		V	C	U	U	N
White-fronted chat *E. albifrons*	S,H,F	C	C	29	R		U	?	N
Mistletoe bird *Dicaeum hirundinaceum*	W	U	*		N	Hi,P,S,B, Mu,Dr,E	U	U	R
Spotted pardalote *Pardalotus punctatus*	W,M,S	*	R	Ø	N				—
Red-browed pardalote *P. rubricatus*					—	Dr,E(?)	*	R	?
Striated pardalote *P. striatus*	W,M	U	C	22	R	S,E	U	U	Nb
Silvereye *Zosterops lateralis*	W,M,S,H	U	R	Ø	N				—
Star finch *Neochmia ruficauda*					—	E,Aq		R	N
Zebra finch *Poephila guttata*	W,M,S,F	U	U	13	R	P,B,Mu, Dr,E,C	C	VC	N
Spotted bowerbird *Chlamydera guttata*					—	Mu,Dr		U	N
Australian magpie lark *Grallina cyanoleuca*	W,T,F	C	C	92	R	P,S,B, Mu,Dr,E	U	C	R
Masked woodswallow *Artamus personatus*	W	U	R		?	P,S,Mu, Sky	C	C	N
Black-faced woodswallow *A. cinereus*	W,S,F	C	C	60	R	Hi,P,S, Mu,C	R	C	R
Little woodswallow *A. minor*					—	Hi,B,C		C	R
White-browed woodswallow *A. superciliosus*					—	Sky	R	R	N
Dusky woodswallow *A. cyanopterus*	W	*	*	Ø	V				—
Grey butcherbird *Cracticus torquatus*	W,M,F,T	C	C	31	R	P,S,B, Mu,Dr,E	U	U	R
Pied butcherbird *C. nigrogularis*	W,F,T		U	69	R	Hi,P,S,B, Mu,Dr,E	U	C	R
Australian magpie *Gymnorhina tibicen*	W,M,T,F	C	C	93	R	Hi,P,S, Mu,E	*	U	R
Grey currawong *Strepera versicolor*	W,M,S	U	R	Ø	R		U	R	—
Australian raven *Corvus coronoides*	W,M,T,F	VC	VC	97	R				—

Table 1 — *continued.*

	Wheatbelt Kellerberrin					Pastoral Murchison River Catchment			
Name	Habitat	Past status	Present status	1987 observers	Season-ality	Habitat	Past status	Present status	Season-ality
Little crow *C. bennetti*	W,M,F	U	U	6	M	Hi,P,S,B, Mu,Dr, E,C		C	N
Torresian crow *C. orru*					—	P,S,B,Mu, Dr,E,C		C	R

STATUS: VC=Very Common; C=Common; U=Uncommon; R=Rare; E=Extinct; *=Mentioned – not enough information to gauge status. HABITAT: *Wheatbelt* — D=Dams; F=Farmland; H=Heath; M=Mallee; S=Shrubland; Sky=Aerial; T=Towns; W=Woodland. *Pastoral* — B=Breakaways; C=Chenopod "steppe" shrublands; Dr=Shrubby creeklines and drainage thickets; E=Riverine woodland; Hi=Hills; Mu=Mulga shrublands on wash plains; P=Footslopes, pediments and colluvial plains. *Aquatic* — Aq=Lakes, swamps, floodwaters, riverine pools and lake beds. SEASONALITY: R=Resident; M=Breeding Migrant; Nb=Nonbreeding Migrant; N=Nomad; V=Vagrant.

TABLE 2. Number of species which have changed in distribution and/or abundance since the turn of the century in the Kellerberrin district of the wheatbelt and in the Murchison River catchment of the pastoral region.

	Change in avifauna				
	Extinct	Decreased	No change or evidence for change	Increased	Total
Wheatbelt: Kellerberrin district					
Non-passerines	6(10%)	8(13%)	32(52%)	15(25%)	61
Passerines	8(11%)	16(23%)	43(62%)	3(4%)	70
Total	14	24	75	18	131
Pastoral region: Murchison River catchment					
Non-passerines	2(4%)	2(4%)	38(74%)	9(18%)	51
Passerines	1(1.5%)	3(4.5%)	51(76%)	12(18%)	67
Total	3	5	89	21	118

The pastoral area

Methods. Some historical information on the pre-1920 status of birds in the southern parts of the Murchison catchment exists from three key accounts of expeditions made before pastoral grazing became a broadscale land use throughout the region. Even so, assessment of post-settlement changes in either the abundance or the range of species in the Murchison River catchment is difficult. There are no census data from past ornithological explorations; quantitative data are limited to explorers' comments on numbers of particular species encountered at different points during their journeys and remarks about absences or first encounters with species. Qualitatively, Austin's journal (1855) provides only glimpses of what he and Kenneth Brown must have encountered during that expedition into the area in 1854. Specimens were taken but only opportunistically. The larger land birds were those most remarked upon, while small bush birds and water birds were evidently largely ignored.

Keartland's observations from the earlier stages of the Calvert expedition in 1896 were published by North (1897). Unlike Austin and Brown, Keartland was a skilled ornithologist and he acted specifically as a naturalist for the expedition. Some 300 bird skins were collected by Keartland between Mullewa and Johanna Springs but these had to be abandoned when the party ran into difficulties. Keartland's notes survived intact and include many general comments on the occurrence, abundance and habitats of species, including a variety of water-birds.

F. Lawson Whitlock's published account of birds found between Lake Austin and the Wiluna area (Whitlock 1910) provides an extraordinarily detailed record. Other records and diary entries survive from his visit to the Yalgoo district, on the southern edge of the study area, in 1904. The thoroughness of Whitlock's knowledge and his coverage of land birds can be gauged by the fact that he recorded virtually every species that he could reasonably have been expected to encounter. By 1909 his visits to the region had effectively completed today's catalogue of arid zone land-birds known from Western Australia, illustrating just how far ahead ornithological exploration of the zone has been in comparison with other groups of fauna and flora. The same degree of thoroughness does not apply to his notes on water-birds, which he (like Austin's expedition) did little to seek out. Undoubtedly, the lack of reference material he would have had available on those species, in combination with a lack of suitable optical instruments with which to observe them, were disincentives for spending time on the aquatic flocks that he would presumably have encountered during the wet seasonal conditions he described.

Data on the contemporary status of species, summarized in Table 1, were gathered largely from opportunistic observations made during a rangeland survey programme conducted during 1985-88. The geographical coverage achieved during this survey was extensive throughout the whole study area, amounting to approximately 14 000 km of slow vehicular traverses and 1 960 sites selected to sample the entire range of landforms, soils and vegetation. Such a non-ornithologically oriented survey could not provide a comprehensive assessment of status for a variety of cryptic and elusive species. In a few cases of expected diurnal species which were not encountered during routine survey work, *ad hoc* searches were conducted in appropriate habitat at other times. By the end of the survey, cumulative coverage was probably poorest for owls and remained unsatisfactory as a basis for accurately assessing the status of several rare species such as grey honeyeater, field wren and thick-billed grass wren (which has evidently disappeared from the area). Sources of recent published observations from the area include Blakers *et al.* (1983), Storr (1985a, 1985b) and Davies (1970). C. P. de Rebeira kindly made available his extensive unpublished observations which served to complement and clarify many of the enigmas and omissions apparent in PJC's data.

Results. The 168 species reliably recorded in the Murchison River catchment and the assessment of their past and present status are shown in Table 1. Forty-four species are water-birds and six (black-shouldered kite, marsh harrier, grey falcon, marsh crake, fork-tailed swift and star finch) are rare or irregular vagrants, which are not considered further in this chapter. The assessed changes in status of the other 118 species are summarized in Table 2. Two species of non-passerine (mallee fowl and little button quail) have decreased in range and/or abundance over the last 100 years or so and a further two species (night parrot and scarlet-breasted parrot) have almost certainly disappeared from the catchment. At least nine species of non-passerine have increased in range/or abundance. The emu, wedge-tailed eagle, Australian dotterel, crested pigeon and galah were present in the region but have expanded their ranges or abundances since pastoral activity commenced. Bourke's parrot, the banded plover and the red-tailed black cockatoo have invaded this and other parts of the pastoral areas of the arid zone where they were previously unknown (Storr 1985a). Breeding habitats of the banded plover, the Australian dotterel and the banded whiteface are highly correlated with the patchy occurrence of degraded and eroded areas which have been caused through overgrazing by stock.

The results for passerines are similar to those for non-passerines. Three species (field wren, white-winged wren and grey currawong) have clearly declined and one species (thick-billed grass wren) has disappeared from the region. One seasonally nomadic species from the semi-arid zone, the white-fronted chat, has been reported only once in recent years, although it may never have been of frequent occurrence. Twelve species of passerine have apparently increased in abundance and/or range since 1910: welcome swallow, grey-crowned babbler, yellow-rumped thornbill, banded whiteface, spiny-cheeked honeyeater, zebra finch, Australian magpie lark, black-faced woodswallow, pied butcherbird, Western (or spotted) bowerbird, little crow and Torresian crow.

Eighty-nine (38 species of non-passerine and 51 species of passerine) species of bird have either not changed in status or there are no clear indications of any changes since 1910.

Discussion

What have been the consequences of agricultural and pastoral development on the avifauna of these regions? The two industries have developed and used the land somewhat differently. Some of the consequences have been similar but clear differences do emerge.

There is nothing subtle about changes in the availability of native vegetation brought about by development of agriculture. Virtually all of the land was thrown open for agricultural development, with little land set aside for conservation. Conditions placed on purchased land obliged the owner to fence and clear the land of its native vegetation. Changing technology has provided one person with the capacity to clear and work extensive areas. The result is a massively fragmented landscape with the remaining area (less than 7% of the total area) of native vegetation being distributed in thousands of small remnants around the countryside. The changes imposed on the landscape have been extensive, rapid and measurable through statistics available on landclearing or from comparisons of successive aerial photographs.

Many of the severest impacts of agricultural development have been attributed to land clearance practices. Bruce Leake lived in the Kellerberrin district all his life and in 1962 he published his account of the wildlife and the changes he noted during his lifetime (Leake 1962). In a single paragraph he set out these changes and the reasons for them.

"The changes that have occurred among the birds are generally the result of altered conditions, such as the clearing of immense areas, and to a less extent improvement in water supplies to birds that drink. Much of the land originally covered with scrub and forest in the Eastern Wheatbelt has been cleared and cultivated, and this has caused a tremendous reduction in numbers of birds such as western rosellas, white tailed black cockatoos, striated field wrens, black-capped tree runners, restless flycatchers or scissors grinders, yellow and broad tailed thornbills, chestnut backed ground birds, western shrike thrushes, wild turkeys, purple backed wrens, rufous treecreepers, currawongs or squeakers, lunated and other honeyeaters, brown flycatchers, hooded robins, wee bills, boobook owls, spotted pardalotes, frogmouths, nightjars, bell birds, ground and king parrots, white-headed swallows, restless flycatchers or scissors grinders (sic), ground cuckoo shrikes, kestrels, falcons and sparrow hawks".

Five of these species have disappeared from the district, one (Carnaby's cockatoo) disappearing within the period between 1977 and 1985. This species appears not to survive in a heavily fragmented landscape (see Saunders 1980; Saunders & Ingram 1987). Many of the changes to the avifauna of the Kellerberrin district (and the wheatbelt at large) can be attributed to habitat clearance which has been to the noticeable detriment of 38 species or 30% of the district's original avifauna (Table 2). The number of birds which have declined would probably be much larger if we had more accurate measures of abundance and distribution in the past. Many species are now restricted to remnants of native vegetation or to parasitising species which are restricted to native vegetation. All of these species must be considered regionally vulnerable. Species in this category include the brown quail, painted button quail, fan-tailed cuckoo, black-eared cuckoo and the bronze cuckoos, scarlet robin, western yellow robin, golden whistler, grey shrike-thrush, blue-breasted wren, white-winged wren, shy hylacola, redthroat, field wren, southern whiteface, broad-tailed thornbill, chestnut-rumped thornbill, white-fronted honeyeater, black honeyeater and spotted pardalote. This amounts to 21 species and over half of the 131 species occurring in the district have disappeared from the area or are at risk now. Saunders (1989) examined the changes in species occurring on an 81 ha reserve in the Kellerberrin district and showed that there had been a loss of three species (southern scrub-robin, western yellow robin and golden whistler) of passerine over the preceding 10 years and that at least one other species (blue-breasted wren) would probably disappear from the reserve soon.

One aim of conservation agencies is to maintain representative samples of the natural landscape and biota occurring within a region (Wallace & Moore 1987). It is not sufficient to establish whether or not any particular species is present in a region. A species must be conserved in populations which are large enough to survive the stochastic events which may drive them into local exinction. It is obvious that the conservation system of the wheatbelt of Western Australia, as shown by the Kellerberrin district, is not adequate at present to preserve the region's avifauna (see Saunders 1989).

Although over 50% of the district's species have been adversely affected or are vulnerable as a result of the changes in distribution in native vegetation, many other species have benefitted. Some have invaded and others have increased in abundance. The species which have invaded are ones that feed on grasses, are dependant on water and forage on open areas. These species (crested pigeon, red-tailed black cockatoo, galah, mulga parrot and elegant parrot) favour "riverine plains" and have expanded their range into the human-created "riverine plains" of the wheatbelt. Some of these species may compete with residents for some resources and compound the effects of habitat fragmentation. The competition between Carnaby's cockatoo and the invading galah over nest hollows fits within this category [Saunders (in press)].

Although the information available on the avifauna of the Murchison River catchment is considerably less detailed than that on the Kellerberrin district, historical changes in the status of some species are demonstrable or strongly implied. The changes in the flora of the region as a result of grazing pressure brought about by European settlement have received some attention (Mabbutt *et al.* 1963; Beard 1976; Mitchell & Wilcox 1989; Department of Agriculture and Department of Land Administration Rangeland Surveys 1985-1988, unpublished). Widespread changes and patchy degradation of the native vegetation have clearly occurred. However, in the arid pastoral environments of the Murchison River catchment, vegetation clearing has not occurred to any significant extent. The major post-settlement impact upon the avifauna has been the vastly increased availability of permanent water supplies on which the pastoral industry itself depends. The most obvious changes to the avifauna are increases among a variety of mainly resident (or now resident) species which drink, at least during summer and other dry seasons, if not throughout the year (Fisher *et al.* 1972). For other species, no adverse consequences of this provision of water have yet to be demonstrated but some may have occurred through changes in competitive balances. For example, in the *Neophema* group of parrots, the original fauna seems to have consisted of one rare or scarce species (scarlet-chested parrot) at the western edge of its range and one absent to rare species (Bourke's parrot) largely beyond its normal range. Scarlet-chested parrots are now probably absent from the area while Bourke's parrots have become common generally throughout all types of *Acacia* shrublands.

Other species which have clearly suffered reductions or extinctions since settlement may not have been able to withstand the combined pressures of several interacting forces of change and disturbance which were first experienced in the period 1880-1910. For example, former local populations of thick-billed grass wrens inhabiting chenopod thickets in drainages below the auriferous ranges would have experienced a combination of preferential overgrazing by sheep and cattle, invasion by rabbits, decreases in small marsupials, increases in large marsupials, the proximity of a heavily populated mining industry and the introduction of feral predators. A special sensitivity to any of these factors may have been sufficient to drive local populations of grass wrens and night parrots into extinction.

That no changes of status can be implied for 138 of 167 species in the region does not mean that no changes have occurred. For many species (for example, nearly all water-birds) the historical information base is virtually non-existent. For other species, any adverse changes experienced as a result of disturbance or selective overgrazing and deterioration of the vegetation in some areas, may have been counterbalanced by favourable changes in others. Two very common insectivores, the southern whiteface and chestnut-rumped thornbill are both likely to have experienced losses in some degraded areas, but have colonized other areas of river frontage plains where invasions of tall inedible shrubs have occurred. Some other common species (e.g., singing honeyeater, yellow-throated miner) that habitually drink from stock watering points may now be locally or generally more abundant than they once were, but again, without better indications of former abundance one cannot be sure.

What is significant is that there is little to indicate that the present-day status of any species is changing at a rapid rate. All three species that have apparently disappeared from the Murchison River catchment did so before or around 1910. Of the three species involved, only for the thick-billed grass wren is there any real prospect of reintroduction and maintenance of a population through habitat protection. Vegetation management and rehabilitation schemes may yet achieve the degree of success which would be prerequisite to re-establishment, perhaps from the Shark Bay population (Curry 1986). Prospects for the reintroduction of scarlet-chested parrots appear poor, unless a large area free of stock watering points becomes available.

The continuing profitability of pastoral wool production makes it likely that present-day patterns of grazing and water distribution will continue to maintain elevated populations of some species while having little effect on others (A. D. Wilson, this volume). Among those least affected by present-day pastoral usage are a wide variety of nomadic terrestrial and water birds, most of which exploit favourable seasonal conditions wherever they are found, exhibit a wide range of habitat tolerance and habitually avoid whole regions during times of environmental stress when droughts or dry seasons prevail.

The species that have increased in range and/or abundance in the Murchison have all done so because of changes brought about by the pastoral industry. Provision of water is the most obvious change and it has allowed the emu, crested pigeon, red-tailed black cockatoo, galah, Bourke's parrot, spiny-cheeked honeyeater, zebra finch, spotted bower birds, Australian magpie lark, pied butcherbird, little crow and Torresian crow to expand from their original distributions, which were restricted to the vicinity of the major streams and rivers.

In some instances, changes in vegetation composition are likely to have been important determinants of present-day range and abundance (e.g., a correlation between the spread of red-tailed black cockatoos and the exotic doublegee, *Emex australis*). Perhaps surprisingly, the commonest symptom of grazing-induced change in mulga shrublands, that of altered or lost populations of understorey shrubs and perrenial herbs, cannot be linked convincingly to any of the observed changes in bird populations in the Murchison River catchment.

The area now known as the wheatbelt of Western Australia forms a broad zone of semi-arid lands between the wet temperate forest region of the extreme south-west with its essentially Bassian avifauna and the arid zone with its Eyrean avifauna beyond the mulga-eucalypt line. As a result, the wheatbelt avifauna has both Bassian and Eyrean elements (Serventy & Whittell 1976). It is not surprising then to find that, excluding water birds and vagrants, the avifauna of the Kellerberrin district and the Murchison River catchment show a high degree of similarity with 88 species in common (Table 1). This amounts to 67% of the species recorded at Kellerberrin and 75% of those recorded in the Murchison. A comparison with the avifauna of the jarrah, *Eucalyptus marginata*, marri, *E. calophylla*, and karri, *E. diversicolor* forests to the south-west of the wheatbelt would undoubtedly show the same high degree of similarity. The wheatbelt is an important link between the temperate forest region and the arid zone and will have an important role to play in any changes which may result as the climate changes over time.

Under a predicted future pattern of increased (though still irregular) summer rainfall resulting from more southerly penetration of cyclonic activity, vegetation composition and faunal communities, including the birds, of the Murchison area could be expected to undergo a number of changes. Potential barriers to changes in bird distribution are virtually non-existent on the northern and eastern sides of the catchment. It is only at the south-western corner that it abuts upon wheatbelt land. In the event that the present line of equal summer and winter moisture availability (Frahmand & Nulsen 1988) were to shift towards the south and west and into this region, several species currently absent from the Murchison or confined to its northern fringe might expand southwards. Blue-winged kookaburras, peaceful doves, diamond doves, spinifex pigeons, black-tailed treecreepers and star finches are among the obvious candidates for southward extension of range. Other species with mainly wheatbelt distributions might decline or disappear altogether, such as the field wren, white-fronted chat, samphire thornbill and grey currawong.

Conservation implications

Although the Kellerberrin district and the Murchison River catchment differ markedly in the scope of their agricultural development and its impact on their avifaunas, development has had deleterious effects in both areas. A number of species have disappeared from the Kellerberrin district and it is obvious that more species will disappear as the avian community adjusts to the massive changes that have been imposed on the landscape over the past 80 years. Patterns of land use in the Murchison have been comparatively stable since the 1940s. Although this in itself is no guarantee that species may not continue to be lost from the Murchison, the rate of disappearance is certainly much lower than in the wheatbelt.

In the Kellerberrin district, extensive historical and contemporaneous data allow one to assess what changes are taking place in the avifauna and to propose management actions to counter the continuing trend of disappearances of species dependant on native vegetation. Detailed information of this sort is not available for the Murchison River catchment but monitoring of the current biota is needed to ensure that future changes in management or land use do not have adverse effects on the regional avifauna.

For conservation to be successful, all the existing elements of a regional biota should be preserved and protected somewhere within that region. This means retaining representative areas of landforms, vegetation and populations of animals which are large enough to sustain themselves. In the wheatbelt this is not possible as most of the native vegetation has been cleared. The remainder is distributed in many small patches and species are being lost from those patches. In the Murchison River catchment immediate threats to native vegetation are less than they are in the wheatbelt, but large areas should be designated as nature reserves or national parks. Such reservations will need to have stock, feral animals and stock watering points removed if they are likely to have demonstrable conservation importance exceeding that of a well-managed pastoral lease (Curry & Hacker 1989).

In both areas it is necessary to carry out further conservation surveys and produce inventories of biological resources. It will also be necessary to set up monitoring points so that change, both natural and human-imposed, can be measured and assessed. Conservation inventories should form the basis of plans for establishing and managing a network of reserves in both regions. In the wheatbelt, all areas of native vegetation should be assessed for their conservation potential and managed to retain the identified nature conservation values. For example, if two reserved areas of high conservation value occur close together it might be desirable to place a high priority on acquiring land and developing a revegetated link between the two areas to foster the exchange of individuals between them.

Finally, it cannot be emphasized too strongly that there exists within the agricultural and pastoral communities a pool of capable and experienced land managers (farmers and graziers), without whom the possibilities for achieving efficient and objective reserve management are limited. Their increasing practical expertise and concern for land conservation needs to be fostered and supported, particularly in the management of land outside of the designated reserve system.

Acknowledgements

We would like to thank Mr A. J. M. Hopkins, Dr J. F. Lynch, Dr S. R. Morton, Dr H. F. Recher and Mr C. P. de Rebeira who criticized earlier drafts of this paper, Mr C. P. de Rebeira who drew the figure, Mrs M. Schwabe and Mrs C. T. Taplin who typed the various drafts, and 16 observers in the Kellerberrin district who provided us with data during 1987.

References

Anon. (1979) *Western Australia. An atlas of human endeavour 1829-1979.* Western Australian Government Printer.

Austin R. (1855) Journal of Assistant-Surveyor R. Austin, commanding an expedition sent by the Government to explore the interior of Western Autralia, north and east of the settled districts. Government Printer, Perth.

Beard J. S. (1976) Murchison 1:1 000 000 vegetation series. *Vegetation Survey of Western Australia.* University of Western Australia Press, Perth.

Beard J. S. (1981) Swan, 1:1000 000 vegetation series. *Vegetation Survey of Western Australia.* University of Western Australia Press, Perth.

Blakers M., Davies S. J. J. F. & Reilly P. N. (1984) *The Atlas of Australian Birds.* Melbourne University Press, Victoria.

Curry P. J. (1986) Habitat characteristics of the thick-billed grass wren *Amytornis textilis* in grazed shrubland in Western Australia. In: *Rangelands: a resource under seige* (eds P. J. Joss, P. W. Lynch & O. B. Williams) *Proc. 2nd. Int. Rang. Congr.* Australian Academy of Science, Canberra.

Curry P. J. & Hacker R. B. (1989) Can pastoral grazing management satisfy endorsed conservation objectives in arid Western Australia? *J. Environ. Manage.* (in press).

Davies S. J. J. F. (1970) A list of plants and animals found at Mileura, Western Australia. Lat. 26°22′S., Long. 117°20′E. Technical Memorandum **3,** 25 pp. Division of Wildlife Research CSIRO, Australia.

Dell, J. (1978) Birds of Kodj Kodjin and Durokoppin Reserves. *Rec. West. Aust. Mus. Suppl.* **7,** 55-68.

Dell J. (1980) Birds of Yorkrakine Rock, East Yorkrakine and North Bungulla Nature Reserves. *Rec. West. Aust. Mus. Suppl.* **12,** 55-67.

Dell J. (1981) Birds of Billyacatting Hill Nature Reserve. *Rec. West. Aust. Mus. Suppl.* **13,** 35-48.

Fisher C. D., Lindgren E. & Dawson W. R. (1972) Drinking patterns and behavior of Australian desert birds in relation to their ecology and abundance. *Condor* **74,** 111-36.

Ford J. R. & Stone P. S. (1957) Birds of the Kellerberrin/Kwolyin District, Western Australia. *Emu* **57,** 9-21.

Frahmand, M. A. & Nulsen R. A. (1988) Climatic change in Western Australia. Division of Resource Management. Western Australian Department of Agriculture.

Gentilli J. (1958) Is the Western Australian Wheatbelt a natural region? *West. Aust. Nat.* **6,** 157-62.

Leake B. W. (1962) *Eastern Wheatbelt Wildlife.* B. W. Leake, Perth.

Mabbutt J. A., Litchfield W. H., Speck N. H., Soufoulis J., Wilcox D. G., Arnold J. M., Brookfield M. & Wright R. L. (1963) General report on lands of the Wiluna-Meekatharra area, Western Australia, 1958. Lands Research Series No. 7. CSIRO, Melbourne.

Mitchell A. A. & Wilcox D. G. (1989) *Plants of the arid shrublands of Western Australia.* University of Western Australia Press, Perth.

Nix H. (1982) Environmental determinants of biogeography and evolution in *Terra Australis.* In: *Evolution of the flora and fauna of arid Australia* (eds W. R. Barker & P. J. M. Greenslade) Peacock Press, Frewville.

Nixon M. & Lefroy R. F. B. (1988) *Road to the Murchison.* Marion Nixon Pty. Ltd., Perth.

North A. J. (1897) List of birds collected by the Calvert Exploring Expedition in Western Australia. *Transactions and Proceedings and Report of the Royal Society of South Australia.* **21,** 125-92.

Saunders D. A. (1980) Food and movements of the short-billed form of the White-tailed Black Cockatoo. *Aust. Wildl. Res.* **7,** 257-69.

Saunders D. A. (1989) Changes in the avifauna of a region, district and remnant as a result of fragmentation of native vegetation: the wheatbelt of Western Australia. A case study. *Biol. Conserv.* **50,** 99-135.

Saunders D. A. (in press) Problems of survival in an extensively cultivated landscape: the case of Carnaby's cockatoo. *Biol. Conserv.* (in press).

Saunders D. A., Arnold G. W., Burbidge A. A. & Hopkins A. J. M. (eds) (1987) *Nature conservation: the role of remnants of native vegetation.* Surrey Beatty & Sons, Chipping Norton.

Saunders D. A. & Ingram J. A. (1987) Factors affecting survival of breeding populations of Carnaby's cockatoo *Calyptorhynchus funereus latirostris* in remnants of native vegetation. In: *Nature conservation: the role of remnants of native vegetation* (eds D. A. Saunders, G. W. Arnold, A. A. Burbidge & A. J. M. Hopkins) pp. 249-58. Surrey Beatty & Sons, Chipping Norton.

Saunders D. A., Rowley I. & Smith G. T. (1985) The effects of clearing for agriculture on the distribution of cockatoos in the southwest of Western Australia. In: *Birds of Eucalypt forests and woodlands: ecology, conservation, management* (eds A. Keast, H. F. Recher, H. Ford & D. Saunders) pp. 309-21. Surrey Beatty & Sons, Chipping Norton.

Serventy D. L. & Whittell H. M. (1976) *Birds of Western Australia,* 5th. edn. University of Western Australia Press, Perth.

Storr G. (1953) Bird-life at Yoting. *West. Aust. Nat.* **4,** 21.

Storr G. M. (1985a) Birds of the Gascoyne Region, Western Australia. *Rec. West. Aust. Mus. Suppl.* **21.**

Storr G. M. (1985b) Birds of the Mid-eastern Interior of Western Australia. *Rec. West. Aust. Mus. Suppl.* **22.**

Wallace K. J. & Moore S. A. (1987) Management of remnant bushland for nature conservation in agricultural areas of southwestern Australia — operational and planning perspectives. In: *Nature conservation: the role of remnants of native vegetation* (eds D. A. Saunders, G. W. Arnold, A. A. Burbidge & A. J. M. Hopkins) pp. 259-68. Surrey Beatty & Sons, Chipping Norton.

Whitlock F. L. (1910) On the East Murchison. *Emu* **9,** 181-219.

Williams O. B., Suijendorp H. & Wilcox D. G. (1980) The Gascoyne Basin. In: *Desertification* (eds M. R. Biowas & A. K. Biowas) pp. 3-106. *Environmental Sciences and Applications 12.* Pergamon Press, Oxford.

Proc. Ecol. Soc. Aust. 1990 — 16:323-345

State forests in Australia and their role in wildlife conservation

S. M. DAVEY[1] and T. W. NORTON[2]

[1]*Department of Forestry, The Australian National University, G.P.O. Box 4, Canberra City, A.C.T. 2601, Australia*

[2]*Centre for Resource and Environmental Studies, The Australian National University, G.P.O. Box 4, Canberra City, A.C.T. 2601, Australia*

Abstract

Management practices for and legislation relating to wildlife conservation in State Forests in Australia are reviewed and future research directions discussed in light of new concepts, theory and computer-based techniques. While State Forests form a substantive part of Australia's remaining forest estate, research suggests that their role in the conservation of native forest wildlife is not being optimized. Generally, ecological data on most forest wildlife are so limited that it is not possible to determine if current wildlife management practices are viable in the long term. Forest studies of the spatial and temporal use of resources by arboreal marsupials are used as a case study to demonstrate the dynamic nature of these resources, and the types of data required by managers. Wildlife values of forests may change markedly over short time-spans (tens of years) but these changes are shown to be relatively predictable. Such predictability is the key to developing long-term management prescriptions. There is evidence to suggest that logging and the conservation of wildlife may be compatible if habitat manipulation (e.g., selective logging, fire) is carefully planned and supervised. These procedures, however, will be considerably more expensive than those currently practised in State Forests. Habitat manipulation to maintain wildlife values in National Parks also appears desirable in the longer term and could play an important role in wildlife conservation. A major change in current attitudes in society will be required if these practices are to be introduced.

Introduction

In this chapter we review the role that State Forests currently play in the conservation of native, forest wildlife (biota). Moreover, we consider the appropriateness of this role in the context of the broad conservation goals outlined in the World Conservation Strategy (IUCN 1980) and the National Conservation Strategy of Australia (Anon 1984). We consider new strategies which may advance future management of native wildlife in these forests.

The Australian forest estate is dominated by four broad types of native forest. In decreasing order of area these are: eucalypt forest; tropical eucalypt and paperbark forest; cypress pine forest; and rainforest (Australian Heritage Commission 1987). Together these forests occupy some five percent of the land. Plantations of softwood and native hardwoods are also maintained; these are seen as an important resource for wildlife management (Gepp 1985) but are not discussed in this chapter.

On the most recent figures available (1984 — derived from a 1976 estimate; Australian Heritage Commission 1987), approximately one-quarter (11 039 000 ha) of the remaining native forests occur on private land while three-quarters (30 243 000 ha) are on publicly-owned land. The extent of plantations is relatively small (866 000 ha). Of the forests of publicly-owned land, about one-sixth (5 019 000 ha) are maintained as National Parks and Reserves while the rest fall into two categories of roughly equal total area. These are forests in which wood harvesting is carried out under government control but are not reserved for that purpose, and forests that are managed by government agencies for multiple uses including wood production (State Forests).

In terms of area alone, State Forests clearly form a substantial part of Australia's remaining forest estate and should play a significant role in the conservation of native, forest wildlife. Current research suggests, however, that the role of State Forests in such is far from being optimized. Before developing this argument further in relation to the current situation, we provide some historical background.

Historical perspective

European forestry was practised first in Australia in the 1870s when colonial governments realized that the forest estate comprised a valuable resource of timber worthy of conservation (Dargavel 1987). Over the next 50 years successive governments allocated many forest areas more or less solely for timber production. Federation in 1901 did not interrupt this pattern as the States retained control over forests. In parallel, forest services were established by governments to manage the utilization of sawlog and, to a much lesser degree, woodpulp resources by an expanding industry. By and large, this expansion proceeded by trial and error (Carron 1985; Dargavel 1988a) but became somewhat more refined and rapid as technology improved and became more accessible, and markets flourished and matured (Carron 1985). By the mid-1920s all States had forestry Acts empowering departments or commissions to administer their forest resources. As the role of the forest services expanded, their management policies evolved in response to a variety of needs and pressures (Norton 1986). By the late 1970s, the forest services were firmly established and an industry based on the export of woodpulp had developed to the extent that the wealth generated by it contributed significantly to the economies of several States and the Nation.

Notwithstanding the autonomy of the State forest services, the Commonwealth Forestry Bureau was created in 1930 by an Act of the Federal Parliment to provide an advisory service and to undertake forestry research and education. The Act has undergone several amendments over the years and today this facility is a division of the Department of Primary Industries and Energy. This Department is responsible for forestry matters at the national level; in particular, the control of export of unprocessed timber, liaison with State, national and international organizations concerned with forestry, and provision of the Secretariat for the Australian Forestry Council. This Council was established in 1964. Its membership comprises the State and Northern Territory Ministers responsible for forestry and the Commonwealth Minister for Primary Industries and Energy. The charter of the Council is to co-ordinate the development of the forestry resource in the general interest of the community and to guide national programmes for the production, utilization and conservation of Australian forests. More recently, a committee (Directors Of Research Committee: DORC) was formed by the Council to review research directions and recommend emphasis for forest research in Australia. Research directions, emphasis and gaps are outlined by Research Working Groups (RWG) which report to DORC. DORC then reports the findings of working groups to Council. It should be emphasized that RWGs cannot comment on policy. RWG 10 appraises the status of research into forest wildlife and habitat management within Australia.

Despite the creation of seemingly appropriate legislation and formal structures relatively early in the development of the forest industry to oversee the wise management of resources, it is clear that wise management has not been achieved. Several eminent commentators have concluded that, in practice, forestry in Australia has been conducted within a philosophical framework, albeit unwritten, of wood primacy (Westoby 1983; Carron 1985; Leslie 1987). That is, its practitioners believed that the principal function of forestry was wood production. Multiple-use management was considered to be timber production plus other uses. Leslie (1987; p. 14) wrote:

> "This misconception set its proponents on a collision course with people who had other ideas of what the principal objectives ought to be for Australia's forests".

The clash was inevitable and the political developments over the past 15 years are well known. More to the point, this history of resource management has had a significant impact on, at least, the past and current status of forest wildlife. Most State Forests have been cut over and few old-growth stands remain. Wildlife habitats have been removed, modified, fragmented and created.

Concurrent with the founding and growth of a forest industry in Australia after 1788 was the development of the agricultural sector. These activities, combined with the more recent phenomena of a significant urban expansion and the widespread establishment of plantations of trees, contributed significantly to a major reduction in the total area of the forest estate. A review of the remaining forest cover in Australia in 1984, for example, estimated that about half of the forests existing at the time of settlement by Europeans had been cleared (Wells *et al.* 1984).

Norton (1989) considered the effects of the large-scale clearance of forests over the past 200 years on the arboreal marsupial fauna (possums, gliders, koala *Phascolarctos cinereus*), a group of forest-dependent species reaching their highest species richness (number of species per unit area) in the tropical rainforests of north Queensland. While it was apparent that the clearing of forests had reduced the availability of habitat for these fauna, and thus reduced their distribution geographically, Norton considered that the effects were more profound than this general analysis indicated. The pattern of forest clearance was shown to be important. Many of the forests cleared were chosen on their suitability for agriculture.

More recent data suggested that many of these forests almost certainly supported optimal habitat for these fauna. Norton (1989) cited the koala as an example. Most of the lowland eucalypt forests, occurring on fertile soils, in the south-east of New South Wales and supporting forest red gum *Eucalyptus tereticornis* have been cleared. As this eucalypt is one of the most preferred food trees of the koala, this animal is now rare in the region. While Norton (1989) found that no species of arboreal marsupial had become extinct since settlement, he considered that the future of several species, particularly Leadbeater's possum *Gymnobelideus leadbeateri,* was at serious risk.

Lunney and Leary (1988) also indicated a similar pattern of forest wildlife becoming locally rare or extinct following modification or removal of their habitat during the clearing of forests for agriculture and forestry purposes. These authors also demonstrated the major impact that exotic species, in this case hares, rabbits and foxes, may have on native wildlife following fragmentation of the forest through clearing. Similar relationships between the abundance and persistence of forest wildlife and habitat clearance and modification have been shown or alluded to in many other studies (e.g., Tyndale-Biscoe & Calaby 1975; McIlroy 1978; Tyler 1979; Groves & Ride 1982; Winter 1984; Grigg *et al.* 1985; Kikkawa 1985; Pahl *et al.* 1988).

At least 20 forest (closed forest, open forest, woodland) plant species have become extinct since settlement by Europeans (Leigh *et al.* 1984). At least another 100 are considered to be endangered (Leigh *et al.*1984). In reality, this situation is misleading as our knowledge of much of the flora, as with the majority of the fauna (Richardson 1983), is scant. In effect, we can only place some reliability on the data available for the higher order taxa. Webb and Tracey (1981), for example, found that from plant specimens representing 765 species collected at 140 rainforest sites in northern Queensland, 117 specimens could not be readily placed and many of them were yet to be described. This occurrence is by no means unusual, particularly for rainforest surveys (Werren & Kershaw 1984). As indicated earlier when considering the forest fauna, the current status of forest flora, in the majority of situations, relates directly to past and current landuse practices.

It is clear that historical patterns of forest-use have had a significant, often detrimental, impact on the population status of many native, forest wildlife. At the same time, knowledge of the effects of these practices on most wildlife is not available. Many more species may have become extinct than is recognized. These are some of the more obvious ecological constraints that managers of forest wildlife face today.

Current situation

The current situation in terms of the practical role that State Forests play in the conservation of wildlife has had a short gestation. Major developments in this area span just 15 years. Most have evolved this decade.

Legislative requirements

A variety of Commonwealth and State Acts of Parliament directly or indirectly concern, *inter alia,* the conservation and protection of forest wildlife in State Forests. More recently, the Federal Government has been increasingly involved in forestry issues in these areas, particularly concerning forests nominated for World Heritage listing or on the Register of the National Estate. The Register is the national inventory of Australia's National Estate and comprises a list of places defined in the Act as:

> "... those places, being components of the natural environment of Australia or the cultural environment of Australia, that have aesthetic, historic, scientific or social significance or other special value for future generations as well as for the present community."

If, for example, a section of State Forest was listed on the Register, no particular form of land-use would be imposed by this action. Rather, the *Australian Heritage Commission Act* 1975 requires Commonwealth Ministers and bodies to ensure that any proposed action(s) in the forest would not adversely affect its National Estate values unless "... there are no feasible and prudent alternatives."

The Commonwealth, in co-operation with the States and Territories, has recently been involved in the development of long-term strategies for forest management through the Australian Forestry Council. The Council recently published the National Forest Strategy for Australia which is consistent with the objectives outlined in the National Conservation Strategy of Australia. Both strategies aim, *inter alia,* to ensure that forests will be protected and managed scientifically and that the genetic diversity will be maintained.

At the State and Territory level, the forest services are required by legislation to adopt, in general, a multiple-use approach in the allocation of forest resources. As an example, State Forests are required to be managed to conserve and protect native flora and fauna and suitable forests are promoted for use for recreational activities.

Both Victoria and Western Australia have recently reassessed their forestry legislation. In 1983, The Report of the Ministerial Review Team in Victoria recommended that the Forests Act should be reviewed and rewritten to take account of several matters of concern, one being to ensure clear objectives for forest management. More recently, the Victorian Forests Commission has been subsumed into a larger department called Conservation, Forests and Lands and the *Flora and Fauna Guarantee Act* has been passed. This Act was written to formalize the government's commitment to the long-term conservation of wildlife on both private and public lands. The *Conservation and Land Management Act* was passed by the Parliament of Western Australia in 1984. This Act aims to make better provision for the use, protection and management of public lands, water, flora and fauna and applies to a variety of lands reserved under the *Land Act,* including State Forests.

Departmental restructuring appears to be in progress in New South Wales and Queensland but specific details are not yet available. The New South Wales *Forestry Act* has been subject to considerable criticism recently (Boer & Preston 1987) but it remains to be seen if this Act will be reformed. In Tasmania, the National Parks and Wildlife Service and the Lands Department were subsumed recently into one agency but the Forestry Commission retained its autonomy.

Planning and management

The public perception of the role of State Forests changed, in general, from one of solely timber production pre 1960s to that of a need for a balance between wood and non-wood values in the 1970s and 1980s. Today, many sections of the public question whether this change in perception is reflected in current management of their State Forests. The National Estate and World Heritage issues, which surfaced in the important debate on forest landuse in the early 1980s, polarized the public's perception of nature conservation in forests. Currently, one point of view totally excludes logging from native forests (see Wootten 1986; Australian Conservation Foundation 1987) as its proponents see logging as being incompatible with the objectives of nature conservation. An opposing view, generally represented by the forest industry, considers the present reserve system (National Parks and associated forest reserves) more than adequate for conservation and that areas not zoned for reservation should be managed intensively for timber production (Gibson 1988).

As a consequence, decision-making on the allocation of forest resources (wood, non-wood) appears to favour a perspective that is subjective and political rather than scientific. Political decision-makers have not, in general, adopted an approach in the allocation of these resources based on ecological and environmental principles. Their response to a most difficult resource problem has been to zone the forest estate into landuse categories, either wood utilization or conservation, it not being widely appreciated or acknowledged that these categories need not be mutually exclusive. Clearly, zoning is an easy way to meet short term objectives. However, the longterm consequences for management flexibility need to be addressed as the zoning practice may limit future options for landuse managers (Walther 1986). Over the period 1987-89 some $120m was offered to Queensland and Tasmania as compensation for losses incurred, socially and economically, in reserving their forests (Sydney Morning Herald, 6th August 1988, p.31). Unfortunately, little support has been given to or appears in prospect for research and/or initiatives directed to finding long-term solutions. Yet, some forests already have been managed successfully for both wood production and wildlife conservation. These include some National Estate forests (e.g., Kioloa State Forest, New South Wales; Perup Forest, Western Australia; Murray River red gum forests, Victoria-New South Wales). Presently, they are managed, using a multiple-use principle, at a reduced logging intensity and retain high wildlife values. Rarely, however, are they cited in the public debate.

There are two underlying principles that are recognized by wildlife managers associated with the forestry sector. The first is that the responsibility of wildlife managers lies in the maintenance of viable populations for the long term just as the timber manager is responsible for ensuring forest products for future generations. The second principle is that properly controlled logging operations pose no threat to the long-term survival of native fauna or flora (Recher 1985) and may well enhance their survival. By properly controlled logging operations, wildlife managers mean that the levels of design, frequency and intensity of operations are pursued in such manner as to meet wildlife conservation objectives in the long term. It is recognized that, by identifying the key requirements of species, and by managing and planning operations around these requirements, this will improve the opportunities for their survival and/or recolonization of the operation area.

Planning for the conservation of forest wildlife presently centres around the principles of island biogeography, minimum viable population and optimal habitat. Current thought on the design procedures for forest wildlife management are discussed more fully in Recher (1985), Recher *et al.* (1987) and Davey (1989). In general, present design involves the linking of reserves, drainage lines (filter strips) and areas unable to be logged for logistic or economic reasons (refuge areas) with unlogged forest through which wildlife can disperse (wildlife corridors). All these features within the design area remain unlogged. It is assumed by managers that wildlife populations retained within the design area are viable and that optimal habitat for all species is represented (Fig. 1).

For planning purposes, managers identify localities rich in fauna or flora, particularly localities supporting sensitive, rare or endangered

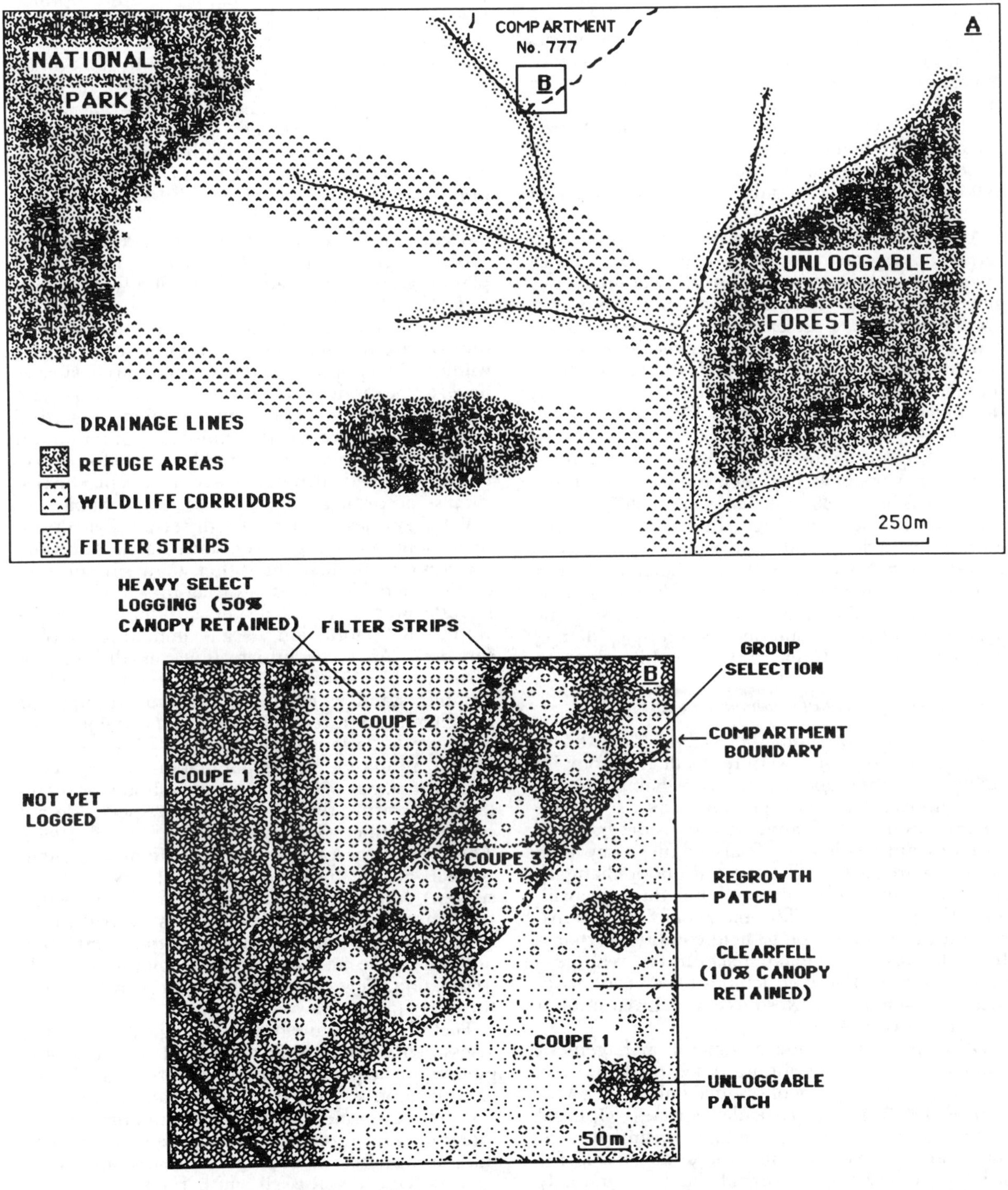

FIG. 1. A hypothetical forest illustrates the design components of present forest wildlife management. (A) Refuge areas, wildlife corridors and filter strips (not drawn to scale) remain unlogged whereas white areas can be logged. (B) Different logging intensities: clearfell, heavy selective and group selection logging affect different wildlife species. Group selection enables forest-dependent species to be maintained and habitat trees retained in logging areas can enable recolonization of areas once the regrowth forest has suitably developed.

species. These localities are currently known as wildlife priority areas (WPA). As WPAs are identified the typical response by forest managers to date has been either to reduce the intensity of harvesting operations, or more frequently, to reserve them from logging for the present time. A number of State Forest services (New South Wales, Victoria, Tasmania, Western Australia) are known to be working actively to identify the location and extent of WPAs in a range of forest types and regions.

As mentioned above, the current approach to forest management is more likely than not to involve a zoning approach, with wildlife zones generally left with minimal disturbance. Zones that are logged have some existing resources retained with long-term recolonization by wildlife in mind. An example of such a resource is the habitat tree; that is, a tree that provides nesting sites for fauna (Recher *et al.* 1980; Mackowski 1984; Inions *et al.* 1989).

The appropriateness of zoning for environmental management has been questioned (Davey 1978; Walther 1986; Recher *et al.* 1987). It has been argued that zoning cannot conserve the infinite variation and corresponding habitats found within the forest estate. Furthermore, it may isolate or fragment populations and lead to genetic degradation in the long term, thus increasing the probability of population decimation by epidemics or predation (see Hopper & Coates this volume).

Development of a wildlife management strategy

Current limitations on the viable (long term) management of forest wildlife result from our inability to comprehend and measure habitat, determine appropriate management units, compare accurate estimates of population densities between localities and predict wildlife abundance away from sampling areas, through time and under changing sets of circumstances. Further, when data are available, the problem wildlife managers face is transferring this information to field conditions, that is from the abstract (habitat space) to the real (geographic space). These difficulties have been recognized for some time (Resource and Environment Consultant Group 1981).

Over the past decade a general management strategy has been developing based on an understanding of the concept of optimal habitat (see Van Horne 1983) for a particular species. Optimal habitat and habitat approaching optimum condition are regarded as the basic units utilized for wildlife conservation. Optimal habitat is potentially the most effective manipulative unit for the management and conservation of wildlife populations. To maintain this habitat in the forest estate requires an understanding of the environment of a species and, in particular, the amounts, arrangements and interactions of resources that make up its optimal habitat (Lyon 1978). Optimal habitat needs to be identifiable and measurable.

Forest wildlife managers require information on the life-histories of species. In the case of fauna this includes behavioural traits such as migrations for food (e.g., honeyeaters) and seasonal thermal-microenvironments (e.g., feathertail glider *Acrobates pygmaeus*; Davey 1984), social interactions, home range, reproduction and dispersal of young. For flora, a knowledge of reproductive and growth patterns is required. In addition, responses to forest fragmentation will influence decisions in planning forest operations for wildlife conservation (Norton 1986).

Clearly, there is a need for the accurate prediction of the distribution and abundance of forest wildlife. The population dynamics of target species have to be understood to a point where precise figures could be incorporated into mathematical functions used for the determination of minimum viable populations (Davey 1989). The information required would include variation in reproductive success according to: (i) the range of habitat in which a species is found; (ii) differing silvicultural operations occurring in the same habitat; and (iii) different manipulations (other than silvicultural) of a single habitat. This information would need to include longevity and dispersal ability of groups within a population (e.g., immature males, females). As well, information on the use of resources by wildlife (niche parameters) is required to understand which resources limit populations or trigger migration to sub-optimal habitats, or sustain populations during periods of stress.

The minimum viable population concept concerns the size at which populations should be maintained to ensure that populations or species do not become extinct (Lehmkuhl 1984; Davey 1989). The crux of the problem for wildlife management is that habitat needs to be made available to support viable populations of taxa. Currently, there are only a very few species of Australian vertebrates for which there is adequate information to enable an estimate of a minimum viable population. Few studies have attempted estimates for other taxa.

Techniques for integrating ecological and timber assessments need to be developed. In these, the guiding principle in sampling design should be the identification of relevant information for each resource component (both wildlife and timber). An integrated inventory strategy may be more efficient than a number of separate resource inventory surveys (Davey & Russell-Smith 1984).

In order to meet the requirements of foresters to be able to predict changes in wildlife habitat resulting from the implementation of various management strategies, it is essential that they be provided with reliable and adequate information bases. These could include mathematical models of the habitat requirements of species so that the models may be incorporated into simulation packages of forest stand dynamics. Also, accurate models for predicting the distribution of wildlife and their potential habitats outside the area under study are required. To achieve this it will be necessary to classify the forest estate by forest sites (Barnes 1984), which will provide a base unit for silvicultural and timber management planning as well as for wildlife management. This base unit is fundamental to sound planning and forest practice (Gessel 1986) and a major consideration for forest management at the moment should be how best to approach forest site classification in State Forests and elsewhere.

Once an inventory of an area has been made, significant localities should be denoted on maps. Significance could be determined in a number of ways; for example: (i) high value of optimal habitat reflecting high population density of a select species; (ii) high habitat-unit value (habitat-unit equates with a high diversity of habitats in a geographic area; following Coulombe 1978); and/or (iii) high diversity or population density of select species. Select species are those that are rare, endangered or sensitive (detrimentally) to forest operations. Management of wildlife, in the first instance, should concentrate on these species.

Forests surveys are a necessity for decision-making as they provide basic but essential information on the distribution and abundance of wildlife. Currently, something of a Catch 22 exists in conducting surveys as they are, normally, very expensive to undertake. Often their costs may prove or appear to be disproportionate to the scale of (proposed) operations in train (Ferguson 1987) and, as a consequence, they are perceived not to be cost effective; something required increasingly by administrative agencies and politicians. Clearly, forest surveys remain a fundamental component of any viable strategy for wildlife management but, at the same time, they need to be made cost effective. This can be achieved by moving, whenever possible, from one-off inventories towards surveys which record, at least, parameters readily adapted to Geographic Information Systems (GIS) and process models. Some of these problems were addressed at a workshop in Adelaide in 1983 (Myers *et al.* 1984) and more recently at a meeting in Sydney in 1988 (M. P. Austin, pers. comm.).

Research

Research on forest wildlife for purposes of management has only been underway for about the last 15 years. Emphasis has been placed upon segments of the avifauna (Keast *et al.* 1985, in particular Christensen *et al.* 1985; Kavanagh *et al.* 1985; Smith 1985), arboreal marsupials (Smith & Hume 1984), habitat identification (e.g., Braithwaite 1984; Davey 1984; Kavanagh 1984; Peters & Cordy 1986) and strategic planning for wildlife management (Recher *et al.* 1980, 1987; Stockwell *et al.* 1989).

Apart from the work done by Webb (1985) on skinks in the Eden region (New South Wales) and Christensen (1980) on endangered macropods in the Perup Forest (Western Australia), few studies of direct relevance to forest wildlife management have been conducted on ground mammals (excluding *Antechinus stuartii, Rattus fuscipes* and *R. lutreolus)*, owls, bats, reptiles, frogs, invertebrates not considered to be pests or members of other animal phyla.

Studies of the flora have generally been inventorial or physiological. Few studies have been approached in such a way as to be easily applied in forest management. Exceptions include the development of successional models (Noble & Slatyer 1980; Shugart *et al.* 1980; Gill 1981; Shugart 1984), derivation of environmental envelopes of taxa (Austin *et al.* 1983), classifications of vegetation types and studies of forest site types.

Much of the emphasis upon forest wildlife research has centred around arboreal marsupials as these were identified by Tyndale-Biscoe and Calaby (1975) as being under direct threat by forest operations. Recently some emphasis has been given to the use of habitat trees and nest boxes by fauna (Golding 1979; Mackowski 1984; Menkhorst 1984; Smith *et al.* 1985; Tidemann & Flavel 1987; Inions *et al.* 1989).

Few studies have addressed satisfactorily the meshing of an ecological understanding of wildlife into a forest planning framework. Such an approach, while considered necessary (Ferguson 1985; Turner 1986), is still in its infancy. Research into arboreal marsupials, however, has progressed further in this direction than that into any other wildlife guild and is now examined as a case study.

Arboreal marsupials: a case study. Several studies on arboreal marsupials have been undertaken with forest planning in mind and have centred around three levels of sampling intensity: regional, forest and stand.

At a coarse scale, as used in Braithwaite's study in the Eden Woodchip Area in southeastern New South Wales (Braithwaite 1984), factors important in influencing the distribution of arboreal marsupials were identified at a regional (Region or District) scale. He found that, for regional planning purposes, forest areas harbouring arboreal marsupials could be readily predicted from the nutritional status of a site, the nutritional status being identified from the eucalypt species found on a site (Braithwaite *et al.* 1983) or geology (Braithwaite 1984). While such a scale was too coarse to satisfactorily determine fauna-vegetation-environmental relationships, it could identify suitable conservation areas. This approach was ideal for differentiating landuse options. His findings were the precursor to the habitat sampling strategy of several more recent studies on arboreal marsupials.

Studies conducted at a State Forest level enable fauna-vegetation-environmental habitat relationships to be determined (Davey & Russell-Smith 1984). At a smaller scale (stands), habitat definition within the context of determining niche differences can be studied. Such scales were used in the arboreal marsupial studies of Henry (1985), Kavanagh (1987a), Norton (1988) and Pahl *et al.* (1988).

The study of Lunney (1987) centred on forests in the Bega district within the Eden Woodchip Area. From observations made in both logged and unlogged coupes, he reported significant declines in populations of arboreal marsupials in coupes that were logged. Fire and drought accentuate the decline. From this study the importance of refuge areas, particularly sheltered unlogged gullies, was apparent. Lunney's data were consistent with those reported by Braithwaite (1984). Further, an important finding was that both drought and fire needed to be considered within a wildlife planning framework, particularly in forests subject to logging. This finding could impinge on the scheduling of logging operations as long-term predictions of periods of increased fire and drought are able to be made (Skidmore 1987). Lunney's study supports our contention that process models need to be incorporated into a wildlife planning framework.

The study of Davey was centred on Kioloa State Forest situated on the south coast of New South Wales and utilized a scale envisaged to be useful for the purposes of planning at the level of the individual forest (see Davey & Russell-Smith 1984). He addressed the question of how to define habitat for the purposes of forest wildlife planning. This required an ability to identify and understand parameters that were useful descriptors of habitat and could readily be coupled to forest planning tools such as GIS or expert systems and, in addition, be recognized and/or understood by forest managers. This was viewed as an extension of the ideas expressed by Davey (1978), Florence (1983) and Davey and Russell-Smith (1984). These authors suggested that native forest management should be developed on ecologically-oriented concepts and emphasized a need for an ecological appreciation and knowledge of the forest resource.

Davey employed two general strategies in his study. In the first of these he examined the habitat of nine species of arboreal marsupials as defined by 70 environmental variables that were ordinated by canonical variate analysis using 41 forest site types. The development of the forest site types involved complex classification procedures. Two canonical variates (CVs) were identified as representing nutrient gradients. Simplistically, CV2 represented the effect of soil or foliage phosphorus (site fertility), whilst CV3 was a function of moisture and/or foliage nitrogen. These two canonical variates best defined habitat space. The habitat of each species was awarded a population category (expressed as animals/hectare) with core areas identified and defined by the habitat space (Fig 2a). Core areas were then identified as hot spots (following Diamond 1975) in geographical space, thereby facilitating the identification and management of important forest stands (Fig. 2b) and habitat units of high value (e.g., insert in Fig. 2b) and the design of wildlife corridors. For example, the distribution of population levels for the mountain brushtail, *Trichosurus caninus*, in this habitat space is illustrated in Figure 3. Using this technique, a site, found in the study area and not sampled for arboreal marsupials, can be placed in the habitat space, once the environmental variables of the site are known. Variables can be read from maps, with minimal sampling on site. Once the position of a site is determined in the habitat space, expected population levels can be obtained. An important finding was that this habitat space could be used to measure the contraction in habitat of the mountain brushtails in response to drought (cf. Morton, this volume; Fig. 3). Equally important was the ability of habitat space to identify phenological changes in important eucalypt species (e.g., *Eucalyptus maculata;* Fig. 4). This would be useful in predicting changes in habitat quality for arboreal marsupials (see Kavanagh 1984, 1987a,b; Norton 1985, 1988), the consequence of which can determine the seasonal scheduling of logging operations.

The second strategy employed by Davey to discriminate habitat involved the use of two structural indices (Davey & Russell-Smith 1984). A foliage biomass index was used to discriminate habitat on the basis of foliage density at different height levels (Davey 1984). The projected foliage index identified forest successional stages, thereby enabling the preferred habitat of possums and gliders to be defined in relation to successional stage. This enabled the prediction of successional changes in a forest stand as well as changes resulting from a particular disturbance. Because population changes of arboreal mammals responded to successional stage (Fig. 5), population levels could be predicted on the basis of successional stage. While these levels changed markedly over relatively short periods of time (decades), the changes were predictable. Such predictability is the key to developing long-term management strategies and prescriptions for these species.

Davey's study identified the dynamic nature, both spatially and temporally, of the habitat and showed that populations levels of arboreal marsupials may fluctuate in a predictable manner. An important finding was identifying a decline in population

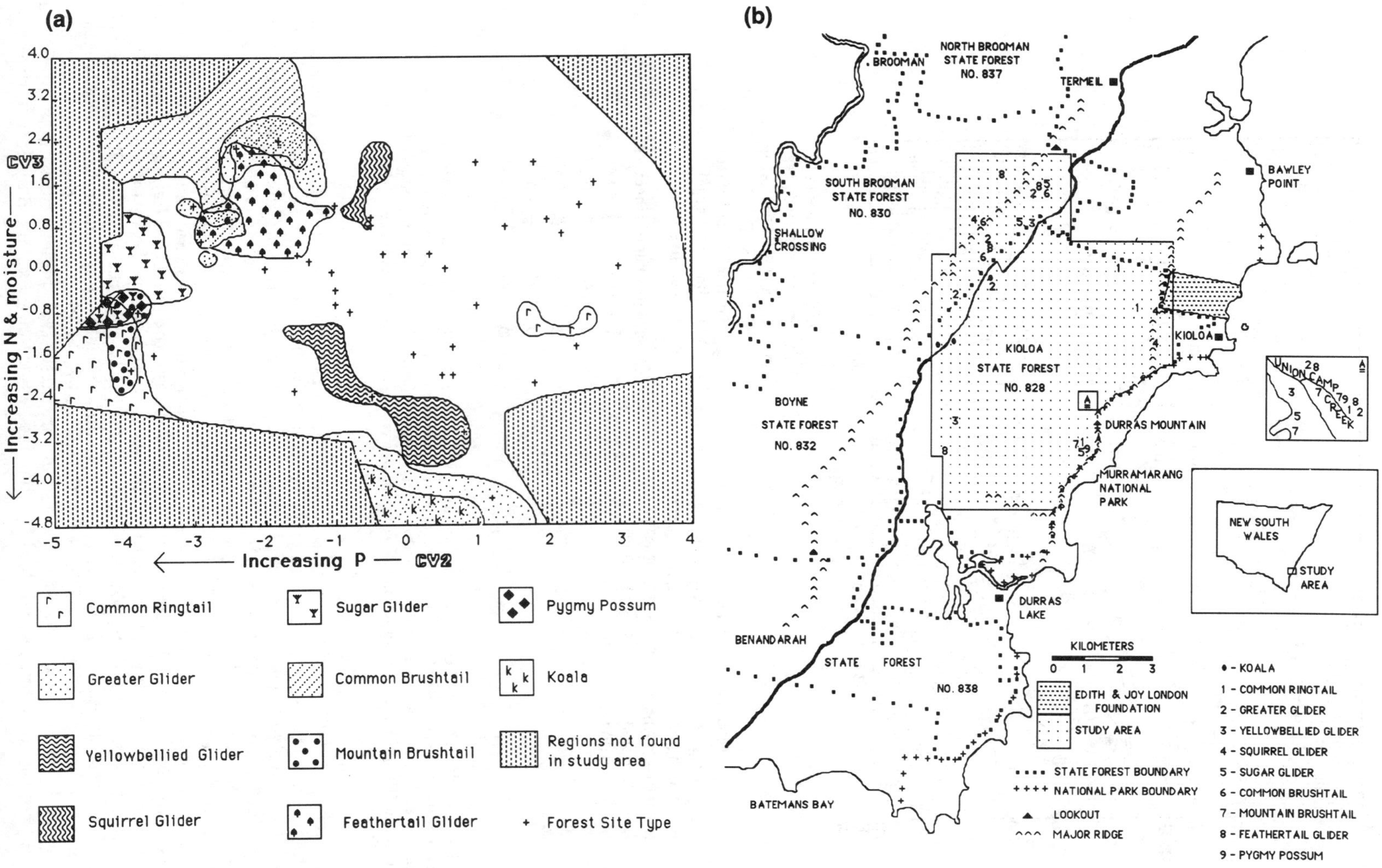

FIG. 2 (a). Core areas of optimum habitat are identified in habitat space, described by two canonical variates (CV), for 10 arboreal marsupials using a forest site classification developed by Davey. Moisture is a measure of a site's soil moisture, N refers to foliar nitrogen of trees, while P can refer to either foliar phosphorus of trees or soil phosphorus. As CV2 becomes more positive site fertility (N, P, K) of tree foliage and soils declines. The study area referred to in (a) is shown in (b). Habitat space can be transformed into geographic space (b) enabling hot spots for each species to be identified on a map for planning purposes. It should be noted that optimum habitat for the koala (obtained from historic records) has been converted from *Eucalyptus tereticornis* forest to agricultural land (refer to comments made in Historical Perspective).

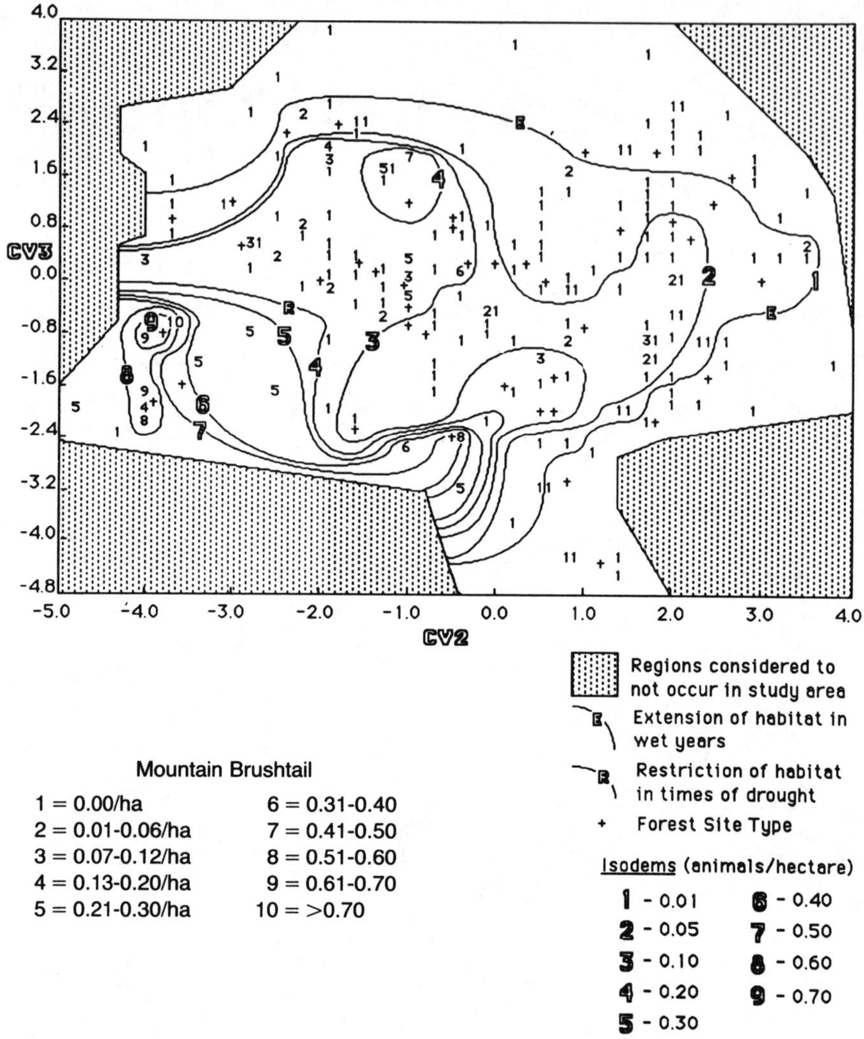

FIG. 3. Isodems (lines of equal population density) for the mountain brushtail were determined from observational data of 171 plots during the period 1980-86 using a contouring package. The restriction in habitat, within habitat space (as in Fig. 2a), during times of drought (R) and expansion in habitat (E) during wet years is presented. Axes are as shown in Fig. 2.

levels of certain species in late to climax successional stages of the forests in question (Fig. 5), thus indicating the need to actively manage forest areas set aside specifically for the conservation of certain arboreal marsupials (Norton & Davey 1986). Equally important is to realize the length of time (see Fig. 5) for habitat to become optimal for most arboreal marsupials after major disturbance (clearfell, intense wildfire). Such a length of time has to be considered in the determination of the rotation length of a particular forest stand.

Kavanagh's study centred on 100 ha of New South Wales tableland forest (Waratah Creek, Coolangubra State Forest) that supported high numbers of arboreal marsupials. His research concentrated on the foraging behaviour of the two largest gliders, the greater glider *Petauroides volans* and the yellow-bellied glider *Petaurus australis*. He found that the patterns of foraging in these two species were related to the phenological changes of the eucalypt species (Kavanagh 1984, 1987a,b). He also found the preferred habitat of both species to be defined best by the resources they used in winter. Clearly, such resources need to be managed to maintain viable populations of these species. Furthermore, he found that the patchy distribution of these two gliders and their preferred habitats, while being explained by floristic diversity,

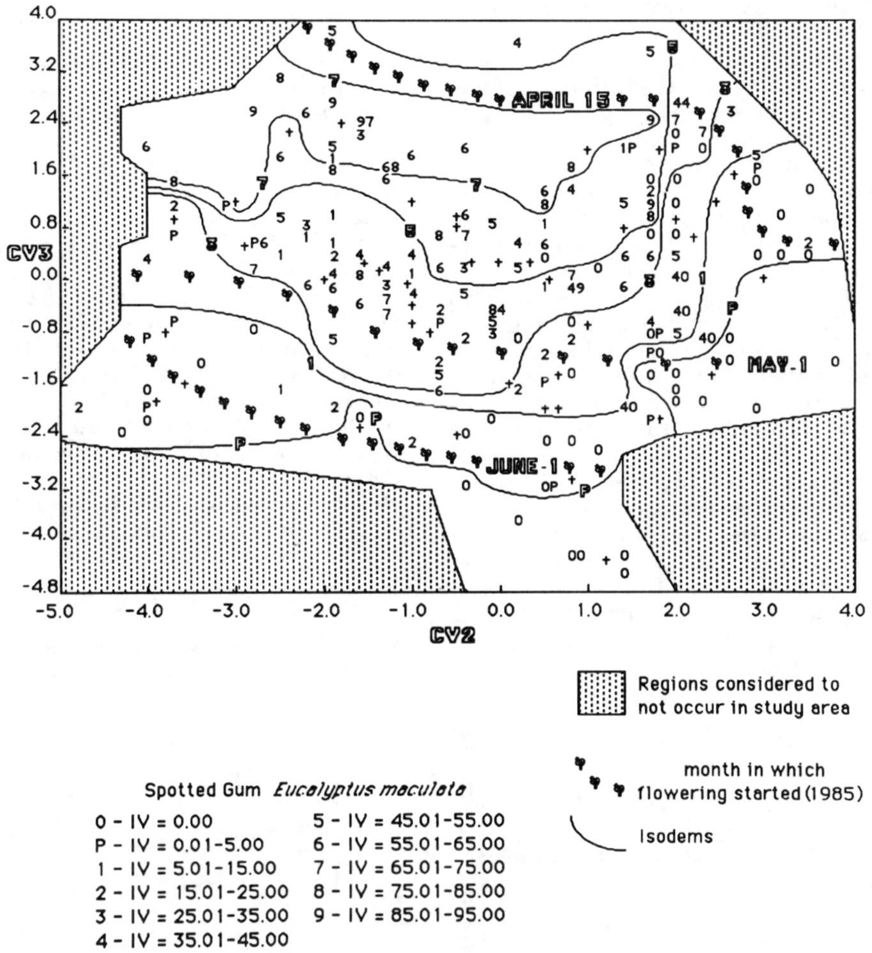

FIG. 4. Importance values (after Davey & Russell-Smith 1984) and the onset of the flowering of *Eucalyptus maculata* are presented in habitat space. The flowering of this species is an important determinant of habitat for the feathertail glider (Davey 1984). Axes are as shown in Fig. 2.

was more a function of eucalypt species richness, the availability of winter resources and the general complexity of the habitat mosaic. Similar findings were reported in the studies of Henry (1985) and Goldingay (1986).

The implications of the data presented by Braithwaite and co-researchers (Braithwaite *et al.* 1983, 1984) for the management of forest wildlife are significant. Clearly, if their hypothesis that the nutritional status of a forest site is a major determinant of the distribution and abundance of arboreal marsupials was correct then this knowledge offered great scope for managers. Given information on the nutritional status of forests in the Eden region, predictions on the abundance of these fauna in different forests could be generated. Once validated, this knowledge would enable more ecologically sound strategies to be developed for the conservation of these fauna. These strategies could then be incorporated into the management plans for the forests of the Eden region. The recognition of this and other potential applications of the findings of Braithwaite *et al.* (1983, 1984) stimulated the study of Norton (1988).

The nutrient hypothesis considered above was derived primarily from the positive correlation reported between the numbers and diversity of species of arboreal marsupials and the quality of eucalypt foliage as defined by concentrations of nitrogen, phosphorus and potassium (Braithwaite *et al.* 1983). Norton (1988) examined this relationship statistically for a local (site) scale in south-east New South Wales. In summary, he identified some of the ecological processes which may effect this correlation. For example, the fecundity of populations of the folivorous greater glider, which demonstrated the strongest positive correlation in the study by Braithwaite *et al.* (1983), was shown to be affected by the availability of relatively high-quality food. Thus, forests in the south-east region, while occurring as a mosaic geographically, could be considered to span a continuum in terms of their abilities to provide high-quality food for this species. At the lower end of the continuum are forests which cannot supply adequate levels of food for this glider to survive. At the higher end are forests which provide large quantities of high-quality food. These resources enable this glider to not only survive but also to maximize its fecundity. Hence, these habitats support the highest population densities of animals. This pattern also appears to hold for many of the other arboreal marsupials (Fig. 2a) and may account for the observations of Braithwaite *et al.* (1983).

Habitat trees (hollow-bearing trees, nest trees) are a critical resource for hollow-dependent arboreal marsupials (Tyndale-Biscoe & Calaby 1975; Recher *et al.* 1980). Studies by Mackowski (1984), Smith *et al.* (1985), and Inions *et al.* (1989) show that these fauna favour particular hollow sizes and types. In a sense, they partition habitat trees on the basis of hollow characteristics. Smith *et al.* (1985) reported that the population density of Leadbeater's possum failed to reach its potential maximum when the number of habitat trees per hectare fell below five (depending upon spatial pattern of the trees). Similar findings have been obtained in forests in south-east New South Wales supporting high-quality habitat for several species of arboreal marsupial (the authors, unpublished data). A most important management implication derived from the latter three studies is that forest managers have to actively manage for future habitat trees. Mackowski (1984) considered that eucalypts have to be at least 140 years of age before hollows are suitable as shelter. However, the rate of hollow formation varies between tree species (Inions *et al.* 1989).

This present case study illustrates how integrated research can provide the basic but strategic information that forest managers require for planning and managing for wildlife at both a regional and local scale. Furthermore, it demonstrates the necessity for accurate resource data.

Upgrading of Biological Surveys

Many faunal and vegetation surveys have focussed on wildlife distribution, enabling the identification

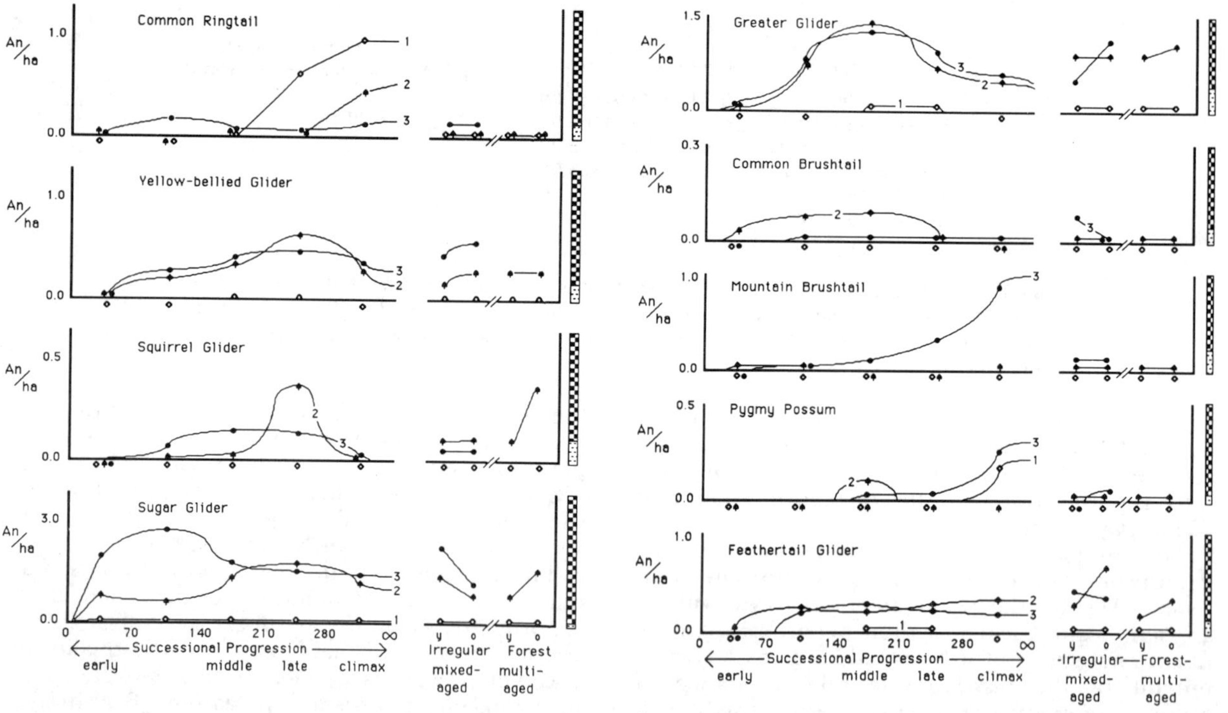

FIG. 5. Estimates of population density of arboreal marsupials along a vegetation successional progression and for irregular (mixed-aged, multi-aged) eucalypt forest. Estimates are for health (sequence 1), stands that remain as sclerophyll forest (sequence 2) and stands that can convert to simple-complex rainforest (sequence 3). Sequences 1 and 2 attain multi-aged stands described as either young (y) or old (o). For mixed-aged forest, stands can be dominated by young (y) or old (o) age swarms. Curves for each species are derived from structural profiles representing 190 plots and over 1 200 hours of spotlight survey. Optimal habitat — vertical dark hatched line; marginal habitat — vertical light hatched line.

of unique, rare or fragile ecological units which may require conservation management (e.g., Emison *et al.* 1984; Macfarlane & Chesterfield 1984). However, these surveys do not facilitate an understanding of the ecological processes that determine associations. Such an understanding is necessary for viable forest planning. Gullan and Norris (1981) commented on this inadequacy, recommending that surveys should be raised to a stage of, at least, a preliminary ecological investigation.

Davey and Russell-Smith (1984) recognized that ecological information useful in predicting and/or identifying ecological units was what forest managers required. They described a methodology for obtaining resource (ecological) information that would be useful in forest resource management and conservation. Their study focussed on a procedure that could be executed by a single operator and was both cost and time effective. Further, the parameters chosen for assessment needed to be readily recognized by forest managers.

An alternative approach to ecological surveys is the gradsect technique (Gillison 1984). This follows the principle that the greatest biological richness is associated with the maximum perceivable environmental gradient in a defined area. Gillison and Brewer (1985) reported that the technique has advantages in cost efficiency and effectiveness in obtaining environmental information over transect and random design techniques. Sampling by gradsect is scale-free, intensity being determined by the steepness of the gradient; the steeper the gradient, the more intense the sampling. A problem exists in situations where a gradient that determines vegetation is not readily discernible or sampling is at a fine scale (Gillison & Brewer 1985); such gradients would be associated with changing edaphic features, particularly changing nutrient-moisture relationships, soil texture, parent material, water tables and drainage.

Wildlife management and information systems

Approaches to gathering resource data that can be incorporated into GIS data bases include inventory and vegetation mapping based on floristics (Gullan 1984) and modi or functional attributes (Fox & Fox 1981; Gillison 1981, 1984; Nix & Gillison 1985). Other approaches being investigated define the forest estate as ordinated gradients, thus enabling communities or forest site types to be defined as subsets of the gradient (Kessell 1981; Kessell *et al.* 1984) or attempt to define the ecotope of specific species as environmental envelopes (Austin 1985; Nix & Gillison 1985).

Two recent developments in computer-based techniques have considerable potential for the management of forest wildlife. BIOCLIM is the acronym for computer software which, using a bioclimatic model, predicts the potential distribution of taxa on the basis of bioclimate (Nix 1986). It requires as input a file of geocoded (latitude, longitude, altitude) point data for the taxa of concern (target taxa). The point data typically represent those localities where a species, for example, has been observed or collected. From this information, BIOCLIM can define the climatic envelope of these localities with the aid of meteorological data represented as climatic surfaces (Hutchinson *et al.* 1984). The package can then generate a map of Australia or a smaller region where localities with similar climatic attributes occur; that is, where a taxon is likely to occur. In the case of target taxa for which few ecological data are available, this technique can be used as the basis for developing a field sampling strategy. Thus, BIOCLIM offers an objective and efficient means to acquire basic but nonetheless essential information, in terms of management, on the distribution of wildlife. To date, a variety of forest wildlife have been investigated in this manner (e.g., Busby 1984; Smith *et al.* 1985; Nix 1986).

Building on the BIOCLIM approach is a proposal to establish an Australian Environmental Geographic Information System (AEGIS) (Nix & Gillison 1985; Mackey *et al.* 1988). The emphasis with AEGIS concerns primary environmental data that are determinants of biotic responses. Research has focussed on methods for deriving reliable estimates of climate, terrain, geology, soil, regolith and vegetation cover. When combined with process knowledge of forest wildlife these primary attribute data have the potential, amongst other things, to predict the distribution of the habitat of taxa. Moreover, they could predict potential changes in habitat (type, location) given known or estimated changes in environmental attributes. Such a scenario may arise, for example, if changes in a forest regime are proposed that will effect changes in the existing environmental attributes.

A technique developed in Tasmania, known as TASFORHAB (Peters 1984; Peters & Cordy 1986), measures habitat variables (floristic, structural habitat and resource information) and produces maps of habitat similarity. Inasmuch as wildlife habitats are related to eco-spatial relationships, the technique affords useful information for the conservation of forest wildlife in production forests. The information is useful to timber managers as well. Its current limitations include an inability to define and determine the habitat of individual species. This can be attributed, in part, to a lack of appropriate research into the habitat requirements of species in that State (Norton 1987). In certain cases the critical habitat of taxa could be identified. TASFORHAB's application to more complex forest and its usefulness in identifying the habitat of individual taxa, however, needs to be tested further.

Two studies in Coolangubra State Forest (New South Wales) have investigated the use of computer planning techniques for wildlife management in

Australian forests. Both used arboreal marsupials as their examples. A study by Stockwell (1987) was conducted in forests surrounding Kavanagh's study area (see above) and considered the usefulness of expert systems in wildlife management. The second study, by Church (1987), appraised the application of an American forest planning programme (FORPLAN), using linear programming optimization (Taha 1976). Both studies found the techniques studied to be valuable tools for management. The limitations of the techniques were not a function of the techniques themselves but more the inadequacy of the resource information for this forest and the limitations of the models used to predict distributions of arboreal marsupials.

Stockwell (1987) developed two expert systems. The first was developed for the purpose of answering common enquiries by forest managers regarding the identification and habitat requirements of arboreal marsupials. The second system was concerned with predicting the distributions of animals within his study area. Predicted estimates were validated and found to be acceptable for short term planning purposes.

Stockwell found that a major difficulty limiting the development of these systems was that the available published information was in a form not readily usable. His study identified gaps in current knowledge and highlighted the need to clarify and formalize, as a set of rules, what is known about the ecology of arboreal marsupials.

Wildlife was one of many components included in the FORPLAN model developed by Church (1987). He set target values for populations of arboreal marsupials, timber and water production, sediment load and recreation visitor-days as constraints and then maximized present net worth in seven different formulations. A problem associated with this type of optimization was an inability to apply an economic value to wildlife species. While this prejudiced the solutions obtained, the study was able to achieve a balance between timber production and wildlife management (conservation) objectives. As an example, guidelines concerning the types and intensity of logging strategies that should be used in this forest were determined. Overall, Church's study supported the findings of earlier overseas studies that FORPLAN is a useful and robust tool in forest and wildlife management (Hoekstra *et al.* 1987).

In their appraisal of the use of FORPLAN in North American forests, Shugart and Gilbert (1987) reported that this package was very useful in examining ecological problems and management implications. Potentially, FORPLAN can be used to manage forests over a variety of scales, both geographic and temporal, as well as to tap into information from other sources (e.g., GIS, expert systems, habitat models, Landsat imagery). In Australia, the potential applications of FORPLAN for use in the management of forest resources are being evaluated in a collaborative project in the Otway Ranges, Victoria, between the Centre for Resource and Environmental Studies and the Department of Forestry at ANU, and the Victorian Department of Conservation, Forests and Lands (*The Colac Herald,* 18th September 1987, p.1).

Future directions

Presently, the planning and management strategies used by forest agencies for wildlife conservation in State Forests in Australia lack integration and are typically framed in generalities. This situation results, in large part, from historical constraints such as the legislative restrictions imposed within the forest estate by administrative boundaries; the tendency of forest agencies to adopt passive or reactionary strategies to wildlife conservation; and insuffficent knowledge due to the long-term lack of support for relevant ecological research. With the benefit of hindsight and the knowledge gleaned from other nations that have experienced similar problems (see e.g., Thomas *et al.* 1988), it is clear that we need a better approach to the conservation of forest wildlife in this country. The approach we envisage is one of greater integration and dynamism and one that will be pro-active rather than passive or reactive. Such an approach would enable planners and managers to make reliable predictions and prescriptions in both the short and longer term. Importantly, this would permit forest agencies to actively plan and manage for novel situations (e.g., climatic change, wildfire) rather than to be surprised by them (see also Hopkins & Saunders 1987). However, this new approach cannot be conceived and implemented in isolation from other demands for forest resources, particularly for wood and water.

A future role envisaged for forestry services is to develop a responsibility in foresters to guide the management of Australia's forests in ways which ensure the survival of forest wildlife (Davey 1978; Shaw 1983; Suckling 1983; Loyn 1985; Recher 1985, 1986). This responsibility is identified in several forest strategy documents (e.g., Australian Forestry Council 1986; Kirner 1986). Recher (1985) considered that this reponsibility should extend beyond forests managed for wood production to include those in National Parks, particularly as there is increasing evidence that habitat manipulation is required within, at least, some National Parks to ensure the long-term conservation of several forest-dependent fauna (see also Norton & Davey 1986). Clearly, wildlife management must become an integral part of forest management and wildlife need to be managed regionally without regard to administrative boundaries and land tenure. It is important that all sources of professional expertise be consulted and encouraged to participate in decision-making processes.

In the latter part of the previous section we considered the current state of some of the research developments for improving our approach to forest wildlife management. Further, we described how the data obtained from these studies were being coupled to computer-based procedures (e.g., GIS, expert systems) that potentially maximize the availability and accessibility of biologically useful information. Below, we discuss the way these developments might evolve.

Approach to a 'Forest Planning Framework'

The basic management unit for wildlife planning is the habitat of taxa. Wildlife habitat is often quantified by a range of biophysical parameters, with faunal habitat, for example, typically being readily determined by the characteristics of the vegetation present (Nix & Gillison 1985; Shugart & Gilbert 1987). As components of the vegetation are what forest industries harvest for timber, the unit that is common to both harvesting and wildlife is, therefore, some measure of vegetation. As a basic unit, the forest site type, which identifies sites characterized by (at least) similar vegetation, is an ecological unit that is most useful to both forestry and wildlife planning (Barnes 1984; Gessel 1986). Thus, the forest site type is the unit required to be identified for planning purposes.

For the purpose of wildfire management, managers must be able to relate a taxon to its biophysical environment and identify the taxon's environmental envelope in habitat space. They also need to be able to relate the dynamics of these envelopes to changes resulting from natural processes (e.g., drought, seasonality, succession) and human-induced disturbances (e.g., logging, fire, fertilizer, pesticide). Therefore, the manager must be able to understand, define and determine habitat. Because a measure of habitat must be incorporated into a planning framework, the question to be addressed is "how best to do this given limited, available knowledge?". The most efficient procedure appears to be by using expert systems. These systems are relatively easily developed once appropriate information on habitat is forthcoming (Stockwell 1987).

Another feature required in the planning framework concerns the designation of appropriate scales for planning. For wildlife planning purposes these are defined best by the forest administrative units. In decreasing order of magnitude these are usually called Region, District, State Forest, Compartment and Coupe. As a planning unit, individual State Forests are seen to be the best scale (1:25 000 map scale; grid $\approx$ 4 ha) to meet wildlife management objectives in the first instance. However, this unit needs to be considered in a regional and district context (1:100 000 map scale; grid $\approx$ 100 ha) for overall regional planning purposes as well as at the finer scale of the compartment and coupe (1:10 000 map scale; grid $\leq$ 1 ha) when considering harvesting operations (Fig. 1). The planning framework must be flexible enough to cope with, at least, these three scales, as different management questions will be asked of managers operating at each scale. Ideally the framework should be scale free.

The planning framework also requires features that enable planners and managers to choose between alternative management strategies based on their economic and environmental effect(s) (Shugart & Gilbert 1987). Hence simulation or optimization models, that provide information on simple effects as well as complex interactions, need to be made available to both planners and managers. To achieve this, we suggest a combination of robust mathematical models and user-friendly information systems. Therefore our framework includes a linear-programming (LP) facility linked into a GIS with supporting models.

Managers need also to monitor planning decisions so they can be updated and modified when necessary. Monitoring would best be undertaken by remote sensing techniques using processors like the microBRIAN image analysis system. However, these techniques have yet to be successfully applied at a fine scale (e.g., habitat level) in native eucalypt forests, although the work of Skidmore (1989) shows considerable promise.

Overall, the features necessary for wildlife planning are habitat models, GIS, LP processors, monitoring systems and expert systems. These should be interactive and linked together into the framework. Critically, the data used to develop and validate many of these features must come directly from or be derived from forest site measurements.

A conceptual model of the integral features of a wildlife management system within our proposed forest planning framework is provided in Figure 6. Illustrated are the factors that determine forest policy, and through this, the overall objectives for forest planning and management. To fulfil these objectives we envisage a planning framework which comprises two systems, a Forest System and a Ecological System, that interact. The technology for the features found in the Forest Planning Framework is available; refinements to features will occur with future technological developments. These features will be interactive once compatible linkages are developed. At some point in the future, forestry, fire behaviour and habitat models will be coupled as part of the LP processor (see Shugart & Gilbert 1987) and the procedures currently offered by AEGIS, the forest and ecological GIS and the monitoring system (see Burrough 1986) will be managed and processed by a single GIS. An expert system will be required to run, manage and implement the Framework.

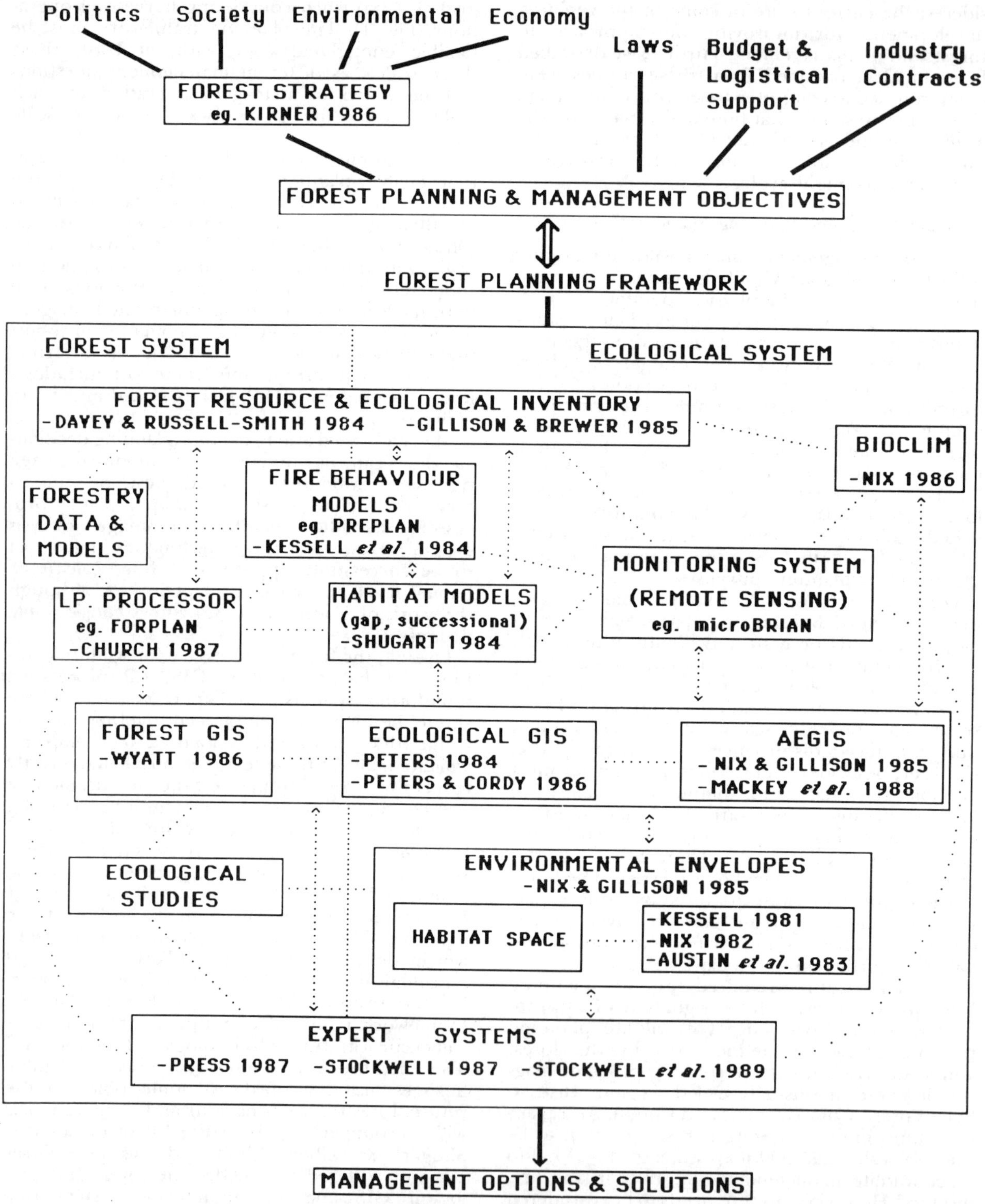

FIG. 6. Conceptual model of the features comprising our proposed forest planning framework and the factors that determine planning and management objectives. Non-arrowed vertical dashed-line separates the two systems.

TABLE 1. General areas of gaps in current knowledge which limit the rate of evolution of our envisaged approach (see Future Directions). Note that all areas warrant immediate attention. However, three relative levels of priority are shown as a guide in the allocation of research effort.

Priority	Area	Topics
1	Status of the Forest Estate	– baseline data including status, quality and health of forest types, ownership.
	Biological Inventories	– comparative techniques, forest site typing.
	Sensitive Species	– priorities vary with forest region.
2	Effects of Different Forestry Practices	– experimentally designed studies of the effects on taxa.
	Processes affecting the Distribution and Abundance of Taxa	– determining key environmental processes and correlates that are transferable to GIS.
	Population Dynamics	– fecundity, dispersal and the inputs for calculating minimum viable populations of taxa.
3	Developments in Information Technology	– tying in Expert Systems and Remote Sensing with habitat and ecological GIS.

Gaps in Knowledge

Figure 6 illustrates how our proposed approach could be formalized and implemented. It indicates the general types and degree of transferability of data, based primarily on forest site measurements, that are required to maximize the strengths of our approach. Clearly, information should be collected with its utility in GIS and mathematical models of forestry and ecological processes in mind. Information should also be presented in ways easily incorporated into expert systems (Stockwell 1987). Further, it is important that the geographic location of study plots be described accurately, particularly their geocode and altitude.

We believe that seven general areas of gaps in our current knowledge exist which limit the rate of evolution of our envisaged approach (Table 1). All of these areas warrant immediate attention. However, within this list we have defined three relative levels of priority as a guide in the allocation of research effort. Further, within each area we have identified topics worthy of initial investigation.

Receiving the highest level of priority are: status of the forest estate, biological inventories and sensitive species. Of paramount importance is an inventory of existing forest types and the status of these types across Australia. This will provide a baseline for an integrated approach to planning and management in forests. Clearly, we need to have accurate data on the general status of the forest estate; for example, the extent of different types of forest, their spatial arrangements, their ownership and their current and planned uses. With this information the identification and management of functional ecosystems can be undertaken. Remote sensing techniques appear to represent part of the solution to this problem, particularly as data would need to be updated regularly.

Biological inventories involving vegetation/forest site/habitat typing and the determination of WPAs are required. However, for this to develop in an integrated manner, biological inventory must be raised from purely a survey that lists wildlife species to one that is ecologically (process) orientated. This will require comparative studies to be undertaken to compare the utility of different inventories, particularly the gradsect approach (Gillison & Brewer 1985) with that proposed by Davey and Russell-Smith (1984). Both techniques can derive forest site types and both produce ecological inventories for an area. For the purposes of inventory design we need to determine whether one approach is superior, whether superiority changes with differing gradients or combinations of gradients and whether both approaches can and should be combined. While not tested, the latter seems most desirable. Furthermore, the methodologies for measuring habitat proposed by Davey and Russell-Smith (1984) need to be tested against such techniques as Gillison's modi or functional attributes (Fox & Fox 1981; Gillison 1981, 1984; Nix & Gillison 1985), Peters' TASFORHAB (Peters 1984; Peters & Cordy 1986) and Walker's crown separation (Walker 1984; Walker & Hopkins 1984; Walker *et al.* 1988).

Sensitive species are species that are adversely affected by forest operations. Lists can be developed for only some segments of the avifauna and mammalian fauna. At this stage, our knowledge is so limited for the other taxa that lists cannot be prepared for reptiles, amphibians, fishes, members of other avial phyla or the flora. Research is only just starting on these taxa. This lack of knowledge is of direct relevance to the three areas listed as second priority (effects of different forestry practices, processes affecting the distribution and abundance of taxa and developments in information technology; Table 1).

Areas awarded second priority (Table 1) are interrelated. Before specific habitat models can be built, much more information on the key processes affecting the distribution and abundance of taxa or groups of taxa are required. Ideally these processes would be tied directly and/or indirectly to generic, resource attributes contained within a GIS data-base. Thus, prediction and, given appropriate validation, prescription could be undertaken on a broadscale. To understand population dynamics, data are required on the life histories (e.g., fecundity, mortality, dispersal) of, at least in the first instance, forest-dependent taxa or species considered to be indicators of biological diversity, guilds or ecosystem stability (e.g., owls, native marsupial carnivores). These types of information are critical for the viable management of functional forest ecosystems and wildlife in fragmented habitats or those subject to recurrent disturbance (see Hopper & Coates, this volume). Such data are a prerequisite for the determination of minimum viable population size or the planning and management of wildlife corridors.

Developments in information technology are given the lowest relative priority of the six mentioned as their rate of expansion is significantly determined by many factors outside our immediate sphere of influence. Moreover, many developments in this field are dependent on feedback from other areas. For example, as new empirical data and/or theory become available on the ecological processes influencing the distribution and abundance of taxa, new expert systems can be tailored to maximize the applicability of this knowledge. Nonetheless, it is clear from Figure 6 that compatible, interactive linkages are required before our proposed wildlife management system can become fully operational. Of immediate importance is the development of remote sensing procedures for typing forest site and habitat.

Finally, it is imperative that representative examples of different functioning forest ecosystems be retained across the forest estate as benchmarks or controls.

New forestry proposals

We support the statement of Hobbs and Hopkins (this volume) that " . . . the history of forestry in Australia has often been one of over-exploitation rather than sustainable yield." While forest managers can identify individual forests that are not being over-cut and are managed on a sustainable yield, the information supporting this general contention for the majority of native forests in Australia is lacking. In fact, many of the research priorities concerning the biological resources, ecosystem processes, ecological stability and stand management regimes of forests that were proposed by the Resource and Environment Consultant Group (1981) are still to be addressed, as are the research priorities identified in Table 1 and discussed previously. Clearly, a more scientific approach to the allocation of forest resources is warranted, particularly given the significant changes in climate forecast within the next 20-50 years (Pittock 1987, 1988a,b; Pearman 1988).

The Forestry and Forest Products Industry Council (FAFPIC; Anon 1988), which represents Australia's second largest manufacturing industry with members from industry and State and Federal Governments (Australian Department of Primary Industry 1987), has recently published a plan advocating a major expansion of the forest products industry in Australia. They contend that significant potential exists for the industry to make major investments totalling some $11 billion for the production of pulp and paper, timber and reconstituted wood products over the next 43 years to the year 2030 (Forestry and Forest Products Industry Council 1987). They believe the major constraint on such an investment is the suitable availability of and access to sufficient forest resources. The FAFPIC document, while highlighting economic and social/employment advantages, does not consider the effect of such a proposal on wildlife conservation, the environment and future forest management flexibility. Such considerations are required by legislation for forests on public lands. Unfortunately, this document is not informative enough for us to appraise the impact of such a plan on the conservation status of forest wildlife.

More recently, the Australian Conservation Foundation has proposed a significant increase in plantations, mostly of native species of hardwoods, as a means of providing wood resources for a reduced domestic industry and enabling logging, generally, to be phased out of native forests (Cameron & Penna 1988). This proposal, as with the FAFPIC plan, has significant implications for the conservation of wildlife in State Forests (see Dargavel 1988b). While it will provide an increased resource for much wildlife, it also has the potential to lead to a reduction in the level of forest wildlife research currently undertaken in State Forests.

Conclusion

The history of planning and management for wildlife conservation in State Forests in Australia has, in general, been dominated by passive or, at best, reactionary strategies. Conservation of wildlife has been achieved indirectly in the past simply through the reservation of forests from timber production or the regulation of the harvesting of trees. More recently, reactionary strategies have been adopted by forest agencies in reponse to an increasing environmental awareness of the general public flowing from significant changes in forest technologies and practices (e.g., a significant increase in the amount of intensive forestry) and increasing fragmentation and reduction of the forest estate. Ecological data on most forest wildlife are so limited that, in general, it is not possible to determine if current wildlife management practices are viable in the long term.

It is clear that a much better approach to forest wildlife management in Australia is required if State Forests are to remain an important resource for wildlife conservation. The approach we have outlined is more integrative and dynamic than is presently in use. It needs to be because we live in a large country supporting a widespread, diverse and dynamic forest environment. As new ecological knowledge and technologies emerge the system we described earlier could evolve. However, we believe that this should not happen passively. Rather it should be promoted actively.

In the longer term, we see an integrative system providing a variety of solutions to small or big problems across a number of spatial and temporal scales within the forest estate. In contrast to the past this system could provide an explicit framework and guidelines for the planning and management of individual forests for the conservation of wildlife. It would facilitate better multiple-use of forests.

In conclusion, decision-making on the allocation of forest resources has become increasingly subjective and politicized. Moreover, the proponents of competing views rarely co-operate to seek viable solutions or to plan integrated strategies for the future. Clearly, this situation is untenable. The need for a rational, scientifically-based approach to optimize timber production and wildlife conservation has never been greater. The challenge for scientists will be to integrate new data and technologies in a manner that will enable this.

Acknowledgements

The ideas presented in this paper have evolved from discussion with both undergraduate and post-graduate students of the Department of Forestry, ANU, academics from this department and CRES, and a wide group of professional colleagues working in the field of forest wildlife management in Australia. We particularly thank L. W. Braithwaite, M. Brown, P. E. S. Christensen, R. L. Cordy, J. Dargavel, R. C. Davey, M. R. Fleming, R. G. Florence, B. C. Gepp, R. P. Kavanagh, J. C. Kehl, R. H. Loyn, M. Macfarlane, B. G. Mackey, O. Nichol, H. A. Nix, D. G. Peters, G. C. Suckling, M. T. Tanton, C. R. Tidemann, B. J. Turner, G. W. Wardell-Johnson and J. E. Williams for their assistance and D. Dunne for typing the manuscript. We should emphasize that the views expressed in the paper do not necessarily coincide with those of the people who have been consulted or have read it during its preparation.

References

Anon. (1984) *A National Conservation Strategy for Australia.* Department of Home Affairs and Environment. Australian Government Printing Service, Canberra.

Anon. (1988) Australian Forestry and Forest Products Industry Council. *Forest and Timber Industries Australasian Bulletin* **10-114**, 8.

Austin M. P. (1985) Continuum concept, ordination methods and niche theory. *Ann. Rev. Ecol. Syst.* **16,** 39-61.

Austin M. P. Cunningham R. B. & Good R. B. (1983) Altitudinal distribution in relation to other environmental factors of several eucalypt species in southern New South Wales. *Aust. J. Ecol.* **8,** 169-80.

Australian Conservation Foundation (1987) *Forestry in Australia.* Policy statement No. 36. Australian Conservation Foundation, Hawthorn, Victoria.

Australian Department of Primary Industry (1987) *Handbook to the Australian Forestry Council.* Canberra.

Australian Forestry Council (1986) *National Forest Strategy For Australia.* Australian Government Publishing Service, Canberra.

Australian Heritage Commission (1987) *Australia's native forests — an information kit.* Department of The Arts, Sport, The Environment, Tourism and Territories. Australian Government Publishing Service, Canberra.

Barnes B. V. (1984) The ecological approach to ecosystem classification. In: *Symposium on Site and Productivity of Fast Growing Plantations* (eds D. C. Grey, A. P. G. Schonau and C. J. Schutz). IUFRO, Pretoria, South Africa.

Boer B. & Preston B. (1987) Forestry — reform and regeneration. *Environ. Plan. Law J.* **4,** 80-100.

Braithwaite L. W. (1984) On identifying important habitat characteristics and planning a conservation strategy for arboreal marsupials within the Eden Woodpulp Concession area. In: *Possums and Gliders* (eds A. P. Smith & I. D. Hume) pp. 501-8. Australian Mammal Society and Surrey Beatty & Sons, Chipping Norton.

Braithwaite L. W., Dudzinski M. L. & Turner J. (1983) Studies on the arboreal marsupial fauna of eucalypt forests being harvested for woodpulp at Eden, New South Wales. II. Relationships between faunal density, richness and diversity, and measured variables of habitat. *Aust. Wildl. Res.* **10,** 231-47.

Braithwaite L. W., Turner J. & Kelly J. (1984) Studies on the arboreal marsupial fauna of eucalypt forests being harvested for woodpulp at Eden, New South Wales. III. Relationships between faunal densities, eucalypt occurrence and foliage nutrients, and soil parent materials. *Aust. Wildl. Res.* **11,** 41-8.

Burrough P. A. (1986) *Principles of Geographical Information Systems for Land Resources Assessment.* Monographs on Soil and Resource Survey No. 12, Clarendon Press, Oxford.

Busby J. R. (1984) *Nothofagus cunninghamii (Southern Beech) Vegetation in Australia.* Australian Government Publishing Service, Canberra.

Cameron J. I. & Penna I. W. (1988) *The Wood and the Trees: Preliminary Economic Analysis of a Conservation Orientated Forest Industry Strategy.* Australian Conservation Foundation, Hawthorn, Victoria.

Carron L. T. (1985) *A history of forestry in Australia.* Australian National University Press, Canberra.

Christensen P. (1980) *The biology of* Bettongia pencillata *Gray, 1837, and* Macropus eugenii *(Desmarest, 1817) in relation to fire.* Forests Department of Western Australia, Bulletin 91.

Christensen P. E. S., Wardell-Johnson G. & Kimber P. (1985) Birds and fire in southwestern forests. In: *Birds of Eucalypt Forests and Woodlands: Ecology, Conservation, Management* (eds A. Keast, H. F. Recher, H. Ford & D. Saunders) pp. 291-9. Surrey Beatty & Sons, Chipping Norton.

Church R. (1987) *Applying FORPLAN to an Australian native forest.* Honours Thesis, Department of Forestry, The Australian National University, Canberra.

Coulombe H. N. (1978) Towards an integrated ecological assessment of wildlife habitat. In: *Integrated Inventories of Renewable Natural Resources: Proceedings of the Workshop* (Co-ordinators — H. G. Lunds, V. J. LaBau, P. F. Ffolliott and D. W. Robinson) pp. 5-23. USDA, GTR, RM-55.

Dargavel J. (1987) Prospects present and preferred. In: *Prospects for Australian Hardwood Forests* (eds J. Dargavel and G. Sheldon) pp. 3-12. Centre for Resource and Environmental Studies Monograph 19, The Australian National University, Canberra.

Dargavel J. (1988a) *Sawing, Selling and Sons.* Centre for Resource and Environmental Studies Monograph 20, Canberra.

Dargavel J. (1988b) Forest industries and forest policies: two new proposals. *Aust. For.* **51,** 8-13.

Davey S. M. (1978) Forestry for the Birds. *Forestry Log* **11,** 13-5.

Davey S. M. (1984) Habitat preferences of arboreal marsupials within a coastal forest in southern New South Wales. In: *Possums and Gliders* (eds A. P. Smith & I. D. Hume) pp. 509-16. Australian Mammal Society and Surrey Beatty & Sons, Chipping Norton.

Davey S. M. (1989) Thoughts towards a forest wildlife management strategy. *Aust. For.* **52,** 56-67.

Davey S. M. & Russell-Smith J. J. (1984) A case study in collecting resource information for forest managers. In: *Survey Methods for Nature Conservation, Vol. 2* (eds K. Myers, C. R. Margules & I. Musto) pp. 263-74. CSIRO Division Water and Land Resources, Canberra.

Diamond J. M. (1975) Assembly of species communities. In: *Ecology and Evolution of Communities* (eds M. L. Cody & J. M. Diamond) pp. 342-444. Belknap Press of Harvard University Press, Cambridge.

Emison W. B., Menkhorst P. W., Mansherg I. M., Norman F. I., Robertson P., Corrick A. H. & Alderson J. A. (1984) Wildlife surveys conducted by the Fisheries and Wildlife Division in Victoria, 1972-1983. In: *Survey Methods for Nature Conservation Vol. 2* (eds K. Myers, C. R. Margules & I. Musto) pp. 166-94. CSIRO Division of Water and Land Resources, Canberra.

Forestry and Forest Products Industry Council (1987) Forest Growth Plan: a submission to the Australian Government, Canberra.

Ferguson I. S. (1985) *Report of the Board of Inquiry into the Timber Industry in Victoria.* Victorian Government Printer, Melbourne.

Ferguson I. S. (1987) Problems and prospects in Victoria. In: *Prospects for Australian Hardwood Forests* (eds J. Dargavel & G. Sheldon) pp. 27- 46. Centre for Resource and Environmental Studies Monograph 19, The Australian National University, Canberra.

Florence R. (1983) Silvicultural practice in eucalypt forests. In: *Proceeding Institute of Foresters of Australia* pp. 115-20. 10th Triennial Conf. 29th August-2nd September, Melbourne.

Fox B. J. & Fox M. D. (1981) A comparison of vegetation classifications as descriptors of small animal habitat preference. In: *Vegetation Classification in Australia* (eds A. N. Gillison & D. J. Anderson) pp. 166-80. CSIRO and Australian National University Press, Canberra.

Gepp B. (1985) Values of Wildlife in Forest Plantations in the Tropics and Southern Hemisphere. In: *Wildlife Management In The Forests And Forestry-Controlled Lands In The Tropics And The Southern Hemisphere* (ed J. Kikkawa) pp. 56-60. University of Queensland Printery, Brisbane.

Gessel S. P. (1986) *Forest Site and Productivity: Forest Sciences edition.* Martinus Nijhoff, Netherlands.

Gibson B. F. (1988) Unlocking the forests. *Appita* **41,** 266-8.

Gill A. M. (1981) Adaptive responses of Australian vascular plant species to fire. In: *Fire and the Australian Biota* (eds A. M. Gill, R. H. Groves & I. R. Noble) pp. 243-72. Australian Academy of Science, Canberra.

Gillison A. N. (1981) Towards a functional vegetation classification. In: *Vegetation Classification in Australia* (eds A. N. Gillison & D. J. Anderson) pp. 30-41. CSIRO and Australian National University Press, Canberra.

Gillison A. N. (1984) Gradient-oriented sampling for resource surveys. In: *Survey Methods for Nature Conservation, Vol. 2* (eds K. Myers, C. R. Margules & I. Musto) pp. 349-74. CSIRO Division of Water and Land Resources, Canberra.

Gillison A. N. & Brewer K. R. W. (1985) The use of gradient-directed transects or gradsects in natural resource surveys. *J. Environ. Manage.* **20,** 103-27.

Golding B. G. (1979) *Use of artificial hollows by mammals and birds in the Wombat Reserve, Daylesford, Victoria.* M.Sc. Thesis, Monash University, Clayton.

Goldingay R. L. (1986) Feeding behaviour of the yellow-bellied glider *Petaurus australis* (Marsupialia: Petauridae) at Bombala, New South Wales. *Aust. Mammal.* **9,** 17-25.

Grigg G., Shine R. & Ehmann H. (1985) *Australasian Frogs and Reptiles.* Royal Zoological Society of New South Wales and Surrey Beatty & Sons, Chipping Norton.

Groves R. H. & Ride W. D. L. (1982) *Species at Risk: Research in Australia.* Australian Academy of Science, Canberra.

Gullan P. K. (1984) Methods using floristics. In: *Survey Methods for Nature Conservation, Vol. 1* (eds K. Myers, C. R. Margules & I. Musto) pp. 57-73. CSIRO Division of Water and Land Resources, Canberra.

Gullan P. K. & Norris K. C. (1981) Floristic classifications, small mammals and birds. In: *Vegetation Classification in Australia* (eds A. N. Gillison & D. J. Anderson) pp. 181-91. CSIRO and Australian National University Press, Canberra.

Henry S. R. (1985) *The diet and socioecology of gliding possums in southern Victoria.* Ph.D. Thesis, Monash University, Clayton.

Hoekstra T. W., Dyer A. A. & Le Master D. C. (1987) *FORPLAN: An Evaluation of a Forest Planning Tool.* USDA Forest Service, GTR/RM—140.

Hopkins A. J. M. & Saunders D. A. (1987) Ecological studies as the basis for management. In: *Nature Conservation: The Role of Remnants of Native Vegetation* (eds D. A. Saunders, G. W. Arnold, A. A. Burbidge & A. J. M. Hopkins) pp. 15-28. Surrey Beatty & Sons, Chipping Norton.

Hutchinson M. F., Booth T. H., McMahon J. P. & Nix H. A. (1984) Estimating monthly mean values of daily total solar radiation for Australia. *Solar Energy* **32,** 277-90.

Inions G. B., Tanton M. T. & Davey S. M. (1989) The effect of fire on the availability of hollows in trees used by the Common Brushtail Possum, *Trichosurus vulpecula* Kerr, 1792, and Ringtail Possum, *Pseudocheirus peregrinus* Boddaerts, 1785. *Aust. Wildl. Res.* **16,** 449-58.

IUCN (1980) *World Conservation Strategy. Living Resource Conservation for Sustainable Development.* Gland, Switzerland.

Kavanagh R. P. (1984) Seasonal changes in habitat use by gliders and possums in south-eastern New South Wales. In: *Possums and Gliders* (eds A. P. Smith & I. D. Hume) pp. 527-43. Australian Mammal Society and Surrey Beatty & Sons, Chipping Norton.

Kavanagh R. P. (1987a) *Floristic and phenological characteristics of a eucalypt forest in relation to its use by arboreal marsupials.* M.Sc. Thesis, Department of Forestry, The Australian National University, Canberra.

Kavanagh R. P. (1987b) Forest phenology and its effect on foraging behaviour and selection of habitat by the yellow-bellied glider, *Petaurus australis* Shaw. *Aust. Wildl. Res.* **14,** 371-84.

Kavanagh R. P., Shields J. M., Recher H. F. & Rohan-Jones W. G. (1985) Bird populations of a logged and unloggd forest mosaic at Eden, New South Wales. In: *Birds of Eucalypt Forests and Woodlands: Ecology, Conservation, Management* (eds A. Keast, H. F. Recher, H. Ford & D. Saunders) pp. 273-81. Surrey Beatty & Sons, Chipping Norton.

Keast A., Recher H. F., Ford, H. & Saunders D. (1985) *Birds of Eucalypt Forests and Woodlands: Ecology, Conservation, Management.* Surrey Beatty & Sons, Chipping Norton.

Kessell S. R. (1981) Application of gradient analysis concepts to resource management modelling. *Proc. Ecol. Soc. Aust.* **11,** 163-73.

Kessell S. R., Good R. B. & Hopkins A. J. M. (1984) Implementation of two new resource management information systems in Australia. *Environ. Manage.* **8,** 251-70.

Kikkawa J. (1985) *Wildlife Management In The Forests and Forestry-Controlled Lands In The Tropics And The Southern Hemisphere.* University of Queensland Printery, Brisbane.

Kirner J. E. (1986) *Victoria Timber Industry Strategy.* Government Printer, Melbourne.

Lehmkuhl J. F. (1984) Determining size and dispersion of minimum viable populations for land management planning and species conservation. *Environ. Manage.* **8,** 167-76.

Leigh J., Boden R. & Briggs J. (1984) *Extinct and Endangered Plants of Australia.* Macmillen, Melbourne.

Leslie A. J. (1987) International logic of a high-value policy. In: *Prospects for Australian Hardwood Forests* (eds J. Dargavel & G. Sheldon) pp. 13-26. Centre for Resource and Environmental Studies Monograph 19, The Australian National University, Canberra.

Loyn A. (1985) Strategies for conserving wildlife in commercially productive eucalypt forest. *Aust. For.* **48,** 95-101.

Lunney D. (1987) Effects of logging, fire and drought on possums and gliders in the coastal forests near Bega, New South Wales. *Aust. Wild. Res.* **14,** 263-74.

Lunney D. & Leary T. (1988) The impact on native mammals of land-use changes and exotic species in the Bega district, New South Wales, since settlement. *Aust. J. Ecol.* **13,** 67-92.

Lyon L. J. (1978) Information requirements for wildlife management. In: *Integrated Inventories of Renewable Natural Resources: Proceedings of the Workshop* (Co-ordinators — H. G. Lund, V. J. LaBau, P. F. Ffolliott & D. W. Robinson) pp. 45-52. USDA, GTR, RM-55.

Macfarlane M. A. & Chesterfield E. A. (1984) Methods and principal findings of a survey of fauna and flora in a high elevation mixed species eucalypt forest in Eastern Australia. In: *Survey Methods for Nature Conservation, Vol. 2* (eds K. Myers, C. R. Margules & I. Musto) pp. 195-205. CSIRO Division of Water and Land Resources, Canberra.

McIlroy J. C. (1978) The effects of forest practices on wildlife in Australia: a review. *Aust. For.* **41,** 78-94.

Mackey B. G., Nix H. A., Hutchinson M. F. & MacMahon J. P. (1988) Assessing representativeness of places for conservation reservation and heritage listing. *Environ. Manage.* **12,** 501-14.

Mackowski C. M. (1984) The ontogeny of hollows in Blackbutt (*Eucalyptus pilularis*) and its relevance to the management of forests for possums, gliders and timber. In: *Possums and Gliders* (eds A. P. Smith & I. D. Hume) pp. 517-25. Australian Mammal Society and Surrey Beatty & Sons, Chipping Norton.

Myers K., Margules C. R. & Musto I. (1984) *Survey Methods for Nature Conservation, Vols. 1 and 2.* CSIRO Division of Water and Land Resources, Canberra.

Nix H. A. (1982) Environmental determinants of biogeography and evolution. In: *Evolution of the Flora and Fauna of Arid Australia* (eds W. R. Baker & P. J. M. Greenslade) pp. 47-66. Peacock Press, Frewville.

Nix H. A. (1986) A biogeographic analysis of Australian elapid snakes. In: *Atlas Of Elapid Snakes of Australia* (ed. R. Longmore) pp. 4-15. Australian Flora and Fauna Series No. 7, Australian Government Publishing Service, Canberra.

Nix H. A. & Gillison A. N. (1985) Towards an operational framework for habitat and wildlife management. In: *Wildlife Management In The Forests And Forestry-Controlled Lands In The Tropics And the Southern Hemisphere* (ed J. Kikkawa) pp. 39-55. University of Queensland Printery, Brisbane.

Noble I. R. & Slatyer R. O. (1980) The use of vital attributes to predict successional changes in plant communities subject to recurrent disturbances. *Vegetatio* **43,** 5-21.

Norton T. W. (1985) Arboreal marsupials and foliar nutrients in eucalypt forests of Australia. *Comm. For. Rev.* **Dec.** 385.

Norton T. W. (1986) Towards understanding the (eucalypt) forest — dependent wildlife of southeast Australia. *Tigerpaper* (FAO: Bangkok) **13,** 24-29.

Norton T. W. (1987) The ecology of small mammals in north-eastern Tasmania. I. *Rattus lutreolus velutinus. Aust. Wild. Res.* **14,** 415-33.

Norton T. W. (1988) *Ecology of Greater Gliders,* Petauroides volans Kerr *1792, in relation to variations in habitat quality in eucalypt forests in south-east New South Wales.* Ph.D. Thesis, Department of Forestry, The Australian National University, Canberra.

Norton T. W. (1989) Australia's arboreal marsupial fauna: past and present. In: *Australia's Ever Changing Forests* (ed K. Frawley) pp. 99-117. Australian Defence Force Academy, University of New South Wales, Canberra.

Norton T. W. & Davey S. M. (1986) Philosophy of wildlife management in national parks. *Ascent* **11 Dec.** 28.

Pahl L. I., Winter J. W. & Heinsohn G. (1988) Variation in responses of arboreal marsupials to fragmentation of tropical rainforest in north eastern Australia, *Biol. Conserv.* **46,** 71-82.

Pearman G. I. (1988) *Greenhouse: planning for climate change.* CSIRO Australia, Melbourne.

Peters D. (1984) Tasforhab. In: *Surveys Methods for Nature Conservation Vol. 2* (eds K. Myers, C. R. Margules & I. Musto) pp. 47-66. CSIRO Division of Water and Land Resources, Canberra.

Peters D. & Cordy R. L. (1986) *Tasforhab Douglas-Apsley Project Report 1,* Australian Heritage Commission, Forestry Commission of Tasmania and National Parks and Wildlife Service (Tas.), Hobart.

Pittock A. B. (1987) Forests beyond 2 000 — effects of atmospheric change. *Aust. For.* **50,** 205-15.

Pittock A. B. (1988a) Economic priorities and environmental research: Part 1 — the context. *Search* **19,** 23-5.

Pittock A. B. (1988b) Economic priorities and environmental research: Part 2 — their importance to Australia. *Search* **19,** 82-4.

Press A. J. (1987) Fire management in Kakadu National Park: the ecological basis for the active use of fire. *Search* **18,** 244-8.

Recher H. F. (1985) A diminishing resource: Mature forest and its role in forest management. In: *Wildlife Management In The Forests And Forestry-Controlled Lands In The Tropics And The Southern Hemisphere* (ed. J. Kikkawa) pp. 28-33. University of Queensland Printery, Brisbane.

Recher H. F. (1986) Integrating national park and forestry objectives for wildlife conservation: the forester's responsibility. *Forestry Log* **18,** 5-7.

Recher H. F., Rohan-Jones W. & Smith, P. (1980) *Effects of the Eden Woodchip Industry on Terrestrial Vertebrates with Recommendations for Management.* Forestry Commission of New South Wales, Research Note 42, Sydney.

Recher H. F., Shields J., Kavanagh R. & Webb G. (1987). Retaining remnant mature forest for nature conservation at Eden, New South Wales: a review of theory and practice. In: *Nature Conservation: The Role of Remnants of Native Vegetation* (eds D. A. Saunders, G. W. Arnold, A. A. Burbidge & A. J. M. Hopkins) pp. 174-94. Surrey Beatty & Sons, Chipping Norton.

Resource and Environment Consultant Group (1981) *Environmental effect of clearcutting native forest: an evaluation of current research and research priorities.* Department of Forestry, The Australian National University, Canberra.

Richardson B. J. (1983) Identifying the Australian fauna: What remains to be done? *Search* **14,** 320-3.

Shaw W. W. (1983) Integrating wildlife conservation and timber production objectives in Australian forests. *Aust. For.* **46,** 132-5.

Shugart H. H. (1984) *A Theory of Forest Dynamics: The Ecological Implications of Forest Succession Models.* Springer-Verlag, New York.

Shugart H. H. & Gilbert B. J. (1987) An ecological appraisal of FORPLAN in National Forest Planning. In: *FORPLAN: An evaluation of a forest planning tool* (eds T. W. Hoekstra, A. A. Dyer & D. C. Le Master) pp. 105-21. USDA Forest Services, GTR/IRM-140.

Shugart H. H., Hopkins M. S., Burgess I. P. & Mortlock A. T. (1980) The development of a succession model for subtropical rain forest and its application to assess the effects of timber harvest at Wiangaree State Forest, New South Wales. *J. Environ. Manag.* **11,** 243-65.

Skidmore A. K. (1987) Predicting bushfire activity in Australia from El Nino/Southern Oscillation events. *Aust. For.* **50,** 231-5.

Skidmore A. K. (1989) A rule based expert system classifies eucalypt forest types using Landsat thematic mapper data and a digital terrain model. *Photog. Eng. Remot. Sen.* **55,** 1449-64.

Smith A. P. & Hume I. D. (eds) (1984) *Possums and Gliders.* Australian Mammal Society and Surrey Beatty & Sons, Chipping Norton.

Smith A. P., Lindenmayer D. & Suckling G. (1985) *The Ecology and Management of Leadbeaters Possum.* WWF (Australia) Project 51, University of New England, Armidale.

Smith P. (1985) Woodchip logging and birds near Bega, New South Wales. In: *Birds of Eucalypt Forests and Woodlands: Ecology, Conservation, Management* (eds A. Keast, H. F. Recher, H. Ford & D. Saunders) pp. 259-71. Surrey Beatty & Sons, Chipping Norton.

Stockwell D. R. B. (1987) *Expert systems and wildlife management: rule-based estimation of the population density of arboreal marsupials in Coolangubra State Forest.* Honours Thesis, Department of Forestry, The Australian National University, Canberra.

Stockwell D. R. B., Davey S. M. & Davis J. R. (1989) Designing an expert system for wildlife management. A. I. *Applications in Natural Resource Management* **3,** (in press).

Suckling G. C. (1983) Future forest directions in fauna conservation. In: *Facing Forestry's Future* (convenor P. J. O'Shaughnessy) pp. 96-9. Proceedings Institute of Foresters of Australia 10th Triennial Conference, Melbourne.

Taha H. A. (1976) *Operations Research: an introduction.* Macmillan, New York.

Thomas J. W., Ruggiero L. F., Mannan R. W., Schoen J. W. & Lancia R. A. (1988) Management and conservation of old-growth forests in the United States. *Wildl. Soc. Bull.* **16,** 252-62.

Tidemann C. R. & Flavel S. C. (1987) Factors affecting choice of diurnal roost site by tree-hole bats *(Microchiroptera)*in south-eastern Australia. *Aust. Wildl. Res.* **14,** 203-9.

Tyler M. J. (1979) *The status of endangered Australasian wildlife.* Royal Society of South Australia, Adelaide.

Tyndale-Biscoe C. H. & Calaby J. H. (1975) Eucalypt forests as refuge for wildlife. *Aust. For.* **38,** 117-33.

Van Horne B. (1983) Density as a misleading indicator of habitat quality. *J. Wildl. Manage.* **47,** 893-901.

Walker J. (1984). Description and classification of Australian non-rainforest vegetation. In: *Survey Methods for Nature Conservation Vol. 1* (eds K. Myers, C. R. Margules & I. Musto) pp. 18-36. CSIRO Division of Water and Land Resources, Canberra.

Walker J., Crapper P. F. & Penridge L. K. (1988) The crown-gap ratio (C) and crown cover: The field study. *Aust. J. Ecol.* **13,** 101-8.

Walker J. & Hopkins M. S. (1984) Vegetation. In: *Australian Soil and Land Survey Field Handbook* (eds R. C. McDonald, R. F. Isbell, J. G. Speight, J. Walker & M. S. Hopkins) pp. 44-67. Inkata Press, Melbourne.

Walther P. (1986) The meaning of zoning in the management of natural resource lands. *J. Environ. Manage.* **22,** 331-43.

Webb G. A. (1985) Habitat use and activity patterns in some south-eastern Australian skinks. In: *Biology of Australasian Frogs and Reptiles* (eds G. Grigg, R. Shine & H. Ehmann) pp. 20-30. Royal Zoological Society of New South Wales and Surrey Beatty & Sons, Chipping Norton.

Webb L. J. & Tracey J. B. (1981) The rainforests of northern Australia. In: *Australian Vegetation* (ed R. H. Groves) pp. 67-101. Cambridge University Press, Cambridge.

Wells K. F., Wood N. H. & Laut P. (1984) *Loss of forests and woodlands in Australia: a summary by State based on rural local government areas.* CSIRO Division of Water and Land Resources Tech. Mem. 4.

Westoby J. C. (1983) Facing the forester's future. *Aust. For.* **46,** 246-50.

Werren G. L. & Kershaw A. P. (1984) *Australian National Rainforest Study Report.* WWF (Australia) Project 44, Sydney.

Winter J. W. (1984) Conservation studies of tropical rainforest possums. In: *Possums and Gliders* (eds A. P. Smith & I. D. Hume) pp. 469-81. Australian Mammal Society and Surrey Beatty & Sons, Chipping Norton.

Wootten H. (1986) Conservation objections. In: *Prospects for Australian Hardwood Forests* (eds J. Dargavel & G. Sheldon) pp. 65-77. Centre for Resource and Environmental Studies Monograph 19, The Australian National University, Canberra.

Wyatt S. W. (1986) *The role of a Geographic Information System in Forest Level Management Planning.* Honours Thesis, Department of Forestry, The Australian National University, Canberra.

Proc. Ecol. Soc. Aust. 1990 — 16:347-357

The use of ant communities to evaluate change in Australian terrestrial ecosystems: a review and a recipe

A. N. ANDERSEN

Tropical Ecosystems Research Centre, CSIRO Division of Wildlife and Ecology, PMB 44, Winnellie, N.T. 0821

Abstract

Ecologists need to develop formal systems of assessing Australian ecosystems that can be used to evaluate land degradation and restoration processes. Ants are likely to be bio-indicators par excellence *in this context because they: (1) are ubiquitously diverse and abundant; (2) interact in many ways with other parts of ecosystems and are functionally important at all trophic levels, such that they integrate a wide variety of biotic and abiotic variables; (3) are readily sampled and processed; (4) are highly sensitive to environmental variables; and (5) respond rapidly to environmental change. Most Australian ants are poorly known at the species level, but provide interpretable results when classified into functional groups whose relative abundances vary predictably with climate, vegetation type, and level of disturbance. The relative abundance of these functional groups is not only sensitive to environmental change, but provides insights into the processes underlying such change. Guidelines are provided for using ants in a standard, environmental evaluation system.*

Introduction

A problem with land degradation and land restoration is that it is difficult to know if land has been degraded or restored. Protagonists such as miners and militant conservationists will always have different perceptions of what levels of degradation or restoration are acceptable, but the real problem is that we have no formal system of assessing Australian ecosystems that can be used to evaluate degradation and restoration processes objectively.

The notion of land degradation typically evokes images of ecological catastrophes such as extensive over-grazing, soil erosion, salination and species extinctions. In reality, of course, the effects of most land management activities are much more subtle and the ecological consequences of them are not immediately apparent. What, for example, are the ecological consequences of burning forests at different frequencies or different times of the year? Which timber extraction option is the most environmentally sensitive? Have mining companies successfully rehabilitated their minesites? Moreover, it is important to be able to recognize ecological change well before it becomes dramatic and possibly irreversible.

Answers to the above questions depend upon the availability of effective environmental evaluation systems. In this regard, Australia lags far behind the United States (Pearsall *et al.* 1986) and Europe (van der Ploeg 1986), where management agencies have developed various habitat evaluation models and data bases to help monitor and understand the consequences of land management practices.

It is generally recognized that these evaluation procedures must adopt a systems approach, rather than deal with only one or a few species (Salwasser & Tappeiner 1981; van Horne 1983), and be sufficiently standardized to be applicable in a broad range of habitats (Pearsall *et al.* 1986).

Adoption of a systems approach to habitat evaluation does not necessarily mean that a large and diverse array of taxa must be studied in detail. Given that ecosystems comprise complex networks of closely interacting organisms, knowledge of a small number of well-chosen taxa should give a strong indication of the status of the ecosystem in general (van der Ploeg 1986). To be successfully incorporated into a standard evaluation system, ideal indicator taxa must meet a special set of criteria (Majer 1983; Greenslade & Greenslade 1984a). First, they must be functionally important and closely interrelated with other parts of the ecosystem, so that they are reliable indicators of the general environment in which they occur. Second, they must be sufficiently abundant, diverse and widespread to be used in a wide variety of habitats. Third, they must be practical to use in the sense of being readily sampled and processed. Finally, they must provide interpretable results, by not only being sensitive to environmental change, but by giving insights into the nature of, and mechanisms underlying, such change.

In this chapter I review the evidence suggesting that ants represent ideal indicator taxa, and argue that an analysis of ant community organization would make a valuable contribution to an objective environmental evaluation system. I then provide guidelines for using ants in this context.

Ants as bio-indicators

Invertebrates best meet the above requirements for indicator taxa. Invertebrates have a strong history of use as bio-indicators in aquatic systems (Hellawell 1978; James & Evison 1979), but their potential for use in terrestrial environments has only become widely appreciated in the last decade or so (Mahoney 1976; Puszkar 1979; Coyle 1981; Majer 1983; Greenslade & Greenslade 1984a; Disney 1986; Eyre 1987). In this period, a wide range of terrestrial invertebrate groups has been suggested for use as bio-indicators in Australia. Notable examples include ants (Table 1), spiders (Mawson 1986), springtails (Greenslade 1985a; Greenslade & Greenslade 1984a, 1987), termites (Nichols & Bunn 1980; Greenslade 1985a) and beetles (Greenslade 1985a; Yen 1987). However ants have several important advantages over other invertebrate groups for use as bio-indicators in a standard environmental evaluation system.

Abundance, diversity and functional importance of ants

Ants are unquestionably Australia's dominant faunal group throughout the arid zone (Greenslade 1979; Greenslade & Greenslade 1984b; Matthews & Kitching 1984), including semi- and seasonally-arid regions, and are one of the most important faunal groups elsewhere. Ant diversity, and apparently also abundance, in Australia are exceptional by world standards. Diversity is especially high in semi- and seasonally-arid regions of southern and northern Australia, where up to 150 species can occur within a single hectare (Andersen 1983a, unpublished data; Andersen & Yen 1985), but even in cool and wet regions well over 50 species per hectare commonly occur (Andersen 1986a, 1988a).

Ants have important influences on soils, vegetation and other faunal groups, both directly, and indirectly through their intimate involvement with a wide range of key ecological processes. As scavengers they are important agents of energy turnover and nutrient cycling and patterning (Brown & Taylor 1970; Greenslade & Greenslade 1984b; Matthews & Kitching 1984; Andersen 1988b). Through their nesting activities they often have an important influence on soil structure and formation (Lyford 1963; Green & Askew 1965; Humphreys 1981; Greenslade & Greenslade 1983; Greenslade 1985a). They have many close associations with plants (Bequaert 1922; Buckley 1982; Beattie 1985), but in Australia seed predation (Ashton 1979; Briese 1982; O'Dowd & Gill 1984; Andersen and Ashton 1985; Andersen 1987a, 1988a) and seed dispersal (Berg 1975, 1981; Davidson & Morton 1981a,b; Majer 1982; Westoby *et al.* 1982; Andersen 1988b,c) appear to be particularly important. Finally, as predators, prey items and competitors, ants have a major influence on the distribution and abundance of other faunal groups, including vertebrates. This can also have important indirect effects; for example, by regulating the composition and abundance of phytophagous insects, ants can play a key role in herbivory (reviewed by Buckley 1982; Beattie 1985).

The diverse array of invertebrates that mimic ants, such as spiders, hemipterans, beetles, mantids and other ants, attests to their functional importance in

TABLE 1. Suggested uses of ants as bio-indicators in Australia.

Bio-Indicator Of	Location	Reference
Minesite rehabilitation	Western Australia	Majer *et al.* 1982
		Majer *et al.* 1984
		van Schagen 1986
	Northern Australia	Majer 1984
	Queensland	Majer 1985
Fire management	Jarrah forests, W. A.	Majer 1980
		Whelan *et al.* 1980
	Kakadu NP, N.T.	Andersen 1987b
Pesticide contamination	N.S.W.	Weir 1978
Habitat disturbance	Mt Lofty Ranges, S.A.	Yeatman & Greenslade 1980
	Rotamah Is., Vic.	Andersen & McKaige 1987
Reserve design	Wog Wog, N.S.W.	Margules 1985

Australian ecosystems (e.g., Greenslade & Halliday 1983). Indeed, it is difficult to think of any other invertebrate group that challenges ants in terms of either their overall importance, or the variety of ecological roles they play. In terms of abundance, diversity and functional importance, ants are therefore ideal candidates for use as bio-indicators.

Although few detailed studies have been conducted, there is evidence of links between the responses of ants to environmental change on the one hand, and the responses of other faunal groups on the other. For example, a study of 30 rehabilitated bauxite mines showed that ant species richness was positively correlated with the richness of other major invertebrate groups such as termites and springtails (Majer 1983). Ant species richness was also correlated with total invertebrate richness on vegetation, as well as with the abundance of many individual taxa. Similarly, ant succession following sand mining in Myall Lakes National Park was matched by small mammal succession (B. J. Fox, pers. comm.).

Another advantage of using ants as bio-indicators is the sensitivity and speed at which they can respond to environmental change. The so-called "relative abundance" of an ant species actually only refers to the numbers of foraging workers, and not necessarily to total colony size. Only a proportion of total workers in an ant colony typically forage at any given time, and this proportion can vary markedly according to environmental conditions. Changes in the relative abundance of different species can therefore be made extremely rapidly through adjustments in the proportion of total workers foraging, without requiring any mortality or (population) recruitment (Andersen & Yen 1985; Andersen 1988a).

Practicality of using ants

No matter how functionally important a potential bio-indicator group is, it will not be effective unless it can be readily sampled and processed. In this sense, ants are also superior to other insect groups.

In the first place, ants are readily sampled with minimal effort and without resort to specialized techniques or equipment. The vast majority of Australian ant species nest in the ground and forage actively on the soil surface (Greenslade 1979), which makes them ideally suited to capture in pitfall traps (Greenslade 1973). Pitfall trapping has been used in almost all Australian studies where information on the relative abundance of ant species has been sought. Some insect taxa, such as beetles, are attracted differentially to the alcoholic preservatives used in traps, so that trapping results can be unreliable for them (Briggs 1961; van der Drift 1963; Luff 1968). This, however, is not a problem with ants (Greenslade & Greenslade 1971). Some ant species are differentially susceptible to capture in traps because of their behaviour: in particular, fast-moving species are more "trappable" than slow-moving ones (Greenslade 1973; Andersen 1983a; Marsh 1984). Nevertheless, pitfall trapping generally provides a good indication of the relative abundance of different species on the ground (Greenslade 1973; Andersen 1983a, 1986a,b) which, I will argue later, is all that is necessary for the effective use of ants as bio-indicators. In addition to epigaeic species, some habitats support large numbers of cryptic (living predominantly within soil and litter) and, to a lesser extent, arboreal species. These can also be readily sampled if necessary (see later section on "guidelines for using ant communities").

Once sampled, specimens must be sorted, preferably to the species level. This is where ants also have advantages over many other invertebrates, as there are no problems with matching different sexes, or juveniles with adults since all workers are adult females. This can pose serious problems with other major invertebrate groups, such as spiders. Moreover, different ant species are separated by simple morphological features, without resort to the difficult taxonomic characters (e.g., reproductive apparatus) so typical of many other insect groups. This also applies to the separation of ant genera (Greenslade 1979), most of which are very distinctive and, with experience, can be recognized at a glance, often even in the field. The genera of other insect groups, on the other hand, are often only recognized after considerable specialist training.

The taxonomy of most Australian insects at the species level is extremely poorly known (Taylor 1983), and ants are no exception. This has led to the convention of distinguishing different species within a genus by number or letter codes, rather than by name. Although this can be frustrating at times, especially when dealing with the larger genera, as I will argue later it really is not an impediment to the use of ants as bio-indicators. Moreover, even though most species cannot be confidently named, many can be readily assigned to species-groups that have distinctive morphologies, habits and distributions. Determination to the species-group level often can provide important ecological information (e.g., Andersen 1986b), but, in the bio-indicator context, in most cases even this is unnecessary.

Compared with most other invertebrate groups, ants are therefore easy to sample, sort and identify. This makes them amenable to study by workers with little or no specialist entomological training, and consequently practical to use in a standard environmental evaluation system.

Interpreting ant data

There is obviously a potential problem with interpreting quantitative data on diverse communities of poorly known species, as is the situation with Australian ants. One approach has been to characterize ant communities by diversity, evenness and species-similarity indices (Majer 1983, 1984, 1985), but this alone is of limited usefulness. Knowledge that communities are more or less diverse than each other, for example, contributes little to an understanding of the processes underlying their differences, and fundamentally different communities can have similar diversities (e.g., a suburban garden might have similar plant diversity to that of a pristine forest). Clearly, in order to provide interpretable results something must be known of the nature of the species themselves. But how is this possible if the species have never even been collected before?

One of the more powerful tools used to advance our understanding of ecological systems is the "functional group" concept, which recognizes species by their ecological behaviour rather than by their taxonomic affinity. The classification of many species into a small number of functional groups can substantialy reduce the apparent complexity of ecological systems, and has the special advantage of allowing comparisons between communities having little species overlap. The functional group concept is therefore especially suited to environmental evaluation systems, and is now being adopted in them. Functional groups of vertebrates, for example, now serve as a basis for evaluating and predicting environmental change in the Blue Mountains of Oregon and Washington (Thomas *et al.* 1975; Hirsch *et al.* 1979; Thomas 1979, 1982), and elsewhere in the USA (Short & Burnham 1982).

P. J. M. Greenslade has developed a functional group classification of Australian ants that is ideal for interpreting ant data in the context of environmental evaluation (Andersen 1987b). The classification is based primarily on studies of the arid zone (e.g. Greenslade 1978; Greenslade & Halliday 1983) but is widely applicable throughout the country (Greenslade & Thompson 1981; Greenslade 1985b; Andersen 1986a,b; Andersen & McKaige 1987). It: (1) bypasses our ignorance of the biology of individual species because, in almost all cases, it operates at the generic level; (2) simplifies complex ant communities by condensing large numbers of species into a small number of groups; (3) provides valuable insights into the major processes operating within ant communities and their associated habitats; and (4) enables ant communities from throughout Australia to be meaningfully compared.

The classification comprises seven functional groups (Table 2), which can be further classified into three major components (Fig. 1). The most important group consists of dominant Dolichoderinae (group 1), mostly species of the ubiquitous genus *Iridomyrmex*, which, together with associated subordinate Camponotinae (group 2), have a major influence on the distribution and abundance of

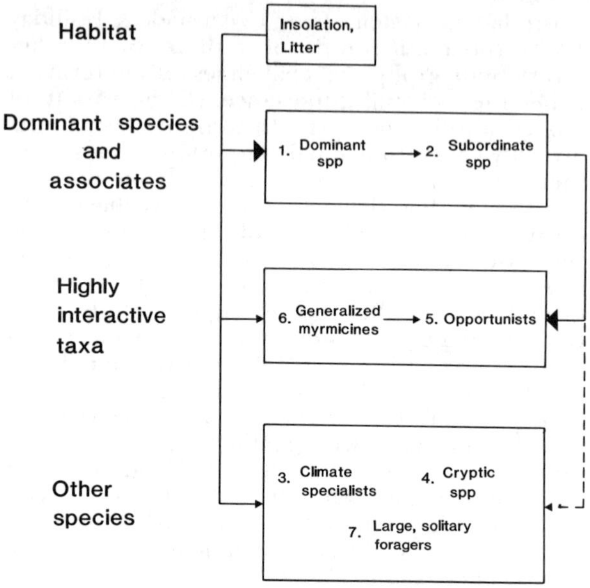

FIG. 1. Simplified Greenslade (1978) model of ant community organization in Australia, showing interactions between habitat variables and functional groups of ants (Table 2). Arrows indicate direction and magnitude of interaction (see text for details) (from Andersen 1987b).

unspecialized taxa ("opportunists" (group 5) and "generalized myrmicines" (group 6)). Species from the remaining groups possess various specializations relating to their body size, distribution and behaviour, reducing their interactions with other ants (Fig. 1).

Many habitat variables potentially influence ants, but the most important appear to be related to vegetation structure (e.g., Majer 1972; Greenslade & Greenslade 1977; Andersen 1982, 1983a, 1986a). As most Australian ants nest and forage on the ground, the structure of the ground layer is particularly important, especially the degree of insolation of the soil surface, and the amount of litter. All ants are influenced to some degree by habitat variables, but, according to Greenslade's model, these variables have a particularly important effect on the abundance of dominant species, which in turn regulates the abundance of highly interactive (unspecialized) taxa (Fig. 1).

Whereas habitat regulates the distribution and abundance of different functional groups within any given region, the regional patterns of ant community organization are broadly related to climate. *Iridomyrmex* is especially prominent in arid, semi-arid and seasonally-arid regions where the high temperatures and open habitats are especially favourable for the high activity and rapid recruitment upon which their dominance depends. Most generalized myrmicines also prefer these conditions, with species of *Monomorium* being especially successful in them despite fierce competition from

TABLE 2. P. J. M. Greenslade's functional groups of Australian ants (with modifications to groups 3, 4, and 7; see text for details).

Group	Common taxa*	Relevant features
1. Dominant Dolichoderinae	*Iridomyrmex*	Highly abundant, active and aggressive; able to monopolize resources
2. Associated subordinate Camponotinae	*Camponotus* *Polyrhachis*	Always co-occurring with *Iridomyrmex*, to which they are competitively subordinate
3. (a) Hot climate specialists	*Melophorus* *Meranoplus*	Behavioural and morphological specializations which enable co-existence with *Iridomyrex*
(b) Cold climate specialists	*Prolasius* *Notoncus*	Restricted to cool and wet regions where the influence of *Iridomyrmex* is reduced.
4. (a) Cryptic species	Many small ponerines and myrmicines	Foraging exclusively within soil and litter; eyes minute or absent
(b) Sub-cryptic species	Many small formicines and dolichoderines	Foraging mostly within soil and litter; eyes larger
5. Opportunists	*Rhytidoponera* *Paratrechina*	Extremely unspecialized behaviour; poor competitors
6. Generalized myrmicines	*Monomorium* *Pheidole* *Crematogaster*	Unspecialized behaviour, but successful competitors due to rapid recruitment and effective defences
7. Large, solitary foragers and/or specialist predators	*Myrmecia* *Leptogenys*	Unlikely to interact much with other ants because of large body size, low population densities and/or specialized diet

*See Table 5 for a more comprehensive list.

Iridomyrmex. This is because they too are capable of rapid recruitment, and are able to defend resources by virtue of their topically-applied venom alkaloids (Jones *et al.* 1988). In hot and dry regions, species of *Iridomyrmex* and generalized myrmicines (mostly *Monomorium*) combined, commonly comprise about 70% of total pitfall catches (Andersen 1983a, 1984.).

Iridomyrmex is less diverse, abundant and influential in the cooler and wetter regions of southern Australia, where cold-climate species and opportunists are typically the most common epigaeic ants (Andersen 1983b, 1986a,b, 1988a). Cryptic species are often also common in these regions (Andersen 1986a,b; Greenslade 1986), and are increasingly a major component of the forest ant fauna with decreasing latitude.

Within any climatic region, the relative abundance of different funtional groups varies predictably with vegetation type (Fig. 2) and level of disturbance (Tables 3 and 4). These two factors determine the habitat variables which influence ant abundance both directly, and indirectly through their effects on *Iridomyrmex* (Fig. 1). A notable feature of habitat disturbance in mesic regions of southern Australia is that it often appears to cause a breakdown in dominance by *Iridomyrmex*, leading to an increase in abundance of unspecialized taxa. For example, opportunists and/or generalized myrmicines consistently increase in relative abundance following fire, an effect that can persist for several years (Table 3). Opportunist species of *Rhytidoponera* appear to be especially sensitive indicators of disturbance in these regions (Table 4).

Fire can have different effects on patterns of ant community organization in other climatic regions. Frequent burning of tropical dry forests in northern Australia, for example, increases the abundance of *Iridomyrmex* because it perpetuates an open habitat favourable for *Iridomyrmex* activity (A. N. Andersen unpub. data). However the abundance of *Iridomyrmex* is reduced at more severely disturbed sites, where unspecialized species of *Monomorium, Tetramorium, Cardiocondyla, Rhytidoponera* and *Odontomachus* often predominate. These sites are also frequently colonized by introduced tramp species such as *P. aratrechina longicornis, Monomorium pharaonis, Solenopsis geminata, Tetramorium*

TABLE 3. Changes in relative abundance of opportunist species (O, belonging to the genera *Rhytidoponera* and *Paratrechina*) and generalized myrmicines (GM, mostly *Monomorium, Pheidole* and *Crematogaster*) following fire in mesic southern Australia.

	% Abundance unburnt	burnt	Time since fire	Site (Reference)
O	12	27		Heathland; Wilson's Promontory
GM	<1	15	4-24 months	Vic. (Andersen 1988a)
O+GM	12	42		
O	14	14		Woodland, Wilson's Promontory,
GM	3	17	9 weeks	Vic. (Andersen 1988a)*
O+GM	17	31		
O	16	29		Woodland; Rotamah Is. Vic.
GM	37	42	7 years	(Andersen & McKaige 1987)
O+GM	53	71		
O	19	56		Woodland; Kojonup W.A.
GM	33	33	2 weeks	(Majer 1980, unpub. data)**
O+GM	52	89		
O	0	5		Woodland; Jandakot,
GM	1	2	2 days	W.A. (Whelan *et al.* 1980)
O+GM	1	6		

* Results of pitfall trapping for a week immediately before the fire, compared with a week 9 weeks after.

** Results of trapping 6 weeks before fire, compared with 2 weeks after.

bicarinatum, Tapinoma melanocephalum and *Pheidole megacephala* (A. N. Andersen, unpub. obs.).

Disturbance can also have more subtle effects by altering the relative abundance of different species within a functional group. This especially applies to *Iridomyrmex*, where one species can be replaced by another following disturbance (*cf.* Fox & Fox 1982).

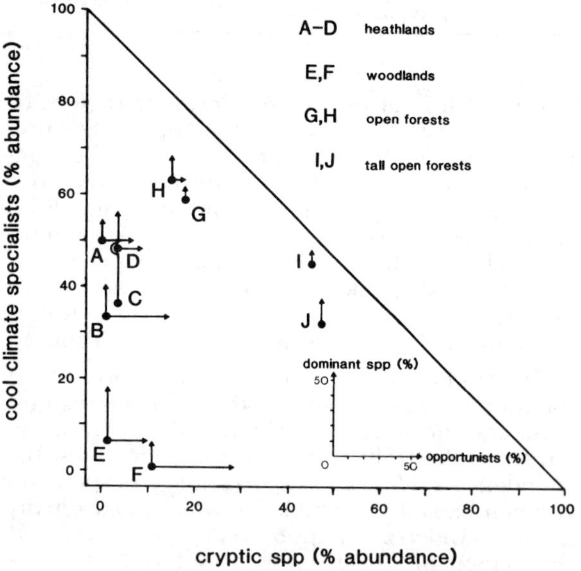

FIG. 2. Ordination of different habitats at Wilson's Promontory in southeastern Australia, using a vector diagram (Lamacraft 1979) based on the relative abundance of the major functional groups of ants (data from Andersen 1986b). The relative abundance of these groups varies predictably with vegetation type.

Guidelines for using ants as bio-indicators

I have argued that ants are ideal bio-indicators of environmental change, with a functional group classification providing an ecologically credible system for habitat evaluation. However, if any habitat evaluation system is to be widely adopted by land management agencies, it must be easy to use as well as ecologically credible. With this in mind, the following guidelines should enable workers with little or no specialist training (but in collaboration with expert advice) to make successful use of ants as bio-indicators. The basic experimental design is assumed to be one where nearby sites have experienced different land management practices (burnt at different frequencies, rehabilitated for different periods of time, etc.), including at least one control site which is as undisturbed as possible. The task is to take one-off measurements at these sites in order to assess the ecological consequences of the land management practices. This is the most simple design possible; obviously a much more robust and preferable approach is to assess the sites before the land management practices are implemented, then to monitor their changes afterwards.

Site selection

Because ants are highly senstive to environmental variables, different species are notoriously patchily distributed within any habitat (Andersen & Greenslade, unpub. data), even when the habitat is apparently homogeneous (Andersen 1982). This makes it highly advisable to have at least two control sites. It is, of course, always preferable to have

TABLE 4. Responses of opportunist species of *Rhytidoponera* to habitat disturbance in southern Australia. Species relative abundance at disturbed sites, or following disturbance (B), is compared with that at undisturbed sites, or prior to disturbance (A).

Species	Type of disturbance	Location	% Abundance A	% Abundance B	Reference
tasmaniensis	Fire	Wilson's Prom. Vic.			
		heathland	4	23	Andersen 1988a
		woodland*	<1	10	
		Rotamah Is., Vic.	16	24	Andersen & McKaige 1987
	Grazing	Rotamah Is., Vic.	16	56	Andersen & McKaige 1987
metallica	Recreation	Mt Lofty Ranges, S.A.	***		Yeatman & Greenslade 1980
inornata	Fire	Kojonup, W.A.	17	56	Majer 1980, unpub. data**
inornata + *violacea*	Fire	Jandakot, W.A.	0	5	Whelan *et al.* 1980

*,** as in Table 3.
*** data not provided, but relative abundance claimed to be correlated with level of disturbance.

TABLE 5. Classification of Australian ant genera into functional groups, following P. J. M. Greenslade's generalized model of ant community organization in Australia (see Table 2 and text). 1 Dominant Dolichoderinae. 2 Associated subordinate Camponotinae. 3a Hot climate specialists. 3b Cold climate specialists. 4a Cryptic species. 4b Sub-cryptic species. 5 Opportunists. 6 Generalized myrmicines. 7 Large, solitary foragers and/or specialist predators. The list excludes many (mostly tropical) genera that are poorly known, but which are rarely common.

MYRMECIINAE		DORYLINAE		DOLICHODERINAE	
Myrmecia	7	*Aenictus*	4a	*Bothriomyrmex*	4b
				Dolichoderus	3b
PSEUDOMYRMECINE		LEPTANILLINAE		*Froggattella*	1
Tetraponera	3a	*Leptanilla*	4a	*Iridomyrmex*	1
				I. darwinianus gp.	4b
PONERINAE		MYRMICINAE		*Leptomyrmex*	1
Amblyopone	4a	*Adlerzia*	6	*Tapinoma*	4b
Anochetus	7	*Anisopheidole*	6	*Technomyrmex*	?5
Bothroponera	7	*Aphaenogaster*	?5	*Turneria*	1
Brachyponera	?4b	*Cardiocondyla*	?5		
Cerapachys	3a	*Colobostruma*	7	FORMICINAE	
Cryptopone	4a	*Crematogaster*	6	*Acropyga*	4a
Diacamma	7	*Epopostruma*	7	*Calomyrmex*	2
Discothyrea	4a	*Glamyromyrmex*	4a	*Camponotus*	2
Heteroponera	?4b	*Mayriella*	?5	*Echinopla*	2
Hypoponera	4a	*Meranoplus*	3a	*Melophorus*	3a
Leptogenys	7	*Mesostruma*	7	*Myrmecorhynchus*	3b
Mesoponera	4a	*Monomorium*(part)	6	*Notoncus*	3b
Mystrium	4a	*M. ('Chelaner')*	3a,b	*Notostigma*	2
Odontomachus	5	*Oligomyrmex*	4a	*Oecophylla*	1
Onychomyrmex	4a	*Orectognathus*	7	*Opisthopsis*	2
Platythyrea	7	*Pheidole*	6	*Paratrechina* (part)	5
Ponera	4a	*Pheidologeton*	4a	*P. minutula* gp.	4b
Proceratium	4a	*Podomyrma*	3a,b	*Plagiolepis*	4b
Rhytidoponera	5	*Quadristruma*	4a	*Polyrhachis*	2
Sphinctomyrmex	4a	*Solenopsis*	4a	*Prolasius*	3b
Trachymesopus	?4b	*Strumigenys*	4a	*Pseudonotoncus*	3b
		Tetramorium	5	*Stigmacros*	?4b

the various treatments replicated. The activity of most ant species is strongly seasonal, at least in southern Australia, with ant activity in general being low during the cool periods of the year (Briese & Macauley 1980; Andersen 1983a, 1986a). It is therefore preferable for sampling to occur during the warmer parts of the year, when most ants are active.

Sampling

As outlined previously, most Australian ant species forage on the ground and are therefore readily sampled using pitfall traps. I consider the ideal trap diameter to be about 4 cm, but smaller sizes are also effective (Majer 1978; Fox & Fox 1982). Traps are partly filled with an alcoholic preservative, and buried with their rims flush with the soil surface. The suitable time period to leave traps in operation varies with the level of ant activity, and is of course constrained by the time available for field work. In hotter regions, sufficient samples can often be obtained after 24 hours, whereas in cool regions a one-week period might be more appropriate. At least ten traps, and preferably 20, should be used at each site. They should be spaced 5-10 metres apart, and can be arranged in a grid or along transect lines depending on the size and shape of the site.

If arboreal and/or cryptic species are also abundant, it might also be useful to sample them. Arboreal species can be collected directly by hand, or sampled by beating or sweeping. Cryptic species can be sampled using one of the several litter extraction techniques outlined in Southwood (1978).

Processing

Ants should be sorted to the species level, and the numbers of individuals of each species in each trap recorded. Numbers are summed across traps to give total species abundance for a site. Sometimes unusually large numbers of a species are collected in one or two traps simply because the traps happened to be placed near a nest entrance or foraging trail. This can obviously lead to a serious bias in the results, which the user must be alert to. One way to circumvent this problem is to transform raw numbers into abundance classes. Species frequency (i.e., percentage of traps in which each species is recorded) is also a useful measure of relative importance.

Species should be identified to genus using Greenslade (1979), or by seeking expert advice. A complete reference collection of voucher specimens should always be seen by a specialist, who can check the generic identifications and species demarcations, and then deposited in an appropriate institution such as the Australian National Insect Collection (CSIRO Division of Entomology, Canberra) or relevant state museum. In the absence of species names, this is the only way future workers will be able to know what species are being dealt with.

Interpretation

The raw data consist of total numbers of each species recorded in traps at each site. These numbers should be converted to relative abundances, because actual numbers of ants caught in traps can vary from site to site because of differences in litter and ground vegetation, and do not necessarily reflect differences in ant abundance (Williams 1959; Greenslade 1964, 1973). The genera can be assigned to functional groups using Table 5, and Table 6 illustrates a useful and convenient format for presenting the data. A vector diagram based on the relative abundances of the major functional groups (as in Fig 2) is an efficient way of illustrating site relationships, and can provide results comparable to those from sophisticated, computer-based ordination

TABLE 6. Suggested format for presenting data on the relative abundance of different ant species. Species are classified into functional groups (Tables 2 and 5), and for each taxon (usually genus) both its total relative abundance and (in brackets) number of species are given. The data are from Andersen and McKaige (1987).

	Land use		
	Control	Burnt	Grazed
1. Dominant Dolichoderinae			
Iridomyrmex	7(3)	5(1)	3(3)
2. Associated subordinate Camponotinae			
Camponotus	21(1)	4(1)	1(1)
3a. Hot climate specialists			
Melophorus	< 1(1)	1(1)	0
b. Cold climate specialists			
Notoncus	< 1(1)	< 1(1)	0
Monomorium (Chelaner)	0	0	< 1(1)
4a. Cryptic species			
Hypoponera	< 1(1)	0	0
Solenopsis	13(2)	15(1)	1(1)
b. Sub-cryptic species			
Heteroponera	0	< 1(1)	< 1(1)
Iridomyrmex darwinianus gp.	2(2)	1(1)	0
Plagiolepis	< 1(1)	< 1(1)	0
Stigmacros	< 1(1)	0	0
Tapinoma	< 1(1)	1(1)	0
sub-total	4(5)	2(4)	< 1(1)
5. Opportunists			
Rhytidoponera	16(2)	29(3)	70(3)
Paratrechina	0	0	4(1)
sub-total	16(2)	29(3)	74(4)
6. Generalized myrmicines			
Monomorium	6(1)	15(2)	1(1)
Pheidole	31(4)	27(4)	20(4)
sub-total	37(5)	42(6)	21(5)
7. Large, solitary foragers			
Myrmecia	0	< 1(1)	0
TOTAL ANTS	187	273	306
TOTAL SPECIES	21	19	16

techniques using species-level data (Andersen, unpub. data). For environmental evaluation purposes, the best two primary axes are likely to be dominant species on the one hand, and generalized myrmicines and/or opportunists (depending on their abundance) on the other.

The data are now ready for interpretation. As outlined previously, special attention should focus on the opportunist and generalized myrmicine groups, as changes in their relative abundance are often correlated with level of disturbance. In the example given in Table 6, for instance, opportunist species comprised 16% of total ants in traps at a relatively undisturbed site, but this figure was 29% and 74% respectively at nearby burnt and grazed sites. These data demonstrate that fire and in particular grazing have had profound influences on the ant communities. This in turn provides a more general indication of the extent to which the environment has been affected by fire and grazing.

Acknowledgements

I am extremely grateful to P. J. M. Greenslade, A. Kinnear, J. D. Majer and D. Saunders for their comments on the draft manuscript, and to J. D. Majer for permission to use his unpublished data in Tables 3 and 4. This is TERC contribution number 638.

References

Andersen A. N. (1982) Seed removal by ants in the mallee of north-western Victoria. In: *Ant-Plant Interactions in Australia* (ed. R. C. Buckley) pp. 31-44. W. Junk Press, The Hague.

Andersen A. N. (1983a) Species diversity and temporal distribution of ants in the semi-arid mallee region of north-western Victoria. *Aust. J. Ecol.* **8,** 127-37.

Andersen A. N. (1983b) A brief survey of ants in Glenaladale National Park, with particular reference to seed-harvesting. *Victorian Nat.* **100,** 233-7.

Andersen A. N. (1984) Community organization of ants in the Victorian mallee. *Victorian Nat.* **101,** 248-51.

Andersen A. N. (1986a) Diversity, seasonality and community organization of ants at adjacent heath and woodland sites in south-eastern Australia. *Aust. J. Zool.* **34,** 53-64.

Andersen A. N. (1986b) Patterns of ant community organization in mesic south-eastern Australia. *Aust. J. Ecol.* **11,** 87-97.

Andersen A. N. (1987a) Effects of seed predation by ants on seedling densities at a woodland site in S. E. Australia. *Oikos* **48,** 171-4.

Andersen A. N. (1987b) Ant community organization and environmental assessment. In: *The Role of Invertebrates in Conservation and Biological Survey* (ed. J. D. Majer) pp. 43-52. Western Australian Department of Conservation and Land Management Report.

Andersen A. N. (1988a) Immediate and longer-term effects of fire on seed predation by ants in sclerophyllous vegetation in south-eastern Australia. *Aust. J. Ecol.* **13,** 285-93.

Andersen A. N. (1988b) Nest mound soil of the seed dispersing ant *Aphaenogaster longiceps* enhances seedling growth. *Aust. J. Ecol.* **13,** 469-71.

Andersen A. N. (1988c) Dispersal distance as a benefit of myrmecochory. *Oecologia* (Berl.) **75,** 507-11.

Andersen A. N. & Ashton D. H. (1985) Rates of seed removal by ants at heath and woodland sites in south-eastern Australia. *Aust. J. Ecol.* **10,** 381-90.

Andersen A. N. & McKaige M. E. (1987) Ant communities at Rotamah Island Victoria, with particular reference to disturbance and *Rhytidoponera tasmaniensis*. *Proc. Roy. Soc. Vict.* **99,** 141-6.

Andersen A. N. & Yen A. L. (1985) Immediate effects of fire on ants in the mallee of northwestern Victoria. *Aust. J. Ecol.* **10,** 25-30.

Ashton D. H. (1979) Seed harvesting by ants in forests of *Eucalyptus regnans* F. Muell in central Victoria. *Aust. J. Ecol.* **4,** 265-77.

Beattie A. J. (1985) *The Evolutionary Ecology of Ant — Plant Mutualisms.* Cambridge University Press, Cambridge.

Bequaert J. (1922) Ants in their diverse relations to the plant world. *Bull. Amer. Mus. Nat. Hist.* **45,** 333-584.

Berg R. Y. (1975) Myrmecochorous plants in Australia and their dispersal by ants. *Aust. J. Bot.* **23,** 475-508.

Berg R. Y. (1981) The role of ants in seed dispersal in Australia lowland heathland. In: *Heathlands and Related Shrublands: Analytical Studies* (ed. R. L. Specht) pp. 41-50. Elsevier, Amsterdam.

Briese D. T. (1982) Relationship between the seed-harvesting ants and the plant community in a semi-arid environment. In: *Ant-Plant Interactions in Australia* (ed. R. C. Buckley) pp. 11-24. W. Junk Press, The Hague.

Briese D. T. & Macauley B. J. (1980) Temporal structure of an ant community in semi-arid Australia. *Aust. J. Ecol.* **5,** 121-34.

Briggs J. B. (1961) A comparison of pitfall trapping and soil sampling in assessing populations of two species of ground beetles (Col: Carabidae). *Rep. E. Mall. Res. Sta.* 1960, 108-12.

Brown W. L. & Taylor R. W. (1970) Hymenoptera: superfamily Formicoidea. In: *The Insects of Australia* pp. 951-9. CSIRO Division of Entomology, Melbourne.

Buckley R. C. (1982) Ant-plant interactions: a world review. In: *Ant-Plant Interactions in Australia* (ed. R. C. Buckley) pp. 111-41. W. Junk Press, The Hague.

Coyle F. A. (1981) Effects of clearcutting on the spider community of a southern Appalachian forest. *J. Arachnol.* **9,** 285-98.

Davidson D. W. & Morton S. R. (1981a) Myrmecochory in some plants (f. Chenopodiaceae) of the Australian arid zone. *Oecologia (Berl.)* **50,** 357-66.

Davidson D. W. & Morton S. R. (1981b) Competition for dispersal in ant-dispersed plants. *Science* **213,** 1259-61.

Disney R. H. L. (1986) Assessments using invertebrates: posing the problem. In: *Wildlife Conservation Evaluation* (ed. M. B. Usher) pp. 271-93. Chapman and Hall, London.

Erye M. D. (1987) Classification of carabid (Coleoptera) habitats in north-east England. *Bulletin of the British Ecological Society* **18,** 94-7.

Fox M. D. & Fox B. J. (1982) Evidence for interspecific competition influencing ant species diversity in a regenerating heathland. In: *Ant-Plant Interactions in Australia* (ed. R. C. Buckley) pp. 99-110. W. Junk Press, The Hague.

Green R. D. & Askew G. P. (1965) Observations on the biological development of macropores in soils of Romney Marsh. *J. Soil. Sci.* **16,** 342-9.

Greenslade P. & Greenslade P. J. M. (1971) The use of baits and preservatives in pitfall traps. *J. Aust. Ent. Soc.* **10,** 253-60.

Greenslade P. & Greenslade P. J. M. (1987) Ecological strategies in collembola: a new approach to the use of terrestrial invertebrates in environmental assessment. In: *Soil Fauna and Soil Fertility* (ed. B. R. Striganova) pp. 245-52. Nauka, Moscow.

Greenslade P. J. M. (1964) Pitfall trapping as a method for studying populations of Carabidae (Coleoptera). *J. Anim. Ecol.* **33,** 301-10.

Greenslade P. J. M. (1973) Sampling ants with pitfall traps: digging-in effects. *Ins. Soc.* **20,** 345-53.

Greenslade P. J. M. (1978) Ants. In: *The Physical and Biological Features of Kunnoth Paddock in Central Australia* (ed. W. A. Low) CSIRO Division of Land Resources Management Technical Paper No. 4.

Greenslade P. J. M. (1979) *A Guide to Ants of South Australia.* South Australian Museum, Adelaide.

Greenslade P. J. M. (1985a) Pterygote insects and the soil: their diversity, their effects on soils and the problem of species identification. *Quaest. Ent.* **21,** 571-85.

Greenslade P. J. M. (1985b) Preliminary observations on ants *(Hymenoptera: Formicidae)* of forests and woodlands in the Alligator Rivers Region, N.T. *Proc. Ecol. Soc. Aust.* **13,** 153-60.

Greenslade P. J. M. (1986) Ants. In: *Natural History of South Australia's Woodlands and Forests* (ed. E. H. Wallace) pp. 154-69. Government Printer, Adelaide.

Greenslade P. J. M. & Greenslade P. (1977) Some effects of vegetation cover and disturbance on a tropical ant fauna. *Ins. Soc.* **24,** 163-82.

Greenslade P. J. M. & Greenslade P. (1983) Ecology of soil invertebrates. In: *Soils: An Australian Viewpoint* pp. 645-69. CSIRO Division of Soils.

Greenslade P. J. M. & Greenslade P. (1984a) Invertebrates and environmental assessment. *Environment and Planning* **3,** 13-5.

Greenslade P. J. M. & Greenslade P. (1984b) Soil surface insects of the Australian arid zone. In: *Arid Australia* pp. 153-76. Australian Museum, Sydney.

Greenslade P. J. M. & Halliday R. B. (1983) Colony dispersion and relationships of meat ants *Iridomyrmex purpureus* and allies in an arid locality in South Australia. *Ins. Soc.* **30,** 82-99.

Greenslade P. J. M. & Thompson C. H. (1981) Ant distribution, vegetation and soil relationships in the Coolola-Noosa River area, Queensland. In: *Vegetation Classification in Australia* (eds A. N. Gillson & D. J. Anderson), pp. 192-207. CSIRO and Australian National University Press, Canberra.

Hellawell J. M. (1978) *Biological Surveillance of Rivers: A Biological Monitoring Handbook.* Water Research Centre, Stevenage.

Hirsch A., Krohn W. B., Schweitzer D. L. & Thomas C. H. (1979) Trends and needs in federal inventories of wildlife habitat. *Transactions of the North American Wildlife and Natural Resources Conference* **44,** 340-59.

Humphreys G. S. (1981) The rate of ant mounding and earthworm casting near Sydney, N.S.W. *Search* **12,** 129-31.

James A. & Evison L. (eds) (1979) *Biological Indicators of Water Quality.* John Wiley, Chichester.

Jones J. H., Blum M. S., Andersen A. N., Fales H. M. & Escoubas P. (1988) Novel 2-ethyl-5-alkylpyrrolidines in the venom of the Australian ant of the genus *Monomorium. J. Chem. Ecol.* **14,** 34-45.

Lamacraft R. R. (1979) A method of displaying differences in botanical composition. *Aust. J. Ecol.* **4,** 407-79.

Luff M. L. (1968) Some effects of formalin on the numbers of Coleoptera caught in pitfall traps. *Entomologist Mon. Mag.* **104,** 115-6.

Lyford W. L. (1963) Importance of ants to Brown Podzolic soil genesis in New England. *Harvard University. Harvard Forest Papers,* No. 7.

Mahoney C. L. (1976) Soil insects as indicators of use patterns in recreation areas. *J. For.* **74,** 35-7.

Majer J. D. (1972) The ant mosaic in Ghana cocoa farms. *Bull. Ent. Res.* **62,** 151-60.

Majer J. D. (1978) An improved pitfall trap for sampling ants and other epigaeic invertebrates. *J. Aust. Ent. Soc.* **17,** 261-2.

Majer J. D. (1980) Report on a study of invertebrates in relation to the Kojonup Nature Reserve Fire Management Plan. *Western Australian Institute of Technology School of Biology Bulletin No. 2.*

Majer J. D. (1982) Ant-plant interactions in the Darling Botanical District of Western Australia. In: *Ant-Plant Interactions in Australia* (ed. R. C. Buckley) pp. 45-61. W. Junk Press, The Hague.

Majer J. D. (1983) Ants: bio-indicators of minesite rehabilitation, land-use, and land conservation. *Environ. Manage.* **7,** 375-83.

Majer J. D. (1984) Recolonisation by ants in rehabilitated open-cut mines in northern Australia. *Reclam. Reveg. Res.* **2,** 279-98.

Majer J. D. (1985) Recolonization by ants of rehabilitated mineral sand mines on North Stradbroke Island, Queensland, with particular reference to seed removal. *Aust. J. Ecol.* **10,** 31-48.

Majer J. D., Day J. E., Kabay E. D. & Perriman W. S. (1984) Recolonization by ants in bauxite mines rehabilitated by a number of different methods. *J. Appl. Ecol.* **21,** 355-75.

Majer J. D., Sartiori M., Stone R. & Perriman W. S. (1982) Recolonization by ants and other invertebrates in rehabilitated mineral sand mines near Eneabba, Western Australia. *Reclam. and Reveg. Res.* **1,** 63-81.

Margules C. R. (1985) The Wog Wog habitat patch experiment: background, objectives, experimental design and sample strategy. Technical Memorandum 85/18, CSIRO Division of Water and Land Resources.

Marsh A. C. (1984) The efficacy of pitfall traps for determining the structure of a desert ant community. *J. Ent. Soc. Sth. Afr.* **47,** 115-20.

Matthews E. G. & Kitching R. L. (1984) *Insect Ecology.* University of Queensland Press, Brisbane.

Mawson P. R. (1986) A comparative study of arachnid communities in rehabilitated bauxite mines. *Western Australian Institute of Technology School of Biology Bulletin No. 14.*

Nichols O. G. & Bunn S. (1980) Termite utilization of rehabilitated bauxite mined areas. *Alcoa of Australia Limited Environmental Research Bulletin No. 9.*

O'Dowd D. J. & Gill A. M. (1984) Predator satiation and site alteration: mass reproduction of alpine ash *(Eucalyptus delegatesis)* following fire in south-eastern Australia. *Ecology* **65,** 1052-66.

Pearsall S. H., Durham D. & Eagar D. C. (1986) Evaluation methods in the United States. In: *Wildlife Conservation Evaluation* (ed. M. B. Usher) pp. 110-33. Chapman and Hall, London.

Puszkar T. (1979) The effect of sulphur industry on epigeic and soil fauna. *Memorabilia Zoologica* **32,** 101-18.

Salwasser H. & Tappeiner J. C. (1981) An ecosystem approach to integrated timber management and wildlife habitat management. *Transactions of North American Wildlife and Natural Resources Conference* **46,** 473-87.

Short H. L. & Burnham K. P. (1982) Technique for structuring wildlife guilds to evaluate impacts on wildlife communities. United States Department of Industry Fish and Wildlife Service, Fort Collins, Colorado, Special Scientific Report, 244.

Southwood T. R. E. (1978) *Ecological Methods* (second edition). Chapman and Hall, London.

Taylor R. W. (1983) Descriptive taxonomy: past, present and future. In: *Australian Systematic Entomology: A Bicentenary Perspective* (eds E. Highly & R. W. Taylor) pp. 93-134. CSIRO, Melbourne.

Thomas J. W. (ed.) (1979) *Wildlife Habitats in Managed Forests — The Blue Mountains of Oregon and Washington.* United States Department of Agriculture Handbook No. 553.

Thomas J. W. (1982) Needs for and approaches to wildlife habitat assessment. *Transactions of the North American Wildlife and Natural Resources Conference* **47,** 35-46.

Thomas J. W., Miller R. J., Black H., Rodiek J. E. & Maser C. (1975) Guidelines for maintaining and enhancing wildlife habitat in forest management in the Blue Mountains of Oregon and Washington. *Transactions of the North American Wildlife and Natural Resources Conference* **41,** 452-76.

van der Drift J. (1963) A comparative study of the soil fauna in forests and cultivated land on sandy soil in Suriname. In: *Studies on the Fauna of Suriname and other Guyana* (eds D. C. Geijskes & P. W. Hummelink) Vol. 6. Uitg. Natuurw. Stud-dring Suriname **32,** 1-42.

van der Ploeg S. W. F. (1986) Wildlife conservation evaluation in the Netherlands: a controversial issue in a small country. In: *Wildlife Conservation Evaluation* (ed. M. B. Usher) pp. 161-80. Chapman and Hall, London.

van Horne B. (1983) Density as a misleading indicator of habitat quality. *Wildl. Mgmt.* **47,** 893-901.

van Schagen J. (1986) Recolonization by ants and other invertebrates in rehabilitated coal mine sites near Collie, Western Australia. *Western Australian Institute of Technology School of Biology Bulletin No. 13.*

Weir J. S. (1978) The ant, *Iridomyrmex,* as a biological indicator of pesticide contamination. Report to the New South Wales State Pollution Control Commission.

Westoby, M., Rice B., Shelley J. M., Haig D. & Kohen J. L. (1982) Plant's use of ants for dispersal at West Head, New South Wales. In: *Ant-Plant Interactions in Australia* (ed. R. C. Buckley) pp. 75-87. W. Junk Press, The Hague.

Whelan R. J., Landedyk W. & Pashby A. S. (1980) The effects of wildfire on arthropod populations in jarrah — *Banksia* woodland. *West. Aust. Nat.* **14,** 214-20.

Williams G. (1959) The seasonal and diurnal activity of the fauna sampled by pitfall traps in different habitats. *J. Anim. Ecol.* **28,** 1-13.

Yeatman E. M. & Greenslade P. J. M. (1980) Ants as indicators of habitats in three conservation parks in South Australia. *S. A. Nat.* **55,** 20-6, 30.

Yen A. L. (1987) A preliminary assessment of the correlation between plant, vertebrate and Coleoptera communities in the Victorian mallee. In: *The Role of Invertebrates in Conservation and Biological Survey* (ed. J. D. Majer) pp. 73-88. Western Australian Department of Conservation and Land Management Report.

Proc. Ecol. Soc. Aust. 1990 — 16:359-367

The value of road reserves to the avifauna of the central wheatbelt of Western Australia

P. CALE

260 Selby Street, Floreat Park, Western Australia 6014

Abstract

The value of road reserve vegetation to the bird communities of the Kellerberrin district of Western Australia, was investigated in 28 sections of nine road reserves. These sections were censused for bird species over a 12-month period, and measures of their physical and biotic attributes were recorded. A total of 60 species representing 88% of the Kellerberrin district's terrestrial avifauna was recorded in road reserves during this study.

The results of this study show that the species complement of a road reserve is determined by the vegetation type present and that species richness is influenced by different factors in each vegetation type. Multiple regression analysis showed that the density of vegetation below 1 m, the mean width of the vegetation, and the density of the vegetation above 8 m, were positively correlated with the average bird species richness per section in woodland road reserve sections. In mallee/shrubland road reserve sections only the mean width of the vegetation significantly influenced bird species richness.

Observations indicate that management actions designed to enhance the conservation value of road reserves, based on maximizing species richness, are insufficient. The temporal and spatial requirements of individual species must also be considered.

Introduction

The Kellerberrin district of the central wheatbelt of Western Australia (31-32°S; 117-118°E) was settled in the 1860s, but it was not until after the opening of the railway in the 1890s that major clearing of the native vegetation began. By 1930 approximately 70% of the area was cleared, this percentage had risen to 90% by 1960, and 94% by 1968. Since that time little major clearing has occurred. This pattern of development is representative of much of the central wheatbelt (Saunders & Curry, this volume).

In an area of 1760 km^2 to the north of Kellerberrin, only 5.7% of the land is uncleared and only 20% of this uncleared land is vested in the Crown, the rest being privately owned (Saunders *et al.* 1987). There are 600 km of roads in this area and the road reserves often contain strips of native vegetation that form verges along the roads (Saunders *et al.* 1987). All of these reserves are vested in the local shire councils or the Main Roads Department. Those managing the road reserves try to maintain the native vegetation, and many still maintain the general physiognomy of the original vegetation, but it is often of poor quality, being heavily invaded by weeds and highly disturbed by livestock and human activities. The conservation value of this roadside vegetation has rarely been considered.

Road reserves provide two important contributions to the conservation of the avifauna. They may act as corridors linking the remaining remnants of native vegetation and they may provide areas of suitable habitat in their own right. Many of the road reserves do physically link remnants of native vegetation but little is known about their use by birds as corridors. However, it seems likely that at least some species are using them for this purpose (Saunders 1977; Newbey & Newbey 1987).

There are few studies in Australia on the value of road reserves as habitat for native animals. Arnold *et al.* (1987) showed that road reserves in the Kellerberrin district were used by a large number of bird species (41) during winter. They showed that species richness was influenced mainly by the width of the strip of native vegetation and that the number of individuals of a given species was influenced by the structure of the vegetation and the width of the reserve. Small passerines also increased in numbers in sections close to remnants of native vegetation, indicating some form of relationship between these remnants and the road reserves.

In a 2 km long section of road reserve near Ongerup, in the wheatbelt, Newbey and Newbey (1987) showed that a large turnover of bird species occurred throughout the year. This they attributed to two factors. Some species used the reserve as a corridor and other species used the short term food resources of the reserve during the flowering season.

This chapter presents the preliminary findings of a study of the avifauna of the Kellerberrin district, with respect to the conservation of road reserves. It also proposes possible future directions in management research that may lead to the enhancement of road reserves as conservation areas.

Study area

The study area is situated to the north of the wheatbelt town of Kellerberrin approximately 250 km east of Perth. It is a 40 km×40 km area with the Great Eastern Highway forming the southern boundary and the Bencubbin/Kellerberrin Road bisecting it from north to south (Saunders & Curry, this volume: Fig. 1).

The area is part of the Mount Caroline vegetation system described by Beard (1980). It is typical of much of the wheatbelt region in that it shows an intricate mosaic character, with many vegetation types mingled together to form broad divisions. Four broad vegetation units are represented in the study area: woodlands dominated by *Eucalyptus salmonoploia*, *E. wandoo* and *E. salubris;* kwongan (heath, thicket and scrub) represented by many different formations; mallee which is generally mixed with woodlands and kwongan or forms transitions between the two; and granite outcrops which are patchily vegetated and generally small in area.

The road reserves in the district vary in width from 20 m to 60 m and generally contain a 10 m wide road with 5-25 m of vegetation on either side. Some reserves (3.6% of censused sections) have vegetation on only one side of the road. Frequently (32% of sections under study) the vegetation of the road reserve is widened by strips of vegetation bordering the edge of private properties. This can produce a strip of vegetation up to 100 m in width which is bisected by the road.

The vegetation on the road reserves is typical of the vegetation in the area, and shows a high degree of heterogeneity. In most reserves the vegetation is highly disturbed by weed invasion and grazing. This is especially prevalent in woodland sections where, due to this disturbance, the understorey is of poor quality or completely absent. The vegetation of many road reserves has been removed to provide access roads and drainage lines. In addition some road reserves have been used for gravel extraction.

Methodology

Twenty-three sections in eight road reserves were each censused 10 times, and five sections in a single reserve were censused eight times between May 1987 and May 1988. Sections within a given road reserve were at least 300 m apart and the vegetation between sections was generally of a different type. Only sections of homogeneous vegetation were selected for censusing, because of the difficulties in quantifying the degree of heterogeneity of the vegetation as a whole. Twenty of the sections were 500 m long, seven were 300 m long and one was 400 m long. The varying lengths were a result of short reserves between two remnants being censused as a unit, as well as the difficulties of finding sections that had homogeneous vegetation.

Each section was censused at mid-afternoon by walking slowly along it's length. For narrow reserves (<30 m) this generally involved walking along the road observing both sides, but for the wide reserves each side was traversed separately. Particular care was taken to prevent multiple sightings, by noting any individuals that crossed the road.

During each census a hand-held tape recorder was used to record all sightings, noting the species, the numbers of individuals, and their location in the section. Locations were determined with the aid of markers set every 100 m along the censused sections. All of this information was later transferred to a census sheet and a map indicating any particular features of the reserve section (e.g., any damage caused by human activity).

The presence of any plant species in flower, with a subjective evaluation of the quantity of flowers, and their potential availability as a food source (i.e., early bud, full flower, or old flowers) was also recorded, as was any indication of nesting.

The vegetation in each section of corridor was evaluated by recording a subjective estimate of the density of vegetation in each of five height classes (<1 m, 1-2 m, 2-4 m, 4-8 m and 8 m+). Density estimates were divided into the four categories defined by Specht (1970) (<10%, 10-30%, 30-70% and 70-100%). Although some sections had minor variations in vegetation along their length, these were averaged to give a single estimate for analysis. Two sections had areas of such divergent vegetation that they could not be averaged and so have been excluded from any analysis involving vegetation estimates.

Stepwise multiple regression analysis was used to evaluate the effect of changes in the mean width of vegetation, the density of vegetation in each height class, the density of vegetation in height classes <1 m, 1-2 m, and 2-4 m combined (UNST), and the distance to the nearest remnant on the average bird species richness per census of woodland, mallee/shrubland and shrubland-heath (the latter combined due to insufficient sections for individual analysis) sections of road reserve.

Five shrubland and heath sections of road reserve that contained a high density of plants with flowers that were used by honeyeaters as a nectar source, were combined for a comparison of temporal changes in honeyeater numbers, with the abundance of flowers. Spearman's rank correlation coefficient (Zar 1974), was used for these comparisons.

To determine the degree of temporal utilization of the road reserve by individual species, an index of utilization was created. This index is the percentage of the possible censuses during which the species was recorded. The number of possible censuses was based on the number of censuses made in sections where the species was found. This value was modified for migratory species by only including censuses made during the period they were present in the region.

The index was calculated for each species in the five major vegetation types found in the road reserves, woodland, mallee/shrubland, and three types of Kwongan; heath, *Allocasuarina* shrubland, and shrubland. The woodland and mallee/shrubland sections were divided: woodland into those with an understorey (value of UNST>4) and those without; and mallee/shrublands into open mallee (value of height class 4-8 m<2) and closed mallee.

Results

Comparison with district avifauna

A total of 56 species of bird was recorded in the 28 road reserve sections. Eight species were recorded only once and a further seven species were recorded only twice. The remaining 41 species were recorded in at least five different sections (Appendix 1). An additional four species (wedge-tailed eagle, brown falcon, spotted harrier and white-backed swallow) were seen on many occasions in road reserves other than those censused and so are included in the following comparison.

A comparison of this species complement with the avifauna recorded for the Kellerberrin district (Saunders & Curry, this volume), shows that 88% of species are represented in road reserves. This comparison excludes vagrants in the district, rare species that have not been recorded in the local area and are presumed locally extinct and all waterbirds and nocturnal species as these were not considered in censuses. If the species that were seen only once or twice in road reserves are removed, the representation of the district's avifauna is reduced to 65%. When the total species complement is broken down on the basis of the relative abundance of species as determined by Saunders and Curry (this volume), the rare species are not well represented (54%), while uncommon (96%), common (96%) and very common species (100%) are almost all represented.

Species utilization of vegetation types

The degree of utilization of the road reserves by individual species is highly dependent on vegetation type (Appendix 1). Only a few species are ubiquitous (Port Lincoln parrot, singing honeyeater, yellow-throated miner, black-faced woodswallow, mulga parrot, crested pigeon and red-capped robin), and even these show some variation in their use of different vegetation types.

None of the species found in woodlands is restricted to those without an understorey, but a number of species (little falcon, Australian magpie, Australian magpie lark, galah, striated pardalote, crested pigeon and pallid cuckoo) were recorded in these woodlands to a greater extent than those with an understorey. Only the crested pigeon made use of the open ground; all of the others were seen almost exclusively in the canopy.

The addition of an understorey to woodland sections adds a new complement of species (willie wagtail, grey fantail, yellow-rumped thornbill and Western warbler). Willie wagtails and yellow-rumped thornbills are common in all vegetation types with a shrub layer, while grey fantails and Western warblers appear to use the middle layers of vegetation. Two species of honeyeater (singing and white-eared honeyeaters) were also seen more frequently in these woodland sections, than those without an understorey, but only the singing honeyeater appeared to actually use the shrub layer.

Many species that were observed occupying mainly the canopy and middle vegetation layers in the road reserve sections (weebill, yellow-throated miner, rufous whistler, grey butcherbird, grey fantail and Western warbler) showed a marked increase in the use of closed mallee as compared to the open mallee sections.

Multiple regression

In woodland sections three factors had significant partial correlation coefficients when analysed by stepwise multiple regression analysis. These formed the following relationship with average bird species richness per census *(S)*:

$$S=1.0V1(0.15)+0.07W(0.01)+1.38V5(0.40)-2.57$$

where *V1* is the density of vegetation in the height class <1 m, *W* is the width of vegetation in the road reserve, and *V5* is the density of vegetation in the height class 8 m+ (standard errors shown in parentheses). This regression explained 96.9% of the variation in *S* (F=72.8, P<0.002).

For the mallee/shrubland sections the following regression was obtained:

$$S=0.04W(0.01)+3.81$$

where *W* is the width of vegetation in the road reserve (standard error shown in parentheses). This regression explained 65.6% of the variation in *S* (F=18.2, P<0.01).

For the combined shrubland-heath sections no significant correlations were obtained. This may have been a result of combining the different vegetation types.

Differences in the length of censused sections of road reserve appears to have little effect on the bird species richness per census.

Temporal changes in species numbers

Red-capped robins showed a cycle in the number of individuals in 23 sections from a high in May 1987 (average 6.5 individuals), to a low in September/October 1987 (average one individual), and returned to a high in May 1988 (six individuals) (Fig. 1).

The temporal change in the mean number of singing honeyeaters per shrubland-heath section, was significantly correlated with the quantity of flowers in these sections (r_s=0.788, P<0.05) (Fig. 2a). This was also the case for the mean number of brown honeyeaters per section, in the same shrubland-heath sections (r_s=0.942, P<0.005). Insufficient numbers of individuals of other species of honeyeaters prevented evaluation of their responses to changes in the availability of flowers.

The correlation between singing honeyeater numbers and the availability of flowers results in a cyclical pattern in shrubland-heath sections of road reserve. A peak in numbers occurred in August 1987, and lows occurred in May 1987 and January/February 1988 (Fig. 2a). A reversed cyclical pattern occurred over the same period in mallee/shrubland sections of road reserve, with peaks in May 1987 and January/February 1988, and a low in September 1987 (Fig. 2b). The correlation between the cylical patterns of singing honeyeater numbers, in shrubland-heath and mallee/shrubland sections, is not significant however (r_s=0.53, P<0.1).

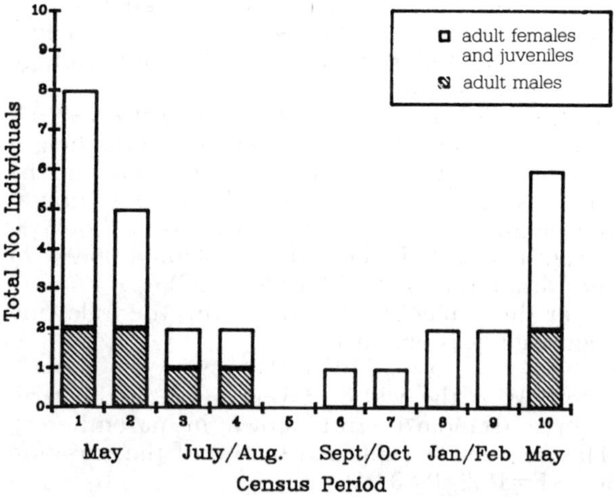

FIG. 1. Temporal changes in the total number of red-capped robins in 23 road reserve sections, from May 1987 to May 1988.

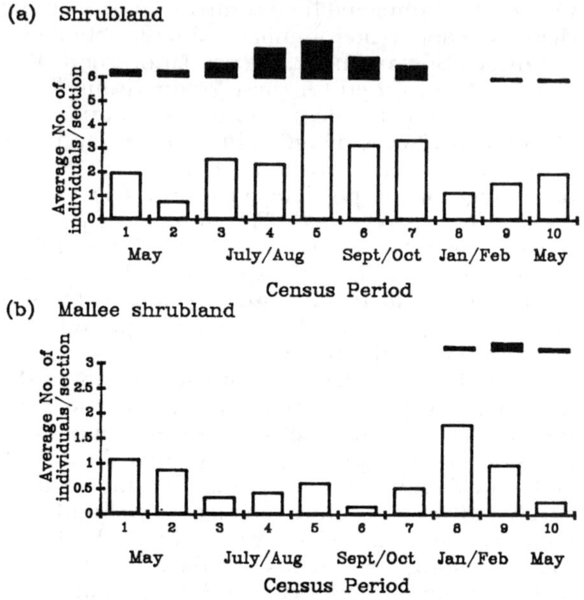

FIG. 2. Temporal changes in the mean number of singing honeyeaters per section, and the availability of flowers in; (a) five shrubland-heath sections, which contain plant species utilized by honeyeaters for nectar, and (b) 10 mallee/shrubland sections, from May 1987 to May 1988. The black bars are a measure of the availability of flowers to honeyeaters as a nectar source.

A cyclical change in foraging activity corresponding to the availability of flowers occurs in singing honeyeaters. During September/October the majority of foraging records are of nectar foraging (60% nectar/40% insects, n=15 records), while nectar foraging and insect foraging are approximately equal in July/August (53% nectar/47% insects, n=15 records) and January/February (50% for both, n=12 records). In May 1987 (20% nectar/80% insects, n=15 records), and May 1988 (33% nectar/67% insects, n=9 records), the majority of foraging records were of insect foraging. A Chi-square contingency test shows that the differences in the foraging records from September/October and from May 1987 are statistically significant (X^2=5, P<0.05), but the difference between September/October and May 1988 are not (X^2=1.6, P<0.25). This cyclical change is not observed in the brown honeyeater where all foraging records throughout the year were exclusively on nectar.

Discussion

Despite the poor quality of the native vegetation present in the majority of road reserves, this network supplies suitable habitat for 88% of the terrestrial species of birds of the Kellerberrin district. One factor that is not considered in this figure is the stability of each species' residence in the road reserve. Determining this for individuals is impossible without some form of individual identification but indirect evidence can be obtained from measures such as the utilization index. This index measures temporal representation of a species and so can be used to eliminate species that were not likely to have been residents of the road reserves. If an arbitrarily chosen value (33) of this index is used, and all species that fail to show a value greater than this in at least one vegetation type are ignored, the results are quite different. Only 38% of the district's avifauna are represented. This suggests that the road reserves do not supply the resources required for many species to reside permanently but are instead used as temporary habitats.

One difficulty with the utilization index results from the very small sample sizes for many species. This is a direct consequence of the low density of birds throughout the area and can only be rectified by greatly increasing the area of road reserve that is censused, or by increasing the number of censuses made. Small sample sizes create a problem in that they reduce the reliability of the utilization index because each recorded sighting has a proportionately greater influence on the value of the index. The species most affected by this are the rare species which are the most important species when considering conservation aspects of road reserves. For these species intensive studies are needed to evaluate their use of road reserves.

What factors influence the number of species that use road reserve vegetation? The values of the utilization index for a given species indicate that vegetation type is of great importance in determining the species complement of a given section of road reserve. Each species has a preference for a given range of vegetation types. It has been found in other bird communities that each species utilizes a range of heights and substrates within the vegetation (Gilmore 1985; Recher & Holmes 1985) and this is probably the case in the present study. Each species' choice of vegetation types may be based on the presence of vegetation within their preferred range of heights.

A problem with using species richness to evaluate the effects of different variables on utilization of different areas is that it does not take into account the identities of the species present. This becomes a problem when comparing areas of different vegetation type because the species in question are not necessarily the same. For this reason the road reserve sections have been divided into major vegetation types for a community-based analysis.

Species richness in woodland sections of road reserve is influenced by the density of vegetation in the understorey and the canopy, and by the width of the vegetation in the road reserve. The increase in species richness with the addition of an understorey is the result of the addition of new species to this vegetation type, and not just an increase in the likelihood of a species being present in a given area of this vegetation type (i.e., not an increase in the level of use of the vegetation type by a species). This was also found by Arnold *et al.* (1987), who recorded that the absence of vegetation under 2 m resulted in the virtual exclusion of small insectivorous passerines.

In theory the enhancement of the vegetation for some species might be to the detriment of others, but this does not appear to be so in the case of woodland road reserves. No species are restricted to woodlands without understorey (Appendix 1) and of those species that utilize such woodlands to a greater extent, only the crested pigeon appears actually to use the open ground. The others are recorded in the canopy and their greater use of this habitat appears to be related to the type of woodland, more than to the absence of an understorey. The woodlands with no understorey that were censused tended to be of mature *Eucalyptus wandoo* and *E. salmonophloia* which supply more of the favoured nesting sites: hollows for galahs and striated pardalotes and a high open canopy for the stick nests of Australian magpies and Australian ravens. These old stick nests are also used by little falcons (Beruldsen 1980).

Only reserve width significantly influenced species richness in mallee/shrubland sections. The vegetation in these sections showed little variation with the coarse measures used in the analysis, and so a true evaluation of vegetation structure on species richness in this vegetation type may not be possible without finer scale analyses. Increases in the utilization index for some species with an increase in canopy density in mallee/shrubland sections of road reserve (Appendix 1) tend to support this view.

Distance to the nearest remnant of native vegetation did not significantly influence species richness in any of the road reserve sections. Arnold *et al.* (1987) found that the area of remnant vegetation around a road reserve section influenced the number of small passerines present. However, the road reserve sections that they studied were selected on the basis of vegetation type alone and some were in road reserves that did not connect to remnants. This was not the case with the present study where all sections are part of road reserves that link two remnants. This may reduce the influence of distance to the remnant because the movement of birds may be facilitated by the continuous vegetation link.

This community-based analysis indicates that the width of the road reserve which is a measure of the area of available habitat, appears to be the most important factor influencing species richness in the road reserve system as a whole. However, the structure of the vegetation in these road reserves also appears to be of importance at least in some vegetation types, but investigations based on finer scale analyses of this factor will be required to determine it's importance in other vegetation types.

This community-based analysis has one major problem. The majority of species using road reserves do so on a short-term basis and the temporal changes occurring in the bird community of a given road reserve are not considered. Temporal changes in utilization can occur as the result of changes in the available resources of the habitat, or by changes in the requirements of a species. They can also be the result of a species using the road reserve only as a corridor for movement, and so will be correlated with the movement patterns of that species. All of these possibilities appear to occur in the road reserve system at Kellerberrin. They are best demonstrated by two examples; the red-capped robin which appears to use the road reserves mainly as corridors, and the singing honeyeater and the brown honeyeater, which appear to shift between habitats, due to changes in the availability of food resources.

Red-capped robins show a yearly cyclical change in the number of individuals in the road reserves, with the lowest numbers being recorded during their breeding season (Fig. 1). The decline prior to the breeding season was also recorded by Arnold *et al.* (1987) for the period between June 3 and July 21, 1986. Newbey and Newbey (1987) also recorded an absence of this species in a single road reserve during the breeding season, and an influx of individuals in the months of April and May 1978 and February and March 1979. I believe that this cyclical pattern in red-capped robin numbers arises from this species mainly using the road reserves as corridors for dispersal and that the increase in numbers prior to the breeding season is the result of dispersal by the previous years juveniles. This hypothesis is supported by the fact that the majority of individuals recorded were not recorded in the next census, indicating that they are in an area for only a short time, and that the majority of recordings were of solitary females or juvenile males (Fig. 1). Since adult and juvenile females cannot be distinguished in the field, many of these individuals could have been juveniles. If this hypothesis is correct then the importance of the road reserves to this species is their physical link to remnants of native vegetation, and a relationship between recordings and vegetation type would not be expected. The observed values of the utilization index support this expectation, with the exception of a marked increase in utilization in *Allocasuarina* shrubland. This high index value, however, is the result of a single reserve section connected to a large remnant.

The second example of temporal changes in the utilization of road reserves is that of two species of honeyeater, the singing honeyeater and the brown honeyeater. These species show differing responses to changes in the availability of nectar in the road reserves. In shrubland-heath road reserve sections, the singing honeyeater shows a cyclical change in numbers that is related to the availability of nectar, as measured by flower abundance. This species also shows a cyclical change in numbers in mallee/shrubland road reserve sections and, though not statistically significant, a negative correlation between the cyclical changes in the two vegetation types is apparent. Corresponding to these observed cyclical changes is a shift in the foraging behaviour from mainly nectar foraging in September/October to foraging mainly on insects in the two May periods. The lack of statistical significance between September/October and May 1988 is probably a result of the low sample size for this May period. Though these data are limited, and so must be considered with caution, their consistency with the other observations should not be ignored.

The population of singing honeyeaters in the Kellerberrin district appears to be comprised of two components; permanent residents which defend territories (Saunders, pers. comm.), and nomads which appear to move throughout the local area temporarily residing in suitable habitats. A similar situation was recorded for New Holland honeyeaters by Paton (1985), and he found that the majority of transients as he termed them, were juvenile birds.

I suggest that the observed relationship between road reserve sections of shrubland-heath and mallee/shrubland vegetation is the result of a seasonal movement by the nomadic sub-population of this species. The territorial residents of road reserves appear to occupy areas that have a high density of suitable flowering plants (i.e., shrublands/heath sections). When nectar supplies are abundant in these areas their territories appear to either breakdown completely or reduce in size, possibly due to the increased abundance of nectar and/or the pressure of increased numbers of individuals resulting from the influx of nomadic singing honeyeaters. A similar situation was observed by Paton (1985) with New Holland honeyeaters. When nectar supplies have been depleted the territorial singing honeyeaters can continue to reside in these road reserve sections, because they can oust the nomadic individuals so reducing the density of birds, and they shift their foraging activities to insect foraging. The ousted nomads move to unoccupied areas (i.e., mallee-shrubland sections) and can reside there temporarily because they are at low densities, and they too shift their foraging activities to insects. This adaptability by singing honeyeaters enables them to utilize the road reserve vegetation throughout the year and to maximize their use of the available habitats.

This is not the case with the brown honeyeater which forages predominantly on nectar throughout the year, a fact also noted by Collins and Briffa (1982). Their numbers increase dramatically in road reserve sections when nectar is available, but when this resource is depleted they move out of the road reserve system entirely. This results in the brown honeyeater being locally nomadic in the Kellerberrin district. This species' use of road reserve habitat is limited to areas of high nectar production (mainly shrubland-heath sections), when that nectar is available.

These two honeyeater species demonstrate two responses to a temporally variable resource, nectar in this case, and indicate the importance of studying a system over an extended period and considering individual species instead of communities or guilds of species, when evaluating the system for management purposes.

The question of management of these road reserves is a complicated one. This study indicates that the species complement of a given area of road reserve is dependent on the type of vegetation present. The species richness can be increased by maximizing the width of the vegetation in the road reserve and, for at least some vegetation types, by increasing the structural diversity of the vegetation. However, maximizing species richness should not be considered the only management objective for these road reserves. The majority of species utilizing the reserves do so only on a temporary basis and an evaluation of the road reserves based on species richness alone fails to consider this. The example of the red-capped robin indicates that the physical position of the reserves can be important, and road reserves that link remnants of native vegetation might be considered more valuable to conservation than other road reserves, because of their use as corridors. The example of the singing honeyeater and the brown honeyeater demonstrate the importance of considering both temporal and spatial changes in individual species requirements. There is also the consideration of the relative importance of different species in conservation efforts. Management that maximizes species richness is of little value if only the common species benefit from the changes. Maximizing the benefits to rare species must be given greater importance in management considerations.

What is required in addition to management based on maximizing species richness, are investigations of individual species. These investigations should concentrate on those species at risk of decline to determine their use of road reserves, so that management can also be based on maximizing the resources required by these species so that they may remain in the road reserve system.

Acknowledgements

This research was funded by the University of Western Australia, the CSIRO Division of Wildlife and Ecology, and a travel grant from the Ecological Society of Australia. I would like to thank Dr C. Dickman, Dr D. Saunders, Dr J. Lynch and Mr D. Cale for their critical review of earlier drafts of this paper.

References

Arnold G. W., Algar D., Hobbs R. J. & Atkins L. (1987) A survey of vegetation and its relationship to vertebrate fauna present in winter on road verges in the Kellerberrin District, W. A. *Department of Conservation & Land Management Technical Report:* **18.** Department of Conservation & Land Management, Perth.

Beard J. S. (1980) The vegetation of the Kellerberrin area Western Australia. 1:250 000 series. *Vegetation Survey of Western Australia.* Vegmap Publications, Perth.

Beruldsen G. (1980) *A Field Guide to Nests and Eggs of Australian Birds.* Rigby, Sydney.

Collins B. G. & Briffa P. (1982) Seasonal variation of abundance and foraging of three species of Australian honeyeaters. *Aust. Wildl. Res.* **9,** 557-69.

Gilmore A. M. (1985) The influence of vegetation structure on the density of insectivorous birds. In: *Birds of Eucalypt Forests and Woodlands: Ecology, Conservation, Management* (eds A. Keast, H. F. Recher, H. Ford & D. Saunders) pp. 21-31. Surrey Beatty & Sons, Chipping Norton.

Newbey B. J. & Newbey K. R. (1987) Bird dynamics of Foster Road Reserve, near Ongerup Western Australia. In: *Nature Conservation: the Role of Remnants of Native Vegetation* (eds D. A. Saunders, G. W. Arnold, A. A. Burbidge & A. J. M. Hopkins) pp. 341-3. Surrey Beatty & Sons, Chipping Norton.

Paton D. C. (1985) Food supply, population structure, and behaviour of New Holland Honeyeaters *Phylidonyris novaehollandiae* in woodland near Horsham, Victoria. In: *Birds of Eucalypt Forests and Woodlands: Ecology, Conservation, Management* (eds A. Keast, H. F. Recher, H. Ford & D. Saunders) pp. 219-30. Surrey Beatty & Sons, Chipping Norton.

Recher H. F. & Holmes R. T. (1985) Foraging ecology and seasonal patterns of abundance in a forest avifauna. In: *Birds of Eucalypt Forests and Woodlands: Ecology, Conservation, Management* (eds A. Keast, H. F. Recher, H. Ford & D. Saunders) pp. 79-96. Surrey Beatty & Sons, Chipping Norton.

Saunders D. A. (1977) The effect of agricultural clearing on the breeding success of the White-tailed Black Cockatoo. *Emu* **77,** 180-4.

Saunders D. A., Arnold G. W., Burbidge A. A. & Hopkins A. J. M. (1987) The role of remnants of native vegetation in nature conservation: future directions. In: *Nature Conservation: the Role of Remnants of Native Vegetation* (eds D. A. Saunders, G. W. Arnold, A. A. Burbidge & A. J. M. Hopkins) pp. 387-92. Surrey Beatty & Sons, Chipping Norton.

Specht R. L. (1970) Vegetation. In: *The Australian Environment* (ed. G. W. Leeper) pp. 44-67. CSIRO, Melbourne.

Zar J. H. (1974) *Biostatistical Analysis.* Prentice-Hall Inc., New Jersey.

APPENDIX 1. The species-specific utilization of different vegetation types in road reserves. The columns are the values of the utilization index for each species in the various vegetation types. Numbers in parentheses are the number of sections of road reserve the species was recorded in. Woodland sections of road reserve have been divided into those with an understorey (UNST>4), and those without an understorey (UNST<4). Mallee/shrubland sections have been divided into closed mallee (density of vegetation in height class 4-8 m>2), and open mallee.

Name	Woodland UNST>4 (3)	Woodland UNST<4 (5)	Mallee/Shrubland Closed (5)	Mallee/Shrubland Open (5)	Shrubland-Heath Heath (2)	Shrubland-Heath *Allocasuarina* Shrubland (3)	Shrubland-Heath Shrubland (3)
Black-shouldered kite *Elanus notatus*	0	13(1)	0	0	0	0	0
Little falcon *Falco longipennis*	0	38(1)	0	0	0	0	0
Nankeen kestral *F. cenchroides*	20(1)	25(2)	0	10(1)	0	0	0
Stubble quail *Coturnix novaezealandiae*	10(1)	0	0	0	0	0	0
Banded plover *Vanellus tricolor*	10(1)	0	0	0	0	0	0
Domestic pigeon *Columba livia*	0	25(1)	0	0	0	0	0
Common bronzewing *Phaps chalcoptera*	0	0	40(1)	0	0	13(1)	13(1)
Crested pigeon *Ocyphaps lophotes*	15(2)	35(5)	13(3)	25(5)	25(2)	11(3)	33(2)
Galah *Cacatua roseicapilla*	20(3)	46(3)	30(2)	11(3)	0	10(1)	11(2)
Port Lincoln parrot *Barnardius zonarius*	90(3)	78(5)	84(5)	69(5)	25(2)	68(3)	57(3)
Mulga parrot *Psephotus varius*	30(3)	30(1)	15(4)	17(5)	15(2)	18(3)	22(2)
Elegant parrot *Neophema elegans*	40(1)	0	0	0	20(1)	0	0
Pallid cuckoo *Cuculus pallidus*	13(1)	29(2)	0	17(1)	0	17(1)	13(1)
Horsfield's bronze cuckoo *Chrysococcyx basalis*	0	0	13(1)	0	0	13(1)	0
Rainbow bee-eater *Merops ornatus*	10(1)	0	0	0	0	0	10(1)
Welcome swallow *Hirundo neoxena*	0	0	0	0	12(1)	0	0
Tree martin *Cecropis nigricans*	45(2)	0	10(1)	25(1)	0	10(1)	40(1)
Richard's pipit *Anthus novaeseelandiae*	0	10(1)	10(1)	25(3)	15(2)	10(1)	13(1)
Black-faced cuckoo-shrike *Coracina novaehollandiae*	10(3)	18(3)	10(1)	10(3)	0	0	25(1)
White-winged triller *Lalage sueurii*	10(1)	0	0	0	0	10(1)	0
Red-capped robin *Petroica goodenovii*	10(2)	13(1)	18(4)	13(3)	15(2)	39(3)	20(1)
Western yellow robin *Eopsaltria griseogularis*	0	0	0	0	0	25(1)	0
Jacky winter *Microece leucophaea*	10(1)	0	0	0	0	0	0
Rufous whistler *Pachycephala rufiventris*	10(1)	0	27(3)	10(2)	0	30(2)	10(1)
Grey shrike-thrush *Colluricincla harmonica*	0	0	15(2)	0	30(1)	39(3)	25(3)
Crested bellbird *Oreoica gutturalis*	0	0	0	0	0	30(1)	0
Grey fantail *Rhipidura fuliginosa*	25(2)	0	56(3)	30(2)	0	56(3)	25(1)
Willie wagtail *R. leucophrys*	55(2)	0	46(5)	52(5)	40(2)	54(3)	68(3)

APPENDIX 1. continued.

Name	Woodland UNST>4 (3)	Woodland UNST<4 (5)	Mallee/Shrubland Closed (5)	Mallee/Shrubland Open (5)	Shrubland-Heath Heath (2)	Shrubland-Heath *Allocasuarina* Shrubland (3)	Shrubland-Heath Shrubland (3)
White-browed babbler *Pomatostomus supercilliosus*	10(1)	20(1)	75(2)	36(3)	40(1)	43(3)	17(2)
Rufous songlark *Cinclorhamphus mathewsi*	10(1)	0	0	0	0	0	0
Blue-breasted wren *Malurus pulcherrimus*	0	0	0	10(1)	0	30(1)	0
Redthroat *Sericornis brunneus*	0	0	30(2)	0	0	0	0
Weebill *Smicrornis brevirostris*	57(3)	33(2)	88(5)	32(4)	20(1)	13(1)	10(2)
Western warbler *Gerygone fusca*	38(2)	0	31(4)	17(1)	0	29(2)	0
Broad-tailed thornbill *Acanthiza apicalis*	0	0	20(4)	0	0	10(1)	0
Chestnut-rumped thornbill *A. uropygialis*	10(1)	0	40(3)	11(2)	20(2)	11(2)	0
Yellow-rumped thornbill *A. chrysorrhoa*	30(3)	0	44(5)	29(5)	20(2)	39(3)	11(2)
Varied sittella *Daphoenositta chrysoptera*	10(1)	0	0	0	0	0	0
Red wattlebird *Anthochaera carunculata*	0	10(1)	10(1)	10(1)	20(1)	10(1)	0
Spiny cheeked honeyeater *Acanthagenys rufogularis*	0	0	0	0	0	10(1)	0
Yellow-throated miner *Manorina flavigula*	37(3)	54(5)	32(5)	18(3)	10(1)	30(1)	56(2)
Singing honeyeater *Lichenostomus virescens*	33(3)	15(2)	48(5)	54(5)	90(2)	71(3)	68(3)
White-eared honeyeater *L. leucotis*	30(1)	0	23(3)	0	10(1)	18(3)	10(1)
Brown-headed honeyeater *Melithreptus brevirostris*	10(1)	0	25(2)	0	0	15(2)	30(1)
Brown honeyeater *Lichmera indistincta*	30(1)	0	15(2)	13(1)	35(2)	45(2)	17(2)
White-fronted chat *Ephthianura albifrons*	40(1)	0	20(1)	25(2)	10(1)	20(1)	0
Striated pardalote *Pardalotus striatus*	20(3)	44(2)	20(3)	20(2)	0	11(2)	17(2)
Zebra finch *Peophila guttata*	0	10(1)	0	27(3)	0	30(1)	25(1)
Australian magpie lark *Grallina cyanoleuca*	33(3)	67(2)	22(5)	21(4)	0	0	11(2)
Black-faced woodswallow *Artamus cinereus*	10(2)	30(1)	17(3)	20(3)	20(2)	36(3)	62(1)
Grey butcherbird *Cracticus torquatus*	20(2)	10(1)	27(3)	10(1)	0	13(1)	20(1)
Pied butcherbird *C. nigrogularis*	10(2)	19(4)	60(1)	0	0	0	0
Australian magpie *Gymnorhina tibicen*	23(3)	72(4)	13(3)	10(2)	0	20(1)	0
Grey currawong *Strepera versicolor*	10(1)	0	0	0	0	0	0
Australian raven *Corvus coronoides*	37(3)	22(4)	60(1)	15(5)	10(1)	10(1)	17(2)
Little crow *C. bennettii*	10(1)	10(1)	0	0	0	0	0

Proc. Ecol. Soc. Aust. 1990 — 16:369-377

The Numbat *Myrmecobius fasciatus* (Myrmecobiidae): history of decline and potential for recovery

J. A. FRIEND

Department of Conservation and Land Management, P.O. Box 51, Wanneroo, W.A. 6065

Abstract

At the time of European settlement of Australia, numbats occurred across much of the southern half of the continent. This distribution began to contract during the mid-1800s, but the most rapid phase of extinction appears to have taken place in the arid zone during the 1940s and 1950s. The timing of this crash coincides both with the establishment of the introduced red fox in the region and with the depopulation of the deserts by Aboriginal people that resulted in the breakdown of their traditional land management practices.

Today numbat populations are limited to a few areas of forest and woodland in south-western Australia. Even these passed through a period of decline during the 1970s, and the species disappeared from several isolated habitat remnants. At Dryandra Forest, the pattern of the decline and subsequent recovery of the numbat population did not suggest that the decline was due solely to low rainfall between 1975 and 1979 or to any increase in the frequency of prescribed burning. Selective poisoning of foxes since 1983 has been followed by a substantial increase in numbat numbers and the higher population levels have been maintained despite further years of low rainfall from 1984-1987.

Re-establishment of populations in areas in which numbats formerly occurred commenced in 1985. Animals from Dryandra Forest and from a captive breeding colony have been used to re-stock Boyagin Nature Reserve, where the species became extinct in the 1970s. Fox control has been carried out continuously since the introductions. Breeding of numbats and establishment of their young have occurred each year. Present indications are that the new population is viable. The potential now exists for translocations to further sites within the former range of the species.

Introduction

The numbat, *Myrmecobius fasciatus,* is a medium-sized marsupial belonging to the monotypic family Myrmecobiidae. It is the only marsupial that feeds exclusively on termites and ants, and many of its distinctive anatomical features are related to this diet (Friend 1989). Its diurnal habit, unusual for marsupials, is also related to its primary dependence on termites, which are principally active during the day.

The numbat's range has contracted greatly since European settlement (Friend & Kinnear 1983), but its local abundance in south-western Australia (Calaby 1960) until the 1970s meant that its status was only recently recognized as endangered by the Western Australian fauna conservation authorities (Anon. 1983). Prior to this, only the eastern subspecies, *M. f. rufus,* was listed as endangered, after its disappearance from South Australia and New South Wales (Calaby 1960).

This chapter details the history of the decline of numbats, speculates on the causes of their decline in different geographical regions and reports on the results of re-introductions of numbats to two reserves where it had recently become extinct.

History of the numbat's decline

It is possible to reconstruct the approximate timing of the contraction in the numbat's range by compiling records of occurrence associated with dates from:

a) specimens held in museums,
b) published first-hand records by reliable observers [Finlayson (1961) and Krefft (1866)],
c) results of interviews in central Australia with Aboriginal people who had lived as nomadic hunter-gatherers (Burbidge & Fuller 1979; Friend *et al.* 1982; Burbidge *et al.* 1988), and

d) information on current distribution collected through publicity campaigns in Western Australia (1981-1985) requesting reports of recent sightings. Reports were verified by investigation (Connell *in litt.*; Connell & Friend 1985).

The numbat's declining range is shown on Figure 1. The information on which it is based is patchy. Few records are available from most of the semi-arid zone, such as the Goldfields and pastoral districts of Western Australia. In the east, Krefft (1866) reported that the species was already on the decline near the confluence of the Murray and Darling Rivers in the south-western corner of New South Wales in 1856-57. Numbats were present near Adelaide at the time of European settlement, but disappeared soon after (Wood Jones 1923). A survey carried out in South Australian country schools prior to 1923 resulted in no reports of recent sightings at all (Wood Jones 1923). The species was clearly present and numerous in the north-western parts of South Australia in the early 1930s (Finlayson 1933). Recent information from Aboriginal people living in the area suggested that it survived near the Mann, Musgrave and Everard Ranges at least until the 1940s (Friend *et al.* 1982).

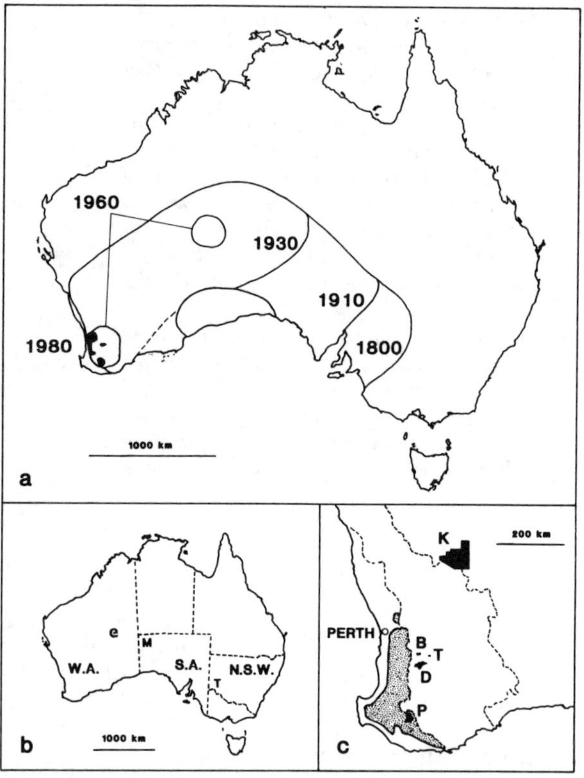

FIG. 1. a. Map of Australia showing the approximate distribution of the numbat, *Myrmecobius fasciatus*, in 1800, 1910, 1930, 1960 and 1980, reconstructed from historical evidence from various sources (see text). Distribution in 1980 shaded black. b. Location of places mentioned in the text. T: Tapio; M: Mann, Musgrave and Everard Ranges; G: Gibson Desert. c. South-west Western Australia with the locations of reserves mentioned in the text (shaded black). P: Perup Nature Reserve; D: Dryandra Forest; B: Boyagin Nature Reserve; T: Tutanning Nature Reserve; K: Karroun Hill Nature Reserve. Dashed lines enclose the most heavily cleared agricultural area (most of which is termed "the wheatbelt"); the stippled area is the portion of main forest belt dominated by jarrah or wandoo.

In Western Australia, there are no records of numbats from the Goldfields region after 1930, although they may have persisted there unrecorded somewhat longer. There is, however, clear evidence of numbats surviving in the central deserts until the 1950s and 1960s (Friend *et al.* 1982), and in the eastern wheatbelt as late as the 1970s (G. Connell *in litt.*). Calaby (1960) recorded the numbat as occupying a series of discontinuous woodland areas in the south-western part of the wheatbelt in 1958. He stated that the species was among the more abundant of the small mammals of the southwest, and that the only real threat to its persistence was the wholesale clearing of its habitat. By 1980, however, numbats were restricted to parts of the main jarrah, *Eucalyptus marginata,* forest belt and a few large reserves on the western fringes of the wheatbelt (Connell & Friend 1985). This was largely due to substantial clearing between 1958 and 1970, but also to a sudden collapse of all known populations in the late 1970s (Perup: Christensen 1980; Dryandra: Turner & Borthwick *in litt.*; Tutanning: L. Marshall, pers. comm.; Boyagin: see below; eastern wheatbelt: G. Connell *in litt.*). A number of populations died out altogether, and even in those areas where numbats were thought secure, numbers dropped to alarmingly low levels (Christensen 1980; Friend 1987).

The pattern of decline of the numbat is shared by many other Australian mammals. The arid zone, in particular, has lost almost all mammal species, including the numbat, in the intermediate body weight range (35 g to 4 200 g, Burbidge & McKenzie 1989). This appears to have occurred over a short time-period in the 1940s and 1950s in Western and central Australia (Burbidge *et al.* 1988). Several species which were lost from the arid zone persisted in the south-west of Western Australia. These included the brush-tailed bettong, *Bettongia penicillata,* the western quoll, *Dasyurus geoffroii* and the red-tailed phascogale, *Phascogale calura* (Strahan 1983) as well as the numbat.

Possible causes of the numbat's decline

The decline of the numbat has not been a smooth process. It occurred rapidly in the arid zone, for instance, while in the Western Australian wheatbelt, populations persisted well after environmental change wrought by Europeans. Clearly, these changes varied in their nature, effect and timing in different parts of the numbat's former range. The recent decline of numbats in the south-west of Western Australia, 20-30 years after that in the deserts, suggests that a new combination of factors had begun to operate there.

Two geographical regions will be examined in an attempt to identify possible causes of the numbat's disappearance from most of its previous range. These are the central deserts, from which numbats have disappeared, and the south-west of Western Australia, where isolated populations persist.

The central deserts

One feature of the loss of fauna from the arid and semi-arid zones is that it has affected mammal species of all major trophic groups. Herbivorous species, such as the hare-wallabies, fungivorous/omnivorous species such as the potoroines, generalist insectivores like the bandicoots, and carnivores like the western quoll have declined, often to extinction, and apparently during the same time-period (Finlayson 1961; Burbidge *et al.* 1988). This is of particular interest with regard to the numbat, because the only other Australian mammal species remotely similar in dietary preference, the echidna, *Tachyglossus aculeatus,* has shown no signs of a contraction in range, despite being well within the weight range susceptible to decline. The echidna persists throughout the previous range of the numbat.

Both numbats and echidnas feed on termites and ants. Although phylogenetically distant, both ants and termites exist in colonies, and in arid climates generally build underground nests and utilize food storage to persist through unfavourable periods (Watson *et al.* 1973). In arid areas both ants and termites are likely to be extremely resilient to environmental fluctuations and to bestow this resilience on their predators.

The sudden disappearance of numbats, but not echidnas, from these areas suggests that it was not related to a decline in the abundance of ants and termites. It is more likely, because of the similar timing, that the numbat's disappearence was due to a factor other than food that also caused the collapse of populations of other medium-sized mammals.

Burbidge *et al.* (1988) reviewed three possible causes for the loss of mammals in central Australia. These were: changes in fire regimes; effects of exotic predators; and competition from exotic herbivores. These three factors appear to have intensified near the time of the faunal collapse in the arid zone. Aboriginal people moved from their traditional land to missions and stations in the 1940s and 1950s. This caused a major widespread change in fire regime, from small patch burns at different times of year to extensive summer wildfires (Bolton & Latz 1978; Johnson & Roff 1982). Secondly, the numbers of the red fox, *Vulpes vulpes,* built up in the region after its arrival prior to 1930 (Finlayson 1961). Thirdly, the rabbit, *Oryctolagus cuniculus,* which had arrived in the 1890s, continued to have plagues of locally devastating effect for 40-50 years thereafter (Finlayson 1935).

The comparison between numbats and echidnas suggests that the decline of the former was not due to changes in the abundance of food. It is unlikely that exotic herbivores would affect numbats and not echidnas. A clear difference between numbats and echidnas is the relative vulnerability of numbats to predation. Numbats rely on their speed and knowledge of refuges within their home range to escape predation by birds and mammals. Echidnas dig into the ground and expose only their spiny coat to the predator. It would seem that cover is much more important to numbats than to echidnas in avoiding predation.

H. H. Finlayson was virtually the only mammalogist active in the arid zone both before and after the collapse of the mammal fauna. He believed that the rapidly growing presence of the red fox in central Australia was responsible for the demise of many native mammals between his major trips during 1931-35 and 1950-1956 to that area (Finlayson 1961). While this, and any other speculation about causes of the loss of mammal species from the region can now be based only on circumstantial evidence, it seems that the effects of the two introduced cursorial predators (red fox and cat, *Felis catus*) on native mammal populations would be greatly exacerbated were cover afforded by vegetation reduced. As old hummocks in the widespread hummock grasslands ("spinifex") provide cover for small vertebrates against predators (Burbidge 1985), extensive wildfires which remove this cover for a short time at least (Johnson & Roff 1982) are likely to increase the success of cursorial predators in securing their prey. Recent work in south-west Western Australia (see below) has indicated that predation by foxes can regulate numbat population numbers there. The most recent reported sighting of a numbat in the arid zone was in about 1965 near the Clutterbuck Hills in the central Gibson Desert (Friend *et al.* 1982), one of the last areas occupied continuously by the traditional owners before the late 1960s when they joined the movement to settlements. It appears that the species persisted longest where the traditional fire regime was most recently in place. The timing of the disappearance of the numbat from the arid zone is consistent with the hypothesis that the changed fire regime reduced cover over large areas to the extent that it allowed predation by the introduced red fox and cat, added to that of natural predators, to exceed the numbat's reproductive capacity. While the fox was only newly arrived in the central deserts at this time, the cat was already well established there in the late 1800s

(Carnegie 1898). Clearly the mammal species that persisted in the central deserts until the 1940s coexisted with the cat before the change in burning regimes.

It is likely that in some areas, rabbit plagues acted to build up fox and cat populations. When rabbit populations crashed, through food shortage, disease, or whatever cause, the large populations of carnivores may then have switched to native mammal species. Wallis and Brunner (1987) showed that between 1973 and 1985 as rabbits became scarce at Sherbrooke Forest, Victoria, they changed from the most frequent to one of the least frequent prey items in the diet of foxes there, and that native mammals assumed a correspondingly greater importance in the diet.

The south-west of Western Australia

It might be expected that the persistence of the numbat at the south-western extremity of its range would give some clue as to the cause of its disappearance elsewhere. Numbats are still found in parts of the jarrah forest where annual rainfall, at over 700 mm, is the highest in the recorded range of the species. Burbidge and McKenzie (1989) speculated that the combined effect of changed fire regimes and introduced herbivores and predators on medium-sized mammals was analogous to the effect of increased aridity. The survival of certain species of medium-sized mammals was therefore better in the south-west than in the deserts, where those species were living closer to their physiological thresholds. There are features other than higher rainfall, however, that distinguish this part of the numbat's former range. Compared with numbat habitat in the arid and semi-arid zones, the forests and woodlands of the south-west have much higher densities of hollow logs (Calaby 1960; Friend 1982), which numbats use for refuge from predators as well as for nesting. In addition, poisonous shrubs of the genera *Gastrolobium* and *Oxylobium* occur widely in the south-west (Aplin 1973). These plants may have bestowed some protection on native mammal species, by reducing fox densities through secondary poisoning (Anon. 1980; Christensen 1980).

The localities in which the numbat persists today are either areas where the incidence of these poisonous shrubs is high (Dryandra Forest and Perup Nature Reserve), or are in jarrah forest, where the density of foxes is relatively low (Christensen *et al.* 1985). In both cases, numbat populations persist only in large reserves (over 15 000 ha), containing large sections over 3 kilometres from the forest-farmland interface, where fox activity is concentrated (Ables 1975; Henry 1986).

In the mid- to late 1970s, the populations of numbats in Dryandra Forest and Perup Nature Reserve crashed to below 10% of their former levels (Dryandra — Fig. 2; Perup — Christensen 1980). Population declines of other medium-sized native mammals were recorded at the same time at Perup (Christensen 1980). King *et al.* (1981) documented an increase in the abundance of foxes at two widely separated sites in the south-west of Western Australia between 1970 and 1977. They attributed this to the cessation in the early 1970s of broadscale rabbit control using 1080 (sodium monofluoroacetate), a practice which, they reasoned, caused significant mortality of foxes through eating poisoned rabbits.

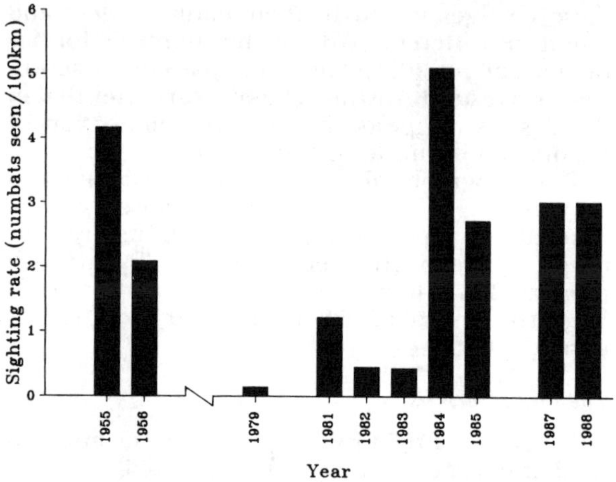

FIG. 2. Rate of numbat sightings (nos./100 km) each year during surveys at Dryandra Forest. Sources of data as in Table 1. Baiting of foxes commenced in September 1982.

There is therefore circumstantial evidence supporting the hypothesis that increased predation by foxes was a major cause of the numbat's decline in the 1970s. The timing of the decline, however, suggested other causes as well, and the most plausible mechanisms are listed below (Anon. 1979):

a) decreased availability of termites, due either to
   — climate change, in particular to a period of several years of low rainfall in the 1970s, or
   — an increased fire frequency following changes in forest management policy, or
b) increased predation, due either to
   — foxes, following the rise in their numbers during the 1970s, or
   — all predators, following the loss of shrub cover and log refuges due to an increased fire frequency.

The security of the species requires an understanding of the processes likely to cause dangerous population crashes like that of the late 1970s, so that numbat populations may be managed to prevent or reduce the severity of such events.

Studies at Dryandra Forest

Studies on the biology of the numbat were carried out in Dryandra Forest between 1981 and 1985, and provided an opportunity to examine the operation of the factors listed above, and to determine which were most likely to have caused the decline.

Following Calaby (1960) the method employed to find numbats was to drive at 15-20 km/h around the forest tracks during the day until a numbat was sighted. The sighting rate over a large distance (at least 200 km) each year was used to provide an index of population size (Table 1). All of the surveys listed in Table 1 were carried out in the central portion (approximately 6 000 ha) of the main block of Dryandra Forest.

During the early part of the study, evidence of fox predation of numbats was found. The remains of two radio-collared numbats were found, chewed and buried. Two scats amongst forty collected along the edge of the forest contained numbat remains.

At the commencement of the study in 1981, the sighting rate was very low (Table 1). In September 1982, a monthly fox-baiting procedure (Kinnear *et al.* 1989) was implemented in the northern section (Area 2: 2 500 ha) of the study area. The southern section (Area 1: 2 000 ha) was left unbaited, as was a buffer strip of 1 km width between the two areas. The two areas contained approximately the same amount of wandoo, *Eucalyptus wandoo*, woodland, the numbat's preferred habitat (Calaby 1960). Baits of fresh meat (100 g), each containing an oat impregnated with 4.5 mg of 1080, were laid at 100 m intervals along forest boundaries and major internal tracks. Between April and October, chickens' eggs, each containing a 1080 oat, were laid at 300 m intervals as well. In 1984 and 1985, surveys carried out in 1981 before baiting commenced were repeated using the same routes (110 km in each area). In 1985, a further 130 km survey was driven in each area (Table 2).

TABLE 1. Distance driven, numbat sightings made, rate of sightings during surveys at Dryandra Forest, and annual rainfall at Narrogin (Bureau of Meteorology). Survey data for 1954-56 provided by J. H. Calaby, 1979 data from J. Turner. Baiting of foxes commenced in September 1982.

Year	Km driven	Sightings	Sightings/ 100 km	Rainfall
1954	53	5		330
1955	385	16	4.16	741
1956	385	8	2.08	419
1979	718	1	0.14	371
1980	—			539
1981	413	5	1.21	513
1982	1 103	5	0.45	549
1983	704	3	0.43	662
1984	275	14	5.09	462
1985	590	16	2.71	435
1986	—			387
1987	200	6	3.00	453
1988	200	6	3.00	534

TABLE 2. Results of surveys in baited and unbaited areas in Dryandra Forest to assess the effect of baiting foxes on numbat sighting frequency. Baiting commenced in September 1982. The first three surveys comprised set routes 110 km long in each area. In 1985, a further 130 km route was driven in each area.

Year	Baiting	Survey distance in each area (km)	Sightings Area 1	Area 2
1981	None	110	1	0
1984	Area 2	110	2	6
1985	Area 2	110	1	2
1985	Area 2	130	2	5

The results of these surveys (Table 2) indicate that recovery of the numbat population in the baited area was more rapid than that in the unbaited area, if it is assumed that the unbaited area could support a similar or higher mean density of numbats compared with the baited area. This implies that fox predation reduces the rate of growth of the numbat population at Dryandra.

Survey data from 1954-56 (J. Calaby, pers. comm.: Table 1) revealed a mean sighting rate of 3.5 per 100 km. Calaby's greatest number of sightings in one day, in a total of 39 days, was five. On a September day during 1977 a group of naturalists saw four numbats in different parts of Dryandra Forest (I. Edgar & B. Fremlin pers. comm.). In the same year, D. Thomas of Taronga Zoo saw three numbats in Dryandra in four days in late October (pers. comm.). By 1979, however, numbat numbers had dropped to the point where J. Turner's survey for the Department of Fisheries and Wildlife (J. Turner & J. Borthwick *in litt.*) resulted in only one sighting in 20 days, at a rate of 0.14 per 100 km (Table 1: data held by the Department of Conservation and Land Management). The recovery of the numbat population in the central portion of Dryandra Forest occurred in 1984, and since then, sighting rates have been similar to those recorded in the 1950s (Table 1 and Fig. 2).

The rainfall record at Narrogin, 30 km south-east of the study area (Fig. 3) shows that the periods 1975-79 and 1984-87 were drier than average, although only the 1979 total was more than 15% below average. The years 1980-83 saw average or above average rainfall. Significant events in summer rainfall since 1975 were the dry year in 1979 despite average summer figures and the high summer rainfall in 1982, but for which the year's total would have been low.

The sustained recovery of the numbat population since 1984 despite another dry period from 1984-87 does not support the hypothesis that low rainfall between 1975 and 1979 alone caused the observed population decline of the late 1970s.

On the other hand, the sighting data (Table 1) support the contention that the recovery in the central portion of the forest was assisted by the

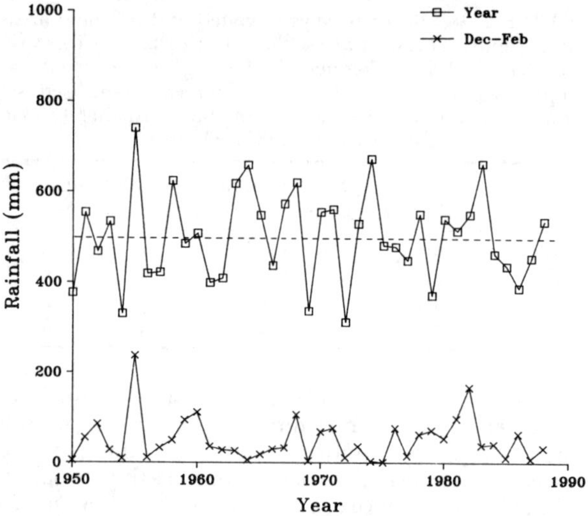

FIG. 3. Annual rainfall and rainfall during the summer months (December of previous year to February) each year from 1950 to 1988 at Narrogin, 30 km south-east of the study site at Dryandra. The dashed line indicates the 97-year average annual rainfall (506 mm).

control of foxes in its northern section, especially since the recovery has been more pronounced in that section (Table 2).

It was not possible to carry out an experiment of sufficient scale to test the hypothesis that predation rates are higher in recently burnt areas. Monitoring radio-collared animals in fires revealed a high rate of survival through the fire and immediately afterwards. It is likely that rates of predation rise after fire due to the decreased understorey cover and log availability. However, in the 10 years from 1970 to 1980, only 21% of Dryandra Forest was burnt (data from Fig. 2 in Burrows *et al.* 1987). It is unlikely, therefore, that prescribed burning alone caused the observed crash, either through food shortage or increased predation.

The indications of the study at Dryandra were that the numbat population crash of the late 1970s was probably caused by increased predation due to an increase in fox numbers.

Clearly, control of fox numbers is necessary to lessen the likelihood of crashes in numbat populations such as that experienced in the late 1970s. Management of the species should involve the protection of strategic populations by continual action to keep fox numbers below the level at which they can exert significant predation pressure on numbats. Control of foxes must also be used in re-establishing numbat populations, firstly to maintain the growth of source populations and secondly to protect newly-established populations, particularly in the critical early stages. A commitment to ongoing control programmes after the initial stages must also be made if new populations are to be set up.

Population re-establishment at Boyagin Nature Reserve

Numbats disappeared from Boyagin Nature Reserve, 40 km north of Dryandra, in the 1970s. They were well known in the reserve by one of the neighbouring landholders in the 1960s (G. C. Matthews, pers. comm.) and there is a reliable report of a sighting of a numbat on the reserve in 1970 (S. White, pers. comm.). W. H. Butler did not record the species in his brief survey of the reserve in 1972 (Butler *in litt.*). In surveys during the present study in 1981, 1983 and 1985, all significant areas of wandoo woodland were searched intensively for numbat diggings. No evidence was found to indicate that numbats were still present.

The re-establishment of numbats at Boyagin was proposed in 1985, as part of a strategy to increase the security of the species by increasing the number of populations. As this was the first attempt to translocate numbats, the approach used was to try to optimize conditions rather than to set up trials to test success under a range of different circumstances. Boyagin Nature Reserve comprises two adjacent uncleared blocks totalling 5 000 ha, surrounded by farmland. It was chosen because it contains the largest remnants of woodland close to the source population at Dryandra, because its vegetation is similar to that of Dryandra, and because of its status as a nature reserve.

As the earlier studies had indicated that fox predation reduces the rate of numbat population growth at Dryandra, a programme of fox control by the use of 1080 in fresh meat baits was commenced on Boyagin Nature Reserve in October 1985.

Young numbats disperse from their maternal home ranges in November or December (Friend 1987). First-year females, unlike males, mate in January immediately following their dispersal (Friend, unpubl.). Hence it is likely that the least disruptive time to carry out a translocation would be during November or December, when animals would normally be dispersing. Young numbats are dependent on their mothers from birth in January or February until October or November, so the process of moving females or dependent young at any other time of the year would interfere directly with the nurturing process.

Seventeen numbats were released in wandoo woodland in two separate valleys in the eastern section (*ca.* 2 000 ha) of Boyagin Nature Reserve during November and December 1985 (Fig. 4). Two adult males were from a captive colony but the other animals were captured at Dryandra Forest and moved directly to Boyagin. The numbats released comprised five first-year and four older males and seven first-year females and one older female. All were fitted wih radio-collars (SM1 transmitters, AVM Instrument Company, California) before release to allow their survival and breeding to be monitored.

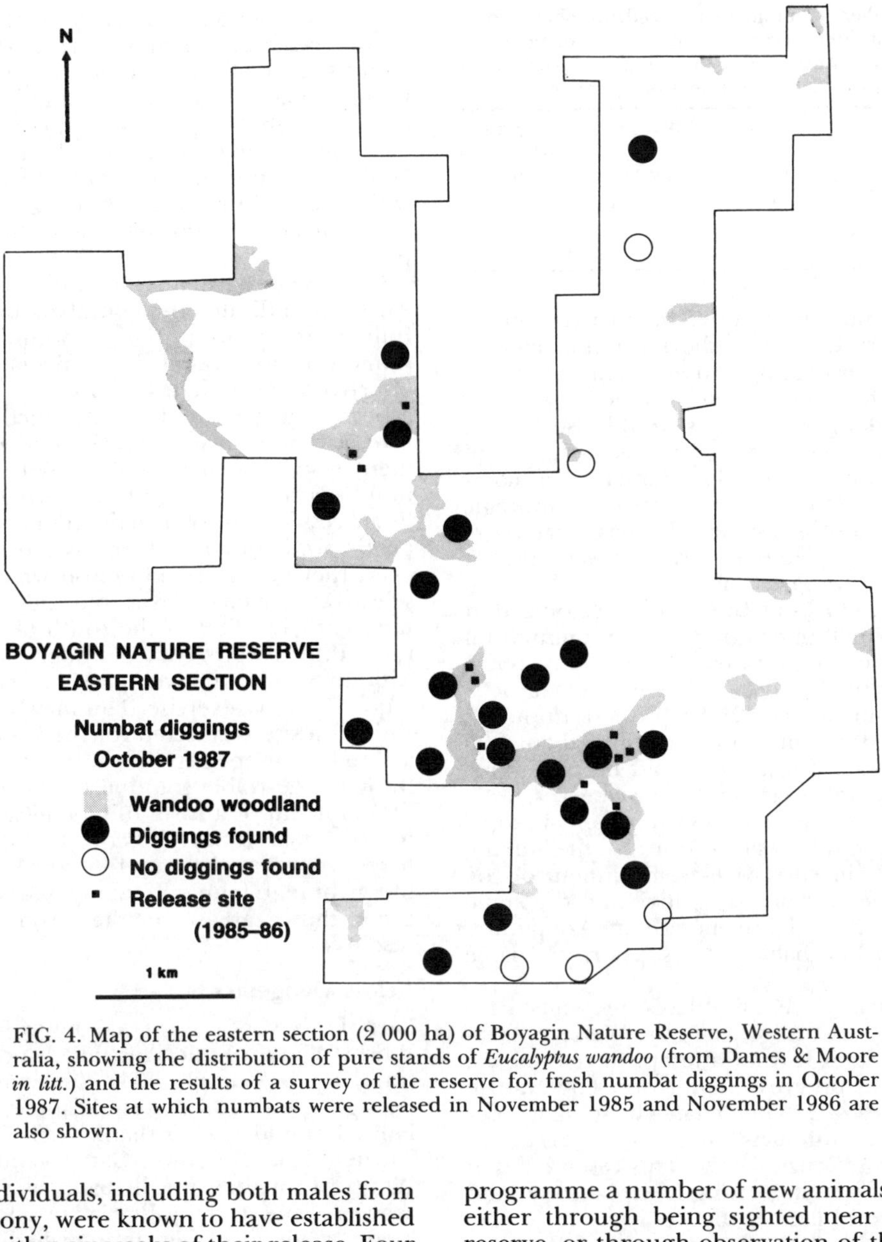

FIG. 4. Map of the eastern section (2 000 ha) of Boyagin Nature Reserve, Western Australia, showing the distribution of pure stands of *Eucalyptus wandoo* (from Dames & Moore *in litt.*) and the results of a survey of the reserve for fresh numbat diggings in October 1987. Sites at which numbats were released in November 1985 and November 1986 are also shown.

Thirteen individuals, including both males from the captive colony, were known to have established home ranges within six weeks of their release. Four of the 17 animals removed their collars or their transmitters failed. All but one of the 13 numbats settled in the release valleys, but some established themselves well away (up to 2 km) from their release sites. Four were known to have died by June 1986. These included two that were taken by raptors, one that was taken and buried by a fox and one that died from unknown causes. Recapture of six of the females revealed that five had produced young by February 1986.

During the following two years, eighteen more numbats were released in Boyagin Nature Reserve (Table 3). During the second and third years of the programme a number of new animals were caught, either through being sighted near tracks in the reserve, or through observation of the burrows of radio-marked females with litters (Table 3).

As a measure of the extent to which the reserve had been recolonized, a search for diggings was carried out, mainly in tracts of wandoo woodland, in October 1987, before the latest release (Fig. 4).

Present indications are that the population is now large enough to be self-sustaining while fox control is maintained. No animals will be introduced in 1988-89, but monitoring will continue. Given the high rate of production of young by newly-moved females, there is little doubt that translocation at the time of dispersal has been a successful strategy.

TABLE 3. Numbers of numbats released, number known dead, number of litters recorded and number of new captures at Boyagin during each of the first three years (July-June) of the re-establishment programme.

	1985-86	1986-87	1987-88
Released	17	8	10
Known deaths	4	5	2
Known litters	5	4	6
New captures	0	6	7

Future translocations could be carried out by management staff after a short training course, and monitored by diggings surveys. The major cost would be the fox control, but evidence is growing that other endangered species would also benefit (e.g., Kinnear *et al.* 1989). The Department of Conservation and Land Management has commenced a programme of fox control on strategic reserves in the south-west, and this has created the opportunity to implement cheaply other numbat re-establishment programmes.

Even if both blocks of Boyagin are colonized, it is a relatively small reserve in terms of a minimum viable population of numbats in the long term. Given that numbats occupy their habitat at a density of about one per 25 ha (Friend, unpubl.), there is room for a maximum of only about 200 established adults. Clearly it would be preferable to re-establish the species on larger reserves.

It appears that, in vegetation remnants in agricultural land in south-western Australia, the limiting factors to the widespread re-establishment of the numbat are the cost of fox control and the small number of remnants large enough and containing sufficient suitable habitat to support a viable population.

The majority of the numbat's previous geographical range occurred east of the Western Australian wheatbelt. A further attempt to re-establish the species is in progress at Karroun Hill Nature Reserve, an area of 300 000 ha in which the former occurrence of numbats has been recorded (Youngson & McKenzie 1977). This reserve lies at the junction of wheat-growing land, pastoral land and vacant Crown land at the western edge of the Goldfields district, 400 km north-east of Perth. A predator control programme of low intensity is in place on this reserve. Early results indicate that the newly introduced animals have no difficulty finding sufficient food, but that predation by introduced carnivores constitutes the greatest hurdle to the establishment of a viable population.

The early stages of re-establishment of numbats at Karroun Hill may present problems not encountered in small isolated vegetation remnants. These include the logistical difficulties of fox control in very large areas as well as the problem of translocated animals dispersing widely after release. In Boyagin Nature Reserve, released animals and dispersing young generally stay within the confines of the reserve, which is surrounded by cleared land. The small area of the reserve ensures that these animals encounter one another. In a reserve large enough to support a population of thousands of numbats, or in an area unbounded by cleared land, dispersing animals could move out of contact with others. Current monitoring of the colony at Karroun Hill should determine the extent of this problem.

Our experience in the south-west of Western Australia indicates that numbats may be successfully re-introduced to sites within their former range if exotic predators, particularly the fox, are effectively controlled. Perhaps the most important aspect of a proposed re-introduction site, therefore, is that it is within reach of adequate management resources. It is most likely that parts of the numbat's former range in eastern Australia could be successfully recolonized under a strict regime of exotic predator control. In eastern Australia, however, there is as yet no poison which can be used selectively against foxes but not against native fauna as can 1080 in the south of Western Australia. Effective fox control under those conditions, without placing native species at risk, becomes a more expensive exercise. This may be judged worthwhile if a site is selected for the re-introduction of a number of species. In the long term, however, the most desirable solution to this problem is the development of a form of biological control of the fox (Kinnear 1989). Effective biological control of foxes will open the way for the reconstruction of the mammal fauna of many parts of Australia, even those areas remote from management resources.

Acknowledgements

Part of this work was carried out with support from World Wildlife Fund Australia, under Projects 50 (captive breeding), 94 (re-establishment at Boyagin) and 108 (re-establishment at Karroun Hill). I would like to thank Neil Thomas, Bruce Turner, Rod Burrows, Garry Connell and Dick Whitford for their invaluable assistance in various aspects of the study. John Calaby and John Turner kindly made available unpublished data. Andrew Burbidge, Jack Kinnear, Graeme Smith and an anonymous referee read the manuscript and made helpful suggestions which greatly improved the final version.

References

Ables E. D. (1975) Ecology of the red fox in America. In: *The Wild Canids* (ed. M. W. Fox). Van Nostrand Reinhold Co., New York.

Aplin T. E. H. (1973) Poison plants of Western Australia. *Western Australian Department of Agriculture Bulletin* **3772,** 1-66.

Anon (1979) Our diminishing heritage — the Numbat. *SWANS* **9,** 40-1.

Anon. (1980) Have poisonous plants helped save some of our native wildlife from extinction? *SWANS* **10,** 10-2.

Anon. (1983) Changes to list of rare and endangered fauna in W.A. *SWANS* **13,** 26-30.

Bolton B. L. & Latz P. K. (1978) The Western Hare-wallaby *Lagorchestes hirsutus* (Gould) (Macropodidae) in the Tanami Desert. *Aust. Wildl. Res.* **5,** 285-93.

Burbidge A. A. (1985) Fire and mammals in hummock grasslands of the arid zone. In: *Fire Ecology and Management of Western Australian Ecosystems* (ed. J. Ford) pp. 91-4. Western Australian Institute of Technology, Perth.

Burbidge A. A. & Fuller P. J. (1979) Mammals of the Warburton region, Western Australia. *Rec. West. Aust. Mus.* **8,** 57-73.

Burbidge A. A., Johnson K. A., Fuller P. J. & Southgate R. I. (1988) Aboriginal knowledge of the mammals of the central deserts of Australia. *Aust. Wildl. Res.* **15,** 9-39.

Burbidge A. A. & McKenzie N. L. (1989) Patterns in the modern decline of Western Australia's vertebrate fauna: causes and conservation implications. *Biol. Conserv.* **50,** 143-98.

Burrows N. D., McCaw W. L. & Maisey K. G. (1987) Planning for fire management in Dryandra Forest. In: *Nature Conservation: the Role of Remnants of Native Vegetation* (eds D. A. Saunders, G. W. Arnold, A. A. Burbidge & A. J. M. Hopkins) pp. 305-12. Surrey Beatty & Sons, Chipping Norton.

Calaby J. H. (1960) Observations on the banded ant-eater *Myrmecobius f. fasciatus* Waterhouse (Marsupialia), with particular reference to its food habits. *Proc. Zool.. Soc. (Lond.).* **135,** 183-207.

Carnegie D. W. (1898) *Spinifex and Sand.* C. Arthur Pearson, London.

Christensen P. E. S. (1980) A sad day for native fauna. *Forest Focus* **23,** 3-12.

Christensen P., Annels A., Liddelow G. & Skinner P. (1985) Vertebrate fauna in the southern forests of Western Australia. *Forests Department of Western Australia Bulletin* **94,** 1-109.

Connell G. W. & Friend J. A. (1985) Searching for numbats. *Landscope* **1,** 21-6.

Finlayson H. H. (1933) On the eremian representative of *Myrmecobius fasciatus* (Waterhouse). *Trans. Roy. Soc. South Aust.* **57,** 203-5.

Finlayson H. H. (1935) *The Red Centre. Man and Beast in the Heart of Australia.* Angus & Robertson, Sydney.

Finlayson H. H. (1961) On central Australian mammals. Part IV. The distribution and status of central Australian species. *Rec. South Aust. Mus.* **14,** 141-91.

Friend J. A. (1982) The numbat: an endangered specialist. *Australian Natural History* **20,** 339-42.

Friend J. A. (1987) Local decline, extinction and recovery — relevance to mammal populations in vegetation remnants. In: *Nature Conservation: the Role of Remnants of Native Vegetation* (eds D. A. Saunders, G. W. Arnold, A. A. Burbidge & A. J. M. Hopkins) pp. 53-64. Surrey Beatty & Sons, Chipping Norton.

Friend J. A. (1989) Myrmecobiidae. Chapter 22. In: *Fauna of Australia. Volume IB. Mammalia* (ed. D. Walton) pp. 583-90. Bureau of Flora and Fauna, Canberra.

Friend J. A., Fuller P. J. & Davis J. A. (1982) The numbat in central Australia. *SWANS* **12(3),** 21-6.

Friend J. A. & Kinnear J. E. (1983) Numbat *Myrmecobius fasciatus.* In: *Complete Book of Australian Mammals* (ed. R. Strahan) p. 85. The Australian Museum and Angus & Robertson, Sydney.

Henry J. D. (1986) *Red Fox.* Smithsonian Institution Press, Washington, D.C.

Johnson K. A. & Roff A. D. (1982) The Western Quoll *Dasyurus geoffroii* (Dasyuridae, Marsupialia) in the Northern Territory: historical records from venerable sources. In: *Carnivorous Marsupials* (ed. M. Archer) pp. 221-6. Royal Zoological Society of New South Wales, Sydney.

King D. R., Oliver A. J. & Mead R. J. (1981) *Bettongia* and fluoroacetate: a role for 1080 in fauna management. *Aust. Wildl. Res.* **8,** 529-36.

Kinnear J. (1989) Outfoxing the fox. *Landscope* **4,** 12-7.

Kinnear J. E., Onus M. L. & Bromilow R. N. (1989) Fox control and rock wallaby population dynamics. *Aust. Wildl. Res.* **15,** 435-50.

Krefft G. (1866) On the vertebrated animals of the lower Murray and Darling, their habits, economy and geographical distribution. *Transactions of the Philosophical Society of New South Wales* **1862-1865,** 1-33.

Strahan R. (Ed.) (1983) *Complete Book of Australian Mammals.* The Australian Museum and Angus & Robertson, Sydney.

Wallis R. & Brunner H. (1987) Changes in mammalian prey of foxes, *Vulpes vulpes* (Carnivora: Canidae) over 12 years in a forest park near Melbourne. *Aust. Mammal.* **10,** 43-4.

Watson J. A. L., Lendon C. & Low B. S. (1973) Termites in mulga lands. *Tropical Grasslands* **7,** 121-6.

Wood Jones F. (1923) *The Mammals of South Australia.* Volume I. South Australian Government Printer, Adelaide.

Youngson W. K. & McKenzie N. L. (1977) The wildlife of the proposed Karroun Hill Nature Reserve, Western Australia. *Western Australian Department of Fisheries and Wildlife Report* **30,** 1-32.

Proc. Ecol. Soc. Aust. 1990 — 16:379-396

Factors affecting small mammal distribution and abundance in the Eastern Otway Ranges, Victoria

B. A. WILSON[1], D. ROBERTSON[2], D. J. MOLONEY[1], G. R. NEWELL[1] and W. S. LAIDLAW[1]

[1]*Biological Sciences, Deakin University, Geelong, Victoria 3217*

[2]*School of Science, Riverina Murray Institute of Higher Education, Wagga Wagga, New South Wales 2650*

Abstract

In the Eastern Otway Ranges 15 species of small mammals are distributed among a complex of vegetation communities including shrublands, coastal complex, heathlands, woodlands and forests. The effects of vegetation and disturbance factors including fire, mining and the plant pathogen Phytophthora cinnamomi *on the distribution, abundance and diversity of small mammals were investigated. Species rich small mammal communities occurred in vegetation with high numbers of sclerophyll plants while the highest abundance of small mammals was recorded in non-sclerophyll sand dune communities.* P. cinnamomi *was associated with small mammal communities of low diversity and abundance. After wildfire the abundance of native small mammals at burnt sites was low for the first three years.* M. musculus *reached peak abundance from 2-3 years and then declined,* P. novaehollandiae *peaked at 3-4 years and exhibited a preference for floristically diverse vegetation. Species such as* R. fuscipes, R. lutreolus *and* A. stuartii *increased from years 3 to 5 while* I. obesulus *remained at low abundance until years 5 to 6. There was no recolonization by native small mammals five years after revegetation on mine rehabilitation sites. Conservation of small mammals will require implementation of suitable fire regimes, limitation of the spread of* P. cinnamomi *and improved regeneration procedures after mining.*

Introduction

In Victoria, 150 years of European settlement has resulted in extensive environmental changes leading to land degradation and loss of indigenous flora and fauna (State Conservation Strategy 1987). At least 24 species of mammal have disappeared and presently some 691 vertebrates and vascular plants are thought to be threatened (State Conservation Strategy 1987; Flora and Fauna Guarantee Act 1988; Ahern 1982; Ahern *et al.* 1985a,b; Menkhorst *et al.* 1987). In 1988 the Flora and Fauna Guarantee Bill was passed. The objective of this legislation is "to ensure that all Victoria's native flora and fauna, its native ecosystems and communities can survive, flourish and retain their potential for evolutionary development". It includes procedures for listing threatened species and threatening processes and the preparation of action statements and management plans. Management and planning decisions to implement such legislation effectively will depend on accurate information on species distribution, habitat preferences, population dynamics and responses to threatening processes. In this chapter we examine aspects of such information relating to small mammal communities in the eastern Otway Ranges.

Research directed at elucidating the habitat preferences of small mammals has concentrated on vegetation characteristics (Braithwaite & Gullan 1978; Braithwaite *et al.* 1978; Cockburn 1978, 1981; Cheal 1979; Fox & Fox 1984; Friend & Taylor 1985). Although some studies have produced variable results with regard to the relative importance of floristic and structural attributes, much light has been shed on the complexity of animal habitat relationships. *Pseudomys* species respond to floristic factors (Cockburn 1978, 1981; Posamentier & Recher 1974; Braithwaite *et al.* 1978; Fox & Fox 1978, 1984), while species such as *Rattus fuscipes* respond mainly to structural factors (Barnett *et al.* 1978; Braithwaite *et al.* 1978). Other species e.g., *Rattus lutreolus, Antechinus swainsonii* are affected by both structural and floristic factors (Braithwaite *et al.* 1978; Stewart 1979; Cockburn 1981). Factors such as habitat diversity, productivity, structural diversity and particularly floristic diversity have also been identified as being of major importance in explaining mammal species diversity and density within different major habitats (e.g., heathlands, forests, the tropics, arid zone) (Watts 1974; Posamentier 1976; Braithwaite *et al.* 1978; Taylor *et al.* 1984; Braithwaite *et al.* 1985; Morton & Baynes 1985; Fox 1983, 1985).

Investigations of disturbances such as fire, mining and logging have also contributed to our understanding of species distribution, habitat relationships and the structure of small mammal communities (Newsome *et al.* 1975; Fox & McKay 1981; Fox & Fox 1978, 1984; Cockburn 1978, 1981; Braithwaite *et al.* 1984; Lunney *et al.* 1987). Studies of the post-fire succession of small mammals after wildfire have shown that reponses vary between species. Some species reach peaks in abundance shortly after a wildfire and are described as early successional species *(Mus musculus, Pseudomys novaehollandiae),* while others have been identified as mid *(Antechinus stuartii, Sminthopsis murina)* or late *(Rattus fuscipes, R. lutreolus)* successional species (Newsome *et al.* 1975; Fox & McKay 1981; Fox 1982, 1983). Fox (1982) proposed that species which prefer habitat of different successional age achieve maximal population densities in response to the vegetation regeneration and leave the succession or decline in numbers as optimal conditions change (habitat accommodation model).

Studies of the recolonization of small mammals following sand mining showed that although there was a similar pattern to that found after fire there was an extended time axis (Fox & Fox 1978, 1984). This was attributed to the more complete disruption of the substrate caused by mining procedures. Few studies have investigated the impact of open-cut mining on the ecology of small mammals in Australia (Thomson 1980; Kentish 1983; Bourne *et al.* 1989).

The fungus *Phytophthora cinnamomi* can cause substantial disturbance to Australian vegetation communities. This water-borne pathogen induces a collection of symptoms called "dieback" (Weste & Marks 1987). The devastating effects of *P. cinnamomi* on sclerophyll communities have been well documented (Weste 1974, 1981; Dawson & Weste 1985; Kennedy & Weste 1986; Dawson *et al.* 1985). A decrease in canopy cover and eventual death of eucalypts, as well as alterations to the diversity and density of understorey species are symptomatic of the disease. There is little known however about the resultant effects on vertebrate communities, although there has been some work on soil and litter invertebrates in infected areas (Postle *et al.* 1986).

In the eastern Otways, Victoria (Fig. 1) 15 species of small mammals are distributed among a complex mosaic of plant communities (Wilson & Moloney 1985a,b). This chapter summarizes the results of studies on the major factors affecting the distribution, diversity and density of the small mammal communities (Kentish 1983; Wilson *et al.* 1986; Laidlaw 1987). We examine; the relationship of small mammals to the floristic and structural vegetation communities, the effects of disturbance factors including wildfire, mine rehabilitation activities and the presence of *P. cinnamomi* infection, and the autecology of the rare *Pseudomys novaehollandiae* (New Holland mouse) and its response to such factors. The implications of the results of these studies with regard to conservation of native small mammal communities and species are discussed and future management plans are proposed.

The study area

The study area (Fig. 1) occurs on a dissected plateau and is characterised by spurs and ridges with steep slopes and valleys to the west, deeply dissected hills to the north and flat hills to the east. Geologically the area consists of Tertiary sediments overlying older Cretaceous strata and generally the soils (e.g., sandy podzols, podzolic soils, lateritic podzols) are of low fertility (Abele 1971; Singleton 1973; Walbran 1971). The study area has predictable, reliable rainfall throughout the year, hot summers and cool winters.

More than 690 native vascular plant species, 25 native mammals (excluding bats) and 150 native bird species have been recorded (White, pers. comm.; Wilson & Moloney 1985a,b; Reilly 1985) and the area is one of the richest orchid areas in Victoria (79 species) (Carr, pers. comm.).

The vegetation communities are predominantly fire prone dry sclerophyll communities (open forest, woodland, heathland, shrubland) with less extensive areas of riparian open forest, fern gullies and damp open forest (Land Conservation Council 1985; Meredith 1986; Wark *et al.* 1987). The coastal fringe of this region consists of sand dune vegetation (Land Conservation Council 1985). Several vegetation communities in the study area have been identified as significant. These include heathy woodland which appears to be floristically the richest in Victoria; closed *Melaleuca squarrosa* shrubland which is now rare in western Victoria, and the Bald Hills heathland which represents the only remaining coastal dry heathland in western Victoria (Land Conservation Council 1987).

Previous studies have indicated that the diversity of the small mammal communities in the area is high (Kentish 1983; Wilson *et al.* 1986; Wilson & Moloney 1985a,b; Conole & Baverstock 1983). Several small mammal species which are rare or uncommon in Victoria occur in the area including *Antechinus minimus* (swamp antechinus), *Pseudomys novaehollandiae* and *Sminthopsis leucopus* (white-footed dunnart) (Land Conservation Council 1985; Ahern 1982; Ahern *et al.* 1985a,b; Menkhorst *et al.* 1987; Kentish 1982; Wilson 1986).

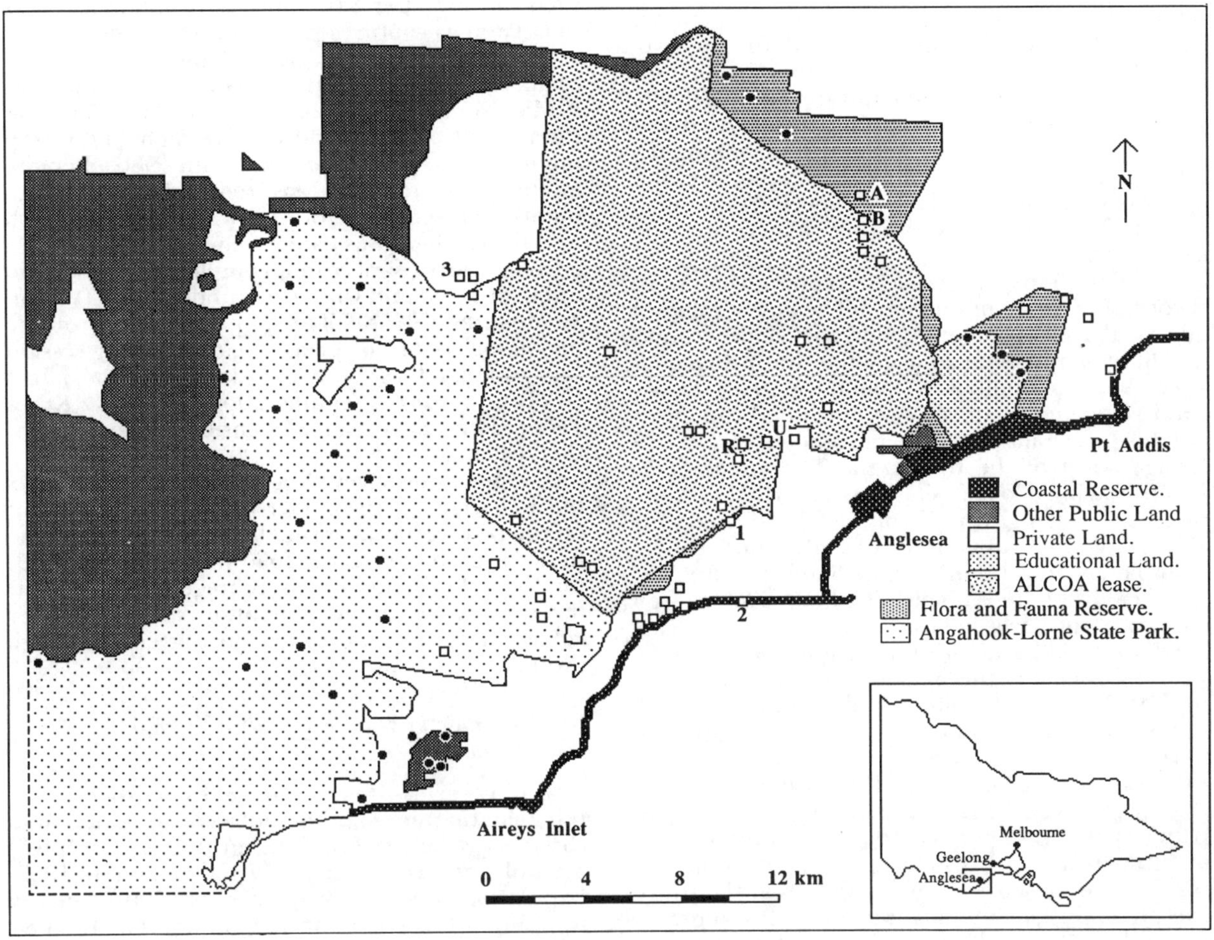

FIG. 1. Land tenure map of the study area and location of the study sites. (●) Trapping sites. (□) Vegetation/Trapping sites. (1, 2 and 3) Fire study grids. (A and B) *P. novaehollandiae* grids. (U and R) Mining study grids.

The area is predominantly public land and a major proportion (7 350 ha) has been leased to Alcoa Australia Pty Ltd since 1961 for brown coal mining (Fig. 1). The location of the mine, its operation and the methods of rehabilitation used are described by Kentish (1983) and Bourne *et al.* (1989). The western limit of the study area falls in the eastern section of the Angahook-Lorne State Park. The northern boundary of the study area is on land owned by International Harvester and used for road testing of vehicles.

The area has been subjected to clearing, logging, mining and other forms of human disturbance but remains dominated by native vegetation. In 1983 wildfires severely burnt approximately 40 000 ha of the Otway Ranges, mostly in the eastern ranges.

Methods

Detailed accounts of animal mark-recapture methods and vegetation analyses are described in Wilson and Moloney (1985a,b), Wilson *et al.* (1986); Laidlaw (1987); Bourne *et al.* 1989). The methods employed are briefly outlined below.

Distribution of small mammals within the land tenure areas

The distribution of small mammal species and their abundance within land tenure areas was assessed by combining the results of all trapping studies (n=56 sites) carried out in the study area during 1986-1987. The abundance was calculated as an average of captures per 100 trap nights at sites where the species occurred.

Relationship of small mammals to vegetation communities

Thirty-three sites trapped between September 1983 and April 1987 were selected for analysis. Twenty-nine of the sites were trapped (30 traps for three nights) in autumn and spring. Four were grid study sites where trapping was carried out every second month (100-158 traps for 3-5 nights) (Wilson & Moloney 1985b). The abundance and species richness (number of species) of small mammals at each site was calculated for the sampling period.

In 1986 floristics at each trapping site were recorded in 10×2 m quadrats. The cover value of all vascular plants together with rock, bare ground and litter were estimated using the Braun-Blanquet scale (Mueller-Dombois & Ellenburg 1974). Structural vegetation data recorded included the mean heights of strata and the percentage projective foliage cover of the tree strata. The status of *P. cinnamomi* infection at sites was recorded visually on a scale ranging from 1 (completely affected with <10% of the area free of infection) to 10 (completely free of infection). Signs of infection were chlorosis and finally death of understorey species particularly indicator species such as *Xanthorrhoea australis* and *Isopogon ceratophyllous*. The sites were assessed for the extent of burn suffered in the 1983 wildfires using a scale ranging from 1 (large area >1 ha unburnt) to 5 (area completely burnt).

Multivariate analyses (CSIRO Taxon P3 programmes; Ross 1982) were used to investigate patterns of variation in two data sets; data set 1, a quadrat data set consisting of binary information (presence or absence) of plant species at the study sites; data set 2, containing numeric measurements of small mammal species richness, abundance of small mammals, *P. cinnamomi* effects, fire effects, quadrat number per site, number of vascular plant species, tree height, tree cover, shrub height, herbaceous plant height.

Polythetic agglomerative classification was performed by the programme MACINF (Ross 1982). Attributes important in delimiting groups were defined by a diagnostic programme (GCOM). The hierarchies were expressed as dendrograms by a plotting programme DENDRO. For data set 2, ordination was based on principal co-ordinates analysis (PCOA). BACRIV, a diagnostic programme, was used to determine which characters were most highly correlated with principal axes. Ordination diagrams were drawn by hand to display two axes.

Recolonization by small mammals after wildfire

Following the wildfires an extensive trapping survey was initiated in the spring of 1983. Twenty-three sites were trapped in spring and autumn (1983-1988). These sites were located in burnt, partially burnt and unburnt habitats and were set over a three night period (i.e., 90 trap nights/site/trapping period). Species abundance was calculated as captures per 100 trap nights.

In May 1984 three grids (4.14 ha each) were established at three of the survey sites for intensive studies of recolonization (Fig. 1). Trapping was carried out on each grid for five nights (790 trap nights) at two month intervals until 1986 when the intensity of trapping was reduced to 300 trap nights per season. Grid 1 was located in heath woodland which was extensively burnt, grid 2 in a partially burnt sand dune complex and grid 3 in unburnt woodland with heath understorey (Wilson & Moloney 1985b). The population size of the major species on the grids at each trapping session was estimated as numbers "known to be alive" (KTBA) a direct enumeration method (Krebs 1966). The yearly average density per hectare was then calculated.

A vegetation survey to record floristic and structural changes in the understorey was carried out on the three grids in spring for four years after the fire. For each species the following were recorded: number of individuals; percentage cover; height; flowering time and regeneration strategy. Cover values were also recorded for litter and bare ground.

Recolonization by small mammals after mine rehabilitation

Trapping grids (Fig. 1) were established on a rehabilitating area and an adjacent unmined area (Bourne *et al.* 1989). The grids were trapped monthly between March 1981 and March 1982. Population sizes were estimated as KTBA. Survey trapping was carried out on both areas in 1983-84 to determine if further changes had occurred.

A vegetation survey was carried out on the rehabilitated and unmined grids (Bourne *et al.* 1989). Floristic and structural data were recorded in quadrats on the rehabilitated areas (area A- revegetated in 1979, area B- revegetated in 1981) and the unmined area (area C). The relative constancy (proportion of quadrats in which a species occurs) was calculated for each community and species having high (>50%) values were selected as character species. The percentage composition of shrubs, herbs, graminoids and ferns on the three areas was calculated and compared. Further vegetation surveys were carried out in 1983-84.

Population ecology and habitat preferences of P. novaehollandiae

After the 1983 wildfire *P. novaehollandiae* was absent from the one known prefire location. Between 1983 and 1988 survey trapping (30-50 traps for three nights) was carried out at 10 sites in areas of suitable habitat. Trapping grids were established at two sites east of the Anglesea River

Fig. 1). Trapping was carried out for three nights every second month on Grid A (5×20 sites with 20 m trap spacing) from July 1985 until April 1988. On Grid B (3×20 sites) trapping was carried out every second month between May 1986 and April 1988. The population size of *P. novaehollandiae* on the two grids was estimated as the minimum number of animals alive (MNA) during a trapping session.

Floristics were described at every trap site in 5×5 m quadrats. The cover/abundance (Braun-Blanquet scale) of vascular plant species was recorded and the data were sorted (Magic programme) into clusters of quadrats having similar species composition (Gullan 1978). Vegetation groups were defined by major discontinuities in the gradient of variations evident in the synthesis table. Species having high relative constancy values (character species) were selected to represent major differences between vegetation groups.

Vegetation structure was measured at five regular intervals on a 2 m transect through each trap site. A pole (1.8 m height, 1 cm diameter) was dropped into the vegetation and the number of contacts (maximum of five) of vegetation in 20 cm intervals recorded. Presence and absence only of vegetation above 1.2 m was recorded. An index of vegetation volume was estimated at each height class as the total number of contacts for the five samples. An index of vegetation volume was calculated at each height class and indices of vertical and horizontal diversity were calculated for each site (see Rosenzweig *et al.* 1975).

The dispersion of *P. novaehollandiae* (captures per 100 trap nights per trap site) was analysed to determine habitat preferences. One-way analysis of variance was used to determine whether animals selected vegetation floristic groups at random. The Student-Newman-Keuls (Sokal & Rohlf 1969) multiple comparison test was used to determine overall floristic preferences.

The effect of vegetation structural parameters (vegetation volume at height classes, vertical and horizontal diversity of vegetation, percentage of bare ground and litter) on *P. novaehollandiae* captures were analysed by stepwise multiple regression.

Results

Distribution of small mammals within the land tenure areas

The distribution and abundance of species together with the species richness of communities in land tenure areas are shown in Table 1. Species not captured during 1986-1987 but captured previously are indicated, as are species now thought to have disappeared from the study area. Species such as *A. stuartii*, *R. fuscipes* and *R. lutreolus* are found in all land tenure areas and occur with high abundance at a large proportion of sites. *I. obesulus* and *S. leucopus* occur in most land tenure areas. The rarer species *C. nanus*, *P. nasuta*, *A. swainsonii*, *A. minimus*, *P. tridactylus* and *P. breviceps* are restricted in their distribution and occur in less than three land tenure areas. *P. novaehollandiae* is found only in the Alcoa lease and flora and fauna reserves. The highest number of species was recorded within the Alcoa lease area.

Relationship of small mammals to vegetation

The results of the MACINF classification of 90 quadrats are summarized in Figure 2 which shows the significant species contributing to major

TABLE 1. Mean richness, distribution and abundance of species in land tenure areas. (n=number of sites, a=mean abundance-captures/100 trap nights).

	Land Tenure Units											
	Alcoa lease (n=16)		Angahook-Lorne State Park (n=22)		Flora and Fauna Reserves (n=9)		International Harvester (n=2)		Coastal Reserve (n=8)		Roadside Reserve (n=1)	
Native Species	n	a	n	a	n	a	n	a	n	a	n	a
A. minimus	2	*#									3	*#
A. stuartii	9	2.716	21	7.453	4	2.266	2	1.050	1	1.113	1	4.444
A. swainsonii							1	*#				
C. nanus	2	*			2	0.185						
I. obesulus	6	0.740	4	1.177	1	*			1	1.113	1	2.222
P. breviceps			1	*			2	*				
P. nasuta	1	*	1	1.100			1	0.080				
P. novaehollandiae	1	1.787			2	1.471						
P. tridactylus	1	*#					1	*				
R. fuscipes	8	5.833	21	12.662	3	0.380	1	0.100	6	5.926	1	*
R. lutreolus	9	4.444	19	8.200	4	3.817	1	0.200	6	11.481		
S. leucopus	5	0.444	1	1.496	3	1.757	1	0.010	2	1.113		
Number spp.	10		7		7		8		6		3	
Mean richness	2.938±2.144		3.590±0.908		1.666±2.500		5.000±2.828		2.833±0.753		3.000±0.000	

* –species captured prior to 1986/87.
#–species possibly locally extinct.

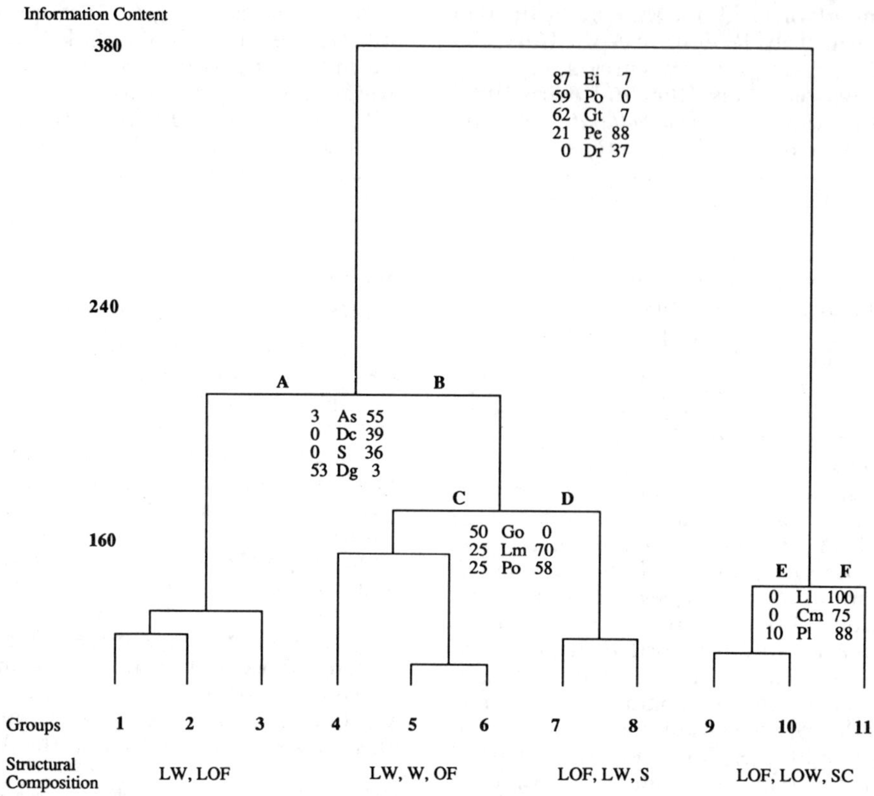

FIG. 2. Dendogram showing hierarchial classification of the floristic quadrat data into 11 groups. Species associated with the main divisions and their percentage frequency of occurence in each division are shown. The structural components include: OF=open forest; LOF=low open forest; LOW=low open woodland; W=woodland; LW=low woodland; S=shrubland; SC=scrub. Species abreviations are: *Acrotriche serrulata,* As; *Clematis microphylla,* Cm; *Dantrhonia caespitosa,* Dc; *Dichrondra repens,* Dr; *Dillwynia glaberrima,* Dg; *Epacris impressa,* Ei; *Gonocarpus tetragynus,* Gt; *Goodenia ovata,* Go; *Lepidosperma laterale,* Ll; *Leptospermum myrsinoides,* Lm; *Platylobium obtusangulum,* Po; *Poa labillardierii,* Pl; *Pteridium esculentum,* Pe; *Stipa* spp. *S.*

divisions. The classification was truncated at the 11 group level which comprises six major divisions A-F. "Group" and "division" refer here to artificial sets of quadrats defined by the computer. "Community" will be used to refer to vegetation types in the field.

At the highest level of the classification hierarchy, quadrats from sclerophyll communities (groups 1-8) were separated from those of fern gully, swamp and sand dune (groups 9-11). The sclerophyll quadrats are characterized by high frequencies of sclerophyllous shrubs, while groups 9-11 are characterized by high frequencies of the fern *Pteridium esculentum* and other herbaceous species. The sclerophyll quadrats of the hierarchy are divided into four major divisions A-D. Quadrats belonging to division A contain higher frequencies of sclerophyllous shrubs than those of divisions B-D. At the next level of the hierarchy, quadrats from forest (division C) are separated from those of woodland (division D). The groups containing quadrats from sand dune, fern gully and swamp are divided to form divisions E (fern gully, swamp and sand dune) and division F (sand dune) (Fig. 2).

The composition of the groups in terms of structure are shown in Figure 2. Groups 1-3 contain quadrats from low woodland and low open forest. These communities have the highest species richness (species number per quadrat) of the sclerophyll communities. They are dominated by low, often multistemmed trees of *Eucalyptus obliqua* and *E. willisii.* Groups 4-6 contain quadrats from woodland and forest with relatively low species diversity. Groups 7 and 8 contain quadrats from low open forest, low woodland and shrubland which contain intermediate species diversity with respect to the other sclerophyll groups. Groups 9-11

TABLE 2. MACINF groups vs abundance, species richness and occurence of native small mammals. Abundance=number of captures/100 trap nights (1983-1987).

	MACINF Groups										
	1	2	3	4	5	6	7	8	9	10	11
Mean abundance	3.100 ±2.300	0.200 ±0.500	3.800 ±2.000	3.200 ±1.500	0.200 ±0.000	4.400 ±0.600	1.600 ±1.900	0	5.000 ±2.600	2.400 ±3.100	9.500 ±2.100
Mean nos. species	4.400 ±2.400	0.500 ±1.000	4.500 1.378	3.000 ±0.577	1.000 ±0.000	2.000 ±0.000	2.600 ±1.689	0	3.000 ±0.577	1.000 ±1.414	3.500 ±0.707
A. minimus	–	–	–	–	–	–	–	–	+	–	+
A. stuartii	+	+	+	+	+	–	+	–	+	–	–
A. swansonii	+	–	–	–	–	–	–	–	+	–	–
S. leucopus	+	–	+	–	–	–	+	–	+	–	+
I. obesulus	+	–	+	+	–	–	+	–	–	–	–
P. nasuta	+	–	+	–	–	–	–	–	–	–	–
P. breviceps	+	–	–	+	–	–	–	–	–	–	–
C. nanus	+	–	+	–	–	–	–	–	–	–	–
P. tridactylus	+	–	+	–	–	–	–	–	+	–	–
R. fuscipes	+	–	+	+	–	+	+	–	+	+	+
R. lutroelus	+	–	+	+	–	+	+	–	+	+	+
P. novaehollandiae	–	+	+	–	–	–	+	–	+	–	–
Total Nos. Species	10	2	9	5	1	2	6	0	8	2	4

are from low open forest, low open woodland and scrub all of which have low species diversity and few if any species typical of sclerophyll communities.

The abundance of small mammal species and the species richness of communities in relation to the MACINF groups are summarized in Table 2. The highest number of native small mammal species were found in quadrat groups 1 (Mean=4.4 species) and 3 (Mean=4.5 species). Species richness was lowest in quadrat groups 2 (Mean=0.5 species) and 8 (Mean=0.0 species). The highest numbers of native small mammal species were recorded in quadrat groups 1 and 3 (10 and nine respectively). The highest mean abundance of native small mammals occurred in quadrat group 11 (9.50) and group 9 (5.00). Quadrat groups 2 and 5 had the lowest abundance of native mammals (0.2) and no native species were captured in group 8.

Several native species (*Antechinus stuartii*, *R. fuscipes* and *R. lutreolus*) occurred in seven to eight of the quadrat groups. Seven native species (*A. minimus*, *A. swainsonii*, *Perameles nasuta*, *Petaurus breviceps*, *Cercartetus nanus*, *Potorous tridactylus* and *Pseudomys novaehollandiae*) were found in only three or fewer quadrat groups. *P. nasuta*, *P. breviceps*, *C. nanus* and *P. tridactylus* did not occur in any of quadrat groups 5 to 11.

The results of the ordination analysis of the numerical data are summarized in Figure 3. The total information content was vector 1 (23%), vector 2 (17%) and vector 3 (13%). Data from the first two vectors are presented here. One of most obvious divisions occurred along axis 1, where sites from groups A and B which were mainly sand-dune communities were separated from others on the basis of higher native mammal abundance and lower proportion of *P. cinnamomi* infestation. Another important pattern involved sites belonging to group E (mainly burnt scrub and woodland sites) which were separated from all other groups on the basis of lower plant species richness and abundance of small mammals, higher fire and *P. cinnamomi* effects and lower tree cover.

Recolonization of small mammals after wildfire

The recovery of the native small mammals on the burnt survey sites in the first three years post-fire was slow compared to the partially burnt sites where numbers increased steadily over the whole survey period (Fig. 4a,b,c). Abundance of native small mammals also increased with time on the unburnt sites. *Mus musculus* (house mouse) bred and increased to extremely large numbers at all sites until three years post-fire when the population declined rapidly. It achieved the greatest abundance on partially burnt sites in the third year after the fire (Fig. 4b).

The relative abundance at burnt sites and the number of sites occupied by the major species captured in the survey trapping are shown in Figures 5a and b. *M. musculus* reached its peak abundance and distribution (maximum number of sites occupied) by the third year. The number of sites occupied by *R. fuscipes* increased until the fifth year. Three species *A. stuartii*, *R. lutreolus* and *Isoodon obesulus* continued to occupy more sites until six years post-fire. *I. obesulus* was slower to move into new sites than the other species and did not show a substantial rate of recolonization until the sixth year (Fig. 5b). *R. fuscipes* was the first

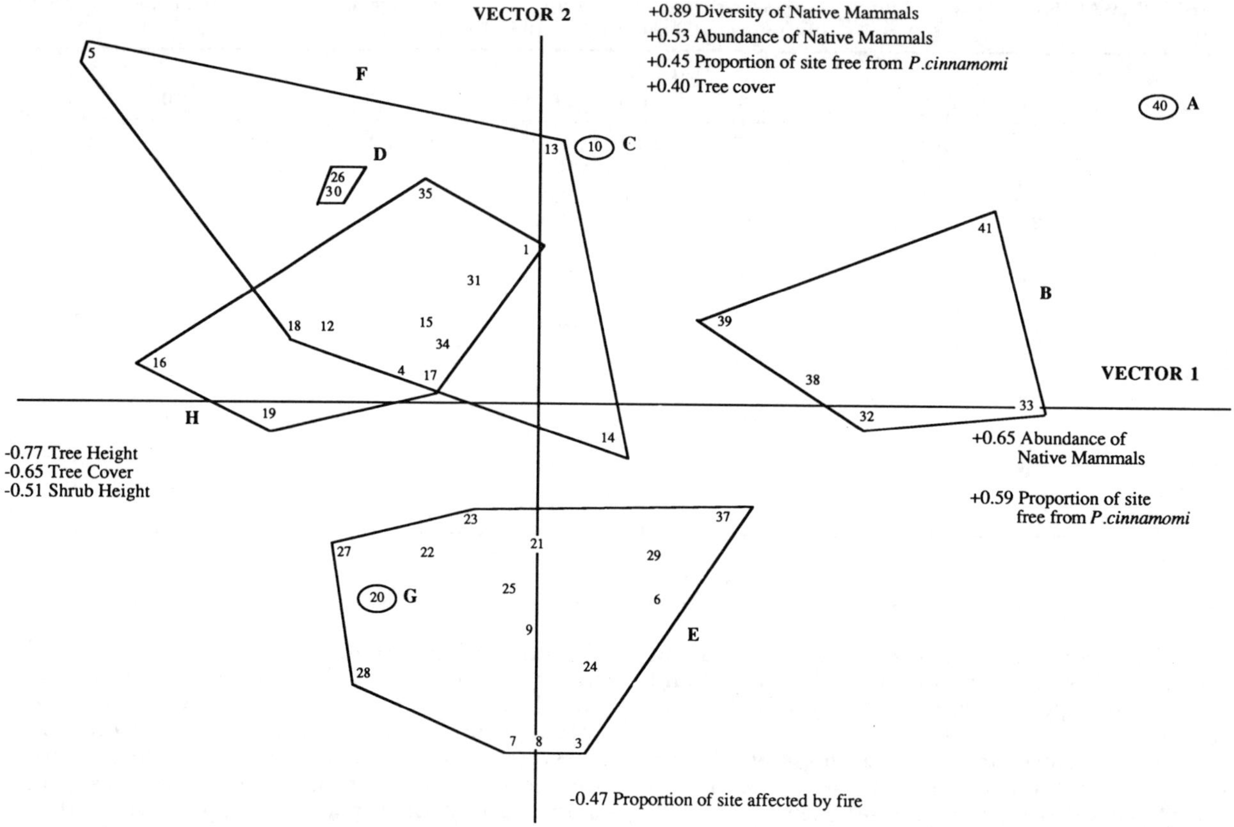

FIG. 3. Ordination of the numeric data (small mammal species diversity; abundance of small mammals; proportion of site free from *P. cinnamomi*; proportion of site affected by fire; quadrat number per site; number of vascular plant species; tree height; tree cover; shrub height; ground stratum height) at study sites, showing vectors 1 and 2. (n=sample number of site).

native species to show a substantial population increase (2-3 years post-fire). It achieved maximal abundance in the fifth year. *A. stuartii* reached maximum abundance in the fifth year and then declined. *I. obesulus* exhibited a slow recovery rate and may well have been increasing in abundance in the sixth year post-fire.

The changes in average yearly densities of the major small mammal species captured on the three study grids are presented in Figures 6a,b and c. Species captured in low numbers included *C. nanus*, and *I. obesulus* (grid 1); *A. stuartii* (grid 2); *A. swainsonii, S. leucopus, P. nasuta, P. tridactylus,* and *P. breviceps* (grid 3). Changes in species densities on the burnt and partially burnt grids showed similar trends. *M. musculus* was the dominant species in the second year post-fire and exhibited a high population density (>30/ha). *R. fuscipes* and *R. lutreolus* increased in density from the third to fifth years post-fire. The populations of *S. leucopus* were low but stable throughout the study. The population densities of all species on the unburnt grid were extremely low compared to those found on the burnt and partially burnt grids.

There was substantial vegetation regeneration on the burnt and partially burnt grids in the first three years. Percentage cover of ground and mid-storey vegetation increased from sparse (mean±SD), (5.0±4.0) in the first year to 57.4±22.6 (burnt grid) and >72.4±5.2 (partially burnt grid) in the third year. The vegetation on both grids reached pre-fire heights by the third year. Vegetation regeneration was mainly by regrowth (approximately 50% of species) on both grids.

All small mammal species present in the area prior to the 1983 wildfire have been captured after the fire (Table 3). However, species such as *A. minimus, P. tridactylus, A. swainsonii, P. novaehollandiae, P. nasuta, C. nanus* and *P. breviceps* were captured at few sites and were present in low numbers. Species losses at particular sites were recorded (e.g., *A. minimus, P. novaehollandiae* and *S. leucopus*). Three species *A. minimus, P. tridactylus* and *A. swainsonii* have not been captured in the area for several years and may be locally extinct.

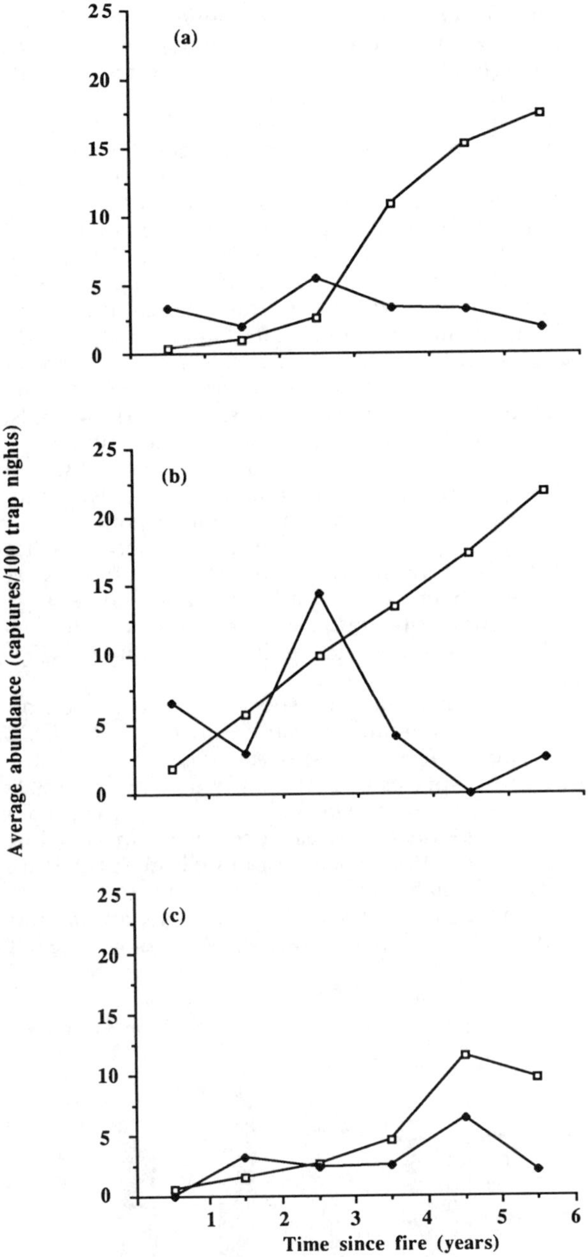

FIG. 4. Yearly average abundance of native small mammals (□–□) and *M. musculus* (⧫–⧫) at the survey sites. (a) burnt, n=22; (b) partially burnt, n=5; (c) unburnt, n=4.

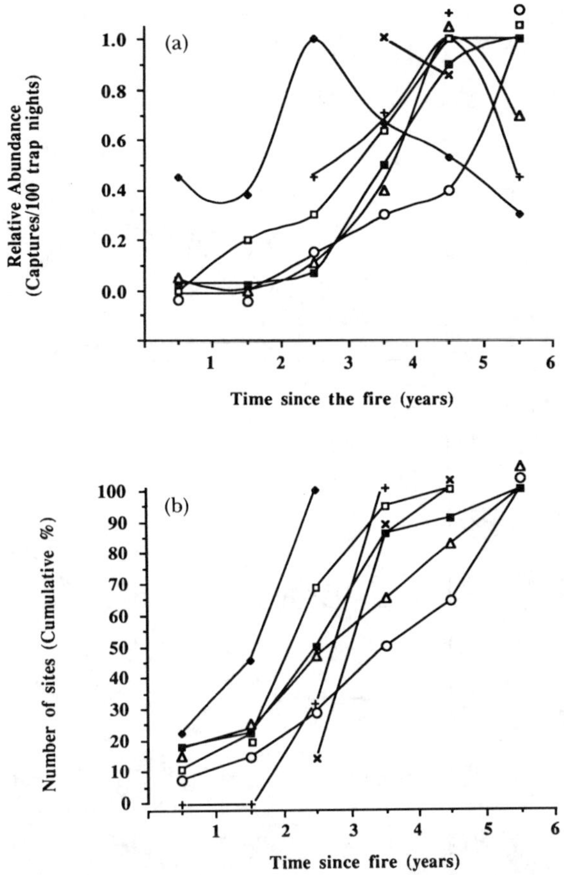

FIG. 5. (a) Relative abundance of small mammals after fire. Values for each species are shown as a proportion of the species' maximum, abundance. (b) Recolonization of small mammals after fire as represented by the cumulative number of sites occupied. △ *A. stuartii;* ○ *I. obesulus;* ⧫ *M. musculus;* + *P. novaehollandiae;* □ *R. fuscipes;* ■ *R. lutreolus;* × *S. leucopus.*

TABLE 3. Small mammal species captured in prefire and postfire studies. *possibly extinct locally (n=no. of sites).

Species	Prefire #	Postfire
A. minimus	+	+ (3)*
A. stuartii	+	+ (17)
A. swainsonii	+	+ (1)*
S. leucopus	+	+ (7)
I. obesulus	+	+ (14)
P. nasuta	+	+ (1)
P. tridactylus	+	+ (1)*
R. fuscipes	+	+ (19)
R. lutreolus	+	+ (22)
P. novaehollandiae	+	+ (4)
C. nanus	+	+ (2)
P. breviceps	−	+ (1)
R. norvegicus	+	+ (13)
M. musculus	+	+ (22)
R. rattus	+	+ (4)

#Conole & Baverstock (1983); Kentish (1983); Wilson (1986); Wilson & Bourne (1984); Wilson *et al.* (1986).

Recolonization by small mammals after mine rehabilitation

At the completion of the survey in 1982 the vegetation on area A (revegetated in 1979) consisted of a sparsely scattered shrub/tree layer (*Eucalyptus viminalis, Acacia pycnantha, A. verticillata*) and a ground cover of *Juncus pallidus, Poa sieberiana* and *Agrostis avanaecea* (Kentish 1983; Bourne *et al.*

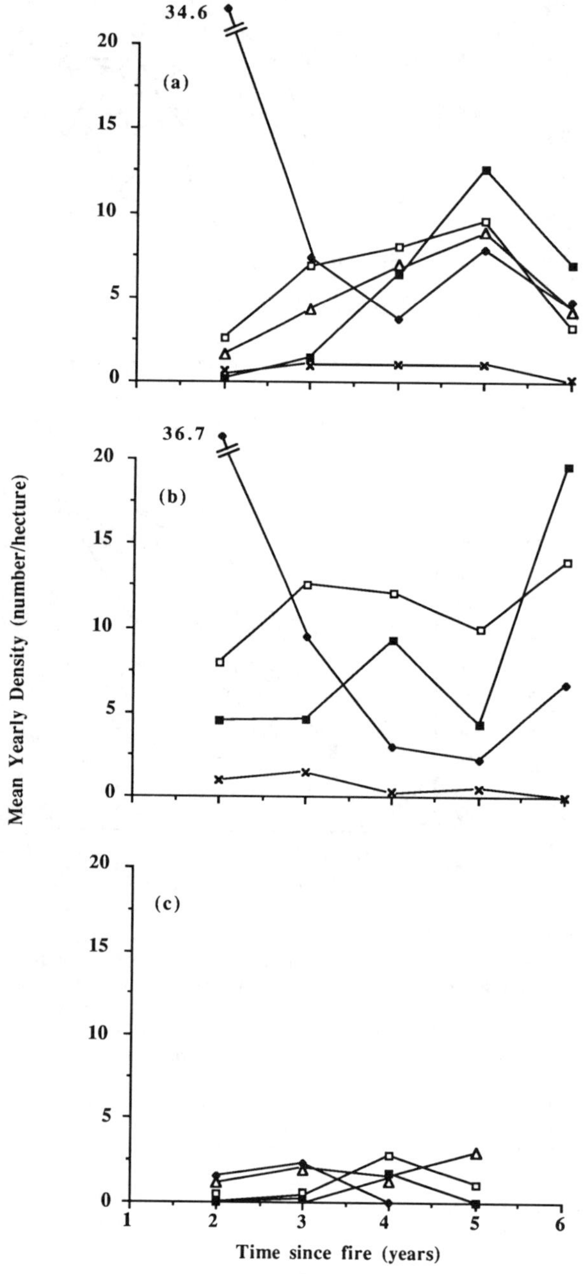

FIG. 6. Mean yearly densities of small mammal species on the fire study grids. (a) burnt; (b) partially burnt; (c) unburnt (symbols as for Fig. 5).

1989). Area B (revegetated in 1981) had a developing tree layer *(E. viminalis),* no shrub layer and a cover crop of *Secale cereale, J. pallidus* and *Senecio* spp. The vegetation on the adjacent unmined area C consisted of two recognizable major communities. Community I was located on dry soils and consisted of a sparse, low tree layer *(E. willisii, E. baxteri)* with a dense sclerophyllous shrub layer *(Leptospermum myrsinoides, L. juniperinum, Banksia marginata).* Community II occurred on a natural drainage area and had a more dense tree canopy and a less prominent shrub layer.

Comparison of the characteristic understorey species (relative constancy) of the revegetated and unmined sites showed that both rehabilitated areas had lower proportions of woody dicotyledons and relatively high proportions of characteristic graminoids and introduced species (Kentish 1983; Bourne *et al.* 1989). Characteristic species present in the unmined area that were absent from the revegetated sites included *B. marginata, Casuarina pusilla, Hibbertia stricta, Goodenia geniculata* and *Burchardia umbellata.* The percentage composition of shrubs, graminoids, ferns and herbaceous species for the areas are shown in Figure 7. The revegetated areas differed markedly from unmined areas with respect to these categories where the percentage composition of graminoids was much greater than that of shrubs. On the unmined area community II differed from community I. The percentage composition of herbaceous species was higher and that of shrubs lower in community II (Fig. 7).

Three small mammal species were captured on grid 1 (revegetation site) during 1981-1982 (Bourne *et al.* 1989). *Mus musculus* was the most numerous species and ranged in density from 2.4-12.5/ha. Two native species *A. stuartii* (one capture) and *R. lutreolus* (three captures) were in very low densities. On the adjacent unmined site (grid 2) six species of small mammals were captured, the most common being *S. leucopus, P. novaehollandiae* and *R. lutreolus. A. minimus* and *I. obesulus* were each

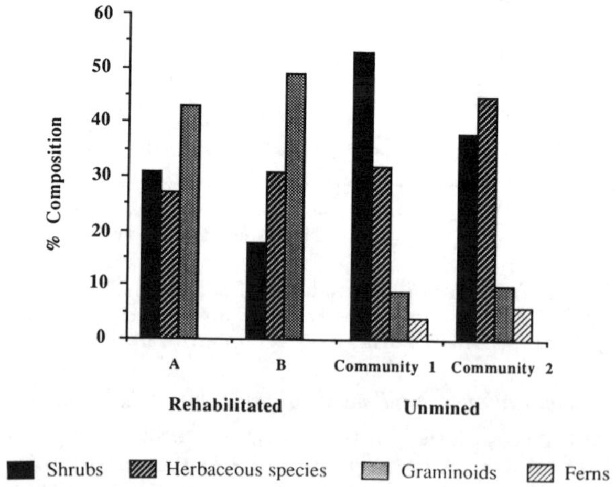

FIG. 7. Percentage composition of vegetation groups on the mining study grids in 1983-84. A — revegetated 1979; B — revegetated 1981.

captured on one occasion and *M. musculus* was trapped infrequently. The KTBA of *P. novaehollandiae* was low throughout the study. The range of population densities for *S. leucopus, P. novaehollandiae* and *R. lutreolus* were 0.3-2.3, 0.3-1.0, 0.8-1.8/ha respectively.

The vegetation on the revegetation sites (A and B) in 1983-84 consisted of a low tree layer (up to 3 m) of *E. viminalis* and *Acacia* spp., and a sparse lower layer of sedges, shrubs and graminoids (up to 60 cm), e.g., *Juncus* spp., *Stipa* spp., *Danthonia setacea, L. juniperinum, Platylobium obtusangulum.* The percentage ground cover was 1-5%, and introduced species constituted 7% of total species number. Total species numbers for A and B were 63 and 50 respectively. The vegetation on area C was burnt in the 1983 fire. When surveyed 18 months later the dominant eucalypts (*E. obliqua* and *E. willisii*) were regrowing. The percentage cover of the understorey ranged from 5-50% and a total of 94 species were recorded. Major understorey species included *C. pusilla, B. marginata, P. obtusangulum, Xanthosia pusilla, L. myrsinoides, X. australis* and *Hypolaena fastigata.*

The trapping surveys (1983-1984) resulted only in the capture of 10 *M. musculus* on both the revegetated and unmined areas. Further survey trapping on the unmined area from 1986 to 1988 (4-6 years post-fire) has resulted in the capture of low numbers of several native species (*R. lutreolus, R. fuscipes, I. obesulus* and *S. leucopus*). No trapping was carried out on the rehabilitated sites over this time period.

Population ecology and habitat preferences of P. novaehollandiae

Prior to the 1983 wildfire the population of *P. novaehollandiae* studied on ummined habitat close to the open cut mine (see above) was the only known location of the species in the area. This population did not survive the fire. The species was captured at four of the ten sites survey trapped between 1983 and 1988. The sites were located on areas of flat to undulating topography with predominantly sandy soils. The vegetation consisted of woodland or open forest with a heathy understorey. Smaller patches of *L. juniperinum* scrub were also present.

The population (minimum numbers alive) of *P. novaehollandiae* on the two study grids was low. On grid A it ranged from one to a maximum of nine in March 1987. The mean (±SD) monthly MNA was low (2.2±1.3) from July 1985 to May 1986. It increased to 5.7±2.9 from July 1986 to May 1987 and decreased to 4.4±2.3 between July 1987 and April 1988. On grid B *P. novaehollandiae* was present in low numbers and one individual only was captured between November 1987 and April 1988.

Five groups representing the major floristic types were identified on the two trapping grids (Table 4). Floristic Group 1 occurs on flat dry areas of Grid A and has dense vegetation to 80 cm, high percentage cover of litter and bare ground, and is floristically rich. Group 2, located predominantly on the sloping and wetter areas, is similar to Group 1 but may be distinguished by the higher cover/abundance of *L. juniperinum, L. myrsinoides* and *P. obtusangulum* and lower species richness. In swampy areas a mid-storey (1.2 m) of *L. juniperinum* was present. The percentage cover of litter is high while that of bare ground is low. Floristic Group 3 is mainly restricted to the elevated areas of dry gravelly soil on grid B. The character species of this group are rarely recorded in Groups 1 and 2. This group has a similar number of species per quadrat to Group 1. The vegetation is dense to 100 cm, litter cover is low and bare ground cover is high. Floristic Group 4 occurs in scattered patches on both grids and is characterized by low species richness. The vegetation cover is dense to 100 cm and there is low litter cover and a high percentage of bare ground. Group 5 located in a wet depression is characterized by low species numbers and high frequencies and cover/abundances of *Gahnia radula, L. juniperinum* and *Tetrarrhena distichophylla.* Dense vegetation occurs to 120 cm, litter cover is medium and bare ground is low. Three sites on Grid A were unclassified and occurred in a disturbed area.

One-way analysis of variance showed that *P. novaehollandiae* did not select the floristic groups at random (Table 5). The species was captured more frequently in floristic group 1 (grid A) and 3 (grid B). These vegetation groups differed in species composition, however, both had high floristic diversity compared to the other groups. A stepwise multiple regression analysis of structural parameters against mean captures at all sites (n=160) showed that structural parameters were not highly significant. However, within the preferred floristic groups (1 and 3) vegetation volumes at 0-20 cm and 101-120 cm did contribute to the dispersion of the species (Table 6). This indicates that within these floristic groups *P. novaehollandiae* prefers dense low cover.

Discussion

The relationship of small mammals to vegetation

The vegetation of the Eastern Otway Ranges is composed of a mosaic of predominantly sclerophyllous communities. The MACINF classification identified a large number of forest, woodland and heathland communities which represent a continuum of vegetation types. These were separated from other plant communities such as fern gully, swamp and sand dune on the basis of the higher species diversity (particularly of sclerophyllous

TABLE 4. Floristic groups identified on grids A and B. (n=number of sites).

	Floristic Group				
	1(n=33)	2(n=57)	3(n=17)	4(n=13)	5(n=33)
Upperstorey	*E. obliqua* *E. willisii*	*E. obliqua* *E. willisii*	*E. obliqua*	*E. obliqua*	*E. obliqua*
Characteristic understorey species	*E. impressa* *H. stricta* *A. myrtifolia* *B. marginata* *C. pusilla* *L. juniperinum* *G. radula* *L. semitieres* *P. sieberiana*	*L. juniperinum* *L. myrsinoides* *P. obtusangulum*	*D. glaberrima* *H. fastigiata* *T. distichophylla* *A. xiphoclada* *E. minus*	*L. juniperinum* *P. esculentum* *G. radula* *T. distichophylla*	*G. radula* *L. juniperinum* *T. distichophylla*
Mean number of species per quadrat	14.30	11.70	13.00	7.15	7.08

shrubs). The range of plant communities appears to reflect soil type, climatic and edaphic factors.

The highest species rich small mammal communities were recorded in two sclerophyllous communities (MACINF groups 1 and 3) which have the highest plant species diversity of the sclerophyll communities. The parallel between plant species richness and small mammal richness has been found in other studies and is likely to be a reflection of the increased variability of food resources (Braithwaite *et al.* 1985).

Although the study showed that species-rich sclerophyll vegetation has high numbers of small mammal species, no native small mammals were present in two MACINF groups (2 and 8) which had high plant diversity. The absence of native species in these groups appears to be related to previous site disturbances. Sites in group 2 have been logged and the one site in group 8 had been cleared. Although the natural vegetation at the latter site has regenerated, the area is isolated from extensive native vegetation by farmland which precludes recolonization by native mammals. The absence of native small mammals in group 5 was predictable as the site is disturbed by the presence of an old plantation of *Pinus radiata* and has a substantially modified understorey.

The greatest abundance of native small mammals was recorded in the non-sclerophyllous MACINF groups 11 and 9. Both of these groups have a high proportion of grasses and sedges in their understorey and occur in wet areas (sand dune swale and wet fern gully). These communities have richer soil and do not suffer water stress as compared to the sclerophyll communities. In a study of the heathlands of Australia, California and South Africa, Fox *et al.* (1985) found that the abundance of small mammals was lower in the Australian communities and concluded that this was due to the lower soil nutrient levels which result in lower plant biomass. The abundance of small mammals in the Eastern Otways thus appears to be related to the productivity of the plant communities in a similar way.

The effects of fire and P. cinnamomi *infection on small mammals*

The ordination analysis identified two disturbance factors, fire and *P. cinnamomi* infection, as being associated with low diversity of small mammal species. Thus low open woodland and low open forest that suffered least in the 1983 fire or had low *P. cinnamomi* infection had the highest plant and small mammal species richness. The analysis also indicated that the presence of *P. cinnamomi* was a major factor affecting small mammal abundance.

Phytophthora cinnamomi was first recorded in the Eastern Otways in the 1970s (Weste & Marks 1974; Parsons *et al.* 1977). It can have a devastating effect on sclerophyll vegetation, eliminating up to 60% of the plant species present (Weste 1974; Dawson *et al.* 1985; Kennedy & Weste 1986). There are few

TABLE 5. Mean captures and habitat preferences of *P. novaehollandiae* in floristic groups. (100 trap nights/ Site±S.D.).

Floristic Group				
1	2	3	4	5
3.39 ±5.3	1.21 ±1.9	2.67 ±4.1	2.10 ±3.6	0.82 ±2.3

$F=3.39^*$, $p<0.05$

Student-Newman-Keuls
Group 5<Groups 2,4<Groups 1,3

TABLE 6. Stepwise multiple regression coefficients for *P. novaehollandiae* dispersion in relation to structural parameters within floristic groups 1 and 3.

Floristic vegetation group	Coefficient and % variance step 1	Coefficient and % variance step 2	Cumulative %
1	Vegetation volume 101-120 cm.	Vertical diversity	29.28
	B= 1.81 r^2=16.38	B=−3.3 r^2= 12.9	
3	Vegetation volume		24.77
	0-20 cm. B= 0.40 r^2=24.77		

studies of the effect of the pathogen on faunal communities; however, Postle *et al.* (1986) have shown that the soil and litter fauna in infected areas is markedly reduced. Although there are no studies of the effects of *P. cinnamomi* on small mammal communities, factors such as the reduction of plant species diversity, elimination of specific species, reduction in cover and a decrease in plant productivity would be expected to ultimately affect small mammals.

The study sites were chosen largely on the basis of their post-fire status and prefire data. Thus site variation was considerable particularly with respect to infection rates of *P. cinnamomi.* Studies are presently being undertaken to map the distribution of the pathogen in the area, to select sites with less variability and to examine the distribution of small mammal species and *P. cinnamomi* in more detail.

Although the immediate effects of the 1983 wildfire were not studied, the extremely low numbers recorded in spring 1983 indicate that few animals survived the fire or the immediate post-fire period. A similar response was reported after the 1972 Nadgee fire where native mammal populations were severely depleted (Newsome *et al.* 1975; Newsome & Catling 1983). Populations of native small mammals at our burnt sites remained at extremely low levels for three years after the fire. They survived and recovered at a greater rate at partially burnt sites. The most significant small mammal species in the early post-fire years was the introduced *M. musculus.* Prior to the fire populations of this species in the study area were low (Kentish 1983; Wilson *et al.* 1986). The large peak in the population in the second to third year after fire and its subsequent decline is comparable to patterns observed in previous post-fire studies (Newsome *et al.* 1975; Fox 1982; Catling 1986; Lunney *et al.* 1987). The abundance of the species during the first to second years in partially burnt survey sites indicates that it survived in these sites and moved into regenerating burnt sites. Unburnt sites did not support large *Mus* numbers.

Mus musculus is known for its ability to survive adverse conditions and exploit opportune changes, especially in Australian cereal crops where its populations periodically increase to plague proportions (Berry & Jackson 1975; Newsome 1969 a,b; Bomford & Redhead 1987). The response of *Mus* to fire in Australian communities is considered to be opportunistic but has not been fully elucidated. It has been suggested that *Mus* fills vacant niches in the post-fire environment when native species are rare or non existent (Catling & Newsome 1981). The results of the present study support such an interpretation as the population numbers of *Mus* and native species show reciprocal trends. The large increase in *Mus* populations resulted from increased breeding but the factors which induce breeding activity in the early post-fire environment are not clear. Wilson and Moloney (1985b) proposed that new growth of vegetation early in the post-fire period may be such a factor. Studies of *M. musculus* on cereal farms have shown that food quality, in particular protein, can increase breeding success (Bomford & Redhead 1987).

The responses of native species recorded on the grid and survey study sites were similar. *A. stuartii* reached its maximum abundance in burnt sites five years after the fire and continued to recolonize new sites up until the sixth year. Similar responses have been found at Nadgee and Myall Lakes (Newsome *et al.* 1975; Fox & McKay 1981; Fox 1982). Of the two *Rattus* species in our study, *R. fuscipes* increased earlier (years 2-4). It achieved maximum colonization of sites and maximum abundance in the fifth year. The response to fire was similar to that described at Nadgee by Newsome *et al.* (1975) but differed from that at Myall Lakes where the species was present only in low numbers on a burnt plot and did not breed successfully until the fourth year (Fox 1982). *Rattus lutreolus* was slower to recolonize sites and to increase abundance; however, it was still increasing both in abundance and number of sites occupied in year six. In contrast at Myall Lakes the species was absent from a burnt plot for 3.5 years despite its presence prior to the fire and its presence on adjacent unburnt vegetation (Fox 1982). In the present study both *Rattus* species were present in low abundance at unburnt sites.

Sminthopsis leucopus was captured in low numbers and at few sites both after and prior to the fire (Kentish 1983; Wilson *et al.* 1986). In other studies the species has also been recorded sporadically and in low numbers (Morton *et al.* 1980; Woolley & Ahern 1983; Braithwaite *et al.* 1984; Lunney & Ashby 1987; Lunney *et al.* 1987). In the Mumbulla State Forest, New South Wales *S. leucopus* preferred sparsely covered sites and it was suggested that dense regenerating areas (after logging and fire) were not suitable for the species (Lunney & Ashby 1987; Lunney *et al.* 1987). Recently *S. leucopus* has been recorded soon after fuel reduction burning in the study area and further analysis of the species habitat preferences in the Eastern Otways is being investigated.

Isoodon obesulus was present in low numbers until years 4 to 6 and was slow to colonize new sites. At Cranbourne on the Mornington Peninsula, Victoria, a preference for heathland regenerating after clearing was reported but this was mainly restricted to juvenile animals (Braithwaite & Gullan 1978; Braithwaite *et al.* 1978; Stoddart & Braithwaite 1979).

There was substantial regeneration of vegetation within the first three years. Studies of regeneration at other heath and heath woodland sites in the Eastern Otways by Wark *et al.* (1987) showed a similar recovery rate. Bare ground decreased from 98% to 39% in the third year and there was a rapid increase in the total biomass of the above ground shrub stratum from 1% to 60%. The recovery of native small mammal populations thus followed the pattern of return of vegetation cover but was slower. The populations appear to be reaching peak abundance five to six years after the fire. Further surveys are required to determine whether populations retain maximum abundances or decline. These results show clearly that *M. musculus* is an early colonizing species. Together with other evidence, the results indicate that *S. leucopus* may also prefer early stages after fire. *R. fuscipes, R. lutreolus* and *A. stuartii* all increased in years 3-5 and could be considered as mid to late succession species. The slow colonization and population increase of *I. obesulus* would indicate that it is a late colonizer. Although no data were available for *P. novaehollandiae* populations for years 1 to 2, the grid studies showed that the species increased in abundance from years three to five and then declined. The pattern of recolonization does not appear to differ significantly from that proposed by Fox (1982) where species achieve maximal abundance as habitat changes meet their requirements. The actual time taken to reach the highest abundance may however differ according to the regeneration rates of the vegetation and the fire regimes in the area.

Recolonization of small mammals after mine rehabilitation

The study of small mammal recolonization after mining showed that *M. musculus* was the only small mammal resident on the revegetated areas A and B up to five years after revegetation. During this period native small mammals were trapped only rarely and were assumed to be transients from surrounding undisturbed areas. The presence of *M. musculus* on these mine rehabilitation sites confirms the opportunistic behaviour of the species after habitat disturbance (Newsome *et al.* 1975; Fox & Fox 1978; Fox & McKay 1981). Prior to the 1983 fire the adjacent unmined area (area C) had three resident native species, *S. leucopus, R. lutreolus* and *P. novaehollandiae.* The study indicates that the vegetation on the regenerating mined areas had not developed to a suitable successional stage for permanent occupation by these native small mammals.

The revegetated areas had a higher proportion of graminoids and introduced species, lower species numbers and lower percentage ground cover than the unmined area. Species characteristic of the unmined areas such as *B. marginata, C. pusilla, H. stricta, G. geniculata* and *B. umbellata* were absent from the revegetated areas. The regeneration of these mined areas was considerably retarded compared with the revegetation after wildfire, where ground cover was from 50-75% three to four years after fire. In comparison there was less than 5% ground cover five years after artificial regeneration of mined areas.

It is possible that vegetative ground cover and species diversity of the revegetated areas may increase over time. However, the revegetated areas are located on a raised, exposed plateau. There is evidence of considerable erosion and the conditions are likely to result in species loss and a decrease in ground cover.

Population ecology and habitat preferences of P. novaehollandiae

From 1983 to 1988 *P. novaehollandiae* was captured at only four sites, all located east of the Alcoa mine. Prior to this the species occurred at a site west of the mine, but this population did not survive the 1983 fire. *P. novaehollandiae* thus exhibits a patchy distribution in the study area, a feature of the species which has been noted in previous studies (Fox & Fox 1978, 1984). The sites where *P. novaehollandiae* was found were flat to undulating in nature, with sandy to gravelly soils. The structural vegetation types, woodland and open forest with heathy understorey are similar to those described in New South Wales and Tasmania (Posamentier & Recher 1974; Hocking 1980) and in Victoria (Seebeck & Beste 1970; Braithwaite & Gullan 1978; Opie, pers. comm.).

The population of *P. novaehollandiae* on grid A reached a maximum in the third year after the 1983 fire (July 1986-May 1987) and then declined. On grid B the population increased in 1987, but only one individual was captured between November 1987 and April 1988. Previous studies in New South Wales have shown that *P. novaehollandiae* maintains low populations after fire and reaches a maximum abundance at 2-3 years (Fox & McKay 1981; Fox 1982). Such responses to successional ages of vegetation have been documented after fire and clearing in New South Wales (Fox & McKay 1981; Fox & Fox 1978, 1984) and Victoria (Braithwaite & Gullan 1978; Opie, pers. comm.). The studies show that *P. novaehollandiae* is absent from early stages (0-2 years), its numbers peak between two and three years after the disturbance and decline in later stages (6-8 years). *Pseudomys novaehollandiae* exhibited microhabitat floristic preferences for two groups which had higher floristic diversity than the other groups. Previous studies have found a similar response to floristic diversity (Braithwaite & Gullan 1978; Cockburn 1978; Fox & Fox 1978, 1984). It is not possible to determine if this response is to the diversity of plant species *per se* or to the presence of particular species. An analysis of *P. novaehollandiae* scats and of potential plant food sources is being undertaken to resolve this question. Structural parmeters were not highly significant for overall habitat preferences, but they did contribute to habitat preference within the preferred floristic groups 1 and 3. These structural factors reflect the species' requirement for low vegetation cover described by Fox and Fox (1978, 1984).

Conservation and management of small mammals

These studies in the Eastern Otways demonstrate the complexity of interactions between vegetation, disturbance factors and small mammal communities. They confirm the importance of floristically diverse vegetation communities for small mammal species richness and for particular species such as *P. novaehollandiae* (Braithwaite & Gullan 1978; Fox & Fox 1978,1984; Cockburn 1978; Taylor *et al.* 1984; Braithwaite *et al.* 1985). The studies show that wildfire and mining have major effects on small mammal communities and highlight the differences in regeneration and recolonization that follow these disturbances (Fig. 8). In general, recolonization by native species in the six years after the wildfire has been substantial. However, native species did not recolonize mine rehabilitation sites over a 5-6 year period. The major factors contributing to this were sub-optimal rehabilitation procedures and the absence of suitable corridors for animal movements. This contrasts with previous studies where small mammals recolonized sites

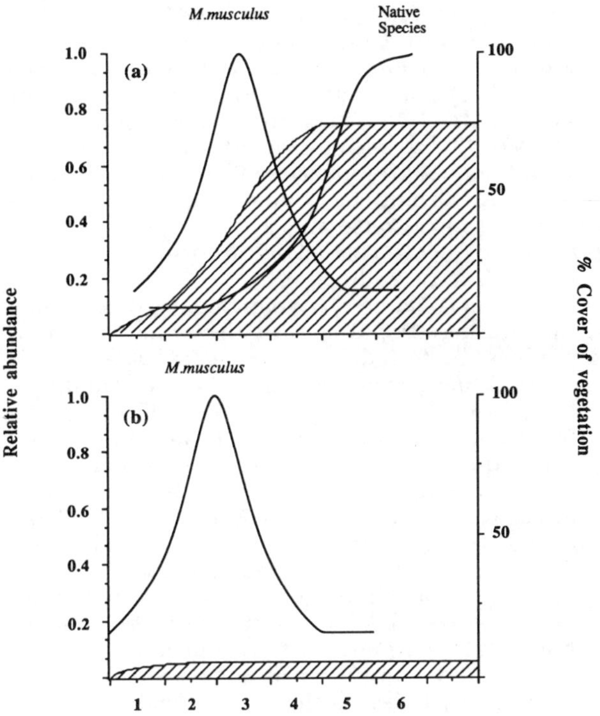

FIG. 8. Diagrammatic comparison of ▨ vegetation recovery and small mammal succession following a disturbance (a) fire (b) mining.

rehabilitated after sandmining (Fox & Fox 1978, 1984). The effects of *P. cinnamomi* on small mammals are yet to be confirmed but would appear to be more subtle but longer lasting than those of fire and mining.

It appears that several species (*A. minimus, A. swainsonii* and *P. tridactylus*) may now have disappeared from the study area as they have not been captured for two to three years. *Pseudomys novaehollandiae,* which is rare in Victoria (Ahern *et al.* 1985a,b; Menkhorst *et al.* 1987) is also at significant risk. These species are vulnerable to the combined effects of disturbance factors due to their low population numbers and restricted distributions.

The results of the study have implications for management of the area to ensure the maintenance of the small mammal communities and species. The importance of floristic diversity indicates that the effect of *P. cinnamomi* should be investigated further and monitored. Although the vegetation in the area has maintained a high level of diversity this could be at risk. In the Eastern Otways the spread of *P. cinnamomi* is relatively slow and it would be possible to identify areas of infection and to conserve the remaining areas of vegetation which are not yet affected (Weste & Marks 1987). To facilitate this, uninfected areas should be isolated and vehicle access restricted.

Fire regimes in the area are also likely to affect small mammal communities and species. Fuel reduction burning practices should be well planned and take into account ecological factors such as the frequency, area burnt, timing and intensity. It is recommended that 1) burning should not be at frequencies of less than four years, 2) patch burning be undertaken to produce a mosaic of successional ages within vegetation types, 3) burning not be carried out in spring when small mammals are breeding.

Burning could be used to maintain and promote suitable successional ages of the vegetation communities preferred by *P. novaehollandiae.* It is essential that burning within these communities is not carried out on a large scale. It is recommended that burning be done in small patches (1-3 ha) within the preferred communities to maintain a variety of successional ages from 2-8 years. This would also ensure that refuge areas are available during and soon after burning. If specific plant species are identified as major food sources then the fire regime should be aimed at maintaining these species. Temporary removal of animals should be considered when burning is being carried out due to the small population size.

The study indicates that mine rehabilitation in the area has been sub-optimal. Alternative and more appropriate techniques are recommended (Bourne *et al.* 1989) including burning before clearing, reducing soil stockpiling and seed burial to optimize seed germination rates, using organic mulch and indigenous species. Mapping and maintenance of corridors for small mammals around the mine area should also be carried out.

The Flora and Fauna Guarantee legislation in Victoria aims to protect native flora, fauna and communities and enable them to retain their potential for evolutionary development. Studies such as those considered in this chapter will be necessary to provide a solid basis for management plans to effectively implement such legislation.

Acknowledgements

We thank M. Menheere, A. Rothwell, D. Benellack, S. Saunders, J. Milgate, and I. Jamieson for providing technical assistance in the field. We pay tribute to the technical work of Alexia Everett, our colleague and friend who died recently as the result of a car accident. The studies were supported by grants to B. A. Wilson from the Department of Conservation, Forests and Lands, Alcoa of Australia Pty Ltd, Deakin Foundation, Ingram Trust and Deakin University. D. J. Moloney was supported by a Deakin University Postgraduate Scholarship. The work was carried out under scientific permits issued by National Parks and Wildlife, Department of Conservation, Forests and Lands, Victoria.

References

Abele C. (1971) Physiography, geology and mineral resources. In: *Barwon Region Resources Study.* Central Planning Authority, Barwon Regional Committee, Government of Victoria.

Ahern L. D. (1982) Threatened wildlife in Victoria and issues related to conservation. *Fisheries and Wildlife Paper Victoria* No. **27,** 1-33.

Ahern L. D., Brown P. R., Robertson P. & Seebeck J. H. (1985a) Application of a taxon priority system to some Victorian vertebrate fauna. Department of Conservation, Forests and Lands. *Arthur Rylah Institute Technical Report* Series No. **30,** 1-48.

Ahern L. D., Brown P. R., Robertson P., Seebeck J. H., Brown A. M. & Begg R. J. (1985b) A proposed taxon priority system for Victorian vertebrate fauna. Fisheries and Wildlife Service Victoria. *Arthur Rylah Institute Technical Report* Series No. **30,** 1-26.

Barnett J. L., How R. A. & Humphreys W. (1978) The use of habitat components by small mammals in eastern Australia. *Aust. J. Ecol.* **3,** 277-85.

Berry R. J. & Jackson M. E. (1975) Adaptation and adaptability in wild-living house mice *(Mus musculus). J. Zool. Soc. Lond.* **176,** 391-402.

Bomford M. & Redhead T. (1987) A field experiment to examine the effects of food quality and population density on reproduction of wild house mice. *Oikos* **48,** 304-11.

Bourne A. R., Kentish K. M., Moloney D. J. & Wilson B. A. (1989) Small mammal recolonization of revegetated heath woodland after open-cut coal mining at Anglesea, Victoria. *Aust. Wildl. Res.* **16,** (in press).

Braithwaite R. W. & Gullan P. K. (1978) Habitat selection by small mammals in a Victorian heathland. *Aust. J. Ecol.* **3,** 109-27.

Braithwaite R. W., Cockburn A. & Lee A. K. (1978) Resource partitioning between communities of south-east Australia. *Aust. J. Ecol.* **3,** 423-45.

Braithwaite R. W., Clayton L. & Parker B. S. (1984) Vertebrate fauna of a 144 ha water catchment within eucalypt forest being harvested for woodpulp at Eden, south-eastern New South Wales. *CSIRO Division Wildlife Rangeland Research, Technical Paper* No. 35.

Braithwaite R. W., Winter J. W., Taylor J. A. & Parker B. S. (1985) Patterns of diversity and structure of mammalian assemblages in the Australian tropics. *Aust. Mammal.* **8,** 171-86.

Catling P. C. & Newsome A. E. (1981) Responses of the Australian vertebrate fauna to fire: an evolutionary approach. In: *Fire and the Australian biota* (eds R. H. Groves, I. R. Noble & A. M. Gill) pp. 273-310. Australian Academy of Science, Canberra.

Catling P. C. (1986) *Rattus lutreolus,* colonizer of heathland after fire in the absence of *Pseudomys* species. *Aust. Wildl. Res.* **13,** 127-39.

Cheal P. (1979) Vegetation analysis — an essential part of small mammal studies. *Bull. Aust. Mamm. Soc.* **6,** 46-7.

Cockburn A. (1978) The distribution of *Pseudomys shortridgei* and its relevance to that of the other heathland *Pseudomys* (Muridae:Rodentia). *Aust. Wildl. Res.* **5,** 213-9.

Cockburn A. (1981) Population regulation and dispersion of the smoky mouse *Pseudomys fumeus.* I Dietary determinants of microhabitat preference. *Aust. J. Ecol.* **6,** 231-54.

Conole L. & Baverstock G. A. (1983) Mammals of the Angahook-Lorne Forest Park, Victoria. *Vic. Nat.* **100,** 224-31.

Dawson P. & Weste G. (1985) Changes in the distribution of *Phytophthora cinnamomi* in the Brisbane Ranges National Park between 1970 and 1980-1981. *Aust. J. Bot.* **33,** 309-15.

Dawson P., Weste G. & Ashton D. H. (1985) Regeneration of vegetation in the Brisbane Ranges after fire and infestation by *Phytphthora cinnamomi. Aust. J. Bot.* **33,** 15-26.

Flora & Fauna Guarantee Act (1988) Acts 1988 No. 47. The Parliament of Victoria.

Fox B. J. (1982) Fire and mammalian secondary succession in an Australian coastal heath. *Ecology* **63,** 1332-41.

Fox B. J. (1983) Mammal species diversity in Australian heathlands: the importance of Pyric succession and habitat diversity. In: *Mediterranean Type Ecosystems — the Role of Nutrients* (eds F. J. Kruger, D. J. Mitchell & J. U. M. Jarvis) pp. 473-89. Springer-Verlag, Berlin.

Fox B. J. (1985) Small mammal communities in Australian temperate heathlands and forests. *Aust. Mammal.* **8,** 153-8.

Fox B. J. & Fox M. D. (1978) Recolonization of coastal heath by *Pseudomys novaehollandiae* (Muridae) following sand mining. *Aust. J. Ecol.* **3,** 447-65.

Fox B. J. & Fox M. D. (1984) Small mammal recolonization of open-forest following sand mining. *Aust. J. Ecol.* **9,** 241-52.

Fox B. J. & McKay G. M. (1981) Small mammal responses to pyric successional changes in eucalypt forest. *Aust. J. Ecol.* **6,** 29-41.

Fox B. J., Quinn R. D. & Breytenbach G. J. (1985) A comparison of small mammal succession in shrublands of Australia, California and South Africa. *Proc. Ecol. Soc. Aust.* **14,** 179-97.

Friend G. E. & Taylor J. A. (1985) Habitat preferences of small mammals in tropical open-forest of the Northern Territory. *Aust. J. Ecol.* **10,** 173-85.

Gullan P. K. (1978) Vegetation of the Royal Botanic Gardens Annexe at Cranbourne, Victoria. *Proc. R. Soc. Vict.* **90,** 225-40.

Hocking G. J. (1980) The occurrence of the New Holland mouse, *Pseudomys novaehollandiae* (Waterhouse) in Tasmania. *Aust. Wildl. Res.* **7,** 71-8.

Kennedy J. & Weste G. (1986) Vegetation changes associated with invasion by *Phytophtora cinnamomi* on monitored sites in the Grampians, Western Victoria. *Aust. J. Bot.* **34,** 251-79.

Kentish K. M. (1982) A new record of the New Holland mouse *(Pseudomys novaehollandiae)* from Anglesea, Victoria. *Vict. Nat.* **99,** 128-9.

Kentish K. (1983) Mine rehabilitation: A study of revegetation and fauna return at Anglesea, Victoria (1980-82). M.Sc. Thesis. Deakin University, Victoria.

Krebs C. J. (1966) Demographic changes in fluctuating populations of *Microtus califonicus. Ecol. Monogr.* **36,** 239-73.

Land Conservation Council (1985) Report on the Melbourne Area, District 1, Review. Land Conservation Council, Victoria.

Land Conservation Council (1987) Final recomendations, Melbourne Area, District 1. Land Conservation Council, Victoria.

Laidlaw W. S. (1987) Factors affecting the distribution and habitat preference of small mammal species in the Angahook-Lorne State Park. Honours Thesis, Deakin University, Victoria.

Lunney D. & Ashby E. (1987) Population changes in *Sminthopsis leucopus* (Gray) (Marsupialia: Dasyuridae), and other small mammal species, in forest regenerating from logging and fire near Bega, New South Wales. *Aust. Wildl. Res.* **14,** 275-84.

Lunney D., Cullis B. & Eby P. (1987) Effects of logging and fire on small mammals in Mumbulla State Forest, near Bega, New South Wales. *Aust. Wildl. Res.* **14,** 163-71.

Menkhorst P., Bennett A. & Lumsden L. (1987) Conservation status of mammals in Victoria. In: *Nature Conservation in Victoria* (eds D. Frood & M. Calder) pp. 103-22. Report to the Victorian National Parks Association Incorporated.

Meredith C. (1986) The vegetation of the Anglesea Lease Area. Report prepared for the Land Conservation Council, Government of Victoria.

Morton S. R. & Baynes A. (1985) Small mammal assemblages in arid Australia: a reappraisal. *Aust. Mammal.* **8,** 159-69.

Morton S. R., Wainer J. W. & Thwaites T. P. (1980) Distributions and habitat of *Sminthopsis leucopus* and *S. murina* (Marsupialia: Dasyuridae) in southeastern Australia. *Aust. Mammal.* **3,** 19-30.

Mueller-Dombois D. & Ellenberg H. (1974) *Aims and methods of plant ecology.* John Wiley, New York.

Newsome A. (1969a) A population of house-mice temporarily inhabiting a South Australian wheatfield. *J. Anim. Ecol.* **38,** 341-60.

Newsome A. (1969b) A population of house-mice permanently inhabiting a reed-bed in South Australia. *J. Anim. Ecol.* **38,** 361-77.

Newsome A. & Catling P. C. (1983) Animal demography in relation to fire and shortage of food: some indicative models. In: *Mediterranean Type Ecosystems — the Role of Nutrients* (eds F. J. Kruger, D. J. Mitchell & J. U. M. Jarvis) pp. 490-505. Springer-Verlag, Berlin.

Newsome A., McIlroy K. & Catling P. (1975) The effects of an extensive wildfire on populations of twenty ground vertebrates in south-east Australia. *Proc. Ecol. Soc. Aust.* **9,** 107-23.

Parsons R. F., Kirkpatrick J. B. & Carr G. W. (1977) Native vegetation of the Otway region. *Proc. R. Soc. Vict.* **89,** 77-88.

Posamentier H. (1976) Relation of habitat structure to small mammal diversity. *Aust. Mamm. Soc. Bull.* **3,** 15-6.

Posamentier H. & Recher H. F. (1974) The status of *Pseudomys novaehollandiae* (the New Holland Mouse). *Aust. J. Zool.* **18,** 66-71.

Postle A. C., Majer J. D. & Bell D. T. (1986) Soil and litter invertebrates and litter decomposition in Jarrah *(Eucalyptus marginata)* forest affected by Jarrah dieback fungus *(Phytophthora cinnamomi). Pedobiologia* **29,** 47-69.

Reilly P. (1985) Victorian Group — Regeneration survey. *Royal Australian Ornithologists Union Newsletter,* No. 64. **8.**

Rosenzweig M. L., Smigel B. W. & Kraft A. (1975) Patterns of food space and diversity. In: *Rodents in desert environments* (eds I. Brakash & P. K. Gosh) pp. 241-68. Dr W. Junk, The Hague.

Ross D. (1982) Taxon Users Manual. Ed. P.3. CSIRO Division of Computing Research.

Seebeck J. H. & Beste H. J. (1970) First record of the New Holland Mouse *(Pseudomys novaehollandiae)* (Waterhouse 1843) in Victoria. *Vict. Nat.* **87,** 280-7.

Singleton O. D. (1973) Mesozoic and tertiary stratigraphy of the Otway Region. In: *Regional guide to Victorian geology* (eds J. McAndrew & M. A. H. Marsden) pp. 114-28. University of Melbourne.

Sokal R. R. & Rohlf F. J. (1981) *Biometry: The principles and practice of statistics in biological research.* 2nd Ed. W. H. Freeland & Co., New York.

State Conservation Strategy (1987) *Protecting the Environment.* Government of Victoria.

Stewart A. P. (1979) Trapping success in relation to trap placement with three species of small mammals; *R. fuscipes, A. swainsonii* and *A. stuartii. Aust. Wildl. Res.* **6,** 1-8.

Stoddart D. M. & Braithwaite R. W. (1979) A strategy for utilization of regenerating heathland habitat by the brown bandicoot (*Isoodon obesulus:* Marsupialia, Peramelidae). *J. Anim. Ecol.* **48,** 165-79.

Taylor J. A., Friend G. R. & Dudzinski M. L. (1984) Influence of sampling strategy on the relationships between fauna and vegetation structure, plant life-form and floristics. *Aust. J. Ecol.* **9,** 281-7.

Thomson P. C. (1980) Population dynamics and feeding ecology of the New Holland Mouse and the house mouse. M.Sc. Thesis, Macquarie University, New South Wales.

Walbran W. I. (1971) Soils. In: *Barwon Region Resources Survey* pp. 200-18. Central Planning Authority, Victoria.

Wark M. C., White M. D., Robertson D. J. & Marriott, P. F. (1987) Regeneration of heath and heath woodland in the north-eastern Otway Ranges following the wildfire of February 1983. *Proc. R. Soc. Vict.* **99,** 51-88.

Watts C. H. S. (1974) The native rodents of Australia: a personal view. *Aust. Mammal.* **1,** 109-15.

Weste G. (1974) *Phytophthora cinnamomi* — the cause of severe disease in certain native communities in Victoria. *Aust. J. Bot.* **22,** 1-8.

Weste G. (1981) Changes in vegetation of sclerophyll shrubby woodland associated with invasion by *Phytophthora cinnamomi. Aust. J. Bot.* **29,** 261-76.

Weste G. & Marks G. C. (1987) The biology of *Phytophthora cinnamomi* in Australasian forests. *Ann. Rev. Phytophathol.* **25,** 207-29.

Wilson B. A. (1986) Reproduction in the female dasyurid *Antechinus minimus maritimus* (Marsupialia: Dasyuridae). *Aust. J. Zool.* **34,** 189-97.

Wilson B. A. & Bourne A. R. (1984) Reproduction in the male dasyurid *Antechinus minimus maritimus* (Marsupalia: Dasyuridae). *Aust. J. Zool.* **34,** 311-8.

Wilson B. A. & Moloney D. J. (1985a) Small mammals in the Anglesea-Aireys Inlet area of southern Victoria — a post fire survey. *Vict. Nat.* **102,** 65-70.

Wilson B. A. & Moloney D. J. (1985b) An investigation of small mammal recolonization and vegetation regeneration in fire affected areas of the Anglesea-Airey's Inlet region. Report to the Ministry of Conservation, Forests and Lands, Victoria.

Wilson B. A., Bourne A. R. & Jessop R. E. (1986) Ecology of small mammals in coastal heathland at Anglesea, Victoria. *Aust. Wildl. Res.* **13,** 397-406.

Woolley P. A. & Ahern L. D. (1983) Observations on the ecology and reproduction of *Sminthopsis leucopus* (Marsupialia: Dasyuridae). *Proc. R. Soc. Vict.* **95,** 169-80.

Proc. Ecol. Soc. Aust. 1990 — 16:397-409

Restoration of biological scenarios: the role of museum collections

BARBARA YORK MAIN

Zoology Department, The University of Western Australia, Nedlands, W.A., Australia 6009

Abstract

This chapter draws attention to the fact that biological collections (albeit fragmentary) from Australia preceded European settlement and major collections varying for different taxa have subsequently accrued (overseas and in Australia) through concerted planning and fortuitous acquisition. Distribution of the biota is now fragmented and in some instances museum collections represent the only material evidence of former distributions. Most places lack complete records of the fauna and flora present at the time of European settlement and at later periods. Thus the cumulative value of natural history collections as a reference source is increasing inversely with the diminution of natural landscapes and habitats. I propose here that biologists can theoretically reconstruct former biological scenarios through reference to these archival collections. I examine for south-western Western Australia, a regional collection of spiders, two species of spider Idiosoma sigillatum *and* Chasmocephalon neglectum *and the frog* Heleioporus albopunctatus, *to show how museum collections of these species document former distributions of greater scope than is apparent at the present time. In turn, a knowledge of the biology of the respective species indicates what the now modified habitats were like. Information similarly obtained by examining other species of the fauna can serve as a guide for reconstruction of habitats in particular localities of former distributions. Justification for restoration of selected sites in contrast to reclamation is addressed as an educational and cultural obligation.*

Introduction

". . . museums and herbaria are still, of course, and always will be, absolutely necessary for the pursuit of biology, not least as adjuncts to the study of ecology . . ." (Tansley 1951)

In earlier periods of scientific curiosity and endeavour, collection of natural history specimens was a *cause celebre*. Biological novelties from around the world were a focus for intellectual investigation of the pattern in the then known world. Today, some of these old collections are the only material evidence of the former existence of some plants and animals; species relegated now to "pseudo-fossil" status through human modification of natural landscapes.

Biological collections (albeit fragmentary) from Australia preceded European settlement. Later collections were made in association with eighteenth and nineteenth century explorations. On a regional basis using south-western Australia as an example, bench mark collections began with Dampier, were continued with the French and British explorations (Alexander 1914, 1916, 1918; Whittell 1954) and culminated with the Berlin and Hamburg Museums' expedition of 1905 (Michaelsen & Hartmeyer 1907-1930). Sizeable collections varying for different taxa have subsequently accrued through concerted planning and opportunistic acquisition. Notable in recent years have been the biological surveys of the Western Australian Museum and the Department of Conservation and Land Management (and its relevant former counterpart within the Department of Fisheries and Wildlife) particularly in the Wheatbelt, Goldfields and Kimberleys. All such surveys resulted in comprehensive collections of selected taxa.

One hundred and fifty nine years of settlement in Western Australia has made a travesty of the natural landscape. Along with denudation of much of the indigenous vegetation there has been a diminution of faunal distributions and extinction of some species. Large scale modification of landscape has been most dramatic in the metropolitan region and agricultural areas, particularly the Wheatbelt. Thus it can be assumed that the reduction and/or loss of species has been proportionately greater in these regions than in other parts of the state. What evidence do we have of the diversity or richness of species 160 years ago compared to today, and what is the abundance of remaining species? How can we go about restoring or at least simulating habitats which have been modified or removed? Assuming that is, that we might want to, if only to create "specimen" areas for the cultural education of our own and future generations. Indeed, the aesthetic and educational purpose for ecological restorations is viewed by many people as desirable and the justification is argued cogently by Atkinson (1989).

In this context I believe we should be looking at old (and even recent) natural history collections in museums and herbaria and re-evaluating them as the markers on a fading print of a prehistoric landscape.

The setting and its biota

I have discussed elsewhere, by using examples from spiders, the significance of extant natural geographic relics in reconstructing theoretical biological scenarios of earlier geological periods with different climatic regimes to the present (Main 1987a). Here I shall demonstrate that from data on distributions of selected species, obtained from museum collections, combined with a knowledge of the biology and habitat requirements of such species or closely related species, we can reconstruct ecological scenarios on a shorter time scale, i.e., within the time span of European settlement. I shall restrict the discussion to the southwest of Western Australia from roughly 26°S at Shark Bay south-eastward to the south coast, taking in the Kalgoorlie area. This region comprises several diagonal belts or natural physiographic/vegetation regions roughly paralleling the rainfall isohyets and ranging from coastal plains with heath and open forest, through low ranges with forest to sclerophyll woodlands, heath and mallee on the plateau. About a third of this area comprises the Wheatbelt which has been severely modified through clearing and cultivation and with its eastern margin giving way to pastoral or open range country largely grazed by sheep and further downgraded by feral animals, predominantly rabbits and goats.

Since European settlement a considerable number and variety of habitat types have contracted or vanished in southern Western Australia, many without prior documentation or observation by biologists. This is clear from some described taxa which are presumed extinct, now rare or have reduced or fragmented ranges. We can only guess at the great unknown category; the unspecified "phantom" species. The biological and habitat requirements of fugitive vertebrates, particularly birds and mammals and some reptiles are more readily ascertained than those of invertebrates. Attempts have and are being made to understand vertebrate needs. Studies include those on the noisy scrub bird, *Atrichornis clamosus* (Smith 1976, 1985; Smith & Robinson 1976), the rock wallaby, *Petrogale lateralis* (Kinnear *et al.* 1984), cockatoos (Saunders *et al.* 1985; Saunders & Ingram 1987), the gecko, *Oedura reticulata* (How & Kitchener 1983), the short necked tortoise, *Pseudemydura umbrina* (Burbidge 1967, 1981), the splendid wren, *Malurus splendens* (Rowley 1981), the numbat, *Myrmecobius fasciatus* (Calaby 1960; Friend 1987; Friend *et al.* 1982) and others (Christensen 1980). While some field studies aimed at understanding habitat needs of terrestrial invertebrates are being made, there are few comprehensive studies on the enduring biology of vulnerable invertebrates *in situ* except possibly those on trapdoor spiders (Main 1978, 1987b) and social spiders (Main 1988).

The fugitive status of species can be recognized from (a) verbal assertions of people (b) written and/or published observations and records and (c) museum collections. The authenticity of such records increases from (a) to (c) i.e., the dated and provenanced museum specimens rate the highest value.

If we determine the biological needs of fugitive species (or of closely related species) in sanctuary habitats and note from museum collections the former geographic localities of such fugitive species, then we can argue retrospectively that such localities (if now alienated) must have presented a habitat at least simulating to some extent that where the respective species now occur.

Thus the present exercise is akin to working from fossils to theoretically reconstruct earlier landscapes but of course within a shorter time frame. Nevertheless we are dealing to some extent with "pseudofossils". I will now illustrate the above premise by examining the "lost" distributions of selected species with fugitive or "pseudofossil" status, i.e., species with contracting distributions or which are possibly extinct.

Rationale for selection of sample species for analysis

A representative invertebrate group-taxon is examined to assess diversity (species richness) at an early period of settlement in relation to present diversity. Spiders were selected as this representative group- taxon (see later).

The status of one selected vertebrate and one invertebrate taxon named during the last century and for which the following data are available is examined: sequential locality records paralleling large scale environmental modification; biological information such as life history, habitat requirements and seasonal behaviour; evidence of, or prediction of a dramatic diminution of distribution and population decline; and accessible museum collections.

Using these criteria I selected the burrowing frog *Heleioporus albopunctatus* (Gray) and the mygalomorph trapdoor spider *Idiosoma sigillatum* (Cambridge) as well as *Chasmocephalon neglectum* Cambridge, a species named in the last century and not recorded again.

From consideration of these species it is believed that extrapolations can be made about the past ecological scenarios of other species also.

Group-taxon: Araneae (spiders)

A large collection of spiders was made in 1905 by the Berlin and Hamburg Museums' expedition to south-western Australia and described shortly afterwards by Simon (1908, 1909). Of the 121 genera and 228 species recorded, 21 and 160 respectively were described as new taxa. The collection, including type material was distributed between the Paris, Hamburg, Berlin and Western Australian Museums. This named collection represents the most comprehensive regional taxonomic collection of spiders from anywhere in Australia. However, identified and provenanced specimens of only 14 of Simon's nominal species have been subsequently both deposited in vouchsafed collections and recorded in the literature. Although additional species are cited in natural history accounts, the distribution of only about 10 species has ever been documented. Because many of the original collection sites are now severely modified or the natural habitats destroyed, it is probable that some species with narrow geographic ranges are now extinct and the distribution of others fragmented. Simon's collection represented 53 type localities, including those of syntypes (Table 1). I can only emphasize that the museum collections representing Simon's species, both the type material and later provenanced specimens, have a very special archival value. It is also these museum specimens which can give us clues to an understanding of the earlier ecological scenarios.

Specific taxon: the burrowing frog Heleioporus albopunctatus

This species was selected because of its early nomenclatural establishment, my long familiarity with its natural history, a well documented biology (Main 1965; Lee 1967) and the Western Australian Museum has comprehensive collections.

TABLE 1. Number of nominal genera and species of spiders recorded by Simon (1908, 1909) collected by the Berlin and Hamburg Museums' expedition to south-western Australia in 1905. The number of newly described taxa are indicated in parentheses. The number of type localities (including syntype localities) are listed according to their arbitrarily graded ecological status. (1) relatively intact; (2) modified but potentially capable of maintaining species in limited sites; (3) destroyed or so modified that species totally absent

		Type localities (53) Current ecological status		
Genera	Species	(1)	(2)	(3)
121 (21)	228 (160)	5	33	15

Background Information — Described from a specimen in the British Museum (Natural History) (Gray 1841) and with the locality given only as Western Australia, *Heleioporus albopunctatus* (Fig. 1a) has long been recognized as widely distributed and occurring abundantly throughout the Wheatbelt (Fig. 2).

FIG. 1. (a) *Heleioporus albopunctatus*, a burrowing frog of south-western Western Australia; (b) (c) *Idiosoma sigillatum*, a trapdoor spider endemic to south-western Western Australia; (b) female (c) male. Scale bars=10 mm.

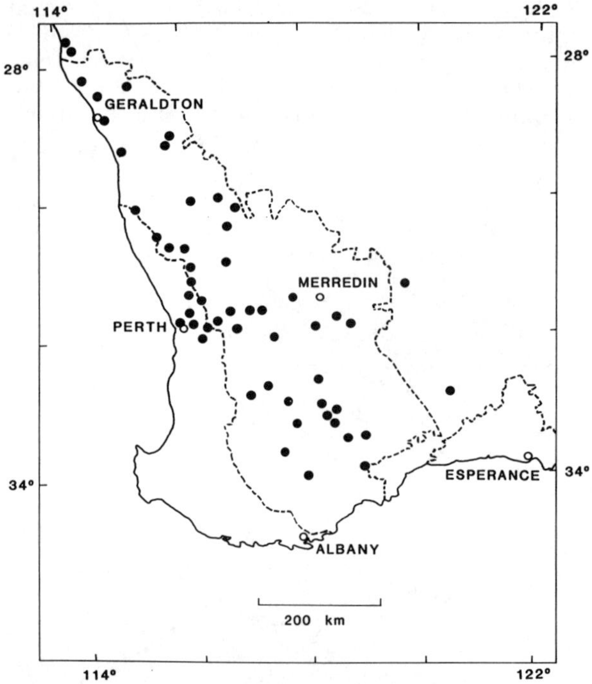

FIG. 2. Geographic distribution of *Heleioporus albopunctatus* (from Tyler *et al.* 1984) and approximate boundary of the Wheatbelt. See also Appendix 1.

Biology — During autumn frogs emerge from aestivation in sandy banks of low lying, non-saline drainage courses and creeks which skirt granite outcrops (Main 1965). Males call from open-holed, deep breeding tunnels to which females are attracted and where the eggs are deposited in frothy masses. Following flooding of the burrows in autumn/winter the tadpoles are flushed into water courses where they metamorphose in ephemeral ponds formed by the subsiding flow of water or in pools at the edge of swamps. Although adult and juvenile frogs aestivate in burrows, individual specimens frequently surface and forage on humid nights throughout the summer.

Frogs and tadpoles are sensitive to salinity as deduced from observations on preferred breeding habitats. Curiously however, Lee in his excellent study of the biology of *Heleioporus* (Lee 1967) made no mention of salinity tolerance although other aspects of water quality as it might affect tadpole development were analysed. The study was conducted in the mid fifties, and just prior to the "million-acres-a-year" clearing syndrome in the Wheatbelt, and thus before salinity had developed as a major environmental crisis. The effect that increasing salinity was to have on frog habitats was not foreseen.

Evidence of diminution of range — A species so widespread, and well represented from its entire range in the Western Australian Museum collections, might appear "safe" in terms of future survival. However, it is dependent for its breeding grounds on the integrity of habitats in low lying watercourses which are now largely destroyed by agriculture (Figs 3a & 3b). Increased salination in long settled parts of the Wheatbelt has been recognized for decades and as a consequence the Western Australian Department of Agriculture in collaboration with the Federal Bureau of Statistics conducted its first salt land survey in 1955 (Burvill 1956). This was followed by further surveys in 1962 (Lightfoot *et al.* 1964), 1974 (Malcolm & Stoneman 1976), 1979 (Henschke 1980) and 1984. Results from the first three surveys are summarized by Stoneman (1978). All those surveys were based on farmers' estimations of total area of salt affected land and percentage of salt affected land previously used for cereal growing. Finally, Malcolm (1983) summarized the underlying factors of salinity in salt-prone soils and topography and in 1988 a Parliamentary Select Committee on Salinity reported on summary findings, projections and recommendations (Anon. 1988). Figure 4 shows the area of salt affected land in 1955, 1962 and 1974 in the three statistical divisions of which two, the Northern and Central are the most relevant to the distribution of *Heleioporus albopunctatus*. Deposition of salt in the soil and decay of the vegetation has destroyed many breeding sites of frogs (Fig. 3).

The last 10 year period during which frogs were collected from respective shires together with percentages of salt affected land in relation to cleared land in the 1974 and 1979 estimates are shown in Table 2. The Select Committee on Salinity Report (Anon. 1988) indicates that some shires in the Central Wheatbelt already have percentages of salt/arable (=cleared) land higher than 4.0 (e.g., Dowerin, Goomalling and Tammin). This means that all the valley floors are probably salt affected and so the main breeding grounds of the frogs have been destroyed in those shires except for possibly a few granite outcrops (e.g., Yorkrakine Rock in the Tammin Shire). More drastically still, it is estimated in the same report that in the next 30 or so years the percentage of salt affected land is likely to increase to 20 percent in some shires. Such an increase would take in higher ground as well as the valley floors and affect many localities (where frogs possibly still occur) listed in Appendix 1.

With reference to the collection localities for the ten year periods (Appendix 1 and Table 2) and noting those shires not represented since 1945, 1955, 1965 and 1975 respectively, it is apparent that Goomalling is not represented at all, Broomehill not since 1945 and Tammin not since 1975. Tammin is one of the most severely salt-affected shires in the State and is situated in the middle of the Central Division so it is geographically in the very core of the distribution of the

FIG. 3. Salt affected farmland formerly inhabited by *Heleioporus albopunctatus* ("Fairfields", Tammin Shire); (a) pools in creek bed, a former breeding site; (b) upslope from the same creek course, trees and shrubs now killed by salt.

species. The outlook for the frog in this shire is ominous. The last collection date (in parentheses) for frogs from shires in the Northern Statistical Division listed as most severely salt-affected in the 1984 survey presents a gloomy prognosis for the frogs: Carnamah (no records), Coorow (1976-85), Dalwallinu (1976-85), Moora (1955), Morawa (1965), Mullewa (1934), Perenjori (no records) and Wongan-Ballidu (1976-85). Although specimens have been collected relatively recently from some severely affected shires e.g., Coorow, Dalwallinu and Wongan-Ballidu they could well be the last. In fact it seems that almost without realizing it we may be witnessing the demise of a species while inadvertently documenting

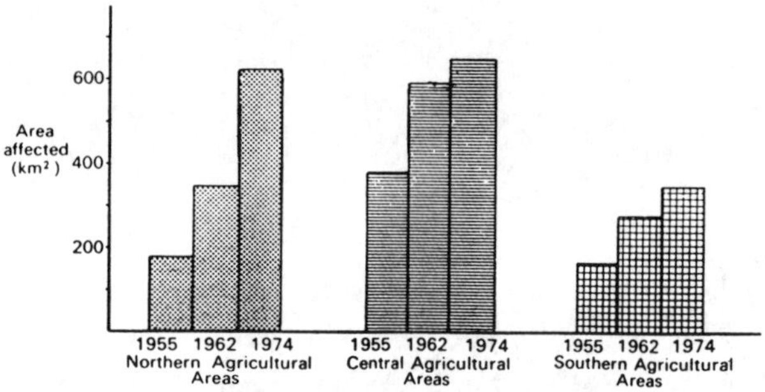

FIG. 4. Area of salt affected land previously used for cereal growing and pasture (from Stoneman 1978); such areas previously supported the breeding habitats of *Heleioporus albopunctatus,* particularly in the northern and central agricultural areas.

TABLE 2. Collection dates of *Heleioporus albopunctatus* specimens in the Western Australian Museum indicating relationship between *last* collection date and rising salinity in respective shire. Each year of collection for the period between 1924-45 and the following 10 year collection periods between 1946-75 are presented. Salinity is listed as percentage of salt affected/cleared land as at the 1974 and 1979 surveys. Note that in some shires salinity has almost doubled in the five years. Statistical divisions: S, Southern; C, Central; N, Northern, P, Perth; Y, Yilgarn.

Years 1924-45 & sequential 10 year periods	Shire	Stat division	% salt/ cleared land 1974	1979
1924-1945	Dowerin (/28)	C	2.9	4.0
(last year of	Kulin (/31)	C	0.7	0.5
collection in	Nyabing-Pingrup	S	0.7	1.4
parentheses)	=Kent (/33)			
	Corrigin (/34)	C	2.2	3.0
	Broomhill (/45)	S	1.6	1.7
1946-1955	Bruce Rock	C	1.0	2.1
	Irwin	N	1.0	0.1
	Moora	N	2.3	3.9
	Wanneroo	P	1.0	0.1
	(Gnangara)			
1956-1965	Gnowangerup	S	0.5	0.4
	Mingenew	N	0.8	0.1
	Morawa	N	2.0	3.9
	Mt Marshall	C	0.5	0.5
	Victoria Plains	N	2.1	2.7
	Yilgarn	Y	0.1	0.2
1966-1975	Chapman Valley	N	0.3	0.7
	Cuballing	C	1.2	1.2
	Lake Grace	S	1.6	2.1
	Kellerberrin	C	1.6	2.5
	Katanning	S	2.7	4.5
	Mukinbudin	C	0.4	0.5
	Narrogin	C	1.4	2.2
	Quairading	C	2.5	4.5
	Tammin	C	3.3	6.1
	York	C	1.8	1.4

its distribution by depositing specimens in a Museum. It is doubtful whether even now, let alone in the future, a deliberate search would as fully record the geographic range as do the collections, obtained sometimes fortuitously at other times by survey workers or in association with research. A corollary to the Museum collection would be documentation of the present distribution simply by driving through the Wheatbelt at night in early winter and recording localities of calling frogs. But what of the continuous habitat that formerly supported these frogs? In most areas, for the present generation, it is not even a memory.

Specific invertebrate taxon: the mygalomorph trapdoor spider Idiosoma sigillatum *(Cambridge)*

History of collections — This species (Figs 1b, c) was described in 1870 from a male specimen collected from the Swan River in 1864, and deposited in the Hope Museum, Oxford (Cambridge 1870). Pocock (1897) described a female from Perth in the British Museum (Natural History) collection. No further specimens appeared in collections until 1928 when a specimen from Perth was deposited in the Western Australian Museum. Between 1929 and 1939 nine more specimens were placed in the Museum. In 1952 when a country naturalist, C. G. Jessup, found what turned out to be a new inland species, *Idiosoma nigrum* Main, I set about searching for specimens of what appeared to be a very rare genus (Main 1952). *I. sigillatum* was susequently located as far north as Mogumber and in the valleys of the northern jarrah, *Eucalyptus marginata,* forest whereas *I. nigrum* was found along the Avon River Valley and eastward into the central and northern Wheatbelt. A third species *I. hirsutum* was recognized from Museum specimens as occurring *only* in a triangular or peninsula area bounded by the confluence of the Swan and Canning Rivers (Fig. 5). I now also recognize two unnamed species; one in the Darling Ranges south of the *sigillatum* range and one east of the *nigrum* distribution. *Idiosoma* is of particular biological significance because it represents the evolutionary adaptive extreme of the genera of the endemic tribe Aganippini. Burrows of these spiders have a conspicuous entrance of twig-lines attached in a fan-like arrangement which act as sensory lines when the spiders are foraging.

Biology of I. sigillatum and hirsutum — Spiders mature to a large size, females with carapace length 11.50-14.00 mm, males 8.40-10.50 mm (Main 1957). The males leave their trapdoor burrows and wander or "run" during the mating season in autumn and early winter. They are long legged and aggressive if disturbed or provoked and hence collected by the general public because they are thought to be funnel web spiders (i.e., *Atrax* which is venomous). Females are sometimes disturbed from their burrows by activities ranging from

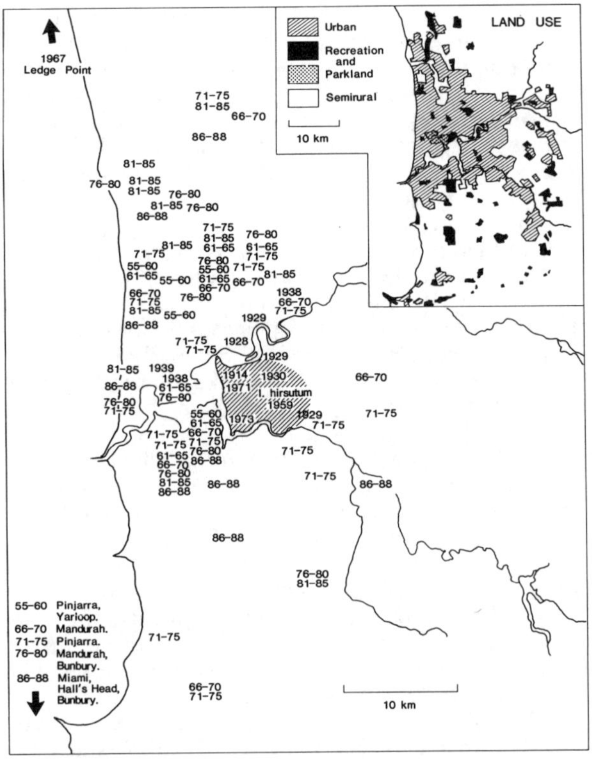

FIG. 5. Distribution of *Idiosoma sigillatum* based on specimens in the Western Australian Museum. Collection (acquisition) dates given in five year groupings (1955-85); actual year for pre-1955; and 1986-88 for post 1985 acquisitions. *Idiosoma hirisutum* acquisitions indicated by actual year. Distributions (see also Appendix 2) were plotted in the relevant grid squares on an enlarged copy of the key map in the *UBD Street Directory, Perth,* 30th edition (Universal Press, Perth). (Inset): Land use of metropolitan Perth environs as at 1970 (adapted from *Atlas of Natural Resources Darling System of Western Australia (1980)* (Department of Conservation and Environment, Western Australia). Note that the recreation and parkland areas range from natural bushland to playing fields, cemeteries and cultivated areas.

agriculture and gardening to road works and building and are likewise collected because of the spiders' aggressive behaviour and striking appearance, particularly the sclerotized, corrugated abdomen (Figs 1b and c).

It is noteworthy that the Western Australian Museum collections of *sigillatum* (which are all fortuitously acquired specimens) are nearly all from the suburban or surrounding areas or southern built up localities. However, I have collected the species from further north and north-east and also have specimens from Rottnest Island (Fig. 6). Nevertheless, as I will show, the species is at risk over the whole of its range, and neither collection i.e. that of the Museum or mine are likely to be extensively augmented in the future.

I. sigillatum occurs in sandy soil and in vegetation dominated by *Casuarina fraseriana* or *Acacia rostellifera* on which they are dependent for the branchlets or phyllodes which are used for twig-lines. Presumably *I. hirsutum* is likewise associated with *Casuarina* and *Acacia.* Dispersal takes place by juveniles when they leave the mother's burrow and search for a burrow site. Once established in a burrow, spiders are sedentary and remain in it throughout life except for adult males which die after mating. Spiders are long lived, probably taking about eight years to mature after which females may go on living for up to twenty or more years. Obviously for a population to persist, stability of the habitat is essential.

Evidence of diminishing geographic range — The urban distribution of Western Australian Museum specimens of *I. sigillatum* and *hirsutum* are shown in Fig. 5 as is land use of this area. From an examination of this figure while bearing in mind the habitat requirements of the spiders, and that they are long lived, sedentary and may take up to eight years to mature, some generalizations can be made regarding past, present and future distribution.

Up until 1939 the only records in the Museum are for Perth, Mt Lawley, Claremont, Nedlands, Bayswater and Cannington; the two nineteenth century specimens (Oxford and British Museum (Natural History)) were from Perth and the Swan River. No collections for the years 1940-1955 exist. From 1956 acquisitions have been made at an average rate of about 3 per year but ranging from six in one year to nine in three years.

There is an absence of specimens from (a) the "old" inner areas north of the Swan River e.g., Subiaco and Leederville where one could expect

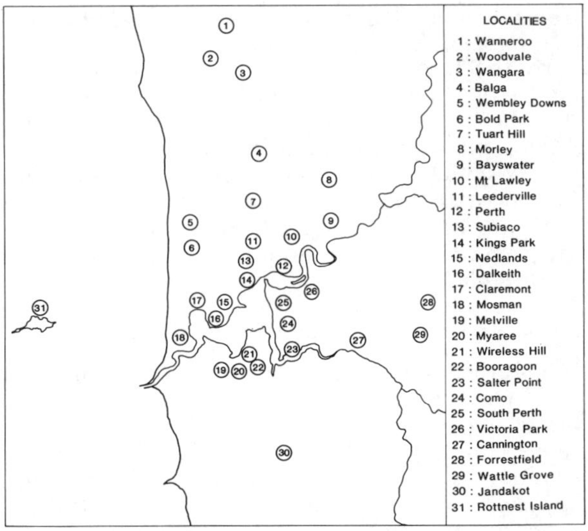

FIG. 6. The coastal plain surrounding the Swan River with locations mentioned in the text.

the species to have occurred and (b) the triangular segment between the junction of the two rivers, the Swan and the Canning, bounded to the east at about Cannington, Forrestfield and Wattle Grove. Significantly this peninsula area comprising South Perth, Victoria Park and Como and Salter's Point is occupied by the allopatric *hirsutum* which is possibly almost extinct. This species has a hirsute abdomen of advantage in sites liable to water logging in that such hairs could retain an air bubble over the respiratory openings (see Rovner 1986). Presumably the species evolved allopatrically by being cut off in a swampier area by an earlier configuration of the wetlands around the rivers.

Museum acquisitions have been made progressively in newer suburbs (a) south of the Swan River e.g., comprising those developed after 1945, (most records dated between 1956-1975 with a few later records e.g., Booragoon and Melville, which retain bushland plots) and (b) northern suburbs from Morley, Tuart Hill, Balga (acquisitions from 1965 onwards) to (c) finally records from as far north as Wangara and Wanneroo and south from Myaree and Jandakot in the 1980s.

Acquisitions are being made still from some older suburbs notably Nedlands and Dalkeith and Mosman Park where vacant lots, University land and river slopes have retained occasional clumps of *Casuarina* with undisturbed litter and soil. Recent building and landscaping have generally exposed female specimens while males have been intercepted during wandering. As these spiders are long lived, we have been probably getting the last survivors.

Finally tracts such as Kings Park, Wireless Hill, Woodvale Research Centre reserve, possibly edges of some golf course fairways (Wembley Downs), Bold Park environs and a few other plots of bush will remain as the last precarious viable refuges for a few years.

By the next century it will almost certainly be only from museum specimens that we will know that the species ever occurred on the coastal plain. At present we can only interpret the occurrence on the coastal plain as being an earlier extension from the north and north-east during an arid period at the time when Rottnest Island was connected to the mainland i.e., about 7-8 000 years ago (Churchill 1959).

Species with a single nineteenth century record

In contrast to the foregoing species of which there are many specimens in collections representing broad geographic distributions, there are species represented by single specimens in museums. An example is the minute spider *Chasmocephalon neglectum* named in 1889 (Cambridge 1889) from a specimen which had been overlooked in the collection at Oxford University for nearly a quarter of a century. I quote from Cambridge:

"Many years ago (1864) I received a large spider from the Swan River, and described and figured it, but until a day or two since have never had occasion to look at it. Examining it, however, now closely, I found on the inner side of one of the folded legs, amongst its numerous hairs, the very minute spider (thus until now wholly overlooked) which forms the type of the present genus and species."

Cambridge's paper was received by the journal in November 1888 so we are talking about an exciting discovery and taxonomic decision made a hundred years ago! The species has not been subsequently collected from the Swan River or anywhere else but a species from Tasmania has been attributed to the genus (Hickman 1944) and another doubtful species from Kuringai National Park, New South Wales (Wunderlich 1976). An unequivocal but unnamed species from Walpole, Western Australia has also been collected and recently attracted a taxonomist from the American Museum of Natural History to visit Western Australia in search of new material of the genus.

A little detective work clearly implicates the 1864 "large spider from the Swan River" as the holotype of *Idiops sigillatus* (now *Idiosoma sigillatum).* The label on the *Idiosoma* specimen also carries the locality "Perth" in addition to "Swan River". Davies (1985) incorrectly attributed the "host" spider to be *Idiommata blackwalli* (Cambridge). The *Idiosoma* could have come from various sites wherever there were *Casuarina* trees, but from the known biology of other species of *Chasmocephalon* the actual site can be further restricted. All *Chasmocephalon* spiders are small, less than 2.0 mm and build horizontal orb webs about 1 to 2 cm in diameter close to the ground, in litter, moss and debris, often in the shelter of rotting logs. The Walpole species occurs in the tingle (*Eucalyptus jacksoni* and *guilfoylei*) forest in permanently wet situations. The most likely situation for the Perth *Chasmocephalon,* mid-nineteenth century, would have been in the shelter of a rotten tuart *(E. gomphocephala)* or jarrah log in the drainage area surrounding the historic spring at the foot of Mt Eliza.

From that single minute spider picked up accidentally 124 years ago by a wandering male trapdoor spider (collected by a Mr Stevens) we can now deduce (but only because the spider is safely preserved in a museum and thus provides a reference for attribution of other species whose biology can still be studied) that along the lower slopes and at the bases of the limestone cliffs along the Swan River between Perth and Fremantle there were pockets of permanently wet and continuously shady clumps of tuart, jarrah and peppermint *(Agonis flexuosa)* trees with an abundance of wet, rotten, mossy logs and debris which was rarely burnt. That the two spiders are also indicative of two distinct microhabitats does not present a fortuitous association. Both are male specimens which is the volent phase for the *Chasmocephalon* and also the "wandering" stage for the *Idiosoma*. Either the *Chasmocephalon* could have drifted from the wet to a dryer habitat or the *Idiosoma* while wandering in search of a female ran "off course" into the adjacent wetter microhabitat.

Discussion

I return now to my premise that "lost" faunal distributions and ecological scenarios of diminished fauna can be reconstructed from specimens in museums and that, in turn, this can give us a guide for restoration of degraded or changed landscapes. Returning to the frog *Heleioporus albopunctatus* it must be emphasized that the geographic distribution is representative of the whole biota of the swales or valley floors of the Wheatbelt. Of the lost invertebrates we know very little. The current proliferation of *Phytophthora cinnamomi* in the jarrah forest and other areas is progressing on a scale of comparable magnitude to the regional landscape degradation (salination) of the Wheatbelt, southwest and south coast of Western Australia. One can foresee a similar cataclysmic loss of bird and mammal species dependent on certain species of susceptible flora; a trend already apparent (see Saunders & Curry, this volume). What is not so obvious in either community is a precursor effect; the disappearance of the unknown invertebrate fauna predominantly arthropods. In a telling essay "The little things that run the world", Wilson (1987) has shown in masterly fashion the absolute dependence of higher plants and animals on invertebrates. Museum collections can guide us from former biological scenarios to restoration of landscapes. In the near future those drums, jars and cabinets in museums and herbaria will be the only material evidence of a once rich habitat in south-western Australia pulsing with a diversity of species [unparalleled elsewhere outside tropical rainforests (Lamont *et al.* 1977)].

Perhaps there is no possibility of restoring the major part of the Wheatbelt. The Agriculture Department of Western Australia, farmers and politicians are involved currently in an enormous problem of arresting and ameliorating salination of the soil. Salination is on a scale unknown elsewhere in non-irrigated areas and there is no comprehensive precedent to follow to restore the landscape. But regardless of a solution, it has been recognized that farmers are not interested generally in large scale *reversion* of a landscape to a semblance of presettlement condition but rather to rebuilding a landscape which will support agriculture of some kind. Accordingly *reclamation* of landscape with salt tolerant forage or holding plants is the farmers' pragmatic ideal. Nevertheless, theoretical reconstruction based on a knowledge of the requirements of former species (e.g., *Heleioporus albopunctatus*) and the species' distributions as ascertained from museum collections and herbaria can be used as a guide to restoration of sample landscapes. Restoration of such sample landscapes can be viewed as a cultural obligation in much the same way as are presentations of social history. There are commendable precedents in eastern Australia viz. restoration of cliffs below the Botanic Gardens in Sydney, using as a guide the botanical lists and collections of Maiden (T. C. Chambers, pers. comm.). Similarly the restoration of Eneogga Creek near Brisbane has been undertaken through reference to old plant lists of the area (R. W. Johnson, pers. comm.).

It surely is not outside engineering capacity, coupled with biological expertise to manipulate a selected terrain, to plant and repopulate it. While restoration of a selected habitat *for the sake* of a frog or even spider, may seem novel or fanciful to some people, it is not so for others. In Suffolk, England a group of volunteers are currently assisting the salvation of the great raft spider, *Dolomedes plantarius*. Water levels in the pools which the spiders inhabit have dropped due to nearby agricultural drainage and volunteers are digging the pools deeper to maintain the habitat for the spiders (Bee 1989).

Reclamation, that is the return of what might be viewed as wasteland into usable land, has generally been the preferred practice. Witness the Wheatbelt being rehabilitated to a so called equilibrium of salt tolerant plants for stock in place of cereal crops on former non saline arable land. However, imbued as we are now with the conservation ethic then surely we should in some instances be thinking of restoration rather than simply reclamation. Allen (1988) has documented the restoration of a mixed forest and cultural landscape in tropical central America. Morrison (1987) has reviewed the philosophy and practicalities of landscape restorations which attempt to simulate visually, selected presettlement landscapes mainly in North America. Guidelines for mimicking original botanical communities there have been derived generally from observations on intact areas with physiographic similarities. Recently Atkinson (1989) has discussed restoration of biotic communities in New Zealand and elsewhere.

It might seem that the magnitude of a salt blighted landscape is beyond complete restoration even in this technological age. Perhaps a makeshift solution will come through "grass roots" perseverance. Indeed some Wheatbelt farmers are attempting what appears to be a combination of restoration/reclamation in their tree planting programmes. Yet, if with the manipulative power of today's gargantuan machinery, whole coastlines can be restructured into harbours, marinas and sophisticated motu-like high density dwelling areas and if mining can turn a landscape inside out, it is surely not beyond the combined vision and skills of biologists and engineers *to reconstruct and restore*, rather than reclaim and rehabilitate sample or specimen sites of our lost landscapes. If we are to think and plan in these terms then we need to know more about the ecological history of the landscapes than just their topography and today's veneer of life forms. We must discover what plants clothed and what animals formerly populated the respective landscapes, and in some cases these clues are only to be found in archival collections in museums and herbaria.

It is appropriate to parody the adage that a bird in the hand is worth two in the bush by saying a frog in the bottle yesterday, will be worth more tomorrow than two frogs in the pond today!

Acknowledgements

I thank the following staff of the Western Australian Museum; Dr L. E. Koch and Dr W. F. Humphreys for access to the Arachnology collections and Julianne Waldock for assistance with sorting *Idiosoma* specimens; Dr R. A. How, R E. Johnstone, L. A. Smith and J. White (Department of Ornithology and Herpetology) for access to records on distribution, collection and acquisition dates of frogs. I am grateful to C. P. de Rebeira, CSIRO, Division of Wildlife and Ecology for preparing Figures 2, 5 and 6 from my drafts.

References

Alexander W. B. (1914) The history of Zoology in Western Australia, Part I, Discoveries in the 17th Century. *Journ. Nat. Hist. and Sci. Soc. W. Aust.* **5,** 49-64.

Alexander W. B. (1916) History of Zoology in Western Australia, Part II, 1791-1829. *J. Roy. Soc. W. Aust.* **1,** 83-149.

Alexander W. B. (1918) History of Zoology in Western Australia, Part III, 1829-1840. *J. Roy. Soc. W. Aust.* **3,** 37-69.

Allen W. H. (1988) Biocultural restoration of a tropical forest. *Bioscience* **30**(3), 156-61.

Anon (1988) Select Committee on Salinity, First Report. Parliament of Western Australia, Legislative Council.

Atkinson I. (1989) Presidential address: Opportunities for ecological restoration. *N. Zeal. J. Ecol.* **11,** 1-12.

Bee L. (1989) Help save the Great Raft Spider. *Br. arachnol. Soc. Newsl.* **54,** Supplement.

Burbidge A. A. (1967) The biology of South-western Australian tortoises. Ph.D. Thesis, The University of Western Australia, Nedlands.

Burbidge A. A. (1981) The ecology of the western swamp tortoise *Pseudemydura umbrina* (Testudines: Chelidae). *Aust Wildl. Res.* **8,** 203-23.

Burvill G. H. (1956) Salt land survey, 1955. *J. Agric. W. Aust.* **5,** 113-9.

Calaby J. H. (1960) Observations on the banded anteater *Myrmecobius fasciatus* Waterhouse (Marsupialia), with special reference to its food habits. *Proc. Zool. Soc. (Lond.)* **135,** 183-207.

Cambridge O. P. (1870) Monograph of the genus *Idiops*, including descriptions of several species new to science. *Proc. Zool. Soc. (Lond.)* **1870,** 101-8.

Cambridge O. P. (1889) On some new species and a new genus of Araneidea. *Proc. Zool. Soc. (Lond.)* **1889,** 34-46.

Christensen P. E. S. (1980) The biology of *Bettongia penicillata* Gray, 1837, and *Macropus eugenii* (Desmarest, 1817) in relation to fire. Forests Department of Western Australia, Bull. **91,** i-ix, 1-90.

Churchill D. M. (1959) Late Quarternary eustatic changes in the Swan River district. *J. Roy. Soc. W. Austr.* **42,** 53-5.

Davies V. T. (1985) Araneomorphae. In: *Zoological Catalogue of Australia, Vol. 3, Arachnida.* (ed. D. W. Walton) pp. 49-125. Australian Government Publishing Service, Canberra.

Friend J. A. (1987) Local decline, extinction and recovery: relevance to mammal populations in vegetation remnants. In: *Nature Conservation: The Role of Remnants of Native Vegetation.* (eds D. A. Saunders, G. W. Arnold, A. A. Burbidge & A. J. M. Hopkins) pp. 53-64. Surrey Beatty & Sons, Chipping Norton.

Friend J. A., Fuller P. J. & Davis J. A. (1982) The numbat in Central Australia. *Swans* **12,** 20-6.

Gray J. E. (1841) A catalogue of the species of reptiles and amphibia hitherto described as inhabiting Australia, with a description of some new species from Western Australia, and some remarks on their geographical distribution. In: *Journals of two expeditions of discovery in North-west and Western Australia, Vol. II* (G. Grey) pp. 422-49. T. & W. Boone, London.

Henschke C. J. (1980) Saltland in statistics . . . the 1979 saltland survey. *J. Agric. W. Aust.* **21,** 116-9.

Hickman V. V. (1944) On some new Australian Apneumonomorphae with notes on their respiratory system. *Pap. Proc. R. Soc. Tas.* **1943,** 179-95, pls l-v.

How R. A. & Kitchener D. J. (1983) The biology of the gecko *Oedura reticulata* Bustard, in a small habitat isolate in the Western Australian Wheatbelt. *Aust. Wildl. Res.* **10,** 543-56.

Kinnear J., Onus M. & Bromilow B. (1984) Foxes, feral cats and rock wallabies. *Swans* **14,** 3-8.

Lamont B. B., Downes S. & Fox J. E. D. (1977) Important value curves and diversity indices applied to a species-rich heathland in Western Australia. *Nature* **265,** 438-41.

Lee A. K. (1967) Studies in Australian Amphibia II. Taxonomy, ecology and evolution of the genus *Heleioporus* Gray (Anura: Leptodactylidae). *Aust. J. Zool.* **15,** 367-439.

Lightfoot L. C., Smith S. T. & Malcolm C. V. (1964) Report of a survey of soil salinity in the agricultural areas of Western Australia. *J. Agric. W. Aust.* **5,** 396-410.

Main A. R. (1965) *Frogs of southern Western Australia.* Western Australian Naturalists' Club, Perth.

Main B. Y. (1952) Notes on the genus *Idiosoma*, a supposedly rare Western Australian trapdoor spider. *West. Aust. Nat.* **3,** 130-7.

Main B. Y. (1957) Biology of aganippine trapdoor spiders (Mygalomorphae: Ctenizidae). *Aust. J. Zool.* **5,** 402-73.

Main B. Y. (1978) Biology of the arid-adapted Australian trapdoor spider *Anidiops villosus* (Rainbow). *Bull. Br. arachnol. Soc.* **4,** 161-75.

Main B. Y. (1987a) Ecological disturbance and conservation of spiders: Implications for biogeographic relics in southwestern Australia. In: *The role of invertebrates in conservation and biological survey* (ed. J. D. Majer) pp. 89-97. Western Australian Department of Conservation and Land Management Report.

Main B. Y. (1987b) Persistence of invertebrates in small areas: Case studies of trapdoor spiders in Western Australia. In: *Nature conservation: the Role of Remnants of Native Vegetation* (eds D. A. Saunders, G. W. Arnold, A. A. Burbidge & A. J. M. Hopkins) pp. 29-39. Surrey Beatty & Sons, Chipping Norton.

Main B. Y. (1988) The biology of a social spider. In: *Australian Arachnology* (eds A. D. Austin & N. Heather) pp. 55-73. Miscellaneous Publication No. 5, Australian Entomological Society, Brisbane.

Malcolm C. V. (1983) Wheatbelt Salinity. Western Australian Department of Agriculture Technical Bulletin No. 52.

Malcolm C. V. & Stoneman T. C. (1976) Salt encroachment — the 1974 saltland survey. *J. Agric. W. Aust.* **17,** 42-9.

Michaelsen W. & Hartmeyer R. (1907-30) *Die Fauna Südwest-Australiens.* Fischer, Jena.

Morrison D. (1987) Landscape restoration in response to previous disturbance. In: *Landscape Heterogeneity and Disturbance* (ed. M. G. Turner) pp 159-72. Springer-Verlag, New York.

Pocock R. I. (1897) On some trapdoor spiders of the family Ctenizidae from South and West Australia, contained in the collection of the British Museum. *Ann. Mag. Nat. Hist.* **6**(19), 109-16.

Rovnor J. S. (1986) Spider hairiness: Air stores and low activity enhance flooding survival in inland terrestrial species. In: *Actas×Congr. Int. Aracnol. Vol. 1* (ed. J. A. Barrientos) pp. 123-29 Juvenil, Barcelona.

Rowley I. (1981) The communal way of life in the splendid wren *Malurus splendens. Z. Tierpsychol.* **55,** 228-67.

Saunders D. A., Rowley I. & Smith G. T. (1985) The effects of clearing for agriculture on the distribution of cockatoos in the south-west of Western Australia. In: *Birds of Eucalypt Forests and Woodlands: Ecology, Conservation, Management* (eds A. Keast, H. F. Recher, H. Ford & D. Saunders) pp. 309-21. Surrey Beatty & Sons, Chipping Norton.

Saunders D. A. & Ingram J. A. (1987) Factors affecting survival of breeding populations of Carnaby's cockatoo *Calyptorhynchus funereus latirostris* in remnants of native vegetation. In: *Nature Conservation: The Role of Remnants of Native Vegetation* (eds D. A. Saunders, G. W. Arnold, A. A. Burbidge & A. J. M. Hopkins) pp. 249-58. Surrey Beatty & Sons, Chipping Norton.

Simon E. (1908) Araneae, 1re partie. In: *Die Fauna Südwest-Australiens* (eds W. Michaelsen & R. Hartmeyer) pp. 360-446. Fischer, Jena.

Simon E. (1909) Araneae, 2me partie. In: *Die Fauna Südwest-Australiens* (eds. W. Michaelsen & R. Hartmeyer) pp. 155-212. Fischer, Jena.

Smith G. T. (1976) Ecological and behavioural comparisons between the Atrichornithidae and Menuridae. In: *Proceedings of the 16th International Ornithological Congress* (eds H. J. Frith & J. H. Calaby) pp. 125-36. Australian Academy of Science, Canberra.

Smith G. T. (1985) The noisy scrub-bird *Atrichornis clamosus.* Does its past suggest a future? In: *Birds of Eucalypt Forests and Woodlands: Ecology, Conservation, Management* (eds A. Keast, H. F. Recher, H. Ford & D. Saunders), pp. 301-8. Surrey Beatty & Sons, Chipping Norton.

Smith G. T. & Robinson F. N. (1976) The noisy scrub-bird: An interim report. *Emu* **76,** 37-42.

Stoneman T. C. (1978) Soil salinity in Western Australia. *J. Agric. W. Aust.* **19,** 99-100.

Tansley A. (1951) What is ecology? *Council for the Promotion of Field Studies.* (Reprinted 1987) *Biol. J. Linn. Soc.* **32,** 5-16.

Tyler M. J., Smith L. A. & Johnstone R. E. (1984) *Frogs of Western Australia.* Western Australian Museum, Perth.

Whittell H. M. (1954) *The Literature of Australian Birds.* Paterson Brokensha, Perth.

Wilson E. O. (1987) The little things that run the world. *Conserv. Biol.* **1,** 344-6.

Wunderlich J. (1976) Spinnen aus Australien. 1. Uloboridae, Theridiosomatidae und Symphytognathidae (Arachnida: Araneida). *Senckenberg. biol.* **57,** 113-24.

APPENDIX 1. *Heleioporus albopunctatus*, localities and 10 year period of collection (acquisition) of specimens in Western Australian Museum. Year of collection given for specimens collected before 1945. Shire in which particular locality occurs given in parentheses. See also Table 2 and Figures 2 and 3.

Year or 10 year period of acquisition	Locality and shire (if different from locality name)
1924	Clackline (Northam)
1927	Waeel (Cunderdin)
1928	Ejanding (Dowerin)
1930	Northam
1931	Kulin
1933	Pingrup (Nyabing/Pingrup=Kent) Seabrook (Northam)
1934	Bullaring (Corrigin), Burnup (Lake Grace), Kulin, Lake Grace
1936	Kukerin (Dumbleyung)
1943	Wamenusking (Quairading), White Peaks (Northampton)
1944	Broomehill, Yellowdine (Yilgarn)
1945	Broomehill
1946-1955	Beechina at Chidlows Turnoff (Mundaring), Bruce Rock, Coorow, Dongara (Irwin), Dongara/Mingenew Rd (Mingenew), Ebano Creek (Morawa), Gillingarra (Victoria Plains), Gnangara (Wanneroo), Jerramungup (Gnowangerup), Kukerin (Dumbleyung), Meckering (Cunderdin), between Miling/Pithara (Moora), Mingenew, Mt Misery (Dandaragan), Murchison River (Northampton), Northam Army Camp (Northam), Ringa (Northam), Waeel (Cunderdin), Wooroloo (Mundaring), Wundowie (Northam), Yorkrakine Rock (Tammin)
1956-1965	Beechina (Mundaring), Dandaragan, Dryandra (Narrogin), Ebano Creek (Morawa), Gillingarra (Victoria Plains), Greenough, Hutt River (Northampton), Hill River Ford (Dandaragan), Kalguddering (Wongan/Ballidu), Mingenew, Pallinup River (Gnowangerup), Pearce (Swan/Guildford), Mt Peron (Coorow), Wialki (Mt Marshall), Wundowie (Northam), Yellowdine (Yilgarn)
1966-1975	Badjaling (Quairading), Benderine (Kellerberrin), Billycatting Hill 17.5 km NE (Mukinbudin), Buntine (Dalwallinu), Dalwallinu, Dongolocking (Dumbleyung), Dryandra (Narrogin), Elachbutting 11.3 km NE (Mukinbudin), Katanning, Kellerberrin 17 km NW, 25 km N (Kellerberrin), Kalbarri (Northampton), Mt Peron (Coorow), Lake Grace, Northampton, Quairading, Lake Magenta (Lake Grace), Stockyard Gully (Dandaragan), Tarin Rock (Dumbleyung), Yuna Reserve (Chapman Valley), Julimar Forest (Toodyay), Yornaning (Cuballing), York, Yorkrakine Rock (Tammin)
1976-1985	Badgingarra (Dandaragan), Bendering (Kondinin), Bruce Rock 57 km E (Narembeen), Lake Dumbleyung (Dumbleyung), Cunderdin, Lake Cronin (Kondinin), Dalwallinu 20 km NE (Dalwallinu), Fitzgerald River National Park (Ravensthorpe), Frank Hann National Park (Ravensthorpe), Julimar Forest (Toodyay), Kalbarri (Northampton), The Lakes (Mundaring), Marchagee (Coorow), Merredin 67.5 km SE, 91.5 km E (Merredin), Mokine Hill (Northam), Midland 20 km NNE (Swan/Guildford), Pearce (Swan/Guildford), Lake Toolibin (Wickepin), Walkaway (Greenough), Mt Rupert 10 km W (Wongan/Ballidu)

APPENDIX 2. Localities and year of collection (acquisition) of *Idiosoma sigillatum* and *hirsutum* specimens in the Western Australian Museum. Males and females pooled. Number of specimens not indicated.

I. sigillatum	1928	Perth
	1929	Cannington, Mt Lawley
	1938	Bayswater, Nedlands
	1939	Claremont
	1955	Pinjarra
	1956	Applecross
	1957	Osborne Park
	1958	Floreat Park
	1959	Yarloop
	1960	Mt Pleasant, Scarborough, Tuart Hill
	1961	Melville, Nedlands
	1962	Dalkeith, Nollamara
	1963	Nedlands
	1964	Dalkeith, Mt Pleasant, Scarborough
	1965	Claremont, Morley Park, Tuart Hill
	1966	Bayswater, Mt Pleasant
	1967	Ledge Point, Melville Heights
	1968	Applecross, Forrestfield, Kwinana, Mt Yokine, Wembley Downs
	1969	Gnangara, Mandurah
	1970	Tuart Hill
	1971	Attadale, Balga, Cannington, Canningvale, Karrinyup, Morley, Mosman Park, Pinjarra, Wanneroo, Wattle Grove
	1972	Dianella, Karrinyup, Morley, Mt Pleasant
	1973	King's Park, Lynwood, Myaree, Orelia, Shenton Park, Wembley Downs
	1974	Bayswater
	1975	Mt Pleasant, Naval Base
	1976	Greenwood, Mandurah, Sorrento
	1977	Cottesloe, Forrestdale, Girrawheen, Joondanna, Sorrento
	1978	Booragoon, Melville, Nedlands
	1979	
	1980	Bunbury, Eaton, Tuart Hill
	1981	Duncraig (South), Padbury
	1982	
	1983	Balga, Dianella, Forrestdale
	1984	Forrestdale, Myaree
	1985	Balcatta, Duncraig, Greenwood, (Perth suburb), Swanbourne
	1986	City Beach, Gosnells, Murdoch, Wangara
	1987	Gwelup?, Hall's Head & Miami (Mandurah)
	1988	Booragoon, Bunbury, Jandakot, Melville
I. hirsutum	1914	South Perth
	1930	Victoria Park
	1959	Perth 6.4 km SE in Pine Plantation
	1971	Como
	1973	Salter's Point

Proc. Ecol. Soc. Aust. 1990 — 16:411-418

Naturalness: the concept and its application to Australian ecosystems

S. G. TAYLOR

Department of Geography, University of Adelaide, G.P.O. Box 498, Adelaide, South Australia 5001

Abstract

The concept of naturalness is implicit in all attempts to assess the extent to which Australian ecosystems have been modified by 200 years of European settlement, as well as by 40 000 years of Aboriginal occupation. Yet, failure to recognize that naturalness is a culturally constructed concept, rather than a universal one, has produced such inconsistency and ambiguity in the terminology used for these assessments that the terminology lacks the precision and ecological rigour required for incorporation in theoretical models of landscape change, and for application to the management of landscape change.

The primary aim of this chapter is to provide a conceptual framework for Australian studies of the modification of natural ecosystems by human intrusion or influences. The meaning of the concept of naturalness in Western technological societies is examined in an attempt to clarify the uncertainty and ambiguity currently associated with the use of the terms "natural landscape", "natural ecosystem" and "natural vegetation". A revised terminology and definitions are presented based on the biocentric concept that places humanity within nature, and treats human intrusion or influences as only one of the many natural agents of landscape change.

Introduction

Naturalness is a culturally constructed concept expressing the perception of Western technological societies that the world consists of the binary opposites, humanity and nature, and that humanity has dominion over nature. This concept, which is both anthropocentric (human-centred) and technocentric, results in formal definitions of natural phenomena that appear to exclude all human beings and their works, but actually exclude only the intrusion and influences of societies seen as being in control of nature rather than under its control (Passmore 1980; Birch & Cobb 1981; O'Riordan 1981).

It might seem that this is a purely semantic issue with little relevance to ecology. Yet, the concept of naturalness is implicit in all attempts to assess humanity's role in landscape change. Failure to recognize that naturalness is a culturally constructed concept, rather than a universal one, has produced such inconsistency and ambiguity in the terminology used for these assessments that the terminology lacks the precision and ecological rigour required for incorporation in theoretical models of landscape change, and for application to the management of landscape change. This point is demonstrated by the confusion associated with the use of the terms "natural landscape", "natural ecosystem" and "natural vegetation".

Natural landscapes, natural ecosystems and natural vegetation

The earliest scientific discipline to study humanity's role in landscape change was not ecology but geography. This focus of geographical studies developed towards the end of the 1800s when German geographers began to define their discipline as *landschaftsgeographie* (i.e., the geography of landscapes). According to this definition, geography was the science concerned with explaining the character of the landscapes found in different environmental regions (Hartshorne 1939).

The German school of landscape geography was the first discipline to make an explicit distinction between natural landscapes *(naturlandschafts)* and landscapes subject to human modification. Using modern terminology, a natural landscape was considered to be the concrete and characteristic product of the biophysical processes operating in a given environmental region, without human intrusion or influences; hence a landscape in the state of naturalness. Based on this definition, many of the early studies in landscape geography were theoretical attempts to identify the geographic agents responsible for shaping the character of the landscapes found in those environmental regions affected by human occupation (Hartshorne 1939). These attempts produced the concept of the cultural landscape *(kulturlandschaft)*.

As implied by the term itself, a cultural landscape was regarded as something more than simply a natural landscape modified hy human intrusion or influences. It was a natural landscape transformed by the cultural preferences and possibilities of its human occupants. In the words of the geographer, Carl Sauer (cited by Leighly 1974), "culture is the agent, the natural area is the medium, the cultural landscape the result".

In Western Europe, human impacts on the landscape were so substantial and pervasive, even in the 1800s, that geographers considered natural landscapes to exist only in Europe's prehistoric past. In the New World, however, European geographers encountered what they perceived to be natural landscapes in the "wilderness" beyond the frontiers of colonial settlement.

This perception, of course, was inexplicable to the New World Aborigines whose cultural homelands were viewed by the European invaders as untamed wilderness, a view based on the notion that a cultural landscape can only be one that has been "civilized" by the consciously creative environmental activities of a technological society (Hendee *et al.* 1978). During the 1800s, pre-technological people were seen as inhabiting natural landscapes because it was assumed that they lived like animals, passively responding to nature, rather than like civilized people, actively dominating nature.

Following the development of the ecosystem concept in the 1930s, natural landscapes were redefined as consisting of natural ecosystems; that is, of biotic communities and their abiotic environments unmodified by human intrusion or influences. Further, the presence of a natural vegetation cover began to be used as the primary indicator of naturalness for terrestrial ecosystems, where natural vegetation was defined as vegetation unmodified by human intrusion or influences (Westhoff 1983). With these minor refinements, the current ecological literature is still reflecting the 19th century idea that technological societies inhabit cultural landscapes and create cultural ecosystems, while pre-technological societies inhabit natural landscapes and are a part of natural ecosystems. Thus Usher (1986) can make the following statement, apparently without embarrassment, in a modern ecological text.

> *In areas of the world, such as Western Europe, where there is a long history of modification of ecosystems by man, it is likely that there are no fully natural ecosystems. In a country like Australia, it is relatively simple to define a natural ecosystem [by regarding the natural condition of the vegetation cover] as that at the time of European settlement.*

Nowhere is there more confusion associated with the use of the terms "natural landscape", "natural ecosystem" and "natural vegetation" than in Australia.

Naturalness in Australia

Inconsistencies and ambiguities

As exemplified by the preceding quotation, it is commonly assumed that Australian landscapes were in a natural state at the time of European settlement and that these natural landscapes consisted wholly of natural ecosystems supporting a natural vegetation cover. This assumption would seem to suggest either that 40 000 years of Aboriginal occupation did not affect the natural character of Australian landscapes or that pre-contact Aborigines were not human, at least in terms of their landscape effects.

In reality, it is widely recognized that traditional Aboriginal land use practices had profound effects on the structure and composition of Australia's pre-European vegetation cover and on its animal inhabitants (Groves 1981; Smith 1982). Despite this recognition, the ecological literature does not restrict the use of the descriptor "natural" to Australian landscapes of the late Pleistocene before the beginning of Aboriginal migration to this continent.

Turning to European landscape effects, if a natural vegetation cover is indicative of a natural ecosystem, and if natural vegetation is conceived to be in the same condition that it was at the time of European settlement (cf. Laut *et al.* 1977), it is arguable that Australia no longer has any natural ecosystems. Virtually the whole of Australia's vegetation cover has heen modified, in some way, by such direct and indirect effects of European settlement as logging, grazing by both domesticated and feral exotic animals, exotic weed and predator invasion, alterations in fire regime, atmospheric pollution, etc. (Groves 1981; Smith 1982).

Nevertheless, it seems somehow inappropriate to treat ecosystems that still support Australian native vegetation, however modified from its pre-European state, as anything other than natural when compared with the kinds of ecosystem characteristic of developed rural and urban landscapes. This dilemma has encouraged the contradictory practice of defining natural ecosystems as "unmodified by human intrusion or influences" and then proceeding to define an ordinal sequence of ecosystems that differ in their degree of naturalness as a function of their degree of modification by human intrusion or influences; *viz.* "natural", "sub-natural", "semi-natural", "near-cultural" and "cultural" (Davis & Glick 1978, p.303). Such ordinal sequences appear to imply that natural ecosystems and cultural ecosystems are not binary opposites, but rather the end points of a progressive continuum produced by the extreme states of a single continuous variable, namely degree of naturalness. If this were true it would make the delimitation of class boundaries within the continuum merely a matter of convenience.

A source of further confusion in the use of the terms "natural landscape", "natural ecosystem" and "natural vegetation" is the redefinition of the humanity/nature relation by environmentalists in recent years. This redefinition is an attempt to replace the anthropocentric/technocentric concept of naturalness with a biocentric (life-centred) one. The biocentric concept envisions the whole of humanity as one global community, within nature and inextricably linked to the biophysical systems of its planetary habitat (Taylor 1986). By placing humanity within nature, the biocentric concept precludes the use of the descriptor "natural" as the antonym of the descriptor "human" in any context, thereby suggesting that all of Australia's ecosystems, past and present, are equally natural.

A biocentric view of naturalness

Those ecologists who have accepted the biocentric concept of naturalness argue that humanity should be treated as only one of the many natural agents of landscape change. This is not to deny that there are points, in both space and time, where humanity becomes a special agent of landscape change, but this special agent is still a natural one (Dasman 1972; Dansereau 1973; Maarel 1980; Taylor 1984; Evernden 1985). When it becomes a special agent, humanity begins to function as though it were an exotic invader of its own habitats, creating landscapes consisting of entirely new kinds of ecosystems that can properly be called "human-dominated", rather than simply "human-modified".

This argument is clearly illustrated by the Holocene history of landscape change in Britain, which provides the most extensively documented model available of the creation of human-dominated landscapes by European land use practices. According to Peterken (1981), Britain last had an essentially continuous cover of native woodland during the Atlantic Period (7 500-5 000 years B.P.) when its terrestrial ecosystems were in a state of dynamic equilibrium with a range of environmental controls, including the effects of nomadic Mesolithic hunting and gathering societies. In pollen profiles, the end of the Atlantic Period is marked by the "elm decline", a sharp and widespread decrease in *Ulmus* spp. pollen. The elm decline is believed to reflect a *qualitative* change in the character of human impact on Britain's native woodland due to the spread of Neolithic, and later Bronze Age, societies (Peterken 1981). In addition to hunting and gathering, these societies cleared areas for cultivation, depasturing their stock in the remnant woodland surrounding the clearings. For some time, the impact of these new land use practices was restricted, both in areal extent and geographic distribution, since native woodland remained widespread, particularly in the Highland Zone, until the end of the Sub-Boreal Period (5 000-2 700 years B.P.). Then Iron Age societies accelerated the rate of woodland clearance until a critical point was reached (probably 2 000 to 1 000 years B.P.) where the remaining native woodland was restricted to one-third of its Atlantic Period distribution area. At this point, remant native woodland occurred only as scattered islands within landscapes consisting of a sea of human-dominated ecosystems and associated human constructions.

Even after a 5 000 year history of fragmentation, some remnants of native woodland still survive within Britain's human-dominated landscapes. This native woodland is known not to have been cleared for cultivation since about 1 600 A.D. when maps of Britain's land cover first became available, although some remnants may have been selectively logged or even clear-felled in historic times, and allowed to regenerate without an intervening period of cultivation. The remnants classed as "native woodland" are clearly recognizable as descendants of the Atlantic Period woodlands, despite millennia of direct and indirect human modification (Peterken 1981).

This example demonstrates that humanity's role in landscape change can not be modelled either as a simple dichotomy (i.e., natural landscape/ecosystem versus cultural landscape/ecosystem), or as a simple progression in degree of modification by human intrusion or influences through space and time (e.g., natural, sub-natural, semi-natural, near-cultural, cultural landscape/ecosystem) (Taylor 1984). The effects of human intrusion and influences do differ in degree through space and time, but they also differ in kind. Thus human-generated landscape changes are marked by discontinuities where humanity becomes a special agent within nature due to qualitative differences between the character of its impact on the landscape and the character of other natural impacts (Dansereau 1973; Maarel 1980).

In Britain, a discontinuity in the Holocene sequence of landscape change occurred at the end of the Atlantic Period (5 000 years B.P.) when Neolithic societies created the first human-dominated ecosystems by clearing areas of native woodland for cultivation. These clearings initiated the fragmentation of the native woodland cover that led to the formation of human-dominated landscapes throughout lowland Britain some 2 000 to 1 000 years B.P. (Peterken 1981).

In Australia, a similar discontinuity occurred 200 years ago when the European invaders began to settle, clear and cultivate land for productive and amenity purposes. In contrast to the British situation, human-dominated landscapes were created locally within the first decade of colonization, and spread throughout Australia's relatively high rainfall, agricultural environments in tens, rather than thousands of years, a fragmentation process Adamson and Fox (1982) have described as "a cataclysm":

> *The European invasion was a watershed beyond which Australian ecosystems are permanently changed. A revolution has occurred and no new steady state is in sight. Completely new ecosystems are being created from the interaction between human activities, the original organisms, the many successful immigrant organisms, and the changed environmental factors.*

Towards a biocentric terminology

The anthropocentric/technocentric concept of naturalness is so deeply embedded in the English language that there seems to be no convenient alternative to the continued use of the adjective "natural" for landscapes that support ecosystems having a native vegetation cover, and for such ecosystems themselves. Nevertheless, it is suggested that the ecological literature, particularly in Australia and other New World countries with surviving Aboriginal populations, should use a biocentric terminology that does not imply human beings and their works are unnatural, or with even less justification, that culture is the prerogative of Western technological societies and especially of the men in these societies. "Human-dominated landscape" and "human-dominated ecosystem" (cf. "man-dominated", Burgess & Sharpe 1981) are suggested as non-racist and non-sexist substitutes for "cultural landscape" and "cultural ecosystem". Together with "natural landscape" and "natural ecosystem", these terms may be acceptable as the basic descriptors of a biocentric terminology.

Simply substituting one set of descriptors for another, however, will not resolve the inconsistency and ambiguity in the terminology presently being used for assessments of humanity's role in landscape change if the new descriptors are not precisely defined, and if their definitions do not rigorously reflect current understanding of the ecology of disturbance. The definitions provided hereafter attempt to satisfy the requirements of precision and ecological rigour.

Natural landscapes, natural ecosystems and native vegetation

The nature and impact of disturbance can only be defined relative to a given level of ecological organization, and relative to the scale (or degree of resolution) at which this level of ecological organization is being observed (Rykiel 1985). In the present context, landscape change can be considered a consequence of disturbance operating at the ecosystem level of ecological organization, and at the scale normally used to identify, delimit and classify biotic communities and their abiotic environments (i.e., at the scale of vegetation structure and dominant plant species composition) (McDonald *et al.* 1984). Following Rykiel (1985), ecosystem disturbance at this scale can be defined as a physical or biological (including human) agent (i.e., a cause) that creates a perturbation (i.e., an effect manifested as a detectable change) in the vegetation structure and dominant plant species composition of the ecosystem. The occurrence of an effect having sufficient magnitude to constitute a perturbation is indicated by either temporary or permanent deviation of the ecosystem from its predisturbance identity, as determined by vegetation structure and dominant plant species composition.

Ecosystem disturbance of a natural landscape may cause two different kinds of change, succession and fragmentation. In the classic model of succession, ecosystem disturbance initiates a perturbation in the structure and dominant plant species composition of the native vegetation cover at a specific site in the landscape. By destroying (or substantially altering the architecture of) one or more vegetation layers, disturbance creates a new native vegetation-type, with a different vegetation structure and/or dominant plant species composition, on the same site (Miles 1979). When this occurs in a natural landscape, regenerative processes normally restore the predisturbance structure and composition of the native vegetation, through an intermediate sequence of (successional) native vegetation-types, over a period of time which may vary from a few years to a few hundred years, depending on the severity of the initial perturbation and on ecosystem attributes. This kind of succession is termed "autogenic" (Miles 1979).

A second kind of succession, termed "allogenic", occurs as a result of changes in the abiotic environments and/or disturbance regimes of entire natural landscapes, usually over periods of hundreds to thousands of years. Instead of a regenerative return to the original (predisturbance) ecosystem state, allogenic succession involves the development of new ecosystem states in equilibrium with a new abiotic environment/disturbance regime, or an ongoing change in ecosystem state in response to a continuously changing abiotic environment/ disturbance regime (Miles 1979).

In reality, the distinction between autogenic and allogenic succession is merely a matter of convenience, since essentially all natural landscapes are subject to more or less local and short-term fluctuations about a notional mean state that is itself subject to more or less widespread and long-term variation. Thus Adamson and Fox (1982) comment that, if a mid-term view (over hundreds of years) is taken, Australia's natural landscapes apparently were in a steady state for at least several millenia before the European invasion. With a short-term view (one to tens of years), however, this steady state can be seen as series of sudden and extreme local fluctuations about a notional mean state as the natural landscapes responded to and recovered from human-generated (i.e., Aboriginal) and other kinds of natural disturbance (e.g., fire, flood, drought, etc.). With a long-term view of thousands of years, the steady state disappears, to be replaced by a pattern of relatively slow but major landscape changes on a continental scale caused by global climatic events and Aboriginal land use practices.

Since the arrival and spread of Western technological societies, allogenic succession has greatly accelerated, so that virtually all natural landscapes are experiencing relatively rapid and ongoing changes in mean state; ". . . the whole continent is in flux because the new impact is open ended and because not all changes instituted earlier have worked through the system" (Adamson & Fox 1982). In addition, European land use practices have introduced a new kind of landscape change to Australia, namely fragmentation.

It is argued here that there is essentially no *qualitative* ecological difference (i.e., difference in kind) between the successional processes initiated in Australia's natural landscapes since the late Pleistocene by Aboriginal occupation, European settlement and other forms of natural disturbance, although there are very considerable and highly significant *quantitative* ecological differences (i.e., differences in magnitude, frequency, areal extent, etc.) in these successional processes. The points in space and time where natural landscapes begin to give way to human-dominated landscapes are marked not by succession but by fragmentation of the native vegetation cover as natural ecosystems are *removed* and *replaced* by human-dominated ones supporting exotic vegetation.

To be useful in practice, the distinction between natural ecosystems supporting native vegetation and human-dominated ecosystems supporting exotic vegetation must be much more carefully defined than has been the case with other attempts to develop similar classification systems (cf. Duvigneaud 1974; Laut *et al.* 1977; Davis & Glick 1978; Brady *et al.* 1979; Westhoff 1983). In this chapter, the distinction between native vegetation and exotic vegetation (except for rainforest vegetation which presents special definitional problems not addressed here) is based on the origin of the *dominant plant species* in the canopy layer and understorey. Native vegetation has *both* a canopy layer and an understorey dominated by plant species which are indigenous to the locality and spontaneous in occurrence (i.e., "native", according to the generally accepted use of this term). In contrast, exotic vegetation has *either* a canopy layer and/or an understorey dominated by plant species which may be:

(a) indigenous to the locality but cultivated (where cultivated is defined as deliberately established for productive/amenity purposes);
(b) alien to the locality and cultivated, or;
(c) alien to the locality and spontaneous (i.e., adventive, according to the generally accepted use of this term).

The emphasis given to dominant plant species by this definition follows from the phytosociological concept that vegetation structure and dominant plant species composition determine the ecological identity of a natural ecosystem (i.e., its formation and association). The replacement of the dominant plant species in the native vegetation canopy layer of a natural ecosystem by other dominant plant species having a different population architecture (i.e., different life forms and/or spatial distribution) produces a new plant community formation and association (Mueller-Dombois & Ellenberg 1974). Native vegetation, therefore, can be said to lose its identity as a native vegetation at the point where its canopy-dominant native plant species are replaced by exotic (i.e., cultivated or adventive) dominants. This is the point at which a natural ecosystem becomes a human-dominated one.

Using this criterion, Australia's natural ecosystems can be defined as those supporting native vegetation in either an autogenic successional (i.e., regenerative) or an equilibrium state. It is accepted that the equilibrium state of Australia's native vegetation is highly dynamic, since this vegetation is experiencing relatively rapid and ongoing allogenic succession as a response to the direct and indirect impacts of European settlement and other natural disturbances. Thus the present equilibrium vegetation of a natural ecosystem may not be the same as the vegetation supported by the site of this ecosystem during different historic phases of European settlement in Australia, or indeed during different prehistoric phases of Aboriginal occupation. Nevertheless, the present equilibrium vegetation is still recognizable as native vegetation by simple field observation using accepted phytosociological criteria (McDonald *et al.* 1984). Moreover, the present equilibrium vegetation has not been "isolated in time" from the pre-Aboriginal native vegetation of the late Pleistocene. It has descended directly from this late Pleistocene native vegetation through an unbroken sequence of autogenic and allogenic successional responses to human-generated disturbance and other natural agents of landscape change.

Human-dominated ecosystems, in contrast, support exotic vegetation that is isolated in both space and time from Australia's native vegetation. This exotic vegetation originated with the introduction of European land use practices only two centuries ago, but it is now the main vegetation-type of the human-dominated landscapes that occur throughout Australia's relatively high-rainfall, agricultural environments.

Human-dominated landscapes, human-dominated ecosystems and exotic vegetation

In Australia's relatively high rainfall, agricultural environments, European land use has involved the extensive clearing of native vegetation and its replacement by exotic vegetation deliberately established for productive/amenity purposes, or accidentally established as a by-product of the settlement process. During the early stages of native vegetation clearance, exotic vegetation occurs as isolated patches and strips in natural landscapes. However, as these occurrences of exotic vegetation grow and coalesce, the natural landscapes are progressively fragmented until small (mainly less than 100 ha) areas of remnant native vegetation occur as isolated patches and strips scattered within a matrix of exotic vegetation and human constructions. It is this landscape pattern that is diagnostic of a human-dominated landscape (Burgess & Sharpe 1981; Taylor 1987).

As a consequence of the settlement process, human-dominated landscapes are the most prevalent landscapes in Australia's relatively high rainfall, agricultural environments. Settlement, however, is still exceedingly sparse throughout Australia's tropical, arid and semi-arid environments. Moreover, the settlement that has occurred in these environments is mainly associated with mining, tourism, Aboriginal social service centres, and extensive livestock grazing using palatable native plant species rather than sown pastures. Thus, in these environments, the exotic vegetation characteristic of human-dominated landscapes is confined to a few scattered points of permanent human occupation, and the entire interior and north-west of the continent can be viewed as a mosaic of natural landscapes with a virtually continuous native vegetation cover (Taylor 1987).

Human-dominated landscapes are by no means unique to Australia. On the contrary, this landscape pattern is commonly associated with the settlement of temperate wooded environments by sedentary agriculturists, and the consequent development of urban population centres dependent on extensive areas of rural land for primary production (Taylor 1984). The result of urbanization is a human-dominated landscape where the land cover is largely an artifact of the settlement process (Burgess & Sharpe 1981). In addition to remnant native vegetation and human constructions, this land cover consists of three different kinds of exotic vegetation; adventive, cultivated and relict.

Adventive vegetation, often described as weedy vegetation, is the typical vegetation of severely disturbed and derelict sites in human-dominated landscapes (e.g., excavations, tips, construction sites, industrial lands, vacant lots, fence lines in rural areas, routeway reserves, etc.). This vegetation has a canopy layer dominated by adventive plant species (alien to the locality and spontaneous), and usually lacks an understorey, or has an understorey that is also dominated by adventive plant species. Management of adventive vegetation often includes periodic mowing, burning or spraying with herbicides, and these practices usually prevent the development of tree and tall (greater than 2 m) shrub layers.

Cultivated vegetation is the most common type of vegetation found in human-dominated landscapes. Such vegetation has a canopy layer dominated by plant species that are either indigenous or alien to the locality, but deliberately introduced to the locality (cultivated) for productive or amenity purposes. Where this type of vegetation has been established for productive purposes, it is often a mono-culture (i.e., an even-aged stand of a single plant species), and lacks a distinct understorey, or has a poorly-developed understorey of indigenous or alien "weed" species. Cultivated vegetation established for amenity (and particularly ornamental) purposes, in contrast, often contains cultivated tree, shrub and herb layers. Moreover, cultivated tree and tall shrub vegetation (whether established for productive or amenity purposes) can also have a well-developed understorey dominated by adventive plant species (e.g., fence row shelter belts in rural land areas) or by indigenous plant species that are relicts of the original native vegetation cover (e.g., tree plantations established for timber production in rural areas). It is therefore necessary to distinguish cultivated vegetation composed primarily of cultivated plant species from cultivated vegetation that has a distinct and well-developed understorey dominated by either adventive or relict plant species.

Relict vegetation is a common type of vegetation found in Australia's human-dominated landscapes, particularly those of its rural areas. Such vegetation is often described as "degraded native vegetation" (cf. Laut *et al.* 1977) because it has a canopy layer dominated by native (spontaneous and indigenous) plant species that are relicts of the original native vegetation cover, but an understorey dominated by adventive or cultivated plant species. When canopy-dominant individuals of relict vegetation die as a result of senescence or of disturbance operating at the population level (e.g., disease, insect attack, windthrow, etc.), they are unlikely to be replaced by natural regeneration for several reasons. In some cases, the seedlings of the canopy species are excluded from or suppressed by the exotic (adventive or cultivated) understorey, and in other cases continuing disturbance operating at the ecosystem level (e.g., grazing) or changes in the abiotic environment (e.g., fertilizer effects on soil nutrients) prevent canopy species regeneration. Because the canopy species of relict vegetation normally do not regenerate, the native canopy layer can be regarded as a transient feature of this essentially human-dominated vegetation-type.

Note that Bridgewater (this volume) uses the term "synthetic" to refer to both adventive and relict vegetation. It is suggested that the distinction between these two kinds of vegetation made in this chapter is a useful one, even though it is acknowledged that relict vegetation is only a transitional stage in the transformation of native vegetation to either metastable cultivated or metastable adventive vegetation. Depending on the life span of the canopy-dominant species, relict vegetation may persist as a significant component of human-dominated landscapes for tens to hundreds of years.

The distinction made in this chapter between exotic vegetation and native vegetation that has been modified from its pre-European (as well as its pre-Aboriginal) state by human intrusion or influences is based on the extreme ecological isolation in time of human-dominated ecosystems from the natural ecosystems they have replaced (Peterken 1981; Taylor 1984). In Australia, the ecological isolation of human-dominated ecosystems has always been high because of the endemism of the Australian biota and the northern hemisphere origin of most of the species deliberately or accidentally introduced by the early European settlers (Groves & Burden 1986). Moreover, the ecological isolation of human-dominated ecosystems is increasing due to the intensification of rural/urban land use, and the consequent radical modification of the abiotic environment in rural/urban areas, particularly at the chemical level (Douglas 1983).

Conclusions

The primary aim of this chapter has been to provide a conceptual framework for Australian studies of the modification of natural ecosystems by human intrusion or influences. The meaning of the concept of naturalness in Western technological societies has been examined in an attempt to clarify the uncertainty and ambiguity currently associated with the use of the terms "natural landscape", "natural ecosystem" and "natural vegetation". The revised terminology and definitions proposed by this chapter overcome some, but not all, of this uncertainty and ambiguity. A source of conceptual and semantic confusion that still requires clarification is succession. As Rykiel (1985) notes with respect to autogenic and allogenic succession, "the difficulty [is] that we seek to define in static terms relativistic concepts that change with our frame of reference". Moreover, the terms "autogenic" and "allogenic" are a source of confusion in themselves since they reflect their origin in the outmoded concept of succession to climax by means of the "relay floristics" or "facilitation" process (Shugart 1984).

Because of the complexity of the matter, no attempt has been made here to classify natural ecosystems with respect to the *degree* of their allogenic successional modification by human intrusion or influences. The definitions of terms "natural ecosystem" and "native vegetation" given in this chapter merely provide a starting point for such a classification.

Similarly, the classification of human-dominated ecosystems into cultivated, adventive and relict types requires further refinement. There is no accepted procedure for grouping exotic vegetation into categories equivalent to native vegetation formations and associations. Thus there is a need to develop structural and compositional subcategories of the exotic vegetation-types defined here.

Finally, although all the definitions given in this chapter can be extended to wetland ecosystems, aquatic ecosystems require radically different approaches to naturalness assessment. To date, very little work has been done on this problem in Australia.

Acknowledgements

I thank Dr Rosemary Purdie for comments on the draft of this chapter which improved the published version.

References

Adamson D. A. & Fox M. D. (1982) Changes in Australasian vegetation since European settlement. In: *A History of Australasian Vegetation* (ed. J. M. B. Smith) pp. 109-46. McGraw-Hill, Sydney.

Birch C. & Cobb J. B. Jr. (1981) *The Liberation of Life from the Cell to the Community.* Cambridge University Press, Cambridge.

Brady R. F., Tobias T., Eagles P. F. J., Ohrner R., Micak J., Veale B. & Dorney R. S. (1979) A typology for the urban ecosystem and its relation to larger biogeographical landscape units. *Urban Ecol.* **4,** 11-28.

Burgess R. L. & Sharpe D. M. (eds) (1981) *Forest Island Dynamics in Man-Dominated Landscapes.* Springer-Verlag, New York.

Dansereau P. (1973) *Inscape and Landscape: the Human Perception of Environment.* Columbia University Press, New York.

Dasman R. F. (1972) *Environmental Conservation* (3rd edn.). John Wiley and Sons, New York.

Davis A. M. & Glick T. F. (1978) Urban ecosystems and island biogeography. *Env. Cons.* **5,** 299-304.

Douglas I. (1983) *The Urban Environment.* Edward Arnold, London.

Duvigneaud P. (1974) L'ecosystème 'urbs'. *Mèmoires de la Sociètè de Botanique de Belgique* **4,** 5-35.

Evernden N. (1985) *The Natural Alien: Humankind and Environment.* University of Toronto Press, Toronto.

Groves R. H. (ed.) (1981) *Australian Vegetation.* Cambridge University Press, Cambridge.

Groves R. H. & Burdon J. J. (eds) (1986) *Ecology of Biological Invasions: an Australian Perspective.* Australian Academy of Science, Canberra.

Hartshorne R. (1939) *The Nature of Geography: a Critical Survey of Current Thought in the Light of the Past.* Association of American Geographers, Lancaster, Pennsylvania.

Hendee J. C., Stankey G. H. & Lucas R. C. (eds) (1978) *Wilderness Management*. USDA Forest Service Miscellaneous Publication No. 1365.

Laut P., Heyligers P. C., Keig G., Löffler E., Margules C. & Scott R. M. (1977) *Environments of South Australia Handbook*. CSIRO Division of Land Use Research, Canberra.

Leighly J. (ed.) (1974) *Land and Life: a Selection from the Writings of Carl Ortwin Sauer*. University of California Press, Berkeley, California.

Maarel E. van der (1980) Towards an ecological theory of nature management. *Okologie* **8,** 13-24.

McDonald R. C., Isbell R. F., Speight J. G., Walker J. & Hopkins M. S. (1984) *Australian Soil and Land Survey Field Handbook*. Inkata Press, Melbourne.

Miles J. (1979) *Vegetation Dynamics*. Chapman Hall, London.

Mueller-Dombois D. & Ellenberg H. (1974) *Aims and Methods of Vegetation Ecology*. John Wiley, New York.

O'Riordan T. (1981) *Environmentalism*. (2nd edn). Pion Ltd, London.

Passmore J. (1980) *Man's Responsibility for Nature: Ecological Problems and Western Tradition* (2nd edn). Duckworth, London.

Peterken G. (1981) *Woodland Conservation and Management*. Chapman and Hall, London.

Rykiel E. J. Jr (1985) Towards a definition of ecological disturbance. *Aust. J. Ecol.* **10,** 361-5.

Shugart H. H. (1984) *A Theory of Forest Dynamics: the Ecological Implications of Forest Succession Models*. Springer-Verlag, New York.

Smith J. M. B. (ed.) (1982) *A History of Australasian Vegetation*. McGraw-Hill, Sydney.

Taylor J. A. (1984) The man/land paradox. In: *Themes in Biogeography* (ed. J. A. Taylor) pp. 254-308. Croom Helm, London.

Taylor P. W. (1986) *Respect for Nature: a Theory of Environmental Ethics*. Princeton University Press, Princeton, New Jersey.

Taylor S. G. (1987) Conservation strategies for human-dominated land areas: the south Australian experience. In: *Nature Conservation: the Role of Remnants of Native Vegetation* (eds D. A. Saunders, G. W. Arnold, A. A. Burbidge & A. J. M. Hopkins) pp. 313-22. Surrey Beatty & Sons, Chipping Norton.

Usher M. B. (1986) Wildlife conservation evaluation: attributes, criteria and values. In: *Wildlife Evaluation* (ed. M. B. Usher) pp. 3-44. Chapman and Hall Ltd, London.

Westhoff V. (1983) Man's attitude towards vegetation. In: *Man's Impact on Vegetation* (eds W. Holzner, M. J. A. Werger & I. Ikusima) pp. 7-24. W. Junk, The Hague.

Proc. Ecol. Soc. Aust. 1990 — 16:419-425

Major issues in land management in Australia with particular reference to the role of educators

V. R. SQUIRES

National Key Centre for Teaching and Research in Dryland Agriculture and Land Use Systems, Roseworthy Agricultural College, Roseworthy, South Australia 5371

Abstract

Two major land management issues in Australia have emerged. Land degradation (erosion, dryland salting, build up of residues) and competition for water will dominate the agenda over the next two decades.

The future for agriculture is uncertain because of changing circumstances and societal pressures for the adoption of more sustainable practices. The new generation of land managers and their advisors needs to adopt a new approach and embrace a land ethic.

The resolution of Australia's major land management problems demands a review of the undergraduate programmes offered by our Universities and Colleges. The reasons for this are outlined.

Introduction

In 1988 Australians celebrated the bicentennial of European settlement in Australia. In considerably less than 200 years Australia had gone from a country with no crops and no domestic livestock to one of the world's major producers and exporters of meat and food grains. This is truly a remarkable achievement for a country where rainfall is variable and uncertain; soils generally infertile and often poor in physical structure; costs of transport and materials high; and the often harsh environment is not conducive to comfortable living for humans.

Australian agriculture is based on a low person-to-land ratio and the farming systems which have developed employ land-using and labour-saving technological innovations. As a nation with seemingly bountiful resources, Australia has rarely faced up to the truth about natural resource limitations. In the short history of European occupation there has always been the view that there were more lands and more resources to develop and a belief that technology could supplement natural resources when needed. This view prevails even though the limits of settlement were reached about 1900.

Resource limitations

Increasingly, however, some regions are experiencing acute resource limitations related to water use and distribution and land allocation which challenge the full capacity of existing social and technical institutions (Bennett & Thomas 1982). The land management problems to be faced in Australia into the third century of European settlement are likely to expand and intensify, especially for agriculture. Concerted Federal, State and local action will be needed to help maintain a sustainable agriculture that is profitable for the farmer and grazier and that effectively addresses the complex and interrelated problems surrounding the agricultural use of land and water resources in Australia.

Distribution of water among users responds to two major institutional forces: the legal system and the market system. For example, the use of water for irrigated agriculture is now the subject of particular controversy and concern. Until recently, government sponsorship of many development activities (and absorption of capital costs) had provided water at well below its cost. Competition for limited water resources (both surface and groundwater) is increasing from industries and municipal users who can afford to pay more for their water. Increasing demands are creating conflicts between agricultural, mining, industrial, municipal, environmental and other users and values and suggest that water in Australia will become much more expensive.

Depletion of ground water resources threatens some agricultural producers and rural communities and diminishes the possibility for using this resource in the future. The greatest potential conflict over groundwater is between mining and agriculture. Examples from South Australia illustrate the problem. Dewatering of coal mines in the Kingston lignite deposit posed threats to irrigation farmers in the south-east of the State (Kinhill Stearns 1983). The development of the Roxby Downs gold/copper/uranium mine in the far north could involve drawing down of the Artesian Basin with serious consequences for pastoralists who rely on bores for stock water. Also, as demand for water for most purposes increases and as scarcity of high quality water is recognized, pressures will mount to shift water to new uses and users. Transferring substantial amounts of water to a new use will have a profound effect on the people and the supporting resources that are left behind. Whether the water is used for irrigation, recreation or industrial purposes as the primary use, that water also generates secondary and tertiary incomes to local economies. This is the so-called multiplier effect. Without the maintenance of a strong and committed Federal role it is conceivable that irrigated agriculture in some areas of Australia may go out of production because of water-quality degradation, rather than loss of supplies. Because of Australia's heavy reliance on a few perennial rivers and the poor and unreliable rainfall, it is not possible to absorb the levels of industrial, municipal and agricultural pollution that might be possible in more water-abundant regions.

Lasting and equitable settlement of conflicts over allocation of land and waters must involve current users, those whose rights have yet to be developed, those where communities and lifestyles might be affected by major policy shifts, and potential users with economic power who might seek to buy land and/or water. There is a clear need to involve both on-site and off-site users. It is imperative that those concerned about short-term economic solutions, rather than long-term issues, not hold too much power to the detriment of sustainable agriculture. Increasing water demands from non-agricultural users plus growing problems of groundwater depletion, salt build-up in agricultural soils, and water-quality deterioration in the major rivers is causing heightened concern about the sustainability of some forms of Australian agriculture.

Long term prospects for agriculture

The long term future of Australia's agriculture is uncertain, especially if commodity prices fall and the cost/price squeeze continues. The elements that have shaped its past (natural resources, people, science and technology, economics and government) will invariably affect its future. New considerations, unknown or discounted as unimportant at present, may influence the future direction of agriculture. About the only certainty is that the future conditions will be different from those of today. Change is inevitable and in some areas likely to require severe adjustment if current trends continue. Many farmers and their advisors are poorly prepared for the long term needs brought about by these problems (Roberts 1985). Whether change ultimately produces a sustainable agriculture that strengthens the farmer and grazier, the region, the State and the Nation depends, in part, on the role that educational institutions play in the coming few years. However, it would be foolish to ignore the role of the State and Federal governments in the area of policy. For example, Australia has neither a comprehensive national water (surface and groundwater) policy nor a national land use policy. As limits of production are reached and long-term productivity is threatened, the government must decide whether it will, acting for the national interest, develop an effective national water and land use policy (Cocks 1985). But first, government must be convinced of the problem in order to develop the will to act. The actions chosen will depend on the level of Australia's commitment to protecting the long-term productivity of its renewable resources.

The fundamental cause of this particular land management problem is the divergence between private and public interests. The short-term gains to private land users are frequently in conflict with long-term resource conservation which is essentially a societal goal. It is difficult for the property owner to see the long-term consequences of an action when there is preoccupation with short-term profitability. The private land user often has a shorter time horizon than society at large. A long-term perspective is needed to guide land use, to make the necessary investments, conserve resources and maintain sustainable yields. People may try always to maximize their incomes in the presence of uncertain conditions, succeeding in good seasons but causing irreversible damage to the resource during drought because of overstocking or overcropping. The responsibility for ensuring that land is not overexploited cannot be left entirely to individuals. Government must play a major support role but the Federal government alone cannot act effectively in this complex area. Federal, State and local governments are all involved in the regulation of land and water use, for agricultural and other uses. In recent years, awareness has increased that most of Australia's water-resource problems transcend State boundaries and are extremely difficult to comprehend, involving a complex web of physical, chemical, biological, economic, legal and socio-political issues. Often they go beyond the ability of a single agency to address effectively (Bennett & Thomas 1982).

Policy issues for land management in Australia

Land management comprises many fields and subjects in the social, biological and physical sciences. Clearly it is not a single-subject discipline. Sustaining any natural system requires knowledge of natural sciences and laws governing it; to ignore or to be ignorant of fundamentals of ecosystems, or ecological units in nature is to risk irreversible environmental deterioration. The issue of sustainability has been identified as the most critical to be faced over the next few years (Anon. 1989).

Agriculture is a human activity system with a number of interacting and interdependent physical and biological components (Fig. 1). Farmers operate at the interface between the biophysical and the socio-economic system. We need a way, therefore, of looking at agriculture as a whole system. Agricultural land and the people who farm it should be regarded together as an entity. Agroecology is such an approach. It is study which is concerned with understanding how farms (and farmers) function. Agricultural systems interact with adjoining systems and with the social systems of the people in them (Fig. 2). The way ecosystems function, including their production of goods and services for human welfare, is a consequence of these interactions. Each farm function can be fully understood only in the context of the whole, that is, how that part fits in with the rest of the farm and with the broader system of which it is a part. Such a way of thinking provides the holism needed to comprehend the interaction between agroecosystems and the human social systems (Wilson 1988). This is the same kind of holism that farmers live with in managing their individual farm enterprises, and it has long been a part of farm management literature, as reflected in terms such as "integrated farming" and "farming systems".

The significance of this approach is that it emphasizes the numerous ecological processes within an agroecosystem that determine how the agroecosystem functions as a whole and emphasizes the numerous points of interaction between agroecosystems and human social systems.

Interactions between agroecosystems and human social systems involve exchanges of energy, materials and information between the two systems (Conway 1985). The passage of energy, material and information from a human social system to an agroecosystem takes the form of human activities and inputs to the agroecosystem that shape its structure and ultimately how it functions. The passage of energy, materials and information from the agroecosystem to the social system consists primarily of the products and services produced by the agroecosystem.

Understanding these exchanges of matter, energy and information between the two systems makes it possible to appreciate how the behaviour of each is shaped by the other through repercussion of events or actions in the coupled system. A comprehension of how the two systems interact also makes it possible to appreciate how agroecosystems

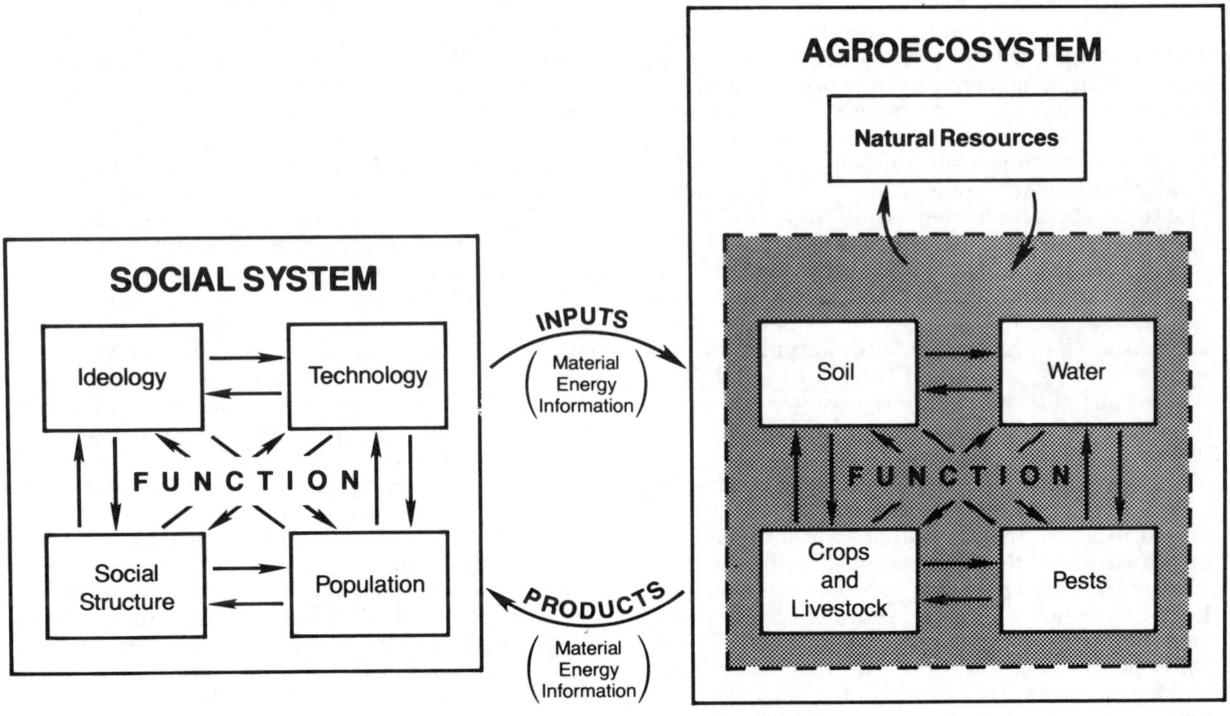

FIG. 1. Interactions between Agroecosystems and the Human Social System. Source: Modified from Rambo 1982. "Natural Resources" in the diagram are soil, water and biological resources from which agroecosystems are constructed.

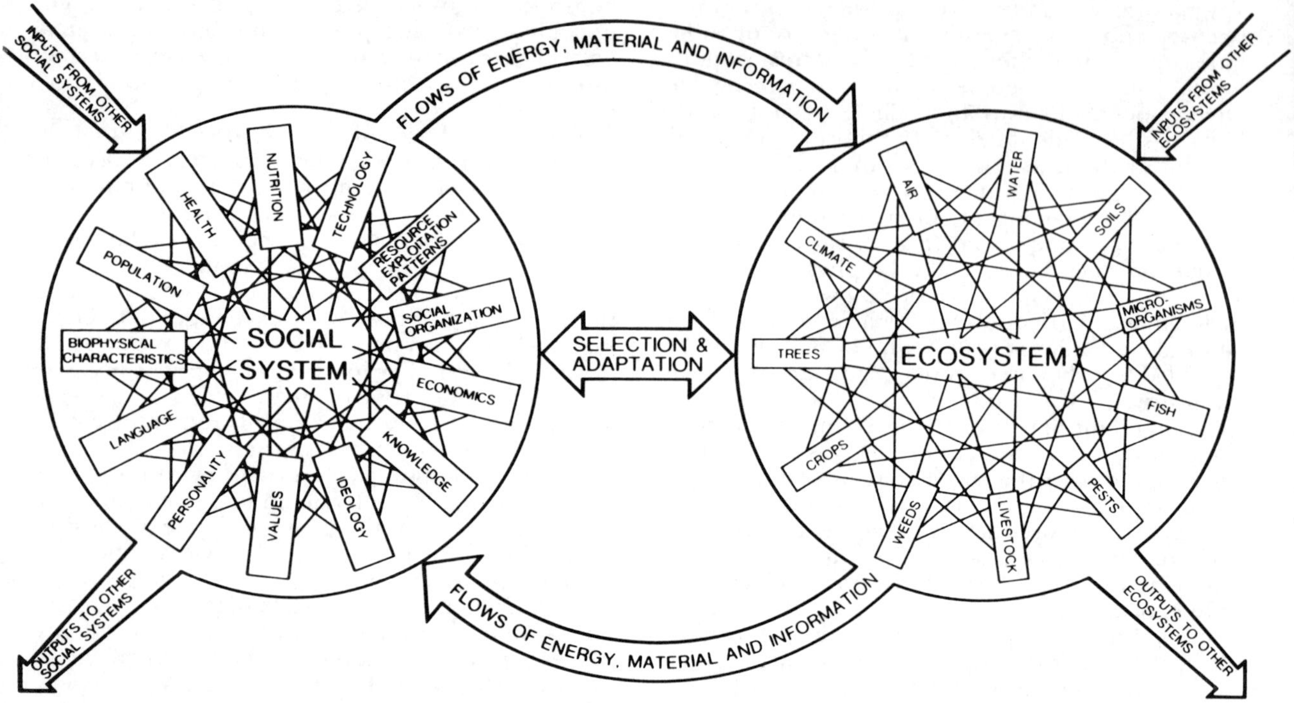

FIG. 2. Social System — ecosystem interactions. Land managers work at the interface of people with their environments.

and human social systems adjust to one another. Studying agroecosystems can be valuable to agriculturalists by helping them design new agricultural systems with *ecological viability* based on an appreciation of agroecosystem processes, and with *social viability* based on an appreciation of the combined agroecosystem-social system. Land managers have to work with people as much as, or even more than, plants, other animals and soils. People use land to meet perceived needs and these needs have to be understood to develop reasonable, realistic management options. Effective land management programmes are often complex and involve land use changes and new resource management practices (Anon. 1989). They also impinge on the income earning capacity of the land user. Failure to recognize this can lead to frustration and disappointment from those who are trying to implement better land management practices.

The principal challenges are to protect and maintain the long-term productivity of dryland agricultural resources, safeguard water quality and ensure equitable distribution for all land users.

Protecting the renewable resource base for productive dryland agricultural resources is a growing concern; especially since irrigated production in some areas is likely to decrease because of water problems. Two areas, in particular, have received inadequate government attention. First, the problem of continued cultivation of marginal or unsuited lands has become critical in the semi-arid lands in most States. Semi-arid land is highly vulnerable to erosion. Second, research and technology into management of marginal cropping areas, and the adjacent semi-arid rangelands, have received scant government support despite the importance of this large region for expanding the range and mix of opportunities for productive agriculture in the drier interior over the long term. Agricultural activity can change the natural resource base in ways which have significant consequences for agricultural sustainability. Perhaps the most important impact is on soil fertility and rates of soil erosion. An agricultural system that progressively reduces its resource base cannot be sustainable. Knowing how agricultural practices affect the land, water and other resources on which crop and livestock production depends is the key to understanding agriculture's ability to function on a sustainable basis.

In the past, exploitation and development (land clearance, tapping underground water, etc.) were the primary objectives. Future productivity was assumed but practical measures to ensure it were not promoted (Spurling 1983). This was partly because of a lack of awareness that productivity was at risk, and partly because technological solutions (such as superphosphate, legume leys, etc.) had masked the productivity decline (Fig. 3).

Technology has enabled Australian farmers to do much with their land and water resources, despite the limitations inherent in both. But

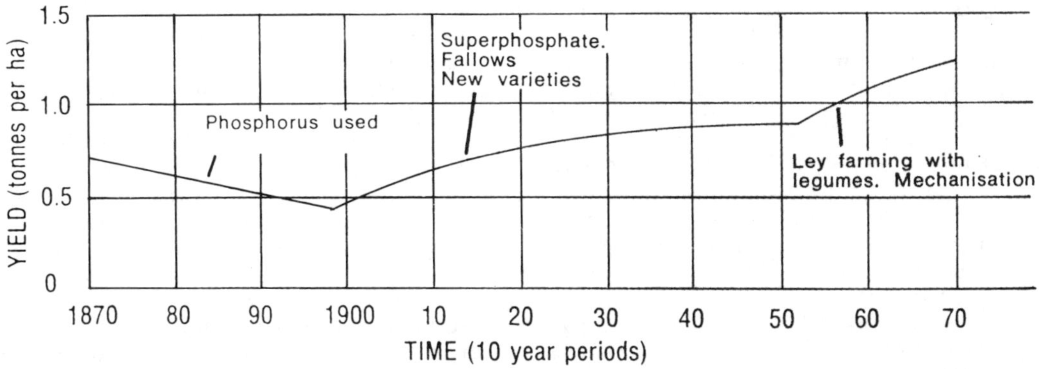

FIG. 3. Resource productivity and technological adjustments in commercial farming: an example from South Australia. Nutrient exhaustion and expansion of cropping into marginal areas led to declines in output. The trend was revised when fallowing, new technologies such as super phosphate and new varieties were released. A further phase of decline in resource productivity followed due to erosion, nutrient depletion, depression and war. More recently, newer technologies such as mechanization, rotations and better legumes have allowed some gains to be made. However, current trends are downward.

technology can be both a blessing and a curse. Clearly, without technology productivity would be low. However, because of it there is a capacity to dramatically modify, even destroy, the natural systems on which agriculture depends.

Technological "solutions" may not be the only appropriate response to ensure an Australian agriculture that is sustainable and profitable over the long term. Also, there was more land "out back" which could be brought into production to compensate. The "pioneer mentality" prevailed. Concern for the land was once considered to be a private responsibility, not a social issue. Today we have entered a new era, and an increasing number of Australians want our land and water resources to be husbanded in a way that encourages most land to be efficiently used on a sustainable basis, retains future land use options, and maintains and enhances their natural environmental and aesthetic qualities.

The sometimes conflicting nature of these aims requires a long-term view of management and the clear definition of managerial and administrative responsibilities. A long-term view will bring maximum benefits both to the community and to industries dependent on land resources. These aims leave no room for exploitative practices that cause lasting environmental damage as the price of short-term profits. They seek to integrate conservation with development; in short, to foster a land ethic (Roberts 1986).

A role for higher education?

There is a clear role for universities and colleges in training people to recognize the interrelatedness within the rural system and to develop a land ethic within them (Reeve *et al.* 1988). The question arises, how can educational institutions be better prepared to meet the changing needs for management of land and water resources in Australia?

It has long been held by most agricultural faculties that, to understand real farming systems, students need to get mud on their boots. I would argue that the farming systems perspective suggests that they also need to get more involved in understanding practical technology and assessing its *impact.* There is a distinction between what Anderson and Hardaker (1979) call *in vitro* and *in vivo* analyses, according to whether the assessment is performed in abstraction or in relation to actual farm circumstances. Most agricultural faculties have been long on the *in vitro* approaches and too little involved in *in vivo.* Students are not merely recipients of data but they must get involved as part of the agroecosystem. Only then can they understand the biological, physical, economic and sociological factors which are the milieu of farmers (Bawden *et al.* 1985).

Educational institutions have to take the lead in education and training in the important field of land management (Reeve *et al.* 1985). They need to recognize the processes involved in moving from a poorly managed (exploitative) system to a well managed (sustainable) system. A poorly managed system is characterised by use in such a way that capital stocks (such as soil or perennial forage plants) are being consumed. The managed system represents a more desirable state and could ensure that the natural resource is being used sustainably.

Education of land managers is the process of communicating skills for changing poorly managed land to well managed land. Education must be concerned not only with the techniques but also with communication systems, processes of information transfer, behaviour modification, institutional structure of government and realities and hierarchies of decision making. The flow diagram (Fig. 4) is a basis for a broad concept of professional education and training for natural systems management. The two sides of the diagram differ in use of knowledge. On the left, the process involves methodologies. It says, in short, that managed ecosystems such as agriculture and forestry are technically complex; but, since not enough is known about them, more information must be collected before advancing. The right-hand route recognizes that only information to answer specific questions is required by those who request it; this information must be in a form meeting the problem-solving needs of the user. The diagram shows that changes in behaviour should result from implementation of strategies and that improved behaviour leads to better managed systems.

We, in the education of land managers, need a dramatic change of tack. We must review our curricula and re-align the allocation of students' time from a didactic to a more experiential approach (Bawden *et al.* 1985).

There are currently some very serious educational problems in land management, especially in agriculture, which need immediate attention. There is evidence that undergraduate agricultural curricula have not changed enough in recent years. We require well-educated and trained people to provide and disseminate knowledge and technology but there is another important role for our graduates. Because our world and our social and ecological systems are changing, so must our educational institutions. If they do not change, their product (the graduates) will be prepared to assume a role which is no longer appropriate or relevant. This is particularly critical because higher education in agriculture is entering a new era for which it is not prepared. This new era will be quite different from anything experienced in the past.

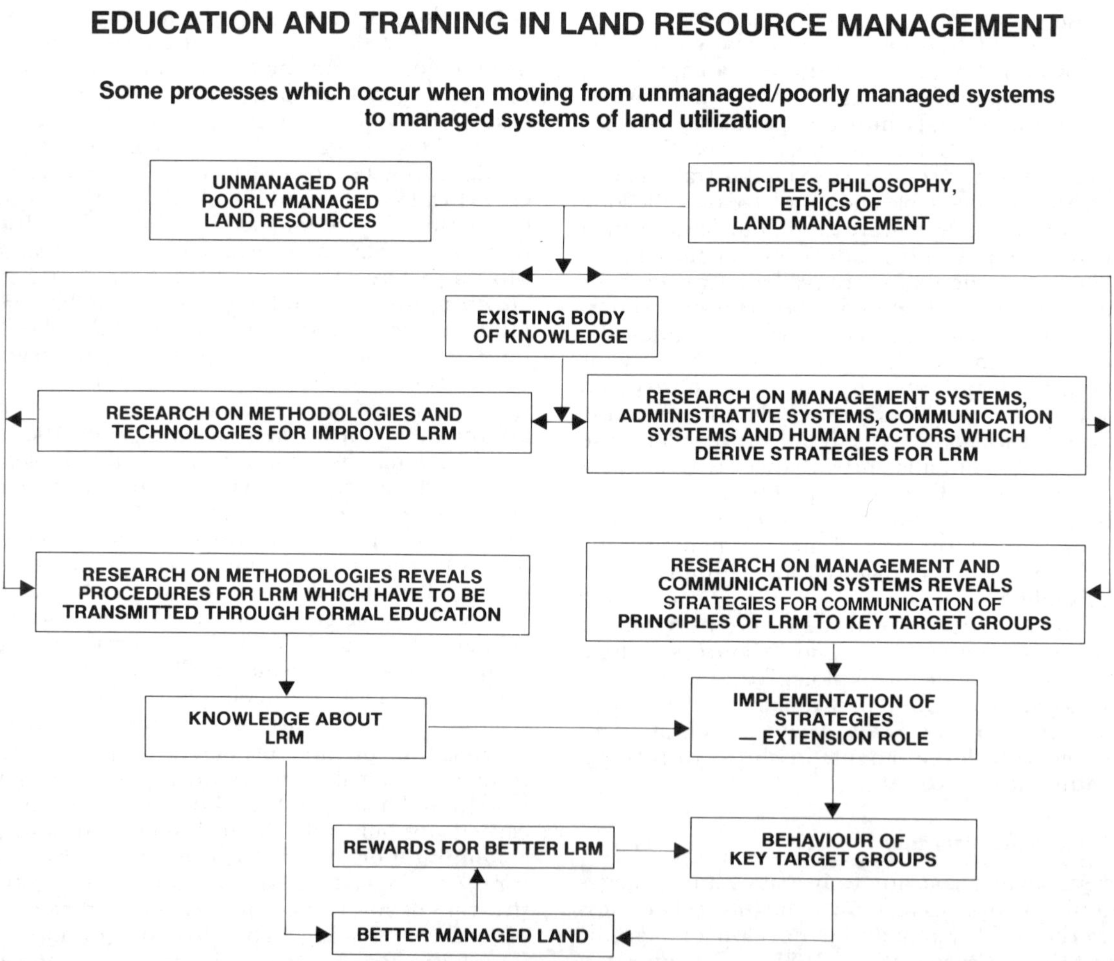

FIG. 4. A badly managed system is one where it is used in such a way that the capital stocks of the resource are being consumed. The managed system represents a more desirable state and should ensure that the natural resources are being used in a sustainable way.

Educational conditions will continue to change rapidly during the new era. If Australia is to have an informed agricultural leadership at the turn of the century at a time in history when it is expected that critical agricultural policy decisions will have to be made, the required educational plans must be drafted and implemented now. Our childrens' food and the nation's agricultural resource base is at stake.

The nature of the problem

Let us examine the nature of the problem for a land management training institution. The mission is clear enough — to secure our future food and agricultural resource base and conserve the environment. To do this we must find some means of continuing to maintain and, hopefully, to improve our agricultural output during a time when the population of the earth will have doubled and both land and water resources will become more scarce. Our land management training institutions must not make the mistake of using out-dated curricula to train tomorrow's leaders. We need to enhance our training programmes in land management with greater attention to the future sociological, ecological, as well as economic implications of the dwindling resource base.

Any proposal to improve the quality and/or effectiveness of education in land management must begin with basic philosophy, mission, objectives and goals. The general mission of land management training institutions could probably be agreed upon but there is little agreement among agricultural educators (in particular) with regards to basic philosophy, objectives and goals.

There seems to be agreement, in U.S.A. at least (Merritt 1984), that graduates should possess six basic competencies: professional and technological; communications; ethics (values) problem solving; inter-personal and leadership.

Educators have been reasonably successful in providing most graduates with beginning-level professional and technical competencies but most graduates are lacking in some of the other listed competencies (Reeve *et al.* 1988). Partly this is because educators have not systematically attempted to assess the degree to which their graduates possessed desired competencies. It is infrequent for teaching staff responsible for a specific area of study to assess carefully the content of the subject/units which formulate the curriculum to ensure that there is adequate opportunity for students to develop the minimum basic competencies. Nor do teaching staff give regular attention to the learning activities included in major subjects/units to ensure that students are engaged in activities which lead to the specified competencies. It is clear to me that the curricula need to be adjusted more systematically and more frequently to meet the *future* needs of the graduates. Most teaching staff perceive their educational task to be one of subject matter building, as opposed to one of planning educational activities to meet human needs in a changing society (see Fig. 2). This condition in particular must be reversed if the quality of education programmes in land management is to improve. Educators must first reach some agreement, in so far as they possibly can, on basic philosophy, mission, objectives, and goals. They must then develop a strategy for reaching their goals.

A better approach to higher education in land management, one which will help graduates learn to live peacefully in a rapidly changing and interdependent global society must be developed. In addition to teaching our young the basic scientific knowledge in their chosen career fields, it is essential that they also be taught basic life-related competencies. Educational programmes in land management should be designed to serve the present and future needs of society and of individuals interested in pursuing careers relevant to food production, natural resource stewardship, and rural living.

References

Anderson J. R. & Hardaker J. B. (1979 Economic analyses in design of new technologies for small farmers. In: *Economics and Design of Small Farmer Technology.* Iowa University Press, Ames.

Anon. (1989) Planning for sustainable agriculture — Why plan? Roseworthy Agricultural College. 127 pp.

Bennett D. & Thomas J. F. (1982) *On rational grounds: systems analysis in catchment land use planning.* Elsevier, Amsterdam.

Bawden R. A., Macadam R. D., Packham R. J. & Valentine I. (1985) Systems thinking and practices in the education of agriculturalists. *Agric. Systems* **13,** 205-25.

Cocks D. (1985) Getting land use policy onto the political agenda. *Search.* **16,** 60-1.

Conway G. (1985) Agroecosystem analysis. *Agric. Admin.* **20,** 31-55.

Kinhill Stearns (1983) Kingston Lignite project: draft E.I.S. 210 pp. Kinhill Stearns, Adelaide.

Merritt R. H. (1984) Challenges for undergraduate education in agricultural sciences. *NACTA Journal,* (Sept), pp. 9-15. National Association of Colleges and Teachers of Agriculture.

Rambo T. (1982) Human ecology research on tropical ecosystems in South-east Asia. *Singapore, J. Trop. Geog.* **3,** 87-99.

Reeve I., Patterson R. & Lees J. (1988) *Land resources: training toward 2000.* Rural Development Centre, University of New England. 386 pp.

Roberts B. R. (1985) Land Stewardship — Our attitudes toward the land. Mulga Lands Symposium Royal Society of Queensland. Brisbane.

Roberts B. R. (1986) Land Care. The politics and theology of Australian land ethics. In: *From Here to Where? Australian Christians owning the past — embracing the future.* pp. 120-34 Uniting Church Press, Melbourne.

Spurling M. B. (ed.) (1983) Agriculture — exploitive or permanent? Roseworthy Agricultural College Centenary Symposium, Roseworthy Agricultural College, South Australia.

Webber G. D., Cocks P. S. & Jefferies B. C. (1976) Farming Systems in South Australia. South Australian Govt. Printer, Adelaide. pp. 104.

Wilson J. (1988) *Changing agriculture: an introduction to systems thinking.* Kangaroo Press.

Proc. Ecol. Soc. Aust. 1990 — 16:427-440

Some management implications for the conservation of vegetation remnants and associated fauna in the central western region of New South Wales

D. C. GOLDNEY and I. J. S. BOWIE

School of Applied Sciences, Mitchell College of Advanced Education, Bathurst 2795

Abstract

The Central Western Region of New South Wales is one of the oldest and most disturbed people-dominated agricultural regions in Australia, European settlement commencing in 1815. The magnitude and intensity of this occupation is described. Particular emphasis is focused on the loss of species and habitat and the measures necessary to conserve and protect the remaining areas of natural vegetation and the associated fauna. Recommendations are made as to how conservation objectives may be achieved.

Introduction

During the 1970s a growing concern was being expressed about the extent of soil and water degradation in the rural areas of Australia and the related losses of species and their habitats.

In the Central Western Region of New South Wales (Fig. 1), this concern was stimulated by population growth in the east, expansion of cereal production in the west and pressures for rural recreation throughout the region from both urban centres within the region and the adjacent metropolises of Sydney and Canberra.

Until the 1970s studies in the Central West of conservation issues in rural areas were few and unrelated. Often such studies were site specific and provided little contribution to understanding the relationships between human activities and the emergence of conservation issues. Nor did they contribute to the development of management policies at a regional level. To rectify this situation a conservation study of the region was prepared and reported to the Australian Heritage Commission (Goldney & Bowie 1987) on behalf of the National Trust (NSW).

That study attempted to consider both the biophysical and socioeconomic dimensions of conservation issues in a regional perspective. As well as providing overviews of both natural and cultural features of the region, it presented assessments in detail of vegetation cover, land tenure, the status of species and habitats and a record of sites of scenic and scientific significance.

The present chapter, while drawing on the data from that study, introduces new information and reflects more comprehensively on the implications for conservation in the region. It attempts to achieve three aims; firstly, documenting the magnitude and intensity of the human occupation of the region; secondly, demonstrating how human impacts have affected species and habitats; and thirdly, spelling out implications of these impacts for environmental management.

The region

The Central Western Region of New South Wales has characteristics which are similar to many areas of Australia inland of the Great Divide. With an area of about 63 000 km^2, extending for some 350 km westwards from the Blue Mountains, the region embraces three of Anderson's (1961) climate-based ecological regions of New South Wales; the Central Tablelands in the east, the Central Slopes and the Central Plains to the west (Fig. 1). There is considerable biotic diversity within the region reflecting variations in climate, geology and soils, especially between the elevated Tablelands and the lower Slopes and Plains. This is reflected very strongly in the geographical patterns of human use and occupation (Fig. 2).

Although the region is not an ecological unit (its boundaries being defined by those of local government areas), there are important unifying features which justify it being treated as a region: the area includes the upper catchments of four New South

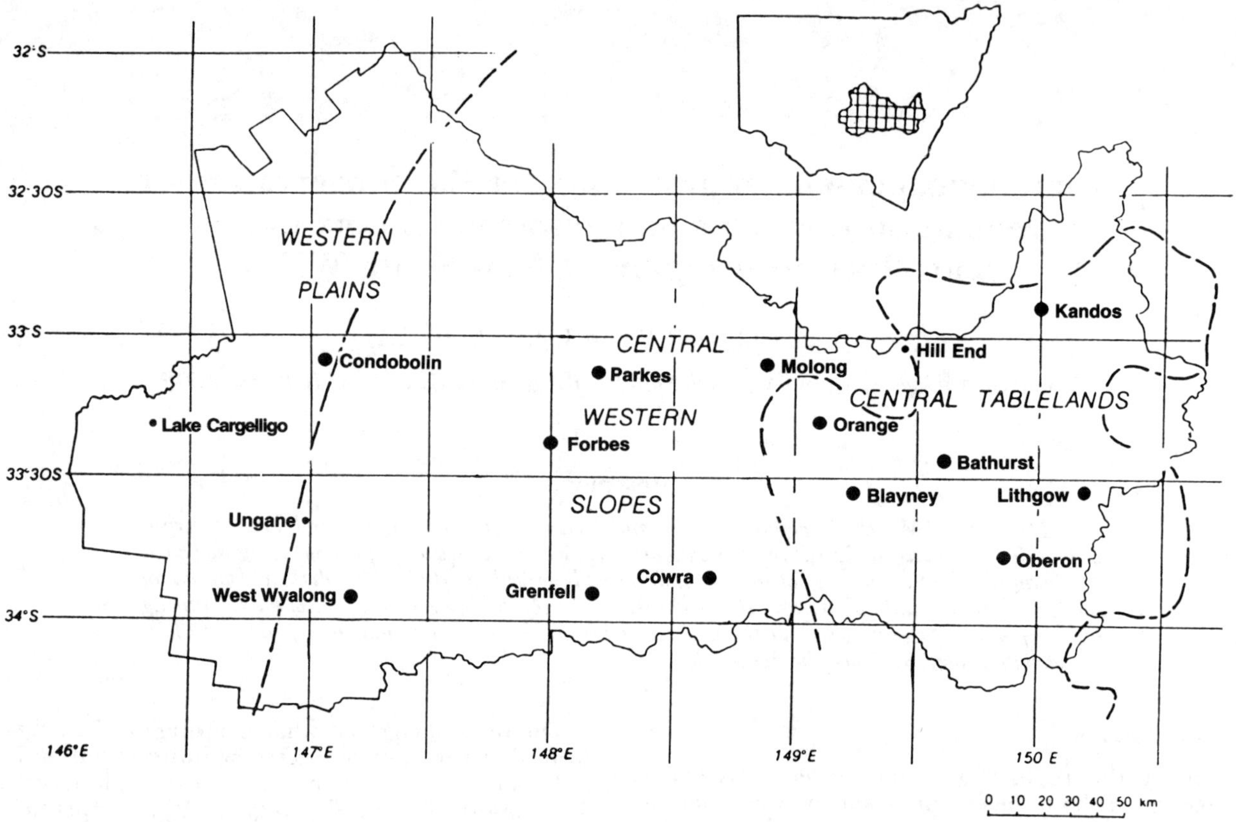

FIG. 1. Central Western Region of New South Wales. Botanical Divisions and Subdivisions after Anderson (1961).

Wales river systems; woodland vegetation forms are predominant and the area has spring and autumn flushes in growth reflecting limitations on plant growth of winter cold, especially in the Tablelands, and the area has summer drought. The region is sparsely settled with an average of 0.8 rural dwellers per km^2. Some 70% of the region's 1986 census population of 162 000 lives in urban centres but significant growth in the rural population is taking place in the tablelands.

Human occupation

The region has been occupied by humans for some 40 000 years. The Aboriginal population was probably no greater than 12 000 when the first Europeans appeared in 1813 (Conyers 1987). Aboriginals used fire extensively for food production (Conyers 1987), thereby creating seral successions which were possibly favourable to selected small mammals (Allan 1983). European settlement from 1815 brought dramatic alterations to regional environments. Since population and agricultural data have been enumerated in censuses for most of the period since European settlement, it is possible to quantify many of the influences of Europeans on these environments. From the 1890s the census data can be adjusted to fit aggregates of today's local government areas which correspond broadly to the Tablelands, Slopes and Plains (Figs 3 to 6).

The first decades of European occupation involved only modest disturbance of regional environments. By the census of 1851, the European population of rural areas in the region amounted to less than 12 000 with another 2 200 in Bathurst, the only urban centre (Table 1). After relaxation of the "limits of settlement" in 1839 grazing activities spread to affect most of the region but the focuses of these activities remained in the small areas of natural grasslands around Bathurst, and in areas of more open woodland and gentle relief. Sheep grazing in particular reduced populations of more edible grasses and succulents but livestock populations would not have reached a million sheep equivalents in this period.

TABLE 1. Central Western Region, New South Wales. Urban Non-Urban Populations 1851-1986.

	1851	1861	1901	1933	1947	1971	1986
Urban Population	2 252	5 688	39 364	72 074	79 151	102 380	112 748
Non-urban Population	11 848[a]	26 312[a]	67 236[a]	78 407	64 481	49 063	48 849
Total	14 100[a]	32 000	106 600	150 481	143 632	151 443	161 597

Note[a]: Approximated from prorated county data.
Source: Population censuses 1851-1986, New South Wales Statistical Registers, Australian Bureau of Statistics, Goldney and Bowie (1987).

The discovery of gold in 1851 provided a catalyst for a second wave of European immigration which was one of dramatic spread and intensification of impacts upon the region. The impacts of mining and processing of gold itself, and later of other minerals as well, had only localized effects, except around the major gold-mining centres of Sofala and Hill End, through clearing of woodland for domestic and industrial fuel. But gold mining attracted to the region large numbers of people, many of whom settled on the land, as well as others who established the foundations of most of the urban settlements of today. Between the censuses of 1851 and 1901, the rural population of the region expanded sixfold.

The dramatic growth in European population stimulated a widespread occupation of rural land (both freehold and Crown leases), after the 1861 Land Act. Between 1861 and 1901 the area of land in rural holdings expanded fourfold to reach over 80% of the area of the region, a percentage which has not changed greatly since. The improved security of tenure which freeholds and leaseholds implied, along with the general

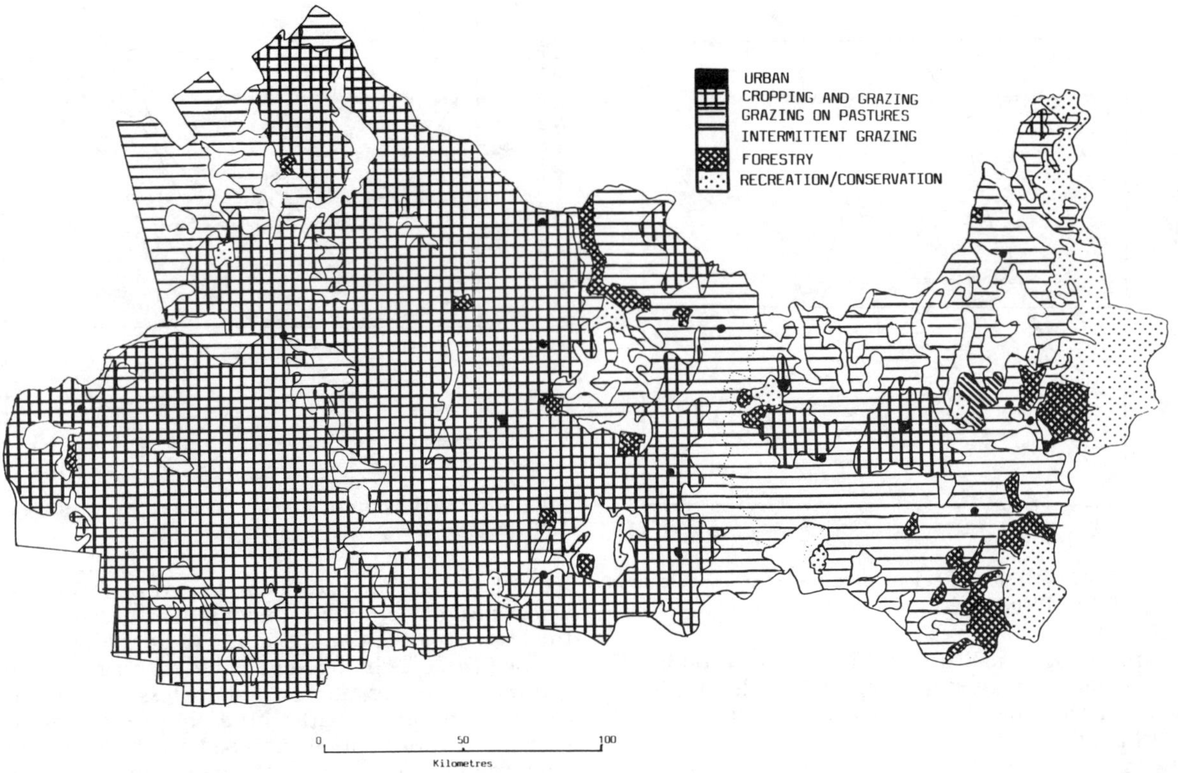

FIG. 2. Adapted from National Mapping 1:1 million land use series.

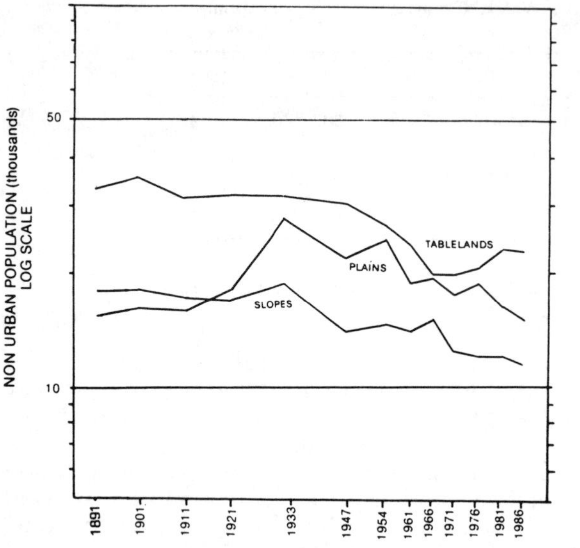

FIG. 3. Central Western Region: Non urban population. Selected years 1891 to 1986.

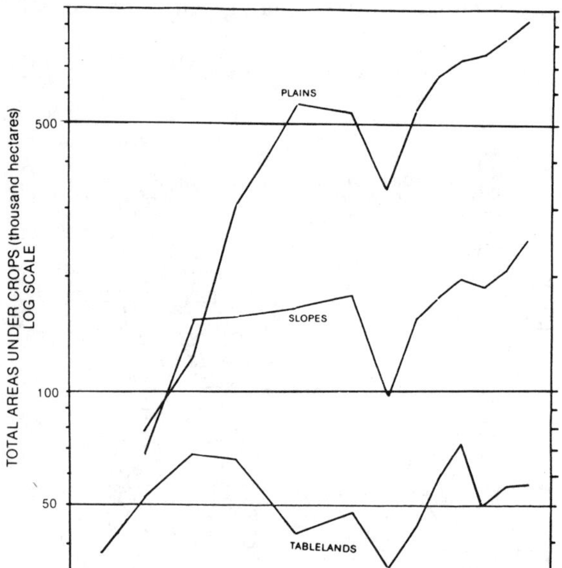

FIG. 4. Central Western Region: areas under crops. Selected years 1891 to 1986.

prosperity of agriculture in this period, encouraged spectacular increases in sheep and cattle numbers between 1861 and 1891, particularly in the Slopes and Plains (Fig. 6).

The expansion of agriculture was at a considerable cost to natural environments. By 1901, 86 years after the first European arrived in Bathurst, nearly

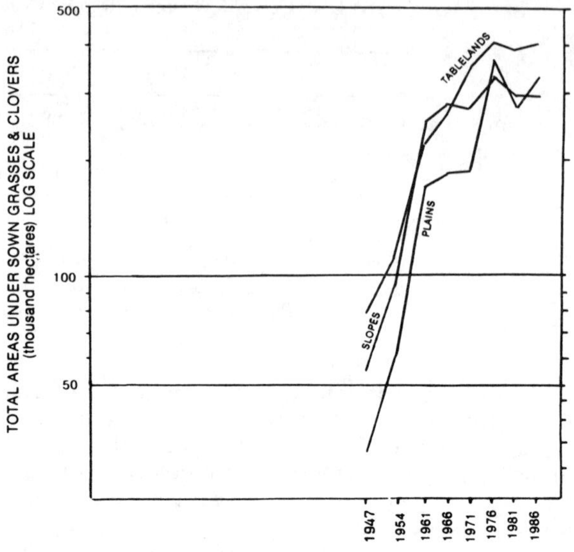

FIG. 5. Central Western Region: areas under sown grasses and covers. Selected years 1947 to 1986.

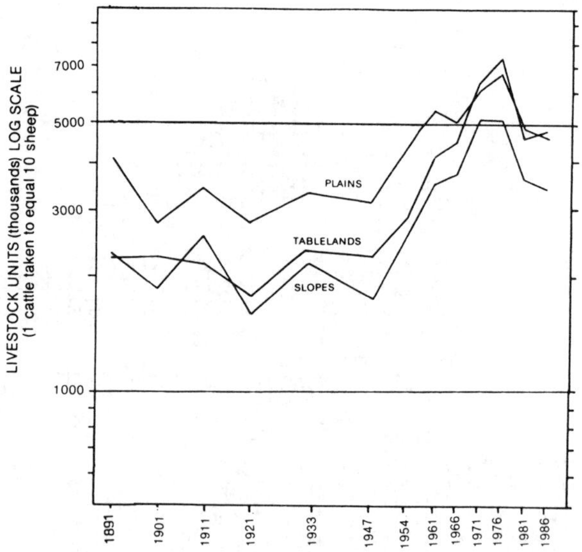

FIG. 6. Central Western Region: livestock. Selected years 1891 to 1986.

half of the region had been cleared of forest and woodland, although a rather higher percentage of the less rugged and more lightly wooded slopes and plains (Table 2) had been cleared. Overgrazing of ground storey vegetation led to loss of habitats, to erosion and, with little improvement of pasture, a rapid spread of weed and pest species. The fragile ecosystems of a generally subhumid environment began to break down. In the 1880s a series of summer droughts led to a collapse of

TABLE 2. Central Western Region. Percentages of total area modified for farming in various years.

	Central Tablelands	Central W Slopes/Plains	Total Region	Comment
1901/2	25.3	56.6	48.4	Areas sown, cropped, ready for cultivation, partly cleared, ringbarked
1919/20	56.7	80.2	73.4	As above
1952/53	59.4	84.4	76.9	Areas of rural holdings less area unused for agricultural purposes
1980/81	49.9	83.2	73.5	Areas in rural establishments other than natural forest/woodland
1981	50.4	84.8	75.7	Areas other than forest/woodland

Source: New South Wales Statistical Registers, Australian Bureau of Statistics, Goldney and Bowie (1987).

the pastoral economy and dramatic reductions in stock numbers, at least in the Slopes and Plains.

The period between Federation and the Second World War was one of consolidation and adjustment in rural occupation. Rural population in the Tablelands and Slopes began to fall as holdings were consolidated and some marginal grazing was withdrawn from agriculture; in the Plains it continued to grow strongly with the employment opportunities opened up by cereal production. Urban populations continued to grow throughout the region to the point where they equalled rural populations.

A particularly important trend to be noted is in the doubling of the area held under freehold and near-freehold tenures (Table 1). By 1940 more than 70% of the region had been alienated. Improved security of tenure for many landholders encouraged considerable investment into the clearing of woodlands and into ploughing of arable land. With these, sheep and cattle numbers grew and stabilized, and the region saw a spectacular expansion of cropping, especially in the Plains and, to a lesser extent, the Slopes. By the late 1930s more than 12% of the region was under crops (Fig. 4).

Thus consolidation and adjustment brought extensive permanent changes to the natural environment. Feral species, along with cultivation, fertilizers and extensive use of fire for pasture improvement led to reductions of habitats and the reduction or extermination of indigenous species (Goldney 1987a).

The war years of the 1940s were years of difficult economic conditions and severe summer drought but they ushered in a period of renewed expansion and a general intensification of rural occupation. Rural population, falling generally from the 1940s as a result of farm consolidation and mechanization had begun to show the effects of a new migration onto the land in the east in and around all larger urban centres which were now growing strongly. Populations more than doubled in the neighbouring metropolitan centres of Sydney and Canberra. Hence there were unprecedented impacts on rural environments through demands for recreation, investment and natural resources for industries; hallmarks of a new affluence which has been largely confined to the cities.

Agriculture also changed in this period. Some increases in areas in rural holdings occurred in the 1950s and 1960s (and some more recent declines, which are largely the result of a changed basis for enumeration in agricultural censuses). These, plus a renewed growth in crop areas to over 20% of the region's area, and the establishment of significant areas of *Pinus radiata* plantations, both made possible by mechanization, led to further clearing of woodlands and forest. This clearing has been largely compensated for by the withdrawal from agriculture of much marginal land (often for reserve or forestry purposes) and extensive reversion to secondary growth in rugged terrain.

Associated with these has been a one hundred fold increase in the area of sown (and essentially "improved") pasture (about 16% of the region's area). This has encouraged a very considerable increase in sheep and cattle numbers (especially in the tablelands), although numbers in the 1980s have fallen as a result of drought and unfavourable economic circumstances. The concomitants of modern land use particularly mechanization and the use of weedicides and pesticides has probably had some serious effects on the remaining vestiges of natural environments. Loss of native species, almost complete in many cleared areas, has extended into native forests and woodlands. There have been extensive invasions of weeds and pests, soils have suffered increasing erosion (Stewart 1968) and water regimes have been greatly altered by pollution and eutrophication. A greater part of the regional environment has been irreversibly changed from its natural condition.

The effects on species and habitats

The effects of human occupation on the natural environments of the region can be seen most directly in the alteration of habitats of indigenous species and the consequent changes to species represented in both the flora and fauna of the region. They have been less directly expressed in changes to soil and water regimes.

The effect of Europeans on flora

Essentially, the vegetation of the region falls into three main groupings: the sandstone complex in the region of Lithow and the Bathurst-Orange-Carcoar-Portland area together comprise the Central Tablelands; further west lie the Central Western Slopes; and a small area forms part of the Western Plains division of the state (Cardale 1987).

The sandstone vegetation ranges in form from open forest to open shrub communities dominated by *Angophora* spp., *Eucalyptus* spp., *Allocasuarina* spp., pockets of rainforest and various heath species. About 12 major alliances, ranging from open forest to mallee woodland, dominate the Tablelands, Slopes and Plains (Table 3), in addition to the mixed Eucalyptus communities of the sandstone complex.

The major changes to the ecology of the sandstone complex which have occurred in recent times probably relate to the nature and frequency of fires. Though most species of the sandstone complex show a remarkable ability to regenerate from seeds, underground rootstocks or by coppicing, the wet sclerophyll communities show a relatively slow recovery after fire, generally passing through a sequence of successional stages. More frequent fires favour the fire tolerant portions of the mosaic thus causing shrinkage of already restricted wetter communities.

Restricted community types of the sandstone complex occupy swamps and cliff faces, which as rare habitat types are of particular significance for occurrence of rare species. Benson (1984) has recently highlighted the importance of such communities in the Newnes Plateau area.

In the Tablelands *Casuarina cunninghamiana* alliances have undergone high degrees of disturbance on many sites due to episodes of stream clearance, invasion of willow species, grazing and extensive weed invasion. The *Eucalyptus fastiata-E. viminalis* alliance, the tallest and densest community type of the region, occupies the most favourable sites and has been considerably depleted to make way for pine plantations and grazing. In many areas it occurs as remnants, or with an understorey extensively modified by grazing. In contrast, the *E. macrorhyncha-E. rossii* alliance is characteristic of less favourable sites and appears to have survived more frequently in a less modified condition. Extensive areas show subtle differences in both combinations of dominant species and in understorey composition, which range from grass tussock dominance to low shrub dominance. Two common Tablelands alliances, *E. melliodora-E. blakelyi* and *E. albens* have a wider range than those preceding, as both extend to the Slopes sub-region. Both types have been extensively modified in structure, due to extensive land clearing, and in understorey composition due to grazing and associated practices, such as pasture improvement and superphosphate application. Such changes have been well documented by Moore (1966).

Extensive clearing in the Slopes and Plains have occurred particularly associated with the wheatbelt. The *Eucalyptus melliodora-E. blakelyi* and *E. albens* alliances represent a westward extension of Tablelands alliances and together with *E. microcarpa* communities have also been considerably modified by grazing. The effects of various grazing regimes on these understories have also been described by Moore (1959).

The remaining alliances extend beyond the Slopes to the Plains with significant changes in species composition. The associations in which cypress pine, *Callitris glaucophylla,* is dominant (Lindsay 1967) have survived in a somewhat less disturbed state due to their importance to the forestry industry. The effects of people on many of the Eucalyptus associations in the slopes and plains are undescribed.

A number of plant species in the region are either extinct, rare or endangered. The greatest concentration of these species is in the north-east corner of the region, with the Newnes Plateau being of particular significance (Benson 1984). Fewer species are listed for the more westerly portions of the region. However, the patterns of vegetation disturbance, particularly related to agriculture and grazing which have been identified by Leigh *et al.* (1984) as the largest single cause of plant species extinctions in Australia, suggest that this may change. Known extinctions within the region include *Grevillea divaricata* collected by Alan Cunningham in 1823, and *Euphrasia argua* collected from plains near Bathurst in 1869. A further 95 plant species are regarded as rare/threatened; 56% of these are conserved in the large National Parks on the eastern border of the region and 44% of these are found elsewhere, but they may not be adequately conserved (Bower *et al.* 1987). Since the vegetation of the region is inadequately mapped (Hayden 1971) then additions may be added to the categories of species and plant associations at risk. The status of the major alliances is indicated in Table 3.

The extent of the remnant forest/woodland cover

Table 4 illustrates the distribution of forest/woodland cover in various size classes in the region. Some 3 484±10% remnants ranging in class size

TABLE 3. Major vegetation alliances* in the Central Western Region.

Alliance	Structure	Location	Status
Mixed communities of sandstone	Rainforest, forest, woodland, heaths	Sandstone only in tablelands	3
E. fastigata *E. viminalis*	Tall open forest/ open forest	Tablelands	2
E. macrorhyncha *E. rossii*	Open forest/ woodlands	Tablelands	3
Casuarina cunninghamiana	Woodland forest/ fringing forest	Tablelands/ watercourses	3
E. pauciflora ssp. *niphophila*	Low woodland/ tall shrubland	Tablelands	3
E. pauciflora *E. stellulata*	Woodland/ low woodland	Tablelands	3
E. melliodora *E. blakelyi*	Woodland/ low woodland	Tablelands, slopes, plains	1
E. albens	Woodland/ low woodland	Tablelands, slopes, plains	1
E. microcarpa	Woodland/ low woodland	Slopes, plains	3
E. dealbata *E. sideroxylon*	Forest/ woodland	Slopes, plains	3
E. populnea *Callitris* sp.	Low woodland/ tall shrubland	Plains	2
E. socialis *E. dumosa*	Low woodland/ tall shrubland	Plains	2
E. camaldulensis	Woodland/forest, fringing forest	Slopes and plains along rivers	2

* Associations vary considerably across slopes and plains as do other prominent species.
1 Poorly conserved in region and elsewhere
2 Poorly conserved in region but conserved elsewhere
3 Conserved in the region and elsewhere

Sources: Hayden (1971); Jacobs and Pickard (1981); Cardale (1987); Bell (1978) and Specht *et al.* (1974).

from 0.04 km^2 to 5 000 km^2 are to be found in the region (Goldney 1987a). Ninety-three percent of these islands are less than 5 km^2 in area. Seventy-six percent of the total land has been substantially cleared since European settlement (Goldney 1987b). The greatest deforestation has occurred in the slopes (86% cleared) followed by plains (83%) and then tablelands (50%). The remnant forest/woodland is confined mainly to land unsuited to agricultural use and is often primarily secondary regrowth. Twenty-nine percent of the forest/woodland is on freehold or near freehold land, with significant differences occurring between slopes, plains and tablelands (Goldney 1987b).

The impacts on vertebrate fauna

The extant vertebrate fauna of the region is surprisingly diverse. Six hundred and nineteen species inhabit part or all of the region, representing about 36% of the terrestial and freshwater vertebrates of Australia (Goldney 1987a).

It is not known how vertebrate populations have changed quantitatively in the region since 1813; little base-line data exists. It appears that for some species there have been significant increases in both numbers and range with the creation of favourable habitat in the form of edges, dams and irrigated areas and the extensive increase in grassland-type habitat.

TABLE 4. Distribution of forest/woodland cover in size classes in Central Western Region (CWR).

Size Classes km²	CWR Region Number	Area km²	Percent of region
0.4– 0.5	2 652	918	1.4
0.5– 5	590	1 518	2.3
5 – 25	164	1 845	2.8
25 – 100	66	2 868	4.4
100 – 500	7	1 217	1.9
500 –1 000	2	1 393	2.1
1 000 –2 000	1	>1 597*	2.4
2 000 –5 000	1	>4 518*	6.9
Totals	3 483	15 874	24.2

* Continuous with woodland/forest cover outside of region.
Source: Modified from Goldney (1987a).

A number of species have become extinct, with the ground dwelling mammals most affected. Whilst the process of extinction can be viewed as a natural process, it is clear that in this region, as elsewhere, the major cause is the destruction of habitat. The pressures on faunal populations continue to be clearing of habitat for agricultural use, drainage of wetlands and the damming of rivers, competition from feral introductions, past rabbit plagues, the creation of remnant vegetation, the widespread road system, land use management practices (including the use of toxic chemicals, control burning, mining, forestry practices, etc.), illegal hunting, trapping and fishing, changing fire patterns, severe droughts, fluctuating stock rates, diseases and feral animal predation. The species which are believed to have succumbed to these pressures and have become extinct in the region since the European invasion are detailed in Table 5.

The extinction of significant numbers of smaller mammals from the region is of considerable interest since they are predominantly medium-sized ground dwelling mammals of grasslands and open forest. In a working paper on the possible factors inducing faunal changes in western NSW and north-western Victoria in the 19th century, Allan (1983) notes that the extinction of medium sized ground dwelling mammals predated the major land clearance programmes of the Europeans, the periods of gross overstocking of sheep and cattle, the arrival of feral animals, particularly rabbits and foxes and the crippling droughts of the late 19th century. Allan goes on to argue that the extinction of this group of mammals in north-western Victoria and western New South Wales is rather related to the rapid demise of the Aboriginal population particularly along waterways. Allan argues that it was the rapid elimination of Aboriginal land management by fire which initiated the sudden and dramatic extinction of the medium sized mammals. Conversely the increase in a small number of larger mammals may also be explained by these changed conditions.

Many species have become or are in danger of becoming locally extinct but may be relatively secure in a regional sense. A number of species are endangered at regional level, and will probably

TABLE 5. Vertebrate species which have become extinct in the region since 1815.

Scientific Name	Common Name	Category	Comment
Macullochella macquariensis	Trout Cod	ER	
Prototroctes maraena	Australian grayling	ER	
Pedionomus torquatus	Plains wanderer	ER	Some uncertainty as to status
Petrophassa scripta	Squatter pigeon	ER	Not seen for more than a decade in the region but highly nomadic in western part of state.
Neophema bourkii	Bourkes parrot	ER	
N. chrysostoma	Blue-winged parrot	ER	
Dasyurus viverrinus	Eastern quoll	ERM	
D. geoffroii	Western quoll	ER	+Some uncertainty if ever in region
Antechinomys laniger	Kultarr	ER	+Some uncertainty if ever in region
Perameles bougainville	Western barred bandicoot	ERM	+Some uncertainty if ever in region
Macrotis lagotis	Bilby	ER	
Bettongia penicillata	Brush-tailed bettong	ER	+Some uncertainty if ever in region
B. gaimardi	Tasmanian bettong	ERM	
B. lesueur	Burrowing bettong	ERM	+Some uncertainty if ever in region
Aepyprymnus rufescens	Rufous bettong	ER	+Some uncertainty if ever in region
Lagorchestes leporides	Eastern hare wallaby	Ex	+Some uncertainty if ever in region
Onychogalea fraenata	Bridled nail-tail wallaby	ER	+Some uncertainty if ever in region
Conilurus albipes	White-footed rabbit-rat	Ex	+Some uncertainty if ever in region
Pseudomys gouldii	Gould's mouse	Ex	+Some uncertainty if ever in region
P. oralis	Hastings River mouse	ER	+Some uncertainty if ever in region
Phascogale tapoatafa	Brush-tailed phascogale	ER	

ER=Extinct in the region; ERM=Extinct in the region and on the mainland; Ex=Extinct; +Based on records north and south of the region.
Source: Modified from Goldney (1987a).

become extinct in the region if their habitat is not adequately conserved. It is difficult to precisely identify these species since, for example, a once common species now observed as uncommon or rare, may have reached an equilibrium with its altered environment, and not be in any real danger of extinction. It is also difficult to make regional pronouncements when some species are widely distributed, others are western or eastern species with some at the extremity of their range distribution. In spite of these difficulties, 38 species appear on balance, to be endangered at the regional level and another 52 species are regarded as rare in the region but probably not endangered (Goldney 1987a).

Implications for environmental management

These effects and their implications for environmental management are of course not unique to this region. The fact that a situation has been documented in some detail does make it possible to suggest implications for environmental management which may have a more general relevance.

From the conservation study of the Central West (Goldney & Bowie 1987) it has emerged that just over 70% of the remaining forested or wooded land in the region is held by the Crown under various tenures. The bulk of this land is held, in roughly equal areas, by the National Parks and Wildlife Service, the Crown Lands Office of the Department of Lands and the Forestry Commission. The other 30% of woodland forest is situated on freehold land. Whilst there are large parcels of Crown lands in the eight national parks which fall within or across the boundaries of the region, the greater part of the forested/wooded lands in these Crown tenures is in smaller parcels, physically separated one from one another. This is the case particularly in the slopes and plains where the percentage of total area in forested/wooded unoccupied Crown lands is lower than in the tablelands.

The implications of this are important. Firstly, particularly in view of the sometimes conflicting objectives of these three and other Crown agencies, there is no co-ordination of the agencies' management activities. Secondly, many of the parcels are probably too small to provide effective and protected habitats for many species, particularly the larger species. Thirdly, there are few opportunities for integrating the management of many of the parcels because the remaining links between them are generally in private hands under freehold or leasehold tenures.

This raises the question of how important remnants and corridors of relatively natural areas on privately occupied lands may be protected. There are no easy answers to this question. In New South Wales the agencies responsible for land use planning of private land holdings are the Department of Planning (with responsibilities for matters of State and regional significance) and local government councils (for matters of local significance).

In the decade since the Environmental Planning and Assessment Act 1979, there has been an impressive improvement of land use planning of rural lands in the State with the employment of qualified planning staff, and the gazetting of local environmental plans to control rural land use.

The scope for local environmental plans and related instruments (such as Wildlife Refuges and Tree Preservation Orders to control natural vegetation) is limited even with the support of the powers to control clearing of other agencies such as the Soil Conservation Service. Many council planners are facing intense opposition from rural landholders towards any restriction of private land use activities and, as a consequence, there is a general reluctance on the part of councils to bring agriculture and forestry within development control, and to use "environmental protection" zoning over private lands.

This reluctance extends to the Department of Planning which, although it has in the past shown some concern to protect the amenity of rural lands, appears to be reluctant now to continue with a proposed State policy to cover development in rural areas.

In this situation there are few instruments by which remaining islands and corridors of natural vegetation on private land in particular can be managed at local or regional level.

Of the many environmental issues in need of urgent attention in the Central Western Region we draw particular attention to three only. We suggest possible approaches to their solution. These are flora and fauna at risk, the remaining uncommitted public land and the problem of remnant forest woodland cover. To our knowledge nearly all rare and endangered species in the region and (self-evidently) the plant alliances and associations that are inadequately conserved are associated with the remnant woodland/forest. There may be rare and endangered plant species associated with cultural features in the region such as cemeteries, agricultural grasslands, old mine sites etc., but this possibility has not been systematically investigated. The uncommitted public land is a subset of the remnant woodland/forest. Hence all three concerns are intimately linked.

Approaches to conservation in the region

The traditional approach to conservation of "buying up" and "locking up" appears an inadequate philosophy when faced with the realities outlined in this chapter. Obviously there are economic limitations to the ability of the Public Sector to both purchase and effectively administer remnant patches of such diverse tenures. Furthermore, the few large areas left do not effectively sample the areas regarded as needing conservation reserves.

The possible approaches to developing effective conservation policies for residual lands in a agricultural region are limited but not without creative options. These options are listed below.

(a) Education programmes are needed to inform primary producers, the public, planners and bureaucrats of the importance of native vegetation remnants to conservation of flora and fauna in the region. Such programmes are based on the premise that the attitudes of individual landowners and the community as a whole will play a decisive role in determining the eventual fate of remnants. Key groups to target in such a programme would be primary producers particularly those with remnant forest/woodland, major farmer organizations, shire and city councils, the Regional Education network including the one Field Study Centre, local media, government land managing agencies, statutory authorities such as electricity distributing councils and political parties and political representatives.

The senior author has been funded by the World Wildlife Foundation (Australia) to lead a group of regional social and environmental scientists in a three-pronged investigation into the region's remnants. Part of the programme will be directed towards education and extension concurrent with an investigation into flora and fauna in remnants (see "j" below). Over a period of 2-3 years, the aims relevant to this section follow.

(1) Survey the attitudes of primary producers about the value of remnant forests on their lands.
(2) Develop media-advertising programmes to help create a climate favourable to remnant conservation as an integrated component of farm management.
(3) Conduct extensive community education programmes to foster awareness about the state and role of remnants in the overall environmental equation through media reports, making audio-visuals and/or speakers available to service clubs, community groups, etc.
(4) Develop a range of resource packs suitable for infant, primary, secondary and tertiary students which would investigate:
   i) the conservation value of the region,
   ii) the remnant woodland/forest problem,
   iii) basic ideas in island biogeographical theory and its application to remnant conservation,
   iv) the fate of species in small remnants,
   v) rare and endangered species in the region,
   vi) value clarification exercises, and
   vii) managing remnants

Such programmes assume that in the long term, fundamental changes in attitudes about and behaviour towards remnants must occur in the society as a whole if remnants are to be conserved. A secondary belief is that students may help to influence the attitudes of the community in which they live.

(5) Develop resource packs and professional displays for regional shows, agricultural field days, etc.
(6) Host a series of regional seminars for primary producers and other land managers with particular emphasis on managing remnants.
(7) Develop strategies for the management of remnant woodland/forest for individual farm units, farm groups, shires and the region as a whole.
(8) Reassess the attitudes of landowners at the close of the programme to determine how successful it may have been.

Whilst we are still determining criteria which could be used to judge the success or otherwise of such an education-extension programme it would need to achieve at least some of the outcomes listed below.

i) Influence State politicians and bureaucrats as to the importance of remnant conservation and management.
ii) Change rural landowners' attitudes which lead to specific behavioural responses which can be objectively measured: e.g., fencing off remnants for reservation or protection, development of farm conservation strategies, tree planting programmes to provide corridors between existing remnants; co-operative conservation programmes between neighbours, the development of covenant agreements to protect remnants, cooperative conservation programmes between private landowners and neighbouring government managed lands.
iii) Influence shire and city councils to address the issues of remnant vegetation, roadside verges, corridors, tree preservation orders, etc., in updated Local Environmental Plans.
iv) A raised community awareness of the importance of remnants in regional conservation.
v) The development of a suitable checklist to aid landowners and planners to assess the conservation values of remnants and to suggest future management strategies which could be appropriate.
vi) The purchase of resource materials by landowners, particularly information on formulating conservation management strategies.
vii) Attendance of farmers at regional seminars on "Disappearing Bushland".
viii) The development of farm/shire based management plans incorporating conservation strategies.
ix) The redirection of some of the energy of regional conservation groups in the larger cities of the region towards regional issues such as remnant conservation, rather than issues outside of the region.

(b) Continuous acquisition or transfer of selected Crown land and freehold lands into Nature Reserves, National Parks, Historic Sites, Forest Preserves and Flora Reserves. The option for further additions of major areas to the conservation reserve system is rapidly diminishing. There are significant opportunities for consolidating existing parks and reserves by enlarging boundaries. There is also the desirability to include at least one riverine National Park along the upper catchment of the Abercrombie River (part of the Lachlan system) where significant continuously forested areas of Crown land remain. Omissions of specific alliances and associations from the reserve system are apparent. The strategy of the National Parks and Wildlife Service in the future, and to a lesser extent, the Department of Lands, will almost certainly focus on the addition of smaller reserves to the system. There is also potential for the Forestry Commission of New South Wales to greatly increase the number of Forest Preserves (2) and Flora Reserves (6) in the region. Some alliances (e.g., *Eucalyptus fastigata* as a dominant) are not well conserved in the region, however they are well conserved elsewhere in New South Wales. Some attention may therefore need to be given to the "rights" of regions to expect where possible, a representative sample of extant native vegetation alliances. Overall, however, it seems likely that relatively few of the uncommitted remaining remnants will be transferred to the reserve system, and probably fewer still from remnants on freehold land.

(c) Retention of other significant areas in a full range of Crown land holdings. In this regard, the Central Western Region Crown lands office in Orange is developing a significant public reserve system, the Central Tablelands Heritage Lands (Crown Lands Office 1986). It is the only division in the State with such a scheme operating. This regional land enterprise whilst encouraging multiple use of land for recreation and access has a strong conservation component and seeks to "maximize . . . flora and fauna habitats . . . by providing a network of Crown land corridors between national parks and native forests". The enterprise replaced the previous incremental approach to allocating land to different uses. A management-structure exists and is funded by leases, licences and concessionaires. The area incorporated into these Heritage lands is currently of the order of 17 000 ha. Unfortunately, as a counter to this optimistic scenario, the current Minister for Lands, the Honorable Mr I. Causly MP, has made it considerably easier for the range of lease land controlled by the Department of Lands to be converted to freehold, even though such land may have significant conservation value.

(d) The greater utilization of existing legislation contained in the Environmental Planning and Soil Conservation Acts, and in the use of exclusions open to the Crown Lands Office when conversion to freehold is applied for. No State Environmental Planning Policy (SEPPS) relating to rural land and particularly vegetation clearance has been published and only a few Regional Environmental Plans relating to rural areas have been gazetted.

In New South Wales, ten environmental protection zones are possible within Local Environmental Plans (LEPs). Those relevant to the CWR are wetlands, water catchment areas, scenic areas, escarpments, archaelogical sites, historical and scientific sites and wildlife refuges. Giblin and King (1987) give anecdotal evidence of the reluctance of councils to include Environmental Protection Zones in LEPs.

It would appear a useful procedure to make it mandatory for LEPs to address significant conservation issues including remnants, roadside verges and conservation corridors, based on meaningful resource inventories. The Department of Environment and Planning has been reluctant to spell out requirements for LEPs. The Soil Conservation Further Amended Act of 1986 No. 142 now allows a wider definition of protected land previously defined under section 21B to include streams and land with slope over 18°. The amendment reconstructed Section 21B to provide extra conditions on what constitutes protected land in the opinion of the Catchment Protection Board. These now include environmentally sensitive land, land subject to erosion or potential erosion, siltation and land degradation.

Existing legislation does in theory have a significant capacity to be used for conservation purposes. Additionally the Crown Lands Office has the well defined right to remove strips of land for conservation, recreation or access when applications for alienation are received. However the advent of the Greiner Government in New South Wales may see a halt to this process.

(e) New Legislation. In the last decade South Australia has passed two pieces of legislation of relevance to the conservation of fragments (Taylor 1987). The SA 1980 Heritage Agreement Scheme provides incentives for private landowners who agree to manage their remnant areas for nature conservation. The legislative basis for the scheme is by means of an amendment to the South Australian Heritage Act. The SA 1985 Native Vegetation Management Act seeks to control clearance of remnant vegetation where it contributes to amenity, water quality, soil erosion control, flood prevention, salinity control and stock shelter.

There has been some discussion of the former in New South Wales and similar Acts could be enacted here but obviously political sensitivities are involved. An additional education programme as to the importance of remnants would need to be undertaken.

As an interim measure, given the extent of land clearance that has occurred in the region, legislation to prohibit any further clearing for a period of five years, with provision made for genuine cases of hardship, could be considered. In this time rational and realistic regional and/or shire-based management plans could be developed.

(f) The recognition by planners and land managing authorities of the importance to fragment management of the theories and practices of island biogeography, landscape ecology, and total catchment management philosophy.

The theory of island biogeography (Macarthur & Wilson 1976) is increasingly being recognized as providing some important insights into the management of island conservation fragments in Australia (Saunders *et al.* 1987). There is however conflicting research as to the optimum strategy for allocating remnant land. Empirical support for the conservation strategy of using small reserves has come from plant studies (Lynch 1987; Ratcliffe 1984). The concept of critical minimum habitat size is however highly relevant to many vertebrate species (Lynch 1987). Nevertheless several small reserves may be found to hold more species than does a single large reserve of the same total area (Humphreys & Kitchener 1982). Corridors appear crucial as a means to inter-patch movement at least for some species (Saunders 1982), even though they may be significantly degraded. Since island biogeography theory is just that, a theory, care needs to be exercised in applying it in practical situations (Hobbs 1988).

Landscape ecology, the applied discipline of reconstruction of degraded landscape units in accordance with sound ecological principles, is commonly practised in parts of Europe and North America (Naveh & Lieberman 1984) but is in its infancy in Australia. As attempts are made to rehabilitate the large areas of Australia's degraded lands (including the Central Western Region) to link existing remnants through corridors and contrived landscaping, then greater attention will be given to the Northern Hemisphere experience in landscape ecology.

Total catchment management (TCM) (Soil Conservation Service of New South Wales, 1986) the philosophy which recognizes that all land management practices require co-ordination within catchment areas particularly in the control of soil erosion, complements the landscape ecology movement and recognizes the importance of remnants in stabilizing catchments. In New South Wales, whilst TCM is the policy of the government, little has been achieved in the way of any realistic implementation.

(g) The recognition that no single authority has the legislative power, time, personnel or expertise to effectively manage the remnant natural areas of the region.

There is an urgent need for government departments including Environment and Planning and the Heritage Council, National Parks and Wildlife Service, Lands, Tourism and Education, Forestry Commission and Soil Conservation Commission and Agriculture to co-operate more actively with each other and with local councils and landholders in promoting information and understanding of the environments of local areas.

(h) The recognition of a major role in conservation management for local landholders. Since farmers and graziers effectively manage most of the land in the region (including 30% of the remnant woodland/forest in the CWR), particular attention should be directed towards their conservation education and to encouraging a key role for them as managers in future regional or shire based management plans.

Training and guidance in the preparation of brief but meaningful management plans would need to be available to landowners through videos, resource packs and on specialized field days/seminars. Emphasis should be placed on the importance of seeing remnant conservation as part of total farm management plans. Some issues which should be addressed in the preparation of such plans at an appropriate, non-technical level include; determining the balance between productive farm land and conservation areas; criteria for allocation of land primarily for conservation purposes; grazing in remnant areas; developing a wildlife inventory; maximising diverse habitats; setting conservation objectives; developing a management strategy; using topographic maps and aerial photographs; cooperating with neighbouring landowners to maximize conservation areas and to develop wildlife corridors; co-operating with landowners within a shire-based/regional management context; access to additional resources; monitoring wildlife populations and assessing rates of change; management of pest species (native and exotic); the use of fire in conservation areas; the economics of wildlife/nature conservation and the role of government, local, state and federal in developing farm conservation policy.

(i) The establishment of Regional and/or shire based Conservation Councils should be encouraged with a brief to draw up objectives for the effective management of remnant vegetation and their associated fauna. Members of such councils could be drawn from relevant government departments, shires and councils, equal numbers of elected landowners and should include at least one ecologist. It would be most effective to target those shires which are likely to co-operate and establish model schemes which can be emulated elsewhere. Day to day management of various tenure types would be left in the main to individual landholders and the relevant administrations in the case of Crown land, backed by overall management plans with agricultural, landscape, catchment and wildlife objectives; the use of economically reasonable incentives such as Heritage Agreements and tax relief; a regional education programme based on the importance of remnant lands; and tax incentives for reafforestation programmes.

(j) The provision of more adequate funding for regional wildlife research should be sought from private and government sources for both pure and applied projects. Funding should be directed towards both amateur and professional groups. The World Wildlife Fund programme described in (a) above is also directing significant funds towards assessing the status of flora and fauna in a stratified sample of the remnants in the region. Some critical components of the island biogeography theory will be examined. These include the relation between species diversity and numbers, and shape and area of remnants as well as the distance of remnants from larger "supply" islands. It may also be possible by inference to comment on a minimal population size critical to a population's survival. Additionally, the more precise identification and location of (regionally) endangered species and their remaining habitat will also be examined.

(k) The mapping of vegetation alliances and associations of the region needs to be urgently carried out and conservation priorities determined for the various alliances and associations.

(l) Some experimentation with reintroductions of animals which are believed to be extinct in the region, but extant elsewhere, should be considered. Likewise the reconstruction of some plant associations and alliances could be attempted in small scale.

(m) Experimentation with fire as a management tool should be undertaken on both private and public lands.

The following are some questions which would be useful for land managers to ask and which should stimulate appropriate investigations. What were pre-European Aboriginal burning practices in particular areas? Is it possible or desirable to emulate such practices today in remnants and corridors in particular? In larger conservation areas? Should fire be used to create mosaics of seral succession in remnants/larger conservation areas, in an attempt to increase both the number of species and habitat diversity? What is the optimal frequency of burning for conservation purposes? For property protection purposes? Are the objectives for the use of fire for conservation and property protection mutually exclusive? At what time of the year should burning be undertaken to achieve optimal conservation objectives? What is/should be the role of local rural fire units, if any, in conservation management? Can property protection policies be implemented in rural areas to minimize destruction/damage to natural areas? Should conservation education be included as part of the fire fighting training of local volunteers and permanent staff? What is the role of hot fire in initiating local extinctions in remnant areas?

(n) A stringent ongoing monitoring capacity to identify what is happening within the rural environments of the region needs urgently to be developed. This process should particularly monitor lands affected by designated developments and land clearance for afforestation and marginal farming. For example, whilst relatively few species of plants and vertebrate have become extinct in the region since 1815, almost certainly the fragmentation of woodland/forest and the assorted impacts on the residual fragments has caused significant local extinctions. The threat of widespread and imminent extinctions may be much greater than is currently recognized. In economic terms alone it may not be realistic to monitor such processes in any but a superficial manner. The World Wildlife Fund programme in the region (see sections a, j), will probably only be able to investigate in detail 100-150 of the 3 500 remnants in the region and to perhaps include some others on a less intense basis. The programme however does envisage the development of a checklist of environmental factors including reporting on appropriate indicator species, which may make some long term indirect monitoring possible. In this regard amateur field naturalists and local landowners may be called on to shoulder greater responsibility and indeed to be given greater recognition by professional scientists in Australia for their contributions. Only two active field naturalist groups are operating effectively in the region.

Conclusion

The disappearance of species through the reduction of their habitats by human activities is a theme which occurs repeatedly in the ecological literature of Australia. Despite this, there have been few successes in the development of tools for the integrated management of natural habitats on public and private lands. Perhaps ecologists have been too concerned with their research on the problems to become active in the formulation and implementation of public policy in this area! But this chapter as well as signalling the problem, has identified possible avenues for policy development and it is to be hoped that these can be followed in the future.

Acknowledgements

We wish to acknowledge Dr Sylvia Cardale for reading and commenting on this manuscript and for allowing us to use material from her chapter "The Vegetation of the Central Western Region" in Goldney and Bowie (1987).

References

Allan H. (1983) *19th Cent. Faunal Changes in Western N.S.W. and N.W. Victoria.* University of Auckland Working Paper in Anthropology. Dept. of Anthropology.

Anderson R. W. (1961) Introduction *Contr. N.S.W. Natl. Herb. F1* Nos 1-18, pp. 1-15.

Australian Bureau of Meteorology (1975) *Climatic Averages (NSW).* Australian Bureau of Meteorology, Melbourne.

Bell F. (1978) *Forest Ecosystems in NSW.* Total Environment Centre, Sydney.

Benson D. G. (1984) *Wallerawang 1: 100 000 Vegetation Sheet.* Central Mapping Authority, Bathurst.

Bower C., Cardale S. & Goldney D. C. (1987) Flora and Fauna sites. In: *The National Trust of Australia (NSW) Scenic and Scientific Survey of the Central Western Region. A Report to the Australian Heritage Commission* (eds D. C. Goldney & I. J. S. Bowie) Vol. II, pp. 204-25. Mitchell College of Advanced Education, Bathurst.

Cardale S. (1987) Vegetation. In: *The National Trust of Australia (NSW) Scenic and Scientific Survey of the Central Western Region. A Report to the Australian Heritage Commission* (eds D. C. Goldney & I. J. S. Bowie) Vol. 1, pp. 111-35. Mitchell College of Advanced Education, Bathurst.

Conyers B. (1987) Aboriginal History and Anthropology. In: *The National Trust of Australia (NSW) Scenic and Scientific Survey of the Central Western Region. A Report to the Australian Heritage Commission* (eds D. C. Goldney & I. J. S. Bowie) Vol. 1, pp. 31-46. Mitchell College of Advanced Education, Bathurst.

Crown Lands Office (1986) *Launch of the Lithgow Macquarie-Turon and Canobolas Heritage Lands. A Regional Land Management Enterprise.* Department of Lands, NSW.

Goldney D. C. (1987a) The Vertebrate Fauna of the Central Western Region. In: *The National Trust of Australia (NSW) Scenic and Scientific Survey of the Central Western Region. A Report to the Australian Heritage Commission* (eds D. C. Goldney & I. J. S. Bowie) Vol. 1, pp. 136-60. Mitchell College of Advanced Education, Bathurst.

Goldney D. C. (1987b) Land Tenure, Forest and Woodland Cover. In: *The National Trust of Australia (NSW) Scenic and Scientific Survey of the Central Western Region. A Report to the Australian Heritage Commission* (eds D. C. Goldney & I. J. S. Bowie) Vol. 1, pp. 252-71. Mitchell College of Advanced Education, Bathurst.

Goldney D. C. & Bowie I. J. S. (eds) (1987) *The National Trust of Australia (NSW) Scenic and Scientific Survey of the Central Western Region. A Report to the Australian Heritage Commission.* Vols. I and II. Mitchell College of Advanced Education, Bathurst.

Giblin W. B. & King S. (1987) The Viability of Planning Control and Renovation as Options in the Conservation of Remnant Vegetation. In: *Nature Conservation: The Role of Remnants of Native Vegetation* (eds D. A. Saunders, G. W. Arnold, A. A. Burbidge & A. J. M. Hopkins) pp. 295-304. Surrey Beatty & Sons, Chipping Norton.

Harriman R. J. & Clifford E. A. (1988) *Atlas of New South Wales.* Central Mapping Authority, Bathurst.

Hayden E. J. (1971) *Natural Vegetation of New South Wales.* Map in Resources of NSW series. Department of Decentralization and Planning, NSW.

Hobbs R. J. (1988) What is the Ecological Theory and Is It of Any Use to Managers? In: *Ecological Theory and Biological Management of Ecosystems* (eds D. A. Saunders & A. A. Burbidge) pp. 15-27. Occasional Paper 1/88. Western Australian Department of Conservation and Land Management, Perth.

Humphreys W. F. & Kitchener D. J. (1982) The Effect of Habitat Utilization on Species-area Curves: Implications for Optimal Reserve Area. *J. Biogeogr.* **9,** 391-6.

Jacobs S. W. L. & Pickard J. (1981) *Plants of New South Wales. A Census of Cycads, Conifers and Angiosperms.* National Herbarium of New South Wales, Royal Botanic Gardens, Sydney.

Leigh J., Boden R. & Briggs J. (1984) *Extinct and Endangered Plants of Australia.* Macmillan, South Melbourne.

Lindsay A. D. (1967) *Forest Types of NSW Cypress Pine Zone.* Forestry Commission of NSW, Tech. Paper No. 8., Sydney.

Lynch J. F. (1987) Responses of Breeding Bird Communities to Forest Fragmentation. In: *Nature Conservation: The Role of Remnants of Native Vegetation* (eds D. A. Saunders, G. W. Arnold, A. A. Burbidge & A. J. M. Hopkins) pp. 123-40. Surrey Beatty & Sons, Chipping Norton.

MacArthur R. G. H. & Wilson E. O. (1976) *The Theory of Island Biogeography.* Princeton University Press, Princeton.

Moore R. M. (1959) Ecological Observations on Plant Communities Grazed by Sheep in Australia. In: *Biogeography and Ecology in Australia* (eds J. A. Keast, R. L. Crocker & C. S. Christian) pp. 500-12. Junk, The Hague.

Moore R. M. (1966) Man and Plant Communities. *Proc. Ecol. Soc. Aust.* **1,** 106-10.

Naveh Z. & Lieberman A. S. (1984) *Landscape Ecology: Theory and Application.* Springer-Verlag, New York.

Ratcliffe D. R. (1984) Post-medieval and Recent Changes in British Vegetation: the Culmination of Human Influence. *New Phytol.* **98,** 73-100.

Saunders D. A. (1982) The breeding behaviour and biology of the White-tailed Black Cockatoo. *Ibis* **124,** 422-55.

Saunders D. A., Arnold G. W., Burbidge A. A. & Hopkins A. J. M. (eds) (1987) *Nature Conservation: The Role of Remnants of Native Vegetation.* Surrey Beatty & Sons, Chipping Norton.

Soil Conservation Service of NSW (1986) *Total Catchment Management. A State Policy.* NSW Government Printer, Sydney.

Specht R. L., Roe E. M. & Boughton V. H. (eds) (1974) Conservation of Major Plant Communities in Australia and Papua New Guinea. *Aust. J. of Bot.* Supplement No. 7.

Stewart J. (1968) Erosion Survey of New South Wales and Eastern and Central Divisions: Reassessment 1967. *Journal of the Soil Conservation Service of NSW* **24,** 139-54.

Taylor S. G. (1987) Conservation Strategies for Human-Dominated Land Areas: The South Australian Example. In: *Nature Conservation: The Role of Remnants of Native Vegetation* (eds D. A. Saunders, G. W. Arnold, A. A. Burbidge & A. J. M. Hopkins) pp. 313-22. Surrey Beatty & Sons, Chipping Norton.

Proc. Ecol. Soc. Aust. 1990 — 16:441-447

Satellite monitoring of soils, vegetation and grazing impacts in the semi-arid region of western New South Wales

A. L. O'NEILL[1] and D. J. ELDRIDGE[2]

[1]*Geography Department, University of Wollongong, P.O. Box 1144, Wollongong, N.S.W. 2500*
[2]*Soil Conservation Service of New South Wales, P.O. Box 198, Chatswood, N.S.W. 2067*

Abstract

Satellite remote sensing provides a relatively rapid and cost effective means of mapping the physical and biological resources of large areas. Landsat Multispectral Scanner digital imagery was used to map soils, vegetation and erosion in a semi-arid area of south-western New South Wales. The classified image showed strong agreement with independently mapped soil boundaries. The technique developed is suitable for identifying and mapping areas affected by fire, overgrazing, erosion, land clearing and cultivation. With an increased awareness of ecologically-based landuse practices, the use of remote sensing for monitoring changes in land use and land degradation is likely to increase.

Introduction

Since the arrival of European pastoralists in western New South Wales in the 1850s, many areas of the vegetation and land resources have been adversely affected by sheep grazing. Records from as early as 1886 indicate that the introduction of domestic stock brought about major changes in the structure of the native vegetation as well as accelerated soil erosion (Moore 1962). By the mid-1880s rabbits had reached the area (Robertson *et al.* 1987). The subsequent competition for forage between rabbits, sheep and native herbivores together with drought conditions which persisted from 1895 until 1903, ruined many pastoralists and led to the disappearance of many edible shrubs and grasses and a marked deterioration in the condition of the land. Devastating droughts around the turn of the century, combined with widespread overgrazing in many parts of the pastoral country led to the holding of a Royal Commission in 1901 into the condition of Crown tenants. In 1982 a Parliamentary Select Committee was established by the New South Wales Government to enquire into landuse, conservation and utilization of natural resources of the Western Division of New South Wales (New South Wales 1983). Arising from the inquiry, various recommendations were made including the need for more objective ecologically-based data in order to make sound management decisions, and the need to adequately monitor condition and trend of the resources.

To manage adequately the physical and biological resources of the region, information on landforms, soils, soil erosion, vegetation type and condition, and present and past land use is required. In order to determine what changes are occurring to the land resources, and whether or not these changes are deleterious, information must be gathered on a regular basis.

Changes such as clearing for cultivation and gross changes in land use are relatively easy to detect, as are areas now suffering from accelerated erosion. The subtle changes in plant community structure or floristics however, are much more difficult to identify and record, and detailed long-term monitoring is required.

The use of satellite information can be cost effective over large areas of inaccessible land (Graetz & Gentle 1982) and the data can be obtained on a systematic and regular basis. Further research is still required to understand better various aspects of reflectance and analysis of Australian soils and land cover, and to assess the accuracy and precision of the classified information produced from the remotely sensed data.

The aim of this study was to investigate the use of Landsat Multispectral Scanner (MSS) digital data to classify and map soil types, soil erosion, vegetation cover and the effects of different grazing pressures over an extensive area of semi-arid western New South Wales.

Methods

Study area

The Willandra Lakes Region covers an area of 6 000 km^2 in central-western New South Wales and is located approximately 1 000 km west of Sydney (Fig. 1). It has a mean annual rainfall of 275 mm with evaporation in the order of 1 500 mm per year. The region, like much of the rest of the semi-arid region of New South Wales, has been extensively grazed since the 1850s (Moore & Perry 1970).

In 1979 the former grazing station "Mungo" was acquired by the New South Wales National Parks and Wildlife Service and was gazetted as Mungo National Park. The adjoining property "Zanci" was acquired in 1984 and, like "Mungo", was subsequently destocked.

Most of the region is pastoral leasehold land used primarily for sheep grazing and some areas have been cleared for opportunistic dryland cropping. In other areas, trees have been selectively ring-barked to increase the growth of palatable pasture species. Extensive bushfires have also affected the structure, density and distribution of the mallee, *Eucalyptus* spp., belah, *Casuarina cristata* and white cypress pine, *Callitrus glaucophylla* woodland.

The perennial vegetation of this area is relatively drought tolerant and is well adapted to the highly irregular rainfall (Graetz 1987). Since it is the perennial vegetation which ultimately determines landscape and soil stability, it is important to monitor perennial cover in order to identify areas susceptible to soil erosion.

The region contains extensive chenopod shrublands dominated by saltbushes, *Atriplex* spp. and bluebush *Maireana* spp. a vegetation type poorly conserved in Australia especially in south-western New South Wales (Specht 1981). Although mallee areas are proportionally better conserved in southern Australia, there may be particular sites of ecological significance, which, because of their protected position between channel and lake systems, have had a significantly different fire history from the mallee woodlands of the region.

The Willandra Lakes Heritage Area is rich in aeolian and lacustrine landforms, providing evidence of early human occupation of south-eastern Australia (Bowler & Magee, unpub. report). The area has a series of interconnected large playa lakes with vegetated floors. Dry for about 16 000 years (Bowler 1978), the lakes are bordered on the eastern and north-eastern sides by sand, clay or gypsum lunettes. The *Walls of China* lunette of Lake Mungo is the most spectacular, being 33 km long and 36 m high. Much of the lunette has been eroded, exposing the red calcareous core on the western side. On the eastern side are extensive mobile white quartz sand dunes. The plains and dunefields surrounding the dry lakes are the result of an intensive period of aeolian activity, being formed by the predominantly westerly winds of the Pleistocene (Bowler 1978). These plains and dunefields are stabilized predominantly by mallee.

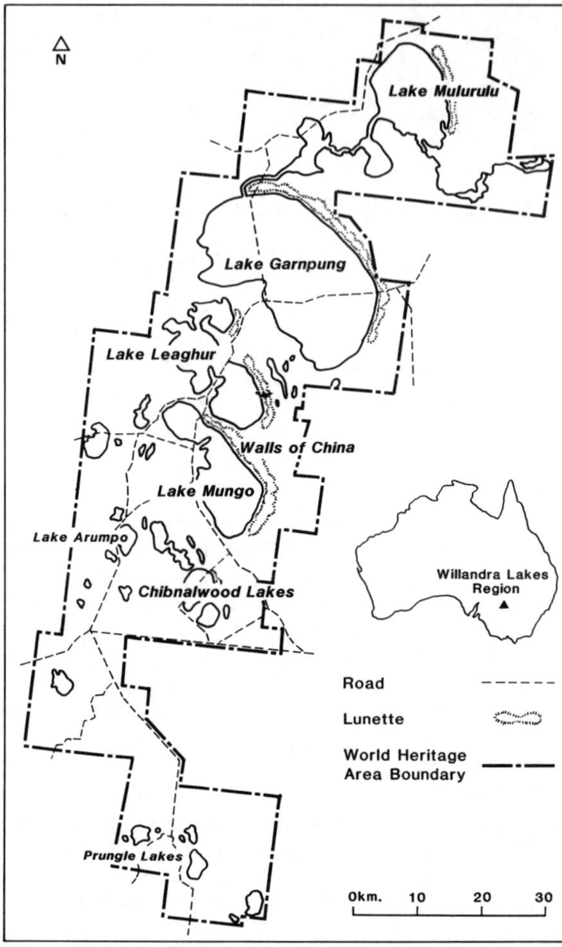

FIG. 1. The Willandra Lakes region showing the boundaries of the World Heritage Area.

The area can be considered as four major systems:

1. Playa lake beds of grey and yellow-grey clay soils and some duplex soils with a dense to sparse cover of chenopods, *Atriplex, Maireana* and *Scleroleana* spp.
2. Lake-bordering lunettes of sand or gypsum either bare or sparsely vegetated.
3. Plains and isolated rises of red-brown calcareous earths with a medium to sparse cover of belah, rosewood, *Heterodendrum oleifolium,* white cypress pine, *Callitris glaucophylla,* and wilga, *Geijera parviflora* with an understorey of perennial grasses, low growing chenopods or bare soil.
4. Dunefields of parallel, parabolic and unaligned dunes of deep, red, siliceous sands with a dense to sparse cover of mallee and an understorey of porcupine grass, *Triodia irritans,* copperburrs or bare sand.

Image analysis

LANDSAT MSS digital data from November 15, 1984 were used to analyse the spectral response of the surface features of Willandra Lakes World Heritage Area. This coincided with the period of initial field data collection.

Analysis of the digital reflectance data was simplified considerably by acquiring the data in late spring. At this time, few of the late winter or spring ephemerals survived the onset of the hot summer in November 1984 thus their specific reflectance was avoided.

The classification of Landsat MSS image data is the process of assigning individual pixels (picture elements) on the basis of their spectral reflectance characteristics in the four spectral bands. These are band 4 (0.5-0.6 um, green), band 5 (0.6-0.7 um, red), band 6 (0.7-0.8 um, near infrared) and band 7 (0.8-1.1 um, near infrared). The Dipix Image analysis system at the University of New South Wales was used for the analysis of the imagery.

Because of the size and diversity of the area, and the lack of vegetation information over much of the region, a migrating means clustering procedure (Kelly 1982; Richards & Kelly 1984) was selected to produce signatures to represent the variety of land surfaces. Details of the methodology are discussed by O'Neill (1987). The signatures of each land surface were used in a maximum likelihood classification procedure (Richards 1986). This whole procedure is often referred to as an unsupervised classification. The image was registered to the 1:100 000 scale map, rectified to the Australian Map Grid and resampled using 50 metre pixels.

Results

Image analysis

The unsupervised clustering procedure generated 13 classes, which, together with two classes derived from training areas, i.e., scalded soil and belah, are shown in Figures 2a and 2b. The complexity of the area is illustrated by the differences in spectral responses of the different classes. In the classified image (Fig. 3), the chenopod classes 5 to 9 were all found to cover very small areas. Although small, each of these areas is ecologically significant. These areas represent cane grass and black box, *Eucalyptus largiflorens* swamps.

The unsupervised clustering procedure relies purely on spectral reflectance data to classify pixels in the image. To describe the spectral classes in terms of ground features, it was necessary to validate these classes in the field.

Sampling was undertaken along roads and property tracks because of time constraints. Sixty field sites were selected and at each site two 100 m transects were surveyed. For each transect the soil type and colour were recorded, the spacing between trees or shrubs was measured, vegetation cover estimated and a plant species list was compiled. This sampling procedure provided information for a provisional classification label. Table 1 describes the vegetation and soils of each of the 15 classes shown in Figure 2a and 2b.

Dense belah/cypress pine woodlands on calcareous earths had a low reflectance in all four bands. The mallee woodlands on red sand had the same reflectance as belah in the green band, slightly higher in the red and much higher in the infrared bands. The chenopod shrublands generally had reflectances

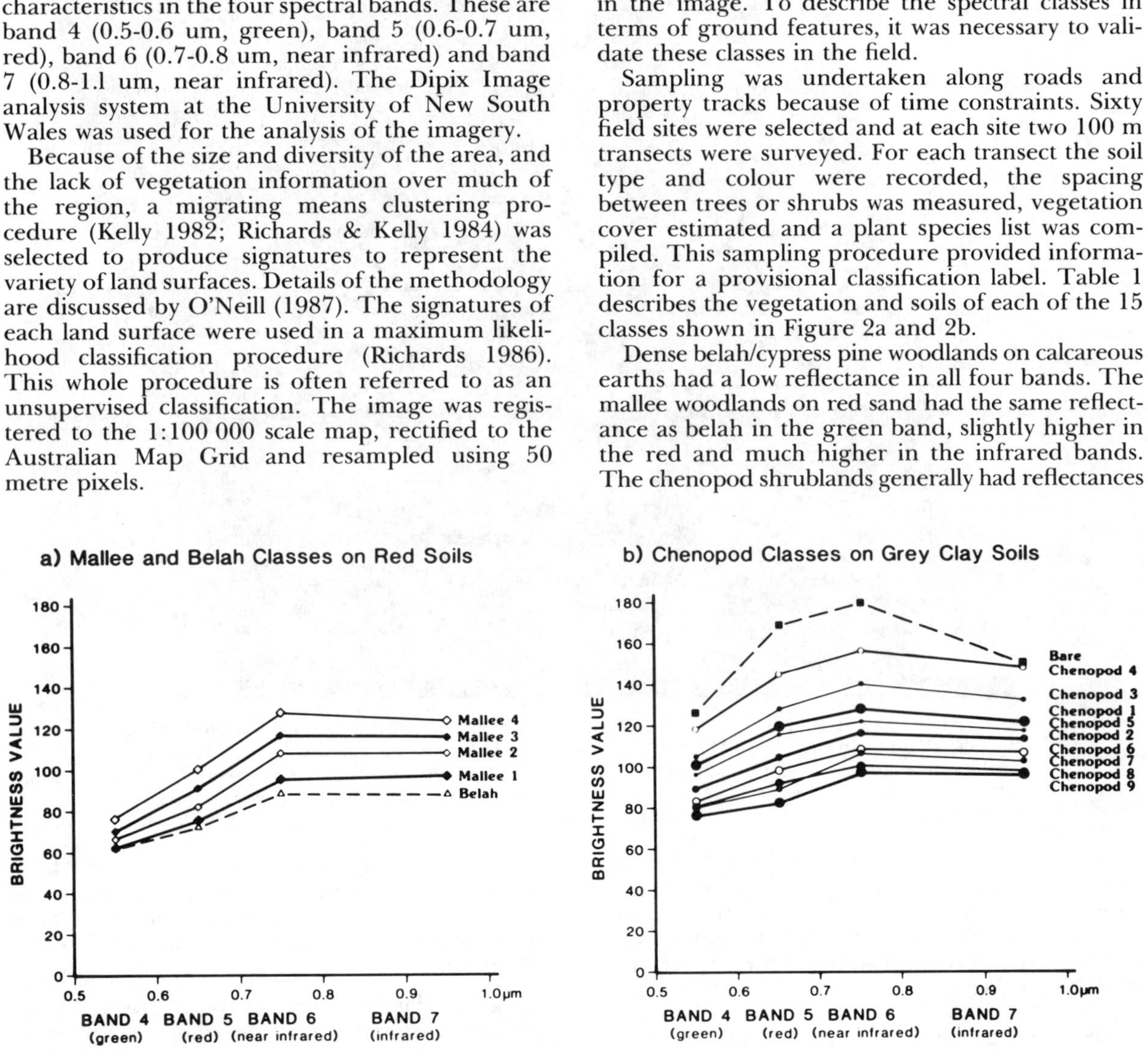

FIG. 2a and 2b. Spectral signatures for a) the mallee and belah classes on red soils and b) the chenopod classes and bare soil on grey clay soils.

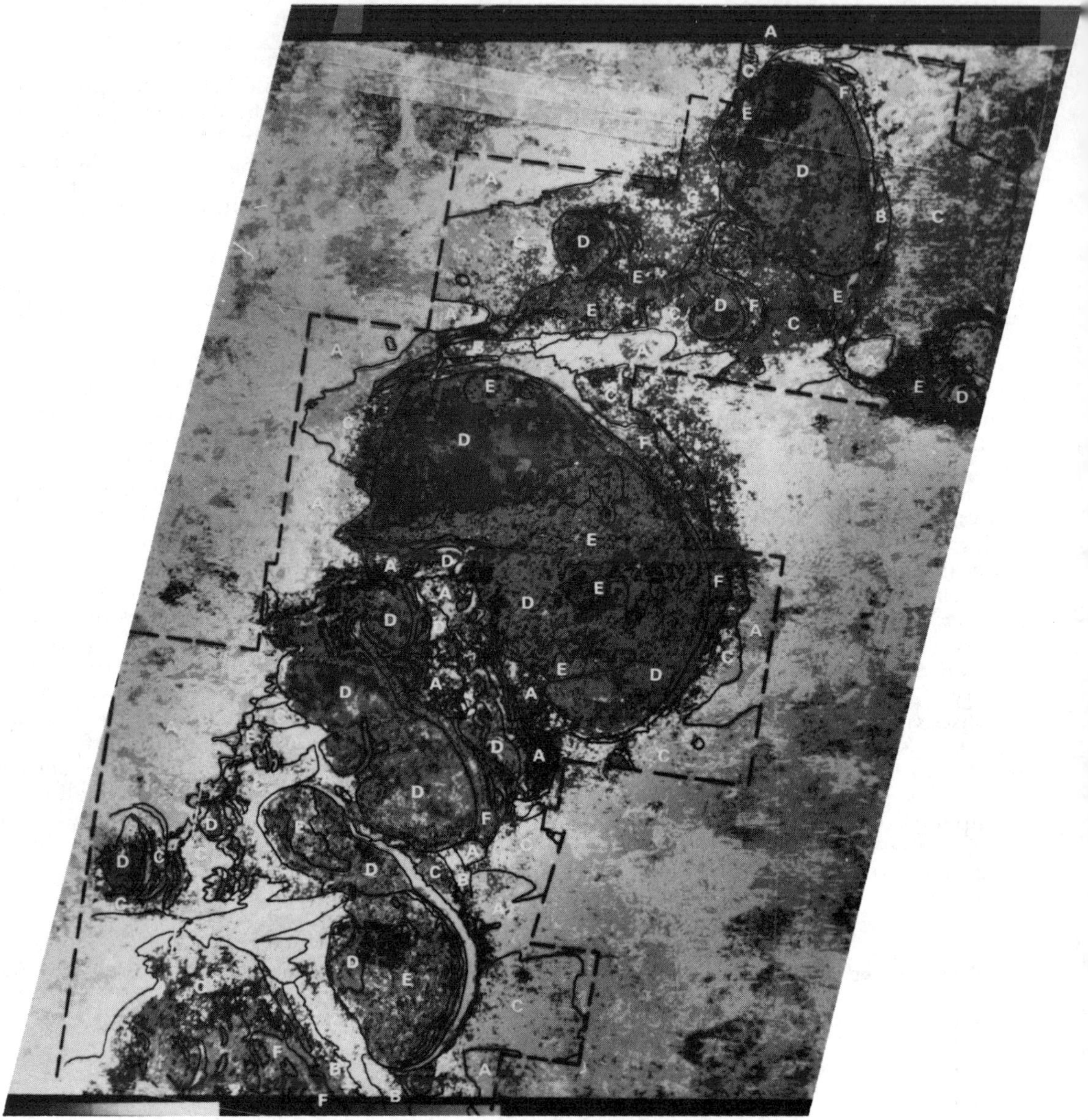

FIG. 3. The Landsat MSS image, classified into 15 classes. The overlay of mapped soil classes shows the following soil groups: A, Red earths with brownish, calcareous and siliceous sand; B, Calcareous earths with siliceous sands; C, Calcareous earths with brown, red, yellow and grey earths; D, Grey self-mulching clays; E, Self-mulching clays with assorted duplex soils; F, Non-cracking clays with brownish and siliceous sands and brown sandy earths.

as high or higher than the belah or mallee woodlands. All had very similar shaped spectral reflectance curves and differed only in the relative amounts of brightness, due to the amount of vegetation cover. Bare soil had the highest reflectance in all bands (Figs 2a and 2b).

Comparison of the classified image and soil maps

The percentage difference between the soil classification and the image is an important way of assessing the classified image. A more important aspect however, is the pattern of error. In order to assess this error, soil class boundaries independently mapped by Eldridge (unpub. report) were superimposed on the image.

A grid 5 km by 5 km was placed over the soil map and the Landsat image, and 122 points on both maps were systematically compared. For each grid point, the soil class and the classified image class were recorded and the results are presented

TABLE 1. Results of field information for each labelled unsupervised class. Willandra Lakes Region, New South Wales.

Class	Vegetation type	Vegetation cover	Soil type
Chenopod 1	Annual chenopod shrubland Dominant – *Dissocarpus paradoxus*	Dense – up to 1 m spacing	Yellow-grey clays
Chenopod 2	Perennial chenopod shrubland. Dominants – *Atriplex vesicaria, Maireana pyramidata, Nitraria billardieri, Sclerolaena divaricata*	Medium-dense approx. 1-2 m spacing	Yellow-grey & grey-brown clays
Chenopod 3	Chenopod shrubland Dominants – *Atriplex vesicaria, Scerolaena divaricata, S. diacantha*	Sparse – 5 m plus spacing	Grey-brown clays
Chenopod 4	Sparse mixed chenopods	Very sparse	Grey clays
Chenopod 5,6,7,8,9	Chenopod shrublands, canegrass and lignum *(Muehlenbeckia cunninghamii)* swamps, isolated box swamps	Dense Less than 1 m spacing	Grey clays with high moisture
Bare soil	Bare sand and scald areas	Bare	Red-brown siliceous sands
Belah	Belah rosewood woodlands Dominants – *Casuarina cristata, Callitris glaucophylla, Heterodendrum oleifolium, Stipa* spp., chenopod or litter understorey	Medium 5-10 m spacing	Red-brown calcareous earths
Mallee 1	Mallee with *Stipa* grasslands or chenopod understorey. Sparse belah and rosewood clumps	Medium 10-100 m tree spacing	Sandy red or red-brown calcareous earths
Mallee 2	Mallee with understorey of small porcupine grass *Triodia irritans* sparse chenopod shrubs or *stipa* spp.	Medium 5-10 m tree spacing	Red sand or red-brown calcareous earths
Mallee 3	Mallee with porcupine grass and bare soil	Medium to sparse 10-100 m tree spacing	Red sand
Mallee 4	Mallee with annual grasses, annual chenopods and bare soil	Very sparse 50-100 m tree spacing	Red sand

in Table 2. There was no clearly defined image class which matched a particular soil class. There were however broad areas of coincidence between the two maps.

Coincidence between the mapped soil class boundaries and the Landsat image boundaries was very high. The six major soil classes outlined in Eldridge (unpub. report) were identified but the classified image divided each of the soil classes into a number of sub-classes (Figure 3). Some features were delineated on the classified image but not on the soils map and vice versa.

In the north west, the boundary between red sand dunes supporting mallee vegetation and the red calcareous earth plains supporting belah woodlands is apparent. Areas in the southern section of Lake Mungo where linear shaped scalds occur on irregular aeolian ridges formed during the final stages of drying of the lake were delineated on both the soils map and the image. These areas are particularly susceptible to erosion and form a mosaic of scalds on duplex soils. The highly mobile and bare dunes on the western margins of Lake Garnpung were easily delineated in both mapping systems.

Some features present on the image but not on the soil map include the small but ecologically important black box swamps within the mallee and belah areas, and the cane grass swamps on many of the lake beds. At the scale of soil mapping, it was not feasible either to describe the soils of these areas nor to map them.

Discussion

Incorrect classification on the image may have occurred on Lake Garnpung as a result of the effects of past cropping on the lake bed. Cropped

TABLE 2. Results of 122 systematic grid sample sites relating image classification class to mapped soil class.

Mapped soil classes	No. of points	Categories on image	% of points in image category
A. Red earths with brownish calcareous and siliceous sands	22	Mallee 1	36.4%
		Mallee 2	31.8%
		Mallee 3	13.6%
		Chenopod 3	13.6%
		Bare soil	4.5%
B. Calcareous earths with brownish sands	22	Chenopod 3	50.0%
		Bare soil	50.0%
C. Calcareous earths with brown, red, yellow or grey earths	34	Mallee 1	32.4%
		Mallee 2	11.8%
		Casuarina	23.5%
		Chenopod 3	20.6%
		Bare Soil	11.8%
D. Grey self mulching clays	21	Chenopod 2	42.9%
		Chenopod 3	33.3%
		Chenopod 4	23.8%
E. Grey self mulching clays with assorted duplex soils	16	Chenopod 2	6.3%
		Chenopod 3	50.0%
		Chenopod 4	6.3%
		Bare soil	31.3%
F. Non cracking clays with brownish and siliceous sand and brown sandy earths	7	Chenopod 2	14.3%
		Chenopod 3	57.1%
		Bare soil	28.6%

areas showed up with higher soil reflectance than surrounding areas due to lack of vegetation. This occurred across a homogeneous soil type.

Soils were mapped independently of their erosional status. The Landsat image was unable to distinguish uneroded duplex soils but delineated eroded duplex soils on the basis of the differences in reflectance between the two erosional states. Duplex soils with a good cover of chenopods could not be distinguished on the image. Furthermore, linear features within the southern lake beds indicating the eroded duplex soils are described on the soil maps but, because of their sporadic distribution in the landscape, were not mapped separately.

Features found on the soil map but not on the image may have resulted for a number of reasons such as matching errors incurred in rectifying the satellite image to the map and field mapping errors and boundary smoothing during the drafting of the soil maps.

Soils were often grouped according to their position in the landscape i.e., as soil landscapes rather than as separate soil types, due to the difficulty of mapping and the small scale of occurrence. Furthermore, the image depicted land cover features such as differences in ground cover, species composition, amount and type of cryptogam crusts and other features largely due to differential management and/or grazing pressures. These often occurred on the same soil type so were not differentiated on the soil map.

Conclusion

The classified image described the major soil types found in the area but was also affected by different cover types which in turn were affected by present and past land use. Although the soil has the strongest influence on the sensor signal, the classified image is delineating broader categories more akin to land systems and land use categories.

This classification technique should be regarded as a systems approach for mapping and monitoring ecological systems. It is a particularly important tool for mapping and monitoring environmental changes over extensive, inaccessible areas such as the arid and semi-arid regions of Australia. With an increase in concern for the managment of arid and semi-arid areas, and, as the areas involved are so vast, the use of satellite data is the only way in which we can assess and monitor the impacts of humans upon the ecosystem (Graetz 1987).

To make the most efficient use of image data, they should be integrated with other available data sets in a computerised, image based Geographic Information System. This allows land managers to access individual sets of data or to examine inter-relationships. Satellite imagery has the advantage of being a non-static data set which can be constantly updated. The Landsat V satellite covers the same area every 16 days.

Future ecological surveys should include site reference co-ordinates in order to facilitate the interpretation of ground truthing of satellite imagery that could be used as a monitoring tool in the future. Expensive field based surveys can then be more cost effective.

Acknowledgements

We are grateful to the New South Wales National Parks and Wildlife Service for allowing us to publish the results of the remote sensing and soils studies which were undertaken on behalf of the New South Wales Department of Environment and Planning. We thank Dr Dawn Williamson and two anonymous referees for their comments on an earlier draft of this paper.

References

Bowler J. M. (1978) Quaternary climate and tectonics in the evolution of the Riverine plain, south-eastern Australia. In: *Landform Evolution in Australia* (eds J. L. Davies & M. A. J. Williams) pp. 70-112. Australian National University Press, Canberra.

Graetz R. D. (1987) Satellite remote sensing of Australian rangelands. *Remote Sens. Envir.* **23,** 313-31.

Graetz R. D. & Gentle M. R. (1982) A study of the relationships between reflectance characteristics in the Landsat wavebands and the composition and structure of an Australian semi-arid shrub rangeland. *Photogram. Eng. & Rem. Sens.* **48,** 1721-30.

Kelly D. J. (1982) The concept of spectral class. A comparison of clustering algorithms. M. Eng. Sci. Thesis. University of New South Wales, Sydney.

Moore R. M. (1962) Effects of the sheep industry on Australian vegetation. In: *The Simple Fleece — Studies in the Australian Wool Industry* (ed. A. Barnard). Melbourne.

Moore R. M. & Perry R. A. (1970) Vegetation. In: *Australian Grasslands* (ed. R. M. Moore) pp. 59-73. Australian National University Press, Canberra.

New South Wales Parliament (1983) *Joint Select Committee of the Legislative Council and Legislative Assembly to Enquire into the Western Division of New South Wales.* D. West, Government Printer, Sydney.

O'Neill A. L. (1987) Landsat, landcover classification using migrating means clustering — Willandra Lakes Region, New South Wales. *Proceedings Fourth Australasian Remote Sensing Conference* pp. 239-49. Adelaide.

Richards J. A. (1986) *Remote Sensing Digital Image Analysis.* Springer-Verlag, Berlin.

Richards J. A. & Kelly D. J. (1984) On the concept of spectral class. *Int. J. Rem. Sens.* **5,** 987-91.

Robertson G., Short J. & Wellard G. (1987) The environment of the Australian sheep rangelands. In: *Kangaroos, their ecology and management in the sheep rangelands of Australia* (eds G. Caughley, N. Shepherd & J. Short) pp. 14-34. Cambridge University Press.

Specht R. L. (1981) Major vegetation formations in Australia. In: *Ecological Biogeography in Australia* (ed. A. Keast). Dr W. Junk, The Hague.

Proc. Ecol. Soc. Aust. 1990 — 16:449-457

Integration of satellite imagery and geographic information systems for habitat mapping and the prediction of fauna distributions: preliminary results and future directions

J. K. HIBBERD[1], A. L. O'NEILL[2], J. MARTHICK[2] and R. L. SIM[2]

[1]*New South Wales National Parks and Wildlife Service, P.O. Box 733, Queanbeyan, NSW 2620*
Present address: Kestel Research, P.O. Box 2546, Canberra, A.C.T. 2601
[2]*Geography Department, University of Wollongong, P.O. Box 1144, Wollongong, NSW 2500*

Abstract

Natural resource managers have a requirement for accurate and easily revised information on vegetation, habitat and fauna distribution. This chapter describes some of the techniques that are being examined currently to integrate satellite imagery with existing environmental data. The suitability of various platforms and sensors is evaluated and results from preliminary use of the LANDSAT 5 satellite's Thematic Mapper sensor to map vegetation structure are described.

This work presents major opportunities and potential benefits for the improved management of vegetation communities and native fauna by conservation authorities, particularly in the development of predictive models for fauna distribution and especially where management agencies have a lack of detailed survey information. This chapter outlines a number of new approaches that the authors wish to see addressed by the ecological community.

Introduction

For 200 years since first European settlement, Australia's flora and fauna has been steadily, and at times, rapidly, reduced in distribution, leading inevitably to a series of national and regional extinctions (Ovington 1978; Recher *et al.* 1979). Natural ecological systems throughout the continent have been modified by a variety of agencies, including pastoralism, clearing for intensive agriculture, forestry, mining and urban expansion.

One major problem facing Australian ecologists is how to maintain (or enhance) the existing diversity and range of biota. Many of the approaches that have been employed in Australia to resolve this problem have centred around creation of suitable land allocation and management systems (Hitchcock 1984). These depend, in the main, on the dedication of representative samples of natural systems in secure and viable reserves; and on improved management techniques for lands of conservation significance outside the reserve system (Hopkins & Saunders 1987).

The problem

The acquisition of a representative reserve system is one of the major objectives of conservation agencies in Australia in achieving their goal of flora and fauna conservation. Many authors (Specht *et al.* 1974; Benson 1985; Australian Heritage Commission 1985) believe that the reserves established to date are, in fact, not secure, viable, or representative.

There are considerable problems inherent in placing undue dependance upon reserve systems to provide for all our conservation needs, not the least of which are difficulties implicit in attempts to manage ecological islands within modified landscapes (Diamond 1975; Slatyer 1975; Saunders *et al.* 1987).

In addition, Soulé (1987) considers that projected climatic changes have the potential for exerting a significant effect upon the viability of species' populations. If extant biological systems are to be maintained across their full environmental range through the next 200 years, government agencies and land managers will need to utilize new ecological techniques and make bold landuse decisions based upon research in this field.

Information systems

Information systems play a major role in maximizing ecological knowledge and its application to land and resource management problems. The traditional information system employed in tackling the problems outlined above is the field-based flora and fauna survey but this has a major associated disadvantage, that of high cost. There is a range of other problems associated with relationships derived from such data because of variability in weather conditions, seasonality or trap-shyness and extreme care is needed in the planning and execution of such work (Emison *et al.* 1984; Lunney & Barker 1986).

Furthermore, examined on a regional basis, the coverage of existing fauna surveys is often extremely patchy (Fig. 1) (Hibberd 1986), and is rarely suitable for evaluating regional faunal population distributions.

New approaches

The potential of remotely sensed imagery

The use of satellite imagery is finding increased use in vegetation mapping particularly in remote and/or inaccessible areas or where it is required to frequently monitor changes in vegetative cover (Woodgate *et al.* 1987; Hall & McCloy 1987). Although mapping derived from remotely sensed products does require the input of detailed field information both for accuracy and for validation, there is potentially considerably less fieldwork required than for the traditional approach.

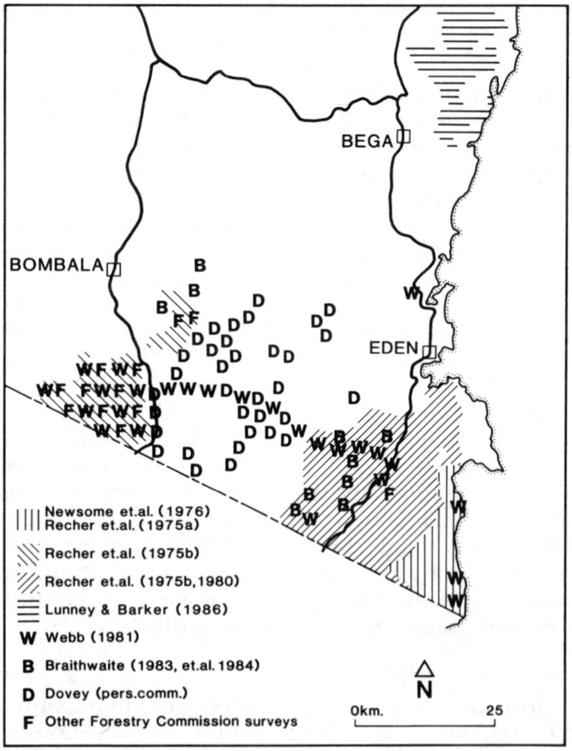

FIG. 1. Coverage of native fauna surveys in south-eastern New South Wales (after Hibberd 1986).

Vegetation communities are rarely static over medium to long periods of time and this is particularly true for fire-prone south-eastern Australia (Recher *et al.* 1975; Fox 1978). As the LANDSAT platform passes over the same area of land every 16 days, mapping can be carried out following fires or other disturbances and both long and short term changes in vegetation cover and structure monitored. The advantages of satellite imagery are that a range of spectral bands in both visible and infra-red wavelengths are recorded and the information is available in digital formats. This assists considerably when combined with other data in geographic information systems (GIS). Computer analyses of individual or combinations of spectral bands can be applied to allow clearer discrimination of relevant features.

Some work in the field of integrating GIS with remotely sensed data has been undertaken in arid Australia (Graetz *et al.* 1984; Milne & Zhou 1987) and in tropical Australia [Nix, Mackey & Hutchinson's work with AEGIS (pers. comm.)]. However, in the rugged forested landscapes of eastern Australia, topographic and canopy closure effects, coupled with the low resolution of the earlier sensors, have severely limited the usefulness of satellite imagery (Good & Hibberd, unpubl.). The availability of higher resolution imagery offering a wider and more definitive spectrum of wavebands, and of more accessible computer databases containing faunal distribution and habitat information, provide new opportunities to refine methods of mapping forest vegetation and fauna habitat.

The contribution of Thematic Mapper (TM) data

Until late 1986 the LANDSAT satellites only provided MSS (multispectral scanner) data with four spectral bands and a resolution of 80 m pixels. However, since then, Australia has been able to receive data from the TM (Thematic Mapper) sensor. Despite its considerable cost (up to $5 000 per 185 km wide scene for a computer compatible tape), TM has become used increasingly for vegetation mapping because of greater resolution (30 m pixels) than the MSS sensor and the provision of greater range of spectral wavebands (seven as compared to four for MSS).

The French SPOT satellite provides 20 m pixels in its MSS sensor (and 10 m pixels in panchromatic mode) but the bandwidths would appear to be less useful for native forest vegetation mapping. Additionally, Australia does not receive continual transmissions from the spacecraft platform and specific requests for data have to be made in advance to France.

The research described in this chapter evaluates Landsat TM spectral bands, and combinations of bands, for the classification of eastern Australian forests and woodlands.

Fauna distribution and modelling

The need for fauna prediction modelling

In order to maximize opportunities for maintaining ecological diversity in the remaining wildlife populations, ecologists need to redress the paucity of information on fauna species distributions and their habitat requirements on a regional basis. Predictive tools which provide measures of change in structural vegetation classes and wildlife populations as a result of climatic or other disturbances could prove valuable keys to successful resource management.

Various workshops of the Australian Mammal Society between 1985 and 1988 have emphatically supported the value of large scale faunal databases, particularly those linked to GIS and modelling systems for correlative and predictive work for environmental management of specific areas. The creation of a potentially Australia-wide system that can be easily accessed and updated by land managers (along the lines of research undertaken at the Arthur Rhylah Institute in Victoria (Emison *et al.* 1984; Macfarlane & Chesterfield 1984) has significant implications for environmental policy making, landuse planning and resource management.

Further research in this field will need to address the formulation of models describing the geographic distribution of fauna species by analysing relationships with their environments along resource gradients. This can be achieved either as single species habitat suitability models (US Fish and Wildlife Service 1981) or community level models [cf. Salwasser *et al.* WHR models (1980)].

An important component of such work, vegetation gradient modelling, has been undertaken in Australia by Austin *et al.* (1983) who analysed eucalypt distributions in south-east New South Wales based on a range of environmental variables; while researchers at the Australian National University Research School of Biological Studies have been evaluating the contribution of expert systems to the modelling of vegetation dynamics (Noble 1987). Predictive fauna modelling in Australia has been largely confined to BIOCLIM (Busby 1985; Nix 1986; Busby 1986), the Resource Mapping System developed by Ferrier (1987), and to research workers at CSIRO Division of Wildlife and Ecology (Austin & Braithwaite, pers. comm.). Figure 2 depicts the BIOCLIM predictive model for *Potorous longipes* which was instrumental in discovering the first evidence of this species in New South Wales. Casperson (1984) in South Australia has undertaken some development work for fauna species on the HABITAT model initiated by Thomas (1979) in the USA. We believe that this approach deserves wider evaluation by the ecological community.

The HABITAT model develops life-form assignments for each species based upon their utilization of habitat niches for feeding and breeding and enables probabilities of association of each life-form group with particular habitat components to be predicted. Other workers in the USA have developed forest-systems dynamics simulators such as Boyce's DYNAST (1980) and Smith *et al.* FORHAB (1981) which combine resource models with models of natural succession. Christensen and Davis (1986) clearly state that habibat suitability forecasts require an understanding of habitat-vegetation relationships, and of vegetation change through time. Taylor and Friend (1984) have recommended the use of stratified random sampling to ensure that biological relevance is not ignored in survey. Similarly, James (1971) coined the term "habitat gestalt" to refer to how an organism perceives its environment. Our models need to ensure that the modellers' "gestalt" is as close to the organism's as possible, that is, we need to describe habitat from the organism's point of view.

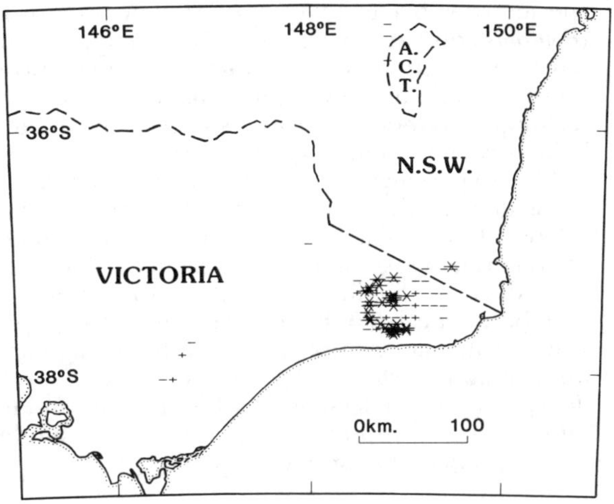

FIG. 2. *Potorous longipes* site locations * and predicted distribution, suitable climates +, and marginal climates – (after Busby 1988).

Future habitat mapping research

The clearing, modification of and continuing pressure upon coastal eucalypt forests has resulted in dramatic reductions in range and abundance of native fauna, especially for ground dwelling mammals within a critical weight/range (Recher *et al.* 1987; Burbidge & Mackenzie 1989). We suggest that priority in research should be given to this species assemblage. Those species such as potoroos (*Potorous* spp.) and bandicoots (*Isoodon* spp.) which have restricted habitat requirements may prove more useful in developing habitat modelling techniques rather than those species with broader habitat requirements e.g., quolls (*Dasyurus* spp.).

The experience of the US Forest Service in constructing the Californian Wildlife Habitat relationship database, which contains information on the relative suitability of various habitats for reproduction, feeding and cover for some 600 vertebrate species (Ohmann & Meyer 1987) provides a valuable example of a wildlife management information system. Approaches need to be developed for assessing wildlife habitat linking sample-based inventory work with fauna models within predefined confidence levels.

The Nadgee project — preliminary study

The study area

The initial work reported on in this chapter was carried out in Nadgee Nature Reserve in far south-eastern New South Wales. Nadgee Nature Reserve is a wilderness area of about 17 000 ha located on the New South Wales South Coast adjoining Croajingalong National Park in Victoria (37°25′S; 149°52′E) (Fig. 3). It stretches about 22 km north from Cape Howe, the most south-easterly point of Australia, and 6 to 8 km inland from the sea reaching elevations of up to 600 m in the Coastal Range. The climate is mild to temperate marine with temperatures 14-23° in January and 7-14° in July and an average rainfall of 972 mm.

Nadgee Nature Reserve was chosen because it has been subject to a wide variety of studies, including vegetation mapping by Groves and Totterdell in 1971, Scobie and Fox in 1972, and Gilmour in 1983 (all unpublished); and research on the long-term impact of fire on wildlife (Newsome *et al.* 1975; Posamentier *et al.* 1981). In particular, the effect of fire upon birds (Fox 1978; Recher 1981) and small mammals (Recher *et al.* 1975) and the subsequent colonization of heaths by mammals after fire (Catling & Newsome 1980; Catling 1986) have been extensively studied. These studies, together with resource information collected by the National Parks and Wildlife Service for fire management purposes, have yielded extensive data on mammal distribution and habitat. Much of the information for Nadgee Nature Reserve is digitized and suitable for use in a geographic information system.

Open forests of *Eucalyptus sieberi, E. smithii, E. gummifera, E. longifolia* and *Angophera floribunda* predominate in both the coastal and upland areas, while tall open forests *(E. cypellocarpa, E. smithii)* occur in the wetter areas along watercourses and in gullies. Some of the more sheltered gullies contain

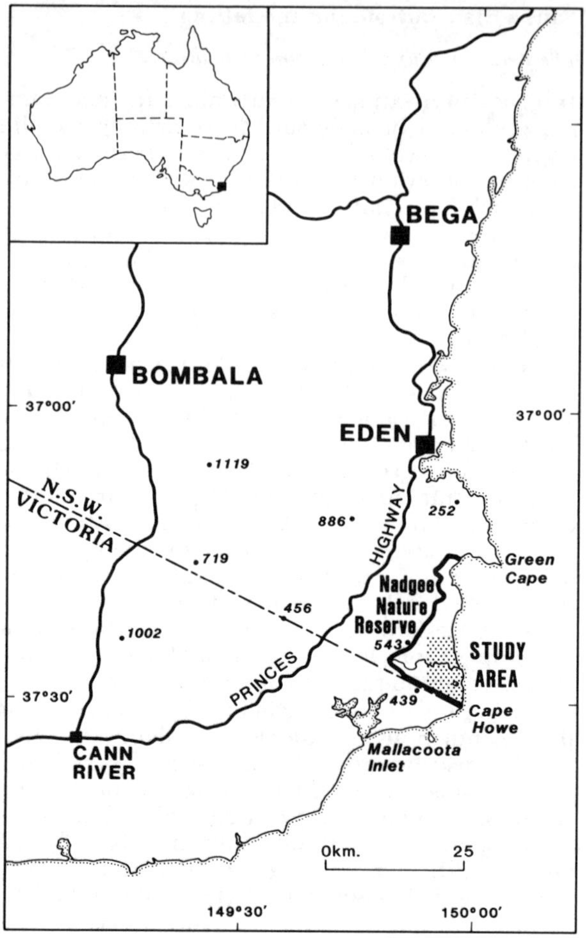

FIG. 3. Location map for study area.

remnants of closed forest. There are major heathland areas behind the 20-50 m high sea cliffs and larger more floristically complex inland heaths (Newsome & Catling 1979) on the plateaux of the Coastal Range. Closed scrubs are found along the dunes and in the swales behind the beaches.

The research of Newsome and Catling (1979) and Recher (1981) has shown that the vertebrate fauna in Nadgee Nature Reserve is closely associated with forested environments, often with heathy understoreys or with grassland ecotones. Their suggestion that habitat mapping seemed more satisfactory in predicting mammal species diversity than vegetation structure mapping alone led us to examine the applicability of integrating satellite imagery and GIS information to improve our ability to predict species' occurrence.

Analysis of Thematic Mapper data

Initial work for this study was carried out to evaluate the suitability of satellite imagery and, in particular the contribution of the various spectral bands in TM data, for mapping coastal forest and heathland communities.

Field inventory. In order to be able to correlate habitat data with categories delineated on satellite imagery, detailed field work was carried out in April and September 1988. Study sites of approximately 90 m×90 m in size were selected within each vegetation type as differentiated from aerial photograph interpretation. Percentage cover and full floristic data for each stratum within each of the study sites were recorded using the point quadrat method.

Results. Analysis of the Thematic Mapper image was undertaken on a MicroBRIAN image processing system and revealed the following.

1. While reflectances in the visible Bands 1-3 are highly correlated for most vegetation types, Band 3 seems preferable since it is less affected by atmospheric scattering.
2. The near infra-red band (Band 4) and middle infra-red bands (Bands 5 and 7) appear extremely useful in discriminating between forest vegetation types (closed forest/open forest/woodland) and heath vegetation types (heathland/sedgeland). While the forested areas have higher reflectances in the near infra-red (Band 4), the heath vegetation has higher reflectances in Bands 5 and 7.
3. Those closed forest areas which are dominated by temperate rainforest species have a higher Band 4 reflectance than any of the other eucalypt dominated forests or woodlands.
4. As the mid infra-red bands (5 and 7) are sensitive to the water content of plants, these bands were also useful in discriminating between sedgelands, *Casuarina* heath, and graminoid-sedge swamps which become waterlogged after heavy rain.
5. All the infra-red bands (4, 5 and 7) proved useful in distinguishing between woodlands with shrub understoreys and those with grass and sedge understoreys; the latter giving higher reflectances in Bands 5 and 7.
6. The near infra-red Band 4 and the mid infra-red Band 5 depicted good discrimination between tall open forests and open forests. Classification of tall open forest proved difficult at times due to shadow effects reducing the radiance values of Band 4.
7. Band 5 also exhibited a tendency to be influenced by the amount and type of exposed soil. Lighter coloured soils appeared to reflect more highly in Band 5 than darker soils, and reflectance decreased with increasing ground cover.

Utilizing the above findings, preliminary results indicate that six heathland vegetation types (as compared to the seven mapped by Groves & Totterdell 1971, unpub.) can be mapped from TM image data on Nadgee Moor (which has a complex assemblage of heath communities). Within the forest areas, results show that Gilmour's detailed level of forest type mapping cannot, at present, be matched by remotely sensed imagery, but the forests can at least be mapped to a structural level of classification. Figure 4 depicts the final classification achieved with Thematic Mapper data.

Implications. The implication of these results is that it is now possible to use satellite imagery to map vegetation structure to a greater degree of complexity than as described by Specht *et al.* (1974). Although the technology therefore would seem to have considerable potential for providing information essential for habitat mapping, a problem which still needs to be overcome is that of scale of resolution. However, the increasing resolution of satellite sensors (now down to 10 m pixels) may provide adequate habitat mapping abilities to be integrated with other geographic datasets and field-derived species habitat requirements — at least on a broad biogeographic scale.

Future directions

Habitat mapping

The physical environment data from the GIS will be integrated as additional bands in the image in order to improve our ability to map vegetation patterns and consequently to predict and map fauna habitats. Further refinement of the vegetation/habitat classification will then be undertaken. The integrated vegetation and fauna models will be run using the final classification output from the image processing. Figure 5 shows the flow and integration of information required to produce such models. At this stage the applicability of knowledge-based information systems (after Noble & Slatyer 1980; Marcot 1986) to wildlife habitat-succession models will be evaluated.

The use of the above data and techniques should enable a set of habitat maps to be drawn for south-eastern Australia that will provide a basis for the description of mammal distributions. The calculation of confidence limits for the prediction of individual species should give an indication of their likelihood of occurrence in each specific mapped habitat class.

The remotely sensed data could then be used as a medium to frequently update and monitor environmental changes.

Testing the models

There is a need for Australian wildlife ecologists to collect more information on the distribution of and habitat preferences of native fauna to aid testing of these models. Such research could draw upon the existing work of, amongst others, Braithwaite and Gullan (1978), Barnett *et al.* (1978), Fox and Fox (1981), and Fox (1984).

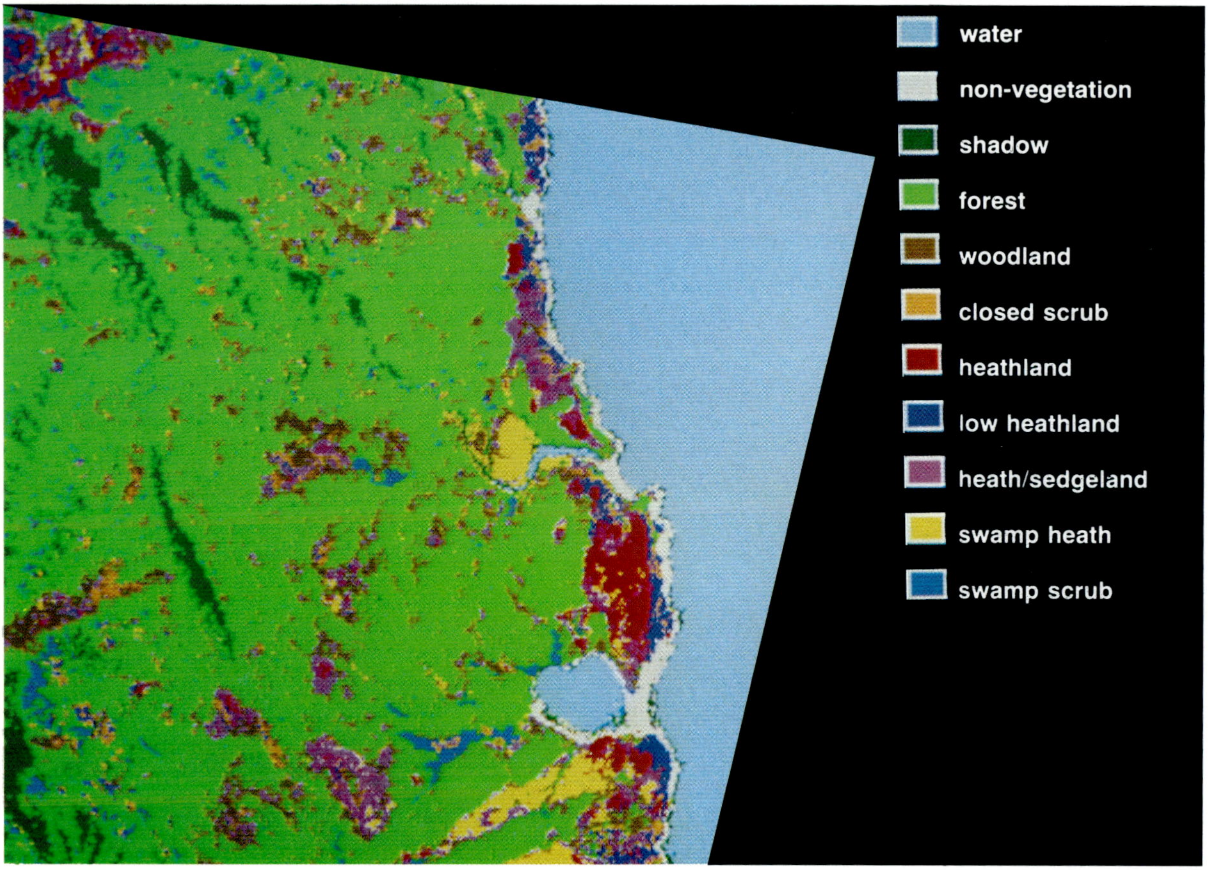

FIG. 4. Vegetation classification of study area from Thematic Mapper LANDSAT imagery.

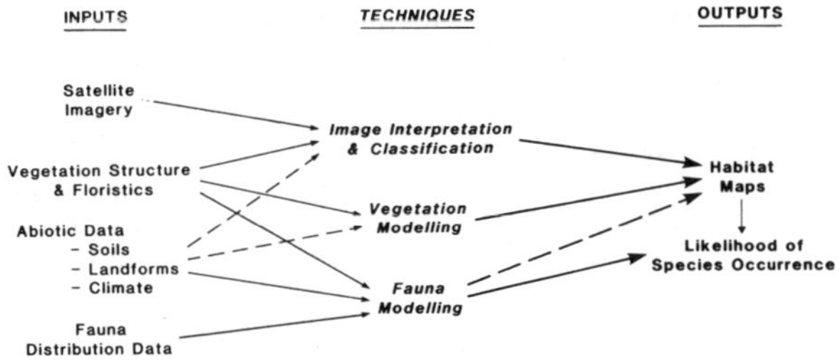

FIG. 5. Integration of information for fauna prediction models.

Re-evaluation of adequacy of reserve systems

If reasonably accurate predictions of faunal distributions can be achieved, together with vegetation mapping, landuse planners and decision-makers will be in a much stronger position in the future to make informed decisions on the representativeness of reserve systems. For example, the proposed expansion of intensive forestry practices to hithertoo lightly modified native forests in south-eastern Australia requires urgent decisions relating to the setting aside of a well-sampled reserve system and improved forest management practices if the loss of important heritage values is not to continue.

An important facet of future management of conservation reserve systems needs to be identified here. Australian land management agencies appear to have given little attention to date to the implications of potential climatic change upon conservation reserves (Chittleborough 1985). While the vast majority of Australian plant and animal species would probably easily cope with small shifts in mean annual temperatures and rainfall, those species with highly limited distributions may well be seriously affected.

If one heeds the warnings of some climatologists (Pearman 1989; Nix, pers. comm.) that an important characteristic of climatic change could be major shifts in the seasonality (as well as absolute amounts) of precipitation, some faunal habitats could well shift dramatically. Such impacts would be heightened by the fragmentation of potential habitats that has been caused by the modification and utilization of natural ecosystems in Australia since European settlement. Unless reserve systems are well designed and incorporate long distance wildlife corridors, there is a very real likelihood of local extinctions.

It is hoped that the techniques developed in this research will assist in assessing the implications of a range of climatic scenarios. The predictive faunal models would then be run iteratively to determine the likely shifts in habitat availability for selected fauna species/assemblages.

Conclusions

Even after 200 years of utilization and degradation, we believe that techniques are becoming available to assist us in improving our management of Australia's natural biotic resources, particularly through a better knowledge and understanding of species distributions and habitat requirements. There should thus be opportunities for halting the loss of further species and, perhaps, even for reconstruction of some severely degraded and species-poor communities.

We would like to reaffirm some of Recher and Lim's (this volume) comments. *We need to act now. We do not have time, or the money, for many of the traditional survey and research techniques. We need accurate baseline population information. We need to improve our management of ecosystems outside the reserve system. We need to urgently consider fauna more adequately than in the past. We need to utilize new technologies which will permit cost effective and accurate data sets to be examined simultaneously.*

From our preliminary research, we believe that geographic environmental data, natural systems models, and remotely sensed information can be integrated to provide predictions of occurrence and abundance within predefined confidence limits for some fauna species. Results gained from these implications are likely to generate further questions on the implications of climatic perturbations upon faunal habitat distributions, and these questions will drive further development of this project.

Most importantly, the information so derived must be communicated to politicians, landuse decision-makers and land managers to minimize the chance that the next 30 years will see the possible extinction of a substantial proportion of our indigenous fauna (particularly small mammals) .

Acknowledgements

We thank the National Parks and Wildlife Service for the provision of facilities and the use of their helicopter in this research and to the Geography Department of Wollongong University for the use of image processing facilities. We are also grateful to Dr Harry Recher, Drs Barry and Marilyn Fox, and to Angus Hopkins for valued critical comments on drafts of this manuscript. We are particularly indebted to Ms Liz Dovey, of the New South Wales National Parks and Wildlife Service, for permission to reproduce her work in Figures 1 and 2, and for her inspiration for this research programme.

References

Australian Heritage Commission (1985) Australia's National Estate: the role of the Commonwealth. Special Australian Heritage Publication Series. No. 1. Australian Government Publishing Service, Canberra.

Austin M. P., Cunningham R. B. & Good R. B. (1983) Altitudinal distribution of several eucalypt species in relation to other environmental factors in southern New South Wales. *Aust. J. Ecol.* **8,** 169-80.

Barnett J. L., Howe R. A. & Humphries W. F. (1978) The use of habitat components by small mammals in eastern Australia. *Aust. J. Ecol.* **3,** 277-85.

Benson J. (1985) Review of the conservation status of vegetation communities in New South Wales. National Parks and Wildlife Service, Sydney.

Boyce S. G. (1980) Management of forests for optimal benefits (DYNAST-OB). United States Department of Agriculture, Forest Service Research Paper SE-204.

Braithwaite L. W. (1983) Studies on the arboreal marsupial fauna of eucalypt forests being harvested for woodpulp at Eden, New South Wales. I. The species and distribution of animals. *Aust. Wildl. Res.* **10,** 219-29.

Braithwaite L. W., Clayton M., Maclean L. & Parker B. S. (1984) Vertebrate fauna of a 144 ha water catchment within eucalypt forest being harvested for woodpulp at Eden, south-east New South Wales. CSIRO Div. Wildlife & Rangelands Res. Tech. Paper No. 35.

Braithwaite R. W. & Gullan P. K. (1978) Habitat selection by small mammals in a Victorial heathland. *Aust. J. Ecol.* **3,** 109-27.

Burbidge A. A. & Mackenzie N. L. (1989) Patterns in the modern decline of Western Australian vertebrate fauna: causes and conservation implications. *Biol. Conserv.* **50,** 143-90.

Busby J. R. (1985) Bioclimate prediction system. Users manual version 1.2. Bureau of Flora and Fauna, Canberra.

Busby J. R. (1986) A biogeoclimatic analysis of *Nothofagus cunninghamii* (Hook.) Oerst. in S.E. Australia. *Aust. J. Ecol.* **11,** 1-7.

Busby J. R. (1988) Potential impacts of climate change on Australia's flora and fauna. In: *Greenhouse planning for climate change* (ed. G. I. Pearman). CSIRO, Melbourne.

Casperson K. D. (1984) Biological survey of the south-east coast of South Australia. In: *Survey Methods for Nature Conservation* (eds K. Myers, C. R. Margules, & I. Musto) pp. 319-30. CSIRO Division of Water and Land Resources, Canberra.

Catling P. C. (1986) *Rattus lutreolus,* colonizer of heathland after fire in the absence of *Pseudomys* species? *Aust. Wildl. Res.* **13,** 127-39.

Catling P. C. & Newsome A. E. (1980) Response of the Australian vertebrate fauna to fire: an evolutionary approach. In: *Fire and the Australian biota* (eds R. H. Groves, I. R. Noble & A. M. Gill). Australian Academy of Science, Canberra.

Chittleborough R. G. (1985) Towards a State conservation strategy: I Planning to meet climatic changes. Western Australian Department of Conservation and Environment. Bulletin No. 207, 1-8.

Christensen N. L. & Davis L. S. (1986) Introduction: Linking wildlife models with models of vegetation succession. In: *Wildlife 2000 — Modelling Habitat Relationships of Terrestrial Vertebrates* (eds J. Verner, M. L. Morrison & C. J. Ralph) pp. 337-8. University of Wisconsin Press.

Diamond J. M. (1975) The island dilemma: lessons of modern biogeographic studies for the design of natural reserves. *Biol. Conserv.* **7,** 129-46.

Emison W. B., Menkhorst P. W., Mansberg I. M., Norman F. I., Robertson P., Corrick A. H. & Alderson J. A. (1984) Wildlife Surveys conducted by the Fisheries and Wildlife Division in Victoria 1972-1983. In: *Survey Methods for Nature Conservation* (eds K. Myers, C. R. Margules & I. Musto) pp. 166-94. CSIRO Division of Water and Land Resources, Canberra.

Ferrier S. (1987) Resource Mapping System. Users Manual. National Parks and Wildlife Service of New South Wales., Sydney.

Fox A. M. (1978) The '72 fire of Nadgee Nature Reserve. *Parks and Wildlife* **2,** 5-24.

Fox B. J. (1984) Small scale patchiness and its influence on our perception of animal species' habitat requirements. In: *Survey Methods for Nature Conservation* (eds K. Myers, C. R. Margules & I. Musto) pp. 162-78. CSIRO Division of Water and Land Resources, Canberra.

Fox B. J. & Fox M. D. (1981) A comparison of vegetation descriptors of small mammal habitat preference. In: *Vegetation Classification in Australia* (eds A. N. Gillison & D. J. Anderson). CSIRO/Australian National University Press, Canberra.

Graetz R. D., Pech R. P., Hindley N. L. & Gentle M. R. (1984) The application of LANDSAT to rangeland management: LIBRIS five years on. *Proc. 3rd Aust. Remote Sensing Conference* pp. 322-31. Gold Coast.

Hall K. A. & McCloy K. R. (1987) Tree and shrub cover using LANDSAT MSS digital data. In: *Proc. 4th Australasian Remote Sensing Conference,* pp. 325-32, Adelaide.

Hibberd J. K. (ed.) (1986) Eden (New South Wales) export woodchip operation 1989-2009. National Parks and Wildlife Service of New South Wales, Sydney.

Hitchcock P. P. (1984) The survey and allocation of land for nature conservation in New South Wales. In: *Survey Methods for Nature Conservation* (eds K. Myers, C. R. Margules & I. Musto) pp. 220-47. CSIRO Division of Water Resources, Canberra.

Hopkins A. J. M. & Saunders D. A. (1987) Ecological studies as the basis for management. In: *Nature Conservation: the Role of Remnants of Native Vegetation* (eds D. A. Saunders, G. W. Arnold, A. A. Burbidge & A. J. M. Hopkins) pp. 15-28. Surrey Beatty & Sons, Chipping Norton.

James F. C. (1971) Ordinations of habitat relationships among breeding birds. *Wilson Bull.* **83,** 215-36.

Lunney D. & Barker J. (1986) Mammals of the coastal forests near Bega, New South Wales. I. Survey. *Aust. Zoologist* **23,** 19-28.

Macfarlane M. A. & Chesterfield E. A. (1984) Methods and principal findings of a survey of flora and fauna in a high elevation mixed species eucalypt forest in eastern Australia. In: *Survey Methods for Nature Conservation* (eds K. Myers, C. R. Margules & I. Musto) pp. 195-205. CSIRO Division of Water and Land Resources, Canberra.

Marcot B. G. (1986) Use of expert systems in wildlife-habitat modelling. In: *Wildlife 2000 — Modelling Habitat Relationships of Terrestrial Vertebrates* (eds J. Werner, M. L. Morrison & C. J. Ralph) pp. 145-50. University of Wisconsin Press.

Milne A. K. & Zhou Q. (1987) Manipulating geographic and remotely sensed data for land cover mapping in arid Australia. *Proc. 4th Aust. Remote Sensing Conference* pp. 544-55. Adelaide.

Newsome A. E. & Catling P. C. (1979) Habitat preferences of mammals inhabiting heathlands of warm temperate coastal, montane and alpine regions of south-eastern Australia. In: *Ecosystems of the world 9A: Heathlands and related shrublands* (ed. R. L. Specht). Elsevier Scientific, Amsterdam.

Newsome A. E., McIlroy J. & Catling P. (1975) The effects of an extensive wildfire on populations of twenty ground vertebrates in south-eastern Australia. *Proc. Ecol. Soc. Aust.* **9,** 107-23.

Nix H. A. (1986) A biogeographic analysis of Australian elapid snakes. In: *Atlas of Elapid Snakes of Australia* (ed. R. Longmore). Australian Flora and Fauna Series No. 7, Australian Government Publishing Service.

Noble I. R. (1987) The role of expert systems in vegetation science. *Vegetatio* **69,** 115-21.

Noble I. R. & Slatyer R. O. (1980) The use of vital attributes to predict successional changes in plant communities subject to recent disturbances. *Vegetatio* **43,** 5-21.

Ohmann J. L. & Meyer K. E. (1987) Wildlife habitats of California's hardwood forests — linking extensive inventory data with habitat models. Proceedings of Symposium: *Multiple Use Management of California's Hardwood Resources.* United States Department of Agriculture General Technical Report PSW-100.

Ovington J. D. (1978) Australia's Endangered Species — mammals, birds and reptiles. Cassell, Australia.

Pearman G. I. (ed.) (1989) Greenhouse: planning for climatic change. CSIRO, Melbourne.

Posamentier H. G., Clark S. S., Hain D. L. & Recher H. F. (1981) Succession following wildfire in coastal heathland (Nadgee Nature Reserve New South Wales). *Aust. J. Ecol.* **6,** 165-75.

Recher H. F. (1981) Bird communities of heath and their management and conservation requirements. In: *Heaths in New South Wales* (ed. B. Haigh). New South Wales National Parks and Wildlife Service, Sydney.

Recher H. F., Lunney D. & Posamentier H. (1975a) A grand natural experiment: the Nadgee Wildfire. *Aust. Nat. Hist.* **18,** 154-63.

Recher H. F., Clark S. S. & Milledge D. (1975b) An assessment of the potential impact of the woodchip industry on ecosystems and wildlife in south-eastern Australia. Dept. Env. Studies. Aust. Mus. Tech Report 75/1.

Recher H. F., Lunney D. & Dunn I. (1979) *A natural legacy: ecology in Australia.* Pergamon Press, Rushcutters Bay, New South Wales.

Recher H. F., Rohan-Jones W. & Smith P. (1980) Effects of the Eden woodchip industry on terrestrial vertebrates with recommendations for management. Forestry Commission of New South Wales Research Note No. 42.

Recher H. F., Shields J., Kavanagh R. & Webb G. (1987) Retaining remnant mature forest for nature conservation at Eden, New South Wales: a review of theory and practice. In: *Nature Conservation: the Role of Remnants of Native Vegetation* (eds D. A. Saunders, G. W. Arnold, A. A. Burbidge & A. J. M. Hopkins) pp. 177-94. Surrey Beatty & Sons, Chipping Norton.

Salwasser H., Capp J. C., Black H. & Hurley J. F. (1980) The California wildlife habitat relationships program: an overview. In: *Management of Western Forests and Grasslands for Non-game Birds* (ed. R. M. DeGraff). Proceedings of Workshop. United States Department of Agriculture, Forest Service, General Technical Report INT-86.

Saunders D. A., Arnold G. W., Burbidge A. A. & Hopkins A. J. M. (eds) (1987) *Nature Conservation: the Role of Remnants of Native Vegetation.* Surrey Beatty & Sons, Chipping Norton.

Slatyer R. O. (1975) Ecological Reserves: size, structure and management. In: *A National System of Ecological reserves in Australia* pp. 22-38. Australian Academy of Science, Report No. 19. Canberra.

Smith T. M., Schugart H. H. & West D. C. (1981) FORHAB: A forest simulation model to predict habitat structure for non-game bird species. In: *The Use of Multivariate Statistics in Studies of Wildlife Habitat* (ed. D. E. Capen). United States Department of Agriculture, Forest Service, General Technical Report, RM-87.

Soulé M. E. (1987) *Viable Populations for Conservation.* Cambridge University Press.

Specht R. L., Roe E. M. & Boughton V. H. (1974) Conservation of major plant communities in Australia and Papua New Guinea. *Aust. J. Bot.* Supplement No. 7.

Taylor J. A. & Friend G. R. (1984) Sampling strategies for fauna surveys. In: *Survey Methods for Nature Conservation* (eds K. Myers, C. R. Margules & I. Musto) pp. 179-92. CSIRO Division of Water and Land Resources, Canberra.

Thomas J. W. (ed.) (1979) Wildlife habitats in managed forest: the Blue Mountains of Oregon and Washington. United States Department of Agriculture, Forest Service, Agriculture Handbook No. 553.

United States Fish and Wildlife Service (1981) Standards for the development of Suitability Index Models. Ecological Services Manual 103. US Department of Interior, Fish and Wildlife Service, Division of Ecological Services. Government Printing Office, Washington D.C.

Webb G. A. (1981) Geographical distribution of reptiles and amphibians in the southern Eden forestry region. Forestry Commission of New South Wales unpubl. rep. no. 783, pp. 1-115.

Woodgate P. W., Parkes D. M. & Roberts J. A. (1987) Vegetation mapping in Victoria's Big Desert using LANDSAT MSS imagery. In: *Proc. 4th Australasian Remote Sensing Conference* pp. 305-24, Adelaide.

Proc. Ecol. Soc. Aust. 1990 — 16:459-465

Reconstruction of South Australia's arid lands: the conservation option

B. COHEN*

National Parks and Wildlife Service, P.O. Box 902, Norwood, S.A. 5067

Abstract

Current Government policy in South Australia aims to create the world's largest network of arid lands national parks. In 1988, land holdings by the National Parks and Wildlife Service in the arid zone totalled 5 913 081 ha in 21 parks. A further 7.5 million ha are proposed as additions to the system.

These impressive achievements in arid land conservation are not as straight forward as they might seem. There is a trend away from the "traditional" national parks model and a move towards multiple use of reserves, albeit with conservation as the stated prime objective. These other uses of reserves include mineral and hydrocarbon exploration, mining, grazing, tourism developments and management for Aboriginal use. This chapter documents the expansion in the arid parks system of South Australia, discusses new trends in park management and speculates on the future implications of these trends for the conservation of South Australia's arid lands.

Introduction

"Australian ecosystems: 200 years of utilization, degradation and reconstruction". Within this framework of land use in Australia, where does management for conservation stand? Conservation reserves may be seen as fitting into the reconstruction phase (or at least, in a saving-from-degradation phase). Reserves are proclaimed over land which has either not been used by non-Aboriginal Australians, or over land which has been used and is subsequently reserved in the name of nature conservation. If conservation reserves represent a form of reconstruction, then South Australia has made spectacular gains in land reconstruction in the State's arid zone over the past six years.

This chapter looks at the reasons behind the expansion in South Australia's reserves system in the 1980s, examines recent trends in management of these reserves, and poses the question of whether conservation reserves in South Australia's arid zone really fit quite so neatly into the reconstruction phase of land management.

Definition of South Australia's arid zone

Definitions of aridity are often linked to growth indices and are devised to be guides to the limits of successful agricultural enterprise or pasture improvement (Squires 1981). For the purposes of this chapter, the precise limits of aridity are less critical. South Australians often use Goyder's line which was drawn across the State to designate the southern extent of the drought of 1865-66 by the Surveyor-General, G.W. Goyder (Fig. 1). Goyder's line is still used generally to demarcate the northerly limit of reliable agriculture in South Australia (Williams 1977). Today there is some agricultural activity north of Goyder's line, although Goyder was correct in stating that agriculture could not be guaranteed to be successful north of his line. In this chapter, I have extended the limit of the arid zone northwards to make for a more realistic boundary (Fig. 1). This places the boundary at the southern edge of the State's pastoral country (South Australian Department of Lands 1986). In effect whether the boundary of the arid zone is pegged at Goyder's line or whether it is drawn around the edge of the pastoral country does not materially affect this appraisal of arid zone parks and their conservation in South Australia.

Political background

It is the policy of the State's Labor Party to work towards creating the world's largest network of arid lands national parks (Australian Labor Party n.d.). Since the election of a Labor Government in South

* The opinions expressed in this chapter are those of the author, and are not intended to represent the views of the National Parks and Wildlife Service or the Minister for Environment and Planning.

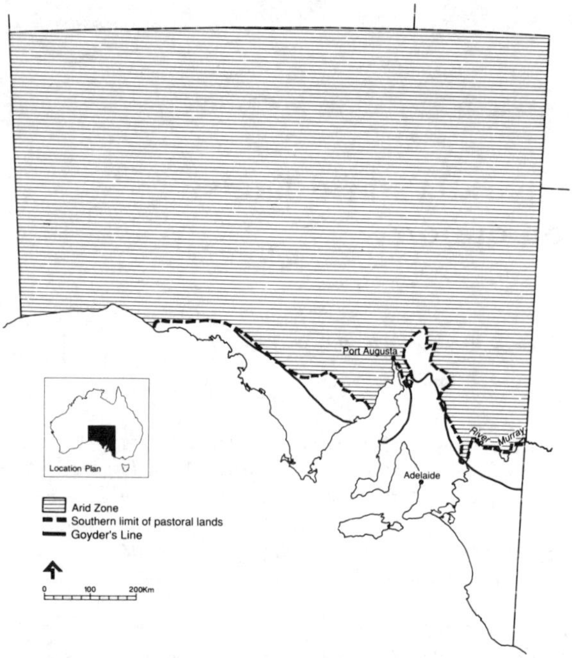

FIG. 1. Extent of South Australia's Arid Zone.

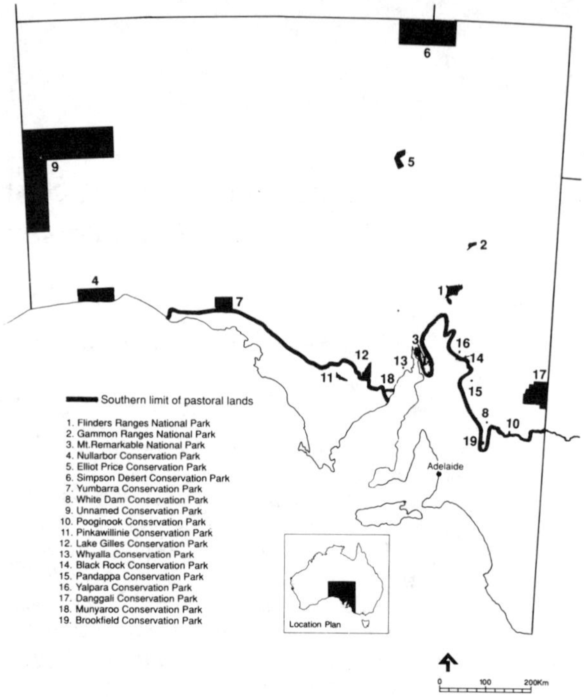

FIG. 2. National Parks and Wildlife Act Arid Zones Reserves 1982.

Australia in November 1982, this policy has provided the impetus for the establishment of reserves in several locations which have long been talked about or wished for within the conservation movement. This has led to a substantial expansion in the reserves system in South Australia's arid zone (Figs 2 and 3). In 1988, total land holdings under the National Parks and Wildlife Act in the arid zone reached 5 913 081 ha. Of this, 2 248 797 ha have been added since 1982 in the form of three new parks and additions to three others. Consequently, since 1982, there has been a 61.4% increase in conservation reserves in the arid zone. The picture looks even more impressive if we project ahead and include those areas proposed for inclusion in the reserves system (Fig. 4). With the addition of these areas, a total of 13 476 946 ha of the arid zone will be managed under the National Parks and Wildlife Act, and arid zone parks will then constitute 13.7% of South Australia's land mass, or 16.0% of the State's arid zone (Fig. 5).

These figures demonstrate that conservation has become a major land use in South Australia's arid zone and that the National Parks and Wildlife Service has become a major land manager. There is no doubt that the 1980s have brought great gains in the reservation of land for conservation in South Australia's arid zone. However, the nature of these land tenure decisions and the management of the land affected by them deserve further investigation.

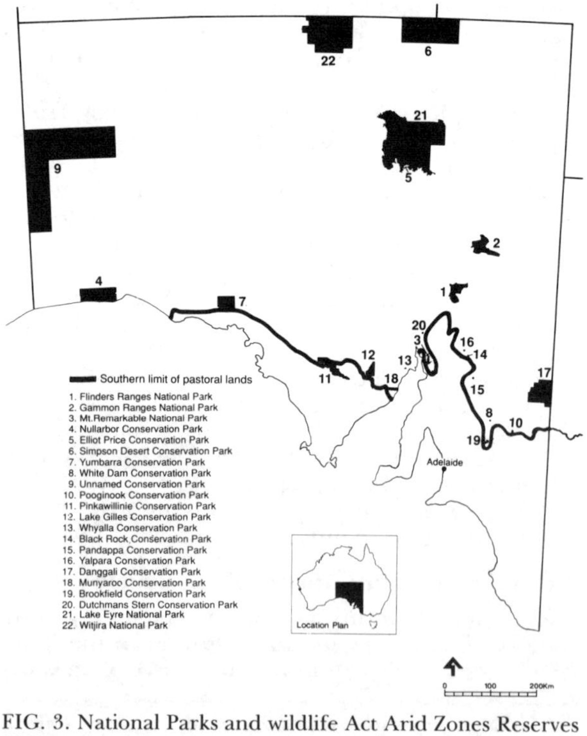

FIG. 3. National Parks and wildlife Act Arid Zones Reserves 1988.

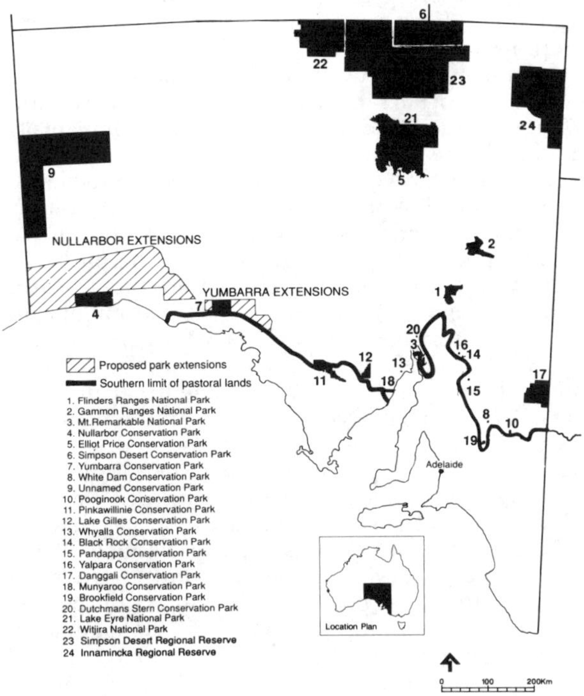

FIG. 4. National Parks and Wildlife Act Arid Zones Reserves. Projected to 1990.

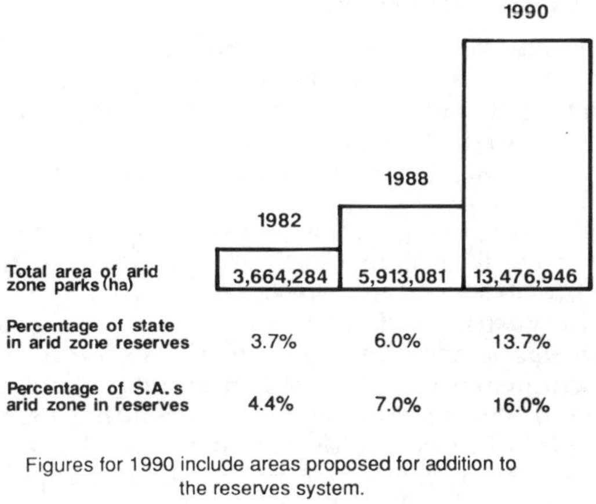

FIG. 5. Growth in South Australia's Arid Zone Reserves System 1982-1988-1990.

The traditional national parks model

The International Union for the Conservation of Nature (IUCN) definition of a national park which was drawn up in 1969 considers a national park to be a relatively large area where ecosystems are not materially altered by human exploitation and occupation, where steps are taken to prevent or eliminate exploitation or occupation and where visitors are allowed to enter under special conditions "for inspirational, educative, cultural and recreative purposes" (McMichael 1980). In Australia, the Council of Nature Conservation Ministers (CONCOM) has based its own definition on the IUCN definition and regards a national park as a

> "relatively large area set aside for its features of predominantly unspoiled natural landscape, flora and fauna, permanently dedicated for public enjoyment, education and inspiration, and protected from all interference other than essential management practices, so that its natural attributes are preserved." (McMichael 1980).

In general, national parks or conservation reserves are seen to be areas where natural processes can take place without undue human interference. They are places where management aims to achieve the highest degree of "naturalness", and, where possible, to maintain the highest level of wilderness values. They are seen to be places where any developments which might take place are small-scale, unobtrusive and relate to, and are compatible with, the perception of the park as a natural, untramelled place, where nature is paramount and substantial human activity is incompatible with the prime purpose of such a reserve (Figgis & Donald 1986). Such facilities as are provided relate to interpretation of the park, or are for the comfort of visitors, or for the protection of the park's assets. In short, a park or reserve is seen to be a place where nature rules supreme and humans are merely guests in the landscape.

It is these perceptions of a national park which I refer to as the traditional model and which, in the case of South Australia, refers to both national parks and conservation parks. In 1988, the foundations of this model were looking decidedly shaky and the very basis of the model was being challenged in South Australia. It was suggested that this idealistic model was no longer viable within the economic and political context of Australia in the 1980s. This process may have taken place somewhat earlier and somewhat faster in other States but, in South Australia it was essentially a phenomenon of the 1980s. South Australia's arid zone parks are as much challenged by these changes as other parks but are given special consideration in this chapter because they are the focus of current government policy and because the issues under consideration are most clearly demonstrated in the State's arid zone reserves.

The challenges to the traditional model arise from a shift towards multiple use parks, albeit with conservation as the stated prime use. Other uses are now being superimposed upon land whose prime tenure is designated for conservation.

What are these other uses which look like changing and challenging the traditional national parks model?

Tourism developments

There has been a boost in tourism and the marketing of tourism throughout Australia. There are claimed to be great economic and employment benefits to be derived from tourism, particularly if it is possible to attract overseas tourists with their foreign currency.

The tourism panacea seems to be particularly attractive to a State like South Australia, with its poorly developed industrial base and where previous industrial initiatives, such as the whitegoods, vehicle and shipbuilding industries have not been as enduring as the State's decision-makers had hoped. The tourism industry in South Australia targets two major kinds of tourism: firstly, special events which draw large crowds over a short period of time such as motor racing, arts festivals and wine festivals; and secondly, less intensive tourism which relies on the natural beauty of the landscape and the native wildlife. South Australia is not unduly blessed with places of conventional great beauty and spectacular scenery. We do not have an Ayers Rock or a Great Barrier Reef, rainforests or great free-flowing rivers. However, there are a few areas which are attractive to the tourism industry and overwhelmingly they are located within the State's reserves system. It is not surprising therefore, that tourism developments are proposed for construction in reserves. At the time this chapter was written, two major developments were proposed for parks in the south of the State, and one for an arid zone park.

The proposed arid zone development is to be built in the Flinders Ranges National Park in the vicinity of Wilpena Pound. The development is a resort comprising an international standard hotel, cottages, bungalows, hostel accommodation, a caravan park, camping grounds, shops and a service station. In addition, there will be a swimming pool and tennis courts and more than 50 flats, units and houses for accommodating staff (Michael Williams and Associates 1988; South Australian Department of Environment and Planning 1989). This development will, in part, replace an existing smaller hotel, camping ground and commercial outlet in the park. The new development will be the largest of its kind in South Australia.

A resort of this scale will create some problems such as the demand for large quantities of water in an arid environment, the impacts of large numbers of visitors on the park, management of increased numbers of people generally, meeting interpretation needs and management of the lease conditions. But the resort does solve some management problems, and importantly, is expected to generate a substantial income for the park.

Whatever the benefits of such a resort development may be, it constitutes a major challenge to the traditional national parks model described above. A resort of this nature extends the limits of what might have been seen as appropriate development in a park.

What is proposed is more than a facility to house people in order for them to enjoy, learn and appreciate the values of a scenic and environmentally significant area. The resort is an attraction in its own right and, in providing its own recreation facilities, provides an additional dimension to the park visitor's experience, a dimension which, it can be argued, is unrelated to the prime purpose of the park.

It is not only the impetus from the tourism industry that drives tourism developments in parks. Conservation groups, park rangers, unions, opposition politicians, farmers' organizations and others have, over the years, claimed that national parks are understaffed and underfunded. In line with many other areas of government funding, there has been little improvement in either staffing or funding and indications are that this will not change substantially. It appears, therefore, that the only avenue now open, is for the National Parks and Wildlife Service to become entrepreneurial and to generate its own income, which can then be put back into improving park management. Encouraging tourism developments in parks is one way of doing this.

Mining

The mining vs. conservation battle has been played out in many arenas in Australia. In South Australia, the expansion of the arid zone parks system and the concerted search for hydrocarbons and minerals has inevitably led to a meeting of these two land uses.

Under Section 43 of the National Parks and Wildlife Act, rights to explore and mine in a reserve may be acquired in two ways: either by resolution of both houses of State Parliament, or by proclamation at the time the park is established (South Australia 1972-87). This means that if mining or exploration rights are sought through resolution of Parliament, the matter comes up for debate in a public forum and is therefore subject to the usual lobbying and media publicity. It generally would come under the spotlight in the same way as any other matter of public interest that is debated in Parliament. By taking the route of what are called joint proclamations (that is, allowing for exploration or mining by means of the proclamation which establishes a new park), the glare of the public and political spotlight is reduced. However, joint proclamations do have the advantage of setting out at the time of proclamation the conditions under which exploration and mining may take place and these proclamations usually make provision for environmental protection.

All but one of the arid zone parks which have been proclaimed since the beginning of this decade have been proclaimed with mining and/or exploration rights provided within the proclamation. This means that parks such as Lake Eyre, Witjira, extensions to the Gammon Ranges and Pinkawillinie, as well as the Simpson Desert Conservation Park (previously dedicated with a joint proclamation) have provision for exploration and/or mining rights. To date no mining has taken place in these parks but substantial seismic exploration work has been undertaken.

Grazing

Grazing by domesticated animals has long been regarded as an unacceptable activity in South Australia's parks, and the usual practice has been to remove stock from all newly dedicated parks. In 1987, Section 34a was added to the National Parks and Wildlife Act to create a new category of reserve, known as a Regional Reserve, which allows for the utilization of the natural resources in the reserve in addition to management for nature conservation (South Australia 1972-87). Creation of this new category of reserve was designed to overcome the special problems for conservation in the Innamincka pastoral lease. Innamincka (Fig. 4) contains part of Cooper Creek and the creek flood-out into the Coongie Lakes system. This wetland system is one of South Australia's most significant wetland habitats and drought refuges for waterbirds (Reid & Gillen 1988). Innamincka pastoral lease has for many years been a vital link in a network of cattle grazing properties managed by a large company. And as geological coincidence would have it, the area overlies part of the Cooper Basin — source of South Australia's natural gas supplies.

The cost of purchasing Innamincka for conservation purposes was prohibitive. But to meet the urgent need to protect wetland habitats and to manage the ever growing visitation to the area, as well as the need to accommodate exploration and production activities of the petroleum industry and continued use by the pastoral industry, the concept of a multiple use park came into being. At no real cost to government, the National Parks and Wildlife Service became the lessor through an administrative arrangement with the Department of Lands, and offered a long term lease to the pastoral company.

What came to be known as a Regional Reserve was seen as the only way to accommodate existing extractive and pastoral uses, while providing some protection to the wetland ecosystems of the area. This multiple-use type of reserve has subsequently been used in extensions to the Simpson Desert Conservation Park and is likely to be invoked in proposed extensions to the Nullarbor National Park. How extensively the Regional Reserve concept is applied in the future remains to be seen, but it might turn out to be the only politically acceptable way to introduce some of the elements of conservation management and to provide protection for specific features, while at the same time to continue pastoral and mining activity in the arid zone. It should also be noted that there is nothing in the legislation to prevent a Regional Reserve from being re-dedicated as a National or Conservation Park in the future, if that is the way government chooses to go.

The Regional Reserve concept is clearly at odds with the traditional national parks model as described above. It explicitly legitimizes extractive and other uses in such reserves. Conservation as a form of land management has to accommodate other uses which, in the past, have been seen to be incompatible with the prime purpose of conservation.

Aboriginal involvement in park management

Since the early 1980s South Australia has been working towards involvement of Aboriginal people in the management of parks. There have been two training programmes for Aboriginal rangers which have resulted not only in the employment of Aboriginal rangers but also in the development of closer links between the National Parks and Wildlife Service and those rangers' own communities. In recent times, contact with Aboriginal communities has developed to include broader aspects of land management in areas of high conservation value. The Service is working towards accommodating Aboriginal aspirations for land management and land ownership, within a broader context of management for conservation. In 1988, for the first time, a conference was held to bring together representatives from Aboriginal communities across the State to discuss issues relating to Aboriginal involvement in conservation management (Kean *et al.* 1988).

As part of this programme, South Australia is considering working towards restitution of ownership of the Unnamed Conservation Park to its traditional Aboriginal owners. This will be the first time in South Australia that a National Parks and Wildlife reserve has had its title invested in any body or organization other than the Crown. The details of the return of the land are still to be worked out but will probably take the form of a lease-back arrangement based on the Uluṟu National Park model (Anon. 1986), where title is vested in the traditional owners who have leased the land to the Australian National Parks and Wildlife Service who, in turn, jointly manage the park with the traditional Aboriginal owners. This model of compromise and accommodation appears to be working well at Uluṟu and should be equally successful in the South Australian situation. It is quite likely that joint agreements of various kinds for the management of other lands may follow later.

There is no doubt that Aboriginal perceptions of the land and Aboriginal relationship to the land are very different from the perceptions of the

Anglo-Australian conservationist. Both groups profess a love for the land, but when Aboriginal people talk about caring for the land or "growing up the country" they may be saying something very different about land management (Figgis & Donald 1986; Anon. 1986; Toyne & Vachon 1984). Aboriginal caring for the land is primarily a commitment to maintain the land within an Aboriginal religious framework. While the Anglo-Australian conservationist may also claim a spiritual commitment to the land, the caring is more to do with physical wellbeing and generally is concerned with visual and ecological integrity of the landscape. In some cases, these different views of the land may lead to compatible land management practices and in others, to different practices. For example, we know that Aboriginal use of fire in land management is very different from Anglo-Australian use, or non-use, of fire.

In working towards accommodation with Aboriginal people in the management of land for conservation, there must be a recognition that Aboriginal cultures are not static. Aboriginal people have in the past taken, and will continue to take on those aspects of the dominant culture's systems and technologies which they see as being useful, relevant or attractive, and will mould them to fit into their own cultural framework. They will, like all people, make decisions about their land and their lives which they consider are to their benefit. These decisions may well result in land use decisions which are at variance with the traditional national parks model.

Implicit in vesting title of a park in Aboriginal communities and implicit in establishing a decision-making system with an Aboriginal majority as in the Uluru model, is an acceptance that Aboriginal land management practices in such a park may prevail and that these land management practices may not necessarily conform to traditional Aboriginal practice or to the traditional national parks model.

Also implicit in these decisions is a willingness to accommodate a different approach to land management, even if this means that the passive approach to land management previously practised by conservation agencies is replaced with something different. So what we are looking at here are conservation reserves which might start looking very different and which might be managed very differently from what has been the case in South Australia in the past.

Implications for the future

What does all this mean for the future of conservation in South Australia's arid zone?

The proposed changes to the reserves system in South Australia as described in this chapter have considerably broadened the spectrum of conservation reserves, so that the traditional Australian model of a national park now fits the real world rather poorly. There is now a much wider range of reserve types accommodating a greater variety of land uses. The traditional model now fits only one end of the spectrum.

The realities of politics and economics in the 1980s have changed the nature of conservation management in South Australia. If conservationists are to be realistic and to save themselves from forever fighting rear-guard actions, some adjustment of the model is needed.

Management of national parks and reserves in Australia is generally the responsibility of State Governments and each State Government will set its conservation programmes within the social, environmental and political context of that State. The advantage of this decentralized system is that it allows progressive and innovative directions in conservation management to take place in those States with innovative and progressive governments. The disadvantage is that it allows for "small decisions" on conservation management to be made rather than taking a holistic approach on a national level. A series of small decisions, even if taken in good faith, may ultimately result in enviromental degradation (Odum 1982).

The regional reserve concept is certainly innovative, and the proposals for joint management with Aboriginal commmunities are certainly progressive. The development of tourist resorts in parks may be considered to be imitative, although not without financial benefits to the management agency. The question arises, however, as to whether these decisions may ultimately lead to environmental degradation, or whether they will stand as monuments to innovative problem solving of the 1980s.

There are several other unanswered questions. How much change to the model is acceptable? How many mines? How many seismic lines? How many resorts? How many people? How much grazing? In other words, what are the acceptable limits to change? If conservation reserves are seen to be reference points against which the effects of change and the effects of perturbation can be measured, can they still perform this function if they are subject to major change themselves? If it is agreed that conservation reserves are part of the reconstruction phase of ecosystem management in Australia, can they continue to be regarded as examples of reconstructed land?

There is a challenge here for research ecologists to start providing managers of conservation reserves with answers to these questions. There is a challenge for conservationists and land managers to establish the limits of the newly expanded spectrum. There is a challenge for all concerned people to ensure that land managers and decision-makers resist the temptation to chip away at the resource lest it drift off into a new cycle of utilization, followed by degradation. Because if this does happen, we leave the land managers of the year

2088 with a legacy of having to reconstruct a second time around. And if we do this, I predict that their task will be very much more complex and very much more difficult than is ours in 1988.

Acknowledgements

I wish to thank Ross Johnston, Bruce Leaver, Mary Lou Morris, Garry Richardson and Tony Robinson for comments on the draft of this manuscript.

References

Anon. (1986) *Uluru (Ayers Rock — Mount Olga) National Park, Plan of Management.* Commonwealth of Australia.

Australian Labor Party (n.d.) *Our Environment, Our Future.* A policy statement from John Bannon, Premier of South Australia.

Figgis P. & Donald B. (1986) *Conservation and Aboriginal Land Rights, the Northern Territory law as a model,* National Environmental Law Association Annual Conference, Canberra, September 1986.

Kean J., Richardson G. & Trueman N. (eds) (1988) *Aboriginal role in nature conservation. Emu Conference June 7-9, 1988.* South Australian National Parks and Wildlife Service and Australian Department of Aboriginal Affairs, Adelaide.

McMichael D. F. (1980) An international perspective. In: Messner, J. & Mosley G. (eds) *The value of National Parks to the Community.* Proceedings of the Second National Wilderness Conference, University of Sydney, November 23-25, 1979. Australian Conservation Foundation, Hawthorn, Victoria.

Michael Williams and Associates (1988) *Proposed Wilpena Station Resort, Flinders Ranges National Park, South Australia.* Draft amendment to the Flinders Ranges National Park Management Plan and Draft Environmental Impact Statement, National Parks and Wildlife Service, Adelaide.

Odum W. E. (1982) Environmental degradation and the tyranny of small decisions, *Bioscience* **32,** 728-9.

Reid J. & Gillen J. (1988) *The Coongie Lakes Study.* South Australian Department of Environment and Planning, Adelaide. (Wildlife Conservation Fund Research Grant Project Report).

South Australia. Legislation. *National Parks and Wildlife Act, (1972-87).*

South Australian Department of Environment and Planning (1989) *Proposed Wilpena Station Resort, Flinders Ranges National Park. Assessment of the potential environmental impacts.* Major Projects and Assessments Branch, Adelaide.

South Australian Department of Lands (1986) *South Australia showing pastoral areas 1:2,000,000 map.* Government of South Australia, Adelaide.

Squires V. (1981) *Livestock management in the arid zone.* Inkata Press, Melbourne.

Toyne P. & Vachon D. (1984) *Growing up the country. The Pitjantjatjara struggle for their land.* McPhee Gribble/ Penguin Books, Fitzroy, Victoria.

Williams M. (1977) *The changing rural landscape of South Australia,* Heinemann, Richmond, Victoria.

Proc. Ecol. Soc. Aust. 1990 — 16:467-473

Natural regions of western New South Wales and their use for environmental management

G. MORGAN and J. TERREY

Landwater Management, Pinnacle West, Bundarra Road, Armidale 2350

Abstract

Western New South Wales is classified into twelve natural regions. Each has a particular climate, geological origin and other biophysical characteristics, and hence has distinctive land use options, constraints and management needs.

By indicating major differences in land processes, energy budgets and in species distributions and patterns of movement, regions can be used as a framework for land use planning, land management and environmental monitoring.

While catchments are a basic land management unit for gravity related processes, natural regions provide a more rational basis for environmental analysis and planning, particularly at state and national levels.

To illustrate the utility of natural regions two basic uses are outlined: the regional classification developed for western New South Wales is used to determine the major regional land conservation objectives for that part of the state; and the regions are used as a framework for assessing the adequacy of the existing conservation reserve system, and to identify additional key conservation areas.

A geographical framework of functional land areas is urgently required in Australia. Natural regions are of a suitable scale and can be delineated readily in most areas through the re-interpretation of existing information.

Introduction

If sustainable patterns of land uses are to be established then land management must be based on a clear understanding of the environment and of environmental processes, at all scales.

At the continental scale, a basic need is to identify the broad functional patterns (Forman & Godron 1984; Zonneveld 1988; Rowe 1988) of the terrestrial environment. These patterns reflect the origins of the landscape, its key biophysical characteristics, and the distinctive land use features that have developed in response to natural and cultural constraints. If these patterns are readily recognizable they provide a strong basis for the development and implementation of regional land management strategies.

Natural patterns, including physiographic, phytogeographic and zoogeographic, have been identified by a number of workers in Australia (see Beard 1981; Heatwole 1987) but the scale of these studies and the consequent heterogeneity of their subdivisions limits their usefulness. Statewide studies, although having artificial boundaries, are a more suitable scale for planning land use.

The ecological survey of South Australia (Laut *et al.* 1977) and a regional study of Queensland (Stanton & Morgan 1977) are two statewide land classifications that have attempted to integrate climate, geology, landform and vegetation into regional divisions of a scale suitable for use by State Government instrumentalities with a responsibility for land use planning and land management. The latter study is now used as a basis for conservation surveys in Queensland (Wilkinson & Gordon 1984; Purdie *et al.* 1986; Sattler 1986) by the Forestry Commission when assessing the conservation significance of plant communities in State Forests (Prineas 1984) and by the Archaeology Branch of the Department of Community Services in their survey and analysis of archaeological sites (Rowland 1984).

In New South Wales there is no single integrated survey that classifies the State into similar broad landscape units although the Soil Conservation Service (1978-88) is nearing completion of its land system mapping for the Western Division, using a scale of 1:250 000. While this will provide a comprehensive basis for land management (Christian

1959) in the western-most 40% of New South Wales, statewide and national perspectives are more readily obtained from a higher order of land classification. Such a classification would provide a framework for interpretation of the land system mapping and be a basis for more detailed land research in other parts of the state.

The study outlined in this chapter was initiated by the National Parks Association of New South Wales in response to the increasing development of marginal agricultural lands in the western part of the State. Its primary aim was to identify areas with significant nature conservation values.

Due to the lack of a suitable regional scheme for most of western New South Wales, land classification formed a large part of the study. Besides providing a framework for the selection of a representative system of conservation areas, the land classification was to provide a basis for ensuring that regional nature conservation areas were complemented by sympathetic and sustainable regional land uses.

The natural regions of western New South Wales

Study area

The study area is that part of the State west of the tablelands of the Great Dividing Range (Fig. 1). It covers an area of approximately 600 000 km^2, representing over 60% of the catchment of the Murray and Darling Rivers. This corresponds approximately with the Central and Western Divisions of the New South Wales Lands Department, the former being dominantly freehold land and the Western Division being leasehold. The Central Division has been largely cleared and undeveloped vegetation is restricted to the more rugged areas and to small parcels of Crown land. Within the Western Division clearing is restricted to the eastern and southern margins, the great proportion of the Division being used for grazing undeveloped native pastures.

Delineation of regions

The land system mapping for the Western Division (Soil Conservation Service 1978-88) covers approximately 50% of the study area. As the higher order of land classification required for this study had to be compatible with the land system mapping, and the regional mapping as a whole required a relatively uniform level of accuracy, the delineation of regions was undertaken at a scale of 1:1 000 000.

The regional base map is the geology map of New South Wales (Pogson 1972) since geology is a fundamental and major influence on landform, soils, local climates, vegetation and fauna (Hughes 1980; Rowe 1988).

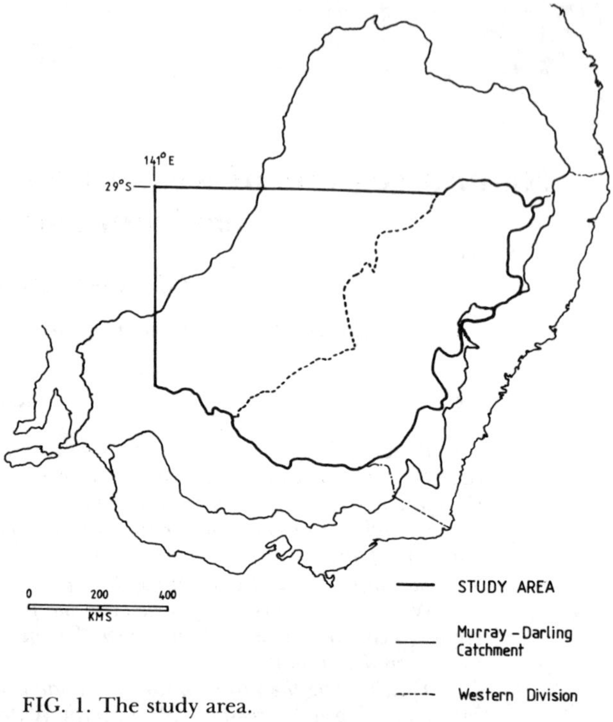

FIG. 1. The study area.

Regions were initially delineated on the basis of major changes in lithology, the primary divisions being unconsolidated Quarternary deposits, horizontally-bedded Mesozoic deposits, and deformed and intrusive rocks of Palaeozoic age and older. Extensive areas of relatively uniform geology were first identified, concentrating initially on bedrock areas, and their landform, soil and vegetation patterns determined from existing information sources and 1:250 000 hardcopy Landsat images. In this way boundaries were more accurately defined and areas not readily subdivided were studied in more detail. Tertiary and Quarternary landscapes were then analyzed, with the catchment characteristics of alluvial areas, and source characteristics of aeolian and colluvial deposits, being major considerations. On more detailed analysis, most regions were further subdivided.

The natural regions

Twelve regions were delineated in western New South Wales (Fig. 2). These were further divided into fifty-six provinces containing a total of one hundred and fifty-seven major regional ecosystems.

The diagnostic characteristic of each region are outlined in Table 1. In the same way that basic functional differences between the regions are indicated by fundamentally different geologies,

TABLE 1. Western New South Wales: Diagnostic key to regions*.

Quarternary deposits (younger than 1.8 million years). Unconsolidated.	alluvial sediments; extensive plains; clay texture-contrast soils.	catchments dominated by bedrock.	summer rainfall dominant; woodlands and grasslands.	DARLING RIVERINE PLAINS	1
			winter rainfall dominant; shrubland, grasslands and riverine forests.	MURRAY RIVERINE PLAINS	2
		catchments dominated by coarse unconsolidated terrestrial sediments.		WARREGO FAN	3
	wind-sorted sediments; dunefields, plains, occasional low stony hills, lakes; sands and loams.	residual deposits from deep weathered sediments.		NORTH WESTERN SANDS	4
		sediments from Murray Basin; marine and alluvial origin.		MURRAY-DARLING DEPRESSION	5
Mezozoic bedrock (younger than 235 million years). Consolidated, horizontally bedded.	rocks fine grained.	stony plateau; rainfall 200-250 mm pa; woodlands and shrublands.		WHITE CLIFFS PLATEAU	6
		undulating plains; rainfall 150-200 mm pa; shrublands and herbfields.		TIBOOBURRA DOWNS	7
	rocks coarse grained.	rainfall 500-600 mm pa; forests and woodlands.		NORTHERN SANDSTONES	8
Palaeozoic and Pre-Cambrian (older than 235 million years). Igneous rocks and deformed sediments.	New England Fold Belt 235-435 m.y.	rainfall 600-850 mm pa; summer dominant.		NORTH WESTERN SLOPES	9
	Lachlan Fold Belt 345-570 m.y.	rainfall 450-700 mm pa; winter dominant.		SOUTH WESTERN SLOPES	10
		rainfall 300-450 mm pa.		COBAR PENEPLAIN	11
	Central Craton Older than 570 m.y.	rainfall 150-200 mm pa.		BROKEN HILL COMPLEX	12

*Sources of biophysical information are listed in Morgan and Terrey *(in litt.)*.

climate, landform, soils and vegetation the provinces and regional ecosystems also differ, at a finer scale. Generally speaking, the provinces delineate significant changes in geology or geomorphology, while the major regional ecosystems are suites of related land systems. Regions correspond with the regions delineated in Queensland (Stanton & Morgan 1977) and the provinces of Laut *et al.* (1977) in South Australia.

The correlation between the provinces and suites of mapped land systems (Soil Conservation Service 1978-88) in the Western Division indicates the value of this rapid approach to land classification, and the strong basis it provides for predicting ecosystem similarities.

The relationship between regions and their major regional ecosystems is illustrated below by a brief description of the Tibooburra Downs Region.

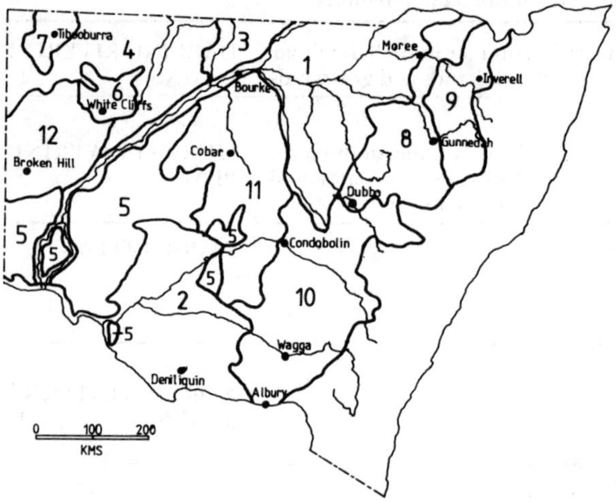

FIG. 2. The Natural Regions of western New South Wales: (1) Darling Riverine Plains; (2) Murray Riverine Plains; (3) Warrego Fan; (4) North Western Sands; (5) Murray-Darling Depression; (6) White Cliffs Plateau; (7) Tibooburra Downs; (8) Northern Sandstones; (9) North Western Slopes; (10) South Western Slopes; (11) Cobar Peneplain; (12) Broken Hill Complex.

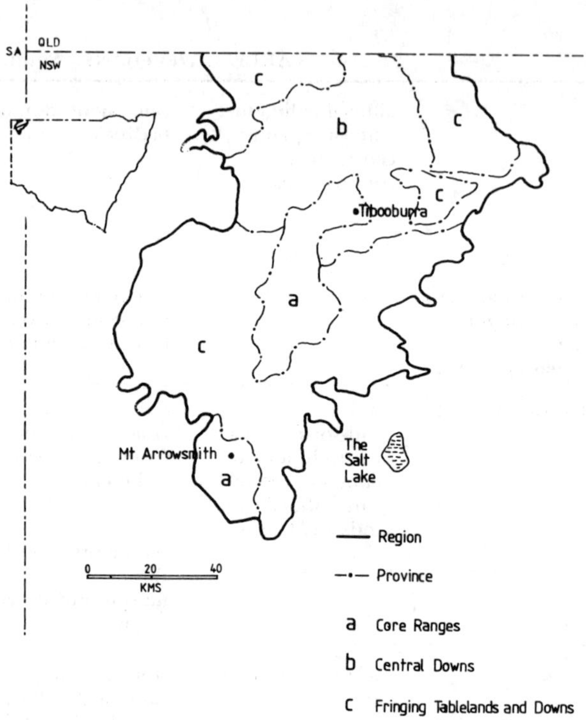

FIG. 3. Region 7: Tibooburra Downs.

The Tibooburra Downs

This is a small bedrock region in the arid north-western corner of the State, rising above the dunes of the surrounding North Western Sands Region. It is largely formed on Cretaceous sediments of the Great Artesian Basin, gently domed over older rocks that now outcrop in central areas (Fig. 3).

The Tibooburra Downs has elements in common with the White Cliffs Plateau to the south-east. Unlike the latter region, the silcrete capping has been largely removed leaving a landscape dominated by gently undulating downs. The Tibooburra Downs also has a lower annual rainfall, and its vegetation is characterized by shrublands and herbfields (Table 1).

Three provinces can be recognized, based primarily on differences in geology.

The Core Ranges Province contains the oldest rocks of the Region, as well as its areas of highest relief. It has two major geologies which are spatially intimately related: the Pre-Cambrian and Cambrian rocks that caused the anticlinal structure of the remainder of the region, and the Jurassic sediments that partly mantle it. Three major ecosystems can be identified in the Core Ranges Province: ranges and ridges on the oldest rocks, with lithosols and low bluebush, *Maireana* spp., shrublands; upper slopes and dissected uplands on sediments, with desert loams and sparse mulga, *Acacia aneura,* communities; and undulating plains on sediments, with finer soils and saltbush, *Atriplex* spp., and bluebush communities.

The overlying Cretaceous sediments, which have been stripped of their silcrete capping, form the second province, the Central Downs. This is characterized by undulating plains divided by broad alluvial flats, these systems forming the two major ecosystems identified. The downs have a variable community of Mitchell grasses, *Astrebla* spp., and forbs, while the finer soils of the alluvials carry a fringing community of gidgee, *Acacia cambagei,* coolabah, *Eucalyptus microtheca,* or river red gum, *E. camaldulensis,* on upper tributaries, and coolabah, black box, *E. largiflorens,* and cooba, *A. salicina,* on lower distributaries.

The third province, the Fringing Tablelands and Downs, is formed on younger Cretaceous sediments and has moderate areas of silcrete plateau remaining. Three major ecosystems were identified: silcrete areas and dissected plateaux, with sparse mulga communities; fringing down areas with Mitchell grasses and forbs; and tributaries and distributaries with river red gum fringing channels and coolabah and gidgee on larger tributaries.

The use of natural regions in environmental management

With the exception of the Queensland examples mentioned previously (Sattler 1986; Prineas 1984; Rowland 1984), most land use planning and environmental analysis in Australia is undertaken on the basis of administrative regions. This is due largely to the long standing of these divisions and the vast amount of data collected and analyzed on that basis. It is also partly due to the absence of generally accepted alternative regional classifications. There has been an attempt to partly rectify this absence by grouping local government areas having similar biophysical characteristics (Department of Home Affairs and Environment 1983).

In 1986 Environment Canada undertook its first comprehensive report on the condition of the national environment using ecozones and ecoregions as the framework for statistical analysis (Wiken 1986; Bird & Rapport 1986). Ecoregions are equivalent to the regions defined in this study. Acknowledging the difficulties involved in changing the regional basis of data analysis, Bird and Rapport (1986) state: "in many cases . . . it has not been possible to convert existing data to suit the new geography . . . These inconsistencies are the inevitable result of the introduction of a new geographical basis for environmental reporting, but they should disappear gradually as the ecozone system framework becomes more widely known and accepted." Similar studies have been initiated in the United States of America (Anon. 1987).

In New South Wales the trend has been towards catchment planning, notably responding to national concerns over the condition of the Murray-Darling catchment. While being the obvious management unit for gravity-related processes, catchments do not provide a biophysical basis for assessing most types of environmental change. In particular, habitat destruction, soil erosion, eucalypt dieback and many other forms of land degradation occur independently of watersheds but these can usually be related directly to fundamental land characteristics and the land use patterns that have developed as a consequence. Provinces provide a rational framework for collecting and analyzing environmental data, for extrapolating information, for devising land use policies and for implementing land management programmes. They enable the preparation of regional plans within which environmental considerations can be comprehensively examined.

Natural regions and their provinces also provide the basis for an inventory of major ecosystems and enable the prediction of ecological differences in the absence of more detailed information. For example genotypes of widespread species will characterize each province, and each province will have distinctive aquatic environments.

Regional objectives

The scale at which natural regions are mapped, usually 1:1 000 000, is particularly useful for determining national and state perspectives of land types and their planning significance. For example they allow primary land conservation objectives to be derived for each region that reflect dominant biophysical attributes, existing land uses, and long-term environmental needs.

On the basis of an analysis of existing biophysical information for western New South Wales the following primary regional objectives can be derived.

- The Darling Riverine Plains and the Murray Riverine Plains regions are dominated by riverine processes. They contain the major aquatic ecosystems in inland New South Wales and the major perennial aquatic ecosystems of inland Australia. They also contain the most extensive irrigated areas in the State. The protection of waterways and their associated wetlands, and the maintenance of the fertile cropping areas on the upper fans, are the primary land conservation objectives.
- The Warrego Fan Region is the southernmost part of an extensive region that lies largely within Queensland. In that State clearing has been widespread and New South Wales contains the region's most extensive areas of undeveloped vegetation. The protection of large areas of these communities is the major land conservation objective.
- The North Western Sands Region is composed largely of loose sands and is easily eroded. The maintenance of ground cover by reducing grazing pressures and controlling shrub regrowth is the major conservation objective. This may be partly achieved by the management of particularly sensitive areas as wilderness.
- The Murray-Darling Depression Region contains extensive areas of saline soils. Deep rooted native vegetation should be retained to the maximum extent, including the protection and management of large areas as wilderness.
- The White Cliffs Plateau, Tibooburra Downs and Broken Hill Complex regions are predominantly fine-textured bedrock areas with shallow soils. They are important grazing regions and provide the run-off for water supplies and alluvial pastures within the adjacent deep sand regions. The major objective is to maintain ground cover by reducing grazing pressures.
- The Northern Sandstones Region contains the major intake areas in New South Wales for the Great Artesian Basin. Maintaining soil permeability and preventing the chemical contamination of ground water are the major land conservation objectives.
- The North Western Slopes and the South Western Slopes regions have shallow soils, steep slopes and relatively high rainfall. They contain most of the catchments of a number of large dams and contribute the greater part of the flow of the Murray-Darling river system. These regions should be managed primarily for catchment protection, large areas of the steeper and

more erodible areas being left undeveloped for multiple use as rough grazing, wildlife corridor and water catchment.

- The Cobar Peneplain Region has shallow loamy soils that are readily eroded: overgrazing is leading to widespread soil loss and dense shrub regrowth. The regeneration of pastures through destocking and the judicious use of fire is the major conservation objective.

More specific objectives can be determined for each province to complement these regional objectives.

Nature conservation assessment

By defining the major regional ecosystems of western New South Wales the regions and provinces provide a strong basis for assessing the adequacy of the existing system of conservation reserves (see Specht *et al.* 1974). The regions are also a framework for selecting suitable areas for additional reserves. These processes are outlined below.

One of the twelve regions delineated, the Warrego Fan, has no conservation reserve, nor do 33 of the 56 provinces. Only 20 (13%) of the 157 major regional ecosystems identified could be considered adequately represented in existing reserves (protected in a large reserve or in several smaller reserves), while 112 (71%) are not represented at all. While these gross deficiencies in western reserves can be partly explained by the parochial concerns of a predominantly coastal population, they are also the inevitable consequence of an inadequate biophysical framework for conservation assessment.

The New South Wales National Parks and Wildlife Service is now in the process (Pressey 1987) of using computer-aided analysis of the land systems maps of the Western Division (Soil Conservation Service 1978-88) to establish a systematic basis for reserve selection. This follows similar work undertaken in Queensland (Purdie *et al.* 1986) and includes the use of natural regions and provinces as indicators of major changes in land system patterns (Pressey, pers. comm.). This is essentially a long-term project and the present study was perceived as providing an interim basis for making pressing land management decisions.

Natural regions and provinces, by defining the boundaries between major land patterns, allow the rapid selection of representative areas containing a wide variety of ecosystems. Ecosystem representation will be optimized by locating reserves astride these significant ecological boundaries, or near boundaries so that they can be linked with reserves in adjacent provinces. Potential conservation areas can be compared within each province and assessed according to accepted criteria such as size, shape, diversity, condition, catchment integrity, tenure and potential for connection to other conservative land uses, initially using visual interpretation of Landsat images, followed by photo-interpretation and field verification. In this way conservation priorities are rapidly determined at very low cost.

A system of conservation reserves for western New South Wales developed on this basis is described in detail by Morgan and Terrey *(in litt.)*. The key areas identified are considered the core nature conservation areas, and field work is now required to verify their significance. An adequate regional conservation plan would build on these core areas and include additional reserves to protect rare species or restricted vegetation communities, as well as multiple use corridor and buffer zones (Noss 1983).

Conclusion

The minimum requirement for nature conservation in Australia may be considered to be the protection of representative areas of all major regional ecosystems in large and viable conservation reserves. If this is to be achieved then there is an urgent need for the broad functional patterns of the continent to be defined and the most significant near-natural areas of each of their major ecosystems to be identified.

The proposed environmental survey of Australia is now primarily concerned with co-ordinating existing environmental data bases (Department of Home Affairs *et al.* 1985). A geographical framework of functional land areas is required however, if these data are to be meaningfully assessed and extrapolated. Existing regional classifications of Australia are not functionally based and are too heterogeneous to provide a rigorous framework for predicting ecosystem differences and similarities or for directing further data collection.

At a time in which options for nature conservation are being rapidly reduced and funding for investigations is miniscule, and in which an understanding of ecosystems and their responses to climatic change may be urgently required, the use of natural regions is a rapid, efficient and cost effective basis for ecological survey, land use planning, land management and environmental monitoring.

Acknowledgements

The work on which this chapter is based was made possible by a bequest from the late Stella Gottleib.

Many individuals and organizations assisted by providing access to unpublished information and through informal discussions. In particular we would like to thank John Pickard of the Western Lands Commission and Peter Walker of the New South Wales Soil Conservation Service for their generous help and constructive advice.

References

Anon. (1987) Ecoregions and resource quality. *Land* **8,** 7.

Beard J. S. (1981) The history of the phytogeographic region concept in Australia. In: *Ecological Biogeography of Australia* (ed. A. Keast) pp. 337-53. Dr W. Junk, Hague.

Bird P. M. & Rapport D. J. (1986) *State of the Environment Report for Canada.* Environment Canada, Ottawa.

Christian C. S. (1959) The eco-complex and its importance for agricultural assessment. In: *Biogeography and ecology in Australia* (eds A. Keast, R. L. Crocker & C. S. Christian) pp. 587-605. Dr W. Junk, Hague.

Department of Home Affairs and Environment (1983) *Environmental Regions of Australia.* Australian Government Publishing Service, Canberra.

Department of Home Affairs and Environment & South Australian Department of Environment and Planning (1985) *Environmental Survey of Australia.* Australian Government Publishing Service, Canberra.

Forman R. T. T. & Godron M. (1984) Landscape ecology principles and landscape function. In: *Methodology in Landscape Ecological Research and Planning* (eds J. Brandt & P. Agger) Volume V, pp. 4-15. Roskilde Universitiesforlag, GeoRuc, Denmark.

Heatwole H. (1987) Major Components and Distributions of the Terrestrial Fauna. In: *Fauna of Australia* (ed. G. R. Dyne) Volume IA, pp. 101-35. Australian Government Publishing Service, Canberra.

Hughes K. K. (1980) The Application of Geology in Land Resource Assessment and Land Management. *Division of Land Utilization Report* Number 80/3. Queensland Department of Primary Industries, Brisbane.

Laut P., Heyligers P. C., Keig G., Löffler E., Margules C. & Scott R. M. (1977) *Environments of South Australia Handbook.* CSIRO Division of Land Use Research, Canberra.

Noss R. F. (1983) A Regional Landscape Approach to Maintain Diversity. *Bioscience* **33,** 700-6.

Pogson D. J. (1972) *Geological Map of New South Wales, Scale 1:1 000 000.* Geological Survey of New South Wales, Sydney.

Pressey R. L. (1987) *The Mulga Lands of New South Wales: Background Information and Trial Reserve Selection Method.* New South Wales National Parks and Wildlife Service, Sydney.

Prineas A. (1984) *Guidelines for mapping vegetation types on State Forests and Timber Reserves, Part 1: Surveying poorly conserved vegetation types.* Land Use and Information Branch, Queensland Department of Forestry, Brisbane.

Purdie R. W., Blick R. & Bolton M. P. (1986) Selection of a Conservation Reserve Network in the Mulga Biogeographic Region of South-Western Queensland, Australia. *Biol. Conserv.* **38,** 369-84.

Rowe J. S. (1988) Landscape Ecology: The Study of Terrain Ecosystems. In: *Landscape Ecology and Management* (ed. M. R. Moss) pp. 35-42. Polyscience Publications, Montreal.

Rowland M. J. (1984) *Archaeology Branch Site Files, 1:250 000 Series with Queensland Biogeographical zones.* Queensland Department of Community Services, Brisbane.

Sattler P. (1986) Nature Conservation in Queensland: Planning the Matrix. *Proceedings of the Royal Society of Queensland,* **97,** 1-21.

Soil Conservation Service, N.S.W. (1978-88) *Land Systems Series, Western Division. 1:250 000.* Sydney.

Specht R. L., Roe E. M. & Broughton V. H. (1974) Conservation of major plant communities in Australia and Papua New Guinea. *Australian Journal of Botany* Supplementary Series Number 7.

Stanton J. P. & Morgan M. G. (1977) *The rapid selection and appraisal of key and endangered sites: the Queensland case study.* Report Number PR4. School of Natural Resources, University of New England, Armidale.

Wiken E. (1986) Terrestrial Ecozones of Canada. *Ecological Land Classification Series No. 19.* Lands Directorate, Environment Canada.

Wilkinson C. G. & Gordon G. (1984) Review of Surveys for conservation purposes in Queensland. In: *Survey Methods for Nature Conservation* (eds K. Myers, C. R. Margules & I. Musto) Volume 2, pp. 118-43. CSIRO Division of Water and Land Research, Canberra.

Zonneveld I. S. (1988) Landscape Ecology and its Application. In: *Landscape Ecology and Management* (ed. M. R. Moss) pp. 3-15. Polyscience Publications, Montreal.

Proc. Ecol. Soc. Aust. 1990 — 16:475-490

Land evaluation: the Victorian land capability assessment approach

A. LINDSAY and K. ROWE

Department of Geography and Environmental Science, Monash University, Clayton, Victoria 3168

Abstract

For prevention of ecological degradation the ability of land to sustain various uses without undue adverse effects must be assessed. The information resulting from the assessment must then be presented in a form which can readily be understood and used by land use planners and managers. A number of assessment systems have been developed in various countries. The land capability assessment procedure developed by the Victorian Land Protection Division is described in this chapter. This system is based on a holistic and interactive model which relates production *and* management *(land use) to* land features *and* impact *on the land. Land features considered to date are usually those which are seen to be most permanent and difficult to alter, such as climate, land form and soil physical properties. An example of land capability assessment using land capability rating tables developed in Victoria is presented. Some difficulties and strengths of the system are discussed. Possible extensions of the approach to assess new land uses such as wildlife conservation are considered.*

Introduction

> *"Moreover, there were tall cultivated trees in abundance, and the mountains afforded pasture for countless herds . . . What is now left . . . is like the bones of a body wasted with disease: the fertile soil has fallen away, leaving only the skeleton of the land."*

Thus wrote Plato two and a half thousand years ago of the deforestation of Attica (quoted in Douglas & Hart 1976).

Production from land involves modifying ecosystems. The benefits are usually obvious; the negative effects, including long-term failure of the land to sustain production, and off-site ecological degradation, may be unforeseen and, for a while, not evident.

The problem is not new, but rising world population and demand are placing increasing strains on the world's ecosystems. Maintenance of these ecosystems must be reconciled with use if humanity is to survive (Anon. 1987a).

Within the non-arid areas of Australia, around 1 800 000 km^2 of land is being used for some form of primary production. It has been estimated that almost half of this area needs some sort of reclamation treatment to maintain or restore productive capacity (Chartres 1987). Considerably more than this has experienced some degree of degradation (Table 1).

The Victorian State Government, (in line with both the *World Conservation Strategy* and the *National Conservation Strategy)* in its *Conservation Strategy for Victoria* (Anon. 1987a) recognizes the need for sustainable development in terms of *(inter alia)*;

— matching type and intensity of use to characteristics of the ecosystem,
— using renewable resources within their ability to maintain the resource base,
— using non-renewable resources wisely,
— controlling direct and indirect environmental impact of development.

The strategy clearly recognizes that the type and intensity of land use imposed must allow the land to continue to be used without reducing its productive capacity over a long period. That is, if production is to be sustainable, then planning and management must ensure that the type and intensity of land use is matched to the capability of the land. The magnitude of land degradation shown in Table 1 indicates a widespread failure to do this. Both imperfect understanding of our ecosystems and failure to communicate what is known to land use planners and managers has been the cause of this failure (Anon. 1987b).

Today knowledge of the relationships between land use and the attributes or features of land is certainly still incomplete. However, enough is

TABLE 1. Forms of degradation in non-arid areas of Australia assessed at 1975 Source Chartres (1987:7).

	Area	
	(000 km^2)	(%)
Area in use	1804	—
Area not requiring treatment	987	55
Form of degradation		
Water erosion	577	32
Wind erosion	57	3
Combined wind and water erosion	55	3
Vegetation degradation	92	5
Dryland salinity-sometimes in combination with water erosion	10	<1
Irrigation area salinity	9	<1
Other	14	<1
Total area requiring treatment	815	45

All values are approximate and have been rounded.

known to provide valuable guidance to land planners and managers (Burch *et al.* 1987). An effective means of communicating this knowledge is essential for sound decision making about land use. It is proposed in this chapter that a formalized system for providing this knowledge, that is a land evaluation system, provides a mechanism for information transfer and can contribute substantially to both land use planning and management.

For a land evaluation system to be successful it must meet several criteria. In the first place it must properly reflect the processes involved in ecosystems and the way these interact with land use. Secondly, it must be acceptable to, and usable by land planners and managers. To this end the method should be explicit and produce consistent results which conform with known performance of land under given land uses. For planners particularly, the information provided should be fairly simple and systematic, and in a form which is readily integrated with the other information which must be considered in planning decisions (Rowe *et al.* 1988). It should, in addition, be in a form which the planners can use in communicating with those affected by their plans. Ideally it should also be usable at a range of planning scales, from broad regional to detailed local determinations (Rowe *et al.* 1988).

Land managers also require a resonably simple and systematic assessment, in a form which can be integrated with other relevant information. However, more detail on factors which influence management decisions is required, and the scale needed is usually large and precise and the information also more precise (Rowe *et al.* 1988). Finally, the system must be one which can be applied using survey personnel available, or for which new personnel can readily be trained. It should make the best use of existing data, be explicit in defining what additional information is required and how it is to be obtained, and provide opportunity for new knowledge obtained by survey personnel to be incorporated into the system.

A tension between the first and the remaining two requirements appears inevitable because of the complexity of the inter-relationships in both natural and managed ecosystems and the need for the relationships to be presented in easily used forms. A variety of approaches to land evaluation or land capability assessment have been developed in countries throughout the world. The United Nations Food and Agricultural Organization has attempted to produce a system which has international application (Food and Agricultural Organization 1976, 1984; Dijkerman 1988). However, the great extent to which land use objectives and management practices are influenced by culture suggests that some tailoring of land evaluation systems to particular countries would usually be necessary. Certainly the Food and Agricultural Organization's system has not been universally adopted and there are workers in land evaluation who consider its complexity also restricts its usefulness (Howe, pers. comm.).

The following discussion outlines the land evaluation procedure most commonly used by the Land Protection Division of the Victorian Department of Conservation, Forests and Lands. Some variations of the system will be considered and its strengths and weaknesses discussed.

In evaluating the system it is assumed that, while it should reflect accurately ecological processes the appropriate basis for judgment is not so much does it include all this information ("is it correct")?; but does it include the information important for making decisions about the use being considered ("is it useful")? Does it lead to planning and management which helps to prevent ecological degradation and retain productive capacity?

What is land capability assessment?

Land capability assessment, that is the ability of land to achieve specified use objectives, including kinds and levels of production and sustainability, evaluates land in terms of its physical and biological attributes and management inputs required (Lindsay 1979; Rowe *et al.* 1988). It involves both production of an adequate land resource inventory and interpretation of this to provide useful information for planners and managers (Van de Graaf 1988). The principles of sustained use and containment of impacts within accepted limits are inherent in the assessment (Rowe *et al.* 1988; Anon. 1987a; Anon. undated; Van de Graaf 1988).

The suitability of land for a given use may be seen to depend on many considerations. These could include proximity to centres of population, land tenure and legal constraints, attractiveness of scenery, transport systems and consumer demand, as well as the biophysical features of the land itself. Attempts to incorporate all relevant factors into a single suitability rating of land for a specified use was popularized in the 1970s by McHarg (1969) (e.g., Burrough 1978; Austin *et al.* 1977). Such broad suitability ratings suffer from several serious drawbacks (Lindsay 1979, 1981). In particular, value judgments concerning the relative importance of various contributing factors such as those mentioned above, are built into the ratings. These values may be neither overt nor explicit and therefore not easily varied in line with changing values in the community. Generally, the ratings do not present details of how various factors are evaluated and integrated to produce the final rating. Mathematically untenable procedures such as adding values on different, or even ordinal scales may be hidden within the assessment. The single rating approach also involves the loss of a great deal of useful information. Planners and managers usually need to know, not only how suitable land is for a use, but why is it so. Finally, assessing all relevant factors at once is likely to be a vast and difficult job, requiring inputs from many disciplines. It is usually best to subdivide the task and work within confined parts of the issue.

Basic concepts

Two concepts are central to practical land capability assessment. The first is the land unit. This is an area of land which has a sufficiently consistent association of land features to respond uniformly and predictably to a particular use. The land features considered when mapping such land units include all the biophysical aspects of the environment; climate, soil, geology, topography, hydrology and vegetation (Rowe *et al.* 1988). The potential of the land for sustained use and its susceptibility to degradation depends on the interplay of all these factors. However, because of the interrelatedness of land features it is also assumed that land units can be defined in terms of a smaller number of features (usually land form, soils, parent material and climate) (Lindsay 1979; Rowe *et al.* 1988).

The second concept concerns the interrelatedness of land features, land use and the effects of use (impacts) on the land. A *conceptual model* developed by Gibbons (1976) illustrates these interactions (Fig. 1). This model forms the basis of the Victorian land capability assessment system and several others which will be mentioned below (Lindsay 1979; Rowe *et al.* 1988; King *et al.* 1986; Semarat 1986). The land capability model proposes that land use should be considered in terms of both management and production. Each may vary in both kind and degree. While *production* relates best to agriculture or forestry it can refer to other types of land use. For example *production kind*

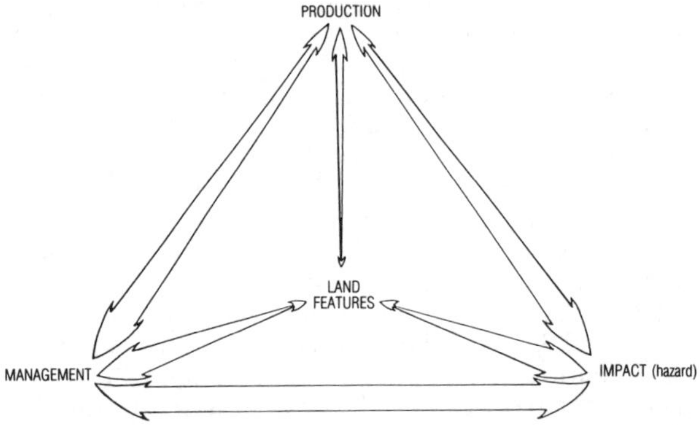

FIG. 1. The Conceptual Model.

could be bush walking on trails and the *degree* could be the number of walkers per day. *Land features* and *impact (or hazard)* also vary in kind and degree; for example the land feature may be soil dispersion and the degree percentage of dispersive clays; the impact may be sheet erosion and the degree be rate of soil loss. Thus there are four interacting variables, each of which be defined in kind and degree.

Interactions between these variables is indicated by two-way arrows (Fig. 1). These emphasise that change to one variable will lead to compensating changes in one or more of the other variables. Where changes are undesirable, management inputs are required to compensate for or overcome these changes. Conceptually these feedbacks allow solutions to the problem of matching land type to land use in all cases. Given sufficient management inputs it is possible to imagine any land being used for any form of production. The restriction then becomes one of cost. Consequently land capability assessment may be used to indicate the kind and degree, and therefore the costs of management required for land to be used to achieve use objectives without unacceptable impacts on the land. Conversely the assessment may be presented in terms of the degradation which would result if the land were used in a given way under defined management.

Because of its complexity, when using the system it is usual to consider the interactions between two variables while the other two are held constant (Van de Graaf 1988). For example management required to achieve a specified level of wheat production on a range of land types, may be given on the assumption that soil erosion (impact) will be confined below a specified limit (a management rating) (King *et al.* 1986). Alternatively the erosion loss which could be expected with a given standard of management and level of crop production on a given area may be assessed (a hazard rating). This would allow judgment to be made as to whether the predicted soil loss would be within tolerable limits or whether changes to management or production type should be made to reduce the hazard.

Methods of integrating information on land and its reponse to use into a capability assessment system vary. Each approach has its deficiencies. If the response functions between the relevant land features and a given land use were known fully then it would be possible to describe land capability in precise mathematical terms. Such a *parametric* approach has been attempted; for example de la Rosa *et al.* (1981) have used linear regression techniques to develop a reasonably simple formula for predicting crop yields from soil properties. The Universal Soil Loss Equation (Appendix 1) relating land features and management to sheet erosion has been used by several workers in an attempt to produce hazard ratings (e.g., for crop production and extractive industries in King *et al.* 1986). The parametric approach has three serious drawbacks. In the first place the production functions are almost never all adequately known. Second, while it may be possible to model some ecological relationships in the form of reasonably simple equations this is often not the case (Lindsay 1987). Third, because the relationships are complex, parametric rating systems are likely to prove difficult to apply for both the land assessor and the user of the rating produced.

An alternative approach adopted by the Victorian Land Protection Division and other workers (e.g., Austin *et al.* 1977; King *et al.* 1986) is to consider separately the limits imposed by various land features on a particular land use. The overall capability rating then attributed to the land is determined by the most limiting land feature (Rowe *et al.* 1988). The principle underlying this *limitations approach* is that there are a number of necessary conditions which must be fulfilled before it is possible to use the land to achieve the particular use objectives. If one of these is inadequately represented no abundance of the others can make up for the deficiencies of the limiting land feature. Distinction can be made between more permanently limiting land features (such as slope or soil texture) and those which may be regarded as temporary (e.g., poor drainage or lack of vegetative cover), but this is generally not done (Van de Graaf 1988). Moreover, the assumption can be made that most limitations may be overcome given appropriate management, the chief concern being whether or not it is economic to do so.

There are several advantages in using the limitations approach. The underlying principle reflects the risk of land degradation and costs of production where a single land feature has a severe effect on a particular land use. It is consistent with the commonly and long observed occurrence of ecological thresholds in biological response systems (Odum 1971). It also avoids the trap which some workers have fallen into of adding quantities measured on different scales in an attempt to achieve a parametric approach (e.g., King *et al.* 1986; Lindsay 1987; Anon. 1986). Perhaps most important, it has permitted the development of systems which are fairly easy to apply and produce information which its users can understand and put into practice. The disadvantages of the approach include its rather simplistic assumptions concerning ecological responses to disturbance. It generally implies that land features are independent in their effects on and responses to land use. Interactions between land features (e.g., slope and soil properties) can only be considered in a fairly superficial way.

Who uses land capability ratings?

The Land Protection Division uses land capability information when providing advice to land managers and planners where changes to land use could lead to land degradation (Rowe *et al.* 1988).

By the late 1970s the system of land capability assessment developed by the then Soil Conservation Authority (now incorporated into the Land Protection Division) had received widespread acceptance by various authorities for land use planning and management (Lindsay 1981; Crooks 1981). For example, the Town and Country Planning Board of Victoria (now incorporated into the Ministry for Planning and Environment) incorporated land capability assessments as bases for planning controls (Lindsay 1981). Both the Melbourne and Metropolitan Board of Works and the Victorian National Parks Service have used land capability assessments when planning development of their parks (Jeffrey & Howe 1976; Guy 1978).

Requirements that land capability assessment be made for certain land uses has now been incorporated into government policy and so has become a statutory requirement for these uses. Specifically the *State Environment Protection Policy* (Gazetted 26 February 1988) requires that forest management "shall be developed in accordance with the Victorian code of Forest Practices" (Anon. 1988a). This Code specifies that assessment of land capability shall provide a basis for specific Forest Management Plans (Anon. 1987b). Hazard ratings for impact of logging within the Central and East Gippsland management areas have already been undertaken (Anon. 1988b).

Protection of the State's wetlands (as required by the Environmental Protection Policy) necessitates that their conservation significance be assessed. This is a type of land capability assessment where the natural features of wetlands are assessed in terms of the influence they have on the ability of the wetland to support wildlife (the use objective). In some instances special management e.g., fencing to exclude stock or special water management to maintain water levels) will be necessary to protect conservation values. An attempt to develop a suitable evaluation procedure has been undertaken by the Department of Conservation, Forests and Lands (Anon. 1986).

More generally, in *A Conservation Strategy for Victoria* the Government has undertaken to:

> "encourage the improvement of land management practices and the matching of type and intensity of use to the capability of the land" (Anon. 1987a).

At a more detailed level of planning, land capability assessment continues to be incorporated into management plans for public parks (e.g., Gellibrand Hill Nature Reserve — Ferrier, Lewis & Mill, unpublished reports 1987; Malmsbury Reservoir — Mathews & Austin, unpublished report 1988) and used by other land developers (Howe, pers. comm.). It is also being used to provide standards and guidelines for rehabilitation of areas being open-cut mined for brown coal (Lindsay, unpublished reports 1987) and to assess potential for new types of land use (e.g., Semarat 1986).

How the Victorian system is applied

The assessment procedure involves serveral stages outlined in a manual entitled *Guidelines for Land Capability Assessment* (Rowe *et al.* 1988). These stages are as follows:

- for the land being assessed determine the land uses of interest for the particular assessment;
- choose appropriate land capability rating tables which satisfy the use objectives (production) and activities (management inputs) involved or, if none is available develop rating tables for these specific activities;
- survey the land to define land units and determine values of relevant land features in each land unit. The latter are those land features listed for use in the rating tables because of their substantial effect on the use; and
- assess the capability of the land by ranking the values for each land feature according to the rating tables and determining the most limiting factors.

The ratings may then be integrated with other information to prepare land use plans for the area, or provide a basis for management decisions. Monitoring the effectiveness of the capability information as the land is used can provide information on the accuracy of the evaluation. This should then be used to improve and update the rating tables. Figure 2 represents the relationships between these stages.

Development of rating tables

Rating tables for a broad range of land uses have already been developed by the Land Protection Division (e.g., in Rowe *et al.* 1988). It should be emphasized here that, as new information comes to hand these tables are updated, and so are never regarded as final. Examples of the tables are given in Table 2. It can be seen that the range of possible capability ratings is usually divided into five discrete classes. Broad definitions of these classes are given in Table 3. Land features which have been found to be of major importance in determining capability are listed in the left hand column of the rating tables as the diagnostic criteria. These land features are also the ones which tend to be difficult or costly to alter. Values of each criterion which correspond to a given level of capability are given in the body of the table. Land features which are limiting but can be rectified easily and at modest cost (e.g., a nutrient deficiency for crop growing) are not seen as major determinants of capability although they will still influence land management.

Use of rating tables — an example

The assessment described here was made largely as a standard for land rehabilitation. The State Electricity Commission of Victoria planned to use an area of land in the Latrobe Valley as a site for dumping open cut mining spoil. Community representatives demanded that any land used for spoil dumping be rehabilitated so that its capability be at least that of the land prior to disturbance. For this reason it was necessary that the capability of the undisturbed land be known. The value of the assessment procedure as a source of guidance for management of the land after severe disturbance (spoil dumping) was also information requested by the State Electricity Commission.

The first step in making this assessment was to choose appropriate land uses for the study area. Land uses considered were those which were common in the district in the past and/or for which the land might be required during the next 50 years. These included informal outdoor recreation, residential development without reticulated sewerage, grazing of improved pastures, intensive horticulture, conservation of wildlife and commercial forestry. Each of these broad land use categories contains several activities or more specific land uses which must be assessed independently as they interact differently with the land (see Fig. 2). Therefore it is important that specific land uses be defined precisely. For example, residential development without reticulated sewerage would require consideration of the following specific land uses; construction of foundations for a light building of one or two stories, provision of septic waste disposal by conventional all-waste septic tank and soil absorption system (or alternative system), construction of secondary roads, and provision of an access track. Residential development without reticulated sewerage requires that the land have acceptable capability for all of these specific land uses.

It may also be necessary to specify the specific impacts to be considered (e.g., soil erosion, pollution of groundwater) if an assessment of hazard is required. In the example given, hazard of soil compaction during forestry operations is assessed (Table 2b).

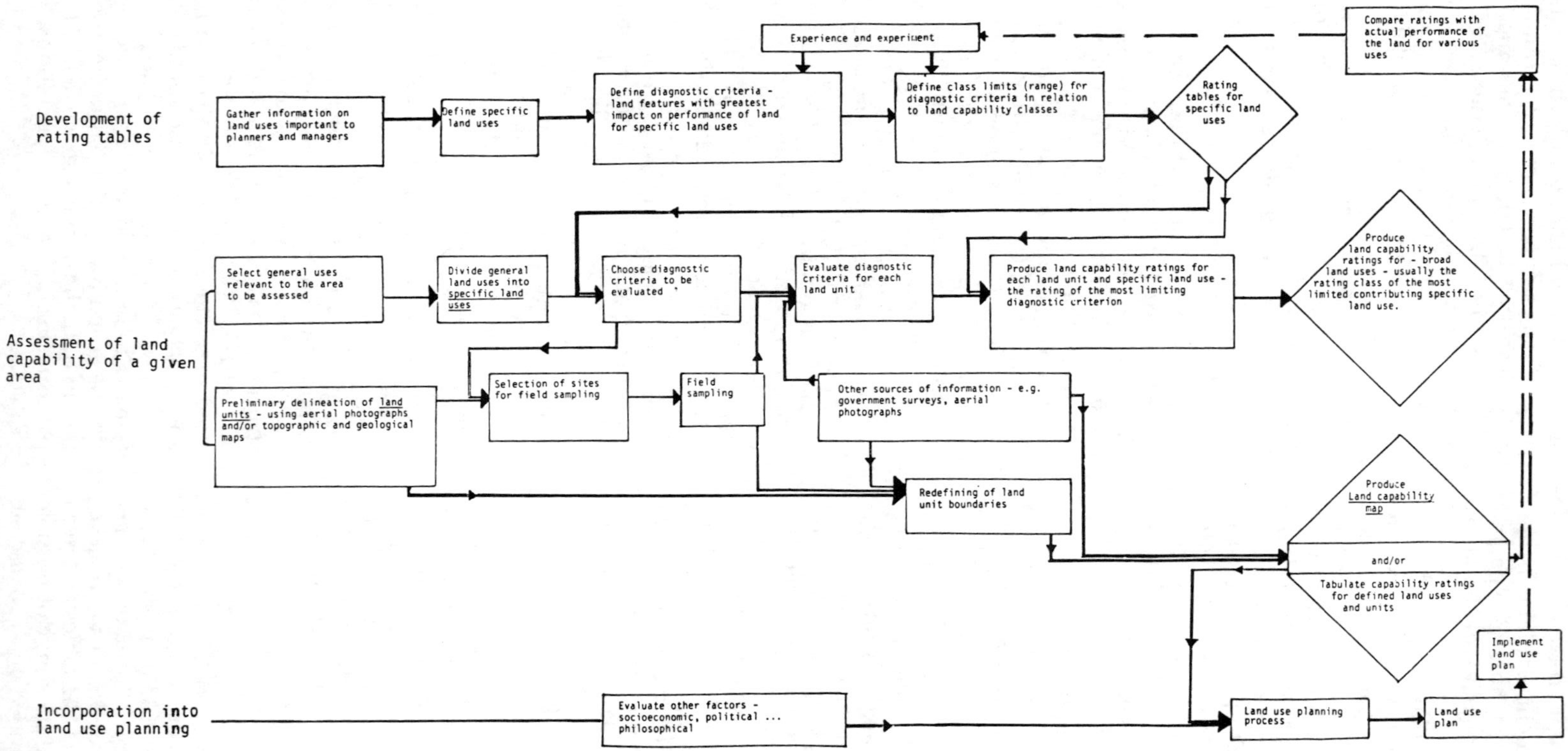

FIG. 2. Steps in land capability assessment and its incorporation into land use planning.

TABLE 2. Selected rating tables used in evaluating the study site.
a) Rating table for on-site effluent disposal
Adapted from Rowe *et al.* (1988).
Areas capable of being used for on-site soil absorption of all waste septic tank effluent from a single family dwelling.

Land features (diagnostic criteria)	Capability Class 1	2	3	4	5
Slope %[1]	0-15			15-30	30-200
Site drainage[2]	EW,W	MW	I	P	VP
Flood frequency[3] (% of years in which floods occurs)	0			0-5	5-200
Depth to seasonal water-table (cm)	500-150	150-120	120-90	90-60	60-0
Shallow permeability[4] (mm/day)	3 000-1000	1 000-300	300-150	150-100	100-0
Depth to hard rock (cm)	500-150	150-125	125-100	100-75	75-0
Gravel (%)	0-5	5-20	20-40	40-75	75-100
Stones (%)	0-2	2-10	10-30	30-60	60-100
Boulders (%)	0-0.02	0.02-2	0.2-2	2-10	10-100
Rock outcrop (%)	0-0.01	0.01-0.1	0.1-1	1-5	5-100
Shrink swell (%)	0-5	5-15	15-20	20-30	—
Slope failure risk	nil			low	high

[1]Reduce class limits by half in slope failure hazard areas.

[2]EW=excessively well drained.
W=well drained.
MW=moderately well drained.
I=imperfect drainage.
VP=very poor drainage.

Site drainage is assessed in terms of the rate at which water moves off a site as a result of runoff and drainage through the profile. It depends on both whole profile permeability and surface slope.

An excessively well drained site is one from which water moves very quickly. The water table does not approach the soil surface even after prolonged rainfall and ponding of water at the surface does not occur at any time.

A moderately well drained site is one from which water moves only moderately quickly. The soil profile is seldom water logged and ponding of water at the surface occurs for less than two days at a time after thorough wetting.

A poorly drained site has the water table within 20 cm of the surface for up to three months at a time. Some ponding of water at the surface may occur for up to one month.

A very poorly drained site has water ponded at the surface for more than a month at a time (Jeffery & Howe 1976, p.5).

[3]Upgrade one class if floods are low velocity, shallow and easily diverted with banks.

[4]Based on determination of hydraulic conductivity, "K" where K exceeds 6.0 m/day risk of pollution of water bodies must be considered.

The second step was selection of rating tables for the specific uses to be considered. These were obtained from two Department of Conservation, Forests and Lands publications (Rowe *et al.* 1988 and Anon. 1988b). Examples are given in Tables 2a-d.

Next, land units were defined and relevant land features (diagnostic criteria) were assessed for each land unit. For the four land uses described in Table 2 the land features listed in Table 4 had to be evaluated. Further land features were required for other uses; e.g., grazing of improved pasture, intensive horticulture and construction of farm dams. Several more were added as being of importance in determining detailed land management in this situation. The last were land features which could be modified readily at moderate cost but would influence the success of a rehabilitation programme (e.g., soil pH and nutrient status).

Seven land units were delineated using aerial photographic interpretation, topographic maps and site inspection (Fig. 3) and values of the land features of interest were determined. These were tabulated in the format shown in Table 4.

It should be recalled here that land capability assessment is not a total assessment of the suitability of land for a given use but rather an assessment of its potential for the use resulting from the biophysical resources. In this case precise definition of the management aspect of land use was required by the client. The Victorian Land Capability assessment procedure is, from time to time, used to

TABLE 2 (b) Rating table for hazard of soil compaction under forestry operations. Adapted from Anon. (1988b). The first four criteria are used in the rating table presented in this report; the following three have been added by the authors. As this table is still in an early stage of development only three classes are used.

Land features (diagnostic criteria)	Capability Class Low (L)	Medium (M)	High (H)
Horizon differentiation[1]	Graduational, Uniform	Uniform, Duplex-No A2 horizon	Uniform, Duplex-A2 horizon bleached or unbleached
Texture[1]	Loams, organic loams	Clay loams, light clays, loamy sands	Silt clays, sandy clays, grey sands, bleached sand
Profile permeability[2] (mm/day)	High (3 000-500)	Moderate (500-100)	Low (100-0)
Site drainage	Well drained	Moderately drained	Poorly drained
Additional criteria[3]			
Shrink-swell potential	High-moderate	Low	
Organic matter content	High (>10%)	Moderate (3-10%)	Low (<3%)
Structure[4]	Strong crumb or SAB	Moderate crumb SAB or AB	Weak or apedal

[1]Notation as in Northcote (1979).
[2]The original table did not give the figures in brackets. These were added by the authors.
[3]Relevance of these land features needs to be tested.
[4]SAB=sub-angular blocky
AB=angular blocky
Weak structure — natural aggregation is almost imperceptible or not clearly seen.
Moderate structure — natural aggregation can be seen clearly but peds still cohere to each other
Strong structure — aggregates are clearly seen and do not cohere to each other

TABLE 2 (c) Land capability ratings for paths and trails Rowe *et al.* (1988). Areas capable of being used for walking trails and bridle paths. Refers to the ability of the natural material to withstand a moderate to high intensity of foot traffic without deteriorating through water or wind erosion.

Land features (diagnostic criteria)	Capability Class 1	2	3	4	5
Slope (%)	0-5	5-8	8-15	15-25	25-200
Site drainage[1]	W	MW EW	I	P	VP
Flooding[2]	0-20			20-100	100-200
Topsoil texture[3]	SL L CL FSL	LS LFS SCL	LC	C MC S	ORGL ORGC
Stones (%)	0-10	10-20	20-30	30-100	100
Boulders (%)	0-1	1-5	5-10	10-100	100
Rock outcrop (%)	0-1	1-5	5-10	10-100	100
Dispersive Clay (%)	0-6	6-10	10-16	16-20	20+

[1]See footnote 2 to Table 2a.
[2]Downgrade to Class 5 if flooding occurs during season of use.
[3]See footnote 1 to Table 2 (b).

provide a context for such detailed management prescriptions. In these cases it would be usual to collect additional information (as here). Though originally developed for broader scale assessment, the system does appear to be providing a useful framework for detailed management evaluation (D. Howe, pers. comm.).

Determination of the capability ratings completed the assessment. Values of the diagnostic criteria found in each land unit were compared with those in the rating tables and capability ratings were assigned to the land unit on the basis of each criterion separately. The overall rating was then taken to be that resulting from the most limiting

TABLE 2 (d) Forestry Radiata Pine for commercial production. Adapted from Rowe *et al.* (1988). Preliminary screening of the area to determine climatic suitability is assumed to have already taken place.

Land features (diagnostic criteria)	Capability Class 1	2	3	4	5
Slope (%)	0-12	12-25	25-35	35-50	50-200
Depth to hard rock	500-100	100-70	70-75	45-35	35-0
Flooding — flash	0-1		1-10	10-20	20-200
— inundation	0-1		1-10	10-20	20-200
Topsoil texture	SL L FSL SCL CL	LS LFS ORGL		S LC	MC C ORGC
Drainage	W	MW	I	EW P	VP
Stones (%)	0	0-2	2-15	15-20	20-100
Boulders (%)	0	0-2	2-15	15-20	20-100
Slope Failure Risk	N			L	H
Organic Matter (%)	40-6	6-4	4-2.5	2.5-0	—

TABLE 3 Meanings of the rating classes (from Rowe *et al.* 1988).

Capability Class	Degree of Limitation to Development	General Description and Management Guidelines
1 (Very good)	The limitations of long term instability, engineering difficulties or erosion hazards do not occur or they are very slight.	Areas with high capability for the proposed use. Standard designs and installation techniques, normal site preparation and management should be satisfactory to minimize the impact of the environment.
2 (Good)	Slight limitations are present in the form of engineering difficulties and/or erosion hazard.	Areas capable of being used for the proposed use. Careful planning and the use of standard specifications for site preparation, construction and follow-up management should minimize development impact on the land.
3 (Fair)	Moderate engineering difficulties and/or moderately high erosion hazard exist during construction.	Areas with fair capability for the proposed use. Specialized designs and techniques are required to minimize development impact on the environment.
4 (Poor)	Considerable engineering difficulties during development and/or a high erosion hazard exists during and after construction.	Areas with poor capability for proposed use. Extensively modified design and installation techniques, exceptionally careful site preparation and management are necessary to minimize the impact on the environment.
5 (Very poor)	Long term severe instability, erosion hazards or engineering difficulties which cannot be overcome practically with current technology.	Areas with very poor capability for the proposed use. Severe deterioration of the environment will probably occur if development is attempted in these areas.

land feature. The final land capability rating included the numerical rating value plus one (or more) letter indicating the limiting land feature which led to the rating. Tables 5a and b show the way in which ratings for the specific uses "on-site effluent disposal" and hazard "soil compaction" were derived using rating Tables 2a and 2b with land features listed in Table 4. For on-site effluent disposal the most limiting land features in land unit GSL.1 were percentage stones and gravel. These gave the unit a rating of 3-4 Ss, Sg for this specific use with the letters Ss and Sg indicate that the limiting land features were stones and gravel. The land unit St2 had a rating of 4G with the letter G indicating that excessive slope was the limiting feature. The other specific land uses (building foundations and provisions of secondary roads and tracks) which contribute to the more general land use "residential development" were less limited than effluent disposal in land unit GSL.1 and so the overall rating for residential development was determined by effluent disposal to be 3-4 Ss, Sg. That is, the most limited specific land use determines the rating of the general land use to which it contributes. For land unit St2, provision of secondary roads was also limited to 4G by excessive slopes and so the overall rating for residential development for this unit was 4G as a result of slopes restricting both effluent disposal and provision of roads.

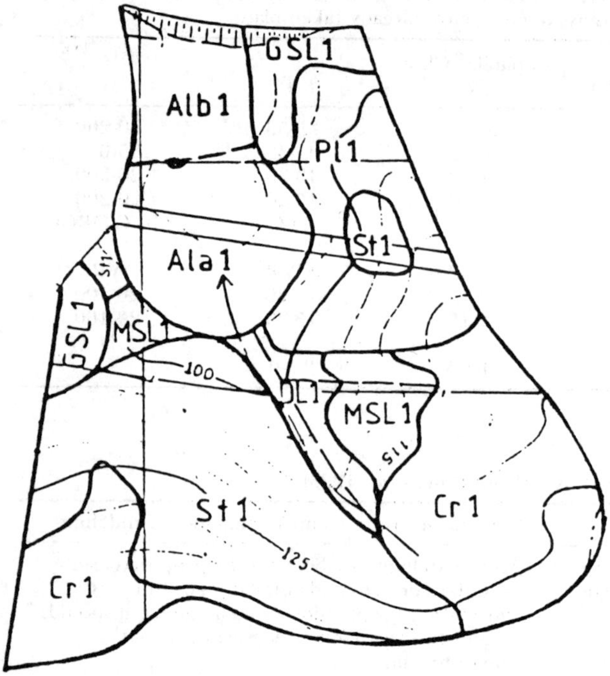

GSLl — gentle lower slope
AL — alluvial areas
ALa1 — better drained
ALb1 — poorer drained
PI1 — medium slope with lateritic layer
MSL1 — medium slope without shallow laterite
DL1 — drainage line
Cr1 — crests, ridges, gentle upper slopes
ST1 — steep exposed slope

FIG. 3. Land Units of the study site.

The rating table for soil compaction is in rather an early stage of development at present (hence the use of only three classes), and so three additional land features which may be relevant were added for consideration. Ratings obtained were compared with observed performance of the land. In a similar way draft rating tables have been developed in the past for a range of land uses. These draft tables have then been used to rate land and the actual performance of the land under specified land uses has then been compared with the land capability assessment as a basis for modifying and upgrading rating tables.

In this particular study the capability ratings obtained for the spoil disposal area were combined with an assessment of wildlife habitat potential to provide a standard for rehabilitation. The State Electricity Commission had also prepared draft plans for the completed spoil dump and details of the type of spoil material to be dumped. An evaluation of the likely capability of the land once the spoil dump was formed was also required. Results were used to provide guidelines for appropriate management of the dump and some modification of its slope, aimed at reducing the proportion of steep slopes.

This example illustrates some key features of the Victorian procedure; namely the division of diagnostic criteria and hence capability into discrete classes, use of complex land qualities, (land features which depend on a number of interacting characteristics of the land — e.g., site drainage), and the limitations approach. Evaluation of a site in isolation from surrounding influences or effects is also usual. If erosion hazard, for example, is high, this will, in all probability, lead to off-site impacts, as well as reduced production of the site, unless management is increased. However, the information on off-site effects is incidental to the primary focus of the assessment. In the same way off-site factors which may impact on the potential of the area for a particular use are not considered. For example, land uses elsewhere which increase accessions to ground water may lead to salinity in a discharge area. This is not dealt with in current land capability assessment procedures. Staff of the Land Protection Division are however proposing to develop salinity hazard maps which will include delineation of recharge areas for use by planners (P. Brewin, pers. comm.). The tendency to use land features which are familiar to soil scientists trained in plant production is a further feature of the procedure. This allows use of extensive existing data including that contained in the Land Protection Division's land systems surveys.

The next section will consider some problems which arise from these and other features of the system.

Evaluation of the Victorian system of land capability assessment

As indicated earlier, the Land Protection Division's land capability assessment system is receiving wide acceptance by planners and administrators. Acceptance is sometimes less whole-hearted amongst land users and those closely involved with them (D. Howe, pers. comm.). Many of these people feel that they already have an adequate understanding of their land. This may be true but does not overcome problems of communication when different sites must be compared or when new land managers and new land uses are involved.

Some difficulties with the present system do exist however. The rather simplistic nature of the procedure and discrete classes do not reflect the complexity of ecological systems. It is not, as mentioned earlier, possible to take full account of interactions between land features or land uses using a limitations approach (Van de Graaf 1988; Lindsay 1979). A few rating tables do make some allowance for interactions between land features;

TABLE 4. Diagnostic criteria and their values for land units of the proposed soil disposal site.

Land feature (diagnostic criteria)[1]	Land Units Gentle lower Slopes (GSL1)	Steep N,W Slopes (St1)
Boulders and rock outcrops (% of area)	0	0
Depth to seasonal water table (cm)	100	200
Depth to rock or impervious layer (cm)	200	300
Dispersive clay (%)	5-10	15
Flood frequency — inundation	0	0
— flash flood	0	0
Horizon differentiation[2]	duplex with no A2 horizon	duplex with bleached A2 horizon
Profile permiability (mm/day)	shallow 400-2000 deep 30-1000	shallow 400-2000 deep 30-1000
Organic matter in top soil (%)	5	2-5
Shrink-swell potential (%)	<5	<5
Site drainage	well drained	excessively well drained
Slope (%)	5-10	25-30
Stones (%)	A horizon, 0 B horizon, 20-50	0 throughout
Gravel (%)	A horizon, 0 B horizon, 20-50	10-20 throughout
Boulders (%)	0 throughout	0 throughout
Unified soil group[3] — A horizon	SM, ML	SM, ML
— B horizon	SC, SM, CH	SC, SM, CH
A horizon — texture[2]	loamy sand sandy loam, loam	silty loam sandy loam
— depth (m)	0.1-0.2	0.1-0.15
B horizon — texture[2]	medium clay	medium clay
Slope failure risk	nil	nil

[1]Criteria needed to assess capability for:
a) commercial forestry using radiata pines
b) on site effluent disposal
c) paths and trails
d) soil compaction hazard with forestry

[2]Northcote's notation (Northcote 1979).

[3]The Unified Soil Classification is based on the soil's particle size distribution and pasticity index. It is correlated with bearing capacity, permeability and shrink-swell characteristics of the soil.

for example, the rating table for intensive cropping allows for the interacting influence of soil structure and slope on erosion hazard by providing different tables for three major soil structure types (Rowe *et al.* 1988). Footnotes such as those used in Table 2a to modify values for flooding and slope also take some account of interactions. The use of land qualities makes some allowance for interactions as well, but land qualities are often rather vaguely defined and assessed subjectively (Rowe *et al.* 1988).

The limitations approach may be inappropriate where effects of different land features can be shown to be multiplicative or additive. The Universal Soil Loss Equation illustrates the former possibility; factors influencing production of water from a catchment (King *et al.* 1986) demonstrate the latter. Where an area has several kinds of production at the same time interactions between land uses can pose special problems; for example where water is one use in a catchment which is also used partly for grazing and partly for forestry or as a wildlife reserve. Clearly using land for grazing or horticulture would impact on the potential of land for water production in a different way than would its use as a wildlife reserve or forest. Land features such as slope or drainage would pose different hazards to many uses if land were forested rather than cultivated. However, interactions between uses are almost never considered (an exception is a rating system for water supply developed by King *et al.* 1986). A problem is that systems which consider such interactions are likely to become excessively cumbersome and difficult to apply.

Apart from difficulty in defining complex land qualities there can also be problems with definition of land uses. Lewis (1988) has pointed out that trails for walking do not have to withstand the same impact as those intended for horse riding,

TABLE 5 (a). Land capability ratings for on-site effluent disposal.

Diagnostic criteria	Gentle lower slope (GSL1) value	rating	Steep N,W slope (St2) value	rating
Slope (%)	5-10	1	25-30	4
Site drainages	W	1	EW	1
Flooding	0	1	0	1
Depth to water table (cm)	100	3	200	1
Shallow permeability (mm/day)	400-2000	1-2	400-2000	1-2
Depth to hard rock (cm)	200	1	300	1
Gravel (%) at depth of excavation	20-50	3-4	10-20	2
Stones (%)	20-50	3-4	0	1
Boulders (%)	0	1	0	1
Rock outcrop (%)	0	1	0	1
Shrink swell (%)	<5	1	<5	1
Slope failure risk	nil	1	nil	1
Overall rating	3-4 due to stones and gravel. Depth to water table is also somewhat limiting.		4 due to slope.	

TABLE 5 (b). Hazard rating for soil compaction with forest operators.

Diagnostic criteria rating	Gentle lower Slope (GSL1) value	rating	Steep N,W Slope (St1) value	rating
1. Horizon differentiation	duplex No. 2	M	duplex with A2	H
2. Texture[1] A horizon	LS, SL, L	L-M	SL, SL	L
B horizon	MC	H	MC	H
3. Profile permiability — shallow (mm/day)	M (400-2000)	L-M	M (400-2000)	L-M
— deep	LM (30-1000)	M-H	LM (30-1000)	M-H
4. Site drainage	W	L	EW	L
5. Shrink-swell (%)	low	M	low	M
6. Organic matter content (%)	5	M	2-5%	M-H
7. Structure[1] A-horizon	weak SAB	H	weak SAB	H
B-horizon	moderate AB	M	moderate AB	M
overall rating based only on current LPD table. Anon 1988b) (criteria 1-4)	M in A horizon due to various soil factors H in B horizon as this is shallow and texture is MC		H due to horizon differentiation and texture of shallow B horizon	
rating with extra criteria 5-7	H in A horizon due to weak structure		H in A horizon due to weak structure	
observe hazard	H		H	

[1]The risk of compaction of the B horizon would depend, in part, on the depth of the A horizon.

yet both uses are dealt with in the one rating table in Rowe *et al.* 1988. Because management is part of land use great variation, even within a clearly defined production type, is possible. this can be dealt with to some extent by defining the rating classes for a given type of production in terms of increasing standards of conservation management required with the poorer capability ratings as in Table 3.

For some land uses the problem of land use definition is more profound. For uses like water production or wildlife protection the concept of what the desired product or use objectives should be is debatable. Various indices have been devised to represent such products (King *et al.* 1986; Kikkawa 1976; Anon 1986). The reasons for choice of production factors and the way they are weighted and combined to produce a single index is often not clear. Again variables measured on dissimilar scales are often added to produce the index. Unless the weightings of each factor transforms these variables to a common scale such additions are improper. Such a transformation would require that a quantitative relationship between each variable and some single measure of production be known. For quantity of water production this could be volume of water per year and the nature of product could be defined precisely. For water quality contributing factors might be turbidity, presence of chemicals or organisms which threaten health, unacceptable taste or smell, or salts which corrode pipes. In other words there is not a single water quality but water qualities, some relating to health, others to aesthetics and still others to maintenance costs of the supply system. Each could be transformed to a single monetary scale by considering the cost of purifying the water to a standard which is acceptable to the community. The acceptable community standard would then become the definition of the product or land use. This could vary with the community involved. In fact such careful definition of land use to permit transformation of contributing factors to a common scale is seldom done. In the case of a land use like wildlife protection, weighting of the factors may largely reflect assumptions concerning the relative importance of such variables as species diversity, rarity of an individual species, or the relative contribution of various aspects of habitat to these variables. In such cases the production index appears to be more a definition of the land use than a transformation of the variables included to a common scale. If there is confusion concerning the nature of the land use an evaluation of the land's potential for this use must also contain confusion. Rowe *et al.* (1988) attempt to give precise definitions for those uses for which they provide ratings (e.g., Table 2a, 2c) as does the Food and Agricultural Organization (Food and Agricultural Organization 1976, 1984; Dijkerman 1988). Where rating tables for new land uses are being developed it is important that the land use or product being considered be precisely and fully defined. It may also be necessary to redefine land uses as management practices or community perceptions change in the future (King *et al.* 1986).

Definition of the rating classes is also less precise than is desirable (refer to Table 3). As indicated above, rating classes can be defined for individual land uses in terms of required management. This is a useful approach. However, even within one rating table, various diagnostic criteria may impose different kinds of limitations on a land use. Consider Table 2a for example. Presence of boulders would restrict water absorption and increase costs of excavation, while flooding would pose a problem of water pollution. Stating the nature of the limitations, as well as giving the simple numerical rating, overcomes this problem to some extent. This is also the approach recommended by the Food and Agricultural Organization (1976, 1984).

Defining land units to which capability ratings apply presents other kinds of problems; for example there may be significant variation in diagnostic criteria over very small distances in the landscape which make mapping them impractical. This problem is particularly severe where the scale of the land use is much smaller than that of a practicable map. Sites for effluent disposal, paths, roads and houses all need only very small areas of acceptable land. For small areas of high value land intensive surveys may be justified.

Where variation occurs within a land unit (such as that in shallow permeability in Table 5), the only appropriate approach may be to point out this variability to potential land users and, where possible, the proportions of the different kinds of land within a land unit. It may even be possible to indicate the pattern of variability. Users can then be instructed to undertake site inspection to establish where there is an adequate site for their use. Indeed this approach is being adopted by the Land Protection Division when carrying out land capability surveys for some clients. An initial sieving process is used to allocate land to three classes. These are areas with little hazard, areas where hazards are most likely to be unacceptably high and areas where further detailed evaluation is needed (D. Howe, pers. comm.; King *et al.* 1986).

The tendency to use as diagnostic criteria those land features of interest for plant production has already been mentioned. Lewis (1988) is particularly critical of the choice and weighting of land features where engineering uses (such as roads and trails) are being considered. For other uses, such as informal outdoor recreation and grazing, the propensity of the area to be reafforested by natural regeneration, or to grow weeds, might be appropriate additional diagnostic criteria. However, rating tables of the type currently used do not easily handle such dynamic processes. Indeed ecological succession is a process usually not considered in any land evaluation systems, even where description of vegetation forms a major part of the assessment.

Part of the difficulty with choice and use of diagnostic criteria is a failure to obtain adequate feedback of knowledge gained by applying the assessment system in the field. No formal procedure has been developed to achieve this and it is becoming clear that an informal approach is not adequate (Rowe *et al.* 1988; D. Howe, pers. comm.). Development of a formal feedback procedure is desirable.

Possibly the greatest problems in the application of any land capability assessment systems are that they must rest on limited knowledge of both the ecological interactions involved and an incomplete data base of land characteristics (Van de Graaf 1988; King *et al.* 1986). In recognition of this problem the Victorian Government has made a commitment to carry out research into processes of soil degradation and the effects of timber harvesting, promote research into how ecosystems function and react to various activities, and continue to develop a data base for natural resource management (Anon. 1987a).

A final point should be made in relation to development of any land evaluation system including land capability assessment. As stated earlier, it is the usefulness of the system which should be the key consideration. To be useful the system must produce reasonably accurate statements concerning the potential of the land to sustain specified use. However, no amount of effort spent on producing an ecologically correct system will be of much value if its potential users are not able or willing to use the system. For this reason the Land Protection Division has made a conscious effort to avoid complexity in the application of its system. In spite of this it appears that potential users need some training in the basis of the approach and its use. When only a few people are involved, training on the job is possible. However, with a decentralized organization or where people from a number of different organizations are involved, specially designed training courses are required. The Department of Conservation, Forests and Lands has recently run brief introductory courses for staff concerned with planning, to heighten awareness of the land capability assessment procedure and encourage its use. However, there would be value in providing more comprehensive training in these skills. This should contribute to further development and better application of the system. The training necessary for those who will apply a land evaluation procedure should therefore be seen as an integral part of the system and not an after thought.

In spite of its various problems the Land Protection Division's land capability assessment procedure has proved useful to both planners and land managers. This is shown by its quite wide acceptance and use. Its simplicity ensures that it is not difficult for its users to understand. Its format allows easy amalgamation of land capability information with other relevant data. It provides, furthermore, a framework for further research into the interactions between land features and land use which is essential for the continued improvement of the rating tables.

The underpinning Gibbons (1976) model (Fig. 1) also provides a conceptual framework for the extension of land evaluation beyond that usually considered by the Land Protection Division. A few of these extensions will be considered next.

Extensions

Land capability assessment has tended to deal mainly with commercial production such as crops or forestry or readily definable uses such as effluent disposal or road construction (Rowe *et al.* 1988; Food and Agricultural Organization 1976, 1984; Grant 1973). A number of possible extensions of the land capability assessment concept have already been alluded to.

Assessment of wildlife production was attempted by King *et al.* (1986). Land features used were vegetation type [using Specht's (1970) classifications], soil type (Northcote *et al.* 1975) and land management practices. Water harvesting is defined as a land use in *Guidelines for Catchment Management*, prepared by the Rural Water Commission of Victoria (Anon. 1987c). King *et al.* (1986) have proposed a land capability rating system which evaluates potential for water production in terms of both quality and quantity.

In *Victoria's Wetlands* (Anon. 1986) conservation of wetland ecosystems can be seen as a land use with various features of the land (particularly those which influence water quality and fluctuations in water depth) affecting the use.

One of the most important tasks in extending the land capability model to new types of land use is defining the land use adequately.

Use of land capability assessment to provide information which will assist with rehabilitation of mined areas has already been mentioned. Two studies by Lindsay (unpublished reports 1987) indicated that the system had value in this area. The sponsoring organization, the State Electricity Commission of Victoria, is continuing to use land capability assessment as a tool in rehabilitation. In this case several extra land features were found to be important (e.g., for building foundations the extent of settling of spoil materials). The potential for assessment of material as it is being exposed and formed into new landscapes was also considered. An on-going evaluation along this line could be built into many land capability assessments, particularly where engineering works are involved.

Several benefits derive from using a conceptual model such as that developed by Gibbons as the basis for a broad range of land evaluations. The model leads to a holistic and systematic procedure and various assessments produced are likely to be compatible and so more readily integrated into planning and management proposals. It is essential for conservation of ecological diversity that they are assessed and represented in a form which gives them weight in decision making processes. As land capability assessment has already become familiar to and accepted by many people, use of a similar procedure is likely to give them such weight.

References

Anon. (1986) *Victoria's Wetlands: a draft wetland conservation statement.* Department of Conservation, Forests and Lands, Department of Water Resources and the Ministry for Planning and Environment, Victoria.

Anon. (1987a) *Protecting the Environment: A Conservation Strategy for Victoria.* Government Printer, Victoria

Anon. (1987b) *Draft Code of Practices — code of Forest Practices for Timber Production.* Department of Conservation, Forests and Lands, Victoria.

Anon. (1987c) *Guidelines for Catchment Management.* Standing Consultative Committee on River Management, Rural Water Commission of Victoria.

Anon. (1988a) State Environment Protection Policy, Waters of Victoria, *Victorian Government Gazette.* No.S13 Friday 26 February 1988, Government Printer, Melbourne.

Anon. (1988b) *Environmental Effects Statement — Proposed Trial of the Value Adding Utilization System Central Gippsland and East Gippsland Forest Management Area.* Department of Conservation, Forests and Lands, Victoria.

Anon. (undated) *Victorian Timber Industry Strategy — Some questions answered.* Government Printer, Melbourne.

Austin M. P., Basinski J. J., Cocks K. D. & Ive J. R. (1977) *South Coast Project.* CSIRO, Technical Memorandum 77/19, Canberra.

Burch G., Graetz D. & Noble I. (1987) Biological and physical phenomena in land degradation. In: *Land Degradation* (eds A. Chisholm & R. Dumsday) pp. 27-48. Cambridge University Press, Cambridge.

Burrough P. A. (1978) Overseas national programme of land resource survey and evaluation. In: *Commonwealth and State Government Collaborative Soil Conservation Study* 1975-77. Report 6, Australian Government Publishing Service, Canberra.

Chartres C. (1987) Australia's Land Resources at risk. In: *Land Degradation* (eds A. Chisholm & R. Dumsday) pp. 7-26. Cambridge University Press, Cambridge.

Crooks M. C. (1981) Planning for recreation and conservation in the Mornington Peninsula. In: *Outdoor Recreation: An Australian Perspective* (ed. D. Mercer) pp. 136-47. Sorret Publishing, Melbourne.

de la Rosa D., Cardona F. & Almorza J. (1981) Crop yield predictions based on the properties of soils in Sevilla, Spain. *Geoderma* **25,** 267-74.

Dijkerman J. C. (1988) An Ustult-Aquult. Tropept Catena in Sierra Leone, West Africa, 11. Land qualities and land evaluation. *Geoderma* **42,** 26-49.

Douglas J. S. & De J. Hart R. A. (1976) *Forest Farming.* Watkins, London.

Food and Agricultural Organization (1976) *A framework for land valuation.* Food and Agricultural Organization, Soils Bulletin No. 32. Food and Agricultural Organization, Rome.

Food and Agricultural Organization (1984) *Guidelines: Land evaluation for rainfed agriculture.* Food and Agricultural Organization, Soils Bulletin No. 52. Food and Agriculture Organization, Rome.

Grant K. (1973), *Terrain Analysis for engineering purposes of the Queenscliff Area,* Victoria CSIRO Division of Applied Geomechanics, Tech. Paper No. 12 Victoria.

Gibbons F. R. (1976) *A study of overseas land capability ratings.* Soil Conservation Authority, Victoria.

Guy G. L. (1978) *Land Capability survey for the You Yangs Forest Reserve.* Soil Conservation Authority, Victoria.

Jefferey P. J. & Howe D. J. (1976) *A study of land in the proposed Yarra Valley Metropolitan Park and interpretation for land based recreation and activities.* Soil Conservation Authority, Victoria.

Kikkawa J. (1976) *The fauna of Wilderness areas as a biological resource.* 47th ANZAAS Congress, Hobart.

King P. N., Carr W. M. B., Allinson J. F. & Boundy K. (1986) *Land capability assessment in Victoria.* Evironmental Report No.28. Graduate School of Environmental Science, Monash University, Melbourne.

Lewis B. (1988) *An assessment of land evaluation methods used for recreational land use.* Master of Environmental Science Thesis, Monash University, Melbourne.

Lindsay A. (1979) *Land capability and land use planning: a case study.* Master of Environmental Science Thesis, Monash University, Melbourne.

Lindsay A. (1981) Recreational land suitability assessment at the urban fringe. In: *Outdoor Recreation: Australian Perspectives* (ed. D. Mercer) pp. 92-106. Sorrett Publishing, Melbourne.

Lindsay A. (1987) *Mathematical Models — their use and limitations.* Environmental Discussion Paper No. 7. Graduate School of Environmental Science, Monash University, Melbourne.

Mc Harg I. L. (1969) *Design with nature.* National History Press, Garden City, New York.

Northcote K. H., Hubble G. O., Isbell R. E., Thompson C. H. & Bettenay E. (1975) *A description of Australian Soils,* CSIRO Division of Soils, Australia.

Northcote K. H. (1979) *A factual key for the recognition of Australian soils.* CSIRO Division of Soils, Australia.

Odum E. P. (1971) *Fundamentals of Ecology* (3rd edition) W. B. Saunders Co. Philadelphia.

Rowe R. K., Howe D. F. & Alley N. F. (1988) *Soil Conservation Practice, Manual of Guidelines for Land Capability Assessment in Victoria.* Department of Conservation, Forests and Lands, Melbourne.

Semarat P. (1986) *Potential for Yabby Farming in Victoria.* Master of Environmental Science Thesis, Monash University, Melbourne.

Specht R. L. (1970) Vegetation. In: *The Australian Environmental* (ed. G. W. Leeper) pp. 44-66 CSIRO. Melbourne University Press, Melbourne, Australia.

Van der Graaf R. H. M. (1988) Land Evaluation. In: *Australian soil and land survey handbook: Guidelines for conducting surveys* (eds R. H. Gunn, J. A. Beattie, R. E. Reid & R. H. M. Van der Graaf) pp. 258-80 Indata Press, Melbourne.

Wishmeier W. H. & Smith D. J. (1965) *Predicting rainfall — erosion losses from cropland east of the Rocky Mountains — Guidelines for selection of pracitices for soil and water conservation.* Agriculture Research Service, U.S.D.A. Handbook No. 282.

APPENDIX 1

The Universal Soil Loss equation relates soil loss from sheet erosion to features of the land and management as follows:

$$e=R\ K\ L\ S\ C\ P$$

where e, is the soil loss in tonnes/ha/yr.

R, the rainfall erosivity, is the number of erosion index units in a normal annual rainfall. The erosion index is a measure of the erosive force of a specific rainfall.

K, the soil erodibility factor, is the erosion rate per unit of erosion index for a specific soil horizon in cultivated continuous fallow on a nine percent slope 22.1 m long.

L, the slope length factor, is the ratio of soil loss from the field slope length to that from a 22.1 m length on the same soil type and gradient.

S, the slope gradient factor, is the ratio of soil loss from the field gradient to that from a nine percent slope.

C, the cover index, is the ratio of soil loss from a field with specified artificial or natural vegetation cover to that from the fallow condition on which the factor K is evaluated. C varies from 1 for ground with no cover to 0.01 for a 90% cover of permanent grass.

P, the erosion control practice factor, is the ratio of soil loss with various surface conditions (e.g., ripped or scraped smooth) to that with straight rows up and down slope. Usually management consists of ripping across the slope which has a P value of 0.8.

(after Wishmeier & Smith 1965).

Proc. Ecol. Soc. Aust. 1990 — 16:491-504

Identifying the National Estate: issues and directions in the assessment of significance of places

M. O'BRIEN and R. PURDIE

Australian Heritage Commission, G.P.O. Box 1567, Canberra, A.C.T. 2601

Abstract

The Australian Heritage Commission is a Commonwealth statutory authority, one function of which is to compile the Register of the National Estate. Early criteria and assessment methods used to identify places of national estate significance were seen to be arbitrary and lacking precision. The range of legislative, technical and political constraints which the criteria and assessment methodology must address, such as consistency, accountability, context, condition and integrity, are briefly discussed. Work aimed at improving the assessment system is described. This includes reformulation of criteria for significance, development of a more explicit interim assessment methodology and improved documentation. Proposed long-term improvements described include the development of a typology of environmental elements comprising the national estate, the preparation of type profiles describing the attributes of each type and the way they relate to the criteria, and implementation of a decision support system for assessment.

Introduction

The Australian Heritage Commission was established by act of the Commonwealth Parliament in 1975 and commenced operation in 1976. Its major function is to compile a register of places which comprise the National Estate. The National Estate is defined as "those places, which are part of the natural environment of Australia or of the cultural environment of Australia, which have aesthetic, historic, scientific or social significance, or other special value to future generations as well as the present community". Since this is the extent of the legal definitions and guidelines for the contents of the Register, it has devolved to the Commission to define criteria and methodologies.

The purpose of the Register is to identify those places which form the National Estate, to serve as an information resource to planners and decision makers and to educate Australians about their heritage. The Commission's role is to *decide* which places are of national estate significance and, subsequently and as the need arises, to *advise* on Commonwealth actions and decisions affecting those places. It should be emphasized that, by its registrations, the Commission *does not* take on land use planning responsibilities, create or manage a reserve system or change land use, ownership or management.

The Commission is one of a variety of government bodies providing advice about a number of registered places which have been the subject of intense public debate about conservation or resource development. A few of these are: sandmining of Shelburne Bay (north-east Cape York), intensive forestry operations in Southern Forests (Tasmania), Coolangubra and Tantawangalo Areas (Eden Region, New South Wales), mining in Kakadu National Park, Stage III (Northern Territory) and World Heritage nomination of the Wet Tropics Region of north-east Queensland. In such circumstances, the Commission is under an obligation to be accountable, rigorous and consistent in its registration decisions.

How the Commission goes about its task of assessing places is influenced by a number of factors, including the purpose of the Register, as established under its legislation, the level of resources available to address this and other tasks, and community expectations of the Commission and of the National Estate. The central issue for the Commission is to establish a consistent means of assessing significance against a defensible threshold, freed of resource constraints conferred by obligations to acquire or manage identified properties. The purpose of this chapter is to give a brief history of the development of the assessment system, to identify the major issues influencing assessment and to indicate directions for further development of criteria and methodology.

Compilation of the Register: 1977-86

The Commission commenced compilation of the Register early in 1977 employing very general criteria which were a product of their time (comparable to the practices then employed by Parks Services). These criteria, set out below, were considered appropriate to the restricted powers of the Commission under its Act.

Criteria in Operation from 1976-86

100 Presence of Endangered Species
101 Scientific Importance
102 Educational Importance
103 Social Importance
104 Aesthetic Importance
105 Historic Importance
106 Recreational and Tourist Importance
107 Diversity of Species and Communities
108 Naturalness
109 Rarity
110 Fragility
111 Representativeness
112 Position in Ecological or Geographic Unit
113 Other . . . *(specify)* . . .

No attempt was made during this period to arrive at a formal or precise specification of thresholds for listing. The registration threshold was established empirically and was reinforced by Commissioners with directly relevant experience, and by the very low turnover of staff and of the members of the Commission's advisory assessment panels. The principal criticisms of the initial criteria were that there was no clear separation of the concepts of "criterion" and "threshold", that the criteria and de facto thresholds lacked precision and were subjective and arbitrary, and that the assessment methodology was ill-defined and poorly documented. Nevertheless, relative consistency in decision-making was achieved despite these inadequacies, principally because of the experience and continuity of consultants, staff and Commissioners. Many of the places assessed under these criteria were those in the initial wave of nominations of very high significance and had been subject to earlier assessment processes. In retrospect, very few places would be likely to be removed as a result of a review applying improved criteria.

Several factors affected the public perception of the Commission's assessment processes. A larger proportion of later nominations were close to the threshold level of significance or were contentious areas which had not been assessed by other recognized assessment procedures. In recent years, the Commission has been drawn increasingly into intense community and political debate surrounding resource utilization proposals affecting places of national estate significance. In this context, action to upgrade criteria and assessment methodology became imperative.

Issues associated with assessment for the Register

Criteria and assessment methodologies must address a range of legislative, technical and political issues. Some of these issues are discussed in the following sections.

Relevance to the definition under the Act

The criteria and assessment methodology must address the terms of the definition of the National Estate under the Australian Heritage Commission Act. They must thus apply to "aesthetic, historic, scientific or social significance, or other special value", and to those categories of place which are encompassed by the term "natural environment of Australia and the cultural environment of Australia". Both prehistoric and historic places are included within the "cultural environment".

Identification of places appropriate to the Register

The issue of how significant a place must be to warrant registration is a vexed one, as the Act is not explicit on the matter. The clear intention of the "Report of the Commission of Inquiry into the National Estate" (the Hope Report) is that the Register would be a comprehensive listing of places of national estate significance.

Some listings or protective measures based essentially on "heritage" values are carried out at State or local government level where there are greater powers to take direct conservation action. Since such powers imply a financial burden for compensation, acquisition or works, the freedom to make an independent listing is much less than at the Commonwealth level. The Register of the National Estate offers the only likely opportunity for a comprehensive listing of significant places in Australia free of such constraints. Any criteria must therefore be consistent with the objective of the Register being an independent, comprehensive listing of National Estate places.

Facts, values and assumptions

Assessment of national estate significance for any reason involves analysis of facts and the attachment of value. Hitherto, there has been inadequate recognition of the distinction, either in public debates or in assessment methodology, between fact and value. This distinction underlies the need for following an explicit values-base for the assessment process, separation of "criteria" (distinguishing attributes) from "thresholds" (cut-off value), and separation of the data analysis step from the attachment of value.

More difficult and fundamental questions arise from acknowledgement of the values component. Such questions include what is the basis of those values? Are they based on a body of literature, an intellectual tradition or on an abiding community concern? Do these values reflect an identifiable and/or measurable quality inherent in a place? Can criteria and an assessment methodology be designed to give effect to these values and if so, how — quantitatively, qualitatively, objectively, rigorously, explicitly? To what extent are these values dependent on the observer? Who holds those values; are they the values of the community, professionals, developers, conservationists, politicians? How should the Commission reflect the community values and acknowledge the values component of its assessment methodology? And how does the assessment of values relate to methods of scientific, aesthetic or historic analysis?

It is unlikely that there will ever be final answers to such questions, since, in part, they reflect changing community values. However, the assessment process must include both elements which conform to the observed norms of scientific method, and elements which involve interpretation and attachment of value or significance. The Commission in its assessment activities must maintain intellectual and professional standards, responsiveness to the Australian community and political credibility. These factors constitute the basis for the Commission to be an independent statutory authority, composed of individuals with experience and expertise in heritage values and charged with the responsibility for making and accounting for those decisions.

Objectivity and assessment of value

It is obviously unacceptable for the Commission to assess places subjectively on the basis of individual preferences of Commissioners, staff or nominators. However, it is unlikely that absolutely objective assessment of value is theoretically possible, let alone practically achievable. Furthermore, it is not possible for any individual or organization to establish universal perception of their "objectivity". That is not to say that rigorous methodology is impossible, but it does underline the requirement to separate consideration of fact from consideration of value.

It is commonly, but erroneously, believed that quantitative methods are inherently objective. In reality, quantitative methods do provide analytical tools which may be extremely useful if they measure consistently and accurately what they purport to measure. However, even prudent and appropriate use of quantitative methods will not establish significance. For example, a landscape analysis based on a quantification of attributes such that running water is always assigned a high landscape value than still water, is simply quantifying one's prejudices, not establishing significance. Assessment of significance is the outcome of an interpretation of the results of various analyses.

In many subject areas, there are no "standard" quantitative models available as yet for analysis of data. For example, there are a number of methods for assessing diversity of species or of communities, but the Commission is not aware of any methodology suited to the assessment of places against thresholds for significance. Frequently even where appropriate quantitative models do exist, there are inadequate data to drive them, forcing a fallback to qualitative methods and expert opinion. The outcome will be a judgement which, although lacking some precision, is no less objective than that produced by quantitative analysis. While it is desirable to know *all* about the values of a place and to be able to analyze data precisely, very often the significance of a place against the criteria *can* be established unambiguously on incomplete data. Many conservation reserves have been established on such a basis.

Whatever the method of analysis, it is important to make explicit all elements of the assessment process from the broad fundamentals to the practical details. These elements include a comprehensive and coherent conservation rationale in sufficient detail to provide an explicit basis for significance criteria, guidelines for applying criteria and thresholds, and the detailed assessment methodology. Although much of the underlying rationale exists in various forms, a detailed synthesis is required of the scientific, philosophical and ethical arguments for conservation of various attributes, e.g., rarity, representativeness, natural processes, etc.

Diverse reasons for significance

The scientific value of places in the natural environment may derive from the degree to which landforms, geology, soils, flora, vegetation, fauna and/or landscapes demonstrate evolutionary patterns and processes, richness or diversity, or rarity. Places may also be important for their representative landscapes, environments or ecosystems, or for their value as research and teaching sites, type localities, and reference or benchmark sites. Natural areas also may be of significance for aesthetic values or historic, prehistoric or social associations. The criteria and assessment methodology must thus encompass the wide range of possible reasons for the significance of a place, and also enable both individual assessment of the diverse elements as well as their aggregation into an overall assessment of the place, recognizing that the significance of the whole may be more than just the sum of the parts.

Consistency of assessment

The Commission has a responsibility to assess the significance of places consistently across the diverse categories of "the National Estate", across State borders, and as community values change over time. While allowing such consistency, the assessment methodology must also be sufficiently flexible to allow for the fact that any single place may include natural and cultural elements. Kakadu National Park, for example, includes significant wetlands, rainforests, woodlands, habitats of rare and endangered flora and fauna and geological features as well as Aboriginal sites of cultural significance. For places with a diversity of elements, any two assessors may disagree in matters of interpretation of, or emphasis on, various data, and on attachment of value. As far as possible, therefore, the assessment process should minimize individual differences between assessors by informative and detailed guidelines and by review processes in decision-making.

Accountability

As a government agency, the Commission has a public responsibility to account for its procedures, including the registration criteria, and decisions. Public debate on conservation/development issues over this decade has resulted in the need for greater accountability of Commission decisions, particularly the assessment procedures and the specific basis for determining the significance of places. Increasingly the Commission has also been called on to publish and justify its advice to the Commonwealth on major resource development issues affecting National Estate places. It is thus important that the Commission develops sound practices and procedures for its activities, consistent with the Act, and arrives at defensible decisions. Not only must the Commission act in this way but also it must be seen to do so.

Professional and community opinion

The Act requires the Commission to identify what might be of significance "for future generations as well as the present community". Thus the Register must reflect the current concern of Australians for their natural and cultural heritage and, at the same time, anticipate changes in that concern. In the Comission's experience, there is, in the wider community, a low level of *understanding* of environmental processes and relationships and of the management requirements to guarantee natural heritage values for future generations. There is probably a larger and growing level of *concern* about the future of the Australian environment, although significant sections of the community appear to remain indifferent to that future.

Changes in community attitudes to the environment in the last 25 years have been dramatic, and are reflected in increased membership of voluntary conservation organizations, stronger opposition to development proposals both within and outside conservation reserves, and more intense lobbying at all political levels by both development interests and conservation organizations. It is also reflected in the increasing area set aside for conservation reserves in Australia (e.g., see Hopkins & Saunders 1987). These changes are likely to accelerate in the coming years in response to major environmental shifts evident at the national and/or global level, including climatic changes, deforestation, desertification, soil erosion and salinization, extinction of flora and fauna species, pollution, etc.

As scientists gain a better understanding of ecological, geological, geomorphological and historical processes, a greater understanding is gained of the ecological, geological, geomorphological and human history of Australia. This growth in understanding must also be reflected in the assessment of significance of places under the Act. The Commission must therefore respond not only to the current state of knowledge and understanding of Australia's natural and cultural heritage and current community concerns and values, but also likely long-term understanding, concerns and outlook for our heritage. These requirements reinforce the obligation for the Commission to be explicit about the underlying values, criteria and assessment procedures for its registration.

Context

The assessment of an individual place for the Register must be in an appropriate thematic and regional context. The requirement for precision in the assessment process demands that the place (and its constituent elements) be assessed in the light of places of a similar type(s). For example, a new nomination for a coastal inlet can only be assessed with reference to other coastal inlets. Such comparisons require a predefined classification system to establish types of places, detailed information about the population of places falling within each type, and, for each place of the type and for related types, detailed information of previous assessments against each criterion.

Condition and integrity

Condition is used in the sense of state of repair or management. In the natural environment, integrity is used predictively, in the sense of viability or resilience. Condition and integrity are used to assess the relative significance of places which may have similar inherent characteristics (soils, landforms, flora, fauna, etc.), and the likelihood of national estate values being maintained in light of natural processes. The acceptable levels of condition and integrity to establish the significance of a place are relative to both the values and the individual type of the place. Thus a rainforest community may have a wide range of values, some of which (e.g., wilderness, benchmark for particular forest type) are affected seriously by disturbance, and others of which (e.g., landscape backdrop) are affected to a lesser degree by the same disturbance. In some regions, plant communities are reduced to linear remnants, e.g., along rivers, roads, or railways; even the best examples might be in a parlous state, but might still be significant examples of those communities because of their scarcity value. In the latter situation, poor condition and integrity may detract from National Estate significance, but the values may still be above the registration threshold. Assessment of such a place may carry a recommendation that the place be reassessed at some later date.

Size and boundaries

Places are listed in the Register irrespective of their size, size being the outcome of assessing the significance of a place against the various criteria. Registered places vary in size from a site such as a single rock outcrop covering less than a hectare to a region such as the Great Barrier Reef which extends over some 250 000 km^2. For a value such as a rare species habitat or a reference catchment, the limits of the habitat or catchment define the boundary of the area having little or no value against the criterion. For other values, e.g., wilderness and representativeness, size influences the degree to which the place meets the criteria for significance, with values declining below threshold levels for areas outside the boundary.

Boundaries of areas to be registered are drawn appropriate to the values of the place from interpretation and generalization of available information. In regions with very low ecological gradients, arbitrary boundaries such as properties are commonly used to provide a reasonable and approximate sampling of environments or communities. In more closely settled areas, cultural elements such as roads, railway lines and property boundaries, are commonly used for registration boundaries, as they frequently represent the present limits of natural vegetation cover or delimit areas subject to particular forms of management such as logging, grazing, cropping, etc. Such cultural elements may also approximate natural boundaries. Commonly, statutory descriptions of boundaries are a combination of natural and cultural features.

Consistency with past decisions

Politically, it is essential that all registered places demonstrably meet revised criteria and improved procedures. It is also essential that existing registrations are reviewed to provide the contextual information against which future nominations will be assessed. Periodic reviews of registered places are essential to reflect current scientific, aesthetic and historic information and theory, current assessment criteria and methodology, current information about the place and about similar places and prevailing community attitudes.

Revised assessment system

In 1986, in response to the complex requirements described earlier, the Commission embarked on a process of reformulating its criteria and developing a more detailed assessment system. The parameters used for the development of criteria and assessment systems are set out in Table 1. This work is being carried out in two stages: an interim method currently being implemented is expected to operate until about 1990-91; longer-term development is expected to be introduced progressively from 1990 onwards.

Criteria and assessment: interim method

The main steps involved in the interim method involve reformulation of general criteria applying to all National Estate places, the development of more explicit assessment methodology and procedures, and improved documentation of methodology, procedures and assessment of places.

The criteria have been redrafted in two forms, firstly as a summary statement and, secondly, a more detailed application guide, including explanatory notes, and inclusion and exclusion guidelines. Examples of both formats are shown at Appendix I and II. The general criteria can be used as a framework for data collection and for establishing a rating of relative significance across the full range of aesthetic, historic, scientific and social values. This is achieved by assembling all relevant data for the place, assessing each data item against the relevant criteria and in the context of similar features elsewhere and giving it a qualitative rating, and then making an overall assessment of the place based on the assessments against individual criteria.

This qualitative analysis is carried out by a simple tabular arrangement of data in a "Values Table" (Appendix III). Assessment ratings are made using a scale from nil, very low, low, moderate, high, very high to outstanding. Rating is not carried out on a numerical or parametric basis. There is no grading of registered places, either by their relative significance or by their date of entry in the Register. The relative significance of places is reflected in the Values Table and in a Statement of Significance prepared at the end of the assessment process. It should be noted that the Values Table procedure is not in itself an assessment. Rather, it is intended that the rating step provide a consistent means of analyzing data so that a professional judgment of significance can be made. The "assessment of significance" step should be seen as subsequent to the various analyses which might be carried out. Assessment of significance will remain a matter of judgment rather than formula.

TABLE 1. Parameters for developing and evaluating criteria and assessment methodology.

Parameters	Evaluation of 1976-86 method	Evaluation of interim method	Evaluation of proposed method
1. Give effect to definition under the Act: "aesthetic, historic, scientific or social significance, or other special value".	Yes	Yes	Yes
2. Encompass "natural environment of Australia and the cultural environment of Australia", i.e., natural, Aboriginal and non-Aboriginal.	Very poorly	Yes, method deals better with the complexity of the natural environment	Yes, but proposed method will achieve further improvements for complex places and environments
3. Respond to changing community values	Not explicitly	Not explicity	Yes
4. Have a sound intellectual basis	Not explicitly	Not explicitly	Yes, see Type Profiles
5. Provide a detailed rationale for conservation objectives, criteria and thresholds	Not explicitly	Partly, in guidelines to criteria; more specific rationale leads to Stage 3	Yes, proposed overall conservation rationale
6. Achieve consistency in assessments both for similar places and between categories	Quite good in practice but not intrinsic to method	Expected to improve, but still not intrinsic to method	Explicit in Type Profiles; detailed comparison procedures to be applied and documented by DSS
7. Achieve consistency in assessments across categories and throughout Australia	Quite good in practice dependent on staff, not intrinsic to method	Expected to be greatly improved by criteria guidelines	Detailed comparison procedures to be applied explicitly
8. Minimize variability between assessors	Quite good in practice because of staff, rather than method	Expected to be greatly improved by criteria guidelines	Detailed comparison procedures to be applied explicitly
9. Documented values base	Not available	Slightly improved by criteria guidelines	Detailed rationale proposed
10. Apply consistently the criteria, assessment methodology and administrative procedures	Dependent on stability and expertise of staff and Commission	Improved documentation of procedures and methodology	Procedures as for interim method; methodology executed by DSS
11. Enable the reasons for registration decisions to be communicated fully	Communication very difficult beyond conservation practitioners	Improved but lacking underlying rationale	Improved rationale and method provide better basis for communication; improving database will also assist
12. Achieve and maintain political and community credibility	System subject to severe criticism; difficult to defend without explicit rationale and procedures	Marked improvement expected with more explicit procedures and documentation	New formula will be: sound rationale and procedures; more specific criteria; improved databases; detailed documentation of assessments
13. Achieve and maintain professional standing for criteria and methodology	Standing was achieved by personal standing of panel members, etc.	Yes, by virtue of expertise of members; also improved and more explicit methodology	Greater recognition expected with more rigorous methodology
14. Achieve and maintain public confidence	AHC standing poor, especially on controversial issues	Marked improvement expected, if public relations effort made	Greater recognition expected, if the improvements can be communicated

TABLE 1 — *continued*

Parameters	Evaluation of 1976-86 method	Evaluation of interim method	Evaluation of proposed method
15. Separate data analysis steps from attachment of value	No — decision-making processes confused	Yes	Yes
16. Utilize rigorous techniques for scientific, aesthetic or historic analysis, (using qualitative or quantitative methods, as appropriate)	On an opportunistic basis, if such work has been done; method relied on expertise of panels	Within limits of data; still emphasis on expert advice	More explicit and appropriate methodologies; development of GIS and other techniques can be incorporated into overall assessment framework
17. Facilitate integration of qualitative and quantitative analyses	Decision-making processes vague	Not explicitly	Yes, significant improvement
18. Develop explicit methodology for attachment of value	None, decision steps not differentiated well	Yes, significantly improved	More explicit criteria and thresholds for each type of place
19. Allow assessment of diverse natural and/or cultural places (or elements within places), each with reference to its own context	Considerable difficulties in methodology and procedures	Greatly improved methodology; procedures probably need further work	Further improvements possible
20. Enable assessments to be made in the context of similar assessments made (a) against each criterion for the type (b) for places as a whole	Done in very rudimentary way	Room for considerable improvement; problems of availability of data in a useful form	Yes, important part of the proposed DSS; intended to be an intelligent system, i.e., refining data, rules, etc.
21. Allow compilation of a comprehenive Register	Yes, but very haphazard when dependent on external nomination process	Sound improvement, but requires better application of contextual data	Yes, better methodology based on environmental typology; further improvements with GIS-based techniques
22. Define sufficiently precise thresholds	Impossible	Some improvement with criteria guidelines, but further improvement required	Yes, improved criteria and thresholds specific to attributes and population of similar places
23. Incorporate detailed knowledge of the attributes and distribution of the type	Not possible; complete dependence on panel experts	No real improvement with criteria guidelines; must be specific to type	Yes, this is the main rationale for type profiles system

Various methods may be employed to conduct such data analysis, according to the particular subject matter and the data available about the place and about similar places. For example, assessment of representation of species or associations in a natural area may be based on numerical analysis, or it may be based on an expert interpretation of remotely sensed data or of output from a geographic information system. Other grounds for significance may be based on qualitative analysis or on a mixture of qualitative and quantitative techniques. The step of assessing significance ascribes a value to the results of the various analyses.

It is essential that the assessment take into account the condition and integrity of a nominated place. Poor condition or artificial modification may have resulted in a loss of value, which may or may not have serious consequences for the National Estate significance of the place. The key questions are whether the area previously disturbed has sufficient National Estate value to warrant registration, or is important to a larger area proposed for registration, or is inconsequential in extent and/or affects significantly the values of the whole area. Figure 1 shows how these factors are taken into account.

The interim scheme described above remedies the major shortcomings of the initial criteria used for the Register and provides a consistent and defensible means of assessing significance. Table 1 shows an evaluation of the interim scheme and identifies those parameters which must be addressed

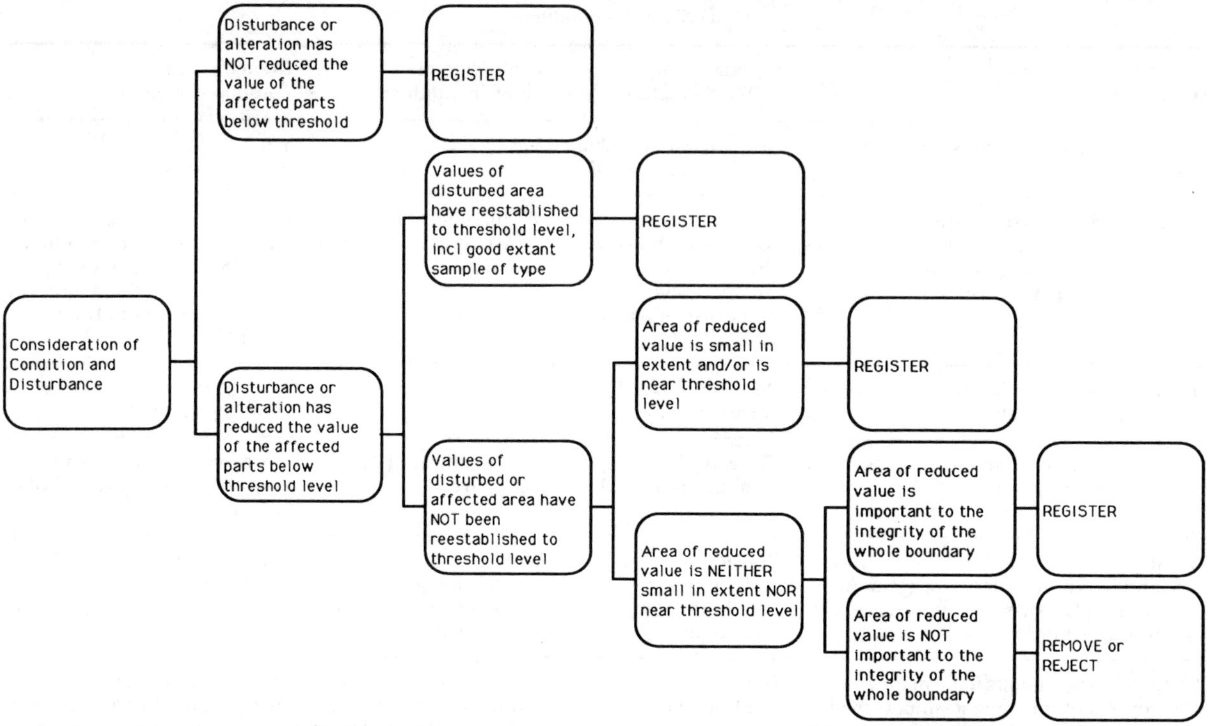

FIG. 1. Decision tree for consideration of condition in assessment of National Estate significance.

in further development work. Aspects of the interim method requiring further development include the need for more precise thresholds and the requirement to defend the methodology and particular decisions to critics and other external observers.

Criteria and assessment: long-term development

The major steps intended for the long-term development of criteria and assessment systems include the development of a detailed conservation rationale, and the development of a type classification and type profiles which provide, for each type, refined criteria for its characteristics, specification of its thresholds, and guidelines on appropriate methodology. Other steps include the development of a database of all judgments against each criterion for each type, continued refinement of appropriate analysis techniques, including use of Geographic Information Systems, review of all previous registration assessments, and the incorporation of type profiles, reference materials, database referrals and report generation capability into a decision support system. Some of these aspects are considered in more detail below.

Development of typology and type profiles

It is proposed to develop a classification tree (typology) for environmental elements (Figure 2), *not* in order to classify places or consituent elements, but rather to assist in the identification of which of one or more models (type profiles) might apply when assessing the significance of a particular place. Accordingly, the choice of models will not necessarily be mutually exclusive, as a particular place might be described by several models (e.g., a wetland may exhibit characteristics of both a saltwater and a freshwater wetland at various times, and so may be assessed with reference to each wetland type). It is anticipated that the typology will function more as a "lattice" than as a "tree", so that an assessor will arrive at the same type profile (e.g., for an Anabranch System) independent of which initial entry point and path is chosen based on a particular interest (e.g., geomorphology, wetlands, rivers or wildlife habitats). A prototype of the classification tree (or lattice) to identify types is under development on the Macintosh computer using HyperCard.

The type profile is intended to provide a model of the type, and will include all its attributes and the way in which they relate to individual criteria. Individual type profiles will thus include a definition of the type and its attributes, a review of the abundance and distribution of the type and of its variability or homogeneity, a review of the condition and integrity of the type, specific application of criteria to the attributes of the type, the scale of significance and registration threshold for the type, guidance on methodologies for assessing the type, and a directory of further resources for additional advice or information.

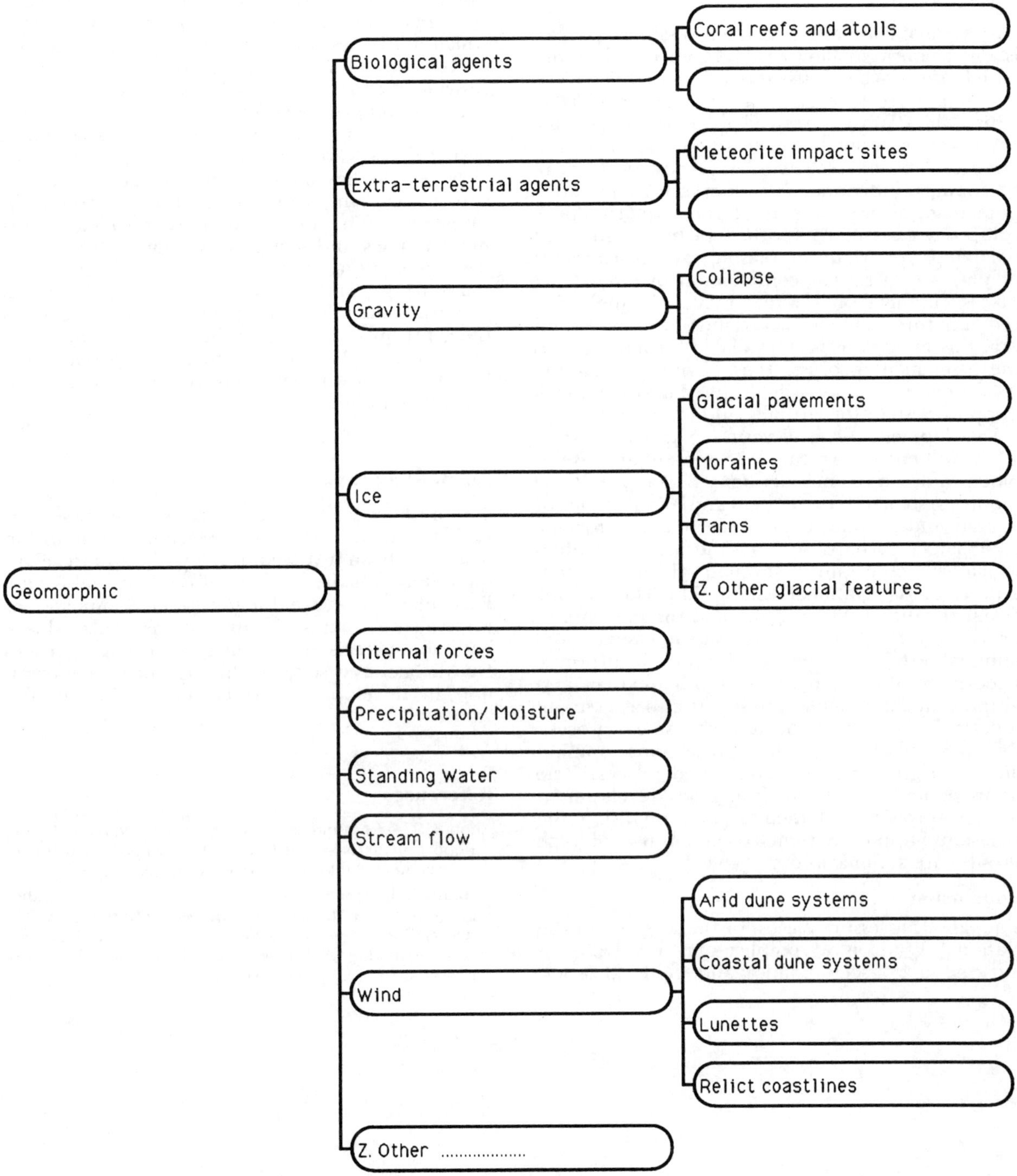

NOTE: All branches of the tree will be subdivided until the derived class describes an homogenous group of places which can be compared and which can be assessed using the same specific criteria.

FIG. 2. Draft typology for some geomorphological places.

Development of Assessments Database and Decision Support System

An essential component of the assessment process is a mechanism to allow easy reference to decisions based on earlier assessments, whether those decisions were to register a place or to reject it. This will require improved systems to facilitate reference to individual assessments against each criterion for the type (e.g., rare aquatic flora in billabongs), individual assessments for each attribute (e.g., all assessments of rare aquatic flora), composite assessments for the type (e.g., significant billabongs for whatever reason), and assessments of places with similar combination of types (e.g., places with billabongs, a major river channel, river red gum forests, mallee woodlands). It is important that this material is both available for and used in the assessment process. It is envisaged that the Assessments Database will be linked into a Decision Support System for virtually automatic access.

Development of a Decision Support System (DSS) will encompass the development and use of type profiles in the manner described above. It will require systematic compilation and maintenance of textual and graphical reference material for the anticipated perhaps a thousand or so profiles which will be required for natural environment aspects of the National Estate. The DSS will be designed so that type profiles and the database of previous assessments are easily accessible, and minimal effort will be required to incorporate selected reference material (both textual and graphic) and assessment comments into assessment reports. By providing a detailed assessment process, the DSS will also help elicit professional opinion on the significance of nominated places. The improvements expected from the development of type profiles and their application using the Decision Support System based on use of type profiles are summarized in Table 1.

Conclusion

The task of assessing places for the Register of the National Estate is a complex one, involving an intricate mix of socio-political and technical issues. Experience has demonstrated the need for explicit discussion of values, clarity about fundamental concepts and assessment steps, a methodology which addresses deduced design parameters, and a decision support system to control process, provide reference material, elicit judgments and generate reports. The interim assessment system currently being implemented provides a sound basis for assessing places for the Register. However, it still has shortcomings in demonstrating consistency and improving the precision of thresholds. The proposed more detailed assessment models and computerised procedures offer the opportunity for greater consistency, while minimizing conflict over assessment of natural resource values. The task is enormous and requires the contribution and co-ordination of a wide range of expertise. It also has many potential beneficiaries and would seem a fertile ground for fruitful collaboration.

Acknowledgements

The authors would like to thank Commissioners and staff of the Australian Heritage Commission who have been involved in the process of developing criteria for the Register of the National Estate. Particular thanks are due to former Commissioner Mr Angas Hopkins, for discussions and feedback both about Register criteria and this paper, and to Dr Michael Pearson, for his significant contribution in developing criteria relating to cultural values.

References

Anon. (1974) Committee of Inquiry into the National Estate, Report of the National Estate (Hope Report). Australian Government Publishing Service, Canberra.

Hopkins A. J. M. & Saunders D. A. (1987) Ecological Studies as the Basis for Management. In: *Nature Conservation: The Role of Remnants of Native Vegetation* (eds D. A. Saunders, G. W. Arnold, A. A. Burbidge & A. J. M. Hopkins). Surrey Beatty & Sons, Chipping Norton.

APPENDIX I

General criteria (November 1988) for assessment of places for the Register of the National Estate.

CRITERION 1:

Importance in the evolution and pattern of Australia's natural and cultural history

1.1 Importance in the evolution of Australian flora, fauna, landscapes or climate.
1.2 Importance in maintaining existing processes or natural systems at the regional or national scale.
1.3 Importance in exhibiting unusual richness or diversity of flora, fauna, landscapes or cultural features.
1.4 Important for their association with events, developments or cultural phases which have had a significant role in the human occupation and evolution of the nation, state, region or community.

CRITERION 2:

Importance as rare, endangered or uncommon aspects of Australia's natural or cultural history

2.1 Importance for rare, endangered or uncommon flora, fauna, communities, ecosystems, natural landscapes or phenomena or wilderness areas.
2.2 Importance in demonstrating a distinctive way of life, custom, process, land-use, function or design no longer practised, in danger of being lost, or of exceptional interest.

CRITERION 3:

Significant in demonstrating the principal characteristics of a class of Australia's natural or cultural places and environments

3.1 Importance in demonstrating the principal characteristics of the range of landscapes, environments or ecosystems, the attributes of which identify them as being characteristic of their class.
3.2 Importance in demonstrating the principal characteristics of the range of human activities in the Australian environment (including way of life, philosophy, custom, process, land-use, function, design or technique).

CRITERION 4:

Importance for associations with lives or works of significant persons in the past

4.1 Importance for their close associations with individuals whose activities have been significant within the history of the nation, state, region.

CRITERION 5:

Importance for aesthetic values or high creative or technical achievement

5.1 Importance in exhibiting particular aesthetic characteristics valued by a community or cultural group.
5.2 Importance in demonstrating a high degree of creative or technical achievement for the time.

CRITERION 6:

Importance for cultural or societal associations or as the focus of cultural or social sentiment

6.1 Importance as places highly valued by the Australian community (or an identifiable or regional segment of it) for reasons of religious, cultural, educational or social associations.

CRITERION 7:

Importance for potential in yielding information contributing to an understanding of Australia's natural or cultural history

7.1 Importance for information contributing to wider understanding of Australian natural history, by virtue of their use as research sites, teaching sites, type localities, reference or benchmark sites.
7.2 Importance for information contributing to a wider understanding of the past human occupation of Australia.

APPENDIX II

Extract from Guide to Application of General Criteria for the Register of the National Estate (November 1988).

CRITERION 1:

Importance in the evolution and pattern of Australia's natural and cultural history

1.1 Importance in the evolution of Australian flora, fauna, landscapes or climate.

Explanatory notes

This criterion applies generally to places which contain evidence and products of *past* climatic, geological, geomorphological, ecological or biological processes (in contrast to Criterion 1.2 below, reflecting continuing processes).

Scope

Climate:

Evidence of former climates, e.g.,
coastlines associated with former sea levels;
soils sequences characteristic of former climates (loess).

Geological:

Evidence of past geological processes, e.g., mineralization, sedimentation, vulcanism;
Erosional and depositional features;
Fault lines;
Sedimentary rock strata;
Fossil localities.

Landform:

Evidence of significant past landform processes, e.g., volcanic landforms, glacial landforms, stratigraphic sequences.

Soils:

Fossil soils;
Lateritic soils;
Sequences of dune creation and burial:

Biological/Ecological:

Relict species or communities;
Disjunct (or outlying) populations;
Refugia;
Atypical locations, substrates, climatic niches;
Sympatric or parapatric relationships;
Limit of range of species or community;
Species with Gondwanic or Indo-Malay affinities.

Inclusion guidelines

Places with fossil or other record of identifiable past climates or environments.
Places with remains of flora and fauna now extinct.
Places with primitive or relict flora or fauna surviving from earlier times.

Exclusion guidelines

Places in which the evidence of past processes is not clearly established.
Places which are replicated by places with clearer evidence or in better condition, although such replicates in other regions may not preclude registration.

Examples of places which are or might be registered using this criterion

Climate:

Willandra Lakes (NSW) — evidence of changes in climate and hydrological regimes over the last 40 000 years.
Quincan Crater (Qld) — evidence of climatic oscillation reflected in deposits containing rainforest pollen and former vegetation types.
Lake George (NSW) — evidence of drying climate and of increased fire frequency.

APPENDIX II — *continued*

Geological

Beecroft Peninsula (NSW) — sedimentary sequences.
Geikie Gorge (WA) — fossil reef.

Landform:

Byaduk Caves (Vic) — volcanic caves system.
Exit Cave (Tas) — prime example of limestone cave formation.
Mt Warning (NSW) — volcanic plug and caldera.
Central Plateau (Tas) — glacial landforms.
Glasshouse Mountains (Qld) — volcanic landforms.
Hallett Cove (SA) — glacial pavement.

Biological/Ecological:

Riversleigh (Qld) — animal fossil.
Naracoorte Caves (SA) — animal fossil.
Dinosaur Cave (Vic) — animal fossil.
Nullabor Caves (SA/WA) — animal fossil.
Quincan Crater (Qld) — see Climate above.
Lake George (NSW) — see Climate above.
Shark Bay (WA) — stromatolites (also a *current* process — see Criterion 1.2).
Wet Tropics (Qld) — primitive flowering plants, e.g., *Idiospermum australiense.*
Wet Tropics (Qld) — refugia for rainforest plants.
Palm Valley (NT) — relict species *(Livistona mariae)*.
Kroombit Tops (Qld) — disjunct occurrence of Sydney Blue Gum *(Eucalyptus saligna)*.
Investigator Island Group (SA) — Stick Nest Rat — relict population and possibly limit of range.
Western Tasmania — endemic species of Gondwanan origin.
Sir Edward Pellew Islands — western and eastern limit of various faunal species.

APPENDIX III

Examples of a values table used for assessment.

No.	Criterion	Value	Rating	Comment
1.1	Evolutionary significance	Palaeontologically significant. The coral reefs of the Abrolhos marine area are the only known examples in the Indian Ocean where currently active reef building processes can be traced back in uplifted fossil reefs of the Pleistocene.	Very high	Geraldton Mid-west Region Development Commission, 1982, The Houtman Abrolhos.
1.2	Continuing natural processes	Internationally important example of a high lattitude coral-algal ecosystem. The most southerly coral complex in the Indian Ocean.	Very high	Hatcher (in press), IUCN Directory of Coral Reefs of International Importance: WA.
1.3	Richness and diversity	A luxuriant and diverse coral algae ecosystem rich in coral, algal, fish, echinoderm and seabird species. Flora and fauna diversity and richness because of presence of both tropical and temperate species.	Very high	Hatcher (in press).
1.4	Phases in cultural development	Historic shipwrecks have important associations with colonial maritime history of Australia. Gazetted shipwrecks in the area are listed individually in the Register. *Batavia* is the second oldest wreck in Australia and the earliest of four wrecks of the Dutch East India Co.	Very high	Hatcher (in press).
2.1	Rare natural phenomena	Habitat of the rare Lesser Noddy Tern which has only one other known breeding site (in the Seychelles). Endangered Green Turtle and Loggerhead Turtle have been sighted in the area.	High	Hatcher (in press).
3.1	Representative natural phenomena	Important representation of aquatic systems, especially in view of diversity of organisms. Important habitat for whale, turtle and sea lion species. Representative terrestrial ecosystems already listed as the Houtman Abrolhos Nature Reserve.	Very high	Hatcher (in press).
5.1	Aesthetic values	Though not as extensive as the Great Barrier Reef, the reefs have been described as some of the most beautiful coral reefs in the world.	High	Hatcher (in press).
7.1	Scientific sites and areas	Important research site for reef ecology, geology and biology, providing examples of both tropical and temperate systems.	Very high	
7.2	Historic sites and areas	The 16 historic shipwrecks found in the area, of which only five have been researched, have significant research potential for marine archaeologists and historians.	High	

Proc. Ecol. Soc. Aust. 1990 — 16:505-507

Institutional processes affecting the management of natural resources

R. D. McKELLAR

Darradup Environmental Services, P.O. Box 13, Nannup, Western Australia 6275, Australia

Abstract

Natural systems have required increasingly intense management from the Government agencies responsible for them. This management has sought to improve or continue their productivity or to ensure their persistence in the face of growing external threats. In general, the approach to managing these complex, dynamic and open natural systems has been adapted from management methods which were developed for much simpler, closed systems such as business organizations or engineering works. However, to be successful, a management institution must reflect the system to be managed. Management of our natural systems has suffered from the structure and style of the management institutions themselves. Therefore, a new form of management institution is required for natural systems.

Because of their knowledge of the dynamics of natural systems, ecologists should engage in shaping natural resource management institutions to conform better with natural systems they are established to manage.

Introduction

Natural systems are complex and dynamic, reacting to a wide variety of external forces including exploitation, newly introduced predators or competitors and climate changes (e.g., Odum 1986). Very often, the internal structure and dynamics of a natural system are not well understood, and the reaction of the system to external forces cannot be predicted.

These diverse factors make the management of natural systems extremely difficult, particularly where an objective of management is to maintain specific but perhaps ephemeral features such as diversity or resilience. If production and harvesting of biotic materials are also desired, management becomes yet more complex and difficult.

This is a very different situation from managing a system which is less complex and less affected by external forces and which has a clear and defined function. For instance, a transport system, whether road, rail or air, is much less complex than a corridor between reserves, even though their functions are in many ways comparable. Similarly, a national food production, storage, distribution and export system is less complex than a forest ecosystem.

Successful traditional engineering and business management systems are usually highly ordered, and structured in a hierarchical framework of information and responsibility. They are generally developed and designed to accomplish specific and defined purposes. Management approaches for them have traditionally reflected their order, structure and hierarchy. Where the complexity of the systems has grown, further levels of order and hierarchy have been implemented, often permitting static management approaches to be successfully continued.

Such traditional management approaches have often been applied to natural systems. The results are now evident throughout Australia and the rest of the world: the systems are failing to cope with new external changes and demands, and their complexity is daily being reduced through extinctions and loss of genetic variety. Successful management of natural systems must recognize the special features of those systems.

Complexity

Beer (1974) has claimed that the "Law of Requisite Variety" is the dominant management law in complex dynamic systems. This law states that the variety (i.e., degree of freedom) present in a management system must match the variety present in the system to be managed. Thus, an institution which is to manage a complex system must have a complexity equal to the system which is to be managed.

There are a number of strategies by which the complexities of the systems concerned can be matched. The management institution can be made more complex, through the establishment of specialized service departments (such as planning, protection, typing, mapping etc.), or through regionalization. This complexity can be enhanced with increased regional or service group autonomy and with a diversification of personnel background and belief. Alternatively, the managed system can be simplified, conceptually or in reality. For instance, management may concentrate on only certain components of the system, such as a valuable species of tree or an indicator species, the success of the management of the entire system being measured in terms of these components. The system itself could also be simplified through the removal of some components by harvesting, spraying, burning or other means.

Thus, steps can be taken towards achieving "requisite variety" by reducing the complexity of natural systems and by increasing the complexity of the management system. However, it is clear that both strategies have a price. Reducing the complexity of natural systems often renders them less capable to respond successfully to surprises, and complex administrative systems are more expensive to operate.

Dynamics

Natural systems are in a continual state of adaptation and evolution. On short time scales, natural systems respond to weather, predator pressure and other forces. On longer time scales, natural systems have evolved through interaction with climate changes, geological processes, soil development and other factors. Where management of natural systems seeks to maintain such a capacity for adaptation and evolution it must therefore also address the openness of natural systems to external forces.

Holling (1978) has written "In order to live successfully with uncertainty, our enviromental management institutions must maintain their responsiveness to change. . . Institutions, like biological systems, learn to handle change by experiencing change. . . . Insulation from small disasters leaves one ill-prepared and vulnerable to larger ones". This implies the need for a "Law of Requisite Openness", in which the openness of a management institution to external stimuli and responsive change needs to match the openness of the system to be managed.

It seems an innate property of our current institutions that they insulate themselves from external disturbance as much as possible. This can result in management lagging far behind clear stimuli for change which arise from within the managed system or the social context. This lag effect is accentuated where there is a personal attachment by senior management to particular management policies. Where a defensive attitude arises when management practices are questioned the lag time becomes further increased.

Organization

The structure of a system governs the flow of information and resources between its components. In natural systems, the exchange of information and resources is greatest between adjacent elements, and weakest between the most distant. The major resource imputs, solar energy and rainfall, are relatively constant (within boundaries) and other inputs such as groundwater and soil nutrients may be substantially determined by the local or adjacent elements of the natural system. Thus each element of a natural system receives inputs of defined types, and produces outputs which it defines itself and passes on mainly to adjacent elements.

Resource management agencies normally have a substantially different organisational framework. The major resources, financial, material and personnel, are provided to the various elements of the management system through decisions made at head office. Some resource sharing between elements may be possible in emergencies such as forest fires, but the outputs expected of each element are defined by the source of the inputs (head office) rather than by each element and its neighbours. This organizational form could be successful where the elements using the resources and generating the products have direct communication links and influence on the decision makers. Usually, however, the management personnel in the field, who are in direct contact with the natural system and so are well aware of its responses to management, are separated by many layers of bureaucracy and some physical distance from the decision makers in head office.

Conclusions

The structure of our natural resource management institutions appears to be contributing to the resource crises we are increasingly experiencing. The institutions do not appear to be able to provide for the complexity, change and self-defining aspects of the natural systems with which they are concerned.

A major challenge for managing our natural systems therefore lies in balancing the relative complexities of the natural and institutional systems, finding an open approach to nature management, and developing appropriate institutional frameworks for resource allocation decisions.

There are no obvious simple answers in this search. The term "institutional bricolage" has been proposed to suggest one approach — "bricolage" referring to a "handyman" approach to management of the institution itself. What would the Public Service Commission and long-time (small and large) empire building bureaucrats think of this? And yet, as each government finds it necessary to establish a Ministry or Unit of Government Management or Public Service Productivity Improvement, new approaches are being sought.

It would be unwise to permit this search to be undertaken only by those who have contributed to the generation and expansion of the problems. It is perhaps time for those who have an understanding of open, complex natural systems to turn their knowledge to the institutions within which so many of them work.

It is perhaps flattering to suggest that ecologists as a group are more aware of system dynamics, openness and structural properties than, say, engineers, foresters or economists. The ecological literature is, after all, well supplied with reports of studies which avoid consideration of unpredictable environmental changes. There is, however, an increasing number of longer-term ecological studies in which the complexity of natural environments is appreciated.

Those ecologists who are willing and able to address natural systems in their full complexity and openness, and are committed to maintaining these and associated attributes could well consider engaging in processes of institutional evolution.

Given the limited success of our past management of Australia's natural systems (see elsewhere in this volume) and the difficulty commonly experienced in integrating new information into management policies and programmes, Australia's ecologists should not lightly shrink from management opportunities or shirk management responsibilities.

Acknowledgements

Bert Main, Angas Hopkins and Denis Saunders provided useful comments on early drafts of this paper. Denis remained patient beyond all reasonable expectation. Many thanks to them all.

References

Beer S. (1974) *Planning for Freedom.* Canadian Broadcasting Corporation, Toronto.

Holling C. S. (ed.) (1978) *Adaptive Environmental Assessment and Management.* John Wiley and Sons, Chichester.

Odum E. P. (1986) Introductory review: Perspective of Ecosystem Theory and Application. In: *Ecosystem Theory and Application* (ed. N. Polunin) pp 1-11. John Wiley and Sons, London.

Proc. Ecol. Soc. Aust. 1990 — 16:509-519

Rehabilitation of disturbed land: long-term prospects for the recolonization of fauna[1]

J. D. MAJER

School of Biology, Curtin University of Technology, P.O. Box U1987, Perth, W.A. 6001, Australia

Abstract

This review looks at the long-term recolonization of fauna on land which has been drastically disturbed by activities such as mining and industry. Whether or not particular animals return, and also the pattern and rate of recolonization, are influenced by a diversity of factors including the nature of the pre-disturbance environment, the size and degree of isolation of the rehabilitated area, the methods used for rehabilitation, climatic factors, competitive effects between recolonizing species, the taxonomic group concerned and by intrinsic properties of the particular ecosystem.

Wide differences exist between the rates and patterns of animal recolonization in different regions and between different taxa. The complexity of factors which operate during animal recolonization highlights the need for more studies on the ecological characteristics of animals in areas which are about to be disturbed. This should enable us to understand better the factors which influence their ability to recolonize disturbed land. For the same reasons, there is also a need for long-term monitoring of animals recolonizing disturbed lands which are situated within a variety of vegetation communities and climatic zones. The acquisition of such data should enable the agencies which are responsible for the design, execution and monitoring of rehabilitation to develop sound guidelines for planning and monitoring of rehabilitation programmes.

Introduction

Rehabilitation of land which has been degraded by mining, industry, salination, grazing and some other types of land-use is often carried out in the hope of restoring the original components and processes of the ecosystem. Sometimes a particular project obtains approval on the basis of the proponent undertaking to restore the environment to its former condition. In other cases, the regulations which are imposed by the relevant authority may require the proponent of a particular project to restore the biota to its former condition.

Whatever the reason for rehabilitation, the practitioner generally considers the plant community while animals generally receive secondary, if any, consideration. There is an increasing realization that the recolonization of animals is of importance to the sustainability of rehabilitation and also for the political acceptance of many projects. In certain situations a company may seek to enhance its reputation by demonstrating that certain species of animals have recolonized their rehabilitated areas.

In this chapter I review some of the factors which influence the patterns and rates of fauna recolonization and even whether particular species will ever return. The answers to these questions are important because they may tell us whether the restoration of a particular community is attainable and they will also assist us in drawing up time-scales for measuring the success of a particular rehabilitation programme.

These questions may only be answered by obtaining long-term data on fauna recolonization but few studies have yet been performed. It is therefore necessary to speculate on the answers by drawing upon ecological theory and then looking for data

[1]This chapter is a shortened version of a chapter in Majer (1989) pp. 143-74 and is printed with permission of Cambridge University Press.

which may illustrate or refute these predictions. It is only possible to mention a selection of factors which influence the process of fauna recolonization. The reader seeking more information is referred to a more complete account given in Majer (1989).

Influence of the type of pre-disturbance Habitat

In order to follow the succession of fauna from the initial to the later stages of rehabilitation it is important to characterize the nature of the pre-disturbance habitat. The originial "habitat templet" proposed by Southwood (1977) provides a useful framework for this. The templet condenses the variety of habitats onto two axes equivalent to their favourableness (particularly for breeding) and predictability (Fig. 1). Within the zones of the habitat templet there is selection for organisms with particular ecological strategies. For instance, highly fecund pioneer or opportunistic species, more correctly termed r-strategists, are favoured in less predictable "r-selecting" environments such as temporary pools, transient decomposition habitats and arable land. On the other hand more numerically constant K-strategists, whose numbers remain near their carrying capacity, are selected for in more predictable and favourable "K-selecting" environments such as oligotrophic lakes, rainforest leaf litter and climax forest communities. An adversity, or "A-selection" strategy is selected for in habitats which are predictable but very unfavourable such as hot springs, montane vegetation and certain deserts. Selection here is for adaptation to the physical environment, for low energy cost resting stages and mechanisms to ensure synchrony of activity when favourable conditions resume.

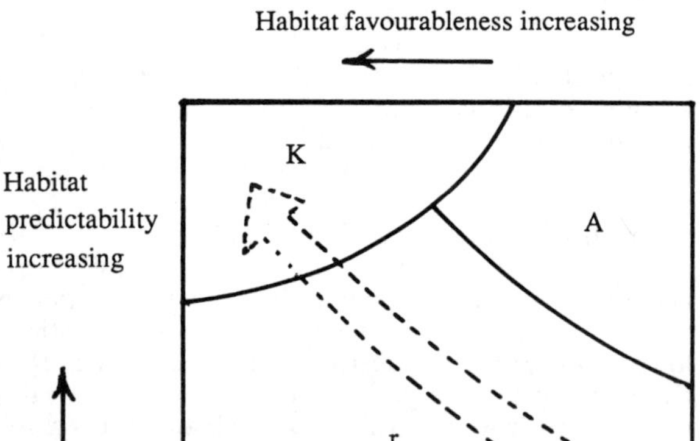

FIG. 1. The habitat templet: relationships between habitat predictability, habitat favourableness, biological unpredictability and selection processes. The zones of K, r- and A-selection are areas in which each type of selection tends to predominate. The dotted line indicates the change in characteristics of a rehabilitated area as it matures (adapted from Greenslade & Greenslade 1983).

Where does an area of recently disturbed land fall on the habitat templet? Prior to rehabilitation being carried out the degraded land is usually characterized by a low spatial complexity, an absence of biological diversity, a harsh microclimate and, in biological terms, an extremely unpredictable future. These conditions place the degraded area within the r-selecting zone of the habitat templet (Fig. 1). However, since the frequency of the disturbance is low (usually once) and its duration short, there is little opportunity for *in situ* selection to take place. The pool of organisms available to colonize the rehabilitated area must therefore come from the surrounding environment in which the disturbed land is located. Disturbed land which occurs within a r-selecting environment should have a large pool of suitably adapted species which can colonize the area, while more predictable and favourable surroundings may provide fewer candidate colonizers.

I am unaware of any studies which have been designed to test this hypothesis although work on ant colonization in Western Australia indicates that recolonization of iron ore mines in the harsh semi-desert Pilbara region is considerably more rapid than in the bauxite mines situated within the more equable mediterranean region (Table 1). Since the Pilbara climate is often unfavourable and the summer rainfall is unpredictable, the assumed presence of a wide range of r-strategists may explain the rapid colonization of ants. Other explanations based on the apparent ease of restoring a structurally simple desert environment may be advanced so, until further studies are performed, we may only speculate on this hypothesis.

TABLE 1. Comparison of the number of ant species in unmined and 12-year-old rehabilitated areas in the mediterranean and harsher, semi-desert regions of Western Australia.

Locality	Number of species present		Percentage return
	Unmined area	Rehabilitated area	
Mediterranean*	27-33	13-20	39-74
Pilbara**	16	16	100

*from Majer *et al.* 1984; **Dunlop *et al.* 1985.

As the rehabilitation matures, the vegetation buffers the microclimate and the area moves towards the K-selecting zone of the habitat templet (Fig. 1). It then tends to become colonized by K-strategists which occur in the locality. The speed with which this happens may depend on the closeness, in a structural and floristic sense, of the rehabilitation to that of the original habitat.

Once again it is difficult to find comparable data with which to test this hypothesis. Comparison of prairie in the western United States, which is relatively easy to reproduce in the structural sense, with forested country in the east provides a clue. Small mammal colonization of rehabilitated coal mines in the Wyoming prairie is relatively rapid in comparison to that in coal mines situated within forested parts of Illinois (Hingten & Clark 1984; Wetzel 1958) (Fig. 2). This could well be explained by the fact that five-year-old rehabilitation in prairie may cater for the habitat requirements of many of the original prairie species, while that in Illinois forested country still lacks the structural and floristic complexity of the original vegetation and therefore does not.

The examples given in this section are not conclusive; other studies may provide data which contradict what has been said here. There is a need for further investigations, using comparable sampling techniques, on the colonization rates of certain taxa in both r- and K-selecting environments and in habitats of differing structural and floristic complexity.

Influence of size and isolation of the rehabilitation

Since the rehabilitated area may differ in its physical and botanical structure from that of the surrounding vegetation it may function as an ecological "island". Therefore, according to MacArthur and Wilson's (1967) theory of island biogeography, the number of species in the island of rehabilitation is associated with the dynamic

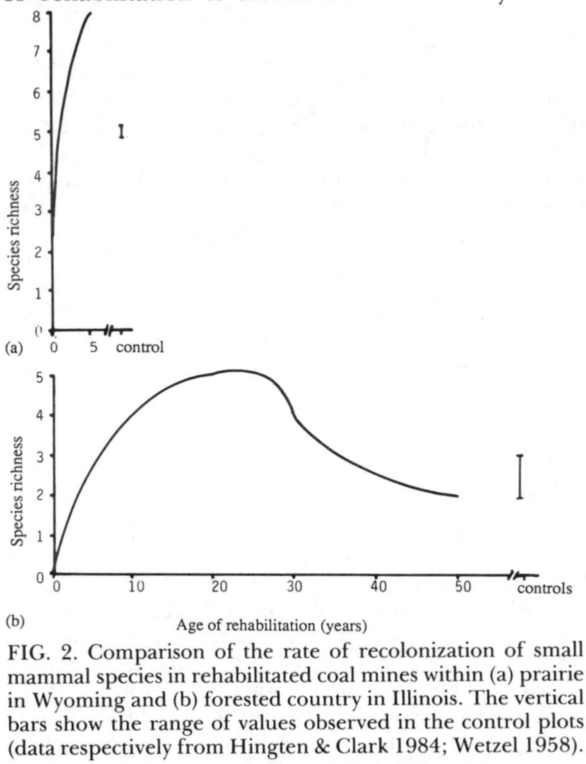

FIG. 2. Comparison of the rate of recolonization of small mammal species in rehabilitated coal mines within (a) prairie in Wyoming and (b) forested country in Illinois. The vertical bars show the range of values observed in the control plots (data respectively from Hingten & Clark 1984; Wetzel 1958).

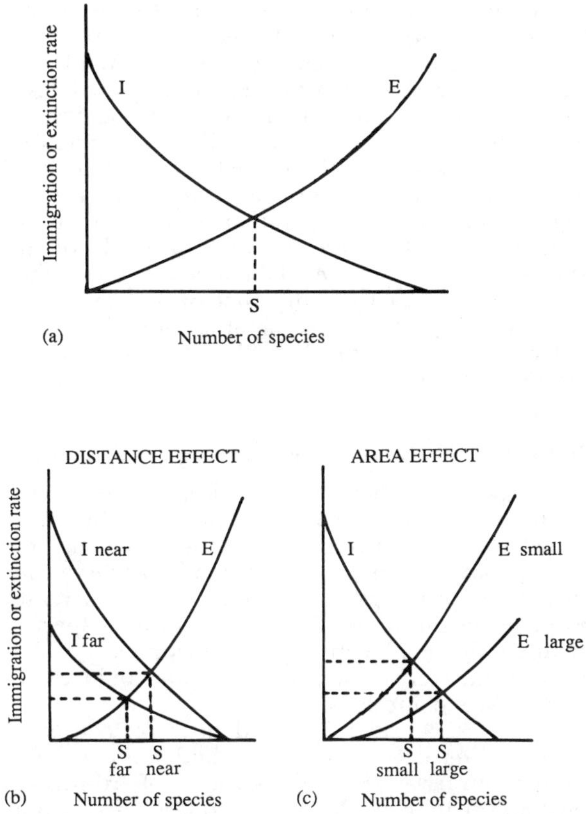

FIG. 3. Equilibrium model of island biogeography in which the rate of immigration (I) and extinction (E) interact to describe the number of species in equilibrium (S*). b) and c) describe how the rate of immigration is influenced by distance from colonizing source and the rate of extinction is influenced by area of the island (adapted from Simberloff 1974).

equilibrium between immigration and extinction of colonizing species (Fig. 3a). The major factors which control the resulting species richness are the degree of isolation and also the area of the island as fewer immigrants are likely to reach isolated areas (Fig. 3b) and more extinctions are likely in smaller areas (Fig. 3c).

There is considerable evidence from the reclamation literature that the degree of isolation influences the number of species occurring in an area. For instance, Majer (1980) found that the number of ant species in Australian bauxite mines, which had been recently topsoiled, declined with increasing distance from the adjacent forest margin (Fig. 4). The most striking evidence comes from the study of Dutch polders which have been reclaimed recently from the sea. The colonization by moles, *Talpa europaea*, can clearly be rated to the proximity of the original mainland, with animals spreading at the rate of about 2 km yr^{-1} along dykes, roads and banks of ditches (Haeck 1969).

I am not aware of any comparative studies of the influence of size of rehabilitated area on the rate of immigration or extinction of colonists. While the rate of immigration may be enhanced by the larger area and longer edge of the site, the degree of isolation of its centre is also likely to increase. These two factors are therefore likely to operate in antagonism. There is however, some evidence for high extinction rates operating in small areas of rehabilitation. In a recent two-year study of an isolated 7 ha area of rehabilitated iron ore waste, Walker *et al.* (1986) found that of 10 vertebrate species immigrating into the area, only four persisted as stable populations; the other six species became extinct.

Influence of habitat changes during rehabilitation

A number of physical and botanical changes occur as the rehabilitation matures. These include some of the following; a reduction in the percentage of bare ground, a more buffered microclimate, an increase in plant cover in certain strata, an increase in plant height, an initial increase in plant species richness, an increase in structural complexity of the vegetation, an increased litter layer and an increase in dead wood and logs.

There is considerable evidence for coupling between the developing plant and animal communities in reclaimed lands. A survey of arthropod, bird and mammal colonization studies carried out by Majer (1989) indicated numerous positive correlations between animal biomass, abundance, richness and diversity with plant cover, richness and diversity and also structural diversity. Brenner and Kelly (1981) and Brenner *et al.* (1982) among others, have also demonstrated relationships between the density of particular plant and particular bird or mammal species. It seems that the presence of certain plants provides opportunities for the establishment of animal species which benefit from the plant in some way. Although the facilitation pathway of succession was conceived to describe the succession of plant communities (Connell & Slayter 1977), these relationships suggest plant facilitation of animal invasions. Furthermore, animal succession is also linked with changes in the nature of the physical environment and the litter layer. For instance, Fox (1982) noted a positive correlation between both ant abundance and small mammal *(Pseudomys)* biomass with a more penetrable substrate while Majer *et al.* (1984) found that ant species diversity is positively associated with percentage of litter cover and the presence of larger logs.

The outcome of this coupling between animals with changes in the physical and botanical environment can be seen with the ant fauna in rehabilitated manganese mines in northern Australia (Fig. 5). Here, the developing ant community tracks the changes in vegetation, litter, dead wood and bare ground. Rehabilitation personnel should be aware of this coupling since the provision of adequate plant cover is not an adequate measure of the success of rehabilitation. In view of the fact that plant cover is obviously of importance to the developing fauna, the provision of adequate floristic and structural diversity as well as logs and litter is important if the full range of native animals are to colonize the area. Although the full range of species may eventually colonize the area without these measures being taken, the rate of recolonization would certainly be slower.

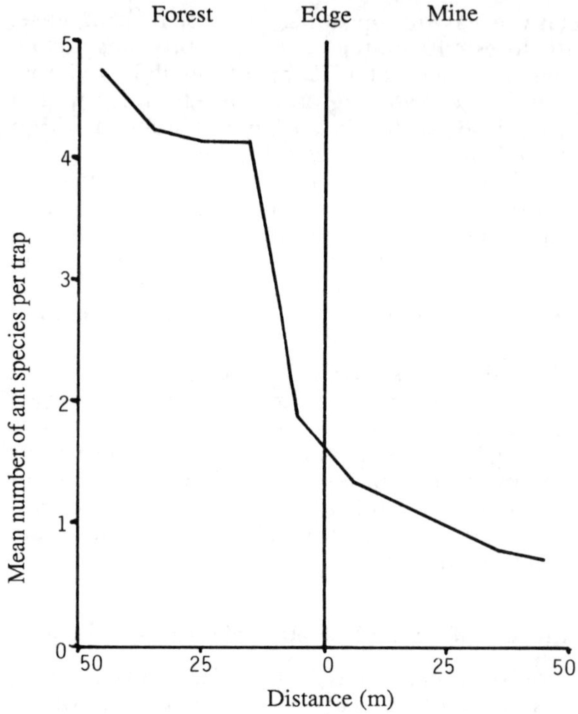

FIG. 4. Variation in number of species of ants caught in pitfall traps extending 50 m into a newly topsoiled bauxite mine from the adjacent forest (from Majer 1980).

Influence of competition between colonizing animals

The importance of interspecific competition in moulding animal community structure has long been recognized (Cody & Diamond 1975). Until recently, its importance in reclaimed land has rested on suspicion. A series of papers on ants and small mammals in Australian sand mined areas provides the first experimental evidence that interspecific competition influences the fauna replacement process in rehabilitated land. The

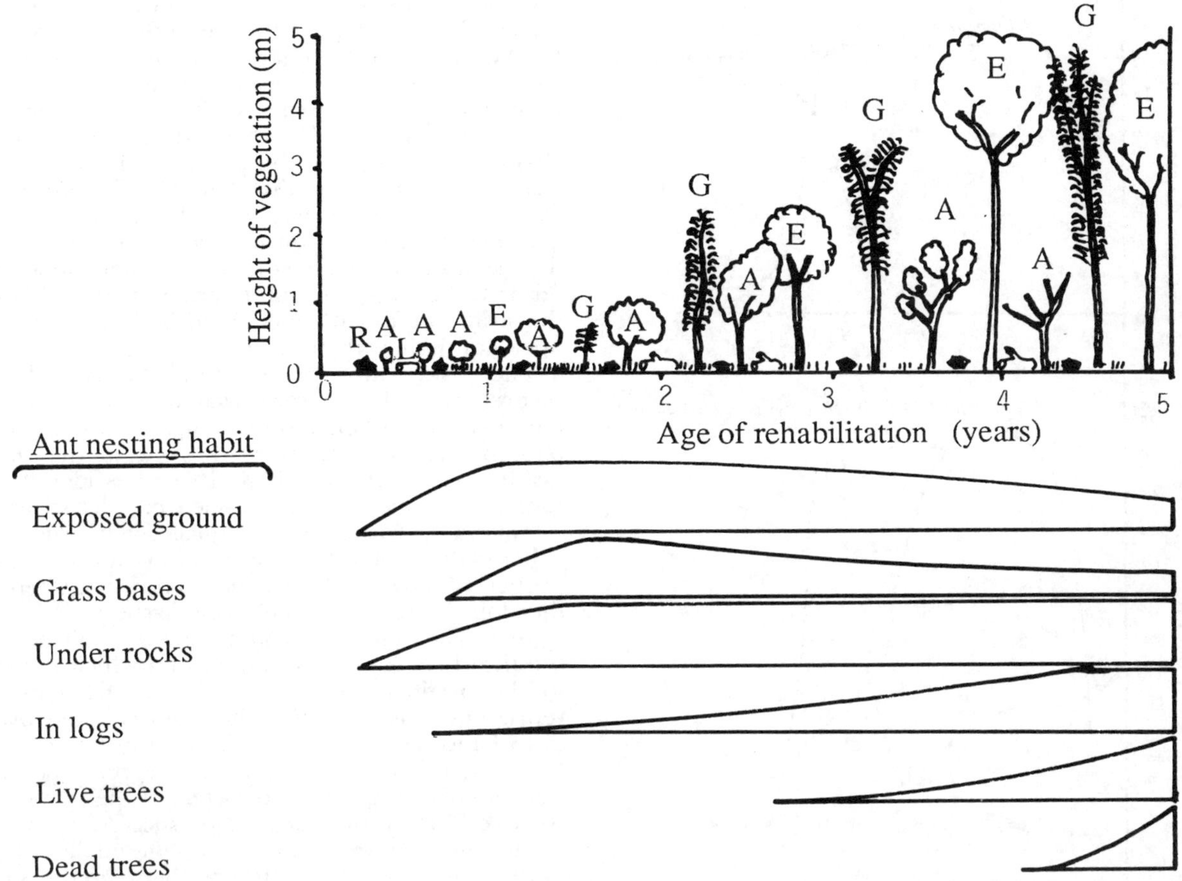

FIG. 5. Schematic representation of ant colonization in rehabilitated manganese mines of Groote Eylandt, northern Australia. Key: A, *Acacia* spp; G, Grevillea *sp.; E. Eucalyptus* spp.; L, logs; R, rocks. (Adapted from data in Majer 1983).

surveys of rehabilitation of increasing age revealed that the ant, *Iridomyrmex* sp. C, initially dominated the rehabilitated sites. After 8-9 years of rehabilitation, *Iridomyrmex* sp. C was replaced by *Iridomyrmex* sp. A (Fox & Fox 1982) (Fig. 6a). Also, the introduced mouse, *Mus musculus,* was replaced by the native *Pseudomys novaehollandiae* as the rehabilitation aged (Fox & Fox 1984) (Fig. 6b).

When the abundance of *Iridomyrmex* sp. A was experimentally reduced, the other ant increased in abundance and spread into the vacated space (Fox *et al.* 1985). The researchers concluded that species C has its foraging area, and hence abundance, reduced by the presence of species A. Similarly, when *Pseudomys* was experimentally removed from, or added to, 1 ha plots of rehabilitation, *Mus* respectively increased or declined in abundance (Fox & Pople 1984). Once again this outcome was considered to result from competition between the two species of mouse. Unfortunately the reciprocal experiments were not performed, but it is likely that the early stages of rehabilitation may favour competitive superiority by one species while later on this may change to advantage the other species. The competitive advantages of one species over the other may in turn be associated with changes in resource availability and in the nature of the habitat.

The existence of competitive interactions is of considerable importance to the fauna colonization process in rehabilitated areas since they may dictate the type of animal present at a particular time and also the replacement sequence of the species. It may be possible to manage the rehabilitation procedure to provide resources for a particular animal species or to accelerate the successional sequence. In the latter case it will be necessary to manage the area for the entire community rather than for populations of individual species. However, in many cases it may be necessary to wait for the area to pass through a series of replacements before a

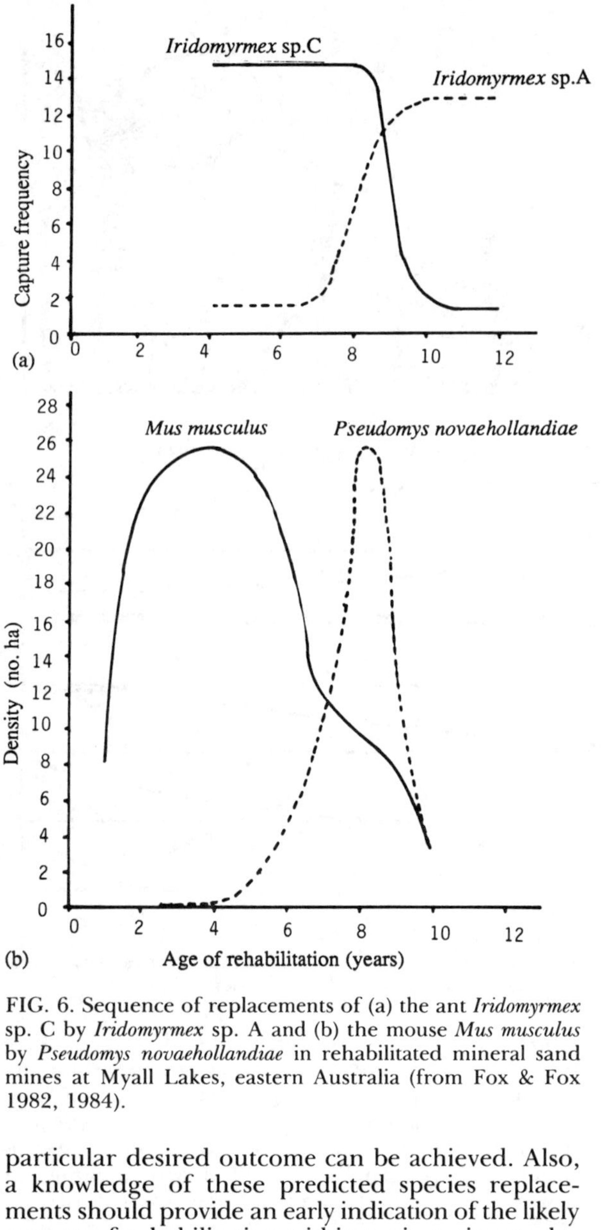

FIG. 6. Sequence of replacements of (a) the ant *Iridomyrmex* sp. C by *Iridomyrmex* sp. A and (b) the mouse *Mus musculus* by *Pseudomys novaehollandiae* in rehabilitated mineral sand mines at Myall Lakes, eastern Australia (from Fox & Fox 1982, 1984).

particular desired outcome can be achieved. Also, a knowledge of these predicted species replacements should provide an early indication of the likely success of rehabilitation within a given time scale.

Components of ecosystem resilience

So far I have largely considered succession at the species level. I will now look at the pattern of disturbance and recovery at the broader levels of the community and ecosystem. This is of practical importance because success of rehabilitation is often measured in terms of community parameters such as species diversity or proportion of original species present. It is therefore desirable to be able to predict, or at least understand, how these parameters may change over time.

There is a profusion of conflicting terminology relating to this subject so, to avoid confusing the reader, I have adopted that recently put forward by Westman (1986). The vertical axis of Figure 7 could describe the population density of an individual species, the species richness of the entire community or it could be an index which integrates characteristics of both the physical and biotic environment. Prior to disturbance the ecosystem exists at a pre-disturbance level. The ecosystem becomes stressed as a disturbance is initiated. The degree of strain inflicted on the ecosystem depends on both the magnitude and period of stress. Thus the strain on a heathland would be larger during open-cut mining than during burning.

The disturbance process and the period of recovery may be characterized by a number of parameters. Resilience is the degree, manner, and pace of recovery of the ecosystem to the predisturbance level. Westman (1986) has identified four components of resilience: "elasticity", "amplitude", "hysteresis" and "malleability" (Fig. 7).

Elasticity is the rate of recovery of an ecosystem property following disturbance. Thus, the steeper the slope of recovery, the more elastic is the ecosystem. The amplitude component of resilience is the threshold level of strain beyond which return to the original state no longer occurs. This is represented by the difference between the predisturbance level and the solid horizontal line on Figure 7. Hysteresis concerns the degree to which the path of change in an ecosystem property under stress differs from the path of succession upon immediate removal of the stress. The malleability of the ecosystem is the difference between the ecosystem's final recovery level and the predisturbance level; the greater the difference, the more malleable is the ecosystem.

Fox and Fox (1986) described a fifth property of resilience: "damping". This is the degree and manner by which a path of restoration is altered by any forces that affect the restoring force. Figure 7 shows the path of recovery of overdamped, critically damped and underdamped ecosystems. The first mentioned ecosystem will take a long time to reach equilibrium rapidly but will overshoot and oscillate for some period of time while the critically damped ecosystem will approach equilibrium more slowly, overshoot slightly and then approach equilibrium level. This ecosystem equilibrates most rapidly and is therefore the most resilient of the three examples. All of the five parameters mentioned above are relevant to the rehabilitation process, but I will here give examples which relate to three of them.

Elasticity

I have already given examples where ecosystems may be more elastic than others as a result of their being r-selecting (Table 1) or structurally simple

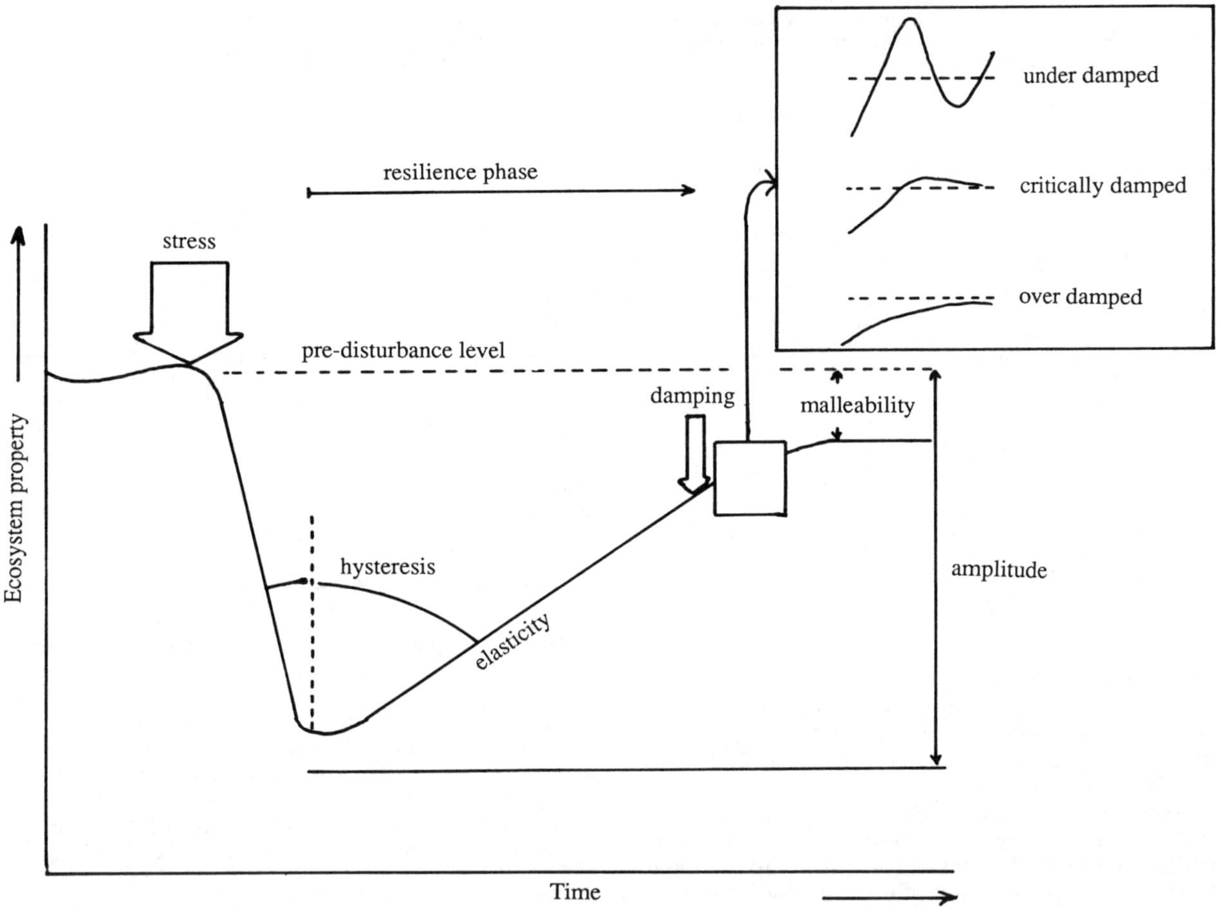

FIG. 7. Model illustrating the response of an ecosystem to disturbance showing its five properties of resilience following cessation of the stress (elasticity, amplitude, hysteresis, malleability and damping).

(Fig. 2a) environments. Climatic pattern may also influence elasticity. Agronomists have long recognized the importance of climate to the growth and succession of plant communities. Although it is well known that rehabilitation of disturbed areas is easier in areas with long growing seasons (National Research Council US 1981), there has been little comment on how climate influences the resilience of animal communities. Mitchell's (1961) work on breeding bird populations in old field succession indicates that the length of growing season may influence the rate of succession of birds. He found only two breeding species and 1.2 breeding pairs per hectare in 18 year old fields on the Anoka Sand Plain, Minnesota compared with Odum's (1950) observation of nine breeding species and 3.2 pairs per hectare in 20 year old fields on the moister and warmer Highlands Plateau of North Carolina.

My own work on ant succession in rehabilitated Australian mines provides data which are in line with this trend (Majer 1984). Comparison of species richness of ants in three-year-old rehabilitation throughout the continent reveals a cline in ant return rate, from lowest in the drier Mediterranean climatic zones to highest in the tropical monsoonal areas (Table 2). Variations in return rate are not correlated with the species richness of animals in the original ecosystem. At this stage it is not certain whether this trend results from ant communities responding to differences in the length of growing season or whether it is because some Australian ecosystems are adapted to higher levels of disturbance than others.

There is some evidence that certain taxa may also be more elastic than others. For instance, the rate of return of species over the first five years in Western Australian bauxite mines was greater for birds than for reptiles and frogs (Fig. 8). Admittedly the sites used in these studies had been subjected to a range of rehabilitation procedures so other explanations are possible. However, it is quite possible that the differences in rates of reptile/frog and bird recolonization may be related to the dispersal abilities of these animals or to the ease with which their habitat requirements may be catered for.

TABLE 2. Estimated number of ant species found in three-year-old rehabilitated plots within mined areas situated throughout Australia (Adapted from Majer 1984).

Location	Mine type	Climate type	Annual rainfall (mm)	Number of ant species
Eneabba, W.A.	Sand	Mediterranean	550	6
Collie, W.A.	Coal	Mediterranean	973	8
Jarrahdale/ Del Park, W.A.	Bauxite	Mediterranean	1 155-1 287	10
Myall Lakes, N.S.W.	Sand	Warm-temperate	1 205-1 362	11
Nth Stradbroke Island, Qld.	Sand	Sub-tropical	1 645	18
Gove, N.T. Groote Eylandt N.T., and Weipa, Qld.	Bauxite and Manganese	Tropical monsoonal	1 277-2 083	22

Amplitude

It is highly probable that major disturbances such as surface mining may stress some ecosystems beyond the amplitude threshold. However, without long-term recolonization studies, coupled with adequate base-line data, it is not possible to distinguish between an ecosystem which has been stressed beyond the amplitude threshold and one which is merely taking a long time to recover. Data on thrip (Thysanoptera) recolonization of Illinois surface-mine spoils indicate that total recovery has not occurred after 32 years (Stannard 1967) (Fig. 9). The Sorensen's index of similarity between thrip species in the 32-year-old spoils to those in the forest was 0.32 compared with an inter-forest value of 0.67. This, and the levelling off in thrip species richness after 23 years (Fig. 9), indicates that the amplitude threshold of the original ecosystem may have been exceeded. In order to confirm this suggestion it would be desirable to determine which species have not yet colonized the mined areas and determine the reasons why they have not.

Damping

The existence of the damping property has important implications to the ways in which rehabilitation success is assessed. Firstly, a

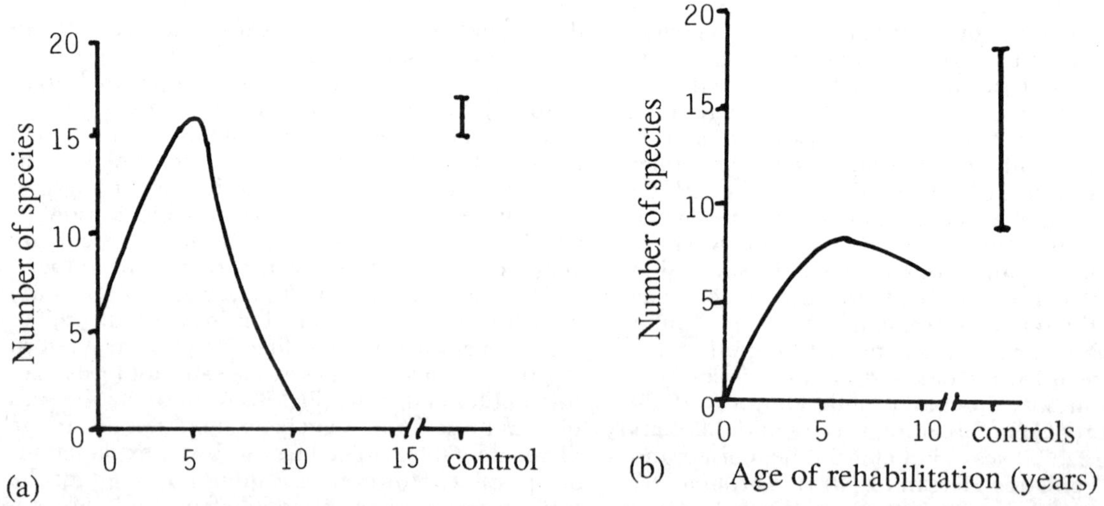

FIG. 8. Comparison of the rate of recolonization of (a) birds and (b) reptiles and frogs in rehabilitated bauxite mines in Western Australia. The vertical bars show the range of values observed in the control plots (data respectively from Nichols & Watkins 1984; Nichols & Bamford 1985).

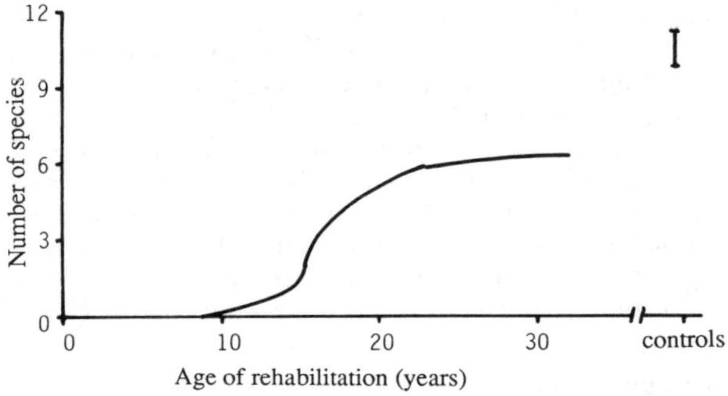

FIG. 9. Rate of recolonization of thrips (Thysanoptera) on Illinois surface-mine spoils. The vertical bar shows the range of values observed in the control plots (data from Stannard 1967).

regenerating ecosystem which exhibits high elasticity may give the impression of being highly resilient. However, it may be an underdamped ecosystem which takes longer to equilibrate than a critically damped ecosystem of lower elasticity. Secondly, the existence of oscillations in ecosystem properties creates problems in selecting the interval after rehabilitation when performance standards should be assessed. There any many examples of oscillations in species richness of insects (Fig. 10a), birds (Fig. 10b) and small mammals (Fig. 2b) following rehabilitation so the problem of when to assess performance standards is widespread.

The end product of succession

It is obviously desirable to know how similar the final community will be in terms of species composition to that of the original community. The answer to this question is hampered by the paucity of long-term data and also by changes in rehabilitation technology. Older types of rehabilitation are generally less sophisticated than those carried out today and are therefore not representative of current rehabilitation practices.

Table 3 shows some similarity index values for fauna in older rehabilitation compared with pre-disturbance vegetation and, as a yardstick of variability, between two plots of pre-disturbance vegetation. Species composition differs between the control plots which have been used in all of these studies, indicating the great variation in species composition that occurs between areas. With the exception of the >56-year-old bird study plots in Illinois, all rehabilitated areas exhibit lower similarity with control plots than do the control

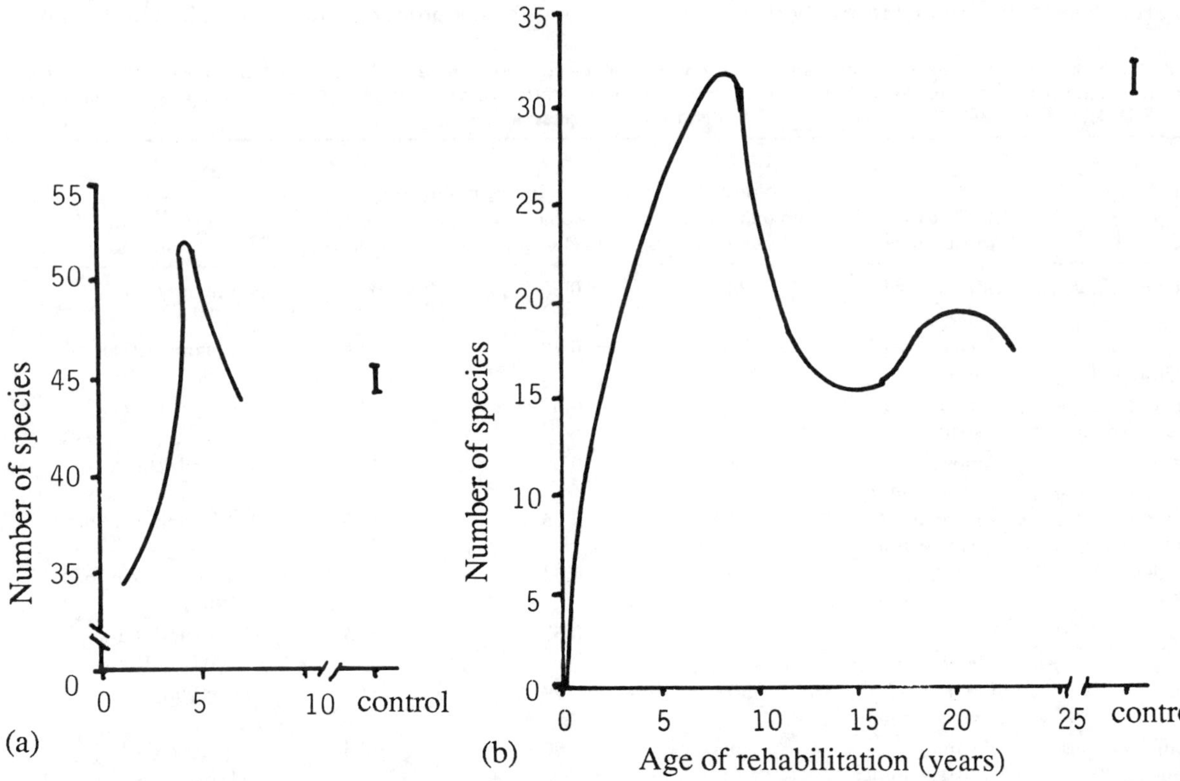

FIG. 10. Rate of recolonization of (a) beetles in rehabilitated Wyoming surface-mines and (b) birds in rehabilitated Virginian surface-mines showing oscillations in species richness that can occur during the course of succession. The vertical bars show the range of values observed in the control plots (data respectively from Parmenter & MacMahon 1989; Crawford *et al.* 1978).

plots with each other. Thus, the fauna in all but one of these rehabilitated areas has not converged with that of the original vegetation in the 5-32 year time scale represented by the data. The similarity index of the >56-year-old plot falls within the same range as the inter-control value although this itself is a low value due to the high variability in fauna composition between control plots.

These data suggest that a considerable period of time is needed for the composition of the fauna of the rehabilitated areas to approach that of the pre-disturbance vegetation. As already mentioned, if the threshold amplitude has been exceeded, the community may never recover to its pre-disturbance composition.

Various writers have pointed out the possible existence of multiple stable equilibria in which the developing community may equilibrate at more than one possible combination of species. Demonstration of this possibility in rehabilitated land is not possible due to the absence of long-term studies on paired plots which have been subjected to identical rehabilitation treatments. Neumann (1971) studied the colonization of slaters (Isopoda), millipedes (Diplopoda) and ground beetles (Carabidae) in 28-year-old forested coal mine spoils. If we allow for the fact that the most isolated area had become colonized by fewer species than the area close to the original forest, both areas were populated by a remarkably similar array of species. Indices of similarity for each taxon in the paired rehabilitation plots are respectively 0.66, 0.87 and 0.74. Brenner and Kelly's (1981) work on birds in 15-year-old surface mines suggests a similar trend (index of similarity between paired rehabilitation plots is 0.76). Thus, based on the limited examples available, it appears that rehabilitated areas of similar age and rehabilitation treatment will tend to become colonized by a similar range of animals.

Conclusions

This review has shown that in both the long- and short-term, the recolonization of fauna is influenced by a wide range of factors which relate to the intrinsic properties of the ecosystem, the taxa which are being considered and the effectiveness of the rehabilitation programme. Therefore, wide differences exist between the rate and patterns of animal recolonization in different regions and also between different taxa within a particular region.

These complexities point to the need for more baseline surveys of the characteristics of the fauna in areas which are about to be disturbed. This will enable us to understand the factors influencing their ability to recolonize rehabilitated land. Secondly, there is a need for carefully monitored long-term studies of fauna in rehabilitated lands in

TABLE 3. Comparison of the Sorensen's index of similarity for selected taxa for which long-term data are available. Comparisons are made between fauna in oldest rehabilitation and pre-disturbance vegetation and also between two areas of pre-disturbance vegetation.

Taxon	Type of disturbance	Age of oldest rehabilitation (yrs)	Index of Similarity: Rehab. vs Control	Index of Similarity: Control vs Control	Reference
Slaters (Isopoda)	German coal spoils	28	0.66	0.91	Neumann (1971)
Millipedes (Diplopoda)	German coal spoils	28	0.72	0.90	Neumann (1971)
Beetles (Carabidae)	German coal spoils	28	0.62	0.91	Neumann (1971)
Thrips (Thysanoptera	Illinois surface-mines	32	0.32	0.67	Stannard (1967)
Predatory invertebrates	Australian bauxite mines	13	0.25	0.39	Nichols and Burrows (1985)
Reptiles and frogs	Australian bauxite mines	10	0.54	0.64	Nichols and Bamford (1985)
Birds	Australian bauxite mines	11	0.57*	0.75	Nichols and Watkins (1984)
Birds	Illinois surface-mines	>56	0.59	0.59	Karr (1968)
Small mammals	Wyoming surface-mines	5	0.67	0.75	Hingten and Clark (1984)

*There is now evidence that the rehabilitation vs forest value is increasing towards forest vs forest levels (L Campeotto unpublished data).

a variety of vegetation communities and climatic zones. These studies should adopt identical monitoring techniques so that valid comparisons may be made between areas and generalizations constructed on the recolonization characteristics in areas of particular vegetation and climatic type.

The acquisition of such data should enable the agencies responsible for the design, execution and monitoring of rehabilitation to develop sound guidelines for planning and monitoring rehabilitation programmes.

References

Brenner F. J. & Kelly J. (1981) Characteristics of bird communities on surface mine lands in Pennsylvania. *Environ. Manage.* **5,** 441-9.

Brenner F. J., Kelly R. B. & Kelly J. (1982) Mammalian community characteristics on surface mine lands in Pennsylvania. *Environ. Manage.* **6,** 241-9.

Cody M. L. & Diamond J. M. (eds) (1975) *Ecology and Evolution of Communities.* Belknap Press, Cambridge, Massachusetts.

Connell J. H. & Slayter R. O. (1977) Mechanisms of succession in natural communities and their role in community stability and organisation. *Am. Nat.* **111,** 1119-44.

Crawford H. S., Hardy D. M. & Abler W. A. (1978) A survey of bird use of strip-mined areas in southern West Virginia. In: *Surface Mining and Fish/Wildlife Needs in Eastern United States* (ed. D. E. Samuel) pp. 241-6. West Virginia University, Morgantown.

Dunlop J. N., Majer J. D., Morris C. J. & Walker K. J. (1985) A preliminary assessment of minesite rehabilitation in the Pilbara iron ore province using ant communities as ecological indicators. *Mulga Research Centre Journal* **8,** 25-31.

Fox B. J. (1982) The influence of disturbance (fire, mining) on ant and small mammal species diversity in Australian heathland. In: *Proceedings of a conference on Dynamics and Management of Mediterranean Type Ecosystems,* pp. 213-9. USDA Forest Service, Riverside, California.

Fox B. J. & Fox M. D. (1984) Small-mammal recolonization of open-forest following sand mining. *Aust. J. Ecol.* **9,** 241-52.

Fox B. J. & Fox M. D. (1986) Resilience of animal and plant communities to human disturbance. In: *Resilience in Mediterranean Ecosystems* (eds B. Dell, A. J. M. Hopkins & B. B. Lamont) pp. 39-64. Dr W. Junk, The Hague.

Fox B. J., Fox M. D. & Archer E. (1985) Experimental confirmation of competition between two dominant species of *Iridomyrmex* (Hymenoptera:Formicidae). *Aust. J. Ecol.* **10,** 105-10.

Fox B. J. & Pople A. R. (1984) Experimental confirmation of interspecific competition between native and introduced mice. *Aust. J. Ecol.* **9,** 323-34.

Fox M. D. & Fox B. J. (1982) Evidence for interspecific competition influencing ant species diversity in a regenerating heathland. In: *Ant-Plant Interactions in Australia* (ed. R. C. Buckley) pp. 99-110. Dr W. Junk, The Hague.

Greenslade P. J. M. & Greenslade P. (1983) Ecology of soil invertebrates. In: *Soils, an Australian Viewpoint* (ed. K. Lee) pp. 645-69. CSIRO Melbourne.

Haeck J. (1969) Colonization of the mole *(Talpa europaea)* in the Ijsselmeer polders. *Ned. J. Zool.* **19,** 145-248.

Hingten T. M. & Clark W. R. (1984) Small mammal recolonization of reclaimed coal surface-mined land in Wyoming. *J. Wildl. Manage.* **48,** 1255-61.

Karr J. R. (1968) Habitat and avian diversity on strip-mined land in east-central Illinois. *Condor* **70,** 348-57.

MacArthur R. H. & Wilson E. O. (1967). *The Theory of Island Biogeography.* Princeton University Press, Princeton.

Majer J. D. (1980). The influence of ants on broadcast and naturally spread seeds in rehabilitated bauxite mined areas. *Recl. Rev.* **3,** 3-9.

Majer J. D. (1983). Recolonization by ants in rehabilitated open-cut mines in Northern Australia. *Recl. Rev. Res.* **2,** 279-98.

Majer J. D. (1984). Ant return in rehabilitated mines — an indicator of ecosystem resilience. In: *Proceedings of the 4th International Conference on Mediterranean Ecosystems* (ed. B. Dell) pp. 105-8. University of Western Australia, Perth.

Majer J. D. (1989). Long-term recolonization of fauna in reclaimed land. In: *Animals in Primary Succession — the Role of Fauna in Reclaimed Lands* (ed. J. D. Majer) pp. 143-74. Cambridge University Press, Cambridge.

Majer J. D., Day J. E., Kabay E. D. & Perriman W. S. (1984). Recolonization by ants in bauxite mines rehabilitated by a number of different methods. *J. Appl. Ecol.* **21,** 355-75.

Mitchell M. J. (1961). Breeding bird populations in relation to grassland succession on the Anoka Sand Plain. *Flicker* **33,** 102-8.

National Research Council US (1981). *Surface Mining: Soil, Coal and Society.* National Academy Press, Washington.

Neumann U. (1971). Die Sukzession der Bodenfauna (Carabidae:Coleoptera, Diplopoda und Isopoda) in den forstlich rekultivierten Gebieten des Rheinischen Braunkohlenrevier. *Pedobiologia* **11,** 143-226.

Nichols O. G. & Bamford M. J. (1985). Reptile and frog utilization of rehabilitated bauxite minesites and dieback-affected sites in Western Australia's jarrah *(Eucalyptus marginata)* forest. *Biol. Conserv.* **34,** 227-49.

Nichols O. G. & Burrows R. (1985). Recolonization of revegetated bauxite mine sites by predatory invertebrates. *Forest Ecol. Manage.* **10,** 49-64.

Nichols O. G. & Watkins D. (1984). Bird utilization of rehabilitated bauxite minesites in Western Australia. *Biol. Conserv.* **30,** 109-31.

Odum E. P. (1950). Bird populations of the Highlands (North Carolina) Plateau in relation to plant succession and avian invasion. *Ecology* **31,** 587-604.

Parmenter R. R. & MacMahon J. A. (1989). Arthropod recolonization of reclaimed strip-mines in southwestern Wyoming: 1. The ground-dwelling beetle fauna (Coleoptera). *Environ. Ent.* (in press).

Simberloff D. S. (1974). Equilibrium theory of island biogeography and ecology. *Ann. Rev. Ecol. Syst.* **5,** 161-82.

Southwood T. R. E. (1977). Habitat, the templet for ecological strategies? *J. Anim. Ecol.* **46,** 337-65.

Stannard L. J. (1967). The redispersal of Thysanoptera (Insecta, thrips) into spoiled strip-mined lands in southern Illinois. *Illinois State Acad. Sci. Trans.* **60,** 130-3.

Walker K. J., Osborne J. M. & Dunlop J. N. (1986). Vertebrate colonization after revegetation of a rehabilitated iron ore waste dump, Newman, Western Australia. In: *1986 Australian Mining Industry Council Environmental Workshop* (ed. D. Whitrow) pp. 113-50. Australian Mining Industry Council, Canberra.

Westman W. E. (1986). Resilience: concepts and measures. In: *Resilience in Mediterranean Ecosystems* (eds B. Dell, A. J. M. Hopkins & B. B. Lamont) pp. 5-19. Dr W. Junk, Dordrecht.

Wetzel R. M. (1958). Mammalian succession in midwestern floodplains. *Ecology* **39,** 262-71.

Proc. Ecol. Soc. Aust. 1990 — 16:521-529

Two hundred years of disturbance: how has it aided our understanding of succession in Australia?

BARRY J. FOX

School of Biological Science, University of New South Wales, P.O. Box 1, Kensington, N.S.W. 2033, Australia

Abstract

The impact of the European invasion of the Australian continent over the last two hundred years has produced many natural laboratories to examine the responses of natural systems to human disturbance. Exogenous disturbances, such as mining, clearing or intensive grazing, produce changes not encountered by the plants and animals in previous natural (endogenous) disturbances such as fire or grazing by native species. This chapter analyses what can be learnt about succession from this Australian experience and examines how current research is contributing to an improved understanding of the mechanisms of animal succession. When comparisons can be made validly, studies of response to exogenous disturbance can provide further insight for the response to endogenous disturbance, while knowledge of response to endogenous disturbance may provide an ability to predict the response to some exogenous disturbances.

A habitat accommodation *model presents animal succession as a response to external changes in the continuing vegetation succession. Animals enter the succession when their specific habitat requirements are met and are replaced when conditions move outside their optimal range reducing their competitive ability. The existence of two components of the model: threshold levels in habitat requirements and the role of interspecific competition in the replacement of species, have been demonstrated by both transplants and manipulations in field experiments.*

Introduction

The Australian landscape is one in which disturbance has long played an important role. It is widely accepted that fire is the natural disturbance most widely implicated in affecting the continent's biota. This is exemplified by the review volume, "Fire and the Australian Biota" (Gill *et al.* 1981). The view espoused there that fire is a natural part of the Australian environment is commonly held, hence identifying fire as an *endogenous disturbance* in the sense of Westman 1978 (see also Westman 1986; Fox & Fox 1986). The importance of fire, both in Australia and other fire-prone regions (Trabaud 1987) has been recognized and identified by effects such as the very rapid responses of plant and animal communities following fire, which are a feature of communities with high resilience (Fox & Fox 1986; Fox *et al.* 1985b). Other features of these highly resilient communities, that have evolved with frequent fire, are the changes in relative abundances of species following fire that have been termed auto-successional by Hanes (1971).

The great age of the Australian continent means that the forces of erosion have long ago reduced any mountains so that we inhabit a vast flat continent with the eastern seaboard as the only area with any significant relief. The result is a lack of reliable surface water over most of the continent which exacerbates the low and irregular rainfall that is an inevitable consequence for a continent at this latitude, lacking mountain ranges to induce orographic precipitation. Hence, for much of the interior, the infrequent and irregular precipitation events brought by thunderstorms, or the more prolonged flood-producing rains from the monsoon latitudes, can be considered as a disturbance endogenous to this environment.

Ironically, it is the burst of growth and reproduction in response to this disturbance which this environment depends on for its continued existence.

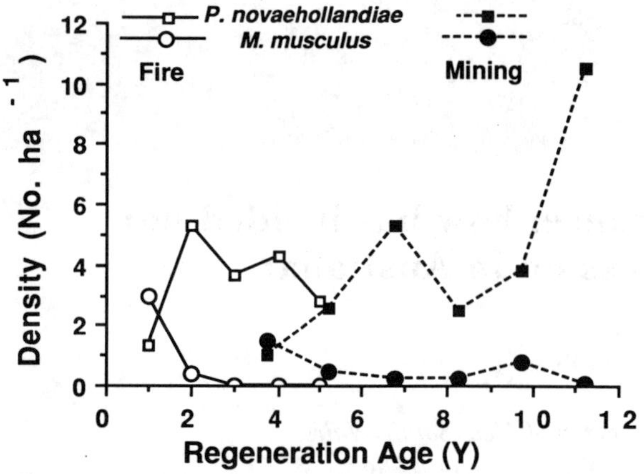

FIG. 1. The response of the introduced *Mus musculus* (house mouse, circles) and the native *Pseudomys novaehollandiae* (New Holland mouse, squares) to fire (open symbols) and sand-mining (solid symbols) for heathland in the Myall Lakes National Park. The data, density of mice (No. ha^{-1}) as a function of regeneration age, are taken from Fox and Fox (1978) for mining and Fox (1982a) for fire.

It is better to consider the unusual precipitation events as disturbance, against a background of drought which is the more usual or common occurrence, rather than consider drought as the disturbance event. However, this sharp dichotomy becomes blurred as one moves along a gradient of increasing precipitation towards the east coast where more regular precipitation is punctuated by both drought and flood events. The removal of topsoil by erosion processes completes a trilogy of disturbances: fire, water and earth which are such an integral part of the Australian environment and constitute its principal endogenous disturbances.

The invasion of the continent by Aboriginal peoples more than 40 000 years ago brought with it the first externally imposed (exogenous) disturbance to the environment. Changed fire regimes are the principal, widespread disturbance associated with Aboriginal people. Their main use of fire was as a tool, to aid in hunting or to promote the growth of some types of plants, and to produce mosaics of relatively small patches of many different regeneration ages. The two hundred year old European invasion greatly increased the impact of exogenous disturbance in magnitude, frequency and type. European use of fire was also primarily as a tool, more often aimed at clearing away vegetation, but also as a means of promoting new growth to be consumed by hard-hooved grazing animals. However, with vast increases in the size of this largely European population, there has been an increased incidence of accidental fires, and those purposefully lit by arsonists. To this must be added the continuing use of fire as a management prescription, to control fuel accumulation in our forests, which has spawned cool fires over vast

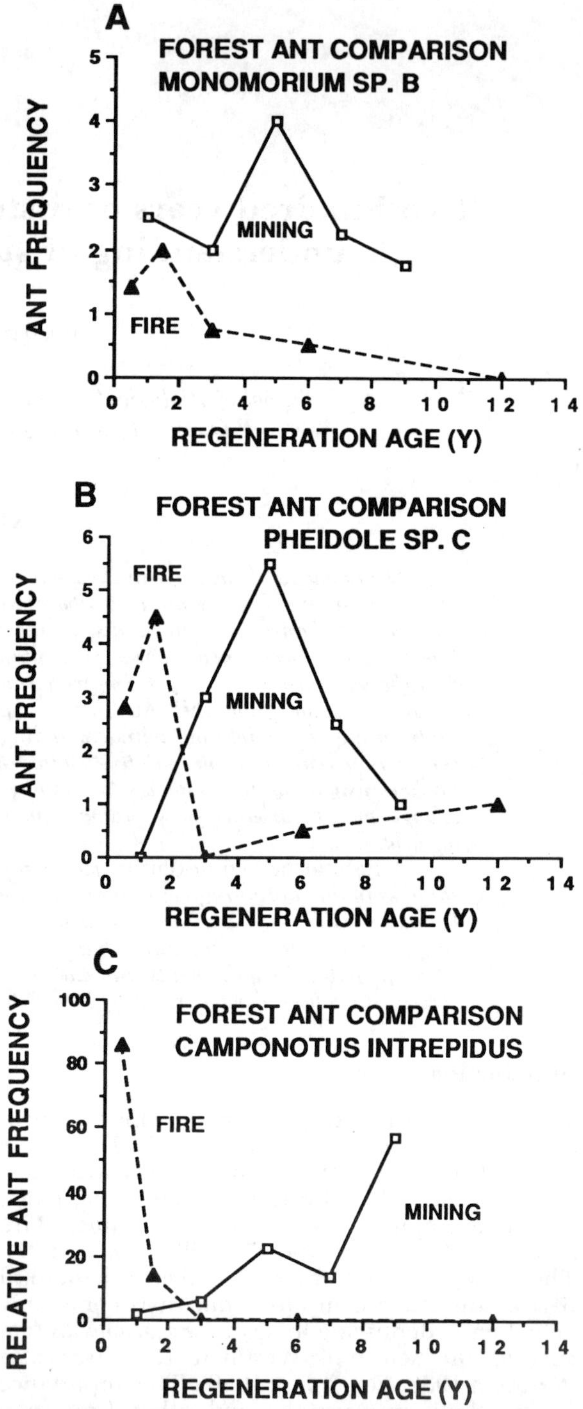

FIG. 2. The response of ant species (ant frequency as a function of regeneration age) to the disturbance of forest fire (solid triangle symbols) and sand-mining (open square symbols) for open forest in Myall Lakes National Park (a) *Monomorium* sp.B, (b) *Pheidole* sp.C, and (c) *Camponotus intrepidus*. The relative ant frequency shown in (c) is the proportional capture in each two year regeneration age class.

tracts of land producing areas of relatively uniform regeneration age from the too frequent burning resulting from such programmes.

In addition to such altered use of fire there have been other massive disturbances that accompanied the European invasion. The removal of two thirds of the forest cover came partly from an expanding need for the timber but more often the clearing was to produce land for agriculture and grazing (Routley & Routley 1975). The disturbance caused by such substantial alienation of habitat inevitably produces marked contractions in range and even extinction of some species of animals and plants. The increased incidence of soil erosion highlights the potential for even more drastic disturbance.

This pressure to clear continues, particularly in the Western Division of New South Wales in an effort to convert the semi-arid shrublands and woodlands, particularly mallee, to wheat farms as has occurred in such a spectacularly rapid fashion in the wheatbelt of Western Australia (see Saunders *et al.* 1985). Although the evidence for range contractions and extinctions is even more apparent in these semi-arid regions there is already in train a much more serious level of disturbance, the degradation of these lands. These areas can produce crops and graze stock successfully during the good years only. But, for much of the time they are not economically viable, so there is overgrazing or the previously cropped land is left fallow, in both cases the result is that topsoil begins to be blown away. This process is desertification and represents an extreme disturbance from which there is little chance for recovery (Messer & Mosley 1983).

However, I will not attempt to document or review the effects of these and all the other disturbances inflicted over 200 years. Rather, I will take an example of endogenous and exogenous disturbance affecting the same systems and ask the questions: How does the regeneration following these two disturbances compare, and what can the study of the regeneration following those two disturbances tell us about their underlying processes? In this chapter I will make direct comparisons of the responses observed at the individual species levels using plots, from open forest on an identical substrate in Myall Lakes National Park, that were subjected to mining or forest fire.

Comparisons of disturbances

The preliminary question to be asked in any such comparison is whether the examples are appropriate. To illustrate this point, I will first consider the example of receding water levels in the Canning Dam Reservoir (Woodruff & Majer 1981). Ants colonized the exposed banks and, although clearly exogenous, it is almost identical with what one might expect following the receding

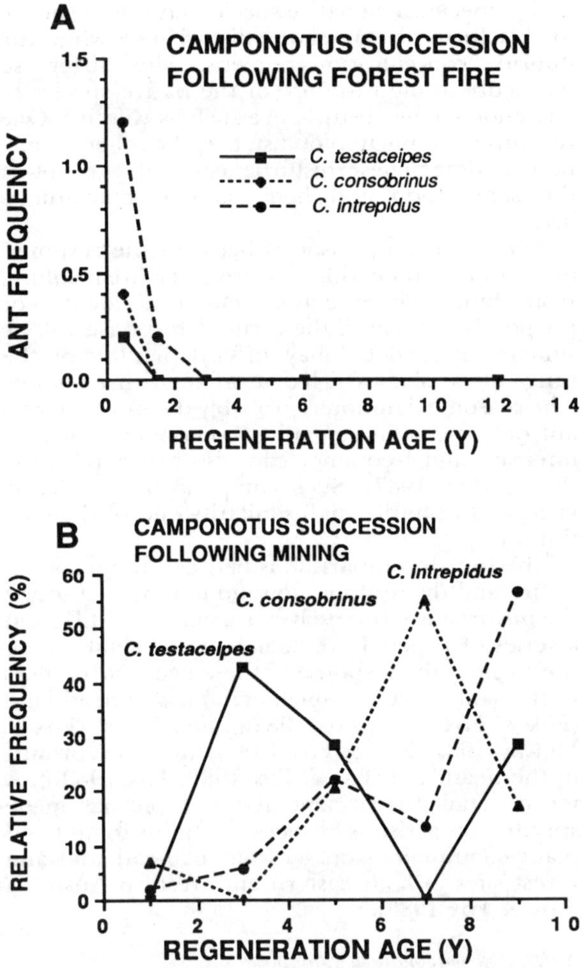

FIG. 3. The succession of three species of *Camponotus; C. testaceipes* (square symbols), *C. consobrinus* (triangular symbols) and *C. intrepidus* (circular symbols) as functions of regeneration age following disturbance by (a) forest fire and (b) sand-mining.

waters of a natural flood (the analogous endogenous disturbance). Such an example is clearly at one end of a continuum of similarity; as it has the same form it certainly is a valid comparison.

A second example, at the other extreme, is a disturbance such as the effect of fluoride fallout on surrounding bushland, caused by the presence of an aluminium smelter (Fox & Andrew 1982). The effects on the native and introduced small mammals are quite opposite. The native species are absent up to 3 km from the smelter and then rise rapidly to the level in the surrounding countryside, while the number of introduced species decreased over the same distance, as one moved away from the smelter. These introduced species are commensal and relate to the human-induced disturbance in a manner that is the opposite of the native species. The native species have no response to this form of exogenous disturbance while the fluoride concentration remains. This differs so markedly to the response of the native species to any endogenous disturbance such as wildfire. One response is a function of distance, the other a function of time. These disturbances then are not of the same form and hence any comparison is invalid.

The third comparison is between the response to a naturally occurring wildfire and a fire resulting from human intervention (either accidental or purposeful). Clearly, the form of response will be similar but the detail may differ depending on the range of variables which determine the fire regime. The season of ignition is probably the most important of these, although other components such as intensity and frequency can also affect response (Fox & Fox 1987). Such comparisons are clearly valid and lie at the high similarity end of the continuum.

The fourth comparison is between the response to fire and the response to sand mining. Although the disturbances themselves are markedly different a series of papers have demonstrated similarity of the form of the response to these two disturbances in the same area: for small mammals in heathland (Fox & Fox 1978; Fox 1982a), and forest (Fox & McKay 1981; Fox & Fox 1984), and also for ants in the heathland (Fox & Fox 1982; Fox 1982b). A more detailed treatment of this topic has made specific comparisons of species richness for ant and plant communities on a variety of heathland and forest sites in both eastern and western Australia (Fox & Fox 1986).

A detailed mammalian example

From previous studies, a summary of data on the small mammal response to fire and mining for sites from the same heathland area in Myall Lakes National Park (see Fox & Fox 1978) can be used to illustrate a similar response to each disturbance (Fig. 1). However, the response to mining is effectively *stretched* along the time axis by a factor of at least four, although it is difficult to establish accurately the magnitude of the *stretching* from these data its existence is clear. In both cases, the declining populations of the introduced house mouse, *Mus musculus,* are replaced by the native New Holland mouse, *Pseudomys novaehollandiae.* A similar set of responses has also been reported from forest sites in Myall Lakes National Park (Fox & Fox 1984), demonstrating again that fire and mining responses may be validly compared and the mining response is *stretched* to be almost fourfold that for fire.

The replacement of the house mouse by the New Holland mouse is a direct result of interspecific competition between these two species, and has been demonstrated by controlled experimental manipulations in the field (Fox & Pople 1984; Fox & Gullick 1989) and controlled behavioural manipulation experiments in enclosures (McLeod 1987). The inability of the New Holland mouse to colonize the earlier successional stages has also been demonstrated by transplant experiments on two separate mining paths at different times (in 1982, Fox & Pople 1984; and again in 1988, Fox & Twigg, unpubl. data).

Forest ants as a model system

The open forest in Myall Lakes National Park grows on a uniform Holocene sand substrate forming a dune system, and makes a natural laboratory within which to examine ecological processes. The forest demonstrates the impact of the 200 year old European invasion. The presence of shell middens identified the Aboriginal impact, this was most noticeable by the presence of bloodwood, *Eucalyptus gummifera,* adjacent to the midden sites in a forest otherwise dominated by blackbutt, *E. pilularis,* and *Angophora costata,* a legacy of the locally increased fire frequency (Fox 1981). The first European impact was also detectable from cut stumps, as limited selective logging had begun in the forest by the 1880s. However, it has remained relatively undisturbed.

Two sets of data were collected from the open forest, one set of data was collected along two mining paths (Big Gibber and Bridge Hill Ridge) that are described in Fox and Fox (1984). Samples were collected in June and December in 1982. The regeneration age of the sixteen plots sampled ranged from 0.35 to 10.3 y and were pooled into five two year age classes to allow the data from the two mining paths to be presented together.

The second set of data was collected from the same forest with the same heath understorey. Three sites of known age (1 month, 3 years 1 month and 13 years 1 month) occurred within a 200 m radius. Four replicates were collected in December 1980 and an additional set was collected in December 1981 when all were one year older. In December 1976 three nearby sites with ages 0.4, 1.8 and 7.6 y had been sampled and these data were included also. Because of the concentration of sites towards the early time period, and the reduced number of sites of later regeneration age, the age classes used were 0-1, 1-2, 2-4, 4-8 and 8-16 y.

For both data sets, the same sampling procedures were used, fifteen 2.5 cm diameter glass vials were buried flush to the ground surface in a 3×5 grid with 20 m spacing. The traps were opened for a 24-hr period immediately they were buried to maximize the capture rate caused by the digging-in effect (Greenslade 1973). To overcome any anomalies of ant abundance introduced because of the techniques used, all data were analysed as frequency of occurrence (i.e., the

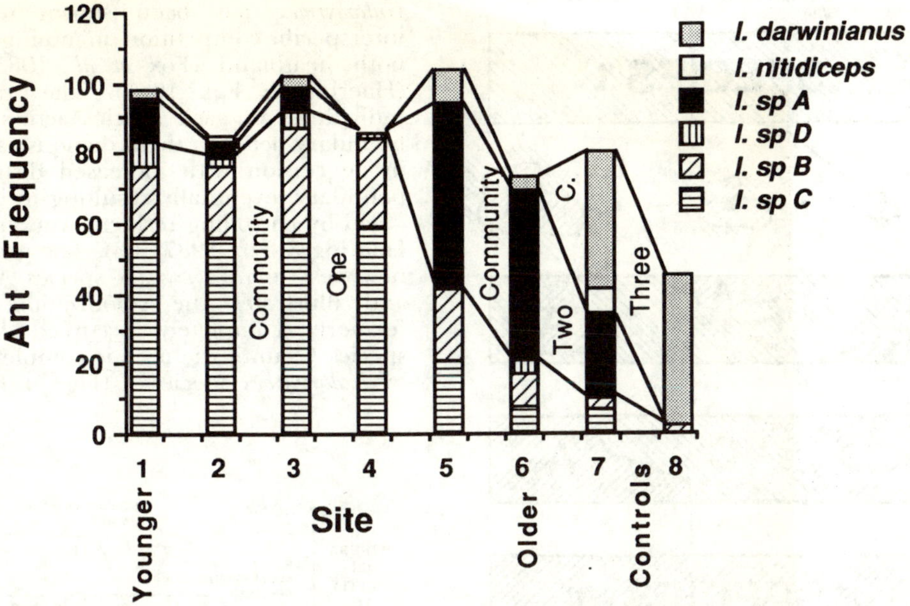

FIG. 4. Species composition for the dominant functional group of *Iridomyrmex*. The data were collected from 24 hr pitfall samples taken in April and December for each of two years (1976 and 1977). The maximum frequency for any one species is 4×15=60 with the summation frequency for the six species shown on the figure. There are two unmined control sites, 7 and 8, with the mined sites arranged from youngest to oldest with respect to the time since topsoil replacement; 1) March 1972; 2) November 1971; 3) February 1971; 4) January 1969; 5) October 1967 and 6) April 1966. The data are taken from Fox and Fox (1982) and the figure is amended from Fox (1982b).

number of vials in which a species is found, maximum 15) on any plot. From this data base I will identify the type of information each form of disturbance provides and relate this to an understanding of the underlying successional processes.

The first comparison to be made is at the species level. Here I have used one species from each of three major genera *Monomorium, Pheidole* and *Camponotus* to illustrate how they differ in their response to the respective disturbances of fire and mining (Fig. 2).

For *Monomorium* sp B, the peak is well defined for both fire and mining (Fig. 2a). The highest ant frequency occurs after fire in the 1-2 y age class, plotted here as 1.5 y, while the highest frequency following mining disturbance occurs in the 4-6 y age class, plotted here as 5 y. A similar situation is demonstrated for *Pheidole* sp C (Fig. 2b). Both show a more than three-fold shift, from fire to mining, in the age at which the successional wave peaks for that species, clearly demonstrating how the time axis has been stretched for the mining disturbance.

The greatest shift is shown by *Camponotus intrepidus,* the response to fire has its highest value at the youngest age class and rapidly decreases to zero while the response to mining increases to its highest relative frequency at the oldest age class (Fig. 2c). This shift should be regarded as a minimum as it is possible that the actual peak values lie outside the range monitored here.

The forest ant succession

To demonstrate that the effect observed is not confined to the individual species selected I illustrate the succession of species of *Camponotus* for the two disturbance effects (Fig. 3). For the succession following fire it is difficult to obtain sufficient resolution on this scale, as the ants are found in the first two age classes, but were not recorded in any other age class up to the 8-16 y age class (Fig. 3a). The results are consistent with the tail-end of the wave for each of these species. A much clearer picture emerges for the same species following sand mining, then it is apparent that the order of the succession is *Camponotus consobrinus, C. testaceipes* and *C. intrepidus* (Fig. 3b) as was implied from the results in Fig. 3a. This demonstrates how observation of the succession following sand mining can assist in understanding the succession following fire by clarifying the successional order of replacement that could not otherwise be resolved with any degree of certainty. The conclusions resulting

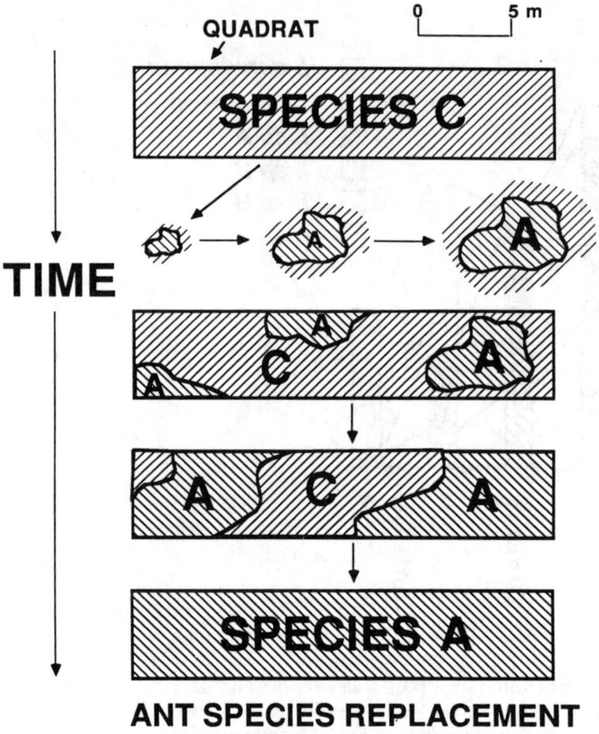

FIG. 5. A diagrammatic representation of the way in which the area controlled by two species of *Iridomyrmex* shifts with time. Beginning with complete domination by *I.* species C and colonization by nests of *I.* species A which expand their territories at the expense of *I.* species C until there is complete domination by *I.* species A. Details of the competitive mechanisms involved are presented by Haering and Fox (1987).

from the comparisons of these ant studies are consistent with those already presented for small mammals (Fox & Fox 1978, 1984) and summarised in this chapter.

The greater resolution available from the sand mining data also demonstrates the successional replacement series for the species of *Pheidole,* and the species of *Iridomyrmex,* found on the regenerating mining path through the forest.

The *Iridomyrmex* response to sand mining through heathland at the southern end of Myall Lakes National Park (Fig. 4) has demonstrated a changeover of communities controlled by the dominant *Iridomyrmex* species (Fox & Fox 1982; Fox 1982b). These results provide strong support for regarding *Iridomyrmex* as keystone species forming the dominant functional group, as first proposed by John Greenslade for semi-arid regions (see Greenslade & Halliday 1983). Functional groups have been shown to be an important component of ant community organization in mesic southeastern Australia (Andersen 1986), and can also serve as important bio-indicators (Andersen this volume).

The replacement of the dominant species of *Iridomyrmex* has been shown to result from interspecific competition on mining paths, through both heathland (Fox *et al.* 1985a) and forest (Haering & Fox 1987). The transplanting of individuals of each species across the territorial boundary between them demonstrates an increase in aggression with increased distance from the boundary, eventually resulting in intruders being killed by patrolling resident workers (see Fig. 4 in Haering & Fox 1987:503). The growth in area of new nests of *Iridomyrmex* species A was recorded and illustrates the colonization of the area, formerly completely occupied by *Iridomyrmex* species C, until the area is completely dominated by *Iridomyrmex* species A (Fig. 5). Given the high

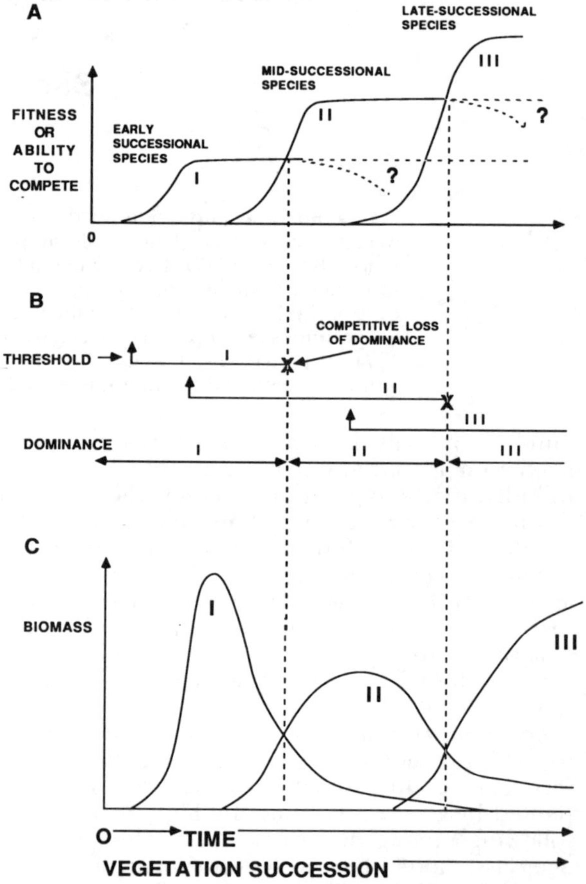

FIG. 6. a) A diagrammatic representation of the fitness or competitive ability as a function of time for three animal species: early successional (I), mid-successional (II), and late successional (III). b) The periods of dominance for each species defined by either the threshold for the entry of a species or the loss of dominance by the preceeding species and the competitive loss of dominance by the species itself. c) A diagram representing biomass (or other measure of abundance or productivity) for three animal species as functions of time and vegetation succession.

mobility and colonizing ability of alates, the fact that no new nests of *Iridomyrmex* species A were observed on the earlier seral stages has been taken to mean that these stages are not suitable habitat, as had been demonstrated by transplant experiments for the New Holland mouse in the mammal studies.

The habitat accommodation model

From the information available from the ant studies presented here, and the small-mammal studies already presented, it is apparent that these data support the *habitat accommodation* model for animal succession which was developed for the small mammal succession following fire on heathland (Fox 1982a). The term *habitat* is included to emphasize the fact that local physical conditions change in response to the external factors produced by the vegetation succession. The animal species enter the succession when changing local physical conditions meet these species specific habitat requirements. In this respect, when comparing this succession to the tolerance and inhibition models of Connell and Slatyer (1977), one might consider the entry to equate to a *habitat facilitation* model and think of such a threshold determining the entry for each species. Species are then competitively replaced in the succession when local physical conditions move out of their optimal range, so reducing their competitive ability. This aspect of the succession might be equated with a *habitat tolerance* model. The detail of the whole *habitat accommodation* model, including both of these aspects, is set out in Fox (1982a).

An explanation for the operation of this successional model is displayed in Figure 6. The fitness or competitive ability for each species is shown on the ordinate as a function of regeneration time or the more apposite variable the progress of the vegetation succession, which could be represented by any appropriate vegetation or habitat parameter (Fig. 6a). The important point here is not so much the detailed shape of these curves, but rather that the curve for species II intersects the curve for species I such that species II becomes competitively superior and the process is then repeated for species II and III. The period of dominance for each species is thus defined from the thresholds to the loss of competitive ability (Fig. 6b) and can then be translated into some measure of abundance or biomass as is demonstrated in Figure 6c.

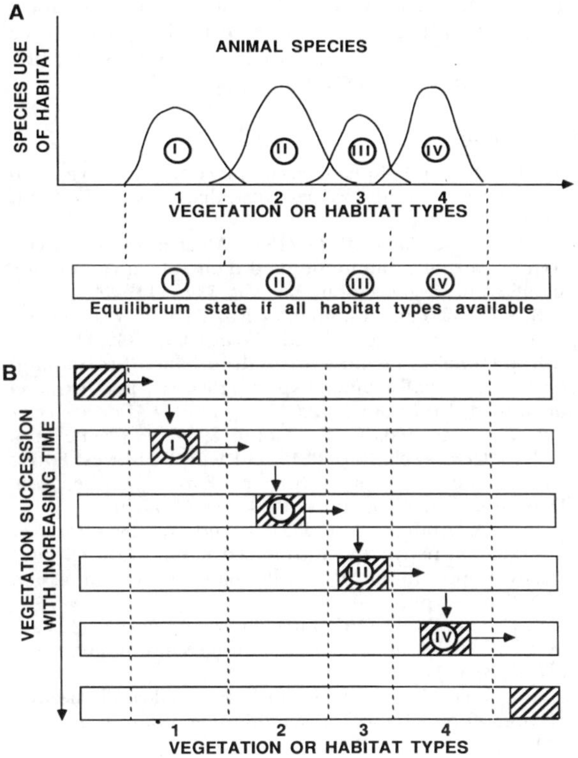

FIG. 7. a) A diagrammatic representation of the use of habitat by four animal species (I, II, III, & IV) for an equilibrium state when all habitat types (1, 2, 3 and 4) are available. b) An *escalator* model for animal succession showing a succession of vegetation types with increasing time and illustrating how they represent habitat suitable for the different animal species which enter the succession when their habitat requirements are met. The result is observed to be an animal succession as a function of time, but, is in fact responding to some habitat parameter which itself is a function of time.

A better understanding of the operation of the model can be obtained from Figure 7. Here we can consider the equilibrium response of the animal species (I, II, III and IV) to the four vegetation or habitat types (1, 2, 3 and 4). The equilibrium situation reflects a gradient analysis for the distribution of each species habitat requirements (or habitat use) along the gradient of habitat availability. The lower part of Figure 7 demonstrates that as the vegetation succession proceeds each habitat type in turn becomes available for colonization by the animal species concerned. I call this model the *escalator model* with each species *stepping on* as the appropriate habitat requirement becomes available, then to be competitively displaced as the next species requirements become available. While the figure is diagrammatic, it clearly explains how the model operates and demonstrates that the model is consistent with both the ant story presented here and the small-mammal story which has already been described (Fox 1982a) and is also summarized here.

Discussion

Confidence in the validity of disturbance comparisons is greatly enhanced when sites are from the same area and have the same substrate. Hence, I have drawn heavily on my own and colleagues work from Myall Lakes National Park to meet this criterion. While this allows me to be confident in the conclusions I draw from these data it should be pointed out that the strength of the inference to be drawn will decrease as the distance between comparative sites increases, and also as they exhibit differences in other factors such as substrate or habitat type.

The ant data used for my main comparisons are based on 24 hr collections from pitfall traps of the ground-active species. So they do not in any way represent a complete census of all species present at the site. However, they do represent an ecologically based sampling methodology that can be replicated and provides a reliable basis for comparison as the number of potentially variable factors is minimized.

The fact that it is possible to demonstrate a similar response to fire and sand-mining for taxa as disparate as ants and small mammals encourages one to generalize. An increased generalization for the conclusions is further supported when both taxa appear to fit the same model for succession. While the conclusions reached here may not be universal, it is indeed encouraging to find such a degree of comparability, which is even further enhanced when it can be shown to hold for both heath and forest habitats.

The concept of the animal succession responding to external factors, in this case the changes in vegetation structure and composition resulting from the plant succession, means that the animal succession is not tied directly to a time axis. While there will be a relationship with time, it is unlikely to be linear. The bell-shaped curves illustrated here for the animal succession imply that species are much more likely to be tied directly to changes in some other factor, such as the amount of vegetation cover or the level of insolation reaching the soil surface. However, perhaps the most important point to be considered in this model for animal succession is the necessity to recognize that the mechanisms that operate for animal succession cannot be the same as those for plant succession, because the vegetation changes themselves are the dominant factor to be considered when examining the animal succession.

The findings of this chapter demonstrate the benefits to be gained from the integration of applied and theoretical aspects of research and show that information exchange can occur in both directions.

Conclusions

When exogenous disturbances (such as mining) are similar in form to endogenous disturbances (such as fire) the time scale for regeneration is *stretched.*

Study of the endogenous disturbances in an area can provide useful information about the potential effects of exogenous disturbances, if the responses are similar in form so that valid comparisons can be made.

Study of exogenous disturbance can provide valuable insight into the processes involved in the endogenous disturbance in an area, if the responses are similar in form so that valid comparisons can be made.

The *habitat accommodation* model for animal succession does appear to describe the form of succession observed in ant communities regenerating after sand-mining.

Acknowledgements

I wish to thank Elizabeth Archer for identifying and counting the ants in the pitfall traps and Bob Taylor for confirming the identification of voucher specimens. Students in the General Ecology course in 1980 and 1982 assisted with the collection of some of the pitfall traps. Marilyn Fox read and commented on an earlier version of the paper which was word processed by Ann Lonergan and Annette McGreavy. The studies reported in this paper have been supported by a number of A.R.G.S. grants, most recently by D18316065 and A18616300.

References

Andersen A. N. (1986) Patterns of ant community organization in mesic south-eastern Australia. *Aust. J. Ecol.* **11,** 87-97.

Connell J. H. & Slatyer R. O. (1977) Mechanisms of succession in natural communities and their role in community stability and organization. *Am. Nat.* **111,** 1119-44.

Fox B. J. (1982a) Fire and mammalian secondary succession in an Australian coastal heath. *Ecology* **63,** 1332-41.

Fox B. J. (1982b) The influence of disturbance (fire, mining) on ant and small mammal species diversity in Australian heathland. In: *Dynamics and Management of Mediterranean-Type Ecosystems* (eds C. E. Conrad & W. C. Oechel) pp. 213-19. Gen. Tech. Rep. PSW-58 Pacific Southwest Forest and Range Experiment Station, Forest Service, U.S. Department of Agriculture, Berkeley, California.

Fox B. J. & Andrew D. (1982) A study of the effects of fluorides on native and introduced fauna in the Hunter Valley. Report to the State Pollution Control Commission of N.S.W. 70 pp.

Fox B. J. & Fox M. D. (1978) Recolonization of coastal heath by *Pseudomys novaehollandiae* following sand mining. *Aust. J. Ecol.* **3,** 447-65.

Fox B. J. & Fox M. D. (1984) Small-mammal recolonization of open-forest following sand-mining. *Aust. J. Ecol.* **9,** 241-52.

Fox B. J. & Fox M. D. (1986) Resilience of animal and plant communities to human disturbance. In: *Resilience of Mediterranean-Type Ecosystems* (eds B. Dell, A. J. M. Hopkins & B. Lamont) pp. 39-64. Tasks for Vegetation Science Series, Dr W. Junk, The Hague, The Netherlands.

Fox B. J. & Gullick G. (1989) Interspecific competition between mice: a reciprocal field manipulation experiment. *Aust. J. Ecol.* **14,** 357-66.

Fox B. J. & McKay G. M. (1981) Small mammal response to pyric successional changes in eucalypt forest. *Aust. J. Ecol.* **6,** 29-42.

Fox B. J. & Pople A. R. (1984) Experimental confirmation of interspecific competition between native and introduced mice. *Aust. J. Ecol.* **9,** 323-34.

Fox B. J., Fox M. D. & Archer E. (1985a) Experimental confirmation of competition between two dominant species of *Iridomyrmex* (Hymenoptera: Formicidae). *Aust. J. Ecol.* **10,** 105-10.

Fox B. J., Quinn R. D. & Breytenbach G. J. (1985b) A comparison of small-mammal succession following fire in shrublands of Australia, California and South Africa. *Proc. Ecol. Soc. Aust.* **14,** 179-97.

Fox M. D. (1981) *Coexistence between eucalypts in a coastal open forest.* Ph.D. Thesis, Macquarie University.

Fox M. D. & Fox B. J. (1982) Evidence for interspecific competition influencing ant species diversity in regenerating heathland. In: *Ant-plant interactions in Australia* (ed. R. C. Buckley) pp. 99-110. Dr W. Junk, The Hague, The Netherlands.

Fox M. D. & Fox B. J. (1987) The role of fire in the scleromorphic forests and shrublands of eastern Australia. In: *The Role of Fire in Ecological Systems* (ed. L. Trabaud) pp. 23-48. SPB Publishing, The Hague, The Netherlands.

Gill M., Noble I. R. & Groves R. (1981) *Fire and the Australian Biota.* Australian Academy of Science, Canberra.

Greenslade P. J. M. (1973) Sampling ants with pitfall traps: digging in effects. *Insectes Sociaux* **20,** 343-53.

Greenslade P. J. M. & Halliday R. B. (1983) Colony dispersion and relationships of meat ants *Iridomyrmex purpureus* and allies in an arid locality in South Australia. *Insectes Sociaux* **30,** 82-99.

Haering R. & Fox B. J. (1987) Short-term coexistence and long-term competitive displacement of two dominant species of *Iridomyrmex:* the successional response of ants to regenerating habitats. *J. Anim. Ecol.* **56,** 495-507.

Hanes T. L. (1971) Succession after fire in the Chaparral of southern California. *Ecol. Monogr.* **41,** 27-52.

Messer J. & Mosley G. (1983) *What future for Australia's arid lands?* Australian Conservation Foundation, Melbourne.

McLeod S. R. (1987) Mechanisms of competition between the murid rodents *Pseudomys novaehollandiae* (Waterhouse) and *Mus musculus* Linneaus. B.Sc. (Hons) Thesis, University of New South Wales.

Routley R. & Routley V. (1975) *The fight for the forests.* Research School of Social Sciences, Australian National University, Canberra, 3rd Edition.

Saunders D. A., Rowley I. & Smith G. T. (1985) The effects of clearing for agriculture on the distribution of cockatoos in the southwest of Western Australia. In: *Birds of Eucalypt Forests and Woodlands: Ecology, Conservation, Management* (eds A. Keast, H. F. Recher, H. Ford & D. Saunders) pp. 309-21. Surrey Beatty & Sons, Chipping Norton.

Trabaud L. (ed.) (1987) *The Role of Fire in Ecological Systems.* SPB Publishing, The Hague, The Netherlands.

Westman W. (1978) Measuring the inertia and resilience of ecosystems. *Bioscience* **28,** 705-10.

Westman W. (1986) Resilience: concepts and measures. In: *Resilience in Mediterranean-type Ecosystems* (eds B. Dell, A. J. M. Hopkins & B. B. Lamont) pp 5-19. Dr W. Junk, Dordrecht, The Netherlands.

Woodruff S. & Majer J. D. (1981) Colonization of ants on the exposed banks of the Canning Dam reservoir Australia. *Aust. Entomol. Mag.* **8,** 41-6.

Proc. Ecol. Soc. Aust. 1990 — 16:531-535

Seed ecology in relation to reclamation: lessons from mined lands in Western Australia

D. T. BELL, S. VLAHOS and S. M. BELLAIRS

Department of Botany, The University of Western Australia, Nedlands, W.A. 6009

Abstract

Complete restoration of regions of the Eucalyptus marginata *forest of Western Australia following bauxite mining is limited by topsoil which contains predominantly seed of annuals and biennials rather than the long-lived perennial shrubs and trees of the pre-mining ecosystem; the ability to collect and respread viable, germinable seed of many of these woody species; and limited amounts of bradysporous species seed. The restoration of regions of Northern Sandplain kwongan shrubland following heavy minerals sand mining has been limited in part by an inability to induce many of the soil-borne seed of the transported topsoil to germinate. However, significant shrubland reclamation has been provided using a shrub mulch which carries large numbers of bradysporous seed to post-mining habitats, the broadcasting of hand collected seed and the planting of nursery stock. Data on community store of seed, seed viability and germinability, and the vegetation of post-mining communities have been provided to illustrate aspects important to minesite reconstruction in Western Australia.*

Introduction

Mining drastically disturbs natural ecosystems and the time and effort to rebuild the structure and function of such ecosystems can be considerable (Collins 1984). However, the ecological lessons learned in the process of disturbed land reconstruction can be most valuable in understanding natural Australian ecosystems. Reconstruction implies that the ecosystem is rebuilt in the image of what it was in its original form. As defined by Allen (1988) *reconstruction* includes restoration, which suggests that the post-mining ecosystem would be exactly as before degradation, and reclamation, which conveys the idea that the post-mining ecosystem would be similar after disturbance, and would be habitable for similar, but not necessarily the same organisms. In North America, *rehabilitation* is a broader term, not necessarily included in reconstruction, meaning that the land would be made useful, but with a different land use and usually composed of different species (Allen 1988). In Australia, the term rehabilitation is used rather loosely as a synonym of reconstruction. In Western Australia, the mining industry is generally charged with the reconstruction of their mining leases by the Government and the larger companies employ professional staffs to carry out associated research and to implement the reconstruction process. This chapter describes research carried out on the seed ecology of ecosystems associated with the reconstruction of the jarrah *(Eucalyptus marginata)* forest of the Darling Range by Alcoa of Australia Ltd. and the kwongan shrublands near Eneabba by Associated Minerals Consolidated Ltd.

Aspects of community seed store, seed germination and seedling establishment are especially important in Australian ecosystems because the developmental process in these systems tends to follow the *initial floristic compositional* model (see Noble & Slatyer 1981) and the composition of the vegetation in the first few years of reconstruction controls ecosystem function for many years (Bell 1988). If inappropriate species are initially established in post-mining ecosystems, particular organisms of the pre-mining habitat may not be able to survive in the post-mining community.

Pre- and post-disturbance site floras

Centred Principal Components Analysis (PCA) was used to ordinate a set of qualitative data on the floristic composition of native jarrah forest sites, the floristic composition of reconstructed bauxite mine sites, and the floristic composition of their respective soil seed banks (Fig. 1). The sites of the native forest grouped together at one end of the first ordination axis, indicating considerable floristic similarity between these sites relative to the variation within the entire matrix of data. The bauxite mining sites grouped near the mid-point of the first axis and the samples representing the floristic composition of the seed banks grouped at the low end of the primary PCA axis. Bauxite mine rehabilitation utilizes topsoil transfer to provide propagules as well as soil organic matter and as a source of mycorrhizal fungi associates and *Rhizobium* for legume innoculation (Tacey 1979). However, the large difference in ordination hyperspace location indicates that the floristic content of the topsoil from native forest regions is very different from the present mature vegetation. An earlier study indicated that the soil seed bank of the northern jarrah forest is dominated by species of annuals and subshrubs despite a domination of the mature vegetation by woody shrubs and trees (Vlahos & Bell 1986). The percentage of the perennial species which occurred both in the existing vegetation and in the soil of these sites as seed was surprisingly low and ranged only from 11 to 29% (Bell 1988). The location of the rehabilitation sites midway between the seed bank and the native forest samples indicates that the provision of additional species as broadcast seed has improved the similarity of the bauxite mine site communities to the native forest regions, but the ordination of the information on floristic composition highlights the fact that the technology of reconstruction process still has a considerable way to go to achieve full restoration of the jarrah forest ecosystem.

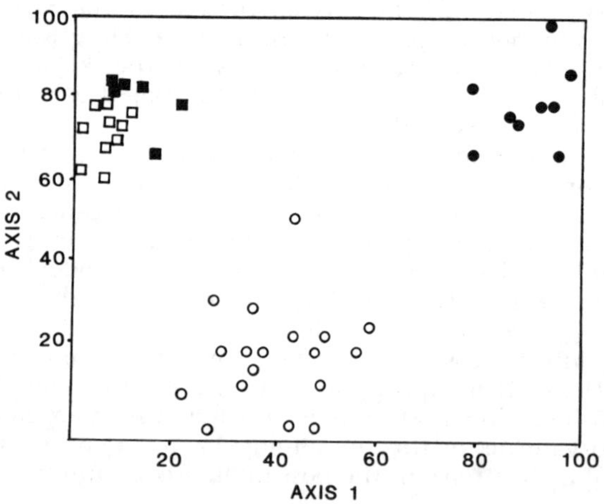

FIG. 1. Centred principal components analysis stand ordination of presence/absence data from 12 native jarrah forest sites (solid circles), the soil seed store flora of six of the 12 jarrah sites (solid squares), the existing components of 18 bauxite mine rehabilitation sites (open circles) and the soil seed bank components of 10 of the rehabilitation sites (open squares), (from Bell 1988).

Jarrah forest seed viability

Jarrah forest legumes have provided the majority of species of the broadcast seed mix used to supplement the bauxite site flora since 1977 (Ward 1988). These native legumes quickly establish a ground cover and supplement the nitrogen content of the nutrient depauparate post-mining ecosystem (Ward *et al.* 1985). Seeds of jarrah forest legume species are easily collected, generally have high viability, and can be stimulated to germinate with a boiling treatment (Glossop 1980). Most of the other species of the jarrah forest understorey, however, are more difficult to collect and often produce very few seeds or mostly inviable seed (Bell *et al.* 1987).

Tests using tetrazolium chloride to judge percentage viability (Moore 1973) and glasshouse plantings of freshly collected seed to provide an estimate of percentage germinability were performed on a wide range of non-leguminous jarrah forest understorey species. The results of this series of trials also indicated that a large portion of the species produce mostly inviable seed (Table 1). Out

TABLE 1. Jarrah forest species tested by tetrazolium chloride for viability and for germination in jarrah forest topsoil under glasshouse conditions after 80 days. Life form codes: SP — resprouting following fire; OS — obligatory re-seeding species following fire.

Species	Life form Code	Percentage Viability	Germination
Agrostocrinum scabrum	SP	52	4
Boronia fastigiata	SP	10	0
B. molloyai	SP	0	0
B. spathulatum	SP	30	2
Clematis pubescens	SP	16	0
Dianella revoluta	SP	10	0
Dryandra carduacea	OS	64	59
D. nivea	SP	5	4
D. polycephala	OS	95	80
D. sessilis	OS	50	82
Eriostemon spicatus	SP	8	0
Glischrocaryon aureum	SP	0	0
Grevillea quercifolia	OS	95	4
Hakea trifurcata	OS	90	87
H. undulata	OS	44	68
Hibbertia hypericoides	SP	28	0
Hybanthus floribundus	SP	20	0
Leptomeria cunninghamii	SP	0	0
Leucopogon oxycedrus	SP	0	0
L. verticillatus	SP	0	0
Lomandra sonderi	SP	56	2
Opercularia echinocephala	SP	10	0
Phyllanthus calycinus	SP	6	7
Pimelea rosea	SP	2	0
P. suaveolens	OS	0	0
Platysace tenuissima	SP	2	0
Sollya heterophylla	SP	0	0
Tetratheca hirsuta	OS	52	2
Thomasia pauciflora	SP	13	8
Trymalium ledifolium	OS	42	1
T. spathulatum	OS	64	0
Xanthosia atkinsoniana	SP	0	0

of the 32 species tested in these trials, 18 had levels of viability below 20%. Predominantly it is the long-lived, resprouting species which have the low viability percentages. Species which must produce seed to re-establish following fire tend to produce a much higher proportion of viable seed. Although these re-seeding species produce a high proportion of viable seed, there are examples of low rates of germination. In a response similar to the jarrah forest legume species, germination in some of these species could be induced by a heat treatment. Bell *et al.* (1987) showed that a dry heating treatment stimulated the germination of *Trymalium ledifolium* and *T. spathulatum*. Good candidates for further heating trials from this study would be *Agrostocrinum scabrum*, *Grevillea quercifolia*, *Lomandra sonderi* and *Tetratheca hirsuta*.

Species showing high germination percentages in the glasshouse soil conditions tended to be species with woody bradysporous fruits (e.g., *Hakea undulata* and *H. trifurcata*) or species that protect seed with numerous bracts (e.g., the *Dryandra* species). These species delay the dehiscence of seed until a fire destroys the parent plant and seeds drop into a post-fire habitat more likely to allow establishment and survival. The chance of a subsequent fire at this time is minimal. Therefore, the evolutionary pressure for a heat stimulus of these seed is also minimal. Once bradysporous fruits open to release the seed to the ash bed there is no advantage of seed requiring a further fire stimulus to germinate. There is evidence, however, of a temperature imposed dormance mechanism which prevents germination in summer (Specht *et al.* 1958).

Community seed store

Soil-borne seed in the jarrah forest is generally plentiful (Table 2) and the transfer of topsoil provides a significant source of seed to the rehabilitation regions. However, as stated earlier, the floristic composition of the soil-borne seed is markedly different from the mature forest ecosystem, but *numerically* this rich source of seed provides an economical method of revegetating bauxite mine sites. Collection of vegetation with the bradysporous fruits to supplement the broadcast seed mix would provide additional species to the rehabilitation areas, but the jarrah forest contains relatively few bradysporous species (Christensen & Kimber 1975) and the mean density of bradysporous seed is also relatively low (Table 2).

In contrast to the jarrah forest, the bradysporous seed density of kwongan shrubland vegetation is quite high (Table 2) and these canopy-borne seeds can provide a significant source of seed for the reconstruction of disturbed lands in areas dominated by this form of plant community. Gravel-mining scars are currently being rehabilitated in the vicinity of Kulin using the canopy-borne seeds of adjacent areas (K. J. Atkins, pers. comm.) and the reclamation procedures employed by Associated Minerals Consolidated Ltd. at Eneabba utilize a mulch obtained by clipping off the tops of the kwongan shrubs as a top dressing to the transported topsoils in areas mined for mineral sands (Carey & Jefferies, pers. comm.). This mulch, in addition to providing significant quantities of canopy-borne seed, provides organic matter and protects the rehabilitation surfaces from wind and water erosion (Bell *et al.* 1986).

TABLE 2. Community seed store in the jarrah forest near Jarrahdale and kwongan vegetation near Kulin and Eneabba.

	Seeds per m^2	Reference
Jarrah Forest		
Soil Seed Store		
Native Forest Areas	769±436	Vlahos & Bell 1986
Stored Topsoil	134± 47	Bell 1988
Canopy Seed Store		
Native Forest Areas	7± 14	Bellairs & Bell unpubl.
Kwongan Shrubland — Kulin		
Canopy Seed Store	1 121±900	Bellairs & Bell unpubl.
Kwongan Shrubland — Eneabba		
Canopy Seed Store	305±202	Bellairs & Bell unpubl.

Eneabba reclamation vegetation

Permanent sampling quadrats have provided further important information on patterns of early establishment in mining reclamation sites. In 1984 permanent quadrats were marked in a newly rehabilitated sand mine at the Eneabba mine site. In November 1985, the twelve replicated 4 m×7 m plots averaged 30 individuals per plot, but more than 93% of these individuals were from species with canopy-borne seed storage (Table 3). In subsequent seasons, the site has added new species; predominantly species of the soil seed bank and consequently the proportion of the individuals contributed by the bradysporous species has dropped from 93% to 74%. In comparison to other years, the 1984 rehabilitation sites are depauperate in soil-borne seedlings especially, but the dominance of the bradysporous species in the first years following revegetation treatments has continued.

Discussion

The ultimate goal of mine site reconstruction should be to provide a post-mining habitat which includes all the structural and functional aspects of the natural ecosystem. The goal of complete restoration is clearly an ideal that can be achieved only

TABLE 3. Vegetation information from rehabilitated Eneabba sand mining site initially treated in 1984. Mean values ± standard deviation from 12 replicate plots of 4 m×7 m.

Sample	Plants	Species	% Bradysporous Plants
Nov. 1985	30±22	7±4	93± 6
Nov. 1986	42±25	11±4	79±10
Nov. 1987	41±22	12±4	74± 8

with exceptional knowledge of the components of the natural ecosystem and significant financial inputs. Reclamation of mine sites is a more realistic goal for the Western Australian mining industry.

In the reclamation of disturbed regions of the jarrah forest, topsoil provides an initial increment of species, but the composition of the soil seed flora requires supplementation with germinable seed and nursery stock to be more similar to the flora of the native forest (Fig. 2). Lack of viable seed from many species of the understorey will be difficult to overcome. The long-lived species which resprout following fire tend to produce few seeds, most of which are inviable. These species, characterized as S- (stress) or C- (competition) tolerant species in Grime's model of plant strategies, expend most of their energy reserves in growth and maintenance rather than reproduction (Grime 1979). The resprouting species contribute approximately three-quarters of the understorey species flora in the jarrah forest (Bell *et al.* 1988). Collection of large quantities of seed from these species for reclamation seed mixes is expensive and many will not be transferred to the post-mining habitat as these species tend to be in very low numbers in the soil seed store. Species requiring re-seeding following fire produce more abundant seed and collections tend to have reasonably high proportions of viable seed. These species tend to have shorter life histories [characterized as R- (ruderal) strategy species by Grime 1979] and energy allocation to reproduction is high in relation to growth and maintenance and the S- and C-strategy species of the jarrah forest. Jarrah forest obligate re-seeding species tend to produce innately dormant seed which in the case of the legumes requires heat to break dormancy (Shea *et al.* 1979). More recent information indicates that some of the non-legume species of the jarrah forest understorey also break dormancy after treatments which simulate the temperature effects of a fire (Bell *et al.* 1987). This study has revealed that species of the jarrah forest which protect their seed from fire in fruit retained on the plant tend to produce highly viable seed which germinates without pretreatment once freed from the fruit. These bradysporous species, however, do not provide a significant source of propagules for reclamation and, therefore, the spread of a

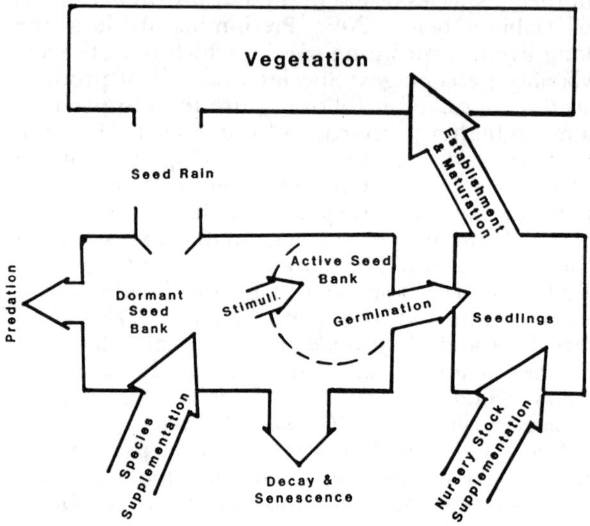

FIG. 2. Representation of the transfers and components of the seed bank, seedling bank and mature vegetation of reconstructed ecosystems (from Bell 1988).

mulch of jarrah forest understorey shrubs would not add greatly to the number of individuals and species in post-mining habitats.

Reclamation after heavy mineral sand mining at Eneabba takes advantage of the large reserves of canopy-borne seed of the kwongan shrublands of the region. The mulch cut from the native vegetation areas contributes strongly to the reclamation site flora in the first years following mining. It is apparent that the species subsequently being added to the flora are species originally transported in the topsoil which slowly break dormancy over following years. In both the reclamation operations described here, it would be advisable to burn the area just prior to stripping the topsoil. This would break the innate dormancy of many of the soil-borne seed and the flora of an area of topsoil return should see the results in the first year of re-establishment of the vegetation. It is recognized, however, that practical problems would make such fires difficult to implement. Also, a potential problem is the longevity in stored topsoil of seed which have had their dormancy broken by heating.

A thorough working knowledge of natural ecosystems and the biology of component species is vital to maximize the ecological reconstruction of Australian ecosystems following mining disturbance. Much of the information on seed store, germination and establishment presented in this chapter has resulted from the opportunity provided by working in co-operative research projects with mining companies. Although some of the native vegetation of the region has been cleared for mining and some of the mining sites will never be fully restored, the knowledge gained from this research has provided valuable insights to the appropriate management of all regional ecosystems, both native and reconstructed (Bell *et al.* 1984). Continued research will ensure that the quality of the reconstruction of Australian ecosystems will be raised along the reconstruction continuum from rehabilitation to reclamation to restoration.

Acknowledgements

Funds for the research were provided by Alcoa of Australia Ltd., Associated Minerals Consolidated Ltd. and the Minerals & Energy Research Institute of Western Australia. We would like to acknowledge the assistance of Mr D. R. Brooks, Mr L. J. Carey, Mr M. Jefferies, Mr M. Ryall, Mr E. D. Kabay, Mr D. F. Waldie, Dr J. H. Gardner and Dr W. A. Loneragan during various stages of the research.

References

Allen E. B.(1988) Introduction: definitions and purpose. In: *The Reconstruction of Disturbed Arid Ecosystems — An Ecological Approach* (ed. E. B. Allen) pp. 1-4. Westview Press, New York.

Bell D. T. (1988) Seed-related autecology in restoration of minesites in Western Australia. In: *The Reconstruction of Disturbed Arid Ecosystems — An Ecological Approach* (ed. E. B. Allen) pp. 5-33. Westview Press, New York.

Bell D. T., Carter D. & Hetherington R. (1986) Experimental assessment of wind erosion after soil stabilization treatments at Eneabba, Western Australia. *Environ. Geochem. Health* **8,** 99-104.

Bell D. T., Hopkins A. J. M. & Pate J. S. (1984) Fire in the kwongan. In: *Kwongan — Plant Life of the Sandplain* (eds J. S. Pate & J. S. Beard) pp. 178-204. University of Western Australia Press, Perth.

Bell D. T., McCaw W. L. & Burrows N. D. (1988) Influence of fire on jarrah forest vegetation. In: *The Jarrah Forest — A Complex Mediterranean Ecosystem* (eds B. Dell, J. J. Havel & N. Malajczuk) pp. 203-15. Kluwer Academic Press, Dordecht.

Bell D. T., Vlahos S. & Watson L. E. (1987) Stimulation of seed germination of understorey species of the northern jarrah forest of Western Australia. *Aust. J. Bot.* **35,** 593-9.

Christensen P. E. & Kimber P. (1975) Effect of prescribed burning on the flora and fauna of south-west Australian forests. *Proc. Ecol. Soc. Aust.* **9,** 85-106.

Collins B. G. (ed.) (1984) *Mine Land Rehabilitation — Is it Sustainable?* Western Australian Institute of Technology School of Biology Bull. No. 7, Bentley, Western Australia.

Glossop B. L. (1980) Germination response of thirteen legume species to boiling. *Alcoa of Australia Ltd. Environ. Res. Bull.* No. 5.

Grime J. P. (1979) *Plant Strategies and Vegetation Processes.* John Wiley & Sons, New York.

Moore R. O. (1973) Tetrazolium staining for assessing seed quality. In: *Seed Ecology* (ed. W. Heydecker) pp. 347-66. Butterworths, London.

Noble I. R. & Slatyer R. O. (1981) Concepts and models of succession in vascular plant communities subject to recurrent fires. In: *Fire and the Australian Biota* (eds A. M. Gill, R. H. Groves & I. R. Noble) pp. 311-35. Australian Academy of Science, Canberra.

Shea S. R., McCormick J. & Portlock C. C. (1979) The effect of fires on regeneration of leguminous species in the northern jarrah (*Eucalyptus marginata* Sm.) forest of Western Australia. *Aust. J. Ecol.* **4,** 195-205.

Specht R. L., Rayson R. & Jackman M. E. (1958) Dark Island heath (Ninety-Mile Plain, South Australia) VI. Pyric succession: Changes in composition, coverage, dry weight, and mineral nutrient status. *Aust. J. Bot.* **6,** 59-88.

Tacey W. (1979) Landscaping and revegetation practices used in rehabilitation after bauxite mining in Western Australia. *Reclam. Rev.* **2,** 123-32.

Vlahos S. & Bell D. T. (1986) Soil seed bank components of the northern jarrah forest of Western Australia. *Aust. J. Ecol.* **11,** 171-9.

Ward S. (1988) Biomass and nutrient accumulation by low-growing understorey on rehabilitated bauxite mines. *Alcoa of Australia Ltd. Environ. Res. Note* No. 13.

Ward S., Pickersgill G. E., Michaelsen D. V. & Bell D. T. (1985) Responses to factorial combinations of N, P, K fertilizers by sapling *Eucalyptus saligna* Sm. and the prediction of response by DRIS indices. *Aust. For. Res.* **15,** 27-32.

Proc. Ecol. Soc. Aust. 1990 — 16:537-543

Weed management and revegetation programmes

F. D. PANETTA[1] and R. H. GROVES[2]

[1]*Department of Agriculture, Baron-Hay Court, South Perth, W.A. 6151*
[2]*CSIRO Division of Plant Industry, G.P.O. Box 1600, Canberra, A.C.T. 2601*

Abstract

The magnitude of weed problems encountered in revegetation programmes is largely a function of the availability of plant propagules. Areas which have been mined represent an extreme case where there can be a good degree of control over the composition of the soil seed bank. For example, invasive species are generally a minor, relatively transient component of the vegetation which establishes following sand mining. When areas have been subjected to long periods of more or less continuous disturbance, annual and perennial weeds often regenerate profusely from soil-stored seed. Strategies which may be adopted to reduce interference from weeds include methods of site preparation which reduce the sizes of weed seed banks, the sowing of large-seeded native species with the potential for rapid, early growth and the use of selective herbicides. When fertilizer is used, its composition should encourage both the establishment and long-term persistence of native species. We argue that if species with the appropriate vital attributes are selected, resilience of the restored community may be achieved, even when species richness is only a fraction of that which was present originally.

Introduction

The restoration of native plant communities on disturbed land is considered by some to be more essential to the long-term survival of Australian plants and animals than is the conservation of small existing remnants (Westoby 1984). However, attempts to restore native communities often fall short of their objectives. The problems which are encountered in revegetation programmes may differ considerably as a consequence of divergent site histories. For instance, the physical and chemical characteristics of mined sites usually require some form of amelioration before any vegetation can be established (Bradshaw & Chadwick 1980), but weed problems are minimal unless weeds are inadvertently introduced during revegetation. Where land has been cleared for long periods of either cropping or grazing by domestic stock, however, exotic species may dominate both the above-ground vegetation and the soil seed bank. Competition from weedy regrowth commonly hinders or prevents establishment of native species, even when nursery stock is planted (Pryor & Clarke 1964; Copley & Venning 1983). Owing to the excessive cost of hand-planting, current methods of broadscale revegetation rely upon techniques involving direct-seeding and natural regeneration, which are most susceptible to problems arising from weed interference.

In this chapter we consider the approaches to revegetation which have been adopted to reduce interference from weeds, using examples from both low and high rainfall environments in Australia. We identify key biological constraints which operate in the revegetation process and suggest possible improvements in methods of restoration.

Replacement of sclerophyll communities after sand mining

Sclerophyll shrub communities, with or without a tree cover of eucalypts, occur along the eastern and southern coastlines of Australia. Because of their occurrence on low nutrient soils (Specht & Rayson 1957; Groves 1981), some of these sclerophyll shrublands have remained relatively undisturbed by human activities until mining for heavy minerals commenced about 30 years ago in eastern Australia and more recently in south-western Western Australia. These vegetation types are now being cleared for open-cut mining of the heavy minerals rutile and zircon. As with other types of mining, the main ecological problem for land management of these areas is not one of weed invasion *per se*. It is more a matter of stabilizing re-shaped sand masses and returning a range of native species from each vegetation stratum.

Soil instability is initially overcome by the planting of a cover crop — *Secale cereale, Avena sativa* or a hybrid *Sorghum* — which rarely persists beyond the first year after planting. At one site on the east coast, north of Newcastle, Clark (1975) reported an increase in the percentage cover of invasive species on fertilized sites in the second year after revegetation, but this increase was short-term (Fig. 1). The invasive species not present before mining were all native to the area, however, and included *Acacia longifolia* var. *sophorae, Entolasia stricta, Gahnia sieberana, Hakea* sp. and *Philotheca salsolifolia.* Results of a later study from Myall Lakes National Park show that of the 133 species which regenerated after mining, 12% were introduced, including the shrub *Chrysanthemoides monilifera* and several species of the herb genus *Conyza* (Fox 1988). Although no data were presented for the abundance of *C. monilifera* at Myall Lakes, this species is often a major invasive weed following sand mining on the north coast of New South Wales (R. H. Groves, pers. obs.).

Revegetation programmes after mining may result in the deliberate introduction of Australian plants not indigenous to the area rather than an increase in weeds, although each outcome of revegetation will depend very much on the proximity of the site to farmland and the species used as initial cover in the programme. In particular, it is essential that weed-free seed be used when cover crops are sown. Specht (1975) grouped native heath species into several categories, based upon seed size and response to fertilizer. We shall return to the latter characteristic subsequently.

Because the pre-mining vegetation usually occurs on soils very low in nutrients, complete fertilizer is commonly applied at rates high enough to ensure adequate early growth of the cover crop. Such high rates, appropriate to the establishment

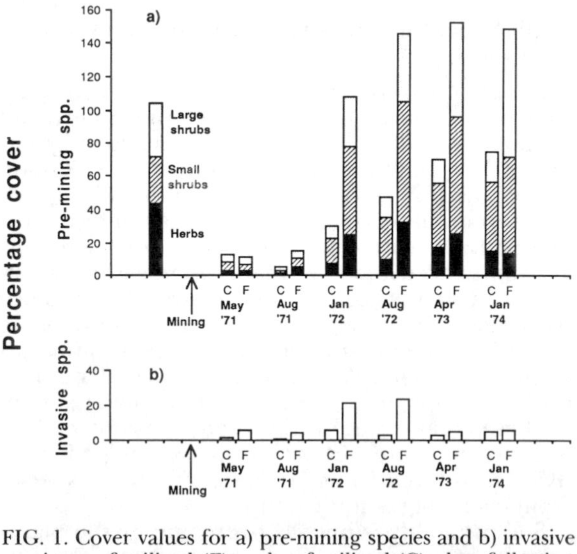

FIG. 1. Cover values for a) pre-mining species and b) invasive species on fertilized (F) and unfertilized (C) plots following sand mining at Bonny Hills, N.S.W. (data of Clark 1975).

of pasture or crop species, may discriminate against regeneration of some of the dominant species present before mining (e.g., the proteaceous element) and favour others (e.g., *Acacia* spp. and weeds). A more discriminating approach to the use of different fertilizer mixes, based on a better knowledge of the different responses of native species to individual elements (Groves & Keraitis 1976) may redress the imbalance in species composition.

Restoration of woodland and mallee communities

In a study examining vegetation succession on abandoned mallee farmland in north-western Victoria, Onans and Parsons (1980) found that newly abandoned land was dominated by the exotic annuals *Brassica tournefortii* and *Asphodelus fistulosus* (sometimes a short-lived perennial), and the native annual *Myriocephalus stuartii.* In most cases, native *Acacia, Dodonaea* and *Cassia* shrubs regenerated from soil-stored seeds and replaced the annuals within a few years. A major finding of this study was that recolonization of cleared land by the original *Eucalyptus* species occurred extremely slowly, if at all. This was partly because of the absence of soil-stored seeds of the mallee eucalypts and partly because of limited seed dispersal. Onans and Parsons (1980) concluded that only by very intensive management could reversion to mallee communities be brought about. In particular, self-perpetuating shrublands dominated by *Dodonaea attenuata* and *Acacia brachybotrya* would have to be manipulated in order to permit the establishment of eucalypts, although no suggestions were made as to how such a change could be brought about.

Dryland agricultural areas, currently the subject of revegetation studies in South Australia, have changed to the stage where there is little if any regeneration of native species from soil-stored seeds (Copley & Venning 1983; Venning & Croft 1984). Although some weedy perennial herbaceous species (e.g., *Phalaris aquatica*) interfere with the establishment of native species, the major impediment to revegetation is the proliferation of exotic annual grasses, mainly *Lolium rigidum* and *Avena fatua.* Winter dominance in the annual pattern of rainfall necessitates autumn sowing of native species. Subsequent vigorous growth of the exotic annual grasses suppresses native seedlings, leading to markedly increased mortality during the first summer following planting (Venning & Croft 1985).

A direct and most successful method to reduce interference from exotic species was to remove physically the major part of the soil seed bank by grading away the top few centimetres of the soil profile (Table 1). This method, however, may be unnecessarily destructive and is unsuitable for sites with shallow soils.

TABLE 1. The effect of site preparation and chemical treatments on the emergence (a) and establishment (b) of direct-seeded native species at Salter Springs, South Australia in 1983 (after Venning & Croft 1984) (a) and 1984 (after Venning & Croft 1985) (b).

	Mechanical weed control		Chemical weed control Pre-emergent			Post-emergent
Species type	Cultivated	Graded	Treflan®	Vorox AA®	Avadex®	Roundup®
(a)						
Large-seeded*	1.07**	32.4	1.87	14.0	2.43	—
Small-seeded	0.00	0.09	0.01	0.22	0.005	—
(b)						
Large-seeded	0.60*** (0.00)	4.82 (3.12)	5.25 (0.00)	4.75 (2.95)	—	1.52 (1.70)
Small-seeded	0.00 (0.00)	0.02 (0.04)	0.00 (0.00)	0.05 (0.02)	—	0.00 (0.00)

* Large-seeded species had seed weights of >5 mg.
** Figures represent percentages of sown viable seed.
*** First set of figures corresponds to end-of-winter counts taken in November 1984; figures in brackets are end-of-summer counts taken in March 1985.

Because most pre-sowing weed control measures had only short-term effects, Venning and Croft's approach was to extend the period of control, initially by using herbicides with both knockdown and residual effects. Application of Vorox AA® (a mixture of amitrole, a knockdown herbicide and atrazine, a residual component) enhanced the survival of large-seeded species, namely *Acacia* spp., *Allocasuarina verticillata* and *Callitris preissii*, but had little effect upon the generally poor establishment of small-seeded species such as *Eucalyptus* spp. and *Melaleuca lanceolata* (Table 1a).

Pre-sowing weed control methods could be supplemented by applying a selective herbicide at 6-8 weeks after seeding, with the aim of removing annual grasses (McMurray 1985). In order to assess the feasibility of this approach, potted plants of both small- and large-seeded species were sprayed after eight weeks' growth with various rates of grass-selective herbicides. The relative toxicity of Fusilade®, Hoegrass® and Sertin® depended upon the species concerned, but Dalapon® had a more consistently toxic effect. When averaged over all species, Sertin® was the least phytotoxic of the herbicides tested. Generally, seedlings of the larger-seeded species (*Dodonaea viscosa, Cassia nemophila, Callitris preissii, Allocasuarina verticillata* and the acacias) were more tolerant of the four herbicides than were seedlings of the smaller-seeded species (*Callistemon macropunctatus, Melaleuca lanceolata* and the eucalypts) (Table 2). When post-emergent applications of Sertin® and Fusilade® were made in field trials, however, the removal of annual grasses simply permitted a number of broad-leaved annual weeds (e.g., *Brassica tournefortii* and *Medicago* sp.) to dominate treated areas (Venning 1986). Recent work in Western Australian woodland and mixed sclerophyllous heath has indicated that a number of native species tolerate applications of diuron, a herbicide commonly used to control broad-leaved annuals in agricultural systems (J. Dodd, pers. comm.). This suggests a potential for selective control of such weeds, provided appropriate species are utilized for revegetation.

Another approach to post-emergent weed control exploited the height differential between weeds and native seedlings. A translocated, non-selective herbicide (Roundup®) was applied by a rope-wick applicator when the weeds were 30 cm tall and native seedlings only 2-5 cm. Results from this method were not encouraging (Table 1).

A method of site preparation which should minimize weed interference may be derived from agricultural methods of managing grass weeds. Because seeds of the annual grasses are relatively short lived, soil seed populations can be depleted fairly rapidly by methods which either prevent recruitment or reduce seed production (Medd 1987). We suggest that interference by grass weeds may be reduced more effectively by the application of appropriate control measures on sites for one (or more) season(s) prior to sowing native species. The lowered densities of grasses in the year of sowing should then postpone the need to apply a selective herbicide until the second year, when native plants would be more advanced and better able to tolerate phytotoxic effects.

Marked year-to-year variation in rainfall appears to be a major impediment to the revegetation of dryland sites. Although the trends in the South Australian experiments were similar in 1983 and 1984, the level of establishment of native species was much lower during the latter year (Table 1). In this year, the break of season was late, resulting

TABLE 2. Toxicity effects induced by post-emergent herbicides commonly used for annual grass control (after McMurray 1985).

Species type	Control	Herbicide			
		Fusilade®	Hoegrass®	Sertin®	2,2-DPA
Large-seeded*	3.7**	9.8	10.2	7.7	25.2
Small-seeded	3.9	13.7	15.9	10.7	33.9

* Large-seeded species had seed weights of >5 mg.
** Evaluation of herbicide effects based upon a visual rating system, utilizing a linear scale where 0=no phytotoxic effects and 100=complete mortality.

in lower soil temperatures at the time of germination, and an early onset of the dry season limited the amount of growth which could occur prior to summer (Venning & Croft 1985). Is it realistic to expect successful revegetation in a "once-off" attempt, when recruitment of native species may be highly episodic under natural conditions? Temperature and rainfall patterns are particularly critical for the successful establishment of small-seeded species (Venning & Croft 1985). Venning (1985) has suggested that an assessment of the suitability of a particular species could be made by using climatic data to determine the frequency with which conditions favouring establishment have occurred in the past. It is highly probable that species with very small seeds will require planting as reasonably advanced individuals, even in the absence of weed competition.

Restoration of rainforest communities

Rainforest sites which have been cleared for long periods are often dominated by exotic shrubs such as *Lantana camara* (Webb *et al.* 1972) and *Baccharis halimifolia* (Westman *et al.* 1975) and/or the perennial grasses *Imperata cylindrica* (Hopkins & Graham 1984), *Stenotaphrum secundatum* and *Pennisetum clandestinum* (Floyd 1985). Seeds of pioneer rainforest species may be present in the soil, but are usually absent or sparingly represented in sites with a long, continuous history of disturbance. For example, weeds comprised 80-90% of the soil seed store under degraded *Imperata* grasslands in northern Queensland (Fig. 2) (Hopkins & Graham 1984). Examination of the composition of the seed bank provides a basis for the prediction of weed problems in the revegetation process, in addition to revealing the presence of desirable species (Panetta 1982).

The use of mechanical methods, such as ripping or rotary hoeing, for controlling the perennial grasses is patently ineffective, since these methods only stimulate vegetative reproduction. Application of Roundup® at 9 L ha^{-1} is recommended for controlling *I. cylindrica* (Anon. 1987). In New South Wales, Floyd (1985) reported that complete coverage by Roundup® (1:100 in water) gave excellent control of *P. clandestinum,* leaving a dead mulch of grass on the surface to conserve soil moisture. Where the grass cover was particularly thick, as with stands of *S. secundatum,* a two-stage spraying programme was advocated, entailing a November application of Roundup® (1:100), burning the dead grass in December and spraying the regrowth in February (Floyd 1985).

Woody exotic species, when present, can be used to facilitate the restoration of rainforest. Provided they are thinned, stands of *Lantana camara* may serve as a "nurse crop" to protect slow-growing late successional rainforest species (Williams 1984). Otherwise, *L. camara* can effectively arrest succession

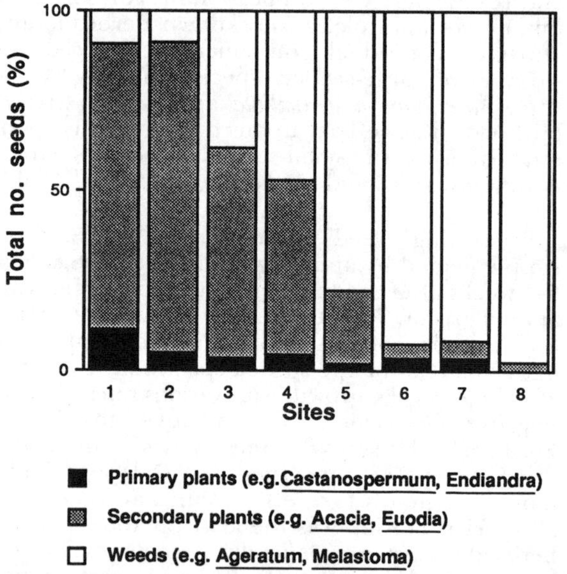

FIG. 2. Histograms showing the proportions of seed banks constituted by primary, secondary and weed species at each of eight sites (viz. 1=primary tropical lowland rainforest; 2 and 3=advanced secondary forest successions after one cycle of clearing and burning; 4=secondary forest succession after several cycles of clearing and burning; 5=deflected succession resulting from invasion by *Eucalyptus* spp. following clearing and burning; 6=*Imperata* grassland after repeated cycles of clearing and burning; 7=initial stage of secondary woody succession on *Imperata* grassland following exclusion of fire; 8=plantation of *Pinus caribaea* established on *Imperata* grassland) (from Hopkins & Graham 1984).

(Webb *et al.* 1972). *Solanum mauritianum*, another native of South America, has been planted to provide protection for seedlings of species which dominate later stages of rainforest succession in New South Wales (Table 3) (Floyd 1985). Individuals of *S. mauritianum* live for only 20-35 years, and the canopy of a stand of this species begins to thin as early as three years after planting, increasing the availability of light to species below. Floyd (1985) commented that as a result of the proximity of undisturbed rainforest to his sites, species of Stages 3 and 4 (Table 3) were dispersed by wind and birds to the regenerated areas. However, where the areas to be restored are either very large or at a distance from a source of propagules, such species must be deliberately introduced.

From what little information is available on rainforest regeneration, then, the general approach is to assist natural succession by planting Stage 2 and Stage 3 species in order to suppress grasses and other weeds, prior to the introduction of the slow-growing species which will eventually dominate the community. Where long periods of site occupation by species such as *I. cylindrica* (indicative of regular burning of the site) have resulted in a loss of soil fertility (Hopkins 1981), improvement of the total soil nutrient status or of individual nutrients may be necessary before native rainforest species can be established. At present very little is known about the nutrient requirements of these.

Even though remnant stands of rainforest may possess higher species richness than can be achieved in restored communities, their perpetuation will depend upon proper management. At Wingham Brush on the lower north coast of New South Wales, a small remnant of lowland subtropical rainforest has been subject to considerable disturbance. Two natural disturbance factors are flooding of the adjacent Manning River and the continual input of propagules by bats. Survival of the tall forest of *Ficus macrophylla* and other rainforest trees is at risk from the growth of introduced herbs and of three climbers introduced from South or Central America — *Anredera cordifolia, Cardiospermum grandiflorum* and *Macfadyena unguis-cati.* The long-term viability of this and other small remnants will only be maintained by continued hand-weeding and application of herbicide (Fox 1988).

Management considerations in the long term

Thus far, we have emphasized the short-term objective of establishing native species, largely through circumventing competition from weeds. However, we recognize that it is as important to ensure that the restored community is resilient (Grubb & Hopkins 1986) in the long term, that it does not degrade to a weed-dominated condition. In all cases, it will be necessary to protect the new community from either major or repeated disturbance, e.g., fire or grazing. It is not realistic to expect the restored community to be insulated from disturbance indefinitely, nor would this necessarily be a desirable condition (see Kirkpatrick 1986).

Since it is unlikely that the diversity of the original community will be achieved during revegetation, it is essential that the species which are re-introduced possess attributes which confer resilience (Panetta 1981). Noble and Slatyer (1980) devised a scheme which utilized vital attributes of species in order to predict successional changes in plant communities following disturbance of variable frequency and intensity. The two major attributes they dealt with related to mechanisms of recovery from disturbance (i.e., whether a species recovered vegetatively, from seed or combinations thereof) and the degree to which a regenerating species was tolerant of competition. Species with certain combinations of vital attributes were considered to be resilient regardless of disturbance regime (Table 4). From Table 4 it can be observed that in almost all cases, species which cannot recruit in the presence of competition for resources are prone to local extinction, as are those for which propagule store depletion accompanies adult death in the course of a disturbance event.

Sclerophyll and rainforest communities differ markedly in relation to the basic patterns of succession which follow disturbance. Early succession of

TABLE 3. Stages in regeneration of lowland subtropical rainforest in northern New South Wales (from Floyd 1985).

Stage 1 Annual herbs	Stage 2 Short-lived shrubs (10-15 years)	Stage 3 Fast-growing secondary trees	Stage 4 Slow-growing eventual dominants
**Tagetes minuta*	*Duboisia myoporoides*	*Alphitonia excelsa*	*Agyrodendron trifoliolatum*
**Bidens pilosa*	*Trema aspera*	*Polyscias elegans*	*Sloanea woollsii*
Solanum nodiflorum	*Omalanthus populifolius*	*P. murrayi*	*Dysoxylum fraseranum*
	**Solanum mauritianum*	*Commersonia bartramia*	*Syzygium crebrinerve*
		C. fraseri	
		Euodia micrococca	

*Introduced species.

TABLE 4. Species types likely to persist (+) under any disturbance regime (after Noble & Slatyer 1980).

	Mechanism of recovery from disturbance						
	Seed-based				Vegetatively-based		
	Seeds readily dispersed	Seed store survives disturbance	Seed store lost at disturbance: Adults killed	Seed store lost at disturbance: Adults survive	Delay before reproduction	No delay before reproduction	Juveniles killed
Tolerant of competition	+	+	—	+	+	+	+
Intolerant of competition	+	—	—	—	—	—	—
Requires vegetative cover	+	+	—	+	+	—	+

vegetation after fire in sclerophyllous communities tends to follow the initial floristic composition model (Egler 1954), with species either regenerating solely from seed, by resprouting or a combination of these (Purdie & Slatyer 1976). By way of contrast, recovery from major disturbance to rainforest generally occurs from buried propagules and follows the relay floristic model (Egler 1954). Seed banks accumulate only for species which occur in the early stages of rainforest succession (Stages 2 and 3 in Table 3) (Hopkins & Graham 1987).

Late-successional species (Stage 4 in Table 3) are good examples of the type which requires vegetative cover to establish. These must reach the site either by natural means of dispersal (e.g., birds) or be deliberately reintroduced following a major disturbance event in a previously restored community.

Westoby (1984) has suggested that if the structure of dominant species is restored, then the majority of other species will return of their own accord. It would appear that we can maximize the resilience of the restored community by selecting species whose vital attributes ensure persistence in the presence of a variety of possible disturbance regimes. Where problem weeds rely upon populations of buried propagules for persistence, it is possible that the selection of native species with a similar strategy will shift the balance towards native species in the long term.

Conclusions

1. Since their seedlings remain small for extended periods, small-seeded native species are disadvantaged with regard to competition with introduced species and tolerance of herbicides which may be employed to control the latter. Where weed competition is a problem or climatic conditions may not be favourable, it will be necessary to introduce small-seeded species as hand-planted nursery stock.
2. Owing to climatic vagaries in low rainfall environments, the revegetation attempt may not succeed in any particular season. The use of large-seeded species should increase the chances of establishment, as suggested by Baker (1972).
3. Fertilizer should be used only in exceptional circumstances. Generally, exotic species respond more vigorously to added nutrients. Where fertilizer is utilized, its composition may favour particular groups of species.
4. Because sclerophyll and rainforest communities follow different models of succession, all species should be introduced simultaneously in sclerophyll revegetation, whereas rainforest requires separate introductions of early- and late-successional species. When exotic shrubs are present, these can be managed as nurse plants to facilitate the introduction of late seral species to rainforest areas.
5. For the purposes of community resilience, it is important to manipulate the soil seed store in order to shift the balance away from exotic species. Introducing plants which form long-lived seeds early in their lives will facilitate the achievement of this objective. Site preparation in the year(s) preceding a restoration effort in a dryland environment should reduce the challenge posed by exotic annual grasses.
6. Members of each stratum of the restored community should possess vital attributes (Noble & Slatyer 1980) which confer resilience in the long term.

Acknowledgements

We thank Gwen Harden for providing us with references on rainforest regeneration. The manuscript was improved by the comments of Julianne Venning and Patrick Pigott. Rod Randall assisted in figure preparation.

References

Anon. (1987) *Australian Weed Control Handbook.* 8th edn. Inkata Press, Melbourne.

Baker H. G. (1972) Seed weight in relation to environmental conditions in California. *Ecology* **53,** 997-1010.

Bradshaw A. D. & Chadwick M. J. (1970) *The Restoration of Land.* Blackwell Scientific Publications, Oxford.

Clark S.S. (1975) The effect of sand mining on coastal heath vegetation in New South Wales. *Proc. Ecol. Soc. Aust.* **9,** 1-16.

Copley P. B. & Venning J. (1983) Rural revegetation in South Australia. Department of Environment and Planning, Adelaide.

Egler F. E. (1954) Vegetation science concepts. I. Initial floristic composition, a factor in old field vegetation development. *Vegetatio* **4,** 412-7.

Floyd A. G. (1985) Management of small rainforest areas. Part II. *Natl Parks J.* **29,** 23.

Fox M. D. (1988) The ecological status of alien plant species. In: *Weeds on Public Land — an Action Plan for Today* (ed. R. G. Richardson) pp. 42-5. The Weed Science Society of Victoria, Melbourne.

Groves R. H. (1981) Heathland soils and their fertility status. In: *Ecosystems of the World.* Vol. 9B, *Heathlands and Related Shrublands* (ed. R. L. Specht) pp. 143-63. Elsevier Scientific Publishing Company, Amsterdam.

Groves R. H. & Keraitis K. (1976) Survival and growth of seedlings of three sclerophyll species at high levels of phosphorus and nitrogen. *Aust. J. Bot.* **24,** 681-90.

Grubb P. J. & Hopkins A. J. M. (1986) Resilience at the level of the plant community. In: *Resilience of Mediterranean-type Ecosystems* (eds B. Dell, A. J. M. Hopkins & B. B. Lamont) pp 21-38. Dr W. Junk Publishers, Dordrecht.

Hopkins M. S. (1981) Disturbance and change in rainforests and the resulting problems in functional classification. In: *Vegetation Classification in Australia* (eds A. N. Gillison & D. J. Anderson) pp. 42-52. Australian National University Press, Canberra.

Hopkins M. S. & Graham A. W. (1984) Viable soil seed banks in disturbed lowland tropical rainforest sites in North Queensland. *Aust. J. Ecol.* **9,** 71-9.

Hopkins M. S. & Graham A. W. (1987) The viability of seeds of rainforest species after experimental soil burials under tropical wet lowland forest in north-eastern Australia. *Aust. J. Ecol.* **12,** 97-108.

Kirkpatrick J. B. (1986) The viability of bush in cities — ten years of change in urban grassy woodland. *Aust. J. Bot.* **34,** 691-708.

McMurray S. K. (1985) Post-emergent herbicides for weed control with direct-seeded native species. Department of Environment and Planning, Adelaide.

Medd R. W. (1987) Weed management on arable lands. In: *Tillage. New Directions in Australian Agriculture* (eds P. S. Cornish & J. E. Pratley) pp. 222-59. Inkata Press, Melbourne.

Noble I. R. & Slayter R. O. (1980) The use of vital attributes to predict successional changes in plant communities. *Vegetatio* **43,** 5-21.

Onans J. & Parsons R. F. (1980) Regeneration of native plants on abandoned mallee farmland in south-eastern Australia. *Aust. J. Bot.* **28,** 479-93.

Panetta F. D. (1981) The rehabilitation of weed-dominated communities. *Vict. Nat.* **98,** 31-7.

Panetta F. D. (1982) *In situ* potential for rehabilitation of sites dominated by blackberry (*Rubus polyanthemos* Lindeb.) *Weed Res.* **22,** 1-6.

Pryor L. D. & Clarke B. (1964) Reforestation of former farm sites on the north coast of New South Wales. *Aust. For.* **28,** 125-35.

Purdie R. W. & Slatyer R. O. (1976) Vegetation succession after fire in sclerophyll woodland communities in south-eastern Australia. *Aust. J. Ecol.* **1,** 223-36.

Specht R. L. (1975) The effect of fertilizers on sclerophyll (heath) vegetation — the problems of revegetation after sand-mining of high dunes. *Search* **6,** 459-61.

Specht R. L. & Rayson P. (1957) Dark Island heath (Ninety-Mile Plain, South Australia). I. Definition of the ecosystem. *Aust. J. Bot.* **5,** 52-85.

Venning J. (1985) Natural regeneration in the mid-north of South Australia: a case study. *S. Aust. Nat.* **60,** 36-41.

Venning J. (1986) 1985 revegetation trials. Department of Environment and Planning, Adelaide.

Venning J. & Croft T. S. (1984) Preliminary assessment of 1983 revegetation trials. Department of Environment and Planning, Adelaide.

Venning J. & Croft T. S. (1985) 1984 revegetation trials. Department of Environment and Planning, Adelaide.

Webb L. J., Tracey J. G. & Williams W. T. (1972) Regeneration and pattern in the subtropical rainforest. *J. Ecol.* **60,** 675-95.

Westman W. E., Panetta F. D. & Stanley T. D. (1975) Ecological studies on reproduction and establishment of the woody weed, groundsel bush (*Baccharis halimifolia L.: Asteraceae*). *Aust. J. Agric. Res.* **26,** 855-70.

Williams G. (1984) Rainforest regeneration. II. *Natl Parks J.* **28,** 12-3.

Westoby M. (1984) Constructive ecology: how to build and repair ecosystems. Australian Environmental Studies Working Paper 1/84. Griffith University, Brisbane.

Proc. Ecol. Soc. Aust. 1990 — 16:545-549

Saltland reclamation: selection of superior Australian tree genotypes for discharge sites

P. G. VAN DER MOEZEL and D. T. BELL

Department of Botany, The University of Western Australia, Nedlands, W.A. 6009

Abstract

The salinization of previously productive agricultural land in Australia has led to research on selecting trees for saltland reclamation. Salt and waterlogging tolerance screening trials were conducted in the glasshouse on 101 species of Acacia, Casuarina, Eucalyptus *and* Melaleuca. *A total of 410 seedlings were selected for micropropagation, representing 7% of seedlings tested under saline waterlogged conditions. Between provenance variation was large for some species and within provenance variation was also significant. Survival in salt waterlogging was related to salt exclusion in* Casuarina *and* Eucalyptus *species. Salinization reduced stomatal conductance in a range of* Eucalyptus *species but there was no clear correlation between salt uptake and stomatal conductance. Guidelines for future research in selecting trees for saltland reclamation are discussed.*

Introduction

The extent of salt-affected land in Australia has been estimated at about 32.4 million ha, or 4.3% of the Australian landmass (Standing Committee on Soil Conservation 1982). Eighty-seven percent of this area is natural salinity, the result of climatic and geological history. The remaining 13%, or 4.2 million ha, is a direct result of 200 years of non-Aboriginal land use. The most serious form of induced non-irrigated salinity is seepage salting which mostly affects the wheat and sheep agricultural zones of Western Australia, South Australia and Victoria. In 1982, seepage salting affected 420 000 ha of previously productive land in Australia. At the average rate of 2% compound increase per year this figure in 1988 would be almost 480 000 ha and increasing by almost 10 000 ha for 1989.

Amelioration of salt-affected seepage sites includes treatment of both discharge and recharge sites. Both land form engineering approaches, which are designed to reduce soil salinity levels and the length of waterlogging periods, and revegetation with salt tolerant plants which use excess saline water can be used to assist rehabilitation of discharge sites. Planting of trees or other deep rooted perennial species on upslope groundwater recharge areas also results in the reduction of waters discharging downslope and will be essential for the permanent reclamation of saline seepages (Morris & Thompson 1983). Discharge areas are inherently saline and periodically waterlogged where the saline groundwater table meets the surface. Reclamation of discharge areas using trees or shrubs requires selection of highly salt tolerant plants, and for extreme conditions, salt and waterlogging tolerant species.

Salt tolerance studies of Australian native woody species have been conducted on *Acacia* (Aswathappa *et al.* 1986), *Casuarina* (El-Lakany & Luard 1982), *Eucalyptus* (Blake 1981; Thomson *et al.* 1987; Van der Moezel & Bell 1987; Van der Moezel *et al.* 1988) and *Melaleuca* (Van der Moezel & Bell 1987). In most instances, ranking of species tolerance was based on growth and/or survival in saline conditions. Van der Moezel *et al.* (1988) exposed seedlings to both salinity and waterlogging since these conditions in the field prove more difficult for tree establishment than salinity alone (Mathur & Sharma 1984; Biddiscombe *et al.* 1985; Thomson *et al.* 1987).

A programme to select woody species tolerant of saline and waterlogging conditions began in 1986 under the National Biotechnology Research Grants Scheme (Kabay *et al.* 1986). Eight screening experiments were conducted on *Acacia, Casuarina, Eucalyptus* and *Melaleuca* species to select the most tolerant individuals which would then be mass produced by micropropagation. This chapter presents a summary of the three-year research programme, highlighting species performances, intraspecific variation, physiological data and research needs for the future.

Methods

A full description of the screening procedure is given in Van der Moezel *et al.* (1988). Seed was collected from plants growing in or around saline sites throughout Australia. Glasshouse experiments were conducted on four-month-old seedlings in non-saline drained control conditions (C), and three treatments which were saline drained (S), non-saline waterlogged (W) and saline waterlogged (SW). A mixture of NaCl, $MgSO_4$ and $CaCl_2$ was used for the S and SW treatments (Na:Mg:Ca ratio of 10:2:1 by weight). Salinities in the S and SW treatments were increased weekly by 7 mS cm^{-1} until most plants in SW had died. Maximum salinity obtained varied according to the particular experiment. A detailed examination of the treatment effects on Na, Cl and K levels in the leaves, stems and roots of salt-stressed and salt waterlogging-stressed seedlings was done in several experiments. Effects of salt and waterlogging on photosynthesis, trans-piration and stomatal conductance was studied on two *Eucalyptus* species using a portable ADC open system photosynthesis apparatus (see Van der Moezel *et al.* 1989) while stomatal conductance measurements were done on five *Eucalyptus* species in a separate experiment to investigate the effects of transpiration on salt uptake under saline conditions.

Results and discussion

Species performance

The total number of Australian species, provenances and seedlings screened in the eight experiments was 101, 323 and 16 736, respectively (Table 1). The number of seedlings selected for micropropagation from the salt waterlogging treatment was 410 which represented approximately 7% of the seedlings in that treatment. Selection pressure varied according to experiment and also to species and provenances within experiments. Seedlings for micropropagation were selected from approximately half the species studied.

The most tolerant species were able to survive in saline waterlogged conditions up to a salinity of 35-63 mS cm^{-1} (Table 2). *Eucalyptus wandoo* was the only species in experiment 2 and was considered very salt waterlogging sensitive, but the other trials provided individuals, provenances and species with high levels of tolerance. The maximum salinity to which the seedlings were exposed should not be compared too strictly between experiments since the stepwise salinization procedures were not identical for each experiment and glasshouse conditions were not controlled throughout the eight sequential experiments. Although experimental variation occurred, the most tolerant *Casuarina* and *Melaleuca* species listed were considerably more tolerant than the best *Acacia* and *Eucalyptus* species. Some other trial species not listed in Table 2 had certain provenances which were salt waterlogging tolerant, e.g., *Eucalyptus socialis, Melaleuca bracteata, M. linariifolia* and *Acacia aulacocarpa*. *Eucalyptus kondininensis* was one species with poor species and provenance tolerance, but had individuals of high salt waterlogging tolerance.

Provenance variation

Variation at the provenance level was large for some species. Sixteen provenances of *Eucalyptus microtheca* had a percentage survival range between 44 and 94% after salt waterlogging for five weeks. In the same experiment and treatment, seven provenances of *E. tereticornis* had survival between 0-50%. In the salt waterlogging treatment of experiment 3, nine *E. sargentii* provenances and 12 *E. occidentalis* provenances had percentage survival ratings of 38-94% and 12-75%, respectively. For *Melaleuca quinquenervia* (experiment 5) survival in salt waterlogging ranged between 0-53%

TABLE 1. Number of species, provenances and seedlings screened for salt and waterlogging tolerance in eight separate experiments. The number of seedlings selected for micropropagation were from the salt waterlogging treatment only.

Experiment	Genus	Screening Species	Screening Provenances	Screening Seedlings	Micropropagation selection Species	Micropropagation selection Seedlings
1	*Eucalyptus, Casuarina*	7	13	1 584	7	31
2	*Eucalyptus*	1	16	1 488	1	3
3	*Eucalyptus*	19	51	2 304	8	39
4	*Casuarina*	5	27	2 168	2	38
5	*Melaleuca, Callistemon*	26	49	1 632	17	146
6	*Acacia*	6	52	2 256	5	56
7	*Eucalyptus*	19	65	2 976	9	45
8	*Acacia*	22	50	2 328	12	52
	Total	101*	323	16 736	58**	410

*4 species included in two separate experiments.
**3 species selected from two separate experiments.

TABLE 2. Listing of the most tolerant species from salt waterlogging treatment in each of the eight experiments and the maximum salinity to which species were exposed.

Experiment	Species	Maximum Salinity ($mS\ cm^{-1}$)
1	*Casuarina obesa*	42
1	*Eucalyptus camaldulensis*	42
3	*Eucalyptus occidentalis*	35
3	*E. sargentii*	35
3	*E. spathulata*	35
4	*Casuarina equisetifolia*	56
4	*C. glauca*	56
4	*C. obesa*	56
5	*Melaleuca acuminata*	63
5	*M. eleuterostachya*	63
5	*M. glomerata*	63
5	*M. halmaturorum*	63
5	*M. sp. aff. lanceolata*	63
5	*M. lateriflora*	63
5	*M. thyoides*	63
6	*Acacia stenophylla*	49
7	*Eucalyptus intertexta*	35
7	*E. microtheca*	49
7	*E. raveretiana*	35
7	*E. striaticalyx*	49
7	*E. tereticornis*	35
8	*Acacia acuminata*	42
8	*A. coriacea*	42
8	*A. cyclops*	42

for three provenances. Variation within and between *Casuarina obesa* provenances was found in experiment 1. Fifteen seedlots of *C. obesa,* consisting of collections from three trees per provenance for five provenances, were screened for salt waterlogging tolerance. Variation in height growth was significantly different between provenances. For three of the five provenances, height growth was significantly different depending upon the mother tree from which the seedlings were grown.

In an experiment using a salt tolerant and salt sensitive provenance of *Eucalyptus camaldulensis,* growth and survival of the sensitive provenance were less than the tolerant provenance after 11 weeks in 10 $mS\ cm^{-1}$ NaCl, fully waterlogged conditions. Seedling growth from both provenances was related to leaf chloride concentrations ($r=0.92$, $p<0.01$). Although mean growth was higher and mean chloride levels were lower ($p<0.01$) in the tolerant provenance, there was considerable overlap in individual seedling reponses (Fig. 1).

The programme to select highly tolerant seedlings for rapid multiplication by micropropagation has found significant variation not only between species, which was expected, but also between

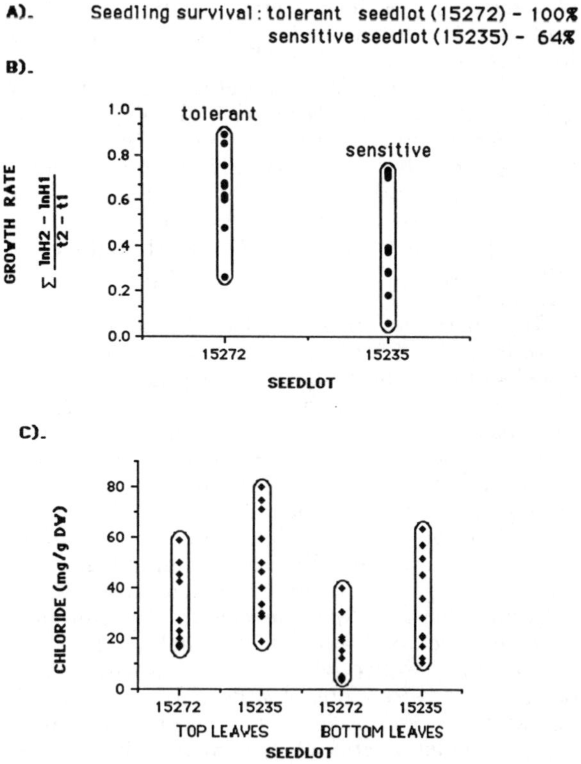

FIG. 1. Responses of two provenances of *Eucalyptus camaldulensis* after 11 weeks in saline (100 mM NaCl) waterlogged conditions. Provenance 15 272 was previously determined to be more salt tolerant the provenance 15 235. (a) Seedling survival; (b) Growth rates, as the sum of the weekly growth rates, where H is height in cm, t is time in weeks; (c) Chloride concentration of top and bottom leaves after 1 week of saline waterlogging exposure.

provenances and within provenances. Choosing the most tolerant individuals to clone ensures the best material will be used for reclamation of saline seepage discharge areas.

Physiology

Salt exclusion from the shoot is the main mechanism of salt tolerance in non-halophytes (Greenway & Munns 1980). The most salt tolerant *Casuarina* and *Eucalyptus* species in these experiments excluded Na and Cl from young growing shoots (Van der Moezel *et al.* 1988). The site of salt exclusion appears to be in the root since all species, regardless of leaf salt concentrations, accumulated salt in root tissue. The combination of salinity and waterlogging greatly increased the Na and Cl levels throughout the whole plant. In the freely draining salt treatment, where salt uptake was less than in salt waterlogging, low seedling survival occurred only in the most sensitive species. Salinization not only increased Na and Cl levels in many species but also decreased K levels. High relative growth of salinized *Eucalyptus* seedlings was related to low foliar Na/K ratios.

Not all seedling deaths and/or reduced relative growth rates were the result of high salt accumulation. For example, *Eucalyptus viminalis* seedlings were found to have low survival in saline and saline waterlogged conditions yet had low tissue Na and Cl levels, comparable with highly tolerant species. Seedling death for *E. viminalis* was thought to be due to dehydration in the low water potential root environment.

Growth reduction of salinized *Eucalyptus camaldulensis* and *E. lesouefii* seedlings was related to lower stomatal conductance, transpiration and net photosynthesis (Van der Moezel *et al.* 1989). *Eucalyptus lesouefii* maintained a higher growth rate and a smaller reduction in stomatal conductance, transpiration and photosynthesis than *E. camaldulensis*. However, *E. lesouefii* accumulated more Na and Cl in plant tissues. The question therefore arose: was salt uptake directly related to transpiration under saline conditions? To answer this question, five *Eucalyptus* species in experiment 7 were selected for bi-weekly stomatal conductance measurements throughout the step-wise salinization procedure. The results showed that salt uptake and stomatal conductance were not related. The most tolerant species, *E. microtheca,* had higher relative growth, stomatal conductance, percentage survival and lower foliar chloride content than the most sensitive species, *E. robusta* (Fig. 2). The apparently conflicting results from the two *Eucalyptus* experiments can be explained by the different effects of salt on species. Three different salt effects have been identified in eucalypts: 1) rapid stomatal closure and salt exclusion (*E. camaldulensis*); 2) gradual stomatal closure with salt uptake (*E. robusta, E. lesouefii*); and 3) gradual stomatal closure and salt exclusion (*E. microtheca*). In type 1, stomatal closure preceeds salt uptake whereas in type 2 stomatal closure is the result of salt uptake. The ultimate aim in selecting salt tolerant eucalypts for maximum growth and water use in saline discharge zones would be to find one that retains high stomatal conductance under saline conditions but excludes salt. This would then allow the plant to survive while transpiring substantial amounts of water and so reducing groundwater levels.

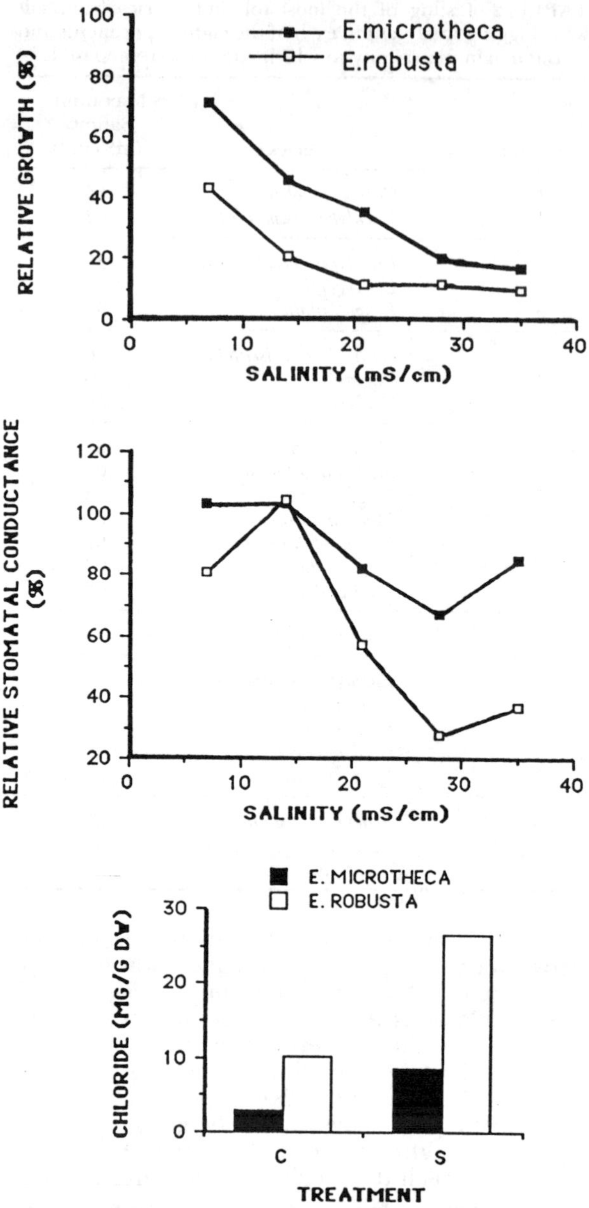

FIG. 2. Salinity treatment effects of *Eucalyptus microtheca* and *E. robusta.* (a) Relative growth in saline treatment as a percentage of control; (b) Relative stomatal conductance in saline treatment compared to control; (c) Chloride concentration of top leaves in non-saline drained control (C) and saline (S) treatment conditions after five weeks.

Future research

The next vital stage in the development of salt tolerant clones for saltland reclamation is to test in the field the genotypes screened in the glasshouse. Trials using *Eucalyptus camaldulensis* and *Casuarina obesa* clones are underway already with a broader selection of species to be planted in the near future. Trial sites range from mildly saline to highly saline and waterlogged, encompassing a broad climatic range. Trials need not be limited to Australia. Many other countries experience land degradation through salinity and highly salt tolerant Australian trees may be useful in those situations. Clones should be tested in the field alongside seedlings from the same seedlot to demonstrate and quantify the suspected superior tolerance of clones. The water use characteristics of the various genotypes should be studied in a range of environmental conditions. The results of water use would help the design of saltland plantations so that maximum tree survival and water use are obtained. Secondary uses (e.g., fuelwood, pulpwood, timber and essential oil potentials) of tolerant genotypes should also be studied.

More screening of genotypes is needed. Although 323 provenances were screened in the eight experiments, some of the most tolerant species were represented by very few provenances, e.g., *E. striaticalyx* and *Acacia cyclops* (one provenance) and *E. raveretiana* (two provenances). Furthermore, species with suspected tolerances were not included in the first series, e.g., *Melaleuca cuticularis, M. tamariscina* and *M. strobophylla*. An extensive seed collection of provenances and families of the best species should be undertaken and rapid screening methods employed for selection. Breeding between closely related tolerant species e.g., *E. sargentii* and *E. spathulata* is a possible way of improving salt tolerance.

From a physiological viewpoint, methods of salt tolerance should be studied further. In particular, research using microanalytical techniques and membrane physiology could elucidate the mechanisms of salt exclusion, thus assisting in improving salt screening procedures. The use of clonal material (both tolerant and sensitive clones) in experiments will reduce variation to the level imposed by the stress alone. Ecologically, understanding mechanisms of salt tolerance and documenting the intraspecific variation in salt tolerance will benefit the study of adaptations of populations to environmental stresses.

References

Aswathappa N., Marcar N. E. & Thomson L. A. J. (1986) Salt tolerance of Australian tropical and sub-tropical acacias. Proceedings Australian Centre for International Agricultural Research Workshop, Gympie, Queensland.

Biddiscombe E. F., Rogers A. L., Greenwood E. A. N. & De Boer E. S. (1985) Growth of tree species near salt seeps, as estimated by leaf area, crown volume and height. *Aust. For. Res.* **15,** 141-54.

Blake T. J. (1981) Salt tolerance of eucalypt species grown in saline solution culture. *Aust. For. Res.* **11,** 179-83.

El-Lakany M. H. & Luard E. J. (1982) Comparative salt tolerance of selected *Casuarina* species. *Aust. For. Res.* **13,** 11-20.

Greenway H. & Munns R. (1980) Mechanisms of salt tolerance in nonhalophytes. *Ann. Rev. Plant Physiol.* **31,** 149-90.

Kabay E. D., Thomson L. A. J., Doran J. C., van der Moezel P. G., Bell D. T., McComb J. A., Bennett I. J., Hartney V. J. & Malajczuk N. (1986) Micropropagation of forest trees selected for salt tolerance. *Aust. Salinity Newsl.* **14,** 60-2.

Mathur N. K. & Sharma A. K. (1984) *Eucalyptus* in reclamation of saline and alkaline soils in India. *The Indian Forester* **110,** 9-15.

Morris J. D. & Thomson L. A. J. (1983) The role of trees in dryland salinity control. *Proc. R. Soc. Vict.* **95,** 123-31.

Standing Committee on Soil Conservation (1982) *Salting of non-irrigated land in Australia.* Report of Working Party on Dryland Salting in Australia for Standing Committee on Soil Conservation.

Thomson L. A. J., Morris J. D. & Halloran G. M. (1987) Salt tolerance in eucalypts. Paper for International Symposium of Afforestation of Salt-affected Soils, Karnal, Feb. 16-20.

Van der Moezel P. G. & Bell D. T. (1987) Comparative seedling salt tolerance of several *Eucalyptus* and *Melaleuca* species from Western Australia. *Aust. For. Res.* **17,** 151-8.

Van der Moezel P. G., Watson L. E. & Bell D. T. (1989) Gas exchange response of two *Eucalyptus* species to salinity and waterlogging. *Tree Physiol.* **5,** 251-7.

Van der Moezel P. G., Watson L. E., Pearce-Pinto G. V. N. & Bell D. T. (1988) The response of six Eucalyptus species and *Casuarina obesa* to the combined effect of salinity and waterlogging. *Aust. J. Plant Physiol.* **15,** 465-74.

Proc. Ecol. Soc. Aust. 1990 — 16:551-556

Rehabilitation agronomy — guidelines for revegetating degraded land

C. V. MALCOLM

Resource Management Division, Department of Agriculture, 3 Baron-Hay Court, South Perth, W.A. 6151

Abstract

Those who benefit from activities that cause land degradation are not required to pay for rehabilitation. Rehabilitation aims to stop soil erosion and improve land for its allotted use. Successful rehabilitation depends on adequate site characterization leading to adequate prescription of site preparation, species and ecotype selection, establishment niche engineering, and protection from predators. Attempts at rehabilitation of arid and semi-arid degraded land have often been unsuccessful. Western Australian experience in rehabilitation of salt-affected farmland has provided techniques that are promising for degraded rangeland and mine dumps.

Introduction

Agricultural development, rangeland use, mining and fuel collection are recognized as major causes of land degradation in many parts of the world. The symptoms of vegetation, soil and land abuse include water and wind erosion, vegetation decline, species loss, salt encroachment, deterioration of land improvements (e.g., fences, buildings, roads) and a decline in economic production. Degraded land provides an unfavourable environment for plant establishment and growth (Springfield 1978; Malcolm 1989). Commonly used agricultural techniques for establishing plant cover (e.g., disc drill, combine seeder) are rarely successful in establishing non-domesticated plants on degraded land. Research in Western Australia aimed at developing revegetation techniques for salt-affected land, degraded rangeland and mine dumps in arid and semi-arid areas, has been termed rehabilitation agronomy and will be reviewed in this chapter. Consideration will be limited to those areas that are degraded to the point where plants must be sown because changes in management will not achieve revegetation in an acceptable time.

Aims of rehabilitation

The most important aim of rehabilitation is to stop the problem getting worse; by preventing an increase in the area of erosion or salt encroachment or by stopping the soil becoming more eroded or saline at any one place.

The second aim is to improve the affected area for the use for which the land is allocated.

The aims are frequently complex. A farmer may wish to revegetate salt-affected land to control erosion, produce forage, minimize treatment cost, increase the value of the farm, improve the appearance of the land, provide shelter for sheep, lower groundwater levels, improve the ecology of the area and silence complaints about wind-blown dust.

The aims of the community and landholder must be compatible or at least amenable to a compromise. For example, an area of salt affected land adjacent to the trans-Australian railway was generating salt laden dust responsible for shorting the railway signals west of Merredin, Western Australia in the early 1980s. The area was fenced, sown to halophytic forage shrubs and protected from grazing for six years. In 1988 it was grazed in autumn without destabilizing the area. Controlled autumn grazing will now be possible each year and will give the farmer a profitable return. In this instance, the costs of seeding were met by the Department of Agriculture as a demonstration.

Site characterization

To determine which species and rehabilitation methods to use it is necessary to know the site characteristics (Malcolm 1986). Factors influencing rehabilitation may be grouped under climate, soil, landscape and biota. The factors are important for germination, establishment, survival and production. In Table 1 an attempt has been made to list all of the environmental factors influencing rehabilitation and highlight the aspects of plant growth most affected.

TABLE 1. Site factors affecting rehabilitation.

Factor	Aspects of plant growth most affected			
	Germination	Establishment	Survival	Production
Climate				
precipitation				
– total			+	+
– seasonal pattern	+	+	+	+
– intensity	+	+	+	
– reliability	+	+	+	
temperature				
– maxima	+			+
– minima	+			+
frost		+	+	
Soil				
physical				
– texture	+	+		
– water retention	+	+	+	+
– crusting	+	+		
– infiltration	+	+	+	+
– slaking, dispersion	+	+	+	
– stability when dry	+	+	+	
– bulk density, penetrability		+	+	+
chemical				
– pH		+	+	+
– salinity	+	+		+
– toxicity	?	+	+	+
– fertility		+		+
Landscape				
slope		+	+	+
aspect	+			
hydrology				
– run-on	+	+	+	+
– run-off	+	+	+	+
– waterlogging		+	+	+
Biota				
weeds		+	+	
diseases			+	+
insects		+		
grazing animals		+	+	+

For plant establishment it is important to have sufficient precipitation in a season of temperatures suitable for germination. The soil water regime needs to remain favourable for radicle growth and seedling establishment. Too little water results in seedling death due to drought; too much water limits root growth of most species due to O_2 deficiency. The water supply to an establishing seedling may be ensured by regular precipitation, or by good soil water storage or by measures designed to concentrate a limited amount of water (e.g., niches, Malcolm & Allen 1981; ponding banks, Jones 1967) and prevent its evaporation (e.g., mulches, Malcolm *et al.* 1982).

Water may also be required to leach inhibitors from seeds (Jones 1968) or salts from the soil and to satisfy the water requirements of the soil matric potential. These conditions must be satisfied for a few weeks in the top few centimetres of soil to allow seedlings to establish (Malcolm *et al.* 1982).

Survival of established plants is affected by water supply at greater depths and is assisted by enabling water to infiltrate preferentially near establishing plants (e.g., furrows, pits).

Soils which are unstable, either to water or wind action, may nullify planting strategies by modifying the surface contours or burying seeds and treatments.

Mine dumps are often built with sloping sides which are unstable, have limited infiltration, and are extremely difficult to treat.

Attempts at rehabilitation may be hampered by competing weeds or predators.

Adequate site characterization will identify the factors likely to affect plant establishment and growth and enable sound rehabilitation techniques to be designed and followed.

Rehabilitation measures

Attempts to rehabilitate degraded land may be compared with growing a crop. There must be measures to prepare the site, to sow the crop, protect the crop from pests and eventually reap the benefit. Crop failure may result from the operation of any one of a range of factors working in different ways at different times. Results are optimized by following what may be called a "critical pathway" (see Table 2 which is discussed in more detail under "Western Australian Programme").

The site characteristics determine the species that may be grown and the measures that are prescribed for preparation, planting and management. Currently, a data bank on species adaptation is being developed and research is aimed at providing an expert system for prescribing site specific establishment methods.

Off-site measures should be applied first as they may ameliorate site conditions. Planning should precede the building of waste dumps so that the materials most amenable to plant growth are on the outside and reshaping is minimized. Soil and water conservation measures must precede sowing.

TABLE 2. Critical pathway for rehabilitation of degraded land.

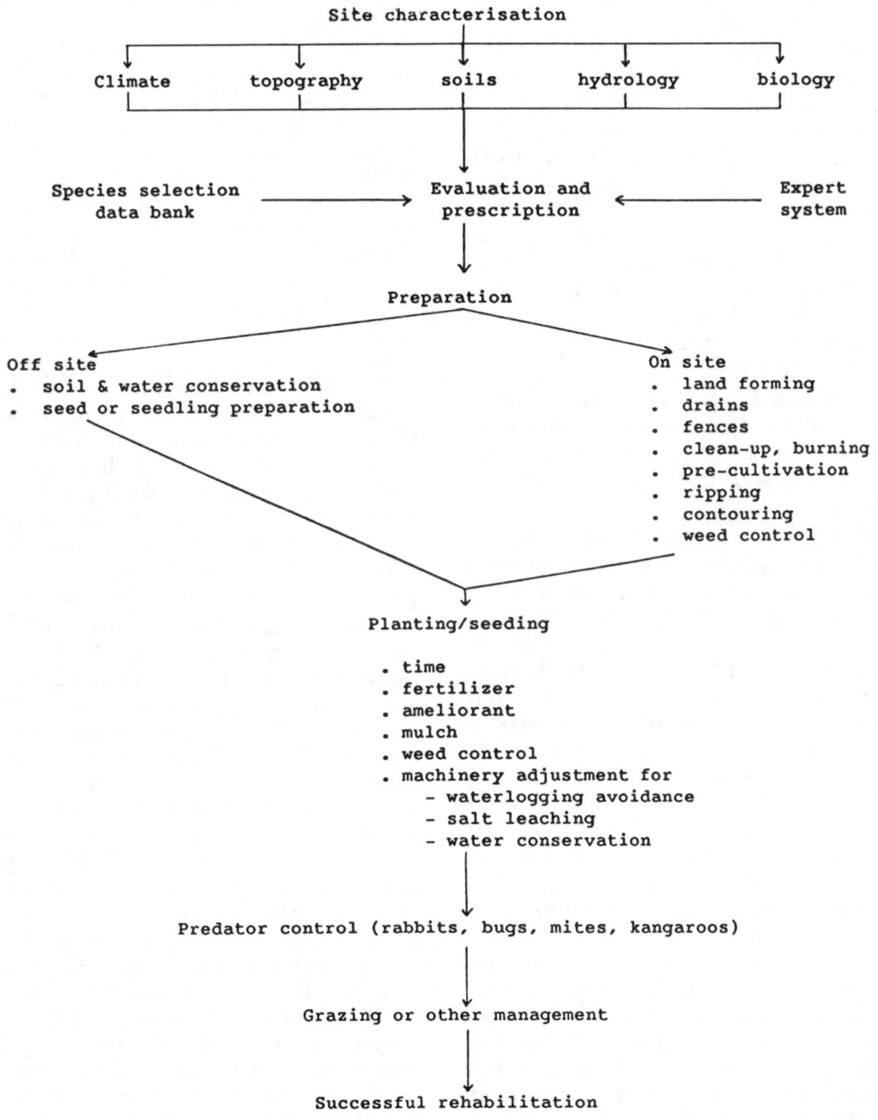

The success of revegetation depends on accurately prescribing the treatment measures on the basis of good site characterization. Neither site characterization nor treatment prescription for revegetation of arid and semi-arid degraded land has received sufficient attention. Consequently, the record of success is poor.

Policy problems

The unfortunate record of failure in rehabilitation of degraded arid and semi-arid land is symptomatic of several basic policy problems.

As a nation (and world) we have failed to give sufficient priority to land rehabilitation. We have continued to allow those who degrade land to escape the consequences.

The actions causing land degradation are sometimes undertaken on land not affected by the consequences and frequently persons causing degradation do not pay for the damage they cause. Land degradation is usually an externality which is not accounted for in the market place. Damaging activities are usually rewarding both to the individual and the community. For example, a farmer may, with community encouragement (or even coercion) clear natural vegetation to grow profitable agricultural crops, but in the process contribute to erosion, salt encroachment and waterlogging on other land. The costs of damage are suffered by other landholders, the community (e.g., road damage) and future generations.

Therefore, it is beneficial to affected landholders *and* the community for damage to be prevented and those engaging in damaging actions *and* the community as a whole should accept some of the responsibility for rehabilitation.

Successful rehabilitation requires treatment of both the cause and the effect and often involves off-site and on-site measures. A redistribution of costs and benefits is involved. As with pollution control the problem is to design economic or regulatory measures that will minimize the damaging effects of externalities. To be successful such measures require the political commitment which results from community awareness and concern. The National Soil Conservation Programme (NSCP) in Australia and the Man and Biosphere Programme and Desertification Conference of UNESCO overseas have contributed greatly to increasing community concern. NSCP funding has provided an incentive for group co-operation and demonstration of techniques. Unfortunately, individuals who have in some cases suffered many years of economic disadvantage due to the actions of others (e.g., farmers in salt-affected valley floors) are still required to pay for rehabilitation. Allowing greater than the present 100% deductibility for taxation purposes would assist the disadvantaged and accelerate adoption of rehabilitation measures. Relief could also be considered for those who undertake to sacrifice productivity by adopting measures designed to prevent land degradation.

As a scientific community we have given inadequate attention to rehabiliation research. The proceedings of the Second International Rangeland Congress were titled "Rangelands: a resource under siege" (Joss *et al.* 1986) but in 634 pages there were only two papers on methods of shrub establishment. For the International Research and Development Conference "Arid Lands Today and Tomorrow" (Whitehead *et al.* 1988) there was only one paper on establishment of plants by reseeding in 1 435 pages. There are two possible reasons for workers to shun this research. The yield of results suitable for good papers in refereed journals is perceived to be low and research on establishment in arid environments is assumed to give a poor return for effort.

The CSIRO has given little attention to rehabilitation research. Rehabilitation research does not have a national focus; it does not rate a national conference; it is not a recognized discipline. Despite the National Soil Conservation Programme (NSCP) we have failed to deal with land degradation adequately because NSCP emphasizes implementation of known technology and has failed to fund significant research on rehabilitation agronomy. State departments have received little NSCP national component research funding and are the only research bodies conducting significant work in the field.

Research on rehabilitation of degraded land is usually accorded a lower priority for funding than production, marketing, and other research because it does not offer a quick dividend. It receives this low rating in all organizations. We have failed to recognize that by combining the requirements to rehabilitate saline agricultural land, degraded rangeland and mine dumps, the justification for a programme of national significance is substantial.

The long-term nature of rehabilitation agronomy research using shrubs militates against its progress when reliance is placed on short-term special fund projects and there is a strong emphasis on the production of scientific publications. For example, in May 1985 an experiment on rangeland rehabilitation was sown at Sturt Meadows Station, near Leonora (Western Australia). Outstanding differences in establishment occurred between different selections of *Atriplex amnicola.* Plant counts up to August 1988 have shown major changes in the relative performance of the selections as a result of emergence from seeds sown up to three years previously (R. Shepherd, pers. comm.). It is difficult to get meaningful results in a short time under these circumstances. Hence it is not well suited for university degree programmes. These difficulties may be the reason for the failure of the world scientific community to develop a centre of excellence in this field. Australia could fill the gap.

Research areas that have been neglected as a result of the above policy problems include plant resources and establishment methods.

Plant resources under-researched

There has been no systematic national effort to explore the Australian flora or introduce exotics for revegetation of arid and semi-arid degraded land. The work that is required is long term and difficult, characteristics that have discouraged its implementation at national and state levels.

In Canberra there is a centre responsible for exploring Australia's tree species resources, maintaining a germ plasm bank and providing seeds and information on the species to overseas countries (S. J. Midgley, pers. comm.). Over 80 countries have been served. There is no similar centre for forage shrub germ plasm, although these resources are of relevance to extensive areas of Australian rangeland and vast areas of degraded land in other countries. Even for species such as *Atriplex vesicaria* and *A. nummularia* which occupy millions of hectares of southern Australian rangelands there has been no significant attempt to evaluate the germ plasm for variation in factors such as ability to recover from grazing, palatability, ability to establish from direct seeding, colonizing ability, and adaptation of ecotypes to particular soil or climatic zones.

In general, in southern Australia, programmes of research aimed at rehabilitation of degraded land in arid and semi-arid areas have failed to explore the available germ plasm and include a reasonably wide range of genetic material. The tendency has been to run a programme using seed commercially available or obtained by canvassing other researchers. Recent research on rehabilitation of overburden dumps at Kalgoorlie, Western Australia, has shown that using locally collected ecotypes of species which colonize readily improves direct seeding results (Fletcher *et al.* 1989).

Plant establishment

Attempts to revegetate degraded environments in arid and semi-arid areas around the world have frequently involved the use of inexact methods of soil disturbance (e.g., pits, furrows, banks, rips) and haphazard distribution of seeds. They have frequently been rewarded by failure. Consequently, there are no reliable seeding methods available for the majority of environments and where extensive areas are being planted to forage shrubs (e.g., Iran) the operation is based on nursery raised plants, a slow, laborious and costly alternative.

Australian attempts at establishment of forage shrubs have shared the same shortcomings as those overseas. There has been: inadequate understanding of the factors affecting plant establishment and survival on degraded land; failure to recognize the importance of establishment niches in hostile environments; and failure to design establishment methods on the basis of species requirements and site characteristics.

There has been an adequate demonstration of many approaches that do not give good results (e.g., attempts to re-establish shrubs on scalds in western New South Wales in the 1950s and 60s). The exception is the use of ponding banks in western New South Wales (Jones 1967), which have proved an outstanding success in establishing annual cover. Unfortunately ponding banks are not sufficiently successful for shrub establishment and they cannot be used on all degraded land.

Western Australian programme

A long-term research programme on the rehabilitation of salt-affected agricultural land has been conducted at the Western Australian Department of Agriculture. The programme has involved germ plasm exploration overseas and in Australia, field screening programmes (Malcolm & Swaan 1989), development of specialized seeding machinery (Malcolm & Allen 1981), grazing studies (Malcolm & Pol 1986), and commercial development. Saltbush growing is a successful commercial operation in Western Australian agricultural areas. About 4 000 ha of saltbush pasture was contract sown during 1988.

The research methods, species, ecotypes and sowing method used on saltland are now being applied to degraded rangelands and mine dumps with encouraging results (Ward 1989; Fletcher *et al.* 1989). Certain basic principles have been shown to operate in all three environments and are incorporated in the critical pathway illustrated in Table 2.

Site characterization — Problems likely to be encountered in revegetation include high salinity, waterlogging, slaking soil, wind blown soil, weed competition and intractable soil (e.g., heavy clay) and can be identified by careful site characterization.

Site preparation — Measures (off- and on-site) which will facilitate revegetation should be undertaken before sowing. For example, for saltland this may include soil conservation measures on the catchment, for rangelands, destocking and ponding may be used and for mine dumps slope reduction is often necessary.

Species and ecotype selection — Species such as *Atriplex bunburyana* have been found to have a superior ability compared to other *Atriplex* spp. to establish when sown in degraded environments. Within *A. amnicola,* selection of seeds from plants that colonize readily has given significant improvements in establishment from seeds sown in saltland, mine dumps and degraded rangeland. The ability to colonize has proved to be a heritable trait in two separate *A. amnicola* selections.

Niche engineering — Seedbed modification has been shown to create favourable positions for the placement of seeds. Criteria for which niche engineering may be designed include reduced salinity, improved water supply and reduced waterlogging. The Mallen Niche Seeder may be adjusted to fulfil these differing functions.

Spot placement — Because shrubs may be sown at a wide spacing rather than as a sward, it is possible to use spot placements of seed and special treatments to reduce the establishment cost per hectare. Vermiculite mulch and sprayed coatings of black latex paint and bitumen emulsion have given promising results.

Protection from predators — Insects, rabbits, kangaroos and sheep have all attacked *Atriplex* spp. in Western Australian programmes and must be controlled.

Using the above techniques, farmland which has been abandoned due to salinity can be made to support grazing sheep during the critical autumn feed shortage (Malcolm & Pol 1986). On current prices (1988) dollar returns from revegetated saltland may equal or exceed those from wheat on nearby non-saline land and the carrying capacity of farms may be substantially increased, (Salerian *et al.* 1987).

Conclusions

Attempts to rehabilitate degraded land in arid and semi-arid parts of Australia have had limited success. Adequate expertise has not been developed because the long-term difficult work with a perceived low potential economic return has been passed over in favour of other research.

The rehabilitation agronomy programme in Western Australia has shown that attention to the principles of site characterization, site preparation, species and ecotype selection, niche engineering, spot placement and protection from predators can result in success. The resulting economic returns have proved to be attractive.

There is a need to develop a centre or network of excellence in Australia in the field of rehabilitation agronomy. Features of the development would include a forage shrub germ plasm bank, a data bank on site characterization and species performance, a facility for the design, construction and testing of machinery for seeding and seed harvesting, long-term research programmes on species evaluation and establishment methods and an export development section.

References

Fletcher M., Malcolm C., Petersen T., Jennings B. & Burnside D. (1989) Abatement of dust in the Kalgoorlie region by revegetation of overburden stockpiles. Minerals & Energy Research Institute of Western Australia Report No. 44.

Jones R. M. (1967) Scald reclamation studies in the Hay district, New South Wales. Part III Reclamation by ponding banks. *J. Soil Conserv. N.S.W.* **23,** 3-17.

Jones R. M. (1968) Studies on the stimulation of germination and emergence of old man saltbush (*Atriplex nummularia* Lindl.) *J. Soil Conserv. N.S.W.* **24,** 271-8.

Joss P. J., Lynch P. W. & Williams O. B. (eds) (1986) *Rangelands: A resource under seige.* Proceedings of the Second International Rangelands Congress. Adelaide, South Australia. 634 pp.

Malcolm C. V. (1986) Saltland management — selecting forage plants for saltland. Western Australian Department of Agriculture Farmnote 32/86.

Malcolm C. V. (1989) Forage shrub production on salt-affected soils. In: *The biology and utilization of shrubs* (ed. C. M. McKell) pp. 553-74. Academic Press, New York.

Malcolm C. V. & Allen R. J. (1981) The Mallen Niche Seeder for plant establishment on difficult sites. *Australian Rangeland Journal* **3,** 106-9.

Malcolm C. V., Hillman B. J., Swaan T. C., Denby C., Carlson D. & D'Antuono M. (1982) Black paint, soil amendment and mulch effects on chenopod establishment in a saline soil. *Journal of Arid Environments* **5,** 179-89.

Malcolm C. V. & Pol J. E. (1986) Grazing and management of saltland shrubs. *Journal of Agriculture Western Australia* **27,** 59-63.

Malcolm C. V. & Swaan T. C. (1989) Screening shrubs for establishment and survival on salt-affected soils in south-western Australia. Western Australian Department of Agriculture Tech. Bull. 81.

Salerian J. S., Malcolm C. V. & Pol E. (1987) The economics of saltland agronomy. Western Australian Department of Agriculture, Division of Resource Management, Tech. Rept. No. 56.

Springfield H. W. (1978) Using mulches to establish woody chenopods — an arid land example. FAO of United Nations Special Readings in Soil Conservation. FAO Conservation Guide, No. 5, 51-64.

Ward B. H. R. (1988) Shrub evaluation for rangeland. Final Report to Rural Credits Development Fund. Western Australia Department of Agriculture WADA 8783.

Whitehead E. E., Hutchinson C. F., Timmerman B. N. & Varady R. G. (eds) (1988) Arid Lands Today and Tomorrow. Proceedings of an International Research and Development Conference, Tucson, Arizona. 1425 pp.

Proc. Ecol. Soc. Aust. 1990 — 16:557-565

Bauxite mine rehabilitation in the Darling Range, Western Australia

S. C. WARD, J. M. KOCH and O. G. NICHOLS

Alcoa of Australia Ltd, P.O. Box 252, Applecross, W.A. 6153

Abstract

Alcoa of Australia Ltd has mined and rehabilitated mined land in the Darling Range for over 20 years. The aim of the rehabilitation programme is to establish a functional forest ecosystem which meets the designated land use objectives for the area.

Current research programmes are investigating the processes of nutrient accumulation, nutrient cycling and floral and faunal succession. Where a legume understorey has been established, nutrient accumulation rates in the young developing system are high and the above ground nutrient pools approach those of the jarrah forest. Weight and nutrient content of annual litterfall is comparable to that in the jarrah and other Australian dry sclerophyll forests with the exception of nitrogen (N). The amount of litterfall N approaches that of Australian wet sclerophyll forests.

In newly rehabilitated sites, plant species numbers and diversity increase rapidly. After 3-5 years there is a levelling off in these parameters due to interspecific competition. As the early dominating species senesce (6-10 yrs), there is a slow invasion by other jarrah forest species. The initial dominance then senescence of pioneer legume species is similar to that seen in the jarrah forest following hot fires.

Detailed faunal studies have demonstrated which rehabilitation techniques favour the return of communities similar to those occurring in the jarrah forest. These techniques have been incorporated into rehabilitation prescriptions.

An understanding of these ecological processes and what affects them allows informed decisions to be made on the use of management tools such as fire. This is essential if rehabilitation is to meet long-term objectives.

Introduction

Alcoa of Australia has been mining bauxite and rehabilitating mined-out areas on the Darling Plateau of Western Australia for more than 20 years. Over that time, increasing knowledge and experience has led to a continually increasing standard of rehabilitation operations.

Alcoa's mines are at Jarrahdale, Huntly, Del Park and Willowdale in the northern jarrah, *Eucalyptus marginata*, forest. The bauxite occurs as isolated pods on the flanks of gentle hillslopes. Individual ore deposits range from 1-100 ha in area, averaging around 10-20 ha. Mineable quality bauxite occupies less than 5% of the forest region.

Successful open-cut mine rehabilitation requires an understanding of how the ecosystem which is being reconstructed functions, and a commitment by the mine operators to sound environmental management. Successful rehabilitation is most likely to be achieved when mine planning takes the rehabilitation of the area into account from the start.

The aim of Alcoa's mine rehabilitation programme is to establish a functional forest ecosystem which meets the designated land use objectives for the particular area. These objectives include timber and water production, recreation and conservation. This chapter describes some of the research into the long term development of the ecosystems on rehabilitated bauxite mines. We demonstrate how this research is being used to improve both rehabilitation techniques and the long term management of rehabilitated areas. Specific areas of research described are the accumulation of nutrients, nutrient cycling and floral and faunal succession.

Nutrient accumulation

The nutrient status of the jarrah forest is lower than most other Australian eucalypt forest ecosystems (Hingston *et al.* 1981). The lateritic soils are particularly deficient in nutrients, and the proportion of the total nutrients in the ecosystem which are incorporated in the plants and litter is particularly high, when compared with other Australian eucalypt forests (Hingston *et al.* 1981). Mining the jarrah forest necessitates removal of the vegetation prior to mining, which inevitably leads to the loss of some nutrients.

To reach a productivity level in rehabilitated areas equivalent to that of the pre-mining forest, requires an input of nutrients to the system and the accumulation and cycling of these nutrients by the vegetation. Nitrogen (N) is the nutrient most limiting the growth of trees in the jarrah forest (Abbott & Loneragan 1986) and rehabilitated areas (Ward *et al.* 1985). Responses to phosphorus (P) have also been found (Ward *et al.* 1985). Most of the N held in soil is in the organic matter and is released for uptake by plants through mineralization. As mineralization is usually low a large capital of N must be accumulated by the system. Inputs of N fertilizer to replace the N lost is not practical after bauxite mining because of the large amounts required, its ephemeral effect and its high cost. This contrasts to P fertilizer which has a high residual value in these soils and where a single application at planting has proved adequate.

The most practical way to accumulate N in a rehabilitated bauxite mine is by using legumes to fix atmospheric nitrogen. As well as fixing N, fast growing legumes provide a quick plant cover (thereby reducing the erosion hazard), change the micro-climate at the soil surface, provide a resource for fauna, and favourably affect soil properties such as structure, cation exchange capacity and water-holding properties. Alcoa have sown an understorey, which includes a high proportion of legumes (mainly acacias), on all rehabilitation sites since 1977. P fertilizer (about 40-45 kg/ha P) is broadcast after seeding to enhance growth, nodulation and N fixation of the jarrah forest legumes (Hansen & Pate 1987; Hingston *et al.* 1982).

The accumulation of N on rehabilitated areas with a dense legume understorey has been measured by Ward and Pickersgill (1985), Koch (1987) and Ward (1988) (Table 1). After 5½ years, the total amount of N in a rehabilitated bauxite pit was similar to that in a 60-year-old pole-stand jarrah forest (Hingston *et al.* 1981). After 3½ years, the biomass (including litter) on a rehabilitated pit with a tall understorey contained nearly 60% of the N in the biomass of the forest site. Even in areas where only short understorey was established, this vegetation had accumulated 132 kg/ha of N after 3½ years. Where no understorey was planted, the biomass contained only 20% or less of the N in the biomass of the forest site, after 7½ years.

These results show that a legume understorey is a very effective method of accumulating N on rehabilitated bauxite mines. Ward and Pickersgill (1985) estimated that the rate of N fixation was 43 kg/ha/yr while Koch (1987) estimated that N fixation rates could be as high as 207 kg/ha/yr. This latter estimate was at very high rates of P application (150 kg/ha P), while the area assessed by Ward and Pickersgill (1985) received around 35 kg/ha P.

Structurally there is a contrast between nutrient distribution in the jarrah forest and the young rehabilitated mines. The latter have a large proportion of the ecosystem N incorporated into the understorey. The tall, well-aerated understorey represents a fire hazard and some sort of fire management is required. Fuel reduction burning is carried out on a regular rotation in the jarrah forest. However, burning young rehabilitated areas could lead to large losses of the N in the understorey by volatilization. Studies of the effect of burning on the nutrient stores in rehabilitated areas are in progress. A fire protection plan for rehabilitated areas has been developed by the Department of Conservation and Land Management (CALM).

The dense understorey competes for resources with the eucalypts which should ultimately be the dominant species on rehabilitated areas. At the site studied by Koch (1987), the average height of the eucalypts growing on areas with a dense understorey

TABLE 1. Biomass (t/ha) and nitrogen (kg/ha) distributions in the jarrah forest and various rehabilitated bauxite mines.

Site Age Reference	Jarrah forest 60 y.o. pole stand Hingston *et al.* (1981)	Rehabilitation 3½ y.o. tall understorey Ward and Pickersgill (1985)	Rehabilitation 3½ y.o. short understorey Ward (1988)	Rehabilitation 5½ y.o. tall understorey Koch (1987)	Rehabilitation 7½ y.o. no understorey Ward and Pickersgill (1985)
Litter weight (includes dead wood)	141	9	2.5	9	1.5-4[B]
Standing biomass	266	27	18[A]	88	24-42
Tree nitrogen	321	10	n.d.	19	36-81
Understorey nitrogen	22	178	132	540	0
Litter nitrogen	132	88	19	124	6-16
Total biomass nitrogen	475	276		683	42-97
Soil nitrogen [C]	1 802	1 149	n.d.	1 326	1 203
Total nitrogen	2 277	1 425		2 009	1 245-1 300

[a]Understorey only, the mass of the trees was not determined.
[b]Range.
[c]Depth of 0-30 cm for 5½ year old rehabilitation, 0-100 cm for the other sites.
n.d. Not determined.

was only 56% of that of the trees where the understorey was sparse, 20 months after planting (Majer & Sawyer 1981). However, this suppressive effective was reduced with time and in the long term should not be important. After 5½ years the eucalypts were much taller than the understorey and the average height of the trees growing amongst the dense understorey had increased to 91% of the average height of trees on areas with a sparse understorey (Koch 1987).

Nutrient cycling

It is not enough for N and other nutrients to be accumulated within the vegetation. Transfer of the N fixed by the understorey to the trees and non-legume understorey species is important. The first step in the transfer of nutrients, and the easiest to measure, is the amount of nutrients which reach the soil surface in the litterfall. Any transfer of nutrients below ground by the shedding of roots is difficult to measure and has not been considered.

In rehabilitation with a tall dense understorey, the average yearly amount of most nutrients in the litter, four to six years after planting, was broadly similar to that in the litter falling in the jarrah forest and other Australian dry sclerophyll forests except for N (Table 2). The N content of the litterfall in the rehabilitated area was about five times that of the dry sclerophyll forests and approached that of wet sclerophyll forests (Table 2). The N content of the litterfall was still around three times that in the jarrah forest at age 8-9 years when much of the acacia dominated understorey was senescent (Table 2). Where an understorey of shorter growing legumes had been planted to reduce the fire hazard associated with a tall understorey, the N content of the litterfall 3-4 years after planting was nearly four times that in the annual litterfall in the jarrah forest (Table 2). The litterfall on rehabilitated areas consisted primarily of legume leaves and flowers. Litterfall in the jarrah forest consists primarily of eucalypt leaves, twigs and bark which have a lower nutrient status than the legume litter.

The average carbon:nitrogen (C:N) ratio of the litter falling on the rehabilitated areas ranged from 39:1 to 54:1 (assuming a carbon content of 50%). This compares with 127:1 and 141:1 for the jarrah forest, 117:1 for a *Eucalyptus obliqua*/*E. baxteri* forest and 65:1 to 81:1 for wet sclerophyll forests (Table 2). Little mineralization of the N in litter occurs until the C:N ratio is reduced to 30:1 to 20:1 or less (O'Connell 1986). At C:N ratios higher than this, immobilization of N occurs. The much lower C:N ratio of the litter falling on rehabilitation compared to the jarrah forest litter suggests that net mineralization of N from the litter will be much greater in the rehabilitated areas. This will result in an increased supply of N for uptake by the trees and understorey. The early release of N from litter on rehabilitated areas is likely to be very important in the development of the ecosystem, as soil stores of N are initially very low. Research is currently being carried out into the decomposition of litter on rehabilitated areas.

Vegetation development in rehabilitated bauxite mined areas

This section deals with understorey vegetation succession in rehabilitated bauxite mined areas. Tree succession is a much longer term process. Due to the relatively short history of bauxite mining and rehabilitation in W.A. (25 years) information on tree succession is not yet available.

Mined areas, if left untreated, are very slow to revegetate. The rehabilitation treatments are a means of accelerating the revegetation process.

TABLE 2. Litterfall (t/ha/yr) and its nutrient content (kg/ha/yr) on rehabilitated bauxite mines and selected Australian eucalypt forest.

	Weight	N	P	K	S	Ca	Mg	C:N ratio (assumes 50% carbon)
Rehabilitated Mines								
4-6 years old, tall dense understorey	4.1	52.5	1.4	10.5	5.8	39.4	5.8	39:1
8-9 years old, tall dense understorey	3.1	28.7	0.7	7.3	3.6	36.9	5.5	54:1
3-4 years old, short dense understorey	3.2	35.4	0.7	10.9	4.0	25.3	5.8	45:1
Forests								
Jarrah forest[A]	2.4-3.1	9.2-12.2	0.4-0.6	5.0-7.0	1.7	16.3-22.0	3.2-7.9	127:1-141:1
Other dry sclerophyll forests[B]	2.3-5.5	9.8	0.3-1.0	2.7-8.9	17.1-30.4	4.2-9.4	9.9	117:1
Wet sclerophyll forests[C]	6.6-10.0	43.5-77.0	1.3-7.5	4.6-31.7	7.4	39.5-122.3	18.7-22.7	65:1-81:1

[A]Hatch (1955), O'Connell *et al.* (1979).
[B]Attiwell *et al.* (1978), Lee and Correll (1978).
[C]Ashton (1975), Richards and Charley (1977), O'Connell and Menage (1983), Turner and Lambert (1983).

Generally, the more intensive the treatments, the faster the revegetation. Landscaping, ripping, topsoil return, seeding and fertilizing produce rapid revegetation and in most cases promote the return of the mined areas to a vegetation not greatly dissimilar to that of unmined jarrah forest (Nichols *et al.* 1985). A return to a vegetation community compatible with unmined jarrah forest vegetation is seen as an important long term aim in rehabilitated areas.

The important factors which control development of bauxite mined areas towards a jarrah forest type vegetation appear to be:

i) the influence of the surrounding forest (includes pit size, species composition of the forest),
ii) application of seed and competition from seeded *Acacia* understorey,
iii) soil seed store,
iv) tree canopy and density, and
v) site fertility.

Influence of the surrounding forest

Pit size or distance from the surrounding forest is important in succession but its effect is very variable and subject to local influences. Observations have shown that tall species which can overhang the rehabilitation areas often produce many hundreds of plants in the rehabilitation. Examples of this are *Banksia grandis*, *Allocasuarina fraseriana* and *Eucalyptus marginata* (jarrah) itself. The mechanisms of dispersal of the smaller understorey species into the mined areas is not well understood. It is expected there would be more species and higher plant densities near the edge of pits than in the middle which are further from the forest. A study is in progress to determine this effect and to measure the rate of spread of various species. However, no data are yet available.

Application of seed

Application of seed to a new bauxite pit immediately increases the species richness. Depending on the species used, the method forms an integral part of promoting a return to a jarrah forest vegetation. About 100 species native to the northern jarrah forest are currently seeded into the bauxite pits. These species are primarily from the families Mimosaceae, Fabaceae, Myrtaceae and Proteaceae. These are the dominant taxa in the northern jarrah forest. *Acacia* and other fast growing jarrah forest legumes are ideally suited to the new bauxite mine environment because they are natural disturbance colonising species. They produce large amounts of seed which is easy to collect and germinate, and they provide the rapid cover necessary to satisfy the high priority aims of prevention of soil erosion and development of a soil profile. *Acacia* are seeded extensively in rehabilitated areas. In the short term, a dense *Acacia* cover reduces the number of invading small shrub and herb species which slows down the rate of succession (Table 3). However, with time, there is a gradual increase in species numbers which appears to accompany the decline and senescence of the fast growing *Acacia* species (Table 4). The number of species per unit area is comparable after 11 years with the jarrah forest (Table 4, Koch 1978). The initial dominance then senescence of *Acacia* also occurs in the jarrah forest following hot fires (Bell & Koch 1980).

TABLE 3. Mean numbers of small shrub and herb species (plants/m^2) in four 6 m×6 m quadrats in dense (mean cover 94.6%) and sparse (mean cover 19.2%) *Acacia* understorey at Jarrahdale minesite. Data from Davis (pers. comm.).

Year Planted	Age	Dense Acacia	Sparse Acacia
		Mean no. of small shrubs & herbs	Mean no. of small shrubs & herbs
1976	12	.701	6.403*
1978	10	.514	.854NS
1979	9	.076	1.271*
1981	7	.590	1.208NS
1982	6	2.138	7.424NS
1983	5	1.271	5.167*
1984	4	.757	1.771NS
1985	3	.284	1.222*
All Years		.693	2.879*

NS — no significant difference in mean number of small shrubs and herbs for dense versus sparse plots.

* — significant difference ($p < 0.05$) in mean number of small shrubs and herbs for dense versus sparse plots.

Soil seed store

The importance of the soil seed store was reported by Tacey and Glossop (1980) for a pit rehabilitated in 1976. Subsequent studies of the same quadrats have been carried out (Nichols & Michaelsen

TABLE 4. Numbers of understorey plant species per 100 m^2 in the "Jungle" study site from 100 1 m×1 m random quadrats from various studies and from a jarrah forest site.

Year	Number of Species	Study
1976	6	Area rehabilitated and seeded with six understorey species.
1979	18	Majer *et al.* 1984.
1985	52	Buckney (pers. comm.)
1987	52	Alcoa unpublished study October 1987
Jarrah Forest	37	Koch 1978 (100 m^2 contiguous quadrats)

1986) and are continuing. Freshly stripped topsoil, removed and replaced in the summer produced a higher species similarity to the jarrah forest than soil which was stockpiled or stripped and returned as a whole deep soil profile (Fig. 1). The beneficial effect of this soil seed store was still evident even after 13 years.

Although all the rehabilitated sites had similar species numbers and diversity the higher similarity of the double stripped plot to the jarrah forest indicates a more "jarrah forest like" species composition. The rapid drop in similarity, species numbers and diversity for the direct whole return plot from 10 to 13 years is due to the presence of annual weeds. The weed invasion followed fertilizer application to a surrounding area in 1986; there may have been fertilizer drift and in addition there was a large input of weed seed from adjoining farmland in 1987 and 1988 and a heavy weed cover in these two years.

Tree canopy

A dense tree canopy has significant effects on the development of the understorey vegetation. A dense canopy favours shade tolerant species and in extreme cases tends to reduce the number of species and the density of understorey plants. Changes in species composition from high light demanding disturbance pioneers to non-pioneer, slower growing species occurs. Figure 2 shows the relationship between tree canopy cover and understorey cover in the 0-50 cm height stratum which is generally the size of mature jarrah forest understorey. In this 20-year-old location there was an inverse relationship between tree canopy cover and 0-50 cm understorey cover. Transects 1 and 2 do not fit the relationship. They were in an area of very poor topsoil (mostly pit floor clay) which has resulted in low values for both parameters. In the case of these two transects, the influence of topsoil quality on vegetation development has overridden the effect of an open tree canopy.

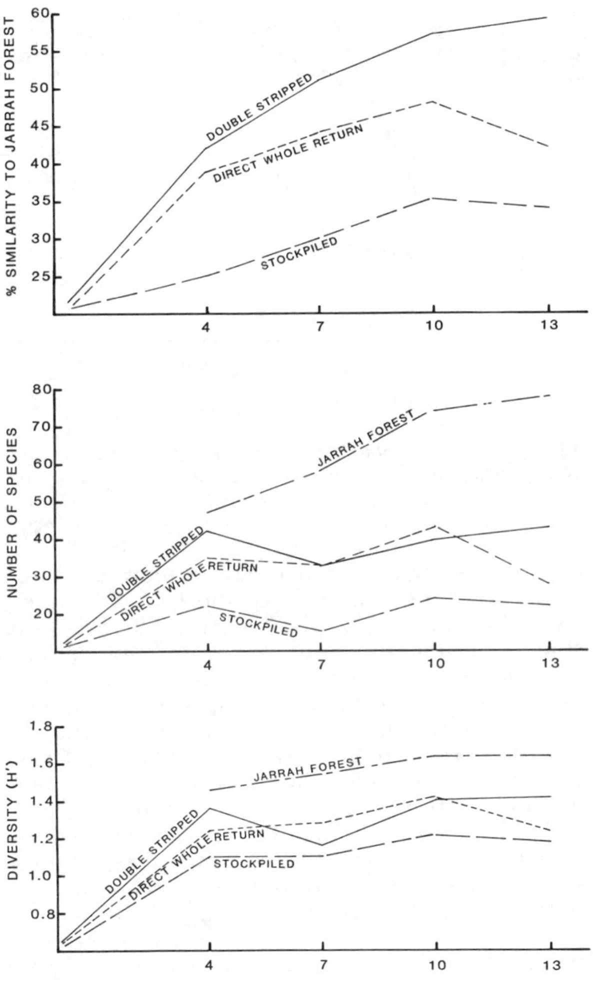

FIG. 1. Successional trends in relation to soil handling techniques in revegetated bauxite mined areas. Four year data were from Tacey and Glossop 1980, seven year data from Nichols and Michaelsen 1986, 10 and 13 year data are new. Data for each treatment and the neighbouring jarrah forest site are from 50 permanent 1 m² quadrats. Percent similarity uses the Sorensens index. Diversity uses the Shannon-Weiner index.

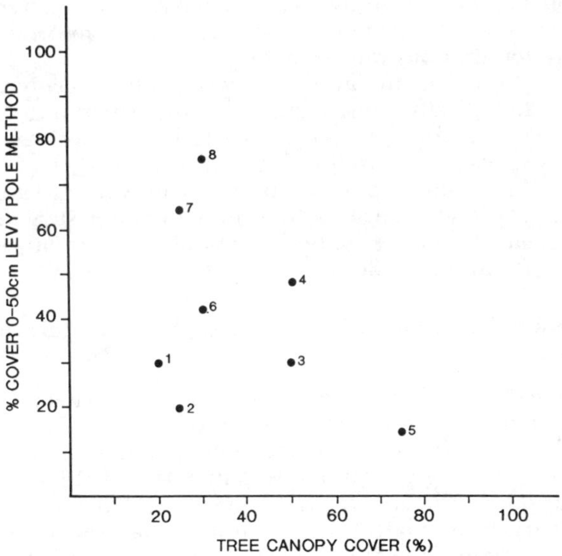

FIG. 2. Percentage cover for the 0-50 cm vegetation stratum in relation to tree canopy cover for 8 transects in 20 year old rehabilitation at Langford Park. There is a negative relationship between tree canopy cover and understorey cover. Transects 1 and 2 are in an area of very poor topsoil (mostly pit floor clay) which has resulted in low values for both parameters. The other transects are in sites of similar quality topsoil.

Site fertility

Site fertility is both a direct and indirect controller of succession in bauxite mines. It directly affects the species composition by favouring species which are suited to the fertility levels of the site. For example *Banksia grandis* does not respond to phosphorus fertilizer (Barrow 1977) and would not be expected to respond to a high fertility site or to application of superphosphate, while *Acacia* species respond to phosphorus (Barrow 1977; Hingston *et al.* 1982) and grow better in rehabilitated pits where phosphorus has been applied (Fig. 3). Indirectly, the ability of certain species to respond to high fertility gives them a competitive advantage in the rehabilitation over species unable to respond to, or which are suppressed by high fertility. This can result in a temporary reduction in species richness in high fertility sites, where, for example a dense *Acacia* stand has developed (Table 3).

There are other factors which control succession in rehabilitated bauxite mined areas. However, the ones described above are the most visible and probably the most important. All of these are based on simple ecological principles and it would seem an easy task to describe vegetation succession in rehabilitated areas. This is not the case. Interaction and combination of these factors has produced a complex range of vegetation types in the rehabilitation and it is very difficult to determine a typical successional pathway. Each site must be considered separately. Overall, however, two important generalizations can be made.

1) With time, the understorey vegetation becomes floristically more similar to the surrounding jarrah forest in terms of species richness and species composition (Table 4, Fig. 1).
2) Fast growing pioneer species (mainly *Acacia*) are gradually replaced by slower growing smaller shrub and herb species typical of the mature jarrah forest flora.

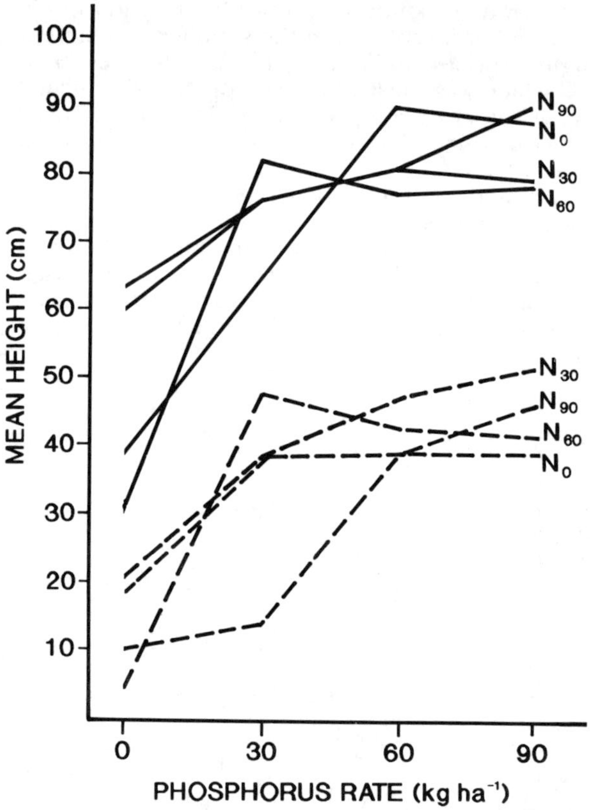

FIG. 3. Height of *Acacia lateriticola* in response to four different phosphorus rates and four different nitrogen rates. Dotted lines are one year old. Solid lines are 2.5 years old. All rates for N and P are kg/ha elemental. From Koch *et al.* (1988).

Faunal succession in rehabilitated bauxite mined areas

The fauna of any forest ecosystem fulfil a vital role in maintaining the functioning of that system. In the jarrah forest, nutrient cycling, pollination, seed and seedling predation are only a few of the processes which are strongly interdependent on the characteristic jarrah forest fauna. Whatever forest ecosystem is established following mining, the fauna must be taken into account. They will colonize from surrounding areas, influence the development of rehabilitation and serve key roles in the resulting ecosystem. Understanding the jarrah forest fauna, and using this understanding to promote their return to mined areas is an important part of successful rehabilitation following bauxite mining.

The ecology of jarrah forest vertebrates is well understood for some species and less well understood for others (Nichols & Muir 1988). Likewise, some invertebrate groups (e.g., ants, spiders and collembola) have been studied whereas others have not. Baseline information on species distribution and ecology is all important in assessing whether the forest ecosystem is being re-established (from the fauna viewpoint) and in developing new methods to further promote fauna return.

To determine whether fauna have successfully returned to revegetated bauxite minesites, we must firstly assess the presence of those parts of the ecosystem on which fauna depend (e.g., plants, suitable soil). Surveys are then needed to determine both the characteristics of fauna return at particular points in time, and ongoing successional trends.

Development of the plants and nutrients are discussed earlier in this chapter. The following section discusses current fauna return, and successional trends in fauna return which are leading towards the re-establishment of a forest fauna community.

A number of studies have shown that a relatively high proportion of most fauna groups return to rehabilitated sites. Reptile (75%), bird (80%), mammal (100% of upland species), ant and collembola (80-90%) studies all show that most species return following rehabilitation (Nichols & Bamford 1985; Nichols & Watkins 1984; Kabay & Nichols 1980; Majer *et al.* 1984). These studies do not confirm that all the habitat requirements of those species returning are fulfilled, but they do indicate that the developing network of mine pods is not forming a barrier to fauna movement. Those species which have not returned have provided information which has been used to modify rehabilitation practices, with the intention of encouraging their return. Examples of this are the increase in return of log and wood material (to encourage ground dwellers such as centipedes) and the change towards more rough barked tree species (to provide more suitable habitat for tree-creepers).

Other parameters must be assessed before it can be concluded that species habitat requirements have been met. For birds, detailed studies (Nichols & Watkins 1984; Collins *et al.* 1985) have shown no significant differences between avian densities in rehabilitated and unmined sites. Similarity indices indicate that the species composition of unmined jarrah forest without dieback is more similar to that of rehabilitated areas than to the bird communities of severely dieback affected forest.

TABLE 5. Arachnid species numbers, diversity and Sorensons similarity index for three rehabilitated and two unmined forest sites. Site 3 had received dense, seeded understorey and Site 4 had been logged three years before the survey. Results are from Mawson (1986).

	Rehabilitated Minesites			Jarrah Forest	
	1	2	3	4	5
Age (yrs)	1	2	8	–	–
Total species	13	17	36	27	47
Diversity	.81	.60	0.98	1.04	1.18
Similarity to Site 5	.31	.35	.48	.41	–

When the same parameters are measured for arachnid populations (Table 5), similar results emerge. For the older rehabilitated site, total numbers of species, and arachnid diversity are similar to the ranges recorded in unmined forest. The most advanced rehabilitated site is the most similar to the better unmined forest site. Studies such as these lead to recommending the establishment of an increased diversity in understorey plant species. This in turn would be expected to result in an increased diversity of insect, bird and other fauna species.

These and other studies performed on different fauna groups provide encouraging indications that the return of forest faunal communities to bauxite minesites is proceeding as well as could be expected, given the relatively young age (up to 12 years) of sites rehabilitated using more advanced methods.

However, it would be useful for future management of existing rehabilitation, and planning of future rehabilitation, if one could obtain some idea of trends in fauna return. Successional studies can indicate what rates of return may be expected in future, given a knowledge of trends in soil and plant community development. Several studies have demonstrated this point.

Casotti (pers. comm.) studied the rates of predatory invertebrate fauna return six years after an earlier study by Nichols and Burrows (1985). Predatory invertebrates serve as useful indicators of whether lower invertebrates (i.e., their food items) are returning, and they occupy a variety of niches (e.g., ground foragers, web-building spiders etc.). Casotti's results (Table 6) demonstrated that sites rehabilitated using more advanced methods (e.g., seeding, fresh topsoil return) develop far better rates of predatory invertebrate return than do sites which are only planted with trees, but which receive no fresh (only stockpiled) topsoil or seeded understorey.

In a similar study, Campeotto (pers. comm.) surveyed bird species return in 1987, six years after the earlier work (Nichols & Watkins 1984). Bird species numbers in rehabilitated areas had increased or stayed the same, as had their similarity to the healthy forest site (Table 7). For most of the parameters assessed, the two rehabilitated sites were more similar to the healthy forest site than were the dieback sites.

As well as these studies, the fauna research programme has studied bird breeding (Curry & Nichols 1986), rare species, and other aspects of fauna ecology. Together, all of the studies help demonstrate the extent to which fauna are returning.

TABLE 6. Species numbers and density of predatory invertebrates in rehabilitated and unmined forest sites. Two surveys were conducted; the first in 1981 (T_1) and the second in 1987 (T_2). Density represents the total number of individuals captured in the survey (Casotti, pers. comm.). Site one received fresh topsoil and was densely seeded, while site two was rehabilitated using methods now outdated; it received stockpiled topsoil and no understorey seed.

	Rehabilitated Sites		Unmined Forest	
			Dieback Affected	Healthy
	1	2	3	4
Age at T_1	5	13		
Species numbers at:				
$-T_1$	37	22	38	43
$-T_2$	46	42	50	49
Density at:				
$-T_1$	551	314	246	467
$-T_2$	1060	214	341	475

TABLE 7. Bird community parameters for rehabilitated and unmined forest sites. Surveys were conducted in 1981 and 1987; 1981 results are in brackets. Data are from Campeotto (pers. comm.).

	Unmined Forest				Rehabilitated Sites	
	Healthy		Dieback Affected			
	1	2	3	4	5	6
Total density Birds/hectare	9.50 (13.21)	13.19 (12.56)	6.16 (13.51)	12.83 (6.9)	7.84 (12.04)	
Species Numbers	18 (16)	21 (16)	13 (9)	10 (14)	16 (16)	19 (12)
Diversity	1.13 (1.09)	1.18 (1.08)	1.06 (0.88)	0.89 (0.99)	1.05 (1.08)	1.05 (0.96)
Similarity to Site 1		0.56 (0.75)	0.71 (0.40)	0.50 (0.47)	0.53 (0.56)	0.65 (0.57)

Future studies will be directed towards further promoting the return of a fauna community more like that found in healthy jarrah forest. This will help ensure that any long term efforts of mining on the forest fauna adjacent to mining areas are minimized.

Concluding comments

The development of a functional ecosystem following bauxite mining is a long term concept and two decades is too short a time to assess the ultimate success of this development.

Our research however has indicated that the processes that are considered important in a forest ecosystem are occurring in the rehabilitated areas and that most parameters that can be measured are proceeding in the desired direction.

Research is continuing, both to improve the quality of our new rehabilitation and to develop management strategies for areas already rehabilitated. The response of rehabilitation areas to fire for example is being studied because it is likely CALM will eventually integrate the rehabilitated mined areas into their control burning programme.

References

Abbott I. & Loneragan O. (1986) Ecology of jarrah *(Eucalyptus marginata)* in the northern jarrah forest of Western Australia. Bulletin No. 1. Department of Conservation and Land Management, Western Australia.

Ashton D. H. (1975) Studies of litter in *Eucalyptus regnans* forest. *Aust. J. Bot.* **23,** 413-33.

Attiwell P. M., Guthrie H. B. & Leuning R. (1978) Nutrient cycling in a *Eucalyptus obliqua* (L'Herit) forest. I. Litter production and nutrient return. *Aust. J. Bot.* **26,** 79-91.

Barrow N. J. (1977) Phosphorus uptake and utilization by tree seedlings. *Aust. J. Bot.* **25,** 571-84.

Bell D. T. & Koch J. M. (1980) Post-fire succession in the northern jarrah forest of Western Australia. *Aust. J. Ecol.* **5,** 9-14.

Collins B. G., Wykes B. J. & Nichols O. (1985) Recolonization of restored bauxite minelands in south-western Australian forests. In: *Birds of eucalypt forests and woodlands: ecology, conservation, management* (eds A. Keast, H. F. Recher, J. Ford & D. Saunders), pp. 341-54. Surrey Beatty & Sons, Chipping Norton.

Curry P. J. & Nichols O. G. (1986) Early regrowth in rehabilitated bauxite minesite as breeding habitat for birds in the jarrah forest of south-western Australia. *Australian Forestry,* **19,** 112-4.

Hansen A. P. & Pate J. S. (1987) Comparative growth and symbiotic performance of seedlings of *Acacia* spp. in defined pot culture or as natural understorey components of a eucalypt forest ecosystem in S.W. Australia. *J. Exp. Bot.* **38,** 13-25.

Hatch A. B. (1955) The influence of plant litter on the jarrah forest soils of the Dwellingup region — Western Australia. Commonwealth of Australia. Forestry and Timber Bureau. Leaflet No. 70.

Hingston F. J., Dimmock G. M. & Turton A. G. (1981) Nutrient distribution in a jarrah (*Eucalyptus marginata* Donn ex Sm.) ecosystem in south-west Western Australia. *For. Ecol. Manage.* **3,** 183-207.

Hingston F. J., Malajczuk N. & Grove T. S. (1982) Acetylene reduction (N_2—fixation) by jarrah forest legumes following fire and phosphate application. *J. Appl. Ecol.* **19,** 631-45.

Kabay E. D. & Nichols O. G. (1980) Use of rehabilitated bauxite mined areas in the jarrah forest by vertebrate fauna. Alcoa of Australia Ltd Environmental Research Bulletin No. 8.

Koch J. M. (1978) Structural changes in the jarrah forest following hot autumn burns. Honours Thesis. Botany Department, University of Western Australia.

Koch J. M. (1987) Nitrogen accumulation in a rehabilitated bauxite-mined area in the Darling Range, Western Australia. *Aust. For. Res.* **17,** 59-72.

Koch J. M., Sudmeyer J. E. & Pickersgill G. E. (1988) Response of seeded eucalypts and understorey to broadcast nitrogen and phosphorus fertilizer. Alcoa of Australia Ltd Environmental Research Bulletin No. 19.

Lee K. E. & Correll R. L. (1978) Litter fall and its relationship to nutrient cycling in a South Australian dry sclerophyll forest. *Aust. J. Ecol.* **3,** 243-52.

Majer J. D. & Sawyer A. (1981) Competition between understorey and planted *Eucalyptus* species in rehabilitated bauxite mined areas. Mulga Res. Centre Ann. Rep. No. 5, pp. 69-70.

Majer J. D., Day J. E., Kabay E. D. & Perriman W. S. (1984) Recolonization by ants in bauxite mines rehabilitated by a number of different methods. *J. Appl. Ecol.* **21,** 355-75.

Mawson P. R. (1986) A comparative study of Arachnid community structure in rehabilitated bauxite mines. Western Australian Institute of Technology Biology Bulletin No. 46.

Nichols O. G. & Bamford M. J. (1985) Reptile and frog utilization of rehabilitated bauxite minesites and dieback affected sites in Western Australia's jarrah *(Eucalyptus marginata)* forest. *Biol. Conserv.* **34,** 227-49.

Nichols O. G. & Burrows R. (1985) Recolonisation of revegetated bauxite minesites by predatory invertebrates. *For. Ecol. Manage.* **10,** 49-64.

Nichols O. G., Carbon B. A., Colquhoun I. J., Croton J. T. & Murray N. J. (1985) Rehabilitation after bauxite mining in south-western Australia. *Landscape Planning* **12,** 75-92.

Nichols O. G. & Michaelsen D. V. (1986) Successional trends in bauxite minesites rehabilitated using three topsoil return techniques. *For. Ecol. Manage.* **14,** 173-5.

Nichols O. G. & Muir B. (1988) Vertebrates of the jarrah forest. In: *The Jarrah Forest* (eds B. Dell, J. J. Havel & N. Malajczuk), pp. 133-53. Kluwer Academic Publishers, Dordrecht, The Netherlands.

Nichols O. G. & Watkins D. (1984) Bird utilization of rehabilitated bauxite minesites in Western Australia. *Biol. Conserv.* **30,** 109-31.

O'Connell A. M. (1986) Effect of legume understorey on decomposition and nutrient content of eucalypt forest litter. *Plant & Soil* **92,** 235-48.

O'Connell A. M., Grove T. S. & Dimmock G. M. (1979) The effects of a high intensity fire on nutrient cycling in jarrah forest. *Aust. J. Ecol.* **4,** 331-7.

O'Connell A. M. & Menage P. (1983) Decomposition of litter from three major plant species of jarrah (*Eucalyptus marginata* Donn ex Sm.) forest in relation to site fire history and soil type. *Aust. J. Ecol.* **8,** 277-86.

Richards B. N. & Charley J. L. (1977) Carbon and nitrogen flux through native forest floors. Proc. Nutrient Cycling in Indigenous Forest Ecosystems Symposium. CSIRO Division of Land Resources Management, Perth, pp. 65-82.

Tacey W. H. & Glossop B. L. (1980) Assessment of topsoil handling techniques for rehabilitation of sites mined for bauxite within the jarrah forest of Western Australia. *J. Appl. Ecol.* **17,** 195-201.

Turner J. & Lambert M. J. (1983) Nutrient cycling within a 27 year old *Eucalyptus grandis* plantation in New South Wales. *Forest Ecol. Manage.* **6,** 155-68.

Ward S. C. (1988) Biomass and nutrient accumulation by low-growing understorey on rehabilitated bauxite mines. Alcoa of Australia Ltd, Environmental Research Note 13.

Ward S. C. & Pickersgill G. E. (1985) Nutrient distribution in two eucalypt plantations growing on rehabilitated bauxite mines. *Aust. J. Ecol.* **10,** 111-24.

Ward S. C., Pickersgill G. E., Michaelsen D. V. & Bell D. T. (1985) Responses to factorial combinations of nitrogen, phosphorus and potassium fertilizers by saplings of *Eucalyptus saligna* Sm., and the prediction of the responses by DRIS indices. *Aust. For. Res.* **15,** 27-32.

Proc. Ecol. Soc. Aust. 1990 — 16:567-577

Conservation of genetic resources in Australia's flora and fauna

S. D. HOPPER and D. J. COATES

Western Australian Wildlife Research Centre, Department of Conservation and Land Management, P.O. Box 51, Wanneroo, W.A. 6065

Abstract

Conservation of the genetic resources of the Australian flora and fauna is advocated as wise management aimed at maximizing the options available to future generations for utilization, study and appreciation of the continent's biodiversity. As a natural resource, the flora and fauna already contribute substantially to the national economy through forestry, the pastoral industry, fisheries, tourism, land reclamation, bee keeping, wildflower harvesting and the kangaroo trade. Moreover, the values of natural vegetation to agricultural and urban existence through ecosystem services such as the provision of clean air and pure water, climatic control, soil conservation and soil salinity control are becoming apparent. The potential of the Australian flora and fauna as sources of food and chemicals useful to humans and as a teaching resource has scarcely been realized.

Effective conservation of genetic resources depends upon an adequate data base and a knowledge of the processes that control and sustain inherited variation. For most of Australia's plants and animals, taxonomic descriptions and geographical information will remain the only data base available for conservation initiatives over the next century. It is, therefore, a matter of national priority that taxonomic research and teaching continue at an accelerated rate.

The application of population genetic studies to conservation is a new field. In this chapter, recent genetic information relevant to reserve design, minimum viable population sizes, selection of priority populations for conservation, endangered species management and revegetation is reviewed. Given the pace of ongoing destruction of native vegetation, the search for general principles in gene resource conservation from a limited sample of the Australian flora and fauna is a key scientific challenge facing the next century.

Introduction

Conservation of genetic diversity is a central tenet of the World and Australian Conservation Strategies (Anon. 1984). This reflects widespread recognition that, without a broad base of biodiversity, present and future options for sustained human existence and a reasonable quality of life will be reduced. Plants and animals provide a natural resource of myriad potential for human utilization, study and appreciation.

In Australia, the native flora and fauna already contribute substantially to the national economy through forestry, the pastoral industry, fisheries, tourism, bee keeping, land reclamation, wildflower harvesting and the kangaroo trade. Moreover, natural vegetation facilitates agricultural and urban existence through ecosystem services such as the provision of clean air and pure water, climatic control, soil conservation and soil salinity control (see, for example, papers by van der Moezel & Bell, and by Malcolm in this volume regarding use of salt-tolerant genotypes in several species to reclaim land degraded by rising saline water tables in Western Australia).

The potential of the Australian flora and fauna as sources of food and chemicals useful to humans, and as a teaching resource has scarcely been realized. For example, research at Macquarie University has demonstrated that extracts from the plant genus *Pimelea* inhibit the development of cancers in mice (Howard & Howden 1975).

While the justification for conserving genetic diversity seems clear, the achievement of this aim is not at all certain. Extinction is a process influenced by many possible factors, some genetic and some demographic (see Soulé 1983). Most of these factors are poorly understood. This lack of knowledge presents a major challenge to the nation.

In this chapter, we wish to focus on the contribution that genetic and biosystematic studies can make to the conservation of Australia's flora and fauna during the next century. Such a contribution depends on knowledge in relation to three key questions.

1. What constitutes the Australian flora and fauna?
2. Where do the component taxa occur in the wild?
3. What genetic management is required to maintain sustained reproduction and recruitment?

The Australian flora and fauna

The Australian vascular flora comprises an estimated 25 000+ species, of which about 18 000 had been described by 1980 (George 1981). In addition there are several thousand non-vascular species. The fauna is much larger, consisting of 200 000-300 000 species, of which about 100 000 have been described (Richardson 1984, 1987).

These statistics signal a major problem facing those who wish to conserve this genetic diversity. A taxonomic description provides the minimal data base on which conservation initiatives depend, and such descriptions do not exist at present for a third of the nation's flora and two-thirds of the fauna (particularly invertebrates).

Moreover, the past few decades have witnessed a decline in taxonomic research and teaching in tertiary educational institutions, coupled with a recent decline in job opportunities at major museums and herbaria. Despite major advances in approaches to systematic biology, the subject has waned in popularity with university administrations and government agencies responsible for museums and herbaria.

This process needs to be reversed as a matter of national priority. Australia urgently needs more taxonomists, and tertiary educational institutions, museums and herbaria have key roles to play in this endeavour. It is vital that taxonomic research and teaching continue at an accelerated rate if conservation efforts are to keep pace with activities that involve permanent destruction of native vegetation, fauna, soil and waters.

In this context, it was heartening at this Symposium to hear Professor McComb, in his presidential address to the Ecological Society of Australia, advocate that Australian universities direct more of their research efforts towards solving uniquely Australian environmental problems (McComb 1989).

The taxonomic description and conservation of the continent's flora and fauna fall squarely within the ambit of such a focus for new research.

In groups where patterns of morphological variation are complex or cryptic, traditional taxonomic methods alone will not be sufficient to define the biological units in nature (Paterson & James 1973). Effective conservation efforts here will require population genetic, morphometric, cytogenetic or biosystematic approaches used, for example, by Shaw (1976) in the grasshopper genus *Caledia*, by King (1979) in gekkos of the *Gehyra variegata-punctata* complex, by Coates (1981) in the scale-leaved trigger plants of the *Stylidium caricifolium* species complex, and by Adams *et al.* (1987) to resolve species boundaries in the bat genus *Eptesicus*.

In some cases, resolution of species boundaries within complexes has clear ecological significance relevant to management, e.g. George's (1981) recognition of the fire sensitive reseeder *Banksia littoralis* var. *seminuda* (now regarded as the species *B. seminuda*) as a distinct taxon from the lignotuberous resprouter var. *littoralis*. Ecologists and managers need to be cautious and aware that accepted species names may well conceal unresolved taxa of this kind.

Although an adequate understanding of species boundaries is essential for conservation, in some (perhaps many) cases a knowledge of genetic variation within species is equally important (Lacy 1988). This is especially the case where breeding systems diverge from the typical diploid sexual outbreeding mode seen in many higher vertebrates and perennial plants. Three well-documented Australian examples are the complex translocation heterozygosity associated with inbreeding in the granite rock herb *Isotoma petraea* (James 1965, 1970), the occurrence of triploid parthenogenetic clones and diploid bisexual individuals within the gekko species *Heteronotia binoei* (Moritz 1983), and the evolution of parthenogenesis in the grasshopper *Warramaba virgo* (White 1980).

Natural hybridization is another major process that generates patterns of variation that need to be understood for effective conservation and management. This process has been well documented in such Australian organisms as eucalypts (e.g., Potts & Reid 1985), kangaroo paws (Hopper 1977), frogs (Woodruff 1981) and grasshoppers (Shaw 1981).

Such descriptions of population genetic structure within species and of natural hybridization exist for very few of Australia's plants and animals. Indeed, our knowledge is rudimentary even for the largest and most common members of many of Australia's ecosystems [e.g., Moran & Hopper's (1987) review of data for eucalypts]. This information is essential for intensive species' management aimed at conservation of representative genetic variation in the wild.

Australian geneticists have made seminal contributions to our knowledge of variation and evolution in native plants and animals in the past (e.g., Smith-White 1959; White 1978; James 1981; Barlow 1982), but this field too has waned in favour of a popular interest in molecular genetics in recent years. We can only hope that the urgency and need for conservation genetic studies will inject new life into this field of inquiry. If so, the challenge will be to develop general principles from a well-chosen but limited sample of the fauna and flora (see below).

Australian biogeography and mapping genetic resources

Mapping species

Mapping the continent's flora and fauna in space and through time has proceeded in close association with taxonomic research, but few organisms have been mapped in sufficient detail to provide data adequate for conservation planning and management. This is nowhere more evident than in comparing initial lists of taxa believed to be rare or threatened with current lists (*cf.* Specht *et al.* 1974; Hartley & Leigh 1979; Leigh *et al.* 1981; Briggs & Leigh 1988; Hopper *et al.* 1990).

Nevertheless, biological survey has made considerable progress over the past few decades and has been the foundation on which some recent national parks and nature reserves have been created (Myers *et al.* 1984). The achievement of representativeness in such reserve systems has proceeded through mapping selected components of the flora and fauna, usually dominants in the case of vegetation mapping (e.g., Beard 1981; Sattler 1986). There remains considerable potential for the use of this approach in extending the reserve network over the next century.

At the same time, detailed mapping of selected taxa needs to be pursued to ensure that the reserve system established to achieve representativeness is complemented by additional reserves that protect taxa left out of the larger reserve network. For example, intensive surveys over the past decade have located 1 386 populations of the 238 taxa of flora declared as endangered in Western Australia (Hopper *et al.* 1990). Of these populations, only 381 (28%) occur on existing reserves, with most occurring on freehold land, road verges, etc. Conservation through reserves alone will not ensure that Australia's genetic resources are protected. Land owners from throughout the community at large each have an important contribution to make (Table 1).

Seeking the assistance of interested volunteers in mapping individual species has been successful recently in the compilation of the Atlas of Australian Birds (Blakers *et al.* 1984) and the Banksia Atlas (Taylor & Hopper 1988). Such an approach has considerable merit for well documented, easily surveyed organisms, particularly in view of the size of Australia and the paucity of professional biologists able to conduct extensive field surveys. Industry too is making contributions to the biogeographic data base through requirements for environmental impact assessment (e.g., Worsley Alumina Pty Ltd 1985).

TABLE 1. Land status for populations of declared endangered flora in Western Australia (from Hopper *et al.* 1990).

Land status	No. of populations
National Parks	207
Nature Reserves	174
State Forest	38
Main Road and Shire Road Reserves	302
Private	216
Other	549
Total	1 386

The application of computer-based Geographical Information Systems to biogeographical data has considerable promise in facilitating conservation and management. The integration of satellite technology with conventional cadastral, topographic, climatic, edaphic and biogeographic mapping will yield interactive capabilities of outstanding value to tomorrow's conservation biologists.

Mapping genetic resources within species

Mapping genetic variation within species is an essential requirement to ensure that populations on reserves provide a representative sample of the existing genetic variation. With current technology it is a time-consuming process and therefore few organisms can be mapped in this way.

How to sample genetic variation efficiently remains problematical. Controlled environment studies of polygenic morphological and physiological variation has been customary in forestry and agriculture for many years (e.g., Green 1971; Grant 1975; van der Moezel & Bell this volume; Malcolm this volume). However, these studies are often unsuitable for rapidly measuring and surveying patterns of genetic variation over the geographic range of a species (Lewontin 1974).

Morphometrics. Multivariate morphometric studies (Reyment *et al.* 1984) provide a powerful and efficient approach to resolving patterns of geographical variation, natural hybridization and sibling speciation (e.g., Campbell & Mahon 1974; Hopper 1977; Coates 1981). A genetic basis to such patterns may be reasonably inferred where consistent morphometric differences are seen between populations in sympatry and/or throughout their geographical ranges.

Morphometric studies are less time-consuming per population sampled than biochemical studies and thus enable better sampling of natural populations. However, the underlying genetic basis of morphometric attributes is often polygenic and requires elaborate breeding programmes to be elucidated. Where such knowledge is required, the simpler inheritance of most isozymes offers a preferable field of study (see below).

Over the past decade we have witnessed a transition from the use of morphometric techniques predominantly on large, expensive main-frame computers to ready availability on modestly priced personal computers. This improvement in accessibility, combined with advances in direct measurement by computer data capture, suggest that multivariate morphometric studies will play an increasingly important role in mapping the genetic resources of those Australian species amenable to such analysis.

Chemosystematics. Cryptic variation in morphologically uniform species may be revealed through assays for secondary compounds such as flavonoids or terpenoids. The occurrence of resinous compounds on the leaves and stems of some native plants has proved useful in studies of geographic variation (e.g., Dell 1975). However, as with morphometric attributes, the underlying genetic basis for secondary compounds may be complex and require sophisticated breeding studies to document.

Chromosome markers. Some groups of native plants, reptiles, native rats, marsupials, insects and spiders display geographic variation in chromosome number or structure that has been mapped by cytogeneticists (e.g., James 1965; Shaw 1976; Baverstock *et al.* 1977; Coates & James 1979; King 1979, 1983; White 1980; Briscoe *et al.* 1982; John & King 1983; Moritz 1984; Rowell 1985; Shaw *et al.* 1988). Such variation can be of particular significance in assessing the genetic resources of a species and in adequately conserving and managing that resource.

In some species complexes, chromosome differences may indicate adaptive differences and/or reproductive isolation of varying completeness between populations. Coates and James' (1979) study of chromosome variation in the triggerplant *Stylidium crossocephalum* illustrates this kind of population differentiation (Fig. 1). Such information would need to be considered carefully in any genetic conservation or management strategies, particularly with regard to translocation or re-establishment programmes (James 1982). Clearly, it would be undesirable to mix individuals of one chromosome variant with those of another if the resultant progeny were likely to have reduced viability or fecundity.

Studies documenting chromosome variation within species of lizards (King 1979, 1983; Moritz 1984), native rats (Baverstock *et al.* 1977), rock wallabies (Briscoe *et al.* 1982) and grasshoppers (Shaw 1976) highlight the need for caution in translocation programmes in many groups of fauna as well.

The data from these studies may also provide useful guides to reserve system design. For example, in *Isotoma petraea* (James 1965, 1970), as many as possible of the chromosomally complex heterozygous populations in the south-western part of the range should be conserved to maximize representation of chromosomally marked genetic resources. In contrast, a smaller selection of the bivalent-forming structurally homozygous populations ranging across arid Australia would need to be represented in reserves if chromosomal variation was the primary focus for conservation [a consideration of isozyme variation suggests that many bivalent-forming populations nevertheless should be reserved despite their chromosomal uniformity to protect other types of detectable genetic variation (James *et al.* 1983)].

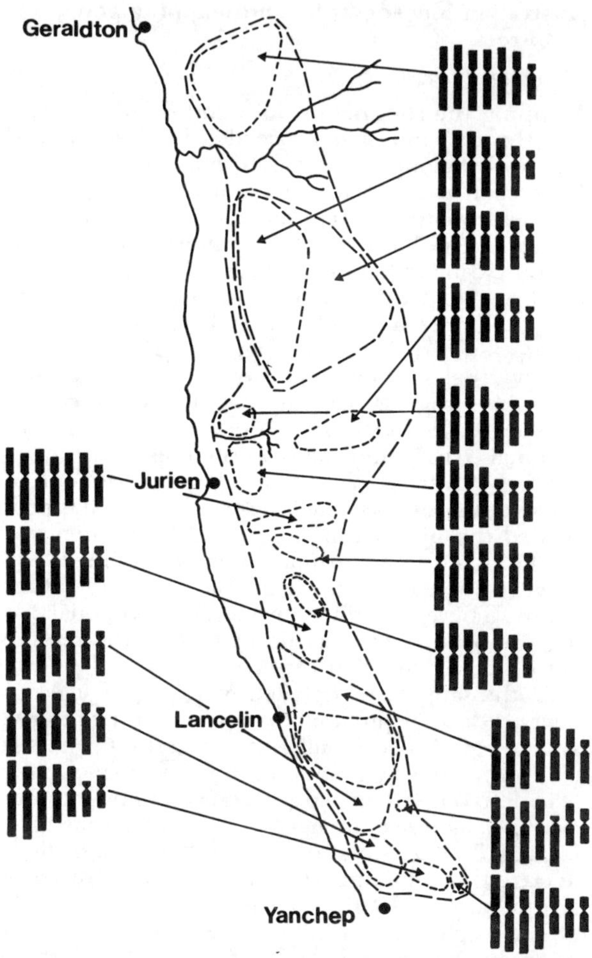

FIG. 1. Distribution in Western Australia of chromosomal variants in the triggerplant *Stylidium crossocephalum* (after Coates & James 1979).

King's (1979, 1983) detailed documentation of chromosomal variation in species complexes of the gekko genus *Gehyra* provides a similar example of the usefulness of cytogenetic studies. It is now clear that dramatically different reserve systems are needed to ensure representativeness of variation in these lizards than would have been predicted on the basis of earlier information.

Isozyme markers. Isozymes currently provide the most efficiently sampled markers for mapping the geographical pattern of genetic variation in wild populations (Lewontin 1974; Brown & Moran 1981; Hamrick 1983). Isozymes, however, provide only a sample of each organism's genetic makeup, so that a lack of isozyme variation in populations need not mean that no genetic variation exists. With this limitation in mind, isozymes nevertheless provide a powerful tool for genetic analysis.

Geographical patterns of genetic variation in populations may be investigated at two levels. The first is to document geographical patterns in genetic differentiation or genetic diversity. These patterns may show a relationship with geographical distance, different biogeographic regions or with marginal versus central areas of a species' distribution. The second level is to investigate associations between geographically varying environmental parameters and genetic variation or specific gene loci.

With regard to correlations between geographic distance and genetic distance, the present evidence is equivocal. Initial indications were that geographic proximity usually is a poor indicator of genetic similarity (Nevo 1978; Levin 1978; Schwaegerle & Schaal 1979; Schaal & Smith 1980; Glover & Barrett 1987). More recently, however, significant positive correlations between geographic and genetic distance were found in several studies (Guries & Ledig 1982; Wendel & Parks 1985), including work on the Western Australian granite rock mallees *Eucalyptus caesia* and *E. crucis* (Moran & Hopper 1983; Sampson *et al.* 1988). These alternative patterns were found within the same species *(Pinus radiata)* in different parts of its distribution by Plessas and Strauss (1986).

Comparisons of levels of genetic variation between geographically marginal populations and central populations have also provided varying results. Theory predicts that marginal populations should exhibit reduced variability but empirical studies indicate that they can be either less, equally or more heterozygous than central populations. Perhaps this is not surprising given the difficulty in distinguishing between what is ecologically marginal and what is just geographically peripheral (Lewontin 1974; Guries & Ledig 1982; Glover & Barrett 1987).

Patterns of isozyme variation in plants are correlated with various life history attributes, such as breeding system, longevity, etc. (Hamrick *et al.* 1979; Loveless & Hamrick 1984). This suggests that it may be possible to derive general principles about genetic variation from a sample of organisms with similar life histories.

Such an approach was investigated by Moran and Hopper (1987) in relation to the conservation of genetic resources in rare and widespread eucalypts. They found that most variation existed within, rather than between, populations in widespread eucalypts. However, the reverse held for *Eucalyptus caesia,* a regionally distributed species occurring in small isolated populations on granite outcrops in the Western Australian wheatbelt (Moran & Hopper 1983). A similar pattern to that in *E. caesia* has been reported recently by Sampson *et al.* (1988) for another granite rock endemic, *E. crucis.*

In some organisms such as marsupials (Baverstock, pers. comm.), isozyme variation is limited within species and provides little data of use to mapping genetic resources. Current studies of mitochondrial DNA variation (Moritz *et al.* 1987) have been successful in documenting genetic differentiation of populations within species of lizards (Moritz & Brown pers. comm.) and fish (Ovenden & White pers. comm). In the future, assaying mitochondrial and other types of DNA may well replace isozyme studies as the most efficient end direct way to sample genetic variation within species (Moritz *et al.* 1987).

Essential genetic management of flora and fauna

Some theoretical considerations

A genetic approach to the management of flora and fauna in the wild is a new field of activity. Much of what we propose below, therefore, is of necessity speculative and needs critical evaluation in future empirical studies.

Firstly, as a fundamental premise, we assume that a primary aim of genetic resource management is the maintenance of genetic diversity.

There is little doubt that the rate of evolution is influenced by the level of genetic variation. Experimental evidence indicates a positive correlation between the amount of heterozygosity and the rate of evolutionary change (Ayala 1965, 1988). Furthermore, a considerable body of knowledge indicates that organisms with high levels of heterozygosity and open recombination systems capable of generating genetic variability are those that constitute evolutionary lineages with the best long-term future (Darlington 1958; James & Hopper 1981).

The selective value of heterozygosity as a fundamental phenomenon in evolution was recognized by Darwin (1859), developed as a concept by Darlington (1958), and empirically explored in Australia at considerable length by James (1981; James & Hopper 1981) and others. The evolution of many unusual breeding systems and cytogenetic attributes may be ascribed to natural selection favouring mechanisms that conserve genetic variation or heterozygosity in the face of enforced inbreeding or harsh environmental circumstances.

If heterozygosity usually is of selective value, it should be possible to demonstrate that heterozygous individuals have higher fitness than homozygous individuals. Such has been the case in many studies (e.g., reviews by Grant 1975; Mitton 1978; Mitton & Grant 1984). An elegant investigation of this hypothesis in a cytogenetically complex situation in *Isotoma petraea* was provided by Beltran and James (1974).

Strong selection for heterozygosity through the life cycle is suggested in many studies of eucalypts and other plant taxa, where heterozygosity in isozyme alleles increases from the seed to the adult stages (e.g., Moran & Bell 1983; Ledig 1986).

Furthermore, individual plants of rare eucalypts that persist in small remnant populations are often highly heterozygous (Moran & Hopper 1987).

On an evolutionary time scale, the high incidence of allopolyploidy and hybrid speciation in plants (Grant 1981) and some animal groups (White 1978) provides additional evidence that selective forces may favour heterozygosity.

To summarise, then, the maintenance of genetic diversity and heterozygosity in natural populations appears to provide the best general strategy for facilitating evolutionary flexibility and persistence for most organisms. However, there are situations where natural selection may favour homozygosity. Such special circumstances would call for different genetic management strategies.

Secondly, the concept of minimum viable population size (MVP) warrants brief discussion because it is a principal issue with regard to both long-term and short-term survival of a species. The concept proposes that there is a minimum effective population size for each organism needed to guarantee a high probability of survival (Soulé 1987).

Understanding the MVP problem is considered to be one of the most important issues in conservation biology (e.g., Lacy 1988). It requires a basic knowledge of the effects of reduction in population size, increased population isolation, inbreeding and the importance of stochastic and deterministic forces in maintaining levels of genetic variation within populations.

Moreover, it is clear that MVP may be influenced by demographic as well as genetic processes (Ewens *et al.* 1987). For example, a single catastrophe may destroy a population irrespective of its genetic makeup, level of inbreeding, etc.

Population genetic theory predicts that large populations should maintain higher levels of genetic variation than small populations because they are less susceptible to loss of variability from stochastic processes. Theory also predicts that the greater the isolation between populations the greater the level of between population variation and that this divergence will be more rapid if the populations are small.

The effects of stochastic processes on genetic variation in populations can be most readily observed when populations are recovering from a bottleneck or new populations are founded by very few individuals. Predictions are a reduction in average heterozygosity and loss of rare alleles although both of these effects depend very much on the size of the bottleneck and the rate of population recovery (see Nei *et al.* 1975).

Although theoretical expectations concerning the effects of population size, bottlenecks and population isolation on levels of genetic variation have been adequately addressed, empirical data on plants and animals remain scarce. A range of situations exist in those cases investigated, perhaps reflecting the diversity of available genetic systems, some of which may conserve heterozygosity, others being neutral, while others favour homozygosity. The eucalypts, for instance, provide contrasting evidence concerning the relationship between population size and levels of genetic variability. In *Eucalyptus pendens*, a localized species with disjunct populations, Moran and Hopper (1987) found no relationship between population size and any single locus diversity measure. For example, the smallest population sampled (North Badgingarra, 27 plants) had higher mean number of alleles per locus, average percentage of polymorphic loci, and observed and expected heterozygosity than one of the largest populations (Coonawarra Downs, 1 000 plants). Similarly in *E. caesia* there was no correlation (Moran & Hopper 1983), whereas in *E. crucis*, a species with a comparable disjunct population structure, Sampson *et al.* (1988) did find a significant correlation between heterozygosity and population size.

Whether small populations can maintain adequate levels of genetic variability depends on the genetic system of the species and the history of the populations. Some species seem to have the capacity to exist in small isolated populations and yet maintain relatively high levels of genetic variability (Moran & Hopper 1987; Coates 1988). It would appear that in these cases selection favouring heterozygosity may be a major factor.

Available evidence suggests that, on genetic criteria, a universally applicable "magic" number of individuals required to constitute a minimum viable population does not exist. Empirical studies of genetic variation and population structure in a diverse range of organisms are needed to ascertain if general guidelines can be established for MVP. In the meantime, we would urge conservation managers to err on the side of caution and to protect populations no matter how small until appropriate genetic and demographic studies can be undertaken.

Genetic management strategies

Where the genetic resources of a species have been adequately documented, a number of management strategies can be implemented to maintain genetic diversity. These strategies include selection of priority populations for conservation, reserve and corridor design, forest tree and fisheries management, and active management of species through propagation, captive breeding and translocation in the wild.

Selection of priority populations for conservation. Managers of wildlife species that are endangered or otherwise in need of special protection are faced often with insufficient resources to look after all known populations. A choice has to be made on the allocation of scarce resources in such circumstances (e.g., for fencing habitat, purchasing habitat, conducting research, etc.). While ecological and political circumstances often dictate the allocation of management priorities, genetic data deserve consideration from a number of perspectives.

Selection of priority populations for genetic conservation can be based on the level of genetic differentiation between populations, the presence of rare but locally common alleles (Brown 1978), and the level of genetic diversity within each population. If we consider just the level of differentiation between populations, the dendrograms from isozyme data that cluster genetic distance values for population divergence in the three species *Eucalyptus caesia* (Moran & Hopper 1983), karri, *Eucalyptus diversicolor,* (Coates & Sokolowski 1989) and *Banksia cuneata* are particularly informative (Fig. 2).

Eucalyptus caesia is characterized by relatively large genetic distance values between populations, while in comparison both karri and *Banksia cuneata* show much lower levels of interpopulation differentiation.

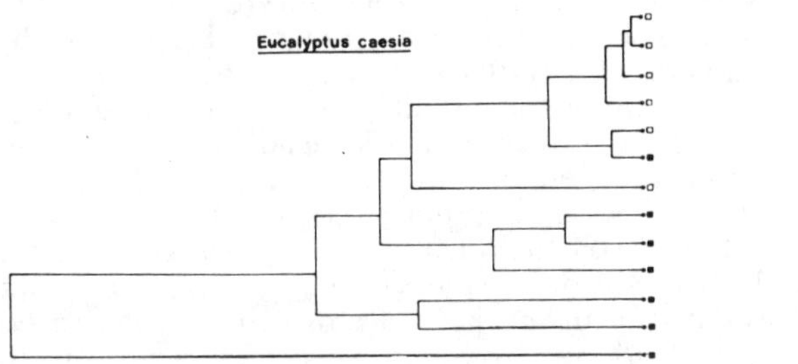

FIG. 2. Dendrograms resulting from cluster analyses based on Nei's genetic distance between 13 populations of *Eucalyptus caesia* subsp. *caesia* (□) and subsp. *magna* (■) (after Moran & Hopper 1983), 13 populations of karri *(E. diversicolor)*, and seven populations of *Banksia cuneata*.

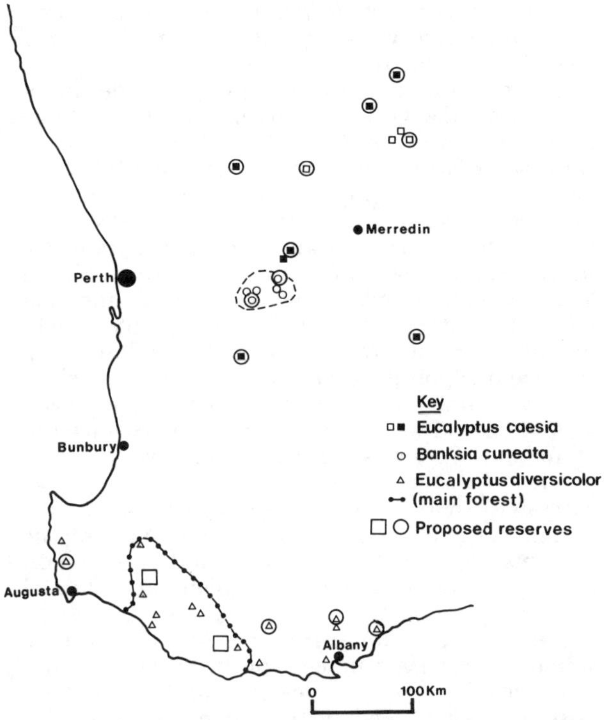

FIG. 3. Distributions and populations used in isozyme studies on *Eucalyptus caesia* subsp. *caesia* (■) and subsp. *magna* (□), karri *(Eucalyptus diversicolor)* and *Banksia cuneata,* and a minimum proposed reserve system for genetic resource conservation in these species. Large reserves are indicated by the open squares.

The dendrograms also indicate that in karri, certain outliers such as the Porongorup populations are noticeably divergent from most other populations while *Banksia cuneata* is characterized by two distinct population groups.

Based on these data, strategies for the conservation of genetic resources in these three species will be quite different (Fig. 3). In *E. caesia* many populations covering its geographic range would need to be protected, while in karri a few large reserves in the main karri belt plus protection of outliers in the Porongorups, Rocky Gully and Leeuwin-Naturaliste block areas would be desirable. In *Banksia cuneata* a minimum of one large population from each of the two population groups would probably ensure adequate protection of genetic resources in this species. Although the information provided in these dendrograms is only a part of the total population data available from isozyme studies it does indicate their value when developing wildlife management programmes.

The work of Johnson and Black (1984) and Johnson (pers. comm.) on genetic resources in marine organisms may be similarly applied to marine conservation programmes.

Reserve and corridor design. Three basic problems concerning reserve design are size, number and spatial arrangement.

There is little doubt that genetic considerations are one of the key concerns in addressing these problems (Soulé & Simberloff 1986). A related concern is the need to determine MVP from a genetic resources point of view. Several factors need to be considered with respect to MVP and reserve design. For instance, the population will need to be sufficiently large to ensure that existing levels of heterozygosity are maintained, and that inbreeding does not reach unacceptable levels. Further, as indicated previously, the long-term survival of a population requires adequate levels of genetic variation, although what these levels should be and what population size is required to maintain these levels are largely unknown for most species. In spite of these difficulties we have already seen that general strategies for reserve design can be obtained from studies on geographic patterns of genetic variation and population genetic structure within both plant and animal species.

The incorporation of corridors in a reserve system has clear advantages, particularly if the aim is to maintain genetic diversity and reduce the possibility of inbreeding by enhancing gene flow and dispersal. Although they may have such benefits, apart from the political and economic difficulties associated with developing a reserve system with corridors, there are other difficulties where corridors could negate the quarantine advantage found in a system of isolated reserves (Soulé & Simberloff 1986).

Where corridors are impractical or do not exist, alternative management strategies are possible to counteract localized extinctions and a lack of gene flow. These strategies may involve translocation of individuals and founding of new populations (see below).

Forest tree and fisheries management. Australian commercial forestry has a strong reliance on wild-growing native hardwoods and so genetic research on wild populations has broken new ground in several key areas. Firstly, provenance trials aimed at establishing the best populations and mother trees for seed for growing at particular locations have been completed on several species (e.g., Green 1971). Secondly, the use of isozymes to investigate mating systems in eucalypts with a view to improving seed orchards was pioneered by A. H. D. Brown and colleagues (Brown *et al.* 1975; Phillips & Brown 1977).

Isozyme surveys of genetic variation in commercially important timber species have proven useful in a number of cases. For example, data on karri, some of which are referred to in the previous sections (Fig. 2), have already indicated priority populations or areas from which seed collections should be made to ensure that any germplasm storage programme will maximize the conservation of genetic resources in this species. These data also show that all major allozyme variants, apart from those in the Rocky Gully population, are adequately represented in conservation reserves and management priority areas. This of course does not necessarily mean that all the genetic variants in karri are adequately reserved. Within the limits of the isozyme technique, however, it is a very useful start.

A knowledge of genetic variation also aids the manager in terms of replanting strategies and in designing seed orchards so as to maximize levels of heterozygosity (Moran pers. comm.).

The application of genetic resource studies to commercial fisheries is a recent development. Pioneering work has been useful in demonstrating the existence of subpopulations or sibling species that may require separate management to maintain sustained yields. For example, Macdonald (1980) showed that the Shark Bay population of the Australian snapper, *Chrysophrys auratus,* had a number of rare alleles not found elsewhere across southern Australia. Further differentiation within the Shark Bay area was shown subsequently by Johnson *et al.* (1986). A second distinctive stock occurred in Spencer Gulf. Depletion of these stocks through overfishing could lead to the loss of significant genetic variation that could not be replaced by recruitment from elsewhere.

By way of contrast, Richardson (1984) found no electrophoretic evidence of population substructuring in the jackass morwong, *Cheilodactylus macropterus,* in south-eastern coastal waters from western Victoria to southern New South Wales and down the east coast of Tasmania. This suggests that heavy local overfishing might be offset by recruitment from elsewhere but Richardson cautioned against management along these lines in view of the limited sampling achieved in his study.

Cultivation, captive breeding and re-establishment or translocation. Intensive genetic management has yet to be carried out on populations of any endangered native Australian plant or animal, although it is an option which may need to be considered in the future if extinction in the wild is to be prevented. This form of management will probably require captive breeding of animals or cultivation of plants followed by their re-introduction or translocation into areas where they were previously known to exist. Alternatively, it may more simply involve translocation of some individuals from one population to another with the aim of enhancing genetic diversity within the recipient population.

Whatever the form of management, it is quite clear that detailed population genetic, breeding system and population ecological studies should be carried out on the species concerned before the programme was initiated. For example, it would be futile to transplant or relocate individuals of species such as *Isotoma petraea, Stylidium crossocephalum,*

Caledia captiva or *Heterontia binoei* without first determining their ability to interbreed and produce viable and fecund offspring. This problem has been addressed by Templeton (1986) in his concept of outbreeding depression.

In the context of this volume, there needs to be some caution in the revegetation of degraded lands if the aim is to re-establish viable breeding populations. In the Western Australian heathlands, for example, remarkable levels of karyotypic and genetic divergence within morphologically uniform widespread species have been documented (e.g., Coates & James 1979; Fig. 1). In such circumstances, it is essential that local seed stocks from as close to the area disturbed as possible be used to revegetate mine sites, road verges, etc. In contrast, species such as karri (Fig. 2) show little genetic differentiation over considerable distances in the main forest and little risk of outbreeding depression would be likely from mixing seed from different geographic sites within the main forest belt. Rather, such management may yield a slight increase in levels of heterozygosity and a more vigorous karri forest.

The difference in likely outcomes of translocation management of species such as *Stylidium crossocephalum* versus karri indicates that a knowledge of population structure, chromosomal and genetic differentiation, and reproductive biology should become an integral part of the expertise of tomorrow's environmental rehabilitation manager.

Philosophical, scientific and economic arguments have been used for and against the intensive management of a species to prevent its extinction in the wild. Some argue that this type of management cannot generally be justified because chances of successful re-introduction are always problematical. Further, they point out that captive breeding or cultivation followed by re-introduction is expensive and that funds going into such programmes are likely to be directed from other programmes which may be more effective in preventing extinction. Although both these points have some merit, the alternative of *ex situ* conservation in botanic gardens or zoological gardens is also plagued by economic difficulties and is likely to be far less efficient in terms of genetic resource conservation. In addition, evolutionary theory predicts that most genetic variation (with the exception of rare variants) can be saved from relatively few individuals if group size is increased rapidly (Nei *et al.* 1975). This suggests that maintaining or restoring adequate levels of genetic variation in a wild population to ensure long-term evolution is a feasible option.

Conclusion

The conservation of genetic resources in Australia's flora and fauna will, of necessity, depend upon the injection of new life in key areas of scientific enquiry. We have highlighted the importance of taxonomic work in defining the biological units which we wish to conserve. We have also noted that a population genetic approach is useful and often essential in resolving complex patterns of variation between and within species, in reserve design for species whose genetic resources are considered important, and in understanding and managing the reproductive biology of such species.

Given the pace of continuing destruction of native vegetation, the search for general principles in gene resource conservation from a limited sample of the Australian flora and fauna is a key scientific challenge facing workers over the next century. The identification of functionally equivalent groups according to their life history, ecology, breeding system and population structure may well enable predictions about genetic resources to be made (e.g., Loveless & Hamrick 1984; Moran & Hopper 1987). We commend this approach to future workers.

Acknowledgements

We are grateful to many colleagues for stimulating debate and collaborative work over the past two decades. A number of ideas presented above had their inception in collaborative studies of SDH and Dr Gavin Moran, to whom we are especially grateful. Useful comments on the manuscript were provided by R. Wyatt, I. Abbott, referees and the senior editor. We thank Jan Imms for assistance with typing, and R. E. S. Sokolowski for preparing the figures.

References

Adams M., Baverstock P. R., Watts C. H. S. & Reardon T. (1987) Electrophoretic resolution of species boundaries in Australian Microptera. I. *Eptesicus* (Chiroptera: Vespertilionidae). *Aust. J. Biol. Sci.* **40,** 143-62.

Anon. (1984) A National Conservation Strategy for Australia. Proposed by a conference held in Canberra in June 1983. 24 pp. Australian Government Publishing Service, Canberra.

Ayala F. J. (1965) Relative fitness of populations of *Drosophila serrata* and *Drosophila birchii. Genetics* **51,** 527-44.

Ayala F. J. (1988) Genotype, environment and population numbers. *Science* **162,** 1453-4.

Barlow B. (1982) Cytogenetic systems in Australian arid zone plants. In: *Evolution of the Flora and Fauna of Arid Australia* (eds W. R. Barker & P. J. M. Greenslade) pp. 161-6. Peacock Publications, Adelaide.

Baverstock P. R., Watts C. H. S. & Hogarth J. T. (1977) Chromosome evolution in Australian rodents. *Chromosoma* **61,** 95-125.

Blakers M., Davies S. J. J. F. & Reilly P. N. (1984) *The Atlas of Australian Birds.* Melbourne University Press, Melbourne.

Beard J. S. (1981) The vegetation of Western Australia at the 1:3,000,000 scale (with map). Forests Department, Perth.

Beltran I. C. & James S. H. (1974) Complex hybridity in *Isotoma petraea* IV. Heterosis in interpopulational heterozygotes. *Aust. J. Bot.* **22,** 251-64.

Briggs J. D. & Leigh J. H. (1988) Rare or Threatened Australian plants. Australian National Parks and Wildlife Service Special Publication No. **14,** 278.

Briscoe D. A., Calaby J. H., Close R. L. Maynes C. M., Mutagh C. E. & Sharman, G. B. (1982) Isolation, introgression and genetic variation in rock-wallabies. In: *Species at Risk: Research in Australia* (eds R. H. Groves & W. D. L. Ride) pp. 73-88. Australian Academy of Science, Canberra.

Brown A. H. D. (1978) Isozymes, plant population genetic structure and genetic conservation. *Theoret. Appl. Genet.* **52,** 145-57.

Brown A. H. D. & Moran G. F. (1981) Isozymes and the genetic resources of forest trees. In: *Isozymes Plant Genetics and Breeding* (eds S. D. Tanksley & T. J. Orton) pp. 219-39. Elsevier, Amsterdam.

Brown A. H. D., Matheson A. C. & Eldridge K. G. (1975) Estimation of the mating system of *Eucalyptus obliqua* L'Herit. by allozyme polymorphism. *Aust. J. Bot.* **23,** 931-49.

Campbell N. A. & Mahon R. J. (1974) A multivariate study of variation in two species of rock crab of the genus *Leptograpsus. Aust. J. Zool.* **22,** 417-25.

Coates D. J. (1981) Chromosome, morphometric and breeding system studies in the *Stylidium caricifolium* species complex (Stylidiaceae). *Aust. J. Bot.* **29,** 397-417.

Coates D. J. (1988) Genetic diversity and population genetic structure in the rare Chittering grass wattle *Acacia anomala. Aust. J. Bot.* **36,** 273-86.

Coates D. J. & James S. H. (1979) Chromosome variation in *Stylidium crossocephalum.* F. Muell. (Angiospermae: Stylidiaceae) and the dynamic co-adaption of its lethal system. *Chromosoma* **72,** 357-76.

Coates D. J. & Sokolowski R. E. (1989) Geographic patterns of genetic diversity in karri (*Eucalyptus diversicolor* F. Muell.) *Aust. J. Bot.* **37,** 145-56.

Darlington C. D. (1958) *Evolution of Genetic Systems.* Oliver and Boyd, London.

Darwin C. (1859) *The Origin of Species.* John Murray, London.

Dell B. (1975) Geographical differences in leaf resin components of *Eremophila fraseri* F. Muell. (Myoporaceae). *Aust. J. Bot.* **23,** 889-97.

Ewens W. J., Brockwell P. J., Gani J. M. & Resnick S. I. (1987) Minimum viable population size in the presence of catastrophes. In: *Viable Populations for Conservation* (ed. M. Soulé) pp. 59-68. Cambridge University Press, Cambridge.

George A. S. (1981) The Genus *Banksia* L.f. (Proteaceae), *Nuytsia* **3,** 239-473.

Glover D. E. & Barrett S. C. H. (1987) Genetic variation in continental and island populations of *Eichornia paniculata* (Pontederiaceae). *Heredity* **59,** 7-17.

Grant V. (1975) *Genetics of Flowering Plants.* Columbia University Press, New York.

Grant V. (1981) *Plant Speciation.* 2nd ed. Columbia University Press, New York.

Green J. W. (1971) Variation in *Eucalyptus obliqua* L. Herit. *New Phytol.* **70,** 897-909.

Griffin A. R., Williams E. R. & Johnson K. W. (1982) Early height growth and frost hardiness of *Eucalyptus regnans* provenances in twelve field trials in south-east Australia. *Aust. For. Res.* **12,** 263-79.

Guries R. P. & Ledig F. T. (1982) Genetic diversity and population structure in pitch pine (*Pinus rigida* Mill). *Evolution* **29,** 213-25.

Hamrick J. L. (1983) The distribution of genetic variation within and among natural plant populations. In: *Genetics and Conservation* (eds C. M. Schonewald-Cox, S. M. Chambers, B. MacBryde & L. Thomas) pp. 343-8. Benjamin-Cummings, London.

Hamrick J. L., Linhart Y. B. & Mitton J. B. (1979) Relationships between life history characteristics and electrophoretically detectable genetic variation in plants. *Ann. Rev. Ecol. Syst.* **10,** 173-200.

Hartley W. & Leigh J. H. (1979) Plants at Risk in Australia. Australian National Parks and Wildlife Service Occasional Paper No. 3 pp. 80.

Hopper S. D. (1977) The structure and dynamics of a hybrid population of *Anigozanthos manglesii* D. Don and *A. humilis* Lindl. (Haemodoraceae). *Aust. J. Bot.* **25,** 413-22.

Hopper S. D., van Leeuwen S., Brown A. P. & Patrick S. J. (1990) Western Australia's endangered flora and other plants under consideration for declaration. Department of Conservation and Land Management, Perth (in press).

Howard H. T. C. & Howden M. E. H. (1975) Antitumor activity in *Pimelea simplex. Cancer Chemotherapy Repts* **59,** 585-6.

James S. H. (1965) Complex hybridity in *Isotoma petraea* I. The occurrence of interchange heterozygosity, autogamy and a balanced lethal system. *Heredity* **20,** 341-53.

James S. H. (1970) Complex hybridity in *Isotoma petraea* II. Components and operation of a possible evolutionary mechanism. *Heredity* **25,** 53-77.

James S. H. (1981) Cytevolutionary patterns, genetic systems and the phytogeography of Australia In: *Ecological Biogeography of Australia* (ed. A. Keast) pp. 763-82. Junk, The Hague.

James S. H. (1982) The relevance of genetic systems in *Isotoma petraea* to conservation practice In: *Species at Risk: Research in Australia* (ed. R. H. Groves & W. D. L. Ride) pp. 63-71. Australian Academy of Science, Canberra.

James S. H. & Hopper S. D. (1981) Speciation in the Australian Flora. In: *The Biology of Australian Plants* (eds J. S. Pate & A. J. McComb) pp. 361-81. University of Western Australia Press, Perth.

James S. H., Wylie A. P., Johnson M. S., Carstairs S. A. & Simpson G. A. (1983) Complex hybridity in *Isotoma petraea* V. Allozyme variation and the pursuit of hybridity. *Heredity* **51,** 653-63.

John B. & King M. (1983) Population cytogenetics of *Atractomorpha similis.* I. C-band variation. *Chromosoma* **88,** 57-68.

Johnson M. S. & Black R. (1984) Pattern beneath chaos: the effect of recruitment on genetic patchiness in an intertidal limpet. *Evolution* **38,** 1371-83.

Johnson M. S., Creagh S. & Moran M. (1986) Genetic subdivision of stocks of snapper *Chrysophrys unicolor,* in Shark Bay, Western Australia. *Aust. J. Mar. Freshw. Res.* **37,** 337-45.

King M. (1979) Karyotypic evolution in Gehyra (Gekkonidae:Reptilia; I. The *Gehyra variegata-punctata* complex. *Aust. J. Zool.* **27,** 373-93.

King M. (1983) Karyotypic evolution in Gehyra (Gekkonidae:Reptilia) III. The *Gehyra australis* complex. *Aust. J. Zool.* **31,** 723-41.

Lacy R. C. (1988) A report on population genetics in conservation. *Conserv. Biol.* **2,** 245-7.

Ledig F. T. (1986) Heterozygosity, heterosis, and fitness in outbreeding plants. In: *Conservation Biology, the Science of Scarcity and Diversity* (ed. M. E. Soulé) pp. 77-104. Sinauer Associates, Sunderland, Massachusetts.

Leigh J., Briggs J. & Hartley W. (1981) Rare or Threatened Australian Plants. Australian National Parks and Wildlife Service Special Publication No. 7, 1-178.

Levin D. A. (1978) Genetic variation in annual Phlox: self compatible versus self-incompatible species. *Evolution* **32,** 245-63.

Lewontin R. C. (1974) *The Genetic Basis of Evolutionary Change.* Columbia University Press, New York.

Loveless M. D. & Hamrick J. L. (1984) Ecological determinants of genetic structure in plant populations. *Ann. Rev. Ecol. Syst.* **15,** 65-95.

MacDonald C. M. (1980) Population structure, biochemical adaptation and systematics in temperate marine fishes of the genera *Arripis* and *Chrysophrys* (Pisces: Perciformes). Ph.D. Thesis. Australian National University, Canberra.

McComb A. J. (1989) Presidential address 1988. After the first 200 years: the future of ecology and ecologists in Australia. *Aust. J. Ecol.* **14,** 1-11.

Mitton J. B. (1978) Relationship between heterozygosity for enzyme loci and variation of morphological characters in natural populations. *Nature* **273,** 661-62.

Mitton J. B. & Grant M. C. (1984) Associations among protein heterozygosity, growth rate and developmental homeotasis. *Ann. Rev. Ecol. Syst.* **15,** 479-9.

Moran G. F. & Bell J. C. (1983) Eucalyptus. In: *Isozymes in Plant Genetics and Breeding, Part B.* (eds S. D. Tanksley & T. J. Orton) pp. 423-42. Elsevier Science Publishers, Amsterdam.

Moran G. F. & Hopper S. D. (1983) Genetic diversity and insular population structure of the rare granite rock species *Eucalyptus caesia* Benth. *Aust. J. Bot.* **31,** 161-72.

Moran G. F. & Hopper S. D. (1987) Conservation of the genetic resources of rare and widespread eucalypts in remnant vegetation. In: *Nature Conservation: The Role of Remnants of Native Vegetation* (eds D. A. Saunders, G. W. Arnold, A. A. Burbidge & A. J. M. Hopkins) pp. 151-62. Surrey Bearry & Sons, Chipping Norton.

Moritz C. (1984) Parthenogenesis in the endemic Australian lizard, *Heteronotia binoei* (Gekkonidae). *Science* **220,** 735-77.

Moritz C., Dowling T. E. & Brown W. M. (1987) Evolution of animal mitochondrial DNA: relevance for population biology and systematics. *Ann. Rev. Ecol. Syst.* **18,** 269-93.

Myers K., Margules C. R. & Musto E. (eds) (1984) *Survey Methods for Nature Conservation.* CSIRO Division of Water and Land Resources, Canberra.

Nei M., Maruyama T. & Chakrabarty R. (1975) The bottleneck effect and genetic variability in populations. *Evolution* **29,** 1-10.

Nevo E. (1978) Genetic variation in natural populations: patterns and theory. *Theoret. Pop. Biol.* **13,** 121-77.

Paterson H. E. & James S. H. (1973) Animal and plant speciation studies in Western Australia. *J. Roy. Soc. West. Aust.* **56,** 31-43.

Plessas M. E. & Strauss S. H. (1986) Allozyme differentiation among populations, stands and cohorts in Monterey pine. *Can. J. For. Res.* **16,** 1155-64.

Phillips M. A. & Brown A. H. D. (1977) Mating system and hybridity in *Eucalyptus pauciflora. Aust. J. Biol. Sci.* **30,** 337-44.

Potts B. M. & Reid J. B. (1985) Analysis of a hybrid swarm between *E. risdonii* Hook. f. and *E. amygdalina* Labill. *Aust. J. Bot.* **33,** 543-62.

Reyment R. A., Blackith R. E. & Campbell N. A. (1984) *Multivariate Morphometrics.* 2nd ed. Academic Press, London.

Richardson B. J. (1984) Identifying the Australian fauna: What remains to be done? *Search* **14,** 320-3.

Richardson B. J. (1987) The animal kingdom in Australia. In: *Fauna of Australia.* General Articles. Vol. 1A (eds G. R. Dyne & D. W. Walton) pp. 294-99. Australian Government Publishing Service, Canberra.

Rowell D. M. (1985) Complex sex-linked fusion heterozygosity in the Australian huntsman spider *Delena cancerides* (Aranae:Sparassidae). *Chromosoma* **93,** 169-76.

Sampson J. F., Hopper S. D. & James S. H. (1988) Genetic diversity and the conservation of *Eucalyptus crucis.* Maiden. *Aust. J. Bot.* **36,** 447-60.

Sattler P. S. (1986) Nature conservation in Queensland: planning the matrix. *Proc. R. Soc. Qld.* **97,** 1-21.

Schaal B. A. & Smith W. G. (1980) The apportionment of genetic variation within and among populations of *Desmodium nudiflorum. Evolution* **34,** 214-21.

Schwaegerle K. E. & Schaal B. A. (1979) Genetic variability and founder effect in the pitcher plant *Sarracenia purpurea* L. *Evolution* **36,** 361-70.

Shaw D. D. (1976) Population cytogenetics of the genus *Caledia* (Orthoptera:Acrididae) I Inter and intra specific diversity. *Chromosoma* **54,** 221-43.

Shaw D. D. (1981) Chromosomal hybrid zones in orthopteroid insects. In: *Evolution and Speciation: Essays in Honour of M. J. D. White* (eds W. R. Atchley & D. W. Woodruff) pp. 146-70. Cambridge University Press, Cambridge.

Shaw D. D., Coates D. J. & Arnold M. L. (1988) Complex patterns of chromosomal variation along a latitudinal cline in the grasshopper *Caledia captiva. Genome* **30,** 108-18.

Smith-White S. (1959) Cytological evolution in the Australian flora. *Cold Spring Harb. Symp. Quant. Biol.* **24,** 273-89.

Soulé M. E. (1983) What do we really know about extinction. In: *Genetics and Conservation* (eds C. M. Schonewald-Cox, S. M. Chambers, B. MacBryde & L. Thomas) pp. 111-24. Benjamin-Cummings, London.

Soulé M. E. (1987) Introduction. In: *Viable Populations for Conservation* (ed. M. E. Soulé) pp. 1-10. Cambridge University Press, Cambridge.

Soulé M. E. & Simberloff D. (1986) What do genetics and ecology tell us about the design of nature reserves? *Biol. Conserv.* **35,** 19-40.

Specht R. L., Roe E. M. & Boughton V. H. (1974) Conservation of Major Plant Communities in Australia and Papua New Guinea. *Aust. J. Bot. Suppl.* 5.

Taylor A. & Hopper S. (1988) *The Banksia Atlas.* Australian flora and fauna series number 8. Australian Government Publishing Service, Canberra.

Templeton A. R. (1986) Co-adaptation and outbreeding depression. In: *Conservation Biology: The Science of Scarcity and Diversity* (ed. M. E. Soulé) pp. 105-16. Sinauer Associates, Massachusetts.

Wendel J. F. & Parks C. R. (1985) Genetic diversity and population structure in *Camellia japonica* L. (Theaceae). *Amer. J. Bot.* **72,** 52-65.

White M. J. D. (1978) *Modes of Speciation.* W. H. Freeman, San Francisco.

White M. J. D. (1980) The genetic system of the parthenogenetic grasshopper *Warramaba virgo.* In: *Insect Cytogenetics, Xth Symposium of Royal Entomological Society of London* (eds R. L. Blackman, M. Ashburner & C. M. Hewitt). Blackwell, Oxford.

Woodruff D. S. (1981) Towards a genodynamics of hybrid zones: studies of Australian frogs and West Indian land snails. In: *Evolution and Speciation: Essays in Honour of M. J. D. White* (eds W. R. Atchley & D. W. Woodruff) pp. 171-97. Cambridge University Press, Cambridge.

Worsley Alumina Pty Ltd (1985) Worsley Alumina Project. Flora and Fauna Studies, Phase Two. pp. 348. Worsley Alumina.

Proc. Ecol. Soc. Aust. 1990 — 16:579-583

Managing Australia's environment: directions for the future

A. J. M. HOPKINS[1], R. A. HOW[2] and D. A. SAUNDERS[3]

[1]*Western Australian Department of Conservation and Land Management, P.O. Box 51, Wanneroo, W.A. 6065*

[2]*Western Australian Museum, Francis Street, Perth, W.A. 6000*

[3]*CSIRO Division of Wildlife and Ecology, LMB No. 4, PO Midland W.A. 6056*

Abstract

The evolutionary history of the Australian landscape has given rise to many unique features which require special management. Degredation has occurred in post-settlement times following disturbance where this management expertise has not existed or has not been provided. There is an urgent need to improve planning and management to halt the decline and to facilitate repair of damaged ecosystems. This chapter suggests ways ecologists can contribute to achieving the goal of sustainable use of our living, natural resources.

Introduction

The theme of the 1988 Symposium of the Ecological Society of Australia was chosen specifically to provide Australian ecologists with the opportunity to take stock of events and consequences of the past 200 years of European settlement and land use and to consider a course or courses of action for the future. All of the contributors were invited to frame their papers in this way and the programme as a whole was structured to facilitate constructive debate and discussion.

The programme was also designed to be comprehensive, touching on as many elements of the Australian environment as possible. These Proceedings reflect that comprehensive approach. Some conspicuous gaps in this volume were covered by papers presented at the Symposium but not submitted for publication.

A key aspect of the Symposium was the concluding plenary session where a number of resolutions were adopted by participants. The resolutions were framed in response to major issues identified throughout the Symposium. They were principally directed at governments and major research institutions and contained recommendations on how those bodies might better deal with the environmental (and related social) problems. A full account of the discussions and resolutions is given in Hopkins (1988).

This final chapter seeks to do three things. Initially it provides a brief précis of the historical events that have shaped the unique Australian landscape and its biota. Secondly, it provides a brief review of what we see as the major issues raised during the Symposium. Finally it draws together a range of those constructive proposals that were advanced including some of the new ideas contained in the revised written papers and those discussed during the plenary session.

The unique Australian landscape

All parts of the world have features that distinguish them from other regions. Australia has a particular palaeohistory that has given rise to the unique assemblage of biophysical features that we see today. It is worthwhile looking at aspects of that palaeohistory to set the scene for subsequent discussion.

Australia is the largest island on Earth. Together with Tasmania and many thousands of continental islands, it covers an area of 7.7M km^2. The isolation of Australia came relatively late in the overall process of plate tectonics: whereas India began to separate from Gondwanaland some 125M years ago (Powell *et al.* 1981), the rifting between Australia and Antarctica was probably not until about 60M years ago (Crook 1981; cf. Frakes *et al.* 1987), a considerable time after the appearance of flowering plants and mammals. It was not until the early

Miocene (25M years ago) that the Australian plate became connected to south-east Asia by a chain of volcanoes, subsequently permitting intermingling of the two biotas (Powell *et al.* 1981). The massive diversification of the plants and animals during the period of isolation gave rise to the progenitors of today's Australian flora and fauna with their many unique and characteristic taxa and features. In terms of the vertebrate fauna, this fascinating evolutionary process seems now to be unravelling in a rapid and spectacular way as a result of the recent discoveries at Riversleigh and the work of Michael Archer (see Williams 1989).

During the time of rifting, Australia moved through some 27 degrees of latitude from a position with its centroid at about 50-60 degrees S latitude to its present position (Crook 1981; Powell *et al.* 1981; Martin 1984; Frakes *et al.* 1987). There were major climatic changes associated with the crustal movements as oceanic circulation patterns changed as well as with global features such as the carbon balance.

Thus, for example, the eastern Gondwanan landmass was characterized by *Nothofagus,* suggesting the widespread existence of temperate rainforest communities. These moist forests persisted throughout much of Australia at least until about 5M years ago (Frakes *et al.* 1987; Archer in Williams 1989) although their distribution became discontinuous as grasslands and sclerophyll dominated vegetation types expanded in drier and more oligotrophic sites at least from the early Miocene.

Two climatic changes, of major significance for the biota of southern Australia particularly, occurred from the end of the Pliocene and into the Quaternary. There was a change in the season of rainfall in the south from summer to winter about 2.5M years ago (Bowler 1982). This was followed by a series (19 or 20) of major fluctuations from warm and wet to cold and dry. These changes must have had profound effects on the biota with the induced stresses giving rise to episodes of extinctions and speciation. The highly endemic flora of south-western Australia is thought to exemplify this process (Lamont *et al.* 1984).

The great antiquity of the Australian landscape also sets this continent apart from the rest of the world. There have been relatively few major episodes of mountain building, volcanism or glaciation. The age and stability, combined with intense weathering, has given rise to a senecent landscape, one that preserves elements that may have existed 300M years ago. The soils are predominantly skeletal, nutrient poor and very susceptible to erosion when the native vegetation is removed.

What we have in Australia, then, is a landscape that is predominantly old and highly weathered, covered by scattered remnants of a modified Gondwanan vegetation with its derived and more complex tropical moist forest types together with the more extensive Australian elements, the sclerophyllous forests, woodlands and shrublands and the grasslands. The vertebrate fauna is predominantly uniquely Australian but with some recently acquired Indo-Malaysian elements. The affinities of the invertebrate fauna are still largely unknown. This is the Australia with which the Aboriginal people came to terms. This is the Australia so severely impacted over the past 200 years by what we might term technological society (*sensu* Taylor, this volume).

Patterns of settlement of technological society in Australia

In January 1788 the first of the Anglo-Celtic settlers began to establish themselves on the Australian continent. This was really only the beginning of a process that we can, in retrospect, see as being of major significance. Individually, developments seemed small and their impacts were barely perceptible to the people involved at the time. We now have a perspective on a great range of small events that together have caused major changes to the Australian environment. As an example, the introduction of rabbits and foxes caused no immediate, conspicuous impacts but we now know that these species have been associated with considerable losses in middle weight range Australian mammals (Morton, this volume; Recher & Lim, this volume).

The actual utilization of Australian ecosystems for agricultural production began in a modest way around Sydney Cove (Benson & Howell, this volume) and subsequently from other epicentres (e.g., Saunders & Curry, this volume) and spread gradually across the continent, with major phases of expansion of dryland farming as late as the 1960s (Saunders & Curry, this volume). Superimposed on the areal expansion has been an intensification; often this has led to changes in land use (e.g., Murray-Murrumbidgee Irrigation Area) while in other areas the level of exploitation has modulated in response to available resources and demand (Pickard, this volume; Freidel *et al.*, this volume; Gardner & McIver, this volume). In the case of the developments around Sydney Cove, degradation was observed and reported quite early (Benson & Howell, this volume); in other cases there are no early records of impressions.

There has been a time lag in the manifestation of effects of many of the land use changes so that we are only now seeing many of the consequences of activities and decisions of 100 years ago (Pickard, this volume). Since the process of change has not finished, we can expect to see further deleterious impacts for some time to come (Recher & Lim, this volume; Saunders & Curry, this volume).

The condition of the Australian landscape prior to settlement

Many authors in this volume stress that Australia was not a pristine wilderness prior to the advent of white settlement. However, the point emphasized by Taylor (this volume) is that settlement brought with it a new suite of attitudes (the Judaeo-Christian ethic) which sought domination over the landscape whereas the Aboriginal people had co-existed within the confines of the environment for a very long period. While there is some debate about the extent of the impact of Aboriginal people on the landscape and biota, it does appear that those people achieved a balance between exploitation and utilization that could be sustained in the long-term, even through periods of major climate flux. As Walsh (this volume) points out, there is still a major opportunity for us to learn from the Aboriginal people about how Australia can be managed — both at a philosophical level and at a level of detail suitable for application to management on a day-by-day basis.

The present state of Australian ecosystems

The chapters presented in this Proceedings provide a comprehensive assessment of the state of major components of the Australian environment. However, it was never intended that this volume should catalogue the gains and losses in a way that might be expected from a State of Environment Report, for example. Rather, the focus of the Symposium and this volume is on processes: it is necessary to be able to adequately describe and understand the processes underlying the changes in order to devise appropriate solutions.

Our knowledge of the Australian environment, either as it was 200 years ago or as it is now, is far from complete. The level of ignorance of the taxonomy of the native invertebrate fauna is particularly critical because they are functionally important in all ecosystems. Most other groups of animals and the flora are also poorly known. The documentation of the biota is proceeding through such programmes as the Australian Biological Resources Survey but many organisms will disappear before they can be properly described or their role in the ecosystems understood (Soulé 1990).

Ecologically, there is a dearth of knowledge about almost every major ecosystem type in Australia in terms of both pattern and process.

To some extent our lack of knowledge about Australian environments reflects our low population base in a large country. However, nowadays that is an insufficient excuse: the nation's economy depends to a large degree on extensive primary production (pastoralism), intensive primary production (agriculture) and intensive mineral production, all activities that must be well supported by ecological research if they are not to permanently degrade the environment.

It is clear from the foregoing chapters that the impact of technological society has been considerable. They document examples of land, water and estuarine degradation, losses of species, losses of productive capacity and so on. Because of lag effects, some of these impacts may become more acute without further alterations to casual factors.

A number of chapters in this volume have focussed on management of Australia's rangelands. Apart from the fact that these cover almost three quarters of the land surface and include many habitat types, they also serve as a useful model for discussion. They have been extensively utilized for grazing by introduced stock. Feral animals, particularly rabbits, have had a major impact. Despite the lack of obvious fragmentation of the vegetation, the indigenous biota is severely depleted (Morton, this volume).

Current estimates are that 1.85M km^2 are in need of repair if pastoralism is to be sustained. Much of the damaged land is also potentially the most productive: the nutrient rich run-on areas.

A comparison with the rangelands of the Middle East is instructive as those lands have been degraded beyond repair by mismanagement. In contrast, Australian rangelands still have the potential to meet a number of land use requirements. The challenge, therefore, is to confront the problems now, to develop management systems that enable them to be utilized on a sustainable basis (see Freidel *et al.*, this volume).

How resilient are Australian ecosystems?

The advent of technological society brought with it major changes to disturbances that affected the landscape. New types of disturbances were introduced (e.g., grazing by ungulates, clearing of native vegetation for agriculture, urban development and so on) and the regimes (frequency, intensity, seasonality, size or patchiness) of pre-existing disturbance types were often changed markedly.

The Australian landscape is often said to be exceedingly fragile because of its great antiquity. Recovery after disturbance is slow, particularly in low productivity environments including those which are nutrient poor and/or arid to semi-arid.

Highly evolved, complex ecosystems such as the moist tropical forests recover slowly too. However, the contrary view seems equally plausible: the antiquity means that the present expressions have experienced a wide range of disturbances and therefore the soils and the plants and animals should be reasonably robust. Ultimately the argument turns on the similarity or difference between the new disturbance regime and that of the past. For most systems we can only speculate on what that pre-European regime might have been although Clark (this volume) has clearly pointed out ways in which appropriate data may be sought. Moreover, our historical records of disturbance are very poor. Thus our capacity to predict effects of new disturbances for management purposes is limited. This deficiency can be overcome to some extent through the study of highly modified ecosystems where some processes continue (B. J. Fox, this volume; Hobbs & Hopkins, this volume).

Given that the recovery process of Australian ecosystems tends to be slow and that many of the new disturbances are very different from those experienced in pre-European times, is it realistic or appropriate to attempt reconstruction (*sensu* Allen 1988)? It can be anticipated that time scales for complete reconstruction would be considerable if indeed it can ever be achieved. As Bell (this volume) points out, reclamation, i.e., returning a functioning ecosystem that is not necessarily identical to the pre-disturbance one, might well be a more realistic and satisfactory goal. Properly designed, the product of a reclamation programme with its synthetic vegetation has much to contribute to production, landscape protection and nature conservation (Bridgewater, this volume; Burke & Youl, this volume).

Public acceptance of reclamation as a goal after a major disturbance such as open-cut mining or even after many lesser disturbances would require land managers to acknowledge that complete reconstruction is seldom feasible. The consequence of this would be an increased demand for large areas to be set aside disturbance-free as national parks or nature reserves.

Future directions

The picture painted in this volume so far is not a pretty one: the aged face of Australia has not weathered well over the past 200 years. However, the picture is not all bleak. The contributors have provided constructive suggestions on how the problems might be addressed.

The lack of knowledge and understanding of Australian environments is a fundamental issue that is hampering our capacity to better plan and manage those environments.

Very substantial areas of Australia have not been surveyed in the level of detail necessary to make good land use planning decisions. There has been discussion of an ecological/environmental survey of Australia but so far the initiatives have been at the State level and therefore quite varied. This proposal should be developed further; some approaches are suggested in this volume (e.g., Morgan & Terrey; Fairweather).

A knowledge of disturbance history is an important factor in interpreting present-day patterns of distribution and abundance of the biota. Techniques for documenting impacts of disturbances over historical time scales are now being developed and warrant further attention. In the meantime we should be keeping better records of on-going disturbances across a range of landscape types.

Studies on the impacts of various disturbances are important to enable us to predict outcomes of land use planning and management decisions. To date these have focused on natural or near-natural environments. However, much can be learned from those environments that are already highly disturbed and often degraded.

Taxonomic studies should continue to be supported at current levels or beyond to provide a foundation for ecological studies. However, there will never be sufficient time to describe all taxa (Soulé 1990). Consequently, attention should be focused on either developing classifications based on functional characteristics (e.g., Anderson, this volume; Hobbs & Hopkins, this volume) or focussing on keystone, vulnerable or indicator taxa (Ehrlich & Murphy 1987; Saunders *et al.* 1987).

It is acknowledged that, whilst all these studies are underway and before results become available to resource managers, many management decisions will need to be made. In making these decisions the opportunity should be taken to increase our knowledge so that the framework for future decisions is enhanced. Procedures for doing this through experimental management and monitoring have been elaborated by Hopkins and Saunders (1987).

The preservation of biological diversity has been identified as a major, global, environmental management objective. The attention focussed on this objective in recent years is a reflection of concern about likely global changes (both climatic and population induced). Australians have a role to play in achieving this objective because we are custodians of a unique biota. In view of the probable scale and rapidity of global changes and our level of ignorance about the biota, a conservative approach to land (and marine and estuarine areas) use decisions must be adopted. Substantial areas should be set aside for nature conservation and remaining areas, including privately owned remnants of native vegetation, should be managed to conserve their biological values.

The present level of degradation of Australian ecosystems and the lack of information about them also points clearly to the need for more judicious management. If, as is suggested in this chapter, reconstruction of degraded ecosystems is seldom possible then greater efforts should be directed towards minimizing disturbances. This is particularly the case in highly fragmented landscapes where the very fact of fragmentation reduces the resilience of the landscape components.

A directly related point is that environmental planning should be based on landscape units rather than on cadastal boundaries as occurs at present. There is a need for a better level of integration of information and expertise in planning and management; this can be facilitated through development of land use plans related to landscape units.

One aspect of landscape planning that is currently being fostered involves incorporating privately owned remnant bushland into the nature conservation reserve system through provision of incentives. Additionally, in some highly fragmented landscapes, areas of farmland are now being replanted in an attempt to produce synthetic vegetation which will functionally complement the remaining natural communities.

There has been much discussion in recent years particularly about sustainability and related concepts such as carrying capacity. Acceptance of the notion that the utilization of natural resources should be on an ecologically sustainable basis is increasing. However, there is considerable uncertainty as to the meaning of the concepts. The challenge, therefore, is to define more clearly what is involved and to develop practical and reliable measures that can permit subsequent validation through monitoring.

If we are to move to a sustainable use of Australia's living, natural resources, it is imperative that the holistic approach of ecological thinking becomes the basis for decision-making at all levels of the community. We need to be more sensitive to the interrelationships and interdependancies rather than focussing on single issues, species or areas. This is the basis of the ecoethic which must become widely accepted and which must be supported by ecologists.

McKellar (this volume) has pointed out that the environment is an open system, subject to many uncontrolled influences. Because of this, traditional, linear thinking patterns often do not provide the best solutions to environmental issues. It is desirable, therefore, the ecologists experienced in dealing with open systems should play a greater role in the decision-making process; as well they should inculcate existing politicians, bureaucrats and land managers with lateral thinking patterns. Institutional arrangements that support the present decision-making process should also be reexamined.

Ecologists must also begin to communicate more effectively. The community must have access to contemporary and pertinent information so that decision-makers can be informed. Advice must be simple enough to be understood but scientific enough to be credible. Educational material must be kept up-to-date (Squires, this volume) and must be taught in a way that ensures its relevance to present day issues. To these ends Saunders *et al.* (1987) have proposed the notion of ecological tything: all ecologists should spend approximately 10% of their time communicating the results of their science. Failure to do this is to abrogate responsibility to the community that supports us.

Ecologists can no longer remain in the background, researching issues in isolation from the community. We must venture into the arena where policies are fought out and see that our knowledge and understanding of the environment is properly used in making the decisions about the world in which we live.

References

Allen E. B. (ed.) (1988) *The Reconstruction of Disturbed Arid Ecosystems — an Ecological Approach.* Westview Press, New York.

Bowler J. M. (1982) Aridity in the late Tertiary and Quaternary of Australia. In: *Evolution of the Flora and Fauna of Arid Australia* (eds W. R. Barker & P. J. M. Greenslade) pp. 35-45. Peacock Publications, Adelaide.

Crook K. A. W. (1981) The break-up of the Australian-Antarctic segment of Gondwanaland. In: *Ecological Biogeography of Australia* (ed. A. Keast) pp. 1-14. Dr W. Junk, The Hague.

Ehrlich P. R. & Murphy D. R. (1987) Monitoring populations on remnants of native vegetation. In: *Nature Conservation: The Role of Remnants of Native Vegetation* (eds D. A. Saunders, G. W. Arnold, A. A. Burbidge & A. J. M. Hopkins) pp. 201-10. Surrey Beatty & Sons, Chipping Norton.

Frakes L. A., McGowran B. & Bowler J. M. (1987) Evolution of Australian environments. In: *Fauna of Australia Vol. 1A. General Articles* (eds G. R. Dyne & D. W. Walton) pp. 1-16. Australian Government Publishing Service, Canberra.

Hopkins A. (1988) B. Report of the Plenary Session of the Geraldton Symposium. *Bull. Ecol. Soc. Aust.* **18(3&4),** 32-9.

Hopkins A. J. M. & Saunders D. A. (1987) Ecological Studies as the basis for management. In: *Nature Conservation: The Role of Remnants of Native Vegetation* (eds D. A. Saunders, G. W. Arnold, A. A. Burbidge & A. J. M. Hopkins) pp. 15-28. Surrey Beatty & Sons, Chipping Norton.

Lamont B. B., Hopkins A. J. M. & Hnatiuk R. J. (1984) The flora — composition, diversity and origins. In: *Kwongan — Plant Life of the Sandplain* (eds J. S. Pate & J. S. Beard) pp. 21-38. University of Western Australia Press, Nedlands.

Martin H. (1984) Australian phytogeography. In: *Vertebrate Zoogeography and Evolution in Australia* (eds M. Archer & G. Clayton) pp. 17-30. Hersperian Press, Carlisle.

Powell C. McA., Johnson D. B. & Veevers J. J. (1981) The early Cretaceous break-up of eastern Gondwanaland, the separation of Australia and India, and their interaction with South East Asia. In: *Ecological Biogeography of Australia* (ed. A. Keast) pp. 15-29. Dr W. Junk, The Hague.

Saunders D. A., Arnold G. W., Burbidge A. A. & Hopkins A. J. M. (1987) The role of remnants of native vegetation in nature conservation: future directions. In: *Nature Conservation: The Role of Remnants of Native Vegetation* (eds D. A. Saunders, G. W. Arnold, A. A. Burbidge & A. J. M. Hopkins) pp. 387-92. Surrey Beatty & Sons, Chipping Norton.

Soulé M. E. (1990) The real work of systematics. *Ann. Missouri Bot. Gard.* **77,** 4-12.

Williams R. (1989) *The Uncertainty Principle.* Australian Broadcasting Commission, Sydney.

Proc. Ecol. Soc. Aust. 1990 — 16: 585-602

Index

A

A-selection 510-1
Aborigines 369-71
  and environmental change 3-4
  animal foods, use of 27-8, 34
  effects on ecosystems 93-4, 96
  fire regimes 27, 31, 35-6
  impact on natural ecosystems 412, 414-5
  land management practices 35-6, 130
  land use, Sydney 120
  movement patterns of 26, 32-4
  perception of wilderness 412
  plant associations, distinction of 32
  plant communities, use of 30-2
  plant foods, use of 28-31, 34
  plant use, Sydney 119-20
  relationship with land 108
  resource management practices 34-5
  seasonal and seasonality, perceptions of 30, 34-5
  settlement of Australia 3
  storage, use of 35
  subsistence strategies 24, 26, 29, 33, 35
  use of fire 120, 222-3, 434
  water sources, use of 27, 32, 34
Abrolhos coral reefs 55-68
Acacia spp.
  A. aneura mulga 222, 239, 470
  A. cambagei gidgee 470
  A. carnei regeneration 267-72
  A. oswaldii regeneration 267-72
  A. rostellifera as habitat for trapdoor spiders 403
  A. salicina cooba 470
Acacias
  cooba 470
  for minesite rehabilitation 558-62
  gidgee 470
  mulga 222, 239, 470
Acanthagenys rufogularis spiny-cheeked honeyeater 306, 317, 319
Acanthiza spp.
  A. apicalis broad-tailed thornbill 318
  A. chrysorrhoa yellow-rumped thornbill 317
  A. iredalei samphire thornbill 320
  A. uropygialis chestnut-rumped thornbill 318-9
Acrobates pygmaeus feathertail glider 328, 331, 333-4
Acts of Parliament 325-6, 338, 340, 429, 435, 437-8
Acts of Parliament — *continued*
  Australian Heritage Commission Act 325
  Conservation and Land Management Act (W.A.) 326
  Flora and Fauna Guarantee Act (Vic) 326
  Land Act (W.A.) 326
Adelaide blue-tongue lizard *Tiliqua adelaidensis* 289
Adventive plants/vegetation 415-7
AEGIS, Australian Environmental Geographic Information System 450
Aegotheles cristatus owlet nightjar 306
Aeolian 442, 445
Agonis flexuosa peppermint 404
Agriculture 430-1
  development of 94, 107-8
  fertilizer use, 46, 48, 50
  in Australia 569
  sustainable 419-23
Agroecology 421-2
Agroecosystem 421-2
Alcoa bauxite mine, Jarrahdale 560
Alexandra's parrot *Polytelis alexandrae* 202
Algal bioassay 42
Algal blooms (see also Phytoplankton) 50
  blue-green 46, 48, 51
  diatom 46, 48
Alice Springs
  fire histories 222
  pastoral district 185
Alien plants 415
Alleles
  isozyme 571, 574
  rare 572
Alliances, vegetation 432-3, 437, 439
Allocasuarina spp. 432
  A. fraseriana 560
Allopolyploidy 572
Alpine
  grevillea *Grevillea australis* 256
  humus soil 256
  mint *Prostanthera cuneata* 256
  orites *Orites lancifolia* 256
  star bush *Asterolasia trymalioides* 256
  vegetation 239, 255-65
Amphibia 340, 515-6
Amplitude resilience of ecosystems 514, 516, 518
Amytornis textilis thick-billed grasswren 202, 317, 319
Anaspides tasmaniae mountain shrimp 86
Angophora spp. 432
  A. floribunda 452
Anigozanthos spp. kangaroo paws
  natural hybridization in 568
Anoka Sand Plain, USA 515
Antechinus spp. 329
  A. minimus swamp antechinus 380, 383, 385-8, 393
Anthochaera carunculata red wattlebird 306
Anthropogenic damage to Abrolhos coral reefs 64, 67
Ants 371, 510-6, 522-8
  as bio-indicators 348-52
  Camponotus spp. 522-3, 525
  community structure 350-2
  ecological importance 348
  effects of disturbance 348
  effects of grazing on 238
  functional groups 350-2
  Iridomyrmex spp. 350-4, 513-4, 525-7
  Monomorium spp. 522, 525
  Pheidole spp. 522, 525
  sampling of 349
Aphelocephala spp.
  A. leucopsis southern whiteface 318-9
  A. nigricincta banded whiteface 317
  A. pectoralis chestnut-breasted whiteface 202
Apiculture
  in W.A. 167
  reserve management for 168-9, 174
Apis mellifera European honey bee 167-73
  competition with native bees 170
Apus pacificus fork-tailed swift 306, 317
Aquatic ecosystems naturalness assessment 417
Aquila audax wedge-tailed eagle 306, 317
Arboreal marsupials 324-5, 329-36
Archaeology in arid zone 24, 26
Arid lands 446
  erosion patterns 196-9
  grazing 195-200
  rangelands 12-3, 18
  sampling methods 195, 198, 200
  satellite imagery 197-8
  spatial patterns 195-200
Arid zone 94, 221-2, 224, 459-64
  archaeology of 24, 26
  climate of 24, 29
  of South Australia 459-64
  plants 267
  vegetation, Aboriginal use of 28-32, 34-5

Arid zone ecosystems
maintenance of 223, 225
prescribed patch burns in 223, 225
structure, composition and dynamics 223
Artamus spp.
A cinereus black-faced woodswallow 317
A. personatus masked woodswallow 306
A. superciliosus white-backed woodswallow 306
Artesian basin 420
Ash forest, Kosciusko, NSW 225
Assessment for nature conservation 472
Asterolasia trymalioides alpine star bush 256
Astrebla spp. mitchell grasses 238, 241, 470
Atlas of Australian Birds 568
Atlas, Banksia 569
Atrichornis clamosus noisy scrub-bird 398
Atriplex spp. saltbush 185, 187, 238, 442, 470
A. amnicola 554-5
A. bunburyana 555
A. nummularia 555
A. vesicaria bladder saltbush 445, 555
defoliation tolerance 236
changes with grazing 238
Attitudes 436
current 107-8, 110
of community to fire 224
of early settlers 106-7
to vegetation, landscapes 94
to revegetation, 138, 140
Australian bass *Macquaria novemaculeata* 81, 85
Australian Conservation Foundation 326, 340
Australian Conservation Strategy 567
Australian dotterel *Peltohyas australis* 317
Australian Forestry Council 424-5, 336
Australian grayling *Prototroctes maraena* 85
Australian Heritage Commission 491
Australian magpie *Gymnorhina tibicen* 293
Australian magpie-lark *Grallina cyanoleuca* 317, 319
Australian Map Grid 443
Australian National Parks and Wildlife Service 223
Australian raven *Corvus coronoides* 291
Australian snapper *Chrysophrys auratus* 574
Avon botanical district 100

B

Banded hare-wallaby *Lagorchestes fasciatus* 289
Banded plover *Vanellus tricolor* 306, 317
Banded whiteface *Aphelocephala nigricincta* 317
Bandicoots *Isodoon* spp. 452
Banksia Atlas 569
Banksias 177-82
co-existence of species 177,180-1
Banksias — *continued*
non-sprouting species 177-9, 181
sprouting species 178-80
Banksia scrub
eastern suburbs of Sydney 117-9, 121-2, 124
Banksia spp.
B. attenuata 178-80
B. burdettii 178-9
B. candolleana 179
B. coccinea 178, 181
B. cuneata 178-9, 181, 573
B. elegans 178-80
B. grandis 560, 562
B. hookeriana 179-80
B. lanata 179
B. leptophylla 179-81
B. littoralis 568
B. menziesii 178-80
B. prionotes 178-9, 181
B. seminuda 568
B. tricuspis 179-80
Basal area 274-5
Bats 329
Eptesicus spp. 568
Bauxite mines 511, 515-6, 518, 560
Beekeepers' Reserve, W.A. 167-9, 172
Beekeeping 567
in W.A. 167
Bees
native 169-72
European honey 167-73
plant species preferences 171, 173-4
Beetles 517-8
Bega, NSW 330
Belah *Casuarina cristata* 442-5
Benchmarks 339-40
Benthic communities
sensitivity of 59, 64-7
Benthic community structure
Abrolhos coral reefs 57
Bettongia spp.
B. gaimardii eastern bettong 289
B. lesueur burrowing bettong 204, 210, 289
B. penicillata brush-tailed bettong 370
Bilby *Macrotis lagotis* 222
change in population 237
BIOCLIM bioclimatic model 335, 338
predictive modelling 451
Biodiversity, conservation of 567
Biogeography, Australian 568
Bio-indicators, ants 348-52
Biological surveys 334-5, 569
collections by W.A. Museum 397
collections by W.A. Department Conservation and Land Management 397
Biosystematics 567-78
Birds 515, 517-8
effects of grazing on 237
Black box *Eucalyptus largiflorens* 443, 470
swamps 445
Black-eared cuckoo *Chrysococcyx osculans* 293, 318
Black-faced cuckoo-shrike *Coracina novaehollandiae* 293, 306
Black-faced woodswallow *Artamus cinereus* 317
Black honeyeater *Certhionyx niger* 318
Black kite *Milvus migrans* 306
Black-shouldered kite *Elanus notatus* 306, 317
Black-tailed treecreeper *Climacteris melanura* 320
Bladder saltbush *Atriplex vesicaria* 236, 445, 555
Blackelyi's red gum *Eucalyptus blakelyi* 150-3, 432-3
Blue-breasted wren *Malurus pulcherrimus* 318
Bluebush *Maireana* spp. 238, 442, 445, 470
Blue-green algae 46, 48
Blue gum high forest 116-7, 121, 124
Blue Mountains National Park, NSW 223
Blue Rock Dam, Vic. 85, 88
Blue-winged kookaburra *Dacelo leachii* 320
Bogong High Plains, Vic. 255-65
Botanic gardens 575
Bothriochloa pertusa 274-6, 281, 283
Bourke's parrot *Neophema bourkii* 317, 319
Box/gum woodland of NSW tablelands 150, 153
Boyagin Nature Reserve, W.A. 370, 375-6
Bradysporous seed store 533
Breeding systems 571, 575
Brisbane Ranges, Vic. 83, 86
Broad-tailed thornbill *Acanthiza apicalis* 318
Broken Hill Complex, NSW 469-71
Bronzeback *Ophidiocephalus taeniatus* 202
Brown falcon *Falco berigora* 306
Brown-headed honeyeater *Melithreptus brevirostris* 306
Brown honeyeater *Lichmera indistincta*, use of road reserves 362, 365
Brown quail *Coturnix australis* 318
Brown trout *Salmo trutta* 83, 86-9
Brushtail possums *Trichosurus* spp.
common *T. vulpecula* 204, 331, 334
mountain *T. caninus* 330-2, 334
Brush-tailed bettong *Bettongia penicillata* 370
Burdekin River, Qld. 279
catchment 279, 282-5
Bureaucracy in pastoral lands 190
Burhinus magnirostris bush stone curlew 293, 306
Burrinjuck Reservoir, NSW 14-8
Burrowing bettong *Bettongia lesueur* 204, 210, 289
Bushfires in semi-arid NSW 246
Bush stone curlew *Burhinus magnirostris* 293, 306

C

Cacatua spp.
C. galerita sulphur-crested cockatoo 291

Cacatua spp. — *continued*
C. pastinator long-billed corella 291, 306
C. roseicapilla galah 291, 293, 306, 317-9
Cainozoic history of Australia 2-7
Calcareous earths 445-6
Caledia spp. grasshoppers 568, 575
California 95
wildlife habitat database 452
Callitris glaucophylla white cypress pine 247, 432, 442, 445
Calluna vulgaris invasion 131-2
Caloprymnus spp.
C. camapicalis desert rat-kangaroo 291
C. campestris plains rat-kangaroo 210
Calyptorhynchus spp.
C. funereus latirostris Carnaby's cockatoo 306, 318
C. magnificus red-tailed black cockatoo 306, 317-9
Camponotus spp. ants 522-3, 525
Cane grass swamps 445
Canis familiaris dingo 203
Canning River, W.A. 404
Cannon ball *Dissocarpus paradoxus* 445
Canonical variate analysis 330-3
Captains Flat, NSW 83-4, 88
Captive breeding 574-5
Carassius auratus goldfish 86
Carnaby's cockatoo *Calyptorhynchus funereus latirostris* 306
disappearance from Kellerberrin 318
Carp *Cyprinus carpio* 86
Carrying capacity
of pastoral lands 185, 191
of rangelands 251
woody plants, effect on 186
Case moth *Plutorectis caespitosa* 259
Casuarina spp.
casuarina-heath 453
C. cristata belah 442-5
C. cunninghamiana 432
C. fraseriana as habitat for trapdoor spiders 403-4
C. obesa, salt tolerance variation in 547
Cat *Felis catus* 203, 209, 371-2
Catastrophe, impact on populations 572
Catchment 46, 50
management 10-8, 79, 81-2, 87, 91, 437
planning 471
Catchment Protection Board (NSW) 437
Cattle 107
effect on vegetation 236, 240
Cattle grazing 255-65
disturbance 260-1
role in restricting woody species 262
role in vegetation change 260-1
selectivity 260, 262-3
Cecropis ariel fairy martin 306
Celmisia sp. silver snow daisy 256
Central Australia 186, 188-90, 221, 224-5, 229, 293
parks and reserves 221-5, 227
ranges 223-4
Central deserts
disappearance of numbats 371
Central Division, NSW 468
Central Western Region, NSW 427-39
Centre for Resource and Environment Studies 336
Cercartetus nanus eastern pygmy possum 331, 334
Certhionyx niger black honeyeater 318
Chaeropus ecaudatus pig-footed bandicoot 291
Chaetomorpha 45
Chasmocephalon neglectum (relic spider) 403-5
Cheilodactylus macropterus jackass morwong 574
Chemosystematics 570
Chenonetta jubata wood duck 306
Chenopod shrublands 443, 445
effects of grazing on 238
Chenopods, pastoral management 187, 190-1
Cherax spp.
C. destructor yabby 83, 89
C. tenuimanus marron 83
Chestnut-breasted whiteface *Aphelocephala pectoralis* 202
Chestnut quail-thrush *Cinclosoma castanotum* 306
Chestnut-rumped thornbill *Acanthiza uropygialis* 318-9
Chlamydera guttata spotted bowerbird 317, 319
Chromosomal differentiation 575
Chromosome markers 570
Chrysococcyx spp.
C. basalis Horsfield's bronze-cuckoo 306
C. osculans black-eared cuckoo 293, 318
Chrysophrys auratus Australian snapper 574
Chthonicola sagittata speckled warbler 293
Cinclosoma spp.
C. castanotum chestnut quail-thrush 306
C. punctatum spotted quail-thrush 293
Cinnamon fungus *Phytophthora cinnamomi* (see also Dieback) 102, 379-80, 382, 385-6, 390-1, 393, 405
Circus aeruginosus marsh harrier 317
Cladophora 45
Classification 443, 446
image 444, 446
Landsat data 443
maximum likelihood 443
of soils 444
unsupervised 443
Clays
non-cracking 446
grey self-mulching 446
Clearfelling 327
Clearing 293, 416, 431-3, 435, 438-9
Climacteris spp.
C. melanura black-tailed treecreeper 320
C. rufa rufous treecreeper 306
Climate change 1-8, 241, 340, 449, 455
Greenhouse effect 72, 76
Climatic
variability, effects on revegetation 539-40, 542
surfaces 335
Climax concept, succession 417
Coal mines 511, 516, 518
Coastal ecosystems 71-4, 76-7
conservation of 71-2, 76-7
impacts on 71-3, 76-7
management of 71, 75-7
research 71-3, 75-6
rocky shores 71-7
sandy beach 71-7
Cobar Peneplain, NSW 469-70, 472
Cockburn Sound, W.A. 40-5
management of 40, 43
nitrogen load, 43, 45
phytoplankton distribution 43
water circulation 42
Collections
made by biological surveys 397
made by Dampier 397
museum (specimens), value of 398
use in interpreting earlier ecological scenarios 399
Colletidae 170-1
Colluricincla harmonica grey shrike-thrush 306, 318
Columba livia domestic pigeon 306
Common brushtail possum *Trichosurus vulpecula* 331, 334
Common ringtail possum *Pseudocheirus peregrinus* 331, 334
Communities
eucalyptus 432-3
responses to fire 229
Community properties
land use impact on 97, 99
responses to fire 224
Competition 262, 293
interspecific 180, 512-3
native pollinators and honey bees 170, 173
Computer applications 446
AEGIS 450
BIOCLIM 335, 338
Geographic Information System 225, 329-30, 335-9, 446, 449-50, 453, 569
microBRIAN 216, 337-8
PREPLAN 223
TASFORHAB 335, 339
Computers
data capture 569
mainframe 569
personal 569
CONCOM 289
Condition
of pastoral lands 189
of rangelands 252
Conflicts among user groups, Abrolhos coral reefs 55-7, 62-3, 66
Conilurus albipes white-footed rabbit-rat 291
Connectivity 132
Conopophila whitei grey honeyeater 317

Conservation 94-5, 97-9, 103, 110, 435-9, 459-61, 463-4
Acacia carnei 267, 271-2
A. oswaldii 267, 271-2
areas 327
arid zone plants 267, 271-2
assessment for 472
biology 572
Department of Forests and Lands 326, 336
ethic 108, 241
ex situ 575
flora and fauna 449, 567-77
genetics 568, 570, 572, 574-5
management 569, 572
marine ecosystems 71-2, 76-7
planning 569
policies 436
reserves in wheatbelt of W.A. 305
small mammals 379, 393-4
status 432
terrestrial vertebrates 287-301
Conservation Commission of N.T. 223-4
Conservation Council 438
Contamination 51
heavy metal 42-3
of seafood 40, 42-3
of water 42-3
Contractions of range
in birds 202, 209
in mammals 201, 207
Control burns 221
Controlled environment studies 569
Cooba *Acacia salicina* 470
Coolabah *Eucalyptus microtheca* 470
Coolangubra State Forest, NSW 332, 335-6
Copperburrs 442
Coracina spp.
C. maxima ground cuckoo-shrike 306
C. novaehollandiae black-faced cuckoo-shrike 293, 306
Coral growth, Abrolhos coral reefs 57, 67
Coral reefs 71-7
anthropogenic damage 64, 67
benthic communities, sensitivity of 59, 64-7
benthic community structure 57
complex karst platforms 58, 60-1, 64, 67
conflicts among user groups 55-7, 62-3, 66
coral growth 57, 67
dissected limestone platforms 58, 60, 63-4
drowned doline fields 60-1, 67
ecological units 58-63, 65-7
fishing effort intensity 57-9, 62-7
geomorphical classes 59-60, 63-4, 66
lagoon patch reefs 57, 60, 63, 67
low energy coral assemblages 61, 64-7
management 56-8
rock lobster fishing patterns 55-6, 58-9, 62-3
Coral reefs — *continued*
shallow reef perimeters 57, 59-60
sheltered leeward reef slopes 57, 60-1, 63, 67
Corridors (see also Road reserves) 435, 437-8
linking conservation reserves 574
wildlife 326-7, 330, 340
Corvus spp.
C. bennetti little crow 317, 319
C. coronoides Australian raven 291
C. orru Torresian crow 317, 319
Cosmopolitan species 142
Coturnix spp.
C. australis brown quail 318
C. novaezealandiae stubble quail 306
County of Cumberland 115-6, 124
Cover crop as aid to revegetation 537, 540-2
Cracticus nigrogularis pied butcherbird 306, 317, 319
Crested bellbird *Oreoica gutturalis* 293
Crested pigeon *Ocyphaps lophotes* 291
expansion of range 306, 317-9
Crested shrike-tit *Falcunculus frontatus* 306
Critical
data set 225
pathway for revegetation 553
weight/range for ground dwelling mammals 451-2
Cropping areas, marginal 422
Crown
agencies 435, 437, 439
Lands Office (NSW) 435, 437
separation 339
Cryptogam crusts 446
CSIRO 224
Cuculus pyrrhophanus fan-tailed cuckoo 318
Cultivated
plants 415
vegetation 416
Cultivation of endangered species 574-5
Cultural ecosystem 412, 414
Cultural landscape 412-4
origin of concept 411-2
Cumberland Plain 115
woodlands 117-21, 123-4
Curriculum development
competency profile for land managers 425
Cyanoramphus novaezealandiae cookii Norfolk Island Green Parrot 133
Cyprinus carpio carp 86
Cytogenetics 568, 570-1

D

Decelo spp.
D. gigas laughing kookaburra 306
D. leachii blue-winged kookaburra 320
Dampier, collections made by 397
Damping, resilience of ecosystems 515-7
Dams
as barriers to fish movements 79, 85-6, 88
ecological effects on streams 79, 84-8
effects on vertebrates 84-5, 88
hypolimnial release of water 84, 85, 87-8
Danthonia caespitosa defoliation tolerance 236
Daphoenositta chrysoptera varied sittella 306
Darling Plateau, W.A. 557
Darling Range, W.A. 531
Darling Riverine Plains, NSW 469-71
Dartmouth Dam, Vic. 84
Dasyurus spp. quolls 452
D. geoffroii western quoll 204, 370-1
D. viverrinus eastern quoll 289
Data collection, time and space scales 55-6, 58, 67
Decision support systems 189, 329-30, 334-41
Register of National Estate 500
Decorticating bark 295
Decreaser species 237
Defoliation of trees in remnant woodland 149, 151, 153-6
Defoliation tolerance
of *Atriplex vesicaria* 236
of *Danthonia caespitosa* 236
of *Thermeda australis* 236
Deforestation 431-3, 435, 438-9
Degradation, land 95, 102, 107, 279, 285
revegetation 575
Demographic processes 572
Dendograms 573
Department of Environment & Planning (NSW) 435, 437, 438, 447
Desert 295
Desertification 523
Desert rat-kangaroo *Caloprymnus camapicalis* 291
Desert, Simpson 295
Destocking lag 101
Devil's Marbles Conservation Reserve, N.T. 224
Diamantina River, Qld. 295
Diamond dove *Geopelia cuneata* 320
Diatoms 15-7
Dieback
Phytophthora cinnamomi 102
rural tree 149-50, 152-6
seagrass 40, 42-5
Dingo *Canis familiaris* 203
Diplopoda (millipedes) 518
Disease as a cause of extinction 204
Dispersal 328
Dispersal agents
land use impacts on 97, 101
role of 106
Dissocarpus paradoxus cannon ball 445
Disturbance 5-6, 8-10, 15-6, 142, 258-63, 414-5, 521-5, 527-8
ecology 109
ecosystems 131-3, 414-5
effects on ants 351-4

Disturbance — *continued*
endogenous 521-4, 528
exogenous 521-4, 528
human 93-103, 521-4, 528
isolation of 511
natural 521-4, 528
regimes 93-5
size of 511
vegetation dynamics 258-63
Diversity 449, 452
effect of grazing on 237
DNA, mitochondrial 571
Doline fields, Abrolhos coral reefs 60-1, 67
Dolomedes plantarius great raft spider conservation of 405
Domestic pigeon *Columba livia* 306
Dominant plant species succession 414-5
Drainage lines 326-7
Dromaius novaehollandiae emu 289, 293, 306, 317, 319
Drought 107, 246, 330, 332, 430, 434, 441
in arid Australia 203, 207
in pastoral lands 185-6, 189
policy 189
Dry sheep equivalent 245, 247
Dryandra forest, W.A. 370, 372-4
Dryandra sessilis 170-2, 174
Drymodes brunneopygia southern scrub robin 306, 318
Dunes 442
parallel, parabolic, unaligned 445
Duplex soils, 442, 445-6

E

Early human occupation, Lake Mungo 442
East Gippsland 289
Eastern barred bandicoot *Perameles gunni* conservation of 133
Eastern bettong *Bettongia gaimardii* 289
Eastern hare-wallaby *Lagorchestes leporides* 291
Eastern Otway Ranges, Vic. 379-81, 389-90, 392-3
Eastern pygmy possum *Cerartetus nanus* 331, 334
Eastern quoll *Dasyurus viverrinus* 289
Echidna *Tachyglossus aculeatus* 371
Ecological
aspects of fire 221, 223, 225
concepts 330, 340
data 221, 223, 225
islands 449
isolation 417
regions of NSW 427
research, historical 253
resilience 131
restoration of landscapes 398
scenarios, theoretical/archival 398, 399
surveys 446, 472
systems 337-8, 446
units, Abrolhos coral reefs 58-63, 65-7
Economics, pastoral management 189, 192
Ecosystem(s)
Australian 221
dynamics 1-18
human-dominated 415-6
management 221-2, 224-5, 229
properties, land use impacts on 94
representation in conservation reserves 472
role of fire in 221, 223
stress, forms of 86-7, 90
Ecotope 335
Edaphic 335
Eden forests 329-30, 333
Education programmes 436
Eductor dredging, ecological effects on streams 83, 88
Eels 85
Effluent 40, 42
industrial 40, 42, 51
urban 40, 42, 51
Elanus spp.
E. notatus black-shouldered kite 306, 317
E. scriptus letter-winged kite 306
Elasticity, resilience of ecosystems 514
Elegant parrot *Neophema elegans* 306, 318
Emu *Dromaius novaehollandiae* 289, 293, 306, 317, 319
Endangered
fauna 326, 329
flora 325
flora in W.A. 569
species 239, 241-2, 289, 432, 434, 439, 569, 572, 574-5
Endogenous disturbance 521-4, 528
Eneabba, W.A. 168, 179, 531-3
Energy inputs, streams 80-1, 88-90
English perch *Perca fluviatilis* 83, 86
Environmental
analysis 470
changes 446
degradation, aquatic ecosystems 39
envelopes 329-35, 338
gradients 256, 258, 263, 335
history 1-18
impact assessment 569
indices problems 477, 487
management 467, 470
monitoring 472
protection 435
protection zones 437
rehabilitation manager 575
Eopsaltria griseogularis western yellow robin 318
Ephemeral species, effect of grazing on 240
Ephthianura spp.
E. albifrons white-fronted chat 317, 320
E. aurifrons orange chat 306
Episodic events 106, 186-8, 190
Eptesicus spp. bats 568
Erosion 98, 103, 110, 239, 295
cells in arid lands 196-9
Escherischia coli water contamination 42
Estuaries 71-7
Peel-Harvey, W.A. 40, 45-51
Estuarine wetlands 117-9, 121, 123-4
Ethnobiology 23
Ethnographic research 26
Ethnotaxonomy 28, 32
Eucalypt(s) 568, 571-2, 574-5
communities 432-3
flowering 333
forests, coastal 451
natural hybridization 568
nutrients 330-1
phenology 330, 332-3
rare 572-3
species richness 333
use by beekeeping 167
woodland
dry tropical rangelands 279, 281
E. melliodora/*E. blakelyi* alliance 150, 153
Eucalyptus spp.
E. albens 432-3
E. baxteri forest, carbon:nitrogen ratio in 559
E. blakelyi Blakelyi's red gum 150-3, 432-3
E. caesia 571-3
E. calophylla 320
E. camaldulensis river red gum 433, 470
salt tolerance of 547-8
E. crucis 571-2
E. cypellocarpa 452
E. dealbata 433
E. diversicolor karri 320, 573-5
E. dumosa 433
E. fastigata 432-3, 437
E. guilfoylei tingle, habitat for relic spider 404
E. gummifera 452
E. jacksoni tingle, habitat for relic spider 404
E. largiflorens black box 443, 470
E. longifolia 452
E. macrohyncha 432-3
E. maculata 330, 333
E. marginata jarrah, 225, 320, 370, 402, 404, 531-4, 557-64
E. melliodora 432-3, 150, 153
E. microcarpa 432-3
E. microtheca coolabah 470
E. obliqua forest, carbon:nitrogen ratio in 559
E. pauciflora 433
E. pendens 572
E. populnea 432
E. rossii 432-3
E. salmonophloia salmon gum 103
E. sideroxylon 433
E. sieberi 452
E. smithii 452
E. socialis 433
E. stellulata 433
E. tereticornis 325, 331
E. viminalis 432-3
E. wandoo 373

European(s)
attitudes of 106
impact on vegetation 93-114
landscapes 106
European honey bee *Apis mellifera* 167-73
European mole *Talpa europaea* 511
European settlement 4, 428, 432
impact on desert-dwelling animals 201-4, 207-9
impact on natural ecosystems 412-7
south-west Australia 303
Eutrophication 431
aquatic ecosystems 39-51
Cockburn Sound 40-5
Peel-Harvey Estuary 45-51
Evapotranspiration 100
Everard Ranges, S.A. 370
Evolution 571, 575
Evolutionary biogeography 287
Exogenous disturbance 521-4, 528
Exotic
introductions, impact on native biota of streams 79, 86-7, 89, 91
species 129-30, 132-3
vegetation 415-7
Experimental
manipulation 513
nature of management 222, 225
Expert systems 451
Exploitation of native vegetation 94
Exploration 462-3
collections made during 397, 399, 402
Extensions of ranges 145
Extinct species 431-2, 434-5, 439
Extinction(s) 289, 291, 431-2, 434-5 439, 449, 455, 511-2, 567, 574-5
and ecosystem management 223
birds 202, 209
causes 203-4, 207-9, 289-91, 293-5
flora 325
ground-dwelling birds 295
in Western Australia 222
local 180
mammals 201-2, 207-9
medium sized, ground dwelling mammals 434
patterns 291-3
rate of, in arid zone 222
rates 291-3
terrestrial vertebrates 287-301

F

Fairy martin *Cecropis ariel* 306
Falco spp.
F. berigora brown falcon 306
F. cenchroides nankeen kestrel 306
F. hypoleucos grey falcon 317
Falcunculus frontatus crested shrike-tit 306
Fan-tailed cuckoo *Cuculus pyrrhophanus* 318
Fantails *Rhipidura* spp. 291
Farm tree programme, Vic. 108
Fauna 433-4
conservation 567-75
Fauna — *continued*
distribution prediction 449, 452, 455
effects of grazing on 237
endangered 326, 329
hollow dependent 334
prediction modelling 451
rare 326, 329
sensitive to forestry 326, 329, 339-40
succession 334
Faunal changes 434
Feathertail glider *Acrobates pygmaeus* 328, 331, 333-4
Felis catus cat 203, 209, 371
Feral species 325, 430, 434
Fertilizers 100
and nutrient accumulation 558
effect on succession 562
effects on native species 538, 542
Field naturalists 439
Field wren *Sericornis fuliginosus* 317-8, 320
Finches 293
Finke George National Park, N.T. 223
Fire 186-7, 261, 330, 332, 337-8, 432, 434, 521-5, 528
Aboriginal use of 120
and grazing 237, 240
as a cause of extinctions 204, 209
as a management tool 124-5, 130-1, 133, 439
behaviour 221, 224-5, 229
biological aspects of 221
changing floristic composition 124
climatic factors 224
control 221, 225
ecological aspects of 221, 223, 225
ecology, application of 221, 223, 225
effect of climatic factors on 224
effect on ants 351-4
effect on Australian ecosystems 222, 225, 229
effect on ecosystem nutrients 558
effect on floristic composition 124
effect on small mammals 379, 381-2, 385-94
in autumn 177, 179-81
in semi-arid NSW 246
in spinifex grasslands 222
in spring 177, 179-81
interval 178, 180-1
patches 227
planning 221, 225
plant recovery after 170-1, 173
practical application of 225
problems of insufficient data, 223, 225
protection from 223
response of hummock grasslands 222
safety 221, 225
short-interval 173
spinifex grasslands 222
Uluru National Park 215, 219
wildfire 186-7
woody plant management 187, 190, 192
Fire-adapted traits 177
Fire ecology
application of 221, 225
Northern Australia 223, 225, 229
research 224, 229
Fire history 3-4, 10-2, 14-5
Alice Springs 222
Northern Territory 222, 224-5
remote sensing 216, 218
spinifex communities 216
Tanami Desert 222
Uluru National Park 216, 218
Fire management 177, 180-2
arid zone 221, 225, 229
manuals 221, 225, 229
models 224, 229
parks and reserves 221-3, 225
plans 225
rules and guidelines 222
strategies 225
Uluru National Park 215-20
urban bushland 124-5
Fire regimes 97, 100-3, 124, 177, 181, 222, 229
and extinctions 222
changes causing numbat decline 372
in Central Australia 371
Fish 340, 571, 574
Fisheries, Australian commercial 567, 574
Fishing 40, 45-6, 48
effort intensity, rock lobster 57-9, 62-7
patterns, rock lobster at Abrolhos coral reefs 55-6, 58-9, 62-3
Fishways 85-6, 89
Flavonoids 570
Flinders Ranges National Park, S.A. 462
Flock bronzewing *Phaps histrionica* 202
Floods, Paroo River 246
Flora 432-3
conservation 567-75
endangered 325
extinctions 325
general 329, 333, 340
reserves 437
Floristic diversity 332
Flycatchers *Myiagra* spp. 291
Foliage biomass index 330
Foraging preferences
European honey bee 169-71, 172
native bees 170-2
Forest(s)
administrative units 337
clearing 325
compartments 327
coupes 327
fragmentation of 325
history 324-5, 331, 340-1
industries 326, 340
logging design 327
management 455
open 522-6
phebalium *P. squamulosum* 256
planning and management 326-8, 335
planning framework 329, 337-40
recurrent disturbance 340

Forests — *continued*
  red gum 325-6, 331
  reserves 326-7, 437
  rotation length 332
  site types 330-3
  survey 328-9, 334
  Sydney area 116-9, 121-4
  system 337-8
  systems dynamics 451
  types 323, 334
  wildlife conservation 323-45
  wildlife research 329-41
Forests and woodlands 430, 432
  prescribed burning 224
  temperate 224
  tropical 224
Forestry
  effects on streams 79, 82, 87-90
  ethics 326, 336
  future directions 336-41
  in Australia 567, 569, 574
  industry 432
  practices in ecosystem management 221
  proposals 340
  strategy 325, 338, 340
Forestry, ANU Department of 336
Forestry Commission (NSW) 435, 437, 439
Fork-tailed swift *Apus pacificus* 306, 317
Fossils, "pseudofossils" 398
Fox *Vulpes vulpes* 203, 209, 289, 291, 293, 295, 325, 371-6
  need for control 374, 376
Fragmentation 293
  of landscapes 97, 100-1, 103
  of vegetation cover 414-6
Freshwater wetlands 117-9, 121-2, 124
Frogs 291, 329
  biology, distribution 400
  Heleioporus albopunctatus 399
  in relation to landscape restoration 397-402, 405-6
  in W.A. Museum collections 400, 402, 408
  natural hybridization in 568
Fuel
  load assessment 224
  production and rainfall 224
  reduction 223
Fugitive species, biological needs of 398
Functional
  attributes 335, 339
  groups in ants 350-2
  land patterns 467, 472
    use in classification 468
  plant types 109

G

Gadopsis marmoratus river blackfish 83, 86, 89
Galah *Cacatua roseicapilla* 291, 293
  expansion of range 306, 317-9
Galaxias spp.
  G. maculatus 85
  G. olidus 80, 86, 88-9
Gambusia affinis mosquito fish 86
Gammon Ranges, S.A. 463
Garden Island, W.A. 40
Gastrolobium sp. 372
Geckos 398, 568, 570, 575
  Heteronotia binoei 568, 575
  Gehyra punctata 568, 570
  G. variegata 568, 570
  Oedura reticulata 398
Gehyra spp. geckos
  G. punctata 568, 570
  G. variegata 568, 570
Geijera parviflora wilga 442, 445
Gene
  dispersal 574
  flow 574
  resource conservation 575
Genetic
  degradation 328
  distance, Nei's 571, 573
  diversity 571-4
  management 572-5
  resources, conservation of 567-77
  studies 567-78, 574
  systems 572
  variation within species 568-71, 574
Geneticists, Australian 568
Geocode 335, 339
Geographic Information Systems 225, 329-30, 335-9, 446, 569
  and satellite imagery 449-50, 453
  for habitat mapping 449-50, 453
Geographic variation
  genetic 570-1, 574
  in species 569-71
Geology as basis for ecosystem survey 468
Geomorphological classes, Abrolhos coral reefs 59, 60, 63-4, 66
Geopelia spp.
  G. cuneata diamond dove 320
  G. placida peaceful dove 320
Geopsittacus occidentalis night parrot 202, 317, 319
George Gill Range, N.T. 224
Germplasm storage 574
Gerygone spp. warblers 291
  G. fusca western warbler 306
Gidgee *Acacia cambagei* 470
Gilbert's whistler *Pachycephala inornata* 306
Gliders (see also Arboreal marsupials)
  feathertail 328, 331, 333-4
  greater 331-4
  squirrel 331, 334
  sugar 331, 334
  yellowbellied 331-2, 334
Global
  climatic change 100, 103
  environmental change 2, 5, 8, 18
Glossopsitta porphyrocephala purple-crowned lorikeet 306
Goals for vegetation and wildlife 241
Golden perch *Macquaria ambigua* 85
Golden whistler *Pachycephala pectoralis* 306, 318
Goldfields (W.A.) 370, 376
Goldfish *Carassius auratus* 86
Gordon River, Tas. 84, 88
Gorse *Ulex europaeus* invasion 130-1, 133
Goulburn River, Vic. 83
Goyder's line 459
Grallina cyanoleuca Australian magpie-lark 317, 319
Granite rock
  herb *Isotoma petraea* 568, 570
  mallees 571-3
Grasshoppers 568, 570, 575
Grassland 255-7
Grazing 463-4
  as a cause of extinctions 204, 207-8
  cattle 255-65
  distribution 237
  domestic animals 245-53
  effect of 235-6, 238
  effect on tropical and subtropical savannahs 241
  industries 240
  interaction with fire 237, 240
  interaction with rainfall 237
  livestock 273, 279, 281
  marginal 431-2
  pressure 273
  rabbits, kangaroos, mammals 267, 269-71
  regimes 103
  sheep and cattle in arid lands 195-200
Great Artesian Basin 470-1
Great raft spider *Dolomedes plantarius* 405
Great Sandy Desert 24, 26, 28, 35
Greater glider *Petauroides volans* 331-4
Greater stick-nest rat *Leporillus conditor* 210, 289
Greenhouse (see Climate change)
Greening Australia 108
Grevillea spp.
  G. australis alpine grevillea 256
  G. divaricata 432
Grey-crowned babbler *Pomatostomus temporalis* 293, 317
Grey currawong *Strepera versicolor* 306, 317, 320
Grey falcon *Falco hypoleucos* 317
Grey honeyeater *Conopophila whitei* 317
Grey self-mulching clays 446
Grey shrike-thrush *Colluricincla harmonica* 306, 318
Ground cuckoo-shrike *Coracina maxima* 306
Ground dwelling mammals 289, 329
  critical weight range 451-2
Ground foragers 293, 295
Ground water resources 419
  conflict over allocation of 420
  stockwaters 420
Guidelines for management of fire 221-2, 225, 229
Gymnobelideus leadbeateri Leadbeater's possum 325, 334
Gymnorhina tibicen Australian magpie 293

H

Habitat 431-5, 429
  accommodation model 521, 527-8

Habitat — *continued*
critical 335
critical minimum 437
evaluation using ants 347-57
favourableness 510
fragmentation 129, 449, 455
general 326, 328-41
"gestalt" 451
heterogeneity 293
identification 329-36, 339
mapping 449, 452-3
marginal 334
models 331-4, 336, 451
mosaic 333
optimal 326, 328, 331, 334
predictability 510
preferences of small mammals 379-80, 382, 385-6, 389
requirements of fauna 451, 453
simplification role in rural tree dieback 150, 152, 154-6
templet 510
Hakea trifurcata 170-2
Hares 325
Harvey River, W.A. 46
Hawkesbury River, NSW 115, 121-2
Hawkesbury sandstone 115
woodlands and heaths of 117, 119-21 123-4
Hazard reduction 223
Heath 255-7, 452-3, 522, 524, 526
Sydney area 117, 119-21, 123-4
Heavy metals, aquatic ecosystems 42
Heidelberg School 107
Heleioporus albopunctatus frog 399
affects of salinity on biology of 400
distribution map 400
distribution of 400, 405
in British Museum (Natural History) 399
in W.A. Museum collections 400, 402, 408
Herbaria 568
collections, relevance to distributions of plants 405
relevance to ecology 397
use in restoration of landscapes 397-8, 405-6
Herbivore preferences 236
Herbivores
extinctions of 202, 207-9
land use impacts on 97, 100
Heritage
agreements 437
lands 437
Heterodendrum oleifolium rosewood 442, 445
Heterogeneity, arid lands 195-200
Heteronotia binoei gecko 568, 575
Heteropogon contortus 273, 281
Heterozygosity 571-2, 574-5
complex translocation, in *Isotoma petraea* 568, 570
Hillend, NSW 429
Hirundo neoxena welcome swallow 293, 317
Historic sites 437
Historical evidence 4-6
History, environmental 1-18
Holism 421
Holland 511
Hollows 328-9, 334
dependent species 334
Holocene Period 4-7
Home range 328
Homozygosity 571-2
Honeyeaters 328
Hooded robin *Melanodryas cucullata* 293, 306
Hope Report 492
Horsfield's bronze-cuckoo *Chrysococcyx basalis* 306
Hotspots for fauna 330-1
House mouse *Mus musculus* 203, 522, 524
Human
disturbance 521-4, 528
-dominated landscape 413, 415-6
occupation 432-4
Humanity/nature relations
anthropocentric concept of 411-2
biocentric concept of 413-4
technocentric concept of 411-2
Hummock grasslands 224, 371
fire effects on 222
Hunter-gatherers 23-37
Hyaleus spp. 170-1
Hybrid speciation 572
Hybridization, natural 568-9
Hydrological characteristics
upland streams 80-1, 89-90
Hydrology
land use impacts on 97-9
timber harvesting and 101
Hypolimnial release of water from dams 84-5, 87-8
Hysteresis, resilience of ecosystems 514

I

Idiosoma spp. trapdoor spider
biology of 402
diminishing range of 403-4
I. hirsutum 402-4, 409
distribution map 403
I. nigrum 402
I. sigillatum 402-4, 409
distribution map 403
museum collection in relation to distribution 402-4, 409
Illinois, USA 511
Image 445
analysis 443
classification 445-6
data 446
Landsat 444-6
satellite 446
Immigration 511-2
Impacts
of pastoral management 149-50, 153-6
on fauna 433-4
of flora 432-3, 435
on habitat 432-3, 435
on populations 572
Inbreeding 571-2, 574
Increaser species 237
Industry 42
effluent 40
impacts 40
Inertia of ecosystems 94, 96
Infertility of soil
effect on mammals 205-6
in arid Australia 205
Information
systems 328-30, 334-41, 449, 453
technology 338-9
Innaminka, S.A. 463
Insects
abundance in remnant woodland 149, 151, 153-6
defoliation role in rural tree dieback 149-50, 152-6
growth and survival 150, 152, 154-6
Integrated survey 467
management 439
Integration of research and management 109-10
Introduced species 141, 143
Inventories
of resources 109, 320, 328, 334-5, 338-40
of streams 79, 87
Inverell, NSW 293
Invertebrates 329
Iridomyrmex spp. ants 350-4, 513-4, 525-7
Iron-ore mines 510, 512
Island biogeography 326, 437, 439, 511
Islands as refuges for mammals 202, 210
Isodems 332
Isolation 293
of disturbance 511
Isoodon spp. bandicoots 452
Isopoda (slaters) 518
Isotoma petraea 568, 570-1, 574
complex translocation heterozygosity 568, 570
Isozymes 569-71, 573-4
Israel 95
agricultural history of 94

J

Jackass morwong *Cheilodactylus macropterus* 574
Jacky Winter *Microeca leucophaea* 289, 293, 306
James Ranges, N.T. 223
Jarrah *Eucalyptus marginata* 320, 370, 372, 557-64
forest habitat of spiders 402, 404
forest 225, 531-4
Jarrahdale, W.A.
Alcoa bauxite mine 560
Jessup, C. G. naturalist 402
Julia Creek dunnart *Sminthopsis douglasi* 289
Juvenile period in plants 180

K

K-selection 510-1
Kakadu National Park, N.T. 7-11, 295

Kangaroo Island, S.A. 297
Kangaroo paws *Anigozanthos* spp.
  natural hybridization in 568
Kangaroo trade 567
Kangaroos *Macropus* spp.
  effects on vegetation 236, 239
  grazing of arid zone plants 269-72
  red *Macropus rufus* 210, 291
Karri *Eucalyptus diversicolor* 320, 573-5
Karroun Hill Nature Reserve, W.A. 370, 376
Karst platforms, Abrolhos coral reefs 58, 60-1, 64, 67
Keartland's observations of avifauna of Murchison district 316
Kellerberrin district, W.A. 303, 305, 320
  avifauna of 306
  description 305
Kinchega National Park, NSW, regeneration of arid zone plants 267-8
King River, Tas. 84, 89
Kings Park, Perth, W.A. 290
Kiola State Forest, NSW 326, 330-4
Koala *Phascolarctos cinereus* 324-5, 331
Kosciusko ash forest 225
Kosciusko National Park 82, 91, 223, 229
Krichauff Ranges, N.T. 223
Kuringai National Park, NSW, habitat for spiders 404
Kwongan 167, 173, 177-80

L

La Trobe River, Vic. 83, 85, 90
La Trobe Valley 293
Labour force 252
Lacustrine 442
Lag effects 95
Lagoon patch reefs, Abrolhos coral reefs 57, 60, 63, 67
Lagorchestes spp.
  L. conspicillatus spectacled hare-wallaby 209
  L. fasciatus banded hare-wallaby 289
  L. hirsutus mala 222
  L. leporides eastern hare-wallaby 291
Lake Burley Griffin, A.C.T. 84, 89
Lake Eyre, S.A. 463
Lake Garnpung, NSW 445
Lake Mungo, NSW 445
Lalage sueurii white-winged triller 306
Lampreys 85
Land
  allocation 419-20
  capability rating tables 479, 481-3
  clearance 100, 107-8, 120-3
  degradation 95, 102, 107, 279, 285, 419, 437, 441, 471, 476
  ethic 419, 423
  evaluation 476-88
  reclamation 567
  suitability assessment, concept and drawbacks 476-7
  systems 224, 446
  tenure of pastoral lands 185
Land capability assessment
  and land planners/managers 476, 484
  concept model of Victoria 476-7, 488
  critique of Victorian system 484-8
  extensions to ecological assessment 488
  parametric v. limitations approach 478
  procedure 479-84
Land classification 468
  functional patterns 467, 472
Land Conservation Council of Victoria 261
Land management 434, 436-7, 467, 472
  regional objectives 471
Land use 431, 434, 446
  Aboriginal 120
  agricultural 120-3
  European 120-3
  extractive industries 123-4
  housing/residential 121-4
  impacts on ecosystems 94, 96-103
  industrial 122-3
  multiple 461, 463
  patterns 94
  planning 106, 109-10, 435, 437, 470, 472
  pressure 39
    of agriculture 46
    of fishing 40, 48
    of industry 40, 42
    of recreation 39
Landsat 449, 450
  imagery 336, 443-5
  Multi Spectral Scanner (MSS) 441, 443-6, 450
  thematic mapper 449-50, 452-3
  V satellite 446
Landscape(s)
  ecology 133, 437
  fragmentation of 97, 100-1, 103
  geography 411
  human modified 397
  human-dominated 413, 415-6
  restoration of 397-8
Landscape change
  Holocene history 413-4
Laughing Kookaburra *Dacelo gigas* 306
Laughing turtledove *Streptopelia senegalensis* 306
Leadbeater's possum *Gymnobelideus leadbeateri* 325, 334
Leaf beetle *Paropsis atomaria* 152, 155-6
Leaf decomposition, upland streams 81, 88
Leake's observations of changes in avifauna 318
Learning, experimental approach to 424
Leeuwin-Naturaliste block, W.A. karri population 573
Legislation (see Acts of Parliament) 325-6, 338, 340
  closer settlement 247
  Commonwealth 325
Legumes for minesite rehabilitation 558-62
Leioproctus spp. 170-1
Leipoa ocellata malleefowl 202, 242, 306, 317
Leonora, W.A. 191
Leporillus spp.
  L. apicalis lesser stick-nest rat 210, 291
  L. conditor greater stick-nest rat 210, 289
Leptospermum laevigatum spread into new habitats 130
Lerderderg River, Vic. 83, 88
Lesser stick-nest rat *Leporillus apicalis* 210, 291
Letter-winged kite *Elanus scriptus* 306
Lichenostomus spp.
  L. leucotis white-eared honeyeater 306
  L. ornatus yellow-plumed honeyeater 306
  L. virescens singing honeyeater 319, 362, 364-5
Life histories 328, 334, 340
Light reduction in aquatic ecosystem
  epiphytes 42
  turbidity 42, 45-6
Lignum *Muehlenbeckia cunninghamii* 445
Limestone platforms, Abrolhos coral reefs 58, 60, 63-4
Lithgow, NSW 432
Litter input into Australian streams 81, 88
Little button quail *Turnix velox* 306, 317
Little corella *Cacatua pastinator* 291, 306
Little crow *Corvus bennetti* 317, 319
Little Sandy Desert 24, 26, 28
Livestock grazing 273, 279, 281
Local government 435-6
Logging 293, 295
  intensity 326-7
  scheduling 330
  selective 327
  types 327
Long-billed corella *Cacatua pastinator* 291, 306
Long-footed potoroo *Potorous longipes* 289, 451
Long-term
  ecological studies 109
  recolonization 517-8
Lophoictinia isura square-tailed kite 306
Lunettes 442

M

MacDonnell Ranges, N.T. 223, 225
Macquaria spp.
  M. ambigua golden perch 85
  M. australasica macquarie perch 81, 85-6, 88
  M. novemaculeata Australian bass 81, 85
Macquarie perch *M. australasica* 81, 85-6, 88
Macroalgae, 46, 48, 51
  accumulation of 45-6, 48, 50
  harvesting of 45-6, 48
  light limitation 48, 50-1
Macropods 210, 291, 329
Macropus rufus red kangaroo 210, 291

Macrotis lagotis bilby 222, 237
Magela Creek, N.T. 7-11
Maintenance of ecosystems 104
Maireana spp. bluebush 442, 470
  M. pyramidata 445
  changes with grazing 238
Mala *Lagorchestes hirsutus* 222
Malleability 514
Mallee 442-4
  woodland in Victoria 225
Malleefowl *Leipoa ocellata* 202, 306, 317
  preservation of 242
Malurus spp.
  M. coronatus purple-crowned wren 293
  M. leucopterus white-winged wren 317-8
  M. pulcherrimus blue-breasted wren 318
  M. splendens splendid wren 398
Mammals (see also Small mammals) 434, 522-4, 526-8
  critical weight range for 451-2
  extinctions of medium-sized 202, 206-8
  grazing of arid zone plants 267, 269-72
  ground 329
  ground dwelling 289
Management 261
  costs 45-6, 50
  experimental nature of 222
  guidelines 221, 222, 225, 229
  impact on eucalypt woodland 149-50, 153-6
  joint, of grazing lands 242
  monitoring 45, 51
  of Abrolhos coral reefs 56-8
  of apiculture reserves 168-9, 174
  of aquatic ecosystems 39-51
  of biological resources 455
  of catchments 48, 50, 79, 81-2, 87, 91
  of coastal ecosystems 71, 75-7
  of Cockburn Sound 40, 43, 45
  of ecosystems 221-33
  of fire 177, 180-2, 221, 223
  of natural resources 221-3, 225, 229
  of pastoral lands 186, 188, 191
  of Peel-Harvey estuary 48
  of rehabilitation 513
  of small mammals 379, 393-4
  of terrestrial vertebrates 287-301
  of urban bushland 124-6
  practices 225
  prescriptions 221, 225
  scale in arid lands 195, 200
  strategies 326-9, 435-9
  wildlife 335-41
Manganese mines 512-3, 516
Mangroves 8-10
Mann Ranges, S.A. 370
Manorina spp.
  M. flavigula yellow-throated miner 319
  M. melanoleuca noisy miner 293
Manual for fire management 221-2, 225, 227, 229
Map Grid, Australian 443
Mapping of species and their genetic resources 568
Mardudjara Aborigines 26
Marine
  conservation 573-4
  ecosystems 71-7
  populations, recruitment of larvae 76
  reserves 71, 76-7
Marri *Eucalyptus calophylla* 320
Marron *Cherax tenuimanus* 83
Marsh crake *Porzana pusilla* 317
Marsh harrier *Circus aeruginosus* 317
Marsupials 570-1
  carnivorous 329-40
  effects of grazing on 237
Martu Aborigines 23-37
Masked woodswallow *Artamus personatus* 306
Mastacomys sp. rat 289
Materialism and sustainable production 110
Mediterranean landscapes 94
Meekatharra, W.A. 190
Melanodryas cucullata hooded robin 293, 306
Melithreptus spp.
  M. brevirostris brown-headed honeyeater 306
  M. lunatus white-naped honeyeater 306
Mercury contamination of streams 83, 88
Metastable ecosystems 94-6, 105
MicroBRIAN 216, 337-8
Microeca leucophaea jacky winter 289, 293, 306
Migrating means clustering procedure 443
Migratory species 295
Miljee *Acacia oswaldii* regeneration 268
Millipedes 518
Milvus migrans black kite 306
Mine(s) 560
  dump rehabilitation 553, 555
  rehabilitation 557-64
    bauxite 531
    gravel 533
    mineral sands 533
  revegetation 575
Minimum viable population 326, 328, 332, 339, 572, 574
Mining 462-3, 521-6, 528
  and exploration 462-3
  effects on streams 79, 83-4, 86-91
  restoration after 103-5, 575
  impacts 429
Mitchell grasses *Astrebla* spp. 470
Mitchell grasslands, effects of grazing on 238, 241
Mitta Mitta River, Vic. 84-5, 88
Modelling
  fauna habitats 451-3
  vegetation 451
  systems, PREPLAN 223
Models
  bioclimatic 335, 338
Models — *continued*
  fire management 224, 229
  habitat 331-4, 336, 451
  habitat accommodation 521, 527-8
  optimization 336-40
  relay floristic, of succession 417
  simulation 336-40
  succession 329
Molecular genetics 568, 570-1
Molonglo River, NSW & ACT 83-4, 87, 89-90
Momba, NSW 245-53
Monitoring 225, 229, 257, 262
  environmental 439
  environmental change 446, 453
  land use 320
  management activities 110
  species change 241-2
  techniques, aircraft 224
Monomorium spp. ants 522, 525
Morphometrics, multivariate 568-9
Morton-Budawang National Park, NSW 223
Mosquito fish *Gambusia affinis* 86
Mountain brushtail possum *Trichosurus caninus* 330-2, 334
Mountain duck *Tadorna tadornoides* 306
Mountain shrimp *Anaspides tasmaniae* 86
MSS 441, 443-6, 450
Mt. Lofty Ranges 293
Muehlenbeckia cunninghamii lignum 445
Mulga *Acacia aneura* 239, 470
  prescribed burning in 222
Mulga parrot *Psephotus varius* 306, 318
Mulga shrublands/woodlands
  pastoral management of 186, 188, 191
Multidisciplinary study 46, 51
Multiple land use 437, 461, 463, 472
Multi Spectral Scanner 441, 450, 443-6
Mungo National Park, NSW 442
Murchison River catchment 303, 320
  avifauna of 317
  changes in avifauna with change in climate 320
  history of bird surveys 316
  land use and description 305
Murray River, W.A. 46
Murray Riverine Plains, NSW 469-71
Murray-Darling
  Depression 469-71
  River System 471
  catchment 468, 471
Mus musculus house mouse 203, 522, 524
Museum collections
  as evidence of geographic distributions 401-4
  of *Heleioporus albopunctatus* 401-2
  of *Idiosoma sigillatum* 403
  use in restoration of landscapes 397-8, 405-6
  Western Australian 397, 399, 403, 408
Museums 568
  Berlin & Hamburg 397, 399
  British (Natural History) 399, 402

Museums — *continued*
Hope (Oxford) 402
Paris 399
relevance to ecology 397-9
Western Australian 397, 399, 403
Musgrave Ranges, S.A. 370
Mycorrhizae 97, 105
Myiagra spp. flycatchers 291
M. inquieta restless flycatcher 306
Myrmecobius fasciatus numbat 204, 369-76, 398
Myrtaceae foraging by bees 169-70

N

N:P ratio, water 42
Nadgee Nature Reserve, NSW 452
Nailtail wallaby *Onychogalea fraenata*, preservation of 241
Nankeen kestrel *Falco cenchroides* 306
Nappa Merrie, S.A. 186
National Conservation Strategy 296, 323, 325
National estate (see also Register of)
Act 325
definition 491
forests 326, 330-6
issues 326
register of 325
National land use policy 420
National parks 295, 323, 326-7, 336, 432, 460-4
biological surveys 569
forests 323
grazing in 463-4
mining and exploration in 462-3
tourism in 462
National Parks & Wildlife Service (NSW) 435, 437, 439, 442, 447
Native
plants 415
vegetation 415-7
clearance 416
destruction of 575
Natural
disturbance 521-4, 528
landscape 411-2, 416
resources, management and documentation 221, 224, 225, 229
selection 572
Natural ecosystem 412, 415-6
management 221, 223, 227
Natural regions 467
means of delineation 468
provinces of 468, 470
use for environmental management 467, 470
Natural vegetation 412
pastoral use of 235, 242
Naturalness 411-2
aquatic ecosystems 417
assessing degree of 412-5, 417
Nature conservation 94, 255, 261
assessment for 472
regional plants for 472
Nature reserves 437
biological surveys 569
Nectar foraging by bees 169-72
Neochmia ruficauda star finch 317, 320
Neophema spp.
N. bourkii Bourke's parrot 317, 319
N. elegans elegant parrot 306, 318
N. spendida scarlet-breasted parrot 317, 319
Nepean-Hawkesbury River, NSW 115, 121-2
Nest boxes 329
New Holland mouse *Pseudomys novaehollandiae* 379-89, 391-4, 512, 514, 522, 524, 527
New South Wales 245-53, 293
central division 468
central western 427-39
forestry 325-6, 328-36
slopes/tablelands 427-8, 432-3, 435
south east 325, 329-34, 451-2
stocking records 245-53
western 427-40, 467
western division 467-8, 472
New Zealand, restoration of landscape 405
Newnes Plateau, NSW 432
Ngatatjara Aborigines 36
Niche 328, 330
engineering 556
Night parrot *Geopsittacus occidentalis* 202, 317, 319
Ningaui sp. 289
Nitraria billardieri chenopod shrubland 445
Nitrogen
in estuarine ecosystems 42, 46
load 43, 45
Nodularia spumigena 46, 48, 51
germination of 48, 51
Noisy miner *Manorina melanoleuca* 293
Noisy scrub-bird *Atrichornis clamosus* 398
Norfolk Island Green Parrot *Cyanoramphus novaezealandiae cookii* 133
North Carolina, USA 515
Northern Australia, fire ecology 223
Northern nailtail wallaby *Onychogalea unguifera* 209
Northern sandplain, W.A. 167-8, 171, 173
Northern Territory, fire histories 222
North-weatern Australia 289
NSW Department of Environment and Planning 447
NSW National Parks and Wildlife Service 442, 447
Numbat *Myrmecobius fasciatus* 204, 369-76, 398
predation by foxes 371-3
translocations of 374, 376
dispersal of 374, 376
home ranges of 375
minimum viable population of 376
Nutrient
accumulation after mining 557-9
dynamics of streams 79-82, 89
gradients 330-3
hypothesis 329, 333
Nutrient — *continued*
load 42-3, 45
recycling 48, 51
Nutrient cycling
after mining 559
land use impacts on 97-8, 100
timber harvesting and 101
Nutrient enrichment
in remnant woodland 150, 152, 154-6, 158
of aquatic ecosystems 39-51
of bushland 124
role in rural tree dieback 150, 152, 154-6
Nutrient input 42, 46, 51, 100
reduction of 43, 50
Nutrients
limiting of 46, 48, 51
nitrogen 42, 46
phosphorus 42, 46, 48, 50-1
soil 48

O

Oceanic exchange 51
Cockburn Sound, W.A. 40-2
Peel-Harvey Estuary, W.A. 50
Ocyphaps lophotes crested pigeon 291, 306, 317-9
Oedura reticulata gecko 398
Olea spp. invasion into native vegetation 131-3
Oncopera spp. swift moth 259
Oncorhynchus mykiss rainbow trout 86
Onychogalea spp.
O. fraenata nailtail wallaby 241
O. unguifera northern nailtail wallaby 209
Ophidiocephalus taeniatus bronzeback 202
Orange chat *Ephthianura aurifrons* 306
Orange, NSW 293, 432
Ordination 531
Oreochromis mossambica 86-7
Oreoica gutturalis crested bellbird 293
Orites lancifolia alpine orites 256
Ormiston Gorge National Park, N.T. 223
Oryctolagus cuniculus rabbit 103, 185-7, 204, 207-8, 236, 240, 246, 248, 250, 269-72, 291, 325, 371, 441
Outbreeding depression 575
Overgrazing 279, 282, 285, 430, 434, 441
and soil erosion 101
Overstocking 252
of pastoral lands 185, 187
Owlet nightjar *Aegotheles cristatus* 306
Owls 329, 340
Oxylobium sp. 372

P

Pachycephala spp.
P. inornata Gilbert's whistler 306
P. pectoralis golden whistler 306, 318
Painted button-quail *Turnix varia* 318
Palaeoecology 1-18
Pardalotus spp.
P. punctatus spotted pardalote 318
P. striatus striated pardalote 306

Parliamentary Select Committee (1982) 441
Paroo River, NSW 246
Paropsis atomaria leaf beetle 152, 155-6
Parthenogenesis 568
Pastoral industry 567
Pastoral management
of remnants 149-50, 153-6
condition 189
decision support systems 189
economics 189, 192
efficiency of production 189-90, 193
of chenopods 187, 190-1
of woodlands 186, 188, 190
overstocking 185, 187
Pastoralism 97
as a cause of extinctions 204, 207-9
Pasture improvement 431-2
Patch burning 223, 227
effect on habitat diversity 219
purpose of 223
spinifex 219
strategy 223-4, 227
Uluru National Park 219, 223-4
Pattern and process 133
Peaceful dove *Geopelia placida* 320
Pedionomus torquatus plains-wanderer 202, 242
Peel-Harvey Estuary, W.A. 40, 45-51
description of 46-7
freshwater inflow 46
nutrient inputs 46
tidal amplitude 46
water circulation 48
Pekina Reservoir, S.A. 18
Peltohyas australis Australian dotterel 317
Peppermint *Agonis flexuosa* 404
Perameles spp.
P. bougainville western barred bandicoot 289
P. gunnii eastern barred bandicoot 133
Perca fluviatilis English perch 83, 86
Perennial herbs, effects of grazing on 236, 240
Perth, W.A. 40, 177, 179
Perup Forest, W.A. 326, 329
Perup Nature Reserve, W.A. 372
Pest species 329
Petauroides volans greater glider 331-4
Petaurus spp.
P. australis yellow-bellied glider 331-2, 334
P. breviceps sugar glider 331, 334
P. norfolcensis squirrel glider 331, 334
Petrogale lateralis rock wallaby 398
Petroica spp.
P. goodenovii red-capped robin 362, 364-5
P. multicolor scarlet robin 318
Petrophassa plumifera spinifex pigeon 320
Phantom species 398
Phaps histrionica flock bronzewing 202
Phascogale calura red-tailed phascogale 370
Phascolarctos cinereus koala 324-5, 331
Phebalium squamulosum forest phebalium 256
Pheidole spp. ants 522, 525
Phosphorus
in estuarine ecosystems 42, 46, 48, 50-1
soil in bushland 124
Phylidonyris spp.
P. albifrons white-fronted honeyeater 318
P. melanops tawny-crowned honeyeater 306
Physical environment 512
Phytophthora cinnamomi cinnamon fungus 102, 379-80, 382, 385-6, 390-1, 393, 405
Phytoplankton
blooms 40, 42-3, 45-6, 48-50
distribution in Cockburn Sound 43
grazing 48
nutrient limitation 45
nutrient uptake 50
Picture elements (pixels) 443
Pied butcherbird *Cracticus nigrogularis* 317, 319
range expansion 306
Pied currawong *Strepera graculina* 291
Pig-footed bandicoot *Chaeropus ecaudatus* 291
Pilbara region, W.A. 510
Pilliga mouse *Pseudomys pilligaensis* 298
Pimelea spp. as anti-cancer agent 567
Pine plantations 291
Pinkawillinie, S.A. 463
Pinus spp.
P pinaster 100
P. radiata 100, 431, 571
Pioneer mentality and land development 422
Pitfall trapping of ants 349
Pittosporum undulatum, spread into new habitats 130
Pixels 443
Plains (NSW) 427-8, 432-3, 435
Plains rat kangaroo *Caloprymnus campestris* 210
Plains wanderer *Pedionomus torquatus* 202
preservation of 242
Planigale spp. 289
Planning 435
for fire 221-2, 225, 227, 229
for parks 222
frameworks 329-30, 337-40
strategic 329
Plant(s)
associations 433, 435
establishment 552
species composition 293
strategies 534
survival 552
vital attributes of 541-2
Plantations 295, 323, 340
Plant communities 224, 537-9, 541-2
mallee 538
rainforest 540-2
sclerophyll shrub 537-9, 541-2
woodland 539
Platycercus icterotis western rosella 306
Playa lakes 442
Plectorhynchus lanceolata striped honeyeater 291
Plectrachne schinzii 224
Pleistocene 442
Plutorectis caespitosa case moth 259
Poa spp.
effects of grazing on 239
P. hiemata snowgrass 256
Podargus strigoides tawny frogmouth 306
Poephila guttata zebra finch 317, 319
Polders 511
Political aspects of rehabilitation 509
Pollen
analysis 8-17
foraging by bees 169-72
Pollinators
attraction of 173-4
land use impacts on 97, 101
Pollution 431
Polygenic variation 569
Polytelis spp.
P. alexandrae Alexandra's parrot 202
P. anthopeplus regent parrot 306
Pomatostomus temporalis grey-crowned babbler 293, 317
Population
bottlenecks 572
distributions, fauna 450-2
dynamics 328, 334, 339-40
ecology 574-5
genetics 568, 572, 574
isolation 572
processes, land use impacts on 97, 99
size, minimum viable 182
structure 569-75
viability 326, 328, 332, 339
Populations
central 571
geographically marginal 571
large 572, 574
minimum viable size 572, 574
priority for conservation 572-3
remnant 572
small 572, 574
Porcupine grass *Triodia irritans* 442, 445
Porongorup Range, W.A. karri population 573
Porzana pusilla marsh crake 317
Possum (see also Arboreal marsupials)
common brushtail 331, 334
common ringtail 331, 334
eastern pygmy 331, 334
mountain brushtail 330-2, 334
Potorous spp. potoroos 451-2
P. longipes long-footed potoroo 289, 451
Prairies 510
Predation 293
Predators, protection from 553, 556
Predator-prey relationships
in urban environments 143-5
Prediction of fauna distributions 449, 451-2, 455
Predictive modelling modules 225
Pre-disturbance habitat 510
PREPLAN modelling system 223

Prescribed burning 223-5, 227, 229, 295
ecological aspects of 224
effect of climatic factors on 224
frequency and intensity of 229
in mulga 222
manipulation of ecological processes 223
objectives 224-5, 227, 229
patch burns 223
temperate forests and woodlands 224
tropical forests and woodlands 224
Primary production in arid lands 195, 199
Private landowners 435
Productivity
role of prescribed burning in 222
decline in 422
Propagule supply
land use impacts on 97
mining and 104
Prostanthera cuneata alpine mint 256
Protea spp. 178, 180
Proteaceae, foraging by bees 169-70
Prototroctes maraena Australian grayling 85
Provenance trials in forestry 574
Provinces 468, 470
Psephotus spp.
P. haematonotus red-rumped parrot 291
P. varius mulga parrot 306, 318
Pseudemydura umbrina short necked tortoise 398
Pseudocheirus peregrinus ringtail possum 289, 331, 334
Pseudomys spp.
P. novaehollandiae New Holland mouse 379-89, 391-4, 512, 514, 522, 524, 527
P. pilligaensis pilliga mouse 289
Public
land, uncommitted 435
policy 439
Purple-crowned lorikeet *Glossopsitta porphyrocephala* 306
Purple-crowned wren *Malurus coronatus* 293
Purple-wood wattle *Acacia carnei*, regeneration 268
Purpureicephalus spurius red-capped parrot 306
Pygmy possum (see Eastern pygmy possum)

Q

Quaternary Period 2-7
Queensland 295, 297
forestry 324-6
north 324-5
south-eastern 293
south-west 186
Quolls *Dasyurus* spp. 452

R

R-selection 510, 514
Rabbit *Oryctolagus cuniculus* 103, 204, 207-8, 291, 325, 371-2, 441
control of, using "1080" 372
dry sheep equivalents 248
Rabbit — *continued*
effects on vegetation 236, 240
grazing of arid zone plants 269-72
in pastoral lands 185-7
on *Momba,* NSW 250
plagues in semi-arid NSW 246
Racist ecological terminology 414
Rainbow trout *Oncorhynchus mykiss* 86
Rainfall
and perennial plant establishment 186-8, 190
and woody plants 187-8, 190
and woody weeds 186, 190
redistribution of 188
unpredictability of 205, 207, 224
variability of 186
Rainforests
fragmentation of 100, 105, 107
history 13-4
species diversity in 405
Ramet recruitment 180
Range distribution 435
Rangeland rehabilitation 551, 554-5
Rangelands
carrying capacity of 251
condition of 252
overstocking 252
regeneration 252
semi-arid 247
vegetation change 252
Rare fauna 288-9, 326, 329
Rare plants
Acacia carnei 268
regeneration of 268
Rare species 239, 241-2, 289, 432, 435, 569
Rats, native 289, 570
Rattus spp. 329
Re-introductions 439
Reclamation 531
landscape 405
saltland 545
Recolonization 326-8
after mining 379-80, 382, 387, 392-4
after wildfire 379, 382, 385, 387, 391-4
of small mammals 379, 382, 385, 387, 390-4
of animals 509-19
Recombinations systems 571
Reconstruction 439, 531
of ecosystems 97-9, 103-8
Recreation 40, 45
impacts of 39
management of 102-3
Recruitment 574
of arid zone plants 267-72
of larvae in marine populations 76
Red calcareous earths 445
Red-capped parrot *Purpureicephalus spurius* 306
Red-capped robin *Petroica goodenovii*
temporal changes in use of road reserves 362, 364
use of road reserves as corridors 364-5
Red data book (ICBP) 288
Red earths 445-6
Red fox (see Fox)
Red kangaroo *Macropus rufus* 210, 291
Red-rumped parrot *Psephotus haematonotus* 291
Red-tailed black cockatoo *Calyptorhynchus magnificus* 306, 317-9
Red-tailed phascogale *Phascogale calura* 370
Redthroat *Sericornis brunneus* 318
Red wattlebird *Anthochaera carunculata* 306
Reefs
shallow perimeters 57, 59-60
sheltered leeward slopes 57, 60-1, 63, 67
Refuge areas 327
for mammals in arid Australia 205-8
Regeneration 239-40, 273, 276, 522-3, 528
niche 181
of *Acacia carnei* 267-72
of *A. oswaldii* 267-72
of arid zone plants 267-72
of rangelands 252
Regent honeyeater *Xanthomyza phrygia* 289
Regent parrot *Polytelis anthopeplus* 306
Regional
ecosystems 468
objectives for land management 471
reserves 463
Regional plans
for land use 468, 471
for nature conservation 472
Regions (see also Natural regions) 467
Register of National Estate 325
as decision support system 500
assessment methodology 496-7
assessment of value 493, 495
compilation of 492
condition of 494
criteria for assessment 492, 495, 501
integrity 494
registration threshold 492-3
size 495
type profile 498
typology 498
Rehabilitation 509-19, 531
after mining 531, 557-64
agronomy 551-6
guidelines 551-6
of ecosystems 97, 104-5
of rangeland 551, 554-5
of streams 79, 87
Reintroduction of species, propagules 105
Relay floristics model of succession 417
Relict vegetation 416-7
Remnant eucalypt woodland
impact of pastoral management 149-50, 153-6
insect abundance and tree defoliation 149, 151, 153-6
nutrient enrichment 150, 152, 154-6 158

Remnant eucalypt woodland — *continued*
soil properties of 149-50, 152-6, 158
species diversity of birds 150, 152, 154-6, 162-5
tree health 150, 153-6
understorey shrubs 150, 152, 154-6, 159-61
Remnant forest/woodland 432-5
Remnant vegetation 100, 293, 305, 318, 416, 432, 435-7
Remote sensing 337-40
fire history 216, 218
microBRIAN 216
Removal of vegetation 94-5, 97-9
Replacement of vegetation 94-5, 97-101
Representative reserve systems 449, 455
Reproduction 328
Reproductive biology 575
Reptiles 291, 329, 340, 515-6
lack of extinctions 202, 209
Research 439
ecosystems 229
fire ecology 224, 229
for conservation 242
forest wildlife 329-41
historical ecological 253
pastoral lands 189
priorities 339-40
wildlife 329-41
working groups 324
Reseeders
abundance of 170-1, 173
bee preference for 171, 173-4
effect of fire on 173-4
Reserve size 291
Reserve systems
conservation of 569-70, 574
representativeness in 569-70
Reserves 241
forest 323
marine 71, 76-7
Reservoir sediments 12-8
Resilience
of ecosystems 514-7
of ecosystems to disturbance 94-7, 103
of restored communities 541-2
Resinous compounds 570
Resource limitations 419-20
Response of community to fire 224, 229
Restless flycatcher *Myiagra inquieta*, reduction in numbers 306
Restoration 531
as cultural ideal 398, 405
as distinct from reclamation 405-6
of ecosystems 103-7
of grazed lands 242
of landscape 397-8, 405-6
of streams 79, 87
Restored communities, resilience of 541-2
Retention of native vegetation 105, 108
Revegetation 105, 108
after mining 537-8
attitudes to 138, 140
Revegetation — *continued*
critical pathway for 553
in Victoria 135-40
methods 551-6
niche engineering for 556
of agricultural land 537-40
of degraded lands 575
site characterization for 551, 553, 555
site preparation for 553, 555
species selection 553
weed management 537-42
Rhipidura spp. fantails 291
Ringtail possum *Pseudocheirus peregrinus* 289, 331, 334
Riparian zone of streams 79-80, 82, 86-90
River blackfish *Gadopsis marmoratus* 83, 86, 89
River continuum concept 80, 87, 91
River-flat forests, Sydney 117-8, 121, 123-4
River red gum *Eucalyptus camaldulensis* 433, 470, 547-8
Road reserves (verges) 437
as corridors for bird movements 359, 363-5
as habitat for bird species 359, 362-3
management research 365
physical characteristics 359, 361, 363, 365
revegetation 575
vegetation characteristics 359, 361, 363-5
Rock lobster
fishing effort intensity 57-9, 62-7
fishing patterns at Abrolhos reefs 55-6, 58-9, 62-3
Rock wallabies *Petrogale* spp. 570
P. lateralis 398
Rocky Gully, W.A., karri population 573-4
Rosewood *Heterodendrum oleifolium* 442, 445
Rotten Swamp, A.C.T. 10-2
Royal Commission NSW (1901) 441
Rudall River, W.A. 24, 26, 33-4
Rufous treecreeper *Climacteris rufa* 306
Rum Jungle, N.T. 84, 90
Rural
dieback 295
landowners 436
population 429-31
Rural tree dieback
extent and severity 149-50
role of habitat simplification 150, 152, 154-6
role of insect defoliation 149, 151-6
role of nutrient enrichment 150, 152, 154-6
role of site stress 149, 155-6

S

Safety, fire, 221, 225
Salinity 295
in relation to frog habitats 400-2
land use impacts on soil 97-100, 103, 110
Salinity — *continued*
of streams 79, 82-3, 89-91
of water column 48-51
photographs of landscape 401
salt land surveys (W.A.) 400, 402
Salmo trutta brown trout 83, 86, 89
Salmon gum *Eucalyptus salmonophloia* 103
Salmonella water contamination 42
Salmonid fish in Australia 80, 83, 86
Salt-affected land
extent of 545
reclamation of 545
Saltbush *Atriplex* spp. 236, 238, 442, 470, 554-5
bladder saltbush *A. vesicaria* 445
pastoral management 185, 187
Saltmarsh 71-4, 76-7
Salt tolerant plants
micropropagation of 545-6
physiology of 547-8
screening trials 545-6, 549
variation in 546-7
Samphire thornbill *Acanthiza iredalei* 320
Sampling methods, arid landscapes 195, 198, 200
Sand mines 512, 516
Sandstone, Sydney area 115, 119, 123, 125-6
Sandstone vegetation 432
Sarcophilus harrisii Tasmanian devil 203
Satellite imagery 224, 441, 446, 569
and GIS for habitat mapping 449-50, 452-3
use on arid landscapes 197-8
Scalds 445
Scarlet-breasted parrot *Neophema spendida*
disappearance from Murchison 317, 319
Scarlet robin *Petroica multicolor* 318
Sclerolaena spp. chenopod shrubland 187, 190-1, 238, 442, 443
S. diacantha 445
S. divaricata 445
Sclerophyll vegetation 104-5
Scrub, Sydney area 117-9, 121-2, 124
Seagrass 72-3, 75-7
dieback 40, 42-5
epiphytes 40, 42
grazing 43
importance of 40, 42
Sea level change 3, 8
Sedimentation of streams 79, 81-2, 86, 89, 91
Seed
banks 177-9, 181, 274, 532, 537-40, 542
dispersal in kwongan flora 173
dormancy 534
Seeling(s)
establishment 178-9
germination, arid zone plants 270-2
inter-fire recruitment 178
post-fire recruitment 177-9, 181-2
recruitment 267-72

Seedling(s) — *continued*
recruitment, arid zone plants 270-2
regeneration 267-72
Seed-safe sites 179-81
ecology 531-5
germination 179-81
orchards 574
post-fire mortality/predation 177, 179-81
release 177-80
set, effect of honey bees on 173
size, constraints on revegetation of 539-40, 542
viability 179, 532
Semi-arid zone 221, 441-2, 446
of NSW 245-53
rangelands 422
woodlands, effects of grazing on 239
Sensitive species 326, 329, 339-40
Sericornis spp.
S. brunneus redthroat 318
S. cautus shy hylacola 318
S. fuliginosus field wren 317-8, 320
Serotiny 177-9
Serpentine River, W.A. 46
Settlement
closer 247
European, affect on landscape 397-8
pre-settlement landscapes 405
Sexist ecological terminology 414
Shale, Sydney area 115-6, 118, 125
Shark Bay, W.A. 574
Sheep
effects on vegetation 236
grazing effects 101
grazing in NSW 441, 446
Shoalhaven River, NSW 85, 88
Short necked tortoise *Pseudemydura umbrina* 398
Shrub(s)
effects of grazing on 236, 239-40
infestation 222
invasion 284-5
species diversity 150, 152, 154-6, 159-61
Shrub-grass dynamics 255-65
Shy hylacola *Sericornis cautus* 318
Sibling
speciation 568-9
species 574
Silvereye *Zosterops lateralis* 306
Silver snow daisy *Celmisia* sp. 256
Silvicultural operations 327-8, 339
Simpson Desert 295, 463
Simpsons Gap National Park, N.T. 223-4
Singing honeyeater *Lichenostomus virescens* 319
changes in food resources of 362, 364
use of road reserves 362, 364-5
Skinks 329
Slaters 518
Slopes (NSW) 427-8, 432-3, 435
Small mammals 512, 514, 518
conservation of 379, 393-4
habitat preferences 379-80, 382, 385-6, 389
Small mammals — *continued*
management of 379, 393-4
recolonization of 379, 382, 385, 387, 390-4
Sminthopsis douglasi Julia Creek dunnart 289
Snowgrass *Pos hiemata* 256
Social interactions, wildlife 328
Sofala, NSW 429
Soil
acidity in remnanι woodland 152-5, 158
Conservation Service (NSW) 435, 438
erosion 10, 12-8, 81, 86, 90, 187-8, 441, 445
fertile islands 188, 192
fertile patches 188, 191
landscapes 446
nitrogen mineralization 150, 152, 154-6, 158
nutrient enrichment 124-5
nutrients 188
organisms 97, 106
phosphorus 124
properties 97-8, 107
reclamation 191
seed banks 177-9, 181, 274, 532, 537-40, 542
surface crusts 101
South Australia 437, 459-64
arid zone 459-64
northern arid 187
national parks 460-4
regional reserves 463
Southern scrub robin *Drymodes brunneopygia* 306, 318
Southern whiteface *Aphelocephala leucopsis* 318-9
South Esk River, Tas. 84, 90
Spartina invasion 130, 133
Spatial heterogeneity in pastoral lands 188-9
Spatial patterns
arid lands 195-200
pastoral lands 190
Spatial scales 6-7, 13-4, 17-8
Species
fugitive 398
management 222
numbers, effects of grazing on 239
phantom 398
responses to fire 229
selection 553
Species diversity 398, 405
effects of grazing on 237
of birds on farms 150, 152, 154-6, 162-5
Species richness 398, 405
diversity 398, 405
fire management 217
maintenance 180
response after fire 217
spinifex communities 217
Uluru National Park 217
upland streams 81, 88
Speckled warbler *Chthonicola sagittata* 293
Spectacled hare-wallaby *Lagorchestes conspicillatus* 209
Spectral reflectance/signatures 443-4
Spencer Gulf, S.A. 574
Spiders
and landscape restoration 398-9, 402-4
social 398
trapdoor 398
Spinifex communities 215-20, 224, 442, 445
and fire 222, 224
dunefields and sandplains 222
fire management of 215-22
porcupine grass *Triodia irritans* 442, 445
T. pungens 215
T. basedowii 215
Uluru National Park 215
Spinifex pigeon *Petrophassa plumifera* 320
Spiny-cheeked honeyeater *Acanthagenys rufogularis* 306, 317, 319
Splendid wren *Malurus splendens* 398
Spot satellite 450
Spotted bowerbird *Chlamydera guttata* 317, 319
Spotted gum 330, 333
Spotted pardalote *Pardalotus punctatus* 318
Spotted quail-thrush *Cinclosoma punctatum* 293
Sprouters, abundance of 170-1, 173
Square-tailed kite *Lophoictinia isura* 306
Squirrel glider *Petaurus norfolcensis* 331, 334
Star finch *Neochmia ruficauda* 317, 320
State Environmental Planning Policy (NSW) 437
State forests
history 324-5
legislation 325-6
management 221, 324, 336
recreational activities 325
wildlife conservation 323-45
Statutary authorities 436
Stewardship of land 108
Stick-nest rat *Leporillus* spp. 210
Stipa spp. 445
Stock numbers 430-1, 434
Stocking
rate 101, 189-92
records, *Momba,* NSW 245-53
Stockyards 251
Streams
dams 79, 84-8
deleterious effects of trout on biota 86, 89, 91
eductor dredging 79, 84-8
energy inputs 80-1, 88-90
forestry 79, 82, 87-90
impact of exotic introductions 79, 86-7, 89, 91
inventories 79, 87
litter input 81, 88
mercury contamination 83-8
mining 79, 83-4, 86-91
nutrient dynamics 79-82, 89
restoration of 79, 87

Streams — *continued*
riparian zone 79-80, 82, 86-90
salinization 79, 82-3, 89-91
sedimentation 79, 81-2, 86, 89, 91
upland and lowland 80-8
Strepera spp.
S. graculina pied currawong 291
S. versicolor grey currawong 306, 317, 320
Streptopelia senegalensis laughing turtledove 306
Stress
in ecosystems 515
role in rural tree dieback 149, 155-6
Striated pardalote *Pardalotus striatus* 306
Striped honeyeater *Plectorhynchus lanceolata* 291
Stubble quail *Coturnix novaezealandiae* 306
Stylidium spp. trigger plants 568, 570
S. caricifolium complex 568
S. crossocephalum 570, 574-5
Subalpine vegetation 239, 255-65
Subdivision of land 245-53
Subtropical savannahs, effects of grazing on 241
Succession 330, 332, 334, 414, 521, 523-5, 527-8, 538, 540-2
allogenic 414-5, 417
autogenic 414-5, 417
climax concept 417
dominant plant species 414-5
end point 517
facilitation model 417
facilitation pathway 512
faunal, after mining 562-4
floral, after mining 560-1
habitat accommodation model 521, 527-8
initial floristic composition model 531
models 329
relay floristics model 417
vegetation structure 414-5
Suckers
recruitment of 270-2
regeneration of 267-72
Sugar glider *Petaurus breviceps* 331, 334
Sulphur-crested cockatoo *Cacatua galerita* 291
Surveys, biological 334-5
Sustainable
agriculture 419-23
land use, framework for 467
land use, regional 468
production 101-2, 106, 110
Sustained
production development 475
yields 574
Swan River, W.A. 402-4
Swamp antechinus *Antechinus minimus* 380, 383, 385-8, 393
Swift moth *Oncopera* spp. 259
Sydney 293
Aboriginal occupation 119-20
climate 116
European utilization 120-3
geology 115-6
Sydney — *continued*
physiographic features 115
vegetation 115-26
loss 124
map 117
Synthetic
ecosystems 103, 105
vegetation 129-32, 417

T

Tablelands (NSW) 427-8, 432-3, 435
Tachyglossus aculeatus echidna 371
Tadorna tadornoides mountain duck 306
Tallowa Dam, Shoalhaven River, NSW 85, 88
Talpa europaea European mole 511
Tanami Desert, fire histories 222, 224
Tanjil River, Vic. 85, 88
Tasmania 289, 297
forestry 326, 328, 335
rainforests 13-4
spiders in 404
Tasmanian devil *Sarcophilus harrisii* 203
Tawny-crowned honeyeater *Phylidonyris melanops* 306
Tawny frogmouth *Podargus strigoides* 306
Taxonomy 568-9, 575
TEC Total Environment Centre 289
Technological solutions and productivity decline 422
Tenure 429, 431, 433, 435, 437
pastoral lands 185
Termite(s) 371-2
effects of grazing on 238
mounds 188
Terpenoides 570
Terrestrial vertebrates
conservation of 287-301
extinction in 287-301
increases in abundance 291
management of 287-301
number of species 288
status 288-9
Tertiary educational institutions 568
Tertiary Period 2
Thematic mapper 449-50, 452-3
Themeda spp.
T. australis defoliation tolerance 236
T. triandra 273, 281
Thick-billed grass wren *Amytornis textilis* 202
disappearance from Murchison 317, 319
Thomson Dam, Vic., 84-5, 88
Thomson River, Vic. 83-4, 88, 90
Threatened species 289, 432, 434, 439, 569
Thrip 516, 518
Thylacine *Thylacinus cynocephalus* 203
Thysanoptera (thrip) 516, 518
Tibooburra Downs, NSW 469-71
Tiliqua adelaidensis Adelaide blue-tongue lizard 289
Timber
harvesting 101
production 326, 336
Time scales 5-6, 13-4, 17-8
Tingle (see *Eucalyptus guilfoylei, E. jacksoni*)
as habitat for relic spider 404
Torresian crow *Corvus orru* 317, 319
Tortoise (see Short necked tortoise)
Tourism 462, 567
Trampling by stock 101
Transfer of information 224
Translocation 570, 574-5
Trapdoor spider *Idiosoma* spp. 402-4, 409
Tree
health in remnant woodland 150, 153-6
hollows 295
preservation orders 435
Treecreepers *Climacteris* spp. 291
Trichosurus spp.
T. caninus mountain brushtail possum 330-2, 334
T. vulpecula common brushtail possum 204, 331, 334
Trigger plants *Stylidium* spp. 568, 570
Triodia spp. spinifex 224
T. basedowii 224
fire management 215-20
species richness 217
Uluru National Park 215
T. delandii 224
T. hubbardii 224
T. irritans porcupine grass 442, 445
T. longiceps 224
T. pungens 224
fire management 215-20
species richness 217
Uluru National Park 215
Tropical
rainforests 324
savannahs, effects of grazing 241
Trout, effects on native stream biota 86, 89, 91
Turnix spp.
T. varia painted button quail 318
T. velox little button quail 306, 317
Turpentine-ironbark forest 116-7, 119, 121-2, 124
Tutanning Nature Reserve, W.A. 370
Type localities 399

U

Ulex europaeus Gorse invasion 130-1, 133
Uluru National Park, N.T. 222, 229
fire management of 215-20
patch burn strategy 223-4
species richness 217
spinifex communities 215
Ulva 45
Umberumberka Reservoir, NSW 12-3, 18
Upland streams
ecological characteristics 79-81, 87
energy inputs 81, 88
hydrological characteristics 80-1, 89-90
leaf decomposition 81, 88
species richness 81, 88

Uranium mining 7-10
Urban
  bushland 124-6
    maintenance and rehabilitation of 125
    management of 124-6
  ecology 141
  population 428-9
Utilitarian view of resource use 107, 109-10
Utilization of native vegetation 95, 97-9, 101-3, 110

V

Vanellus tricolor banded plover 306, 317
Varied sittella *Daphoenositta chrysoptera* 306
Vegetation 432-3
  Aboriginal use of arid zone 28-32, 34, 35
  adventive 416-7
  alliances 432-3, 437, 439
  and prescribed fire 224
  classification 329, 450
  cultivated 416
  European impact 93-114
  exotic 415-7
  gradient modelling 451
  islands 432-5
  land use impacts 97-103
  measurement 224
  native 415-7
  natural 412
  pastoral use of 235, 242
  relict 416-7
  remaining in Sydney 124
  synthetic 417
  types in Sydney 116-9
Vegetation change 1-18, 252
  aesthetic value 241
  long-term measurement 241
  pastoral value 239, 241
  recreational value 240
  water catchment 240
Vegetation clearance 120-3, 136
  for agriculture 100, 107-8
  native 416
Vegetation dynamics 255-65
  cyclic changes 258
  disturbance 258-63
  rare events 260-1
  seedling establishment 258
  shrub invasion 255
  shrub persistence 258-61
  shrub senescence 258-61
  vegetative reproduction 258
Vegetation loss
  agricultural 120-3
  extractive industries 123-4
  industrial 122-3
  suburban development 121-4
  Sydney area 124
  timber cutting 120-2
Vegetation maps/mapping 229, 569
  of Sydney 117
Vegetation structure 293
  succession 414-5
Vegetative suckers
  Acacia carnei 267-78, 270-1
  survival 267-8, 270-1
Vertebrates 433-4
Victoria
  Farm Tree Program 108
  forestry 325-6, 328
  mallee woodland 225
  north west 434
  revegetation 135-40
Vital attributes, plants 541-2
Vulnerable species 229
Vulpes vulpes fox 203, 209, 289, 291, 293, 295, 325, 371-6

W

Walls of China, Lake Mungo 442
Wanaaring floods 246
Wanderrie, pastoral management 191
Wandoo *Eucalyptus wandoo* 373-5
Warblers *Gerygone* spp. 291
Warramaba virgo grasshopper 568
Warrego Fan, NSW 469-72
Watarrka National Park, N.T. 223-4
Watchbox Creek, Vic. 86
Water
  catchments 101
  circulation in Cockburn Sound 42
  production 336
  supply, effect on avifauna 318-9
Water Buffalo 103
Water quality
  degradation 40
  salinity 48
  turbidity 46, 518
Watering points 251
Wedge-tailed eagle *Aquila audax* 306, 317
Weed
  control, herbicide use 539-42
  invasion 124-5, 431, 538
  management 537-42
Weeds 100, 416
  annual 538-9, 542
  in Sydney bushland 124-5
  perennial 538, 540-1
Welcome swallow *Hirundo neoxena* 293, 317
Western Australia 94, 100, 102-3, 295
  environmental degradation in 39-51
  forestry 325, 328
  jarrah forest in 225
  north 289
  northern sandplain 167-8, 171, 173
  south-western 177-80, 182
Western barred bandicoot *Perameles bougainville* 289
Western Desert Aborigines 23-37
Western Division, NSW 246, 467-8, 472
Western NSW 467
Western quoll *Dasyurus geoffroii* 204, 370-1
Western rosella *Platycercus icterotis* 306
Western warbler *Gerygone fusca* 306
Western yellow robin *Eopsaltria griseogularis* 318
Wetlands 72, 74, 76-7
  estuarine 117-9, 121, 123-4
  freshwater 8-10, 117-9, 121-2, 124
Wheatbelt of Western Australia 100, 295, 303, 397, 400, 405
  area cleared 303
  clearing 400
  disappearance of numbats 370, 376
  distribution of *Heleioporus albopunctatus* 399, 402
  modification of landscape 397
  restoration of 405
  salinity 400
White-backed woodswallow *Artamus superciliosus* 306
White Cliffs Plateau, NSW 469-71
White Cliffs, NSW 247
White cypress pine *Callitris glaucophylla* 247, 432, 442, 445
White-eared honeyeater *Lichenostomus leucotis* 306
White-footed rabbit-rat *Conilurus albipes* 291
White-fronted chat *Ephthianura albifrons* 317, 320
White-fronted honeyeater *Phylidonyris albifrons* 318
White-naped honeyeater *Melithreptus lunatus* 306
White-winged triller *Lalage sueurii* 306
White-winged wren *Malurus leucopterus* 317-8
Whitlock's observations of avifauna of Murchison district 317
Whyalla, S.A. 190
Wianamatta Shale 115, 118
Wilcannia, NSW 246
Wilderness 108, 412, 471
  in Australia 108
  in United States 108
  perception by Aborigines 412
Wildfire 186-7, 223-4, 227, 379, 381-2, 385-94
  and coroner's inquests 223
  Uluru National Park 215, 219
Wildflower harvesting 567
Wildlife
  corridors 455, 326-7, 330, 340
  habitat succession modelling 451, 453
  management 335-41
  refuges 435
  research 329-41
  strategy 326-9
Wildlife conservation 323-45
  forests 323-45
Wilga *Geijera parviflora* 442, 445
Willandra Lakes, NSW 442
  World Heritage Region 442-3
Wilpena Pound, S.A. 462
Witjira, S.A. 463
Woodchip areas 329-30, 333, 340
Wood duck *Chenonetta jubata* 306
Woodland 177-8, 453
  dry tropical rangelands 279, 281
  pastoral management 186, 188, 190
  Sydney area 117-21, 123-4

Woody plants
  as weeds 186, 190, 246
  carrying capacity 186
  excessive 187-8, 190
  fire 187, 190, 192
  rainfall 186-8, 190
Woody weeds (see Woody plants)
World Conservation Strategy 323, 567
World Heritage 325-6
World Heritage Area, Willandra Lakes 442-3
World Wildlife Fund 436
Wren (see *Malurus*)
Wyoming, USA 511

X

Xanthomyza phrygia regent honeyeater 289

Y

Yabby *Cherax destructor* 83, 89
Yellow-bellied glider *Petaurus australis* 331-2, 334
Yellow-plumed honeyeater *Lichenostomus ornatus* 306
Yellow-rumped thornbill *Acanthiza chrysorrhoa* 317
Yellow-throated miner *Manorina flavigula* 319

Z

Zebra finch *Poephila guttata* 317, 319
Zoological gardens 575
Zosterops lateralis silvereye 306